*Edited by
Detlef Stolten and
Bernd Emonts*

**Hydrogen Science and
Engineering**

Related Titles

Hirscher, M. (ed.)

Handbook of Hydrogen Storage

New Materials for Future Energy Storage

2010

Print ISBN: 978-3-527-32273-2; also available in electronic formats

Godula-Jopek, A., Jehle, W., Wellnitz, J

Hydrogen Storage Technologies

New Materials, Transport and Infrastructure

2012

Print ISBN: 978-3-527-32683-9; also available in electronic formats

Stolten, D. (ed.)

Hydrogen and Fuel Cells

Fundamentals, Technologies and Applications

2010

Print ISBN: 978-3-527-32711-9; also available in electronic formats

Stolten, D., Scherer, V. (eds.)

Efficient Carbon Capture for Coal Power Plants

2011

Print ISBN: 978-3-527-33002-7; also available in electronic formats

Stolten, D., Scherer, V. (eds.)

Transition to Renewable Energy Systems

2013

Print ISBN: 978-3-527-33239-7; also available in electronic formats

Godula-Jopek, A. (ed.)

Hydrogen Production

by Electrolysis

2014

Print ISBN: 978-3-527-33342-4; also available in electronic formats

Ullmann's Energy: Resources, Processes, Products

3 Volume Set

2015

Print ISBN: 978-3-527-33370-7; also available in electronic formats

Edited by Detlef Stolten and Bernd Emonts

Hydrogen Science and Engineering

Materials, Processes, Systems and Technology

Volume 1

Verlag GmbH & Co. KGaA

Editors

Prof. Dr. Detlef Stolten
Forschungszentrum Jülich GmbH
IEK-3: Fuel Cells
Leo-Brandt-Straße
52425 Jülich
Germany

Dr. Bernd Emonts
Forschungszentrum Jülich GmbH
IEK-3: Fuel Cells
Leo-Brandt-Straße
52425 Jülich
Germany

All books published by **Wiley-VCH** are carefully produced. Nevertheless, authors, editors, and publisher do not warrant the information contained in these books, including this book, to be free of errors. Readers are advised to keep in mind that statements, data, illustrations, procedural details or other items may inadvertently be inaccurate.

Library of Congress Card No.: applied for

British Library Cataloguing-in-Publication Data
A catalogue record for this book is available from the British Library.

Bibliographic information published by the Deutsche Nationalbibliothek
The Deutsche Nationalbibliothek lists this publication in the Deutsche Nationalbibliografie; detailed bibliographic data are available on the Internet at http://dnb.d-nb.de.

© 2016 Wiley-VCH Verlag GmbH & Co. KGaA, Boschstr. 12, 69469 Weinheim, Germany

All rights reserved (including those of translation into other languages). No part of this book may be reproduced in any form – by photoprinting, microfilm, or any other means – nor transmitted or translated into a machine language without written permission from the publishers. Registered names, trademarks, etc. used in this book, even when not specifically marked as such, are not to be considered unprotected by law

Print ISBN: 978-3-527-33238-0
ePDF ISBN: 978-3-527-67429-9
ePub ISBN: 978-3-527-67428-2
Mobi ISBN: 978-3-527-67427-5
oBook ISBN: 978-3-527-67426-8

Cover Design Formgeber, Mannheim
Typesetting Thomson Digital, Noida, India
Printing and Binding Markono Print Media Pte Ltd, Singapore

Printed on acid-free paper

Contents

List of Contributors *xxxi*

Volume 1

Part 1 **Sol–Gel Chemistry and Methods** *1*

1 **Hydrogen in Refineries** *3*
   James G. Speight
1.1 Introduction *3*
1.2 Hydroprocesses *4*
1.2.1 Hydrotreating Processes *6*
1.2.2 Hydrocracking Processes *8*
1.2.3 Slurry Hydrocracking *10*
1.2.4 Process Comparison *10*
1.3 Refining Heavy Feedstocks *11*
1.4 Hydrogen Production *12*
1.5 Hydrogen Management *14*
   References *17*

2 **Hydrogen in the Chemical Industry** *19*
   Florian Ausfelder and Alexis Bazzanella
2.1 Introduction *19*
2.2 Sources of Hydrogen in the Chemical Industry *22*
2.2.1 Synthesis Gas-Based Processes *22*
2.2.2 Steam Reforming *23*
2.2.3 Process Variations *25*
2.2.3.1 Partial Oxidation *25*
2.2.3.2 Autothermal Reforming *25*
2.2.3.3 Pre-reforming *25*
2.2.3.4 Water–Gas Shift Conversion *26*
2.2.3.5 Gasification *26*
2.2.3.6 Other Waste and Coupled Streams *26*
2.2.4 Electrolytic Processes *26*
2.2.4.1 Alkaline Electrolysis *27*

2.2.4.2   PEM Electrolysis   *27*
2.2.4.3   High-Temperature Electrolysis   *27*
2.2.5   Hydrogen Production Steam Reforming versus Electrolysis   *28*
2.2.6   Hydrogen as Coupled Stream in the Electrolytic Production of Chlorine   *28*
2.2.6.1   Membrane Cell Process   *29*
2.2.6.2   Mercury Cell Process   *30*
2.2.6.3   Daphragm Cell Process   *31*
2.2.6.4   New Developments   *31*
2.3   Utilization of Hydrogen in the Chemical Industry   *32*
2.3.1   Ammonia   *32*
2.3.2   Methanol   *34*
2.3.3   Other Uses and Applications of Hydrogen   *36*
2.3.4   Current Developments and Outlook   *37*
          References   *38*

3   Chlorine–Alkaline Electrolysis – Technology and Use and Economy   *41*
           Alessandro Delfrate
3.1   Introduction   *41*
3.2   Production Technologies   *42*
3.2.1   Electrochemistry of Chlorine Production   *42*
3.2.2   Mercury Electrolyzer Technology   *43*
3.2.3   Diaphragm Electrolyzers   *45*
3.2.4   Ion Exchange Membrane Electrolyzers   *46*
3.2.5   Research   *49*
3.2.6   Breakthrough Technologies: Chlorine-Alkali Production with Oxygen Depolarized Cathode (ODC)   *49*
3.3   Use of Chlorine and Sodium Hydroxide   *52*
3.3.1   Chlorine   *52*
3.3.2   Sodium Hydroxide   *53*
3.3.3   Economy of Chlorine and Caustic Soda   *53*
3.3.4   Energy Savings   *55*
           References   *56*

Part 2   Hydrogen as an Energy Carrier   *57*

Part 2.1   Introduction and National Strategies   *57*

4   Hydrogen Research, Development, Demonstration, and Market Deployment Activities   *59*
           Jochen Linssen and Jürgen-Friedrich Hake
4.1   Introduction   *59*
4.2   Germany   *60*

4.2.1	Energy Framework and Relevant Policies	*60*
4.2.2	Hydrogen Related Energy Policy Strategies	*60*
4.2.3	Hydrogen Research, Development, Demonstration, and Deployment Activities	*61*
4.2.3.1	Transportation	*61*
4.2.3.2	Hydrogen Production	*63*
4.2.3.3	Stationary and Residential Applications	*63*
4.2.3.4	Special Markets	*64*
4.2.3.5	Industry Activity	*64*
4.3	Norway	*65*
4.3.1	Energy Framework and Relevant Policies	*65*
4.3.2	Hydrogen Related Energy Policy Strategies	*65*
4.3.3	Hydrogen Research, Development, Demonstration, and Deployment Activities	*66*
4.3.3.1	HyNor Project	*67*
4.3.3.2	ZEG Power	*67*
4.3.3.3	Transnova Hydrogen Projects	*68*
4.4	European Union	*68*
4.4.1	Energy Framework and Relevant Policies	*68*
4.4.2	Hydrogen Related Energy Policy Strategies	*69*
4.4.3	Hydrogen Research, Development, Demonstration, and Deployment Activities	*69*
4.5	Canada	*70*
4.5.1	Energy Framework and Relevant Policies	*70*
4.5.2	Hydrogen Related Energy Policy Strategies	*74*
4.5.3	Hydrogen Research, Development, Demonstration, and Deployment Activities	*74*
4.6	United States of America	*76*
4.6.1	Energy Framework and Relevant Policies	*76*
4.6.2	Hydrogen Related Energy Policy Strategies	*76*
4.6.3	Hydrogen Research, Development, Demonstration, and Deployment Activities	*77*
4.7	Japan	*78*
4.7.1	Energy Framework and Relevant Policies	*78*
4.7.2	Hydrogen Related Energy Policy Strategies	*79*
4.7.3	Hydrogen Research, Development, Demonstration, and Deployment Activities	*80*
4.8	International Networks	*80*
4.8.1	Hydrogen Implementing Agreement of the IEA	*81*
4.8.2	IPHE	*81*
4.8.3	EHA	*81*
4.8.4	International Association for Hydrogen Energy	*82*
	Acknowledgment	*82*
	References	*82*

Part 2.2 Thermochemical Hydrogen Production 85

5 Thermochemical Hydrogen Production – Solar Thermal Water Decomposition 87
Christian Sattler, Nathalie Monnerie, Martin Roeb, and Matthias Lange
5.1 Introduction 87
5.2 Historical Development 88
5.3 Present State of Work 89
5.3.1 Metal/Metal Oxide Thermochemical Cycles 89
5.3.1.1 FeO/Fe_3O_4 90
5.3.1.2 Ferrites 91
5.3.1.3 Hercynite Cycle 93
5.3.1.4 Manganese Ferrite plus Activated Sodium Carbonate 94
5.3.1.5 Zn/ZnO 94
5.3.1.6 CeO_2/Ce_2O_3 95
5.3.2 Sulfur Cycles 96
5.3.2.1 Hybrid Sulfur Cycle (Westinghouse, Ispra Mark 11) 96
5.3.2.2 Mark 13 V2 98
5.3.2.3 Mark 13A 98
5.3.2.4 Sulfur–Iodine or General Atomics Process (ISPRA Mark 16) 99
5.3.3 Other Cycles 100
5.3.3.1 UT3 (Ca/Fe/Br Cycle) 101
5.3.3.2 Hybrid Copper–Chlorine Cycle 101
5.3.3.3 Uranium–Europium Cycle 102
5.4 Conclusion and Outlook 102
Nomenclature 103
References 103

6 Supercritical Water Gasification for Biomass-Based Hydrogen Production 109
Andrea Kruse
6.1 Introduction 109
6.1.1 Hydrothermal Biomass Conversions 109
6.1.2 Properties of Water 111
6.2 Model Compounds 113
6.2.1 Glucose 113
6.2.2 Cellulose 115
6.2.3 Amino Acids 115
6.2.4 Phenols 115
6.2.5 Others 116
6.3 Biomass 116
6.3.1 Influence of Salts 118
6.3.2 Influence of Proteins 118
6.3.3 Influence of Lignin 118
6.4 Catalysts 119

6.5	Challenges *119*
6.5.1	Heating-Up *119*
6.5.2	Heat Recovery *120*
6.5.3	Yields *120*
6.5.4	Salt Deposition *121*
6.5.5	Material Choice *121*
6.5.6	Catalyst Stability *122*
6.6	Scale-Up and Technical Application *122*
6.7	New Developments *122*
6.8	Conclusion *123*
	References *123*

7 Thermochemical Hydrogen Production – Plasma-Based Production of Hydrogen from Hydrocarbons *131*
Abdullah Aitani, Shakeel Ahmed, and Fahad Al-Muhaish

7.1	Introduction *131*
7.2	Non-thermal Plasma *132*
7.2.1	Gliding-Arc Plasma *132*
7.2.2	Microwave Plasma *136*
7.2.3	Dielectric Barrier Discharge (DBD) Plasma *139*
7.2.4	Corona Discharge *142*
7.2.5	Spark and Pulsed Plasmas *142*
7.3	Thermal Plasma *144*
7.3.1	DC Torch Plasma *144*
7.3.2	Three-Phase AC Plasma *145*
7.3.3	DC-RF Plasma *145*
7.4	Concluding Remarks *146*
	Acknowledgment *147*
	References *147*

8 Solar Thermal Reforming *151*
Christos Agrafiotis, Henrik von Storch, Martin Roeb, and Christian Sattler

8.1	Introduction *151*
8.2	Hydrogen Production via Methane Reforming *152*
8.2.1	Thermochemistry and Thermodynamics of Reforming *152*
8.2.2	Current Industrial Status *153*
8.3	Solar-Aided Methane Reforming *154*
8.3.1	Solar Concentration Systems *154*
8.3.2	Coupling Reforming with Solar Energy: Solar Receiver–Reactor Concepts *154*
8.3.3	Worldwide Research into Solar Thermal Reforming of Methane *157*
8.3.3.1	Indirectly Heated Reactors *159*
8.3.3.2	Directly Irradiated Reactors *163*
8.4	Current Development Status and Future Prospects *167*
	References *169*

9	**Fuel Processing for Utilization in Fuel Cells** *173*
	Ralf Peters
9.1	Introduction *173*
9.2	Scope of the Work and Methodical Approach *174*
9.3	Chemical Engineering Thermodynamics *175*
9.3.1	Thermodynamic Property Relations *175*
9.3.2	Chemical Equilibrium *178*
9.4	Unit Operations *180*
9.4.1	Catalytic Reactors *180*
9.4.1.1	Hydrogen Generation *181*
9.4.1.2	Water–Gas Shift Reaction *184*
9.4.1.3	Preferential Oxidation *186*
9.4.1.4	Selective CO Methanation *187*
9.4.1.5	Catalytic Combustion *188*
9.4.2	Separation Devices *189*
9.4.2.1	Adsorption Process *189*
9.4.2.2	Membrane Process *189*
9.4.3	Balance-of-Plant Components *191*
9.4.3.1	Pumps and Compressors *191*
9.4.3.2	Heat Exchangers and Evaporators *192*
9.5	Subsystems of Fuel Processing *192*
9.5.1	Optimization of the Subsystem Reforming *193*
9.5.1.1	Process Selection and Optimization *193*
9.5.1.2	Catalyst Development *200*
9.5.1.3	Current Challenges in Fuel Processing *201*
9.5.1.4	Technical Outlook *204*
9.5.2	Design of Complete Gas Cleaning System *205*
9.6	Conclusion *208*
	Acknowledgments *209*
	References *209*
10	**Small-Scale Reforming for On-Site Hydrogen Supply** *217*
	Ingrid Schjølberg, Christian Hulteberg, and Dick Lieftink
10.1	Introduction *217*
10.2	Definition *218*
10.3	Reforming Technologies *219*
10.3.1	Steam Methane Reforming (SMR) *219*
10.3.2	Partial Oxidation *220*
10.3.3	Hydrogen Purity *221*
10.4	Feedstock Options *223*
10.4.1	Upgraded Biogas *223*
10.4.2	Biomethanol and bioDME *224*
10.4.3	Biodiesel *224*
10.4.4	Biopropane *225*
10.5	Suppliers and Products *225*

10.5.1	Cost Trends	*225*
10.5.1.1	Central Plant Production	*225*
10.5.1.2	On-Site SR	*226*
10.5.1.3	Economics: On-Site versus Central Plant Production	*227*
10.6	Emerging Technologies	*228*
10.6.1	Material Development	*229*
10.6.2	Reactor Improvements and Design Aspects	*230*
10.6.3	Clean-Up Technology	*230*
10.6.4	Small-Scale CO_2 Capture	*231*
10.7	Process Control	*232*
10.7.1	Condition Monitoring	*232*
10.7.2	Control Structures	*233*
10.8	Safety	*234*
10.9	Conclusion	*235*
	References	*235*

11 Industrial Hydrogen Production from Hydrocarbon Fuels and Biomass *237*
Andreas Jess and Peter Wasserscheid

11.1	Options to Produce Hydrogen from Fuels–An Overview	*237*
11.2	Hydrogen Production from Solid Fuels (Coal, Biomass)	*242*
11.2.1	Basic Principles and Reactions of Syngas Production from Solid Fuels	*242*
11.2.2	Hydrogen Production by Gasification of Solid Fuels	*242*
11.3	Syngas by Partial Oxidation of Heavy Oils	*244*
11.4	Syngas by Steam Reforming of Natural Gas	*246*
11.5	Conclusions	*249*
	References	*251*

Part 2.3 H_2 from Electricity *253*

12 Electrolysis Systems for Grid Relieving *255*
Filip Smeets and Jan Vaes

12.1	Introduction	*255*
12.2	Energy Policies around the Globe Drive Demand for Energy Storage	*256*
12.3	The Options for Integration of Intermittent Renewable Energy Sources	*261*
12.4	The Evolution of the Demand for Energy Storage	*268*
12.5	The Role of Electrolyzers in the Energy Transition	*270*
12.6	The Overall Business Case and Outlook	*274*
12.6.1	De-carbonization of all Energy Sectors and the Chemical Industry	*275*
12.6.2	Combination of Low-Cost Power Generation with Low-Cost Gas Infrastructure	*276*

12.6.3	Grid-Scale Renewable Energy Storage *276*
12.6.4	Optimization of Power System Operation with Intermittent Generators *277*
12.7	Conclusions *278*
	References *279*

13 Status and Prospects of Alkaline Electrolysis *283*
Dongke Zhang and Kai Zeng

13.1	Introduction *283*
13.2	Thermodynamic Consideration *285*
13.2.1	Theoretical Cell Voltages *285*
13.3	Electrode Kinetics *287*
13.3.1	Hydrogen Generation Overpotential *288*
13.3.2	Oxygen Generation Overpotential *290*
13.3.3	Cell Overpotential *291*
13.4	Electrical and Transport Resistances *292*
13.4.1	Electrical Resistances *292*
13.4.2	Bubble Phenomena *293*
13.4.2.1	Bubble Departure Diameter Predictions *293*
13.4.2.2	Comparison of Model Predictions with Experimental Observations *296*
13.5	Research Trends *297*
13.5.1	Electrodes *297*
13.5.2	Electro-Catalysts *298*
13.5.3	Electrolyte and Additives *302*
13.5.4	Bubble Management *302*
13.6	Summary *303*
	References *304*

14 Dynamic Operation of Electrolyzers – Systems Design and Operating Strategies *309*
Geert Tjarks, Jürgen Mergel, and Detlef Stolten

14.1	Introduction *309*
14.2	Process Steps and System Components *310*
14.2.1	Alkaline Electrolysis *311*
14.2.2	PEM Electrolysis *312*
14.2.3	Gas Separation *313*
14.2.4	Compression *314*
14.2.5	Gas Drying *316*
14.2.6	Power Supply *317*
14.3	Dynamic Operation of Electrolyzers *317*
14.3.1	Operation Range *318*
14.3.2	Dynamic System Response *320*
14.3.3	Startup and Shutdown *322*
14.4	System Design Criterion *322*

14.4.1	System Efficiency	*322*
14.4.2	Investment Cost and Hydrogen Production Costs	*325*
14.4.3	Lifetime	*326*
14.4.4	Safety	*327*
14.5	Conclusion	*327*
	References	*328*

15 Stack Technology for PEM Electrolysis *331*
Jürgen Mergel, David L. Fritz, and Marcelo Carmo
15.1 Introduction to Electrolysis *331*
15.1.1 History of PEM Electrolysis *333*
15.1.2 Challenges Facing PEM Electrolysis *334*
15.2 General Principles of PEM Electrolysis *335*
15.2.1 State-of-the-Art *335*
15.2.2 Stack Design *338*
15.2.2.1 Catalyst Coated Membranes (CCMs) *338*
15.2.2.2 Porous Transport Layer (PTL) *340*
15.2.2.3 Separator Plates *343*
15.2.2.4 Stack Operation *345*
15.2.3 Future Trends in PEM Electrolysis *349*
15.2.3.1 Cost Reduction *350*
15.3 Summary *355*
References *356*

16 Reversible Solid Oxide Fuel Cell Technology for Hydrogen/Syngas and Power Production *359*
Nguyen Q. Minh
16.1 Introduction *359*
16.2 Reversible Solid Oxide Fuel Cell Overview *359*
16.2.1 Operating Principles *359*
16.2.2 Features *361*
16.3 Solid Oxide Fuel Cell Technology *366*
16.4 Solid Oxide Electrolysis Cell Technology *372*
16.5 Reversible Solid Oxide Fuel Cell Technology *379*
16.6 Summary *383*
References *383*

Part 2.4 H$_2$ from Biomass *391*

17 Assessment of Selected Concepts for Hydrogen Production Based on Biomass *393*
Franziska Müller-Langer, Konstantin Zech, Stefan Rönsch, Katja Oehmichen, Julia Michaelis, Simon Funke, and Elias Grasemann
17.1 Introduction *393*
17.2 Characteristics of Selected Hydrogen Concepts *394*

17.2.1 Concepts under Discussion *394*
17.2.2 Concepts for Further Assessment *398*
17.2.2.1 Concept A – Wood Chip Gasification *398*
17.2.2.2 Concept B – Biogas Reforming *398*
17.2.2.3 Concept for Hydrogen Distribution of Produced Hydrogen *399*
17.3 Concept Assessment of Technical Aspects *401*
17.3.1 Methodical Approach *401*
17.3.2 Overall Efficiencies *401*
17.4 Concept Assessment of Environmental Aspects *402*
17.4.1 Methodical Approach *403*
17.4.2 GHG Emissions *403*
17.4.3 Cumulated Non-renewable Energy Demand *405*
17.5 Concept Assessment of Economic Aspects *406*
17.5.1 Methodical Approach *406*
17.5.2 Total Capital Investments of Hydrogen Production *407*
17.5.3 Hydrogen Production Costs *408*
17.5.4 Hydrogen Distribution Costs *410*
17.6 Summary *411*
Acknowledgment *411*
References *412*

18 Hydrogen from Biomass – Production Process via Fermentation *417*
Balachandar G., Shantonu Roy, and Debabrata Das
18.1 Introduction *417*
18.1.1 Current Energy Scenario *417*
18.1.2 Importance and Applications of Hydrogen as a Fuel *419*
18.1.3 Conventional Hydrogen Production *419*
18.1.4 Biological Hydrogen Production *419*
18.1.4.1 Biochemistry behind Thermophilic Biohydrogen Production via Dark Fermentation *420*
18.1.4.2 Microbial Characteristics of Thermophilic Hydrogen Producing Bacteria *421*
18.2 Hydrogen Production from Biomass as Feedstock *422*
18.2.1 Agricultural Residues *423*
18.2.2 Municipal Solid Waste and Sewage *424*
18.2.3 Industrial Residues *424*
18.2.3.1 Distillery Industry Waste *425*
18.2.3.2 Food Industry Waste *425*
18.3 Reactor Configurations and Scale-Up Challenges *427*
18.3.1 Reactor Configurations *427*
18.3.2 Current Status of Technologies Available on Scale Up *429*
18.4 Economics and Barriers *430*
18.5 Future Prospects *431*
18.6 Conclusion *431*
Acknowledgment *432*
References *432*

Part 2.5 Hydrogen from Solar Radiation and Algae *439*

19 Photoelectrochemical Water Decomposition *441*
Sebastian Fiechter
19.1 Introduction *441*
19.2 Principles of Photoelectrochemical Water Splitting *442*
19.2.1 Photoelectrochemical Cells with a Single Photoelectrode *443*
19.2.2 Photoelectrochemical Cells with Two Photoelectrodes *446*
19.2.3 Electrocatalysts and Overvoltage *448*
19.3 Design of Water Splitting Devices *448*
19.4 Nano- and Microstructured Photoelectrodes *455*
19.5 Economic Aspects *457*
19.6 Concluding Remarks *457*
References *458*

20 Current Insights to Enhance Hydrogen Production by Photosynthetic Organisms *461*
Roshan Sharma Poudyal, Indira Tiwari, Mohammad Mahdi Najafpour, Dmitry A. Los, Robert Carpentier, Jian-Ren Shen, and Suleyman I. Allakhverdiev
20.1 Introduction *461*
20.2 Biological H_2 Production *463*
20.3 Physiology and Biochemistry of Algae and Cyanobacteria for H_2 Production *465*
20.4 Hydrogenase and Nitrogenase for H_2 Production *466*
20.4.1 Hydrogenase and H_2 Production *466*
20.4.2 Uptake Hydrogenase and Hydrogen Production *467*
20.4.3 Nitrogenase and H_2 Production *469*
20.5 Photosystems and H_2 Production *469*
20.6 Factors Affecting Hydrogen Production *470*
20.7 Designing the Photosynthetic H_2 Production *471*
20.8 Leaf and Solar H_2 Production *472*
20.9 Biofuel and Hydrogen Production by Other Organisms *473*
20.10 Available Methods to Enhance Photosynthetic Hydrogen Production *474*
20.10.1 Photolytic H_2 Production by Microorganisms *474*
20.10.2 Photosynthetic Bacterial H_2 Production *475*
20.10.3 Dark Fermentative H_2 Production *475*
20.10.4 Genetic Engineering to Enhance Hydrogen Production *476*
20.11 Application of Biohydrogen *477*
20.12 Conclusion and Future Prospectus *477*
Acknowledgments *478*
Abbreviations *478*
References *478*

Part 2.6 Gas Clean-up Technologies 489

21 PSA Technology for H_2 Separation 491
Carlos A. Grande
21.1   Introduction 491
21.2   Basics of PSA Technology 492
21.3   Selective Adsorbents; Commercial and New Materials 499
21.4   Improving the PSA Cycle 501
21.5   Summary 503
       Acknowledgments 504
       References 504

22 Hydrogen Separation with Polymeric Membranes 509
Torsten Brinkmann and Sergey Shishatskiy
22.1   History 509
22.2   Basics of Membrane Gas Separation 510
22.3   Hydrogen Separation and Fractionation by Gas Permeation 516
22.3.1   Hydrogen/Nitrogen Separation 517
22.3.2   Hydrogen/Carbon Monoxide Separation 517
22.3.3   Hydrogen/Hydrocarbon Separation 518
22.3.4   Hydrogen Separation and Fractionation Applications in Renewable Hydrogen Production 519
22.4   Membrane Materials and Modules 519
22.4.1   Membrane Classification 519
22.4.1.1 Morphological Classification 520
22.4.1.2 Geometrical Classification 523
22.4.2   Membrane Defect Curing 524
22.4.3   Membrane Module Classification 525
22.4.3.1 Hollow Fiber Modules 525
22.4.3.2 Flat Sheet Modules 526
22.4.4   Membrane Material Classification 528
22.4.4.1 Membranes Based on Rubbery Polymers 528
22.4.4.2 Glassy Polymer Based Membranes 530
22.5   Process Examples 531
22.5.1   Hydrogen Separation from Purge Streams in Ammonia Production 531
22.5.2   Hydrogen Separation from Hydrocracker Flash Gas 532
22.5.3   Carbon Dioxide Removal from Biomass Gasification Product Gas 534
22.6   Conclusions 535
         Nomenclature 536
         References 537

23	**Gas Clean-up for Fuel Cell Systems – Requirements & Technologies**	*543*

Matthias Gaderer, Stephan Herrmann, and Sebastian Fendt

23.1 Introduction *543*
23.2 Background *543*
23.3 Fuel and Pollutants *545*
23.3.1 Main Gas Components *545*
23.3.1.1 Hydrogen *546*
23.3.1.2 Carbon Monoxide *546*
23.3.1.3 Methane *546*
23.3.1.4 Carbon Dioxide *547*
23.3.1.5 Nitrogen *547*
23.3.1.6 Steam *547*
23.3.2 Trace Gas Components *547*
23.3.2.1 Tars *547*
23.3.2.2 Particulate Matter (Particles) *549*
23.3.2.3 Alkali Compounds *549*
23.3.2.4 Sulfur Compounds *549*
23.3.2.5 Nitrogen Compounds *550*
23.3.2.6 Halogen Compounds *550*
23.3.2.7 Siloxanes *550*
23.3.2.8 Other Potential Contaminants *550*
23.4 Pollutant Level Requirements *550*
23.5 Technologies to Remove Pollutants *551*
23.5.1 Cold Gas Clean-Up *552*
23.5.2 Hot Gas Clean-Up *552*
23.5.3 Particulate Matter *552*
23.5.4 Alkali Components *555*
23.5.5 Tars and Higher Hydrocarbons *555*
23.5.5.1 Thermal Cracking *556*
23.5.5.2 Catalytic Reduction *557*
23.5.6 Sulfur Components *558*
References *559*

Volume 2

Part 3 Hydrogen for Storage of Renewable Energy *563*

24	**Physics of Hydrogen**	*565*

Carsten Korte, Tabea Mandt, and Timm Bergholz

24.1 Introduction *565*
24.2 Molecular Hydrogen *565*
24.2.1 The H_2 Molecule *565*
24.2.1.1 Covalent Bonding and Molecular Orbitals *565*
24.2.1.2 Natural Isotopes *566*
24.2.1.3 Nuclear Spin-States, Ortho- and Para-H_2 *566*

24.2.2 Thermodynamic Properties 569
24.2.2.1 Pressure–Temperature Phase Diagram 569
24.2.2.2 Pressure–Volume Phase Diagram 571
24.2.2.3 Joule–Thomson Effect 572
24.2.3 Reaction Kinetics 574
24.2.3.1 Reaction with O_2 – Thermodynamics 574
24.2.3.2 Reaction with O_2 – Microscopic Mechanisms 574
24.2.3.3 Reaction with O_2 – Explosion Limits 577
24.2.4 Transport Kinetics 578
24.2.4.1 Thermal Conductivity 578
24.2.4.2 Diffusion in Gasses 579
24.2.4.3 Permeation and Diffusion in Polymers 580
24.2.4.4 Permeation and Diffusion in Metals 583
24.3 Hydrides 588
24.3.1 Classification and Properties of Hydrides 588
24.3.1.1 Ionic Hydrides 588
24.3.1.2 Covalent Hydrides 590
24.3.1.3 Complex Hydrides (Hydrido Complexes) 591
24.3.1.4 Interstitial (Metallic) Hydrides 592
24.3.2 Formation of Hydrides 593
24.3.2.1 Ionic and Covalent Hydrides 593
24.3.2.2 Interstitial (Metallic) Hydrides 594
24.3.3 Clathrates 597
References 598

25 Thermodynamics of Pressurized Gas Storage 601
Vanessa Tietze and Detlef Stolten
25.1 Introduction 601
25.2 Calculation of Thermodynamic State Variables 602
25.2.1 Ideal Gases 602
25.2.2 Real Gases 603
25.3 Comparison of Thermodynamic Properties 606
25.3.1 Compressibility Factor 606
25.3.2 Joule–Thomson Coefficient 607
25.3.3 Isentropic Exponent 609
25.4 Thermodynamic Analysis of Compression and Expansion Processes 610
25.4.1 Isothermal and Isentropic Compression 611
25.4.2 Isenthalpic and Isentropic Expansion 615
25.5 Thermodynamic Modeling of the Storage Process 617
25.5.1 Governing Equations 617
25.5.2 Heat Transfer Equations 619
25.6 Application Examples 620
25.6.1 Refueling of a Vehicle Storage Tank 620
25.6.2 Salt Cavern 622
25.7 Conclusion 624
References 625

26	**Geologic Storage of Hydrogen – Fundamentals, Processing, and Projects** *629*
	Axel Liebscher, Jürgen Wackerl, and Martin Streibel
26.1	Introduction *629*
26.2	Fundamental Aspects of Geological Hydrogen Storage *631*
26.2.1	Physicochemical Properties of Hydrogen and Hydrogen Mixtures *631*
26.2.2	Interaction between Hydrogen and Microbial Inventories *634*
26.2.3	General Types of Geological Storage Option *635*
26.2.4	Hydrogen Storage in Porous Rocks *636*
26.2.5	Hydrogen Storage in Caverns *640*
26.3	Process Engineering *642*
26.3.1	Gas Transport: Pipelines and Tubing *642*
26.3.2	Compressors *643*
26.3.3	Metering *644*
26.3.4	Controlling and Safety Components *644*
26.3.5	Geologic Storages: Survey *645*
26.3.6	Geologic Storages: Construction *646*
26.3.7	Geologic Storages: Initial Testing *647*
26.3.8	Geologic Storages: Operation *648*
26.4	Experiences from Storage Projects *649*
26.4.1	Hydrogen Storage Projects *650*
26.4.2	Pure Hydrogen Storage Projects *650*
26.4.2.1	Clemens Dome, Texas, USA *650*
26.4.2.2	Moss Bluff, Texas, USA *650*
26.4.2.3	Teesside Project, Yorkshire, UK *651*
26.4.3	Town Gas Storages *651*
26.4.3.1	Town Gas Storage at Ketzin, Germany *652*
26.4.3.2	Town Gas Storage at Lobodice, Czech Republic *652*
26.4.3.3	Town Gas Storage at Beynes, Ile de France, France *654*
26.5	Concluding Remarks *654*
	References *655*
27	**Bulk Storage Vessels for Compressed and Liquid Hydrogen** *659*
	Vanessa Tietze, Sebastian Luhr, and Detlef Stolten
27.1	Introduction *659*
27.2	Stationary Application Areas and Requirements *660*
27.3	Storage Parameters *661*
27.4	Compressed Hydrogen Storage *662*
27.4.1	Hydrogen Compression *662*
27.4.2	Hydrogen Pressure Vessels *663*
27.4.2.1	Conventional Small-Scale Bulk Storage *663*
27.4.2.2	Current and Potential Small- to Medium-Scale Bulk Storage *665*
27.4.2.3	New Ideas for Medium-Scale Bulk Storage *669*
27.5	Cryogenic Liquid Hydrogen Storage *670*
27.5.1	Hydrogen Liquefaction *670*

27.5.2	Liquid Hydrogen Storage Tanks *671*
27.6	Cost Estimates and Economic Targets *675*
27.7	Technical Assessment *678*
27.8	Conclusion *683*
	References *684*

28 Hydrogen Storage in Vehicles *691*
Jens Franzen, Steffen Maus, and Peter Potzel

28.1	Introduction: Requirements for Hydrogen Storage in Vehicles *691*
28.2	Advantages of Pressurized Storage over Other Storage Methods *693*
28.3	Design of a Tank System *695*
28.3.1	Flow Diagram and Description of the Components *695*
28.3.2	Container *697*
28.4	Specific Requirements for Compressed Gas Systems for Vehicles *699*
28.4.1	Legal and Normative Requirements *699*
28.4.2	Refueling *700*
28.5	Special Forms of Compressed Gas Storage *704*
28.5.1	Parallel Hydride and Compressed Gas Storage *704*
28.5.2	Metal Hydride-Filled Pressure Containers *705*
28.5.3	Conformable Containers *705*
28.6	Conclusion *707*
	References *707*

29 Cryo-compressed Hydrogen Storage *711*
Tobias Brunner and Oliver Kircher

29.1	Motivation for Cryo-compressed Hydrogen Vehicle Storage *711*
29.2	Thermodynamic Opportunities *714*
29.3	Refueling and Infrastructure Perspectives *717*
29.4	Design and Operating Principles *719*
29.4.1	System Design *720*
29.4.2	Operating Principles *722*
29.5	Validation Challenges of Cryo-compressed Hydrogen Vehicle Storage *725*
29.5.1	Validation Procedure *726*
29.5.2	Validation Challenges and Opportunities *729*
29.5.3	Hydrogen Safety Validation *730*
29.6	Summary *731*
	References *731*

30 Hydrogen Liquefaction *733*
Alexander Alekseev

30.1	Introduction *733*
30.2	History of Hydrogen Liquefaction *734*
30.3	Hydrogen Properties at Low Temperature *735*
30.3.1	Thermodynamic Properties *735*
30.3.2	Ortho and Para Modifications of Hydrogen *735*

30.3.2.1 Underlying Physics 735
30.3.2.2 Ortho-to-Para Conversion and Liquefaction of Hydrogen 737
30.3.2.3 Available Data 738
30.3.2.4 Some Useful Thermodynamics 738
30.4 Principles of Hydrogen Liquefaction 739
30.4.1 Power Requirements 739
30.4.2 General Principle 740
30.4.3 Simple Joule–Thomson Process with Nitrogen Precooling 742
30.4.3.1 Basic Process 742
30.4.3.2 Integration of Ortho-to-Para Conversion 744
30.4.4 Evolution of the Hydrogen Liquefaction Processes 745
30.4.5 Process Design of the Precooling Part 745
30.4.6 Precooling by Nitrogen Brayton Cycle 746
30.4.6.1 Process Description 746
30.4.6.2 Evaluation 748
30.4.7 Mixed Gas Refrigeration for Precooling Purposes 748
30.4.8 Final Cooling 749
30.5 Key Hardware Components 751
30.5.1 Compression 752
30.5.1.1 Impact of the Isentropic Exponent 752
30.5.1.2 Low Density 753
30.5.1.3 Screw Compressor 753
30.5.1.4 Reciprocating (Piston) Compressors 755
30.5.2 Expansion Turbine (or Expander or Turbine) 755
30.5.2.1 Oil Bearing 756
30.5.2.2 Gas Bearing 757
30.5.3 Heat Exchangers 758
30.6 Outlook 760
References 761

31 Hydrogen Storage by Reversible Metal Hydride Formation 763
Ping Chen, Etsuo Akiba, Shin-ichi Orimo, Andreas Zuettel, and Louis Schlapbach
31.1 Introduction 763
31.2 Summary of Energy Relevant Properties of Hydrogen and its Isotopes 764
31.3 Hydrogen Interaction with Metals, Alloys and Other Inorganic Solids 764
31.4 Hydrogen Storage in Intermetallic Compounds 767
31.4.1 AB_5 Type Compounds 771
31.4.2 AB_2 Type Hydrogen Absorbing Alloys 771
31.4.3 AB Type Alloy TiFe 772
31.4.4 Intermetallic Hydrides 773
31.5 Hydrogen Storage in Complex Hydrides 773
31.5.1 Alanates (tetrahydroaluminate) 774

31.5.2 Amides *775*
31.5.3 Borohydrides (Tetrahydroborate) *779*
31.6 Physisorption and High Open-Porosity Structures for Molecular Hydrogen Storage *781*
31.7 Other Energy Relevant Applications of Hydrogen Interacting Materials *784*
31.8 Conclusions and Outlook *785*
References *786*

32 Implementing Hydrogen Storage Based on Metal Hydrides *791*
R.K. Ahluwalia, J.-K. Peng, and T.Q. Hua
32.1 Introduction *791*
32.2 Material Requirements *792*
32.2.1 Operating Temperatures *792*
32.2.2 Material Thermodynamics *794*
32.2.3 Containment Tank *794*
32.2.4 Buffer Hydrogen *796*
32.2.5 Desorption Kinetics *797*
32.2.6 Sorption Kinetics *798*
32.2.7 Material Compaction and Heat Transfer *799*
32.3 Reverse Engineering: A Case Study *800*
32.3.1 MH Refueling: Temperature Profile and Conversion *802*
32.3.2 System Analysis Model *803*
32.3.3 Reference Targets *804*
32.3.4 Sensitivity Study *805*
32.4 Summary and Conclusions *807*
Acknowledgments *808*
References *808*

33 Transport and Storage of Hydrogen via Liquid Organic Hydrogen Carrier (LOHC) Systems *811*
Daniel Teichmann, Wolfgang Arlt, Eberhard Schlücker, and Peter Wasserscheid
33.1 Hydrogen Storage and Transport for Managing Unsteady Renewable Energy Production *811*
33.2 Liquid Organic Hydrogen Carrier (LOHC) Systems *814*
33.3 Development of LOHC-Based Energy Storage Systems *819*
33.4 Applications of LOHC-Based Energy Storage Systems *822*
33.4.1 LOHC Systems for the Storage of Renewable Energy Equivalents, in Particular for Decentralized Storage in Heat–Storage Coupling *823*
33.4.2 Energy Transport over Long Distances with LOHC *824*
33.4.2.1 Import of Solar Energy from Northern Africa *825*
33.4.2.2 Import of Renewable Energy from Iceland *826*
33.4.3 Mobile Applications *827*
33.5 Conclusions *828*
References *829*

Part 4 Traded Hydrogen *831*

34 Economics of Hydrogen for Transportation *833*
Akiteru Maruta
34.1 Introduction *833*
34.2 Hydrogen Transportation System *833*
34.2.1 FCEVs *833*
34.2.2 Hydrogen Infrastructure *835*
34.3 Economics of Hydrogen for Transportation *836*
34.3.1 Hydrogen Cost *836*
34.3.1.1 Two Approaches for Hydrogen Cost Calculation *836*
34.3.1.2 Bottom-Up Approach for Hydrogen Cost *836*
34.3.1.3 Top-Down Approach for Hydrogen Cost *839*
34.3.1.4 Total Cost of Ownership (TCO) Approach *841*
34.3.2 Economics of Social Cost and Benefits *841*
34.3.2.1 Social Costs *842*
34.3.2.2 Social Benefits *843*
34.4 Conclusion *845*
References *846*

35 Challenges and Opportunities of Hydrogen Delivery via Pipeline, Tube-Trailer, LIQUID Tanker and Methanation-Natural Gas Grid *849*
Krishna Reddi, Marianne Mintz, Amgad Elgowainy, and Erika Sutherland
35.1 Introduction *849*
35.2 Variation in Demand for Hydrogen *850*
35.3 Refueling Station Components and Layout *852*
35.3.1 Gaseous Hydrogen Refueling Station *852*
35.3.2 Cryo-Compressed and Gaseous Refueling Station with Liquid Delivery *854*
35.3.3 Refueling Station Challenges *854*
35.4 Distributed Production of Hydrogen *856*
35.5 Central or Semi-central Production of Hydrogen *857*
35.5.1 Gaseous Hydrogen Delivery *857*
35.5.1.1 Pipeline Delivery Pathway *857*
35.5.1.2 Tube-Trailer Delivery Pathway *860*
35.5.1.3 Challenges *863*
35.5.2 Liquid Hydrogen Delivery Pathway *864*
35.5.2.1 Components Layout *864*
35.5.2.2 Cost Estimates *864*
35.5.2.3 Challenges *865*
35.6 Power-to-Gas Mass Energy Solution (Methanation) *866*
35.6.1 Hydrogen Methanation Process *866*
35.6.2 Current Applications *867*
35.6.3 Challenges and Opportunities *869*
35.7 Outlook and Summary *870*
Note *871*
References *872*

36 Pipelines for Hydrogen Distribution *875*

Sabine Sievers and Dennis Krieg

- 36.1 Introduction *875*
- 36.2 Overview *875*
- 36.2.1 Pipelines in Comparison to Other Transportation Possibilities *875*
- 36.2.2 An Overview of Existing Hydrogen Pipelines *876*
- 36.2.3 Some Material Concerns *877*
- 36.3 Brief Summary of Pipeline Construction *879*
- 36.3.1 Planning and Approval *879*
- 36.3.2 Assessment of Preferred Pipeline Routes and Alternative Routes *880*
- 36.3.3 Project Execution, Construction and Commissioning *883*
- 36.4 Operation of an H_2 Pipeline *886*
- 36.4.1 The Control Center *886*
- 36.4.2 Operations *887*
- 36.5 Decommissioning/Dismantling/Reclassification *888*
- 36.6 Conclusion *888*
- References *889*

37 Refueling Station Layout *891*

Patrick Schnell

- 37.1 Introduction *891*
- 37.2 Basic Requirements for a Hydrogen Refueling Station *892*
- 37.2.1 Car Refueling *892*
- 37.2.2 Bus Refueling *894*
- 37.3 Technical Concepts for Hydrogen Filling Stations *895*
- 37.3.1 Hydrogen Production *895*
- 37.3.1.1 Hydrogen Production from Biomass *896*
- 37.3.1.2 Electrolysis *897*
- 37.3.1.3 Steam Reforming *900*
- 37.3.1.4 Byproduct Hydrogen *900*
- 37.3.1.5 Biological Hydrogen Production *900*
- 37.3.2 Hydrogen Delivery *900*
- 37.3.2.1 CGH_2 Delivery *901*
- 37.3.2.2 LH_2 Delivery *901*
- 37.3.2.3 Pipeline Delivery *901*
- 37.3.3 Major Components of Hydrogen Refueling Stations *902*
- 37.3.3.1 Production *902*
- 37.3.3.2 Hydrogen Storage *902*
- 37.3.3.3 Compressors *903*
- 37.3.3.4 Pre-cooling *904*
- 37.3.3.5 Dispenser *905*
- 37.3.3.6 Controls *905*
- 37.3.4 Integration of Hydrogen Refueling Stations *906*
- 37.3.5 Facility Size/Space Requirements *907*
- 37.4 Challenges *907*

37.4.1	Standardization	*907*
37.4.2	Costs	*909*
37.4.3	Reliability	*910*
37.4.4	Approval Processes	*910*
37.4.5	Gauged H_2-Metering	*911*
37.4.6	Refueling According to Technical Guidelines	*911*
37.4.7	Hydrogen Quality	*912*
37.5	Conclusion	*913*
	References	*914*

Part 5 Handling of Hydrogen *917*

38 Regulations and Codes and Standards for the Approval of Hydrogen Refueling Stations *919*
Reinhold Wurster

38.1	Introduction	*919*
38.1.1	Explanation of the term "Regulations, Codes and Standards (RCS)"	*919*
38.1.1.1	Regulations	*919*
38.1.1.2	Codes of Practice	*920*
38.1.1.3	Standards	*921*
38.1.1.4	Referencing of standards	*922*
38.1.1.5	Why Globally Harmonized Standards are Needed	*923*
38.1.2	General Requirements for the Approval of Hydrogen Refueling Stations (HRSs)	*923*
38.2	European Union and Germany	*924*
38.2.1	Europe	*924*
38.2.2	Germany	*928*
	References	*930*

39 Safe Handling of Hydrogen *933*
William Hoagland

39.1	Introduction	*933*
39.2	Hydrogen Safety and the Elements of Risk	*934*
39.2.1	Assessing Risk	*935*
39.2.2	Current Shortcomings of QRA for Hydrogen Safety	*935*
39.3	The Unique, Safety-Related Properties of Hydrogen	*937*
39.4	General Considerations for the Safe Handling of Hydrogen	*938*
39.4.1	Gaseous Hydrogen	*939*
39.4.2	Liquid Hydrogen	*939*
39.4.3	Handling Emergencies	*940*
39.5	Regulations, Codes, and Standards	*940*
39.6	International Collaborations to Prioritize Hydrogen Safety Research	*942*

39.6.1	Survey of Hydrogen Risk Assessment Methods, 2008	942
39.6.2	Knowledge Gaps White Paper, 2008	942
39.6.3	Comparative Risk Assessment Studies of Hydrogen and Hydrocarbon Fueling Stations, 2008	943
39.7	Current Directions in Hydrogen Safety Research [6]	943
39.7.1	Research on the Physical Behavior of Leaked or Leaking Hydrogen	943
39.7.2	Hydrogen Storage Systems Safety Research	944
39.7.3	Research that Supports Early Market Applications	945
39.7.4	Research on Risk Mitigation Measures	945
39.7.5	Simplified Tools to Assess and Mitigate Risk	947
39.8	Summary	947
	References	948
	Bibliography	948

Part 6 Existing and Emerging Systems 949

40 Hydrogen in Space Applications 951
Jérôme Lacapere

40.1	Liquid Hydrogen for Access to Space	951
40.1.1	Liquid Storage	951
40.1.2	Constraints Due to Liquid Hydrogen Use	952
40.1.3	Insulation	953
40.2	To Go Beyond GTO	954
40.2.1	Coasting Phase	954
40.2.2	Re-ignition Preparation Phase	955
40.3	Relevant Tests in Low Gravity Environment	958
40.4	In-Space Propulsion	960
40.5	Conclusion	961
	References	963

41 Transportation/Propulsion/Demonstration/Buses: The Design of the Fuel Cell Powertrain for Urban Transportation Applications (Daimler) 965
Wolfram Fleck

41.1	Introduction	965
41.2	Operational Environment	966
41.3	Requirements	967
41.3.1	Propulsion Power to Drive	967
41.3.2	Auxiliary Power Demand	970
41.3.3	Heating and Air Condition Power Demand	971
41.4	Design Solutions	973
41.4.1	NEBUS	973
41.4.1.1	Design Solution	973
41.4.2	CUTE and HyFleet: CUTE Program	975
41.4.3	Citaro FuelCELL-Hybrid	978

41.5	Test and Field Experience	*982*
41.5.1	NeBus	*982*
41.5.2	CTA, CMBC	*982*
41.5.3	CUTE and HyFleet:CUTE Program	*984*
41.5.4	NaBuz DEMO and CHIC	*984*
41.6	Future Outlook	*986*
41.6.1	Transit Bus Applications	*986*
41.6.2	Truck Applications	*987*
41.6.3	Life Time and Product Costs	*989*
41.6.4	Summary	*989*
	References	*990*

42 Hydrogen and Fuel Cells in Submarines *991*
Stefan Krummrich and Albert Hammerschmidt

42.1	Background	*991*
42.1.1	When it All Began . . .	*992*
42.2	The HDW Fuel Cell AIP System	*992*
42.3	PEM Fuel Cells for Submarines	*993*
42.3.1	Introduction	*993*
42.3.2	The Oxygen/Hydrogen Cell Design	*994*
42.3.2.1	Constructive Features/Cell Design of Siemens PEM Fuel Cell	*994*
42.3.2.2	Results from Fuel Cell Operation	*995*
42.3.3	Constructive Feature of Fuel Cell Module for Submarine Use	*995*
42.3.3.1	Preconditions	*995*
42.3.3.2	Cascaded Fuel Cell Stacks [2]	*996*
42.3.3.3	Pressure Cushion for Uniform Current Distribution [4]	*998*
42.3.3.4	Fuel Cell Module	*999*
42.3.3.5	Results from Fuel Cell Module Operation	*1000*
42.3.3.6	Safety Features of Submarine Fuel Cell Modules	*1001*
42.4	Hydrogen Storage	*1002*
42.5	The Usage of Pure Oxygen	*1004*
42.6	System Technology – Differences Between HDW Class 212A and Class 214 Submarines	*1005*
42.7	Safety Concept	*1006*
42.8	Developments for the Future – Methanol Reformer for Submarines	*1006*
42.8.1	System Configuration	*1006*
42.8.2	Challenges of the Methanol Reformer Development	*1007*
42.8.3	Hydrogen Purification Membranes	*1008*
42.8.4	High Pressure Catalytic Oxidation	*1008*
42.8.5	Integration on Board a Submarine	*1008*
42.9	Conclusion	*1009*
	References	*1010*

43	**Gas Turbines and Hydrogen** *1011*	
	Peter Griebel	
43.1	Introduction *1011*	
43.2	Combustion Fundamentals of Hydrogen relevant for Gas Turbines *1012*	
43.2.1	Ignition Delay *1014*	
43.2.2	Flame Speed *1016*	
43.2.3	Flame Temperature, Stability, and Emissions *1018*	
43.3	State-of-the-art Gas Turbine Technology for Hydrogen *1019*	
43.4	Research and Development Status, New Combustion Technologies *1022*	
43.4.1	New Combustion Technologies *1024*	
43.4.1.1	MILD Combustion or Flameless Oxidation *1024*	
43.4.1.2	Lean Direct Injection, Multi-injection, and Micro-mixing *1026*	
43.5	Concluding Remarks *1028*	
	References *1028*	

44	**Hydrogen Hybrid Power Plant in Prenzlau, Brandenburg** *1033*	
	Ulrich R. Fischer, Hans-Joachim Krautz, Michael Wenske,	
	Daniel Tannert, Perco Krüger, and Christian Ziems	
44.1	Introduction *1033*	
44.2	Description of the Concept of the Hybrid Power Plant at Prenzlau *1035*	
44.2.1	Overview *1035*	
44.2.2	Alkaline Electrolyzer *1036*	
44.2.3	Safety Engineering *1039*	
44.3	Operating Modes of the Hybrid Power Plant *1042*	
44.3.1	Hydrogen Production Mode *1042*	
44.3.2	Base Load Mode *1043*	
44.3.3	Forecast Mode *1044*	
44.3.4	EEX Mode *1045*	
44.4	Operational Management and Experiences *1045*	
44.4.1	Dynamic Load Operation *1045*	
44.4.2	Temperature Influence on Stack Voltage *1047*	
44.4.3	Influence of Activated Electrodes *1049*	
44.4.4	Voltage Efficiency of the Electrolyzer *1050*	
44.5	Outlook *1050*	
	References *1051*	

45	**Wind Energy and Hydrogen Integration Projects in Spain** *1053*	
	Luis Correas, Jesús Simón, and Milagros Rey	
45.1	Introduction *1053*	
45.2	The Role of Hydrogen in Wind Electricity Generation *1055*	
45.2.1	Mini Grids *1056*	
45.2.2	Electricity Storage *1056*	

45.2.3	Fuel Production	*1058*
45.2.4	Comparison of the Three Configurations	*1059*
45.3	Description of Wind–Hydrogen Projects	*1059*
45.3.1	RES2H2 Project	*1060*
45.3.2	Hidrólica Project	*1061*
45.3.3	ITHER Project	*1063*
45.3.4	Sotavento Project	*1064*
45.4	Operation Strategies Tested in the Sotavento Project	*1066*
45.4.1	Peaking Plant Strategy	*1067*
45.4.2	Strategy of Deviation Correction	*1069*
45.4.3	Strategy for Increasing the Capacity Factor of the Wind Farm	*1070*
45.4.4	Load Leveling Enabling Distributed Generation or Island	*1071*
45.5	Conclusions	*1071*
	References	*1072*
46	**Hydrogen Islands – Utilization of Renewable Energy for an Autonomous Power Supply** *1075*	
	Frano Barbir	
46.1	Introduction	*1075*
46.2	Existing Hydrogen Projects on Islands	*1077*
46.3	System Design/Configuration	*1082*
46.4	Key Technologies	*1083*
46.4.1	Electrolyzer	*1083*
46.4.2	Hydrogen Storage	*1085*
46.4.3	Fuel Cell	*1086*
46.5	System Issues	*1087*
46.5.1	Capacity Factor	*1087*
46.5.2	Coupling Efficiency	*1087*
46.5.3	Intermittent Operation	*1088*
46.5.4	Water Consumption	*1088*
46.5.5	Performance Degradation with Time	*1088*
46.6	Sizing	*1088*
46.7	Energy Management	*1090*
46.8	Other Uses/System Configurations	*1092*
46.8.1	Demand Side Management	*1092*
46.8.2	Seawater Desalination	*1092*
46.8.3	Oxygen Use	*1093*
46.8.4	Fuel for Vehicles	*1093*
46.9	Conclusions	*1093*
	References	*1094*

Index *1097*

List of Contributors

Christos Agrafiotis
Deutsches Zentrum für Luft- und
Raumfahrt e.V.
Linder Höhe
51147 Köln
Germany

Rajesh Ahluwalia
Argonne National Laboratory
Nuclear Engineering Division
9700 S. Cass Avenue
Lemont, IL 60439
USA

Abdullah M. Aitani
King Fahd University of Petroleum
& Minerals (KFUPM)
Center of Research Excellence in
Petroleum Refining &
Petrochemicals
University Blvd. 4597
34463 Dhahran
Saudi Arabia

Etsuo Akiba
Kyushu University
International Institute for
Carbon-Neutral Energy Research
(WPI-I2CNER)/Department of
Mechanical Engineering
Fukuoka City, Kyushu
Japan

Alexander Alekseev
Linde AG
Dr.-Carl-von-Linde-Str. 6–14
82049 Höllriegelskreuth
Germany

Suleyman I. Allakhverdiev
Institute of Plant Physiology
Russian Academy of Sciences
Botanicheskaya St. 35
Moscow 127276
Russia

and

Institute of Basic Biological
Problems
Russian Academy of Sciences
Pushchino
Moscow Region 142290
Russia

and

Department of Plant Physiology
Faculty of Biology
M.V. Lomonosov Moscow State
University
Leninskie Gory 1-12
Moscow 119991
Russia

Wolfgang Arlt
Lehrstuhl für Thermische
Verfahrenstechnik der Friedrich-
Alexander-Universität (FAU)
Erlangen-Nürnberg
Egerlandstrasse 3
91058 Erlangen
Germany

Florian Ausfelder
DECHEMA Gesellschaft für
Chemische Technik und
Biotechnologie e.V.
Theodor-Heuss-Allee 25
60486 Frankfurt am Main
Germany

Frano Barbir
University of Split
Faculty of Electrical Engineering
Mechanical Engineering and Naval
Architecture (FESB)
R. Boskovica 32
21000 Split
Croatia

Alexis Bazzanella
DECHEMA Gesellschaft für
Chemische Technik und
Biotechnologie e.V
Theodor-Heuss-Allee 25
60486 Frankfurt
Germany

Timm Bergholz
Forschungszentrum Jülich GmbH
Institut für Energie- und
Klimaforschung
Elektrochemische
Verfahrenstechnik (IEK-3)
Wilhelm-Johnen-Straße
52428 Jülich
Germany

Torsten Brinkmann
Helmholtz-Zentrum Geesthacht,
Centre for Materials and Coastal
Research
Insitute of Polymer Research
Verfahrenstechnik
Max-Planck-Straße 1
21502 Geesthacht
Germany

Tobias Brunner
BMW Group
Knorrstr.147
80788 München
Germany

Marcelo Carmo
Forschungszentrum Jülich GmbH
Institute of Energy and Climate
Research
IEK-3: Electrochemical Process
Engineering
Wilhelm-Johnen-Straße
52425 Jülich
Germany

Robert Carpentier
Groupe de Recherche en Biologie
Végétale, Université du Québec à
Trois-Rivières
C.P.500, Trois-Rivières
QC G9A 5H7
Canada

Ping Chen
Chinese Academy of Sciences
Dalian Institute of Chemical Physics
457 Zhongshan Road
Dalian 116023
China

Luis Correas
Fundación para el Desarrollo de las Nuevas Tecnologías del Hidrógeno en Aragón
Parque tecnológico Walqa
22197 Huesca
Spain

Debabrata Das
Indian Institute of Technology Kharagpur
Department of Biotechnology
Kharagpur 721302
India

Alessandro Delfrate
ThyssenKrupp Uhde Chlorine Engineers (Italia) S.r.l.
Via L. Bistolfi 35
20134 Milan
Italy

Amgad Elgowainy
Argonne National Laboratory
9700 S. Cass Avenue
Lemont, IL 60439
USA

Sebastian Fendt
Technical University Munich
Institute Energy Systems
Boltzmannstrasse 15
85748 Garching
Germany

Sebastian Fiechter
Helmholtz-Zentrum Berlin für Materialien und Energie GmbH
Institute for Solar Fuels
Hahn-Meitner-Platz 1
14109 Berlin
Germany

Ulrich R. Fischer
BTU Cottbus
H2-Forschungszentrum
03013 Cottbus
Germany

Wolfram Fleck
Daimler AG
73230 Kirchheim / Teck- Nabern
Germany

Jens Franzen
Daimler AG
HPC NAB
73230 Kirchheim unter Teck
Germany

David L. Fritz
Forschungszentrum Jülich GmbH
Institute of Energy and Climate Research
IEK-3: Electrochemical Process Engineering
Wilhelm-Johnen-Straße
52425 Jülich
Germany

Simon Funke
Fraunhofer Institute for Systems and Innovation Research ISI
Breslauer Str. 48
76139 Karlsruhe
Germany

Balachandar G.
Indian Institute of Technology Kharagpur
Department of Biotechnology
Kharagpur 721302
India

Matthias Gaderer
Technical University Munich
Institute Regenerative Energy Systems
Schulgasse 16
94315 Straubing
Germany

Carlos A. Grande
Senior Research Scientist
SINTEF Materials and Chemistry
Forskningsveien 1
0373 – Oslo
Norway

Elias Grasemann
Deutsches Biomasseforschungszentrum Gemeinnützige GmbH
Department Biorefineries
Torgauer Str. 116
04347 Leipzig
Germany

Peter Griebel
German Aerospace Center (DLR)
Institute of Combustion Technology
Pfaffenwaldring 38–40
70569 Stuttgart
Germany

Jürgen-Friedrich Hake
Forschungszentrum Jülich
Institute of Energy and Climate Research – Systems Analysis and Technology Evaluation (IEK-STE)
Wilhelm-Johnen-Straße
52425 Jülich
Germany

Albert Hammerschmidt
Siemens AG
Coburger Str. 47a
91056 Erlangen
Germany

Stephan Herrmann
Technical University Munich
Institute Energy Systems
Boltzmannstrasse 15
85748 Garching
Germany

William Hoagland
Element One, Inc.
7253 Siena Way
Boulder, CO 80301
USA

Thanh Hua
Argonne National Laboratory
Nuclear Engineering Division
9700 S. Cass Avenue
Lemont, IL 60439
USA

Christian Hulteberg
Lund University
Chemical Engineering
Box 124
22100 Lund
Sweden

Andreas Jess
Lehrstuhl für Chemische Verfahrenstechnik der Universität Bayreuth
Universitätsstrasse 30, 95440 Bayreuth
Germany

Oliver Kircher
BMW Group
Knorrstr.147
80788 München
Germany

Carsten Korte
Forschungszentrum Jülich GmbH
Institut für Energie- und
Klimaforschung
Elektrochemische
Verfahrenstechnik (IEK-3)
Wilhelm-Johnen-Straße
52428 Jülich
Germany

Hans-Joachim Krautz
BTU Cottbus
H2-Forschungszentrum
03013 Cottbus
Germany

Dennis Krieg
AIR LIQUIDE Advanced
Technologies GmbH
Hans-Güther-Sohl-Str. 5
40235 Düsseldorf
Germany

Perco Krüger
BTU Cottbus
H2-Forschungszentrum
03013 Cottbus
Germany

Stefan Krummrich
ThyssenKrupp Marine Systems
GmbH
Werftstraße 112–114
24143 Kiel
Germany

Andrea Kruse
Universität Hohenheim
Institute of Agricultural
Engineering (44U1)
Garbenstraße 9
70599 Stuttgart
Germany

Jérôme Lacapere
Absolut System
2 Rue des Murailles
38170 Seyssinet Pariset
France

Matthias Lange
Deutsches Zentrum für Luft- und
Raumfahrt e.V.
Linder Höhe
51147 Cologne
Germany

Axel Liebscher
GFZ German Research Centre for
Geosciences
Centre for Geological Storage
Telegrafenberg
14473 Potsdam
Germany

Dick Lieftink
HyGear
P.O. Box 5280
6802 EG Arnhem
Norway

Jochen Linssen
Forschungszentrum Jülich
Institute of Energy and Climate
Research – Systems Analysis and
Technology Evaluation (IEK-STE)
Wilhelm-Johnen-Straße
52425 Jülich
Germany

Dmitry A. Los
Institute of Plant Physiology
Russian Academy of Sciences
Botanicheskaya St. 35
Moscow 127276
Russia

Sebastian Luhr
Forschungszentrum Jülich GmbH
Institute für Elektrochemische
Verfahren
Wilhelm-Johnen-Straße
52425 Jülich
Germany

Tabea Mandt
Forschungszentrum Jülich GmbH
Institut für Energie- und
Klimaforschung
Elektrochemische
Verfahrenstechnik (IEK-3)
Wilhelm-Johnen-Straße
52428 Jülich
Germany

Akiteru Maruta
Technova Inc.
13th Floor Imperial Hotel Tower
Uchisaiwai-cho, Chiyoda-ku
Tokyo 100-0011
Japan

Steffen Maus
Daimler AG
HPC T332
70546 Stuttgart
Germany

Jürgen Mergel
Forschungszentrum Jülich GmbH
Institute of Energy and Climate
Research
IEK-3: Electrochemical Process
Engineering
Wilhelm-Johnen-Straße
52425 Jülich
Germany

Julia Michaelis
Fraunhofer Institute for Systems
and Innovation Research ISI
Breslauer Str. 48
76139 Karlsruhe
Germany

Nguyen Q. Minh
University of California, San Diego
Center for Energy Research
9500 Gilman Drive
La Jolla, CA 92093-0417
USA

Marianne Mintz
Argonne National Laboratory
9700 S. Cass Avenue
Lemont, IL 60439
USA

Nathalie Monnerie
Deutsches Zentrum für Luft- und
Raumfahrt e.V.
Linder Höhe
51147 Cologne
Germany

Franziska Müller-Langer
Deutsches
Biomasseforschungszentrum
Gemeinnützige GmbH
Department Biorefineries
Torgauer Str. 116
04347 Leipzig
Germany

Mohammad Mahdi Najafpour
Department of Chemistry
Institute for Advanced Studies in
Basic Sciences (IASBS)
Zanjan, 45137-66731
Iran

and

Center of Climate Change and
Global Warming
Institute for Advanced Studies in
Basic Sciences (IASBS)
Zanjan, 45137-66731
Iran

Katja Oehmichen
Deutsches
Biomasseforschungszentrum
Gemeinnützige GmbH
Department Bioenergy Systems
Torgauer Str. 116
04347 Leipzig
Germany

Shin-ichi Orimo
Tohoku University
WPI Advanced Institute for
Materials Research (WPI-AIMR)/
Institute for Materials Research
(IMR)
Sendai, Miyagi
Japan

Jui-Kun Peng
Argonne National Laboratory
Nuclear Engineering Division
9700 S. Cass Avenue
Lemont, IL 60439
USA

Ralf Peters
Forschungszentrum Jülich GmbH
Institute of Energy and Climate
Research
IEK-3: Energy Process Engineering
Wilhelm-Johnen-Straße
52428 Jülich
Germany

Peter Potzel
Daimler AG
HPC X456
71059 Sindelfingen
Germany

Roshan Sharma Poudyal
Department of Molecular Biology
Pusan National University
Busan 609-735
Republic of Korea

and

Central Department of Botany
Tribhuvan University
Kirtipir, Kathmandu
Nepal

Krishna Reddi
Argonne National Laboratory
9700 S. Cass Avenue
Lemont, IL 60439
USA

Milagros Rey
Gas Natural, S.A.
Plaza del Gas No. 1
08003 Barcelona
Spain

Martin Roeb
Deutsches Zentrum für Luft- und
Raumfahrt e.V.
Linder Höhe
51147 Cologne
Germany

Stefan Rönsch
Deutsches
Biomasseforschungszentrum
Gemeinnützige GmbH
Department Biorefineries
Torgauer Str. 116
04347 Leipzig
Germany

Shantonu Roy
Indian Institute of Technology
Kharagpur
Department of Biotechnology
Kharagpur 721302
India

Christian Sattler
Deutsches Zentrum für Luft- und
Raumfahrt e.V.
Linder Höhe
51147 Cologne
Germany

Ingrid Schjølberg
Norwegian University of Science
and Technology (NTNU)
IVT-IMT
7491 Trondheim
Norway

Louis Schlapbach
NIMS National Institute for
Materials Science
Tsukuba, Ibaraki
Japan

and

Kyushu University
WPI I2CNER
Fukuoka City, Kyushu
Japan

Eberhard Schlücker
Lehrstuhl für Prozessmaschinen
und Anlagenbau, Friedrich-
Alexander-Universität (FAU)
Erlangen-Nürnberg
Cauerstrasse 4
91058 Erlangen
Germany

Patrick Schnell
TOTAL Deutschland GmbH
Jean-Monnet-Straße 2
10557 Berlin
Deutschland
Germany

Jian-Ren Shen
Photosynthesis Research Center
Graduate School of Natural Science
and Technology/Faculty of Science
Okayama University
Okayama 700-8530
Japan

Sergey Shishatskiy
Helmholtz-Zentrum Geesthacht,
Zentrum für Material- und
Küstenforschung
Verfahrenstechnik
Max-Planck-Straße 1
21502 Geesthacht
Germany

Sabine Sievers
AIR LIQUIDE Deutschland GmbH
H2 Energy Deutschland
Hans-Güther-Sohl-Str. 5
40235 Düsseldorf
Germany

Jesús Simón
Fundación para el Desarrollo de las
Nuevas Tecnologías del Hidrógeno
en Aragón
Parque tecnológico Walqa
22197 Huesca
Spain

Filip Smeets
Hydrogenics Europe NV
Nijverhiedstraat 48c
2260 Oevel
Belgium

James G. Speight
CD&W Inc.
2476 Overland Road
Laramie, WY 82070
USA

Detlef Stolten
Forschungszentrum Jülich GmbH
Institute of Energy and Climate
Research
IEK-3: Electrochemical Process
Engineering
Wilhelm-Johnen-Straße
52425 Jülich
Germany

Martin Streibel
GFZ German Research Centre for
Geosciences
Centre for Geological Storage
Telegrafenberg
14473 Potsdam
Germany

Erika Sutherland
US Department of Energy
Fuel Cell Technologies Office
1000 Independence Avenue, SW
Washington, DC 20585
USA

Daniel Tannert
BTU Cottbus
H2-Forschungszentrum
03013 Cottbus
Germany

Daniel Teichmann
Hydrogenious Technologies GmbH
Weidenweg 13
91058 Erlangen
Germany

Vanessa Tietze
Forschungszentrum Jülich GmbH
Institute für Elektrochemische
Verfahren
Wilhelm-Johnen-Straße
52425 Jülich
Germany

Indira Tiwari
Department of Microbiology
Pusan National University
Busan 609-735
Republic of Korea

Geert Tjarks
Forschungszentrum Jülich GmbH
Institute of Energy and Climate
Research
IEK-3: Electrochemical Process
Engineering
Wilhelm-Johnen-Straße
52425 Jülich
Germany

Jan Vaes
Hydrogenics Europe NV
Nijverhiedstraat 48c
2260 Oevel
Belgium

Henrik von Storch
Deutsches Zentrum für Luft- und
Raumfahrt e.V.
Professor-Rehm-Straße 1
52428 Jülich
Germany

Jürgen Wackerl
Research Centre Jülich
Institute of Energy and Climate
Research (IEK)
52425 Jülich
Germany

Peter Wasserscheid
Lehrstuhl für Chemische
Reaktionstechnik der Friedrich-
Alexander-Universität (FAU)
Erlangen-Nürnberg, Egerlandstrasse
3, 91058 Erlangen
Germany

and

Forschungszentrum Jülich,
„Helmholtz-Institut Erlangen-
Nürnberg" (IEK-11),
Nägelsbachstrasse 49b, 91052
Erlangen
Germany

Michael Wenske
BTU Cottbus
H2-Forschungszentrum
03013 Cottbus
Germany

Reinhold Wurster
Ludwig-Bölkow-Systemtechnik
GmbH
Daimlerstr. 15
85521 Ottobrunn
Germany

Konstantin Zech
Deutsches
Biomasseforschungszentrum
Gemeinnützige GmbH
Department Biorefineries
Torgauer Str. 116
04347 Leipzig
Germany

Kai Zeng
The University of Western
Australia
UWA Centre for Energy
35 Stirling Highway
Crawley, WA 6009
Australia

Dongke Zhang
The University of Western
Australia
UWA Centre for Energy
35 Stirling Highway
Crawley, WA 6009
Australia

Christian Ziems
BTU Cottbus
H2-Forschungszentrum
03013 Cottbus
Germany

Andreas Zuettel
Empa, Swiss Federal Institute for
Material Science & Technology
Überlandstrasse 129
8600 Dübendorf
Switzerland

Part 1
Sol–Gel Chemistry and Methods

1
Hydrogen in Refineries
James G. Speight

1.1
Introduction

A critical issue facing the world's refiners today is the changing landscape in processing petroleum crude into refined transportation fuels under an environment of increasingly more stringent clean fuel regulations, decreasing heavy fuel oil demand, and increasing supply of heavy, sour crude. Hydrogen network optimization is at the forefront of world refineries options to address clean fuel trends, to meet growing transportation fuel demands and to continue to make a profit from their crudes [1]. A key element of a hydrogen network analysis in a refinery involves the capture of hydrogen in its fuel streams and extending its flexibility and processing options. Thus, innovative hydrogen network optimization will be a critical factor influencing future refinery operating flexibility and profitability in a shifting world of crude feedstock supplies and ultra-low-sulfur (ULS) gasoline and diesel fuel.

The chemical nature of the crude oil used as the refinery feedstock has always played the major role in determining the hydrogen requirements of that refinery. For example, the lighter, more paraffinic crude oils will require somewhat less hydrogen for upgrading to, say, a gasoline product than a heavier more asphaltic crude oil. It follows that the hydrodesulfurization of heavy oils and residua (which, by definition, is a hydrogen-dependent process) needs substantial amounts of hydrogen as part of the processing requirements.

In fact, the refining industry has been the subject of the four major forces that affect most industries and which have hastened the development of new petroleum refining processes: (i) the demand for products such as gasoline, diesel, fuel oil, and jet fuel, (ii) feedstock supply, specifically the changing quality of crude oil and geopolitics between different countries and the emergence of alternate feed supplies such as bitumen from tar sand (*oil sand*), natural gas, and coal, (iii) technology development such as new catalysts and processes, especially processes involving the use of hydrogen, and (iv) environmental regulations that include more stringent regulations in relation to sulfur in gasoline and diesel [2–6].

Hydrogen Science and Engineering: Materials, Processes, Systems and Technology, First Edition.
Edited by Detlef Stolten and Bernd Emonts.
© 2016 Wiley-VCH Verlag GmbH & Co. KGaA. Published 2016 by Wiley-VCH Verlag GmbH & Co. KGaA.

Categories (i), (ii), and (iv) are directly affected by the third category (i.e., the use of hydrogen in refineries) and it is this category that will be the subject of this chapter. This chapter presents an introduction to the use and need for hydrogen petroleum refineries in order for the reader to place the use of hydrogen in the correct context of the refinery. In fact, hydrogen is key in allowing refineries to comply with the latest product specifications and environmental requirements for fuel production being mandated by market and governments and helping to reduce the carbon footprint of refinery operations.

The history and evolution petroleum refining has been well described elsewhere [6,7] and there is little need to repeat that work here except to note that it is not the intent of this chapter to ignore the myriad of processes in modern refineries that do not use hydrogen but may be dependent upon hydrogenated products in one way or another.

1.2
Hydroprocesses

The use of hydrogen in thermal processes is perhaps the single most significant advance in refining technology during the twentieth century and is now an inclusion in most refineries (Figure 1.1). Hydrogenation processes for the conversion

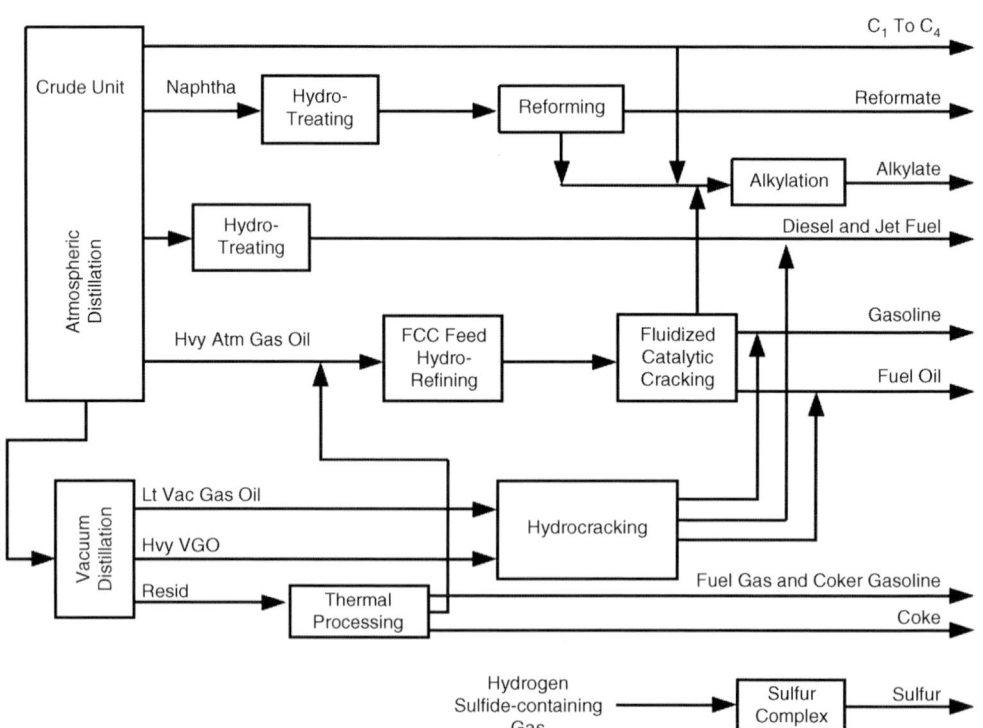

Figure 1.1 Example of the relative placement of hydroprocesses in a refinery.

of petroleum fractions and petroleum products may be classified as *destructive hydrogenation* and *nondestructive hydrogenation*.

Nondestructive hydrogenation (hydrotreating, simple hydrogenation) is generally used for the purpose of improving product quality without appreciable alteration of the boiling range. Mild processing conditions are employed so that only the more unstable materials are attacked. Nitrogen, sulfur, and oxygen compounds undergo reaction with the hydrogen to remove ammonia, hydrogen sulfide, and water, respectively. Unstable compounds that might lead to the formation of gums, or insoluble materials, are converted into more stable compounds.

On the other hand, destructive hydrogenation (hydrogenolysis, hydrocracking) is characterized by the conversion of feedstock higher molecular weight constituents into lower-boiling value-added products. Such treatment requires severe processing conditions and the use of high hydrogen pressures to minimize polymerization and condensation reactions that lead to coke formation.

The process uses the principle that the presence of hydrogen during a thermal reaction of a petroleum feedstock will terminate many of the coke-forming reactions and enhance the yields of the lower boiling components such as gasoline, kerosene, and jet fuel; processing parameters vary depending upon the character and properties of the feedstock (Tables 1.1 and 1.2).

Table 1.1 Summary of typical hydrogen applications and production processes in a refinery.

Hydrogen application	Hydrogen production process
Naphtha hydrotreater: uses hydrogen to desulfurize naphtha from atmospheric distillation; must hydrotreat the naphtha before sending to a catalytic reformer unit.	**Catalytic reformer:** used to convert the naphtha-boiling range molecules into higher octane reformate; hydrogen is a byproduct
Distillate hydrotreater: desulfurizes distillates after atmospheric or vacuum distillation; in some units aromatics are hydrogenated to cycloparaffins or alkanes	**Steam–methane reformer:** produces hydrogen for the hydrotreaters or hydrocracker
Hydrodesulfurization: sulfur compounds are hydrogenated to hydrogen sulfide as feed for Claus plants	**Steam reforming of higher molecular weight hydrocarbons:** produces hydrogen from low-boiling hydrocarbons other than methane
Hydroisomerization: normal (straight-chain) paraffins are converted into iso-paraffins to improve the product properties (e.g., octane number)	**Recovery from refinery off-gases:** process gas often contains hydrogen in the range up to 50% v/v
Hydrocracker: uses hydrogen to upgrade heavier fractions into lighter, more valuable products	**Gasification of petroleum residua:** recovery from synthesis gas (syngas) produced in gasification units
	Gasification of petroleum coke: recovery from synthesis gas (syngas) produced in gasification units
	Partial oxidation processes: analogous to gasification process; produce synthesis gas from which hydrogen can be isolated

Table 1.2 Hydroprocessing parameters.

Parameter			
Conditions	Feedstocks	Products	Variations
Solid acid catalyst (silica-alumina with rare earth metals, various other options)	Distillates	Lower molecular weight paraffins	Fixed bed (suitable for liquid feedstocks)
Temperature: 260–450 °C (500–845 °F) (solid/liquid contact)	Refractory (aromatic) streams	Some methane, ethane, propane, and butane	Ebullating bed (suitable for heavy feedstocks)
Pressure: 1000–6000 psi hydrogen	Coker oils	Hydrocarbon distillates (full range depending on the feedstock)	
Frequent catalysts renewal for heavier feedstocks	Cycle oils	Residual tar (recycle)	
Gas oil: catalyst life up to three years	Gas oils	Contaminants (asphaltic constituents) deposited on the catalyst as coke or metals	
Heavy oil/tar sand bitumen: catalyst life less than one year	Residua (as a full hydrocracking or hydrotreating option). In some cases, asphaltic constituents (S, N, and metals) removed by deasphalting		

1.2.1
Hydrotreating Processes

The commercial processes for treating, or finishing, petroleum fractions with hydrogen all operate in essentially the same manner. The feedstock is heated and passed with hydrogen gas through a tower or reactor filled with catalyst pellets. The reactor is maintained at a temperature of 260–425 °C (500–800 °F) at pressures of 100–1000 psi, depending on the particular process, the nature of the feedstock, and the degree of hydrogenation required. After leaving the reactor, excess hydrogen is separated from the treated product and recycled through the reactor after removal of hydrogen sulfide. The liquid product is passed into a stripping tower where steam removes dissolved hydrogen and hydrogen sulfide and, after cooling, the product is taken to product storage or, in the case of feedstock preparation, pumped to the next processing unit.

Thus, in a typical catalytic hydrodesulfurization unit, the feedstock is deaerated and mixed with hydrogen, preheated in a fired heater (315–425 °F; 600–800 °F), and then charged under pressure (up to 1000 psi) through a fixed-bed catalytic reactor. In the reactor, the sulfur and nitrogen compounds in the feedstock are converted into hydrogen sulfide and ammonia. The reaction products leave the reactor and after cooling to a low temperature enter a liquid/gas separator. The hydrogen-rich gas from the high-pressure separation is recycled to combine with the feedstock, and the low-pressure gas stream rich in hydrogen sulfide is sent to a gas treating unit where the hydrogen sulfide is removed. The clean gas is then suitable as fuel for the refinery furnaces. The liquid stream is the product from hydrotreating and is normally sent to a stripping column for removal of hydrogen sulfide and other undesirable components. In cases where steam is used for stripping, the product is sent to a vacuum drier for removal of water. Hydrodesulfurized products are blended or used as catalytic reforming feedstock.

Hydrofining is a process that first went on-stream in the 1950s and is one example of the many hydroprocesses available. It can be applied to lubricating oils, naphtha, and gas oils. The feedstock is heated in a furnace and passed with hydrogen through a reactor containing a suitable metal oxide catalyst, such as cobalt and molybdenum oxides on alumina. Reactor operating conditions range from 205 to 425 °C (400–800 °F) and from 50 to 800 psi, and depend on the kind of feedstock and the degree of treating required. Higher-boiling feedstocks, high sulfur content, and maximum sulfur removal require higher temperatures and pressures.

After passing through the reactor, the treated oil is cooled and separated from the excess hydrogen, which is recycled through the reactor. The treated oil is pumped to a stripper tower where hydrogen sulfide, formed by the hydrogenation reaction, is removed by steam, vacuum, or flue gas, and the finished product leaves the bottom of the stripper tower. The catalyst is not usually regenerated; it is replaced after about one year's use.

Distillate hydrotreating (Figure 1.2) is carried out by charging the feed to the reactor, together with hydrogen in the presence of catalysts such as tungsten-nickel sulfide, cobalt-molybdenum-alumina, nickel oxide-silica-alumina, and platinum-alumina. Most processes employ cobalt-molybdenum catalysts, which generally contain about 10% of molybdenum oxide and less than 1% of cobalt oxide supported on alumina. The temperatures employed are in the range 260–345 °C (500–655 °F), while the hydrogen pressures are about 500–1000 psi [8].

The reaction generally takes place in the vapor phase but, depending on the application, may be a mixed-phase reaction. Generally, it is more economical to hydrotreat high-sulfur feedstocks prior to catalytic cracking than to hydrotreat the products from catalytic cracking. The advantages are that (i) sulfur is removed from the catalytic cracking feedstock, and corrosion is reduced in the cracking unit, (ii) carbon formation during cracking is reduced so that higher conversions result, and (iii) the cracking quality of the gas oil fraction is improved.

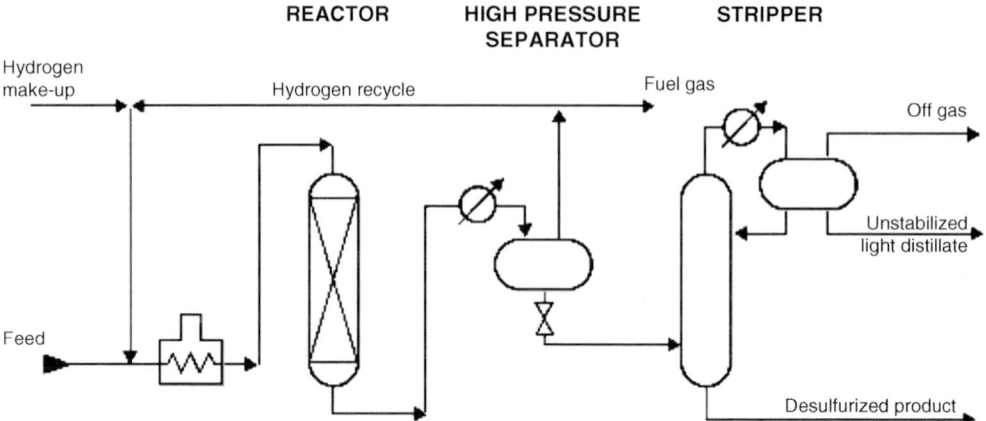

Figure 1.2 A distillate hydrotreater for hydrodesulfurization. Source: US Department of Labor, OSHA Technical Manual, Section IV, Chapter 2: Petroleum Refining Processes. http://www.osha.gov/dts/osta/otm/otm_iv/otm_iv_2.html.

Hydrotreating processes differ depending upon the feedstock available and catalysts used. Hydrotreating can be used to improve the burning characteristics of distillates such as kerosene. Hydrotreatment of a kerosene fraction can convert aromatics into naphthenes, which are cleaner-burning compounds. Lube-oil hydrotreating uses catalytic treatment of the oil with hydrogen to improve product quality. The objectives in mild lube hydrotreating include saturation of olefins and improvements in color, odor, and acid nature of the oil. Mild lubricating hydrotreating also may be used following solvent processing. Operating temperatures are usually below 315 °C (600 °F) and operating pressures below 800 psi. Severe lube hydrotreating, at temperatures in the 315–400 °C (600–750 °F) range and hydrogen pressures up to 3000 psi, is capable of saturating aromatic rings, along with sulfur and nitrogen removal, to impart specific properties not achieved at mild conditions.

Hydrotreating also can be employed to improve the quality of pyrolysis gasoline (*pygas*), a by-product from the manufacture of ethylene. Traditionally, the outlet for pygas has been motor gasoline blending, a suitable route in view of its high octane number. However, only small portions can be blended untreated owing to the unacceptable odor, color, and gum-forming tendencies of this material. The quality of pygas, which is high in di-olefin content, can be satisfactorily improved by hydrotreating, whereby conversion of di-olefins into mono-olefins provides an acceptable product for motor gas blending.

1.2.2
Hydrocracking Processes

Hydrocracking (Figure 1.3) is similar to catalytic cracking, with hydrogenation superimposed and with the reactions taking place either simultaneously or

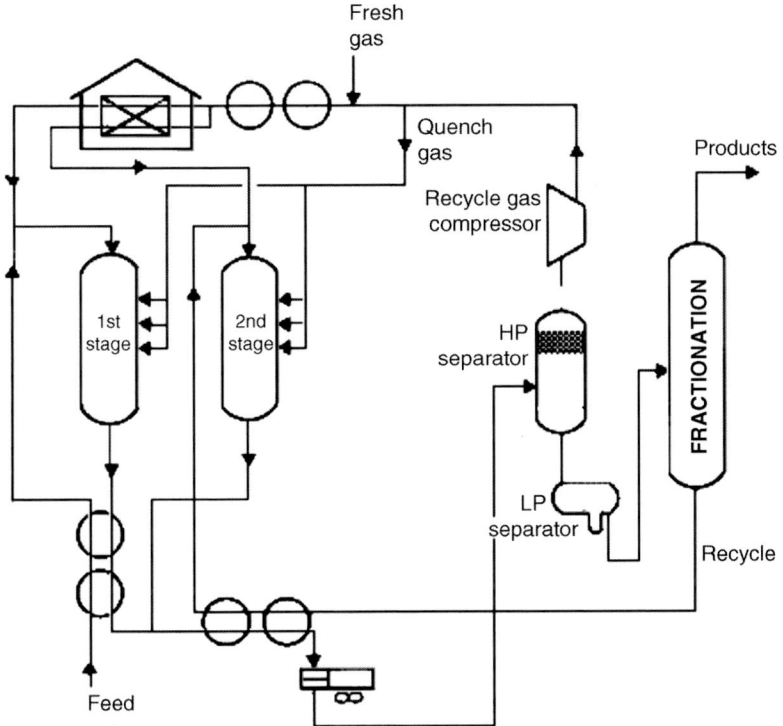

Figure 1.3 A single-stage or two-stage (optional) hydrocracking unit. Source: US Department of Labor, OSHA Technical Manual, Section IV, Chapter 2: Petroleum Refining Processes. http://www.osha.gov/dts/osta/otm/otm_iv/otm_iv_2.html.

sequentially [2–6]. Hydrocracking was initially used to upgrade low-value distillate feedstocks, such as cycle oils (high aromatic products from a catalytic cracker, which usually are not recycled to extinction for economic reasons), thermal and coker gas oils, and heavy-cracked and straight-run naphtha. These feedstocks are difficult to process by either catalytic cracking or reforming, since they are characterized usually by a high polycyclic aromatic content and/or by high concentrations of the two principal catalyst poisons – sulfur and nitrogen compounds.

Hydrocracking employs high pressure, high temperature, and a catalyst. Hydrocracking is used for feedstocks that are difficult to process by either catalytic cracking or reforming, since these feedstocks are characterized usually by high polycyclic aromatic content and/or high concentrations of the two principal catalyst poisons, sulfur and nitrogen compounds. The hydrocracking process largely depends on the nature of the feedstock and the relative rates of the two competing reactions, hydrogenation and cracking. Heavy aromatic feedstock is converted into lighter products under a wide range of very high pressures (1000–2000 psi) and fairly high temperatures (400–815 °C; 750–1500 °F), in the presence of hydrogen and special catalysts. When the feedstock has a high paraffinic content, the primary function of hydrogen is to prevent the formation of polycyclic aromatic compounds. Another important role of hydrogen in the hydrocracking process is to reduce tar formation and prevent buildup of coke on the catalyst. Hydrogenation also serves to convert sulfur and nitrogen compounds present in the feedstock into hydrogen sulfide and ammonia, respectively.

Typically, hydrocracking is a process with options: single stage hydrocracking or two-stage hydrocracking. In the first stage of the process (Figure 1.3), preheated feedstock is mixed with recycled hydrogen and sent to the first-stage reactor, where catalysts convert sulfur and nitrogen compounds into hydrogen sulfide and ammonia. Limited hydrocracking also occurs. After the hydrocarbon leaves the first stage, it is cooled and liquefied and run through a hydrocarbon separator. The hydrogen is recycled to the feedstock. The liquid is charged to a fractionator. Depending on the products desired (gasoline components, jet fuel, and gas oil), the fractionator is run to cut out some portion of the first stage reactor out-turn. Kerosene-range material can be taken as a separate side-draw product or included in the fractionator bottoms with the gas oil. The fractionator bottoms are again mixed with a hydrogen stream and charged to the second stage. Since this material has already been subjected to some hydrogenation, cracking, and reforming in the first stage, the operations of the second stage are more severe (higher temperatures and pressures). Like the outturn of the first stage, the second stage product is separated from the hydrogen and charged to the fractionator.

More often than not, especially with the influx of heavy feedstocks into refineries, hydrocracking is a two-stage process combining catalytic cracking and hydrogenation, wherein heavier feedstocks are cracked in the presence of hydrogen to produce more desirable products. Hydrocracking also produces relatively large amounts of iso-butane for alkylation feedstock and the process also performs isomerization for pour-point control and smoke-point control, both of which are important in high-quality jet fuel.

1.2.3
Slurry Hydrocracking

In terms of slurry hydrocracking processes, metals that have been screened as potential slurry catalysts and include transition metal-based catalysts derived from vanadium, tungsten, chromium, and iron. Homogeneous catalysts based hydrocracking technology has been developed for upgrading of heavy crude and tar sand bitumen [6,9]. In this process the hydrocracking catalyst is homogeneously dispersed as a colloid with particles similar in size to that of asphaltene molecule, which results in high conversion of asphaltene constituents [10,11].

1.2.4
Process Comparison

A comparison of hydrocracking with hydrotreating is useful in assessing the parts played by these two processes in refinery operations. Hydrotreating of distillates may be defined simply as the removal of nitrogen-, sulfur-, and oxygen-containing compounds by selective hydrogenation. The hydrotreating catalysts are usually cobalt plus molybdenum or nickel plus molybdenum (in the sulfide) form impregnated on an alumina base. The hydrotreating process conditions (1000–2000 psi hydrogen and approximately 370 °C, (700 °F)) are such that appreciable hydrogenation of aromatics will not occur. The desulfurization reactions are usually accompanied by small amounts of hydrogenation and hydrocracking.

Hydrocracking is an extremely versatile process that can be utilized in many different ways such as conversion of the high-boiling aromatic streams that are produced by catalytic cracking or by coking processes. To take full advantage of hydrocracking the process must be integrated in the refinery with other process units (Figure 1.1).

1.3
Refining Heavy Feedstocks

Over the past three decades, crude oils available to refineries have generally decreased in API (American Petroleum Institute) gravity [3,6,7]. There is, nevertheless, a major focus in refineries on the ways in which heavy feedstocks (such as heavy oil and tar sand bitumen) can be converted into low-boiling high-value products [2–7,12–18]. Simultaneously, the changing crude oil properties are reflected in changes such as an increase in asphaltene constituents, an increase in sulfur, metal, and nitrogen contents. Pretreatment processes for removing such constituents or at least negating their effect in thermal process would also play an important role.

The limitations of processing these heavy feedstocks depend to a large extent on the amount of higher molecular weight constituents (i.e., asphaltene constituents and resin constituents) that contain the majority of the heteroatom-containing compounds, which are responsible for high yields of thermal and catalytic coke [6]. Be that as it may, the essential step required of a modern refinery is the upgrading of heavy feedstocks, particularly atmospheric and vacuum residua.

Upgrading feedstocks such as heavy oils and residua began with the introduction of hydrodesulfurization processes [8,19]. In the early days, the goal was desulfurization but, in later years, the processes were adapted to a 10–30% partial conversion operation, intended to achieve desulfurization and obtain low-boiling fractions simultaneously, by increasing severity in operating conditions. However, as refineries have evolved and feedstocks have changed, refining heavy feedstocks has become a major issue in modern refinery practice and several process configurations have evolved to accommodate the heavy feedstocks [2,6,7,12].

For example, hydrodesulfurization of light (low-boiling) distillate (naphtha or kerosene) is one of the more common catalytic hydrodesulfurization processes since it is usually used as a pretreatment of such feedstocks prior to deep hydrodesulfurization or prior to catalytic reforming. A similar concept of pretreating residua prior to hydrocracking to improve the quality of the products is also practiced [6,7]. Hydrodesulfurization of such feedstocks, which is required because sulfur compounds poison the precious-metal catalysts used in the hydrocracking process, can be achieved under *relatively* mild conditions. If the feedstock arises from a cracking operation (such as cracked residua), hydro-pretreatment will be accompanied by some degree of saturation resulting in increased hydrogen consumption.

Finally, there is not one single heavy feedstock upgrading solution that will fit all refineries. Market conditions, existing refinery configuration, and available crude prices all can have a significant effect on the final configuration. Furthermore, a proper evaluation, however, is not a simple undertaking for an existing refinery. Evaluation starts with an accurate understanding of the market for the various products along with corresponding product values at various levels of supply. The next step is to select a set of crude oils that adequately cover the range of crude oils that may be expected to be processed. It is also important to consider new unit capital costs as well as incremental capital costs for revamp opportunities along with the incremental utility, support, and infrastructure costs. The costs, although estimated at the start, can be better assessed once the options have been defined, leading to the development of the optimal configuration for refining the incoming feedstocks.

1.4
Hydrogen Production

Hydrogen is generated in a refinery by the catalytic reforming process, but there may not always be the need to have a catalytic reformer as part of the refinery sequence. Nevertheless, assuming that a catalytic reformer is part of the refinery sequence, the hydrogen production from the reformer usually falls well below the amount required for hydroprocessing purposes. Consequently, an *external* source of hydrogen is necessary to meet the daily hydrogen requirements of any process where the heavier feedstocks are involved, which is accompanied by various energy changes and economic changes to the refinery balance sheet [20].

Thus, hydrogen production as a by-product is not always adequate to the needs of the refinery and other processes are necessary [6,7,21]. Thus, hydrogen production by steam reforming or by partial oxidation of residua has also been used, particularly where heavy oil is available. Steam reforming is the dominant method for hydrogen production and is usually combined with pressure-swing adsorption (PSA) to purify the hydrogen to greater than 99% v/v. However, the process parameters need to be carefully defined in order to optimize capital cost. An unnecessarily stringent specification in the hydrogen purity may cause undesired and unnecessary capital cost – an example is the residual concentration of nitrogen that should not be less than 100 ppm.

The most common, and perhaps the best, feedstocks for steam reforming are low boiling saturated hydrocarbons that have a low sulfur content, including natural gas, refinery gas, liquefied petroleum gas (LPG), and low-boiling naphtha.

Natural gas is the most common feedstock for hydrogen production since it meets all the requirements for steam–methane reformer feedstock. Natural gas typically contains more than 90% methane and ethane with only a small percentage of propane and higher boiling hydrocarbons [6,22]. Natural gas may (or most likely will) contain traces of carbon dioxide with some nitrogen and other impurities. Purification of natural gas, before reforming, is usually relatively straightforward. Traces of sulfur must be removed to avoid poisoning the reformer catalyst; zinc oxide treatment in combination with hydrogenation is usually adequate.

However, one of the key variables in operating the steam–methane reforming unit is maintaining the proper steam to carbon ratio, which can be difficult when natural gas containing various hydrocarbons [6,22] is used as the feedstock and if the feedstock is typically a mixture of refinery fuel gas and natural gas, the composition is not constant. If the steam to carbon ratio is too low, carbon will deposit on the catalyst, lowering the activity of the catalyst. In some cases, the catalyst can be completely destroyed, and the unit will need to be shut down to change the catalyst, thereby causing a disruption in the hydrogen supply. On the other hand, if the steam to carbon ratio is run too high, this wastes energy, increases steam consumption, and could also affect throughput.

Light *refinery gas*, containing a substantial amount of hydrogen, can be an attractive steam reformer feedstock since it is produced as a by-product. Processing of refinery gas will depend on its composition, particularly the levels of olefins and of propane and heavier hydrocarbons. Olefins, which can cause problems by forming coke in the reformer, are converted into saturated compounds in the hydrogenation unit. Higher boiling hydrocarbons in refinery gas can also form coke, either on the primary reformer catalyst or in the preheater. If there is more than a small percentage of C_3 and higher compounds, a promoted reformer catalyst should be considered, to avoid carbon deposits.

Refinery gas from different sources varies in suitability as hydrogen plant feed. Catalytic reformer off-gas, for example, is saturated, very low in sulfur, and often has high hydrogen content. The process gases from a coking unit or from a fluid catalytic cracking unit are much less desirable because of the content of unsaturated constituents. In addition to olefins, these gases contain substantial amounts of sulfur that must be removed before the gas is used as feedstock. These gases are also generally unsuitable for direct hydrogen recovery, since the hydrogen content is usually too low. Hydrotreater off-gas lies in the middle of the range. It is saturated, so it is readily used as hydrogen plant feed. The content of hydrogen and heavier hydrocarbons depends to a large extent on the upstream pressure. Sulfur removal will generally be required.

The gasification of petroleum residua, petroleum coke, and other feedstocks such as biomass [6,7,23,24] to produce hydrogen and/or power may become an attractive option for refiners. The premise that the gasification section of a refinery will be the *garbage can* for deasphalter residues, high-sulfur coke, as well as other refinery wastes is worthy of consideration. Other processes such as ammonia dissociation, steam–methanol interaction, or electrolysis are also available for hydrogen production, but economic factors and feedstock availability assist in the choice between processing alternatives.

Hydrogen is produced for use in other parts of the refinery as well as for energy and it is often produced from process by-products that may not be of any use elsewhere. Such by-products might be the highly aromatic, heteroatom, and metal containing reject from a deasphalting unit or from a mild hydrocracking process. However attractive this may seem, there will be the need to incorporate a gas cleaning operation to remove any environmentally objectionable components from the hydrogen gas.

When the hydrogen content of the refinery gas is greater than 50% v/v, the gas should first be considered for hydrogen recovery, using a membrane or pressure swing adsorption unit. The tail gas or reject gas that will still contain a substantial amount of hydrogen can then be used as steam reformer feedstock. Generally, the feedstock purification process uses three different refinery gas streams to produce hydrogen. First, high-pressure hydrocracker purge gas is purified in a membrane unit that produces hydrogen at medium pressure and is combined with medium pressure off-gas that is first purified in a pressure swing adsorption unit. Finally, low-pressure off-gas is compressed, mixed with reject gases from the membrane and pressure swing adsorption units, and used as steam reformer feed.

Various processes are available to purify the hydrogen stream but since the product streams are available as a wide variety of composition, flows, and pressures, the best method of purification will vary. And there are several factors that must also be taken into consideration in the selection of a purification method. These are: (i) hydrogen recovery, (ii) product purity, (iii) pressure profile, (iv) reliability, and (v) cost, an equally important parameter not considered here since the emphasis is on the technical aspects of the purification process.

1.5
Hydrogen Management

In general, considerable variation exists from one refinery to another in the balance between hydrogen produced and hydrogen consumed in the refining operations. However, what is more pertinent to the present chapter is the amounts of hydrogen that are required for hydroprocessing operations, whether these be hydrocracking or the somewhat milder hydrotreating processes. For effective hydroprocessing, a substantial hydrogen partial pressure must be maintained in the reactor and, in order to meet this requirement, an excess of hydrogen above that actually consumed by the process must be fed to the reactor. Part of the hydrogen requirement is met by recycling a stream of hydrogen-rich gas. However, the need still remains to generate hydrogen as makeup material to accommodate the process consumption of 500–3000 scf/bbl depending upon whether the heavy feedstock is being subjected to a predominantly hydrotreating (hydrodesulfurization) or to a predominantly hydrocracking process.

Many existing refinery hydrogen plants use a conventional process, which produces a medium-purity (94–97% v/v) hydrogen product by removing the carbon dioxide in an absorption system and methanation of any remaining carbon oxides. Since the 1980s, most hydrogen plants have been built with pressure swing adsorption (PSA) technology to recover and purify the hydrogen to purities above 99.9%. Since many refinery hydrogen plants utilize refinery off gas feeds containing hydrogen, the actual maximum hydrogen capacity that can be synthesized via steam reforming is not certain since the hydrogen content of the off-gas can change due to operational changes in the hydrotreaters.

Hydrogen management has become a priority in current refinery operations and when planning to produce lower sulfur gasoline and diesel fuels [25–28]. Along with increased hydrogen consumption for deeper hydrotreating, additional hydrogen is needed for processing heavier and higher sulfur crude slates. In many refineries, hydroprocessing capacity and the associated hydrogen network is limiting refinery throughput and operating margins. Furthermore, higher hydrogen purities within the refinery network are becoming more important to boost hydrotreater capacity, achieve product value improvements, and lengthen catalyst life cycles.

Improved hydrogen utilization and expanded or new sources for refinery hydrogen and hydrogen purity optimization are now required to meet the needs of the future transportation fuel market and the drive towards higher refinery profitability. Many refineries developing hydrogen management programs fit into the two general categories of either a catalytic reformer supplied network or an on-purpose hydrogen supply.

Multiple hydrotreating units compete for hydrogen – either by selectively reducing throughput, managing intermediate tankage logistics, or running the catalytic reformer sub-optimally just to satisfy downstream hydrogen requirements. Part of the operating year still runs in hydrogen surplus, and the network may be operated with relatively low hydrogen utilization (consumption/production) at 70–80%. Catalytic reformer off gas hydrogen supply may swing from 75% to 85% hydrogen purity. Hydrogen purity upgrade can be achieved through some hydrotreaters by absorbing heavy hydrocarbons. But without supplemental hydrogen purification, critical control of hydrogen partial pressure in hydroprocessing reactors is difficult, which can affect catalyst life, charge rates, and/or gasoline yields.

More complex refineries, especially those with hydrocracking units, also have on-purpose hydrogen production, typically with a steam–methane reformer that utilizes refinery off gas and supplemental natural gas as feedstock. The steam methane reformer plant provides the swing hydrogen requirements at higher purities (92% to more than 99% hydrogen) and serves a hydrogen network configured with several purity and pressure levels. Multiple purities and existing purification units allow for more optimized hydroprocessing operation by controlling hydrogen partial pressure for maximum benefit. Typical hydrogen utilization is 85–95%.

Over the past four decades, the refining industry has been challenged by changing feedstocks and product slate. In the near future, hydroprocessing options (especially for heavy feedstocks) will become increasingly flexible with improved technologies and improved catalysts. The increasing focus to reduce sulfur content in fuels will assure that the role of *desulfurization* in the refinery increases in importance [29]. Currently, the process of choice is the hydrotreater, in which hydrogen is added to the fuel to remove the sulfur from the fuel.

Because of the increased attention for fuel desulfurization various new process-concepts are being developed with various claims of efficiency and effectiveness. The major developments in three main routes to desulfurization will be (i) advanced hydrotreating (new catalysts, catalytic distillation, processing at mild conditions), (ii) reactive adsorption (type of adsorbent used, process design), and (iii) oxidative desulfurization (catalyst, process design).

However, residuum hydrotreating requires considerably different catalysts and process flows, depending on the specific operation so that efficient hydroconversion through uniform distribution of liquid, hydrogen-rich gas and catalyst across the reactor is assured. In addition to an increase in *guard bed* use [6,7] the industry will see an increase in automated demetallization of fixed-bed systems as well as more units that operate as ebullating-bed hydrocrackers.

For heavy oil upgrading, hydrotreating technology and hydrocracking technology will be the processes of choice. For cleaner transportation fuel production, the main task is the desulfurization of gasoline and diesel. With the advent of various techniques, such as adsorption and biodesulfurization, the future development will be still centralized on hydrodesulfurization techniques.

In fact, hydrocracking will continue to be an indispensable processing technology to modern petroleum refining and petrochemical industry due to its flexibility to feedstocks and product scheme, and high quality products. In particular, high quality naphtha, jet fuel, diesel, and lube base oil can be produced through this technology. The hydrocracker provides a better balance of gasoline and distillates, improves gasoline yield, octane quality, and can supplement the fluid catalytic cracker to upgrade heavy feedstocks. In the hydrocracker, light fuel oil is converted into lighter products under a high hydrogen pressure and over a hot catalyst bed – the main products are naphtha, jet fuel, and diesel oil.

In summary, the modern refinery faces the challenge of meeting an increasing demand for cleaner transportation fuels, as specifications continue to tighten around the world and markets decline for high-sulfur fuel oil. Innovative ideas and solutions to reduce refinery costs must always be considered, including: (i) optimization of the hydrogen management network, (ii) multiple feedstock options for hydrogen production, (iii) optimization of plant capacity, and last but certainly not least (iv) use of hydrogen recovery technologies to maximize hydrogen availability and minimize capital investment.

References

1 Long, R., Picioccio, K., and Zagoria, A. (2011) Optimizing hydrogen production and use. *Petrol. Tech. Q.*, Autumn, 1–12.
2 Speight, J.G. and Ozum, B. (2002) *Petroleum Refining Processes*, Marcel Dekker Inc., New York.
3 Speight, J.G. (2005) Natural bitumen (Tar Sands) and heavy oil, in *Coal, Oil Shale, Natural Bitumen, Heavy Oil and Peat, from Encyclopedia of Life Support Systems (EOLSS), Developed under the Auspices of the UNESCO*, EOLSS Publishers, Oxford, UK.
4 Hsu, C.S. and Robinson, P.R. (eds) (2006) *Practical Advances in Petroleum Processing*, Volumes **1** and **2**, Springer Science, New York.
5 Gary, J.H., Handwerk, G.E., and Kaiser, M.J. (2007) *Petroleum Refining: Technology and Economics*, 5th edn, CRC Press, Taylor & Francis Group, Boca Raton, FL.
6 Speight, J.G. (2014) *The Chemistry and Technology of Petroleum*, 5th edn, CRC Press, Taylor & Francis Group, Boca Raton, FL.
7 Speight, J.G. (2011) *The Refinery of the Future*, Gulf Professional Publishing, Elsevier, Oxford, UK.
8 Ancheyta, J. and Speight, J.G. (2007) *Hydroprocessing of Heavy Oils and Residua*, CRC Press, Taylor & Francis Group, Boca Raton, FL.
9 Kriz, J.F. and Ternan, M. (1994) Hydrocracking of heavy asphaltenic oil in the presence of an additive to prevent coke formation. US patent 5,296,130. March 22.
10 Bhattacharyya, A. and Mezza, B.J. (2010) Catalyst composition with nanometer crystallites for slurry hydrocracking. US patent 7,820,135, October 26.
11 Bhattacharyya, A., Bricker, M.L., Mezza, B.J., and Bauer, L.J. (2011) Process for using iron oxide and alumina catalyst with large particle diameter for slurry hydrocracking. US patent. 8,062,505, November 22.
12 Khan, M.R. and Patmore, D.J. (1997) Heavy oil upgrading processes, in *Petroleum Chemistry and Refining* (ed. J.G. Speight), Taylor & Francis, Washington, DC, ch 6.
13 Rana, M.S., Sámano, V., Ancheyta, J., and Diaz, J.A.I. (2007) A review of recent advances on process technologies for upgrading of heavy oils and residua. *Fuel*, **86**, 1216–1231.
14 Rispoli, G., Sanfilippo, D., and Amoroso, A. (2009) Advanced hydrocracking technology upgrades extra heavy oil. *Hydrocarb. Process.*, **88** (12), 39–46.
15 Stratiev, D. and Petkov, K. (2009) Residue upgrading: challenges and perspectives. *Hydrocarb. Process.*, **88** (9), 93–96.
16 Stratiev, D., Tzingov, T., Shishkova, I., and Dermatova, P. (2009) Hydrotreating units chemical hydrogen consumption analysis: a tool for improving refinery hydrogen management. Proceedings. 44th International Petroleum Conference, Bratislava, Slovak Republic. September 21–22.
17 Motaghi, M., Shree, K., and Krishnamurthy, S. (2010) Consider new methods for bottom of the barrel processing – Part 1. *Hydrocarb. Process.*, **89** (2), 35–40.
18 Motaghi, M., Shree, K., and Krishnamurthy, S. (2010) Consider new methods for bottom of the barrel processing – Part 2. *Hydrocarb. Process.*, **89** (2), 55–88.
19 Speight, J.G. (2000) *The Desulfurization of Heavy Oils and Residua*, 2nd edn, Marcel Dekker, Inc., New York.
20 Cruz, F.E. and De Oliveira, S. Jr, (2008) Petroleum refinery hydrogen production unit: exergy and production cost evaluation. *Int. J. Thermodynam.*, **11** (4), 187–193.
21 Lipman, T. (2011) An Overview of Hydrogen Production and Storage Systems with Renewable Hydrogen Case Studies. A Clean Energy States Alliance Report. Conducted under US DOE Grant DE-FC3608GO18111, Office of Energy Efficiency and Renewable Energy Fuel Cell Technologies Program, United States Department of Energy, Washington, DC.
22 Mokhatab, S., Poe, W.A., and Speight, J.G. (2006) *Handbook of Natural Gas Transmission and Processing*, Elsevier, Amsterdam.

23 Speight, J.G. (2008) *Synthetic Fuels Handbook: Properties, Processes, and Performance*, McGraw-Hill, New York.
24 Speight, J.G. (ed) (2011) *The Biofuels Handbook*, Royal Society of Chemistry, Cambridge, p. 2011.
25 Zagoria, A., Huychke, R., and Boulter, P.H. (2003) Refinery hydrogen management – the big picture, in *Proceedings. DGMK Conference on Innovation in the Manufacture and Use of Hydrogen. Dresden, Germany. October 15–17* (ed. G. Emig), DGMK, p. 95.
26 Davis, R.A. and Patel, N.M. (2004) Refinery hydrogen management. *Petrol. Tech. Q.*, (Spring), 28–35.
27 Méndez, C.A., Gómez, E., Sarabia, D., Cerdá, J., De Prada, C., Sola, J.M., and Unzueta, E. (2008) Optimal management of hydrogen supply and consumption networks of refinery operations, in *Proceedings. 18th European Symposium on Computer Aided Process Engineering – ESCAPE 18* (eds B. Braunschweig and X. Joulia), Elsevier, Amsterdam.
28 Luckwal, K. and Mandal, K.K. (2009) Improve hydrogen management of your refinery. *Hydrocarb. Process.*, **88** (2), 55–61.
29 Babich, I.V. and Moulijn, J.A. (2003) Science and technology of novel processes for deep desulfurization of oil refinery streams: a review. *Fuel*, **82** (2003), 607–631.

2
Hydrogen in the Chemical Industry

Florian Ausfelder and Alexis Bazzanella

2.1
Introduction

The chemical industry is the world's largest producer and consumer of hydrogen. Hydrogen fulfills various roles along the chemical value chain. First and foremost, hydrogen is an integral part of many of the largest volume chemicals, both of organic and inorganic nature. It can be supplied as a reactant either in molecular form or bound within a molecule to produce the desired products. Arguably the worldwide most important process involving hydrogen is the production of ammonia for the production of nitrogen-based fertilizers.

Hydrogen can also be a co-produced stream of a chemical process. Within the chemical industry, the chlorine alkaline process to produce chlorine co-produces stoichiometrically large amounts of hydrogen of very high purity. Side streams with high concentration of hydrogen in combination with other unwanted compounds can be used as fuel to supply process heat, if separation and purification are either impractical or too expensive.

Molecular hydrogen is used as a reduction agent, both within the chemical industry and in other industrial sectors, such as metals, to reduce the precursor molecule, alloy, or ore into the desired product. This reducing function also serves to remove undesired compounds from mixtures, for example in the removal of sulfur-containing compounds or in the cleaning of semi-conductor surfaces. Within the chemical industry, hydrogen is also widely used for catalyst regeneration.

Hydrogen can be a safety hazard if not handled properly. It is therefore not surprising to find hydrogen generation and consumption in most cases taking place on the same industrial site. This, in turn, gives rise to a rather high uncertainty in the overall global production numbers for hydrogen. The overall hydrogen market is estimated to range from about 45 million metric tons [1] to 50 million metric tons [2] or even 65 million metric tons [3]. Data from the international fertilizer association report the global production of ammonia to amount 166 million metric tons in 2012 [4], suggesting a stoichiometric hydrogen production of 29 million metric tons solely for the production of ammonia.

Hydrogen Science and Engineering: Materials, Processes, Systems and Technology, First Edition.
Edited by Detlef Stolten and Bernd Emonts.
© 2016 Wiley-VCH Verlag GmbH & Co. KGaA. Published 2016 by Wiley-VCH Verlag GmbH & Co. KGaA.

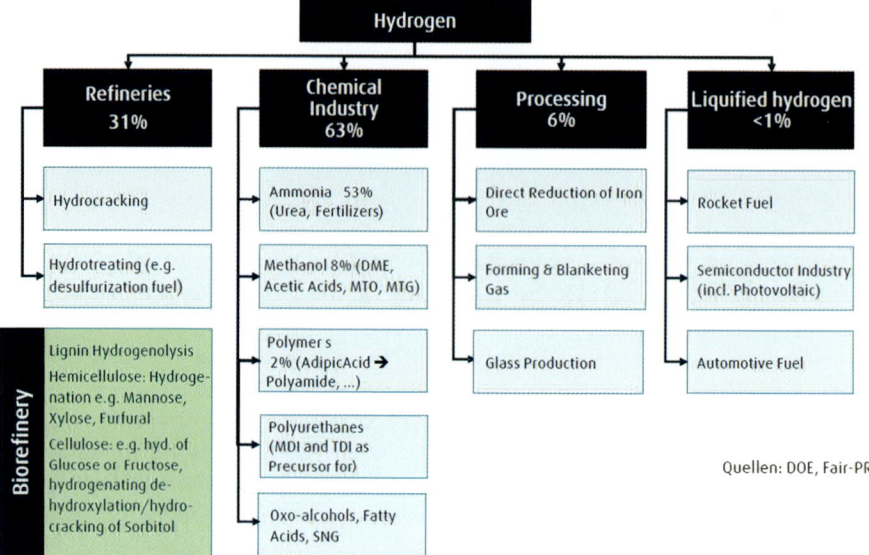

Figure 2.1 Hydrogen utilization across different sectors. Diagram by N. Schödel, Linde AG, used with kind permission.

Figure 2.1 outlines the different applications for hydrogen in different sectors. Two main sectors, refineries and chemical industry, define the hydrogen market, with ammonia being the dominant product. Due to the dominance of ammonia in the overall consumption of hydrogen, it can be used as an indication of worldwide hydrogen production in the future. Ammonia in turn is used to produce nitrogen-based fertilizers, which are linked to population growth and agricultural land use and intensification.

Due to the scarcity of global data, the following discussion is based on European data collected by Eurostat, the European statistics office. Eurostat requests yearly production and trade data from the EU member states for a large variety of products and chemicals, among them different gases such as hydrogen. The data is published in publically available databases [5].

Most European hydrogen production within the chemical industry is used directly for the production of ammonia, a small amount for the production of methanol, and some to supply other processes. Only the latter is reported as hydrogen production to Eurostat. Out of this production, about 750 kt hydrogen are traded as commodity, which amounts to an estimated market size of €1.6 billion. There are slight discrepancies between the data reported by Eurostat [5] on ammonia production and those reported by the international fertilizer association (IFA) [4]. For the following discussions, the IFA values are used for ammonia.

Currently, European chemical industry hydrogen production amounts to 4.8 million tons and is dominated by its use in ammonia production, which takes up around 64% of the overall hydrogen production (based on stoichiometric

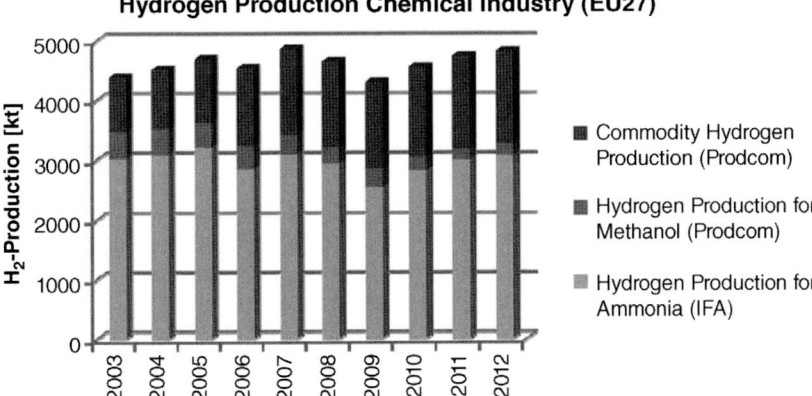

Figure 2.2 Hydrogen production by the European chemical industry. Estimated from data by IFA [4] and Eurostat [5].

calculations on data reported in Reference [5]). Past hydrogen production within the chemical industry of the EU27 member states is displayed in Figure 2.2.

In the future, see Figure 2.3, European ammonia production is expected to moderately decline. In contrast, overall hydrogen production is expected to stay roughly at present levels, due to the increase of commodity hydrogen production for uses other than ammonia or methanol, which is expected to grow in line with the overall growth of the chemical industry in Europe. These estimates, however, have to be treated with caution as they do not assume any significant role of hydrogen in the energy sector.

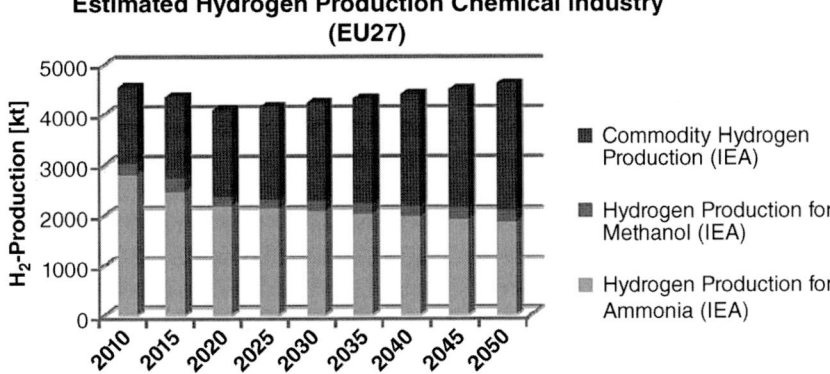

Figure 2.3 Expected development of the European hydrogen production within the timeframe of 2010 to 2050. Ammonia and methanol production data based on IEA-ICCA-DECHEMA catalysis roadmap data [6]. Free hydrogen production scaled to overall growth of chemical production in Europe.

2.2
Sources of Hydrogen in the Chemical Industry

2.2.1
Synthesis Gas-Based Processes

The vast majority, around 96%, of the overall hydrogen production is based on fossil fuels [1]. Refineries cover the main part of their hydrogen requirements via internal refinery processes, especially catalytic reforming. Additional hydrogen requirements are met by gasification of heavy residues. Further details are discussed in Section 1.2. Hydrogen-rich off-gases are also produced as a result of coking processes, but are generally burned to supply process heat.

The chemical industry generally requires significant amounts of on-purpose generated hydrogen, mainly for the production of ammonia and methanol. Hydrogen is generally produced via synthesis gas-based processes that can, in principle, be applied to any carbon-containing material, including waste or biomass. A choice for a specific technology depends on availability of the feedstock and the nature of downstream processes. A generalized reaction scheme is given in reaction (2.1):

$$C_xH_y + xH_2O \rightarrow xCO + \left(x + \frac{y}{2}\right)H_2 \qquad (2.1)$$

Depending on the feedstock's H:C ratio, the synthesis gas reaction translates the reactants into a certain H_2:CO product ratio, which may or may not match the requirements of the downstream processes. However, partial application of the water–gas shift reaction (Reaction (2.2)) adjusts the overall H_2:CO ratio by conversion of CO into and steam into CO_2 and additional H_2:

$$xCO + xH_2O \Leftrightarrow xCO_2 + H_2 \qquad (2.2)$$

Overall, the combination of both process steps, for maximum hydrogen production, results in the generalized reaction (Reaction (2.3)):

$$C_xH_y + xH_2O \rightarrow xCO + \left(x + \frac{y}{2}\right)H_2$$

$$xCO + xH_2O \Leftrightarrow xCO_2 + H_2$$

$$C_xH_y + 2xH_2O \rightarrow xCO_2 + \left(2x + \frac{y}{2}\right)H_2 \qquad (2.3)$$

The subsequent utilization of the generated synthesis gas and therefore the required H_2:CO ratio in combination with the feedstock for synthesis gas generation plays the dominant role in the design and implementation of any given synthesis gas-based process.

The production of urea, that is, ammonia synthesis followed by reaction with CO_2, reaction (2.4), is an extreme case of synthesis gas based processes. Since no

CO but instead CO_2 is required in the downstream process steps (Reaction (2.3)), reaction (2.3) can be driven towards maximum hydrogen yield.

$$2NH_3 + CO_2 \rightarrow (NH_2)_2CO + H_2O \tag{2.4}$$

In contrast, the production of methanol via synthesis gas requires formally a $H_2 : CO$ ratio of $2 : 1$ (Reaction (2.5)) and in its technical implementation an even slightly higher $H_2 : CO$ ratio:

$$2H_2 + CO \rightarrow H_3COH \tag{2.5}$$

The production of higher hydrocarbons via Fischer–Tropsch reactions (Reaction (2.6)) also requires formally a $H_2 : CO$ ratio of $2 : 1$:

$$2xH_2 + xCO \rightarrow (H_2C)_x + xH_2O \tag{2.6}$$

2.2.2 Steam Reforming

Steam reforming of natural gas is by far the most dominant process employed to produce hydrogen in the chemical industry. This process, using natural gas as feedstock, produces the highest $H_2 : CO$ ratio of all synthesis gas processes. Natural gas contains a mix of light hydrocarbons, with methane being the dominant one, and other impurities.

Figure 2.4 shows a generalized process scheme for steam reforming.

The incoming feedstock is conditioned for the steam reforming process. In the case of natural gas, it is beneficial to remove sulfur directly at the beginning of the process chain. Within the desulfurization step sulfur is converted into H_2S, adsorbed, and removed.

The actual steam reforming takes place in heated tubes, where the gas mix of high-temperature steam and desulfurized feed is exposed to the catalyst at high pressures. The amount of steam is adjusted to produce subsequently the $H_2 : CO$ ratio required. The catalysts are generally nickel-based. Typical temperature ranges of 750–900 °C and pressure ranges of 3–25 bar are applied, depending on the individual process design. The reforming reaction (reaction (2.1) for the general reaction, reaction (2.7) for the reaction of methane) converts hydrocarbons and steam into a synthesis gas consisting of a mix of CO and H_2:

$$CH_4 + H_2O \Leftrightarrow CO + 3H_2 \tag{2.7}$$

The reforming reaction is endothermic and requires significant amounts of heat supplied via the high-temperature steam as well as external burning of natural gas and off-gases from the process after hydrogen separation. Efficient heat recovery in the overall process is thus critical for economic operation of the plant.

The resulting synthesis gas leaving the reformer unit is then subjected to a shift-conversion reaction (water–gas shift) with additional steam to adjust the $H_2 : CO$ ratio (Reaction (2.2)):

$$xCO + xH_2O \Leftrightarrow xCO_2 + H_2 \tag{2.2}$$

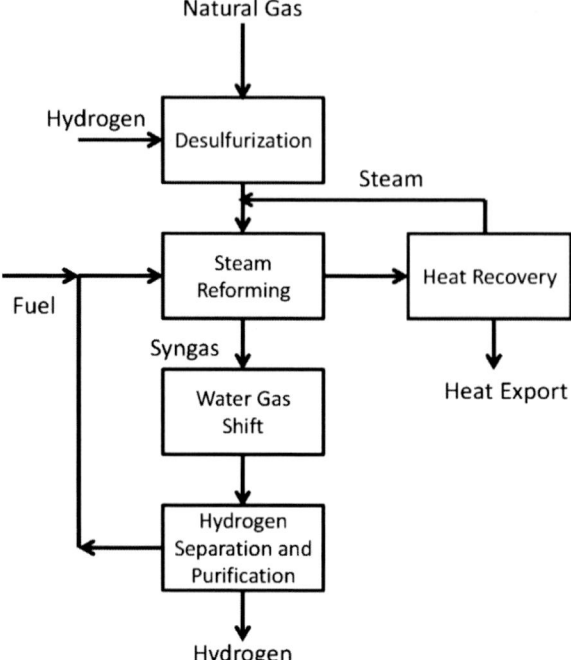

Figure 2.4 Generalized process scheme for hydrogen production via steam reforming.

However, CO_2 also reacts with methane, similarly to steam, in so-called dry reforming, resulting in syngas with a $H_2:CO$ ration of 1. (Reaction (2.8)):

$$CH_4 + CO_2 \Leftrightarrow 2CO + 2H_2 \tag{2.8}$$

These reactions are equilibrium driven and the resulting gas mixture depends on the operational conditions of the plant. The syngas is then subjected to separation process steps depending on its subsequent use.

In the case of hydrogen production (e.g., for downstream ammonia synthesis), CO_2 and water are trapped and hydrogen is separated and purified from the remaining gases by pressure swing adsorption (PSA), membrane processes and/or methanation of left-over CO. Off-gases can also be used as fuel to supply some of the heat required in the reforming stage.

In the case of methane, steam reforming can stoichiometrically produce up to four units of hydrogen gas for each unit of methane, which forms the theoretical upper limit for hydrogen production based on natural gas. These attributes make steam reforming of natural gas a uniquely well-suited synthesis gas process for the production of hydrogen in combination with subsequent ammonia and urea synthesis, for which captured carbon dioxide from the water–gas shift reaction is used.

While natural gas is the most commonly used feedstock for steam reforming it can also be applied to refinery off-gases, LPG (liquefied pressurized gas), and naphtha as feedstocks. Steam reforming is a widely accessible technology and all major chemical engineering companies supply turn-key solutions to their clients with various extra features to improve gas quality and reduce energy usage of the plants.

2.2.3
Process Variations

2.2.3.1 Partial Oxidation
Non-catalytic partial oxidation (thermal partial oxidation) of the feedstock (Reaction (2.9)) is a reforming reaction, where the feedstock reacts with an under-stoichiometric amount of oxygen, resulting in partial combustion of the feedstock. It directly supplies the process heat to the subsequent reforming reaction. Pure oxygen as additional feedstock is available for example from air separation units (ASUs) that, for example, supply nitrogen for ammonia plants. The reaction is carried out at high temperatures beyond 1200 °C:

$$CH_4 + \tfrac{3}{2}O_2 \Leftrightarrow CO + 2H_2O \tag{2.9}$$

If carried out as a catalytic reaction (Reaction (2.10)), sulfur-free feed is required and the reaction proceeds at 800–900 °C. It produces a carbon-free synthesis gas with a $H_2:CO$ ratio of 2 : 1:

$$2CH_4 + O_2 \Leftrightarrow 2CO + 4H_2 \tag{2.10}$$

Partial oxidation allows for a more compact process design and removes the requirement of external heating. The resulting CO is then converted with steam via the shift-reaction (Reaction (2.2)) into CO_2 and H_2. Partial oxidation can be carried out non-catalytically via combustion through a burner, or catalytically via a catalytic bed without the requirement of a flame.

2.2.3.2 Autothermal Reforming
Autothermal reforming is a combination of partial oxidation and classical steam reforming, optimized to benefit from the advantages of both technologies [7]. Oxygen and steam are used to react with the feed (Reactions (2.7) and (2.10) for methane). The overall reaction is exothermic. Process variations based on air have been proposed due to oxygen generation facilities making up to 40% of the total cost of the plant [8]. In principle, air could also be used instead of pure oxygen. However, nitrogen as the dominant gas in air, while non-reactive under the process conditions, would be carried through the process and significantly increases the unit volumes and costs of plant operation.

2.2.3.3 Pre-reforming
Depending on the composition and quality of the feedstock, pre-reforming applies an upstream reforming unit to ensure complete conversion of higher hydrocarbons (Reaction (2.1)).

2.2.3.4 Water–Gas Shift Conversion

The water–gas shift reaction (Reaction (2.2)) adjusts the $H_2:CO$ ratio as required in the downstream process steps. The high-temperature sweet water–gas shift requires a high steam to dry gas ratio to prevent iron-carbide formation on the iron-based catalyst and to keep it active. For downstream low-temperature sweet water–gas shift, the copper-based catalyst requires a completely sulfur-free synthesis gas.

Sour water–gas shift can be applied to synthesis gas compositions that contain sulfur. The metal oxide-based catalyst is active once converted into a metal-sulfide – thus it is kept in its active form by sulfur in the gas mix [9]. Sour water–gas shift is beneficial when initial desulfurization is not possible (e.g., liquid or solid feedstocks) or incomplete. It can also lead to a more compact process design due to reduced steam load relative to sweet water–gas shift. Desulfurization and removal of CO_2 is applied after the sour water–gas shift.

2.2.3.5 Gasification

Gasification can be applied to all solid and liquid carbon-containing feedstocks, including biomass and waste. The dominant feedstock, however, is coal. These feedstocks have a lower $H:C$ ratio than natural gas and in the case of biomass are also composed of significant amounts of oxygen and water. Processes based on gasification of coal have a significant contribution to global greenhouse gas emissions. Coal-based production of ammonia was responsible for around 180 Mt CO_2 emissions in 2010, while coal-based methanol production was responsible for around 30 Mt CO_2 emissions [6]. Gasification of coal is also currently the dominant process for subsequent Fischer–Tropsch reactions to produce liquid fuels [10].

There is a large variety of technologies available from all major technology suppliers, each with their own special features for certain feedstocks or downstream integration strategies.

2.2.3.6 Other Waste and Coupled Streams

Hydrogen-rich off-gases are for example co-produced by coking processes. These gases are made up of complex mixtures and are often used thermally to produce process heat rather than separated and purified. Other off-gases can be rather pure, that is, dehydrogenation reactions in refineries or chemical plants. The dehydrogenation of ethylbenzene for the production of styrene is one of the largest technical dehydrogenation processes. A global production of 21 million tons produced via this route in 2010 co-produced a stream of 400 kt pure H_2.

2.2.4
Electrolytic Processes

While electrolytic production of hydrogen is only responsible for a small amount of the overall hydrogen produced, it contributes to processes where relatively small amounts of pure hydrogen are required. Furthermore, electrolysis is the key technology to attempts to convert electrical energy produced by renewable energy sources like windpower or photovoltaic into chemical energy which then can be easily stored.

2.2.4.1 Alkaline Electrolysis

The chemical industry presently mass-produces hydrogen via the alkaline electrolysis. This process is a well-established production method for hydrogen when high purity is required. Hydrogen is co-generated with oxygen by the electrolysis of a concentrated (about 30% weight) potassium hydroxide in water solution at elevated temperatures between 60 and 90 °C. The overall efficiency of the electrolytic process is of the order of 70–80%. The electrolysis is carried out in stacks, which can be built up to reach the required capacity. An electrolytic stack produces up to $740\,m^3$-$H_2\,h^{-1}$. The largest production facilities produce up to $30\,000\,m^3$-$H_2\,h^{-1}$ with the according number of electrolytic stacks [11]. Most of the production is carried out under atmospheric pressure conditions. However, high-pressure electrolysis plants have been developed. For example, Lurgi developed technology for alkaline electrolysis for up to 60 bar of pressure, which may facilitate the use of the hydrogen in subsequent process steps. This process is well established within the chemical industry and explained in further detail in Chapter 2.4.

2.2.4.2 PEM Electrolysis

A more recent development is based on the proton-exchange membrane (PEM) for separation of the electrolytic cells. Uses water. First, PEM-electrolyzers have been introduced into the market and cover power ranges up to 2 MW. The main advantages of this technology are its very short response time and its large dynamic range, which make it uniquely suitable to transform renewable electricity generated by unsteady power sources to generate hydrogen. Currently, several projects are under construction. Further details on this technology are discussed in Section 2.4.

2.2.4.3 High-Temperature Electrolysis

Electrolytic processes, by their very nature require electricity to formally break chemical bonds. However, the necessary energy does not have to be supplied only as electrical current. High-temperature electrolysis uses the principles of an inverted solid oxide fuel cells (SOFC) to produce hydrogen. Solid oxide electrolysis of water is versatile and can be used to co-electrolyze CO_2 in order to produce a synthesis gas of the desired composition. Due to part of the energy required being supplied as high temperature heat, the electricity consumption for the production of hydrogen is significantly lower than for any of the above electrolytic processes. The stacks can be operated with high pressure and easily expanded to accommodate downstream methanol or methanization reactions [12]. This technology offers advantages in reduced electricity consumption and in the easy combination with follow-up processes. However, it requires a significant high-temperature heat source. Presently, its development is in the laboratory stage. However, several international projects have taken up the challenge to foster further development of the technology, for example, the FP7 projects RELHY [13] (end date 2011). From the point of view of the chemical industry, which is generally more interested in synthesis gas as such than in hydrogen production alone, this technology offers a promising new road of development.

2.2.5
Hydrogen Production Steam Reforming versus Electrolysis

Steam reforming is the dominant process to produce hydrogen in the chemical industry. All processes require substantial amounts of energy that are either supplied by natural gas or by electricity. Therefore, the relative price level of electricity and natural gas largely define the relative competiveness of the processes with respect to each other. A sensitivity analysis of the hydrogen production costs of various processes has been carried out by IEA [3]. It clearly demonstrates the relative cost dependencies on the price of the energy carrier. At present, the production of hydrogen via alkaline electrolysis is about 3–5 times higher than by steam reforming, depending on the local gas and electricity prices. Gas prices and electricity prices are coupled, since natural gas can be used to produce electricity at a competitive price; it is unlikely that electrolytic processes will be cost competitive with large-scale steam reforming in the foreseeable future.

2.2.6
Hydrogen as Coupled Stream in the Electrolytic Production of Chlorine

Chlorine is one of the most important products of the chemical industry. As with hydrogen, it is generally produced and immediately used on site; about a third of the European chlorine production is used for the production of vinyl chloride and another 30% for the production of isocyantes and oxygenates [14]. Only a fraction of chlorine is traded (only around 6% of European production is transported beyond the production facility). Chlorine is mass produced by electrolytic processes that co-produce hydrogen. From the point of view of the chlorine producers, the co-production of hydrogen is both a waste of energy and a safety hazard. Therefore, present developments aim at process modifications that reduce or eliminate hydrogen co-production.

Hydrogen is also produced as a coupled product by chlorine-alkaline electrolysis in the production of chlorine. Chlorine is currently produced by the following processes:

- membrane cell process,
- mercury cell process,
- diaphragm cell process.

All of the above co-produce hydrogen gas. Total European chlorine production capacity amounts to 12 550 kt per year out of which 12 174 kt per year are based on any of the above processes. The membrane cell process is dominant with 7343 kt (68%), followed by the mercury cell process with 3271 kt (26%), and the diaphragm cell process with 1635 kt (13%) of the European chlorine production capacity. France, Germany, and the Benelux countries host about 70% of the European production capacity [15]. Chlorine production is affected by the overall economic

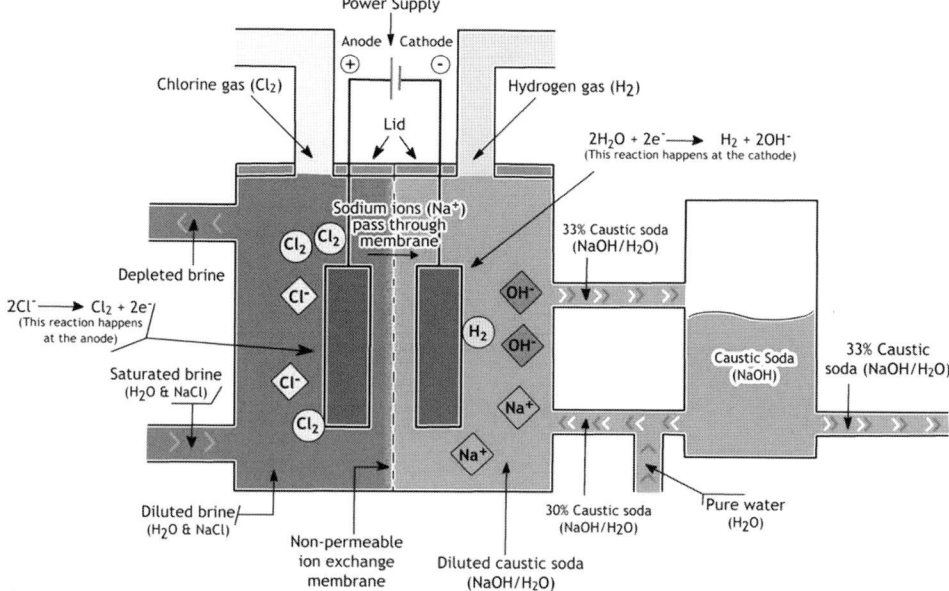

Figure 2.5 Schematic overview of the membrane cell process for the production of chlorine. Source: www.eurochlor.org.

development and effective production reached 9701 kt (77%) [16] of the production capacity. Eurostat estimated the EU27 production in 2012 to be 7998 kt [5].

This amounts to a stoichiometrical hydrogen production capacity of 346 kt H_2 per year (315 m^3 H_2 per t Cl = 3.8 Mm^3 [14]). However, the European chlorine industry only valorizes 88% (2012) of the hydrogen produced, with a decreasing trend over the last two years.

2.2.6.1 Membrane Cell Process

The membrane cell process is based on a saturated brine solution that enters the anode part of the cell (Figure 2.5). The anode directly converts chloride ions of chlorine (Reaction (2.11)):

$$2Cl^- \rightarrow Cl_2 + 2e^- \tag{2.11}$$

Sodium ions can pass through the non-permeable ion exchange membrane to the cathode side of the cell. The solution in this part of the cell is made up of a concentrated caustic soda solution (30%). The reaction at the cathode produces hydrogen directly (Reaction (2.12)):

$$2H_2O + 2e^- \rightarrow 2OH^- + H_2 \tag{2.12}$$

The cathode cell caustic soda solution therefore becomes more concentrated (up to 33%) and needs to be refreshed. The overall products are chlorine, hydrogen, and concentrated caustic soda solution. This process has the lowest electricity demand of the chlorine production processes and a growing in share of the

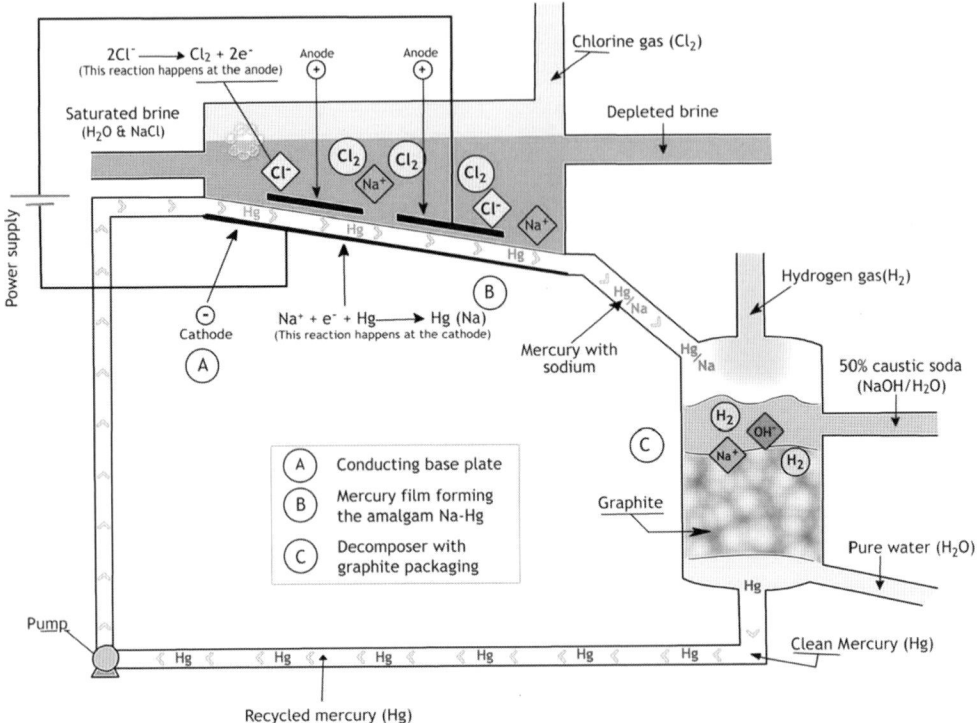

Figure 2.6 Schematic overview of the mercury cell process for the production of chlorine. Source: www.eurochlor.org.

European chlorine production capacity (51% in 2011) [15]. However, the resulting chlorine may contain oxygen in intolerable amounts and may require additional purification.

2.2.6.2 Mercury Cell Process

A saturated solution of brine (NaCl in water) flows over the cathode (Figure 2.6). At the cathode, Na^+ reacts with mercury to form an amalgam (Reaction (2.13)), while the cathode reaction produces chlorine (Reaction (2.11)):

$$Na^+ + e^- + Hg \rightarrow Hg(Na) \qquad (2.13)$$

Mercury and amalgam are transported to the graphite decomposer, where pure mercury is recovered and recycled, hydrogen and a concentrated caustic soda solution are co-produced (Reaction (2.14)):

$$2Hg(Na) + 2H_2O + graphite \rightarrow 2Hg + 2NaOH + H_2 \qquad (2.14)$$

The process produces high purity chlorine; however, the main drawbacks of this process are the use of mercury and its high electricity demand. Presently, European mercury emissions by mercury-cell chlorine plants are 0.81 g per ton of chlorine produced. The remaining 31 plants in Europe will be phased by 2020 [16].

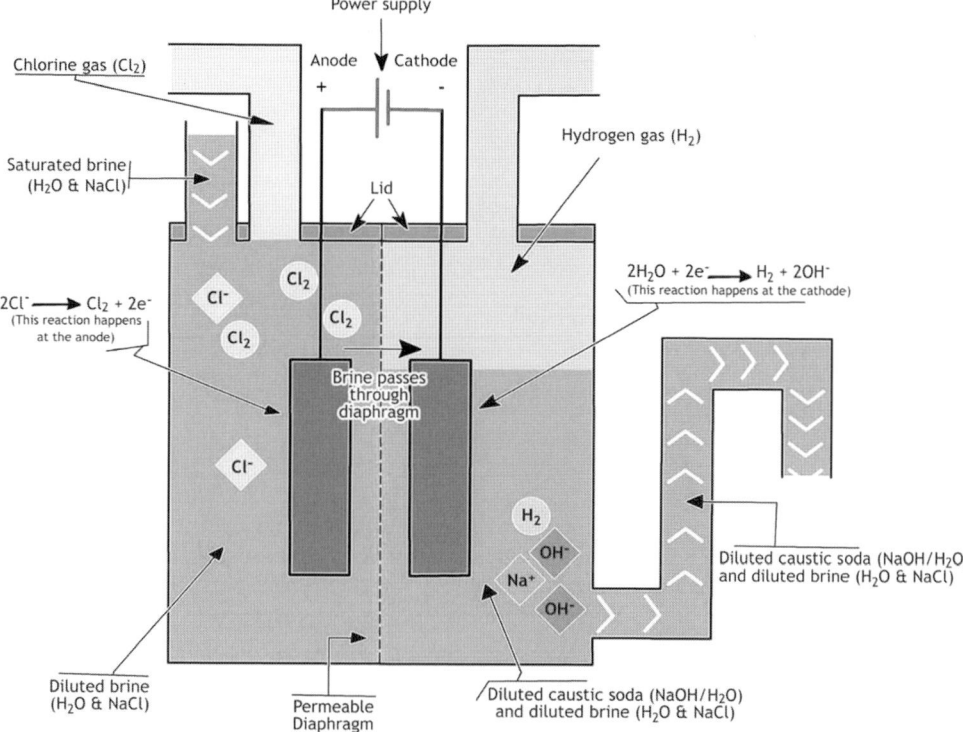

Figure 2.7 Schematic overview of the diaphragm cell process for the production of chlorine. Source: www.eurochlor.org.

Due to environmental concern, this process is globally in terminal decline. From 2002 to 2012 the number of plants decreased from 91 to 50 (−45%) and the production capacity from 9.1 to 4.9 million ton (−46%) [17].

2.2.6.3 Daphragm Cell Process

The diaphragm cell process uses a permeable diaphragm to separate the anode and cathode of the electrolysis cell (Figure 2.7). While it separates the gases, the brine solution can pass from the anode to the cathode part of the cell. The formal reactions at the anode and cathode are identical to reactions (2.11) and (2.12) from the above membrane process. However, the resulting mixed solution, containing caustic soda and brine does require significant amounts of steam for separation, if no other way of site specific process integration is available. The diaphragm cell process currently makes up around 14% of European chlorine production capacity [15].

2.2.6.4 New Developments

New technologies are likely to change the state-of-the-art in the chemical industry in chlorine production. New production methods for chlorine using oxygen depolarized electrodes, essentially a fuel cell at the cathode, reduce the overall amount of energy required to produce chlorine and also eliminates the by-product hydrogen. Even though markets and new applications for hydrogen, for example, in fuel cells, are in development, the chlorine producers are expected to switch to hydrogen-free chlorine production technologies in the mid- and long-term time horizons.

2.3
Utilization of Hydrogen in the Chemical Industry

2.3.1
Ammonia

Ammonia is the largest consumer of hydrogen. Global production of ammonia amounted to 166 million metric tons in 2012 [4], suggesting a stoichiometric hydrogen production of 29 million metric tons solely for the production of ammonia. As ammonia is the precursor compound of nitrogen fertilizers, global production is continuously increasing, corresponding to the ever increasing world population. A large amount of the produced ammonia is directly converted into urea using the CO_2 originating from the steam reforming used for hydrogen supply. The International Fertilizer Organization quantified the world urea production as 162 Mio. tons in 2012 [4]. European production is around 17 Mio. tons. Apart from this, virtually all organic nitrogen compounds are derived from ammonia. Nitric acid is an important derivative generated by oxidation of ammonia with air over a platinum catalyst (Ostwald process).

Ammonia is globally produced almost exclusively via the Haber–Bosch process, a catalytic process in which hydrogen and nitrogen are passed over an iron catalyst at elevated temperature (300–550 °C) and pressure (15–25 MPa, 150–250 bar) according to the following reaction (2.15):

$$N_2 + 3H_2 \Leftrightarrow 2NH_3 \qquad (2.15)$$

Air is directly or indirectly used as the source of nitrogen; nitrogen is usually fed as air, but in some plants (e.g., partial oxidation) it is a purified gas from an associated air separation facility. Hydrogen is usually provided by synthesis gas, and the reformer for syngas production is an integral part of an ammonia plant. Around 75% of the syngas production for ammonia is based on steam reforming of gas, followed by partial oxidation of coal. The latter is strongly increasing, in particular in China. Approximately 1% of global ammonium is still produced from hydrogen from electrolysis, the applied hydrogen being a by-product of chlorine production as described above.

Figure 2.8 schematically depicts the main steps in ammonia production, based on steam reforming of natural gas.

Prior to the reformer, the natural gas is passed over a catalyst in a *desulfurizer* to scrub sulfur traces which would poison the catalysts used in the process. Process steam is added, the mixture is pre-heated and largely converted into hydrogen, carbon dioxide and carbon monoxide in the *primary reformer* at

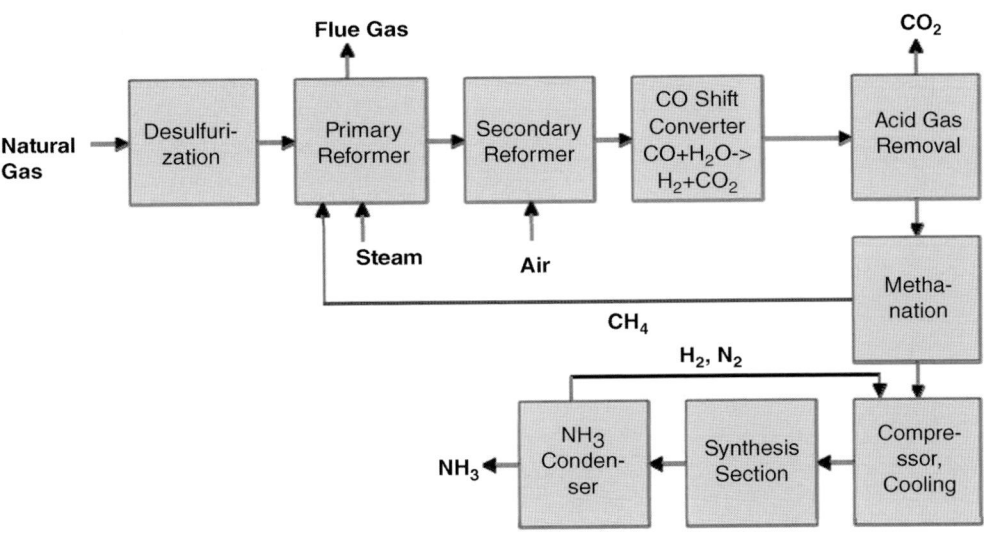

Figure 2.8 Schematic depiction of ammonia synthesis from natural gas.

temperatures around 820 °C and pressures of approx. 40 bar. In the *secondary reformer*, the process gas leaving the primary reformer is contacted with pre-heated air causing partial combustion conversion of the residual methane into hydrogen, carbon dioxide, and carbon monoxide. The process gas is cooled from around 1000 to 370 °C and fed to the *shift converter*, where carbon monoxide is converted with steam into additional hydrogen and carbon dioxide. CO_2 is then removed from the process gas in an absorber (*acid gas removal*) using a solvent such as monoethanolamine or other amines, potassium carbonate, or sulfinol (sulfolane and an alkanolamine). The solvent is regenerated in a stripper column (not shown in the scheme). After the absorber, the gas stream contains mainly hydrogen and nitrogen, but also traces of carbon oxides, which would poison the ammonia synthesis catalyst. In the *methanator*, these species are converted back into methane, which is redirected to the reformer section. After compression, the purified process gas, now consisting of hydrogen and nitrogen in a 3:1 ratio and small amounts of residual methane, is directed to the *ammonia synthesis reactor* containing several beds of catalyst (with iron oxide and ruthenium catalyst) and cooling between each pass. The ammonia synthesis is an exothermic equilibrium reaction involving a volume reduction. The equilibrium is therefore favored by high pressures and low temperatures. Temperatures between 390 and 500 °C and highly active catalysts are used to achieve acceptable reaction rates. Only 15% conversion is achieved in a single pass, meaning that the ammonia must be condensed by refrigeration and the remaining gas mixture is fed back into the reactor. This *condenser* provides liquid ammonia and leaves unreacted hydrogen and nitrogen in the gas phase, which is returned as recycle gas stream to the ammonia synthesis section. Since the residual methane in the process gas would be enriched in the loop, a small amount of purge-gas is taken

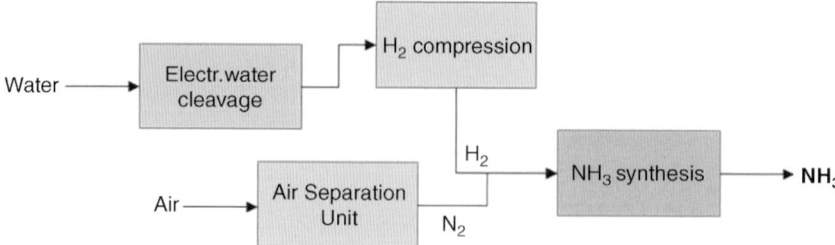

Figure 2.9 Alternative ammonia synthesis based on hydrogen produced from water electrolysis.

from the loop to discharge the methane. The separated liquid ammonia is expanded and stored or processed further, for example, in a urea plant.

For economic reasons, ammonia plants are usually large-scale plants. The practical limitation with manageable equipment size and loop volume is between 2000 and 3000 tons per day. Nevertheless major ammonia process contractors nowadays offer large plant concepts (mega-ammonia technology) with 3000–6000 ton-per-day units. These capacity extensions are realized by producing additional ammonia outside the synthesis loop. In the Uhde Dual Pressure process for instance, a once-through reactor is installed upstream of the conventional synthesis loop, and allows splitting of ammonia production into a low-pressure once-through synthesis and a conventional high-pressure synthesis loop with conventional reactor size.

The described process scheme will represent the state of ammonia synthesis for a long time. Figure 2.9 shows the alternative scheme of an ammonia process based on hydrogen produced from water electrolysis. This requires an electrolyzer for water cleavage, and an air separation unit for the delivery of nitrogen. The electrolysis is particularly energy-intensive and hence the scheme is only realistic when renewable energy is used. Apart from this, this process is not economically viable for the time being.

2.3.2
Methanol

Methanol is the second largest consumer of hydrogen within the chemical industry. It serves as a base chemical to produce formaldehyde, MTBE, acetic acid, as well as fuels via methanol-to-gasoline or other base chemicals as olefins via methanol-to-olefin or methanol-to-propylene routes. Global methanol production amounted in 2010 to around 48 million tons [6], with European production at around 1.6 million tons. The dominant feedstock is natural gas, which is used in about 75% of the production. Most of the remaining production is based on coal with small contributions from other feedstocks like naphtha or heavy oil.

Methanol has been suggested for a central position in the value chains of the chemical industry [18,19] and to replace car fuel. Notwithstanding future developments, methanol currently is central in the production of some high-volume chemicals and has an interesting potential for changing existing value chains within the chemical industry.

Methanol production is based on syngas-processes like steam reforming and coal gasification as discussed above. The following discussion is based on the Lurgi-MegaMethanol Process® [20], currently the state-of-the-art process for high volume methanol production. Synthesis of methanol from syngas requires a stoichiometric number of 2 (Eq. (2.16)):

$$S_r = \frac{(H_2 - CO_2)}{(CO + CO_2)} = 2.0 - 2.1 \tag{2.16}$$

The individual technical implementation will depend on the feedstock available. Conventional steam reforming might be used in small and medium-size applications. However, for large world-scale plants different variations are chosen.

Light feedstock such as light natural gas is desulfurized and reformed with steam and oxygen at around 40 bar and reformer outlet temperatures of around 1000 °C. The resulting syngas is compressed and transferred to the methanol-synthesis unit. To meet the above condition, it is necessary to feed an additional hydrogen stream into the methanol-synthesis unit. This hydrogen can be supplied by recycling hydrogen after separation of methanol from the remaining gases.

Heavier feedstocks are unable to meet the above requirement for optimum synthesis conditions (i.e., stoichiometric number within the range 2.0–2.1) and need an external hydrogen supply. The most economic solution is a combination of steam reforming and autothermal reforming by which a fraction of the desulfurized feed gas is stream reformed at high pressure of 35–40 bar and low temperature of 700–800 °C in order to provide the necessary hydrogen, and then combined with the non-reformed feed gas and together subjected to autothermal reforming.

The reactions of syngas to methanol, reaction (2.17) and reaction (2.18), are highly exothermic processes and the overall economics of a methanol plants is critically dependent on efficient heat recovery and management:

$$2H_2 + CO \Leftrightarrow CH_3OH \tag{2.17}$$

$$3H_2 + CO_2 \Leftrightarrow CH_3OH + H_2O \tag{2.18}$$

The reaction takes place in tubes filled with the catalyst (Cu, ZnO on alumina) and placed in water. The heat of reaction is transferred to the surrounding water and leads to partial evaporation. Control of the steam pressure allows control of the reaction conditions and an isothermal operation. A second, gas-cooled tube reactor is applied after the first isothermal one. The cooling agent consists of the syngas feed for the first reactor that countercurrents the product gas stream. The product gas from the isothermal reactor is cooled down over the catalyst bed and the driving force to achieve the equilibrium is maintained, while the feed gas for the isothermal reactor is heated up and less preheating is required before entering the isothermal reactor. Figure 2.10 summarizes the different process options for methanol synthesis.

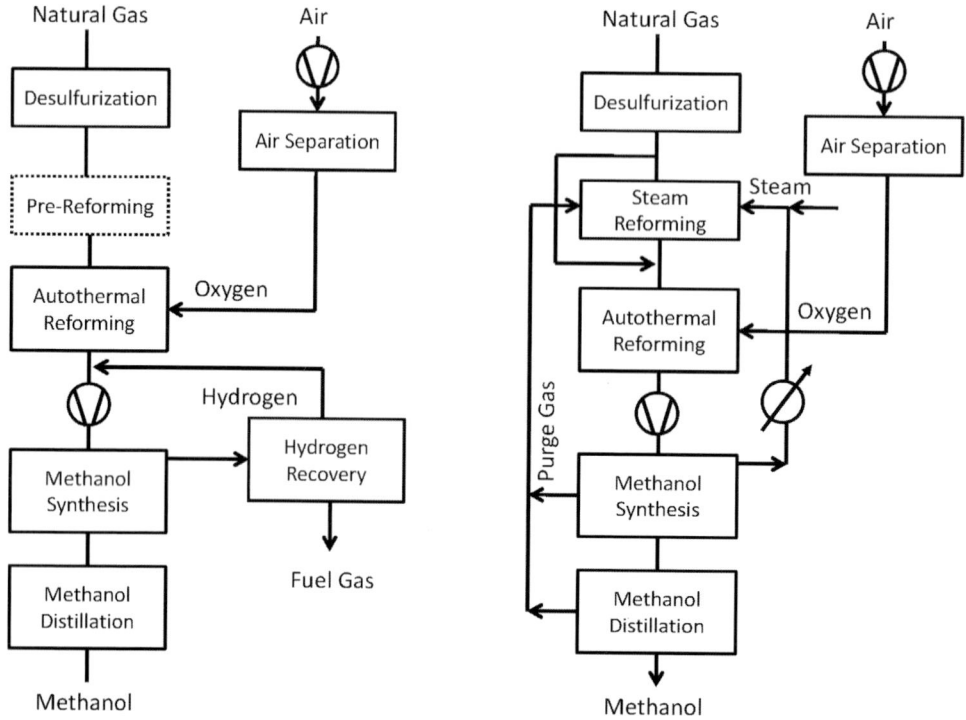

Figure 2.10 Process schematics for methanol production. Adapted from Lurgi [20].

2.3.3
Other Uses and Applications of Hydrogen

While ammonia and methanol production are the principal consumers of hydrogen, many other process are operating in which hydrogen is used as a feed. Another large-scale process is the production of fuels via steam reforming/gasification of natural gas or coal followed by Fischer–Tropsch synthesis. The globally dominant producer of synthetic fuels, SASOL, reported an overall production of synthetic fuels of 7443 Mio. ton for 2013. The technology has been successfully employed on alternative resources like Canadian shale gas [21].

Other processes are hydrogenation reactions that are commonly used and enter the chemical value chain at various positions. Furthermore, hydrogenation is a versatile tool in combination with nitration to introduce functionalities into a molecule, for example, in the production of MDI and TDI, precursors for subsequent polyurethane production [22]. Generally, hydrogenation is also applied to saturate aromatic and unsaturated compounds like BTX or to remove double bonds from chemical systems. Hydrogen is also used in this function to harden animal or vegetable fats in the food industry, for example, in the production of margarine.

2.3.4
Current Developments and Outlook

The current debate on power-to-gas technologies has a strong impact on the chemical industry. Various concepts not only extend beyond power-to-gas towards power-to-fuels but also to power-to-chemicals [23]. In principle, these concepts would open up the opportunity of chemistry beyond fossil fuels and also additional value creation.

However, while power-to-gas might produce hydrogen, chemical transformations in the chemical industry are generally based on carbon, more so than on hydrogen. While power-to-gas technologies might have the potential to decarbonize the transport sector, the chemical industry will continue to require carbon-based feeds. Several interesting processes and concepts have been developed and need to be demonstrated to be economically viable. This represents a significant economic challenge since hydrogen is produced via electrolytic processes in any power-to-chemicals concept.

For the chemical industry, the prospects of an increased role of hydrogen within the energy system offer opportunities and challenges alike. On one side, the chemical industry has elaborate experience working with hydrogen and new branches might open up within the classical, fossil-fuel-based value chains. These developments, however, will not break the dominance of the larger energy sector on the smaller chemical sector, as both continue to utilize the same raw material base for their products and services. The chemical industry could reduce its dependence on fossil fuels if power-to-gas were to be implemented on a large scale.

Furthermore, beside the availability of hydrogen, oxygen would also be available in large quantities as a side-stream of the electrolysis. This might lead to a rearrangement of current value chains as other oxygen-based process options might become more viable.

Biorefineries, another interesting line of current development, could be substantially altered by a large abundance of hydrogen. Biomass feedstocks for biorefineries generally have high oxygen content. Hydrogen could in principle be used in analogy to desulfurization of fossil fuel to deoxygenate biomass feedstocks. Biorefineries in combination with hydrogen from renewable energy sources could be the basis of a fully sustainable chemical industry, albeit probably not on the level of the current chemical industry production scale.

The effect of the possible abundance of renewable hydrogen on the economics of ammonia production is still unknown. Renewable hydrogen is unlikely to be produced at competitive price levels with respect to current steam reforming technology. If, however, large amounts of renewable hydrogen are required as energy/fuel, it is likely to raise the general price level for hydrogen. This in turn might affect ammonia production. Since ammonia is the base chemical for fertilizers, changes in price and cost levels will influence the price and cost base for the agricultural production of crops for food and animal feed, with unforeseeable consequences for global food supply.

References

1. Air Products and Chemicals (2015) Is hydrogen available today? http://www.airproducts.com/industries/Energy/Power/Power-Generation/faqs.aspx (accessed 21 January 2015).
2. Conte, M., Di Mario, F., Iacobazzi, A., Mattucci, A., Moreno, A., and Ronchetti, Ma. (2009) *Energies*, **2**, 150–179. doi: 10.3390/en20100150
3. OECD/IEA (2007) IEA Energy Technology Essentials: Hydrogen Production & Distribution. Available at http://www.iea.org/publications/freepublications/publication/essentials5.pdf.
4. International Fertilizer Industry Association (2002–2015) Statistics: Production and International Trade. http://www.fertilizer.org/en/doc_library/Statistics/PIT/ammonia_public.xlsx (accessed 21 January 2015).
5. Eurostat (2002–2012) PRODCOM. http://ec.europa.eu/eurostat/data/database (accessed 21 January 2015).
6. IEA, ICCA, DECHEMA (2013) Technology Roadmap Energy and GHG Reductions in the Chemical Industry via Catalytic Processes. Available at http://www.dechema.de/dechema_media/Chemical_Roadmap_2013_Final_WEB-p-4584-view_image-1-called_by-dechema2013-original_site-dechema_eV-original_page-124930.pdf
7. Rostrup-Nielsen (2002) Syngas in perspective. *Catal. Today*, **71**, 243–247.
8. Rostrup-Nielsen, J.R., Dybkjaer, I., and Aasberg-Petersen, K. (2000). *Prepr. ACS Petr. Chem. Div.*, **45** (2), 186.
9. Haldor-Topsoe (2009) Sulphur resistant/sour water-gas shift catalyst. http://topsoe.ru/business_areas/gasification_based/Processes/~/media/PDF%20files/SSK/topsoe_SSK%20brochure_aug09.ashx.
10. Dechema (2009) Positionspapier: Kohlenveredlung. http://www.dechema.de/dechema_media/Kohlenveredlung.pdf (in German).
11. Smolinka, T., Günther, M., and Garche, J. (2011) Stand und Entwicklungspotential der Wasserelektrolyse zur Herstellung von Wasserstoff aus regenerativen Energien, NOW-Studie, Fraunhofer ISE, Freiburg 2011.
12. Hansen, J.B., Christiansen, N., and Nielsen, J.U. (2011) *ECS Trans.*, **35** (1), 2941–2948. doi: 10.1149/1.3570293
13. Patyk, A. (21 July 2014) RelHy – Innovative Solid Oxide Electrolyser Stacks for Efficient and Reliable Hydrogen Production, EU FP7 Project. http://www.itas.kit.edu/english/erts_completed_paty08_relhy.php.
14. Eurochlor (2014) Chlorine Industry Review 2012–2013. http://www.eurochlor.org/media/70861/2013-annualreview-final.pdf (accessed 21 January 2015).
15. Eurochlor (2013) Installed chlorine production capacities. http://www.eurochlor.org/media/82824/2013-europeanproductioncapacities.pdf (accessed 21 January 2015).
16. Eurochlor (2015) The mercury cell process. http://www.eurochlor.org/the-chlorine-universe/how-is-chlorine-produced/the-mercury-cell-process.aspx (accessed 21 January 2015).
17. UNEP Global Mercury Partnership Chlor-Alkali Area (June 2012) Conversion from Mercury to Alternative Technology in the Chlor-Alkali Industry. http://www.unep.org/hazardoussubstances/Portals/9/Mercury/Documents/chloralkali/Partnership%20Document%20on%20the%20Conversion%20from%20Mercury%20to%20Alternative%20Technology%20in%20the%20Chlor-Alkali%20Industry.pdf (accessed 21 January 2015).
18. Olah, G.A., Goeppert, A., and Surya Prakash, G.K. (2009) *Beyond Oil and Gas: The Methanol Economy*, Wiley-VCH Verlag GmbH, Weinheim.
19. *Methanol: The Basic Chemical and Energy Feedstock of the Future* (2014) (eds M. Bertau, H. Offermanns, L. Plass, F. Schmidt, and H.J. Wernicke), 1st edn, Springer Verlag, Heidelberg.
20. AirLiquide (2010) Lurgi MegaMethanol®, Lurgi GmbH. http://www.zeogas.com/files/83939793.pdf.
21. SASOL (2013) Integrated Annual Report 2013. http://www.sasol.com/sites/default/files/publications/integrated_reports/downloads/Sasol%20IR%202013lores.pdf

22 Boustead, I. (March 2005) Eco-profiles of the European Plastics Industry Diphenylmethane diisocyanate (MDI), Plastics Europe.
23 Ausfelder, F., Beilmann, C., Bertau, M., Bräuninger, S., Heinzel, A., Hoer, R., Koch, W., Mahlendorf, F., Metzelthin, A., Peuckert, M., Plass, L., Räuchle, K., Reuter, M., Schaub, G., Schiebahn, S., Schwab, E., Schüth, F., Stolten, D., Teßmer, G., Wagemann, K., Ziegahn, K.-F. (2015) Energiespeicherung als element einer sicheren energieversorgung. *Chem. Ing. Tech.*, **87** (1–2), 17–89.

3
Chlorine–Alkaline Electrolysis – Technology and Use and Economy

Alessandro Delfrate

3.1
Introduction

Sodium chloride is the raw material for chlorine production, and almost all of the chlorine produced today comes from the electrolysis of sodium chloride brine solutions. Sources of sodium chloride are sea salt, rock salt from mines, or well brine, where brine is produced directly by injection of water into underground salt deposits. Approximately 2% of chlorine is produced using potassium chloride as raw material, thus producing potassium hydroxide as co-product. Potassium hydroxide is used in the glass industry, fertilizer industry, batteries production, and in food processing. Other technologies for the industrial production of chlorine include the oxidation of hydrochloric acid by electrolysis or by catalytic oxidation: in such processes only chlorine is produced, without any caustic and hydrogen coproduction. Such technologies are becoming important in managing the world market of hydrochloric acid, which is presently unbalanced by the growing polyurethane business.

In the last 100 years, three electrolysis technologies have been used for the production of chlorine: the oldest is based on the use mercury as a cathode for an electrolysis process. The environmental and health problems related to the use of mercury are forcing chlorine producers to phase out this technology, so that today less than 50 plants are still in operation worldwide [1].

The second production method is the so-called diaphragm technology: the anodic and cathodic chambers are physically separated by a porous diaphragm, originally based on asbestos material. The replacement of asbestos with less pollutant materials has prolonged the life of existing plants, while the high power consumption and low quality of caustic soda are the major reasons for plant modernization.

State of the art electrolyzers for production of chlorine are based today on membrane technology: the anodic chamber is separated from the cathodic chamber by a cationic ion exchange membrane, thus granting safe operation of the electrolyzers and high purity of the chlorine, caustic, and hydrogen products.

Hydrogen Science and Engineering: Materials, Processes, Systems and Technology, First Edition.
Edited by Detlef Stolten and Bernd Emonts.
© 2016 Wiley-VCH Verlag GmbH & Co. KGaA. Published 2016 by Wiley-VCH Verlag GmbH & Co. KGaA.

All three industrial technologies have been developed for chlorine production and are based on common principles in order to guarantee the safety of operation, to minimize power consumption, and to guarantee the best possible quality of the products:

a) Chlorine, on the anode side, and caustic with hydrogen on the cathode side must be kept separate because they react violently when coming together.
b) Ohmic losses within the circuit must be reduced to the maximum extent.
c) Overvoltage at the electrodes must be minimized by using specific catalysts.
d) Current efficiency must be maximized, avoiding parasitic reactions like production of oxygen or chlorates.

3.2
Production Technologies

3.2.1
Electrochemistry of Chlorine Production

Chlorine is mainly produced through the electrolysis of sodium chloride solution.

Approximately 2% of total chlorine production is obtained through electrolysis of potassium chloride, with the same electrochemical reaction as with sodium chloride:

$$\text{Anodic reaction}: 2\,Na^+ + 2\,Cl^- \rightarrow Cl_2 + 2e^-$$

$$\text{Cathodic reaction}: H_2O + 2e^- \rightarrow H_2 + 2OH^-$$

$$\text{Overall process } 2NaCl(aq) + 2H_2O \rightarrow Cl_2(gas) + 2\,NaOH(gas) + H_2(gas)$$

Chlorine and caustic soda are produced in a fixed ratio:

anode: 1 ton Cl_2 (gas);
cathode: 1.128 ton NaOH (acq) + 0028 ton H_2 (gas).

Production of chlorine is an energy intensive process, and electricity consumption is the most important parameter in the chlor alkali production. The average DC energy consumption is 3.4 MWh per metric tonne of chlorine produced (Figure 3.1) [2]; electricity can represent over 50% of the cost for chlorine production, depending from the geographical area. Other costs are related to the sodium chloride consumption and the operating cost of the chlorine plant. There are no direct emissions of greenhouses gasses in the chlorine manufacturing process, but there are carbon dioxide emissions related to the fuel used to generate electricity and to generate steam required during production.

Figure 3.1 Main parameters for chlorine production cost.

3.2.2
Mercury Electrolyzer Technology

Figure 3.2 gives a schematic diagram of the mercury electrolyzer process.

In mercury electrolysis, physical separation between chlorine and caustic soda is achieved in a two-step reaction. The saturated sodium chloride solution (brine) is purified by a precipitation process, and is then introduced into the electrolyzer on top of the mercury film. The cathodic circuit is made of a thin film of mercury, about 3 mm thick, flowing on a steel plate from top to the bottom of the electrolyzer.

Electrodes made of mixed metal oxide coated titanium material are positioned close to the mercury film and represent the anodes. Chlorine is produced at the anodes, and is collected in the main chlorine circuit. Together with chlorine, metallic sodium is produced, resulting in a mercury–sodium amalgam that exits the cell from the bottom. In a separate decomposer, amalgam with 0.2–0.4% in weight of sodium reacts with water to produce hydrogen gas and caustic soda as 50% w/w solution.

To reduce power consumption, computer-controlled systems have been developed to automatically keep the distance between anodes and the mercury film at the minimum value possible. This process produces high purity chlorine (98% w/w) that for general utilization does not requires special purification. Caustic soda is produced as a 50% w/w solution, which represents the industry standards for transportation. Both caustic soda and hydrogen must be treated to remove the residual mercury.

The major drawback of this technology is related to the use of mercury (Figure 3.3). Owing to environmental restrictions, mercury technology is going to be progressively abandoned, with a target of complete replacement within 2017 in Europe [3].

3 Chlorine–Alkaline Electrolysis – Technology and Use and Economy

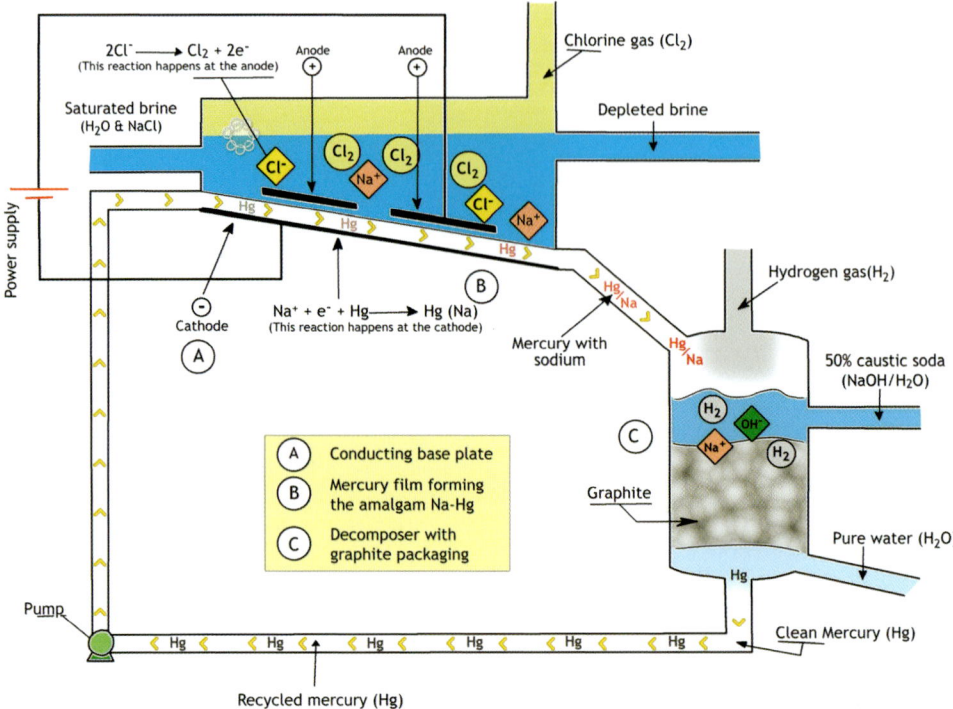

Figure 3.2 Mercury electrolyzer: schematic. Source: EuroChlor (www.eurochlor.org).

Figure 3.3 Mercury cell room.

The amount of mercury still in use in a chlorine plant has had to be reported on a yearly basis to the European Commission since 2009. When a plant is closed or converted into membrane all the mercury is stabilized by conversion into sulfide and stored in safe salt mine disposal sites; an international export ban has been active since 2011 [4].

At present there are about 55 plants worldwide still using mercury technology, with a total capacity of about 5 million tons of chlorine [3].

3.2.3
Diaphragm Electrolyzers

In diaphragm electrolysis, physical separation between chlorine and caustic soda with hydrogen is achieved by using a permeable diaphragm that separates anodic and cathodic chambers (Figure 3.4).

The saturated sodium chloride solution (brine) is purified by a precipitation process, and is then introduced into the electrolyzer. At the anodic side, chloride ions are oxidized to chlorine that is collected in the main chlorine circuit.

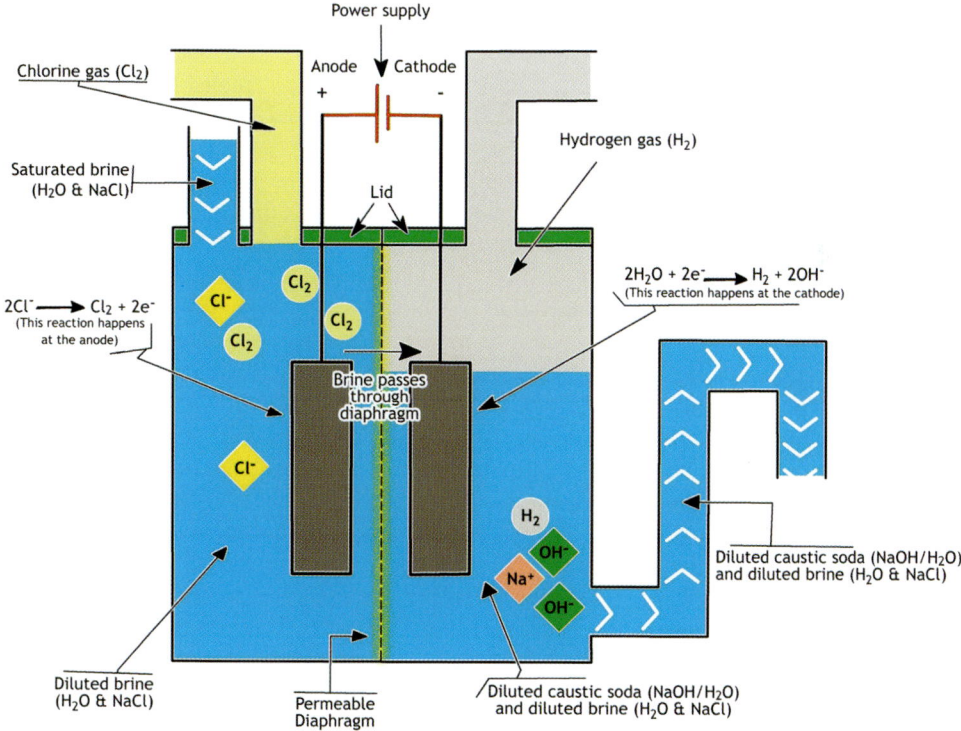

Figure 3.4 Diaphragm electrolysis: schematic. Source: EuroChlor (www.eurochlor.org).

Figure 3.5 Diaphragm cell room.

Mixed noble metal oxide coated titanium electrodes are used as anodes. The diluted brine solution crosses the diaphragm and reacts at the cathode, producing hydrogen and a diluted solution of caustic soda and sodium chloride (Figure 3.5).

The diaphragm has multiple roles in this process: the tangled fiber structure of the asbestos allows liquids to pass through, but not fine gas bubbles, and in addition hinders to a large extent back-diffusion of the cathodically-formed OH^- ions to the anode. This process produces relatively low purity chlorine, with elevated oxygen concentration; further purification steps through chlorine liquefaction are generally required. Caustic soda is produced at low concentration (12–14% w/w) and contaminated by sodium chloride (15–17% w/w). In most plants, caustic concentration units are installed to recover the sodium chloride in solid form, which is re-used to prepare brine, and concentrate caustic to a 50% w/w solution, which is the industry standards for transportation.

The major drawback of this technology is related to the use of asbestos as diaphragm material. Asbestos is recognized as causing lung cancer when breathed with air. Alternative materials for diaphragm have been developed by the industry, mainly based on zirconium based PTFE filled materials to replace asbestos. Owing to the high energy consumption and overall low quality of products, diaphragm plants are being progressively converted into membrane technology.

3.2.4
Ion Exchange Membrane Electrolyzers

Membrane technology represents today the Best Available Technique for chlorine production (Figure 3.6) [3]. Reduced energy consumption when compared

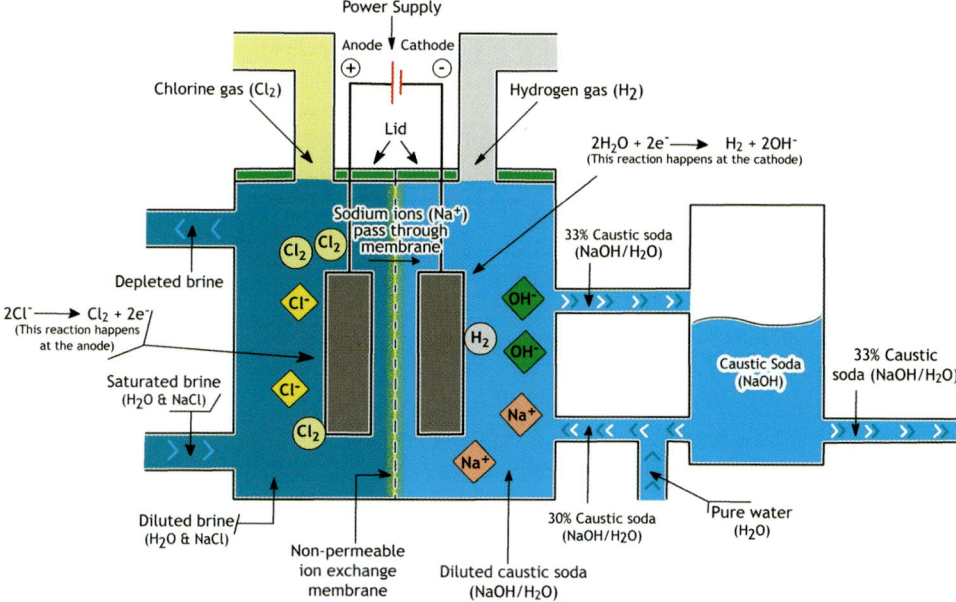

Figure 3.6 Membrane electrolysis process. Source: EuroChlor (www.eurochlor.org).

to mercury and diaphragm technologies, high quality of products, and safe and environmental friendly operation make membrane technology the only one used for new chlorine plants over the last 20 years.

In membrane plants, the anodic compartment is physically separated from the cathodic compartment by an ionic exchange membrane (Figure 3.7). This membrane is a two-layer membrane, with a sulfonic and carboxylic layer with perfluorinated Ionomers, and guarantees the passage of sodium or potassium ions

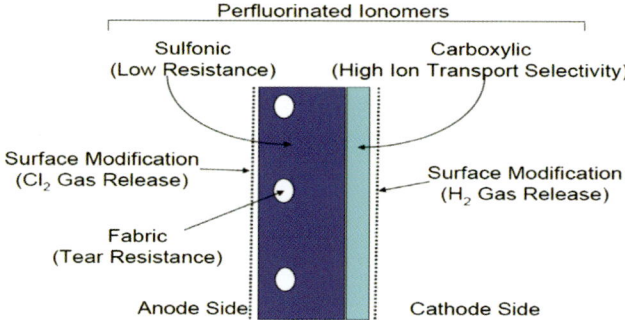

Figure 3.7 Cross section of ion exchange membrane.

with high specificity to other ions [5]. The saturated sodium chloride solution (brine) requires a very high degree of purification, which is achieved by a combination of precipitation process, filtration, and final removal of cathions with ion exchange resins to ppb levels.

Brine purity is a critical operating parameter for the performances of the electrolyzer, influencing the power consumption, purity of products, and operating life of electrodes and membrane. Membrane degradation due to the deposition of impurities on its surface or within its internal structure is strongly dependant on the brine quality. Specific anions, such as fluorine, are responsible for premature anode consumption, while organic contamination results in foam inside the anodic cells with resulting problems in terms of brine distribution and membrane performance. Pure brine at about 310 g/l enters the anodic compartment, where chloride ion oxidation takes place at the anode. Brine circulation inside the anodic compartment is promoted to avoid local gradients of concentration. Titanium is used for the anodic compartment, and anodic electrodes are titanium based and activated with a layer of mixed noble metal oxides, for example, iridium or platinum. Chlorine is collected from the top of the anodic element and sent to the main chlorine circuit. Sodium ions cross the cationic ion exchange membrane to reach the cathodic compartment.

At the cathodic compartment, a diluted solution of caustic soda at about 30% w/w is fed from the bottom of the cell, and leaves as 32% w/w solution at the top of the cell, together with hydrogen. Part of this caustic is sent to the products storage tanks, and part is diluted and recirculated within the cathodic circuit. The cathodic compartment is made entirely of nickel, while the cathodic electrodes are made of activated nickel (Figure 3.8).

Figure 3.8 A 150 000 ton per year Cl_2 100% cell room.

The membrane process produces high quality products; sodium hydroxide concentration unit and chlorine liquefaction systems are installed depending on market requirements. Hydrogen is pure to 99.9%, and can be used as a chemical or sold to the market.

3.2.5
Research

To reduce the power consumption and the capital cost necessary for the electrolysis of ion exchange membrane electrolyzers, the main area of research is focused on the following subjects:

1) The reduction of electrodes overpotential through the development of innovative catalytic coatings.
2) The development of new generation of ionic membranes to further reduce ohmic losses.
3) An increase in operating current density of the electrolyzers. Ion exchange membrane electrolyzers are today operated in the range of 5.5 to over $7\,\text{kA}\,\text{m}^{-2}$, and the trend is to increase the operating current density to over $8\,\text{kA}\,\text{m}^{-2}$. As a consequence, smaller electrolyzers could be installed, with savings on capital cost.

3.2.6
Breakthrough Technologies: Chlorine-Alkali Production with Oxygen Depolarized Cathode (ODC)

After over 20 years of research and development, the production of chlorine and caustic using oxygen depolarized cathode (ODC) technology (known also as gas diffusion electrode) became available to the industry in 2013. The ODC technology represents a breakthrough innovation with a 30% electricity saving over conventional membrane electrolysis.

The electrochemistry of chlorine production with ODC technology is according to the following reactions:

Anode reaction : $2Cl^-(aq) \rightarrow Cl_2(gas) + 2e^-$

Cathode reaction : $H_2O + \frac{1}{2}O_2(gas) + 2e^- \rightarrow 2OH^-(aq)$

Total reaction : $2NaCl(aq) + H_2O + \frac{1}{2}O_2(gas) \rightarrow Cl_2(gas) + 2NaOH^-(aq)$

The anodic reaction of chlorine evolution is the same as for the conventional electrolysis process. At the cathode, oxygen is used to prevent the formation of hydrogen, with the direct consequence of reducing the cell voltage from about 3 V, as in the standard electrolysis, to about 2 V. Figure 3.9 describes the electrochemistry of chlorine caustic with ODC.

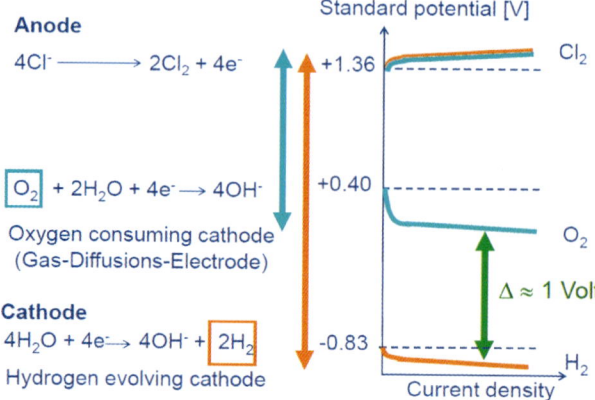

Figure 3.9 Electrochemistry of NaCl electrolysis with an ODC (oxygen depolarized cathode).

Notably, no hydrogen is produced during the electrolysis process. A three-phase reaction – liquid, gas and solid – takes place on an oxygen depolarized cathode, where oxygen, caustic soda, and the catalyst come in contact.

The ODC structure consists of a current distributor, a catalyst, and a binder. The current distributor acts as a support structure for the catalyst and for the binder, and it is made of a woven metal mesh (Figure 3.10).

The reaction proceeds in several steps: the oxygen consumption reaction starts by permeation of oxygen into the porous ODC structure, followed by a second step, in which oxygen dissolves in the caustic electrolyte and diffuses onto the catalyst surface. Oxygen is then chemically reduced, and the reaction products are removed by convective transport.

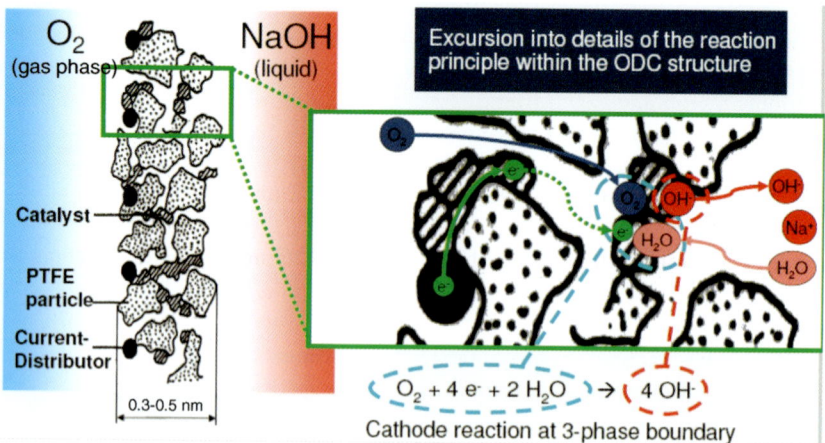

Figure 3.10 ODC structure.

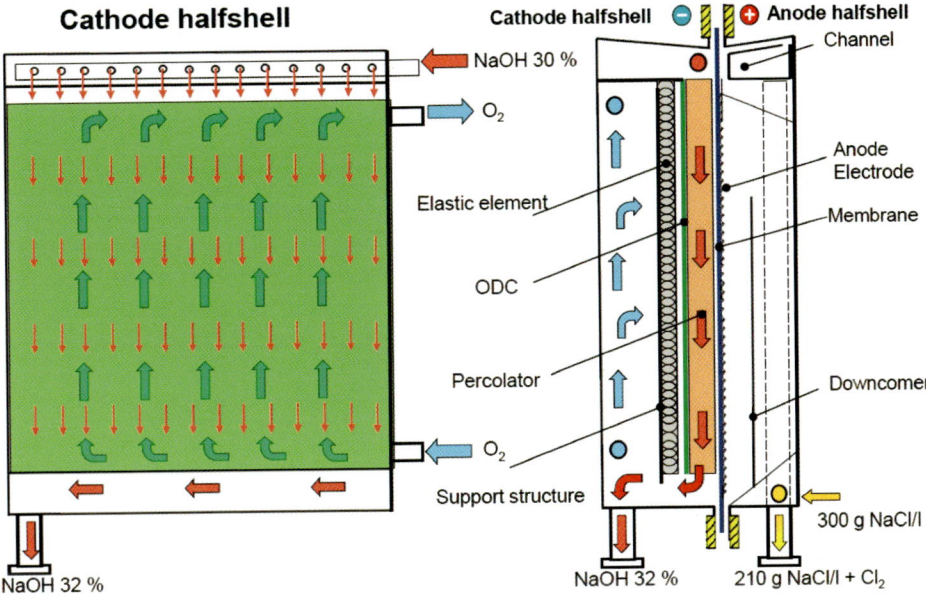

Figure 3.11 Cross section of NaCl ODC based electrolysis halfshell.

The basic electrolyzer configuration is similar to the conventional membrane ones, in order to make this technology compatible for the retrofitting of existing systems. Owing to the three-phase reaction taking place on ODC surface, the electrolyzer design must guarantee that all the specific operating conditions of the ODC are fulfilled (Figure 3.11). The anodic compartment of the electrolyzer, where chlorine evolution takes place, is similar to the conventional membrane configuration.

At the cathode side, a nickel support structure is used for mechanical support and current distribution. The nickel elastic element positioned over the support structure has the function of distributing the current to the ODC and to provide the right pressure to guarantee the mechanical stability of the cathodic array. The ODC is positioned on top of the elastic element. To avoid ODC flooding and to guarantee the even distribution of the catholyte all over the ODC, a special plastic porous media, called a percolator, is positioned between the ion exchange membrane and ODC. The percolator enables the formation of a caustic film between the membrane and the ODC, granting constant oxygen and caustic pressure distribution over the entire height of the compartment [6].

This configuration results in a power consumption of 1400 kWh/metric ton NaOH (100%) at kA/m^2 operating current density.

Technology validation to operate NaCl–ODC electrolyzers at higher current densities is today ongoing.

3.3
Use of Chlorine and Sodium Hydroxide

3.3.1
Chlorine

Chlorine is at the basis of important and fundamental industrial production, either as a part of the final chlorinated product or used only as intermediate [7]. Today chlorine production is about 60.6 million ton. The main use of chlorine is in the production of PVC, which represents over 33% of the overall consumption of chlorine (Table 3.1). PVC is a material widely used for applications such as pipes production, and windows frames and in the automotive industry for car components and in many other products. Isocyanate production, as the basis of polyurethane chemistry, is the second important application for chlorine.

Isocyanates are used in the production of paints and coatings to create weather-resistant surfaces. They are also used to make building materials (e.g., Styrofoam, flexible foams, adhesives, elastomers, and binders) and in the production of manufactured goods (e.g., bedding, furniture, clothing, appliances, electronics, tires, and packaging). Chlorine is an intermediate for the production of epoxy resins, as well as for the production of polycarbonate, widely used in DVD production, glasses, and car components. Chlorine is used for medicine production as well as for crop protection chemicals (almost 50% of crop protection chemicals are based on chlorine chemistry).

Other chlorine uses include pulp and paper industry, titanium production, hydrochloric acid synthesis for the food industry and for industrial use (such as for oil and gas and shale gas extraction).

Chlorine is also the main disinfectant for drinking water: in all industrialized countries drinking water is made safe with chlorine, and the lack of water disinfection is one of the main causes for the death of children in developing countries.

In the chlorine industry, to summarize the main use of chlorine for different applications, it has become common to refer to the so-called "chlorine tree."

Table 3.1 Uses of chlorine. Data from World Chlorine Council.

Application	Proportion (%)
Organics	32
Vinyls	33
Water treatment	6
Chlorine intermediates	5
Pulp and paper	3
Inorganics	2

Table 3.2 Uses of sodium hydroxide. Data from World Chlorine Council.

Application	Proportion (%)
Organics	17
Soaps, detergents, textiles	15
Alumina	14
Inorganics	12
Pulp	12
Water treatment	5
Others	25

This is a graphical representation that uses a tree as a model, where the different branches represent different chlorine use in the different markets, and the ramifications the different products. The chlorine tree is available from the main chlorine associations and chlorine related companies [8].

3.3.2
Sodium Hydroxide

Sodium hydroxide is a basic chemical used in many industrial fields. Sodium hydroxide production today is about 64.7 million ton per year. The most important application of sodium hydroxide is as a reactant in the production of organic chemicals (Table 3.2). The pulp and paper industry is a consumer of sodium hydroxide, as well as the cellulose industry. Aluminum refining from bauxite consumes about 4% of the world production of sodium hydroxide. Sodium hydroxide is a fundamental chemical for the production of soaps, detergents, and for the cosmetic industry. Sodium hydroxide is used for water treatment for the flocculation of heavy metals and as pH regulator. On reacting with chlorine, it is also employed for the production of sodium hypochlorite, which is used for sanitation and disinfection. Sodium hydroxide is also used in various diversified productions, such as in the food industry, synthesis of pharmaceutical compounds, and neutralization of acids. As with chlorine, the so-called "caustic tree," available from the main chlorine associations, summarizes the industrial applications for sodium hydroxide.

3.3.3
Economy of Chlorine and Caustic Soda

The worldwide installed capacity for chlorine is about 76.8 million ton. The chlorine industry has some specific characteristics that has driven its development in different regions of the world.

Chlorine is produced simultaneously to sodium hydroxide and hydrogen, but while caustic soda can be stored in very large quantities, both in liquid and solid

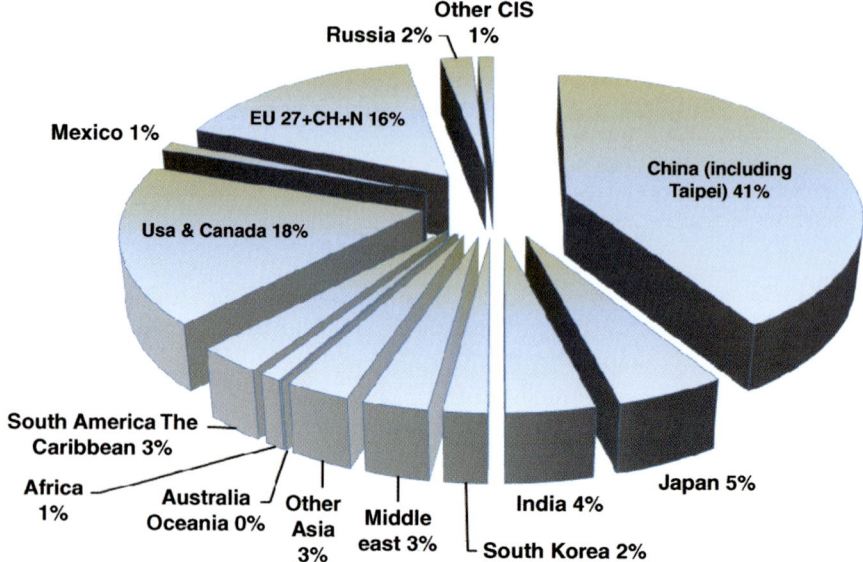

Figure 3.12 World chlorine production. Data from World Chlorine Council.

form, and can be shipped on the sea, chlorine is only stored in limited quantities and is transported generally by rail and by truck. For such reasons, generally, chlorine plants are associated with the local possibility of consuming chlorine rather than caustic. As an example, Australia is a net importer of sodium hydroxide for the aluminum industries; local production is extremely small since no market for chlorine exists there.

China is the biggest producer of chlorine. The percentage of installed chlorine capacity in different countries is reported in Figure 3.12.

In recent years, the chlorine industry has been subject to a consolidation process, especially in Europe and in the USA. In *Europe* the commitment to close or convert existing mercury plants into membrane technology by 2020 is progressing; the percentage of total capacity using mercury technology has fallen from 54% in 2000 to 30.6% in 2011 [2]. In 2011 membrane technology accounted for 51% of European chlorine capacity. In the *Unites States of America*, after a period of market consolidation, with the progressive conversion of mercury and diaphragm plants into membrane plants and a reduction of chlorine capacity, the low price of energy resulting from the use of shale gas is promoting again investments in energy intensive productions, including new chlorine alkali plants. In *China*, after a very rapid increase in the installed capacity, now accounting for about 31 million ton of chlorine, the industry is undergoing a phase of rationalization. China has no mercury plants, and is replacing the existing diaphragm plants with more energy efficient membrane plants. In *South America*, while the demand for caustic soda is raising, the market for chlorine cannot support the increase of local production of chlorine, and so industry needs are supported by caustic imports. *India*, despite its population, is still a low chlorine consumer. With a yearly production of about three million tons of chlorine, India has an average consumption of 1.85 kg of chlorine per capita, compared with 13 kg in China and 7.8 kg in Brazil [2]. Considering that chlorination of drinking water is still not mandatory, there is a high potential for increasing production. The market outlook for the coming years considers the conversion of the still existing mercury and diaphragm plants into membrane technology, and an increase in production with a compound annual growth rate (CAGR) of about 2% per year.

3.3.4
Energy Savings

Energy saving in the chlorine industry has always been at the center of innovation. The adoption of membrane technology afforded an approximately 30% reduction in electricity needed for electrolysis versus mercury technology. The advancement of ion exchange membrane technology is largely based on the development of more efficient low ohmic ion exchange membranes, new families of electrocatalytic coatings for anode and cathode electrodes, and the overall reduction of ohmic losses within the electrolyzer structure. This technology evolution is relatively close to the theoretical limit, and only incremental improvements will be possible for the future.

The recent market introduction of ODC (oxygen depolarized cathode) technology, with 30% energy savings over ion exchange membrane technology, constitutes the next generation of electrolyzer. The market adoption of this technology greatly depends on the specific characteristics of each chlorine plant. The main variables to be considered when evaluating the cost–benefit of ODC include the need for oxygen as part of the production process, which is not needed for conventional membrane electrolysis, the absence of hydrogen, which in many plants has a commercial value as a chemical or as a fuel, and the cost of grid electricity, which makes this technology interesting when the cost is high. Through the analysis of such variables it is possible to identify plants where ODC technology brings advantages versus plants where economics are not yet mature for this application. Another possible way to reduce chlorine plant power consumption is through utilization of the hydrogen that is co-produced during the electrolysis process. The installed plant capacity of 76.8 million ton of chlorine per year results in a hydrogen generation capacity of 2.15 million ton per year.

Hydrogen produced during the electrolysis process has several characteristics that makes it ideal as energy vector for production of electricity: Its quality is extremely pure, over 99.9% v/v and the only impurities are inert gases. Its production is continuous since chlorine plants run 24 h a day for over 340 days per year. Even in the case of problems with the membrane integrity of the electrolyzer, hydrogen is never contaminated by chlorine.

Hydrogen from the chlorine industry is used as a chemical for production of hydrochloric acid and for hydrogenation processes; in many cases it is sold to the market through hydrogen pipelines or compressed in cylinders. As an alternative, hydrogen is used as a clean fuel to produce steam, which is widely used in chemical plants. Only a small fraction of hydrogen is vented to the atmosphere, the last option when no other use is possible. The European chlorine association aims to improve hydrogen utilization, and to further reduce the 10% that is still vented today [2]. Different technologies have been tested to convert industrial by-product hydrogen into clean electricity. In 2009, the first hydrogen powered turbine began operation, close to Venice (Italy), producing 12 MW electricity plus an additional 4 MW from the steam generated by the turbine. Hydrogen was provided by a cracker in the area. Fuel cell developers have considered the chlorine industry as an early market application, and demonstration models have been installed in Europe and in the USA in recent years, evolving from a few kilowatts to megawatt size. The energy reduction potential is about 20% of the electricity consumed for electrolysis, depending on the fuel cell technology considered.

References

1 World Chlorine Council (2011) Reduction of Mercury Emissions and use from the Chlor-Akali Sector Partnership.
2 Eurochlor (2013) The Amazing Chlorine Universe. Available at http://www.eurochlor.org/media/64565/01-the_amazing_chlorine_universe.pdf (accessed 27 November 2014).
3 Joint Research Centre (2014) Best Available Techniques (BAT) Reference Document for the Production of Chlor-alkali, JRC Science and Policy Reports.
4 European Council (22 October 2008) On the banning of exports of metallic mercury and certain mercury compounds and mixtures and the safe storage of metallic mercury. *Official J. Eur. Union*, **51**, 75
5 Grot, W. (2007) *Fluorinated Ionomers*, William Andrew, Norwich, NY.
6 Faita, G. and Fulvio, F. (filed 2002) European Patent Specification EP 1 446 515 B1 Electrochemical cell with gas diffusion electrodes.
7 World Chlorine Council (2002) Chlorine's Chemistry Role in our Daily Life. Available at http://www.worldchlorine.org/wp-content/themes/brickthemewp/pdfs/chlorine.pdf (accessed 27 November 2014).
8 World Chlorine Council (2014) Chlorine: The Element of Surprise. www.chlorinetree.org.

Part 2
Hydrogen as an Energy Carrier

Part 2.1
Introduction and National Strategies

Hydrogen Science and Engineering: Materials, Processes, Systems and Technology, First Edition.
Edited by Detlef Stolten and Bernd Emonts.
© 2016 Wiley-VCH Verlag GmbH & Co. KGaA. Published 2016 by Wiley-VCH Verlag GmbH & Co. KGaA.

4
Hydrogen Research, Development, Demonstration, and Market Deployment Activities

Jochen Linssen and Jürgen-Friedrich Hake

4.1
Introduction

Hydrogen is a flexible energy carrier that can be used as fuel for transportation or other applications as well as chemical feedstock. Together with electrolysis, hydrogen can be an interface between the power sector and other end consumer technologies by offering flexible operation and the option for storing electricity or other energy in chemical form. Especially together with renewable energies and the intermittent generation of wind and photovoltaics (PV), hydrogen can be an important element that supports the integration of renewable electricity by offering controllable consumption and storage. Hydrogen is one of only three potentially net zero-emission energy carriers (along with electricity and biofuels) and is an important part of decarbonization strategies to reduce greenhouse gas emissions (see Reference [1]).

Given these advantages hydrogen technologies are under research, development, and demonstration worldwide, and are on the way to mass market. Because every nation or international association has a different focus on energy policy and what will be the role of hydrogen in the future energy system, this chapter points out national and international strategies to make hydrogen an important element of the energy supply system. Some information for this chapter was developed during the work of the International Energy Agency (IEA) Hydrogen Implementing Agreement (HIA) Task 30 on Global Hydrogen Resource Analysis.

The chapter covers the hydrogen research strategies of selected countries and gives an overview of existing hydrogen funding schemes and activities worldwide. The list of selected countries does not claim to be complete as there are more countries with comprehensive hydrogen support programs. The period up to December 2013 is covered, along with some important updates up to mid-2015.

Hydrogen Science and Engineering: Materials, Processes, Systems and Technology, First Edition.
Edited by Detlef Stolten and Bernd Emonts.
© 2016 Wiley-VCH Verlag GmbH & Co. KGaA. Published 2016 by Wiley-VCH Verlag GmbH & Co. KGaA.

4.2
Germany

4.2.1
Energy Framework and Relevant Policies

The German Federal Government set ambitious targets for energy and climate policy with the introduction of the Energy Concept in September 2010. The energy supply system in Germany therefore will undergo a fundamental transformation. At the same time, a secure and affordable supply of energy will also be imperative in the future if Germany is to remain a competitive business location. The Federal Government laid the foundations for this new energy era in the summer of 2011 with the adoption of a comprehensive legislative package known as the "Energy Package" (six laws and one ordinance). The master plan for the development of the German Energy System towards 2050 contains concrete goals for different energy related sectors:

- phasing out of nuclear energy by the end of 2022;
- reduction of CO_2 emissions to 40% by 2020, 55% by 2030, and 80% by 2050 (reference year for emissions: 1990);
- reduction of the primary energy consumption to 50% by 2050;
- share of renewable energy regarding primary energy to be a minimum 30% by 2030;
- reduction of electricity consumption (gross): 10% by 2020 and 25% by 2050;
- share of renewable energy for electricity generation 50% by 2030 and 80% by 2050;
- reduction of heat demand for space heating 20% by 2020;
- share of renewable energy for transportation fuels 10% by 2020, 1 million electric vehicles (EV) by 2020, and 6 million EV by 2030 (this can be BEV (battery electric vehicle), PHEV (plug in electric vehicle), and fuel cell electric vehicles (FCEVs)).

To achieve the goals of the energy concept the federal government intensifies energy research to identify new technologies which maintain a reliable energy supply at affordable prices in future. On the basis of the sixth Energy Research Program, R,D&D projects are developing technologies for the energy supply of tomorrow. The development of new storage technologies is a central thrust of energy research. Hydrogen is an important part of that research strategy.

4.2.2
Hydrogen Related Energy Policy Strategies

The German research, development and demonstration hydrogen activities are mainly combined in the "National Hydrogen and Fuel Cell Technology

Innovation Programme (NIP)". The activities are added to by industrial R&D activities, programmes of Federal States, and European research cooperation.

The total budget within NIP in the period 2006–2016 is €1.4 billion, coming in equal parts from NIP and the participating industries. Governmental sponsors of the NIP are the Federal Ministry of Transport, Building and Urban Affairs, the Federal Ministry of Economics and Technology, the Federal Ministry of Education and Research, and the Federal Ministry for the Environment, Nature Conservation and Nuclear Safety.

Research, development and demonstration projects within NIP are concentrated on the R&DD topics transportation, production, stationary and residential household applications, special markets, and cross-cutting issues. The financial expenses and goals of the individual topics have been described [2].

The first phase of the NIP ends in 2016. At the end of summer 2013 a proposal was prepared by the representatives of industry and science on the advisory board of NOW GmbH for the future scope of work in the second phase (2014–2023). The defined goals for 2025 are [3]:

- 500 filling stations;
- 500 000 FCEV;
- 2000 buses for public transport;
- hydrogen production from renewable energies: flexibility and storage option;
- integration in the entire energy system: combining electricity, gas, and transport sector;
- 1500 MW installed electrolysis capacity;
- power to gas and hydrogen storage;
- 500 000 micro CHP FC (combined heat and power fuel cell) units;

The financial requirements for the NIP II is about €3.9 billion, of which the industry part is planned to be €2.3 billion.

4.2.3
Hydrogen Research, Development, Demonstration, and Deployment Activities

4.2.3.1 Transportation
The flagship project within the current NIP is the Clean Energy Partnership (CEP). One of the main goals is to develop, build, and operate a safe hydrogen refueling infrastructure with 50 filling stations for 5000 fuel cell electric vehicles (FCEV) by 2015. Therefore, several filling stations were taken into operation within 2012. Table 4.1 lists new filling stations installed in 2012 and some technical data.

TOTAL, Enertrag, and Linde are currently planning another "multi-energy filling station" at Berlin-Brandenburg-International Airport within the CEP.

The industry partners involved in CEP in 2012 were GM/Opel, Berliner Verkehrsbetriebe, BMW, Daimler, EnBW, Enertrag, Ford, Hamburger Hochbahn, Linde, Shell, Siemens, Statoil, TOTAL, Toyota, Vattenfall, and Volkswagen.

Table 4.1 New filling stations in Germany in 2012.

Parameter	Berlin–Heidestraße	Hamburg–Barmfelder Chaussee	Hamburg–Cuxhavener Straße	Düsseldorf–Höherweg
H_2 production	Centralized wind electrolysis and BtH	Byproduct	Centralized wind electrolysis	Byproduct
H_2-transportation	GH_2-trucks	GH_2-trucks	GH_2-trucks	GH_2-trucks
Pressure (bar)	700 (filling)	700–1000 (at storage)	700 (filling)	200 (storage) 350 and 700 (filling)
Filling time (min)	3	3	3	3–5
Filling capacity (kg day^{-1})	200	40	30–35 cars per day	212
Highlights	Advanced cooling system, Ion-compressor	Meets SAE J2601	Upgraded with "H2small"	First of at least seven filling stations in NRW
Partners	Total, Linde, Enertrag	Shell, Air Products, Daimler AG	Total, Linde, Enertrag	Air Liquide, Opel, 3M, e-plus, EnergieAgentur.NRW

Source: References [4,5].

Figure 4.1 Filling station at Fraunhofer ISE in Freiburg. Source: Reference [6].

In addition to the CEP, Fraunhofer ISE in Freiburg (South of Germany) built a filling station and is operating it. There, the hydrogen is produced by a pressurized PEM-electrolyzer operating with wind and solar power. The capacity of the electrolyzer is 13 kg per day and the produces hydrogen with a purity of 5.5 (99.9995%). The hydrogen is compressed in two temporary stages up to a pressure of 700 bar. [6] The filling station (Figure 4.1) is funded outside of CEP by the Federal State of Baden-Württemberg.

In 2012 the HyCologne Initiative celebrated one year of operation of Phileas busses. The buses consume 15 kg H_2 per 100 km and have a storage capacity of 40 kg. Within the project the installation and operation of a safe H_2 infrastructure is another important aspect. This study gathers operating experience and information for standardization of infrastructure, too.

4.2.3.2 Hydrogen Production

The main focus of the Hydrogen Production part of the NIP is hydrogen production from renewables. Coupling of renewable energy with electrolyzers is the main funding topic within NIP. In 2012 NOW published a report on the technical status of electrolyzers and their ability to use fluctuating energies for hydrogen production. Another study evaluates processes and technologies for producing hydrogen from biomass. The study uses technical, economic, and ecological criteria for the concept evaluation and delivers recommendations for the German hydrogen strategy.

4.2.3.3 Stationary and Residential Applications

In the project Callux more than 200 fuel cell heating appliances were in operation in 2012. Baxi-Innotech, Hexis, and Vaillant installed the micro CHP units mainly in one and two-family houses. More than 1 million operating hours had been in reached in 2012. A highlight within the Callux project is the first installation of a wall-mounted FC operating in Karlsdorf. This fuel cell has an electric power of 1 kW and a heating power of 2 kW. However, for fuel cell heating applications the market launch is expected at the end of the Callux project in 2016.

A molten carbonate fuel cell (MCFC) was installed at the new building of the Federal Ministry of Education and Research (BMBF). Its operation started in December 2014 and it is expected to be in continuous operation for ten years [7]. The goals for the electric efficiency are 40% and 60% overall efficiency including heat generation.

Another stationary fuel cell was installed at a Mercedes-Benz dealership in Hamburg. The phosphoric acid fuel cell (PAFC) has an electric capacity of 100 kW and a thermal capacity of 120 kW. Activities on building insulation and the use of the fuel cell will reduce CO_2 emissions by 60%. The payback period of the €1 million investment in the fuel cell will be six to seven years.

4.2.3.4 Special Markets

In the Special Markets programme essential goals are market preparation and launch, the demonstration of practical suitability, and transition to series production. Projects in the Special Markets area are divided into application segments of remote systems, UPS (uninterruptable power supply), independent/hybrid power supply, electricity supply for camping applications, industrial and special transport vehicles, and micro fuel cells. To reach the goals requires are optimization of components and systems, integration developments, and refueling infrastructure as well as meeting approval issues [2].

As a cross cutting activity, the strategic alliance Clean Power Net was established with approximately 70 Germany companies from the information and communication sector. The goal of this network is to accelerate the commercialization of fuel cells in ICT and process automation applications.

4.2.3.5 Industry Activity

In 2009 leading industrial companies initiated the so-called "H2Mobility" group. The goal of the industry group is the establishment of a nationwide network of hydrogen fueling stations in Germany. In 2013 the H2Mobility activity supported by Air Liquide, Daimler, Linde, NOW, OMV, Shell, and Total agreed on a concrete action plan for the construction of a hydrogen refueling network in Germany. The goal of the action plan is to expand the current network (15 filling stations) to a public hydrogen infrastructure of about 400 H2 filling stations in 2023. The hydrogen supply by then will be suitable for everyday use in urban areas and rural areas. The average distance to reach a hydrogen filling station should be no more than 90 km. An intermediate step to this long-term goal is to build 100 hydrogen stations over the next four years. This will be the precondition for the market success of fuel cell electric vehicles (FCEVs) and is combined with an overall planned investment of around €350 million. The launch of FCEVs in German by the first manufacturers is scheduled for 2015 [8a]. In addition, for example, the sale of the Toyota FCEV Mirai started in Japan December 2014. The market launch in the USA and Europe is planned for summer or autumn in 2015 [8b].

4.3
Norway

4.3.1
Energy Framework and Relevant Policies

Norway has the special energy supply situation in that it explores and generates ten times more energy products than it consumes. Most of the exported fossil energy carriers (oil and natural gas) go to the European market. Domestic electricity generation comes nearly exclusive from hydropower with the option to export electricity to Europe. Consequently, about 60% of Norwegian energy consumption is based on renewable energy. A political agreement on climate policy was decided on in 2008, and a white paper from the government was presented in 2012. With this paper the Norwegian Government intends to take a number of steps to reduce greenhouse gas emissions and promote technological advances. Some of the most important are: establishing a new climate and energy fund, raising the CO_2 tax rate for the offshore industry, and improving public transport.

Norway has set a target of reducing CO_2 emissions by 30% by 2020. The Norwegian long-term goal for 2050 will to be carbon neutral. Due to population growth and increasing prosperity CO_2 emissions from transport are increasing the most. To meet the Norwegian climate targets, transport emissions must be reduced. Transport emissions can only be reduced by cutting back on the use of fossil fuels. Therefore, the use of alternative fuels such as hydrogen, electricity, and various types of biofuels are promoted.

4.3.2
Hydrogen Related Energy Policy Strategies

There is no specific hydrogen and fuel cells program in Norway. The R&D activities have mainly been supported under the large RENERGI program, covering renewable energy, energy efficiency, and other clean energy topics. Carbon capture and storage (CCS) has a dedicated R&D&D program in Norway. The RENERGI program period has come to an end, and the new program EnergiX has the same structure and content, but with a new program plan, and is in place. Hydrogen and fuel cells are still part of the R&D program.

To promote and accelerate the deployment of alternative fuels various incentives are installed in Norway. Therefore, the Norwegian agency Transnova was set up. Its main policy instrument was to provide grants for demonstration and pilot projects that reduce CO_2 emissions in the transport sector. Grants were mainly given to projects at a stage between R&D late phase and market introduction. Transnova did not provide support to basic science/research. In January 2015, Transnova was transferred to Enova SF.

The high taxation of conventional cars combined with an attractive incentive scheme for zero emission vehicles (e.g., sales and VAT exemptions, reduced annual road tax, tolls and municipal parking fee exemptions, access to public transport lanes) has made Norway an attractive early market for both battery and fuel cell electric vehicles. In addition, Norway has a large surplus of hydroelectric power and may contribute to developing the use of hydrogen in the European energy system. Norway set the goal of reducing CO_2 emissions by 30% in 2020 and to be carbon neutral in 2050. To meet these climate targets, transport emissions must be reduced significantly notwithstanding national transport demand rising owing to population growth and increasing prosperity. Accordingly, the reductions goals can only be met by replacing fossil with renewable fuels. Consequently, the use of renewable and carbon neutral fuels like hydrogen, electricity, and various types of biofuels is promoted.

4.3.3
Hydrogen Research, Development, Demonstration, and Deployment Activities

Hydrogen research and development activities are performed by the major institutions in Norway. Figure 4.2 gives a survey of projects that Norwegian R&D institutions are involved in.

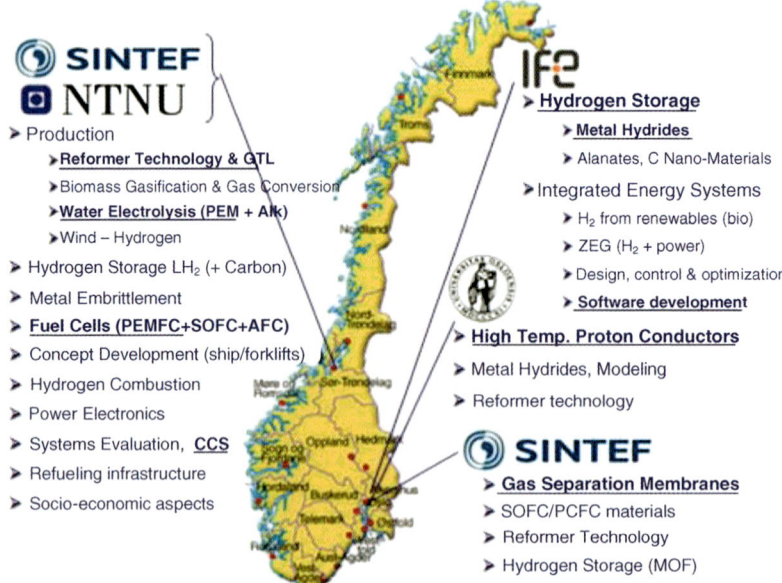

Figure 4.2 Norwegian R&D institutions related to hydrogen. Source: Reference [9].

In Norway in 2013 six hydrogen refueling stations were in operation. The development of this infrastructure began in 2004 with the HyNor project. Financial support for building up the infrastructure was provided by Transnova.

4.3.3.1 HyNor Project

The HyNor project, a joint public–private partnership initiated to demonstrate a real-life implementation of a hydrogen refueling infrastructure, continued in 2011, but with a focus on a denser network in the Oslo region, rather than a "hydrogen highway," which was the motivation at the start of the project. In 2011, one hydrogen refueling station was added to the network, and one was closed down, leaving the number of operating stations by the end of the year at four. The station that closed its operation was the station in Stavanger, and the new station is part of the H2-Moves Scandinavia project, which is a European lighthouse project under the FCH JU (Fuel Cells and Hydrogen Joint Undertaking). The project involves a state of the art filling station in Oslo, which will provide 19 FCEVs in the project with hydrogen, and a mobile refueling system that will be used for a European road tour [10].

The filling stations in operation use hydrogen from different resources and demonstrate various technologies for production and supply of hydrogen. Two additional stations were opened in 2012: one in Lillestrøm, 25 km north of Oslo; and one hydrogen bus station in the city of Oslo. The bus station is part of the HyNor Oslo bus project, which is also part of the FCH JU project CHIC (Clean Hydrogen in European Cities). VanHool supplies the buses, and Air Liquide Norway AS will deliver hydrogen from a filling station at Rosenholm bus facilities in Oslo. The buses will be in daily operation on a regular route for a minimum of five years.

4.3.3.2 ZEG Power

ZEG technology provides the power concept with the highest energy efficiency from hydrocarbon fuels, more than 80% for large-scale plants, and ZEG technology is the only concept with the potential to produce electricity and hydrogen with integrated CO_2 capture less expensively than today. In addition, there are no emissions from a ZEG Power plant, NO_x, particles, or aerosols. In 2013 ZEG Power realized a $20\,kW_{el} + 30\,kW_{H2}$ ZEG plant (BioZEG) based on biomass in connection with the Hynor Lillestrøm hydrogen filling station. The technology also shows great flexibility with respect to fuel, since all types of hydrocarbons can be used: natural gas, biogas, gasified coal, tar, or oil. In addition, the relative amount of electricity and hydrogen (and heat) can be adjusted according to market demand. The technology is developed as a joint cooperation between the Institute for Energy Technology (IFE) and Christian Michelsen Research AS (CMR Prototech) through the development company ZEG Power AS. A prestudy of the plant (50 kW) has shown a potential of more than 70% in total efficiency [11].

4.3.3.3 Transnova Hydrogen Projects

In the project "H2Moves" a hydrogen filling station has been built and 17 vehicles have been made available to professional staff and the general public. "NextMove" stimulates collaboration between Norway, Sweden, and Denmark by coordinating and strengthening the purchase and maintenance of hydrogen vehicles and through education. The project "HYOP" manages and develops hydrogen filling stations in Norway. In addition, Transnova is supporting the development of two small-scale production technologies. In the project "GasPlas" a cold plasma technology is being developed to produce hydrogen from methane without emitting CO_2. The technology is scalable and allows production-on-demand onsite. At "Hynor Lillestrøm" new and efficient ways to produce hydrogen from landfill methane and through electrolysis are being developed [8].

4.4
European Union

4.4.1
Energy Framework and Relevant Policies

In 2009 the European Commission set up the climate and energy package. The package is a set of binding legislations that aim to ensure the European Union meets its ambitious climate and energy targets for 2020. These targets (known as the "20-20-20" targets) set three key objectives for 2020 [12]:

- 20% reduction in EU greenhouse gas emissions from 1990 levels;
- raise share of EU energy consumption produced from renewable resources to 20%;
- 20% improvement in the EU's energy efficiency.

As a further development the European Union published a White Paper on transportation in 2011. The developed roadmap consists of 40 concrete initiatives for the next decade to build a competitive transport system that will increase mobility, remove major barriers in key areas, and fuel growth and employment. At the same time, the proposed measurements will dramatically reduce Europe's dependence on imported oil and cut carbon emissions in transport by 60% by 2050. Some of the key goals include [13]:

- phasing out of conventionally-fueled cars in cities;
- 40% use of sustainable low carbon fuels in aviation;
- 40% cut in shipping emissions;
- 50% shift of medium distance intercity passenger and freight journeys from road to rail and waterborne transport;

All of which will contribute to a 60% cut in transport emissions by the middle of the century.

4.4.2
Hydrogen Related Energy Policy Strategies

At the beginning of 2013 the European Commission published a proposal for a new European Directive on the deployment of alternative fuels infrastructure. The main focus is on evaluation of alternative fuel options available to substitute mineral oil products and the contribution to reducing greenhouse gas emissions from transport. The directive stated the main alternative fuel options: electricity, hydrogen, biofuels, compressed natural gas (CNG), liquefied natural gas (LNG), gas-to-liquid (GTL), and liquefied petroleum gas (LPG). The lack of alternative fuel infrastructure is considered as a major obstacle to the market introduction of alternative fuels and consumer acceptance. The proposal aims at ensuring the build-up of an alternative fuel infrastructure and the implementation of common technical specifications for this infrastructure in the EU [14].

European R,D&D activities related to Hydrogen and Fuel Cells are summarized under the "Fuel Cells and Hydrogen Joint Undertaking" (FCH JU). The FCH JU is a public–private partnership (European Commission, fuel cell and hydrogen industry, and research community) supporting research, technological development, and demonstration (RTD) activities in fuel cell and hydrogen energy technologies in Europe. The goal of the partnership is to accelerate the introduction of hydrogen and fuel cells onto the market as an important element in a low carbon energy supply system. The JU was established in 2007/8, based on a strong priority for this field in EU. The FCH JU supports basic research, technological development, technology demonstration, and support actions to promote public acceptance, education, and standardization (Figure 4.3) [15].

Under the new EU framework program for research and innovation "Horizon 2020"–starting in 2015 for a five years period–hydrogen will remain as a significant research and demonstration topic. New calls for hydrogen and fuels research projects are in preparation.

4.4.3
Hydrogen Research, Development, Demonstration, and Deployment Activities

The funding system for Hydrogen and Fuel Cell related R&DD projects follows open and competitive calls for project proposals and an independent evaluation. About €470 million, over a six year period, have been granted by the European Union under the FCH JU. The projects are classified by the application areas transport and refueling infrastructure, hydrogen production and distribution, stationary power production and combined heat and power, early markets and cross-cutting issues [15]. Table 4.2 gives a survey of projects initiated in the call 2011.

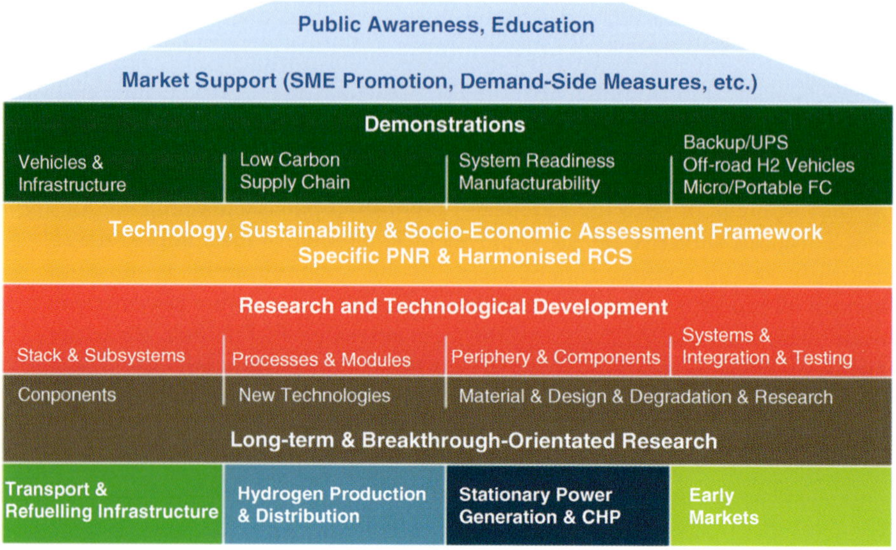

Figure 4.3 Survey of FCH JU support activities. Source: Reference [15].

All supported activities contribute to reducing the time to market penetration of hydrogen and fuel cell technologies and to delivering mature, robust, and cost-effective hydrogen supply and fuel cell technologies.

4.5
Canada

4.5.1
Energy Framework and Relevant Policies

Canada's Energy Policy Framework sets a series of principles and agreements. This leads to an energy policy that refers to market orientation, gives due respect to jurisdictional authority, and takes care of the role of the provinces. Canadian energy policy pays attention to flexibility to ensure an economically competitive and innovative energy sector that sustainably delivers a secure, reliable, and safe supply. To achieve specific policy objectives the Canadian government makes use of targeted intervention in the market process through regulations or other means. Canada and the USA concluded the North American Free Trade Agreement that emphasizes the importance of competitive market behavior and encourages investment in Canadian energy markets.

The National Energy Board of Canada supports the areas of environmental protection and efficient energy infrastructure and markets in the regulation of pipelines, transmission lines, energy development, and trade. The installed

Table 4.2 Survey of FCH JU supported projects of the call for proposals made in 2011.

Acronym	Short description	Stationary and CHP	H$_2$ production/ storage	Transport/ refuelling	Cross-cutting issues	Early markets
CLEARgen Demo	Integration and demonstration of large stationary fuel cell systems for distributed generation	X				
EVOLVE	Evolved materials and innovative design for high-performance, durable, and reliable SOFC cell and stack	X				
EURECA	Efficient use of resources in energy converting applications	X				
TriSOFC	Durable solid oxide fuel cell tri-generation system for low carbon Buildings	X				
ene.field	European-wide field trials for residential fuel cell micro-CHP	X				
FluMaBack	Fluid management component improvement for back up fuel cell systems	X				
T-CELL	Innovative SOFC architecture based on triode operation	X				
ARTIPHYCTION	Fully artificial photo-electrochemical device for low temperature hydrogen production		X			
NOVEL	Novel materials and system designs for low cost, efficient and durable PEM electrolyzers		X			
Don Quichote	Demonstration of new qualitative innovative concept of hydrogen out of wind turbine electricity		X			
BOR4STORE	Fast, reliable, and cost-effective boron hydride based high capacity solid state hydrogen storage materials		X			
ELECTROHYPEM	Enhanced performance and cost-effective materials for long-term operation of PEM water electrolyzers coupled to renewable power sources		X			

(continued)

Table 4.2 (Continued)

Acronym	Short description	Stationary and CHP	H$_2$ production/ storage	Transport/ refuelling	Cross-cutting issues	Early markets
EDEN	High energy density mg-based metal hydrides storage system		X			
UNIfHY	Biomass steam gasification for hydrogen production		X			
MATHRYCE	Material Testing and Design Recommendations for Components exposed to Hydrogen enhanced fatigue		X			
IMPALA	Improve PEMFC with Advanced water management and gas diffusion Layers for Automotive application			X		
SWARM	Demonstration of Small 4-Wheel fuel cell passenger vehicle Applications in Regional and Municipal transport			X		
Phaedrus	High Pressure Hydrogen all electrochemical decentralized refuelling station			X		
CATHCAT	Novel catalyst materials for the cathode side of MEAs suitable for transportation applications			X		
IMMEDIATE	Innovative autoMotive MEa Development – implementation of Iphe-genie Achievements Targeted at Excellence			X		
IMPACT	Improved lifetime of automotive application fuel cells with ultra-low Pt-loading			X		
PUMA MIND	Physical bottom Up Multiscale Modelling for Automotive PEMFC Innovative performance and Durability optimization			X		
STAMPEM	Stable and low cost manufactured bipolar plates for PEM Fuel Cells			X		
ARTEMIS	Automotive PEMFC Range extender with high temperature Improved MEAs and Stacks			X		

Project	Description				
HyTransit	European Hydrogen Transit Buses in Scotland	X			
Stack-Test	Development of PEM fuel cell stack reference test procedures for industry		X		
HyUnder	Assessment of the potential, the actors, and relevant business cases for large-scale and seasonal storage of renewable electricity by hydrogen underground storage in Europe		X		
HyLIFT-EUROPE	Large-scale demonstration of fuel cell powered material handling vehicles			X	
HYPER	Integrated hydrogen power packs for portable and other autonomous applications			X	
LiquidPower	Fuel cell system for back-up-power and telecom applications with methanol reformers			X	
BeingEnergy	Integrated low temperature methanol steam reforming and high temperature polymer electrolyte membrane fuel cell			X	
SAPIENS	SOFC Auxiliary Power In Emissions/Noise Solutions			X	
PURE	Development of auxiliary power unit for recreational yachts			X	

Source: Reference [15].

Program on Energy Research and Development supports the development of efficient and clean energy technologies [16].

4.5.2
Hydrogen Related Energy Policy Strategies

Hydrogen and fuel cell technologies represent a unique opportunity for Canada to reduce the environmental footprint of many sectors, such as transportation, which accounts for approximately 30% of total energy use and 28% of all greenhouse gas emissions in Canada.

Canada is one of the global leaders in hydrogen and fuel cell research, development, and early stage commercialization and is also a large hydrogen producer and user with long experience and extensive expertise in this area. Canada's industry covers nearly all elements within the supply chain. Important industrial companies with locations in Canada are, for instance, Ballard Power Systems, Hydrogenics [17].

Two national groups are active in the hydrogen arena. The Canadian Hydrogen and Fuel Cell Association (CHfCa) provides services and support to Canadian corporations, government, and educational institutions in promoting the development, demonstration, and deployment of hydrogen and fuel cells of hydrogen technologies. The industry group "Industry Canada" aims to enhance the competitiveness of the Canadian industry. This includes hydrogen and fuel cell activities [17].

4.5.3
Hydrogen Research, Development, Demonstration, and Deployment Activities

In 2011, the Canadian hydrogen and fuel cell sector reported employment at approx. 2000, a stable number of demonstration projects (approx. 120), and an increase both in number of strategic alliances and research partnerships. Hydrogen and fuel cell related facilities and activity, RD&D expenditure, and employment were largely concentrated in region of British Columbia.

In 2011 the RD&D expenditure of industry and research institutions was around $(CAD)137 million. Most of the expenditure went on research and development activities (Table 4.3). Expenditure on demonstration projects was lower, at approximately $(CAD)20 million.

Refueling infrastructure was the main area of focus, with 50% of overall demonstration projects, followed by mobile applications for auxiliary power units and drivetrains [17].

Regarding the topics of research and development, Canada is working to increase the viability of hydrogen production and purification technologies through R,D&D that reduces costs and improves safety, durability, and efficiency. In addition, activities to design, develop, and demonstrate affordable, durable, and lightweight hydrogen storage systems for transportation are being undertaken. For hydrogen utilization the focus of R&D is on the increase of

4.5 Canada | 75

Table 4.3 Canadian RD&D expenditures in $(CAD).

	2011 Total RD&D expenditure ($ million)		
	R&D	Demonstration	Total
Corporate	83.0	13.3	96.3
Government	22.2	7.1	29.3
Academic and non-profit	11.0	—	11.0
Total RD&D	116.2	20.4	136.6

Source: Reference [17].

viability for fuel cells in transportation, through cost reduction and improved safety, durability, and efficiency of PEMFCs. Canada is active in the development of codes and standards pertaining to the production, quality, storage, dispensing, and refueling infrastructure for hydrogen and fuel cells in the transportation sector. This includes the development of harmonized codes and standards together with the USA [9]. The progress made in hydrogen and fuel cells is illustrated in Figure 4.4, which gives a timeline of the history in R&DD work from 1980 to 2010.

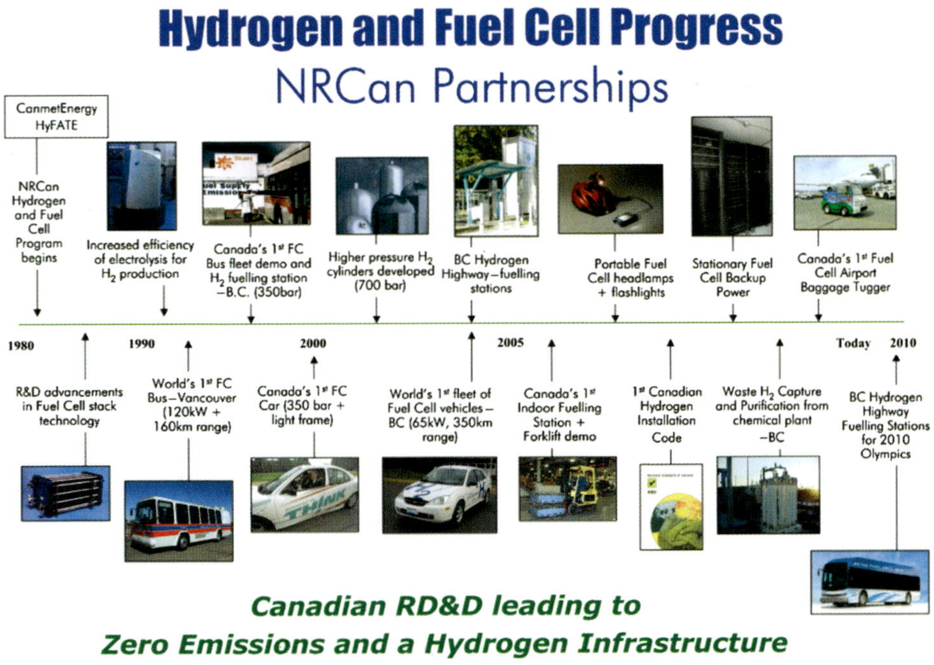

Figure 4.4 Timeline of hydrogen and fuel cells progress in Canada. Source: Reference [18].

4.6
United States of America

4.6.1
Energy Framework and Relevant Policies

The energy policy of the United States of America aims to reduce the dependence on foreign oil and position the USA as the global leader in clean energy technologies. The goals for a secure energy future of the USA are the increase of safe and responsible domestic oil and gas production, new fuel economy standards for cars and trucks to provide efficient transportation, the improvement of energy efficiency in homes and buildings, to provide consumers with choices to reduce costs/save energy, and take steps to make the USA a leader in the clean energy options. For instance, in 2011, American oil production reached the highest level in nearly a decade and natural gas production reached an all-time high [19].

The overall goals in energy policy of the US government are supplemented by State and local government activities to develop and install the next generation of clean energy technologies at the local level.

4.6.2
Hydrogen Related Energy Policy Strategies

The Hydrogen and Fuel Cells Program is coordinated across the Department of Energy, including activities in the offices of Energy Efficiency and Renewable Energy (EERE), Science (SC), Nuclear Energy (NE), and Fossil Energy (FE). The goal of the program is to prepare comprehensive efforts to enable the widespread commercialization of hydrogen and fuel cell technologies in diverse sectors of the economy. The program addresses the full range of challenges facing the development and deployment of hydrogen and fuel cell technologies by integrating basic and applied research, technology development and demonstration, and other supporting activities.

Highlights of the program's accomplishments in hydrogen related RD&D are documented annually in the US DOE Hydrogen and Fuel Cells Program Annual Progress Report (see Reference [20]). In the fiscal year (FY) 2013, the Congressional budget was $98 million for the DOE Hydrogen and Fuel Cells Program. The program is organized into distinct sub-programs focused on specific areas of RD&D, as well as other non-technical challenges.

In addition to the US governmental programmes, states prepare funding or set incentives to establish hydrogen applications on the market. Some states, including California, have policies in place to support the future role of hydrogen in vehicle applications. As a front runner state California established policy mechanisms to improve market entry of zero emission vehicles (ZEVs). Manufacturers are required to increase sales of ZEV (including plug-in hybrid, battery electric, and fuel cell electric vehicles) over time. By 2020 California will have established adequate infrastructure to support 1 million ZEVs and more than 1.5 million ZEVs in 2025. Incentive funding is widely acknowledged as necessary to make the business case for investing in the early commercial stations. Recently, eight state governors have signed an MOU to develop ZEV mandates similar to California.

4.6.3
Hydrogen Research, Development, Demonstration, and Deployment Activities

The mission of the Hydrogen and Fuel Cells Program is to enable widespread commercialization of a portfolio of hydrogen and fuel cell technologies through applied research, technology development and demonstration, and diverse efforts to overcome institutional and market challenges. The key goals are the development of hydrogen and fuel cell technologies for early markets (stationary power, lift trucks, portable power), mid-term markets (CHP, APUs, fleets, and buses), and long-term markets (light duty vehicles). Nearly 300 projects are currently funded at companies, national laboratories, and universities/institutes. The focus will be critical RD&D and in future an increased focus on low-carbon hydrogen pathways. The highest share of funding (61%) goes to cross-cutting topics and transportation (27%). Figure 4.5 points out the structure of the programme and the way to commercialization.

The Hydrogen and Fuel Cells Program sets key targets for fuel cell transport applications at $30\,kW^{-1}$ and a lifetime of 5000 h and for stationary applications at $1500\,kW^{-1}$ and a lifetime of 60 000–80 000 h depending on application. The target for hydrogen production prices is approximately $2–4 per kg hydrogen. All results of governmental funded projects are documented and publicly available (see Reference [20]). Figure 4.6 shows the collaborations within and outside the program.

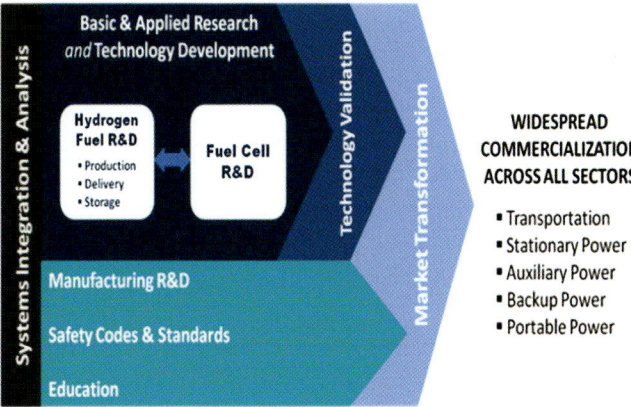

Figure 4.5 US DoE Hydrogen and Fuel Cells Program Overview. Source: Reference [21].

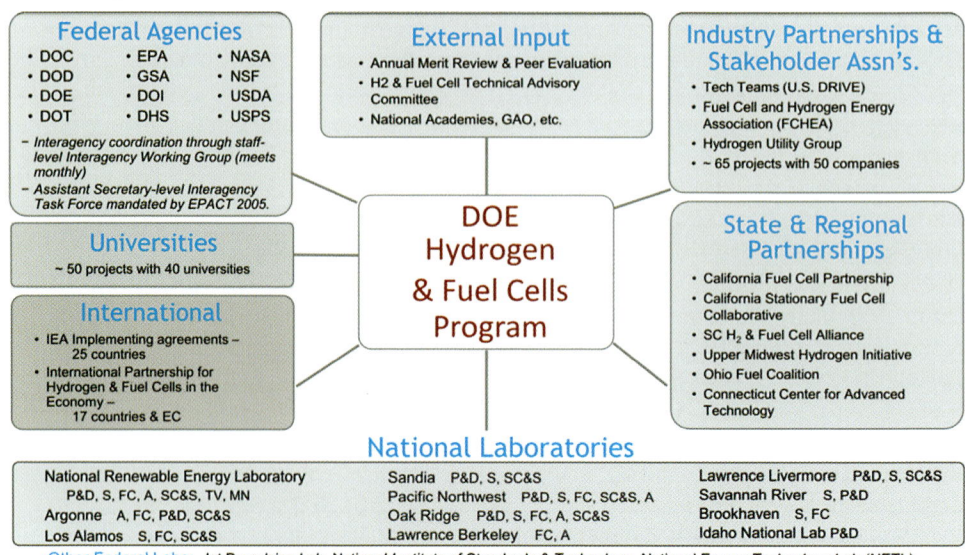

Figure 4.6 Collaborations of the Hydrogen and Fuel Cells Program. Source: Reference [21].

In addition to the governmental programs, a new public–private partnership–H2USA–has been designed to establish infrastructure to support future vehicles throughout the country. H2USA was launched in the USA in March 2013. The partners consist of major automakers, energy companies, component companies, and US Government Departments. The mission of H2USA is to promote the commercial introduction and widespread adoption of FCEVs across America through the creation of a public–private collaboration to overcome the hurdle of establishing a hydrogen infrastructure. The work will be done in working groups: hydrogen refueling station, locations, market support and acceleration and financing.

4.7
Japan

4.7.1
Energy Framework and Relevant Policies

Over the past three decades, Japan has taken significant steps to reduce its dependence on fossil fuels and diversify its foreign sources of supply. Japan's National Energy Strategy of 2006 aimed to *increase* the share of nuclear energy in the fuel mix to at least 40% by 2030. After the Fukushima disaster, the Japanese Cabinet approved in October 2011 an Energy White Paper calling for the *reduction* of nuclear power in the energy mix [22]. The new four pillars of the

Energy Policy are safety of Nuclear Energy, Environmental Challenge of Fossil Fuels, Practical Use of Renewable Energy, and the Potential of Energy Efficiency [23].

4.7.2
Hydrogen Related Energy Policy Strategies

Japan has set up many activities, policy measures, and incentives to bring fuel cells and hydrogen to the market. In Japan, 17 hydrogen refueling stations were already in operation in 2013. Toyota has already started selling its FCEV Mirai, in December 2014. The sales goal in Japan is a couple of hundred in 2015 [8b]. Policy mechanisms for encouraging the purchase of FCVs are expected soon. For refueling these cars 100 refueling stations are required first. Government subsidies for refueling stations enacted in the FY2013. The subsidies cover up to 50% of the construction cost of new stations. The goal for 2025 is to reach two million vehicles and 1000 fueling stations in Japan. Therefore, FCEVs and hydrogen stations should be business and viable by that time [24].

Presently, hydrogen production in Japan comes from fossil fuels and as a by-product of industrial processes. In the longer term H_2 production should come from the reformation of oil or natural gas with carbon capture and storage and renewable resources. This will include imports from other countries. Figure 4.7 gives more details of the Japanese commercialization scenario for hydrogen and fuel cells.

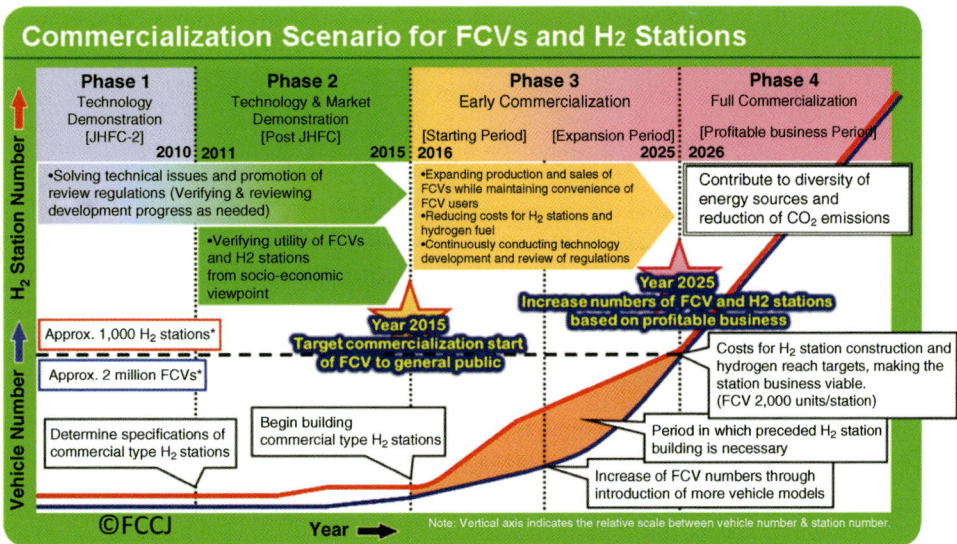

Figure 4.7 Japanese commercialization scenario for hydrogen and fuel cells.
Source: Reference [24].

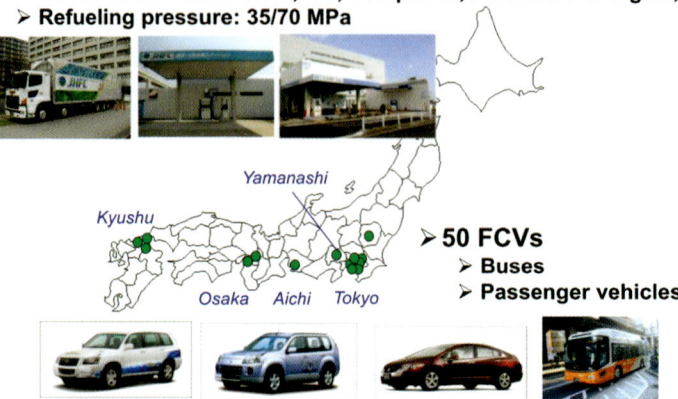

Figure 4.8 JHFC3 outline of a demonstration project. Source: Reference [25].

4.7.3
Hydrogen Research, Development, Demonstration, and Deployment Activities

NEDO is promoting hydrogen and fuel cell technology development projects in basic research, experimental studies, and the establishment of benchmarks and standards in cooperation with industry, academic institutions, and public research institutes. In 2012 NEDO was carrying out seven large demonstration projects in the fields of stationary fuel cell systems, FCVs, and hydrogen infrastructure.

After successful completion of JHFC2, JHFC3 commenced in April 2012. Sixteen hydrogen refueling stations and approximately 50 FCVs are being operated in the Tokyo, Aichi, Osaka, and Fukuoka regions. The project was planned for five years (2012–2016). The funding is ¥3 billion (approx. $30 million) in FY2012. Figure 4.8 shows a map of Japan and the hydrogen refueling demonstration projects and the passenger cars and buses in operation.

METI (Ministry of Economy, Trade and Industry) started a three-year subsidiary program for the construction of hydrogen refueling stations in FY2013. The subsidiary rate is up to 50% of the construction cost. Subsidy is delivered through the Next Generation Vehicle Promotion Center. In FY2013, subsidy will be delivered to 19 hydrogen refueling stations. The hydrogen refueling rates of all station is more than 300 $Nm^3 \ h^{-1}$. Of the 19 stations, 17 are centralized hydrogen refueling stations. The budget in FY2013 was ¥4.6 billion ($46 million) and requested ¥8.25 billion (US$82.5 million) in FY2014 [24].

4.8
International Networks

Worldwide there are many research, demonstration, and deployment activities around hydrogen and its use as a clean fuel and as part of sustainable energy systems. Different associations, institutions, and NGOs formed over recent decades aim to bring this knowledge together and push hydrogen forward on its way to the market.

4.8.1
Hydrogen Implementing Agreement of the IEA

The Hydrogen Implementing Agreement (HIA) is a technical Implementing Agreement of the International Energy Agency of the OECD. The HIA establishes international collaboration in hydrogen research, development, and demonstration. The Agreement's tasks and activities cover research issues in production, storage, conversion, safety, integrated systems, infrastructure, and systems analysis. The strategic framework for the phase 2010 to 2015 is collaborative R,D&D and analysis work that (i) advances hydrogen science and technology, (ii) positions hydrogen for technical progress, and (iii) promotes hydrogen awareness, understanding, and acceptance that fosters technology diffusion and commercialization [26].

4.8.2
IPHE

The mission of the International Partnership for Hydrogen and Fuel Cells in the Economy (IPHE) is to accelerate the transition to a hydrogen economy. It was established in 2003 as an international institution to organize, coordinate, and implement international research, development, demonstration, and commercial utilization activities related to hydrogen and fuel cell technologies. Currently, 18 OECD and non-OECD nations have committed to collaborating to advance the commercialization of hydrogen and fuel cell technologies in an effort to improve their energy supply and reduce environmental impacts. The main mission of the partnership is to develop uniform codes and standards and to educate and inform stakeholders and the general public of the benefits of, and challenges in, establishing a hydrogen economy [27].

4.8.3
EHA

The European Hydrogen Association (EHA) was established in 2000 and started a network to promote the use of hydrogen as an energy vector in Europe. The missions of the EHA are to provide information dissemination, to identify and advertise hydrogen-related capabilities in Europe, and to represent activities, gathering, and mediation of expertise and education/training. Since 2008 the EHA has hosted the Project HyER representing over 35 regions in Europe active in clean hydrogen/electric power and vehicles deployment [28]. Currently, the EHA represents 21 national hydrogen and fuel cell organizations and nine European companies as well universities active in the development of hydrogen technology [29].

4.8.4
International Association for Hydrogen Energy

This non-governmental organization brings together institutes and research worldwide to support hydrogen as one option for an abundant clean fuel. The association stimulates the exchange of information in the hydrogen energy field through its publications and sponsorship of international workshops, short courses, and conferences. An additional goal is to inform the general public of the important role of hydrogen energy in clean energy systems. For information dissemination the *International Journal of Hydrogen Energy* is published by Elsevier on behalf of the International Association for Hydrogen Energy and organizes the biennial World Hydrogen Energy Conferences held around the world.

Acknowledgment

The authors thank the members and secretariat of the Hydrogen Implementing Agreement of the International Energy Agency and the experts of Task 30 "Global Hydrogen Systems analysis" for the input of country specific information.

References

1 IEA (2012) Energy Technology Perspectives 2012. IEA, Paris.
2 NOW (2012) Annual Report 2011. Available at http://www.now-gmbh.de/fileadmin/user_upload/RE_Publikationen_NEU_2013/Publikationen_NOW_Berichte/NOW_Jahresbericht_2011_EN.pdf (accessed 20 June 2012).
3 NOW (2013) Hydrogen and Fuel Cell Technologies–Key Pillars of the Energy Transition 2.0: Further Development of the National Innovation Programme for Hydrogen and Fuel Cell Technology (NIP). Available at http://www.now-gmbh.de/fileadmin/user_upload/Download/hydrogen_and_fuel_cell_technologies_key_pillars_of_the_energy_transition_2_0.pdf.
4 TÜV SÜD & Ludwig Bölkow Systemtechnik (2012) Hydrogen Stations Worldwide. http://www.netinform.net/h2/H2Stations (accessed 4 May 2013).
5 CEP (2013) Clean Energy Partnership, www.cleanenergypartnership.de (accessed 5 May 2013).
6 Fraunhofer ISE (2012) Mit Sonne und Wasser Auto fahren–Fraunhofer ISE weiht solare Wasserstoff-Tankstelle ein. Press release 2 March 2012.
7 Press release Fraunhofer-Institut für Keramische Technologien und Systeme IKTS (2014) http://www.energie.fraunhofer.de/de/presse/presseinformationen-1/fraunhofer-ikts-250-kw-brennstoffzellenkraftwerk-liefert-energie-fuer-bundesministerium-fuer-bildung-und-forschung. Press release 4 December 2014 (accessed 22 May 2015).
8 (a) NOW (30 September 2013) H2 Mobility Initiative. http://www.now-gmbh.de/en/aktuelles-presse/2013/h2-mobility-initiative.html; (b) Voelcker, J. (18 November 2014) 2016 Toyota Mirai priced at $57,500, with $499 monthly lease. Green Car Reports. http://www.greencarreports.com/news/1095511_2016-

toyota-mirai-priced-at-57500-with-499-monthly-lease (accessed 20 May 2015).
9. HIA (2012) Hydrogen Implementing Agreement Annual Report 2012: IEA Agreement on the Production and Utilization of Hydrogen.
10. HYNOR (22 October 2013) Current Hydrogen Projects. http://hynor.no/en/stations.
11. Meyer, J. (2012) Techno-economical study of the Zero Emission Gas power concept. *Energy Proc.*, **4**, 1949–1956.
12. European Commission (2009) The EU climate and Energy Package. http://ec.europa.eu/clima/policies/package/index_en.htm (accessed 22 November 2013).
13. European Commission (2011) White Paper Roadmap to a Single European Transport Area–Towards a competitive and resource efficient transport system. COM(2011) 144 final. http://ec.europa.eu/transport/themes/strategies/2011_white_paper_en.htm.
14. European Comission (2013) Proposal for a Directive of the European Parliament and of the council on the deployment of alternative fuels infrastructure. 2013/0012 (COD). http://www.europarl.europa.eu/oeil/popups/ficheprocedure.do?lang=en&reference=COM%282013%290018#tab-0.
15. FCH JU (2013) Fuel Cells and Hydrogen Joint Undertaking, Brussels. www.fch-ju.eu/ (accessed 21 November 2013).
16. NRC (2013) Overview of Canada's Energy Policy. http://www.nrcan.gc.ca/energy/policy/1352 (accessed 11 May 2013).
17. CHFCA (2012) Canadian Hydrogen and Fuel Cell Sector Profile 2012. Canadian Hydrogen and Fuel Cell Association and PricewaterhouseCoopers LLP. http://www.chfca.ca/media/FINAL%20WEB%202389-01%20CHFC%20Sector%20Profile%202012%20EN.pdf.
18. CanmetENERGY (2013) Hydrogen and Fuel Cell Timeline. http://canmetenergy.nrcan.gc.ca/transportation/2278 (accessed 26 July 2013).
19. The White House (2012) The Blueprint for a Secure Energy Future: Progress Report. http://www.whitehouse.gov/sites/default/files/email-files/the_blueprint_for_a_secure_energy_future_oneyear_progress_report.pdf.
20. DOE (2013) 2013 Annual Merit Review and Peer Evaluation Report. www.hydrogen.energy.gov/annual_review13_report.html.
21. Satyapal, S. (2013) Hydrogen & fuel cells–program overview–IN ENERGY, D. O. (Ed.) 2013 Annual Merit Review and Peer Evaluation Meeting 13 May 2013, Washington D.C. http://www.hydrogen.energy.gov/pdfs/review13/03_satyapal_plenary_2013_amr.pdf (accessed 14 May 2015).
22. Calabrese, J. (2012) Japan's new energy future and the Middle East. American University in Washington, D.C./Middle East Institute. http://www.mei.edu/content/japan%E2%80%99s-new-energy-future-and-middle-east#_ftnref38.
23. METI (2012) Japan's challenges towards recovery. http://www.meti.go.jp/english/earthquake/nuclear/japan-challenges/index.html.
24. FCCJ (2013) Fuel Cell Commercialization Conference of Japan. http://fccj.jp/eg/.
25. Ito, T. (2012) NEDO's fuel cell and hydrogen activities in Japan. Presented at the 19th World Hydrogen Energy Conference, 3–7 June 2012, Toronto, Canada.
26. HIA (2013) Hydrogen Implementing Agreement. Bethesda, Maryland, USA. www.ieahia.org/ (accessed 25 November 2013).
27. IPHE (2013) International Partnership for Hydrogen and Fuel Cells in the Economy. iphe.net/ (accessed 13 October 2013).
28. The European Hydrogen Association (13 October 2013) Hydrogen: From when and where to here and now. www.h2euro.org/.
29. European Hydrogen Association (2015) EHA members. http://www.h2euro.org/2012/eha-members (accessed 20 May 2015).

Part 2.2
Thermochemical Hydrogen Production

Hydrogen Science and Engineering: Materials, Processes, Systems and Technology, First Edition.
Edited by Detlef Stolten and Bernd Emonts.
© 2016 Wiley-VCH Verlag GmbH & Co. KGaA. Published 2016 by Wiley-VCH Verlag GmbH & Co. KGaA.

5
Thermochemical Hydrogen Production – Solar Thermal Water Decomposition

Christian Sattler, Nathalie Monnerie, Martin Roeb, and Matthias Lange

5.1
Introduction

Solar thermal water decomposition for hydrogen production opens up the possibility of very high efficiencies. As the process is technically challenging plants will be especially competitive for large-scale production to serve mass markets [1–3].

The direct thermal dissociation of water (Figure 5.1) is theoretically the simplest reaction to produce hydrogen [4–6]. Therefore, research was performed especially in the 1970s and 1980s to understand and demonstrate the technology. A very comprehensive overview of the technology has been given by Kogan *et al.*, 1997–2000 [7–10]. Thermodynamically, complete dissociation takes place at about 4500 K but reasonable conversion would be achievable at about 2500 K. However, it was shown that the provision of heat at such temperatures is technically very challenging and economically not feasible. In addition, the production of the explosive hydrogen and oxygen gas mixture was a problem that could not be solved with respect to the safety of the process:

$$H_2O \rightarrow H_2 + \tfrac{1}{2} O_2 \tag{5.1}$$

To avoid these problems, so-called thermochemical cycles were developed to reduce the necessary temperature for the thermolysis reaction into two or more consecutive steps. These sets of reactions still offer very high conversion efficiencies, making them possibly the most effective alternative for large-scale hydrogen production. Therefore, nuclear and solar heated thermochemical cycles have been under development since the oil crises of the 1970s. Following the signing of the Kyoto Protocol a new surge in interest in thermochemical cycles took place. Thermochemical cycles avoid the H_2/O_2 separation problem and thereby reduce the temperature level. The cycles were originally developed to use high-temperature heat from nuclear power plants for hydrogen production but since very high temperature nuclear reactors are presently not so much in favor interest has turned to use concentrated solar energy like in solar thermal power plants as a renewable heat source.

Hydrogen Science and Engineering: Materials, Processes, Systems and Technology, First Edition.
Edited by Detlef Stolten and Bernd Emonts.
© 2016 Wiley-VCH Verlag GmbH & Co. KGaA. Published 2016 by Wiley-VCH Verlag GmbH & Co. KGaA.

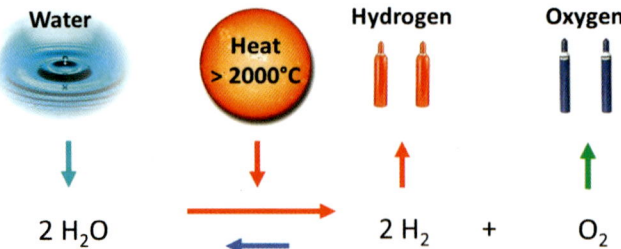

Figure 5.1 Thermal decomposition of water.

5.2
Historical Development

Research into thermochemical cycles started in the 1960s and more than 3000 thermochemical cycles have been reported in the literature [2,11,12]. However, only a few of them are relevant for a bulk production of hydrogen. During the 1970s and early 1980s, with the oil crisis, many studies and comparisons were carried out to identify the most promising cycles based on different criteria such as thermodynamics, theoretical efficiencies, and projected cost [13–16]. Most of the development effort made during those years was promoted by the nuclear energy sector, with the objective of diversifying the use of high temperature thermal energy supplied by nuclear reactors. The main results were reported from the programs developed by the Joint Research Center of the European Union in Ispra, Italy [17], by General Atomics [18] and Westinghouse [19] in USA, as well as by the Japanese Atomic Energy Research Institute. In the late 1980s interest in thermochemical cycles decreased greatly because of low-priced fossil fuels and the mistrust of nuclear energy caused by the accidents at Chernobyl and Three Mile Island. Only Japan continued the work, driven by its energy independence policy. Since then until the late 1990s, few additional findings were reported, mainly on the UT-3 cycle developed by and named after the University of Tokyo [20] and on the sulfur–iodine cycle originally proposed and named after the company General Atomics. A renaissance in the research and development of thermochemical cycles has taken place in the past few years. It was motivated by the production of hydrogen as a greenhouse-gas free energy carrier to fulfill the requirements of the Kyoto Protocol. Prospective studies and comparative analyses of cycles have been made as in the 1970s, but, furthermore, hardware was developed and tested. The integration of a concentrating receiver system (CRS) with these thermochemical cycles is quite new and contrary to the work conducted 30 years ago – the more important heat source is now concentrated solar radiation compared to nuclear heat. Owing to the high temperatures required for solar thermal water decomposition, only concentrating solar technologies working with a point focusing system can provide the necessary temperatures with high efficiency.

5.3
Present State of Work

Thermochemical cycles decompose water into hydrogen and oxygen via a series of chemical reactions at least one of which is highly endothermic. For solar thermal water decomposition the heat is provided by concentrated solar energy in a CRS, consisting of heliostats that concentrate the sunlight onto a receiver located at the top of a tower. Solar thermochemical cycles are still at the development stage and are very promising processes for "green" hydrogen mass production. Today, temperatures of more than 1273 K have been demonstrated on solar towers on the scale of several 100 kW_{th} and temperatures of more than 800 K are demonstrated on the scale of several 100 MW_{th} [21].

Two of the most intensively investigated families of cycles are based on metal oxides and on sulfuric acid. Some cycles employ an electrochemical step, peculiarly for the hydrogen production. As a subclass these variants are named "hybrid thermochemical cycles." The most famous is the hybrid sulfur cycle or Westinghouse cycle [19]. Thermochemical cycles have the potential to reach high efficiencies, which make them very interesting. Several of them offer theoretical thermodynamic efficiencies of 50% or more, but practical values are less due to thermal losses and some irreversible formation of by-products. In any event, the processes are able to use solar energy more efficiently than when converting it into electricity [22].

5.3.1
Metal/Metal Oxide Thermochemical Cycles

One of the most intensively investigated processes for thermochemical hydrogen production is based on metal oxides (MOs). MO cycles are very attractive because they involve fewer and less difficult steps than other processes at lower temperature, having thus the capacity to reach higher efficiency. The reaction scheme is as follows:

$$MO_x + yH_2O \rightarrow MO_{(x+y)} + yH_2 \quad \text{(splitting)} \quad (5.2)$$

$$MO_{(x+y)} \rightarrow MO_x + \tfrac{y}{2}O_2 \quad \text{(regeneration)} \quad (5.3)$$

The reactions include a metal or metal oxide capable of removing the oxygen of the water molecule (water-splitting step) at temperatures significantly lower than the one-step thermal water splitting. In the next step (regeneration), the metal oxide is reduced again, liberating some of its lattice oxygen. A large variety of redox materials consisting either of oxide pairs of multivalent metals (e.g., Fe_3O_4/FeO) [16,23] or systems of metal oxide/metal (e.g., ZnO/Zn) [24,25] have been evaluated for such purposes. In all the cycles, reduction of the oxidized metal oxide to generate its reduced form by setting oxygen free is the thermodynamically limiting step.

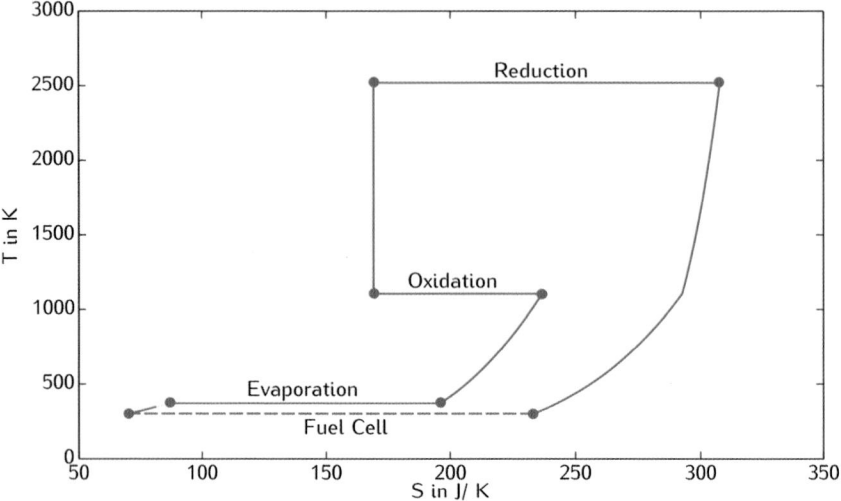

Figure 5.2 T–S diagram of two-step thermochemical cycle to split 1 mol of water.

Significant advantages of these cycles are that pure hydrogen is produced and that oxygen and hydrogen are generated in separate steps, so that no separation of these two gases is needed. Water splitting takes place at temperatures below 1073 K by oxidation of the metaloxide. It is regenearate by thermal reduction above 1073 K. The concept has been proven experimentally in laboratory models as well as in solar reactors operated in solar furnaces on a scale up to 10 kW$_{th}$ [26,27]. The temperature limits for large applications are calculated to be between 1573 and 2273 K. However, annual efficiencies have to be analyzed intensively to compare the very high temperature thermochemical cycles with their rivals like efficient electrolysis processes.

A vivid way of analyzing the thermodynamics of thermochemical cycles is by representation in a TS-diagram (Figure 5.2). The graph shows the entropy change in the gas phase when all gases are at 1 bar. The integral of the closed curve is equivalent to the ideal work that can be generated in a fuel cell, in other words the Gibbs energy of reaction of one mole of hydrogen with oxygen. When reducing the oxygen partial pressure during reduction, the area can be enlarged on the right-hand side, providing the possibility of reducing the upper reaction temperature.

5.3.1.1 FeO/Fe$_3$O$_4$

The production of hydrogen based on metal/metal oxide thermochemical cycles was originally proposed in 1977 by Nakamura using Fe$_3$O$_4$/FeO as redox material [28]. It consists of two steps, which are as follows:

$$Fe_3O_4(l) \rightarrow 3FeO(l) + \tfrac{1}{2}O_2 \qquad > 1873\ K \qquad (5.4)$$

$$3FeO + H_2O \rightarrow Fe_3O_4 + H_2 \qquad \text{exothermal} \qquad (5.5)$$

Using concentrated solar energy at temperatures above 1873 K, Fe_3O_4 can be reduced to FeO. By the subsequent exothermic reaction of FeO and H_2O at much lower temperatures, H_2 and Fe_3O_4 are generated [29]. The overall efficiency of the ideal Carnot system is reported to be between 50.7% and 62.5%. For an operating plant losses would lower it to 20.4–25.1% depending on the concentration of the solar radiation coupled into the reactor. The thermal dissociation of magnetite to wurtzite was experimentally investigated in a solar furnace at 2300 K in 1982 [30]. To avoid re-oxidation, it was necessary to quench the products but this quenching led to an energy penalty of up to 80% of the solar energy input [31], which causes some difficulties, among other things loss of oxide by vaporization. Nevertheless the reaction temperature for the two-step water-splitting process can be reduced with activated iron based redox systems.

5.3.1.2 Ferrites

Recently, research has focused on mixed ferrites to solve the problem associated with pure Fe_3O_4/FeO as redox material. Partial substitution of the iron in Fe_3O_4 by other metals, such as manganese, nickel, zinc, or cobalt, forms mixed metal oxides $(Fe_{1-x}M_x)_3O_4$. These mixed ferrites can be reduced at lower temperatures than pure Fe_3O_4 and the reduced phase $(Fe_{1-x}M_x)_{1-y}O$ is still capable of splitting water in the same way as FeO. These mixed-ferrites are very attractive for hydrogen production using water because they involve solid–gas reactions with a minimum of steps and reactants.

Such use of ferrites was recently investigated and demonstrated within the European projects HYDROSOL 3-D and HYDROSOL-2. In these projects, a two-step thermochemical cycle for solar hydrogen production from water has been developed and can be written as follows:

$$ZnFe_y{}^{2+}Fe_{2-y}{}^{3+}O_{4-y/2} + xH_2O$$
$$\rightarrow ZnFe_{y-x/2}{}^{2+}Fe_{2-y+x/2}{}^{3+}O_{4-y/2+x} + xH_2 \quad \text{(water splitting)} \tag{5.6}$$

$$ZnFe_{y-x/2}{}^{2+}Fe_{2-y+x/2}{}^{3+}O_{4-y/2+x}$$
$$\rightarrow ZnFe_y{}^{2+}Fe_{2-y}{}^{3+}O_{4-y/2} + \tfrac{x}{2}O_2 \quad \text{(reduction)} \tag{5.7}$$

A quasi-continuous operating monolithic solar receiver reactor was built at the DLR in Germany. It consists of two separate chambers with fixed honeycomb absorbers in both chambers (Figure 5.3). The honeycomb structured ceramics were coated with a thin layer of the mixed iron oxide redox system and placed inside a solar receiver. It is a rare example of a closed thermochemical cycle using only solar radiation for heating the reactors.

Since 2008 a 100 kWth pilot plant installed on a solar tower at Plataforma Solar de Almería in Spain has been in operation (Figure 5.4). Thanks to the multi-chamber arrangement of the reactor, quasi-continuous solar hydrogen

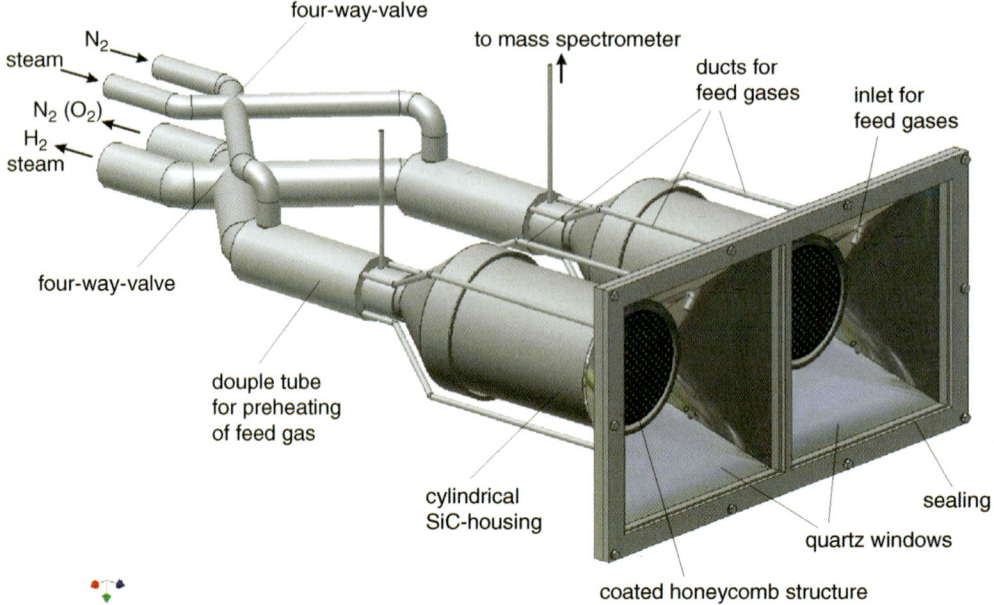

Figure 5.3 Schematic of the solar HYDROSOL reactor.

production is possible. Significant amounts of hydrogen were produced with a steam conversion of up to 30% [32].

Also rotating reactor types were developed to carry out solar hydrogen production by mixed ferrites thermochemical cycles. Two examples are the rotary-type reactor of the Tokyo Institute of Technology [33] and the counter rotating ring receiver reactor recuperator CR5 of the Sandia National Laboratory [34].

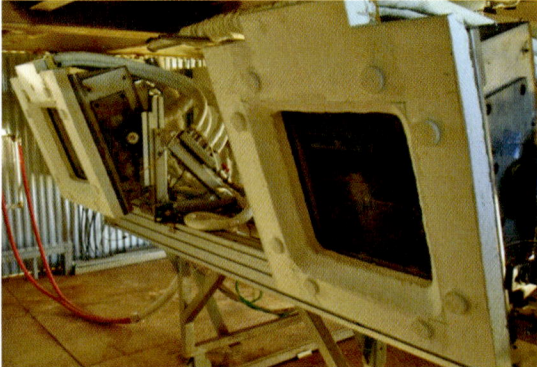

Figure 5.4 Hydrosol pilot plant in operation at Plataforma Solar de Almería in Spain.

For the latter, analyses suggest that thermal efficiencies of up to about 50% are possible [35].

Work on fluidized bed reactors was also started recently. It is still difficult to introduce solar radiation into such a reactor efficiently and separate the reactants and products in parallel [36]. However, a solar thermochemical water splitting reactor using an internally circulating fluidized bed of $NiFe_2O_4/m\text{-}ZrO_2$ particles was developed and demonstrated at laboratory scale. About 45% of the $NiFe_2O_4$ was converted and was then completely reoxidized with steam at 1273 K to generate hydrogen [37].

Finally, the utilization of a rotary kiln as high temperature thermochemical reactor was analyzed very recently [38]. Such a rotating cavity reactor combines relatively low radiation heat losses with a uniform temperature of the reactive material. The reactive material can be irradiated directly and the rotation allows intensification of the heat and mass transfer inside the reactive material. A thermal model describing the behavior of this reactor was developed and validated. For a given incident flux, the optimal aperture depends on the operating temperature.

Cobalt and nickel ferrites are also promising because of their favorable thermodynamic properties and their high melting points. They have been investigated among others by Kodama *et al.* [39].

Recently, hydrogen production via a redox cycle using iron oxide and cobalt ferrites has been investigated [40]. $Co_{0.85}Fe_{2.15}O_4$ films produced without deactivation four times more hydrogen than iron oxide films.

5.3.1.3 Hercynite Cycle

The efficacy of using cobalt ferrites deposited on Al_2O_3 substrates in a ferrite water splitting redox cycle was recently studied at the University of Colorado [41]. It was shown that the reduction temperature of the redox cycle is thus lower (473–573 K) than by using only $CoFe_2O_4$ because of a reaction between the ferrite and the substrate forming hercynite ($FeAl_2O_4$). This can be described as:

$$CoFe_2O_4 + 3Al_2O_3 \rightarrow CoAl_2O_4 + 2FeAl_2O_4 + \tfrac{1}{2}O_2 \qquad (5.8)$$

$$CoAl_2O_4 + 2FeAl_2O_4 + H_2O \rightarrow CoFe_2O_4 + 3Al_2O_3 + H_2 \qquad (5.9)$$

Significant quantities of hydrogen are produced at a reduction temperature of 1473 K while little hydrogen is generated using only $CoFe_2O_4$ until 1673 K.

In the thermochemical cycles used to produce hydrogen for water splitting, the heat recovery is very important and has to be done sufficiently. If not, the heat losses can be damaging for the efficiency and economics of the process [42]. It has, thus, been recently suggested that both reaction steps (oxidation and reduction) be carried out at the same temperature [43] to avoid heat losses due to the temperature difference between the two steps. However, this implies increasing the temperature for the water oxidation and avoiding the production of hydrogen and oxygen at the same time. Muhich *et al.* have studied the isothermal water splitting hercynite cycle. It was shown that this cycle was more efficient under isothermal conditions at 1623 K [43].

5.3.1.4 Manganese Ferrite plus Activated Sodium Carbonate

To further decrease the temperature of the ferrite process a three-step water splitting cycle using the $Na_2CO_3/MnFe_2O_4/Fe_2O_3$ system is under development. Within the context of experiments, $MnFe_2O_4$ and Na_2CO_3 nanostructured materials, H_2 was successfully produced from H_2O at relatively low temperatures (<1073 K):

$$2MnFe_2O_4 + 3Na_2CO_3 + (1+6\delta)H_2O$$
$$\Rightarrow 6\,Na(Mn_{1/3}Fe_{2/3})O_{2+\delta} + 3CO_2 + (1+6\delta)\ H_2 \quad 1073\,K \tag{5.10}$$

$$6Na(Mn_{1/3}Fe_{2/3})O_{2+\delta} + 3Fe_2O_3$$
$$\Rightarrow 2MnFe_2O_4 + 6NaFeO_2 + (1+6\delta)2O_2 \quad 1273\,K \tag{5.11}$$

$$2NaFeO_2 + CO_2 \Rightarrow Na_2CO_3 + Fe_2O_3 \quad 873\,K$$
$$\delta = (n-3)/6 \text{ with } 3 < n < 4 \tag{5.12}$$

The system consisting of manganese ferrite mixed with "mechano-chemically activated" Na_2CO_3 is reported to produce H_2 with kinetics faster than other ferrites [44]. At temperatures higher than the melting point of Na_2CO_3, the reactivity among species rises and the quartz walls of the reactor reacts with the system to form unwanted side-products that abruptly reduce the H_2 production. This is a common problem with high temperature redox reactions. Above 1000 K effects very often take place that were not taken into account when choosing materials for the reactors. Therefore, a proper reactor design has to contain a study on the material behavior under the redox conditions and the interaction with the reactants at the necessary temperatures. In this case the reaction has to be kept at a very constant temperature, which is rather difficult in a solar heated reactor. In addition, for technical apparatus quartz should be replaced by an inert material or protected by an inert layer.

5.3.1.5 Zn/ZnO

Zn/ZnO is also a promising redox system and has been investigated widely. The reaction consists of the solar endothermic dissociation of ZnO into Zn and O_2 at about 2273 K [31] and in the hydrolysis of Zn producing hydrogen at 723 K [45]:

$$ZnO \rightarrow Zn + \tfrac{1}{2}O_2 \quad 2273\,K \tag{5.13}$$

$$Zn + H_2O \rightarrow ZnO + H_2 \quad 723\,K \tag{5.14}$$

The Zn/ZnO cycle can reach a maximum exergy conversion efficiency of 29% and a solar thermal efficiency of 40%. A study for a 90-MW solar reactor operated at 2273 K has been carried out [46]. Solar H_2 production costs for a large-scale plant with an annual production rate of 61 million kWh_{th} were estimated to be between US$0.13 and US$0.15 kWh_{th}^{-1}. This is thermodynamically one of the most efficient cases, but the decomposition of ZnO requires very high temperatures, above 2273 K, and produces a mixture of Zn and O_2, which has to be quenched to avoid their recombination. This influences the efficiency and safety of the process. The gas quenching to minimize Zn recombination constitutes, consequently, the major challenge of this cycle [47].

The solar dissociation of ZnO has been investigated and experimental tests have been performed in a solar furnace since 1977 [48]. The hydrolysis of Zn nanoparticles has been carried out in a tubular reactor at the University of Minnesota [49]. Hydrogen was produced with an experimental conversion of 61–79%. The thermal efficiency of the cycle is estimated to be 36% without heat recovery from the quench process [50]. A solar chemical reactor consisting of a windowed rotating cavity receiver lined with ZnO particles was built and demonstrated at a power level of 10 kW in 1998 [51]. Scaling-up the reactor technology to 1 MW solar thermal power input has the potential of reaching a solar-to-chemical energy conversion efficiency of 56% [52]. Recently, a 100 kW_{th} reactor was tested by the Swiss Paul Scherrer Institute at the 1 MW solar furnace of CNRS-PROMES in Odeillo, France [53].

Because of the severe conditions concerning the temperature and the separation of zinc and oxygen, another process using additional carbon containing raw materials was developed in order to favor the reduction of ZnO. This carbothermal reduction of ZnO was demonstrated in a 300 kW_{th} reactor in the frame of the EU project SOLZINC at the solar tower of WIS in Israel [51,54].

5.3.1.6 CeO_2/Ce_2O_3

Ceria was proposed as water splitting material in 1985 by Otsuka *et al.* [55] and is presently the most prominent redox system. However, the pure CeO_2/Ce_2O_3 two-step cycle needs very high temperatures:

$$2CeO_2(s) \rightarrow Ce_2O_3(s) + \tfrac{1}{2}O_2(g) \quad 2273\ \text{K},\ 100-200\ \text{mbar} \quad (5.15)$$

$$Ce_2O_3(s) + H_2O(g) \rightarrow 2CeO_2(s) + H_2(g) \quad 673-873\ \text{K} \quad (5.16)$$

The feasibility of this process was demonstrated by several research groups. The thermal reduction was performed in the solar furnace of CNRS-PROMES in Odeillo, France at 2273 K under reduced pressure (100–200 mbar). The water splitting was investigated in a fixed bed reactor. Fast kinetics could be shown at 673–873 K as well as the complete conversion of Ce_2O_3 into CeO_2 thanks to the high reactivity of the reduced cerium oxide with water. A remarkable calculated thermal efficiency of 68% was reported [56]. This cycle has also been investigated by the California Institute of Technology and the ETH who have built and tested together a prototype reactor in a high flux solar simulator. A high rate solar fuel production was demonstrated. An efficiency of 16–19% was calculated with the first thermodynamic analysis but some increase is expected by taking into account the recovery of the sensible heat, optimization of the reactor, and system integration.

Recently, cerium-based oxides $M_xCe_{1-x}O_2$, where M = Gd, Y, Sm, Ca, Sr, have been studied, and compared with pure ceria, from the point of view of their thermodynamics concerning their application in the solar thermochemical hydrogen production [57]. It was shown that, at a given temperature, more hydrogen is produced at equilibrium with pure ceria than with any of the cerium based metal oxides considered. Moreover, the solar-to-fuel energy conversion efficiency was shown to be greater for pure ceria.

A new solar thermochemical reactor concept using CeO_2 as a reactive material has been developed recently [58]. It uses a moving packed bed of reactive particles of CeO_2. It was shown that the conversion efficiency of solar energy into hydrogen can exceed 30% at the design point of the fully developed system.

5.3.2
Sulfur Cycles

Several encouraging thermochemical cycles for water splitting are based on sulfur systems. Originally, sulfur-based cycles were developed to produce hydrogen by coupling them with high temperature heat from nuclear very high temperature reactors (VHTRs). Most sulfur cycles have in common the thermal splitting of sulfuric acid into sulfur dioxide and oxygen, which is a high-temperature and endothermic reaction step. This reaction necessitates temperatures above 1123 K (Eq. (5.17b)) even if efficient catalysts are employed. Moreover, the sulfuric acid decomposition reaction occurs in a corrosive atmosphere and thus constitutes the main issue in the coupling with a very high temperature heat source. As VHTRs are at present only available as research installations and only very few for power generation are under construction, it seems to be more appropriate to generate the required temperature level by concentrated solar energy for processes on a large scale. Several sulfur-based cycles have been investigated during recent decades. The principal work has been performed by General Atomics and Westinghouse in the USA, the European Joint Research Centre in Ispra, Italy, and the Japan Atomic Energy Research Institute (JAERI, now Japan Atomic Energy Agency, JAEA). The best known cycles are the hybrid sulfur cycle, the Mark 13 cycle, and the sulfur–iodine cycle.

5.3.2.1 Hybrid Sulfur Cycle (Westinghouse, Ispra Mark 11)
The hybrid sulfur cycle (Figure 5.5), also known as Westinghouse cycle or ISPRA Mark 11 cycle, is a hybrid electrochemical–thermochemical cycle [59]. It was originally proposed by Brecher and Wu in 1975 [60] and developed by Westinghouse Electric corporation.

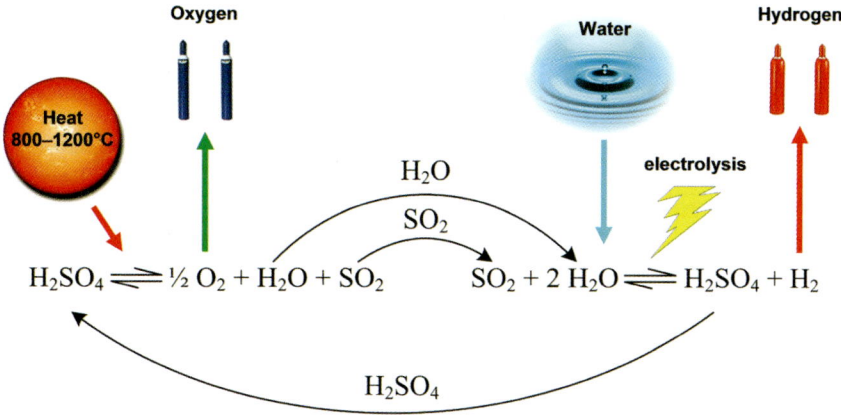

Figure 5.5 Hybrid sulfur cycle.

It is a two-step cycle consisting of the splitting of sulfuric acid, which is divided into two sub-steps (the vaporization (5.17a) and the dissociation of sulfuric acid (5.17b)) and the decomposition of the resulting SO_2 (5.16):

$$H_2SO_4 \rightarrow H_2O + SO_3 \qquad > 723\,K \tag{5.17a}$$

$$SO_3 \rightarrow SO_2 + \tfrac{1}{2}O_2$$
$$> 1123\,K \text{ with catalyst } (1423\,K \text{ without catalysts}) \tag{5.17b}$$

$$2H_2O + SO_2 \rightarrow H_2SO_4 + H_2 \qquad \text{electrolysis, } 353\,K \tag{5.18}$$

This process is called a hybrid cycle because of the combination of the thermal decomposition of sulfuric acid with the electrochemical oxidation of SO_2. This requires electrical power but is more efficient than the electrolytic splitting of water. The voltage needed for the electrolysis is 0.17 V and therefore much lower than the 1.23 V needed for water electrolysis, which reduces consequently the electrical power consumption. Sulfuric acid is vaporized to produce steam and sulfur trioxide, which is decomposed at high temperature in the presence of a catalyst into sulfur dioxide and oxygen. This reaction runs better at higher temperatures, which can be reached with concentrated solar energy. The overall thermal efficiency of the process is calculated to be about 40% [59]. The Savannah River National Laboratory (SRNL) has carried out a lot of work on electrolyzer and component development as well as on flow-sheet modeling [61,62]. Carbon-supported platinum electrodes are used for the SO_2 oxidation. Cells made from ceramics such as silicon carbide, silicon nitrite, and cermets have excellent resistance to corrosion by sulfuric acid at ambient temperature and at low acid concentration. Catalysts principally based on iron oxide are available for accelerating the reaction rate of the SO_3 reduction at temperatures of 1123 K. The kinetics of the reaction are much faster if higher temperatures are available. Therefore, the use of catalysts might be reduced or made unnecessary if the sulfuric acid splitting is coupled to concentrated solar radiation at temperatures above 1123 K. Process design and economic analysis have been performed for this coupling by SRNL and CSIRO [63–65]. Hydrogen production costs with such a process were calculated to be between $US3.19 and $US5.57 kg^{-1}. In the frame of the EU projects HycycleS and HYTHEC, the reliability and the potential hybridization of the hybrid sulfur cycle with solar energy [66] were analyzed and qualified materials were identified for a technical realization of the solar thermal decomposition of sulfuric acid [67]. For this step, a test reactor was developed and tested in the solar furnace of DLR in Cologne at temperatures up to 1223 K. A thermal efficiency of up to 28% was found while re-radiation of the solar absorbers was identified as the main heat loss. This can be reduced by using smaller windows and a better insulation of the housing. The results indicate that the efficiencies can additionally be increased at higher flow rates of sulfuric acid than performed due to the limitations of the equipment used. The multi-chamber concept of the reactor is scalable and mainly uses commercial materials like steel and mass produced silicon carbide structures while only few components are made of quartz especially for the reactor. Thus, this development is an important step in the direction of industrial realization of solar hydrogen production by the sulfur based thermochemical cycle. A project to scale-up the technology into the several 100 kWth scale was recently started (SOL2HY2 Web-Site: https://sol2hy2.eucoord.com/home/body.pe).

5.3.2.2 Mark 13 V2

The Mark 13 V2 is a hybrid cycle and was developed by JRC Ispra, Italy for coupling to a concentrating solar power plant [17]:

$$SO_2(g) + Br_2(l) + 2H_2O(l) \rightarrow 2HBr(g) + H_2SO_4(l) \quad 373 - 413 \text{ K} \quad (5.19)$$

$$2HBr(l) \quad H_2(g) + Br_2(l) \quad \text{(electrochemical step)} \quad (5.20)$$

$$H_2SO_4(g) \rightarrow H_2O(g) + SO_2(g) + \tfrac{1}{2}O_2(g) \quad 1123 - 1393 \text{ K} \quad (5.21)$$

A 1.4×10^6 GJ yr^{-1} solar power system was simulated. Decomposition of the hydrobromic acid can be performed by a chemical sub-cycle or in an electrolytic cell. The energy demand of this reaction is high, since HBr has a high free energy of formation. However, unlike the sulfur–iodine cycle H_2SO_4 and HBr can be obtained at high concentration and in separate phases (Eq. Eq. 5.19).

Although theoretically feasible, the removal of SO_2 from the O_2/SO_2 gas stream (Eq. 5.21) is very difficult in terms of corrosion, efficiency, and cost. The calculated process efficiency of the Mark 13 cycle is 37%, which results in an overall solar H_2 production efficiency of 21%. The production cost for the process is calculated at US$51.57 GJ^{-1}.

5.3.2.3 Mark 13A

The Mark 13A cycle was first designed as a desulfurization process for refineries. It was developed by JRC Ispra, Italy and is based on two of the latter's three

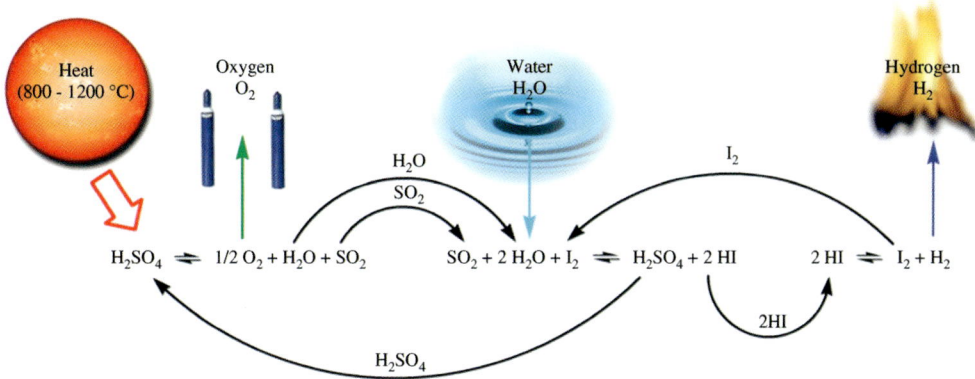

Figure 5.6 Sulfur–iodine process.

reactions of the Mark 13 V2 cycle (Eqs. 5.19 and 5.20). The successful operation of a bench-scale unit was followed by the construction and operation of a pilot plant at the Saras refinery in Sarroch, Sardinia, Italy [68]. This plant was designed for a $32\,000\,\text{Nm}^3\,\text{h}^{-1}$ flue gas flow that emanated from an internal power station that operated on a mixture of heavy fuel oil and refinery gas. After some additional modifications, the plant started regular operation in 1990. Since the sulfuric acid was not recycled, the cycle was not closed but the energy stored in the flue gas was used for hydrogen production, as a by-product of the desulfurization. Nevertheless, this is still the largest applications ever to produce hydrogen by a technology developed for thermochemical hydrogen production.

5.3.2.4 Sulfur–Iodine or General Atomics Process (ISPRA Mark 16)

The sulfur–iodine water-splitting cycle (Figure 5.6), also known as the ISPRA Mark 16 or General Atomics process [69], is a promising cycle for efficient production of hydrogen.

The cycle was originally developed in the USA by General Atomics in the mid-seventies to split water using high-temperature heat from nuclear power plants. In the eighties, the coupling to a solar thermal plant was designed [70]. It was demonstrated on a solar concentrator of the Georgia Institute of Technology in 1984 [70,71]. The three basic reactions are:

$$2H_2O + SO_2 + xI_2 \rightarrow H_2SO_4 + 2HIX \quad (300 - 400\,\text{K}) \quad (5.22)$$

$$2HI_X \rightarrow xI_2 + H_2 \quad (400 - 1000\,\text{K}) \quad (5.23)$$

$$H_2SO_4 \rightarrow H_2O + SO_2 + \tfrac{1}{2}O_2 \quad (1120 - 1200\,\text{K}) \quad (5.24)$$

The first reaction, which is exothermic, is known as the "Bunsen reaction." This reaction requires surplus water and iodine to proceed and to obtain the two acids HI and H_2SO_4. The separation of these two acids is one of the issues of this cycle. Indeed, the removal by distillation requires a certain amount of energy, which impacts directly the cycle efficiency. Thus, one of challenges of this cycle is to find an efficient separation process requiring less energy than distillation as well as to reduce the consumption of water and iodine. The two acids are decomposed following Eqs. (5.23) and (5.24), respectively. Decomposition of H_2SO_4 is carried out in the vapor phase in a catalytic reactor at a temperature above 1123 K, analogous to the processes described above. Each individual section was successfully demonstrated in a glass, quartz, and Teflon lab-scale facility by JRC in Ispra, Italy in the 1980s [72]. Despite the less promising economic calculations and the fact that the H_2SO_4 and HI decomposition caused severe corrosion problems, improvements to this process have continued at diverse locations, like in the technical University of Aachen in Germany [73], in Japan by JAERI, who developed a pilot test plant [74], and by JAEA, who has operated a closed-loop continuous hydrogen production at a rate of $32\,l\,h^{-1}$ for 20 h [75].

In the last decade the sulfur-iodine cycle was back in the focus because of its potentially high efficiency. The sulfuric acid splitting step was thus recently demonstrated in the solar furnace of the German Aerospace Center DLR in Cologne, Germany [76] within the European project "Hydrogen by Thermochemical Cycles" HYTHEC [77] and its successor the HyCycleS in which the effective potential of the sulfur–iodine cycle for massive hydrogen production was also investigated. For industrial application, a $2.4\,GW_{th}$ modular helium reactor (H2-MHR) driven plant was designed. This reactor should be capable of supplying temperatures up to 1173 K [71]. The overall efficiency of the cycle was calculated to be 42% at 1123 K with the assumption that efficiencies of the hydrogen production process of up to 52% might be achievable at a temperature of 1223 K. This seems to be difficult to achieve with a nuclear reactor. General Atomics, SANDIA National Laboratories, and the French CEA have collaborated to set up an electrically heated pilot plant at the General Atomics site in San Diego, USA. The three partners each designed one of the steps. The water splitting test apparatus was designed for a hydrogen production rate of $100–200\,l\,h^{-1}$ [78]. It was unfortunately not possible to solve some of the issues described in the 1980s for this plant. A continuation of this joint venture is consequently presently not planned.

5.3.3
Other Cycles

Several other thermochemical cycles were historically developed to be operated by heat from nuclear power plants. A major aim was reduction of the temperature to make such processes also available for reactor types other than VHTRs providing heat at about 773 K. The first worked on cycle was the iron-chloride cycle [79]. After some drawbacks more complex candidates came into focus like the UT-3 and the hybrid copper chlorine cycle. The latest proposal, to go as low as 573 K, is the uranium–europium cycle.

5.3.3.1 UT3 (Ca/Fe/Br Cycle)

The UT-3 cycle was proposed in 1978 by the University of Tokyo (UT), after which it is named [80]. The process consists of four gas–solid reactions, two reactions based on Ca-compounds and the other two based on Fe-compounds:

$$CaBr_2 + H_2O \rightarrow CaO + 2HBr \quad\quad 950 - 1000 \text{ K} \quad\quad (5.25)$$

$$CaO + Br_2 \rightarrow CaBr_2 + \tfrac{1}{2}O_2 \quad\quad 750 - 800 \text{ K} \quad\quad (5.26)$$

$$Fe_3O_4 + 8HBr \rightarrow 3FeBr_2 + 4H_2O + Br_2 \quad\quad 500 - 550 \text{ K} \quad\quad (5.27)$$

$$3FeBr_2 + 4H_2O \rightarrow Fe_3O_4 + 6HBr + H_2 \quad\quad 950 - 1000 \text{ K} \quad\quad (5.28)$$

The overall cycle was demonstrated first in a laboratory scale test rig and then as a pilot plant at the University of Tokyo. For this pilot plant conventional materials were used [81,82]. In the last decade, a new UT-3 process has been conceived to be coupled with a solar heat source [83]. An overall thermal efficiency of 49.5% was calculated. Since it is easy to separate solid from gas products, it is decisive for the cycle to use solid reactants with high reactivity and durability. Thus, research has focused on the stability of the solids and on the membrane separation processes. According to work performed in the 1980s and 1990s, the UT-3 cycle can potentially be a competing process in hydrogen production, particularly if a membrane separator is used. In this case the thermal efficiency was calculated to be 45%, but recently the overall efficiency of the cycle is reported to be lower, to 13%, which is not enough to be competitive with the water electrolysis.

5.3.3.2 Hybrid Copper–Chlorine Cycle

The copper–chlorine cycle is a four-step cycle and was principally investigated by the Argonne National Laboratory in the USA [84]. It consists of three thermal reactions in which H_2, O_2, and HCl are generated and an electrochemical step in which CuCl is disproportionated to yield copper metal and $CuCl_2$. The oxygen is released from the copper oxychloride between 720 and 800 K, which is the highest temperature limit for this cycle and which represent its main benefit. The heat required for the last step (Eq. (5.32)), producing oxygen, can be provided by a molten salt central solar receiver. Moreover, the electrochemical step requires a low voltage, namely, between 0.4 and 0.6 V [85], which is another advantage:

$$2Cu(s) + 2HCl(g) \rightarrow 2CuCl(l) + H_2(g) \quad\quad 700 - 750 \text{ K} \quad\quad (5.29)$$

$$4CuCl(s) \rightarrow 2CuCl_2(aq) + 2Cu(\text{electrochemical}) \quad\quad 300 - 350 \text{ K} \quad\quad (5.30)$$

$$2CuCl_2(s) + H_2O(g) \rightarrow CuO \cdot CuCl_2(s) + 2HCl(g) \quad\quad 620 - 673 \text{ K} \quad\quad (5.31)$$

$$CuO \cdot CuCl_2(s) \rightarrow 2CuCl(l) + \tfrac{1}{2}O_2(g) \quad\quad 803 \text{ K} \quad\quad (5.32)$$

The idealized efficiency of the CuCl cycle was calculated to be 41–43% (LHV). The Argonne National Laboratory (ANL) has completed an ASPEN simulation to perform a sensitivity studies [84]. The first reaction, Eq. (5.27), producing hydrogen has been tested in a furnace at the ANL [86]. Large laboratory scale reactors have been tested successfully. The University of Ontario Institute of Technology (UOIT) in Canada is in the lead in reporting new developments on this cycle related process and material development as well as economic evaluation [87–89].

5.3.3.3 Uranium–Europium Cycle

The search for even lower temperature cycles has led to interesting proposals like this one. The cycle is based on heavy element halide chemistry to reach a maximum temperature of 573 K:

$$2(UO_2Br_2 \cdot H_2O) \rightarrow 2\text{"}UO_3 \cdot H_2O(s)\text{"} + 4HBr(g) + 2H_2O(g) \quad 573\,K \quad (5.33)$$

$$4EuBr_2 + 4HBr \rightarrow 4EuBr_3 + 2H_2(g) \quad \text{Exothermic} \quad (5.34)$$

$$4EuBr_3 \rightarrow 4EuBr_2 + 2Br_2(g) \quad 573\,K \quad (5.35)$$

$$2\text{"}UO_3 \cdot H_2O(s)\text{"} + 2Br_2 + 4H_2O \\ \rightarrow 2(UO_2Br_2 \cdot H_2O) + O_2(g) \quad \text{Exothermic} \quad (5.36)$$

The exact stoichiometry of "$UO_3 \cdot H_2O_{(s)}$" has not been determined yet. It is uncertain if the major research that would be necessary to be carried out to demonstrate its feasibility will be done. Nevertheless, low temperature cycles are relevant because the outcome of the Generation IV International Initiative might be that no new high temperature nuclear reactors will be developed [89].

5.4
Conclusion and Outlook

The development of thermochemical cycles for water splitting is directly related to the discussion on climate change and the need for CO_2 free production of hydrogen, and it can also be a chemical commodity to upgrade heavy crudes like Canadian oil sands or to produce ammonia for fertilizers. It might also have an important place in the future energy economy as well for powering stationary fuel cells for heat and power production such as for transportation purposes in cars, busses, ships, or even airplanes. The first peak in development of thermochemical water splitting technologies was caused by the oil crises of the 1970s. After a decade of low interest during the 1990s, since 2000 limited fossil resources, rising prices, and the environmental impact of fossil energy carriers have been the main drivers of its comeback. Although research has gone on for nearly 40 years, no thermochemical cycle is in operation at an industrial scale yet. But the efficiency of the cycles makes it very likely that this will happen in the next decade, especially since alternative technologies will have difficulties in reaching the same high efficiencies as well as their low land use. This implies that the calculated production cost will make these technologies competitive. The market introduction of solar-thermal power plants, especially solar towers, makes a heat source available to power a wide range of thermochemical cycles. This reduces the concerns that prevent investors from engaging in these new technologies. If VHTR nuclear reactors do become commercially available, as recent developments in China suggest, it would promote even further the possibilities of thermochemical cycles for hydrogen production by water splitting.

Nomenclature

ANL	Argonne National Laboratory, USA
CEA	Commissariat à l'Énergie Atomique = French Atomic Energy Commission
CRS	Concentrating Receiver System
DLR	Deutsches Zentrum für Luft - und Raumfahrt e.V. = German Aerospace Center
GA	General Atomics
H2-MHR	Modular helium gas cooled reactor
JAEA	Japan Atomic Energy Agency
JAERI	Japan Atomic Energy Research Institute
JRC	Joint Research Centre of the European Union
LHV	Lower heating value, defined as the amount of heat released by combusting a specified quantity (initially at 298 K or another reference state) and returning the temperature of the combustion products to 423 K
MO	Metal oxide
UT3	Thermochemical cycle named after the University of Tokyo
VHTR	Very high temperature reactor

References

1 De Beni, G.M. and Marchetti, C. (1970) Hydrogen, key to the energy market. *Eurospectra*, **9**, 46–50.

2 Marchetti, C. (1973) Hydrogen and energy. *Chem. Eco. Eng. Rev.*, **5**, 7–25.

3 Winter, C.-J. (2000) *On Energy of Changes – The Hydrogen Solution*, Gerling Akademie Verlag, Munich.

4 Russell, J.L. Jr., McCorkle, K.H., Norman, J.-H., Porter, J.-T. II, Roemer, T.S., Schuster, J.R., and Sharp, R.S. (1976) Water Splitting – a progress report, in *World Hydrogen Energy Conference, 1st, Miami Beach, FL, 1-3 March 1976, Proceedings Volume 1* (ed. T. Nejat Veziroğlu), University of Miami, p. 1A-105–1A-124.

5 Brecher, L.E. (1982) Decomposition of water, US Patent 4330523.

6 Etievant, C. (1991) Solar high-temperature direct water splitting – a review of experiments in France. *Sol. Energy Mater.*, **24**, 413–440.

7 Kogan, A. (1997) Direct solar thermal splitting of water and on-site separation of the products I. Theoretical evaluation of hydrogen yield. *Int. J. Hydrogen Energy*, **22**, 481–486.

8 Kogan, A. (1998) Direct solar thermal splitting of water and on-site separation of the products – II. Experimental feasibility study. *Int. J. Hydrogen Energy*, **23**, 89–98.

9 Kogan, A. (2000) Direct solar thermal splitting of water and on-site separation of

the products — IV. Development of porous ceramic membranes for a solar thermal water-splitting reactor. *Int. J. Hydrogen Energy*, **25**, 1043–1050.

10 Kogan, A. (2000) Direct solar thermal splitting of water and on-site separation of the products. III. Improvement of reactor efficiency by steam entrainment. *Int. J. Hydrogen Energy*, **25**, 739–745.

11 Abraham, B.M. and Schreiner, F. (1974) General principles underlying chemical cycles which thermally decompose water into the elements. *Ind. Eng. Chem., Fundam.*, **13**, 305–310.

12 Chao, R. (1974) Thermochemical water decomposition processes. *Ind. Eng. Chem., Prod. Res. Dev.*, **13**, 94–101.

13 Fletcher, E.A. and Moen, R.L. (1977) Hydrogen and oxygen from water. *Science*, **197**, 1050–1056.

14 Ducarroir, M., Bilgen, E., Foex, M., and Sibieude, F. (1977) Use of solar energy for direct and two-step water decomposition cycles. *Int. J. Hydrogen Energy*, **2**, 251–257.

15 Knoche, K.F. (1977) Entropy production, efficiency, and economics in the thermochemical generation of synthetic fuels 1. The hybrid sulfuric acid process. *Int. J. Hydrogen Energy*, **2**, 377–385.

16 Nakamura, T. (1977) Hydrogen production from water utilizing solar heat at high temperatures. *Sol. Energy*, **19**, 467–475.

17 Beghi, G.E. (1986) A decade of research on thermochemical hydrogen at the joint research centre, ISPRA. *Int. J. Hydrogen Energy*, **11**, 761–771.

18 Besenbruch, G.E. and McCorkle, K.H. (1981) Thermochemical Water Splitting with Solar Thermal Energy: Final Report, GA-A16022, General Atomics, San Diego, CA.

19 Spewock, S., Warde, C.J., and Brecher, L.E. (1976) The Westinghouse sulfur cycle for the thermochemical decomposition of water, in *World Hydrogen Energy Conference, 1st, Miami Beach, FL, 1-3 March 1976, Proceedings Volume 1* (ed. T. Nejat Veziroğlu), University of Miami, pp. 9A-1–9A-16.

20 Sakurai, M., Tsutsumi, A., and Yoshida, K. (1995) Improvement of Ca-pellet reactivity in UT-3 thermochemical hydrogen production cycle. *Int. J. Hydrogen Energy*, **20**, 297–301.

21 Hooper, R. (2013) California's heliostat mirrors harness sun's power. *New Sci.*, **218**, 24–25.

22 Diver, R.B. and Kolb, G.J. (2008) Screening Analysis of Solar Thermochemical Hydrogen Concepts, SANDIA National Laboratories, Albuquerque, NM SAND2009-1900, March 2008.

23 Gokon, N. and Kodama, T. (2007) Thermochemical cycles for high-temperature solar hydrogen production. *Chem. Rev.*, **107**, 4048–4077.

24 Steinfeld, A. (2002) Solar hydrogen production via a two-step water-splitting thermochemical cycle based on Zn/ZnO redox reactions. *Int. J. Hydrogen Energy*, **27**, 611–619.

25 Steinfeld, A. (2005) Solar thermochemical production of hydrogen – a review. *Sol. Energy*, **78**, 603–615.

26 Roeb, M., Sattler, C., Klüser, R., Monnerie, N., de Oliveira, L., Konstandopoulos, A.G., Agrafiotis, C., Zaspalis, V.T., Nalbandian, L., Steele, A.M., and Stobbe, P. (2006) Solar hydrogen production by a two-step cycle based on mixed iron oxides. *J. Sol. Energy*, **128**, 125–133.

27 Schunk, L. and Steinfeld, A. (2009) Kinetics of the thermal dissociation of ZnO exposed to concentrated solar irradiation using a solar-driven thermogravimeter in the 1800–2100 K range. *AIChE J.*, **55**, 1497–1504.

28 Nakamura, T. (1977) Hydrogen production from water utilizing solar heat at high temperatures. *Sol. Energy*, **19**, 467–475.

29 Steinfeld, A., Sanders, S., and Palumbo, R. (1999) Design aspects of solar thermochemical engineering – a case study: two-step water-splitting cycle using the Fe3O4/FeO redox system. *Sol. Energy*, **65**, 43–53.

30 Sibieude, F., Ducarroir, M., Tofighi, A., and Ambriz, J. (1982) High temperature experiments with a solar furnace: the decomposition of Fe_3O_4, Mn_3O_4, CdO. *Int. J. Hydrogen Energy*, **7**, 79–88.

31 Steinfeld, A. (2005) Solar thermochemical production of hydrogen – a review. *Sol. Energy*, **78**, 603–615.

32 Roeb, M., Säck, J.P., Rietbrock, P., Prahl, C., Schreiber, H., Neises, M., de Oliveira, L., Graf, D., Ebert, M., Reinalter, W., Meyer-Grünefeldt, M., Sattler, C., Lopez, A., Vidal, A., Elsberg, A., Stobbe, P., Jones, D., Steele, A., Lorentzou, S., Pagkoura, C., Zygogianni, A., Agrafiotis, C., and Konstandopoulos, A.G. (2011) Test operation of a 100 kW pilot plant for solar hydrogen production from water on a solar tower. *Sol. Energy*, **85**, 634–644.

33 Ishihara, H., Kaneko, H., Hasegawa, N., and Tamaura, Y. (2008) Two-step water-splitting at 1273–1623 K using yttria-stabilized zirconia-iron oxide solid solution via co-precipitation and solid-state reaction. *Energy*, **33**, 1788–1793.

34 Siegel, N.P., Diver, R.B., Moss, T.A., Miller, J.E., Evans, L., Hogan, R.E., Allendorf, M.D., Stuecker, J.N., and James, D.L. (2008) Innovative Solar Thermochemical Water Splitting, SANDIA National Laboratories, Albuquerque, NM SAND2008-0878, March 2008.

35 Coker, E.N., Rodriguez, M.A., Ambrosini, A., Stumpf, R.R., Stechel, E.B., Wolverton, C., and Meredig, B. (2008) Sandia Final Report: Fundamental Materials Issues for Thermochemical H2O and CO2 Splitting Final Report.

36 Kondo, N., Gokon, N., Mataga, T., and Kodama, T. (2009) Internally circulating fluidized bed reactor with NiFe2O4 particles for thermochemical water splitting, in *SolarPACES International Symposium, Berlin, 15–18 September 2009* (eds T. Mancini and R. Pitz-Paal), Deutsches Zentrum für Luft- und Raumfahrt e.V, Stuttgart, p. 11.

37 Gokon, N., Takahashi, S., Yamamoto, H., and Kodama, T. (2008) Thermochemical two-step water-splitting reactor with internally circulating fluidized bed for thermal reduction of ferrite particles. *Int. J. Hydrogen Energy*, **33**, 2189–2199.

38 Tescari, S., Neises, M., Oliveira, L.d., Roeb, M., Sattler, C., and Neveu, P. (2013) Thermal model for the optimization of a solar rotary kiln to be used as high temperature thermochemical reactor. *Solar Energy*, **95**, 279–289.

39 Gokon, N., Murayama, H., Nagasaki, A., and Kodama, T. (2009) Thermochemical two-step water splitting cycles by monoclinic ZrO_2-supported $NiFe_2O_4$ and Fe_3O_4 powders and ceramic foam devices. *Solar Energy*, **83**, 527–537.

40 Scheffe, J.R., Allendorf, M.D., Coker, E.N., Jacobs, B.W., McDaniel, A.H., and Weimer, A.W. (2011) Hydrogen production via chemical looping redox cycles using atomic layer deposition-synthesized iron oxide and cobalt ferrites. *Chem. Mater.*, **23**, 2030–2038.

41 Scheffe, J.R., Li, J., and Weimer, A.W. (2010) A spinel ferrite/hercynite water-splitting redox cycle. *Int. J. Hydrogen Energy*, **35**, 3333–3340.

42 Roeb, M. and Satter, C. (2013) Isothermal water splitting. *Science*, **341**, 470–471.

43 Muhich, C., Evanko, B., Weston, K., Lichty, P., Liang, X., Martinek, J., Musgrave, C., and Weimer, A.W. (2013) Efficient generation of H_2 by splitting water with an isothermal redox cycle. *Science*, **341**, 540–542.

44 Padella, F., Alvani, C., La Barbera, A., Ennas, G., Liberatore, R., and Varsano, F. (2005) Mechanosynthesis and process characterization of nanostructured manganese ferrite. *Mater. Chem. Phys.*, **90**, 172–177.

45 Alxneit, I. (2008) Assessing the feasibility of separating a stoichiometric mixture of zinc vapor and oxygen by a fast quench – model calculations. *Solar Energy*, **82**, 959–964.

46 Fahrni, R. (2002) Hydrogen Production, an Overview of Hydrogen Production Methods and Costs Today, Term paper WS01/02 ETH Zürich, Institute of Energy Technologies.

47 Abanades, S., Charvin, P., Lemort, F., and Flamant, G. (2008) Novel two-step SnO_2/SnO water-splitting cycle for solar thermochemical production of hydrogen. *Int. J. Hydrogen Energy*, **33**, 6021–6030.

48 Steinfeld, A. (2005) Solar thermochemical production of hydrogen – a review. *Solar Energy*, **78**, 603–615.

49 Hamed, T.A., Venstrom, L., Alshare, A., Brulhart, M., and Davidson, J.H. (2009) Study of a quench device for the synthesis and hydrolysis of Zn nanoparticles: modeling and experiments. *J. Solar Energy*, **131**, 031018 9.

50 Meier, A. and Steinfeld, A. (2010) Solar thermochemical production of fuels. *Adv. Sci. Technol.*, **74**, 303–312.

51 Palumbo, R., Lede, J., Boutin, O., Ricart, E.E., Steinfeld, A., Möller, S., Weidenkaff, A., Fletcher, E.A., and Bielicki, J. (1998) The production of Zn from ZnO in a high-temperature solar decomposition quench process – I. The scientific framework for the process. *Chem. Eng. Sci.*, **53**, 2503–2517.

52 Schunk, L., Lipinski, W., and Steinfeld, A. (2009) Heat transfer model of a solar receiver-reactor for the thermal dissociation of ZnO – Experimental validation at 10 kW and scale-up to 1 MW. *Chem. Eng. J.*, **150**, 502–508.

53 Villasmil, W., Brkic, M., Wuillemin, D., and Steinfeld, A. (2013) Pilot scale demonstration of a 100-kWth solar thermochemical plant for the thermal dissociation of ZnO. *J. Sol. Energy Eng.*, **136** (1), 011016–011016-11.

54 Wegner, K., Ly, H.C., Weiss, R.J., Pratsinis, S.E., and Steinfeld, A. (2006) In situ formation and hydrolysis of Zn nanoparticles for H2 production by the 2-step ZnO/Zn water-splitting thermochemical cycle. *Int. J. Hydrogen Energy*, **31**, 55–61.

55 Otsuka, K., Hatano, M., and Morikawa, A. (1985) Decomposition of water by cerium oxide of δ-phase. *Inorg. Chim. Acta*, **109**, 193–197.

56 Flamant, G. and Abanades, S. (2006) Thermochemical hydrogen production from a two-step solar-driven water-splitting cycle based on cerium oxides. *Sol. Energy*, **80**, 1611–1623.

57 Scheffe, J.R. and Steinfeld, A. (2012) Thermodynamic analysis of cerium-based oxides for solar thermochemical fuel production. *Energy & Fuels*, **26**, 1928–1936.

58 Ermanoski, I., Siegel, N.P., and Stechel, E.B. (2013) A new reactor concept for efficient solar-thermochemical fuel production. *J. Sol. Energy. Eng.*, **135**, 031002.

59 Bilgen, E. (1988) Solar hydrogen production by hybrid thermochemical processes. *Solar Energy*, **41**, 199–206.

60 Brecher, L.E. and Wu, C.K. (1975) Electrolytic decomposition of water, US Pat. 3888750.

61 Gorensek, M.B. and Summers, W.A. (2009) Hybrid sulfur flowsheets using PEM electrolysis and a bayonet decomposition reactor. *Int. J. Hydrogen Energy*, **34**, 4097–4114.

62 Summers, W.A. and Gorensek, M.B. (2006) Nuclear hydrogen production based on the hybrid sulfur thermochemical process. Presented at the International Congress on Advances in Nuclear Power Plants (ICAPP), Reno, NV USA.

63 Corgnale, C. and Summers, W.A. (2011) Solar hydrogen production by the hybrid sulfur process. *Int. J. Hydrogen Energy*, **36**, 11604–11619.

64 Hinkley, J.T., O'Brien, J.A., Fell, C.J., and Lindquist, S.-E. (2011) Prospects for solar only operation of the hybrid sulphur cycle for hydrogen production. *Int. J. Hydrogen Energy*, **36**, 11596–11603.

65 Steimke, J.L. and Steeper, T.J. (2005) WSRC-TR-2005-00310: Characterization Testing of H2O–SO2 Electrolyzer at Ambient Pressure, Aiken, SC, USA.

66 Monnerie, N., Schmitz, M., Roeb, M., Quantius, D., Graf, D., Sattler, C., and Lorenzo, D.D. (2011) Potential of hybridisation of the thermochemical hybrid-sulphur cycle for the production of hydrogen by using nuclear and solar energy in the same plant. *Int. J. Nucl. Hydrogen Prod. Appl.*, **2**, 178–201.

67 Thomey, D., Roeb, M., Oliveira, L.d., Gumpinger, T., Schmücker, M., Sattler, C., Karagiannakis, G., Agrafiotis, C., and Konstandopoulos, A. (2010) Characterisation of construction and catalyst materials for solar thermochemical hydrogen production. Presented at the Conference on Materials for Energy, Karlsruhe, Germany.

68 Velzen, D.V. (1991) Desulphurization and Denoxing of Waste Gases producing Hydrogen as a By-product.

69 Brown, L.C., Besenbruch, G.E., Lentsch, R.D., Schultz, K.R., Funk, J.F., Pickard, P.S., Marshall, A.C., and Showalter, S.K. (2003) High Efficiency Generation of Hydrogen Fuels using Nuclear Power, Final

Technical Report for the period August 1, 1999 through September 30, 2002, General Atomics GA-A24285.
70 General Atomics (1985) Decomposition of Sulfuric Acid using Solar Thermal Energy, GA-A17573, General Atomics GA-A17573.
71 Schultz, K.R. (2003) Use of the modular helium reactor. Presented at the World Nuclear Association Annual Symposium, London, 3–5 September 2013.
72 Agency, I.A.E. (1999) Hydrogen as an energy carrier and its production by nuclear power, IAEA-TECDOC-1085, International Atomic Energy Agency.
73 Roth, M. and Knoche, K.F. (1989) Thermochemical water splitting through direct Hi-decomposition from $H_2O/HI/I_2$ solutions. *Int. J. Hydrogen Energy*, **14**, 545–549.
74 Kasahara, S., Kubo, S., Okuda, H., Terada, A., Tanak, N., Inaba, Y., Ohashi, H., Inagaki, Y., Onuki, K., and Hino, R. (2004) A pilot test plan of the thermochemical water-splitting iodine-sulfur process. *Nucl. Eng. Des.*, **223**, 355–362.
75 Yamawaki, M., Nishihara, T., Inagaki, Y., Minato, K., Oigawa, H., Onuki, K., Hino, R., and Ogawa, M. (2007) Application of nuclear energy for environmentally friendly hydrogen generation. *Int. J. Hydrogen Energy*, **32**, 2719–2725.
76 Roeb, M., Noglik, A., Sattler, C., and Pitz-Paal, R. (2009) Experimental study on sulfur trioxide decomposition in a volumetric solar receiver-reactor. *Int. J. Energy Res.*, **33**, 799–812.
77 Roeb, M., Noglik, A., Rietbrock, P.M., Mohr, S., de Oliveira, L., Sattler, C., Cerri, G., de Maria, G., Giovanelli, A., Buenaventura, A., and de Lorenzo, D. (2005) HYTHEC: Development of a dedicated solar receiver-reactor for the decomposition of sulphuric acid. Presented at the European Hydrogen Energy Conference EHEC 2005, Zaragoza, Spain.
78 Onuki, K., Kubo, S., Terada, A., Sakaba, N., and Hino, R. (2009) Thermochemical water-splitting cycle using iodine and sulfur. *Energy Environ. Sci.*, **2**, 491–497.
79 Hegels, S., Cremer, H., Knoche, K.F., Schuster, P., Steinborn, G., Wozny, G., and Wüster, G. (1980) Status report on thermochemical iron/chlorine cycles: a chemical engineering analysis of one process. *Int. J. Hydrogen Energy*, **5**, 231–252.
80 Yalcin, S. (1989) A review of nuclear hydrogen production. *Int. J. Hydrogen Energy*, **14**, 551–561.
81 Tadokoro, Y., Kajiyama, T., Yamaguchi, T., Sakai, N., Kameyama, H., and Yoshida, K. (1997) Technical evaluation of UT-3 thermochemical hydrogen production process for an industrial scale plant. *Int. J. Hydrogen Energy*, **22**, 49–56.
82 Aochi, A., Tadokoro, T., Yoshida, K., Kameyama, H., Nobue, M., and Yamaguchi, T. (1989) Economical and technical evaluation of UT-3 thermochemical hydrogen production process for an industrial scale plant. *Int. J. Hydrogen Energy*, **14**, 421–429.
83 Bilgen, E., Sakurai, M., Tsutsumi, A., and Yoshida, K. (1996) Solar UT-3 thermochemical cycle for hydrogen production. *Sol. Energy*, **57**, 51–58.
84 Lewis, M. and Masin, J. (2005) An Assessment of the Efficiency of the Hybrid Copper-Chloride Thermochemical Cycle, Argonne National Laboratory.
85 Naterer, G., Gabriel, K., Wang, Z., Daggupati, V., and Gravelsins, R. (2008) Thermochemical hydrogen production with a copper-chlorine cycle. *Int. J. Hydrogen Energy*, **33**, 5439–5489.
86 Lewis, M.A., Serban, M., and Basco, J.K. (2004) Kinetic study of the hydrogen and oxygen production reactions in the copper-chloride thermochemical cycle. Presented at the AIChE Spring National Meeting.
87 Naterer, G.F., Daggupati, V.N., Marin, G., Gabriel, K.S., and Wang, Z.L. (2008) Thermochemical hydrogen production with a copper–chlorine cycle, II: Flashing and drying of aqueous cupric chloride. *Int. J. Hydrogen Energy*, **33**, 5451–5459.
88 Wang, Z.L., Naterer, G.F., Gabriel, K.S., Gravelsins, K.S., and Daggupati, V.N. (2009) Comparison of different copper–chlorine thermochemical cycles for hydrogen production. *Int. J. Hydrogen Energy*, **34**, 3267–3276.

89 Naterer, G., Suppiah, S., Lewis, M., Gabriel, K., Dincer, I., Rosen, M.A., Fowler, M., Rizvi, G., Easton, E.B., Ikeda, B.M., Kaye, M.H., Lu, L., Pioro, I., Spekkens, P., Tremaine, P., Mostaghimi, J., Avsec, J., and Jiang, J. (2009) Recent Canadian advances in nuclear-based hydrogen production and the thermochemical Cu–Cl cycle. *Int. J. Hydrogen Energy*, **34**, 2901–2917.

6
Supercritical Water Gasification for Biomass-Based Hydrogen Production

Andrea Kruse

6.1
Introduction

Current difficulties in accessing fossil energy sources has already resulted in increased costs of, for example, petroleum, although the resources have not yet been exhausted. This demonstrates that it is necessary to explore alternative energy sources. One of these energy sources is biomass. In addition, the use of biomass as an energy source is "CO_2-neutral," as growing plants take up as much CO_2 as is released during combustion or natural degradation. As both processes, formation and use, take place on the same time scale, contrary to fossil fuels, CO_2 emission is not increased by the use of biomass.

The wet biomass could be dried and combusted or gasified by a "dry process." However, drying is associated with considerable costs at such high water content [1]. For such biomass, "wet" or "green" biomass, hydrothermal biomass conversion methods are superior. The expression "hydrothermal" was originally used in geology to name the reaction in water at increased temperatures and pressures. Depending on the reaction conditions, different fuels can be produced. These fuels are a solid, liquid (with a very high viscosity), or gases. The selectivities towards these products are usually higher and the temperatures lower than in the analogous dry processes. The reason for the lower temperature is the high reactivity of biomass in water. To avoid evaporation of the necessary reaction medium, hydrothermal biomass conversions are high pressure processes, in the range 2–35 MPa.

6.1.1
Hydrothermal Biomass Conversions

Figure 6.1 gives an overview of different biomass conversion processes. At lower temperatures, there are different pretreatment [2,3] methods, for example, simply cooking of biomass. The (exothermic) degradation of biomass starts at around 170 °C [4].

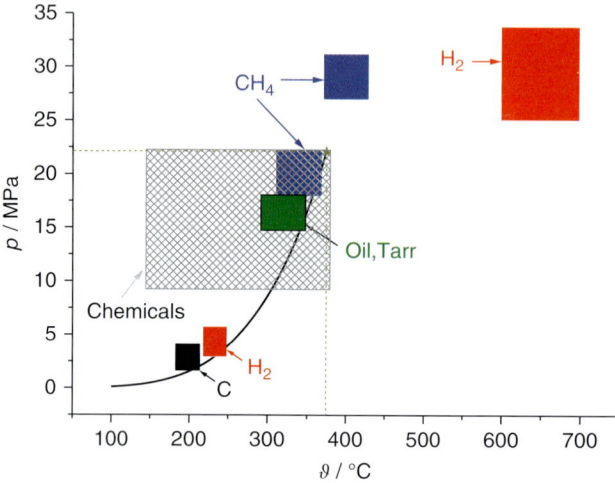

Figure 6.1 Overview of different hydrothermal biomass conversions and the saturation pressure curve of water (simplified) [20].

At around 200 °C, we enter the operation conditions of hydrothermal carbonization (HTC) of biomass. Here artificial coal is produced with very high yield, often in the presence of an acid as catalyst [5–7].

At slightly higher temperatures hydrogen can be produced from compounds originating from biomass using noble metal catalysts. This process, called aqueous reforming [8], only works, for thermodynamic reasons, at very low concentrations and up to now not with real biomass.

At around 300–350 °C biomass liquefaction occurs [9–13]. This process, developed originally by the company Shell as hydrothermal upgrading, leads to a heavy product with a lower oxygen content and higher heating value than the product of dry biomass liquefaction (fast pyrolysis [1]). In some cases catalysts are applied [10,14,15] and recently the focus has been on liquefaction of algae [14–19].

Near the critical point, catalyzed hydrothermal gasification occurs. Methane and CO_2 are the main products and heterogeneous catalysts are necessary [21–24]. The reason for choosing this temperature range is mainly the temperature dependence of the equilibrium composition (Figure 6.2). For the calculation a mixture of glucose and phenol is used as substitute for biomass. The equilibrium calculations predict a significant dependence of the gas composition on temperature (and pressure, not shown). Near the critical point of water (374 °C, 22.1 MPa) methane is the main fuel gas formed in equilibrium. To produce hydrogen, higher temperatures, above the critical point of water, are necessary as shown in equilibrium calculation (Figure 6.2).

For supercritical water gasification to hydrogen no addition of a catalyst is necessary. This process is called supercritical water gasification (SCWG), because the reaction conditions are above the critical point [20,25–27].

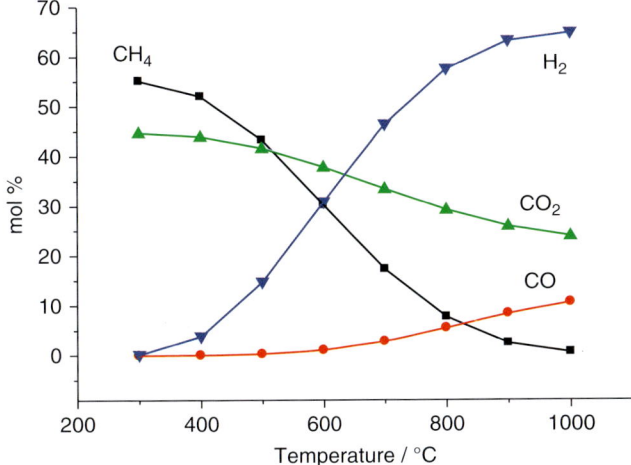

Figure 6.2 Calculated gas yields for a glucose/phenol mixture representing biomass as a function of temperature at 30 MPa with 10 wt% dry matter content (5.4 wt% glucose and 4.6 wt% phenol), calculated with Aspen Plus, RGibbs reactor, "PENG-ROB" method. Feedstock content in equilibrium: glucose <10–20 mol.%, phenol 10–18 to <10–20 mol.%, decreasing with temperature.

In the supercritical range the density of water is highly dependent on the pressure; therefore, the properties of water are dependent on pressure as well.

In addition, in the case hydrogen as preferred product, the biomass has to be heated and the reaction conditions of other hydrothermal processes are crossed. Even in the case of very fast heating, biomass is first hydrolyzed; the formed intermediates degrade and after this gases are formed. In other words: the products of low temperature biomass conversion processes are intermediates or side products of the hydrothermal gasification of biomass.

This chapter focuses on the process of supercritical water gasification to produce hydrogen. In this field some reviews have been published, which have to be considered to get a full overview of the work done [20,25–29].

6.1.2
Properties of Water

All hydrothermal biomass conversion processes, summarized in Figure 6.1, benefit from the special properties of hot compressed liquid or supercritical water (Figure 6.3 [30,31]). The influence of solvent properties on the chemical reactions enables the variety of products possible via biomass conversion. Water is an extraordinary compound because it changes its properties as function of temperature (and above the critical point also as a function of pressure) like no other solvent.

If water is heated at increased pressure (Figure 6.1), the density decreases. As a consequence and because of the increased thermal movement of the water

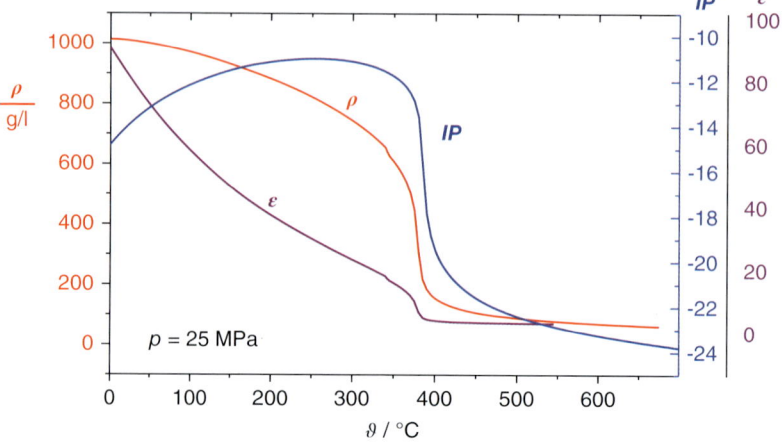

Figure 6.3 Properties of water as function of temperature ([30], data from Reference [33]); ε = relative static dielectric constant, ρ = density, and IP = -log(ionic product).

molecules, the interference of different water molecules is weaker. This leads to less and less stable hydrogen bonds, although clusters by hydrogen bonds exist still at high temperature. The second consequence is that the relative static dielectric constant decreases as a result of the lower density. The distances between the water molecules are larger and they show less interaction with each other, which is directly measureable as low dielectric constant and lower ability to dissolve polar compounds. Ambient water is known to be a polar solvent, being able to dissolve salts. With increasing temperature the polarity of water as solvent decreases, which is shown as decreasing dielectric constant in Figure 6.1. At supercritical conditions, water behaves as a non-polar organic solvent with good solubility for organic components and gases and poor solubility for salts. The single water molecule is still polar and very reactive because of the weaker/fewer hydrogen bonds [30,32]. A reaction medium consisting of polar molecules behaving as a non-polar solvent is unique in chemistry.

At subcritical conditions, the ionic product of water is higher than at ambient conditions. The reason for this is the endothermic character of the self-dissociation of water. The high H^+ and OH^- concentrations enable rather higher reaction rates of reactions that usually require the presence of acids and bases. In other cases, water itself works as catalyst, fulfilling the role usually performed by H^+ or OH^-. On the other hand, solvation energies also influence a chemical reaction, and water changes it solvation properties with temperature. The varied solvation energies of educts, intermediates, intermediate states, and products change the reaction enthalpies and activation energies [30,34]. Because of the high reactivity of and in water, polar compounds in particular show a high reactivity. Therefore, unsurprisingly, biomass shows fast degradation [31]. Recently, Akizuki has given an overview of the role of water in different reaction pathways [35], completing earlier summaries [36].

6.2
Model Compounds

The first step of the reaction of biomass in hot water is the rather fast hydrolysis of cellulose to glucose [37,38] (more detailed: [39]). This is the most important difference between hydrothermal biomass conversion and the dry biomass conversion for which higher temperatures are necessary [4]. In hydrothermal processing a solid–liquid system turns into a reaction in aqueous solution by splitting into smaller intermediates, which are dissolved. These reactions in solution are much faster due to little or no heat and mass transfer limitations.

As a consequence of the fast hydrolysis, glucose gasification should be similar to biomass gasification. On the other hand, glucose is much easier to handle, because it is soluble in water. Feeding a solid, especially into small laboratory plants, is difficult. Therefore, a lot of basic studies were carried out with glucose solutions.

6.2.1
Glucose

The number of degradation products of glucose is much lower than that of biomass. Therefore, study of the reaction of glucose in water is the key to understanding biomass conversion. Knowledge of the chemistry of biomass conversion is mainly based on knowledge of the glucose conversion reaction pathways.

The glucose converts partially into fructose. By water elimination, hydroxymethylfurfural (HMF) from fructose is produced (see Figure 6.4 below) [40–42]. HMF may polymerize easily. This polymerization is the most important reaction of hydrothermal carbonization [6]; it is an unwanted side reaction in view of hydrogen production [43].

By aldol-splitting, smaller compounds are formed from glucose and fructose [45–50], for example, acids. These acids transform the hydrothermal carbonization and the degradation reactions into auto-catalyzed processes because, for example, the HMF formation and other water elimination reactions are catalyzed by acids. At slightly increased temperature the splitting reactions of glucose and fructose are increased and the water behaves like a less polar solvent with increased solubility for the intermediates. This is the temperature range of hydrothermal liquefaction (Figure 6.1). In this temperature range (300–350 °C) the water is still polar enough as solvent to support glucose/fructose degradation reactions via polar or ionic reaction pathways [50]. On the other hand, it is less polar as a solvent, showing a high solvent power for the intermediates, which avoids polymerization reactions.

Near the critical point of water (374 °C, 22 MPa), the solvent shell of the ions from water can no longer survive, because of the low density. Therefore, the ionic product decreases to very low values (Figure 6.3 [30]) as well. As consequence of the lower density, free radical reactions become more important, leading to the formation of gases [51,52].

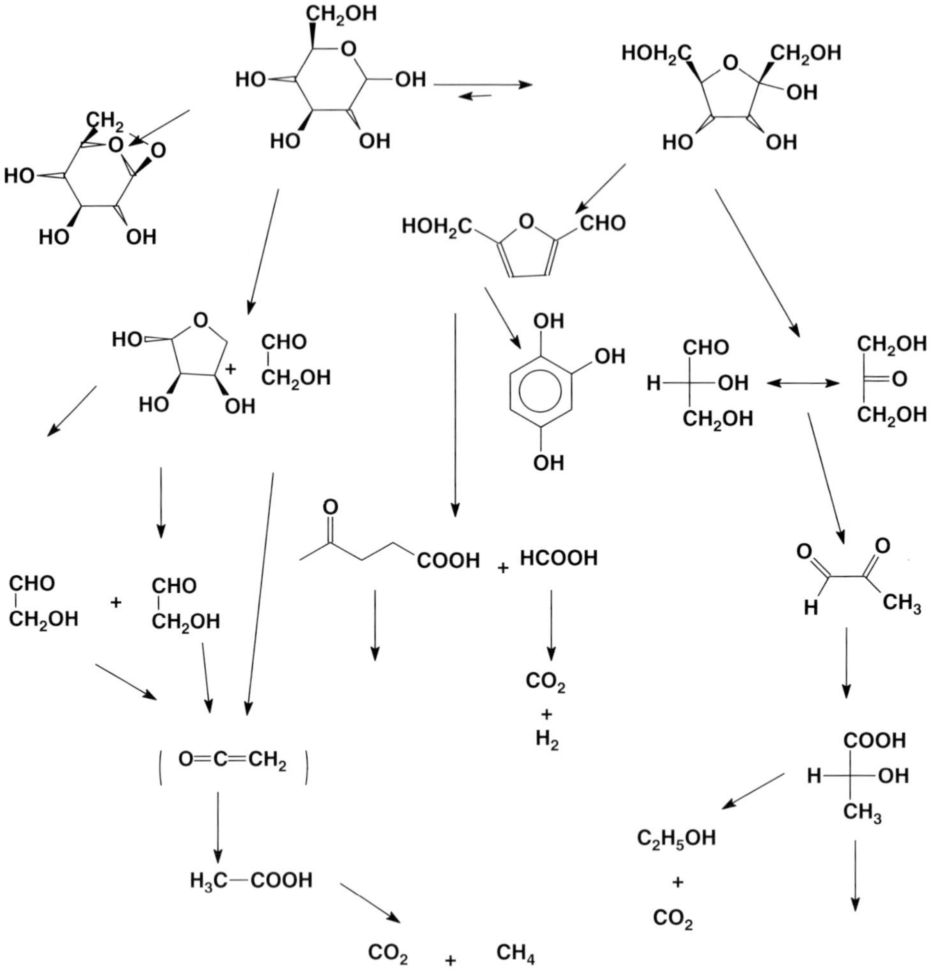

Figure 6.4 Firsts steps of glucose conversion under hydrothermal conditions (simplified) [44].

Although the studies of glucose conversion give an important view of the chemistry of supercritical water gasification, there are significant differences. First, the solubilization of biomass is not always very fast, especially in the case of lignocelluloses the lignin protects the biomass carbohydrates from water attack. Then, for example, char may be formed, which cannot be gasified under conditions at which lignin-free biomass is gasified completely. A second aspect is the gas composition: typically in the case of glucose gasification the CO content is much higher than predicted by thermodynamics (Figure 6.2) and found for biomass conversion. This is a consequence of the presence of salts in the case of biomass (see below). If salts are added to glucose the gasification products are very similar to biomass results. In addition to glucose, other simple sugars like xylose [53] and fructose [54] have been studied.

6.2.2
Cellulose

A typical property of cellulose is the crystallinity forced by hydrogen bonds. A very important aspect of the use of cellulose as model compound for biomass is whether destruction of the hydrogen bonds, by water swelling the cellulose, or the hydrolysis of glycosidic bonds is the faster process [38,55,56]. The answer seems to be that this depends on the temperature. In addition, in the case of cellulose the question of the influence of ingredients like salts is important and the comparability to biomass is limited.

6.2.3
Amino Acids

Amino acids are the product of protein hydrolysis [57–67]. Most studies have focused on the goal of using proteins as feedstock for chemical products or food additives. Amino acids may further react via hydrolysis of the amino group or decarboxylation. Studies with glycine show a rather low reaction time, lower than phenol. In the experiments with glycine, a dark solid product was also visible in the quartz reactors [68].

6.2.4
Phenols

The reactivity of phenols or their precursor lignin in supercritical water is low compared with carbohydrate conversion [69–71]. The reaction rate of conversion of phenols is significantly increased by the presence of nickel [71]. This has to be considered when experiments are carried out in autoclaves consisting of nickel base alloys [71]. The conversion of phenols leads also to higher molecular weight products by reaction of intermediates with each other [72,73].

Phenols may be formed via the degradation of lignin, a natural ingredient especially of wood and herbs. On the other hand, phenols can also be formed from carbohydrates [74]. It was suggested that phenols are formed by rearrangement of HMF [75,76]. This is not in accordance with the found reaction time dependence and the reaction order. For this reaction, a reaction order of around two was obtained [74]. This suggest that phenols from carbohydrates are formed by something like a dimerization, for example, a Diels–Alder reaction, of two smaller intermediates from carbohydrate (here glucose) degradation. This usually occurs in the higher temperature range of 400 °C [74].

6.2.5
Others

If water after a gasification experiment is analyzed, often acetic acid and phenols are found. These two compounds are hurdles for complete gasification [77]. The acetic acid concentration is influenced by the composition [78], the temperature, and slightly by pressure. Although thermodynamics predict complete gasification at low temperature and the formation of hydrogen and carbon dioxide at higher temperature, it is rather inert and without a catalyst it reacts to give methane and carbon dioxide [79]. The concentration can be rather high [80] but is decreased in the presence of $KHCO_3$. Studies of the hydrothermal degradation of sodium acetate, likely formed in the presence of carbonates, show that there is nearly no visible degradation below 400 °C. In addition, at higher temperatures, CH_4 is the main product in the temperature range investigated. For H_2 formation at higher temperatures, the formation of oxalate as intermediate is assumed [81]. If H_2 is the wanted product (SCWG) and acetic acid is formed, which is usually the case, the thermodynamically predicted hydrogen yield cannot be reached because either the acetic acid/acetate does not degrade or CH_4 is also formed at high temperatures because it is produced via decarboxylation of acetic acid.

Comparison of gasification of alcohols and organic acids shows that the alcohols are more easily gasified. Both show an interesting oscillating behavior in terms of gas composition, for example, of the methane content observed. This is explained by simple β-scission mechanisms; here whether the carbon number is even or not makes a difference [82].

6.3
Biomass

Most studies concerning supercritical gasification focus on model compounds. Biomass is much more complex, because it contains different types of chemical compounds. In the case of lignocelluloses plants these are cellulose, hemicellulose, lignin, and inorganic compounds (ash).

Figure 6.4 shows the main reaction pathways of hydrothermal biomass degradation in a simplified way using key compounds. These key compounds are representatives for reactions pathways and are found in conversions of biomass and model compounds. The use of key compounds helps us to understand the chemistry pathway by connecting results from biomass with those from model compounds [77].

The dark gray range in the middle of Figure 6.5 shows the chemical reaction of Figure 6.4 in a simplified way. In addition to the previous reaction from biomass to simple sugars, gas formation at increased temperature and the unwanted reaction to a solid product are included. Lignin is degraded to mainly phenols,

6.3 Biomass | 117

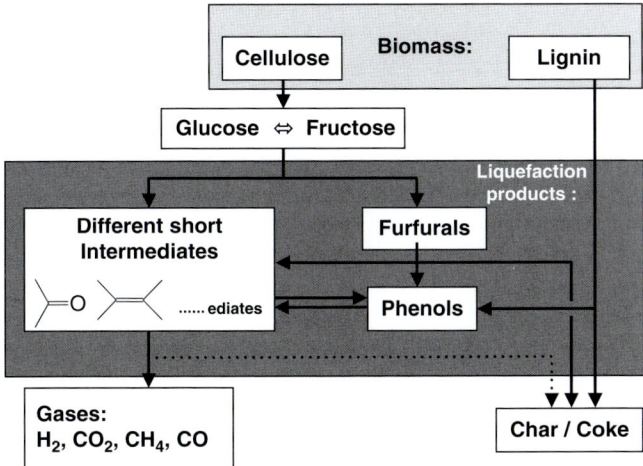

Figure 6.5 Simplified reaction scheme of hydrothermal biomass degradation.

which influences other reaction pathways. Figure 6.6 compares the concentrations of different key compounds, indicating that a very important aspect of the conversion of biomass is the presence of salts. The concentrations of key compounds are more similar to biomass conversion for the mixture of glucose with salt than for pure glucose conversion.

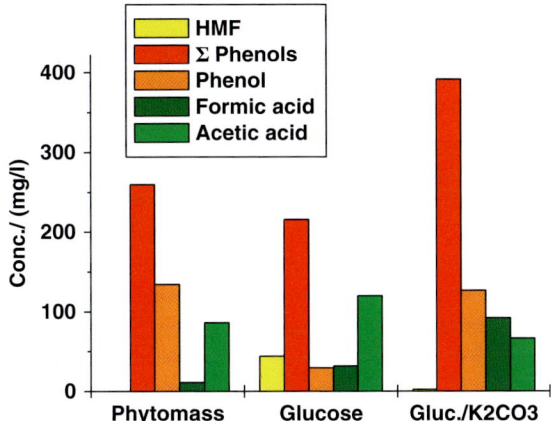

Figure 6.6 Concentration of typical intermediates for the degradation of a phytomass (mixture of carrots and potatoes) glucose and glucose with K_2CO_3 (1 °C min^{-1} to 500 °C, 5% dry matter content; 0.5% (g/g) K_2CO_3). Reprinted with kind permission of Wiley VCH Verlag GmbH [83].

6.3.1
Influence of Salts

In hydrothermal reactions, the addition of salts leads to more gas and less solid yields, and a changed composition of intermediates (Figure 6.6). Detailed studies hint at the assumption that all effects are a direct or indirect consequence of the water–gas shift reaction (Eq. (6.1)) [78]. The water–gas shift reaction is catalyzed by salts, most efficiently alkali salts, which are necessary to reach the calculated equilibrium yields for supercritical water gasification. These alkali salts are usually in the biomass; therefore, it is not necessary to add them.

Water–gas shift reaction:

$$CO + H_2O \leftrightarrow CO_2 + H_2 \tag{6.1}$$

6.3.2
Influence of Proteins

Studies with protein containing biomass and mixtures of model compounds with amino acids show lower gas yields. In the product mixture, nitrogen containing heterocyclic compounds and relatively stable free radicals are found. It is assumed that protein degradation products [12,58,68], namely amines, and hydroxyaldehydes form free radical scavengers by the Maillard reaction. The gas formation is a free radical reaction; therefore, the presence of scavengers reduces the reaction rate of the free radical chain. This is especially the case at short reaction times in a tubular reactor, because the nitrogen containing compounds are consumed and the reaction rate normalizes after a certain reaction time [84,85]. Hints in favor of this assumption are the identification of N-heterocyclic compound products and the existence of rather stable organic free radicals after reaction [84].

6.3.3
Influence of Lignin

The degradation of lignin needs higher temperatures than the degradation of carbohydrates [4]. Therefore, lignin may "protect" the carbohydrates from degradation [43] in the case of rapid heating-up [86] or remain as a "skeleton" [5] in the case of carefully chosen reaction conditions with complete solubility of carbohydrates.

The formation of phenols has consequences in view of the goal of complete supercritical biomass gasification. First, phenols are difficult to gasify (see above). Second, the gas formation of carbohydrates is influenced: investigation of a glucose–phenol mixture in a tubular reactor shows not only the lower reactivity of phenol compared with glucose; in addition, the phenol leads to a decrease of gas formation from glucose, the mechanism of which could not be identified by this study [87]. In any event, phenols are known to be free radical scavengers [88]; therefore, the effect is similar to that described in Section 6.3.2 and the free-radical gas formation rate is decreased.

6.4
Catalysts

To reach high yields of hydrogen, the presence of alkali salts for catalysis of the water–gas shift reaction is necessary [89]. In addition, for the formation of methane, nickel and noble metal catalysts are necessary [22]. The reason for this is that the hydrogenation of carbon monoxide to methane is kinetically inhibited. The same catalysts are also used to produce hydrogen at higher temperature. This should enable the hydrogen formation from, for example, methane and other compounds. An overview concerning the use of heterogeneous catalysts for gasification is given by Azadi and Farnood [23] as well as Guo *et al.* [90].

Based on early work by Antal *et al.* [91], coke produced from biomass was used to as catalyst. In the early work the salts precipitating in supercritical water plugged the solid bed of the carbon catalyst used [92]. Later, the coke particles were pumped through the reactor and recycled. By this process plugging was avoided [93].

6.5
Challenges

If real biomass is used for a technical process of supercritical water gasification to produce hydrogen, first it has to be pumped to feed into the reactor and to reach the necessary pressure. Usually, biomass slurries with a dry matter content of more than 10% or 20%, depending on the type of biomass used, is no longer pumpable. Therefore, the dry mass content is limited. It has to be pointed out that pumping is easier on a large scale with large pumps than on the laboratory scale [94]. Therefore, the challenge of pumping is on a technical scale less pronounced than can be assumed by literature reports.

6.5.1
Heating-Up

Often, unwanted reactions of the intermediates are the reason for reduced hydrogen yields or incomplete gasification. Here an important aspect is heating-up, because temperatures are crossed at which other reaction pathways, which in supercritical water gasification are unwanted reaction pathways, are preferred. It is well known that high heating rates lead to higher gas yields. One study suggests a heating rate faster than 2.2 K s^{-1} [95].

6.5.2
Heat Recovery

Although the evaporation of water is avoided by the pressure, the sensible heat necessary to heat-up a large amount of water is a very important. To reach high efficiencies, which is the usable heat and heating value of product gas relative to the sum of the heating value of biomass as well as all energies consumed to run the process, the heat for temperature increase has to be recovered. In this case, considering the low temperature heat of the products after heat exchanger, separation, and so on the calculated energy efficiencies of the different approaches and process designs vary between 44% and 65%. The exergy efficiencies published are in the range 41–52% [96–100]. Here, only the exergy of product gases and no use of the out-put stream heat is considered; therefore, the efficiency is lower. One possibility is to feed in a mixture with high dry matter content and mix it with hot water, which is recycled. The disadvantage is that pumping of biomass is only possible at rather low dry mass content. The other possibility is to use a heat exchanger [94]. Such a heat exchanger was found to be efficient, but plugging might occur. In addition, a heat exchanger limits the heating rate and therefore the yields might be not optimal.

6.5.3
Yields

Not in all cases and for all biomass are the yields in supercritical water gasification (SCWG) of biomass as predicted by equilibrium calculations. Often side reactions of the intermediates are the reason for reduced hydrogen yields or incomplete gasification. Here an important aspect is heating-up, because the temperatures ranges of hydrothermal carbonization and liquefaction are crossed. Compounds formed here may have a low reactivity at supercritical conditions, which reduces therefore the gas yields (Section 6.5.1).

In a tubular reactor [79,101] and batch reactor [102] a decrease of the relative gas yield (per dry mass in the feedstock) with increasing concentration or feedstock dry matter content, respectively, is usually observed. In a CSTR (continuous stirred tank reactor) the opposite is observed: the relative gas yield increases with feedstock dry matter content [83]. This decrease in relative gas yield can be explained by competition from an adverse high order reaction with the wanted gas formation. In such a case, the concentration is increased; the high order reaction is accelerated more than the low order reaction. Likely, the gas formation has a low reaction order, for example, 1. Kruse *et al.* [83] suggest a reaction order of around 4 for the adverse reaction, because such a reaction order is necessary to predict the observed decease of the relative gas yield observed in tubular reactors by a simple model of competing reaction pathways: the gasification (reaction order ≈ 1) and the polymerization (reaction order 4). HMF easily polymerizes and was found to have a reaction order of above 4 concerning its reaction to tarry compounds [103]. Therefore, the polymerization of furfural is a good candidate for this adverse competing reaction path occurring in parallel to gasification.

Now the question occurs as to what is different in the CSTR. If the polymerization is suppressed in this reactor type this would explain the results. In this case, the relative gas yield would increase at higher dry matter content, as observed. In general, reactions of high reaction order are suppressed in a CSTR, because the back-mixing of the reacting agent with "fresh" biomass and later products leads to a lower concentration of the reacting agent. This is not the only reason. As mentioned above, hydrogenating agents reduce the furfural concentration and back-mixing enables contact of the early degradation product HMF with the late product hydrogen, formed via the water–gas shift reaction [83]. That hydrogenation of furfurals occurs by hydrogen formed via the water–gas shift reaction has been proven by the reaction of deuterated glucose in normal "light" water [104].

6.5.4
Salt Deposition

The properties of supercritical water lead to a reduced solubility of salts. Therefore, plugging by salt deposition is a challenge. This is especially the case if in addition solid catalysts are used. Notably, some salts form a second liquid phase, between a high-density aqueous phase and a melt, depending on reaction conditions. These so-called type I water-salt systems are often found in the case of potassium salts (except sulfate). Other salts contained in biomass, like magnesium salts, show type II behavior and deposit as solid particles [105]. Both types of salts can be separated by a cyclone [94] or by gravitation [105]. Another possibility is to use salts showing type I behavior to solve type II system salts [106]. Or, mainly used for supercritical water oxidation, cold water is added at the bottom of the reactor to dissolve the salts again [107].

6.5.5
Material Choice

As consequence of the relative high temperature and pressures, nickel base alloys are used. Hydrogen formation seems not to be problematic. For example, in gasification experiments with methanol the reactor tube (nickel base alloy 625) can be used for more than 1000 h [108,109]. Anyway, biomass may include many different components leading to corrosion [110]. In the case of zoo mass in a CSTR severe corrosion was found, likely because of the sulfur in the biomass [85]. In addition, for glucose with K_2CO_3, corrosion products (Ni, Cr, and Mo particles in the outflow) were found, indicating weak corrosion [108,111]. The combination of methanol, K_2CO_3, and supercritical water produces severe corrosion of Ni-base alloy 625 and a selective solution of Ni, Mo, and Fe [112]. In the conversion of corn silage, the corrosion observed was described as "a serious problem" [113]. Technical ceramics are not a solution; besides not being mechanically stable, they show strong degradation [114]. Carbon materials are stable enough and might be used as layers inside the reactor [114].

6.5.6
Catalyst Stability

Most studies with catalysts, except for pure carbon, show deactivation [24]. A focus of recent research has been to understand deactivation and to solve this problem (see below). Here, the stability of the support at hydrothermal conditions as well as poisoning of the active layer by sulfur, salts, and carbon are of special interest [24].

6.6
Scale-Up and Technical Application

The test facility VERENA at the Institute for Catalysis Research and Technology at KIT consists of components that may be scaled up [94]. It represents the largest supercritical water gasification facility without catalyst addtion existing to date. In the plant VERENA many different types of biomass, for example, corn silage, are successfully converted into gas. Feeding becomes less of a problem as the size of the plant and the pump increases! The heat transfer works with an efficiency higher than 80% and gas cleaning was successfully demonstrated at this scale ([94] and literature cited there). At the University of Hiroshima, in cooperation with industry, a plant with a carbon catalyst on the scale of around $50\,L\,h^{-1}$ is in operation [115]. For other larger plants, even below the VERENA throughput, see References [25,116].

6.7
New Developments

To increase the yields, pretreatment methods were tested: acid pretreatment increases the hydrogen yield [117] or liquefaction [118]. The idea of liquefying the biomass before gasification is not new. It has been carried out for near-critical methane formation [119] and carbon-catalyzed supercritical gasification to hydrogen [115].

A special focus is on new catalysts, for example, based on Ni-Mg-Al. Here $NiMg_{0.6}Al_{1.9}$ shows the highest stability and hydrogen yield. The Mg-content inhibits poisoning by carbon layer formation [120,121]. In addition, new catalysts based on Ru [122] or with carbon nanotubes have been studied [123]. Well-known catalysts have been studied in more detail to understand efficiency and poisoning [124–131]. Another focus is the gasification of special feedstock, for example, algae [132–137] or pyrolysis products [121,126,138]. In addition, the gasification of sewage sludge and municipal waste has been investigated

[139–142]. This feedstock is of special interest because of the rather high incineration cost.

6.8
Conclusion

Over recent decades numerous studies have been published. The technique of supercritical water gasification of biomass is nearly mature for an industrial application. There are, however, still some hurdles to overcome. Two important aspects are material choice and salt deposition. Some ideas to solve the challenges exist but there has been no opportunity to prove them at industrial-like conditions. The idea of using a catalyst and, if possible, reducing the temperature to avoid some material problems has not been successful yet, because of a lack of catalyst stability. An exception is the fluidized carbon catalyst concept introduced by the University of Hiroshima; here additional interesting results are expected [115].

On the other hand, the chemical process during supercritical water gasification is rather well understood. This opens up the opportunity to handle a wide range of biomass. New efforts are necessary to transfer this knowledge into an industrial process.

References

1 Dahmen, N., Henrich, E., Kruse, A., and Raffelt, K. (2010) *Biomass to Biofuels: Strategies for Global Industries* (eds H.P. Blaschek, N. Qureshi, A. Vertes, and H. Yukawa), John Wiley & Sons, Ltd, Chichester, pp. 91–122.

2 Mosier, N., Wyman, C., Dale, B., Elander, R., Lee, Y.Y., Holtzapple, M., and Ladisch, M. (2005) Features of promising technologies for pretreatment of lignocellulosic biomass. *Bioresour. Technol.*, **96** (6), 673–686.

3 Ingram, T., Wörmeyer, K., Lima, J.C.I., Bockemühl, V., Antranikian, G., Brunner, G., and Smirnova, I. (2011) Comparison of different pretreatment methods for lignocellulosic materials. Part I: conversion of rye straw to valuable products. *Bioresour. Technol.*, **102** (8), 5221–5228.

4 Ibbett, R., Gaddipati, S., Davies, S., Hill, S., and Tucker, G. (2011) The mechanisms of hydrothermal deconstruction of lignocellulose: new insights from thermal-analytical and complementary studies. *Bioresour. Technol.*, **102** (19), 9272–9278.

5 Titirici, M.M., Thomas, A., Yu, S.H., Müller, J.O., and Antonietti, M. (2007) A direct synthesis of mesoporous carbons with bicontinuous pore morphology from crude plant material by hydrothermal carbonization. *Chem. Mater.*, **19** (17), 4205–4212.

6 Titirici, M.M., Antonietti, M., and Baccile, N. (2008) Hydrothermal carbon from biomass: a comparison of the local structure from poly- to monosaccharides and pentoses/hexoses. *Green Chem.*, **10** (11), 1204–1212.

7 Hoekman, S.K., Broch, A., and Robbins, C. (2011) Hydrothermal carbonization (HTC) of lignocellulosic biomass. *Energy Fuels*, **25** (4), 1802–1810.

8 Luo, N., Fu, X., Cao, F., Xiao, T., and Edwards, P.P. (2008) Glycerol aqueous

phase reforming for hydrogen generation over Pt catalyst – effect of catalyst composition and reaction conditions. *Fuel*, **87** (17–18), 3483–3489.

9 Goudriaan, F. and Peferoen, D.G.R. (1990) Liquid fuels from biomass via a hydrothermal process. *Chem. Eng. Sci.*, **45** (8), 2729–2734.

10 Watanabe, M., Bayer, F., and Kruse, A. (2006) Oil formation from glucose with formic acid and cobalt catalyst in hot-compressed water. *Carbohydr. Res.*, **341** (18), 2891–2900.

11 Bhaskar, T., Sera, A., Muto, A., and Sakata, Y. (2008) Hydrothermal upgrading of wood biomass: influence of the addition of K2CO3 and cellulose/lignin ratio. *Fuel*, **87** (10–11), 2236–2242.

12 Toor, S.S., Rosendahl, L., and Rudolf, A. (2011) Hydrothermal liquefaction of biomass: a review of subcritical water technologies. *Energy*, **36** (5), 2328–2342.

13 Akhtar, J. and Amin, N.A.S. (2011) A review on process conditions for optimum bio-oil yield in hydrothermal liquefaction of biomass. *Renew. Sustain. Energy Rev.*, **15** (3), 1615–1624.

14 Hammerschmidt, A., Boukis, N., Hauer, E., Galla, U., Dinjus, E., Hitzmann, B., Larsen, T., and Nygaard, S.D. (2011) Catalytic conversion of waste biomass by hydrothermal treatment. *Fuel*, **90** (2), 555–562.

15 Biller, P., Riley, R., and Ross, A.B. (2011) Catalytic hydrothermal processing of microalgae: decomposition and upgrading of lipids. *Bioresour. Technol.*, **102** (7), 4841–4848.

16 Anastasakis, K. and Ross, A.B. (2011) Hydrothermal liquefaction of the brown macro-alga Laminaria saccharina: effect of reaction conditions on product distribution and composition. *Bioresour. Technol.*, **102** (7), 4876–4883.

17 Qin, L., Wu, Y., Zou, S., Chen, Z., Yang, M., and Chen, H. (2010) Experimental research on direct liquefaction of microalgae in subcritical/supercritical water. *Taiyangneng Xuebao*, **31** (9), 1079–1084.

18 Savage, P. (2010) ACS National Meeting Book of Abstracts, Boston, MA.

19 Yu, G., Zhang, Y., Schideman, L., Funk, T.L., and Wang, Z. (2011) Hydrothermal liquefaction of low lipid content microalgae into bio-crude oil. *Trans. ASABE*, **54** (1), 239–246.

20 Kruse, A. (2009) Hydrothermal biomass gasification. *J. Supercrit. Fluid*, **47** (3), 391–399.

21 Rabe, S., Nachtegaal, M., Ulrich, T., and Vogel, F. (2010) Towards understanding the catalytic reforming of biomass in supercritical water. *Angew. Chem. Int. Ed.*, **49** (36), 6434–6437.

22 Elliott, D.C. (2008) Catalytic hydrothermal gasification of biomass. *Biofuels, Bioprod. Bioref.*, **2** (3), 254–265.

23 Azadi, P. and Farnood, R. (2011) Review of heterogeneous catalysts for sub- and supercritical water gasification of biomass and wastes. *Int. J. Hydrogen Energy*, **36** (16), 9529–9541.

24 Guo, Y., Wang, S.Z., Xu, D.H., Gong, Y.M., Ma, H.H., and Tang, X.Y. (2010) Review of catalytic supercritical water gasification for hydrogen production from biomass. *Renew. Sustain. Energy Rev.*, **14** (1), 334–343.

25 Kruse, A. (2008) Supercritical water gasification. *Biofuels, Bioprod. Bioref.*, **2** (5), 415–437.

26 Matsumura, Y., Minowa, T., Potic, B., Kersten, S.R.A., Prins, W., van Swaaij, W.P.M., van de Beld, B., Elliott, D.C., Neuenschwander, G.G., Kruse, A., and Jerry, A. Jr. (2005) Biomass gasification in near- and super-critical water: status and prospects. *Biomass Bioenerg.*, **29** (4), 269–292.

27 Peterson, A.A., Vogel, F., Lachance, R.P., Froling, M., Antal, M.J. Jr., and Tester, J.W. (2008) Thermochemical biofuel production in hydrothermal media: a review of sub- and supercritical water technologies. *Energy Environ. Sci.*, **1** (1), 32–65.

28 Kong, L., Li, G., Zhang, B., He, W., and Wang, H. (2008) Hydrogen production from biomass wastes by hydrothermal gasification. *Energy Sources, Part A*, **30** (13), 1166–1178.

29 Xia, F., Tian, S., Gu, J., and Ning, P. (2013) Progress of near critical and supercritical water gasification of

biomass. *Chem. Bull./Huaxue Tongbao*, **76** (2), 118–123.

30 Kruse, A. and Dinjus, E. (2007) Hot compressed water as reaction medium and reactant: properties and synthesis reactions. *J. Supercrit. Fluid*, **39** (3), 362–380.

31 Möller, M., Nilges, P., Harnisch, F., and Schröder, U. (2011) Subcritical water as reaction environment: fundamentals of hydrothermal biomass transformation. *ChemSusChem*, **4** (5), 566–579.

32 Ikushima, Y., Hatakeda, K., Sato, O., Yokoyama, T., and Arai, M. (2000) Acceleration of synthetic organic reactions using supercritical water: noncatalytic Beckmann and pinacol rearrangements. *J. Am. Chem. Soc.*, **122** (9), 1908–1918.

33 Meyer, C.A., McClintock, R.B., Silvestri, G.J., and Spencer, R.C. Jr. (1992) *Steam Tables – Thermodynamic and Transport Properties of Steam*, 6th edn, The American Society of Mechanical Engineers, New York.

34 Kruse, A. and Vogel, H. (2008) Heterogeneous catalysis in supercritical media: 2. Near-critical and supercritical water. *Chem. Eng. Technol.*, **31** (9), 1241–1245.

35 Akizuki, M., Fujii, T., Hayashi, R., and Oshima, Y. (2014) Effects of water on reactions for waste treatment, organic synthesis, and bio-refinery in sub- and supercritical water. *J. Biosci. Bioeng.*, **117** (1), 10–18.

36 Hunter, S.E. and Savage, P.E. (2004) Recent advances in acid- and base-catalyzed organic synthesis in high-temperature liquid water. *Chem. Eng. Sci.*, **59** (22–23), 4903–4909.

37 Varhegyi, G., Szabo, P., Mok, W.S.L., and Antal, M.J. (1993) Kinetics of the thermal-decomposition of cellulose in sealed vessels at elevated pressures - effects of the presence of water on the reaction-mechanism. *J. Anal. Appl. Pyrol.*, **26** (3), 159–174.

38 Tolonen, L.K., Zuckerstätter, G., Penttilä, P.A., Milacher, W., Habicht, W., Serimaa, R., Kruse, A., and Sixta, H. (2011) Structural changes in microcrystalline cellulose in subcritical water treatment. *Biomacromolecules*, **12** (7), 2544–2551.

39 Yu, Y. and Wu, H. (2010) Understanding the primary liquid products of cellulose hydrolysis in hot-compressed water at various reaction temperatures. *Energy & Fuels*, **24** (3), 1963–1971.

40 Chuntanapum, A., Shii, T., and Matsumura, Y. (2011) Acid-catalyzed char formation from 5-HMF in subcritical water. *J. Chem. Eng. Jpn*, **44** (6), 431–436.

41 Chuntanapum, A. and Matsumura, Y. (2011) Role of 5-HMF in supercritical water gasification of glucose. *J. Chem. Eng. Jpn*, **44** (2), 91–97.

42 Antal, M.J. Jr., Mok, W.S., and Richards, G.N. (1990) Mechanism of formation of 5-(hydroxymethyl)-2-furaldehyde from D-fructose and sucrose. *Carbohydr. Res.*, **199** (1), 91–109.

43 Kaayildirim, T., Sinag, A., and Kruse, A. (2008) Char and coke formation as unwanted side reaction of the hydrothermal biomass gasification. *Chem. Eng. Technol.*, **31** (11), 1561–1568.

44 Kruse, A. and Vogel, G.H. (2010) *Handbook of Green Chemistry*, Wiley-VCH Verlag GmbH.

45 Bobleter, O. and Pape, G. (1968) Der hydrothermale abbau von glucose. *Monatsh. Chem./Chemical Monthly*, **99** (4), 1560–1567.

46 Kabyemela, B.M., Adschiri, T., Malaluan, R.M., and Arai, K. (1997) Kinetics of glucose epimerization and decomposition in subcritical and supercritical water. *Ind. Eng. Chem. Res.*, **36** (5), 1552–1558.

47 Kabyemela, B.M., Adschiri, T., Malaluan, R.M., and Ohzeki, H. (1997) Rapid and selective conversion of glucose to erythrose in supercritical water. *Ind. Eng. Chem. Res.*, **36** (12), 5063–5067.

48 Kabyemela, B.M., Adschiri, T., Malaluan, R.M., and Arai, K. (1999) Glucose and fructose decomposition in subcritical and supercritical water: detailed reaction pathway, mechanisms, and kinetics. *Ind. Eng. Chem. Res.*, **38** (8), 2888–2895.

49 Williams, P.T. and Onwudili, J. (2005) Composition of products from the supercritical water gasification of glucose:

a model biomass compound. *Ind. Eng. Chem. Res.*, **44** (23), 8739–8749.

50 Srokol, Z., Bouche, A.G., van Estrik, A., Strik, R.C.J., Maschmeyer, T., and Peters, J.A. (2004) Hydrothermal upgrading of biomass to biofuel; studies on some monosaccharide model compounds. *Carbohydr. Res.*, **339** (10), 1717–1726.

51 Bühler, W., Dinjus, E., Ederer, H.J., Kruse, A., and Mas, C. (2002) Ionic reactions and pyrolysis of glycerol as competing reaction pathways in near- and supercritical water. *J. Supercrit. Fluid*, **22** (1), 37–53.

52 Promdej, C. and Matsumura, Y. (2011) Temperature effect on hydrothermal decomposition of glucose in sub- and supercritical water. *Ind. Eng. Chem. Res.*, **50** (14), 8492–8497.

53 Aida, T.M., Shiraishi, N., Kubo, M., Watanabe, M., and Smith, J. (2010) Reaction kinetics of D-xylose in sub- and supercritical water. *J. Supercritical Fluids*, **55** (1), 208–216.

54 Aida, T.M., Tajima, K., Watanabe, M., Saito, Y., Kuroda, K., Nonaka, T., Hattori, H., Smith, J., and Arai, K. (2007) Reactions of D-fructose in water at temperatures up to 400 °C and pressures up to 100 MPa. *J. Supercritical Fluids*, **42** (1), 110–119.

55 Zhao, Y., Lu, W.J., and Wang, H.T. (2009) Supercritical hydrolysis of cellulose for oligosaccharide production in combined technology. *Chem. Eng. J.*, **150** (2–3), 411–417.

56 Zhu, G., Ma, Y., and Zhu, X. (2010) Reactions of cellulose in supercritical water. *Chin. J. Org. Chem.*, **30** (1), 142–148.

57 Lamoolphak, W., Goto, M., Sasaki, M., Suphantharika, M., Muangnapoh, C., Prommuag, C., and Shotipruk, A. (2006) Hydrothermal decomposition of yeast cells for production of proteins and amino acids. *J. Hazard. Mater.*, **137** (3), 1643–1648.

58 Klingler, D., Berg, J., and Vogel, H. (2007) Hydrothermal reactions of alanine and glycine in sub- and supercritical water. *J. Supercritical Fluids*, **43** (1), 112–119.

59 Abdelmoez, W., Yoshida, H., and Nakahasi, T. (2010) Pathways of amino acid transformation and decomposition in saturated subcritical water conditions. *Int. J. Chem. React. Eng.*, **8** (1), article A107. doi:10.2202/1542-6580.1903.

60 Cheng, H., Zhu, X., Zhu, C., Qian, J., Zhu, N., Zhao, L., and Chen, J. (2008) Hydrolysis technology of biomass waste to produce amino acids in sub-critical water. *Bioresour. Technol.*, **99** (9), 3337–3341.

61 Islam, M.N., Kaneko, T., and Kobayashi, K. (2003) Reaction of amino acids in a supercritical water-flow reactor simulating submarine hydrothermal systems. *Bull. Chem. Soc. Jpn*, **76** (6), 1171–1178.

62 Kang, K.Y. and Chun, B.S. (2004) Behavior of hydrothermal decomposition of silk fibroin to amino acids in near-critical water. *Korean J. Chem. Eng.*, **21** (3), 654–659.

63 Li, J. and Brill, T.B. (2003) Decarboxylation mechanism of amino acids by density functional theory. *J. Phys. Chem. A*, **107**, 5993–5997.

64 Li, J. and Brill, T.B. (2003) Spectroscopy of hydrothermal reactions 25: kinetics of the decarboxylation of protein amino acids and the effect of side chains on hydrothermal stability. *J. Phys. Chem. A*, **107**, 5987–5992.

65 Li, J. and Brill, T.B. (2003) Spectroscopy of hydrothermal reactions, part 26: kinetics of decarboxylation of aliphatic amino acids and comparison with the rates of racemization. *Int. J. Chem. Kinetics*, **35** (11), 602–910.

66 Quitain, A.T., Daimon, H., Fujie, K., Katoh, S., and Moriyoshi, T. (2006) Microwave-assisted hydrothermal degradation of silk protein to amino acids. *Ind. Eng. Chem. Res.*, **45** (13), 4471–4474.

67 Rogalinski, T., Herrmann, S., and Brunner, G. (2005) Production of amino acids from bovine serum albumin by continuous sub-critical water hydrolysis. *J. Supercrit. Fluid*, **36** (1), 49–58.

68 DiLeo, G.J., Neff, M.E., Kim, S., and Savage, P.E. (2008) Supercritical water gasification of phenol and glycine as models for plant and protein biomass. *Energy & Fuels*, **22** (2), 871–877.

69 Resende, F.L.P. and Savage, P.E. (2009) Expanded and updated results for supercritical water gasification of cellulose and lignin in metal-free reactors. *Energy & Fuels*, **23** (12), 6213–6221.

70 Resende, F.L.P. and Savage, P.E. (2010) Kinetic model for noncatalytic supercritical water gasification of cellulose and lignin. *AIChE J.*, **56** (9), 2412–2420.

71 DiLeo, G.J., Neff, M.E., and Savage, P.E. (2007) Gasification of guaiacol and phenol in supercritical water. *Energy & Fuels*, **21** (4), 2340–2345.

72 Huelsman, C.M. and Savage, P.E. (2012) Intermediates and kinetics for phenol gasification in supercritical water. *Phys. Chem. Chem. Phys.*, **14** (8), 2900–2910.

73 Huelsman, C.M. and Savage, P.E. (2013) Reaction pathways and kinetic modeling for phenol gasification in supercritical water. *J. Supercrit. Fluid*, **81**, 200–209.

74 Kruse, A., Henningsen, T., Pfeiffer, J., and Sinag, A. (2003) Biomass gasification in supercritical water; influence of the dry matter content and the formation of phenols. *Ind. Eng. Chem. Res*, **42** (16), 3711–3717.

75 Luijkx, G.C.A., Van Rantwijk, F., and Van Bekkum, H. (1993) Hydrothermal formation of 1,2,4-benzenetriol from 5-hydroxymethyl-2-furaldehyde and D-fructose. *Carbohydr. Res.*, **242** (2), 131–139.

76 Luijkx, G.C.A., van der Horst, W., Koskinen, S.O.A., van Rantwijk, F., and Van Bekkum, H. (1994) Hydrothermal formation of hydroxylated benzenes from furan derivatives. *J. Anal. Appl. Pyrol.*, **28** (2), 245–254.

77 Sinag, A., Kruse, A., and Schwarzkopf, V. (2003) Key compounds of the hydropyrolysis of glucose in supercritical water in the presence of K_2CO_3. *Ind. Eng. Chem. Res*, **42** (15), 3519–3521.

78 Kruse, A. and Dinjus, E. (2005) Influence of salts during hydrothermal biomass gasification: the role of the catalysed water-gas shift reaction. *Z. Phys. Chem.*, **219** (3), 341–366.

79 Yu, D., Aihara, M., and Antal, M.J.J. (1993) Hydrogen production by steam reforming glucose in supercritical water. *Energy & Fuels*, **7** (5), 574–577.

80 Jin, F., Zhou, Z., Kishita, A., and Enomoto, H. (2006) Hydrothermal conversion of biomass into acetic acid. *J. Mater. Sci.*, **41** (5), 1495–1500.

81 Onwudili, J.A. and Williams, P.T. (2010) Hydrothermal reactions of sodium formate and sodium acetate as model intermediate products of the sodium hydroxide-promoted hydrothermal gasification of biomass. *Green Chem.*, **12** (12), 2214–2224.

82 Chakinala, A.G., Kumar, S., Kruse, A., Kersten, S.R.A., van Swaaij, W.P.M., and Brilman, D.W.F. (2013) Supercritical water gasification of organic acids and alcohols: the effect of chain length. *J. Supercrit. Fluid*, **74**, 8–21.

83 Kruse, A. and Faquir, M. (2007) Hydrothermal biomass gasification – effects of salts, backmixing and their interaction. *Chem. Eng. Technol.*, **30** (6), 749–754.

84 Kruse, A., Maniam, P., and Spieler, F. (2007) Influence of proteins on the hydrothermal gasification and liquefaction of biomass. 2. Model compounds. *Ind. Eng. Chem. Res.*, **46** (1), 87–96.

85 Kruse, A., Krupka, A., Schwarzkopf, V., Gamard, C., and Henningsen, T. (2005) Influence of proteins on the hydrothermal gasification and liquefaction of biomass. 1. Comparison of different feedstocks. *Ind. Eng. Chem. Res*, **44** (9), 3013–3020.

86 Hashaikeh, R., Fang, Z., Butler, I.S., Hawari, J., and Kozinski, J.A. (2007) Hydrothermal dissolution of willow in hot compressed water as a model for biomass conversion. *Fuel*, **86** (10–11), 1614–1622.

87 Weiss-Hortala, E., Kruse, A., Ceccarelli, C., and Barna, R. (2010) Influence of phenol on glucose degradation during supercritical water gasification. *J. Supercrit. Fluids*, **53** (1–3), 42–47.

88 Jin, F., Cao, J., Kishida, H., Moriya, T., and Enomoto, H. (2007) Impact of phenolic compounds on hydrothermal oxidation of cellulose. *Carbohydr. Res.*, **342** (8), 1129–1132.

89 Elliott, D.C., Sealock, L.J., and Butner, R. S. (1986) Aqueous catalyst systems for the water-gas shift reaction .3. Continuous gas processing results. *Ind. Eng. Chem. Prod. Res. Develop.*, **25** (4), 541–549.

90 Guo, Y., Wang, S.Z., Xu, D.H., Gong, Y. M., Ma, H.H. and Tang. X.Y. (2010) Review of catalytic supercritical water gasification for hydrogen production from biomass. *Renew. Sust. Energ. Rev.*, **14**, 334–343.

91 Xu, X., Matsumura, Y., Stenberg, J., and Antal, M.J. Jr. (1996) Carbon-catalyzed gasification of organic feedstocks in supercritical water. *Ind. Eng. Chem. Res.*, **35**, 2522–2530.

92 Antal, M.J., Allen, S.G., Schulman, D., Xu, X.D., and Divilio, R.J. (2000) Biomass gasification in supercritical water. *Ind. Eng. Chem. Res.*, **39** (11), 4040–4053.

93 Matsumura, Y. and Minowa, T. (2004) Fundamental design of a continuous biomass gasification process using a supercritical water fluidized bed. *Int. J. Hydrogen Energy*, **29** (7), 701–707.

94 Boukis, N., Galla, U., Müller, H., and Dinjus, E. (2007) Biomass gasification in supercritical water. Experimental progress achieved with the Verena pilot plant. Presented at the 15th European Biomass Conference and Exhibition, 7–11 May 2007, Berlin, Germany.

95 Fang, Z., Minowa, T., Smith, R.L., Ogi, T., and Kozinski, J.A. (2004) Liquefaction and gasification of cellulose with Na_2CO_3 and Ni in subcritical water at 350 °C. *Ind. Eng. Chem. Res.*, **43** (10), 2454–2463.

96 Calzavara, Y., Joussot-Dubien, C., Boissonnet, G., and Sarrade, S. (2005) Evaluation of biomass gasification in supercritical water process for hydrogen production. *Energy Conversion Manage.*, **46** (4), 615–631.

97 Gasafi, E., Meyer, L., and Schebek, L. (2007) Exergetic efficiency and options for improving sewage sludge gasification in supercritical water. *Int. J. Energy Res.*, **31** (4), 346–363.

98 Feng, W., van der Kooi, H.J., and Swaan Arons, J. (2004) Biomass conversions in subcritical and supercritical water: driving force, phase equilibria, and thermodynamic analysis. *Chem. Eng. Process.*, **43** (12), 1459–1467.

99 Lu, Y., Guo, L., Zhang, X., and Yan, Q. (2007) Thermodynamic modeling and analysis of biomass gasification for hydrogen production in supercritical water. *Chem. Eng. J.*, **131** (1–3), 233–244.

100 Matsumura, Y. (2002) Evaluation of supercritical water gasification and biomethanation for wet biomass utilization in Japan. *Energy Conversion Manage.*, **43** (9–12), 1301–1310.

101 D'Jesus, P., Artiel, C., Boukis, N., Kraushaar-Czarnetzki, B., and Dinjus, E. (2005) Influence of educt preparation on gasification of corn silage in supercritical water. *Ind. Eng. Chem. Res.*, **44** (24), 9071–9077.

102 Srokol, Z.W. and Rothenberg, G. (2010) Practical issues in catalytic and hydrothermal biomass conversion: concentration effects on reaction pathways. *Top. Catal.*, **53** (15–18), 1258–1263.

103 Chuntanapum, A. and Matsumura, Y. (2009) Formation of tarry material from 5-HMF in subcritical and supercritical water. *Ind. Eng. Chem. Res.*, **48** (22), 9837–9846.

104 Kruse, A., Bernolle, P., Dahmen, N., Dinjus, E., and Maniam, P. (2010) Hydrothermal gasification of biomass: consecutive reactions to long-living intermediates. *Energy Environ. Sci.*, **3** (1), 136–143.

105 Schubert, M., Aubert, J., Müller, J.B., and Vogel, F. (2012) Continuous salt precipitation and separation from supercritical water. Part 3: interesting effects in processing type 2 salt mixtures. *J. Supercrit. Fluid*, **61**, 44–54.

106 Kruse, A., Forchheim, D., Gloede, M., Ottinger, F., and Zimmermann, J. (2010) Brines in supercritical biomass gasification: 1. Salt extraction by salts and the influence on glucose conversion. *J. Supercrit. Fluid*, **53** (1–3), 64–71.

107 Xu, D.H., Wang, S.Z., Gong, Y.M., Guo, Y., Tang, X.Y., and Ma, H.H. (2010) A novel concept reactor design for preventing salt deposition in supercritical water. *Chem. Eng. Res. Des.*, **88** (11), 1515–1522.

108 Habicht, W., Boukis, N., Franz, G., and Dinjus, E. (2004) Investigation of nickel-based alloys exposed to supercritical water environments. *Micochim. Acta*, **145**, 57–62.

109 Boukis, N., Habicht, W., Franz, G., and Dinjus, E. (2003) Behavior of Ni-base alloy 625 in methanol-supercritical water systems. *Mater. Corros.*, **54**, 326–330.

110 Marrone, P.A. and Hong, G.T. (2009) Corrosion control methods in supercritical water oxidation and gasification processes. *J. Supercrit. Fluids*, **51** (2), 83–103.

111 Sinag, A., Kruse, A., and Rathert, J. (2004) Influence of the heating rate and the type of catalyst on the formation of selected intermediates and on the generation of gases during hydropyrolysis of glucose with supercritical water in a batch reactor. *Ind. Eng. Chem. Res*, **43** (2), 502–508.

112 Habicht, W., Boukis, N., Hauer, E., and Dinjus, E. (2011) Analysis of hydrothermally formed corrosion layers in Ni-base alloy 625 by combined FE-SEM and EDXS. *X-Ray Spectrom.*, **40** (2), 69–73.

113 D'Jesus, P., Boukis, N., Kraushaar-Czarnetzki, B., and Dinjus, E. (2006) Influence of process variables on gasification of corn silage in supercritical water. *Ind. Eng. Chem. Res.*, **45** (5), 1622–1630.

114 Richard, T., Poirier, J., Reverte, C., Aymonier, C., Loppinet-Serani, A., Iskender, G., Pablo, E.B., and Marias, F. (2012) Corrosion of ceramics for vinasse gasification in supercritical water. *J. Eur. Ceram. Soc.*, **32** (10), 2219–2233.

115 Nakamura, A., Kiyonaga, E., Yamamura, Y., Shimizu, Y., Minowa, T., Noda, Y., and Matsumura, Y. (2008) Gasification of catalyst-suspended chicken manure in supercritical water. *J. Chem. Eng. Jpn*, **41** (5), 433–440.

116 Gassner, M., Vogel, F., Heyen, G., and Marèchal, F. (2011) Optimal process design for the polygeneration of SNG, power and heat by hydrothermal gasification of waste biomass: process optimisation for selected substrates. *Energy Environ. Sci.*, **4** (5), 1742–1758.

117 Lu, Y., Guo, L., Zhang, X., and Ji, C. (2012) Hydrogen production by supercritical water gasification of biomass: explore the way to maximum hydrogen yield and high carbon gasification efficiency. *Int. J. Hydrogen Energy*, **37** (4), 3177–3185.

118 Sawai, O., Nunoura, T., and Yamamoto, K. (2013) Application of subcritical water liquefaction as pretreatment for supercritical water gasification system in domestic wastewater treatment plant. *J. Supercrit. Fluid*, **77**, 25–32.

119 Douglas, C. and Elliott, G. (2004) Chemical processing in high-pressure aqueous environments. 7. Process development for catalytic gasification of wet biomass feedstocks. *Ind. Eng. Chem. Res.*, **43** (9), 1999–2004.

120 Li, S., Guo, L., Zhu, C., and Lu, Y. (2013) Co-precipitated Ni-Mg-Al catalysts for hydrogen production by supercritical water gasification of glucose. *Int. J. Hydrogen Energy*, **38** (23), 9688–9700.

121 Li, S., Lu, Y., Guo, L., and Zhang, X. (2011) Hydrogen production by biomass gasification in supercritical water with bimetallic Ni-M/+|Al2O3 catalysts (M = Cu, Co and Sn). *Int. J. Hydrogen Energy*, **36** (22), 14391–14400.

122 Zhang, L., Xu, C., and Champagne, P. (2012) Activity and stability of a novel Ru modified Ni catalyst for hydrogen generation by supercritical water gasification of glucose. *Fuel*, **96**, 541–545.

123 DeVlieger, D.J.M., Thakur, D.B., Lefferts, L., and Seshan, K. (2012) Carbon nanotubes: a promising catalyst support material for supercritical water gasification of biomass waste. *ChemCatChem*, **4** (12), 2068–2074.

124 Azadi, P., Afif, E., Foroughi, H., Dai, T., Azadi, F., and Farnood, R. (2013) Catalytic reforming of activated sludge model compounds in supercritical water using nickel and ruthenium catalysts. *Appl. Catal. B-Environ.*, **134–135**, 265–273.

125 Azadi, P., Afif, E., Azadi, F., and Farnood, R. (2012) Screening of nickel catalysts for selective hydrogen production using supercritical water gasification of glucose. *Green Chem.*, **14** (6), 1766–1777.

126 Chakinala, A.G., Chinthaginjala, J.K., Seshan, K., Van Swaaij, W.P.M., Kersten, S.R.A., and Brilman, D.W.F. (2012) Catalyst screening for the hydrothermal gasification of aqueous phase of bio-oil. *Catal. Today*, **195** (1), 83–92.

127 Chowdhury, M.B.I., Hossain, M.M., and Charpentier, P.A. (2011) Effect of supercritical water gasification treatment on Ni/La$_2$O$_3$-Al$_2$O$_3$-based catalysts. *Appl. Catal. A-Gen.*, **405** (1–2), 84–92.

128 Dreher, M., Johnson, B., Peterson, A.A., Nachtegaal, M., Wambach, J., and Vogel, F. (2013) Catalysis in supercritical water: pathway of the methanation reaction and sulfur poisoning over a Ru/C catalyst during the reforming of biomolecules. *J. Catal.*, **301**, 38–45.

129 Osada, M., Yamaguchi, A., Hiyoshi, N., Sato, O., and Shirai, M. (2012) Gasification of sugarcane bagasse over supported ruthenium catalysts in supercritical water. *Energy & Fuels*, **26** (6), 3179–3186.

130 Peterson, A.A., Dreher, M., Wambach, J., Nachtegaal, M., Dahl, S., Nèrskov, J.K., and Vogel, F. (2012) Evidence of scrambling over ruthenium-based catalysts in supercritical-water gasification. *ChemCatChem*, **4** (8), 1185–1189.

131 Yamaguchi, A., Hiyoshi, N., Sato, O., and Shirai, M. (2012) Gasification of organosolv-lignin over charcoal supported noble metal salt catalysts in supercritical water. *Top. Catal.*, **55** (11–13), 889–896.

132 Brandenberger, M., Matzenberger, J., Vogel, F., and Ludwig, C. (2013) *Biomass Bioenerg.*, **51**, 26–34.

133 Faeth, J.L. and Savage, P.E. (2012) The effects of heating rate and reaction time on hydrothermal liquefaction of microalgae. Presented at AIChE 2012 – 2012 AIChE Annual Meeting.

134 Freitas, A.C.D. and Guirardello, R. (2013) Thermodynamic analysis of supercritical water gasification of microalgae biomass for hydrogen and syngas production. *Chem. Eng. Trans.*, **32**, 553–558.

135 Guan, Q., Wei, C., and Savage, P.E. (2012) Kinetic model for supercritical water gasification of algae. *Phys. Chem. Chem. Phys.*, **14** (9), 3140–3147.

136 Miller, A., Hendry, D., Wilkinson, N., Venkitasamy, C., and Jacoby, W. (2012) Exploration of the gasification of Spirulina algae in supercritical water. *Bioresour. Technol.*, **119**, 41–47.

137 Onwudili, J.A., Lea-Langton, A.R., Ross, A.B., and Williams, P.T. (2013) Catalytic hydrothermal gasification of algae for hydrogen production: composition of reaction products and potential for nutrient recycling. *Bioresour. Technol.*, **127**, 72–80.

138 Zhang, L., Champagne, P., and Charles Xu, C. (2011) Supercritical water gasification of an aqueous by-product from biomass hydrothermal liquefaction with novel Ru modified Ni catalysts. *Bioresour. Technol.*, **102** (17), 8279–8287.

139 Chen, Y., Guo, L., Jin, H., Yin, J., Lu, Y., and Zhang, X. (2013) An experimental investigation of sewage sludge gasification in near and super-critical water using a batch reactor. *Int. J. Hydrogen Energy*, **38** (29), 12912–12920.

140 Chen, Y., Guo, L., Cao, W., Jin, H., Guo, S., and Zhang, X. (2013) Hydrogen production by sewage sludge gasification in supercritical water with a fluidized bed reactor. *Int. J. Hydrogen Energy*, **38** (29), 12991–12999.

141 Sawai, O., Nunoura, T., and Yamamoto, K. (2014) Supercritical water gasification of sewage sludge using bench-scale reactor: advantages and drawbacks. *J. Mater. Cycles Waste Manage.*, **16**, 82–92.

142 Zhai, Y., Wang, C., Chen, H., Li, C., Zeng, G., Pang, D., and Lu, P. (2013) Digested sewage sludge gasification in supercritical water. *Waste Manage. Res.*, **31** (4), 393–400.

7
Thermochemical Hydrogen Production – Plasma-Based Production of Hydrogen from Hydrocarbons

Abdullah Aitani, Shakeel Ahmed, and Fahad Al-Muhaish

7.1
Introduction

The majority of hydrogen is currently produced by catalyst-based chemical processes such as steam reforming of natural gas (methane) and other petroleum sources (naphtha) [1–5]. Steam methane reforming (SMR) is by far the most important and widely used process for the industrial production of hydrogen using large manufacturing plants [1]. Other routes for hydrogen production from hydrocarbons include partial oxidation (POX), autothermal reforming (ATR), and decomposition to hydrogen and carbon [1–3]. About 50% of the produced hydrogen is consumed by petroleum refining processes, followed by ammonia synthesis, methanol synthesis, and other miscellaneous uses. The renewed interest in developing more efficient hydrogen processes is due to the rapid development of fuel cell technology for automotive and transport sector applications.

Plasma, which is an ionized gas known as a distinct fourth state of matter, is an attractive alternative route for the production of hydrogen-rich gas from gaseous and liquid hydrocarbons. Plasma replaces catalysts and accelerates chemical reactions due to both a temperature and an active-species effect [4,5]. The role of the plasma is not only to provide energy to the system but only to generate radical and excited species, allowing the initiation and enhancement of chemical reactions. The input power is technically regarded as a key parameter in operating the plasma chemical processing and in sustaining the plasma stability. In the presence of steam or air, hydrocarbons are converted into H•, •OH, and O• radicals, thereby promoting reductive and oxidative reactions. These active species can lead to long-chain reactions of fuel conversion and hydrogen production. Compared to plasma reforming and POX, plasma decomposition of hydrocarbons co-produces carbon, which is isolated in solid form, thus avoiding the cost of sequestering CO_2.

Among the different plasma characteristics, one can distinguish two main categories: non-thermal and thermal plasmas. In non-thermal plasmas, the electrical power is normally low (tens to hundreds watts). Non-thermal plasma

Hydrogen Science and Engineering: Materials, Processes, Systems and Technology, First Edition.
Edited by Detlef Stolten and Bernd Emonts.
© 2016 Wiley-VCH Verlag GmbH & Co. KGaA. Published 2016 by Wiley-VCH Verlag GmbH & Co. KGaA.

accelerates chemical reactions at low temperature as well, through the generation of active species by fast electrons [3,4]. If active species, generated by non-thermal plasma, are capable of promoting many cycles of chemical transformation, then a high specific productivity of plasma can be combined with the low energy consumption of traditional catalysts. The temperature level required to shift chemical equilibrium is relatively low (600–1000 K). It was concluded that non-thermal plasma alone is inefficient for converting hydrocarbons because of the high activation energy required by electron impact processes [6].

In thermal plasma applications, electrical power injected in the discharge is relatively high (e.g., higher than 1 kW) and the neutral species and the electrons interact for longer than microseconds, so that they have approximately the same temperature (around 5000–10 000 K). The use of this technology is therefore not relevant for fuel cell applications, which require an efficient production of hydrogen in terms of energy consumption [7]. The high enthalpy of thermal plasma has been used as the main heat source to decompose hydrocarbons into hydrogen rich gas and solid carbonaceous nano-materials [8]. This technique has turned out to be very effective and clean, without emission of greenhouse gases and other pollutants, as compared to conventional processes. Reaction enthalpies in the process are supplied from external electrical power through arc and/or inductively coupled discharges, and the high reaction temperatures of 2000–10 000 K are easily achievable.

7.2
Non-thermal Plasma

Several plasma discharge methods such as gliding arc, microwave, corona, dielectric barrier discharge, and others have been used for the conversion of hydrocarbons (mainly methane) with the addition of O_2, CO_2, or H_2O into hydrogen. Table 7.1 presents types of non-thermal plasmas used for the production of hydrogen from hydrocarbons.

7.2.1
Gliding-Arc Plasma

A group of researchers has developed a gliding-arc Plasmatron device for hydrocarbon reforming into syngas for automotive (on-board) applications [9–11]. An optional catalyst was used for processing of the hydrogen rich-gas and light hydrocarbons into additional hydrogen and CO. Some of the capabilities of the Plasmatron include fast start time, wide dynamic range of operation (with fuel input rates spanning $0.1–3\,g\,s^{-1}$), small and compact physical dimensions, and low electrical power consumption. A wide variety of hydrocarbon fuels, including methane, propane, and liquid fuels such as gasoline, diesel, bio-diesel, and JP-5, can be reformed in the Plasmatron. In a modified version of the Plasmatron, hydrogen yield reached 50% and conversion rates were as high as 80% in

Table 7.1 Types of non-thermal plasmas used in hydrogen production from hydrocarbons.

Plasma type	Brief description
Gliding arc discharge	High plasma density consisting of two electrodes wherein the arc starts at the shortest distance between electrodes. Characterized by simplicity and stability of discharge
Microwave	High frequency electromagnetic radiation in the GHz range. It is capable of exciting gas discharges without the requirement of electrodes
Dielectric barrier discharge (DBD)	Created by electric discharge between two electrodes separated by an insulating barrier. Low temperature makes DBD attractive in generating plasma at atmospheric pressure
Corona discharge	Its electrical discharge is brought about by ionization of a gas surrounding a conductor. The discharge occurs when the electric strength is high enough to form a conductive region without electrical arcing

diesel reforming [12,13]. Several novel gliding arc plasma reformers were developed and investigated for the partial oxidation of several fuels, including in the POX and ATR regimes. Figure 7.1 presents the schematics of a Plasmatron device that operates as part of an integrated system with a solid oxide fuel cell to create a complete plasma-assisted on-board auxiliary power unit (APU). By recycling heat and a portion of the fuel cell exhaust, it was possible to limit the amount of external air used as an oxidant, promote endothermic steam reforming reactions, and increase the overall quality and heating value of the reformate stream.

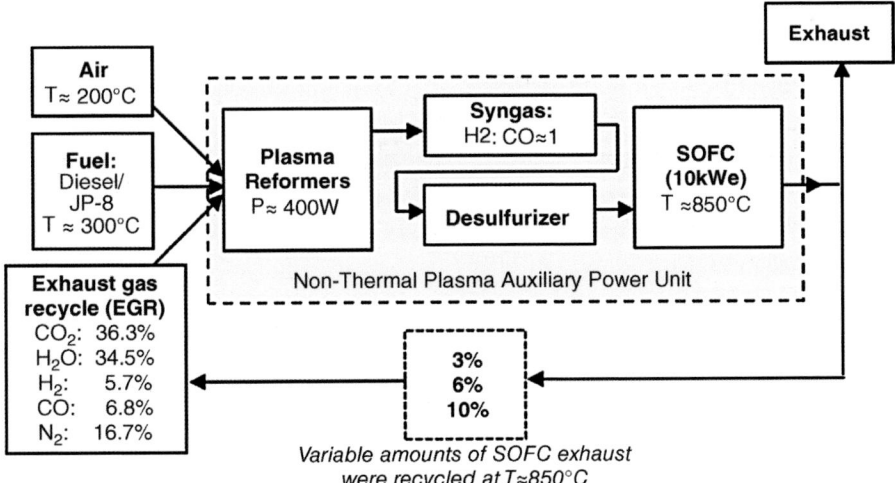

Figure 7.1 Schematic diagram of the major components of a gliding arc plasma auxiliary power unit operated in steam-oxidative diesel reforming mode [13].

In another approach, the technology and design of a plasma reactor composed of two gliding arcs in series were proposed in terms of gasoline conversion into hydrogen-rich gas under ATR or steam reforming conditions for automotive applications [7]. The reformer is flexible in changing the geometry of the electrodes, the reaction volume, inter-electrode gap, and to work with one or two gliding arcs in series. It has a very short time startup (few seconds), a large operating range of fuel power (from 10 to 40 kW), and compactness and robustness. The system could process different hydrocarbons and is tolerant to sulfur content and carbon deposit. The effects of the different operating parameters (pressure, temperature, air and steam ratio, inlet flow rates) on the reformer efficiency and the composition of the product gas were investigated. The reformer efficiency reached 12.3% in ATR and 26% in steam reforming for a pressure equal to 2×10^5 Pa. It was claimed that this efficiency could, however, be increased by working at higher pressure (2.5 or 3×10^5 Pa) and especially at higher temperature (1000 K or higher).

The partial oxidation of propane to syngas has been studied in a gliding arc apparatus with electric consumption at less than 2% of a fuel cell electric output [14]. The 1-liter reactor operated at atmospheric pressure and needed less than 100 W of electric assistance to produce up to $3 \, m^3(N) \, h^{-1}$ of pure syngas corresponding to about 10 kW of electric power of an ideal fuel cell. Propane was totally reformed at more than 70% energy efficiency with total absence of soot.

The effect of four kinds of additive gases (He, Ar, N_2, and CO_2) during methane pyrolysis was investigated in a gliding-arc plasma [15]. Methane conversion produced hydrogen, C_2H_2, and solid carbon as main products. A high frequency AC power supply with a maximum voltage of 10 kV and a maximum current of 100 mA was connected to the electrodes of the gliding arc to generate the plasma. Methane conversion reached 40% at 15 kHz and went up to 45% at 20 kHz. Increasing the frequency increased the total power, resulting in higher dissociation of methane and generated more energetic electrons. The selectivity to hydrogen reached 40%, and C_2H_2 reached 20%. Argon, helium, and nitrogen produced a positive effect on methane conversion and reduced the power consumption. The conversion reached 65% at 10% CH_4 diluted with 90% of argon or helium. Increasing dilute gas concentration produced higher selectivity of hydrogen and reduced the selectivity of C_2H_2. However, when methane was mixed with CO_2, the conversion of methane was lower and the main products were hydrogen, CO, and C_2H_2.

The gliding arc discharge accelerated chemical reactions of methane conversion into syngas at low temperature and power consumption of about 300 W [16]. The apparatus enhanced the interaction of key radicals such as $CH_3^{\bullet}$ and $H^{\bullet}$ followed by chemical reactions involving other hydrocarbons. However, to lower further the cost of hydrogen production, the experimental setup must consider recycling the plasmagen gas (argon). A gliding arc discharge reactor was also used to decompose ethanol into primarily hydrogen and CO with small amounts of CH_4, C_2H_2, C_2H_4, and C_2H_6 [17]. The ethanol concentration, electrode gap, input power (14 W), and Ar flow rate all affected the conversion of

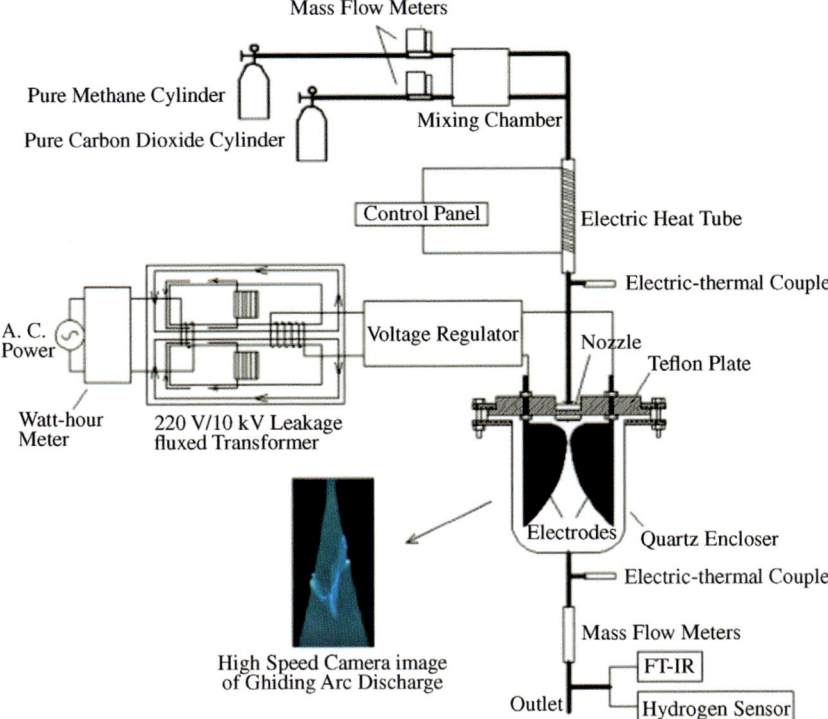

Figure 7.2 Schematic diagram of the experimental procedure for gliding arc discharge plasma assisted methane reforming with CO_2 [18].

ethanol, with results ranging from 40.7% to 58.0% and energy efficiency of 8%. Interestingly, for all experimental conditions the S_{H2}/S_{CO} selectivity ratio was quite stable at around 1.03. The mechanism for the decomposition of ethanol was also described.

The influence of feed gases proportion on the performance of gliding arc discharge plasma assisted methane reforming with CO_2 process has been investigated [18]. Figure 7.2 shows a schematic diagram of the experimental procedure. An additional analysis of the theoretical conditions of equilibrium and its comparison with the experimental results was also included. The gliding arc was effective in converting CH_4/CO_2 into syngas with C_2H_2 and C_2H_4 detected as the main hydrocarbon compounds. Strong dehydrogenation was suggested with no obvious C_2H_6 observed in the effluent and serious coke formation and deposition detected in the relatively high CH_4/CO_2 feed ratio and supply voltage condition.

A new design of a micro-sized gliding arc discharge reactor was investigated for the reforming of methane [19]. Hydrogen and acetylene were dominantly produced, with high selectivities of 75% and 70–90%, respectively. Small amounts of other products (including ethylene, ethane, butadiene, and coke) were also formed. The results showed that the methane conversion strongly depended on input power (4–12 W) and reactor thickness (0.65–4.25 mm). At a methane flow rate of 100 cm^3 min^{-1}, an input power of 6 W, a frequency of 50 Hz, and an electrode gap distance of 4 mm, the temperatures at the outer wall of the plasma reactor were found to be about 125, 130, and 150 °C, respectively, for the three systems: plasma alone, plasma + unloaded catalyst, and plasma + Ni-loaded catalyst. The interaction between plasma and Ni catalyst provided a significant enhancement of methane conversion (50%), while the selectivity to hydrogen remained unchanged.

Hydrogen-rich gas was produced using a gliding arc reformer consisting of plasma and catalytic reactors [20]. Optimum conditions for producing H_2 from methane were also suggested. The conditions consisted of an O_2/C ratio of 0.64, a total gas flow rate of 14.2 L min^{-1}, a catalyst reactor temperature of 672 °C, a reformer energy density of 1.1 kJ L^{-1}, and a steam flow ratio of 0.8. The maximum H_2 concentration of 41.1% was obtained under the optimum conditions. In addition, methane conversion rate, H_2 yield, and reformer thermal efficiencies were 67%, 54%, and 35%, respectively.

The combination of steam reforming with POX of CO_2-containing methane has been investigated in a gliding arc discharge plasma [21]. The results showed an increase in either methane conversion or syngas yield with increasing input voltage and electrode gap distance, whereas the opposite trends were observed with increasing HCs/O_2 feed molar ratio and input frequency. The optimum conditions were found at a HCs/O_2 feed molar ratio of 2 : 1, an input voltage of 14.5 kV, an input frequency of 300 Hz, and an electrode gap distance of 6 mm, providing high methane and oxygen conversions with high syngas selectivity and relatively low power consumptions, as compared with the other processes.

The effect of stage number of multistage gliding arc discharge system on the process performance of the combined steam reforming and POX of simulated natural gas was investigated [22]. The simulated natural gas contained 70% methane, 10% ethane, 10% propane and 10% CO_2. Increasing stage number from 1 to 3 stages at constant feed flow rate enhanced the reactant conversions, and hydrogen yield with the reduction of energy consumption. Under the operating conditions at a HCs/O_2 feed molar ratio of 2 : 1, an applied voltage of 14.5 kV, an input frequency of 300 Hz, and an electrode gap distance of 6 mm, the lowest energy consumption was 3.49×10^{-17} W s per molecule of reactants converted or 2.04×10^{-17} W s per molecule of hydrogen produced obtained from three stages of plasma reactors at residence time and feed flow rate of 4.11 s and 100 cm^3 min^{-1}, respectively.

7.2.2
Microwave Plasma

Microwave plasma accelerates the conversion of hydrocarbon due to the specific influence of the plasma-active particles (ions, radicals), making temperatures as low as 500 °C effective [23]. The preheated methane (400–600 °C) was fed to a

Plasmatron in which the gas was exposed to a pulse-periodic microwave discharge with high-power (250 kW). Methane conversion up to about 30% was shown at temperatures from 250 to 600 °C. The following possible mechanisms of plasma effect on methane conversion were considered: thermal acceleration related to local overheating within microwave streamer channels; radical acceleration provided by possible chain reactions of methane radicals such as $CH_3{}^\bullet$, $H^\bullet$, and $CH_2{}^\bullet$ generated by plasma from matrix gas; autocatalytic methane conversion on the surface of carbon black particles generated in the plasma zone; and ion acceleration driven by ion/molecular chain reactions during methane conversion.

A process utilizing microwave plasma and a catalytic reaction for the production of hydrogen and carbon black from methane has been developed and tested [24]. The reactor was designed to be able to insert the honeycomb type catalyst pellet such as Pd and Pt with a diameter of almost 114 mm. The conversion was carried out with 2.45 GHz microwave-generated plasma, and high conversion yields of hydrogen and carbon black were obtained. It was found that hydrogen and carbon black were produced at the mole ratio of 2 : 1, and carbon black of around 30 nm was produced, which is comparable to classical furnace black. Methane conversion reached 96% at 3 kW of applied plasma power, pressure of 100 torr, and $1.0\,L\,min^{-1}$ CH_4 flow rate. The selectivity to hydrogen ranged from 83% to 95% under the above conditions. Platinum-loaded catalysts showed higher activities than a Pd catalyst, and the productivity of carbon black increased with Pt loading, up to a maximum of 3 wt%.

The use of a single-stage, dry methane plasmolysis process has been demonstrated for producing hydrogen and carbon powders at atmospheric-pressure [25]. A microwave torch plasma system was assembled with a maximum stationary power of 5 kW in continuous-wave mode. A high selectivity to hydrogen (86.0%) and carbon powder and a low required energy consumption of hydrogen were obtained simultaneously at a low-applied power and a high inlet concentration of methane. Large amounts of nano-carbon powders, which consisted of carbon atoms and had a graphite-rhombohedral structure with a particle size of about 50 nm, were produced. A similar study reported the effect of temperature and power source on the synthesis of nanostructured carbon and high-purity hydrogen from microwave plasma cracking of methane [26].

A theoretical research to methane reactivity in a microwave plasma studied the effects of various process parameters on energy efficiency, methane conversion, and selectivity to acetylene, ethylene, and ethane [27]. The results showed a strong correlation between the energy efficiency of the discharge and the specific energy supplied. A high energy efficiency required a high local gas temperature, which played a key role in determining the system reactivity.

Two types of microwave plasma source were used for hydrogen production via methane conversion [28]. The first plasma type was nozzle-less waveguide-supplied coaxial-line-based and the second type was a nozzle-less waveguide-supplied metal-cylinder-based. Both types were operated with microwave frequency of 2.45 GHz and power up to a few kW with high gas flow rates.

Optical emission spectroscopic measurements showed that the high gas temperature (4000–6000 K) made microwave plasma an attractive tool for hydrogen production via hydrocarbon conversion. The experiments were conducted at a methane flow rate up to $12\,000\,L\,h^{-1}$ and additional nitrogen or CO_2 swirl flow was used. The absorbed microwave power was up to 5000 W. For the nozzle-less waveguide-supplied metal-cylinder-based plasma, the best achieved results of hydrogen production rate and energy efficiency of hydrogen production were $1100\,NL(H_2)\,h^{-1}$ and $350\,NL(H_2)$ per kWh of microwave energy used.

In methanol reforming, the selectivity to hydrogen increased from 77.5% to 85.8% when the applied power was elevated from 800 to 1400 W [29]. The selectivities to carbon-containing byproducts were in the order: $CO >$ carbon black $> C_2H_2 > CH_4 > CO_2 = C_2H_4$. In addition, a higher conversion of methanol with a higher selectivity to hydrogen was achieved at a higher applied power. A low required energy consumption of hydrogen (13.2 eV per molecule-H_2) was obtained at a low applied power (800 W) and a higher inlet concentration of methanol (5.0%).

A microwave (2.45 GHz) "tornado"-type plasma with a high-speed tangential gas injection at atmospheric pressure conditions was used for methanol reforming to syngas [30]. The hydrogen production rate dependence on the partial methanol flux was investigated both in Ar and Ar + water plasma environments. Figure 7.3 shows the schematics of experimental set-up. It was found that the hydrogen production rate increased by nearly a factor of 1.5 when water was added into the plasma at 100% methanol conversion.

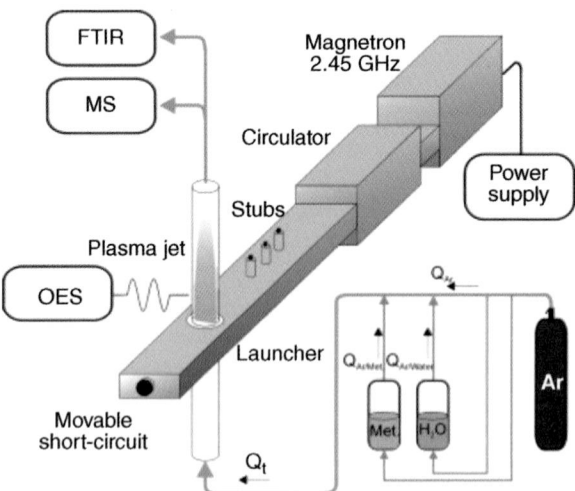

Figure 7.3 Schematic diagram for a microwave (2.45 GHz) "tornado"-type plasma apparatus with a high-speed tangential gas injection for methanol reforming to syngas [30].

7.2.3
Dielectric Barrier Discharge (DBD) Plasma

A dielectric barrier discharge (DBD) plasma apparatus with two concentric elongated electrodes and a dielectric barrier between them has been used to convert methane into hydrogen and carbon [31]. A high voltage pulse generator was connected to the electrodes and, when powered, created the DBD plasma in the gas passing through the gap. Solid carbon was separated from hydrogen by filtration or by using a negatively-charged electrode, to which the carbon was attracted because it carried a positive charge, and the two products were collected and stored in separate cylinders.

Dry reforming of methane reforming has been investigated under the conditions of DBD plasma [32]. The grounded electrode was covered by the dielectric material (quartz). The product contained only gaseous hydrocarbons, syngas, and oxygenates (acetic acid, propanoic acid, ethanol, and methanol). It was concluded that an optimum feed ratio of CH_4/CO_2 exists to obtain the maximum selectivity to the desired oxygenate. The highest selectivity to acetic acid was 5.2%, achieved at methane and CO_2 conversions of 64.3% and 43.1%, respectively.

In another study, dry reforming of methane was investigated in a DBD reactor combined with different $Ni/\gamma-Al_2O_3$ catalysts [33]. Three different catalyst packing methods were used to study their influence on the plasma–catalyst interactions. Compared to the fully packed method, partially packing the $Ni/\gamma-Al_2O_3$ catalyst either in a radial or axial direction into the discharge area showed strong filamentary discharge due to the large void fraction in the discharge gap, which significantly enhanced the physical and chemical interactions between the plasma and catalyst. The synergistic effect of DPD plasma and 10 wt% $Ni/\gamma-Al_2O_3$ resulted in increased methane conversion (56.4%) and hydrogen yield (17.5%), which were attributed to strong plasma–catalyst interactions and high activity of the $Ni/\gamma-Al_2O_3$ catalyst.

The partial oxidation of methane to syngas has been investigated in a DBD reactor. Methane and oxygen conversions depended on temperature and the energy input per unit flow [34]. The product selectivities observed were compared with thermal equilibrium calculations. When a $Ni/\alpha-Al_2O_3$ catalyst was present in the plasma zone, the interactions between nickel catalyst and plasma resulted in oxidation of CO to CO_2 at 300 °C and above, whereas the selectivity to H_2 and H_2O remained stable. The catalyst shifted the distribution of the carbon oxides towards equilibrium and the plasma induced gas-phase processes dominated over surface catalysis.

Plasma-catalytic methanol-steam reforming has been investigated in a tubular quartz reactor with annular-shaped electrodes, in which Cu/ZnO catalysts supported on Al_2O_3 pellets were packed [35]. The methanol conversion under the electric discharge increased as the discharge voltage and frequency increased. The electric discharge provided sufficient energy to break chemical bonds of methanol and steam. At a discharge voltage of 2.0 kV and 220 °C, methanol conversion increased slightly from 40% to 45% on increasing the discharge frequency from 10 to 50 kHz. When an alternating current is used for the electric discharge, a dielectric heating occurs inherently. The dielectric heating is proportional to the permittivity of the dielectric barrier material and the discharge frequency.

A novel plasma process has been proposed for extracting energy from fossil fuels without CO_2 emissions while producing hydrogen and carbon suboxide, $(C_3O_2)_n$, which is a reddish, brown polymer and an important constituent of organic fertilizers [36]. A variable alternating current (AC) power supply was used to generate the DBD plasma. The power supply had an operational frequency of 50 Hz–1.66 kHz and a maximum peak-to-peak voltage range of 20–34 kV. The conversion of n-butane/air and the characterization of the by-products with energy dispersive X-ray spectroscopy were discussed. Thermodynamic results showed an energy efficiency of up to 78% for producing carbon suboxide from various hydrocarbon fuels when compared to the energy efficiency of producing syngas (100%).

The decomposition of hexadecane has been investigated in a nanosecond hybrid catalytic pulsed DBD plasma reactor with methane as the carrier gas [37]. Figure 7.4 shows an overall scheme of the experimental apparatus. The plasma

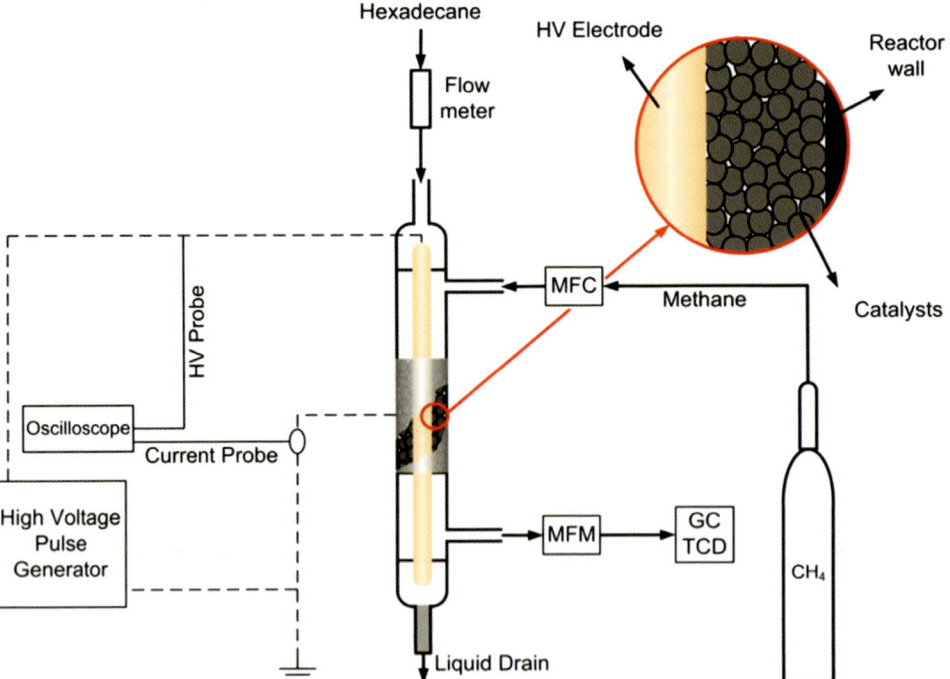

Figure 7.4 Schematic diagram of packed DBD (dielectric barrier discharge) reactor for the cracking of hexadecane [37].

was generated in a tubular reactor made of quartz glass, with an outer diameter of 13 mm and 1.5 mm wall thickness, acting as a dielectric separating inner and outer electrodes. The reactor performance increased significantly when the discharge zone was packed with granules of Mo–Ni/Al_2O_3 catalyst. Energy efficiency and hydrogen product varied between 36.9 and 194.4 L kWh^{-1} and 17.7% and 63.7%, respectively. The highest energy efficiency was achieved at 52.3 W power input. The production rate and concentration of hydrogen were reported at 108.03 mL min^{-1} and 63.7%, respectively. However, the breakdown voltage decreased significantly when the reactor was packed with TiO_2 based catalyst.

The energy efficiency of methane conversion determined by experimentation using DBD was 1%, although theoretical analysis suggested 8%, implying that DBD alone is invariably more inefficient [6]. A synergistic effect between DBD and Ni-catalysts was demonstrated, thereby increasing energy efficiency up to 50%. Methane conversion as well as product distribution were governed by chemical equilibrium at a given catalyst temperature. The authors concluded that not only methane but also water vapor must be excited simultaneously to promote overall methane reforming. The appropriate combination of DBD plasma, catalyst, and mid-temperature thermal energy was crucial to maximization of energy efficiency.

The direct conversion of methane using a DBD reactor has been investigated at different flow rates and discharge voltages [38]. Maximum conversion of methane reached 80% at an input flow rate of 5 mL min^{-1} and a discharge voltage of 4 kV. Optimum conditions occurred at a high discharge voltage and higher input flow rate. No hydrocarbon product was detected using the thermal method, except hydrogen and carbon. The selectivity to ethane was increased when Pt and Ru catalysts were used in the plasma reaction.

Methane conversion into hydrogen and C_2 hydrocarbons has been investigated in four electric discharge systems: (i) a needle-to-plate reactor by pulsed streamer, (ii) pulsed spark discharge, (iii) a wire-to-cylinder DBD reactor by pulsed DC DBD, and (iv) AC DBD at atmospheric pressure and ambient temperature [39]. In the former two electric discharge processes, acetylene was the dominant C_2 product. Pulsed spark discharges gave the highest acetylene yield (54%) and H_2 yield (51%) with 69% methane conversion in a pure methane system at 12 W of discharge power. In the two DBD processes, ethane was the major C_2 product and pulsed DC DBD provides the highest ethane yield.

A nanosecond pulsed DBD system was used to clarify the effect of hydrocarbon feed chain length and carrier gas type (Ar–CH_4 mixtures) on hydrogen production through plasma cracking [40]. To survey the effect of hydrocarbon feed chain length, C_1–C_{16} straight-chain alkanes were employed. Energy efficiency and hydrogen production rate vary between 23.8 and 121.1 L kWh^{-1} and 17.0 and 34.1 mL min^{-1} in the produced gas, respectively. The highest energy efficiency was achieved when *n*-hexadecane was used as a feed with 55.5 W discharge power. Owing to electron energy consumed in dissociation of methane, hydrocarbon cracking was more efficient in argon than in argon–methane mixtures.

7.2.4
Corona Discharge

A corona-based plasma reforming unit has been used to convert methane and propane into CO_x-free hydrogen [41,42]. Argon was used to provide additional electrons and photons for higher reaction rates. A series of experiments was performed for positive corona discharge at a fixed inter-electrode distance (15 mm) to study the effects of discharge power (range 14–20 W) and residence time (60, 120, 180, and 240 s). Discharge power and residence time had a positive influence on methane conversion, hydrogen selectivity, and energy conversion efficiency. Longer discharge gaps favor hydrogen production. In the case of propane conversion, the hydrogen content reached 19% at a power input of 102 W and discharge time of 8.18 min.

A high-frequency pulsed plasma process for methane conversion into acetylene and hydrogen has been used in a co-axial cylindrical type of reactor [43,44]. The performance of this process was compared with conventional arc and POX processes. The pulsed plasma consisted of a pulsed corona discharge and a pulsed spark discharge. Most of the energy was injected over the duration of the pulsed spark discharge. The influences of pulse frequency and pulse voltage on methane conversion rate and product selectivity were investigated.

Combining steam reforming and POX of methane using an AC corona discharge resulted in reducing oxygen requirement and increasing thermal efficiency of the system [45]. The addition of water vapor greatly enhanced the conversions of methane/oxygen and improved the energy efficiency in an oxygen-lean system. The energy consumed to convert one methane molecule decreased dramatically from 68 to 13 eV per molecule converted with an increase in water vapor content from 0 to 50% at a CH_4/O_2 ratio of 5. At these conditions, input power had more influence on the methane and oxygen conversions than applied frequency.

7.2.5
Spark and Pulsed Plasmas

Using a small-scale plasma converter for small engines, hydrogen was produced through the reformation of ionized methane and air mixture by means of a spark discharge [46]. Hydrogen concentration was improved when the intake gas mixture resided longer in the reaction channel, although the hydrogen volume flow rate revealed an inverse trend. In general, under optimal operating conditions, the plasma converter produced a maximum hydrogen concentration of 48% and a hydrogen volume flow rate of 70 mL min^{-1}. In a plasma tubular reactor, it was concluded that in combined low-temperature plasma steam and dry reforming, H•, •OH, and O• radicals help activate more methane molecules and increase hydrogen abstraction from hydrocarbon species [47]. As a result, methane conversion was much higher than that in the pure methane discharge or in the steam reforming discharge under similar experimental conditions. A significant reduction in power consumption at 8 eV per molecule of C converted was also reported.

A continuous production of hydrogen and carbon black from methane has been investigated by decomposition of methane using a system of direct current (DC)-spark discharge plasma [48]. A plasma reactor with a specific design of electrodes was employed to examine the reactor performance regarding operating conditions such as feed flow rate, input power, and electrodes distance. At specified operating conditions (power = 21 W and methane flow rate = 150 mL min^{-1}), the plasma converter produced a hydrogen concentration of 45% at a hydrogen volume flow rate of 75 mL min^{-1}.

Diesel, kerosene, gasoline, and methane have been reformed by spark discharges between needle and plate electrodes at room temperature and atmospheric pressure [49]. The gaseous products from liquid fuels consisted of 65–70% hydrogen and 30–35% light hydrocarbons having two carbon atoms per molecule (i.e., C_2), or three carbon atoms per molecule (i.e., C_3). The product gases were 90% hydrogen and 10% C_2 in the case of methane reforming. The energy efficiency for the production of gaseous products was highest in the case of gasoline at 3.8 mol kWh^{-1}, followed by kerosene, diesel, and methane at 3.2, 3.0, and 2.4 mol kWh^{-1}, respectively.

In another study, liquid hydrocarbon fuels such as 2,2,4-trimethylpentane, n-hexadecane, n-tridecane, and diesel fuel were reformed in a liquid-phase plasma reactor at temperatures of less than 60 °C [50]. To reduce the onset of sparking in insulating liquids, metal chips were inserted in the electrode spacing. When a pulsed square high voltage of 8 kV was applied, transparent hydrocarbon fuels rapidly turned black due to formation of black powders. Hydrogen gas concentration in the produced gas was about 60–70% when the electrodes and chips consisted of aluminum. When 2,2,4-trimethylpentane was used as a test fuel, the generation rate of the hydrogen gas was 52 mL min^{-1} for an input energy of 32 W, with an energy efficiency of 3.9 mol-H_2 kWh^{-1}.

A High-frequency pulsed plasma (HFPP) has been generated by applying a pulsed voltage in an atmosphere of methane with simultaneous production and storage of hydrogen in titanium or a metallic hydride [51]. The efficiency of the system was optimized by adjusting the frequency of the pulsed plasma, with the objective of maximum selectivity to hydrogen. Pyrolytic carbon condensed on the lower walls of the reactor but did not interfere with the kinetics of hydrogen absorption. The authors claimed that the carbon produced in such a manner could be considered a byproduct of the process and can be used as a natural gas adsorbent, once adequately activated. Contents in excess of 50% carbon with respect to the total volume of methane were collected from the plasma chamber structure. In another study, methane and propane were converted into hydrogen and carbon at near ambient conditions using reactors with stationary and moving electrodes for continuous conversion [52]. Different electrode materials (e.g., graphite, stainless steel, Ni, Ni-Cu, Fe) were used in the reactor, and their effect on the reactor performance was evaluated. It was found that the distribution of the reaction products was mostly affected by the residence time in the plasma

Table 7.2 Types of thermal plasmas used in hydrogen production from hydrocarbons.

Type of plasma	Brief description
DC torch	Direct current plasma requires a supply of electrons at the cathode. In an arc plasma, this electron supply is produced by heating the cathode region to temperatures where thermionic emission is rapid
3-Phase AC	A novel three-phase alternating current (AC) plasma process for the production of carbon black and pure hydrogen
DC-RF hybrid	Uses DC-radio frequency (RF) hybrid plasma at high-temperature and longer residence time

reactor. Hydrogen concentration in the effluent gas reached 40–50 vol.% (balance CH_4 and C_{2+} hydrocarbons).

7.3
Thermal Plasma

Thermal plasmas, normally generated by DC (direct current) arc or inductively coupled RF (radio frequency) discharge, can be described as a high enthalpy flame with extremely high temperature fields (1000–20 000 K) and a wide range of velocity fields from several m per s to supersonic values [8]. Table 7.2 presents the types of thermal plasmas developed for the production of hydrogen from hydrocarbons.

7.3.1
DC Torch Plasma

A thermal plasma-assisted pyrolysis process known as the Kvaerner Carbon Black and Hydrogen Process (CB&H) has been developed for the decomposition of a wide spectrum of hydrocarbons into carbon black and hydrogen. The process was based on a specially designed version of plasma torch with coaxial graphite electrodes [53,54]. The advantages of this process include yields of high thermal efficiency (>90%) and purity of hydrogen (98%); however, it is an electrical-energy-intensive process consuming more than 1 kWh per m^3-H_2 due to a high reaction temperature [55]. A wide range of traditional as well as new qualities of carbon black such as carbon cones can be produced by varying the process parameters, that is, the plasma gas flow rate and temperature, the feedstock flow rates and temperature, and the location and number of feedstock injection points.

An advanced combined cycle for power generation and hydrogen production from hydrocarbons has been described [56] using an electric arc of plasma-black reactor. The elemental carbon was fed to a molten carbonate fuel cell to produce electrical power, part of which was fed back to power the hydrogen plasma.

Hydrogen product was used in a solid oxide fuel cell for power generation and the remaining high temperature energy in a back-end steam Rankine cycle for additional power. The combined cycle plant was projected to yield electrical power plant efficiencies ranging from over 70% to exceeding 80% based on higher heating values.

Another study on DC thermal plasma pyrolysis of methane was conducted using special designed reactor geometry for providing supersonic expansion of a plasma jet [57]. The equilibrium products between about 1000 and 2500 °C were carbon and hydrogen. The effects of operating parameters were investigated through detailed chemical reaction simulations to optimize processing conditions and to maximize the yield of carbon products. The kinetics of the decomposition of methane required higher product temperature, necessitating higher inlet temperature or greater power input.

An experimental linear 55–69 kW DC plasma torch operating at atmospheric pressure has been developed and used to convert propane into hydrogen-rich gas [58]. The average temperature of water vapor plasma jet in the exhaust nozzle varied from 2000 to 3000 K, and the plasma jet velocity was 210–350 m s^{-1}. Hydrogen and CO gases were produced as the main products, in addition to small amounts of CO_2 and H_2O and O_2. Propane conversion depended on the plasma jet temperature, total gas flow rate, and stable operating conditions of the plasma torch.

7.3.2
Three-Phase AC Plasma

The three-phase AC plasma technology has been developed at an intermediate scale between the laboratory bench and the industrial pilot plant [59]. Initially focusing on the development of a new environment-friendly process for the production of carbon black and hydrogen, research diversified towards the synthesis of fullerenes and nanotubes. The approach mainly deals with process optimization and with the understanding of the main process and product relationships. Direct thermal conversion of methane, acetylene, and propane into hydrogen and carbon black was conducted using a thermal plasma system [60,61]. Thermodynamic equilibrium compositions were calculated based on the minimization of Gibb's free energy, and conversion experiments were performed on the basis of calculation results. It was observed that the carbon black particles were spherical with diameters mainly in the range 10–80 nm. The size of carbon black synthesized from methane was observed to be smaller than that from propane feed.

7.3.3
DC-RF Plasma

The use of DC-RF hybrid thermal plasma was demonstrated for the production of hydrogen and carbon black from methane [62]. A stainless steel reaction chamber (0.4 m outer diameter, 1.5 m long) was used. A DC-RF hybrid torch (a DC torch with an 8-mm anode nozzle, coupled with an RF torch having a 50-mm-diameter confinement tube) was fabricated on the basis of theoretical considerations. The injected methane was converted mostly into hydrogen with a small volume fraction of C_2H_2, and fine carbon particles of 50–200 nm.

The decomposition of dodecane ($C_{12}H_{26}$) to carbon black and hydrogen was investigated in an inductive plasma reactor [63]. Thermodynamic and kinetic models were compared with experimental results; a kinetic reaction model best represented the experimental results. Morphological analysis of the solid product indicated that a high plasma temperature was the most important factor affecting the final morphology of carbon black formed during the reaction. Various operating conditions, such as plasma power (40 kW) and He used as the plasma torch sheath gas, influenced the production of carbon black. The maximum input carbon recovered, as a solid mass, was 28%. It was possible that hydrogen, both as radical and elemental hydrogen formed in the reaction zone, acted as a reactive species that promoted the recombination and hydrocarbon formation reactions.

7.4
Concluding Remarks

Research efforts are ongoing to enhance and optimize the production of hydrogen from hydrocarbons via various types of plasma systems that have been developed to facilitate POX, ATR, and SMR. To compete with existing commercial thermo-catalytic processes (SMR and POX), researchers worldwide are investigating techniques to enhance the performance of various plasma discharges, especially non-thermal plasma. Various types of plasma reactor systems have been investigated that include gliding-arc, dielectric barrier discharge (DBD), microwave, corona, pulsed discharge, and other hybrid discharges. Table 7.3 presents a comparison of the preferred type of plasma for hydrocarbon conversion into hydrogen and process requirement. Plasma replaces catalysis, and accelerates chemical reactions, mainly because of its high-temperature features.

Non-thermal plasma can accelerate chemical reactions at low temperature through the generation of active species by fast electrons, thereby minimizing energy and material consumption. The advantages of the plasma route include operational flexibility, compactness of the apparatus, low investment, and low operating costs. Plasma conversion overcomes many limitations of the conventional SMR route such as catalyst deactivation and size and utilization of heavy hydrocarbons as feed. Various gaseous and liquid hydrocarbon fuels (including alcohols) have been investigated, with most studies emphasizing methane reforming and POX. In decomposition or pyrolysis processes, the use of plasma results in co-producing carbon in solid form, thus avoiding the cost of sequestering CO_2. Plasma systems have the ability to induce gas phase reactions but with

Table 7.3 Preferred plasma type for hydrogen production from hydrocarbon conversion.

Hydrocarbon conversion route	Process requirements	Preferred type of plasma
Reforming (endothermic reaction)	Steam or CO_2 (dry) reforming of hydrocarbons. The process needs significant energy requirement with efficiency less than 25%. Plasma-catalyst approach is used to reduce heat input	• Microwave • pulsed corona discharge • DBD
Partial oxidation (exothermic reaction)	Well suited for gaseous and liquid HCs. In hybrid systems, plasma can activate catalytic reactions at low temperatures	• Gliding arc discharge
Autothermal reforming (ATR)	Oxidative steam reforming that uses air to provide a supplementary source of reaction heat to assist in the steam conversion reactions	• Gliding arc discharge
HC decomposition (pyrolysis)	High energy requirements similar to reforming process; however, it co-produces carbon instead of CO_2 (plasma acts like catalyst)	• Microwave • DBD

less selectivity to products compared with catalytic routes. To overcome this disadvantage, innovative plasma approaches are being based on the combination of plasma with catalytic routes by using selective catalysts inside plasma reactors. Moreover, the utilization of a hybrid SMR-POX catalytic system makes the plasma route a promising and cost effective approach for the production of hydrogen from hydrocarbons.

Acknowledgment

The author appreciates the support provided by the Ministry of Higher Education, Saudi Arabia for establishing the Center of Research Excellence in Petroleum Refining & Petrochemicals at King Fahd University of Petroleum & Minerals (KFUPM).

References

1 Muradov, N.Z. and Veziroğlu, T.N. (2008) "Green" path from fossil-based to hydrogen economy: an overview of carbon-neutral technologies. *Int. J. Hydrogen Energy*, **33** (23), 6804–6839.

2 Holladay, J.D., Hu, J., King, D.L., and Wang, Y. (2009) An overview of hydrogen production technologies. *Catal. Today*, **139** (4), 244–260.

3 Ahmed, S., Aitani, A., Rahman, F., Al-Dawood, A., and Al-Muhaish, F. (2009) Decomposition of hydrocarbons to hydrogen and carbon. *Appl. Catal. A General*, **359**, 1–24.

4 Fridman, A. (2012) *Plasma Chemistry*, Cambridge University Press, Cambridge, pp. 676–754.

5 Slovetskii, D.I. (2006) Plasma-chemical processes in petroleum

chemistry. *Rev. Petrol. Chem.*, **46** (5), 295–304.
6 Nozaki, T. and Okazaki, K. (2013) Non-thermal plasma catalysis of methane: principles, energy efficiency, and applications. *Catal. Today*, **211**, 29–38.
7 Paulmier, T. and Fulcheri, L. (2005) Use of non-thermal plasma for hydrocarbon reforming. *Chem. Eng. J.*, **106**, 59–71.
8 Seo, J. and Hong, B. (2012) Thermal plasma synthesis of nano-sized powders. *Nucl. Eng. Technol.*, **44** (5), 9–20.
9 Bromberg, L., Cohn, D.R., Rabinovich, A., Surma, J.E., and Virden, J. (1999) Compact plasmatron-boosted hydrogen generation for vehicular applications. *Int. J. Hydrogen Energy*, **24** (4), 341–350.
10 Bromberg, L., Cohn, D., Rabinovich, A., Alexeev, N., Samokhin, A., Ramprasad, R., and Tamhankar, S. (2000) System optimization and cost analysis of plasma catalytic reforming of natural gas. *Int. J. Hydrogen Energy*, **25**, 1157–1161.
11 Green, J., Domingo, N., Storey, J., Wagner, R., Armfield, J., Bromberg, L., Cohn, D., Rabinovich, A., and Alexeev, A. (2000) Experimental evaluation of SI engine operation supplemented by hydrogen rich gas from a compact plasma boosted reformer, SAE Technical Paper Series, 2000-01-2206.
12 Gallagher, M.J., Polevich, A., Rabinovich, A., and Fridman, A. (2009) Characterization of a new glid-arc plasmatron on-board reformer for an integrated solid oxide fuel cell auxiliary power unit. Presented at the AIChE Spring Meeting, Tampa, April 2009.
13 Gallagher, M.J. (2010) Partial oxidation and autothermal reforming of heavy hydrocarbon fuels with non-equilibrium gliding arc plasma for fuel cell applications, PhD Thesis, Drexel University, pp. 105–119.
14 Czernichowski, A., Czernichowski, M., and Wesolowska, K. (2003) Glidarc-assisted production of synthesis gas through propane partial oxidation in *First International Conference on Fuel Cell Science, Engineering and Technology, Rochester, NY, 21-23 April 2003*, The American Society of Mechanical Engineers, pp. 175–179.

15 Indarto, A., Choi, J., Lee, H., and Song, H. (2006) Effect of additive gases on methane conversion using gliding arc discharge. *Energy*, **31** (14), 2986–2995.
16 Garduño, M., Pacheco, M., Pacheco, J., Valdivia, R., Santana, A., Lefort, B., and Rivera-Rodríguez, C. (2012) Hydrogen production from methane conversion in a gliding arc. *J. Renew. Sustain. Energy*, **4** (1021202), 1–8.
17 Wang, B., Ge, W., Lü, Y., and Yan, W. (2013) H_2 production by ethanol decomposition with a gliding arc discharge plasma reactor. *Frontiers Chem. Sci. Eng.*, **7** (2), 145–153.
18 Bo, Z., Yan, J., Li, X., Chi, Y., and Cen, K. (2008) Plasma assisted dry methane reforming using gliding arc gas discharge: effect of feed gases proportion. *Int. J. Hydrogen Energy*, **33** (20), 5545–5553.
19 Rueangjitt, N., Akarawitoo, C., Sreethawong, T., and Chavadej, S. (2007) Reforming of CO_2-containing natural gas using an AC gliding arc system. *Plasma Chem. Plasma Process.*, **27** (5), 559–576.
20 Yang, Y., Lee, B., and Chun, Y. (2009) Characteristics of methane reforming using gliding arc reactor. *Energy*, **34** (2), 172–177.
21 Pornmai, K., Jindanin, A., Sekiguchi, H., and Chavadej, S. (2012) Synthesis gas production from CO_2-containing natural gas by combined steam reforming and partial oxidation in an AC gliding arc discharge. *Plasma Chem. Plasma Process.*, **32** (4), 723–742.
22 Arthiwet, N. and Chavadej, S. (2012) Combined plasma reforming of CO2-containing natural gas with steam and partial oxidation in a multistage gliding arc discharge system: effect of stage number. *Chem. Eng. Trans.*, **29**, 1111–1116.
23 Deminsky, M., Jivotov, V., Potapkin, B., and Rusanov, V. (2002) Plasma-assisted production of hydrogen from hydrocarbons. *Pure Appl. Chem.*, **74**, 413–418.
24 Cho, W., Lee, S., Ju, W., Baek, Y., and Lee, J. (2004) Conversion of natural gas to hydrogen and carbon black by plasma and application of plasma carbon black. *Catal. Today*, **98**, 633–638.

25 Tsai, C.H. and Chen, K.C. (2009) Production of hydrogen and nano carbon powders from direct plasmalysis of methane. *Intl. J. Hydrogen Energy*, **34**, 833–838.

26 Tian, M., Batty, S., and Shang, C. (2013) Synthesis of nanostructured carbons by the microwave plasma cracking of methane. *Carbon*, **51** (1), 243–248.

27 Ravasio, S. and Cavallotti, C. (2012) Analysis of reactivity and energy efficiency of methane conversion through non-thermal plasmas. *Chem. Eng. Sci.*, **84**, 580–590.

28 Jasiński, M., Czylkowski, D., Hrycak, B., Dors, M., and Mizeraczyk, J. (2013) Atmospheric pressure microwave plasma source for hydrogen production. *Int. J. Hydrogen Energy*, **38** (26), 11473–11483.

29 Wang, Y., You, Y., Tsai, C., and Wang, L. (2010) Production of hydrogen by plasma-reforming of methanol. *Int. J. Hydrogen Energy*, **35** (18), 9637–9640.

30 Bundaleska, N., Tsyganov, D., Saavedra, R., Tatarova, E., Dias, F.M., and Ferreira, C.M. (2013) Hydrogen production from methanol reforming in microwave "tornado"-type plasma. *Int. J. Hydrogen Energy*, **38** (22), 9145–9157.

31 Fletcher, D. (2004) Production of hydrogen and carbon from natural gas or methane using barrier discharge non-thermal plasma, US patent application US2004148860, August 5, 2004.

32 Zhang, Y., Li, Y., Wang, Y., Liu, C., and Eliasson, B. (2003) Plasma methane conversion in the presence of carbon dioxide using dielectric-barrier discharges. *Fuel Proc. Technol.*, **83** (1–3), 101–109.

33 Tu, X. and Whitehead, J.C. (2012) Plasma-catalytic dry reforming of methane in an atmospheric dielectric barrier discharge: understanding the synergistic effect at low temperature. *Appl. Catal. B: Environ.*, **125**, 439–448.

34 Pietruszka, B., Anklam, K., and Heintze, M. (2004) Plasma-assisted partial oxidation of methane to synthesis gas in a dielectric barrier discharge. *Appl. Catal. A: General*, **261** (1), 19–24.

35 Lee, D.H. and Kim, T. (2013) Plasma-catalyst hybrid methanol-steam reforming for hydrogen production. *Int. J. Hydrogen Energy*, **38** (14), 6039–6043.

36 Odeyemi, F., Pekker, M., Rabinovich, A., Fridman, A.A., Heon, M., Mochalin, V.N., and Gogotsi, Y. (2012) Low temperature plasma reforming of hydrocarbon fuels into hydrogen and carbon suboxide for energy generation without CO_2 emission. *IEEE Trans. Plasma Sci.*, **40** (2), 1362–1370.

37 Hooshmand, N., Rahimpour, M.R., Jahanmiri, A., Taghvaei, H., and Mohamadzadeh Shirazi, M. (2013) Hexadecane cracking in a hybrid catalytic pulsed dielectric barrier discharge plasma reactor. *Ind. Eng. Chem. Res.*, **52** (12), 4443–4449.

38 Indarto, A., Choi, J., Lee, H., and Song, H. (2006) Methane conversion using dielectric barrier discharge: comparison with thermal process and catalyst effects. *J. Natural Gas Chem.*, **15** (2), 87–92.

39 Li, X., Zhu, A., Wang, K., Xu, Y., and Song, Z. (2004) Methane conversion to C_2 hydrocarbons and hydrogen in atmospheric non-thermal plasmas generated by different electric discharge techniques. *Catal. Today*, **98** (4), 617–624.

40 Taghvaei, H., Jahanmiri, A., Rahimpour, M.R., Shirazi, M.M., and Hooshmand, N. (2013) Hydrogen production through plasma cracking of hydrocarbons: effect of carrier gas and hydrocarbon type. *Chem. Eng. J.*, **226**, 384–392.

41 Aleknaviciute, I., Karayiannis, T.G., Collins, M.W., and Xanthos, C. (2013) Methane decomposition under a corona discharge to generate COx-free hydrogen. *Energy*, **59**, 432–439.

42 Aleknaviciute, I., Karayiannis, T.G., Collins, M.W., and Xanthos, C. (2013) Plasma-assisted decomposition of gaseous propane to produce COx free hydrogen. *Int. J. Low-Carbon Technol.*, **8** (3), 197–202.

43 Yao, S.L., Suzuki, E., Meng, N., and Nakayama, A. (2002) A high-efficiency reactor for the pulsed plasma conversion of methane. *Plasma Chem. Plasma Proc.*, **22** (2), 225–237.

44 Yao, S., Nakayama, A., and Suzuki, E. (2001) Acetylene and hydrogen from

pulsed plasma conversion of methane. *Catal. Today*, **71** (1–2), 219–223.
45 Supat, K., Chavadej, S., Lobban, L.L., and Mallinson, R.G. (2003) Combined steam reforming and partial oxidation of methane to synthesis gas under electrical discharge. *Ind. Eng. Chem. Res.*, **42** (8), 1654–1661.
46 Horng, R., Chang, Y., Huang, H., and Lai, M. (2007) A study of the hydrogen production from a small plasma converter. *Fuel*, **86** (2), 81–89.
47 Le, H., Hoang, T., Mallinson, R., and Lobban, L. (2004) Combined steam reforming and dry reforming of methane using AC discharge. *Stud. Surf. Sci. Catal.*, **147**, 175–180.
48 Moshrefi, M.M., Rashidi, F., Bozorgzadeh, H.R., and Zekordi, S.M. (2012) Methane conversion to hydrogen and carbon black by dc-spark discharge. *Plasma Chem. Plasma Proc.*, **32** (6), 1157–1168.
49 Malik, M.A., Hughes, D., Malik, A., Xiao, S., and Schoenbach, K.H. (2013) Study of the production of hydrogen and light hydrocarbons by spark discharges in diesel, kerosene, gasoline, and methane. *Plasma Chem. Plasma Proc.*, **33** (1), 271–279.
50 Matsui, Y., Kawakami, S., Takashima, K., Katsura, S., and Mizuno, A. (2005) Liquid-phase fuel re-forming at room temperature using nonthermal plasma. *Energy Fuels*, **19** (4), 1561–1565.
51 Da Silva, C., Ishikawa, T., Santos, S., Alves, C., and Martinelli, A. (2006) Production of hydrogen from methane using pulsed plasma and simultaneous storage in titanium sheet. *Int. J. Hydrogen Energy*, **31**, 49–54.
52 Muradov, N., Smith, F., and Bokerman, G. (2009) Non-thermal plasma assisted decomposition of light hydrocarbons to hydrogen-rich gas and carbon. Presented at the AIChE Spring Meeting, Tampa, April 2009.
53 Bakken, J., Jensenb, R., Monsenb, B., Raanessb, O., and Wrernesb, A. (1998) Thermal plasma process development in Norway. *Pure Appl. Chem.*, **70**, 1223–1228.
54 Gaudernack, P. and Lynum, S. (1998) Hydrogen from natural gas without release of CO_2. *Int. J. Hydrogen Energy*, **21**, 1087–1093.
55 Lynum, S. (1996) The Kvaerner CB&H Process. Presented at Carbon Black World 96, March 1996, Nice, France.
56 Steinberg, M. (2006) Conversion of fossil and biomass fuels to electric power and transportation fuels by high efficiency integrated plasma fuel cell (IPFC) energy cycle. *Int. J. Hydrogen Energy*, **31**, 405–411.
57 Fincke, J., Anderson, R., Hyde, T., and Detering, B. (2002) Plasma pyrolysis of methane to hydrogen and carbon black. *Ind. Eng. Chem. Res.*, **41**, 1425–1435.
58 Tamošiunas, A., Grigaitiene, V., Valatkevičius, P., and Valinčius, V. (2012) Syngas production from hydrocarbon-containing gas in ambient of water vapor plasma. *Catal. Today*, **196** (1), 81–85.
59 Fulcheri, L. (2002) Plasma processing: a step towards the production of new grades of carbon black. *Carbon*, **40**, 169–176.
60 Yang, H., Nam, W., and Park, D. (2007) Production of nanosized carbon black from hydrocarbon by a thermal plasma. *J. Nanosci. Nanotechnol.*, **7**, 3744–3749.
61 Lee, T., Nam, W., Baeck, S., and Park, D. (2007) Application of thermal plasma for production of hydrogen and carbon black from direct decomposition of hydrocarbon. *J. Korean Ind. Eng. Chem.*, **18** (1), 84–89.
62 Kim, K. and Takahashi, S. (2005) Production of hydrogen and carbon black by methane decomposition using DC-RF hybrid thermal plasmas. *IEEE Trans. Plasma Sci.*, **33**, 813–823.
63 Merlo-Sosa, L. and Soucy, G. (2005) Dodecane decomposition in a radio-frequency (RF) plasma reactor. *Int. J. Chem. Reactor Eng.*, **3** (A4), 1–23.

8
Solar Thermal Reforming

Christos Agrafiotis, Henrik von Storch, Martin Roeb, and Christian Sattler

8.1
Introduction

Virtually all hydrogen produced today is sourced from fossil fuels, with the principal method employed being the catalytic reforming of methane (CH_4, the principal component of natural gas and other gaseous fuels) [1]. Two different reactions can be distinguished in the methane reforming process: steam methane reforming (SMR) and CO_2 (or dry) methane reforming (DMR), represented by Eqs. (8.1) and (8.2) respectively:

$$CH_4 + H_2O \rightleftharpoons 3\,H_2 + CO \qquad \Delta H^o_{298\,K} = +206\ \text{kJ mol}^{-1} \qquad (8.1)$$

$$CH_4 + CO_2 \rightleftharpoons 2\,H_2 + 2\,CO \qquad \Delta H^o_{298\,K} = +247\ \text{kJ mol}^{-1} \qquad (8.2)$$

Both reactions are highly endothermic, and are, thus, favored by high temperatures; industrial reforming processes are carried out between 800 and 1000 °C [1]. The required energy is supplied by combustion of additional natural gas and process waste gas from the downstream hydrogen purification step. The share of natural gas consumed as fuel varies from 3% to 20% of the total plant's natural gas consumption [2]. The reaction gas product mixture is known as synthesis gas (syngas): a gas mixture that contains varying amounts of CO and H_2 whose exothermic conversion into fuel and other products has long been commercially practiced, for example, via the Fischer–Tropsch technology [3]. In fact hydrogen and syngas are the basic raw materials to produce synthetic liquid fuels (SLFs) and chemicals via industrially available processes.

An entirely "renewable" hydrogen production route would involve renewable energy sources such as solar or wind energy as well as renewable feedstocks like water or biomass. However, at least for a transition period, hydrogen supply at a competitive cost can only be achieved from hydrocarbons – essentially natural gas – using well-known commercial processes like steam reforming. As an intermediate step, considerable effort is being spent on developing a hybrid hydrogen technology in which concentrated solar power (CSP) is used to provide the heat for the high temperature endothermic steam–methane reforming reaction. In doing so, solar energy is embodied thermochemically in the product hydrogen, being thus able to be stored at ambient conditions, transported from the point of collection to where it is required, and used outside daylight hours. Such a transitional technology is considered by many to be an essential stepping stone from current practice to a truly renewable-based hydrogen economy [4,5]. Besides natural gas, there are plenty of other promising methane-containing gases for hydrogen production such as coke oven gas, refinery gas, and biogas [6]. The solar-aided production of hydrogen from methane as well as from such feedstocks has been presented in detail in a recent review by the present authors [7]. The present chapter – based essentially on this review – deals only with solar-aided methane steam/CO_2 reforming reactions, summarizing the up-do-date research in this area.

8.2
Hydrogen Production via Methane Reforming

8.2.1
Thermochemistry and Thermodynamics of Reforming

Both reforming reactions (8.1) and (8.2), due to their increase in number of moles, are favored by low pressures. However, commercial reforming plants are operated at much higher pressure levels (usually above 25 atm) due to process optimization: syngas is merely an intermediate product that is further processed (e.g., to hydrogen for ammonia production or to methanol) via operations requiring high pressures [1]. In the presence of water the reforming reaction is followed by the mildly exothermic water–gas shift (WGS) reaction (Eq. (8.3)) and, thus, the overall reaction of steam reforming followed by WGS is described by Eq. (8.4):

$$CO + H_2O \rightleftharpoons H_2 + CO_2 \qquad \Delta H^o_{298\,K} = -41 \text{ kJ mol}^{-1} \qquad (8.3)$$

$$CH_4 + 2H_2O \rightleftharpoons 4H_2 + CO_2 \qquad \Delta H^o_{298\,K} = +165 \text{ kJ mol}^{-1} \qquad (8.4)$$

As can be seen from reactions (8.1) and (8.2) the H_2/CO ratio in the product can differ between 3 and 1 respectively. Therefore, to provide a high hydrogen yield, steam reforming followed by WGS is most suitable, as a H_2/CO ratio of up to 4 can be achieved. However, the required properties of syngas depend on its subsequent utilization, since for further processing to liquid fuels, for instance, a molar ratio of H_2/CO above 3 is unsuitable [8,9].

Both reforming reactions can generally be catalyzed by metals of group VIII of the periodic system. In commercial applications, Ni-based catalysts supported on mixed oxides of Ca-Al [10] or Mg-Al [11] proved to be the most suitable, due to low cost and high catalytic activity. Other catalysts are not preferred due to high costs (noble metals) or technical issues (Fe and Co). Two main issues in reformer design utilizing Ni-based catalysts are sulfur compounds in the feed and carbon formation at the catalytically active site. While the first can usually be eliminated by use of a hydrogenator followed by a ZnO bed [12], the latter is a more complicated issue. Carbon formation in reforming reactions occurs mainly due to methane decomposition (pyrolysis) or disproportionation of carbon dioxide according to Eqs. (8.5) and (8.6) respectively:

$$CH_4 \rightarrow C + 2H_2 \tag{8.5}$$

$$2CO \rightarrow C + CO_2 \tag{8.6}$$

Thus, in commercial steam reforming plants, most commonly the introduction of excess steam into the feed is used to reduce the risk of carbon formation. Promoters (e.g., alkaline earth metal oxides) are also added to the Ni-based catalyst to inhibit carbon formation [1,4,8,12,13]. After the shift reaction, the product gas ("raw synthesis gas") consists mainly of H_2 and CO_2, and some impurities such as unconverted CH_4 and CO.

8.2.2
Current Industrial Status

The commercial feedstock of choice for syngas production is natural gas and the most widely used process is steam reforming as discussed above. For large-scale chemical processes such as oil refining, steam reformers manufactured by large plant engineering companies are producing hydrogen of the order of 100 000–1 000 000 kg hydrogen per day. Steam reforming is conducted usually inside high-temperature alloy tubes, 10–13 m long × 100–150 mm in diameter, packed with nickel catalyst and heated by radiation and convection from burning natural gas or refinery waste fuel gas (Figure 8.1) [14]. Typical inlet temperatures

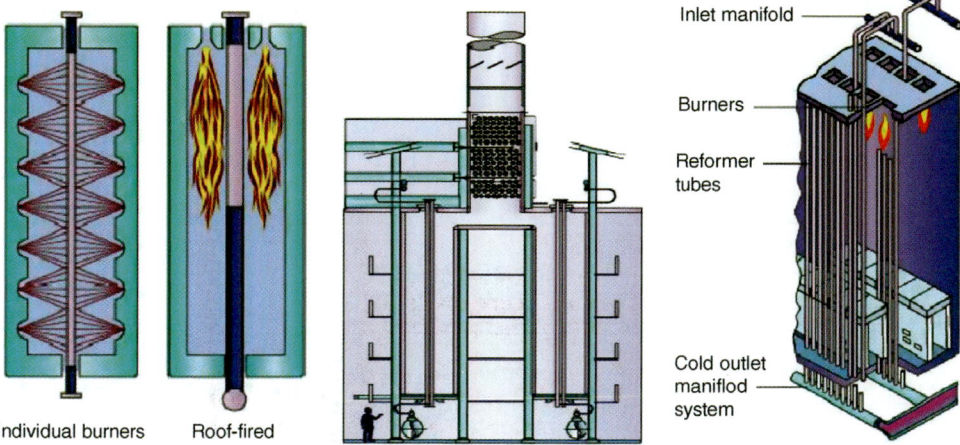

Figure 8.1 Schematics of operation of typical industrial reformers [14].

to the catalyst bed are 450–650 °C and product gas leaves the reformer at 800–950 °C depending on the application. Tubular reformers are designed with various tube and burner arrangements, either side- or top-fired. The capital cost for these large-scale steam reformers is in the range of €200–400 per kg-H_2 d^{-1} [1,4,14,15].

8.3
Solar-Aided Methane Reforming

8.3.1
Solar Concentration Systems

Concentrated solar radiation can be used to provide high-temperature process heat as the necessary energy source for the performance of endothermic chemical reactions in the so-called solar thermochemical processes [4]. Of the four kinds of optical configuration systems that use movable reflectors (mirrors) tracking the sun and achieving large-scale concentration of solar energy at pilot and commercial CSP plants – namely parabolic trough (PT) collectors, linear Fresnel (LF) reflectors, dish–engines (DE), and tower/central receivers (CRs) – the first two currently operate at moderate solar concentration ratios at working temperatures usually below 500 °C [16]. Thus, regardless of the type of reforming reactor and catalyst, the higher temperatures required for steam or CO_2 methane reforming processes effectively limit the solar concentrator choice to the two types of high-solar-concentration, namely, the parabolic dish collector and the central tower receiver.

8.3.2
Coupling Reforming with Solar Energy: Solar Receiver–Reactor Concepts

If solar energy is used to provide the heat of reaction for endothermic reactions (8.1) and (8.2), on the one hand, a reduction of fossil fuel consumption is achieved because by this way fossil fuels are not used to generate high temperature but only as a feedstock. On the other hand, the product gas will contain 26% and 31% of solar energy embodied in chemical form respectively (low heating value basis, assuming water stays as a vapor). As already mentioned, in the presence of water the reforming reaction is followed by the water–gas shift (WGS) reaction according to Eq. (8.3). Because the WGS reaction is exothermic, the conversion of CO into hydrogen and CO_2 actually reduces the amount of energy that can be stored in the products to 21% in both cases [5].

For the efficient design and operation of solar receiver/reactors, concepts from "traditional" chemical reactor engineering should be combined with ways to achieve efficient reactor heating via concentrated solar irradiation. Just as in "traditional" non-solar chemical engineering, the catalytic reactor type can be separated into two broad categories depending on whether the catalyst particles are distributed randomly or are "arranged" in space at the reactor level: the first category includes packed and fluidized catalytic beds; the second includes the so-called "structured" catalytic systems like honeycomb, foam, and membrane catalytic reactors [17].

In solar-aided chemistry applications the reactor concepts above have to be combined with effective absorbance of concentrated solar irradiation achieved via the solar receiver. The receiver's task is to "trap" (absorb) the concentrated solar radiation and ". . . to provide the maximum conversion of the energy of incident radiation to the energy of chemical reaction products . . ." [18]. Solar receivers are divided into two broad categories according to the mechanism of transferring heat to the heat transfer fluid (HTF): directly- and indirectly-heated ones. The common characteristic of the historically precedent indirectly irradiated receivers (IIR) is that the heat transfer to the working fluid does not take place upon the surface that is exposed to incoming solar radiation. Instead, there is an intermediate opaque wall, heated by the irradiated sunlight on one side and transferring the heat to a "working" fluid on the other side [19]. The simplest examples are conventional tubular receivers that consist of absorbing surfaces exposed to the concentrated solar irradiation. The heat transfer fluid (e.g., a gas or a molten salt) is moving in a direction vertical to that of the incident solar radiation (Figure 8.2a, left); heat transfer to the thermal fluid takes place through the receiver opaque walls by conduction.

In the case of receiver–reactors two further options are possible: in the first option the receiver and the reactor are de-coupled. The latter is not located

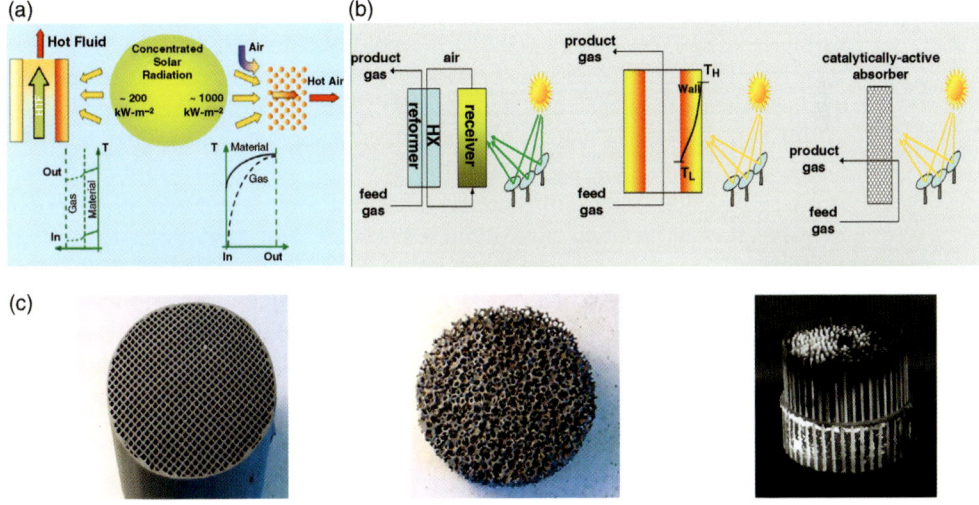

Figure 8.2 (a) Operational principles of indirectly heated tubular (left) and directly heated volumetric (right) solar receivers [25]; (b) operation principles of indirectly irradiated allothermal reformer (left), indirectly irradiated integrated tubular reactor/receiver (middle), and directly irradiated reactor/receiver (right); (c) monolithic ceramic structures explored as volumetric receivers: honeycomb (left), foam (middle) [23], and pin-finned (right) [19].

within the solar absorbing module: it is heated via the enthalpy of the solar-receiver-heated heat transfer fluid, transferred via insulated tubes to the reactor's site ("allothermal" reactor). The operation principle of this concept is shown schematically in Figure 8.2b (left). The same de-coupled configuration can be used when another, non-solar-heated, heat transfer fluid is available (e.g., helium from nuclear reactors). Alternatively, the solar-heated tubular receivers can contain the catalyst material in the form of either packed beds or even liquid beds (when a material is heated slightly above its fusion temperature) or, finally, as structured assemblies – again in this case the catalyst is heated via conduction through the tube walls. The operation principle of this concept is shown schematically in Figure 8.2b (middle). Finally, another version of indirectly heated receivers is the so-called heat pipe receiver, where concentrated sunlight is employed for the evaporation of a liquid substance that in turn condenses on the tubes containing the heat transfer fluid, liberating the heat of vaporization [20].

However, the most direct and therefore potentially the most efficient method would involve direct heating of the heat transfer fluid by the concentrated beam, thus eliminating the wall as the light absorber and heat conductor. Directly irradiated receivers (DIRs) use fluid streams or solid particles directly exposed to the concentrated solar radiation. A key element of all directly irradiated receivers is the absorber: the component that absorbs concentrated sunlight and transports its energy to a working fluid flowing within and over it. Directly irradiated receivers are also called "volumetric" receivers since they enable the concentrated solar radiation to penetrate and be absorbed within the entire volume of the absorber. In different designs, the absorber is either a stationary matrix (grid, wire-mesh, foam, honeycomb, etc.) or moving (usually solid) particles.

Directly irradiated receivers with stationary absorbers are relatively simple and the most common of the volumetric receiver family. Here, the absorbing matrix must be able to absorb highly concentrated radiation, while providing sufficient heat convection to the working gas flow. It is also required to sustain thermal stresses created by large temperature gradients as well as thermal shock caused by rapid heating–cooling cycles. Such receivers, first proposed in the early 1980s [21,22] are essentially compact heat exchangers consisting of high-porosity material structures capable of absorbing the concentrated solar radiation and transferring it to a gaseous heat transfer medium [23,24]. For instance, gas (air) can be driven through the absorber, flowing past the absorbing structures parallel to the direction of the incoming radiation (Figure 8.2a, right) [25] and be heated by convection. Hence, the volumetric receiver concept entails that the solar irradiation and the heat extraction take place on the same surface simultaneously (Figure 8.2b, right). Since the heat transfer surface per unit of incident solar radiation is increased compared to a tube absorber design, higher flux densities can be transferred without reaching the temperature limit of the absorber material. As already mentioned, such structures can be steel or ceramic wire meshes, ceramic foams (Figure 8.2c, middle) – or multi-channeled honeycombs (Figure 8.2c, left). In fact such ceramic honeycomb receivers, developed by the

German Aerospace Center (DLR), have been extensively tested in CSP facilities [24,26,27], having currently reached the level of commercial exploitation in the 1.5 MW$_e$ Solar Tower Jülich (STJ) Thermal Power Plant [28]. Researchers at the Weizmann Institute of Science (WIS), Israel, have introduced an alternative design of a structured volumetric (directly-irradiated) solar absorber, nicknamed "porcupine", constructed with ceramic elongated heat transfer elements (i.e., the porcupine "quills") implanted in a base plate; a schematic view presented in Figure 8.2c (right) [19]. Often the operating conditions demand that the absorber be physically separated from the ambient, for example, when the flow is pressurized, or the working fluid is not air. In these cases the receiver must be equipped with a transparent window, which allows concentrated light to enter the receiver while separating the working gas from ambient air [29].

In a direct analogy with "conventional" catalytic applications [30] it becomes obvious that all three structured porous volumetric solar absorber modules shown in Figure 8.2c can be coated with proper functional materials capable of performing/catalyzing various high-temperature chemical reactions – among them reforming – and thus be "transformed" and adapted to operate as solar chemical receiver/reactors where chemical reactions can take place in an efficient and elegant manner with the aid of the functional materials immobilized upon their porous walls [31]. Such devices consist of an integrated solar receiver–chemical reactor module, in which a catalyst-coated porous matrix volumetrically absorbs concentrated solar (radiant) energy directly at the catalyst sites, promoting heterogeneous reactions with gases flowing through the matrix (absorber) – thus they are frequently referred as directly irradiated volumetric receiver–reactors (DIVRRs).

Each receiver/reactor type has its advantages and drawbacks; it is not therefore surprising that essentially all the above concepts have been used for solar-driven reforming to a greater or lesser extent, having reached different scale-up levels. In fact, natural gas reforming was the first process for solar conversion of hydrocarbons tested at an engineering scale of some hundreds of kilowatts of solar power input. Non-structured, solar receiver/reactors, that is, based on stationary or moving particles, have been primarily developed and employed for solar treatment of solid carbonaceous feedstocks [32] and therefore are not discussed here.

8.3.3
Worldwide Research into Solar Thermal Reforming of Methane

The concept of solar-driven reforming was first coined in 1982 in association with proposing the CO_2/CH_4 reforming–methanation cycle as a mechanism for converting and transporting solar energy via solar receivers [33] and operation of a solar tubular reformer with a ruthenium (Ru) catalyst [34]. Interest in this technology was revived at the late 1980s with the initial research in solar-driven reforming focusing on the concept of a closed loop for storage and transport of solar energy, in analogy to that with high temperature energy being supplied by a

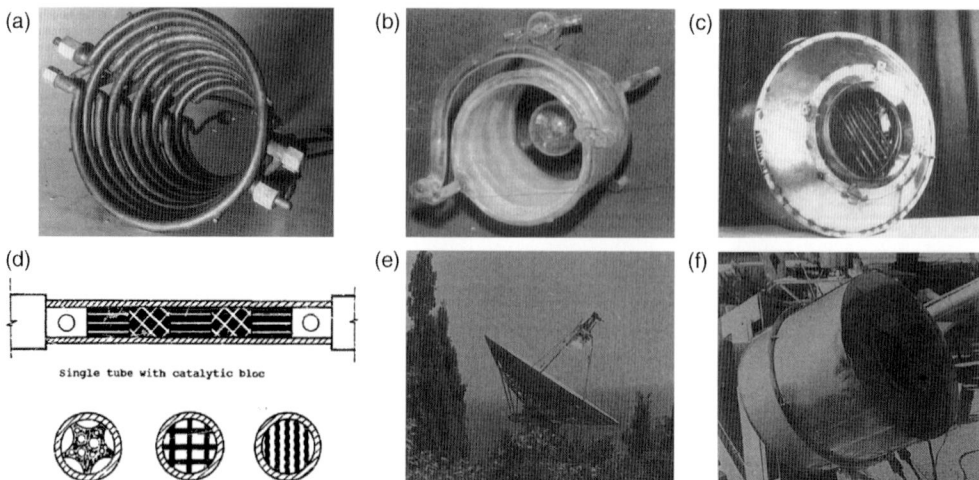

Figure 8.3 Various solar reformers developed and tested in the Soviet Union in the early 1990s: (a), (b) tubular metallic and quartz reactors, respectively, containing catalyst in powder form; (c) tubular reactor containing structured catalysts (i.e., pellets, honeycombs); (d) section of a single catalytic tube and typical structured catalysts [18]; (e) solar reformer in operation at the focal point of a dish concentrator; (f) the solar reformer after operation [36].

nuclear reactor. Parallel work took place during the same period in the former Soviet Union [18,35,36], where various catalytic reforming reactors/receivers were designed and tested on the focal point of a parabolic dish on the same research line (Figure 8.3). The reaction was carried out at levels of total power input of incident solar energy of ca. 5 kW and pressures between 1.8 and 2.8 atm; almost complete methane reformation was reported at gas temperatures at the catalyst bed outlet of between 650 and 700 °C [36]. Between 1988 and 1992 solar-based CO_2 methane reforming experiments on a laboratory scale were started at Sandia National Laboratories (SNL), USA [37], at DLR, and at the WIS, both with indirectly heated tubular reactors as well as with the first windowed receiver–reactors where the catalyst was heated directly by a concentrated solar beam [38–40]. The technical characteristics and results of all these studies have been comparatively summarized [41].

A significant amount of work has been conducted over the last 25 years on the development and scale-up of the technology of solar reforming of methane and other hydrocarbons all over the world, principally by laboratories and research institutes that possess pilot-to-large scale CSP facilities [8]. Notably, considerable work – not included here – has also been conducted on solar-aided methane cracking, or thermal decomposition, to produce hydrogen and elemental carbon. Research efforts worldwide are presented below, grouped on the basis of the solar heating concept employed (indirect or direct) and sub-divided on the basis of the heat transfer medium employed in each case, in order to delineate the technical characteristics of each approach rather than the chronological evolution of the technologies tested [42].

8.3.3.1 Indirectly Heated Reactors

Heat Transfer Medium: Air
This concept was first tested within the Advanced Steam Reforming of Methane in Heat Exchange (ASTERIX) joint Spanish–German project carried out by the Centro De Investigaciones Energéticas Medioambientales Y Tecnológicas (CIEMAT), Spain, the German Company Steinmüller, and DLR in the late 1980s and early 1990s [43]. The project employed the allothermal heating concept, that is, a solar-driven tower system was used to produce hot air (at 1000 °C and 9 bar) to drive a separate, 170 kW, 6 m long, tubular, packed catalyst bed, steam reformer. Typically solar-heated air was supplied to the reformer at 980 °C and exited the reformer at approximately 420 °C. Final CH_4 conversions reported were between 68% and 93% depending on the reformer's temperature, which varied from 702 to 803 °C.

Between 1993 and 1998 the WIS operated a solar central receiver for development of high-temperature technology, including the storage and transport of solar energy via CH_4 reforming. In contrast to the ASTERIX concept, in this case the reformer tubes were exposed to concentrated solar irradiation: a cavity receiver containing eight vertical 2.5-inch-diameter, 4.5 m long reformer tubes was built that was designed to produce syngas at 800 °C. To address scalability issues, WIS researchers transferred the concept to the so-called "beam down" receiver optics that allows a multi-megawatt tubular reformer to be built on the ground in a manner resembling the construction of a conventional, commercial reformer with roof burners. However, even though a complete 50 MW_{th} solar reforming system was conceptually designed [44] such a reformer has not yet been built and tested.

In Australia, solar reforming of methane is particularly attractive in view of the country's enormous areas of favorable insolation and its very large reserves of natural gas and coal bed methane that are co-located in many regions. The Commonwealth Scientific and Industrial Research Organization (CSIRO) began work on solar reforming in 1999, with its 25 kW single coil reformer (SCORE – Figure 8.4a). The catalyst is packed in-between the inner and outer tubes; the inner tube is purely for counter-current heating of the feed water stream. Production of solar-enriched fuels and hydrogen via steam reforming of natural gas at temperatures and pressures up to 850 °C and 20 bar was demonstrated on this Ni-catalyst-based, reformer coupled to a 107 m^2 dish concentrator between 1998 and 2001. In 2004, CSIRO built a single-tower heliostat field of 500 kW capacity (Figure 8.4c) with the objective of demonstrating this solar reforming process on a larger scale, which took place in 2006, over a commercial catalyst at reaction temperatures of 700–800 °C and pressures of 5–10 bar. Subsequently, CSIRO designed and installed in June 2009 a much larger, hexagonal-shape, dual-coil reformer

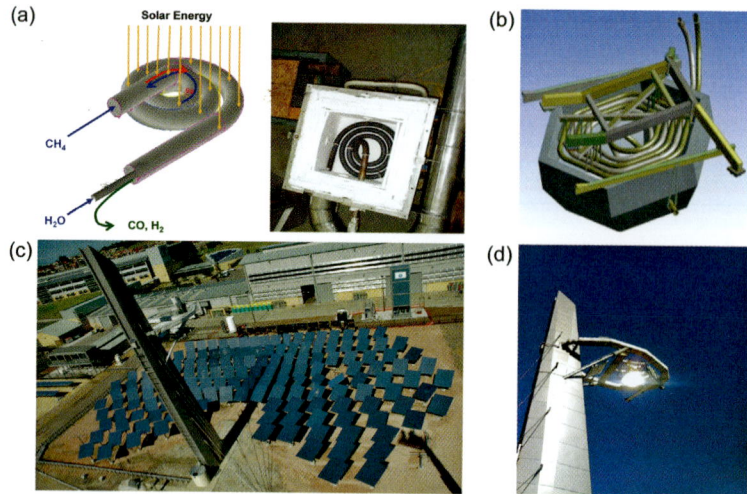

Figure 8.4 CSIRO's reformer technology development: (a) the single coil solar reformer and its operating principle; (b) the double-coil reformer; (c) the single-tower, 500 kW, heliostat field at Newcastle, Australia; (d) the reformer on the tower during solar operation [6,46].

(DCORE) capable of processing 200 kW thermal (LHV) natural gas at 10 bar and 850 °C (Figure 8.4b and d) [45,46].

Heat Transfer Medium: Sodium (Na) Vapors

In this concept, liquid sodium contained in an evacuated chamber evaporates under the effect of concentrated sunlight impinging on one surface of the containment. The sodium vapor condenses on the reactor tubes in the chamber and liberates the heat of vaporization isothermally to the heat transfer fluid through the walls since evaporation–condensation occurs at a constant temperature. Passive techniques (channels, wicks, gravity, etc.) return the liquid sodium to the absorber. The proof-of-concept of such a reactor was first demonstrated for steam reforming under simulated solar conditions [20,47] and subsequently under "real" solar irradiation within a SNL–WIS joint research project in which a 20 kW sodium reflux solar reformer was built and tested at WIS's solar furnace (1983–1984) for the CO_2 reforming of methane [48]. Figure 8.5 shows a schematic of the operation concept and actual reactor photographs. The flammability of Na vapors was considered critically in the solar community and caused the abandonment of this concept for some time. However, recently, sodium – and molten metals in general – as heat transfer fluids have come back into the focus of several research groups [49].

Heat Transfer Medium: Molten Salts

Molten salts such as $NaNO_3/KNO_3$ mixtures have been successfully employed on a large scale not only for the storage of solar heat in CSP plants but also as

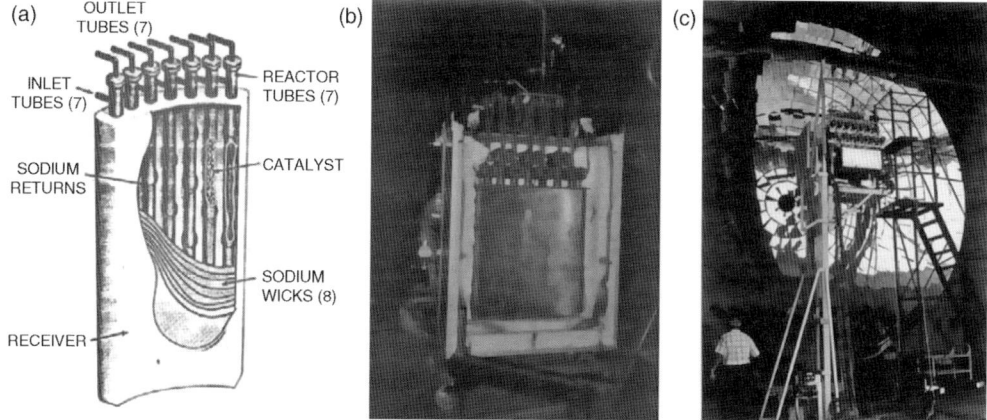

Figure 8.5 SANDIA-WIS's sodium reflux heat pipe solar receiver–reformer, built and tested at WIS's solar furnace (1983–1984) for CO_2 reforming of methane: (a) heating concept principle; (b) photograph of the receiver without the front panel; (c) the receiver installed in the facility at WIS's solar tower [48].

combined heat transfer fluid/heat storage media operating at 565 °C in the recently inaugurated Gemasolar solar tower power plant [50]. Their attributes lead several research groups to propose the utilization of thermal storage with a molten salt in order to provide constant-rate solar heat supply also for an energy demanding industrial chemical process like steam reforming.

The group of Niigata University, Japan, has proposed the concept of dry reforming of methane employing molten salts of the system K_2CO_3/Na_2CO_3 – which has a eutectic melting temperature of 710 °C – as the carrier material for the catalyst powder. In their initial configuration [51] a direct contact bubble reactor (DCBR) – or a "liquid bed" – was proposed where the preheated CH_4/CO_2 reactant gases were introduced through an aperture at the bottom of the reactor passing through the molten salt containing the catalyst powder. In subsequent studies [52] they replaced the "salt bath" with Ni-loaded porous alumina or zirconia spheres impregnated with the salt material to improve upon heat storage/release characteristics. Currently, in cooperation with Inha University, Korea [53] this concept has culminated in a double walled tubular receiver–reactor called MoSTAR (molten-salt tubular absorber/reformer), where the inner tube is filled with catalyst balls and the space between the inner tube and the outer absorber tube with a phase change (carbonate) and a ceramic material (MgO) to increase the composition's thermal conductivity. The next project step is to demonstrate the performance of such "double-walled" tubular absorbers/reformers with molten-salt thermal storage reactors on-sun with a 5 kW$_{th}$ dish-type solar concentrator.

Instead of employing the molten salt inside the reformer reactor, the group at Ente per le Nuove tecnologie, l' Energia e l' Ambiente (ENEA), Italy, has put forward the idea of heating a tubular reformer externally ("allothermally") by a molten salt, more or less in the same way that the industrial classical reformers are heated by hot combustion gases. Materialization of this concept would enable coupling of allothermal reformers to parabolic trough solar plants; however, this also implies that the particular reaction has to be performed at temperatures lower than 565 °C. One idea in this direction is to employ membrane reactors where the selective removal of hydrogen from the products stream enhances the conversion and allows performance of the reaction at lower temperatures [54]. ENEA is working in this direction; they have adopted a heat-exchanger-shaped (shell-and-tube) reactor, optimized its dimensions, and assessed its performance [55,56]. However, due to the small methane conversion, this concept does not yield syngas but enriched methane (EM), that is, methane with 17% hydrogen. Engineering and experimental activities aimed to the development of a prototype apparatus are in progress.

Heat Transfer Medium: Solid Particles

A somewhat different approach to solar-driven CO_2-reforming is being investigated at NREL and the Department of Chemical Engineering at the University of Colorado, USA. In this approach what is heated by concentrated solar radiation is a "target" graphite tube that can thus reach ultra-high temperatures (around 1900 °C) under a non-oxidative atmosphere. The same reactor configuration can be used for either the solar thermal dissociation of methane to hydrogen and carbon black or for non-catalytic CO_2 methane reforming to produce syngas [57]. For the latter case, a triple-tube, solar powered, fluid-wall aerosol flow reactor design has been used to produce syngas with a lower H_2/CO ratio, as required for Fischer–Tropsch synthesis. A schematic and a photograph of the reactor are shown in Figure 8.6a [58]. In this case, carbon particles are produced *in situ* as a by-product: it is though claimed that these are not detrimental to the process but on the contrary, are preferably flowed with reactant gas through the reactor tube

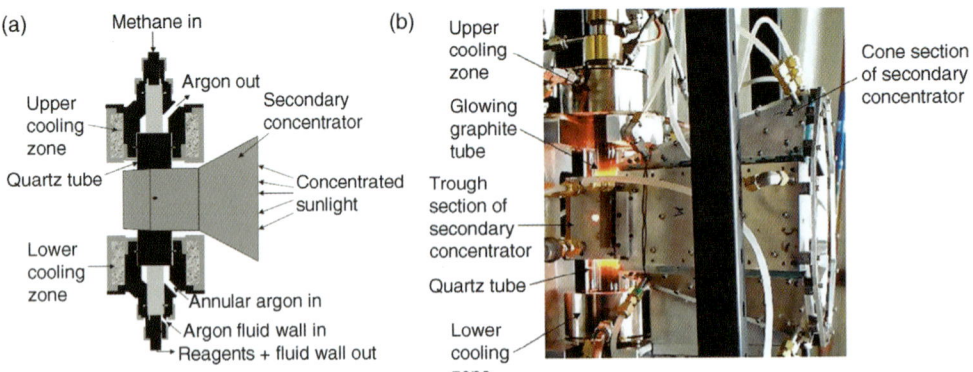

Figure 8.6 University of Colorado's solar-thermal aerosol reactor: (a) schematic and (b) photograph of actual reactor heated to approximately 1200 °C [58].

as radiation absorbers to facilitate heating and reaction – justifying the terminology of "aerosol" reactor. Operating with residence times of the order of 10 ms and temperatures of approximately 1700 °C, methane and CO_2 conversions of 70% and 65%, respectively, were reported in the absence of any added catalysts.

8.3.3.2 Directly Irradiated Reactors

Structured Reactors Based on Ceramic Honeycombs
Researchers at WIS have irradiated Rh-coated alumina honeycombs in their solar furnace through a sapphire window to catalyze the reaction of CO_2 methane reforming [40]. Due to failure issues the sapphire window was substituted by a fused silica bell jar reactor (Figure 8.7 [39]) and the catalysts with cordierite honeycombs coated with a Rh-alumina washcoat. When a temperature of about 800 °C was reached, methane was added, keeping the ratio of CO_2/CH_4 at about 1.3. Reported methane conversions reached 67%. However, no scaled-up reformers based on honeycombs have been reported.

Structured Reactors Based on Ceramic Foams
In fact reactors based on ceramic foams were the first structured reactors to be tested for solar-aided methane reforming and today are the most developed ones tested on a level of a few hundred kilowatts of solar input. Their prevalence over honeycomb reactors for in particular high-pressure application is owing to their inherently higher porosity, which allows solar radiation to penetrate deeper into the volume of the materials, as well as to their sufficient radial heat transfer mechanism, which prevents flow instabilities at the high temperatures encountered during solar operation [59]. The first examples of such solar reactors can be traced back to 1990 when solar reforming of methane with CO_2 in a directly

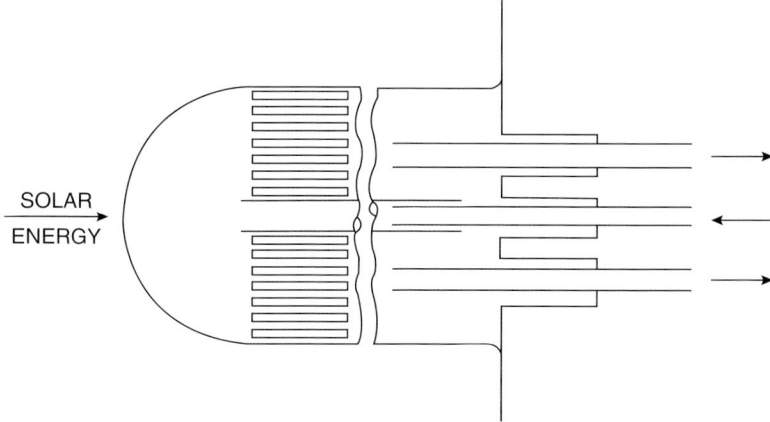

Figure 8.7 Sketch of WIZ's transparent bell jar solar reactor with honeycomb catalyst [39].

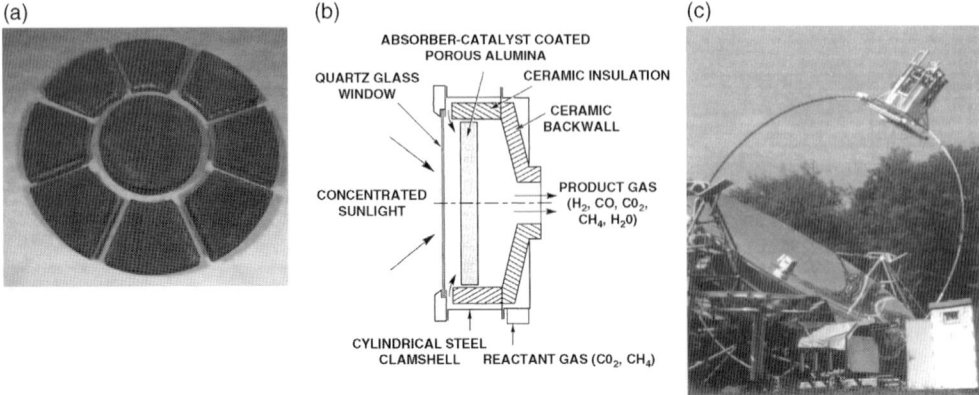

Figure 8.8 CEASAR foam absorber/reactor: (a) photograph of first CEASAR receiver; (b) schematic of receiver/reactor's operation; (c) reactor on parabolic dish test facility [60].

irradiated volumetric receiver–reactor was demonstrated in the "catalytically enhanced solar absorption receiver" (CAESAR) experiment conducted by SNL, USA and DLR, Germany [60]. Reticulated porous α-alumina–mullite foam disks, coated with a Rh/γ-alumina washcoat, were assembled in a multilayered flat-disk configuration within an absorber (Figure 8.8a), located behind a quartz window aperture and directly irradiated by concentrated solar energy (Figure 8.8b) in the focal point of a parabolic dish capable of 1–50 kW solar power (Figure 8.8c). The maximum methane conversion was almost 70%, demonstrating the "proof-of-concept" of CO_2 methane reforming.

A second generation solar chemical receiver–reactor [61,62] was developed by implementing a higher power unit and a parabolic quartz window behind which receiver foam pieces made of silicon carbide (SiC) were arranged in a domed cavity configuration (Figure 8.9a and b). For solar testing, the assembled receiver was installed on top of the tower of the WIS Solar Test Facility Unit in Israel (Figure 8.9c). Typical operating temperatures ranged from 700 to 860 °C, with an absolute pressure of 3.5 bar, reaching methane conversions of 84–88%. These ceramic-foam-based solar reactors were pursued further within Project SOLASYS (partners: DLR, WIS, and Ormat Pty Ltd.) where the technical feasibility of solar reforming with liquid petroleum gas (LPG) as feedstock and combustion of the product gases (syngas mixture) in a gas turbine to generate electricity at the 300 kW$_e$-scale was successfully demonstrated. An advanced and more compact and cost-effective volumetric receiver/reformer has been developed within the successor Project SOLREF with the rationale of operating at a higher power level (400 kW$_e$) as well as at higher pressure and temperature conditions, for example, 950 °C and 15 bar, which would result in higher conversion rate of methane into hydrogen and higher efficiency (Figure 8.9d). The solar reformer was tested and validated under real solar conditions at the WIS (Figure 8.9e) demonstrating the feasibility of the SOLREF technology. Relevant

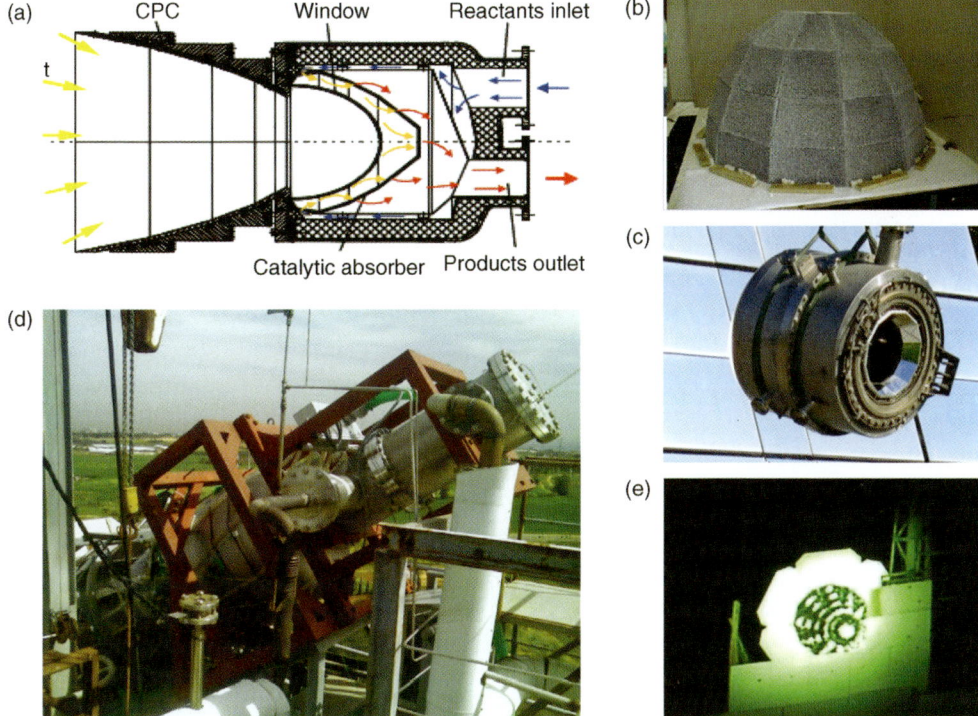

Figure 8.9 "Evolution" of the SOLASYS-SOL-REF ceramic foam-based directly irradiated (DIVVRR) solar steam methane reformer: (a) sketch of operation principle; (b) assembled dome of the solar receiver; (c) actual reactor; (d) reactor installed on the solar tower of the WIS, Israel; (e) reformer in operation [14].

publications claim that the cost of hydrogen produced by solar steam reforming of methane is very close to conventional steam reforming with the break-even point being at a natural gas price of €0.35 m^{-3} [63,64].

The same Japanese group from Niigata University mentioned above has also studied CO_2-reforming on a laboratory scale using a new type of Ru/Al$_2$O$_3$ catalytically activated, metallic (Ni-Cr-Al) foam absorber, directly irradiated by a solar-simulated xenon lamp light and subjected to solar flux levels in the range 180–250 kW m^{-2} [65]. In a series of parallel publications [66,67], the same group tested a series of Ni-based catalysts, trying to establish robust catalytic systems free of expensive metals like Ru or Rh. The current status of this development is that the system of choice for subsequent scale-up is a Ni/MgO-Al$_2$O$_3$ catalyst coated on SiC foams [67].

Researchers from Inha University in Korea are also investigating structured reactors based on metallic foams for CO_2 methane reforming. They have proposed and developed a double-layer metal foam absorber/reactor in which the front part is not catalytically active but realizes the absorbing of solar radiation

Figure 8.10 The metallic-foam-based 5 kW$_{th}$ absorber/reactor of Inha University, Korea, used for CO_2 reforming of methane on the solar dish system (INHA-DISH1) [42].

and heat transfer to the gaseous reactants fluid. The rear part is a Ni metal foam coated with Ni/Al$_2$O$_3$ catalyst. The absorber was tested in several campaigns with respect to solar CO_2 reforming of methane on a parabolic dish capable of providing 5 kW$_{th}$ solar power (INHA DISH1 – Figure 8.10 [42]). The temperature at the center of the irradiated surface of the absorber ranged from 900 to 1000 °C and the maximum CH$_4$ conversion reached 60% [68]. Compared to "conventional" catalytically activated ceramic foam absorbers, direct irradiation of the metallic foam absorber was claimed to exhibit superior reaction performance at relatively low insolation or at low temperatures and better thermal resistance, preventing cracking by mechanical stress or thermal shock.

Structured Reactors Based on Fins
WIS's researchers have designed a reformer based on the directly irradiated annular pressurized receiver (DIAPR) [29] and the "porcupine" concept, for operation at high temperatures and pressures. Figure 8.11a shows a schematic of the reformer design and operating principle [69,70]. Catalytic elements for use in this reformer based on Ru/Al$_2$O$_3$ promoted with Mn oxides to avoid carbon deposition were developed, characterized supported on alumina pins by washcoating, and long-term tested with respect to chemical and thermal stability. With input from computational models and corrective actions taken therein, a prototype "porcupine" reformer using a conical quartz window was built (Figure 8.11b) [29,70] and tested with respect to CO_2 methane reforming on the solar tower of WIS in 2010. The CO_2 to CH$_4$ ratio was about 1 : 1.2. Eight test runs were performed, focusing on the influence of gas pressure (4–9 atm) and flow rate on the CH$_4$ conversion. The maximum absorber temperature was kept below 1200 °C. The conversion of CH$_4$ reached 85%. Further work aimed at improving the total efficiency of the system is in progress.

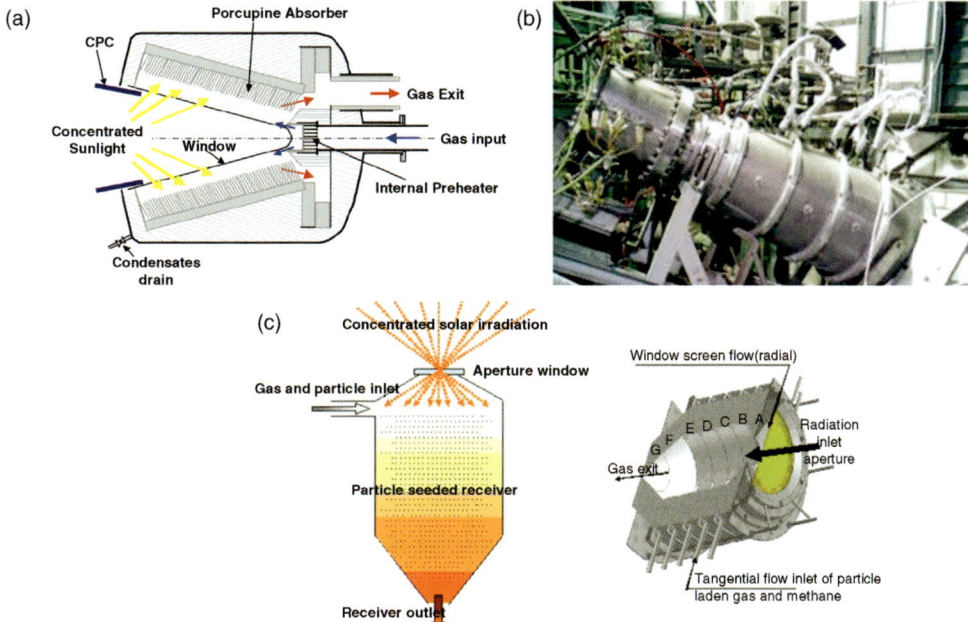

Figure 8.11 WIZ's solar reformers: (a), (b) "porcupine" directly irradiated solar reformer, schematic of design and actual reactor as built [70]; (c) directly irradiated particle solar receiver: schematic of operation principle and of receiver actually used [71].

Non-structured Reactors Based on Solid Particles

Researchers at WIS performed dry methane reforming in a directly irradiated particle receiver seeded with carbon black particles in a molar ratio of C/CO_2 about 0.5/100 and with CO_2/CH_4 inlet ratios varying from 1 : 1 to 1 : 6 [71]. The receiver uses a moving radiation absorber, that is, particles entrained in the reforming gas mixture that have two functions: to absorb the solar radiation and transfer it to the gas and to act as a reaction surface for the reforming reactions (Figure 8.11c). Dry solar reforming was conducted without any catalyst at a temperature between 950 and 1450 °C at the WIS's solar tower. The exit gas temperatures were 1100–1450 °C. Depending on the temperature, a share of the carbon particles reacted with CO_2 to form CO, that is, the particles in the receiver were gradually partially or totally consumed. The reforming experiments indicated that the carbon black particles entrained in the gas flow augmented the reaction of methane cracking.

8.4
Current Development Status and Future Prospects

While the long-term incentives for solar reforming may be apparent, there is a need to search for near-term applications, establish priorities for R&D activities, and attract industrial financial support. In this perspective, for the subsequent scale-up of current solar methane reforming technologies to demonstration level, several choices have to be made with respect to critical issues that will impact the ability of the technology to enter the marketplace.

With respect to the CSP technology to be employed, even though the first solar-coupling step of many of the reforming studies above has been to test the reactors on solar dish receivers, scale-up of the reforming dish technology has many technical problems; it is therefore highly probable that large-scale solar thermal reforming will require the economies of scale offered by central tower receiver heliostat fields that can comfortably generate solar thermal fluxes in the MW capacity.

The technology of choice will be one that can be deployed cost-effectively at large scale. The choice of directly irradiated volumetric receivers versus cavity (tubular) indirectly irradiated ones for solar-driven catalytic reactions will be a key decision. Volumetric absorbers appear to offer better thermal and performance properties than conventional tubular ones. Since solar energy is absorbed directly by the catalyst, temperatures are highest at the reaction sites and the chemical reactions are likely to be kinetically-limited rather than heat transfer-limited as in conventional tubular reactors. A further advantage is that the intensity of the impinging solar radiation can be five times higher than in a tubular receiver – Figure 8.2a). The concurrent flow of solar radiation and chemical reactants reduces absorber temperatures and re-radiation losses.

Indirectly irradiated reactors eliminate the need for a window at the expense of having less efficient heat transfer – by conduction – through the walls of an opaque absorber. Thus, the disadvantages are linked to the limitations imposed by the materials of the absorber, with regard to maximum operating temperature, inertness to the chemical reaction, thermal conductivity, radiative absorbance, resistance to thermal shocks, and suitability for transient operation. In this perspective, perhaps "technically simpler" concepts, like, for instance, the tubular indirectly-heated ("allothermal") reformers employed in the ASTERIX project, might be more attractive for large-scale implementation and demonstration of the technology.

Another set of issues relates to the choice of steam or CO_2 for reforming. There are pros and cons for each option with a clear choice only for certain open-cycle applications. For example, if methanol is the desired end-product, the amount of steam or CO_2 used would give an optimal CO/H_2 ratio in the syngas. If H_2 is the desired product, steam reforming is the choice. A process configuration considered more recently is the mixed reforming of methane where it reacts with a mixture of water and CO_2, a process especially advantageous for methane sources with a high CO_2 content such as biogas (45–70 mol.% CH_4) [6,8].

In conclusion, significant progress has been made through the research efforts described above and the solar driven steam reforming of methane has been demonstrated at pilot scale. The technologies for DIVRR and tubular reactors are both ready for small commercial scale installations, and it is anticipated that plants in the 1–5 MW$_{th}$ range will be constructed in Australia and Europe within the next couple of years. The transitional nature of this technology – coupling renewable solar thermal energy and conventional fossil fuels to make either a synthesis gas or hydrogen – is seen as a key first step on the path to sustainable hydrogen and energy production.

References

1 Liu, K., Song, C., and Subramani, V. (2010) *Hydrogen and Syngas Production and Purification Technologies*, John Wiley & Sons, Inc. Hoboken, NJ.
2 Mueller-Langer, F., Tzimas, E., Kaltschmitt, M., and Peteves, S. (2007) Techno-economic assessment of hydrogen production processes for the hydrogen economy for the short and medium term. *Int. J. Hydrogen Energy*, **32**, 3797–3810.
3 Rostrup-Nielsen, J.R. (2002) Syngas in perspective. *Catal. Today*, **71**, 243–247.
4 Steinfeld, A. and Palumbo, R. (2001) Solar thermochemical process technology, in *Encyclopedia of Physical Science and Technology* (ed. R.A. Meyers), Academic Press, pp. 237–256.
5 Sattler, C. (2010) Hydrogen production by the solar-driven steam reforming of methane, in *Hydrogen and Fuel Cells: Fundamentals, Technologies and Applications* (ed. D. Stolten), Wiley-VCH Verlag GmbH, Weinheim.
6 Turpeinen, E., Raudaskoski, R., Pongrácz, E., and Keiski, R. (2008) Thermodynamic analysis of conversion of alternative hydrocarbon-based feedstocks to hydrogen. *Int. J. Hydrogen Energy*, **33**, 6635–6643.
7 Agrafiotis, C., Von Storck, H., Roeb, M., and Sattler, C. (2013) Solar thermal reforming of methane feedstocks for hydrogen and syngas production – a review, *Renew. Sustain. Energy Rev.*, **29** (January), 656–682.
8 Sun, Y., Ritchie, T., Hla, S.S., McEvoy, S., Stein, W., and Edwards, J.H. (2011) Thermodynamic analysis of mixed and dry reforming of methane for solar thermal applications. *J. Nat. Gas Chem.*, **20**, 568–576.
9 Tomishige, K., Chen, Y., and Fujimoto, K. (1999) Studies on carbon deposition in CO_2 reforming of CH_4 over nickel–magnesia solid solution catalysts. *J. Catal.*, **181**, 91–103.
10 Lemonidou, A., Goula, M., and Vasalos, I. (1998) Carbon dioxide reforming of methane over 5 wt.% nickel calcium aluminate catalysts – effect of preparation method. *Catal. Today*, **46**, 175–183.
11 Gadalla, A.M. and Bower, B. (1988) The role of catalyst support on the activity of nickel for reforming methane with CO_2. *Chem. Eng. Sci.*, **43**, 3049–3062.
12 Sehested, J. (2006) Four challenges for nickel steam-reforming catalysts. *Catal. Today*, **111**, 103–110.
13 Berman, A., Karni, R.K., and Epstein, M. (2005) Kinetics of steam reforming of methane on Ru/Al$_2$O$_3$ catalyst promoted with Mn oxides. *Appl. Catal. A-Gen.*, **282**, 73–83.
14 Epstein, M. (2011) Presented at Solar Thermal Reforming of Methane, 2nd SFERA Winter School at ETH Zürich, Switzerland, 24–25 March 2011.
15 Rostrup-Nielsen, J.R. (1984) Catalytic steam reforming, in *Catalysis: Science and Technology* (ed. J.R. Anderson), Springer-Verlag, Berlin.
16 Richter, C., Teske, S., and Nebrera, J.A. (2009) Concentrating Solar Power Global Outlook 09. Greenpeace International/European Solar Thermal Electricity Association (ESTELA)/IEA SolarPACES Report.
17 Cybulski, A. and Moulijn, J.A. (eds) (2005) *Structured Catalysts and Reactors*, 2nd edn, CRC Press, Boca Raton, FL.
18 Anikeev, V.I. and Kirillov, V.A. (1991) Basic design principles and some methods of investigation of catalytic reactors-

receivers of solar radiation. *Solar Energy Mater.*, **24**, 633–646.

19 Karni, J., Kribus, A., Rubin, R., and Doron, P. (1998) The "porcupine": a novel high-flux absorber for volumetric solar receivers. *J. Solar Energy. Eng.*, **120**, 85–95.

20 Richardson, J., Paripatyadar, S., and Shen, J. (1988) Dynamics of a sodium heat pipe reforming reactor. *AICHE J.*, **34**, 743–752.

21 Bilgen, E. and Galindo, J. (1981) High temperature solar reactors for hydrogen production. *Int. J. Hydrogen Energy*, **6**, 139–152.

22 Olalde, G. and Peube, J. (1982) Etude expérimentale d'un récepteur solaire en nid d'abeilles pour le chauffage solaire des gaz à haute température. *Rev. Phys. Appl.*, **17**, 563–568.

23 Fend, T., Hoffschmidt, B., Pitz-Paal, R., Reutter, O., and Rietbrock, P. (2004) Porous materials as open volumetric solar receivers: experimental determination of thermophysical and heat transfer properties. *Energy*, **29**, 823–833.

24 Fend, T., Pitz-Paal, R., Reutter, O., Bauer, J., and Hoffschmidt, B. (2004) Two novel high-porosity materials as volumetric receivers for concentrated solar radiation. *Solar Energy Mater. Solar Cells*, **84**, 291–304.

25 Ávila-Marín, A.L. (2011) Volumetric receivers in solar thermal power plants with central receiver system technology: a review. *Solar Energy*, **85**, 891–910.

26 Chavez, J.M. and Chaza, C. (1991) Testing of a porous ceramic absorber for a volumetric air receiver. *Solar Energy Mater.*, **24**, 172–181.

27 Agrafiotis, C.C., Mavroidis, I., Konstandopoulos, A.G., Hoffschmidt, B., Stobbe, P., Romero, M., and Fernandez-Quero, V. (2007) Evaluation of porous silicon carbide monolithic honeycombs as volumetric receivers/collectors of concentrated solar radiation. *Solar Energy Mater. Solar Cells*, **91**, 474–488.

28 Hennecke, K., Schwarzbözl, P., Koll, G., Beuter, M., Hoffschmidt, B., Göttsche, J., and Hartz, T. (2009) The Solar Power Tower Jülich—A solar thermal power plant for test and demonstration of air receiver technology, in *Proceedings of ISES World Congress, 2007 (Vol I–Vol V)*, Springer, Berlin, pp. 1749–1753.

29 Karni, J., Kribus, A., Doron, P., Rubin, R., Fiterman, A., and Sagie, D. (1977) The DIAPR: a high-pressure, high-temperature solar receiver. *J. Solar Energy Eng.*, **119**, 74–78.

30 Heck, R.M., Farrauto, R.J., and Gulati, S.T. (1995) *Catalytic Air Pollution Control*, John Wiley & Sons, Inc., New York.

31 Agrafiotis, C.C., Pagkoura, C., Lorentzou, S., Kostoglou, M., and Konstandopoulos, A.G. (2007) Hydrogen production in solar reactors. *Catal. Today*, **127**, 265–277.

32 Piatkowski, N., Wieckert, C., Weimer, A.W., and Steinfeld, A. (2011) Solar-driven gasification of carbonaceous feedstock – a review. *Energy Environ. Sci.*, **4**, 73–82.

33 Chubb, T.A. (1980) Characteristics of CO_2-CH_4 reforming-methanation cycle relevant to the solchem thermochemical power system. *Solar Energy*, **24**, 341–345.

34 McCrary, J.H., McCrary, G.E., Chubb, T.A., Nemecek, J.J., and Simmons, D.E. (1982) An experimental study of the CO_2-CH_4 reforming-methanation cycle as a mechanism for converting and transporting solar energy. *Solar Energy*, **29**, 141–151.

35 Anikeev, V., Bobrin, A., Khanaev, V., and Kirillov, V. (1993) Chemical heat regeneration in power plants. *Int. J. Energy Res.*, **17**, 233–242.

36 Anikeev, V., Parmon, V., Kirillov, V., and Zamaraev, K. (1990) Theoretical and experimental studies of solar catalytic power plants based on reversible reactions with participation of methane and synthesis gas. *Int. J. Hydrogen Energy*, **15**, 275–286.

37 Spiewak, I., Tyner, C.E., and Langnickel, U. (1993) Applications of Solar Reforming Technology. SAND93-1959, UC-237, Albuquerque, New Mexico, USA.

38 Hogan, R., Skocypec, R., Diver, R., Fish, J., Garrait, M., and Richardson, J. (1990) A direct absorber reactor/receiver for solar thermal applications. *Chem. Eng. Sci.*, **45**, 2751–2758.

39 Levy, M., Rubin, R., Rosin, H., and Levitan, R. (1992) Methane reforming by direct solar irradiation of the catalyst. *Energy*, **17**, 749–756.

40 Levy, M., Rosin, H., and Levitan, R. (1989) Chemical reactions in a solar furnace by direct solar irradiation of the catalyst. *J. Solar Energy Eng.*, **111**, 96–97.

41 Kirillov, V.A. (1999) Catalyst application in solar thermochemistry. *Solar Energy*, **66**, 143–149.

42 Meier, A. (2010) Task II Solar Chemistry Research. SolarPaces Annual Report.

43 Böhmer, M., Langnickel, U., and Sanchez, M. (1991) Solar steam reforming of methane. *Solar. Energy Mater.*, **24**, 441–448.

44 Segal, A. and Epstein, M. (2003) Solar ground reformer. *Solar Energy*, **75**, 479–490.

45 McNaughton, R. and Stein, W. (2009) Improving efficiency of power generation from solar thermal natural gas reforming. Presented at the 15th International SolarPACES Conference, September 15th–18th, Berlin, Germany.

46 Stein, W. (2009) Hydrogen production by the solar-driven steam reforming of methane, in *Encyclopedia of Electrochemical Power Sources* (ed. J. Garche), Elsevier, Amsterdam.

47 Paripatyadar, S.A. and Richardson, J.T. (1988) Cyclic performance of a sodium heat pipe, solar reformer. *Solar Energy*, **41**, 475–485.

48 Diver, R.B., Fish, J.D., Levitan, R., Levy, M., Meirovitch, E., Rosin, H., Paripatyadar, S.A., and Richardson, J.T. (1992) Solar test of an integrated sodium reflux heat pipe receiver/reactor for thermochemical energy transport. *Solar Energy*, **48**, 21–30.

49 Hering, W., Stieglitz, R., and Wetzel, T. (2012) Application of liquid metals for solar energy systems. EPJ Web of Conferences: EDP Sciences.

50 Burgaleta, J.I., Arias, S., and Ramirez, D. (2011) GEMASOLAR, the first tower thermosolar commercial plant with molten salt storage. Presented at 17th SolarPACES Conference, 20–23 September 2011, Granada, Spain.

51 Kodama, T., Koyanagi, T., Shimizu, T., and Kitayama, Y. (2001) CO_2 reforming of methane in a molten carbonate salt bath for use in solar thermochemical processes. *Energy Fuels*, **15**, 60–65.

52 Kodama, T., Isobe, Y., Kondoh, Y., Yamaguchi, S., and Shimizu, K.I. (2004) Ni/ceramic/molten-salt composite catalyst with high-temperature thermal storage for use in solar reforming processes. *Energy*, **29**, 895–903.

53 Kodama, T., Gokon, N., Inuta, S., Yamashita, S., and Seo, T. (2009) Molten-salt tubular absorber/reformer (MoSTAR) project: the thermal storage media of Na_2CO_3-MgO composite materials. *J. Solar Energy Eng.*, **131** (4) 04013.

54 Giaconia, A., de Falco, M., Caputo, G., Grena, R., Tarquini, P., and Marrelli, L. (2008) Solar steam reforming of natural gas for hydrogen production using molten salt heat carriers. *AICHE J.*, **54**, 1932–1944.

55 De Falco, M., Giaconia, A., Marrelli, L., Tarquini, P., Grena, R., and Caputo, G. (2009) Enriched methane production using solar energy: an assessment of plant performance. *Int. J. Hydrogen Energy*, **34**, 98–109.

56 DeFalco, M. and Piemonte, V. (2011) Solar enriched methane production by steam reforming process: reactor design. *Int. J. Hydrogen Energy*, **36**, 7759–7762.

57 Dahl, J.K., Buechler, K.J., Weimer, A.W., Lewandowski, A., and Bingham, C. (2004) Solar-thermal dissociation of methane in a fluid-wall aerosol flow reactor. *Int. J. Hydrogen Energy*, **29**, 725–736.

58 Dahl, J.K., Weimer, A.W., Lewandowski, A., Bingham, C., Bruetsch, F., and Steinfeld, A. (2004) Dry reforming of methane using a solar-thermal aerosol flow reactor. *Ind. Eng. Chem. Res.*, **43**, 5489–5495.

59 Pitz-Paal, R., Hoffschmidt, B., Böhmer, M., and Becker, M. (1997) Experimental and numerical evaluation of the performance and flow stability of different types of open volumetric absorbers under non-homogeneous irradiation. *Solar Energy*, **60**, 135–150.

60 Buck, R., Muir, J.F., and Hogan, RE. (1991) Carbon dioxide reforming of methane in a solar volumetric receiver/reactor: the CAESAR project. *Solar Energy Mater.*, **24**, 449–463.

61 Tamme, R., Buck, R., Epstein, M., Fisher, U., and Sugarmen, C. (2001) Solar

upgrading of fuels for generation of electricity. *J. Solar Energy Eng.*, **123**, 160–163.

62 Wörner, A. and Tamme, R. (1984) CO_2 reforming of methane in a solar driven volumetric receiver–reactor. *Catal. Today*, **46**, 165–174.

63 Möller, S., Kaucic, D., and Sattler, C. (2006) Hydrogen production by solar reforming of natural gas: a comparison study of two possible process configurations. *J. Solar Energy Eng.*, **128**, 16–23.

64 Pregger, T., Graf, D., Krewitt, W., Sattler, C., Roeb, M., and Möller, S. (2009) Prospects of solar thermal hydrogen production processes. *Int. J. Hydrogen Energy*, **34**, 4256–4267.

65 Kodama, T., Kiyama, A., and Shimizu, K.I. (2003) Catalytically activated metal foam absorber for light-to-chemical energy conversion via solar reforming of methane. *Energy Fuel*, **17**, 13–17.

66 Kiyama, A., Moriyama, T., and Mizuno, O. (2004) Solar methane reforming using a new type of catalytically-activated metallic foam absorber. *J. Solar Energy Eng.*, **126**, 808–814.

67 Gokon, N., Yamawaki, Y., Nakazawa, D., and Kodama, T. (2010) Ni/MgO–Al_2O_3 and Ni–Mg–O catalyzed SiC foam absorbers for high temperature solar reforming of methane. *Int. J. Hydrogen Energy*, **35**, 7441–7453.

68 Lee, J., Lee, J., Shin, I., and Seo, T. (2011) Solar CO_2-reforming of methane using a double layer absorber. Presented at 17th SolarPACES Conference, 20–23 September 2011, Granada, Spain.

69 Berman, A., Karni, R.K., and Epstein, M. (2006) A new catalyst system for high-temperature solar reforming of methane. *Energy Fuel*, **20**, 455–462.

70 Rubin, R. and Karni, J. (2011) Carbon dioxide reforming of methane in directly irradiated solar reactor with porcupine absorber. *J. Solar Energy Eng.*, **133** (2), 021008.

71 Klein, H.H., Karni, J., and Rubin, R. (2009) Dry methane reforming without a metal catalyst in a directly irradiated solar particle reactor. *J. Solar Energy Eng.*, **131**, 021001.

9
Fuel Processing for Utilization in Fuel Cells
Ralf Peters

9.1
Introduction

Fuel choice plays an important role in most mobile and stationary applications. Although hydrogen is the ideal fuel gas for a fuel cell, it is not widely available in today's infrastructure. In regard to automotive systems, gaseous hydrogen can be stored at 70 MPa as fuel for PEFCs (polymer electrolyte fuel cells). Automotive stacks deliver 50–70 kW$_e$ to an electric motor for propulsion. It is technically feasible to reach a driving range of about 400–500 km. In future, the hydrogen infrastructure and sustainable sources for hydrogen production must be considerably expanded. The mass-specific and volumetric power densities of hydrogen storage systems are not high enough to fulfill the demands of large fuel cell systems in mobile applications such as trucks, ships, and airplanes. In aircraft applications, liquid hydrogen storage systems can act as a bridging technology for medium-sized systems with an electric power of 10–100 kW$_e$. Mass considerations are more important here than volumetric requirements for hydrogen storage systems because additional space demands can be realized on airplanes to a certain extent. Furthermore, mass aspects dominate due to a higher fuel consumption, that is, a delta of 0.16 kg fuel per 1 kg additional mass [1]. In the long term, larger fuel cell systems (>150 kW$_e$) will still require the use of liquid energy carriers [2]. Hydrogen is not feasible as an energy carrier for truck applications because of the long distances involved and the handling of additional fuels by the driver. In addition to the dominant fuel in heavy duty transport (i.e., diesel), LPG, LNG and DME are being discussed as alternate fuel options [3,4]. For maritime applications, LPG is a better alternate fuel to marine gas oil and heavy fuel oils than hydrogen. For ferries and small boats, hydrogen would be the first choice if these ships were operated near the coast and a harbor. For sail boats, methanol is a further option as it could be used directly in a DMFC [5]. For use in the leisure sector, such as for camping gas, a propane–butane mixture (LPG) is foreseen as a useful energy carrier. Finally, stationary house heating systems generally prefer natural gas as a hydrogen source, and to a minor extent kerosene (Japan) or heating oil EL (Europe).

Hydrogen Science and Engineering: Materials, Processes, Systems and Technology, First Edition.
Edited by Detlef Stolten and Bernd Emonts.
© 2016 Wiley-VCH Verlag GmbH & Co. KGaA. Published 2016 by Wiley-VCH Verlag GmbH & Co. KGaA.

9.2
Scope of the Work and Methodical Approach

This chapter deals with the task of fuel processing for fuel cells. Since a large number of book chapters are dedicated to fuel processing for fuel cells, to special aspects and design methods of these technologies, this contribution focuses on process engineering demands. Additional review articles complete the available information on fuel processing for fuel cells [6–12]. Reviews provide a snapshot of state-of-the-art technology at a particular point in time. They focus on specific items, such as heterogeneous catalysis [7,8,13], ethanol and methanol reforming [6,7,12], or multi-fuel reforming for fuel cell vehicles [9]. The latest review by Xu *et al.* published in 2013 considers aspects of thermochemistry and reactor design [11]. Sometimes, a review reflects only a constricted part of complete information – for example, fuel processing papers from a certain number of national groups and companies or from selected areas. In this contribution, a mixture of important basic and current research results will be used. The cited references are only a selection of the huge amount of papers available. For more detailed information, recent papers are cited for further reading. Specific information concerning chemical reaction paths and catalytically active materials, especially answers to questions of heterogeneous catalysis, can be found in Reference [13]. General methods of process engineering and aspects of system design are discussed in Reference [14].

What do we need to know to begin with? Hydrogen can be produced by chemical reactions. A chemical reaction is performed under kinetic, mass transport, or thermodynamic limitations. It is important to note that a specific chemical reaction occurs at favored conditions. In addition, a fuel reacts with air, steam, or with both – these are referred to as educts. These educts must be processed at ideal conditions for the chemical reaction. Gases must be compressed, liquids must be pumped, and all educts must be heated. Liquid fuels and water should be evaporated. Subsequently, they have to be mixed by the most appropriate method. Thermodynamic quantities, such as enthalpy, are used for process analysis. Therefore, calculation methods will be discussed in Section 9.3.

The variety of available fuels, different fuel cell types, and various power classes defined by an application mean that a plurality of possible system configurations exists. Instead of discussing them in different classes of systems, another attempt is made in this contribution, that is, fuel processing is divided into unit operations. The ultimate task is to produce hydrogen from an energy carrier that is well established in today's infrastructure or one that will reach this status in the near future. Some examples are gases or liquefied gases, and energy carriers produced as liquid fuels from fossil and renewable sources. Before tackling the complexity of all of these different systems, the reader should be reminded of the basics of process engineering. A whole process can be split into various unit operations. In Section 9.4, different unit operations in fuel processing will be explained, and subsequently combined into chains, which describe subsystems.

Further unit operations include compressors and pumps, heat exchangers, evaporators, and mixers. Process control instruments are not addressed here. These subsystems can be connected at the end to monitor the whole process.

In Section 9.5, subsystems for reforming and gas cleaning will be discussed in more detail. Optimization is necessary for a chain of single process units in regard to their interaction on a subsystem level. Current research on catalyst development and technically oriented development are presented to provide an outlook on challenges that must be addressed to achieve suitable fuel cell systems for different applications.

9.3
Chemical Engineering Thermodynamics

Fuel processing consists of heating up fluids, separating compounds from each other, and several chemical reactions. To design these processes and the corresponding devices, several thermodynamic calculation methods must be applied. Two of them are discussed in the next sections: (i) the determination of thermodynamic properties, such as density, specific heat capacity, and enthalpy, and (ii) chemical equilibrium.

9.3.1
Thermodynamic Property Relations

Fuel cell systems generally convert an energy carrier into a hydrogen-rich gas, and subsequently hydrogen is consumed by the electrochemical processes in the fuel cell. The latent heat of the fuel cell tail gas, that is, the anode off-gas, is used for recuperation purposes, whereby the flammable compounds in the anode off-gas must be burned completely in a separate device. Possible energy carriers are natural gas and liquid fuels, such as LPG, gasoline, kerosene, and diesel on the fossil side, and biogas, bioethanol, and biodiesel on the renewable side. These energy carriers or fuels have to be mixed with air and/or steam before entering the first stage of fuel processing. Liquid fuels and water, if required by the process, must be evaporated before they enter the mixing device. Owing to the high number of energy carriers, a broad variety of system options exist. Therefore, different methods must be applied to determine compound properties.

The product gas in the first stage of fuel processing, that is, reforming, contains only a few species – mainly hydrogen, steam, nitrogen, carbon monoxide, carbon dioxide, and methane – in different compositions depending on process variation and energy carrier. The species that occur in all subsequent processes are the same, but their concentrations change with the chemical reactions applied stepwise. Figure 9.1 shows a decision tree for determining the thermodynamic quantities of specific volume (v) and enthalpy (H). In advance of the explanations in the next section, a list of process streams, of reaction conditions, and of the pre-selected hardware will be outlined first. If the stream consists of

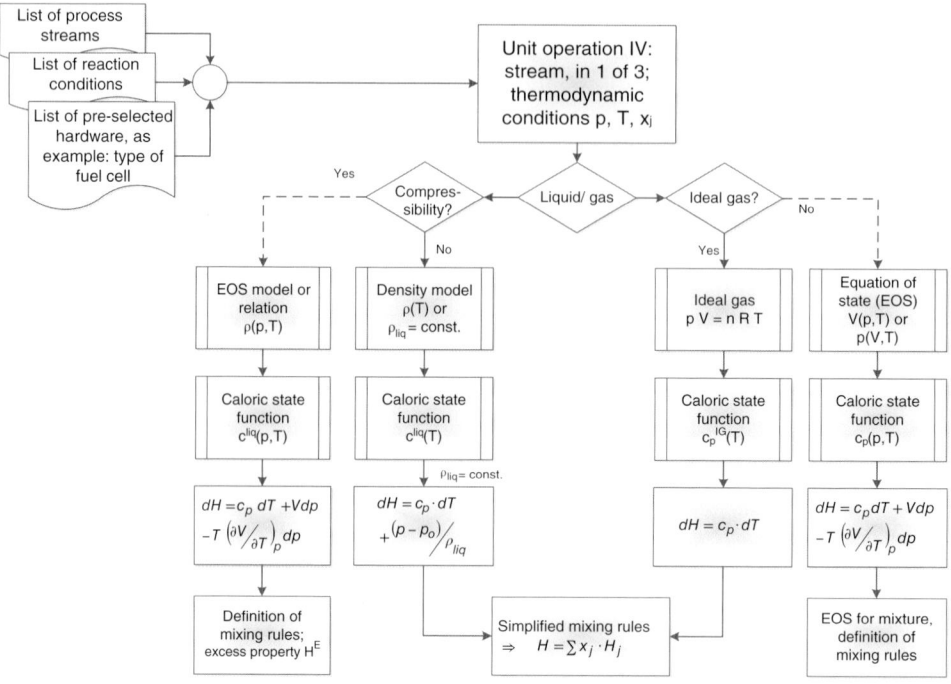

Figure 9.1 Decision tree for determination of the thermodynamic quantities of specific volume (v) and enthalpy (H).

gaseous compounds, it must be decided whether an ideal gas can be assumed for the single species and for the mixture of these species. Otherwise, an equation of states (EOS) must be applied. This leads to a high degree of complexity in all subsequent calculation steps. In the case of the reformate composition, the ideal gas state can be assumed. The same path exists for incompressible fluids, here liquids. Compressible fluids also demand an EOS, whereby specific volumes are often less accurate than those for the gaseous state. The EOS for mixtures, in particular, leads to a high level of complexity in regard to numeric solutions, even for simple mixing rules. The ideal gas state has the opposite effect. Therefore, the ideal gas state and incompressible liquids should be assumed wherever possible.

To verify whether EOS is really required, the deviation from the ideal gas law, expressed by the fugacity coefficient (φ), should be determined. The EOS applied here is taken from a proposal by Redlich–Kwong–Soave [15]. The EOS parameters can be determined from critical data, such as critical pressure and critical temperature of the pure substance. Non-idealities increase with decreasing temperature and increasing pressure. Three gas mixtures should be checked: (i) the product gas of a low-temperature water–gas shift reactor (LT-WGS) at 1.5 bar and 300 °C, (ii) a mixture of steam and dodecane ($C_{12}H_{26}$) ($H_2O/C = 1.8$) after mixing at 400 °C and elevated pressure of 4 bar, and (iii) a methanol–water mixture ($H_2O/C = 1.3$) at 280 °C and 10–20 bar. The calculation scheme for applying the EOS to the fugacity coefficients of pure components and especially to mixtures is quite complex. Detailed information can be found elsewhere [16]. The single species CO, CO_2, H_2, CH_4, N_2, and H_2O in a LT-WGS reach fugacity coefficients in the range 0.997–1.0007 with the highest value for CO and the lowest for steam. The mixture has a fugacity coefficient of 1.00004. Regarding the second mixture, fugacity coefficients of single-species steam and dodecane are 0.9950 and 0.9320, respectively. The mixture studied with a ratio $H_2O/C = 1.8$ has a fugacity coefficient of 0.9937, which is near that of pure steam. This results from the very low molar concentration of dodecane of about 4.45% (vol.), which is typical for such mixtures. Methanol and water are both high-polar liquids, which also leads to molecular interactions in the gaseous phase and to non-idealities. The fugacity coefficient decreases from 0.997 at a pressure of 1 bar, up to 0.972 at 10 bar, and to 0.945 at 20 bar. Nevertheless, the assumption of the ideal gas state can be made for reformate. Even for gaseous methanol–water mixtures, calculations with the ideal gas state lead to good approximations.

An empirical equation to determine the ideal gas heat capacity can be found in *Perry's Chemical Engineering Handbook* [17] (see Eq. (9.1)). For a huge number of organic and inorganic compounds, parameters C_1–C_5 are tabled. These values are an outcome of a DIPPR project (Design Institute for Physical Properties). Enthalpy and entropy can be determined by integrating c_p and c_p/T. Usage of the sinh function enables reliable curve behavior in the complete range of validity, and beyond that to a certain extent. Unfortunately, progressive integration methods must be applied to derive formula for enthalpy and entropy. Equations (9.2) and (9.3) are given to avoid unnecessary mathematical effort:

$$c_p^{IG} = C_1 + C_2 \left[\frac{C_3/T}{\sinh(C_3/T)} \right]^2 + C_4 \left[\frac{C_5/T}{\sinh(C_5/T)} \right]^2 \tag{9.1}$$

$$\overline{H}^{IG} = \overline{H}_o^{IG} + C_1 \cdot T + C_2 \cdot C_3 \cdot \coth(C_3/T) - C_4 \cdot C_5 \cdot \tanh(C_5/T) \tag{9.2}$$

$$\overline{S}^{IG} = \overline{S}_o^{IG} + C_1 \cdot \ln T + C_2 \cdot (C_3/T \cdot \coth(C_3/T) - \log(\sinh(C_3/T))) \\ - C_4 \cdot (C_5/T \cdot \tanh(C_5/T) - \log(\cosh(C_5/T))) - R \cdot \ln\left(\frac{p}{p_o}\right) \tag{9.3}$$

All enthalpy and entropy values demand an integration constant $\overline{H}_o^{IG}$. The constant $\overline{H}_o^{IG}$ does not correspond to the value of standard enthalpy of formation at 25 °C, $\Delta \overline{H}_f(298\,K)$. Tables of standard enthalpies of formation, $\Delta \overline{H}_f(298\,K)$, and standard entropies of formation, $\Delta \overline{S}_f(298\,K)$, are available [17]. Standard enthalpies of formation are usually determined by heat-of-combustion measurements [16] from heating values $\Delta \overline{H}_C(298\,K)$. Klotz and Rosenberg describe an approximation method to calculate standard enthalpies of formation using tabulated bond energies [18]. As an example, a value of −272.33 kJ mol^{-1} was calculated by bond energies for dodecane ($C_{12}H_{26}$), while the exact value is −290.72 kJ mol^{-1}. The integration constant $\overline{H}_{j,o}^{IG}$ for each compound j in a mixture can be determined by setting $\overline{H}_{j}^{IG} = \overline{H}_{j,f}(298\ K)$ in Eq. (9.2) to 298 K. The same method must be applied for $\overline{S}_{o}^{IG}$ in Eq. (9.3).

If a component is not available in standard handbooks, progressive approximation methods can be applied. In the case of oleic acid methyl ester ($C_{19}H_{36}O_2$), which is the main component in biodiesel, there are no heat capacity data available in Reference [17]. The method of Joback and Reid [19] leads to a heat of formation of $\Delta \overline{H}_f(298\ K) = 563.1$ kJ mol^{-1} for the gas phase while a value of $\Delta \overline{H}_f(298\ K) = 726.87$ kJ mol^{-1} can be found for the liquid phase [20]. If a solid material, such as ZnO, is applied as sorbent material, corresponding tables can be found in References [21,22] for ZnO and ZnS. Furthermore, thermodynamic data such as standard enthalpy of formation can also be found in the *NIST Chemistry WebBook* [23].

9.3.2
Chemical Equilibrium

Chemical reactions are restricted by chemical equilibrium. Two limiting cases exist:

1) Chemical equilibrium is on the educt side: no reaction takes place at all.
2) Chemical equilibrium is on the product side: The chemical reaction is not limited and will be stopped only if one educt is consumed completely. Combustion reactions show this behavior at air ratios higher than 1.

In all other cases, chemical equilibrium data must be generated to calculate maximum conversion until equilibrium is reached. The thermodynamic quantity for evaluating phase changes and chemical reactions is Gibbs energy (*G*). In general, it is determined by:

$$G = H - T \cdot S \tag{9.4}$$

It can be calculated from molar specific enthalpy and entropy data of each species j by Eqs. (9.2) and (9.3) applying linear mixing rules (see Figure 9.1). The gradient of chemical reactions is described by conversion or by reaction numbers (λ_i) for individual reactions i. The reaction number is defined by:

$$\lambda = \frac{n''_j - n'_j}{\nu_j} \tag{9.5}$$

whereby ' marks the beginning of reaction, " the end of reaction, and ν_j the stoichiometric coefficient, that is, the number of moles j going into the reaction.

9.3 Chemical Engineering Thermodynamics

Equilibrium is given for a minimization of the change in Gibbs energy (dG) during the reaction:

$$\frac{dG}{d\lambda} = 0 \tag{9.6}$$

In practice, the difference in Gibbs energy between educt mixture and product can be calculated as integral change as:

$$\frac{\Delta G}{n''} = \overline{G}'' - {n'}/{n''} \cdot \overline{G}' \tag{9.7}$$

This rather complex procedure can be simplified by applying the law of mass action [16]. An equilibrium constant K_i at constant pressure is defined for a chemical reaction i. The temperature dependence of K_i is described by an Arrhenius equation. In regard to fuel processing in fuel cells, three reactions are determined by chemical equilibrium: carbon formation, methane formation, and the water–gas shift reaction. A complex mechanism underlies the first reaction, and should be avoided by choosing suitable operation parameters. Equilibrium calculations provide pointers for these parameters. Other effects can also cause carbon formation in a region near the critical line drawn by chemical equilibrium. More information on carbon deposition can be found in References [13,24,25]. If methanation and the water–gas shift reaction both occur, they must be considered in parallel. The law of mass action for an ideal gas can be expressed in this case as:

$$K_{\text{Methane}}(T) = \frac{y_{\text{CH}_4}^{\text{eq}} \cdot y_{\text{H}_2\text{O}}^{\text{eq}}}{y_{\text{CO}}^{\text{eq}} \cdot (y_{\text{H}_2}^{\text{eq}})^3} \cdot \left(\frac{p_o}{p}\right)^2 \tag{9.8}$$

$$K_{\text{WGS}}(T) = \frac{y_{\text{CO}_2}^{\text{eq}} \cdot y_{\text{H}_2}^{\text{eq}}}{y_{\text{CO}}^{\text{eq}} \cdot y_{\text{H}_2\text{O}}^{\text{eq}}} \tag{9.9}$$

Molar concentrations can be determined by two different methods: applying reaction numbers or using balance equations for atoms a: C, O, H, and N. For a given set of species j – for example CO, CO_2, CH_4, O_2, N_2, H_2O, and H_2 – four balance equations should be defined according to:

$$\sum N_{ja} \cdot n'_j = \sum N_{ja} \cdot n''_j \tag{9.10}$$

For completion, $C_m H_n O_l$ should be added to the list of occurring species. The addition of oxygen to the chemical formula of fuel enables an application to hydrocarbons $C_m H_n$, and also to alcohols such as methanol (CH_3OH). If long-chain hydrocarbons $C_n H_m$ are processed, oxygenates $C_q H_p O_o$ can be formed. The educt composition must be known. If eight components are considered, four quantities still must be determined using additional information – for example, that oxygen is consumed completely by oxidation reactions and that fuel is converted without residues into gas. The remaining two quantities could be calculated by Eqs. (9.8) and (9.9) in the case of chemical equilibrium. A more detailed analysis will follow after a discussion of fuel processing reactions.

9.4
Unit Operations

In regard to fuel processing for fuel cells, three groups of unit operations will be discussed in the next section, namely, catalytic reactors, separation devices, and balance-of-plant components. Important equations for their design are given, and the challenges for single units are outlined. This information will then be summarized in the next section to form first subsidiary systems.

9.4.1
Catalytic Reactors

Hydrogen generation from hydrocarbon fuels requires chemical process units. In these units, hydrogen is formed by the core reaction, namely, reforming. In addition to hydrogen, other gases are produced, which must be separated, reduced, or completely eliminated by subsequent process units before the reformate enters the fuel cell. Such units or reactors can be designed as isothermal reactors with cooling or heating devices or as adiabatic reactors with no heat exchange. Chemical reactions can occur completely in a homogenous gas phase. Homogeneous reactions are only possible at high temperatures, which are necessary to overcome thermophysical barriers hindering spontaneous reactions at arbitrary conditions. Catalysts facilitate chemical reactions at moderate or lower temperatures. They do not change the final composition of the product. Owing to their material properties, they influence the reaction path of the chemical reaction. Therefore, catalysts are used in most reactors for fuel processing in fuel cell systems. More information on heterogeneous catalysis and catalytic materials in fuel processing can be found elsewhere [13]. Hydrogen generation must proceed as a continuous process and is mostly performed in tube reactors with or without inner construction elements. The catalyst is implemented in the chemical reactor as a fixed bed, catalyst particles, or by coated devices. These latter devices can be metallic sheets, structured plates, or ceramic monoliths. Before the catalytically active material is added, a washcoat is coated on the device to enable adherence and a good catalyst distribution on and in the structure. Chemical reactions in the gas phase take place in the catalytic reactor at residential times between several milliseconds to seconds. In regard to the compactness of chemical reactors in mobile fuel cell systems, residential times of 50–200 ms must be reached in the single, fuel processing components.

The velocity of a chemical reaction is determined by reaction kinetics. Kinetics are more complex to define when catalysts are used in fuel processing. In principle, a reaction rate only describes the chemical process on the active site of the catalyst. However, the process on a catalyst also consists of mass transport to and from the active side. These transport phenomena include convective and diffusive mechanisms outside and inside of the catalyst. A process rate that is an apparent reaction rate accounts for this. Another possibility for reactor design is the usage of the gas hourly space velocity (GHSV). The GSHV is the reciprocal value of mean residence time. It is the ratio of volume flow to catalyst volume – including the solid materials. Reactor design is possible when GHSVs are known. GHSVs can be determined from experience, from information provided by the catalyst manufacturer, or by applying kinetic models.

9.4.1.1 Hydrogen Generation

Hydrogen can be generated using different conversion technologies for natural gas, alcohols, gasoline or diesel. A hydrocarbon fuel C_nH_m is converted into a hydrogen-rich gas by steam reforming according to:

$$C_nH_m + nH_2O \rightleftharpoons nCO + \left(\frac{m}{2} + n\right)H_2 \qquad (\Delta H > 0 \text{ kJ mol}^{-1}) \qquad (9.11)$$

The reaction enthalpy is positive, leading to a heat requirement for steam reforming, which can be realized by catalytic combustion. This process is described in Section 9.4.1.5. Partial oxidation (POX) uses air as an oxygen source and has a negative reaction enthalpy. A drawback of POX is nitrogen dilution. If the reaction is performed in adiabatic reactors, nitrogen damps the resulting temperature peak of the POX reaction:

$$C_nH_m + \tfrac{n}{2}O_2 \rightleftharpoons nCO + \tfrac{m}{2}H_2 \qquad (\Delta H_R < 0 \text{ kJ mol}^{-1}) \qquad (9.12)$$

Autothermal reforming constitutes a compromise between steam reforming and partial oxidation. POX provides the heat demanded by steam reforming. For example, a diesel–steam mixture of 625 K is mixed with preheated air of 400 K and ignited on the catalyst. POX occurs immediately and the temperature increases to 1150 K. Owing to endothermic steam reforming, the temperature decreases under adiabatic conditions to 995 K. Different reaction kinetics for POX, steam reforming, and the water–gas shift reaction lead to a significant temperature profile in the reformer.

Two side reactions must also be taken into account, that is, the water–gas shift reaction (WGS) in Eq. (9.13) and methane formation Eq. (9.14), which is the opposite of methane steam reforming:

$$CO + H_2O \rightleftharpoons CO_2 + H_2 \qquad (\Delta H_{R,1023.15\text{ K}} = -34.6 \text{ kJ mol}^{-1}) \qquad (9.13)$$

$$CO + 3H_2 \rightleftharpoons CH_4 + H_2O \qquad (\Delta H_{R,1023.15\text{ K}} = -225.3 \text{ kJ mol}^{-1}) \qquad (9.14)$$

If the contact time of the gas molecules at the active sites of the catalyst is long enough, the gas composition in the product gas will be in chemical equilibrium.

The chemical formula in Eq. (9.11) can also be adapted to oxygen-containing fuels, such as biodiesel (FAME: fatty acid methyl ester) or alcohols:

$$C_nH_mO_l + (n-l)H_2O \rightleftharpoons nCO + \left(\frac{m}{2} + n - l\right)H_2 \quad (\Delta H_R > 0 \text{ kJ mol}^{-1}) \quad (9.15)$$

$$C_nH_mO_l + \frac{(n-l)}{2}O_2 \rightleftharpoons nCO + \frac{m}{2}H_2 \quad (9.16)$$

FAME from rapeseed oil consists mainly of oleic acid methyl ester $C_{19}H_{36}O_2$. For example, the steam reforming of oleic acid methyl ester results in:

$$C_{19}H_{36}O_2 + 17H_2O \rightleftharpoons 19CO + 35H_2 \quad (\Delta H_{R,923.15\,K} = 2734.6 \text{ kJ mol}^{-1}) \quad (9.17)$$

and partial oxidation of ethanol (C_2H_5OH) in:

$$C_2H_6O + \tfrac{1}{2}O_2 \rightleftharpoons 2CO + 3H_2 \quad (\Delta H_{R,1073.15\,K} = 30.6 \text{ kJ mol}^{-1}) \quad (9.18)$$

Notably, pure POX of ethanol is weakly endothermic. Simultaneously, methanation occurs leading to an overall reaction scheme with a certain heat release.

Methanol is an exception in regard to CO formation. Carbon monoxide is formed by the decomposition of methanol together with two molecules of hydrogen. Another mechanism for hydrogen production from methanol can be derived by experimental investigations. Methanol decomposition occurs only in the first millimeters of a fixed catalyst bed. The main part of methanol reacts via steam reforming according to:

$$CH_3OH + H_2O \rightleftharpoons CO_2 + 3H_2 \quad (\Delta H_{R,553.15\,K} = 59.8 \text{ kJ mol}^{-1}) \quad (9.19)$$

Carbon monoxide is formed in the downstream part of the catalyst bed by the reverse water–gas shift reaction (WGS in Eq. (9.13)).

The educt mixture of a reformer is defined by the ratios of steam to carbon (S/C) and of oxygen to carbon (O_2/C):

$$S/C = \frac{n_{H_2O}}{n_C} = \frac{\dot{n}_{H_2O}}{n \cdot \dot{n}'_{C_nH_m}} \quad (9.20)$$

$$O_2/C = \frac{n_{O_2}}{n_C} = \frac{\dot{n}_{O_2}}{n \cdot \dot{n}'_{C_nH_m}} \quad (9.21)$$

Several papers use the ratio O/C instead of O_2/C, that is, O/C = $2O_2$/C. Conversion of the original fuel should be as high as possible. During catalytic screening, a catalyst is chosen based on activity and selectivity at incomplete conversion, that is, $\zeta = 0.6$–0.9. The conversion here can be determined by:

$$\zeta_{C_nH_m} = \frac{\left(\dot{n}''_{CO} + \dot{n}''_{CO_2} + \dot{n}''_{CH_4}\right)}{n \cdot \dot{n}'_{C_nH_m}} \quad (9.22)$$

At high conversion, that is, $\zeta > 0.95$, this method fails due to inaccuracies in the experimental values [14]. In such cases, it is better to determine the residual original fuel:

$$\zeta_{C_nH_m} = 1 - \frac{\dot{n}''_{C_nH_m}}{\dot{n}'_{C_nH_m}} \tag{9.23}$$

However, notably, the residual fuel in many experiments differs from the original fuel due to cracking and formation processes. In addition, oxygen-containing hydrocarbons can occur as an intermediate or as a byproduct. The fuel composition C_nH_m then changes to an average number $C_qH_pO_o$. Furthermore, Eqs. (9.15) and (9.16) include oxygen-containing feedstock:

$$\zeta_C = 1 - \frac{q \cdot \dot{n}''_{C_qH_pO_o}}{n \cdot \dot{n}'_{C_nH_mO_l}} \tag{9.24}$$

The residual fuel can be found as a gas component, for example, as propene C_3H_8, as a liquid component in an organic liquid phase ($C_{14}H_{28}$ etc.), or as acetone (C_3H_6O) dissolved in an aqueous phase mainly consisting of excess water from reforming. Instead of analyzing each individual component, the overall composition can be defined for the single thermodynamic phases or for the complete mixture:

$$\dot{n}''_{C_qH_pO_o} = \dot{n}''^g_{C_{q1}H_{p1}O_{o1}} + \dot{n}''^{l,org}_{C_{q2}H_{p2}O_{o2}} + \dot{n}''^{l,aq}_{C_{q3}H_{p3}O_{o3}} \tag{9.25}$$

All occurring species can be evaluated by balance equations for the atoms of carbon (C), oxygen (O), hydrogen (H), and nitrogen (N) (see Eq. (9.10)). Finally, the resulting relations can be used to calculate the concentrations of all components in the gas phase. These balanced values can be compared with the later experiments. Another outcome of balancing is a relation between hydrogen and carbon monoxide in the product gas flow to the carbon in the original fuel flow. Carbon monoxide is taken as a desired product because CO can be converted into the target product of hydrogen by WGS:

$$\frac{\dot{n}''_{H_2} + \dot{n}''_{CO}}{n(\dot{n}'_{C_nH_m} - \dot{n}''_{C_nH_m})} = \zeta_{C_nH_m} \left\{ \frac{m}{2} + n\left[2 - 4\sigma_{CH_4} - 2\left(\frac{O_2}{C}\right) - \sum_j \left(\frac{p}{2} - o\right) \cdot \sigma^{(j)}_{C_qH_pO_o}\right] \right\} \tag{9.26}$$

In process engineering, it is useful to define selectivities for the main products and byproducts. The main product is the modified hydrogen yield (see Eq. (9.26)) and the byproducts are defined in Eqs. (9.27)–(9.29) for methane, carbon dioxide, and higher hydrocarbons, respectively:

$$\sigma_{CH_4} = \frac{\dot{n}''_{CH_4}}{n(\dot{n}'_{C_nH_m} - \dot{n}''_{C_nH_m})} = \frac{\dot{n}''_{CH_4}}{n\zeta_{C_nH_m} \cdot \dot{n}'_{C_nH_m}} \tag{9.27}$$

$$\sigma_{CO_2} = \frac{\dot{n}''_{CO_2}}{n \cdot (\dot{n}'_{C_nH_m} - \dot{n}''_{C_nH_m})} = \frac{\dot{n}''_{CO_2}}{n \cdot \zeta_{C_nH_m} \cdot \dot{n}'_{C_nH_m}} \quad (9.28)$$

$$\sigma^{(j)}_{C_qH_pO_o} = \frac{\dot{n}''^{(j)}_{C_qH_pO_o}}{n \cdot (\dot{n}'_{C_nH_m} - \dot{n}''_{C_nH_m})} = \frac{\dot{n}''^{(j)}_{C_qH_pO_o}}{n \cdot \zeta_{C_nH_m} \cdot \dot{n}'_{C_nH_m}} \quad (9.29)$$

This product yield reaches an optimum at complete conversion and for suppressed formation of byproducts $C_qH_pO_o$:

$$\zeta_{C_nH_m} = 1; \sigma^{(j)}_{C_qH_pO_o} = 0 \quad (9.30)$$

$$\frac{\dot{n}''_{H_2} + \dot{n}''_{CO}}{\dot{n}'_{C_nH_m}} = \frac{m}{2} + n\left[2 - 4\sigma_{CH_4} - 2\left(\frac{O_2}{C}\right)\right] \quad (9.31)$$

A maximum yield can be reached for pure steam reforming ($O_2/C = 0$) and for reduced methane formation. Reaction conditions also contribute to reduced methane formation. Methane reaches its highest value at equilibrium, which can be calculated by the law of mass action (see Eq. (9.8)). A detailed process analysis for hydrogen generation will follow in the next section. Different processes will be evaluated concerning hydrogen content, temperatures, potential for carbon deposits, and subsystem complexity.

9.4.1.2 Water–Gas Shift Reaction

The water–gas shift reaction (WGS) is a well-known basic process in the chemical industry [26]. Equation (9.32) repeats the chemical reaction (9.13) with the calculated heat of reaction for the relevant temperature region. The reaction is only weakly exothermic:

$$CO + H_2O \rightleftharpoons CO_2 + H_2 \quad (\Delta H_{R,573.15\,K} = -39.1\,\text{kJ mol}^{-1}) \quad (9.32)$$

The advantage of WGS is the production of hydrogen and simultaneous decrease in carbon monoxide. As can be seen from Eq. (9.32), WGS demands steam. Therefore, by injecting water into WGS, the partial load of steam can be increased. Conversion is limited due to chemical equilibrium at high temperatures (HTs) and is determined by low catalytic activities at low temperatures (LTs). Figure 9.2 shows CO conversion in dependence on temperature for two different process routes in relation to chemical equilibrium data ($H_2O:C = 1.3-2.5$). The product gas leaves an autothermal reformer for example at nearly 1000 K with a CO concentration of about 7.6% (vol.). The molar ratio of steam to carbon monoxide of the product gas is 1.3. For such a mixture, chemical equilibrium limits CO conversion to 80% at 623 K. To reach a WGS outlet concentration of 1%, a conversion of 87% is required. Lower temperatures, such as 573 K, offer a slightly higher conversion, but the reaction velocity is rather slow near equilibrium. Two process routes are shown in Figure 9.2 that meet the challenges of WGS reactor design. The gray crosses split the chemical

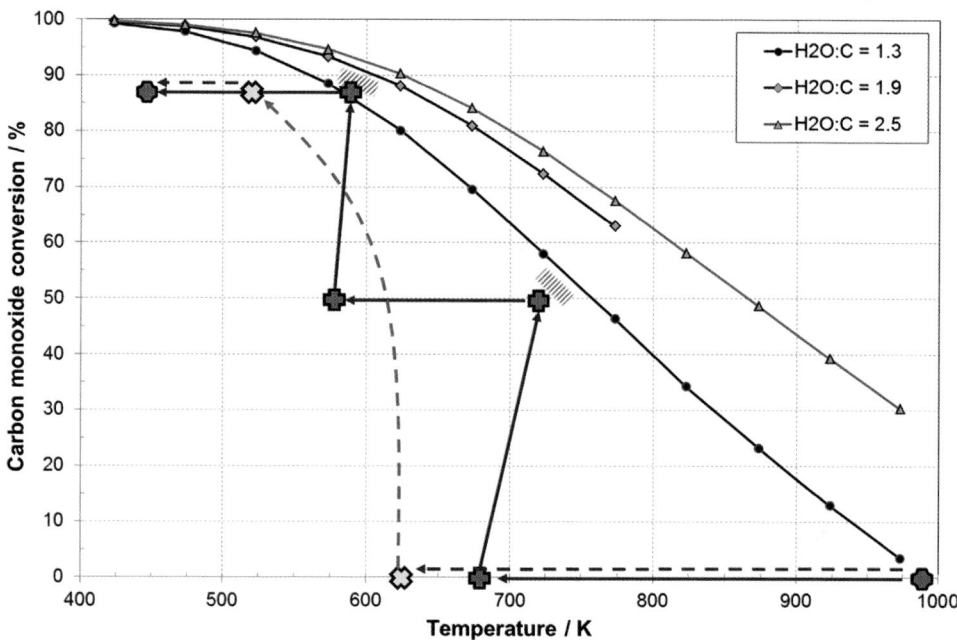

Figure 9.2 CO conversion in dependence on temperature for two different process routes in relation to chemical equilibrium data ($H_2O:C$: 1.3–2.5).

reaction into two adiabatic reactor stages with inlet temperatures of 673 K (HT-WGS) and 573 K (LT-WGS). Owing to the heat of reaction, the temperatures increase by about 50–60 K in HT-WGS and by about 20–30 K in LT-WGS. The shift reaction in HT-WGS is designed for a conversion of 50% to avoid a strong negative influence of equilibrium. GHSVs are in the range 75 000–90 000 h^{-1} at 673–733 K. Intermediate cooling is realized by water injection. This measure increases the molar steam-to-carbon ratio to 1.9. After LT-WGS, the product gas with a CO content lower than 1% (vol.) must be cooled to 453–473 K before entering a HT-PEFC or a PROX reactor in a PEFC system. The GHSV of LT-WGS is chosen as 45 000 h^{-1} to reach an overall GHSV of about 30 000 h^{-1}. As an alternative process, the reformate can be cooled to 623 K. Water injection or steam supply leads to a steam-to-carbon ratio of 2.5. The WGS reactor operates in a cooled mode, leading to an outlet temperature of 523 K. A detailed reactor design will ultimately decide which system is more compact and better to operate.

Commercial catalysts are available at high temperatures above 623 K and at low temperatures below 473 K. Suitable catalysts include those within the group Ib metals, the group VIII metal oxides, and metal sulfides. The demand for sufficient stability restricts the choice of catalysts. The materials used are iron oxide (Fe_3O_4) and iron sulfide (FeS) for the high-temperature range and copper for low temperatures. Fe_3O_4 is a typical high-temperature catalyst for temperatures above 350 °C. In the low-temperature range, metallic copper is still catalytically active below 200 °C, but the finely dispersed copper particles are sensitive to thermal sintering at higher temperatures. In addition, a copper catalyst must be kept under a non-oxidative atmosphere. In regard to automotive fuel cell applications, precious metal catalysts and non-pyrophoric base metal catalysts were developed in the past [27–31]. The temperature range for most fuel cell applications is between 523 K and 673 K.

9.4.1.3 Preferential Oxidation

Carbon monoxide is undesired in low-temperature fuel cells of type PEFC. It blocks the electrochemically active sites on the anode catalysts, and must therefore be removed from the anode fuel gas. Preferential oxidation makes use of catalysts, which suppress undesired hydrogen oxidation:

$$CO + \tfrac{1}{2} O_2 \rightleftharpoons CO_2 \quad (\Delta H_{R, 473\,K} = -283.6 \text{ kJ mol}^{-1}) \quad (9.33)$$

$$H_2 + \tfrac{1}{2} O_2 \rightleftharpoons H_2O \quad (\Delta H_{R, 473\,K} = -243.6 \text{ kJ mol}^{-1}) \quad (9.34)$$

The selectivity for reaction (9.33) is in the range 40–60% [32–36]. It can be defined as:

$$\sigma_{CO} = \frac{\dot{n}'_{CO} - \dot{n}''_{CO}}{(\dot{n}'_{CO} - \dot{n}''_{CO}) + (\dot{n}'_{H_2} - \dot{n}''_{H_2})} \quad (9.35)$$

In this case, a catalyst with high activity and selectivity for CO oxidation is required to suppress undesirable H_2 oxidation. Catalytic materials such as gold reach a selectivity of about 80% at an operation temperature of 80–120 °C, but the highest selectivity is gained at an elevated partial pressure of CO. At a low partial pressure of CO, that is, at small CO concentrations, selectivity drops to 20% [33,36]. Unfortunately, such a performance is disadvantageous in terms of counteracting high concentrations at elevated temperatures of 200–250 °C and residual amounts at stack temperature.

Depending on the performance of the water–gas shift reactor, the inlet CO concentration is 0.5–1% (vol.). The desired CO concentration varies between 10 and 30 ppmV. The process is highly exothermic. Intermediate cooling is required between adiabatic reactor stages or cooled reactor devices. Increased operation temperatures lead to lower selectivity, which forces hydrogen oxidation and further increases the temperature. Temperature control and the avoidance of hot spots are therefore important challenges for PROX reactor design. Although oxidant supply necessitates additional process and regulation expenditure, it also facilitates reaction control by restricting the O_2 educt. Importantly, a good catalyst suppresses methanation even at high temperatures of about 250 °C [35].

Catalyst screening for PROX catalysts often reveals a small operation window in terms of temperature. Candidate catalysts are usually noble metals, such as Pt, Pd, Rh, or Ru, but Cu/Co, Cu/Cr, and Ni/Co/Fe mixtures, as well as Ag, Cr, Fe, and Mn, are also being studied as catalyst materials [37]. Broader operation temperature ranges are possible for Pt catalysts [37] and for Pt-Ru catalysts impregnated on mixed oxides (MnO_2-CuO) [38]. Gold catalysts are active at low temperatures of 80 °C and are well suited for a PROX stage in front of or integrated in a PEFC [36]. Unfortunately, gold catalysts deactivate more rapidly than corresponding Pt catalysts [34]. Ceria-supported platinum or iridium catalysts are particularly suitable for CO removal at high hydrogen concentrations. Ceria-based catalysts are often used as supports for three-way catalysts in exhaust gas cleaning. These supports operate as oxygen buffers, and provide oxygen even during lean combustion operation [34]. Core–shell nanoparticles of Ru-Pt offer a low light-off temperature of 30 °C [39].

The effort and the complexity of a PROX reactor should be minimized. The number of PROX stages varies between one and three [32]. One stage is only possible if the CO concentration is lower than 1% (vol.). Ahluwalia *et al.* [32] have conducted reactor design studies and corresponding experiments for CO inlet concentrations between 1% and 3% (vol.). The adiabatic stages were interrupted by intermediate cooling to remove the heat of reaction. The inlet temperature for each stage was 100 °C. In a one-stage design, a temperature increase up to 280 °C was observed at an inlet concentration of 1.3% (vol.) CO and an air ratio of 2. For a compact reactor design, a GHSV of 30 000–80 000 h^{-1} is necessary for a single stage. More information on the applied catalyst is available in Korotkikh and Farrauto [35]. The overall selectivity can be improved by adding further stages, for example, by changing from a two-stage to a three-stage design. Another option is reducing the inlet CO concentration to 0.5% (vol.). This leads to a larger low-temperature shift stage. Finally, an optimized process design for complete CO removal must be found with regard to reactor volume, catalyst mass, and cost.

9.4.1.4 Selective CO Methanation

An alternative process to PROX is the methanation of carbon monoxide according to Eq. (9.36). It must be selective in relation to the methanation of carbon dioxide. Both processes are exothermic. It is important to cool the reactor. In comparison to PROX, temperature control cannot be realized by controlling an educt flow such as oxygen.

$$CO + 3H_2 \rightleftharpoons CH_4 + H_2O \quad \left(\Delta H_{R,473\,K} = -213.3 \text{ kJ mol}^{-1}\right) \quad (9.36)$$

$$CO_2 + 4H_2 \rightleftharpoons CH_4 + 2H_2O \quad \left(\Delta H_{R,473\,K} = -173.2 \text{ kJ mol}^{-1}\right) \quad (9.37)$$

Considering thermodynamic equilibrium, CO methanation is preferred over CO_2 methanation, which can be expressed as a selectivity for CO methanation:

$$\sigma_{CO} = \frac{\dot{n}'_{CO} - \dot{n}''_{CO}}{\left(\dot{n}'_{CO} - \dot{n}''_{CO}\right) + \left(\dot{n}'_{CO_2} - \dot{n}''_{CO_2}\right)} \quad (9.38)$$

In contrast to the use of methanation in industrial process technology [40,41], Eq. (9.38) displays the challenge associated with gas purification by methanation in fuel processing for PEFCs to exclusively hydrogenate CO into CH_4 without consuming additional hydrogen produced by CO_2 hydrogenation. Methanation of CO_2, in particular, can increase the risk of hotspots or lead to an out-of-control process. Only a few papers have considered methanation for gas cleaning in fuel cells [42–44]. Methanation causes higher hydrogen losses, particularly at partial load. It can be applied in stationary systems [44]. Methane will not affect the electrochemistry of the fuel cell but it must be burned as part of the system tail gas. Owing to the high stability of methane, its concentration in the tail gas should be limited to 1–2%.

Suitable catalysts are to be found, in particular, among the transition metals of group VIII and among the noble metals. Pichler [41] identified Ru, Ni, Co, Fe, and Mo as suitable materials (listed in order of decreasing activity). Vannice [40] reported that Ni/Al_2O_3 catalysts are 85% less active and 35% more selective than Ru/Al_2O_3 catalysts. Methanation of CO on Ru/Al_2O_3 and on Co/Al_2O_3 has been investigated in combination with the WGS and PROX activity of the catalysts [42,44]. The concentrations of CO_2 and H_2O affect catalyst activity [44] and also determine whether WGS or methanation occur [42]. Finally, a product gas concentration of 100 ppmv CO, a GHSV of $13\,500\,h^{-1}$, and H_2 conversion of 10–30% are unsuitable for PEFC application [43].

9.4.1.5 Catalytic Combustion

The tail gas of a fuel cell contains unconverted hydrogen, and, depending on the fuel cell type and the fuel processing, methane. Hydrogen and carbon monoxide oxidation have already been described in Section 9.4.1.3 under the heading of preferential oxidation. Methane is the most difficult species to convert in a burner. Emissions should be limited due to its high greenhouse effect, which is 25-times higher than that of carbon dioxide:

$$CH_4 + 2O_2 \rightleftharpoons CO_2 + 2H_2O \quad (\Delta H_{R,773\,K} = -792.8\,\text{kJ mol}^{-1}) \quad (9.39)$$

$$CO + \tfrac{1}{2}O_2 \rightleftharpoons CO_2 \quad (\Delta H_{R,773\,K} = -283.3\,\text{kJ mol}^{-1}) \quad (9.40)$$

$$H_2 + \tfrac{1}{2}O_2 \rightleftharpoons H_2O \quad (\Delta H_{R,773\,K} = -246.2\,\text{kJ mol}^{-1}) \quad (9.41)$$

Methane oxidation can best be performed on monoliths coated with a catalytically active material, such as Pd or Pt. The required operation temperature is 450–500 °C. Hydrogen ignites on Pt, even at room temperature. The presence of CO could lead to blocking effects and to a shift in ignition temperature to 200 °C [45]. Only a few papers have reported on catalytic burner development for fuel cell systems [46–49]. Thermal power ranges from $230\,W_{th}$ via $6.5\,kW_{th}$ up to $250\,kW_{th}$.

9.4.2
Separation Devices

9.4.2.1 Adsorption Process

One possible physical gas purification process is adsorption. In separating gas mixtures, use is made of the physical effect that large-molecular, heavy, easily liquefiable gases adsorb much more strongly than light not-easily liquefiable gases, such as hydrogen. A high partial pressure of the adsorbate and low temperatures promote adsorption, while a low partial pressure and high temperatures encourage desorption. An adsorbent must be chosen according to capacity and selectivity for the desired adsorbate [50]. Since the capacity of all adsorbents is limited, all adsorption processes are necessarily discontinuous processes. Suitable cyclic variation of the process conditions permits adsorption–desorption cycles, thus ensuring continuous operation. In many industrial applications, the loading and unloading process is most frequently realized by pressure cycling adsorption (PSAPSA), and less frequently by temperature cycling adsorption (TCA) or by cyclically flushing the adsorbers with inert gas.

In addition to numerous other applications [51], hydrogen purification and H_2S removal are two of the main fields using adsorption in fuel cell systems. Majlan *et al.* [52] applied activated carbon to reduce CO from 4000 ppmv to 1.4 ppmv. In parallel, CO_2 was reduced from 5% (vol.) to 7 ppmv. The gap between reformate and inlet PSA must be closed by an enlarged WGS.

Sulfur-containing hydrocarbons, such as thiophene and benzothiophene, are converted in an autothermal reformer into hydrogen sulfide (H_2S). H_2S must be reduced down to the ppmv range for HT-PEFC application and even lower, to several ppbv, for PEFC application. In regard to SOFC systems, uncertainties lead to a broad range of tolerable concentration of 5–100 ppmv [53]. The presence of 1 ppmv H_2S affects WGS in the anode and leads to a remarkable drop in cell voltage [54]. Sulfur poisoning, even at 0.1 ppmv H_2S, causes a performance loss of about 20% at 750 °C within 40 h at a delay of 20 h [55]. Experimentally observed regeneration times are not applicable.

A possible adsorbate for sulfur is ZnO, which reacts with H_2S according to:

$$H_2S + ZnO \rightleftharpoons H_2O + ZnS \quad \left(\Delta H_{R, 623.15\ K} = -173.6\ kJ\ mol^{-1}\right) \quad (9.42)$$

The operation range of a ZnO safeguard bed is between 300 and 400 °C. Its application is thus only reasonable for low-temperature fuel cells.

9.4.2.2 Membrane Process

In principle, membranes can be divided into porous and nonporous membranes. The candidate materials for porous ceramic membranes are silicates, aluminium, or zirconium oxides, as well as active charcoal substrates with a zirconium coating. Industrially produced ceramic membranes have a compact design due to their high flow rates and they are very low-priced. There are, in principle, four different mechanisms of gas separation by microporous membranes: Knudsen diffusion, surface diffusion, capillary condensation with mass transport in the liquid phase, and separation by the molecular sieve effect.

Nonporous polymer membranes are used industrially primarily for the separation of substances with low molar masses by reverse osmosis, for the separation of vapors by vapor permeation, for the separation of liquid mixtures by pervaporation, and for gas separation by gas permeation. Gas permeation is predominantly used to separate and recover hydrogen from synthesis gas loops in the chemical industry and in refineries. In the past, two other processes aside from the electrochemically active membrane were considered for membrane applications in fuel cell technology. Carbon monoxide must be separated from a hydrogen-rich product gas, particularly from the reformate of methanol steam reforming [56,57]. Sulfur-containing hydrocarbons should also be removed from fuel [58].

Hydrogen can be separated from reformate with nonporous metal membranes, such as Pd/Ag membranes [59]. Product purity is very high, that is, 99.9995% (vol.). High pressures of about 10–20 bar are demanded on the feed side of the membrane. Pd/Ag membranes have been tested mainly in combination with methanol steam reforming because compressing liquid energy carriers and water in the liquid state using a fuel pump is rather effective and does not demand much auxiliary power. In the case of gaseous energy carriers, such as natural gas or air supply in POX and ATR reforming, high compression losses prohibit their application in a completed system. A first overview of the variety and progress of manufacturing technologies was given in 1999 by Uemiya [60]. Several factors prevent the widespread application of ceramic-supported Pd/Ag membranes. During activation, process cracks and pinholes are often formed, which lead to an undesired pollution of the product gas. A huge number of research projects are dedicated to producing dense Pd membranes. Tong *et al.* reported on a pinhole-free Pd/Ag membrane on a porous metal support [61]. Hydrogen permeation is affected by several factors such as mass transfer, competitive adsorption, coking, and influence of steam and CO_2, which must be taken into account [62,63]. Carbon dioxide has no inhibition effect on hydrogen permeation at temperatures higher than 220 °C, whereas steam and carbon monoxide demand temperatures higher than 400 °C [62]. WGS plays an important role in these inhibitive effects [63]. Carbon deposits are formed by dehydrogenation of CO_2 or CO. Palladium membranes have also been applied in membrane reactors such as WGS for compactness of the fuel processor design [64]. Hydrogen permeation through Pd membranes can be calculated by Sievert's law. A modification for the inclusion of surface effects has been presented by Vadrucci *et al.* [65]. A review on hydrogen separation membranes including metallic, silica, zeolite, carbon-based, and polymer membranes has been published by Ockwig and Nenoff [59].

Sulfur-containing hydrocarbons can be removed from kerosene by a pervaporation process in the liquid state [66]. Pervaporation can be applied as a one- or two-stage process before fuel spraying and evaporation. Unfortunately, membrane materials are not yet mature. More information can be found in Reference [67].

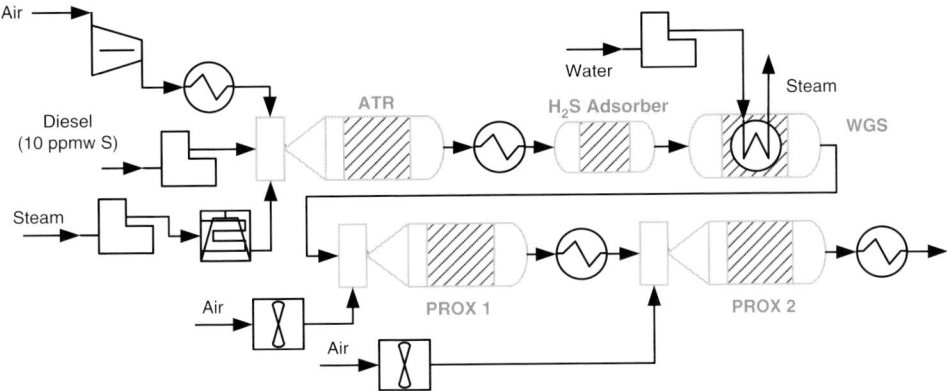

Figure 9.3 Fuel processor subsystem consisting of autothermal reforming of diesel and a gas cleaning unit for PEFC systems.

9.4.3
Balance-of-Plant Components

The balance-of-plant (BOP) components consists of pumps, fans, compressors, and expanders, which charge various media to the chemical reactors, heat exchangers, evaporators, and condensers to prepare these media for ideal reaction conditions. In general, media must achieve the required thermodynamic conditions in regard to pressure and temperature. Single components can be selected only in a system context. Figure 9.3 shows an example of a fuel processor subsystem for the autothermal reforming of diesel and a gas cleaning unit for PEFC systems. This system design is just one of various different designs. Chemical reactors are drawn in gray; important elements are pumps for water, an air compressor, two air fans, and certain heat exchangers. Air is compressed for the air inlet of ATR, while diesel is only charged to the mixing chamber and steam must be delivered by a water pump and a subsequent evaporator. Owing to the different operation temperatures of the chemical reactors, heat exchangers are demanded in-between. Examples are the heat exchangers between the ATR and H_2S adsorber and behind the PROX stages. To present alternative designs, WGS is performed with an evaporative cooling of water.

9.4.3.1 Pumps and Compressors
During the first design phase of basic engineering of a fuel cell system, important decisions must be made. The air supply concept shown in Figure 9.3 is not fixed nor is it the ideal solution for each application. The most important criterion for the development of a supply concept is efficiency. The efficiencies of compressors and expanders are often chosen for process analysis as isentropic efficiencies or they are determined from polytrophic coefficients. Values of 75–80% for isentropic efficiency lead to excellent system efficiencies, but they are rather

challenging to reach in practice. In the next design phase, namely, detailed engineering, pumps and compressors must be selected from procurement lists. Unfortunately, compressors optimized for fuel cell systems are rather rare. Important selection criteria are compactness, reliability, efficiency, power class, load profile, low noise emissions, operating characteristics, and costs.

9.4.3.2 Heat Exchangers and Evaporators

Choosing heat exchangers for a fuel processing unit is often difficult due to their operation temperature. Heat exchangers operating at temperatures between ambient temperatures and 623 K can be bought off-the-shelf. At a range of 623–773 K, heat exchangers must be made of special materials due to certification issues. At higher temperatures above 773 K, heat exchangers are customized products. Therefore, it makes sense to integrate heat exchangers directly into the reactor design and to construct them together with the complete reactor vessel. Instead of a separate diesel evaporator, a nozzle is often used for fuel spraying. More information can be found in specialized papers [68–73]. Equation (9.43) shows the energy balance for a heat exchanger as an open system including heat losses $\dot{Q}_{loss}$ and the option of integrating electric heating P_{el} – for example, for start-up purposes:

$$\sum \dot{n}' \cdot \overline{H}'_i + P_{el} = \sum \dot{n}'' \cdot \overline{H}''_i + \dot{Q}_{loss} \qquad (9.43)$$

9.5
Subsystems of Fuel Processing

In Section 9.4 (Unit operations), all chemical reactors of fuel processing were discussed as single processes. These single processes must be combined in a clever manner to obtain a fuel cell system operated in a well-defined application with well-known fuels. Owing to the variety of systems, only the methods of system design are discussed in this contribution. Aspects of system design are restricted here to the subsystem of fuel processing. Notably, according to the pinch-point method [74], heat exchanger networks are split into two subsystems near the pinch point. These subsystems can be optimized separately below and above the pinch. The pinch of fuel cell systems is usually at the operating temperature of the stack. This is caused by a remarkable amount of heat produced by the electrochemical reaction at an almost constant temperature. At a gross efficiency of 50%, a 10 kW stack delivers 10 kW$_e$ electric power and 10 kW$_{th}$ heat. Low-temperature fuel cell systems such as PEFCs and HT-PEFCs have their pinch at 353 and 433–473 K, respectively. Fuel processing is performed in a temperature range of 423–1273 K for PEFCs and 523–1273 K for HT-PEFCs. Therefore, stack cooling and fuel processing can be optimized separately in regard to heat exchange. SOFC systems have their pinch at the inlet temperature of the stack, that is, 973–1273 K. The heating demands of a gas processor are

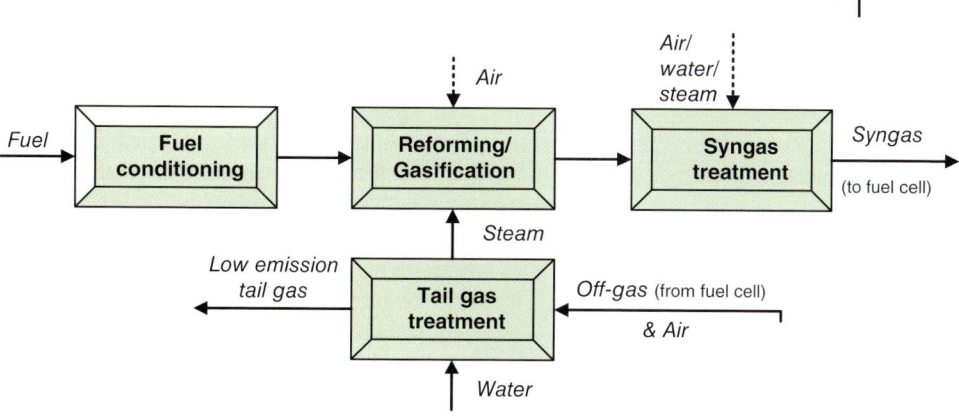

Figure 9.4 Arrangement of fuel processing subsystems with no connecting heat exchangers.

strongly affected by the overall system design. Furthermore, fuel processing for most SOFC systems is carried out as steam reforming of methane or by partial oxidation of diesel. A subsequent gas treatment is not necessary. Therefore, this section is forfeited in low-temperature fuel cell systems.

Figure 9.4 shows an arrangement of fuel processing subsystems with fuel conditioning, reforming, syngas, and tail gas treatment. The subsequent sections here focus on reforming and syngas treatment as the most critical subsystems. Fuel conditioning involves desulfurization, and, for a few applications, rectification of light fuel fractions [58]. Tail gas treatment cleans the fuel cell off-gas, which contains residual amounts of hydrogen, methane, and carbon monoxide. The heat of reaction can be used for steam generation or as a heat source for steam reforming.

9.5.1
Optimization of the Subsystem Reforming

9.5.1.1 Process Selection and Optimization
In the literature, various reformer configurations exist that are modifications of the basic processes of steam reforming (see Eq. (9.10)) and catalytic partial oxidation (CPOX; see Eq. (9.12)). Recent papers reflect the state of the art of the different technologies, including steam reforming [75,76], partial oxidation [77–79], and autothermal reforming [69,80–87]. Autothermal reforming is the most well-known method. It combines the two other basic processes to avoid heat generation by additional combustion processes. Owing to the educt mixture composition, high amounts of nitrogen and steam dilute the product gas.

The advantage of CPOX is its fast response behavior, its short start-up time, its compact design, and the absence of water and thus steam. Pure catalytic partial oxidation, Eq. (9.12), involves the risk of carbon deposits. To avoid a direct water supply and simultaneously improve the process conditions in relation to carbon deposits, CPOX can be modified by anode gas recycling. The steam-to-carbon ratio can be increased to 0.3–1 in this way.

The advantage of steam reforming is its higher efficiency and a higher hydrogen concentration in the product gas. A high risk of carbon deposits still exists at steam-to-carbon ratios of about 4.5, although the minimum ratio should be somewhat higher than 1 based on thermodynamic calculations [88]. Moreover, steam reforming requires a heat source such as a combustion process performed in a separate compartment near the reforming section. Two drawbacks of this configuration are the longer start-up time and slower response behavior. To improve the heat balance and start-up time and to decrease the risk of carbon deposits, small amounts of air can be added to the fuel/steam mixture. Different research groups applied this modification as oxidative steam reforming for methanol [89,90] and ethanol [91]. At oxidative conditions (O/C = 0.2–0.4), the amount of nitrogen will be reduced. The method was also applied for propane [92] and diesel [93,94].

Figure 9.5 shows a map with different types of hydrogen generation processes indicated by their educt mixture, namely, by the H_2O/C and O/C ratios (instead of O_2/C). In addition to those reforming processes discussed earlier, areas with conditions for pyrolysis [95] and supercritical steam reforming are also

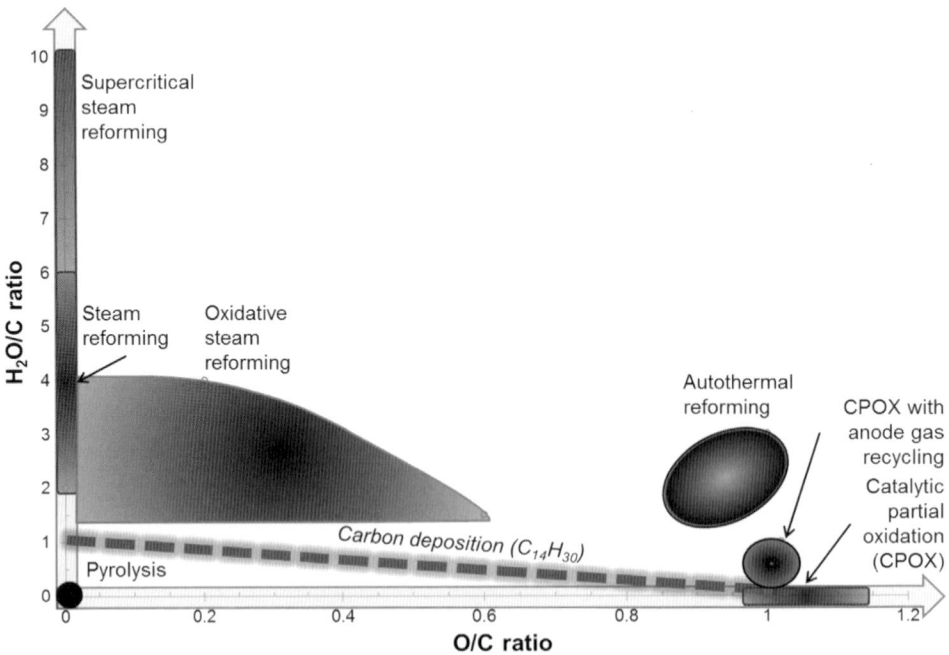

Figure 9.5 Different types of hydrogen generation processes indicated by their educt mixture, namely, by the H_2O/C and O/C ratios.

shown [96,97]. In principle, pyrolysis is an undesired side reaction leading to carbon deposits according to:

$$C_nH_m \rightleftharpoons nC + \frac{m}{2} H_2 \quad (\Delta H_R < 0 \text{ kJ mol}^{-1}) \tag{9.44}$$

Supercritical steam reforming has the advantage of producing a well-suited product composition of hydrogen, carbon dioxide, and steam. Owing to the critical conditions of water (220.5 bar, 647.3 K [16]), process conditions of 250–276 bar and 823–973 K are a challenge for reactor design. A compact and light-weight reactor vessel is difficult to realize.

Autothermal reforming is advantageous in regard to carbon deposition. However, it necessitates the design of air supply and steam generation. Other process routes that have been reported in the literature are hybrid steam reforming [98] and plasma reforming [99,100].

The main difference between the various reforming processes is product gas composition. Figure 9.6 shows the product gas spectra for various hydrogen

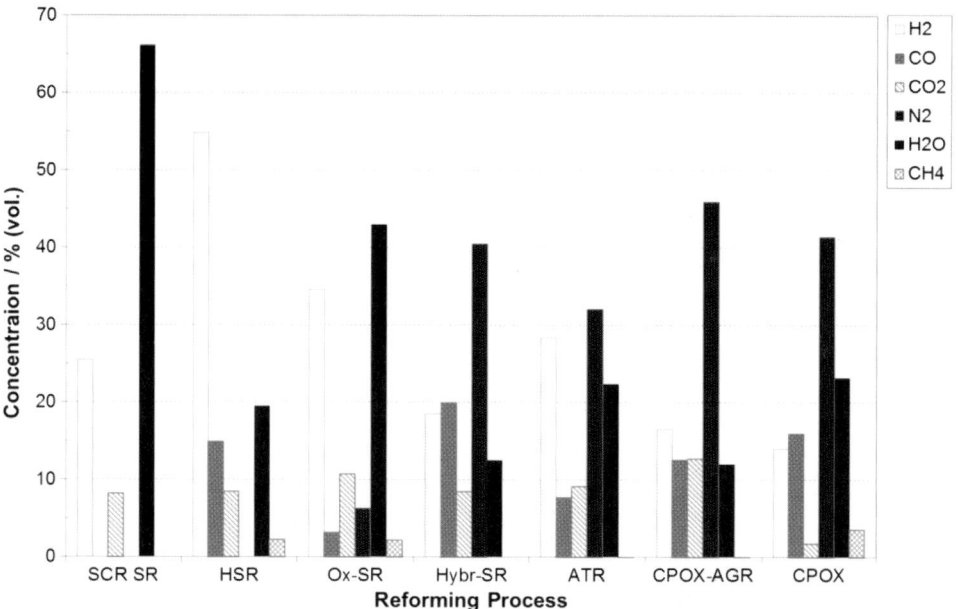

Figure 9.6 Product gas composition for various hydrogen generation processes: supercritical (SCR SR), balance for $H_2O/C = 10$; heated steam reforming (HSR), chemical equilibrium data for $H_2O/C = 2$, 700 °C, 1.5 bar; oxidative steam reforming (Ox-SR), chemical equilibrium data for $H_2O/C = 4$, $O/C = 0.2$, 590 °C, 1.5 bar; hybrid steam reforming (Hybr-SR), data derived from Reference [98]; autothermal reforming (ATR), chemical equilibrium data for $H_2O/C = 1.9$, $O/C = 0.94$, 715 °C, 1.5 bar; catalytic partial oxidation with anode gas recycling (CPOX AGR), chemical equilibrium data for $H_2O/C = 1$, $O/C = 1$, 732 °C, 2 bar; catalytic partial oxidation (CPOX), chemical equilibrium data for $O/C = 1$, 1080 °C, 1.5 bar. Fuel: dodecane ($C_{12}H_{26}$) except for CPOX AGR, which used isooctane (C_8H_{18}). The effect of alkane chain length on composition is minor.

generation processes. Dodecane ($C_{12}H_{26}$) was chosen as fuel for all calculations, except for SOFC with anode gas recycling, which used isooctane (C_8H_{18}). The effect of alkane chain length on composition was minor. Steam reforming produced the highest amount of hydrogen with 54% (vol.) on a water-free basis for supercritical steam reforming. Oxidative steam reforming and autothermal reforming followed with concentrations of 35% (vol.) and 28% (vol.), respectively. Most process routes were evaluated with the aid of chemical equilibrium data under reliable and suitable operation conditions in regard to system aspects such as steam availability. Owing to a somewhat lower reactor temperature and lower oxygen partial pressure, oxidative steam reforming produced a higher methane concentration than autothermal reforming. The burden of nitrogen dilution was reduced, while the excess of steam was quite high. CPOX produced the lowest hydrogen concentration of about 12–13% (vol.). The share of carbon monoxide, carbon dioxide, and methane can be improved by implementing anode gas recycling. Hybrid steam reforming [98] generates a product gas mixture that is more like that of the CPOX route with anode gas recycling (AGR). It integrates total combustion as the heat source for steam reforming by directly using tail gas. This design does not require a heat exchanger, but leads to a dilution of the typical product gas of a steam reformer by nitrogen.

How are these processes considered from the perspective of process engineering? Figures 9.7 and 9.8 show layouts of subsystems as a series of unit operations. The simple and unified configuration of medium charge and heat exchange before all fluids are mixed provide the basis for the scheme in Figure 9.7a. It is not useful to heat, evaporate, and superheat steam and diesel and mix them with hot air at 350 °C. Diesel would pyrolyze under these conditions. Experience in reactor design has shown that a configuration with fuel

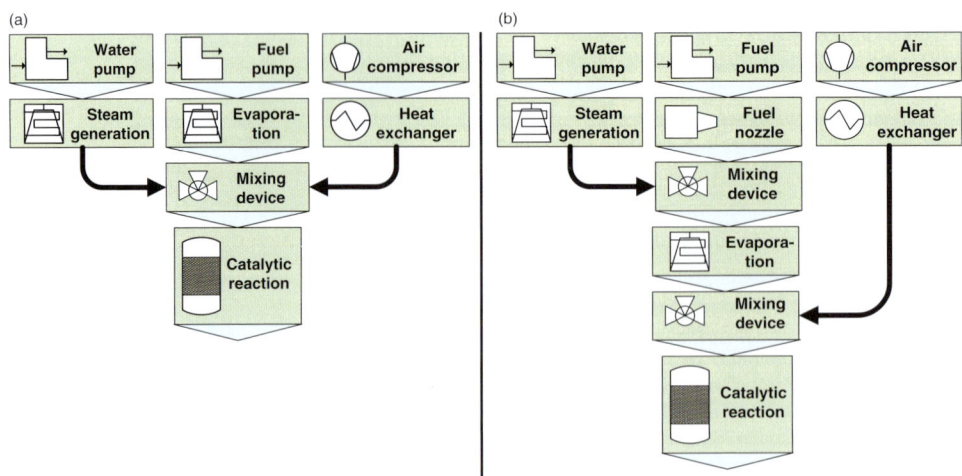

Figure 9.7 Fuel processor subsystem for the autothermal reforming of diesel as a draft (a) and as a successfully tested (b) design. Individual processes are depicted as unit operations.

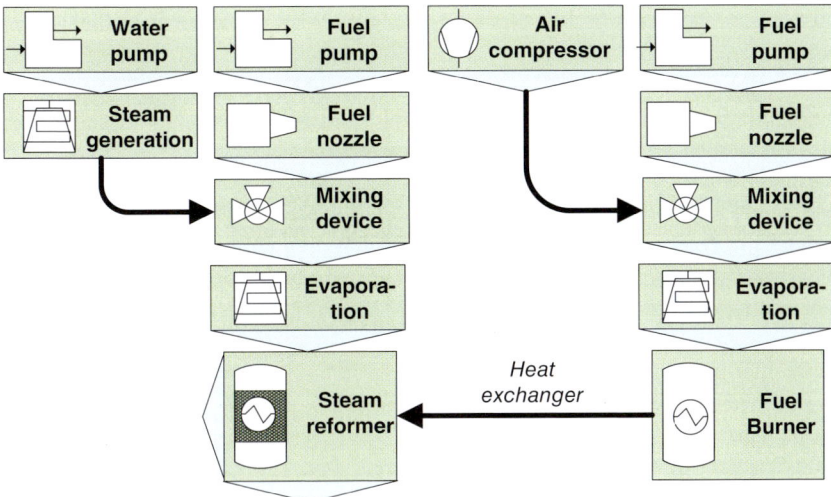

Figure 9.8 Fuel processor subsystem for steam reforming of diesel as a potential design. Individual processes are depicted as unit operations.

injection by a nozzle and the evaporation of the spray in steam is the most appropriate. Heated air is added in a secondary mixing device. This concept, which is illustrated as unit operations in Figure 9.7b, is more complex to some extent. However, it enables operation without fuel pyrolysis and without carbon deposits. Figure 9.8 shows an analogous layout for steam reforming. Obviously, this process is more complex with increased hardware.

Three of the most successful reformer developments should be analyzed in regard to the avoidance of carbon deposition and the effort of steam production. These are CPOX with anode gas recycling [87], autothermal reforming [86], and steam reforming [101]. Figure 9.9 shows the heat of reaction and required heat for steam production with different operation points. Steam reforming of dodecane has a heat of reaction of 1987.5 kJ mol^{-1} while partial oxidation releases −988.4 kJ mol^{-1}, both at 750 °C. Regarding the chemical reaction for steam reforming (Eq. (9.11)), a minimum heat of 849.6 kJ mol^{-1} is necessary to supply superheated steam at 750 °C. Engelhardt *et al.* [101] proposed a molar steam-to-carbon ratio of 6 to avoid carbon deposition. As a consequence, a heat demand of 5097.6 kJ mol^{-1} must be covered. Roychoudhury *et al.* [102] reported on successful experiments with a mixture corresponding to O/C = 0.9 and H$_2$O/C = 0.95. The heat requirements for superheated steam at 750 °C are 807.1 kJ mol^{-1}. In a SOFC system, steam can be supplied by anode gas recycling. Autothermal reforming is discussed in Section 9.4.1.1 as a combination of heated steam reforming and partial oxidation. Under the conditions that dodecane, steam, and air are heated up to 750 °C and the product gas also leaves the ATR at 750 °C, a closed heat balance, that is, $\Delta H_R \approx 0$ kJ mol^{-1}, would be given for

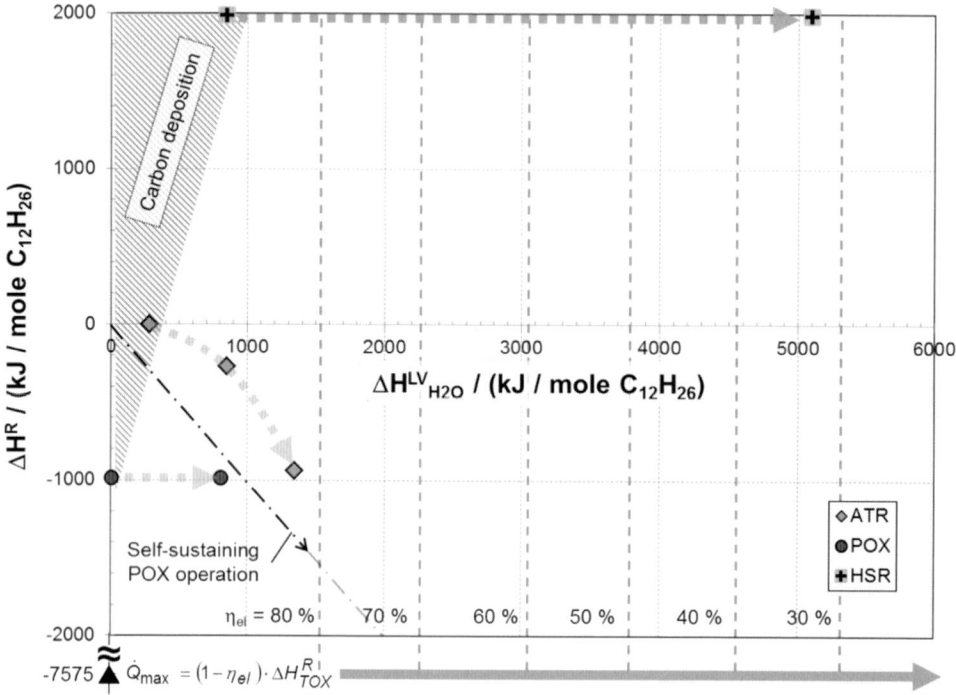

Figure 9.9 Heat of reaction and heat for steam supply for different reforming processes of dodecane ($C_{12}H_{26}$). Process parameters are taken from References [86,87,101].

two-thirds CPOX and one-third steam reforming. This mixture leads to a chemical reaction:

$$C_{12}H_{26} + 4H_2O + 4O_2 \rightleftharpoons 12CO + 17H_2 \quad (\Delta H_{R,1023\,K} = 0 \text{ kJ mol}^{-1}) \tag{9.45}$$

The corresponding ratios of the educt mixture are $H_2O/C = O_2/C = 0.33$. As indicated in Figure 9.9, such an educt mixture is located in the region of carbon deposits [88] with the ratios $H_2O/C = 1.1$ and $O/C = 1.02$. An increase in steam, for example up to $H_2O/C = 1$, leads to the occurrence of the weakly exothermic WGS reaction (Eq. (9.13)), and finally, to the chemical reaction:

$$C_{12}H_{26} + 12H_2O + 4O_2 \rightleftharpoons 4CO + 8CO_2 + 25H_2 \quad (\Delta H_{R,1023\,K} = -269.3 \text{ kJ mol}^{-1}) \tag{9.46}$$

Finally, the ATR process cannot be performed as originally intended. Pasel et al. reported on long-term experiments with ARAL Ultimate diesel, desulfurized kerosene, GTL, and BTL diesel for 1000, 2000, 5000, and 3500 h [85,86] at ideal mixing ratios of $H_2O/C = 1.9$ and $O/C = 0.94$. These ratios led to a share of

94% POX and only 6% HSR. The third rhombus symbol in Figure 9.9 was calculated for similar ratios, that is, $H_2O/C = 1.9$ and $O/C = 0.9$. Reflecting experimental results, about 7 from 12 CO molecules are shifted to CO_2. In addition, the thermal energy that is brought into the mixing chamber must be limited. Ignition occurs at temperatures between 300 and 400 °C. Owing to the heat of reaction, the temperature at the inlet to the reforming catalyst increases rapidly, but should not exceed 1000 °C. At higher temperatures, catalyst activity is affected by sintering. The calculation resulted in an exothermic heat of reaction of $-933\,kJ\,mol^{-1}$ at 1023 K and an enthalpy supply of about $1344\,kJ\,mol^{-1}$ to produce steam at a reduced temperature of 450 °C. An arrow marks a self-sustaining POX process. As indicated in Figure 9.9, the parameters of Roychoudhury et al. [87] fulfilled this condition, while the parameters of Pasel et al. required a further heat supply, which was realized by their catalytic burner [85].

Returning to the balance of the complete fuel cell process, the ultimate energy input in chemical form is the input of $C_{12}H_{26}$ with a lower heating value of $-7575\,kJ\,mol^{-1}$ at 298 K. The main task of a fuel cell system is to produce electricity with a high efficiency. Residual energy is released in the form of thermal heat, which can be used for heating purposes. In addition to air preheating and fuel conditioning, steam production is another use of this residual energy. Figure 9.9 shows six vertical lines corresponding to electrical efficiencies of 30–80%. Even at high efficiencies for partial load operation, residual latent heat can cover the demand of steam production for ATR and POX processes. For steam reforming at high molar ratios of H_2O/C in the range of 4–6, steam production is a challenge. In addition to $2000\,kJ\,mol^{-1}$ for the endothermic reaction, about $5000\,kJ\,mol^{-1}$ are required for steam production under these conditions. A more detailed analysis must include further heat requirements for air and fuel preheating. A thermodynamic analysis, such as a pinch-point analysis [88], considers the quality of available latent heat. Such an analysis may show that the combustion of fresh fuel as a heat source is necessary. The advantage of a higher hydrogen concentration for steam reforming is reduced by increased steam dilution, which is a drawback associated with steam supply.

Process selection is made on the basis of experimental experience gained by the developing group and on the basis of system requirements. After initial selection – for example, in favor of autothermal reforming – a set of operation conditions must be defined. According to Figure 9.7b, fuel is sprayed and evaporated in superheated steam before being mixed with preheated air. The temperature ranges from 400 to 500 °C for superheated steam, and from ambient temperature to 200 °C for air. The temperatures of diesel and jet fuels should not exceed 150 °C to avoid coking in small channels, for example in the nozzle. Under these conditions, hydrogen and carbon monoxide yields vary depending on the conversion, selectivities for methane and residual hydrocarbons, and the O/C ratio (see as O_2/C ratio in Eq. (9.26)). Assuming complete conversion and negligible byproducts, Eq. (9.26) can be simplified to Eq. (9.31). Chemical equilibrium calculations can be performed to determine maximum yields under adiabatic conditions. The addition of air at 120 °C to a dodecane–steam mixture of

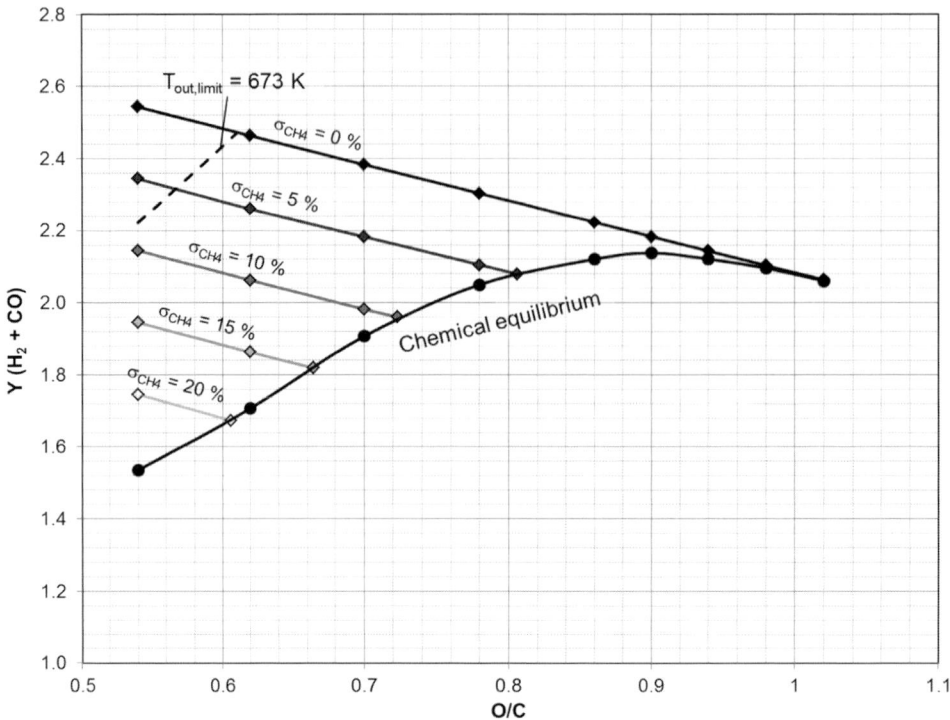

Figure 9.10 Mapping of hydrogen yield for autothermal reforming of dodecane ($C_{12}H_{26}$). Operation conditions: dodecane–steam mixture at 350 °C, air preheated to 120 °C, steam-to-carbon ratio of $H_2O/C = 2.5$.

350 °C leads to an exit temperature of 624 °C and a methane selectivity of 5% for mixing ratios of $O/C = 0.8$ and $H_2O/C = 2.5$. Figure 9.10 maps the hydrogen and carbon monoxide yields as a function of the oxygen-to-carbon ratio (O/C) and selectivity to methane. The yield increases slightly with an increasing O/C ratio due to lower methane selectivity. At $O/C = 0.9$, selectivity reaches its maximum. The addition of more air leads to an enhanced combustion of hydrogen. At constant O/C ratios, methane formation can be suppressed by a suitable catalyst. At low O/C ratios, the yield is limited by the inlet temperature of the shift stage, indicated here by an outlet temperature of 400 °C. In practice, an outlet temperature between 650 and 800 °C is envisaged. To avoid carbon deposition and to suppress methane formation, O/C ratios between 0.9 and 0.95 are recommended. The yield is 2.05 to 2.15 (mole ($H_2 + CO$) per mole C). This value is somehow lower than 2.4 at low O/C ratios, which leads to deposits.

9.5.1.2 Catalyst Development

Although catalyst development belongs to the discipline of heterogeneous catalysis and not directly to process engineering, it should still be discussed briefly here because it plays a highly important role in hydrogen production. Catalysts are tailored to each fuel and the selected processes under the conditions of the intended application. For several fuels, catalyst formulations are well known because processes are already established in the chemical industry. Natural gas or methane reforming is performed with Ni catalysts [25]. For methanol steam reforming, Cu catalysts – mainly Cu/ZnO on γ-Al_2O_3 as a support – are preferred. As a further option, Pd/ZnO catalysts have also been investigated [12,13]. For ethanol steam reforming, the catalyst choice is somewhat larger, that is, Co/ZnO, ZnO, Rh/Al_2O_3, Rh/CeO_2, and Ni/La_2O_3-Al_2O_3 [6].

Catalyst development for fossil fuels such as gasoline, jet fuels, and diesel has been a matter of research in industry and universities over the last decade. While knowledge on industrial catalysts is limited [31], a series of research papers have been published over the last ten years [103–115]. In 2002, Krumpelt et al. [114] developed new catalysts for SOFC technology, whereby transition metals were supported on oxide-conducting substrates. Tested transition metals were Ru, Pd, Fe, Cu, Pt, Ni, Co, and Au. Substrates were ceria, zirconia, or lanthanum gallate, which were later doped with small amounts of Gd, Sm, or Zr. Only Ru resulted in complete conversion for the whole temperature range even at a low GHSV of 3000 h^{-1}. Experiments were performed under autothermal conditions: $H_2O/C = 1.14$ and $O_2/C = 0.46$. Cheekatamarla and Lane reported on the development of an efficient bimetallic catalyst for hydrogen production from diesel fuels [103]. In the current literature, the following catalytically active materials are preferred: Ru [110,115], Rh [106,109,112], and bimetallic Rh-Pt [104,113] or Ni-Rh catalysts [107]. These research studies can also be divided into those for gasoline [106,107,114,115], kerosene [109], diesel [104,110,112,113], and corresponding surrogates. Different washcoats such as CeO_2, ZrO_2, Al_2O_3, TiO_2, SiO_2, La_2O_3, MgO, and Y_2O_3 were investigated by Ferrandon et al. [106,107] and Karatzas et al. [113], and CeO_2-ZrO_2 was found to be most promising.

9.5.1.3 Current Challenges in Fuel Processing

Avoiding carbon deposition and the sulfur tolerance of catalysts play a crucial role in reforming. Residual hydrocarbons should be avoided because they lower efficiency and more importantly they can negatively affect the reactor and the fuel cell. Ethylene (C_2H_4) and aromatics, in particular, are regarded as precursors for carbon deposits.

Kaila et al. [111] reported on the sulfur tolerance of a Zr-supported Rh/Pt catalyst in relation to different sulfur species. The addition of 10 ppm of 4,6-dimethyldibenzothiophene (DMDBT) to commercial low-sulfur diesel led to a slight degradation of an ATR operating at $H_2O/C = 3$, $O_2/C = 0.34$, and 700 °C. DMBT is a typical compound in diesel fuel, whereby 4,6-DMBT is the most challenging for desulfurization [53,116]. The addition of 10 ppm H_2S led to strong performance loss, which was reversible except for some irreversible effects at lower temperatures. According to the results, the ATR of liquid fuels should not be performed below 700 °C to avoid catalyst deactivation when RhPt is used. The close interaction between Rh and Pt prevents bonds with sulfur.

Cheekatamarla and Lane [104] have also reported on a loss in hydrogen yield caused by periodic switching between low-sulfur synthetic diesel and blended 150 ppm H_2S diesel for ATR operation conditions ($H_2O/C = 2.5$, $O/C = 1$, preheating to 400 °C). Analogous effects were observed for 200 ppm SO_2. The hydrogen yield decreased, and CO and CH_4 concentrations in the product gas increased. The SO_2 changed the surface, chemical, and structural characteristics of the catalysts due to migration of SO_3 in the ceria bulk [105]. The catalysts were reactivated by H_2 purging, but some irreversible changes were identified.

Ferrandon et al. [108] also investigated the effects of sulfur in gasoline reformed with Rh/La-Al_2O_3 under autothermal conditions ($H_2O/C = 2.0$, $O/C = 0.9$, $T_{Furnace} = 700$–800 °C). A switch from sulfur-free to 34 ppm S gasoline led to a strong increase in residual hydrocarbons, such as propene, benzene, butene, ethane, and toluene, at 700 °C. At 800 °C, residuals were much lower with benzene as the main component accompanied by some ethane and ethene. Increasing the steam-to-carbon ratio from 2 to 2.5 and then to 3 resulted in a stable and weak loss of hydrogen yield. Adding potassium in the form of Rh-K/La-Al_2O_3 led to increased reaction temperatures accompanied by reduced steam reforming activity. Furthermore, the inhibition of carbon deposits and site blockages was demonstrated. Qi et al. [115] interpreted their activity loss during long-term experiments (1000 h) as a strong etching and corrosive effect of K_2O on the Al_2O_3 support. Roychoudhury et al. [102] also reported on the effect of different sulfur species in fuel on the occurrence of C_2 and C_3 residues. Sulfur species should be added very carefully. During 1100 h of ATR diesel operation ($H_2O/C = 0.9.0$, $O/C = 0.95$) C_3 residues were lower than 10 ppmv. During testing of jet fuel with 400 ppmw, sulfur residues increased after 60 h to C_2 and C_3 concentrations lower than 50 ppmv and 10 ppmv, respectively. Doping low-sulfur jet fuel with 400 ppmv DBT led to much higher concentrations of up to 960 ppmv C_2 and 120 ppmv C_3.

Shamsi et al. [117] investigated carbon deposition on Pt/alumina catalysts under POX ($O/C = 1$), ATR ($H_2O/C = 1.5$, $O/C = 0.6$), and HSR ($H_2O/C = 3$) conditions. They used tetradecane ($C_{14}H_{30}$), decalin ($C_{10}H_{18}$), and methylnaphthalene ($C_{11}H_{10}$) as typical compounds in diesel to reflect the influence of alkanes, naphthenes, and aromatics, latter both binuclear. The experiments clearly showed the sequences $C_{10}H_{18}$, $C_{14}H_{30}$, and $C_{11}H_{10}$, and ATR, POX, and HSR were found to increase carbon deposition. The second sequence was supported by experiments by Yoon et al. with ethylene (C_2H_4) as a model fuel and Pt/CGO (Gd-doped CeO_2) [118].

The formation of undesired byproducts also depends on fuel composition. About half of the cited research groups in this contribution used commercial fossil fuels [69,79,80,82–85,87,94,106–109,111–113], commercial biofuels [86], or fuel surrogates [75,101,110,117] for basic research and technical development. Studies by Kang et al. and Kopasz et al. showed that reactivity in reforming reactions decreased in the sequences of alkanes, olefins, naphthenes, mono-aromatics, and di-aromatics analyzed [82,119]. The species used were hexane, cyclohexane, toluene, and methylnaphthalene, and, secondly, isooctane, octane, methylcyclohexane, toluene, and trimethylbenzene. In most studies, residual hydrocarbons consisted mainly of olefins and aromatics. The importance of proper evaporation and an effective mixing of educts to avoid residuals was pronounced by different groups [69,82,85,120]. Yoon et al. [73] applied an ultrasonic injector and thus reduced ethene concentration in the reformate (via ethene reforming) by a factor of $\frac{1}{3}$ compared to operation without this injector. After 250 h of testing, the amount of ethene in the reformate was 600 ppmv (value was dry and nitrogen-free). Lindström et al. [69] confirmed their mixing chamber development by reducing diesel slip from 800 ppmv to less than 10 ppmv. Ethene, ethane, propane, and propene formation was observed by several research groups. Some examples are: 3108 ppmv C_2 and 1777 ppmv C_3 at 80% load for oxidative steam reforming [94], 426 ppmv C_2H_6 and 176 ppmv C_3H_8 for diesel steam reforming at full load [101], 1000 ppmv C_2H_4 and 7000 ppmv C_3H_6 at the beginning for autothermal kerosene reforming [80], nearly 1.5% vol. (C_2H_4 and C_2H_6) after 1000 h under ATR conditions of octane [115], 150 ppmv C_2H_4 and 100 ppmv C_2H_6 after 30 h catalyst screening [109], and 600–100 ppmv C_2H_4 under ATR conditions ($H_2O/C = 2.0$, O/C = 0.98, diesel) at 630–760 °C.

Another evaluation aspect is the duration of testing. Only a few refereed papers have reported on the longevity of catalysts and reformers [80,85,86,115,120] for operation periods between 300 and 6000 h. High conversion, that is, higher than 99.5%, was shown after 1000 h for premium diesel [120], after 2000 h for low-sulfur kerosene [85], and after 5000 h for gas-to-liquid kerosene [86]. A fuel switch to commercial bio-to-liquid diesel led to a decrease in conversion from >99.99% to 99.67% after a further 1000 h [86]. Residual hydrocarbons in the gas phase were 60 ppmv C_2H_6, 100 ppmv C_2H_4, and 70 ppmv C_3H_6. These values after 6000 h ATR operation are still somewhat smaller than those reported for fresh catalysts. Methane amounts to 0.35% (vol.) CH_4. It is not considered an undesired product because its appearance is forced by chemical equilibrium and it does not harm subsequent reactors or the fuel cell. It must be combusted in the afterburner unit. The heat of combustion from residual methane may be required, depending on the system design.

Depending on the analytic equipment used, not all residual hydrocarbons can be measured in the gas phase. A gas chromatograph, for example, demands a dry gas phase, probably after gas cooling and water condensation. Only a few research groups have analyzed this condensate [85,86,102,120]. They found organic carbon in the range 0.5–130 mg l^{-1}. Pasel et al. [86] reported a constant value of 20 mg l^{-1} during a 5000 h GTL-kerosene ATR test. Trace concentrations of acetic acid, acetone, and butanone were found in the condensed aqueous phase. A conversion into the wet gas phase led to concentrations of about 0.6 ppmv acetic acid and 0.3 ppmv acetone. The required quality for the reforming reaction must be verified in experiments on a system level. Residual hydrocarbons play a decisive role for fuels that contain higher hydrocarbons.

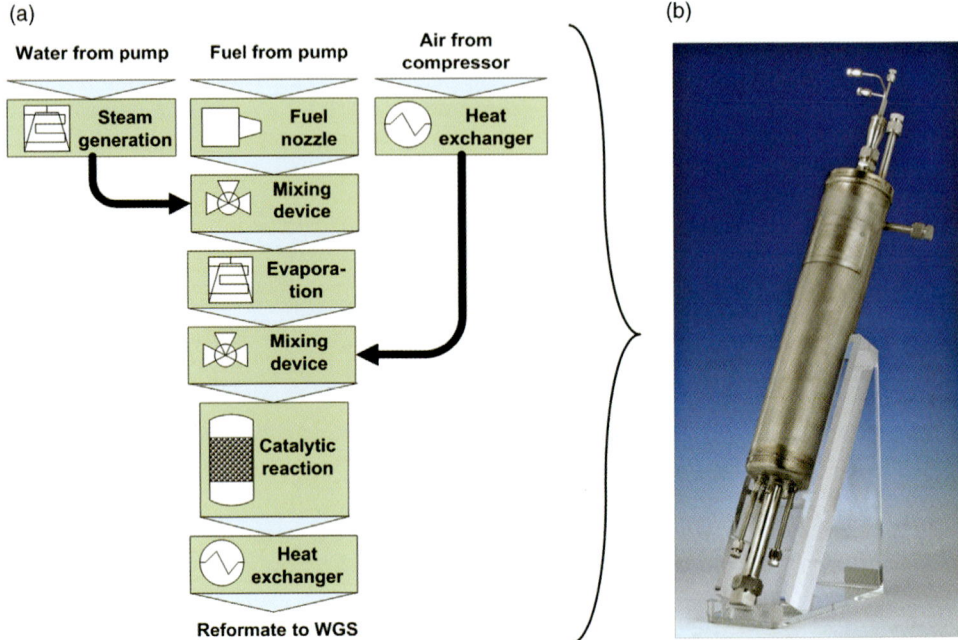

Figure 9.11 Implemented unit operations (a) and autothermal reformer designed and manufactured by Forschungszentrum Jülich (b).

9.5.1.4 Technical Outlook

Within the last two sections, the basics of hydrogen production were discussed in detail. Catalyst development and current testing activities for different process options have reached a level that will allow reactor designs to be transferred from basic developmental research to industrialized process technologies. The integration of different unit operations in one device reflects progressive development. Figure 9.11 shows an autothermal reformer developed and constructed by Forschungszentrum Jülich. The reactor consists of eight unit operations, two of which, that is, steam generator and heat exchanger for reformate cooling, belong together as a coupled unit. A divided mixing chamber accommodates steam/fuel mixing, fuel evaporation, and mixing with air. Heat conduction through the metal jacket functions as the heat exchanger for air. All other components of a fuel cell system should be designed in a similar manner, integrating different unit operations in one device.

Based on the results reported in Section 9.5.1.3, operation parameters must be chosen in a narrow range. Catalysts should not be used at temperatures higher than 1273 K due to sintering effects nor should they be used at temperatures lower than 973 K because of carbon deposition [111]. The lower limit is reached in chemical equilibrium at the ATR exit under the conditions given in Figure 9.10 for an oxygen-to-carbon ratio of $O/C = 0.94$. The ratio is somewhat lower for lower H_2O/C. Therefore, mixture parameters were varied between

O/C = 0.86–0.94 and H_2O/C 1.7–1.9 (see also References [72,86,120]). Finally, integrating the fuel processing system and fuel cell stack in a single system is an ongoing task for future fuel cell system development.

9.5.2
Design of Complete Gas Cleaning System

Different fuel cell types require an adapted fuel gas quality. The main poisons are carbon monoxide and sulfur. A distinction should be made between irreversible and reversible degradation effects. A reversible degradation effect leads to a certain loss in cell voltage, which can be recovered by purging with sulfur-free fuel gas. Elevated CO concentrations can be tolerated for a short time during load changes, even by a PEFC. PEFCs demand the highest fuel quality with concentrations of carbon monoxide lower than 10–50 ppmv and hydrogen sulfide lower than 100 ppbv. Jörissen et al. [121] showed that the amount of accumulated sulfur is decisive for the voltage decrease. Lower H_dS concentrations will only prolong the point in time when a certain degradation effect will occur. The experiments by Schmidt and Baurmeister [122] with HT-PEFCs identified a strong interaction between CO and H_2S. HT-PEFCs can be operated using a fuel gas with 1% CO and 10 ppmv H_2S with a drop in cell voltage of 4%. In contrast to other types of fuel cells, methane and carbon monoxide can be used as fuels in SOFC systems. Sulfur in the form of hydrogen sulfide or thiophene will also lead to a considerable performance loss. A defined tolerance limit for SOFCs has yet to be determined [53]. However, even a hydrogen sulfide content of 0.1 ppm H_2S leads to partially irreversible performance loss in practice [55].

Higher hydrocarbons also affect cell performance. Uniform threshold values are not available. Moore et al. [123] reported a performance loss of 28% at a low current density of 200 mA cm^{-2} measured 15 min after exposure to 50 ppm benzene on the cathode side. Li et al. [124] published the effect of exposing the cathode side to 1, 5, 10, and 50 ppm toluene (C_7H_8). At a current density of 750 mA cm^{-2}, the drop in cell voltage (from 700 mV) was 5%, 6%, 10%, and 22% at the respective C_7H_8 concentrations. Toluene on the anode side affected the cell performance to a lesser extent. Dorn et al. [125] reported a cell voltage drop of 8 mV at 600 mV (1.3%) and 60 °C when 20 ppmv C_7H_8 was added, whereby the effect of 2 ppmV CO was rather strong, that is, 35%. This effect was amplified by 20 ppmv C_7H_8 to 49%.

Figure 9.12 illustrates the required gas cleaning processes for carbon monoxide and hydrogen sulfide in a PEFC system. The product gas exits the ATR at 750–800 °C with CO and H_2S concentrations of 10% (vol.) and 0.8–1 ppmv, respectively. Sulfur must be reduced to 10 ppmw upstream of the reformer due to the limited sulfur tolerance of the reforming catalyst. Diesel fuel contains sulfur in the form of methylated dibenzothiophene [53], which will be converted during reforming into hydrogen sulfide.

Air dilution and the change in mole numbers during reforming reduce the amount of sulfur by a factor of ten. H_2S can be removed from reformate by a

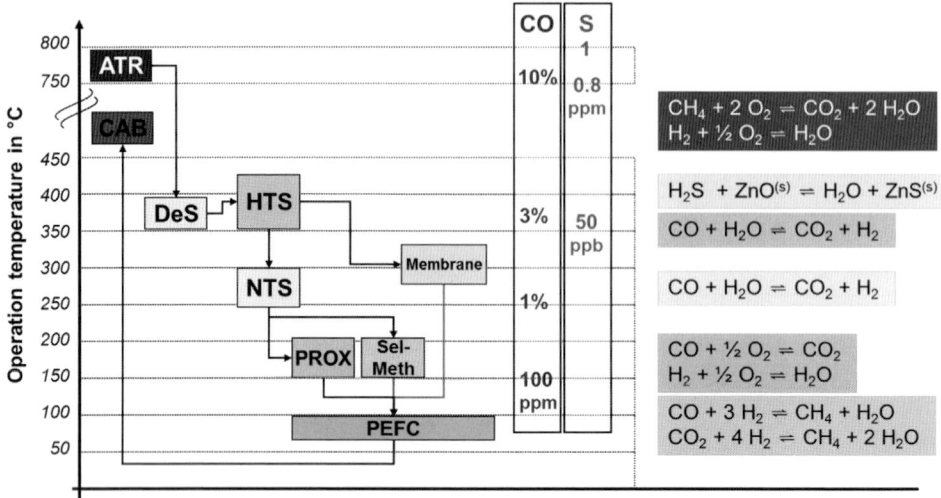

Figure 9.12 Principle scheme of gas cleaning processes for carbon monoxide and hydrogen sulfide in a PEFC (polymer electrolyte fuel cell) system.

ZnO bed at 400 °C. The ZnO bed should be exchanged or regenerated at a threshold of 100 ppmv. Novochinskii *et al.* reported a sorption capacity of 9 mg-S per g-ZnO under the conditions given above [126]. Carbon monoxide can be converted by WGS in two steps at different inlet temperatures to the adiabatic reactors, here 400 and 300 °C. An intermediate CO concentration might be chosen between 3% (vol.) and 4% (vol.). The residual CO must be removed from the fuel gas by a fine-cleaning process, which also causes a loss of hydrogen. Possible cleaning processes are preferential CO oxidation or selective CO methanation. Both reactor types were operated in a temperature range of 150–200 °C. The CO concentration at the stage between WGS and fine cleaning was 1% (vol.). This allowed a compact WGS design despite the fact that PROX would be more challenging. PROX design can be simplified with a decreasing inlet concentration of CO. In regard to the limitations of WGS by kinetics and chemical equilibrium, a reasonable CO concentration would be 0.5% (vol.) at the PROX outlet.

Figure 9.13 shows a gas cleaning subsystem as a draft design for a PEFC. The subsystem contains 16 unit operations. In this proposed system, WGS is designed as a boiling reactor. Water is evaporated in WGS and will be superheated by the heat exchangers behind and in front of the H_2S scrubber. The air supply and the air cooling system of the two PROX stages indicate the complexity of gas cleaning for PEFCs. These unit operations can be omitted for HT-PEFC technology.

In regard to compactness, integrating gas cleaning components in one or two devices is a major task for further development. Kolb *et al.* [127] proposed microreactor technology, similar to the sketch in Figure 9.13. An alternative system design includes adiabatic reactors and intermediate cooling [32]. Detailed

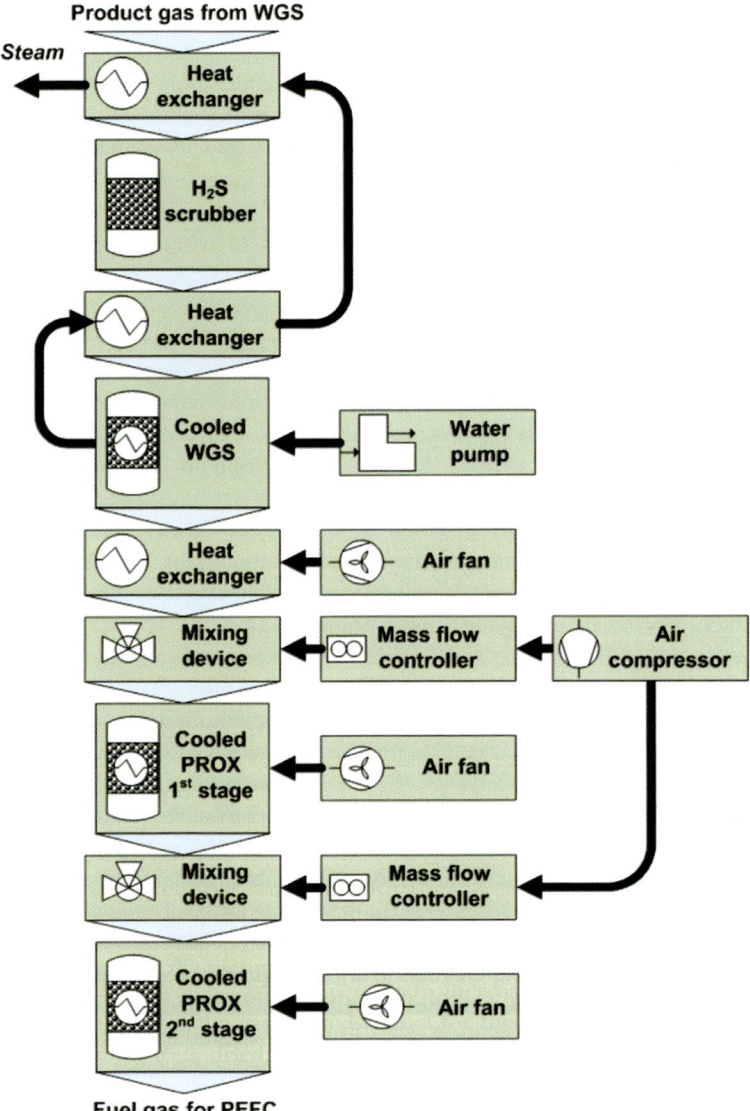

Figure 9.13 Gas cleaning subsystem as a possible design for a PEFC. Individual processes are represented as unit operations.

engineering and practical experience will decide on the suitability of these options.

A further challenge in system development is the interaction of the single components in dynamic operation. Liu *et al.* [128] reported on the deactivation mechanism of WGS catalysts during shutdown in reformate. They found that carbonates cover the CeO_2 surface of a Pt/CeO_2 WGS catalyst and block the Pt metal surface. Sintering effects of Pt did not occur. As a consequence, regeneration with 400 °C hot air must be implemented in the system design. In addition to carbon formation during shutdown, a second mechanism for Au/CeO_2 catalysts is described for steady-state operation in realistic reformate at 250 °C. H_2 and CO led to a reduction of the active metals to Au^0 and Ce^{3+} and weakened the strong $Au–CeO_2$ interaction [129].

9.6
Conclusion

In principle, hydrogen is the ideal fuel gas for a fuel cell. Hydrogen can be produced directly by electrolysis or by chemical conversion from other gaseous or liquid fuels. The chemical conversion of such fuel into a hydrogen-rich gas for use in fuel cells is termed fuel processing. The hydrogen in such systems is referred to as short-lived hydrogen. Different types of fuel cells, such as the SOFC, HT-PEFC, and PEFC, have different operating temperatures and different electrochemical processes, which give rise to different fuel quality requirements. The two most important catalyst poisons for specific fuel processing components and the fuel cell are sulfur and carbon monoxide.

If hydrogen could be produced under economically reasonable conditions and if it were widely available in a future infrastructure, it would be the first choice for operating passenger cars and busses. Liquid fuels will continue to be used in the future in mobile applications with high driving powers, such as trucks, airplanes, ships, and trains, due to their high volume-specific and mass-specific storage densities. The fuels most commonly used at the moment are middle distillates, and include kerosene, diesel, and extra-light heating oil. In the future, biofuels will have to replace fossil sources.

In the short- and medium-term, blends will be used. Research and development work is concentrating on second-generation fuels and their fabrication, which do not compete with the food production industry. This contribution dealt with the task of fuel processing for fuel cells. It focused on general methods of process engineering and aspects of system design. All relevant unit operations of a fuel processing system were discussed. Detailed aspects of process design, catalyst development, and current challenges were analyzed in relation to the reforming of higher hydrocarbons.

While fuel processing for pure fluids, such as methanol and methane, as well as for gaseous mixtures such as liquid-petroleum gas (LPG), natural gas, or biogas, is already an industrialized and mature process, reforming higher hydrocarbons remains challenging today. Current results from catalyst research must be considered in catalyst manufacturing and should lead to improved reformers. The success of reformer development depends on a stable and robust catalyst and primarily on educt preparation. An effective catalyst requires a completely evaporated fuel homogeneously mixed with air or steam or both of them. Some groups have already developed reformers with the desired performance. Process optimization, the integration of all demanded components into a fuel processor and later into a complete system, and the transfer to industrialized technologies are major tasks for the future.

Acknowledgments

The author would like to thank the Department of Fuel Processing and Systems at Forschungszentrum Jülich GmbH, Institute of Energy and Climate Research – Energy Process Engineering (IEK-3) for their excellent cooperation in recent years.

References

1 Pratt, J.W., Klebanoff, L.E., Munoz-Ramos, K., Akhil, A.A., Curgus, D.B., and Schenkman, B.L. (2013) Proton exchange membrane fuel cells for electrical power generation on-board commercial airplanes. *Appl. Energy*, **101**, 776–796.

2 Peters, R. and Samsun, R.C. (2013) Evaluation of multifunctional fuel cell systems in aviation using a multistep process analysis methodology. *Appl. Energy*, **111**, 46–63.

3 Metz, S. (2013) Merchant LNG: natural gas instead of diesel. Available from http://www.the-linde-group.com/en/clean_technology/clean_technology_portfolio/merchant_liquefied_natural_gas_lng/merchant_lng/index.html (accessed 17 July 2013)

4 Arcoumanis, C., Bae, C., Crookes, R., and Kinoshita, E. (2008) The potential of di-methyl ether (DME) as an alternative fuel for compression-ignition engines: a review. *Fuel*, **87** (7), 1014–1030.

5 Dewald, U. (2005) SFC bringt Brennstoffzelle für Segelyachten auf dem Markt. Available from www.innovations-report.de/html/berichte/energie_elektrotechnik/bericht-38537.html (accessed 16 October 2010).

6 Haryanto, A., Fernando, S., Murali, N., and Adhikari, S. (2005) Current status of hydrogen production techniques by steam reforming of ethanol: a review. *Energy & Fuels*, **19**, 2098–2106.

7 Cheekatamarla, P.K. and Finnerty, C.M. (2006) Reforming catalysts for hydrogen generation in fuel cell applications. *J. Power Sources*, **160** (1), 490–499.

8 Ghenciu, A.F. (2002) Review of fuel processing catalysts for hydrogen production in PEM fuel cell systems. *Curr. Opin. Solid State Mater. Sci.*, **6**, 389–399.

9 Pettersson, L.J. and Westerholm, R. (2001) State of the art of multi-fuel reformers for fuel cell vehicles: problem identification and research needs. *Int. J. Hydrogen Energy*, **26**, 243–264.

10 Qi, A., Peppley, B., and Karan, K. (2007) Integrated fuel processors for fuel cell application: a review. *Fuel Process Technol.*, **88** (1), 3–22.

11 Xu, X., Li, P., and Shen, Y. (2013) Small-scale reforming of diesel and jet fuels to make hydrogen and syngas for fuel cells: a review. *Appl. Energy*, **108**, 202–217.

12 Yong, S.T., Ooi, C.W., Chai, S.P., and Wu, X.S. (2013) Review of methanol reforming-Cu-based catalysts, surface reaction mechanisms, and reaction schemes. *Int. J. Hydrogen Energy*, **38** (22), 9541–9552.

13 Peters, R. (2008) Fuel processors, in *Handbook of Heterogeneous Catalysis* (eds G. Ertl, H. Knozinger, F. Schuth, and J. Weitkamp), Wiley-VCH Verlag GmbH, Weinheim, pp. 3045–3080.

14 Samsun, R.C. and Peters, R. (2012) Methodologies for fuel-cell process

engineering, in *Fuel Cell Science and Engineering* (eds D. Stolten and B. Emonts), Wiley VCH Verlag GmbH, Weinheim, p. 49.

15 Soave, G. (1972) Equilibrium constants from a modified Redlich–Kwong equation of state. *Chem. Eng. Sci.*, **27**, 1197–1203.

16 Smith, J.M. and van Ness, H.C. (1987) *Introduction to Chemical Engineering Thermodynamics*, 4th edn, McGraw-Hill, New York.

17 Perry, R.H. and Green, D.W. (1997) *Perry's Chemical Engineering Handbook*, 7th edn, McGraw-Hill, New York.

18 Klotz, I.M. and Rosenberg, R.M. (1994) *Chemical Thermodynamics*, 5th edn, John Wiley & Sons, Inc., New York.

19 Joback, K.G. and Reid, R.C. (1987) Estimation of pure-component properties from group-contributions. *Chem. Eng. Commun.*, **57** (1–6), 233–243.

20 Riddick, J.A., Bunger, W.B., and Sakano, T.K. (1985) *Techniques of Chemistry*, 4th edn, John Wiley and Sons, Inc., New York, vol. **II** Organic Solvents, p. 424.

21 Madelung, O., Rössler, U., and Schulz, M. (eds) (1999) Zinc sulfide (ZnS) heat capacity, in *The Landolt-Börnstein Database*, Springer Materials, Berlin.

22 Madelung, O., Rössler, U., and Schulz, M. (eds) (1999) Zinc oxide (ZnO) Debye temperature, heat capacity, density, melting point, vapor pressure, hardness, in *The Landolt-Börnstein Database*, Springer Materials, Berlin.

23 NIST (2011) NIST Chemistry WebBook. Available from: http://webbook.nist.gov/chemistry/ (accessed 26 June 2013)

24 Rostrup-Nielsen, J.R. (1974) Coking on nickel catalysts for steam reforming of hydrocarbons. *J. Catal.*, **33**, 184–201.

25 Rostrup-Nielsen, J.R. (1984) Catalytic steam reforming, in *Catalysis – Science and Technology* (eds J.R. Anderson and M. Boudart), Springer, Berlin, p. 130.

26 Hinrichsen, K-.O., Kochloefl, K., and Muhler, M. (2008) Water gas shift and COS removal, in *Handbook of Heterogeneous Catalysis* (eds G. Ertl, H. Knozinger, F. Schuth, and J. Weitkamp), Wiley-VCH Verlag GmbH, Weinheim, pp. 2905–2920.

27 Balakos, M.W. and Wagner, J.P. (2002) High performance water-gas shift catalysts for fuel processing. Presented at Fuel Cell Seminar, Palm Springs: Courtesy Associates.

28 Fu, Q., Saltsburg, H., and Flytzani-Stephanopoulos, M. (2003) Active nonmetallic Au and Pt species on ceria-based water-gas shift catalysts. *Science*, **301** (5635), 935–938.

29 Ruettinger, W., Ilinich, O., and Farrauto, R.J. (2003) A new generation of water gas shift catalysts for fuel cell applications. *J. Power Sources*, **118** (1–2), 61–65.

30 Swartz, S.L., Seabaugh, M.M., McCormick, B.E., and Dawson, W.J. (2002) Ceria-based water-gas-shift catalysts. Presented at Fuel Cell Seminar. Palm Springs: Courtesy Associates.

31 Wieland, S., and Baumann, F., and Starz, K.A. (2001) New powerful catalysts for autothermal reforming of hydrocarbons and water-gas shift reaction for on-board hydrogen generation in automotive PEMFC application, SAE Technical Papers, Society of Automotive Engineers, Warrendale, p. 5. doi: 10.4271/2001-01-0234.

32 Ahluwalia, R.K., Zhang, Q., Chmielewski, D.J., Lauzze, K.C., and Inbody, M.A. (2005) Performance of CO preferential oxidation reactor with noble-metal catalyst coated on ceramic monolith for on-board fuel processing applications. *Catal. Today*, **99** (3–4), 271–283.

33 Kahlich, M.J., Gasteiger, H.A., and Behm, R.J. (1999) Kinetics of the selective low-temperature oxidation of CO in H_2-rich gas over Au/γ-Fe_2O_3. *J. Catal.*, **182**, 430–440.

34 Mariño, F., Descorme, C., and Duprez, D. (2004) Noble metal catalysts for the preferential oxidation of carbon monoxide in the presence of hydrogen (PROX). *Appl. Catal. B Environ.*, **54** (1), 59–66.

35 Korotkikh, O. and Farrauto, R. (2000) Selective catalytic oxidation of CO in H_2: fuel cell applications. *Catal. Today*, **62**, 249–254.

36 Schubert, M.M., Kahlich, M.J., Gasteiger, H.A., and Behm, R.J. (1999) Correlation between CO surface coverage and

selectivity/kinetics for the preferential CO oxidation over Pt/γ-Al$_2$O$_3$ and Au/α-Fe$_2$O$_3$: an in-situ DRIFTS study. *J. Power Sources*, **84**, 175–182.

37 Oh, S.E. and Sinkevitch, R.M. (1993) Carbon monoxide removal from hydrogen-rich fuel cell feed streams by selective catalytic oxidation. *J. Catal.*, **142**, 254–262.

38 Dudfield, C.D., Chen, R., and Adcock, P.L. (2001) A carbon monoxide PROX reactor for PEM fuel cell automotive application. *Int. J. Hydrogen Energy*, **26**, 763–775.

39 Alayoglu, S., Nilekar, A.U., Mavrikakis, M., and Eichhorn, B. (2008) Ru-Pt core-shell nanoparticles for preferential oxidation of carbon monoxide in hydrogen. *Nat. Mater.*, **7** (4), 333–338.

40 Vannice, M.A. (1975) The catalytic synthesis of hydrocarbons from H$_2$/CO mixtures over group VIII metals. *J. Catal.*, **37**, 449–461.

41 Pichler, H. (1952) Twenty-five years of synthesis of gasoline by catalytic conversion of carbon monoxide and hydrogen. *Adv. Catal.*, **4**, 272–337.

42 Batista, M.S., Santiago, E.I., Assaf, E.M., and Ticianelli, E.A. (2005) Evaluation of the water-gas shift and CO methanation processes for purification of reformate gases and the coupling to a PEM fuel cell system. *J. Power Sources*, **145** (1), 50–54.

43 Dagle, R.A., Wang, Y., Xia, G.-G., Strohm, J.J., Holladay, J., and Palo, D.R. (2007) Selective CO methanation catalysts for fuel processing applications. *Appl. Catal. A-Gen.*, **326** (2), 213–218.

44 Echigo, M. and Tabata, T. (2003) A study of CO removal on an activated Ru catalyst for polymer electrolyte fuel cell applications. *Appl. Catal. A-Gen.*, **251** (1), 157–166.

45 Federici, J.A. and Vlachos, D.G. (2011) Experimental studies on syngas catalytic combustion on Pt/Al$_2$O$_3$ in a microreactor. *Combust. Flame*, **158** (12), 2540–2543.

46 Dokupil, M., Spitta, C., Mathiak, J., Beckhaus, P., and Heinzel, A. (2006) Compact propane fuel processor for auxiliary power unit application. *J. Power Sources*, **157** (2), 906–913.

47 Lee, S., Ahn, K.Y., Lee, Y.D., Han, J., Im, S., and Yu, S. (2012) Flow uniformity of catalytic burner for off-gas combustion of molten carbonate fuel cell. *J. Fuel Cell Sci. Technol.*, **9** (2), 021006.

48 Meißner, J., Pasel, J., Samsun, R.C., Scharf, F., Wiethege, C., and Peters, R. (2014) Catalytic burner with internal steam generation for a fuel-cell-based auxiliary power unit for middle distillates. *Int. J. Hydrogen Energy*, **39**, 4131–4142.

49 Sarioglan, A., Can Korkmaz, Ö., Kaytaz, A., Akar, E., and Akgün, F. (2010) A 5 kW$_t$ catalytic burner for PEM fuel cells: effect of fuel type, fuel content and fuel loads on the capacity of the catalytic burner. *Int. J. Hydrogen Energy*, **35** (21), 11855–11860.

50 Kumar, R. (1994) Pressure swing adsorption process: performance optimum and adsorbent selection. *Ind. Eng. Chem. Res.*, **33**, 1600–1605.

51 Kohl, A. and Nielson, R. (1997) *Gas Purification*, 5th edn, Elsevier, Amsterdam.

52 Majlan, E.H., Wan Daud, W.R., Iyuke, S.E., Mohamad, A.B., Kadhum, A.A.H., Mohammad, A.W., Takriff, M.S., and Bahaman, N. (2009) Hydrogen purification using compact pressure swing adsorption system for fuel cell. *Int. J. Hydrogen Energy*, **34** (6), 2771–2777.

53 Pasel, J. and Peters, R. (2012) Desulfurization for fuel-cell systems, in *Fuel Cell Science and Engineering* (eds D. Stolten and B. Emonts), Wiley-VCH Verlag GmbH, Weinheim, p. 1011–1044.

54 He, H.P., Wood, A., Steedman, D., and Tilleman, M. (2008) Sulphur tolerant shift reaction catalysts for nickel-based SOFC anode. *Solid State Ionics*, **179** (27–32), 1478–1482.

55 Weber, A., Dierickx, S., Kromp, A., and Ivers-Tiffée, E. (2013) Sulfur poisoning of anode-supported SOFCs under reformate operation. *Fuel Cells*, **13**, 487–493.

56 Emonts, B., Hansen, J.B., Jørgensen, S.L., Höhlein, B., and Peters, R. (1998) Compact methanol reformer test for fuel-cell powered light-duty vehicles. *J. Power Sources*, **71**, 288–293.

57 Lin, Y.-M. and Rei, M.-H. (2000) Process development for generating high purity

hydrogen by using supported palladium membrane reactor as steam reformer. *Int. J. Hydrogen Energy*, **25**, 211–219.

58 Peters, R., Latz, J., Pasel, J., and Stolten, D. (2008) Desulfurization of jet A-1 and heating oil: general aspects and experimental results. *ECS Trans.*, **12** (1), 543–554.

59 Ockwig, N.W. and Nenoff, T.M. (2007) Membranes for hydrogen separation. *Chem. Rev.*, **107**, 4078–4110.

60 Uemiya, S. (1999) State-of-the-art of supported metal membranes for gas separation. *Separ. Purif. Method*, **28**, 51–85.

61 Tong, J., Shirai, R., Kashima, Y., and Matsumura, Y. (2005) Preparation of a pinhole-free Pd–Ag membrane on a porous metal support for pure hydrogen separation. *J. Membr. Sci.*, **260** (1–2), 84–89.

62 Hou, K. and Hughes, R. (2002) The effect of external mass transfer, competitive adsorption and coking on hydrogen permeation through thin Pd/Ag membranes. *J. Membr. Sci.*, **206**, 119–130.

63 Gielens, F.C., Knibbeler, R.J.J., Duysinx, P.F.J., Tong, H.D., Vorstman, M.A.G., and Keurentjes, J.T.F. (2006) Influence of steam and carbon dioxide on the hydrogen flux through thin Pd/Ag and Pd membranes. *J. Membr. Sci.*, **279** (1–2), 176–185.

64 Tosti, S., Basile, A., Chiappetta, G., Rizzello, C., and Violante, V. (2003) Pd–Ag membrane reactors for water gas shift reaction. *Chem. Eng. J.*, **93** (1), 23–30.

65 Vadrucci, M., Borgognoni, F., Moriani, A., Santucci, A., and Tosti, S. (2013) Hydrogen permeation through Pd–Ag membranes: surface effects and Sieverts' law. *Int. J. Hydrogen Energy*, **38** (10), 4144–4152.

66 Wang, Y., Latz, J., Dahl, R., Pasel, J., and Peters, R. (2009) Liquid phase desulfurization of jet fuel by a combined pervaporation and adsorption process. *Fuel Process Technol.*, **90** (3), 458–464.

67 Pasel, J., Wang, Y., Hürter, S., Dahl, R., Peters, R., Schedler, U., and Matuschewski, H. (2012) Desulfurization of jet fuel by pervaporation. *J. Membr. Sci.*, **390–391**, 12–22.

68 Ahmed, S., Pereira, C., Ahluwalia, R., Kaun, T., Liao, H.K., Lottes, S., Novick, V., and Krumpelt, M. (2002) Quick start fuel processor. Presented at the 2002 National Laboratory R&D Meeting, Golden, Colorado, 9–10 May 2002.

69 Lindström, B., Karlsson, J.A.J., Ekdunge, P., De Verdier, L., Häggendal, B., Dawody, J., Nilsson, M., and Pettersson, L.J. (2009) Diesel fuel reformer for automotive fuel cell applications. *Int. J. Hydrogen Energy*, **34** (8), 3367–3381.

70 Mao, C.-P., Short, J., Buelow, P., Caples, M., Siders, R., and Clausen, D. (2005) Innovative injection and mixing systems for diesel fuel reforming. Presented at the SECA 6th Annual Workshop 2005, Department of Energy: Pacific Grove, California.

71 Ogrzewalla, J., Adomeit, P., Severin, C., Pischinger, S., and Patil, P. (2003) Optimized mixture formation for fuel cells. Presented at the Fuel Cell Seminar. 2003. Orlando, Florida: Courtesy Associates.

72 Pasel, J., Meißner, J., Pors, Z., Palm, C., Cremer, P., Peters, R., and Stolten, D. (2004) Hydrogen production via autothermal reforming of diesel fuel. *Fuel Cells*, **4** (3), 225–230.

73 Yoon, S., Kang, I., and Bae, J. (2009) Suppression of ethylene-induced carbon deposition in diesel autothermal reforming. *Int. J. Hydrogen Energy*, **34** (4), 1844–1851.

74 Linnhoff, B. (1989) Pinch technology for synthesis of optimal heat and power systems. *J. Energy Resour. Technol.*, **111–148**, 137.

75 Maximini, M., Engelhardt, P., Grote, M., and Brenner, M. (2012) Further development of a microchannel steam reformer for diesel fuel. *Int. J. Hydrogen Energy*, **37** (13), 10125–10134.

76 Thormann, J., Pfeifer, P., Schubert, K., and Kunz, U. (2008) Reforming of diesel fuel in a micro reactor for APU systems. *Chem. Eng. J.*, **135S**, 74–81.

77 Hartmann, L., Lucka, K., and Köhne, H. (2003) Mixture preparation by cool

flames for diesel-reforming technologies. *J. Power Sources*, **118** (1–2), 286–297.

78 Hartmann, M., Maier, L., and Deutschmann, O. (2011) Hydrogen production by catalytic partial oxidation of iso-octane at varying flow rate and fuel/oxygen ratio: from detailed kinetics to reactor behavior. *Appl. Catal. A-Gen.*, **391**, (1–2). 144–152.

79 Lindermeir, A., Kah, S., Kavurucu, S., and Mühlner, M. (2007) On-board diesel fuel processing for an SOFC-APU – Technical challenges for catalysis and reactor design. *Appl. Catal. B-Environ.*, **70** (1–4), 488–497.

80 Aicher, T., Lenz, B., Gschnell, F., Groos, U., Federici, F., Caprile, L., and Parodi, L. (2006) Fuel processors for fuel cell APU applications. *J. Power Sources*, **154** (2), 503–508.

81 Kang, I. and Bae, J. (2006) Autothermal reforming study of diesel for fuel cell application. *J. Power Sources*, **159** (2), 1283–1290.

82 Kang, I., Bae, J., and Bae, G. (2006) Performance comparison of autothermal reforming for liquid hydrocarbons, gasoline and diesel for fuel cell applications. *J. Power Sources*, **163** (1), 538–546.

83 Karatzas, X., Nilsson, M., Dawody, J., Lindström, B., and Pettersson, L.J. (2010) Characterization and optimization of an autothermal diesel and jet fuel reformer for 5 kWe mobile fuel cell applications. *Chem. Eng. J.*, **156** (2), 366–379.

84 Lenz, B. and Aicher, T. (2005) Catalytic autothermal reforming of jet fuel. *J. Power Sources*, **149**, 44–52.

85 Pasel, J., Meisner, J., Pors, Z., Samsun, R., Tschauder, A., and Peters, R. (2007) Autothermal reforming of commercial Jet A-1 on a 5 kW$_e$ scale. *Int. J. Hydrogen Energy*, **32** (18), 4847–4858.

86 Pasel, J., Samsun, R.C., Peters, R., and Stolten, D. (2013) Fuel processing of diesel and kerosene for auxiliary power unit applications. *Energy & Fuels*, **27**, 4386–4394.

87 Roychoudhury, S., Lyubovsky, M., Walsh, D., Chu, D., and Kallio, E. (2006) Design and development of a diesel and JP-8 logistic fuel processor. *J. Power Sources*, **160** (1), 510–513.

88 Blum, L., Samsun, R.C., and Peters, R. (2012) Principles of system engineering, in *Fuel Cell Science and Engineering* (eds D. Stolten and B. Emonts), Wiley-VCH Verlag GmbH, Weinheim.

89 Basile, A., Gallucci, F., and Paturzo, L. (2005) Hydrogen production from methanol by oxidative steam reforming carried out in a membrane reactor. *Catal. Today*, **104** (2–4), 251–259.

90 Velu, S., Suzuki, K., Okazaki, M., Kapoor, M.P., Osaki, T., and Ohashi, F. (2000) Oxidative steam reforming of methanol over CuZnAl(Zr)-oxide catalysts for the selective production of hydrogen for fuel cells: catalyst characterization and performance evaluation. *J. Catal.*, **194** (2), 373–384.

91 Fierro, V., Akdim, O., Provendier, H., and Mirodatos, C. (2005) Ethanol oxidative steam reforming over Ni-based catalysts. *J. Power Sources*, **145** (2), 659–666.

92 Silberova, B., Venvik, H.J., and Holmen, A. (2005) Production of hydrogen by short contact time partial oxidation and oxidative steam reforming of propane. *Catal. Today*, **99** (1–2), 69–76.

93 Venkataraman, G., Sethuraman, S., Lux, K., Elder, W., Bhalerao, A., and Namazian, M. (2008) Distillate-fuel to power systems using high-temperature PEMFCs. Fuel Cell Seminar. Phoenix, Arizona: Courtesy Associates.

94 O'Connell, M., Kolb, G., Schelhaas, K.P., Schuerer, J., Tiemann, D., Ziogas, A., and Hessel, V. (2009) Development and evaluation of a microreactor for the reforming of diesel fuel in the kW range. *Int. J. Hydrogen Energy*, **34** (15), 6290–6303.

95 Ledjeff-Hey, K., Formanski, V., Kalk, T., and Roes, J. (1998) Compact hydrogen production system for solid polymer fuel cells. *J. Power Sources*, **71**, 199–207.

96 Taylor, J.D., Herdman, C.M., Wu, B.C., Wally, K., and Rice, S.-F. (2003) Hydrogen production in a compact supercritical water reformer. *Int. J. Hydrogen Energy*, **28** (11), 1171–1178.

97 Pinkwart, K., Bayha, T., Lutter, W., and Krausa, M. (2004) Gasification of diesel

oil in supercritical water for fuel cells. *J. Power Sources*, **136** (2), 211–214.

98 Marty, P. and Grouset, D. (2003) High temperature hybrid steam-reforming for hydrogen generation without catalyst. *J. Power Sources*, **118** (1–2), 66–70.

99 Czernichowski, A. (2001) GlidArc assisted preparation of the synthesis gas from natural and waste hydrocarbons gases. *Oil Gas Sci. Technol.*, **56** (2), 181–198.

100 Kim, S.C. and Chun, Y.N. (2008) Experimental study on partial oxidation of methane to produce hydrogen using low-temperature plasma in AC Glidarc discharge. *Int. J. Energy Res.*, **32** (13), 1185–1193.

101 Engelhardt, P., Maximini, M., Beckmann, F., and Brenner, M. (2012) Integrated fuel cell APU based on a compact steam reformer for diesel and a PEMFC. *Int. J. Hydrogen Energy*, **37** (18), 13470–13477.

102 Roychoudhury, S., Walsh, D., Mastanduno, R., Junaedi, C., DesJardins, J., and Morgan, C. (2008) Long term operation of a diesel/JP-8 fuel processor. Fuel Cell seminar. Phoenix, Arizona: Fuel Cell Associates.

103 Cheekatamarla, P. and Lane, A. (2005) Efficient bimetallic catalysts for hydrogen generation from diesel fuel. *Int. J. Hydrogen Energy*, **30** (11), 1277–1285.

104 Cheekatamarla, P.K. and Lane, A.M. (2005) Catalytic autothermal reforming of diesel fuel for hydrogen generation in fuel cells. *J. Power Sources*, **152**, 256–263.

105 Cheekatamarla, P.K. and Lane, A.M. (2006) Catalytic autothermal reforming of diesel fuel for hydrogen generation in fuel cells. *J. Power Sources*, **154** (1), 223–231.

106 Ferrandon, M. and Krause, T. (2006) Role of the oxide support on the performance of Rh catalysts for the autothermal reforming of gasoline and gasoline surrogates to hydrogen. *Appl. Catal. A-Gen.*, **311**, 135–145.

107 Ferrandon, M., Kropf, A.J., and Krause, T. (2010) Bimetallic Ni-Rh catalysts with low amounts of Rh for the steam and autothermal reforming of n-butane for fuel cell applications. *Appl. Catal. A-Gen.*, **379** (1–2), 121–128.

108 Ferrandon, M., Mawdsley, J., and Krause, T. (2008) Effect of temperature, steam-to-carbon ratio, and alkali metal additives on improving the sulfur tolerance of a $Rh/La–Al_2O_3$ catalyst reforming gasoline for fuel cell applications. *Appl. Catal. A-Gen.*, **342** (1–2), 69–77.

109 Harada, M., Takanabe, K., Kubota, J., Domen, K., Goto, T., Akiyama, K., and Inoue, Y. (2009) Hydrogen production by autothermal reforming of kerosene over $MgAlO_x$-supported Rh catalysts. *Appl. Catal. A-Gen.*, **371** (1–2), 173–178.

110 Haynes, D.J., Campos, A., Berry, D.A., Shekhawat, D., Roy, A., and Spivey, J.J. (2010) Catalytic partial oxidation of a diesel surrogate fuel using an Ru-substituted pyrochlore. *Catal. Today*, **155**, 84–91.

111 Kaila, R.K., Gutiérrez, A., and Krause, A.O.I. (2008) Autothermal reforming of simulated and commercial diesel: the performance of zirconia-supported RhPt catalyst in the presence of sulfur. *Appl. Catal. B-Environ.*, **84** (1–2), 324–331.

112 Karatzas, X., Creaser, D., Grant, A., Dawody, J., and Pettersson, L.J. (2011) Hydrogen generation from n-tetradecane, low-sulfur and Fischer–Tropsch diesel over Rh supported on alumina doped with ceria/lanthana. *Catal. Today*, **164** (1), 190–197.

113 Karatzas, X., Jansson, K., González, A., Dawody, J., and Pettersson, L.J. (2011) Autothermal reforming of low-sulfur diesel over bimetallic RhPt supported on Al_2O_3, $CeO_2–ZrO_2$, SiO_2 and TiO_2. *Appl. Catal. B-Environ.*, **106** (3–4), 476–487.

114 Krumpelt, M., Krause, T.R., Carter, J.D., Kopasz, J.P., and Ahmed, S. (2002) Fuel processing for fuel cell systems in transportation and portable power applications. *Catal. Today*, **77**, 3–16.

115 Qi, A., Wang, S., Fu, G., and Wu, D. (2005) Autothermal reforming of n-octane on Ru-based catalysts. *Appl. Catal. A-Gen.*, **293**, 71–82.

116 van Rheinberg, O., Lucka, K., Kohne, H., Schade, T., and Andersson, J. (2008) Selective removal of sulphur in liquid fuels for fuel cell applications. *Fuel*, **87** (13–14), 2988–2996.

117 Shamsi, A., Baltrus, J.P., and Spivey, J.J. (2005) Characterization of coke deposited on Pt/alumina catalyst during reforming of liquid hydrocarbons. *Appl. Catal. A-Gen.*, **293**, 145–152.

118 Yoon, S., Kang, I., and Bae, J. (2008) Effects of ethylene on carbon formation in diesel autothermal reforming. *Int. J. Hydrogen Energy*, **33** (18), 4780–4788.

119 Kopasz, J.P., Applegate, D., Rustic, L., Ahmed, S., and Krumpelt, M. (2000) Effects of gasoline components on fuel processing and implications for fuel cell fuels. Fuel Cell Seminar. Portland: Courtesy Associates.

120 Porš, Z., Pasel, J., Tschauder, A., Dahl, R., Peters, R., and Stolten, D. (2008) Optimised mixture formation for diesel fuel processing. *Fuel Cells*, **2**, 129–137.

121 Jörissen, L., Lehnert, W., Garche, J., and Tillmetz, W. (2007) Lifetime of PEMFCs. *Mater. Sci. Forum*, **539–543**, 1303–1308.

122 Schmidt, T.J. and Baurmeister, J. (2006) Durability and reliability in high-temperature reformed hydrogen PEFCs. *ECS Trans.*, **3** (1), 861–869.

123 Moore, J.M., Adcock, P.L., Lakeman, J.B., and Mepsted, G.O. (2000) The effects of battlefield contaminants on PEMFC performance. *J. Power Sources*, **85**, 254–260.

124 Li, H., Zhang, J., Fatih, K., Wang, Z., Tang, Y., Shi, Z., Wu, S., Song, D., Zhang, J., Jia, N., Wessel, S., Abouatallah, R., and Joos, N. (2008) Polymer electrolyte membrane fuel cell contamination: testing and diagnosis of toluene-induced cathode degradation. *J. Power Sources*, **185** (1), 272–279.

125 Dorn, S., Bender, G., Bethune, K., Angelo, M., and Rochelea, R. (2008) The impact of trace carbon monoxide/toluene mixtures on PEMFC performance. *ECS Trans.*, **16** (2), 659–667.

126 Novochinskii, I.I., Song, C., Ma, X., Liu, X., Shore, L., Lampert, J., and Farrauto, R. J. (2004) Low-temperature H_2S removal from steam-containing gas mixtures with ZnO for fuel cell application. 1. ZnO particles and extrudates. *Energy & Fuels*, **18**, 576–583.

127 Kolb, G., Schürer, J., Tiemann, D., Wichert, M., Zapf, R., Hessel, V., and Löwe, H. (2007) Fuel processing in integrated micro-structured heat-exchanger reactors. *J. Power Sources*, **171** (1), 198–204.

128 Liu, X., Ruettinger, W., Xu, X., and Farrauto, R. (2005) Deactivation of Pt/CeO_2 water-gas shift catalysts due to shutdown/startup modes for fuel cell applications. *Appl. Catal. B-Environ.*, **56** (1–2), 69–75.

129 Liu, X., Guo, P., Wang, B., Jiang, Z., Pei, Y., Fan, K., and Qiao, M. (2013) A comparative study of the deactivation mechanisms of the Au/CeO_2 catalyst for water–gas shift under steady-state and shutdown/start-up conditions in realistic reformate. *J. Catal.*, **300**, 152–162.

10
Small-Scale Reforming for On-Site Hydrogen Supply

Ingrid Schjølberg, Christian Hulteberg, and Dick Lieftink

10.1
Introduction

On-site hydrogen supply is important in the development of hydrogen supply infrastructure both for the transport sector as well as for stationary applications. Both reforming and electrolysis are foreseen as future on-site supply technologies. Producing hydrogen by reforming is a well-proven technology and has been applied in large-scale centralized production plants (i.e., Haldor Topsøe and Mahler AGS) since the 1950s. However, hydrogen can be produced on a small scale using the same technology, by down scaling of equipment, and a large effort is being made by several equipment suppliers to harmonize components and to offer containerized reformer units. Small-scale reforming of hydrocarbons is currently one of the cheapest options for on-site hydrogen production and is in direct competition with the industrial gas market [1].

A large deployment of hydrogen refueling stations is predicted for the period 2020–2030; this is to meet the demand arising from the rollout of a large number of hydrogen fuel cell vehicles. To meet the demand for hydrogen, on-site small-scale reforming of natural gas is one option due to the extensive availability of natural gas through existing distribution networks. The UK H2 mobility program foresees that a hydrogen production mix will consist of 47% steam methane reforming in 2030 [2]. Projections based on high-volume production indicate that reforming natural gas at the fueling station can produce hydrogen at a cost close to $2 per gge [3]. On-site reforming is seen as an intermediate solution due to the inevitable emission of fossil-based CO_2. On a long-term perspective such production must be based on renewables (to achieve low fossil-based CO_2 emissions), as on-site carbon dioxide capture is costly and adds complexity to design and operation.

Large roll-out of small-scale reformer units require standardized and harmonized components to reduce manufacturing costs and it is one of the main challenges to enable mass production of such units. In addition, there is a need for cheaper and more robust materials and in particular for hot parts, sulfur tolerant catalysts, more compact clean-up components, as well as improved safety and

Hydrogen Science and Engineering: Materials, Processes, Systems and Technology, First Edition.
Edited by Detlef Stolten and Bernd Emonts.
© 2016 Wiley-VCH Verlag GmbH & Co. KGaA. Published 2016 by Wiley-VCH Verlag GmbH & Co. KGaA.

control methods. In the following an overview of the technology and suppliers is given as well as a short presentation of future technology development.

10.2
Definition

Reforming is a processing technique where the molecular structure of hydrocarbons is rearranged to alter the gas properties. Steam methane reforming is often applied to natural gas to alter the combustion characteristics. Thermal reforming alters the properties of low-grade naphthas by converting the molecules into those of higher octane number by exposing the materials to high temperatures and pressures.

There is no common definition of the term *small scale* relative to production rate. Production rates may be in the range of 0.48 Nm^3 h^{-1} for small-scale methanol reformers via 50–500 Nm^3 h^{-1} natural gas based hydrogen generators to thousands of Nm^3 h^{-1} for centralized units. A typical steam reforming system may consist of many components in addition to the reformer itself such as compressor, burner, desulfurization, clean-up, purification, heat exchangers, and so on, as illustrated in Figure 10.1. Harmonization of capacities, with the objective to standardize components and size, will reduce costs, minimize engineering hours, and reduce the need for expensive customization. It is suggested that such harmonization is carried out for 100, 300, and 500 Nm^3 h^{-1} [4]. These sizes are chosen to meet the need of neighborhood, small forecourt, and forecourt type production.

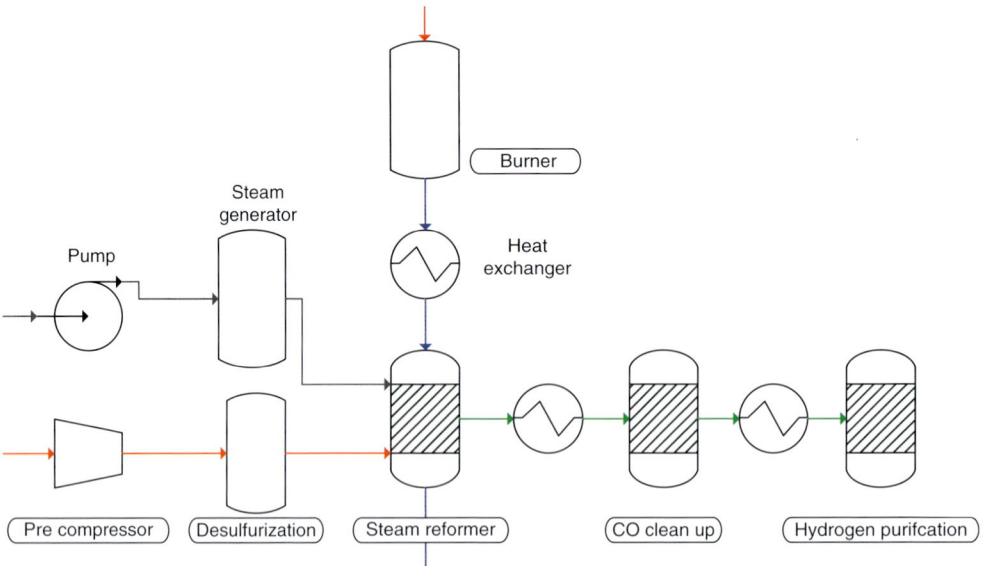

Figure 10.1 Schematics of components in an industrial reformer unit.

The NREL's study on state of the art of electrolytic hydrogen production has categorized four different size ranges [5]. Similarly, available on-site, small-scale reformer systems can be categorized by defining sizes relative to, for instance, the number of hydrogen cars that can be served:

- small neighborhood: serves fuel needs of 5–50 cars with a hydrogen production rate of 1000–10 000 kg-H_2 year^{-1} (1.5–15 $Nm^3 h^{-1}$);
- neighborhood: serves fuel needs of 50–150 cars with a hydrogen production rate of 10 000–30 000 kg-H_2 year^{-1} (15–45 $Nm^3 h^{-1}$);
- small forecourt: serves fuel needs of 150–500 cars with a hydrogen production rate of 30 000–100 000 kg-H_2 year^{-1} (45–150 $Nm^3 h^{-1}$);
- forecourt: serves fuel needs of more than 500 cars with a hydrogen production rate larger than 100 000 kg H_2 year^{-1} (>150 $Nm^3 h^{-1}$).

Stakeholders in IEA Hydrogen Implementing Agreement defined small scale in the range of 50 Nm^3/h to 500 Nm^3/h, and this definition is applied in the following.

Table 10.2 below presents several commercial available reformers indicating the availability of systems in the small-scale range.

10.3
Reforming Technologies

Reforming technologies are based on the blending of a hydrocarbon fuel with steam and potentially with oxygen and air. Strictly speaking, reforming without water is not seen as reforming but in some cases it could be used as a hydrogen production technology. The basic technologies are steam reforming (SR) and partial oxidation (POX) [6].

10.3.1
Steam Methane Reforming (SMR)

Steam reforming is an endothermic process; the required heat can be delivered by burning part of the fuel in a separate burner. The heat is transferred by a heat exchanger system to the reformer catalyst. The steam reforming process is the most commonly considered one for small-scale production of hydrogen from natural gas. The reaction involves the reaction of methane with steam over a catalyst surface at elevated temperatures to yield hydrogen:

$$CH_4 + H_2O \rightarrow CO + 3H_2, \quad \Delta H = 206 \text{ kJ mol}^{-1} \tag{10.1}$$

The reaction is an equilibrium reaction and the production of hydrogen is favored by low pressure and high temperature. The normal operating temperature is from 700 to 800 °C and pressures range from ambient to 30 bar. The reaction is endothermic and therefore heat is required to perform the reaction.

This heat is normally supplied by combustion of waste gas stream in the hydrogen production and supplemental methane, should it be required. The reaction is performed using a catalyst, which is most commonly Ni-based; the cost of noble metal-based catalysts is, however, not prohibitive on the smaller scale. The catalysts employed are prone to carbon formation, coking, under standard operating conditions, which is why an excess of steam is used to suppress this formation. Steam-to-carbon ratios of 2.5 : 1 to 4.5 : 1 are common. Another catalyst poison is sulfur, which has to be removed upstream of the reactor.

The steam reforming reaction is connected to another equilibrium reaction, the water–gas shift:

$$O + H_2O \rightarrow CO_2 + H_2, \quad \Delta H = -41 \text{ kJ mol}^{-1} \tag{10.2}$$

At the elevated temperatures (700–800 °C) used in steam reforming CO is favored and to increase the hydrogen yield the temperature has to be reduced to as low-a-temperature as possible. At reduced temperature it is possible to increase the hydrogen produced using one or several catalytic reactors; the suitable operating temperature is governed by catalyst kinetics. The catalysts traditionally used are Fe/Cr for high temperature water–gas shift (350–450 °C) and Cu/ZnO for low temperature water–gas shift (250–280 °C); medium temperature noble metal catalysts are also available (300–350 °C) [7].

10.3.2
Partial Oxidation

In partial oxidation the reaction heat is provided by the partial combustion of hydrocarbons:

$$CH_4 + 0.5O_2 \rightarrow CO + 2H_2, \quad \Delta H = -73 \text{ kJ mol}^{-1} \tag{10.3}$$

There are three main routes for partial oxidation, namely, non-catalytic [8] (POX), auto-thermal reforming (ATR), and catalytic partial oxidation (CPO) [9]. Catalysts range from base-metals such as cobalt and nickel to noble metals such as iridium, palladium, rhodium, and ruthenium.

Aside from catalyst poisoning and carbon formation problems, the POX reaction is performed at higher temperatures than SR and ATR, making catalyst stability and sintering more of a concern.

In autothermal reforming oxygen (air) is added together with steam to the fuel. Part of the fuel is combusted, providing heat to the reform reaction in the same reactor. ATR reactors are usually simpler than steam reformers, but the mixing of fuel with oxygen needs cautious controls to avoid undesired side effects like pre-combustion, occurrence of hot spots in the reactor, and carbon formation. At elevated pressures the mixing of oxygen and fuel is even more challenging from a safety point of view.

An ATR catalyst is usually a precious metal wash-coated on a monolithic structure. The source of oxygen can be air, which results in addition of nitrogen as well, which increases the flows through the system. The production of

Table 10.1 Comparison of reformer technologies [11].

Technology	Pros	Cons
Steam reforming	Well-proven technology; oxygen supply not required; low operating temperature; highest H_2/CO ratio; handles various feedstocks	Highest air emissions; heavy system; heat source required; slow startup
Partial oxidation	Higher sulfur tolerance; no external heat source; compact system; quick startup	Low H_2/CO ratio; coke formation; high operating temperature; oxygen and air supply required; exothermic reaction
Autothermal reforming	Medium operating temperature; no external heat source; favorable H_2/CO ratio; low coke formation; lower footprint	Limited experience; oxygen or air required

hydrogen usually requires a high pressure separation step, pressure swing adsorption, or membrane separation. When ATR is used as hydrogen production technology the air has to be compressed before the fuel processing or a compression step of the reformate needs to be incorporated. Both compression steps have an impact on the efficiency of the system. In the case of hydrogen production for a fuel cell system, however, ATR at ambient pressure has a benefit, because it is small, easy to modulate and the compression steps is not required. In Reference [10] the sustainability of five alternative processes for hydrogen production by steam reforming of natural gas was investigated, concluding that the integrated reactor process and autothermal process are the most sustainable.

In a steam reformer the energy producing combustion reaction is separated from the high pressure steam reforming, which makes the oxygen control easier. Table 10.1 gives a comparison of the reforming technologies.

10.3.3
Hydrogen Purity

To produce high purity hydrogen, purification technologies are available for separating hydrogen from the others species like CO, CO_2, and CH_4. Such separation technologies need elevated pressures – both pressure swing adsorption (PSA) and membrane separators work at pressures of 6 bar and higher. PSA has been commercially available since the early 1970s. The main target for PSA development is to increase the yield of the systems and reduce the costs of small-scale systems. There are still improvements possible in the realm of adsorbent materials and in materials for the pressure vessels. Since a higher pressure difference favors the separation of the gas species, stronger materials are needed that still are (relatively) light weight and economically affordable.

Conversely, a vacuum can also be used to create bigger pressure differences to improve the yield of the PSA. This pressure requirement makes the use of ATR as hydrogen production technology less favorable. To select the best fuel processing technology one should look at the efficiency which is related to a series of operating conditions [12]. Next to efficiency the technology also heavily depends on the customer demands for temperature, pressure, and purity of the hydrogen product.

Commonly speaking, steam reforming is known as the most economical and efficient technology for producing hydrogen in the small scale range.

The various applications of hydrogen have a range of purity demand covering from 99.999% (5.0 quality) and higher for fuel cell and electronics applications to more relaxed requirements for industrial applications. The exact speciation of the impurities is an important factor. Natural gas often contains traces of noble gasses (He, Ne), which are considerably more difficult to remove than nitrogen, which is present in some L-gasses in, for example, the Netherlands, These species, however are chemically inert. Other impurities like carbon monoxide have to be removed to sub-ppm levels for some applications because it is a poison for instance for fuel cells. The purity of the hydrogen product depends partly on the components (main and impurities) in the feed stream of the fuel processor but mainly on the performance of the PSA or membrane reactor in cleaning up the product. A typical hydrogen production versus purity curve is shown in Figure 10.2 based on HyGear's HGS-L system. These data are generated on multiple days using a steam

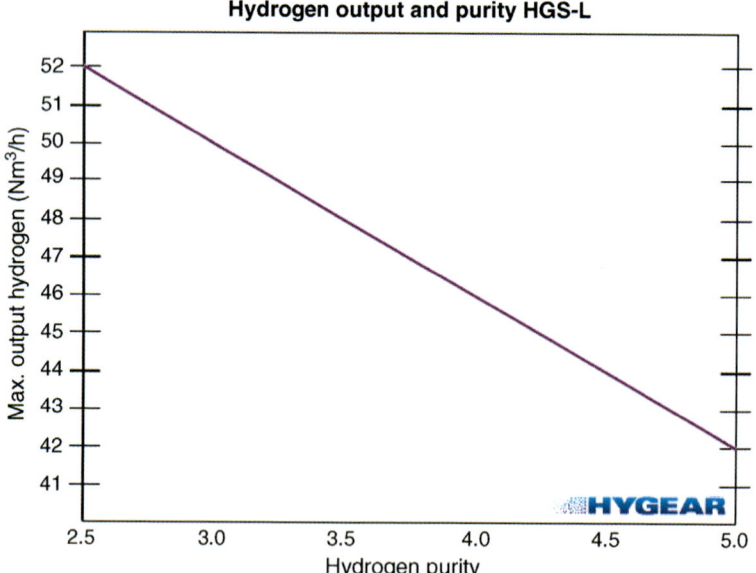

* All data and values are indicative and might differ due to local circumstances and feedstock characteristics.

Figure 10.2 Hydrogen production versus purity for HyGear's HGS-L system.

methane reformer followed by a PSA. These yield–purity curves are a function of the PSA bed size, PSA controls, and adsorption and desorption characteristics of the PSA materials. The HGS-L system produces 42–52 Nm^3 h^{-1} of hydrogen product (99.5–99.999% purity) at a maximum 5 bar(g) delivery pressure. The feedstock is natural gas (950 MJ h^{-1} at 7–8 bar(g)) and it has a 7.1 kW$_e$ in electricity consumption. The weight of the unit is 5500 kg and the dimensions 4056 × 2438 × 2591 mm^3 (length × width × height). The unit has a 30 min start up time from warm and 3 h from cold state and an output of 0–100% H$_2$ product flow.

10.4
Feedstock Options

Small-scale reforming can be performed on any gaseous or liquid hydrocarbon source available as feedstock. In many cases the differences in reforming technologies are in the preparation/clean-up of the hydrocarbon feed stream. Trace contaminants (S, P, Si, Cl, F) have to be removed as they harm the reformer catalyst and other downstream modules. Technologies for the removal will vary with the concentration and speciation of the impurities. A second variable which depends on the type of feedstock is the amount of steam that has to be generated for reforming. This determines the efficiency and reliability of reforming of specifically high molecular feedstocks like diesel.

The fuel choice is made based on logistics and availability of the fuel in the area where the system will be placed. The most common feedstock for steam reformers is based on methane, as natural gas or (upgraded) biogas. In some developing countries like India, natural gas is not the fuel of choice; there is a demand for on-site hydrogen generation system fueled by LPG. LPG is one of the most widely used and available fuels in the world. Due to its low condensation point, it is a highly logistical fuel and can be transported to almost any site.

In some international research projects, the production of on-site hydrogen based on renewable feedstocks like biodiesel (www.nemesis-project.eu) or bioethanol is studied. The reasoning is to use the bio-feedstock as an environmentally neutral technology as transition towards a hydrogen based economy.

Emissions associated with the feedstock can be reduced by using renewable feedstock or using a blend of fossil and renewable feedstock [1]. Upgraded biogas, biomethanol, bioDME, bioethanol, biodiesel, bioFT-diesel, and biopropane are examples of such feedstock.

10.4.1
Upgraded Biogas

Upgraded biogas (CH$_4$ concentrations ranging from 85% to 100%) is, chemically, equal to natural gas and there should be no major problems associated with using upgraded biogas in a commercial reformer system. Biogas from landfills needs to be cleaned not only of the previous mentioned components such as sulfur and others but also from nitrogen [13]. Carbon dioxide and nitrogen can be present in biogas in significant concentrations ranging from 0 to 50 vol.%. Nitrogen is an unwanted impurity in produced hydrogen and should be removed preferably before reforming. Though the nitrogen will not harm the reformer system, it must be heated to reformer temperatures and it increases the gas hourly space velocity in the reformer reactor up to 5% and therefore decreases the efficiency of the system. Downstream of the reformer the nitrogen can also be removed by PSA technologies. Carbon dioxide from the biogas is also preferably separated from the biogas as it drives the chemical equilibrium away from optimal hydrogen production. In addition, it increases the chance of carbon formation in the reformer reactor. Small concentrations of phosphorous and silicon (0–50 mg m^{-3}) in the biogas [14] can poison the reformer catalyst and should be taken out together with sulfur species.

10.4.2
Biomethanol and bioDME

Methanol steam reforming is a way to release hydrogen from methanol and water. In addition, DME can be transformed into hydrogen by steam reforming. The feedstock is mixed with water and catalyst reforming is performed at 240–280 °C at 10–30 bar on a copper containing catalyst. The focus is on the suppression of CO but the gas from the reformer typically contains less than 1 mol.% CO, which simplifies the clean-up process for pure hydrogen. Contamination of methanol fuel with chloride and sulfate may poison the reforming catalyst. The chlorine can accidently be introduced in the distribution chain and will lead to reduced engine performance due to the formation of plugging deposits and cause corrosion in engine components.

10.4.3
Biodiesel

Biodiesel is produced from vegetable oils, animal fats, or recycled restaurant greases. It is safer and cleaner than petroleum-based diesel. Biodiesel can be used in its pure form (B100) or blended with petroleum diesel. Common blends include B2 (2% biodiesel), B5, and B20. In general, biodiesel contains less sulfur than fossil diesel [15]. A requirement for the biodiesel industry is to refine the fuel to remove sulfur to the level of 15 ppm as specified by the US Environmental Protection Agency. Sulfur content in raw materials is typical lower than 30 ppm and, for example, washing of the biodiesel is required to insure minimal sulfur carry into the fuel.

Biodiesels are a promising feedstock for the production of hydrogen by reforming. Hydrogen can be produced by using commercial reforming catalysts and process conditions close to those for steam-reforming of naphtha. However, activities remain mostly at the R&D stage and most of available reformers are laboratory-scale reactors.

The main challenge with biodiesel reforming is related to soot formation. Gradual deactivation of the catalysts by carbon deposition is often observed during long-duration reforming experiments. Current R&D efforts are addressing this problem [16]. The combination of biomass gasification and subsequent Fischer–Tropsch (FT) synthesis results in a sulfur free, pure alkane based diesel fuel that is easier to reform than the fossil counterparts.

10.4.4
Biopropane

Propane reformers are available on a commercial basis and as biopropane is identical to propane there are no reformer issues that need to be addressed.

10.5
Suppliers and Products

A limited number of companies offer small-scale solutions in hydrogen production by reforming, since minimizing the SR-technology is accompanied by challenges related to cost versus energy efficiency. As technology is improving, it may be stated that the number of companies offering energy efficient small-scale reformers is growing. Due to competitive sensitive information, it is difficult to obtain reliable data on which companies offer this technology and how mature their solution is. Table 10.2 gives an overview of public information on companies capable of offering small-scale SR systems between 5 and 1000 $Nm^3 h^{-1}$. Efficiencies defined by the high heating value (HHV) lie in the area of 65–75 [1].

10.5.1
Cost Trends

At a time when clean and green technology serves as a unique selling point and more and more countries impose on themselves (or others) a reduction in the emission of greenhouse gasses, opportunities are created for clean and energy efficient solutions. This also calls for solutions in hydrogen supply: in line with total energy efficiency improvement policies, energy security policies, and local emissions reductions, hydrogen as an energy carrier is starting to be implemented. On-site hydrogen generation improves the efficiency and emission reduction of the total supply chain even further.

10.5.1.1 Central Plant Production
Most of the hydrogen (95%) in the USA is produced in large-scale plants by steam methane reforming [17]. The bulk of such produced hydrogen is immediately used on-site for oil refinery. This production method will remain to supply large amounts of hydrogen or occasional hydrogen requests, but it is inevitable

Table 10.2 Overview hydrogen generator sizes of some equipment suppliers.

Company name	Product capacity[a] (Nm³ h⁻¹)				Specification
	5–50	50–100	100–500	500–1000	
Air Products				x	Up to 1000+
Ally High Tech			x	x	400–20 000
Angstrom Advanced		x	x	x	50–20 000
Caloric			x	x	200–10 000
HyGear	x	x	x		4–250
Linde			x	x	100–1000
Mahler AGS			x	x	100–10 000
Mitsubishi Kakoki		x	x	x	50–1000+
Osaka	x	x			30–100
Haldor Topsøe			x	x	300–10 000

a) The numbers above are found on the companies' web sites and might not represent the actual capabilities of the companies.

that these companies will have to meet CO_2-reduction policies and this will increase production prices by at least 35% [18]. Furthermore, the rising prices of fossil fuels [19] used for distribution of centrally produced hydrogen to end users by road transport will contribute to higher prices of hydrogen supplied by central plants as well. In addition, the firm safety restrictions for hydrogen storage and transportation are more and more difficult to meet. On-site production of hydrogen omits such storage and transportation. Therefore, on-site hydrogen generation is expected to become a more interesting option in many cases and is thus expected to grow in the near future. Moreover, the International Energy Agency predicts that fossil fuels will remain the dominant energy sources until 2035 [20].

10.5.1.2 On-Site SR

Hydrogen production by steam methane reforming (SMR) generates hydrogen from natural gas. On-site SMR has a higher efficiency than electrolysis, at about 70% for SMR against around 60% for electrolysis. When the comparison is made based on natural gas as fuel the efficiency of electrolyzers becomes around 30% taking into account the large-scale production of electricity (efficiency of 52%) and transport losses (7%). Large-scale hydrogen production has a 10–15% higher energy efficiency compared with on-site production due to economy of scale and lower heat losses. However, compression and distribution of the hydrogen is highly ineffective, resulting in 20% lower overall energy efficiency than with on-site SMR. Since natural gas is the direct input for the process and carbon dioxide is emitted on the spot, SMR is often not considered as being a clean green solution. Indeed, fossil fuel reforming does not eliminate CO_2 release into the atmosphere, but reduces the carbon dioxide emissions as compared to the burning of conventional fuels due to increased efficiency. Depending on the feedstock, small-scale steam methane reforming can become cleaner and contribute to the reduction of methane and CO_2 outlet into the atmosphere.

10.5.1.3 Economics: On-Site versus Central Plant Production

Adaptation of on-site production technologies depends heavily on cost-effectiveness of the solutions. Centralized production of hydrogen is most cost-effective as long as the hydrogen does not need to be transported on the road. When the hydrogen needs to be spread across a wider area, on-site hydrogen production is often beneficial as illustrated in Figure 10.3. Depending on the geographical position the distance for break-even for on-site SMR is approximately 400 km. The world energy outlook to 2040 predicts that electricity prices will increase due to higher fuel prices, increased use of renewables, and, in some regions, CO_2 pricing [21]. Natural gas price projections are significantly lower than past years due to an expanded shale gas resource base [22]. This development in electricity and natural gas prices can lead to increased use of small-scale reformers for on-site hydrogen supply. One special driver for costs of reformers is the engineering of custom designed units. Currently units are produced on demand and mass production benefits are not achieved. Another cost driver is design and building of the manufacturing system. Investments in new manufacturing processes will not occur unless the market is stable and growing, but when that happens the cost of production may be reduced by more than 60%. Figure 10.4

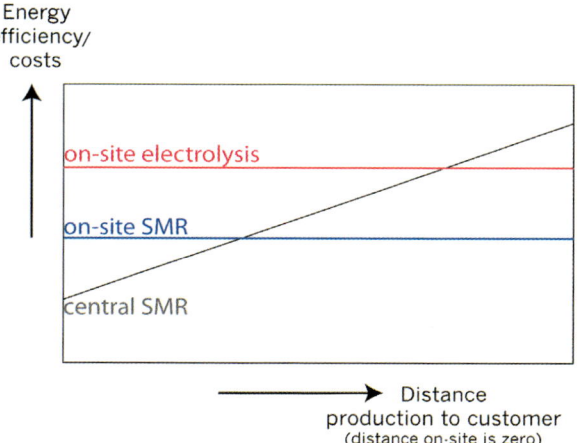

Figure 10.3 Break even points for energy efficiency relative to costs versus production method and distribution distance (0–400 km).

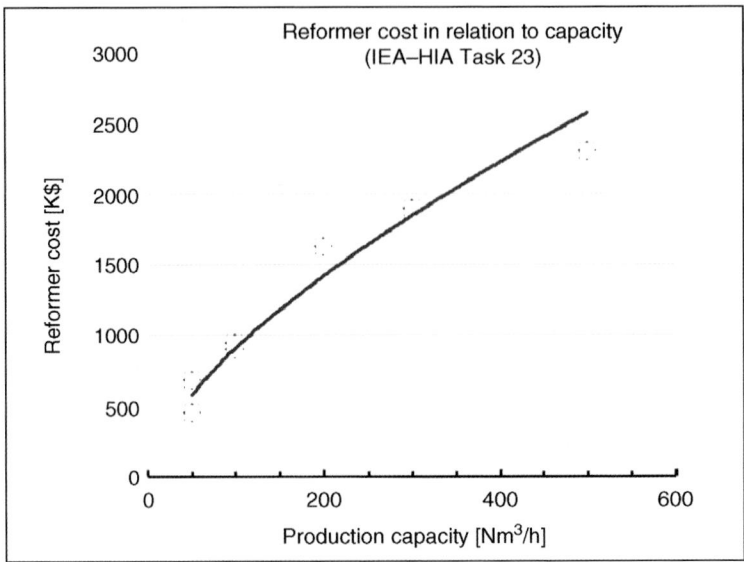

Figure 10.4 Reformer cost versus production capacity [1].

shows the cost development of reformers relative to production capacities. Larger plants are as expected relatively cheaper.

The footprint of a small-scale hydrogen generator based on steam reforming for $100\,m^3\,h^{-1}$ production available in the industry at the moment is between 10 and $15\,m^2$ [1], depending on the required amount of storage. There will always be storage of hydrogen necessary even when hydrogen is produced locally to cover peak demand for instance during rush hours in refueling stations. For refueling stations this footprint is important because of space limitation on the sites. The space is comparable with the area needed for maneuvering trucks with trucked-in hydrogen.

Next to the footprint of the machine itself, safety zoning around the system should also be considered. Usually, there should be a zone of 3–5 m around the system depending on the components (ATEX Directives) and the systems around the hydrogen generator like storage tanks, buildings, and so on. Depending on the site and local and national regulations those safety distances can vary.

10.6
Emerging Technologies

Producing hydrogen by steam-methane reforming is a well-proven technology but other technologies for small-scale production of hydrogen are emerging and are described briefly in the following. These improvements over state-of-the-art technology will be grouped into material improvements (including catalysts), reactor improvements, and hydrogen clean-up improvements; in some cases some solutions are difficult to separate and have been put under the category where it makes the most sense.

10.6.1
Material Development

When designing a small-scale SMR system there are many elements that differ compared to the design of a centralized hydrogen production plant. When looking at the *modus operandi* of the small-scale system it will inherently be prone to more variations in load and many more startup/shut-down cycles can be anticipated. The catalyst used must therefore be chosen to be more robust with respect to air; in addition, pyrophoricity should be avoided. The cost of the catalyst in relation to the overall investment cost is also relatively small, which is why a highly active noble metal catalyst may be considered for the system [23]. Another area of interest is the development and implementation of sulfur tolerant catalysts for the various reactions involved in hydrogen generation. The reason for this is to further enhance the degrees of freedom in system design. The use of noble metal catalysts will further the sulfur resistance of the systems as these may be formulated to be resistant to some level of sulfur [24].

Another interesting development area is the use of *in situ* sorption of a reaction product to further the production of hydrogen, altering the equilibrium position by this phase-transition: sorbent-enhanced reforming (SER). The two potential gas components that are considered for removal are carbon dioxide and hydrogen, with the first being by far the most investigated route [25,26]. There has been much effort spent in this area over recent decades; however, the advent of small-scale systems opens up new innovative solutions. For instance, the cost of regeneration may be reduced significantly by using streams other than steam, which is normally considered for the regeneration. Such streams may be off-gas streams from PSA, flue-gas, or cathode off-gas (when integrated with a fuel cell). There is also significant potential for improvement of the sorbent and combination of sorbent and catalyst. Hydrogen sorption-enhanced reforming (HSER) is another alternative process for hydrogen production. It uses a bed containing a mixture of a reforming catalyst and a selective sorbent to remove H_2 at high temperatures from the reaction zone, shifting the equilibrium limitation of conventional steam methane reforming.

This section describes the requirements for materials used in small-scale reformer technology. The environments in the reformer can vary a lot, ranging from exposed to air at room temperature and at (catalytic) burner temperatures of 800 °C and higher. The other extreme is a hydrogen-containing atmosphere at similar temperature levels. Another complicating factor is the presence of steam, which can create an oxidizing atmosphere relative to some metals even in the presence of hydrogen. Reformers for the production of hydrogen always contain an elevated pressure step like PSA or membrane separation. Consequently, next to the elevated temperatures the material should also withstand pressure up to 20 bar or higher depending on the layout of the hydrogen clean-up process. Typical metals used currently in the high temperature and pressure zones of the reformer are Inconel and FeCr alloy type materials. The processes taking place that degrade the reformer system, for example, hydrogen embrittlement and metal dusting, should be determined by life-time studies of materials. With these studies, one can determine the degradation rate and design the proper wall thickness of the reactors and heat exchangers to reach the desired lifetime of the reactors, guaranteeing the safety aspects of the system as well.

10.6.2
Reactor Improvements and Design Aspects

The most obvious advantage is the compact design of heat-exchanger reactors in comparison to ordinary steam reformer units. Indeed, most of the volume corresponding to the oven and the burners can be removed. In addition, the catalyst volume can be reduced since the thin layers added to the surfaces guarantee an effective utilization of the active material. An additional advantage is that the close coupling between the reactions permits the temperature and the process to be controlled very precisely. In addition, micro-channel reactors may be used where micro-technology is employed to improve heat and mass transfer rates either in reactors or indeed in heat-exchanger reactors [27].

A membrane reformer (MRF) equipped with palladium membrane modules for *in situ* hydrogen separation can offer more compact, simpler, and more efficient hydrogen production than the conventional steam methane reformer. Plasma reforming describes processes where hydrogen production takes place in plasma, with or without the presence of a catalyst. A plasma is an ionized gaseous medium, often described as the fourth state of matter, which is composed of free moving electrons, positive ions, radicals, neutral atoms, and molecules. A plasma can be generated by several methods, but electrically generated plasmas (cold plasmas) appear to be the best suited for hydrogen production.

10.6.3
Clean-Up Technology

Gas clean-up can roughly be divided in two types of technologies, PSA (pressure swing adsorption) and membrane technologies. Developments in hydrogen clean-up are driven not only by cost and performance, but also by the purity requirements of the fuel cell. If the fuel cell requires higher purity more materials and controls have to be developed to meet these requirements. This is valid for both PSA and membrane technologies. This will spur development of new classes of materials to be used as sorbents in a PSA set-up, equilibrium and rate-limited based, as well as improvement of Pd and Pd-alloy membranes to reduce the contamination of CO and other fuel cell poisons in the product gas. This will be done both by improved manufacturing technology and also by new membrane materials.

10.6.4
Small-Scale CO_2 Capture

Technology for small-scale carbon capture and sequestration (CCS) exists but profitability depends on availability of storage. Whenever there is a demand for CO_2 in the neighborhood of the small-scale reformers CCS should be considered.

A comparison of the CO_2 footprint of hydrogen production from fossil fuels was given in Reference [1]. The CO_2 in the off gas of a membrane reformer can be easily separated and captured by direct liquefaction owing to highly concentrated CO_2. Figure 10.5 shows an example of how CO_2 in the PSA off-gas is liquefied and then purified by cryogenic distillation.

There are several examples of prototype reactors with CO_2 capture, one is at a Tokyo gas refilling station, showing good performance combined with compactness [28].

The CO_2 emissions from small-scale reforming are associated with [1]:

- feedstock
- reformer efficiency
- electricity source.

Figure 10.6 shows carbon footprint of reforming natural gas relative to reformer efficiency. The figure shows that the reformer efficiency does not have a significant effect on the carbon footprint, indicating that emissions associated with electricity production and feedstock are significant.

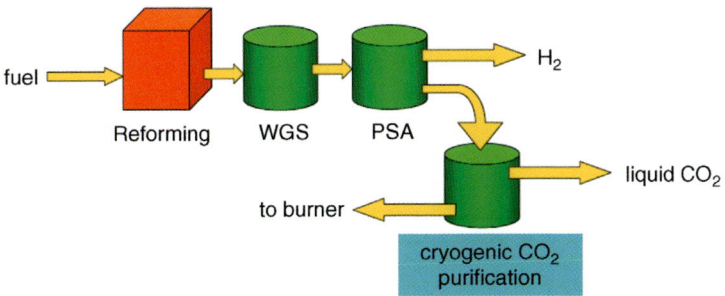

Figure 10.5 Steam methane reforming and water–gas shift reactor (WGS) and liquefied CO_2 (PSA, pressure swing adsorption) [1].

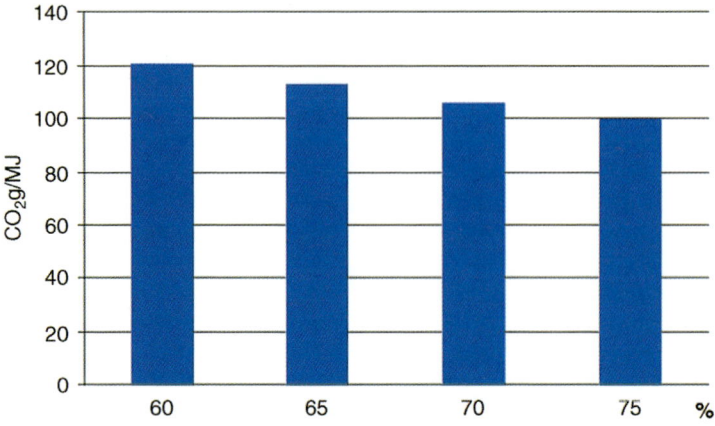

Figure 10.6 Carbon footprint of hydrogen versus reformer efficiency [8].

10.7
Process Control

Process control of small-scale reformers includes control of pressure, temperature, and flow rates. Many sophisticated control strategies have been developed in recent decades, but the PID (proportional-integral-derivative) controller is the most common in industrial use due to its simplicity and robustness. Catalyst deactivation and reformer performance are main control objectives for optimal design of control systems for small-scale reformers. The catalyst may be permanently deactivated by sintering (loss of surface area) during operation, therefore careful control of burners are necessary even though high-temperature materials are used. In addition, temperature control must be tight to avoid overheating of reactor components. Process control is considered at several layers including regulatory control of the units and coordinated control of the whole facility. The latter is often denoted as plant-wide control. High operability and robustness of reformer units can be achieved through robust control structure design [29].

Nevertheless, nonlinear process dynamics can provide challenges with regards to the use of PID controllers [30]. More advanced control schemes are often required to achieve minimal energy consumption, handling variations in plant load, and minimization of steam consumption and optimization of startup time.

10.7.1
Condition Monitoring

Small-scale on-site hydrogen generators are expected to run unattended, without the continuous interference of an operator. This requires a solid and robust control strategy of the system, as is common practice for commercial systems. These controls can only work when there are sensors located in the system for measuring critical variables such as temperature, flow rates, and ratios. Some sensors are critical for operation and safety, for instance thermocouples and oxygen and CO sensors while other sensors are not critical for (safe) operation but can reduce the operating cost.

Temperature control of tubular steam reformation reactors is critical to reactor performance and catalyst life. As shown in Reference [31], temperature sensor locations near the heat source result in reduced temperature fluctuations throughout the catalyst bed with virtually no temperature overshoot at any location. In this control, measuring the lambda value or oxygen content of the flue-gas is also important to control the maximum temperature of combustion in the combustor and avoid reducing conditions in the combustion section.

A YSZ oxygen sensor can be applied as the detector for the real-time sensing of methane steam reforming over a Ni/Al_2O_3 catalyst. The monitoring of reaction progress along the gas flow direction in the catalytic bed has been successfully demonstrated. Furthermore, the degradation behavior of catalytic activity was detected properly during a long-term operation [32].

Another example is the sulfur sensor that gives a warning when the desulfurization adsorbent is (almost) used. Currently, these sensors are not available for commercial operation of a system and indirect measurements and calculations are used to determine the end of life of the desulfurization bed. Diagnosis is needed when the fault is not easily extractable from the sensor signals and depends on what is measured and in what way the signals are utilized.

The answer to the question of what should be measured and what should be estimated is also a matter of cost–benefit evaluation. Factors of influence are possibility/quality and costs of developing sensors versus estimation models. Methods that utilize existing sensors and are based on a simple, though reliable, model are preferred. This assures low development costs, installation costs, and maintenance costs for the monitoring system. In addition, a generic method that is equally applicable, for example, to any control valve increases its value considerably.

10.7.2
Control Structures

Control structures for steam reformers depend on the measures for quality and both direct and indirect alternatives are applied. Commonly, the reformers are controlled by a PID controller manipulating a fuel gas valve to control the coil outlet temperature (COT), that is, the gas outlet temperature, of the reformer, as shown in Figure 10.7. COT control is adequate for control of the hydrogen concentration in the outlet of the reformer under feed composition variations only when the steam-to-carbon ratio (S/C) is controlled.

When instead the steam-to-gas (S/G) ratio is controlled, the hydrogen concentration varies largely under feed composition variations. To compensate for small variations in fuel pressure, the COT is often controlled via cascade control to a fuel flow controller.

One of the main control objectives related to reforming to produce hydrogen is keeping the hydrogen production rate at a set point given by the production

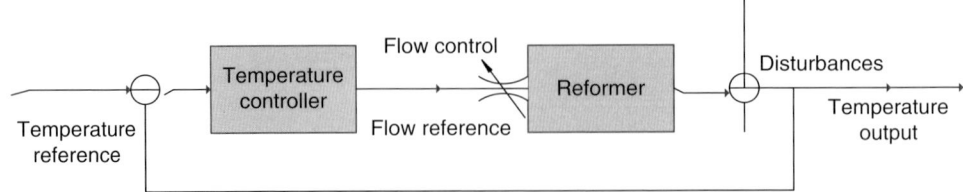

Figure 10.7 Typical temperature control loop for SMR.

demand. There are two sub-objectives for the control: (i) set point tracking and (ii) disturbance rejection. For the control system to handle variations in the disturbances at various operating points, knowledge of the response of the hydrogen production to changes in input variables is crucial. This is also a well-known feature of heat exchangers. Hence, nonlinear control strategies are often suggested for heat exchanger reactors.

In conventional autothermal steam-reformers, the spatial temperature distribution leads to hot-spot challenges [33]. Figure 10.8 shows a classical control structure for ATR with disturbance rejection.

10.8
Safety

Placing small-scale hydrogen supply units in urban areas require that regulations, codes, and standards are fulfilled. In addition, public confidence and acceptance is needed. The risk involved with hydrogen service stations with on-site reforming represents a challenge as major hazards such as fire and explosions arise from the handling of flammable gases [35].

An overview of some relevant standards and material aspects for hydrogen systems are given in Reference [1]. Moreover, primary risks with hydrogen are described in the ISO 15916, which is relevant for both hazard and risk assessment. International standards form the basis in the development of safety systems and

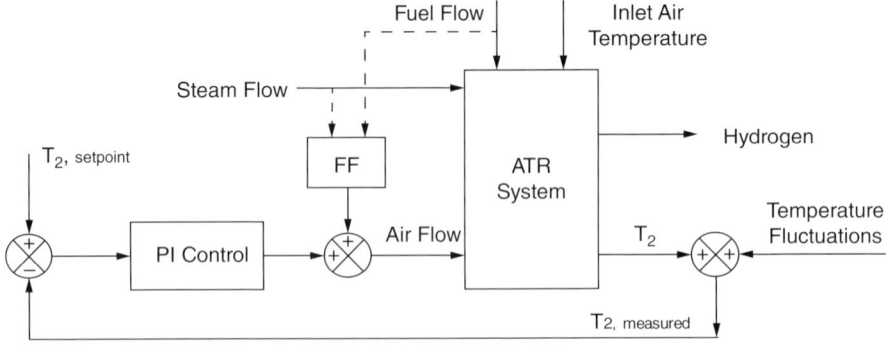

Figure 10.8 Classical feed-forward control structure [34].

the main goal of such systems is to reduce the probability of fatality/process accidents/hazardous situations to an acceptable level.

10.9
Conclusion

Fossil fuel is one of the main sources for hydrogen in the near future, and steam methane reforming is currently one of the cheapest options for on-site hydrogen supply. A large effort is being made by several equipment suppliers to harmonize components and to offer standardized containerized reformers with capacities in the area of $50-100\,\text{Nm}^3\,\text{h}^{-1}$. Improvements of state-of-the-art technology in materials (including catalysts), reactor design, and in hydrogen clean-up components is leading to more sustainable reformer units. Whenever there is a demand for CO_2 in the neighborhood of the small-scale reformers, small-scale CCS could be considered.

References

1 Schjølberg, I. et al. (2012) *IEA-HIA Task 23 Small-scale Reformers for On-site Hydrogen Supply*, IEA-HIA ISBN 978-0-9815041-4-8.
2 UK H2 Mobility (2015) Phase 1 Results, Production & Distribution. Available at http://www.ukh2mobility.co.uk/the-project/production-and-distribution/ (accessed 25 May 2015).
3 DOE (2012) Hydrogen Production, Technical Plan.
4 Schjølberg, I., Hulteberg, C., Yasuda, I., and Nelsson, C. (2012) Small scale reformers for on-site hydrogen supply. *Energy Proc.*, **29**, 559–566.
5 Ivy, J. (2004) Summary of Electrolytic Hydrogen Production. National Renewable Energy Laboratory, NREL/MP-560-36734.
6 Stolten, D. (ed.) (2010) *Hydrogen and Fuel Cells*, Wiley-VCH Verlag GmbH, Weinheim.
7 Hulteberg, C. (2007) *Hydrogen, an Energy Carrier of the Future*, Media-Tryck, Lund. ISBN 978-91-628-7337-0.
8 Brandin, J. and Liljedahl, T. (2011) Unit operations for production of clean hydrogen-rich synthesis gas from gasified biomass. *Biomass Bioenerg.*, **35**, S8.
9 Smith, MW. and Shekawat, D. (2011) Catalytic partial oxidation, in *Fuel Cells*, Elsevier Science Publishers B.V., Amsterdam.
10 Tugnolia, A., Landucci, B., and Cozzani, V. (2008) Sustainability assessment of hydrogen production by steam reforming. *Int. J. Hydrogen Energy*, **33**, 4345–4357.
11 Xu, X., Li, P., and Shen, Y. (2013) Small-scale reforming of diesel and jet fuels to make hydrogen and syngas for fuel cells: a review. *Appl. Energy*, **108**, 202–217.
12 Feitelberg, A.S. and Rohr, D.F. Jr. (2005) Operating line analysis of fuel processors for PEM fuel cell systems. *Int. J. Hydrogen Energy*, **30**, 1251–1257.
13 IEA (2009) IEA Bioenergy Task 37: Energy from biogas and landfill gas. http://www.iea-biogas.net/files/daten-redaktion/download/publi-task37/upgrading_rz_low_final.pdf.
14 Masebinu, S.O., Aboyade, A., and Muzenda E. (2015) Enrichment of biogas for use as vehicular fuel: A review of the upgrading techniques. *Int. J. Res. Chem., Met. Civil Eng. (IJRCMCE)*, **1** (1), 2349–1450.
15 He, B.B., Van Gerpen, J.H., and Thompson, J.C. (2009) Sulfur content in selected oils and fats and their corresponding methylesters. *Appl. Eng. Agric.*, **25** (2), 223–226.

16 Martin, S., Kraaij, G., Ascher, T., Wails, D., and Wörner, A. (2015) An experimental investigation of biodiesel steam methane reforming. *Int. J. Hydrogen Energy*, **40** 95–105.

17 Lipman, T. (2011)An overview of hydrogen production and storage systems with renewable hydrogen case studies. A clean energy state alliance report. Available at www.cesa.org.

18 Molburg, J.C. and Doctor, R.D. (2003) Hydrogen from steam-methane reforming with CO_2 capture. Presented at the 20th Annual International Pittsburgh Coal Conference, 15–19 September 2003, Pittsburgh, PA.

19 U.S. Energy Information Administration (2015) Annual Energy Outlook 2015 With Projections to 2040. http://www.eia.gov/forecasts/ieo/.

20 International Energy Agency (2012) World Energy Outlook 2012, Paris. Available at http://www.iea.org/publications/freepublications/publication/English.pdf.

21 International Energy Agency (2014) World Energy Outlook 2014 Factsheet, Paris. Available at http://www.worldenergyoutlook.org/media/weowebsite/2014/141112_WEO_FactSheets.pdf.

22 U.S. Energy Information Administration (2013) Analysis and Projections: Technically recoverable shale oil and shale gas resources: an assessment of 137 shale formations in 41 countries outside the United States. www.eia.gov/analysis/studies/worldshalegas.

23 Kolb, G. (2008) *Fuel Processing For Fuel Cells*, Wiley-VCH Verlag GmbH, Weinheim, ISBN 978-3-527-31581-9.

24 Hulteberg, C. (2012) Sulphur-tolerant catalysts in small-scale hydrogen production, a review. *Int. J. Hydrogen Energy*, **37**, 3978–3992.

25 Michelsen, F.A., Schjølberg, I., and Lund, B.L. (2007) Dynamic system analysis of a small scale hydrogen production plant. In 8th IFAC: Int. Symp. Dynamics and Control of Process Systems (DYCOPS), Cancun, Mexico, pp. 249–254, doi: 10.3182/20070606-3-MX-2915.00040.

26 Halabi, M.H.M. (2011) *Sorption Enhanced Catalytic Reforming of Methane for Pure Hydrogen Production*, Technische Universiteit Eindhoven. ISBN 978-90-386-2454-9.

27 Kiwi-Minsker, L. and Renken, A. (2005) Microstructured reactors for catalytic reactions. *Catal. Today*, **110** (1–2), 2–14.

28 Kurokawa, H., Shirasaki, Y., and Yasuda, I. (2010) Demonstration of highly-efficient distributed hydrogen production from natural gas with CO2 capture, in *18th World Hydrogen Energy Conference 2010 – WHEC 2010, Parallel Sessions Book 3* (ed. D. Stolten and T. Grube), Forschungszentrum Jülich GmbH, Zentralbibliothek, Verlag. ISBN: 978-3-89336-653-8.

29 Recio-Garrido, D., Ocampo-Martinez, C., and Serra-Prat, M. (2012) Design of optimization-based controllers applied to an ethanol steam reformer for hydrogen production. *Int. J. Hydrogen Energy*, **37**, 11141–11156.

30 Michelsen, F.A., Schjølberg, I., and Lund, B.L. (2007) Dynamic system analysis of a small scale hydrogen production plant. 8th IFAC International Symposium on Dynamics and Control of Process Systems (DYCOPS), 6–8 June 2007, Cancun Mexico.

31 Vernon, D.R., Erickson, P.A., Liao, C.-H., Tang, H.-Y., and Hsu, J. (2009) Implications of sensor location in steam reformer temperature control. *Int. J. Hydrogen Energy*, **34** (2), 877–887.

32 Matsui, T., Saburi, C., Okuda, S., Muroyama, H., and Eguchi, E. (2011) Real-time sensing of methane steam reforming by YSZ oxygen sensor. *Int. J. Hydrogen Energy*, **36**, 2945–2949.

33 Hüppmeier, J., Bar, S., Baune, M., Koch, D., Grathwohl, G., and Thöming, J. (2010) Oxygen feed membranes in autothermal steam-reformers – a robust temperature control. *Fuel*, **89**, 1257–1264.

34 Hua, Y., Chmielewski, D.J., and Papadias, D. (2008) Autothermal reforming of gasoline for fuel cell applications: controller design and analysis. *Power Sources*, **182**, 298–306.

35 Schjølberg, I. and Østdahl, A.B. (2008) Security and tolerable risk for hydrogen service stations. *Technol. Soc.*, **30** (1), 64–70.

11
Industrial Hydrogen Production from Hydrocarbon Fuels and Biomass

Andreas Jess and Peter Wasserscheid

11.1
Options to Produce Hydrogen from Fuels–An Overview

Today approximately 50×10^6 tons of hydrogen are produced industrially. The largest part of this quantity is applied in the petrochemical and the chemical industry, mainly for desulfurization and hydrogenation reactions [1]. Some 95% of the actual industrial hydrogen production is generated from fossil fuels by so-called steam reforming, partial oxidation or gasification processes.

Hydrogen production from hydrocarbon fuels and biomass always proceeds via synthesis gas (syngas), which is a general term for the mixture of hydrogen and carbon monoxide that typically forms first in all hydrogen production processes using carbon containing substances as feedstocks [2]. As shown in Figure 11.1, syngas may be produced from any fossil fuel and from biomass.

For heavy oils like vacuum gasoil and for solid feedstocks like coal and biomass, syngas production is always based on non-catalytic partial oxidation as these feedstocks often contain catalyst poisons like sulfur or heavy metal impurities. For this non-catalytic partial oxidation of coal and biomass the term "gasification" is often used.

In contrast, natural gas or light hydrocarbons are catalytically converted with steam as these feedstocks can be easily desulfurized, and are typically free of contaminants that would lead to catalyst deactivation. This catalytic process is called steam reforming. For natural gas, partial oxidation is usually not an economical option as the investment costs for the required cryogenic air separation are high.

Depending on the final use of the syngas, several additional treatment steps are still needed, primarily to remove unwanted impurities from the raw syngas such as H_2S and CO_2 and to adjust the required H_2-to-CO ratio. For the production of pure hydrogen the initial conversion step that results in a H_2/CO mixture is followed by a water–gas shift step. The latter converts CO with water into CO_2 and additional hydrogen (Eq. (11.1)):

$$CO + H_2O \leftrightarrow CO_2 + H_2 \quad \Delta_R H^0_{298} = -41 \text{ kJ mol}^{-1} \quad (11.1)$$

Hydrogen Science and Engineering: Materials, Processes, Systems and Technology, First Edition.
Edited by Detlef Stolten and Bernd Emonts.
© 2016 Wiley-VCH Verlag GmbH & Co. KGaA. Published 2016 by Wiley-VCH Verlag GmbH & Co. KGaA.

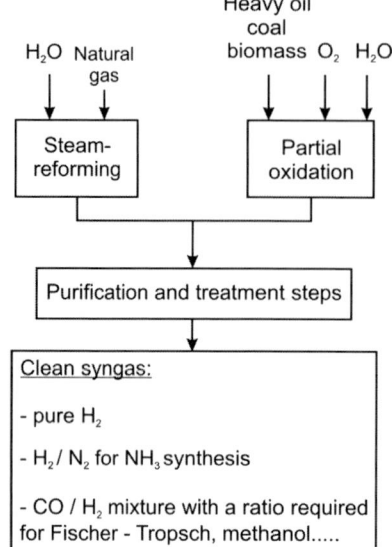

Figure 11.1 Production of syngas based on fossil fuels and biomass.

The water–gas shift step is followed by a pressure-swing-adsorption for hydrogen purification.

Figure 11.2 shows a simplified block diagram of the main processes involved in syngas production from light feedstocks. Moreover, the figure gives an overview of the potential technical applications of the produced syngas besides the production of pure hydrogen.

The feed for steam reforming, typically natural gas, usually has to be desulfurized prior to its conversion into syngas. If only small amounts of sulfur are present in the form of H_2S, reaction with solid ZnO to ZnS is used. For higher contents of H_2S, scrubbing with a solvent is applied, and if the feed also contains organic sulfur compounds, a hydrodesulfurization step is needed.

For coal and heavy oil based syngas production, feedstock purification is up to now not possible. Hence, during partial oxidation, the organic sulfur compounds are converted in the reducing reactor atmosphere into H_2S. Consequently, H_2S and other unwanted impurities like ash, tar, and solid carbon (soot) have to be separated from the raw syngas before its further treatment.

Even if all these unwanted impurities have been removed, extended treatment steps are still required to produce pure hydrogen. These involve water–gas shift reaction (Eq. (11.1)), CO_2 removal, and methanation to remove traces of CO. The sequence of treatment steps is illustrated in Figure 11.3 for hydrogen production in the context of an ammonia plant. Note that in this specific case it is important to reduce the amount of CO in the final ammonia synthesis gas (a mixture of N_2 and H_2 in the ratio 1:3) to avoid deactivation of the iron based ammonia synthesis catalyst. Table 11.1 gives typical compositions of the syngas at the different stages of the syngas treatment.

11.1 Options to Produce Hydrogen from Fuels–An Overview

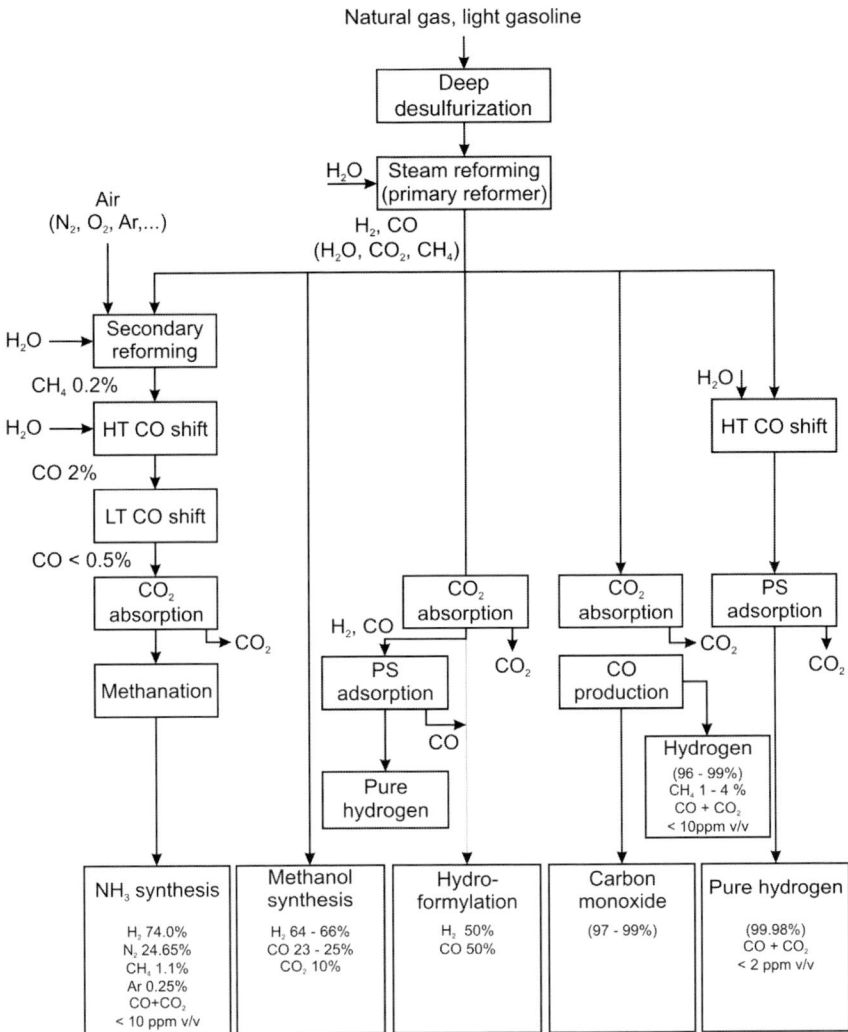

Figure 11.2 Production and treatment of syngas depending on the final use of the syngas for the example natural gas as feedstock. Adapted from Reference [3].

The two primary steps of (ammonia) syngas production from natural gas are steam reforming followed by autothermic (secondary) reforming where the residual methane of the syngas is converted with air by catalytic partial oxidation. The syngas then contains hydrogen and nitrogen, but still also CO, CO_2, H_2O, and – in a very small quantity – CH_4 (Table 11.1). Separation of these compounds is possible in principle, for example, by cryogenic processes, but this is very costly. Therefore, purification of the raw syngas is realized mainly by a sequence of chemical conversion steps: The syngas leaving the secondary reformer contains appreciable amounts of CO and CO_2. Separation of CO_2 by ad- or absorption is

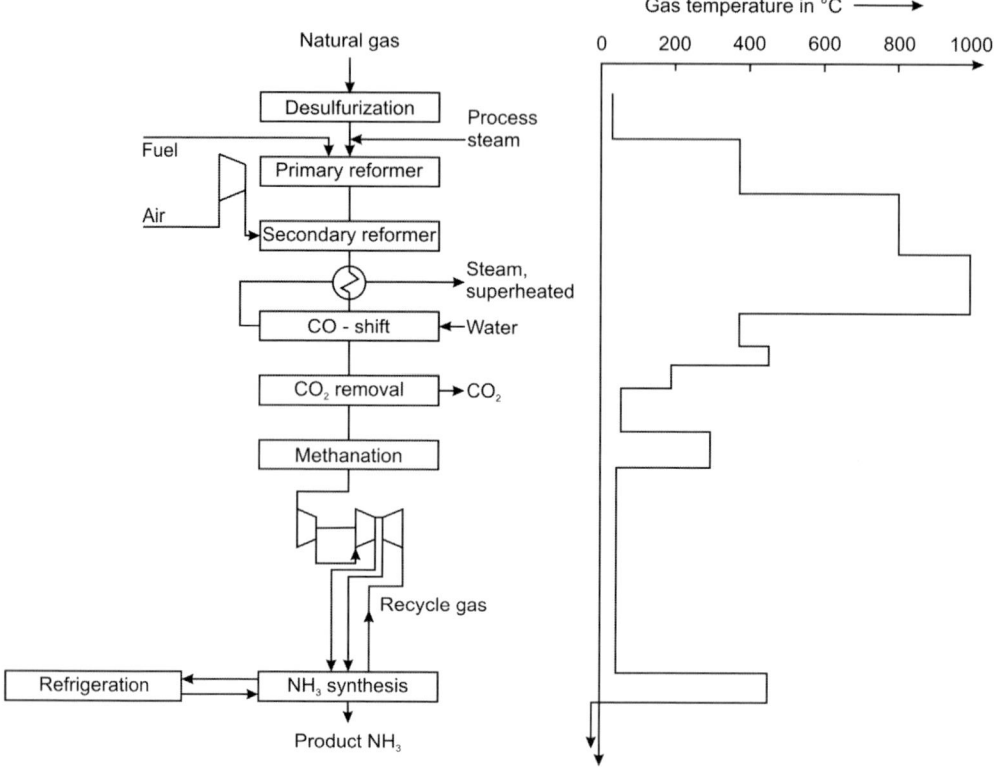

Figure 11.3 Block diagram and temperature profile for an ammonia plant using hydrogen from methane steam reforming. Adapted from Reference [4].

Table 11.1 Typical composition of syngas for NH_3 synthesis at different stages of raw syngas treatment for natural gas with 14% N_2 as feedstock. Data from Reference [5].

Composition (vol.%) at the exit of:	Primary reformer	Secondary reformer	High temp. shift	Low temp. shift	Methanation
H_2	40.8	36.2	42.3	44.2	74
CO	6.4	8.2	2.1	0.2	<5 ppm
CO_2	5.8	4.8	10.9	12.8	—
CH_4	4.6	0.2	0.2	0.2	0.7
H_2O	39.8	36.0	29.9	28.0	—
N_2	2.6	14.3	14.3	14.3	25
Ar	—	0.2	0.2	0.2	0.3

11.1 Options to Produce Hydrogen from Fuels–An Overview

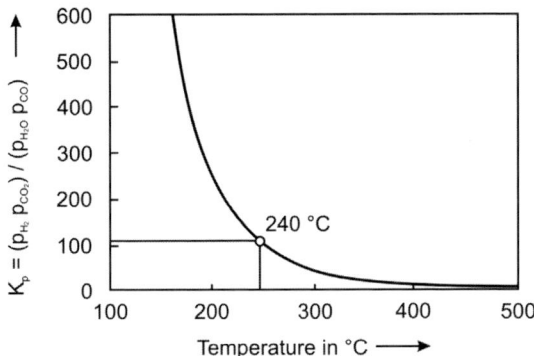

Figure 11.4 Influence of temperature on equilibrium constant of the water–gas shift reaction.

relatively straightforward, but CO separation is not. Thus CO is at first converted into CO_2 by the water–gas shift reaction (Eq. (11.1)). The shift reaction is exothermic and thus the equilibrium is favored by low temperatures (Figure 11.4). Thus, the reaction temperature should be kept as low as possible. However, the choice of reaction temperature is limited by the activity of the catalyst.

The industrially applied Fe-Cr water–gas shift catalyst is sufficiently active only at temperatures above about 300 °C. Other catalysts based on copper and zinc are active enough at a temperature of about 200 °C but these catalysts are very sensitive to poisoning and require extremely pure gases, typically with less than 1 ppm H_2S. In practice, the water–gas shift reaction is carried out in two adiabatic fixed-bed reactors with intermediate cooling between both converters. The first high temperature shift reactor operates with a Fe-Cr catalyst, the second low temperature shift reactor contains the more active Cu-Zn system. At the exit of the second shift reactor, the CO_2 present in the converted syngas is removed in a gas scrubber, usually by chemical absorption in aqueous amine solutions, for example, mono- or diethanolamine.

Although the majority of the CO originally present in the syngas is converted by the water–gas shift reaction into CO_2 there is still too much CO present as the water–gas shift reaction is limited by thermodynamic constraints. For a typical temperature of about 240 °C at the exit of the low temperature shift reactor, the value for K_p ($= p_{H2}\, p_{CO2}/(p_{H2O}\, p_{CO})$) is about 100 (Figure 11.4) which still corresponds to a CO content of about 0.2% (see also Table 11.1). In the case of ammonia synthesis, this remaining CO is converted by methanation, which is the reverse of the steam reforming reaction:

$$CO + 3H_2 \leftrightarrow CH_4 + H_2O \quad \Delta_R H^0_{298} = -206\ \text{kJ mol}^{-1} \quad (11.2)$$

The equilibrium favors methanation at low temperature, and therefore a reaction temperature of about 450 °C is used (whereas steam refoming is performed typically at about 800 °C).

If syngas is produced by partial oxidation of heavy oils or gasification of solid fuels oxygen is needed, which is produced by cryogenic air separation. In this case nitrogen is available and the CO separation is then conducted by a low temperature liquid nitrogen wash at about $-190\,°C$. By this means the N_2 needed for the NH_3 synthesis is also added.

11.2
Hydrogen Production from Solid Fuels (Coal, Biomass)

11.2.1
Basic Principles and Reactions of Syngas Production from Solid Fuels

Coal gasification to syngas is a complex network of parallel and consecutive reactions. Assuming for simplicity that coal is represented by carbon ("C") the following main reactions have to be considered:

$$C + O_2 \leftrightarrow CO_2 \quad \Delta_R H^0_{298} = -394 \text{ kJ mol}^{-1} \tag{11.3}$$

$$C + CO_2 \leftrightarrow 2CO \quad \Delta_R H^0_{298} = +172 \text{ kJ mol}^{-1} \tag{11.4}$$

$$C + H_2O \leftrightarrow CO + H_2 \quad \Delta_R H^0_{298} = +131 \text{ kJ mol}^{-1} \tag{11.5}$$

The direct oxidation of carbon to CO is not shown here as in technical processes the primarily formed CO is rapidly oxidized in a consecutive homogenous gas-phase reaction to CO_2. Thus Eq. (11.3) summarises the oxidation of C to CO_2 (intrinsically only favored at low temperatures) and the homogeneous CO oxidation.

Coal gasification is commonly conducted in an autothermic way, that is, gasification is carried out with a mixture of O_2 and H_2O, so that the combustion generates the heat needed to reach the high temperatures required for the endothermic gasification reactions (Eqs. (11.4) and (11.5)). Beside these heterogeneous reactions, homogeneous reactions like the water–gas shift reaction also occur. In addition, coal is a complex mixture of organic and mineral compounds. Hence, if coal is heated, various gases are evolved by pyrolysis (H_2, CH_4, CO, aromatics, tar, etc.), and the solid residue is transformed into a porous char. Depending on the process and reactor type, the volatile compounds and the char are further converted by thermal cracking and gasification reactions.

11.2.2
Hydrogen Production by Gasification of Solid Fuels

Reactors for syngas production by gasification of solid fuels like coal and probably in future also biomass are moving bed, fluidized-bed, or entrained flow systems. Table 11.2 lists the main characteristics of these reactors. Depending on the temperature of the gasifier the mineral matter is released as a liquid (slag) or a solid (ash).

The *fixed bed gasifier* (*Lurgi* reactor) is operated countercurrently. The coal bed is supported on a rotating grate. Coal enters the gasifier at the top and is

Table 11.2 Typical conditions of coal gasification in different reactor types. Data taken from Reference [6].

Reactor type	Fixed bed	Fluidized bed	Entrained bed	Texaco/Shell
Diameter of coal particles (mm)	6–40	1–8	<0.1	<0.1
Pressure (bar)	35	25	1	35
Suitable coal	Non-caking	Reactive	All	All (coal slurry)
Temperature (max./outlet) (°C)	1100/300	1000/900	2000/1000	
Ash/slag removal	Solid	Solid	Liquid	
O_2 consumption (m^3 (STP) per 10 MJ coal)	0.1–0.13	0.16–0.19	0.19–0.24	
Steam consumption (kg per 10 MJ coal)	0.3–0.5	0.19–0.24	0–0.05	
Typical composition of syngas (dry basis) for hard coal as feedstock in vol.%				
H_2	39	46	31	35
CO	22	30	58	50
CO_2	28	22	10	14
CH_4	10	1	<0.5	—
H_2S, COS, NH_3	Up to 1			
By-products	Tar, oil	None	None	None

slowly heated, dried, and then pyrolyzed on its way down; vice versa, the product gas is cooled before it leaves the reactor. In the gasification zone the coal is partly gasified by steam and CO_2, until a reaction temperature of about 900 °C is reached, and the rates of both reactions become too low. The remaining coal is finally combusted in the lowermost zone where temperatures of 1200 °C are reached. The solid ash leaves the reactor at the bottom. A disadvantage of the fixed-bed reactor is that a large amount of by-products is released as tar and oil. Only non-caking coals can be processed. Caking coals would lead to the formation of agglomerates leading to an increased pressure drop or even plugging. A novel development is a slagging reactor with a higher temperature in the combustion zone of above 1700 °C.

In the *fluidised-bed gasifier* (*Winkler* process) the coal particles are well mixed, which leads to a moderate and uniform temperature. Char particles that leave the reactor together with the product gas are recovered in cyclones and are recycled. The ash is removed at the bottom. The solid phase is well mixed, which leads to a lower carbon conversion compared to a plug flow system. Some carbon is lost via the ash. The coal should therefore be reactive (e.g., brown coal) to permit high conversion.

In the *entrained-flow gasifier* (*Koppers–Totzek* process) the fine coal particles react cocurrently with steam and oxygen. The residence time is only a few seconds and the temperature must be high to ensure a high coal conversion. The mineral matter is removed as molten slag. The entrained-flow reactor can handle all types of coal.

Note that these three reactor types differ significantly with regard to the hydrogen content in their product gas (Table 11.2). These differences originate from the different operating conditions such as, for example, different reaction temperatures and different amounts of oxygen and steam in the reactor.

Besides syngas and hydrogen production, coal gasification has recently also been considered as an attractive option for power generation. The so-called *combined cycle power plant* produces first a fuel gas rich in H_2 and CO by coal gasification. After cleaning, the fuel gas is combusted with compressed air in a gas turbine that drives the electricity generator and also the air compressor. The exhaust gas of the turbine is still hot and therefore used for steam generation and electricity generation in a classical steam turbine cycle, respectively. Compared to a classical power plant based on coal combustion with an efficiency of about 35%, the combined cycle power plant is significantly more efficient (41–43%) [6b].

11.3
Syngas by Partial Oxidation of Heavy Oils

Partial oxidation is the reaction of hydrocarbons with an insufficient amount of oxygen with regard to complete combustion, and is usually conducted at temperatures of up to 1600 °C and pressures up to 100 bar. The reactions of partial oxidation may be simplified as follows:

$$C_nH_m + \tfrac{n}{2}O_2 \rightarrow n\,CO + \tfrac{m}{2}H_2 \tag{11.6}$$

$$C_nH_m + n\,H_2O \rightarrow n\,CO + \left(\tfrac{m}{2} + n\right)H_2 \tag{11.7}$$

CO and H_2 are the main products, but CO_2 and H_2O are also formed by combustion and by the water–gas shift reaction. Moreover, some CH_4 is formed by thermal cracking. Beside oxygen, steam is added, which leads to more hydrogen (Eq. (11.7)) than expected according to Eq. (11.6) only. Sulfur, which is always present in heavy oils, is converted into H_2S (about 95%) and COS (5%).

Under the conditions of partial oxidation, carbon should not exist according to the Boudouard equilibrium (Eq. (11.4)), but nevertheless soot can be present in the raw gas. During partial oxidation of methane soot formation is practically zero, whereas in heavy-oil gasification up to 2% of the feedstock mass is converted into soot.

Two processes are commercially established for heavy oil gasification, the *Shell* and the *Texaco* process (Figures 11.5 and 11.6) but the main steps of both processes are practically identical [7]. Besides partial oxidation of natural gas and residue oil, this process has been developed further for coal slurries (Table 11.3).

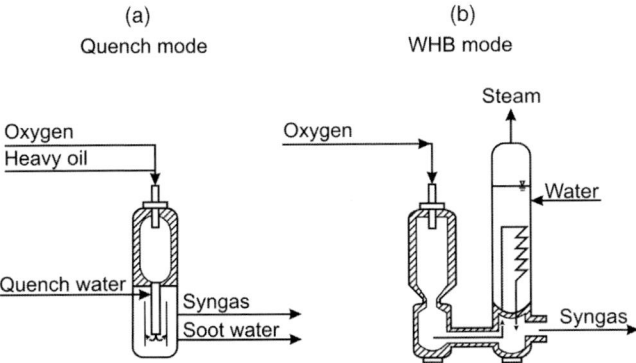

Figure 11.5 Gasification of heavy oil and raw gas cooling in the Texaco gasification process (WHB: waste-heat boiler). Adapted from Reference [4].

Oxygen, pre-heated heavy oil and steam are fed to the refractory-lined reactor where the oil is sprayed into the reactor. The reaction starts in a water-cooled burner, which also houses the atomizing gun and the oxygen and steam inlets.

The product gas mixture leaves the reactor at about 1350 °C and is either cooled in a waste-heat boiler (WHB mode, Figure 11.5b) or by direct quenching with water from the soot water cycle (quench mode, Figure 11.5a).

The *Texaco* process with the quench mode is shown in Figure 11.6. The quench is followed by a two-step soot-scrubbing, a venture-scrubber, and a counter currently operated packed scrubbing column. The soot is extracted from the water with naphtha, and the soot–naphtha suspension is separated in a settler from the

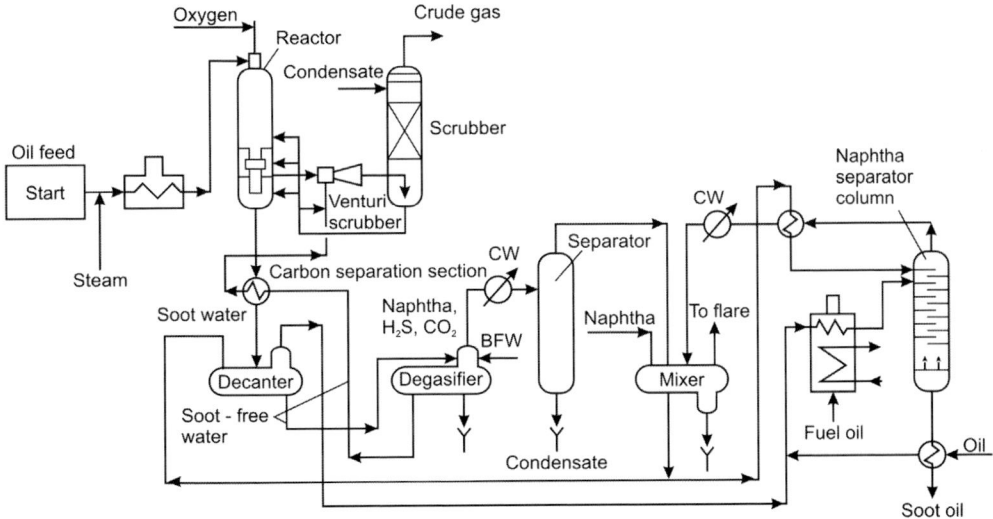

Figure 11.6 Gasification of heavy oil (quench mode) and soot removal in the Texaco gasification (BFW: boiler feed water). Adapted from Reference [4].

Table 11.3 Feedstock and raw syngas composition of the Texaco process. Data from Reference [4].

	Natural gas	Naphtha	Heavy fuel oil	Tar (from coal)
Feedstock composition				
Approximate formula	$CH_{3.71}$	$CH_{2.31}$	$CH_{1.28}$	$CH_{0.75}$
C (wt%)	73.40	83.8	87.2	88.1
H (wt%)	22.76	16.2	9.9	5.7
O (wt%)	0.76	—	0.8	4.4
N (wt%)	3.08	—	0.7	0.9
S (wt%)	—	—	1.4	0.8
Ash (wt%)	—	—	—	0.1
Raw gas composition				
H_2 (vol.%)	61.1	51.2	45.8	38.9
CO (vol.%)	35.0	45.3	47.5	54.3
CO_2 (vol.%)	2.6	2.7	5.7	5.7
N_2 (vol.%)	1.0	0.1	0.3	0.8
H_2S (vol.%)	—	—	0.3	0.2
CH_4 (vol.%)	0.3	0.7	0.5	0.1
Soot (kg 1000 m^3 (STP))	—	1.8	10	6.1
Consumption for 1000 m^3 (STP) syngas				
Feedstock (kg)	262	297	232	356
Oxygen (m^3 (STP))	248	239	240	243
Steam (kg)	—	74	148	186

almost carbon-free water, which is recycled to the quench and scrubbing section. The soot–naphtha phase is mixed with feed oil, after which the naphtha is distilled off in a stripping column and returned to the soot-extraction.

The raw syngas of partial oxidation of heavy oil contains almost equal amounts of carbon monoxide and hydrogen, about 5% CO_2, and small amounts of CH_4, N_2, Ar, and H_2S (Table 11.3, second column from right). For subsequent synthesis with a higher demand in hydrogen (e.g., methanol synthesis, *Fischer–Tropsch* synthesis) or for the production of pure hydrogen, a part or all of the CO is converted with steam into CO_2 and H_2 in a water–gas shift step. By addition of nitrogen from the air separation unit, also syngas suitable for ammonia can be produced.

11.4
Syngas by Steam Reforming of Natural Gas

The main reaction of steam reforming of natural gas (here taken as methane) is:

$$CH_4 + H_2O \leftrightarrow CO + 3\,H_2 \quad \Delta_R H^0_{298} = +206 \text{ kJ mol}^{-1} \quad (11.8)$$

This reaction is favored by a low pressure and a high temperature.

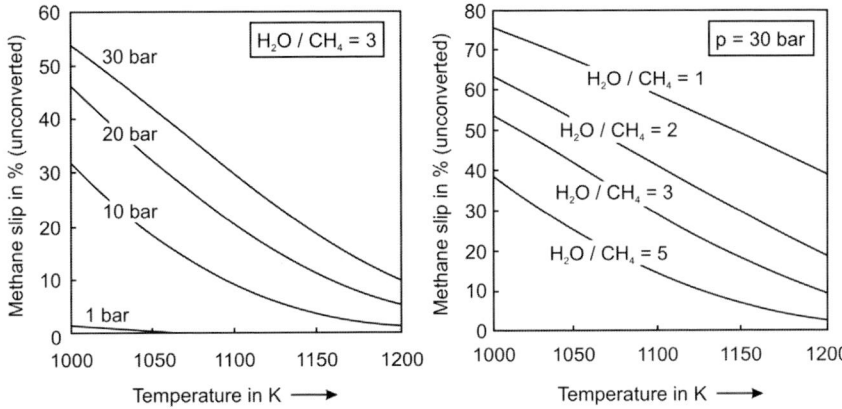

Figure 11.7 Equilibrium content of methane during steam reforming for different pressures, temperatures, and steam-to-carbon ratios. Adapted from Reference [6b].

The equilibrium content of methane for different pressures, temperatures, and steam-to-carbon ratios is shown in Figure 11.7. Thermodynamically, steam reforming is hindered by elevated pressures as the number of molecules increases. Nevertheless a pressure of 30 bar is adjusted in the steam reforming reactor as the syngas is needed in the subsequent processes under pressure (e.g., in the form of compressed hydrogen or for catalytic conversion in, for example, ammonia synthesis or Fischer–Tropsch synthesis). Note that the energy requirement for compression of natural gas is much lower than for syngas compression with a higher volume rate. At 30 bar, the equilibrium conversion of methane into H_2 and CO is only complete at a temperature of over 1100 °C, but temperatures in excess of 900 °C cannot be applied with regard to metallurgical constraints.

Although steam reforming is carried out at high temperature, a nickel catalyst is still required due to the high stability of methane. The catalyst is contained in tubes, which are placed inside a furnace that is heated by combustion of fuel (Figure 11.8).

The desulfurized natural gas feed is mixed with steam and preheated to 500 °C before entering the reformer tubes. The heat for the reforming reaction is supplied by combustion of fuel in the furnace which may contain up to 500 tubes with a length of 10 m and a diameter of 10 cm. Figure 11.9 shows axial profiles of the tube wall temperature and the heat flux.

The results of simulations of the steam reforming process carried out by Froment and coworkers based on kinetic data and the respective heat and mass balances are shown in Figure 11.10 [8].

The *difference* between the external and internal tube skin temperature is about 30 K and is much smaller than the difference between the flue gas temperature (about 1100 °C) and the external tube skin temperature ($\Delta T_{\text{flue gas - external skin}}$: 200–300 K) or the difference of between the internal tube skin temperature

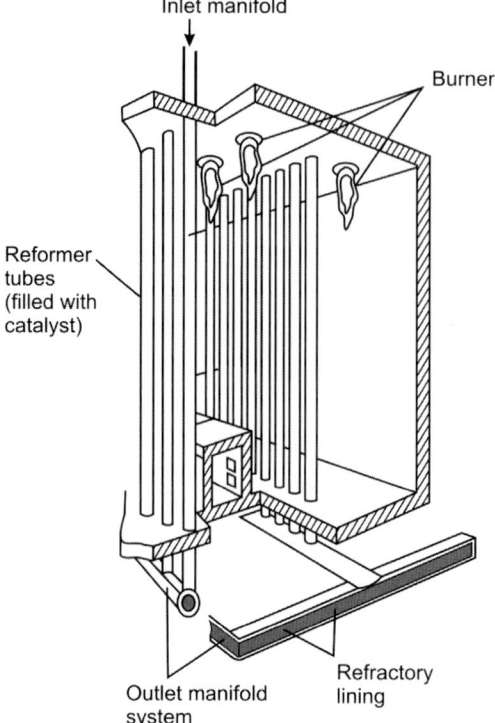

Figure 11.8 Primary reformer. Adapted from Reference [4].

and the average process gas temperature inside the tube ($\Delta T_{\text{internal skin – process gas}}$: 120–210 K). Thus with regard to heat transfer the process is determined by the heat transfer by convection and radiation from the fire box to the tube as well as by the heat transfer from the internal tube skin to the fixed bed. The exit gas

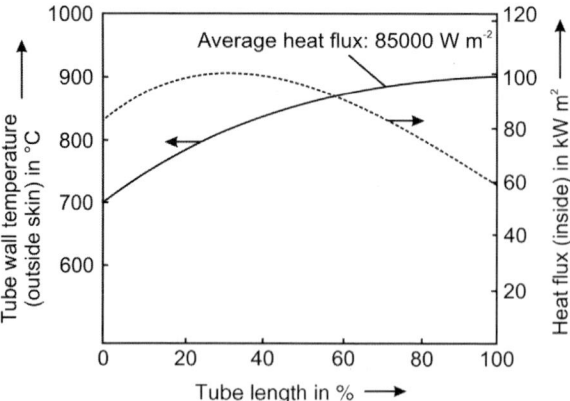

Figure 11.9 Temperature profile and heat fluxes in a side-fired primary reformer. Adapted from Reference [4].

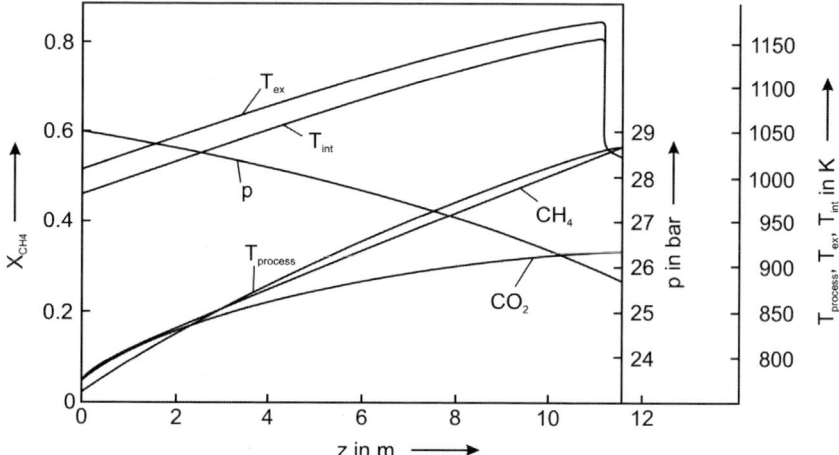

Figure 11.10 Evolution of methane conversion (X_{CH4}), mean radial process gas temperature and external and internal tube skin temperatures ($T_{process}$, T_{ex}, and T_{int}, respectively), and total pressure in a tube of a steam reformer (simulations by a one-dimensional reactor model), internal/external tube diameter: 10.2/13.2 cm, heated tube length: 11.1 m, ring-shaped cat. (height: 1 cm, diameters: 0.8 and 1.7 cm), molar steam to methane ratio: 3.4, average flue gas temperature: 1100 °C. Adapted from from Reference [8].

of the steam reformer may be further converted with air/oxygen by autothermic reforming in the presence of a catalyst. The reactor is a refractory lined vessel (Figure 11.11) and thus higher temperatures can be applied than in steam reforming.

Autothermic reforming of natural gas and light hydrocarbons is usually not applied on its own, due to the high investment and operating costs (oxygen). For ammonia synthesis, an autothermic reformer is frequently used downstream of the steam reformer where conversion of the unconverted methane with air (oxygen) takes place. The advantage of this arrangement is also that the air supplies the required nitrogen, and thus no expensive oxygen plant is needed. For syngas for methanol and Fischer–Tropsch plants, a combination of steam and autothermic reforming is sometimes used to adjust the required CO to H_2 ratio.

Typical process data of steam (primary) reforming and secondary reforming of natural gas for the production of hydrogen or ammonia syngas are summarized in Table 11.4. The product gas of the secondary reformer is then further processed (CO-shift, CO_2 removal, methanation, as shown in Figure 11.2).

11.5
Conclusions

State-of-the-art technologies for the industrial production of hydrogen from hydrocarbon fuels or biomass generate in their first step synthesis gas, a mixture of hydrogen and carbon monoxide, which by itself is an important intermediate

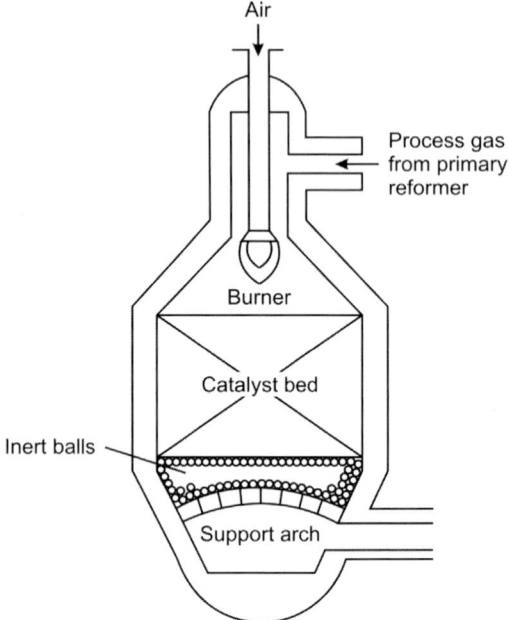

Figure 11.11 Secondary reformer. Adapted from Reference [4].

Table 11.4 Typical process data for steam reforming and secondary reforming of natural gas (total pressure: 40 bar at exit of primary reformer). Data from Reference [2].

	Feedstreams to primary reformer		Product gas of primary reformer	Product gas of secondary reformer
	Natural gas	Steam		
CH_4 (vol.%)	91.2	—	8.5	0.4
C_nH_m (vol.%)	5.8	—	—	—
N_2 (vol.%)	1.0	—	0.2	15.9
CO_2 (vol.%)	2.0	—	6.1	5.0
H_2 (vol.%)	—	—	39.3	36.7
CO (vol.%)	—	—	5.9	9.1
H_2O (vol.%)	—	100	40.0	32.7
Ar (vol.%)	—	—	—	0.2
Total (tonne h^{-1})	30.2	91.0	121.2	199.6
Total (kmol h^{-1})	1714	5054	8817	12 500

Natural gas as fuel for primary reformer: 9.4 tonne h^{-1} (24% of total natural gas)
Air as feedstock for secondary reformer: 78.4 tonne h^{-1}

for the production of many chemicals such as methanol, ammonia, or diesel oil (via Fischer–Tropsch synthesis). Syngas production from natural gas proceeds via steam reforming where natural gas is catalytically converted with steam into hydrogen and carbon monoxide. The reforming step is typically followed by water–gas shift reaction, CO_2 removal, and a range of purification steps (e.g., methanation, pressure swing adsorption) to obtain hydrogen in the required quality. Coal or biomass gasification to syngas is commonly conducted in fixed-bed, fluidized-bed, or entrained-flow gasifiers. Gasification is carried out with a mixture of O_2 and H_2O. The combustion generates the heat needed to reach the high temperatures required for the subsequent endothermic gasification with CO_2 and H_2O. Beside these heterogeneous reactions, homogeneous reactions like the water–gas shift reaction also occur. In addition, various gases are evolved by coal pyrolysis (H_2, CH_4, CO, aromatics, tar). For syngas based on heavy oils, partial oxidation is used, whereby the hydrocarbons react with an insufficient amount of oxygen with regard to complete combustion. The reaction is typically conducted at temperatures of up to 1600 °C and pressures up to 100 bar.

References

1 Armaroli, N. and Balzani, V. (2011) *Energy for a Sustainable World: From the Oil Age to a Sun-Powered Future*, 1st edn, Wiley-VCH Verlag GmbH, Weinheim.

2 Jess, A. and Wasserscheid, P. (2013) *Chemical Technology*, 1st edn, Wiley-VCH Verlag GmbH, Weinheim, pp. 536–558.

3 Baerns, M., Behr, A., Brehm, A., Gmehling, J., Hofmann, H., Onken, U., and Renken, A. (2006) *Technische Chemie*, Wiley-VCH Verlag GmbH, Weinheim.

4 Appl, M. (1999) *Ammonia*, Wiley-VCH Verlag GmbH, Weinheim.

5 Bakemeier, H., Huberich, T., Krabetz, R., Liebe, W., Schunck, M., and Mayer, D. (1997) Ammonia, in *Ullmann's Encyclopedia of Industrial Chemistry*, 5th edn on CD-ROM, Wiley-VCH Verlag GmbH, Weinheim.

6 (a) Onken, U. and Behr, A. (1996) *Chemische Prozesskunde*, Georg Thieme Verlag, Stuttgart; (b) Moulijn, J.A., Makkee, M., and van Diepen, A. (2004) *Chemical Process Technology*, Wiley-VCH Verlag GmbH, Weinheim; (c) Arpe, H.-J. (2007) *Industrielle Organische Chemie*, Wiley-VCH Verlag GmbH, Weinheim.

7 Supp, E. (1997) Noncatalytic partial oxidation and special gasification processes for higher-boiling hydrocarbons, in *Ullmann's Encyclopedia of Industrial Chemistry*, 5th edn on CD-ROM, Wiley-VCH Verlag GmbH, Weinheim.

8 (a) Xu, J. and Froment, G.F. (1989) Methane steam reforming, methanation and water-gas shift: a. intrinsic kinetics. *AIChE J.*, **35**, 88; (b) Xu, J. and Froment, G.F. (1989) Methane steam reforming, methanation and water-gas shift: a. diffusional limitations and reactor simulation. *AIChE J.*, **35**, 97; (c) Plehiers, P.M. and Froment, G.F. (1989) Coupled simulation of heat transfer and reaction in a steam reforming furnace. *Chem. Eng. Techn.*, **12**, 20–26.

Part 2.3
H₂ from Electricity

12
Electrolysis Systems for Grid Relieving

Filip Smeets and Jan Vaes

12.1
Introduction

Increasingly, national and regional energy policies around the globe are driven by concerns around sustainability, security, and competitiveness. This, invariably, leads to policies favoring a continuously growing penetration of renewable energy sources in the total energy mix of states and regions to mitigate the effects of anthropogenic global warming. The two most potent renewable energy sources, wind and solar, while being potentially unlimited are unfortunately intermittent, providing a variable output with limited predictability. This leads to oversupply of renewable electricity at certain moments, and shortages at others. Since the existing electricity network has hardly any storage capacity and conventional sources have limited responsiveness, the grid stability is increasingly being challenged. This already leads to frequent curtailment of wind turbines, reverse flow on the grid in high solar regions, higher operating costs of conventional power plants, and blackouts or power outages.

Relieving the grid of excess renewable electricity can be accomplished by using these free renewable electrons to power a water electrolysis unit producing hydrogen, hence converting renewable power into renewable hydrogen. This renewable hydrogen can be temporarily stored before re-use in industrial, mobility, or heat applications, or re-electrified in the case of electricity shortage. As a storage solution the hydrogen gas can also be injected into the natural gas network either directly or after methanation with CO_2, thus providing a unique seasonal storage solution. This concept is also known as "Power-to-Gas," and is gaining traction in many jurisdictions as a viable solution to meet energy policy objectives and realize the energy transition at the lowest possible cost to society.

In this chapter we will explain the impact of an increased share of renewable energy generation on the electricity grid, the options we have to integrate intermittent renewable energy sources, and the evolution of the demand for energy storage. Additionally the role of electrolyzers in the energy transition is described and the business case and outlook for Power-to-Gas are pictured. Finally, all topics are summarized in the conclusions.

Hydrogen Science and Engineering: Materials, Processes, Systems and Technology, First Edition.
Edited by Detlef Stolten and Bernd Emonts.
© 2016 Wiley-VCH Verlag GmbH & Co. KGaA. Published 2016 by Wiley-VCH Verlag GmbH & Co. KGaA.

12.2
Energy Policies around the Globe Drive Demand for Energy Storage

On 15th December 2011, the European Commission adopted the Communication "Energy Roadmap 2050," revealing its commitment to reducing greenhouse gas emissions by 2050 to 80–95% of the 1990 levels. A policy that was accepted in the context of necessary efforts to be taken by developed countries as a whole. In the Energy Roadmap 2050 the Commission explores the challenges posed by obtaining the EUs de-carbonization objective while ensuring security of energy supply and competitiveness. The Energy Roadmap 2050 is the basis for developing a long-term European framework together with all stakeholders [1]. In any of the possible scenarios, it is clear that the biggest share of energy supply technologies in 2050 comes from Renewables. Hence a large-scale implementation of sustainable and renewable energy sources by 2050 is unavoidable. The European CO_2 emission reduction target of 80% relative to the 1990 emission level, implies that the power production sector should be fully sustainable by 2050 and that other sectors, like industry and mobility, should rely largely on sustainable use of energy sources.

Similar policies promoting the increased penetration of renewables have been communicated in other parts of the world as depicted by Figure 12.1 [2].

Low carbon technologies are needed to reconcile increasing energy demand with the decarbonization challenge. Wind and solar PV are at the forefront of power sector decarbonization and set to expand rapidly with projected and demonstrated growth rates of +29% CAGR from 2005 to 2017 [3,4].

Wind and solar have the highest potential and are both techno-economically and regulatory the most advanced. Unfortunately they come with several drawbacks with respect to conventional carbon based sources and some other renewable energy sources. Their variability is significant (minutes scale) in terms of power system integration because their power output is variable on multiple timescales, depending on daily and seasonal patterns and on local weather conditions. Additionally the predictability of their output is typically limited to a range below 10% of the root mean square error of its power as rated one day ahead. This makes long-term forecasting difficult and certainly less predictable than output from fossil-fuel technologies. Their output is also subject to ramp events. In addition wind and solar have a low degree of plant dispatchability. Their load factor, that is, the % of power they deliver in relation to the installed power, and thus their competitiveness depend on the inherent quality of their natural resources. For instance, high wind load factors are typically found at the north shores of Europe, grid parity for solar installations is found in the south. Typically, good generation centers tend to be far away from major consumption centers. Fortunately their limited invasiveness makes their geographic diversity potential high. These characteristics (variability, unpredictability, low load factor, and low dispatchability) do not fit well with the existing infrastructure of the power system. Two trends arise regarding the implementation of intermittent energy sources: (i) the effects of wind power integration on the energy system,

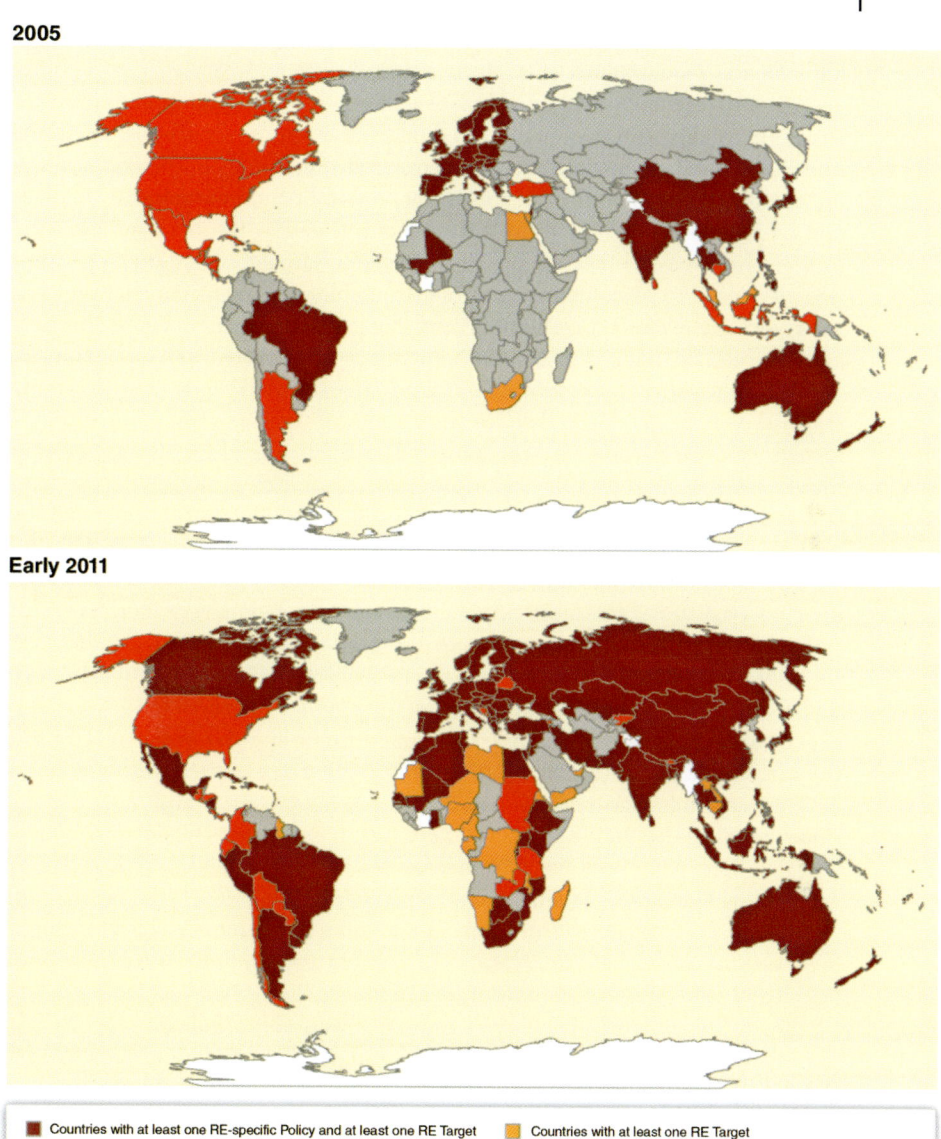

Figure 12.1 Global view of policies promoting the increased penetration of renewables.

mainly causing power transmission scale challenges, and (ii) the effects of photovoltaic solar power integration, which mainly challenges power distribution scale flexibility.

The challenge of operating the power-system operation lies in the precise balancing of supply and demand at all times. The International Energy Agency

(*IEA*) states stability, balance, and being adequate (in providing power) as the fundamental requirements of an electricity grid system [5]. *Stability* refers to frequency and voltage control to comply with the grid's technical limits within seconds – both characteristics are affected by fluctuations in supply and demand. It is not possible to store alternating current and hence supply to the grid has to be matched to the draw from the grid instantaneously and continuously on a second-by-second basis. Matters are further complicated by the fact that both active (resistive) power and reactive power (power that circulates in the grid but does no work) need to be controlled, in order to maintain the set frequency and voltage respectively. *Balancing* refers to load changes over minutes or days that must be balanced. To achieve this, system operators rely on ancillary services, either capacity reserves that can provide more power when needed or flexible customers with controllable loads that are willing to vary their consumption within an agreed window. By doing so the balanced grid has a load-following capability: that is, supply to the grid can be ramped up or down as required to meet demand within minutes. System operators call upon power producers and consumers, depending on their marginal operation costs and degree of flexibility.

Adequacy to provide power refers to the capacity needed to meet peak demand even under the most extreme conditions and on a very long term scale from months to years. To be adequate, the grid must have sufficient production capacity to meet this peak demand with an acceptable level of reliability. System operators rate power plants according to the amount of power that they can be expected to reliably produce at peak demand times and this is translated into a capacity credit granted to each power plant [6,7].

Wind and solar PV systems make demand–supply matching more difficult since they increase the need for flexibility within the system, while not contributing significantly to flexibility. The increased need for flexibility is reflected in the residual load variations (demand minus intermittent output).

Their limited participation in flexibility pool resources is reflected in the low capacity credits that are typically granted to wind and solar by system operators, based on the amount of power that they can be expected to reliably produce at peak demand. The IEA estimates in its New Policies scenario that the capacity credit of wind and solar will range between 5% and 20% by 2035 depending on regional variations and technologies. The 5% lower limit for European wind capacity credit means that, out of 450 GW of installed capacity, only 22.5 GW can be relied upon to meet peak demand, while the average annual output will around 112 GW. From this, it can be estimated that 89.5 GW of additional flexibility capacity (i.e., the difference between average annual output and capacity credit) must be found elsewhere to ensure system adequacy.

According the DG Energy, EU, with levels of intermittent renewable energy sources (*RES*) higher than 25% of the overall electricity consumption, the production has to be curtailed in low consumption periods to avoid grid perturbation and grid congestion, unless the RES excess can be stored [8]. Existing flexible resources may, up to a point, be able to manage the additional residual load variations that result from the increased use of wind and solar, if these energy sources can be used more efficiently.

Intermittent RES penetration thresholds are subject to debate and tend to be system specific, but there are upper limits beyond which a radical change in the way power systems operate will be needed. This is due to three factors: (i) The ability of variable renewables to meet peak demand declines as deployment increases – that is, the higher the penetration rate, the lower the capacity credit. (ii) The average annual power output of variable renewables increases roughly linearly with installed capacity. As a result, the differential between average annual output and the capacity credit, which measures additional flexibility needs, will increase. (iii) Variable renewables have the lowest marginal operating costs because of the absence of fuel costs. They come first in the merit order ranking used by system operators to decide which generator to dispatch.

Consequently renewables will (i) displace base load plants, increasing their cost of production because very high utilization rates are required to amortize the initial investment in baseload plants (e.g., nuclear). They will also (ii) affect the economics of the peak power plants used to back up wind and solar, as back-up plants will be used less often, resulting in increasing prices to maintain profitability. Finally, (iii) production of electricity from renewables may need to be curtailed, if the system cannot accommodate it (Figure 12.2).

Therefore, system costs are likely to increase faster than growth in the penetration of variable renewable in the existing power system operating model [7].

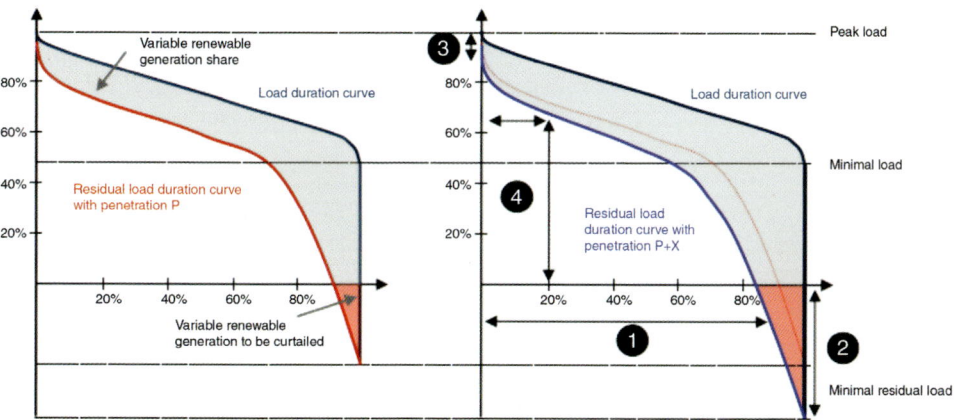

Figure 12.2 Impact of growing RES penetration on load duration curve and residual load duration curve [9]. Impact of increased variable renewable in the power mix: ❶ Reduce the residual base load that is, full load hours utilization of the power plants; ❷ Increase the wasted energy that is, time when renewables need to be curtailed due to excessive production; ❸ Provide smaller capacity credit that is, participate less in system adequacy; and ❹ Reduce the utilization of peak power plants running infrequently (<20% of the time).

Let us look now at some country-specific data. Several studies on the Dutch energy system presented results on the amount of projected curtailed wind power in the future without any storage in place. Ummels reports about 6.2 TWh of curtailed wind power for the Netherlands in a scenario of 12 GW wind capacity and no international exchange [10]. Undoubtedly, international exchange is crucial in energy systems analyses; however, it could be argued that whenever oversupply from wind power in Northern Germany occurs, an oversupply of wind power from the Dutch offshore wind parks is likely, hence limiting the possibilities for international exchange. Consistently, Velthuis [11] reports 2–10 TWh excess wind power per year in 2050 (different scenarios) and Boer [12] reports about 2.4 TWh of power to be available for storage at 12 GW wind power installed. However, a more detailed analysis from KEMA [13] shows that the actual curtailment from wind energy is also highly sensitive to the installed must-run capacity of combined heat and power units, for example, in industry. Depending on the amount of assumed wind and must-run capacities, but excluding contributions of solar power, the authors found a maximum wind curtailment of 0.5 TWh per year in their simulations. Incorporating the added capacity of solar power will result in more curtailment.

The increase in wind capacity can result in excess power production and may require curtailment of wind power generation at times of overproduction. In the UK in 2011, operators of Scottish wind farms were paid to constrain their supply so as not to overload the transmission grid. The total energy lost increased up to 75 000 MWh, for which customers would pay on average 21pence per kWh amounting to a total of £15.8 million [14].

Three studies [15] have investigated this phenomenon in the UK. From Table 12.1 it is evident that an increase in installed wind capacity, and thus a higher penetration of wind power production, leads to an increase in shedding.

In contrast to the Redpoint *EFC* scenario with annual power shedding up to 400 GWh, the high renewables (*HR*) scenario in the same study is characterized by a higher wind penetration and shedding of 2 TWh. The Strbac study [16]

Table 12.1 Wind capacity, penetration and annual electricity shedding in different scenarios [15].

	Wind capacity (GW)	Wind penetration (%)	Shedding (TWh)
Redpoint *EFC* 2030	39.5	32.4[7]	0.4
Rdepoint *HR* 2030	48.4	39[4]	2
Pyöry *High* 2030	59	42	3
Pyöry *Very High* 2030	68	49	6
ICL *Grassroots* 2030	72	N.A.	30
Pyöry *High* 2050	100	55	37.5
Pyöry *Very High* 2050	119	63	40.9
Pyöry *Max RES*	189	79	116.2

models annual renewable curtailment against installed storage capacity for different scenarios. Without measures against curtailment nor storage capacity installed, 30 TWh of wind power would have to be shed in 2030.

Wind farm constraints take place in the UK already. In 2013, 213 000 MWh of wind energy was constrained for a total of £18 868 000 market value up to the point when the report was issued [17]. A linear projection for entire year 2013 indicates constraint generation of 335 000 MWh for £30 000 000. These data take into account the 30 largest wind farms that participate in the balancing mechanism (BM) and that represent about 4 GW of the 10 GW installed wind power [18]. For these wind farms only half of the wind constraints are reported within the BM while the potential other half occurs within bilateral contracts. One can thus assume that in 2013 the total amount of constrained wind from the 30 largest wind farms amounted to 670 GWh with a constraint payment value of £60 000 000. This results in a constraint factor of 2.5% based on the total amount of installed wind power. Wind constraints of ~10% are a reality for certain UK wind farms already.

Surpluses, despite considerable network expansions, are also to be expected in Germany. The amount of surplus energy depends mainly on the expansion of renewables and the local and international network expansion. In the ideal case of a perfectly developed electricity grid over Germany (i.e., the so called "copper plate" model, assuming no bottlenecks in distribution) the BMU scenarios lead to surpluses from 2020 onwards and rising to about 170 TWh$_{el}$ in 2050 despite a European energy market [19]. In the 100% RES scenario of the Federal Environment Agency, which also was calculated with the Fraunhofer IWES model SimEE, the surpluses are, under ideal network conditions and national self-sufficiency, dependent on the meteorological year, about 80–110 TWh$_{el}$. In the SRU study [20] that involves Scandinavia, there is no explicit indication of an actual storage requirement. However, the storage requirement is expected to be of the same order of magnitude as in the UBA IWES study and the BMU scenarios, deduced from the import balance of 76.4 TWh$_{el}$ plus the use of 15.7 TWh$_{el}$ national large-scale storage capacity [20]. In reality, however, the network is not ideal and transmission or distribution grid bottlenecks apply, in the so-called "energy management" many surpluses from wind energy are already curtailed locally. In 2010 alone in Schleswig-Holstein about 0.1 TWh$_{el}$ wind energy was shed.

This, together with the growth of intermittent renewables challenges the integrity of the electricity grid and the load following paradigm of the current system. System solutions need to be developed if we want to avoid frequent curtailment of RES that hinders the growth of their share in the energy mix.

12.3
The Options for Integration of Intermittent Renewable Energy Sources

To mitigate the integration costs of growing intermittent RES, system operators will have to draw upon alternative resources (Figure 12.3). The problems of

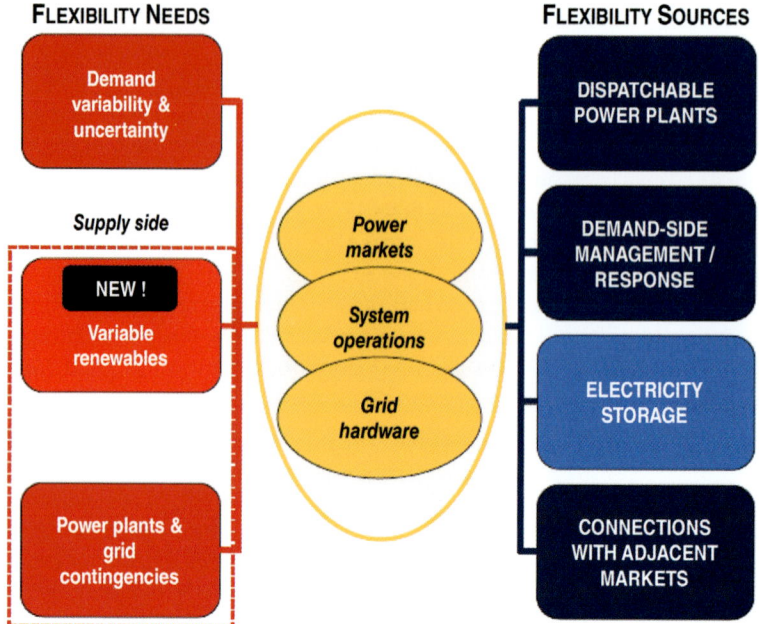

Figure 12.3 Flexibility resources and flexibility needs in the power system [3,7].

variability and uncertainty can be managed – and the disadvantages mitigated – if the power system incorporates sufficient flexibility that can support a larger proportion of variability. Flexibility can be quantified as the power that can be ramped up or down over a given time period: megawatts per minute for instance. This will be particularly crucial to counteract the low capacity credit of wind and solar, and to monetize excess electricity, avoiding curtailment and waste of investment capital. The IEA lists four main categories of flexibility sources: dispatchable power generation (supply-side management), demand-side management (response side), energy storage, and interconnection between grids [6].

Energy storage facilitates the integration of the renewable electricity supply to the energy system, but it is not a stand-alone solution. The main complementary technology options for integrating variable renewables are:

- electrical grid modernization and improved operation schemes at all voltage levels (transmission and distribution grid expansion and upgrade, including interconnections of different smart grid elements; improved planning, operation, and grid management);
- supply-side management (improved flexibility of conventional generation, centralized and decentralized);
- demand-side management (flexibility of important consumers, e.g., metal industry, chemical industry, paper industry, households, e-mobility, etc.).

None of these options provides a simple solution to a complex problem. In the longer term only coordinated interaction between options such as generation, transmission, distribution, storage, and consumption of electrical energy facilitates effective integration of renewable energy generated in the power supply system. These options are to some extent overlapping or interchangeable, but none of them is a silver bullet.

Various studies have analyzed the deployment of renewable energy and the accompanying need for infrastructure expansion in the European power supply and on a regional level [21–23].

Different possible future scenarios for renewable electricity in Europe result in different requirements for the extension and investment needs of the electricity networks. Notably, transmission grid expansion is a very long process (5–10 years, in specific cases up to 20 years) that requires major investments as well as a long-term permission process at European level. Land use and other environmental impacts are playing an important role. However, increased transmission capacity alone will not be sufficient to ensure security of supply in an energy system completely based on renewables [24]. In the past, many interconnector and transmission grid projects were not implemented in time or have not been realized at all [23,25,26]. This may lead to increased storage needs in the next few years.

Supply-side management (*SSM*) means the improvement of flexibility of the conventional generation mix or dispatchable renewable power plants (e.g., biofuel, biomass, solar/wind power with energy storage). Future conventional power plants will have to meet the following requirements to compensate the fluctuating input of renewable electricity and to stabilize the electricity grid:

- frequent startup and shutdown,
- quick response capability,
- higher ramp rates,
- extension of load range (decreasing "must run percentage"),
- improved part load operational efficiency.

Power plants have different characteristics making some more suited to supplying certain functions. Coal-fired power plants are not flexible and also nuclear power plants are inflexible to a large extent and thus mainly suited for base load coverage. Arrhenius [27] concludes that of conventional power plants only the gas-fired power plants can meet all requirements resulting from an increased renewable integration [24].

Regarding demand-side management (*DSM*), it can be stated that [24,28,29]:

- Theoretical load shift potentials exist already today in industrial areas as well as in households and in the small commerce sector. In particular peak loads lasting for minutes can be reduced by the application of DSM.
- Currently, the practical applications of DSM are limited to industry as they allow significant energy cost reduction. Other DSM strategies/applications

in the residential, small-commercial and mobility sectors do practically not exist as of today but could gain momentum in the future.
- All DSM participants, large or small, require additional "smart grid" technology elements, and thus the cost increases. Besides cost, acceptance is also required from both the utilities and the customers to enable the connection to and the interaction with the electricity grid, implying loss of privacy and, possibly, autonomy.
- In most cases the implementation and/or the increased utilization of practical DSM potential require new investments. These investments have to be weighed against those required for alternative solutions such as storage systems or grid extension.
- The potential of DSM for electricity load management will also in the future remain limited to an hourly level.

Even if all aforementioned non-storage options for renewable energy integration are perfectly realized, there still remains a substantial need for energy storage: [24]

- for valorization of excess renewable electricity;
- to match energy supply with demand;
- to provide assured power capacity at "low-wind and low-sun" times;
- for the transition towards flexible conventional power plant operation characteristics;
- to increase grid stability, system black-start capability, and local supply security.

Depending on the regional boundary conditions, energy storage may be a more important axis for solving several issues other than grid extension.

Electricity storage has two primary functions: (i) leveling the demand–supply curve and (ii) providing ancillary services to ensure power quality and providing operating reserves. In addition, electricity storage facility has several operational applications (Figure 12.4 [30]).

To cope with the growing need for energy storage, several electrical and chemical technologies are being developed. Technologies are still constrained by their design limitations and can therefore only compete for applications suited to their technical features (Figure 12.5 [7,31]), allowing a classification according to their power rating and discharge time requirements:

- power rating: the storage device's instantaneous ability to withdraw/inject energy from/into the grid;
- the discharge time: time needed to provide this energy; it corresponds to the energy capacity of the storage divided by the power rating.

Flywheels, supercapacitors, and superconducting magnetic energy storage [*SMES*], as well as some chemical batteries, are competing to provide the desired

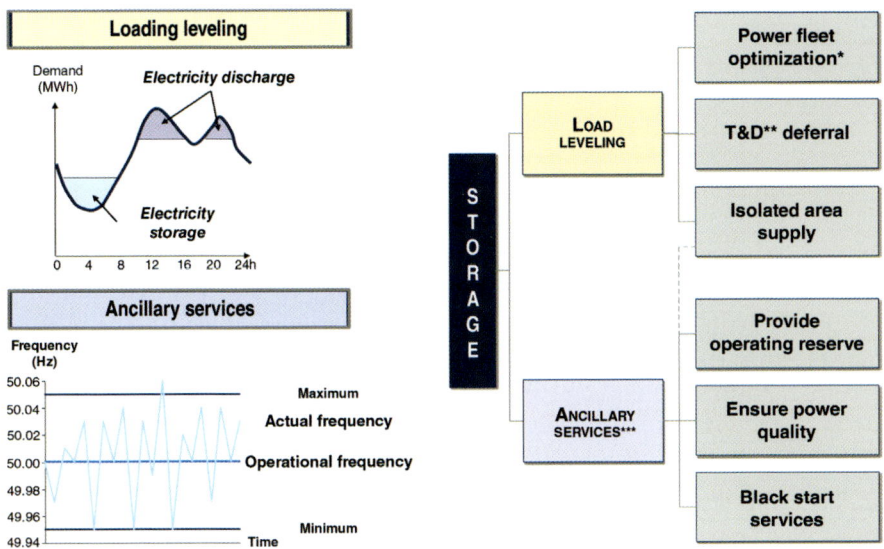

Figure 12.4 Primary functions of storage and main operational applications [30].

power quality and reliability. Batteries with a multitude of chemical compositions – such as nickel metal hydrides [NiMH], sodium sulfur [NaS], lithium ion [Li-ion], and lead acid – are competing with flow batteries – such as vanadium redox [VRB] and zinc bromine [Zn Br] – that allow us to decouple energy and power, for hour-to-day load shifting. Flow batteries are scalable and could compete for applications with wider power and energy ranges. Pumped hydro storage [PHS] and compressed air energy storage [CAES] are the main technologies for power-fleet optimization and large-scale intermittent balancing. Pumped hydro storage is, however, not adequate for the frequency regulation market, as it has a high power and energy rating. Hydrogen-based storage solutions are the only ones that can meet intermittent balancing requirements at very high penetration levels, being able to fulfill small power requirements using fuel cells and also large amounts of power can be provided deploying combustion turbines.

Large-scale electricity storage, apart from hydro storage, has long been perceived as impractical. It has been generally avoided in favor of storing primary energy sources and energy inertia in power generators [3]. PHS is currently the only widespread large-scale technology. Its development was boosted in the 1980s by price arbitrage opportunities arising from the growing spread between low and peak electricity prices, and runs timely in sync with nuclear power capacity expansion. Attempts to develop compressed air energy storage, batteries, or flywheels have, so far, been unsuccessful for large scale applications. Their use has been limited to niche applications and a small number of demonstration plants. Thermal-based electricity storage is emerging in parallel with the

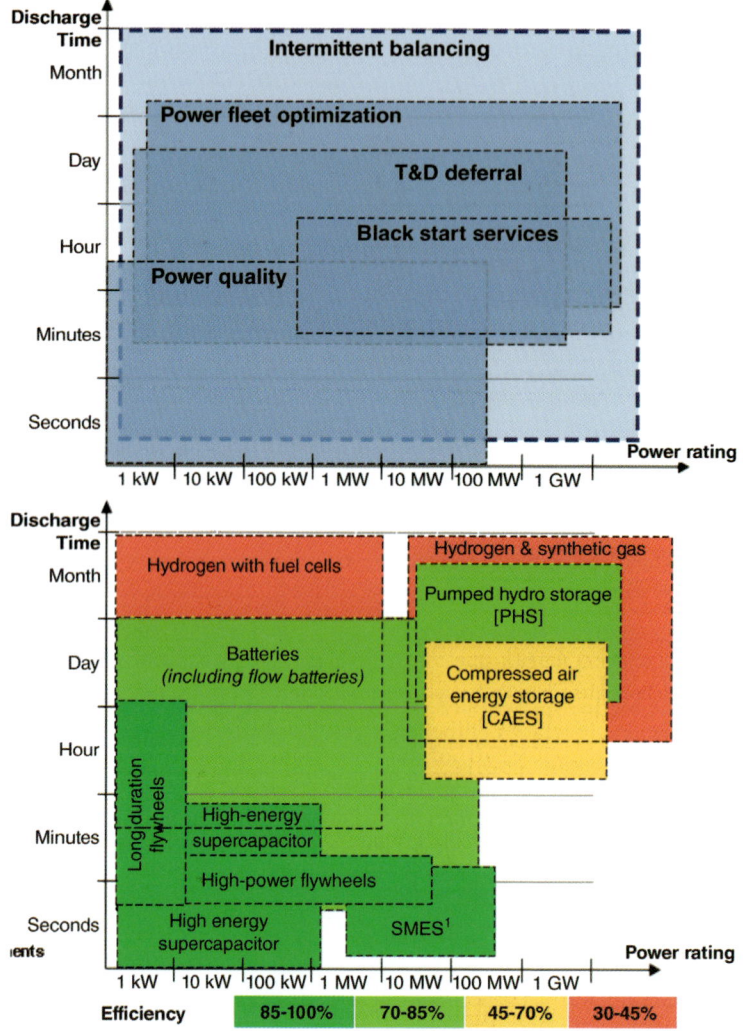

Figure 12.5 Electricity storage application requirements and features of technologies [7,31].

development of concentrating solar power plants. Finally, chemical electricity storage in batteries is getting increasing attention, despite still being at an early stage of development, with around 8 MW of demonstration capacity in existence (Figure 12.6 [7,32]).

The European Association for Storage of Energy (*EASE*) [33] is defining a comprehensive overview of the grid-related services that energy storage solutions can provide (Figure 12.7).

12.3 The Options for Integration of Intermittent Renewable Energy Sources | 267

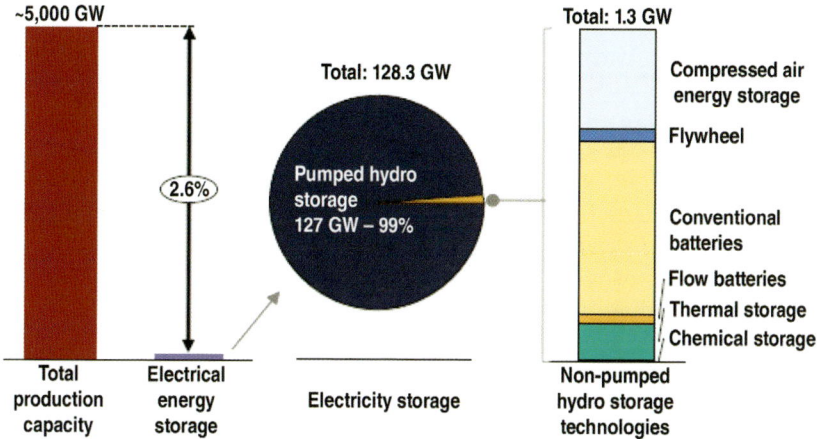

Figure 12.6 Electricity energy storage installed capacities by technology [32].

From these 26 service applications a few contribute to the integration of renewables. The integration of wind or solar energy will increase the need for most of these service applications, and not only those in the Renewable Generation segment (Figure 12.7).

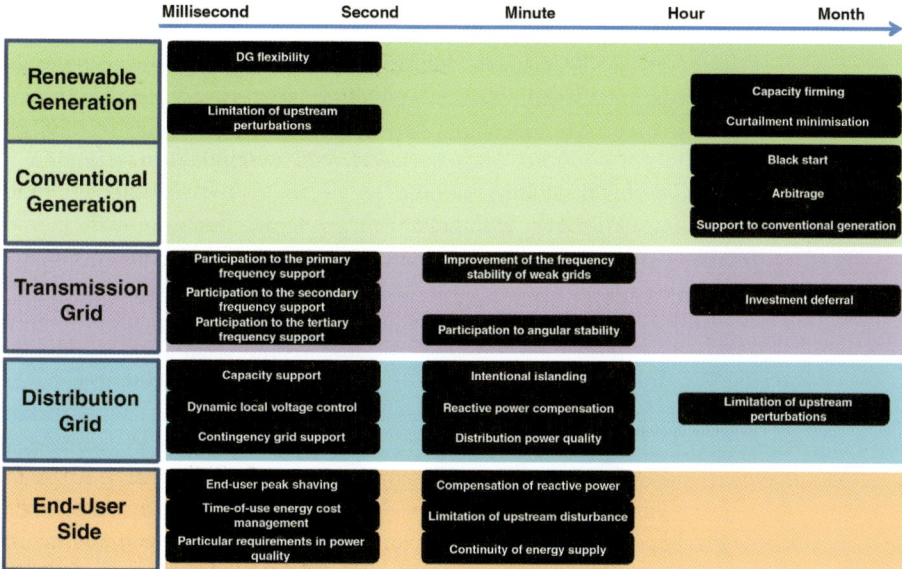

Figure 12.7 Overview of services that energy storage solutions can provide relating to the different aspects of electricity grid operation.

12.4
The Evolution of the Demand for Energy Storage

EASE and EERA have jointly developed a roadmap on energy storage for the EU, the "European Energy Storage Technology Development Roadmap towards 2030" [33]. These policy recommendations can be seen as a starting point to define significant and relevant energy storage aims in the EU towards 2030.

Today, the EU electricity system storage capacity is around 5% of the total, already installed, generation capacity. Pumped hydro-electricity storage (PHES) represents the highest share of all large scale electricity storage capacity in Europe (99% worldwide). As postulated widely in References [1,34], an increase in the global energy and electricity storage capacity will be required in the future. With levels of intermittent RES generation higher than 25% of the overall electricity consumption, the production has to be curtailed in low consumption periods to avoid grid perturbation and grid congestion, unless the RES excess can be stored [8].

New energy storage capacity and technologies will therefore be needed for the RES based energy supply envisaged by the EC policy. Until the development and the implementation of new technologies and capacities, pumped hydrogen as large scale storage technology and natural gas storage, in combination with rapidly responding CCGTs back-up power plants, will play an important role in the medium term transition to a RES based energy system in Europe.

Neither the European Commission, through the European Energy Programs, nor the different energy storage projects realized or being realized in Europe have quantified the energy storage needs in the EU in the medium or long term.

The main results found with regard to the quantification of future EU energy storage needs are presented in scientific articles in the field of energy storage. Of special interest are the studies of Heide *et al.* [35,36]. Both studies analyze a highly or fully developed renewable European electricity system based on wind and solar together with a high amount of energy storage based on hydrogen in underground salt caverns. In Reference [36] a scenario based on 100% wind and solar power generation is studied. The required energy storage capacity is estimated to be of the order of 12–15% of the annual European electricity consumption, corresponding to 400–480 TWh (2007 data), assuming an optimal seasonal mix of 60% wind and 40% solar power. Deviating from this optimum would increase the need for storage, up to doubling the requirements when in wind-only or solar-only scenarios. Some of the assumptions made and conclusions obtained by the authors are:

- "A storage energy capacity of several hundred TWh represents an incredibly large number. For pumped hydro and compressed air storage in Europe this is fully out of reach."
- "Excess wind and solar power generation can be used to significantly reduce the required storage needs for a fully renewable European power system."

- "The combination of hydro storage lakes and hydrogen storage will be able to contribute to solving Europe's search for long term storage."
- "Power transmission across Europe is needed to balance local negative power mismatches with positive mismatches in other regions."

The final results for a fully renewable EU energy power system (assumptions: 1300 GW wind power, 830 GW solar power, 50% excess generation) require a hydrogen storage system of 50 TWh energy capacity and 220 GW discharge power, assuming 60% water electrolyzer efficiency, 60% fuel cell efficiency, and hydrogen underground storage. Taking into account that a typical large cavern field has a volume of $8 \times 10^6 \, m^3$ [37], providing storage capacity of 1.3 TWh, the required amount of energy storage could be covered by 39 large salt cavern fields with hydrogen storage.

Although a more detailed quantification of the energy storage needs in the EU has not been undertaken yet, the necessity of energy storage to compensate the intermittency of RES is obvious. It will play a key role in the future European energy structure, as it is envisaged in the EUs Energy 2020 program and the Energy Roadmap 2050. A commonly accepted roadmap on energy storage for the EU or adequate documents needs to define these goals and will be the basis for establishing a well-defined program. Hydrogen underground storage is believed to be a really promising technology due to its high volumetric energy storage density and the fact that hydrogen storage in caverns does not pose unsurmountable technological hurdles and is already standard practice [36]. Even the most ambitious case of 100% RES integration seems to be feasible by the implementation of a large but realistic amount of hydrogen energy storage. Taking into account that the penetration of RES in the energy mix by 2050 can be less than estimated, and counting also on the development of other storage technologies to contribute to the energy management, together with non-storage solutions such as high voltage transmission lines, currently being defined as priorities, the objectives established in the Energy Roadmap 2050 could be fulfilled without major problems [24].

The increased penetration of variable renewables in the electricity mix is expected to create new long-term (seasonal) and large-scale (TWh) electricity storage requirements. It is likely to reduce the correlation of supply and demand, generating very large deficits of supply at times of high demand, if the weather is unfavorable, and significant surpluses when demand is low but the wind is blowing or the sun is shining. Researchers of the Fraunhofer Institute [38] simulated a 100% renewables-based electricity system for 2050 on behalf of the German Ministry of Environment. With load management (e.g., scheduling charging times for electric vehicles or air-conditioning use), their model predicts 82.7 TWh and 84.7 TWh of surplus and deficit, respectively (Figure 12.8). While only illustrative, the model highlights that, when variable renewables provide an important share of the electricity mix, long-term, large-scale energy storage is essential to avoid massive waste of energy and to complement other flexibility

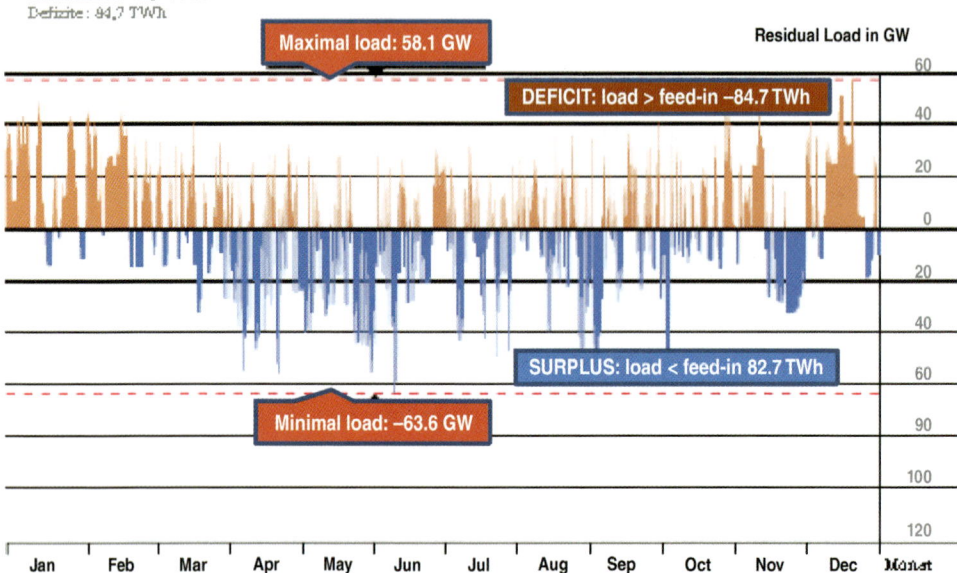

Figure 12.8 German residual basic load for 2050, based on 2009 data [38].

sources (e.g., without load management, the deficit would be 52.8 TWh and the surplus 153.9 TWh). The penetration threshold of renewables that triggers the need for storage is under debate and is, in essence, system specific (from 35% to 100% in the literature), but there is a consensus that above 80% there will be a need for long-term, large-scale electricity storage.

12.5
The Role of Electrolyzers in the Energy Transition

Modern energy systems are characterized by a high penetration of intermittent renewable power generators. Political targets demand ever higher penetration levels. This puts challenges on the power system. In traditional power systems fossil generators adapt to inflexible power demand on top of the base-load provided by fossil and nuclear generators. A higher penetration of non-dispatchable, intermittent generators has a negative impact on this supply-side flexibility. Therefore, modern energy systems must feature energy storage concepts and demand-side flexibility.

Hydrogen-based energy storage solutions are unique, creating a new group of storage technologies – large scale chemical storage (as opposed to small scale chemical storage in batteries). In contrast to peer electricity storage technologies, chemical storage is based on a conversion process from one carrier into another – the charging phase – where electricity produced from primary renewable sources (wind and solar) is converted into a new energy carrier – hydrogen. While the volumetric energy density of hydrogen (kWh l^{-1}) is not as good as those of hydrocarbons, it still offers unrivaled energy density compared to other bulk storage technologies such as pumped hydro storage.

Electricity stored as hydrogen can be converted back into electricity during the discharge phase. Here, the main value of hydrogen solutions lies in their ability to store energy in bulk. Besides providing a re-electrification pathway, chemical storage creates alternative uses for the stored energy. These include blending hydrogen with natural gas, converting it into synthetic methane, converting it into liquid fuel, valorizing hydrogen for use as a feedstock in the chemical and petrochemical industries, or using it as a fuel for mobility. Conversion has downsides associated with its virtues. It can lead to additional efficiency losses and requires investment in conversion (and re-conversion) facilities. The big questions are: Under which circumstances do the virtues of chemical storage outweigh its disadvantages? Can a system operate without storage even when variable renewables are a large part of the electricity mix? And, finally, can the drawbacks be mitigated by technological progress? Hydrogen based storage solutions can, in theory, compete for the whole spectrum of storage applications, as they decouple charging, storage, and discharging, as well as benefiting from the molecule's very high energy density per unit of mass. However, energy losses inherent in the conversion processes (power-to-hydrogen and hydrogen-to-power) are likely to limit the cycling frequency, except in niche applications that leverage the unique properties of hydrogen (discharge time, energy density) or those of its underlying technologies (reliability of fuel cells for back-up power or power quality applications).

In practice, the main applications of hydrogen-based storage solutions are expected to combine the three chief advantages of chemical storage: (i) Very high energy density, allowing extensive storage capacity. A hydrogen molecule has an energy density per volume of 2.7–160 kWh m^{-3} for pressures from 1 to 700 bar, compared with 0.27 kWh per 100 meter elevation differential per m^3 of water for pumped storage, and 2–7 kWh m^{-3} for compressed air storage for pressures from 20 to 80 bar. (ii) Hydrogen solutions benefit from negligible losses during the storage phase, allowing long-term storage, unlike, say, batteries. (iii) Hydrogen based technologies have the ability to react to erratic weather patterns and thus to fluctuations in storage requirements.

Chemical storage, exploiting hydrogen's versatility and energy value per mass, opens up alternatives to the usual approach to electricity storage, due to its three-dimensional shifting ability:

- Time-shifting. If the volumetric energy density of hydrogen is not as good as hydrocarbons, it still offers unrivaled energy density compared to other bulk storage technologies such as pumped hydro storage. It represents the only technology that would be able to bridge several weeks of windless or cloudy conditions or to provide security of supply on the same level as gas or oil stocks.

- Location-shifting. Hydrogen-based technologies could also reduce infrastructure investments for integrating wind and solar PV generators to the grid. Conversion of excess of electricity produced from renewables into hydrogen allows leveraging of existing networks: (i) the power network, and by locating storage facilities at congestion nodes to level the load (as with any storage technology); (ii) the gas network, when it already exists as a legacy natural gas network; (iii) hydrogen for road, rail, and maritime transport options and, in some regions, hydrogen pipelines.
- End-use-shifting. The versatility of hydrogen based storage solutions, compared with other electricity storage technologies, means they are not restricted to producing electricity. Hydrogen can be used in its traditional markets, as an upgrader in refineries for removing sulfur from sour crude and for enriching heavy oil fractions into lighter, more valuable products. It can also be used as a commodity in many industrial processes, such as ammonia production and in the semiconductor industry. As a transport fuel, hydrogen can be used to power fuel cell electric vehicles, but it can also be blended with natural gas in gas networks or in gas refueling stations to fuel compressed natural gas vehicles without major changes in the process. Hydrogen is also commonly used with CO_2 to synthesize liquid and gaseous fuels (e.g., methane, methanol, dimethyl ether) that are essentially chains of hydrogen and carbon molecules. Such green fuels are also building blocks of a CO_2 neutral energy mix.

Power-to-gas plants provide these features. They convert electrical power into a storable gas, hydrogen or methane. This concept is suited to absorb excessive power resulting from the increasing penetration of intermittent renewable power generators. As a gas, this excessive renewable power can be stored within the gas network and made available to all energy sectors and the chemical industry. *Power-to-gas* plants couple the power network to the gas network and thus provide the missing link towards a fully integrated energy system (Figure 12.9).

This creates several opportunities:

- de-carbonization of all energy sectors, the chemical industry and the transport sector;
- combination of low-cost power generation with low-cost gas infrastructure;
- "grid-scale" renewable energy storage;
- optimization of power system operation with intermittent generators;
- pathway towards a Hydrogen Economy.

These opportunities support sustainability, affordability, and security of supply in modern energy systems. They increase the deployment potential for intermittent generators and offer a pathway towards a renewable gas based energy economy through the provision of cost-competitive and renewable hydrogen and/or methane.

12.5 The Role of Electrolyzers in the Energy Transition

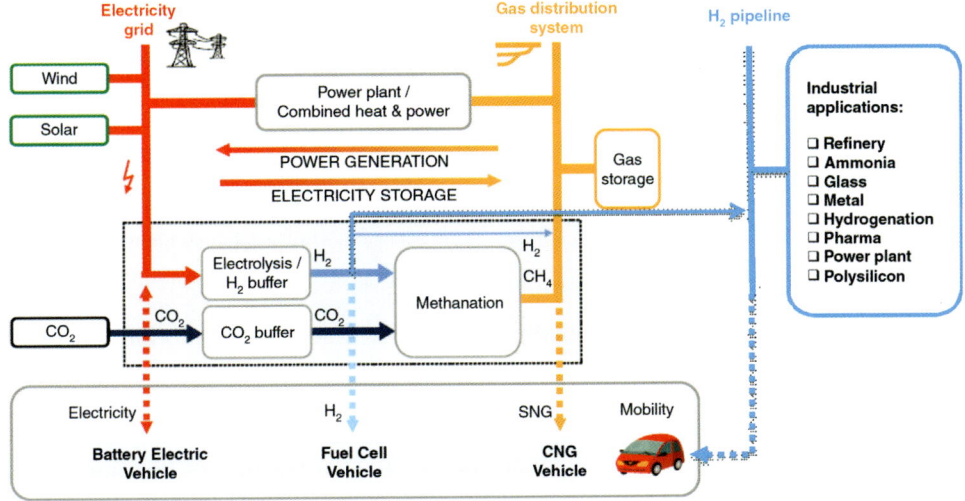

Figure 12.9 Hydrogen-based energy conversion routes.

From an exergetic perspective electricity should always be deposited as electricity on the electricity grid when possible (highest efficiency). However, when problems occur in the electricity sector (such as congestion, negative electricity prices, or physical damage) or an electricity infrastructure is lacking at the production site, then electricity can be converted into hydrogen. This hydrogen can then be accommodated directly in the gas grid, utilized in the chemical industry or mobility sector, stored in a hydrogen buffer in order to be re-converted into electricity at a later moment, or converted into methane. Although parts of the gas grid that do not feed critical appliances can tolerate up to 20% H_2, by volume, setting national limits above 5 vol.% will be difficult [44]. The conversion into methane can be appropriate whenever injection in the gas grid is preferred but limited by a maximum allowable amount of hydrogen.

Another reason for power-to-gas is to overcome continuous transmission capacity constraints (grid specific) or to transport energy over long distances. In this situation, electricity will be converted into hydrogen or hydrocarbons continuously.

The pathways from electricity to one of the end products mentioned above are visualized by Figure 12.9. With technology that is available today, the first step in the conversion of electricity into hydrogen is indispensably water electrolysis. As soon as the hydrogen has been produced different pathways become possible. When hydrogen is converted into methane or other hydrocarbons, a carbon source (e.g., CO_2) is required for synthesis.

The amount of hydrogen that can be added to natural gas in the natural gas infrastructure is limited because of the effects of hydrogen on the Wobbe index of natural gas, on the combustion behavior of the gas mixture, and its effect on material integrity. The allowable amount of hydrogen blending is case specific and depends on the end user sectors. Because of the limitations for hydrogen injection into the natural gas grid, the potential storage capacity of hydrogen in the gas infrastructure is significantly lower than the total storage capacity of methane.

The gaseous energy carrier hydrogen can enable large amounts of renewable electricity to be utilized for other types of consumption. When high levels of penetration of renewable energy sources are to be achieved, then the use of hydrogen (or methane) from renewables will play a key role in re-electrification to cover residual loads, in generation of high-temperature heat, and in replacing fossil fuels for transportation. This is especially true in view of the limited potential of biomass. Both paths of usage, hydrogen and methane, begin with hydrogen generated electrochemically from renewables. In a further process stage (methanation), the hydrogen (and its energy) can be converted into synthetic methane. Since losses are unavoidable in the provision and utilization of hydrogen and methane, it is obvious that all processes for direct utilization of electricity from renewables should be exploited to their fullest in the heat and transportation sectors first.

12.6
The Overall Business Case and Outlook

Power-to-gas (*PtG*) plants convert electrical energy into chemical energy, in the form of a gas. The key technology of this concept is water electrolysis which consumes electricity to split water into its chemical compounds, hydrogen and oxygen. The innovativeness of the PtG concept comes with the idea of using excessive, carbon-neutral electricity and processing hydrogen into synthetic methane via CO_2 methanation. This approach was first formulated by Dr Specht as *Renewable Power Methane* in association with a patent filed in 2009 [39]. It establishes a link between the electricity and the natural gas networks via the injection of renewable synthetic methane into the gas grid. Furthermore, it allows diffusing renewable energy, generated in the power system, to all energy sectors and the chemical industry. Since 2009, the commonly used name for this concept has changed to *power-to-gas* and the idea of injecting hydrogen directly into the gas network or using it in other applications has attracted interest as well.

To enable the efficient deployment of PtG systems, policy makers and investors need to understand their economic and environmental benefits. These are affected by the limitations of the potential product markets (power, heat, transport, industry).

A study by Bünger [40] has shown that the transport sector offers the highest achievable prices for hydrogen (Figure 12.10). This is based on applying current fuel prices per km for a conventional vehicle (e.g., petrol, diesel) as a bench mark to the highly efficient fuel cell electric vehicles (*FCEV*) powered by hydrogen.

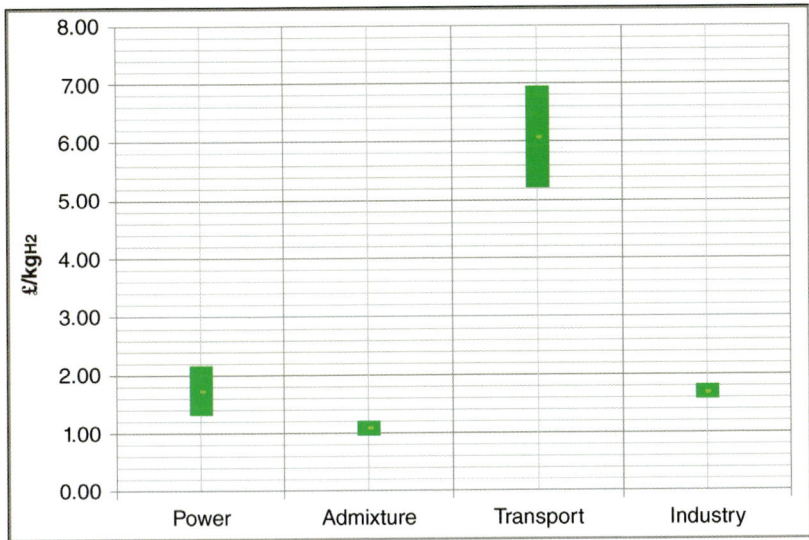

Figure 12.10 Achievable hydrogen prices per sector. (Adapted from Reference [40].)

The cost of hydrogen will vary highly as a function of electricity cost, utilization rate, and unit capacity number.

The results of Reference [41] indicate potential business cases for large-scale PtG projects providing hydrogen for the chemical industry and in the hydrogen mobility sector. They show that product prices are not necessarily the determining factor for economic viability, but that operation strategy, investment costs, power costs, and the ability to commercialize by-products can have significant case-specific impacts as well. These impacts are quantified in this study. The industrial and transport sectors also exhibit the highest de-carbonization potential.

12.6.1
De-carbonization of all Energy Sectors and the Chemical Industry

Figure 12.9 displays the role of PtG within the power and the gas network and possible pathways for renewable hydrogen or methane. In current energy systems, the gas network is linked to the power network via gas-fired power stations (gas-to-power). The heat and transport energy sectors can theoretically be served from the power and the gas network, but due to low electrification degrees the power network is rarely used for energy supply. The chemical industry is provided with feedstock from the gas network only. Renewable energy is mostly generated in the power sector and available in the power network. Thus, linking the power network to the gas network and creating a mutually integrated energy system allows renewable energy generated in the power sector to be accessed by all energy sectors [39].

Table 12.2 Energy storage capabilities of UK power and gas network.

	Power network	Gas network
Consumption (TWh a^{-1})	367.45	857
Average power (GW)	42	98
Storage (TWh)	0.03	40
Theoretical energy supply duration (h)	0.7	408 (= 17 d)

12.6.2
Combination of Low-Cost Power Generation with Low-Cost Gas Infrastructure

Renewable generators (e.g., wind) are characterized by low marginal costs due to the lack of fuel costs. They come first in the merit order. In combination with the intermittent character of wind power generation, this may lead to excessive renewable power generation subject to local power transmission constraints or insufficient demand. The costs for excess power are zero or negative if subsidies for renewable power generation are in place [42]. The conversion of this low-cost power into a gas allows accessing low-cost gas storage and distribution infrastructure. Tables 12.2 and 12.3 compare the storage capabilities and investment costs of the power and the gas network in the UK. The theoretical energy supply duration of the gas network is substantially higher than that of the power network, despite low gas storage capabilities in the UK. Thus, the gas network is a valuable asset to store energy.

12.6.3
Grid-Scale Renewable Energy Storage

The high storage capacity and duration (grid-scale energy storage) of the gas grid is complemented by its geographic size. This combines the temporal dissociation of "charging" and "discharging" (e.g., seasonal storage) with geographic dissociation. Renewable energy can be injected into the gas grid when and where excess production occurs and used when and where

Table 12.3 Investment cost examples for power and gas networks [33].

	Power network (Brit Ned)	Gas network (bbl company)
Length (km)	260	230
CAPEX (€)	600 million	500 million
Capacity (GW)	1	20
	€2.3 kW^{-1} km^{-1}	€0.11 kW^{-1} km^{-1}

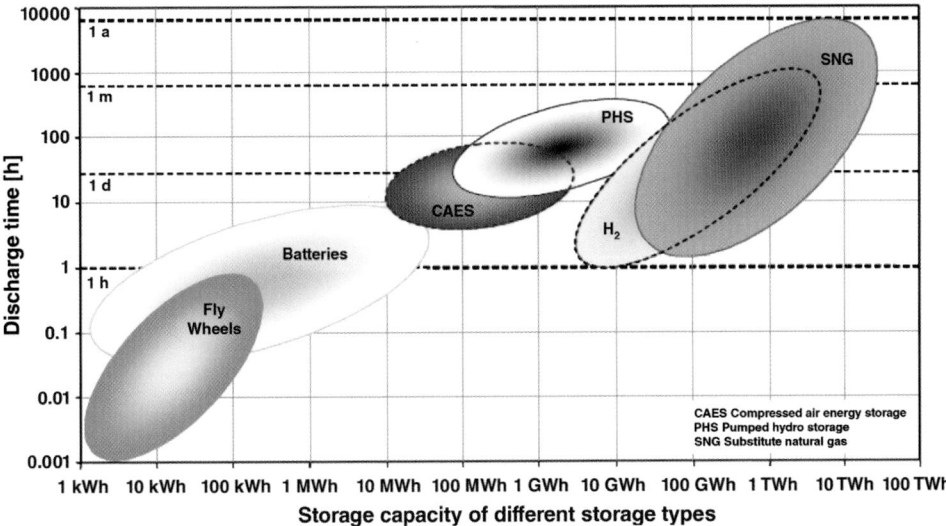

Figure 12.11 Discharge time and storage capacity of different electricity storage systems [43].

demand exists. PtG plants are not subject to geographic constraints (e.g., caverns for CAES) and subsequently exhibit high flexibility allowing optimal location selection, for example, close to wind farms. The limitations on scalability are set by the gas grid. Figure 12.11 depicts the storage capacity versus the discharge time of different energy storage technologies, clearly illustrating the vastness of the PtG (H_2 and SNG) option.

12.6.4
Optimization of Power System Operation with Intermittent Generators

The flexibility of the technology used in PtG plants enables the immediate uptake of excessive power and thereby optimizes the operation of power systems with intermittent generators. Immediate demand management minimizes curtailment and enables higher penetration of intermittent, renewable generators. In addition, PtG plants increase the system balancing capacity through the provision of demand- and supply-side balancing services such as short term operating reserve (*STOR*) and frequency control demand management (*FCDM*). This contributes to minimizing the ramping of thermal generators and associated efficiency losses that translate into higher costs and CO_2 emissions. Shifting renewable energy from the power to the gas network helps to defer capital investments in network reinforcements. It allows determination of an economic optimum for renewable energy transmission. The absorption of increasing price fluctuations helps in stabilizing the power system economically [15].

12.7
Conclusions

Wind and solar PV are at the forefront of power-sector decarbonization and set to expand rapidly. The variable output of wind and solar PV makes demand–supply matching more difficult and increases the need for flexibility within the system. New flexibility resources must be developed in addition to dispatchable power plants. With the exception of pumped hydro storage, the deployment of electricity storage is at an embryonic stage. Hydrogen energy storage solutions are based on the conversion of electricity into a new energy carrier, hydrogen, by means of water electrolysis. Exploiting hydrogen's versatility, chemical energy storage opens up alternatives to the usual approach to electricity storage. The physical properties of hydrogen make it particularly suited, but not limited, to large-scale, long-term re-electrification applications. Hydrogen can be re-electrified in a direct electrochemical process, using fuel cells, but also using conventional thermal combustion turbines. Hydrogen re-electrification results in poor round-trip efficiency, which is likely to impede its development in the short term. Converting electricity into a new energy carrier enables extracted energy to be transported through alternative infrastructure and opens up possibilities for de-carbonizing other sectors such as transportation, heat and industry.

Decentralized hydrogen production is essential in order to minimize hydrogen transportation costs.

Power-to-gas may change the rules of the energy game by linking the natural gas grid with the power grid. Although parts of the gas grid that do not feed critical appliances can tolerate up to 20% H_2, by volume, setting national limits above 5 vol.% will be difficult. Hydrogen blending is an elegant early stage solution for monetizing electricity surpluses in countries with highly developed natural gas infrastructure. Methanation produces synthetic natural gas from H_2 and CO_2, which is not restricted to blending-ratio limits. Several synthetic liquid fuels can also be synthesized from hydrogen and carbon.

Hydrogen is a vital molecule for mobility. FCEVs are back in the spotlight, but successful deployment depends on the cost of fuel cells and on solving the challenge of the hydrogen infrastructure roll-out. Hydrogen-enriched compressed natural gas vehicles may provide solutions to the hydrogen development challenge, and to cleaner mobility.

Hydrogen is an important industrial gas, with 87% of demand coming from the chemicals and petrochemicals industries. Electrolytic hydrogen is unlikely to compete with steam methane reforming in terms of cost, but may provide refineries close to H_2 equilibrium with additional operational flexibility (and a greener image). Small-scale ammonia production for fertilizers, coupled with distributed production, could make economic sense in remote locations.

The main challenge for hydrogen conversion is economic rather than technical: How to find a sustainable business case in an uncertain environment? The versatility of the hydrogen carriers opens the way to end-uses that valorize the power conversion into H_2 as a service or the H_2 produced as a feedstock.

Applications that valorize H_2 as a feedstock and benefit from support mechanisms for low-carbon solutions are likely to drive the first hydrogen energy developments. The business-models of electrolytic hydrogen are inherently system- and application-specific.

The value of hydrogen-based energy solutions lies predominantly in their ability to convert renewable power into green chemical energy carriers. There is no silver bullet for hydrogen-storage solutions: end-use requirements must be matched with the features of individual energy systems. In all cases, the deployment of hydrogen systems requires cost reductions and public support. Many individual, hydrogen-related technologies are technologically mature, but electrolysis and power-to-gas are in the investor's "valley-of-death;" despite the need for large investments in light house projects, hydrogen-based energy-storage projects are still a way from commercial deployment. Public authorities have a wide variety of temporary incentives at their disposal to help transform hydrogen-based solutions into self-sustaining commercial activities.

References

1 European Commission's communication (2011) Energy roadmap 2050, COM(2011) 885 final of 15 December 2011.
2 IPCC (2012) Special Report on Renewable Energy Sources and Climate Change Mitigation (SRREN).
3 International Energy Agency – IEA (2012) Energy Technology Perspectives (2012).
4 International Energy Agency – IEA (2012) Renewable Energy – Medium Term Market Report 2012.
5 International Energy Agency (2011) Harnessing Variable Renewables: A Guide to the Balancing Challenge", OECD/IEA.
6 Fuel Cell Today (May 2013) Water Electrolysis & Renewable Energy Systems.
7 SBC Energy Institute (February 2014) Hydrogen-based Energy Conversion.
8 European Commission, Directorate General for Energy (2013) The Future Role and Challenges of Energy Storage.
9 SBC Energy Institute Analysis (2012) based on Ueckerdt (2012), Variable Renewable Energy in Modeling Climate Change Mitigation Scenarios, Potsdam Institute of Climate Impact Research.
10 Ummels, B. (2009) Wind Integration – Power System Operation with Large-Scale Wind Power in Liberalised Environments. PhD thesis of Delft University of Technology at Tennet.
11 Velthuis, M. (2012) The Role of Large Scale Energy Storage Systems in the Electricity Grid of the Netherlands in 2050. Master Thesis at University of Groningen, EES 2012-136T.
12 Boer, de H.S. (2012) The Application of different types of Large Scale Energy Storage Systems in the Dutch Electricity System at different Wind Power Penetration Levels. An Environmental, Economical and Energetic Analysis on Power-to-gas, Compressed Air Energy Storage and Pumped Hydro Storage. Master Thesis at University of Groningen & DNV KEMA.
13 KEMA (2010) Integratie van Windenergie in het Nederlandse Elektriciteitssysteem in de Context van de Noordwest Europese Elektriciteitsmarkt. Rapportage 30920572-Consulting 10-01-0198, in opdracht van het Ministerie van Economische Zaken.
14 Cooley, G. (presented 11th October 2012) Hydrogen Energy Storage – Nextgen.
15 Grant, P. and Skillings, S. (2009) Decarbonising the GB Power Sector: Evaluating Investment Pathways, Generation Patterns and Emissions through to 2030. Redpoint Energy Ltd.
16 Strbac, G. et al. (2012) Strategic Assessment of the Role and Value of Energy Storage Systems in the UK Low

Carbon Energy Future. Imperial College London.

17 REF (2013) Balancing Mechanism Wind Farm Constraint Payments (19 08.2013).

18 REF (2013) Notes on Wind Farm Constraint Payments (19 08.2013).

19 Nitsch, J., Sterner, M., Wenzel, B. et al. (2010) Langfristszenarien und Strategien für den Ausbau der erneuerbaren Energien in Deutschland bei Berücksichtigung der Entwicklung in Europa und global - Leitstudie 2010. Herausgegeben von Natur-und Reaktorsicherheit BMU Bundesministerium für Umwelt. Berlin.

20 SRU (2010) Sachverständigenrat für Umweltfragen: 100% erneuerbare Stromversorgung bis 2050: klimaverträglich, sicher, bezahlbar, SRU, Berlin.

21 SUSPLAN (2011) EU-project 2008–2011: Efficient integration of renewable energy into future energy systems, Development of European energy infrastructures in the period 2030 to 2050.

22 German Energy Agency (2010) German Energy Agency (dena): power grid study II: Integration Erneuerbarer Energien in die deutsche Stromversorgung im Zeitraum 2015–2020 mit Ausblick 2025, Report, Berlin.

23 Ten-Year Network Development Plan (TYNDP) (2012) European Network of Transmission System Operators for Electricity.

24 HyUnder (8th of August 2013) Assessment of the Potential, the Actors and Relevant Business Cases for Large, Scale and Long Term Storage of Renewable Electricity by Hydrogen Underground Storage in Europe, Deliverable No. 2.1 Benchmarking of large scale hydrogen underground storage with competing options.

25 Bundesnetzagentur für Elektrizität (2011) Gas, Telekommunikation, Post und Eisenbahnen: Monitoringbericht.

26 Bundesnetzagentur für Elektrizität (2012) Gas, Telekommunikation, Post und Eisenbahnen: Monitoringbericht.

27 Arrhenius: Institute for Energy and Climate Policy, Hamburg (October 2011) Kurzstudie: Die künftige Rolle von Gaskraftwerken in Deutschland.

28 European Renewable Energy Network (January 2012) Study, LBST, Hinicio, Centre for European Policy Studies, Technical Research Centre of Finland.

29 Verband der Elektrotechnik Elektronik Informationstechnik (VDE) (June 2012) ETG-Task Force: Demand Side Management: Ein notwendiger Baustein der Energiewende: Demand Side Integration, Lastverschiebungspotenziale in Deutschland, Studie der Energietechnischen Gesellschaft im VDE (ETG), Frankfurt am Main.

30 SBC Energy Institute (September 2013) Electricity Storage Factbook.

31 United Sates Department of Energy – US DoE (2011) Energy Storage Program Planning Document.

32 Electric Power Research Institute – EPRI (2010) "Electricity Energy Storage Technology Options – A White Paper on Applications, Costs, and Benefits".

33 EASE/EERA (March 2013) Joint EASE/EERA recommendations ES Technology Development Roadmap 2030.

34 European Commission (2010) Energy 2020 – A Strategy for Competitive, Sustainable and Secure Energy. European Commission, COM(2010) 639 final, 2010.

35 Heide, H., Greiner, M., Von Bremen, L., Hoffmann, C., Speckmann, M., and Bofinger, S. (2010) Seasonal optimal mix of wind and solar power in a future highly renewable Europe. *Renew. Energ.*, **35** (11), 2483–2489.

36 Heide, H., Greiner, M., Von Bremen, L., and Hoffmann, C. (2011) Reduced storage and balancing needs in a fully renewable European power system with excess wind and solar power generation. *Renew. Energ.*, **36**, 2515–2523.

37 London Research Institute (March 2010) Survey of Energy Storage Options in Europe Report.

38 Fraunhofer Institut für Windenergie und Energiesystemtechnikfor Umwelt Bundes Amt (2010) 2050: 100% – Energie Ziel 2050: 100% Strom aus erneuerbaren Quellen.

39 Specht, M., Sterner, M., Stuermer, B., Frick, V., and Hahn, B. (2009) "Renewable Power Methane – Stromspeicherung durch Kopplung von Strom- und Gasnetz

– Wind/PV-to-SNG." Registered on 09.04.2009. Patent pending no: 10 2009 018 126.1.

40 Bünger, U. (2012) The role of hydrogen in future energy markets. Presented at the 19th World Hydrogen Energy Conference (WHEC 2012), 3–7 June 2012, Toronto, Canada.

41 Schmidt, O. (September 2013) *Power-to-Gas* Business Models in the United Kingdom, Imperial College London.

42 Sensfuß, F., Ragwitz, M., and Genoese, M. (2008) The merit-order effect: a detailed analysis of the price effect of renewable electricity generation on spot market prices in Germany. *Energ. Policy*, **36** (8), 3086–3094.

43 Specht, M. *et al.* (2010) Themen 2009, Forschungsverbund Erneuerbare Energien, S. 69, Berlin.

44 SBC Energy Institute (Feb 2014) Hydrogen-based Energy Conversion Presentation.

13
Status and Prospects of Alkaline Electrolysis
Dongke Zhang and Kai Zeng

13.1
Introduction

Hydrogen is mainly used in petroleum refining, ammonia production [1,2], and, to a lesser extent, metal refining such as nickel, tungsten, molybdenum, copper, zinc, uranium, and lead [3,4]. The global annual hydrogen consumption amounted to more than 449 billion cubic meters (NTP) in 2010 [5]. The large-scale nature of such hydrogen consumption requires large scale hydrogen production to match. As such, hydrogen production is dominated by reforming of natural gas [6] and gasification of coal and petroleum coke [7,8], as well as gasification and reforming of heavy oil [9,10]. Although water electrolysis to produce hydrogen (and oxygen) has been known for some 200 years [11,12] and has the advantage of producing extremely pure hydrogen, its applications are often limited to small scale and unique situations where access to large scale hydrogen production plants is not possible or economical, such as marine, rockets, space crafts, electronic industry, and food industry as well as medical applications. Water electrolysis represents only 4% of the global hydrogen production [13,14].

With ever increasing energy costs owing to the dwindling availability of oil reserves, production, and supply [15] and due to concerns over global warming and climate change attributed to man-made carbon dioxide (CO_2) emissions associated with the fossil fuel use [16], particularly coal [17], hydrogen as an energy carrier has in recent years become very popular for several reasons: (i) it is perceived as a clean fuel, emitting almost nothing other than water at the point of use; (ii) it can be produced using any energy source, with renewable energy being most attractive [18]; (iii) it works with fuel cells [19–21] and, together, they may serve as one of the solutions to the sustainable energy supply and use puzzle in the long run, in the so-called "hydrogen economy" ideology [22,23].

Water electrolysis can work beautifully well at small scales and, by using renewable electricity [24], it can also be considered sustainable. In a conceptual distributed energy production, conversion, storage and use system for remote communities [24], as illustrated in Figure 13.1, water electrolysis may play an

Hydrogen Science and Engineering: Materials, Processes, Systems and Technology, First Edition.
Edited by Detlef Stolten and Bernd Emonts.
© 2016 Wiley-VCH Verlag GmbH & Co. KGaA. Published 2016 by Wiley-VCH Verlag GmbH & Co. KGaA.

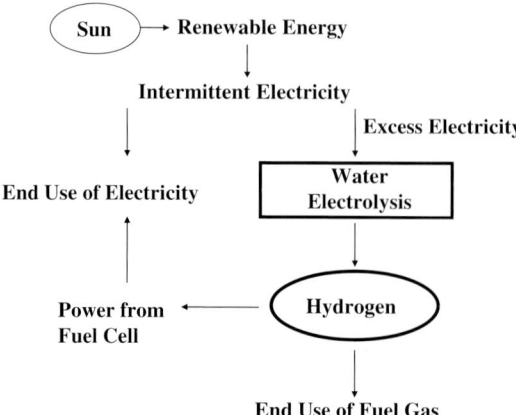

Figure 13.1 Schematic illustration of a conceptual distributed energy system with water electrolysis playing an important role in hydrogen production as a fuel gas and energy storage mechanism [24].

important role as it produces hydrogen using renewable energy as a fuel gas for heating applications and as an energy storage mechanism. When abundant renewable energy is available, excess energy may be stored in the form of hydrogen produced by water electrolysis. The stored hydrogen can then be used in fuel cells to generate electricity or used as a fuel gas.

Remote areas with abundant solar and/or wind resources for renewable electricity can take advantage of water electrolysis to produce hydrogen to meet their energy need for households, such as lighting and heating [25], powering telecommunication stations [26] and small-scale light manufacturing industry applications, electricity peak shaving, and in integrated systems, both grid-connected and grid-independent [27]. Hydrogen produced by renewable electricity can offer a great advantage, being mobile and flexible in scale, which is essential to the energy supply in remote areas away from the main electricity grid.

Small scale water electrolyzers can avoid the need for a large fleet of cryogenic, liquid hydrogen tankers or a massive hydrogen pipeline system. The existing electrical power grid could be used as the backbone of the hydrogen infrastructure system, contributing to the load leveling by changing operational current density in accordance with the change in electricity demand [28]. On a small scale, pure hydrogen and oxygen can find diverse applications including gases in laboratories and oxygen for life-support system in hospitals [29].

While possessing these advantages of availability, flexibility, and high purity, to achieve widespread application hydrogen production using water electrolysis needs improvements in energy efficiency, safety, durability, operability and portability, and, above all, reduction in costs of installation and operation. These open up many new opportunities for research and development leading to technological advancements in water electrolysis. This chapter also aims to identify such new research and development opportunities. We begin with an overview of the thermodynamics and kinetics of water electrolysis in the context of electrochemistry, laying a theoretical basis for scientific analysis of the published electrolysis systems and data. We then examine recent trends in research and publication to identify the gaps for improvements – the prospects of water electrolysis and the need for further research and development.

13.2 Thermodynamic Consideration

13.2.1 Theoretical Cell Voltages

Water is one of the thermodynamically most stable substances in nature and it is always an uphill battle to try to pull water molecules apart into the element hydrogen and oxygen molecules. If we want hydrogen (and oxygen) from water by electrolysis, we have to at least overcome an *equilibrium cell voltage, $E°$*, which is also called the "electromotive force." With established reversibility and the absence of a cell current between the two different electrode reactions, the open cell potential is $E°$, which is defined as the equilibrium potential difference between the respective anode and cathode [30] and is described by Eq. (13.1):

$$E° = E_{anode}° - E_{cathode}° \tag{13.1}$$

Equation (13.2) relates the change in the Gibbs free energy (ΔG) of the electrochemical reaction to the equilibrium cell voltage as follows:

$$\Delta G = nFE° \tag{13.2}$$

where n is the number of moles of electrons transferred in the reaction and F is the Faraday constant. The overall water electrolysis cell reaction $E_{H_2O}°(25\,°C) = 1.23\,V$ and the change in Gibbs free energy of the reaction is $+237.2\,kJ\,mol^{-1}$ [31], which is the minimum amount of electrical energy required to produce hydrogen. The cell voltage at this point is known as the *reversible potential*. Hence, the electrolysis of water to hydrogen and oxygen is thermodynamically unfavorable at room temperature and can only occur when sufficient electrical energy is supplied. In contrast, when the electrolysis process is performed under adiabatic conditions, the total reaction enthalpy must be provided by an electrical current. Therefore, to maintain the adiabatic condition, a *thermoneutral* voltage is required to maintain the electrochemical reaction without heat generation or adsorption [32].

Even when the equilibrium potential is met, the electrode reactions are inherently slow and then an *overpotential* (η) above the equilibrium cell voltage is needed to kick start the reaction due to the activation energy barrier, low reaction rate, and bubble formation [33,34]. According to the resistances mentioned above, the input of additional energy is also essential to drive the ionic migration process and overcome the resistance of the membrane as well as the electrical circuit. This extra energy requirement causes within the cell a *potential drop* (iR_{Cell}, where i is the current through the cell and R_{Cell} is the sum of electrical resistances of the cell, which are a function of electrolyte properties, the form of the electrodes, and cell design). The *cell potential* (E_{cell}) can be written as Eq. (13.3), which is always 1.8–2.0 V at the current density of 1000–3000 A m^{-2} in industry water electrolysis [35]. The total overpotential is the sum of overpotentials or barriers from the hydrogen and oxygen evolution reactions, electrolyte concentration difference, and bubble formation. Under mild conditions, when gas bubbles and the electrolyte concentration differences can be neglected, the sum of the overpotential can be calculated using Eq. (13.4), where j is the current density (current per unit of electrode surface area) at which the electrolysis cell operates. Both the overpotential and the ohmic loss increase with current density and may be regarded as causes of inefficiencies in the electrolysis whereby electrical energy is degraded into heat, which must be taken into account in any consideration of energy balance:

$$E_{cell} = E_{anode} - E_{cathode} + \sum \eta + iR_{cell} \tag{13.3}$$

$$\sum \eta = |\eta_{anode}(j)| + |\eta_{cathode}(j)| \tag{13.4}$$

Figure 13.2 shows the relationship between the electrolyzer cell potential and the operating temperature [15,36]. The cell potential–temperature plane is divided into three zones by the so-called equilibrium voltage line and thermoneutral voltage line. The equilibrium voltage is the theoretical minimum potential

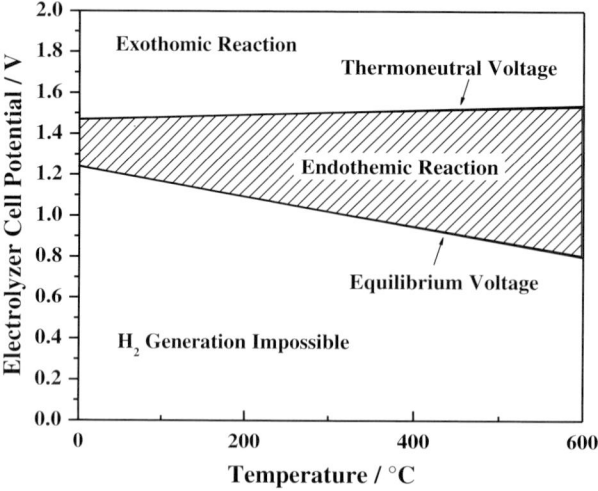

Figure 13.2 Cell potential for hydrogen production by water electrolysis as a function of temperature [24].

required to dissociate water by electrolysis, below which the electrolysis of water cannot proceed. The equilibrium voltage deceases with increasing temperature. The thermoneutral voltage is the actual minimum voltage that has to be applied to the electrolysis cell, below which the electrolysis is endothermic and above which it is exothermic. The thermoneutral voltage naturally includes the overpotentials of the electrodes, which are only weakly dependent on temperature. Thus, the thermoneutral voltage exhibits only a slight increase with temperature. We denote the thermoneutral voltage as $E_{\Delta H}$. If water electrolysis takes place in the shaded area in Figure 13.2, the reaction will be endothermic.

13.3 Electrode Kinetics

The rate of the electrode reaction, characterized by the current density, firstly depends on the nature and pretreatment of the electrode surfaces. Secondly, the rate of reaction depends on the composition of the electrolytic solution adjacent to the electrodes. The ions in the solution near the electrodes, under the effect of the electrode, form layers, known as the *double layer* [37]. Taking the cathode in a KOH solution as an example, the charge layer is formed by hydroxyl ions and potassium ions according to the charge of the electrodes. Finally, the rate of the reaction depends on the electrode potential, which is characterized by the reaction overpotential. The study of electrode kinetics seeks to establish the macroscopic relationship between the current density and the surface overpotential and the composition of the electrolytic solution adjacent to the electrode surface [38].

For a one-step, one-electron reaction, through the relationship between the current and the reaction rate, the dependence of the current density on the surface potential and the composition of the electrolytic solution adjacent to the electrode surface is given by the Butler–Volmer equation [33]:

$$i = i_{\text{cathode}} - i_{\text{anode}} = FAk^0\left[C_O(0,t)e^{-\alpha f(E-E^\circ)} - C_R(0,t)e^{f(1-\alpha)(E-E^\circ)}\right] \quad (13.5)$$

where A is the electrode surface area through which the current passes, k^0 is the *standard rate constant*, α refers to the *transfer coefficient* whose value lies between 0 and 1 for this one-electron reaction, f is the F/RT ratio and t and 0 in the brackets are, respectively, the specific time at which this current occurs and the distance from the electrode. For the hydrogen evolution reaction, $C_O(0,t)$ stands for the concentration of reaction species at the cathode in the oxidized state, the hydrogen ions (H^+), while $C_R(0,t)$ is the concentration of reaction product hydrogen $\frac{1}{2}H_2$, which is in the reduced state.

The Butler–Volmer equation can be simplified as:

$$i = i_0\left(e^{-\alpha f\eta} - e^{(1-\alpha)f\eta}\right) \quad (13.6)$$

where i_0 is known as the exchange current density [39], which is the current of the reversible hydrogen evolution reaction. From the simplified equation above, we can derive the overpotential at each electrode. In the absence of the influence of mass transfer and at large overpotentials (>118 mV at 25 °C), one of the terms in Eq. (13.6) can be neglected. For example, at a large negative overpotential, $e^{-\alpha f \eta} \gg e^{(1-\alpha)f\eta}$, the relationship between i and η $(E - E°)$ can be written as the Tafel equation [40]:

$$\eta = a + b \log i \qquad (13.7)$$

where:

$$a = \frac{2.3RT}{\alpha F} \log i_0$$

and $b = -2.3RT/\alpha F$.

The linear relationship between the overpotential and the logarithm of current density is characterized by the slope b and exchange current density (i_0). The slope is also known as the Tafel slope. Both parameters are commonly used as kinetic parameters to compare electrodes in electrochemistry.

From the above analysis, the rate of electrolysis can be expressed by the current or current density. Furthermore, the current can be reflected by i_0, which is the current associated with the reversible hydrogen evolution reaction on the surface of the electrodes. The rate of the reaction is also directly determined by the overpotential, which depends on several factors. One of the important factors is the activation energy, E_A, which determines the rate constant k^0 in Eq. (13.5), according to the Arrhenius law. The term E_A is strongly influenced by the electrode material and is, thus, a focus of continuing research effort. To reduce the activation energies of the electrode reactions, or reduce the overpotential, it is therefore necessary to consider how they are related to the electrode materials and surface configurations.

13.3.1
Hydrogen Generation Overpotential

The mechanism of the hydrogen evolution reaction is widely accepted [15] to be a step involving the formation of adsorbed hydrogen:

$$H^+ + e^- \rightarrow H_{ads} \qquad (13.8)$$

which is followed by either chemical desorption:

$$2H_{ads} \rightarrow H_2 \qquad (13.9)$$

or electrochemical desorption:

$$H^+ + e^- + H_{ads} \rightarrow H_2 \qquad (13.10)$$

where the subscript "ads" represents the adsorbed state.

The overpotential of the hydrogen evolution reaction is generally measured by the Tafel equation:

$$\eta_{cathode} = 2.3 \frac{RT}{\alpha F} \log \frac{i}{i_0} \qquad (13.11)$$

In this equation, i_0, the exchange current density of the reaction, which can be analogized as the rate constant of reaction, is a function of the nature of the electrode (cathode) material [38]. The overpotential of the hydrogen production means there is an additional energy barrier in the process of hydrogen formation.

The overpotential on the cathode is directly related to the formation of hydrogen in the vicinity of the electrode. The formation of hydrogen is intrinsically determined by the bond between hydrogen and the electrode surface. Palladium has the lowest heat of adsorption of hydrogen of 83.5 kJ mol^{-1}, as compared to 105 kJ mol^{-1} for Ni [41]. In addition, hydrogen formation is also influenced by the electrode properties, the type and concentration of electrolyte, and temperature. By comparing the kinetic data, including the exchange current density and Tafel slope, the relationship between these factors can be revealed. Table 13.1 compares the kinetic parameters, represented by the exchange current density and Tafel slope, of the hydrogen evolution reactions on different metal electrodes.

For the hydrogen evolution reaction, it is necessary to identify the rate determining step. If hydrogen adsorption, (13.8), is the rate-determining step, an electrode material with more edges and cavities in its surface structure, which favor easy electron transfer, will create more electrolysis centers for hydrogen adsorption. If the hydrogen desorption steps, (13.9) and (13.10), are the rate determining steps, the physical properties of the electrode such as surface roughness or perforation will either increase the electron transfer by increasing the reaction area or prevent bubbles from growing, which in turn increase the overall rate of electrolysis.

Increasing the overpotential may lead to a change in the reaction mechanisms. In other words, the rate determining step will alter within different potential ranges. When the potential is low, the electron transfer is not as fast as

Table 13.1 Kinetics parameters of hydrogen production on different electrode metals.

Metal	Heat of H_2 adsorption (kJ mol^{-1})	Electrolyte	Temperature (°C)	i_0 (A m^{-2})	Tafel slope (mV)
Ni [42]	105	1M NaOH	20	1.1×10^{-2}	121
Fe [43]	109	2M NaOH	20	9.1×10^{-2}	133
Pb [44]	N/A	6N NaOH	25	4×10^{-2}	121
Zn [44]	N/A	6N NaOH	25	8.5×10^{-6}	124
Co [45]	N/A	0.5M NaOH	25	4.0×10^{-3}	118
Pt [15]	101	0.1N NaOH	22	4.0	105
Au [15]	N/A	0.1N NaOH	25	4.0×10^{-2}	120

desorption. As such, hydrogen adsorption will be slow and become the rate determining step. Conversely, when the potential is high enough to enable the hydrogen adsorption rate to be greater than the desorption rate, hydrogen desorption will be the rate determining step.

13.3.2
Oxygen Generation Overpotential

The mechanism of oxygen evolution reaction is more complex than the pathways suggested for hydrogen evolution. Several theories have been presented and discussed in the literature and the most generally accepted mechanism involves the following steps:

$$OH^-_{ads} \rightarrow OH_{ads} + e^- \tag{13.12}$$

$$OH^- + OH_{ads} \rightarrow O_{ads} + H_2O + e^- \tag{13.13}$$

$$O_{ads} + O_{ads} \rightarrow O_2 \tag{13.14}$$

One of the charge transfer steps is rate controlling. The Tafel slope variations can be used to identify the rate determining step. For example, a slow electron transfer step (13.12) determines the reaction at low temperatures; in contrast, a slow recombination step (13.14) is controlling at high temperatures on a nickel electrode. The different Tafel slopes between the steps can be used to judge the mechanisms [46,47].

The overpotential of the oxygen evolution reaction is generally measured by the Tafel equation:

$$\eta_{anode} = 2.3 \frac{RT}{(1-\alpha)F} \log \frac{i}{i_0} \tag{13.15}$$

The reaction rate decreases with increasing activation energy; thereby, reducing the activation energy is always favored for more efficient water electrolysis. Furthermore, the activation energy can be lowered by using appropriate electrocatalysts. Table 13.2 compares the kinetic parameters, again represented by the

Table 13.2 Kinetic parameters of oxygen production on different metals.

Metal	Electrolyte	Temperature (°C)	i_0 (A m^{-2})	Tafel slope (mV)
Pt [48]	30% KOH	80	1.2×10^{-5}	46
Ir [49]	1 N NaOH	N/A	1.0×10^{-7}	40
Rh [49]	1 N NaOH	N/A	6.0×10^{-8}	42
Ni [50]	50% KOH	90	4.2×10^{-2}	95
Co [48]	30% KOH	80	3.3×10^{-2}	126
Fe [48]	30% KOH	80	1.7×10^{-1}	191

exchange current density and Tafel slope, of oxygen evolution reactions on different metal electrode materials.

Generally speaking, the overpotential of oxygen evolution is more difficult to reduce than that of hydrogen evolution, owing to the complex mechanism and irreversibility. Alloys of Fe and Ni have been found to be able to reduce the overpotential to some extent [51].

13.3.3
Cell Overpotential

Since the cell potential contains both anode and cathode reactions, it is necessary to identify the contributions of each of them to the cell voltage and factors influencing them in order to understand the overpotential resistance. The characteristic trend of the effect of temperature on the overpotential is summarized by Kinoshita [35]. As shown in the Figure 13.3 an increase in temperature will result in a decrease in the overpotential at the same current density.

The overpotential is not only a function of temperature but also a function of current density [52]. As can be seen from Figure 13.4, the overpotentials from hydrogen and oxygen evolutions are the main sources of the reaction resistances. The other obvious resistance at high current densities is the ohmic loss in the electrolyte, which includes resistances from the bubbles, diaphragm, and ionic transfer. Understanding these resistances can open up opportunities to enhance the efficiency of water electrolysis.

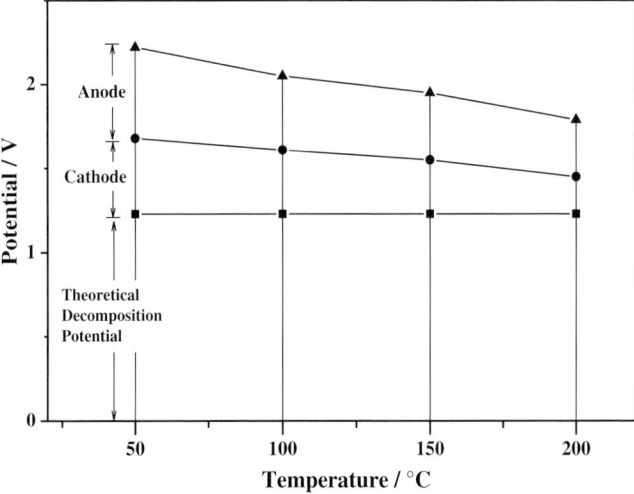

Figure 13.3 Illustration of the contributions of anode and cathode polarization to the cell voltage of an alkaline water electrolysis cell [24].

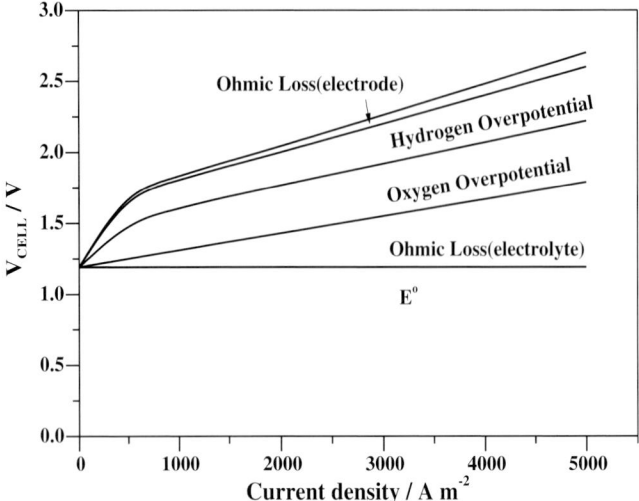

Figure 13.4 Compositions of the typical cell voltage of an alkaline water electrolysis cell [24].

13.4
Electrical and Transport Resistances

13.4.1
Electrical Resistances

By dividing the overpotential due to each of the components identified in Figure 13.4 by the current density, all of the resistances related to these components can be normalized based on the unit of ohm, which makes it possible to compare energy losses caused by different resistances, as illustrated in Figure 13.5, where $E_{loss,\ electrolyte}$ includes energy losses due to the bubbles in the electrolyte and ionic transfer resistances. Figure 13.5 also shows that the energy losses caused by the reaction resistances increase relatively slowly as the current density increases. The energy loss in the electrical circuit is relatively small. However, the energy loss due to ionic transfer resistance in the electrolyte becomes more significant at higher current density. The dotted and dashed lines are the bubble resistance and the total resistance, respectively. The energy loss due to bubble coverage on the electrode surfaces, and thus the total energy loss, is hypothetical, on the basis of 50% of the electrode surface being covered by bubbles.

Although the relationship between current density and energy loss in Figure 13.5 does not specify all of the resistances mentioned above, it approximately presents the relationships among the losses. More interestingly, the energy loss due to bubbles formed on the electrodes should be considered as the major contribution to the total energy loss. Therefore, minimizing the bubble effect is a key to electrolyzer efficiency improvement.

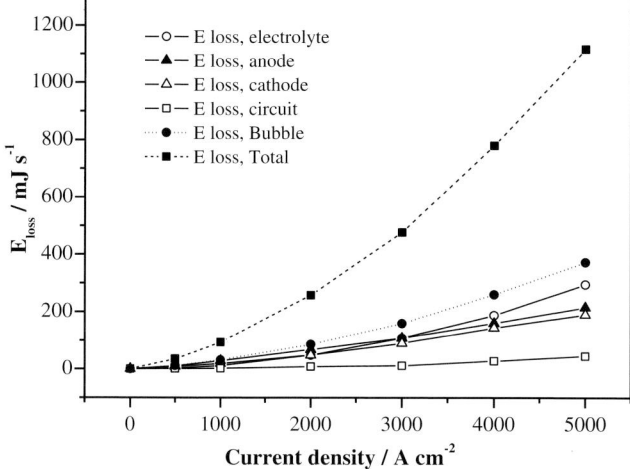

Figure 13.5 Qualitative comparison of the energy losses caused by reaction resistances, ohmic resistance, ionic resistance, and bubble resistance [24].

13.4.2
Bubble Phenomena

Owing to the significant electrical resistance of bubbles, understanding the bubble phenomena is an important element in the development of any water electrolysis systems. A force balance analysis around an electrolytic bubble has been performed on the electrolytic gas evolution model. The model was based on a vertical electrode with an electrolytic gas bubble evolution model that simply involved bubble formation on an electrode and was called the *stagnant model*.

Figure 13.6a shows a gas bubble on an electrode surface in the stagnant model. The x coordinate is in the direction against gravity and the y coordinate is normal to the x coordinate and pointing away from the electrode.

Figure 13.6b shows that the forces acting on a gas bubble originate from various sources. The buoyancy and surface tension exist due to the density difference between liquid and gas and the properties of the solution, respectively [53]. The force incurred by the temperature field effect is neglected by maintaining the temperature of the electrolyte constant.

The forces acting on a gas bubble can be decomposed into components along the x and y coordinates, resulting in possible movements of gas bubbles in the corresponding directions. These movements are noted as departure and lift-off, respectively.

13.4.2.1 Bubble Departure Diameter Predictions
When the gas bubble is attached to the electrode surface as shown in Figure 13.6, both $\sum F_x = 0$ and $\sum y = 0$ are satisfied. Once one of these conditions is broken, the gas bubble departs or lifts off from the electrode, respectively.

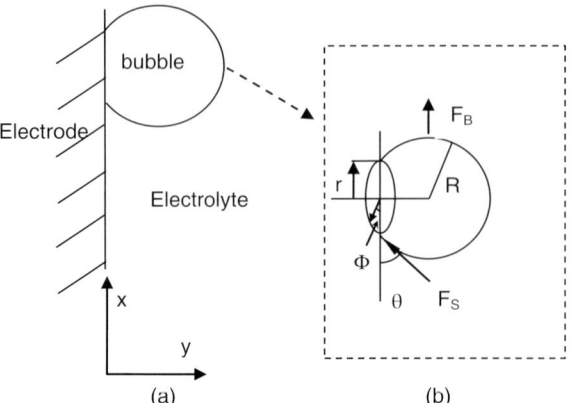

Figure 13.6 Schematic diagram of a gas bubble on an electrode surface (a) and the forces acting on the bubble (b).

In the stagnant model, when the gas bubble attaches on the electrode surface, $\sum F_x = \sum F_b > 0$. Therefore, the buoyancy force is not balanced with the sum of the other forces. This unbalance will force the bubble to tilt upwards to reach a new force balance and then depart when the bubble can no longer tilt after sufficient growth in size.

The upward tilting of the bubble causes so-called advancing and receding angles [54], which are denoted as α and β in Figure 13.7. Therefore, there will be an interfacial tension force in the x coordinate direction. This force will be balanced with the buoyancy force to maintain bubble attachment on the electrode.

Resolving the force balance of buoyance and surface tension [56], yields Eq. (13.16):

$$\tfrac{1}{3}(\rho_L - \rho_B)g\pi R^3(1 + \cos\theta)^2(2 - \cos\theta) - 2r\sigma\frac{\pi(\alpha - \beta)}{\pi^2 - (\alpha - \beta)^2}[\sin\alpha + \sin\beta] = 0$$

(13.16)

where ρ_L and ρ_B are the densities of liquid and bubble, respectively, which can be considered as constants; R is the critical diameter of the bubble; σ is the

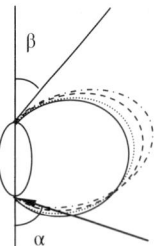

Figure 13.7 Advancing and receding angles of a gas bubble attached to a vertical electrode surface [55].

interfacial tension between the gas and electrolyte liquid; g and π are gravitational and pi constants, respectively.

Therefore, the critical diameter of a gas bubble, above which the bubble departs, in the stagnant model can be approximated as Eq. (13.17):

$$R = \sqrt{\frac{6 \sin\theta \sigma(\alpha - \beta)[\sin\alpha + \sin\beta]}{(\rho_L - \rho_B)[\pi^2 - (\alpha - \beta)^2] \cdot g \cdot (1 + \cos\theta)^2(2 - \cos\theta)}} \qquad (13.17)$$

The critical diameter for gas bubble departure can be expressed in Eq. (13.18) by simplifying Eq. (13.17):

$$R = kf(\theta, \alpha, \beta)\sqrt{\sigma} \qquad (13.18)$$

where:

$$f(\theta, \alpha, \beta) = \sqrt{\frac{\sin\theta(\alpha - \beta)(\sin\alpha + \sin\beta)}{[\pi^2 - (\alpha - \beta)^2](1 + \cos\theta)^2(2 - \cos\theta)}} \qquad (13.19)$$

It can then be predicted that varying the electrolyte concentration and the cell voltage will alter the interfacial tension force and thus the critical diameter for bubble departure. These changes can be estimated in theory and thus the critical diameter for bubble departure can be calculated. Experimental data can be recorded to verify the dependence of the critical diameter on parameters such as cell voltage and electrolyte concentration. The theoretical values can then be compared to validate the predictions.

To calculate the critical diameter for bubble departure, a few parameters in Eq. (13.17) need to be obtained independently and analyzed, such as the contact angle and surface tension. The advancing and receding contact angles are usually measured experimentally [57]. They characterize the flexibility of the gas bubble in distortion. It is determined by the surface tension and, thus, the electrode surface defects [58] and the properties of the gas and liquid [59]. Although the understanding of this phenomenon is still poor, it can be ascertained that the bubble size is a function of θ, α, β, and the interfacial tension, σ.

The contact angles for both hydrogen bubbles and oxygen bubbles have been experimentally determined by Matsushima [57] and adapted for our model predictions. The average contact angles for hydrogen bubbles and oxygen bubbles were 43° and 50°, respectively. The advancing angle and the receding angle can be written as Eqs. (13.20) and (13.21) and $\Delta\theta$ was reported as not more than 10° [60]; the critical diameter of the hydrogen bubble and the oxygen bubble can be calculated according to Eq. (13.17):

$$\alpha = \theta + \Delta\theta \qquad (13.20)$$

$$\beta = \theta - \Delta\theta \qquad (13.21)$$

Take a 0.5M KOH solution as an example, its density is $1.02\,\text{kg m}^{-3}$ at 20 °C, and its surface tension is $72.4\,\text{mN m}^{-1}$. Assuming $\Delta\theta$ is 6°, the critical diameter for the departure of hydrogen and oxygen gas bubbles was predicted to be 0.50 and 0.57 mm, respectively.

Increasing the potential applied to the electrodes enhanced the wettability of the electrodes [61], which led to an increase in the surface tension and $\Delta\theta$. On the other hand, increasing electrolyte concentration led to an increase of electrolyte viscosity and surface tension. The increase in electrolyte viscosity caused not only less coalescence of bubbles [62] but also a reduction in $\Delta\theta$.

13.4.2.2 Comparison of Model Predictions with Experimental Observations

Prediction of the critical diameter for bubble departure was based on Eq. (13.17), where σ, θ, α, and β were unknown parameters. The value or the range of these parameters was sourced from the literature [57,60,63]. Table 13.3 summarizes the values used in the prediction of the critical diameter of bubble departure. The surface tension of the KOH solution was found to be 73.3, 78.5, 81.7, and 86.4 dyne cm^{-1} for the electrolyte solutions of 0.5, 1.0, 2.0, and 4.0 M KOH, respectively [63]. Owing to the effect of the electrode potential, the values of surface tension increased accordingly. The contact angles for oxygen and hydrogen were independent of the electrolyte and electrode potential, and were reported to be 50° and 43° according to Matsushima [57]. The terms α and β were assumed through Eqs. (13.20) and (13.21), where $\Delta\theta$ was chosen in the range between 0° and 10° [60], and the values of $\Delta\theta$ and σ chosen were in agreement with the previous analysis of the effect of current density and KOH concentration [57,60,63].

With the predicted values for the contact angle and surface tension, the critical diameter for bubble departure can be calculated. Figure 13.8 compares the predicted and the measured values of the critical diameters under different conditions. As can be seen from Figure 13.8, generally good agreement is evident for both hydrogen and oxygen gas bubbles.

Table 13.3 Summary of parameters used in the prediction of the critical diameter for bubble departure.

Current density/concentration		σ (dyne cm^{-1})	θ_{H2} (°)	θ_{O2} (°)	$\Delta\theta$ (°)
Current density (mA cm^{-2})	0.30	84.0	50	43	6
	0.45	89.0	50	43	8
	0.60	94.0	50	43	9
	0.75	100.0	50	43	10
Concentration (M)	0.5	94.0	50	43	8
	1.0	94.0	50	43	4
	2.0	94.0	50	43	3
	4.0	94.0	50	43	1.5

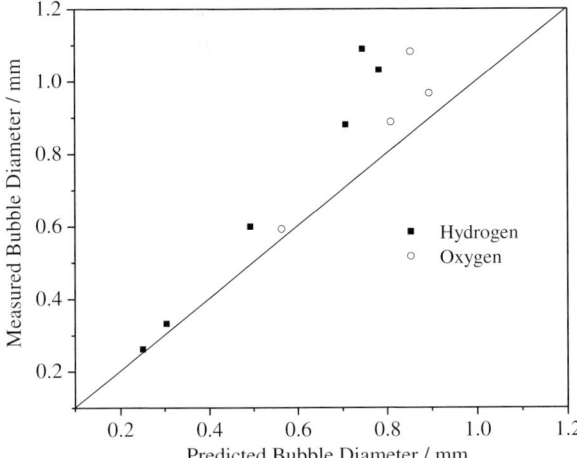

Figure 13.8 Comparison of the predicted and measured critical diameters for hydrogen and oxygen gas bubbles [55].

Although the predicted critical diameters through the detailed force analysis were generally in good agreement with the measured values, the critical diameter for electrolytic gas bubble departure is highly dependent on the electrolyte concentration and cell voltage at low current densities. An increase in the electrode potential would result in an increase in the interfacial tension, while an increase in the KOH concentration could have an adverse effect by reducing $\Delta\theta$. The convection caused by local heating, bubble detachment, and electrolyte flow can also influence the critical diameter of the electrolytic gas bubbles. All these factors are important when considering the critical bubble diameter under complex conditions and require more research effort to understand the bubble phenomena.

13.5
Research Trends

Intensive studies have been performed in the above areas to promote the development of water electrolysis [64]. The current research trends of alkaline water electrolysis are discussed from several aspects including electrodes, electrolytes, ionic transport and bubble formation. These research efforts are not only significant for water electrolyzer development but also for the fundamentals of the electrochemistry.

13.5.1
Electrodes

Metal electrodes are normally adopted in the gas evolving processes. As discussed in the kinetics and the development of electrolyzer sections earlier, the most widely used electrode material is nickel because of its stability and favorable activity. However, deactivation is a major challenge faced by the electrode materials, even for nickel. The mechanism of nickel electrode deactivation is the formation of a nickel hydride phase at the electrode surface due to high hydrogen concentration. Iron coating can prevent the nickel hydride phase from forming and hence prevent the electrode deactivation [65]. Dissolved vanadium species are also found to activate nickel cathodes during hydrogen evolution in the alkaline media [40]. Addition of iron to manganese–molybdenum oxides enhanced the stability of electrodes. The iron addition also enhanced the oxygen evolution efficiency. The formation of nickel triple oxides seems to be responsible for the enhancement of both oxygen evolution efficiency and stability [66]. Therefore, electro-catalysts are the key to enhancing and stabilizing the electrode activity.

Apart from material selection, electrode modifications in the cell design are also important in water electrolysis. The electrode surfaces are commonly modified by slits and holes to facilitate the escape of gas bubbles. The holes must be appropriate to prevent gas trapping. Typical diameters for electrode perforation in alkaline water electrolysis are around 0.1 and 0.7 mm for hydrogen and oxygen, respectively [30]. Louvered, finned, or slotted electrodes are also used to promote the removal of bubbles.

Surface modifications to the electrodes by means of mechanical polishing and electrochemical deposition have been used by the authors to improve their activities [67]. Taking the hydrogen evolution reaction on Ni electrodes as an example, electrode modification by mechanical polishing shows enhanced apparent activities compared with base Ni. The rougher the Ni electrode surface after the mechanical polishing, the better the electrode activity by exhibiting a lower overpotential. The concepts of effective surface area and relative roughness factor have been introduced and their values can be derived from electrochemical impedance studies [67]. When the effective surface area is applied, the Tafel curves of the Ni electrodes polished with different sandpapers collapsed into a narrow band that can be described by the equation $\eta = 0.02 + 0.191 \log(j'')$, where j'' is the exchange current density based on the effective surface area [67]. The equation represents the intrinsic activities of Ni electrodes towards hydrogen evolution reaction and also validates the roughness factor for presenting intrinsic activity. While the electrodes modified by electrochemical deposition of Ni–Co do change their apparent electrode activities, their intrinsic activities are different, depending on their surface Ni–Co compositions as varied with the deposition time [67].

13.5.2
Electro-Catalysts

To some extent, the electrode itself is a catalyst by affecting the activation energy of the electrochemical reaction. However, doping or coating the electrode with a more stable and active layer is often used in electrode design. Similar to catalysts, electro-catalysts facilitate charge transfer or chemical reaction, reducing the activation energy of the reaction. The obvious effect of an electro-catalyst is to reduce the overpotential of either or both of the two half reactions. The role of the electro-catalyst is affected by the electronic structure of the electrodes. In the hydrogen evolution reaction, Ni, Pd, Pt with d^8s^2, $d^{10}s^0$, and d^9s^1 electronic configurations, respectively, exhibit minimum overvoltage values and Zn, Cd, Hg with $d^{10}s^2$ electronic configuration show maximum values. The spill-over theory in electro-catalysis by Bockris and Mchardy enables the understanding of the interaction between substances [15].

Alloys, with different electronic distributions in the metal, are adopted to improve the activity of electrodes. For example, an alloy of Mo and Pt was found to give a significant upgrade in electrolytic efficiency in comparison with its individual components and conventional cathode materials [68]. More examples are listed in Table 13.4. The Tafel slope and exchange current density of the hydrogen evolution reaction in alkaline solutions at temperatures near 70 °C are used to compare the activities of Ni and Ni based alloy.

The doping material could be chosen from a wide range of metals. Noble metals are commonly used as electro-catalysts. Ruthenium dioxide (RuO_2), prepared by pyrolysis and calcination, clearly shows electro-catalytic activity for the oxygen evolution reaction [73]. An anode electro-catalyst with the formula $Ir_xRu_yTa_zO_2$ has been claimed to achieve an overall voltage of 1.567 V at 1 A cm^{-2} and 80 °C, equating to an energy consumption of 3.75 kWh Nm^{-3} H_2 and an efficiency of 94% with a total noble metal loading of less than 2.04 mg cm^{-2} [74]. Non-noble metals also find their electro-catalytic activities. The doping of Li increases the electrical conductivity of these materials. The key to better performance is that the roughness factor increases with the percentage of Li up to 3%, favoring oxygen evolution [75]. CeO_2 is also found to alter the phase composition of a Ni electrode and improved the activity of Ni [76], and a synergetic effect

Table 13.4 Tafel slopes of Ni alloys.

Material and preparation method	Electrolyte	Temperature (°C)	Tafel slope (mV)	i_0 (A cm^{-2})	η_{250}[a] (mV)
Ni (wire) [69]	30% NaOH	70	99	5.5×10^{-5}	362
$Ni_{79}Mo_{20}Cd$ [70]	1M NaOH	70	125	N/A	N/A
$Ni_{70}Mo_{29}Si_5B_5$ (amorphous) [71]	30% KOH	70	118	1.8×10^{-6}	489
$Ni_{60}Mo_{40}$ mixture (ball-milled 20 h) [72]	30% KOH	70	50	1.7×10^{-2}	58

a) Hydrogen evolution reaction overpotential under 2500 A m^{-2}.

towards hydrogen evolution reaction may exist between CeO_2 particle and Ni electrodes [77,78].

Hydrous oxide coated Fe electrodes have good redox properties and electro-catalytic activity towards the oxygen evolution reaction [79]. The growth of hydrous iron oxy-hydroxide film shows excellent electro-catalytic properties on the parent metal. RuO_2/IrO_2 mixed oxides have been examined for oxygen evolution; the key kinetic step in the oxygen evolution reaction process involves the electrochemical generation of an unstable and reactive oxy-metal surface group of high oxidation state [80].

The physical properties of the electrode materials also influence the electro-catalytic activity. A larger BET surface area and porosity of the oxide catalyst powder have been achieved by small La addition by Singh et al. [81]. They observed a reduction in the charge transfer resistance for the oxygen evolution reaction on an electrode made of oxide powder.

Nanostructures have also received much attention as they enlarge the material surface area and enable unique electronic properties. The increased active area of nanostructured electrode reduces the operating current density of the electrolyzer. A 25% reduction in overpotential and 20% reduction in energy consumption were achieved by the use of a Ru nano-rod cathode compared to the planar Ru cathode. The improvement was attributed to the increased active area of the nanostructured electrode, which reduces the operating current density of the electrolyzer [31]. Prashant V. Kamat [82] also proposed different nanostructures for improving the performance of photo-electrolysis facilitating the charge transfer, which has the potential to be applied as electrodes for water electrolysis.

The preparation methods of electrodes are an important factor in terms of improving the electrode surface properties such as roughness. Coatings are another common technique in electrode preparation. For example, a composite of Ni, Fe, and Zn prepared from the electro-deposition showed good stability for up to 200 h under a current density of $1350\,A\,cm^{-2}$. This material showed good activity as well; in 28% KOH at 80 °C its overpotential is about 100 mV, which is significantly lower than that of mild steel (400 mV) [83]. A catalyst-coated membrane (CCM) with a five-layer structure has been developed [84]. The five-layer CCM exhibits the highest performance and stability, attributed to expansion of the triple-phase boundaries for electrochemical reactions and the improvement of contact and mass transfer resistance. Tables 13.5 and 13.6 list several electro-catalysts that have been found helpful to reduce the overpotential or stabilize the electrodes for industrial water electrolysis.

In general, physical modifications of electrodes help the removal of gas bubbles from the electrodes. The electrode material influences the overpotential significantly. The electronic property and the surface property determine electro-catalytic performance of the coating or doping. Alloys, nano-structured materials, transition metals, and noble metals could be used to improve the electrode activity.

Table 13.5 Oxygen overpotential of different electrode materials.

Composition formula	Method	T (°C)	Electrolyte	C (mol dm^{-3})	j (A m^{-2})	η_{oxygen} (mV)	Reference
Ni + spinel type Co_3O_4	Thermal decomposition	25	KOH	1	1000	235 ± 7	[81]
Ni + La doped Co_3O_4	Thermal decomposition	25	KOH	1	1000	224 ± 8	[81]
MnOx modified Au	Electro-deposition	25	KOH	0.5	100	300	[85]
Li10% doped Co_3O_4	Spray pyrolysis	Room temperature	KOH	1	10	550	[75]
Ni	N/A	90	KOH	50 wt%	1000	300	[86]
$La_{0.5}Sr_{0.5}CoO_3$	Spray-stiner	90	KOH	50 wt%	1000	250	[86]
$Ni_{0.2}Co_{0.8}LaO_3$	Plasma jet projection	90	KOH	50 wt%	1000	270	[86]

Table 13.6 Hydrogen overpotential of different electrode materials.

Composition formula	Method	T (°C)	Electrolyte	C (mol dm^{-3})	j (A m^{-2})	$\eta_{hydrogen}$ (mV)	Reference
Ni-Fe-Mo-Zn	Co-deposition	80	KOH	6	1350	83	[87]
Ni-S-Co	Electro-deposition	80	NaOH	28 wt%	1500	70	[88]
Ni50%-Zn	Electro-deposition	N/A	NaOH	6.25	1000	168	[89]
$MnNi_{3.6}Co_{0.75}Mn_{0.4}Al_{0.27}$	Arc melting	70	KOH	30 wt%	1000	39	[90]
Ti_2Ni	Arc melting	70	KOH	30 wt%	1000	16	[91]
Ni50% Al	Melting	25	NaOH	1	1000	114	[92]
Ni75%Mo25%	Co-deposition	80	KOH	6	3000	185	[93]
Ni80%Fe18%	Co-deposition	80	KOH	6	3000	270	[93]
Ni73%W25%	Co-deposition	80	KOH	6	3000	280	[93]
Ni60%Zn40%	Co-deposition	80	KOH	6	3000	225	[93]
Ni90%Cr10%	Co-deposition	80	KOH	6	3000	445	[93]

13.5.3
Electrolyte and Additives

Most commercial electrolyzers have adopted alkali (potassium or sodium hydroxide) solutions as the electrolyte. Energy consumption during the electrolysis of water is significantly reduced by small quantities of activating compounds by the effect of ionic activators [94,95].

Ionic liquids (ILs) are organic compounds. At room temperature, they are liquids solely consisting of cations and anions, thus possessing very good ionic conductivities and stability [96]. Imidazolium ILs have been used as an electrolyte for hydrogen production through water electrolysis. Current densities higher than 200 A m^{-2} and efficiencies greater than 94.5% have been achieved using this ionic liquid in a conventional electrochemical cell with platinum electrodes at room temperature and atmospheric pressure. The catalytic activity of the electrode surface was not affected during the electrolysis mainly due to the chemical stability of the ILs [97]. However, ILs normally have high viscosity and low water solubility, which is not favored for mass transport, resulting in low achievable current densities and thus low hydrogen production rates. Therefore, there is a need to search for more suitable ionic liquids with high conductivity and solubility to facilitate electron transfer and water electrolysis, respectively.

Compared to research effort on the development of electro-catalysts, developmental effort on new electrolytes is relatively low. However, there is still potential to improve the overall efficiency by using electrolyte additives to enhance ionic transfer by reducing the electrolyte resistance. On the other hand, the adoption of electrolyte additives could tune the affinity between electrolyte and electrodes and help to manage the bubble behavior.

13.5.4
Bubble Management

Bubble formation and its transportation are a major reason for extra ohmic losses. Not only the dissolution of gas, but also the interface of gas between electrodes and electrolyte leads to excessive resistances to water electrolysis.

Bubble behavior has been intensively studied [98–100] in the sense of electrochemistry, but no mechanisms or models have been applied to alkaline water electrolysis. The microgravity condition under which the buoyancy effect is eliminated has been used to study the bubble behavior. Water electrolysis under microgravity has been found to result in stable froth layer formation, and the accompanying ohmic resistance increases with the froth layer thickness. The contributions of electrode surface coverage by bubbles and electrolyte-phase bubble void fraction to the ohmic drop have also been studied [101]. Under terrestrial gravity, the bubble sizes are smaller. Therefore, reducing the residence time of bubble staying on the electrodes is the key to minimizing the bubble size and thus reducing the bubble resistance.

According to the theory of surface tension, a hydrophilic electrode prefers water rather than bubbles. It means that bubble sizes are not easily grown. The mass transfer and ionic transfer between electrodes and electrolyte could then be enhanced. Similarly, surfactant additive can be used to reduce the surface tension, which can minimize the bubble size or accelerate the departure of the bubbles and then achieve the same effect of hydrophilic materials. In the meantime, these additives should be inert to the electrochemical reaction [102] and stable during the process.

Fluid mechanical means, by circulating the electrolyte solution to sweep the bubbles off the electrode surfaces, can also be applied. To sweep the bubbles off the electrode surface, the velocity of the fluid should be high enough, which in turn will benefit the mass transfer of the electrolyte and eliminate the concentration difference. Therefore, mechanically forcing the bubble to depart from the electrode surface is an alternative way to eliminate the bubbles formed on the electrode surface.

13.6 Summary

Alkaline water electrolysis combined with renewable energy can be integrated into the distributed energy system by producing hydrogen for end use and as an energy storage media. Compared to the other major methods for hydrogen production, alkaline water electrolysis is simple but currently less efficient. The challenges for widespread use of water electrolysis are also durability and safety. These disadvantages require further research and development effort.

Based on the thermodynamic and kinetic analyses of alkaline water electrolysis, several resistances hindering the efficiency of the alkaline water electrolysis process have been identified. These include resistances due to bubbles, activation energy of the electrode reaction, ionic transfer, and electrical resistances in the circuit. The bubble resistance can be reduced by electrode modification and electrolyte additives. Reaction overpotential can be optimized by electrode material selection and preparation. In addition, transport-related resistances such as bubble resistance and electrolytic resistance can be reduced by improving mass transport such as bubble elimination by electrolyte circulation. By identifying the resistances causing extra energy losses, this study opens up opportunities to minimize the energy consumption in alkaline water electrolysis, especially at high current densities.

Further R&D efforts to improve the efficiency are needed to enhance the efficacy of alkaline water electrolysis. These include developing electro-catalysts to significantly reduce electrochemical reaction resistance, electrolyte additives to facilitate the electron transfer and ionic transfer and to reduce electrode surface tension, electrode surface profile modifications and surface coatings, and more, importantly, managing the gas bubble resistances.

References

1. Ramachandran, R. and Menon, R.K. (1998) An overview of industrial uses of hydrogen. *Int. J. Hydrogen Energy*, **23** (7), 593–598.
2. Lattin, W.C. and Utgikar, V.P. (2007) Transition to hydrogen economy in the United States: A 2006 status report. *Int. J. Hydrogen Energy*, **32** (15), 3230–3237.
3. Eliezer, D. *et al.* (2000) Positive effects of hydrogen in metals. *Mater. Sci. Eng. A*, **280** (1), 220–224.
4. Eliaz, N., Eliezer, D., and Olson, D.L. (2000) Hydrogen-assisted processing of materials. *Mater. Sci. Eng. A*, **289** (1–2), 41–53.
5. Abbasi, T. and Abbasi, S.A. (2011) 'Renewable' hydrogen: prospects and challenges. *Renew. Sustain. Energy Rev.*, **15** (6), 3034–3040.
6. Turner, J.A. (2004) Sustainable hydrogen production. *Science*, **305** (5686), 972–974.
7. Rosen, M.A. and Scott, D.S. (1998) Comparative efficiency assessments for a range of hydrogen production processes. *Int. J. Hydrogen Energy*, **23** (8), 653–659.
8. Trommer, D. *et al.* (2005) Hydrogen production by steam-gasification of petroleum coke using concentrated solar power - I. Thermodynamic and kinetic analyses. *Int. J. Hydrogen Energy*, **30** (6), 605–618.
9. Momirlan, M. and Veziroglu, T.N. (2002) Current status of hydrogen energy. *Renew. Sustain. Energy Rev.*, **6** (1–2), 141–179.
10. Sato, S. *et al.* (2003) Hydrogen production from heavy oil in the presence of calcium hydroxide. *Fuel*, **82** (5), 561–567.
11. Stojic, D.L. *et al.* (2003) Hydrogen generation from water electrolysis —possibilities of energy saving. *J. Power Sources*, **118** (1–2), 315–319.
12. Tarasatti, S. (1999) Water electrolysis: who first? *J. Electroanal. Chem.*, **476**, 90–91.
13. Dunn, S. (2002) Hydrogen futures: toward a sustainable energy system. *Int. J. Hydrogen Energy*, **27** (3), 235–264.
14. de Souza, R.F. *et al.* (2007) Electrochemical hydrogen production from water electrolysis using ionic liquid as electrolytes: towards the best device. *J. Power Sources*, **164** (2), 792–798.
15. Bockris, J.O.M. *et al.* (1981) *Comprehensive Treatise of Electrochemistry*, vol. 2, Plenum Press, New York.
16. Turner, J.A. (1999) A realizable renewable energy future. *Science*, **285** (5428), 687.
17. Mueller-Langer, F. *et al.* (2007) Techno-economic assessment of hydrogen production processes for the hydrogen economy for the short and medium term. *Int. J. Hydrogen Energy*, **32** (16), 3797–3810.
18. Steinfeld, A. (2002) Solar hydrogen production via a two-step water-splitting thermochemical cycle based on Zn/ZnO redox reactions. *Int. J. Hydrogen Energy*, **27** (6), 611–619.
19. Grigoriev, S.A., Porembsky, V.I., and Fateev, V.N. (2006) Pure hydrogen production by PEM electrolysis for hydrogen energy. *Int. J. Hydrogen Energy*, **31** (2), 171–175.
20. Granovskii, M., Dincer, I., and Rosen, M.A. (2006) Environmental and economic aspects of hydrogen production and utilization in fuel cell vehicles. *J. Power Sources*, **157** (1), 411–421.
21. Kreuter, W. and Hofmann, H. (1998) Electrolysis: the important energy transformer in a world of sustainable energy. *Int. J. Hydrogen Energy*, **23** (8), 661–666.
22. Bockris, J.O. (2002) The origin of ideas on a hydrogen economy and its solution to the decay of the environment. *Int. J. Hydrogen Energy*, **27** (7–8), 731–740.
23. Bockris, J.O.M. and Veziroglu, T.N. (2007) Estimates of the price of hydrogen as a medium for wind and solar sources. *Int. J. Hydrogen Energy*, **32** (12), 1605–1610.
24. Zeng, K. and Zhang, D. (2010) Recent progress in alkaline water electrolysis for hydrogen production and applications. *Prog. Energy Combust.*, **36** (3), 307–326.

25 Hollmuller, P. et al. (2000) Evaluation of a 5 kW(P) photovoltaic hydrogen production and storage installation for a residential home in Switzerland. *Int. J. Hydrogen Energy*, **25** (2), 97–109.

26 Varkaraki, E., Lymberopoulos, N., and Zachariou, A. (2003) Hydrogen based emergency back-up system for telecommunication applications. *J. Power Sources*, **118** (1–2), 14–22.

27 Barbir, F. (2005) PEM electrolysis for production of hydrogen from renewable energy sources. *Solar Energy*, **78** (5), 661–669.

28 Oi, T. and Sakaki, Y. (2004) Optimum hydrogen generation capacity and current density of the PEM-type water electrolyzer operated only during the off-peak period of electricity demand. *J. Power Sources*, **129** (2), 229–237.

29 Kato, T. et al. (2005) Effective utilization of by-product oxygen from electrolysis hydrogen production. *Energy*, **30** (14), 2580–2595.

30 Wendt, H. and Kreysa, G. (1999) *Electrochemical Engineering*, 1st edn, Springer-Verlag, Berlin.

31 Kim, S. et al. (2006) Water electrolysis activated by Ru nanorod array electrodes. *Appl. Phys. Lett.*, **88** (26), 263106–263113.

32 Leroy, R.L., Bowen, C.T., and Leroy, D.J. (1980) The thermodynamics of aqueous water electrolysis. *J. Electrochem. Soc.*, **127** (9), 1954–1962.

33 Bard, A.J. and Faulkner, L.R. (2001) *Electrochemical Methods Fundamentals and Applications*, 2nd edn, John Wiley & Sons, Inc., New York.

34 Rossmeisl, J., Logadottir, A., and Norskov, J.K. (2005) Electrolysis of water on (oxidized) metal surfaces. *Chem. Phys.*, **319** (1–3), 178–184.

35 Kinoshita, K. (1992) *Electrochemical Oxygen Technology*, 1st edn, John Wiley & Sons, Inc., New York.

36 Viswanath, R.P. (2004) A patent for generation of electrolytic hydrogen by a cost effective and cheaper route. *Int. J. Hydrogen Energy*, **29** (11), 1191–1194.

37 Oldham, K.B. and Myland, J.C. (1993) *Fundamentals of Electrochemical Science*, 1st edn, Academic Press, San Diego.

38 Newman, J.S. (1991) *Electrochemical Systems*, Prentice Hall, New Jersey.

39 Rieger, P.H. (1987) *Electrochemistry*, 1st edn, Prentice-Hall, New Jersey.

40 Abouatallah, R.M. et al. (2001) Reactivation of nickel cathodes by dissolved vanadium species during hydrogen evolution in alkaline media. *Electrochim. Acta*, **47** (4), 613–621.

41 Trasatti, S. (1971) Work function, electronegativity and electrochemical behaviour of metals: 2. Potentials of zero charge and electrochemical work functions. *J. Electroanal. Chem.*, **33** (2), 351–378.

42 Krstajic, N. et al. (2001) On the kinetics of the hydrogen evolution reaction on nickel in alkaline solution - Part II. Effect of temperature. *J. Electroanal. Chem.*, **512** (1–2), 27–35.

43 de Chialvo, M.R.G. and Chialvo, A.C. (2001) Hydrogen evolution reaction on a smooth iron electrode in alkaline solution at different temperatures. *Phys. Chem. Chem. Phys.*, **3** (15), 3180–3184.

44 Lee, T.S. (1971) Hydrogen overpotential on pure metals in alkaline solution. *J. Electrochem. Soc.*, **118** (8), 1278–1282.

45 Correia, A.N., Machado, S.A.S., and Avaca, L.A. (1999) Studies of the hydrogen evolution reaction on smooth Co and electrodeposited Ni-Co ultramicroelectrodes. *Electrochem. Commun.*, **1** (12), 600–604.

46 Pickett, D.J. (1979) *Electrochemical Reactor Design*, 2nd edn, Elsevier, Amsterdam.

47 Choquette, Y., Menard, H., and Brossard, L. (1990) Electrocatalytic performance of composite coated electrode for alkaline water electrolysis. *Int. J. Hydrogen Energy*, **15** (1), 21–26.

48 Miles, M.H., Huang, Y.H., and Srinivasan, S. (1978) Oxygen-electrode reaction in alkaline-solutions on oxide electrodes prepared by thermal-decomposition method. *J. Electrochem. Soc.*, **125** (12), 1931–1934.

49 Damjanov., A., Dey, A., and Bockris, J.O.M. (1966) Electrode kinetics of oxygen evolution and dissolution on Rh Ir and Pt-Rh alloy electrodes. *J. Electrochem. Soc.*, **113** (7), 739.

50 Miles, M.H. et al. (1976) Effect of temperature on electrode kinetic parameters for hydrogen and oxygen evolution reactions on nickel electrodes in alkaline solutions. *J. Electrochem. Soc.*, **123** (3), 332–336.

51 Potvin, E. and Brossard, L. (1992) Electrocatalytic activity of Ni-Fe anodes for alkaline water electrolysis. *Mater. Chem. Phys.*, **31** (4), 311–318.

52 Leroy, R.L. (1983) Industrial water electrolysis - present and future. *Int. J. Hydrogen Energy*, **8** (6), 401–417.

53 Kulkarni, A.A. and Joshi, J.B. (2005) Bubble formation and bubble rise velocity in gas-liquid systems: A review. *Ind. Eng. Chem. Res.*, **44** (16), 5873–5931.

54 Van Helden, W.G.J., Van Der Geld, C.W.M., and Boot, P.G.M. (1995) Forces on bubbles growing and detaching in flow along a vertical wall. *Int. J. Heat Mass Trans.*, **38** (11), 2075–2088.

55 Zhang, D. and Zeng, K. (2012) Evaluating the behavior of electrolytic gas bubbles and their effect on the cell voltage in alkaline water electrolysis. *Ind. Eng. Chem. Res.*, **51** (42), 13825–13832.

56 Zhang, D. and Zeng, K. (2012) Evaluating the behavior of electrolytic gas bubbles and their effect on the cell voltage in alkaline water electrolysis. *Ind. Eng. Chem. Res.*, **51** (42), 13825–13832.

57 Matsushima, H., Fukunaka, Y., and Kuribayashi, K. (2006) Water electrolysis under microgravity. Part II. Description of gas bubble evolution phenomena. *Electrochim. Acta*, **51** (20), 4190–4198.

58 Drelich, J., Miller, J.D., and Good, R.J. (1996) The effect of drop (bubble) size on advancing and receding contact angles for heterogeneous and rough solid surfaces as observed with sessile-drop and captive-bubble techniques. *J. Colloid Interface Sci.*, **179** (1), 37–50.

59 Tadmor, R. (2004) Line energy and the relation between advancing, receding, and young contact angles. *Langmuir*, **20** (18), 7659–7664.

60 Yeoh, G.H. et al. (2008) Fundamental consideration of wall heat partition of vertical subcooled boiling flows. *Int. J. Heat Mass Trans.*, **51** (15–16), 3840–3853.

61 Brussieux, C. et al. (2011) Controlled electrochemical gas bubble release from electrodes entirely and partially covered with hydrophobic materials. *Electrochim. Acta*, **56** (20), 7194–7201.

62 Craig, V.S.J., Ninham, B.W., and Pashley, R.M. (1993) The effect of electrolytes on bubble coalescence in water. *J. Phys. Chem.*, **97** (39), 10192–10197.

63 Dunlap, P.M. and Faris, S.R. (1962) Surface tension of aqueous solutions of potassium hydroxide. *Nature*, **196** (4861), 1312–1313.

64 Pettersson, J., Ramsey, B., and Harrison, D. (2006) A review of the latest developments in electrodes for unitised regenerative polymer electrolyte fuel cells. *J. Power Sources*, **157** (1), 28–34.

65 Mauer, A.E., Kirk, D.W., and Thorpe, S.J. (2007) The role of iron in the prevention of nickel electrode deactivation in alkaline electrolysis. *Electrochim. Acta*, **52** (11), 3505–3509.

66 Abdel Ghany, N.A. et al. (2002) Oxygen evolution anodes composed of anodically deposited Mn-Mo-Fe oxides for seawater electrolysis. *Electrochim. Acta*, **48** (1), 21–28.

67 Zeng, K. and Zhang, D. (2014) Evaluating the effect of surface modifications on Ni based electrodes for alkaline water electrolysis. *Fuel*, **116**, 692–698.

68 Stojic, D.L. et al. (2006) Intermetallics as advanced cathode materials in hydrogen production via electrolysis. *Int. J. Hydrogen Energy*, **31** (7), 841–846.

69 Huot, J.Y. and Brossard, L. (1987) Time-dependence of the hydrogen discharge at 70-degrees-c on nickel cathodes. *Int. J. Hydrogen Energy*, **12** (12), 821–830.

70 Conway, B.E. and Bai, L. (1986) H2 evolution kinetics at high activity Ni-Mo-Cd electrocoated cathodes and its relation to potential dependence of sorption of H. *Int. J. Hydrogen Energy*, **11** (8), 533–540.

71 Huor, J.Y. et al. (1989) Hydrogen evolution on some Ni-base amorphous alloys in alkaline solution. *Int. J. Hydrogen Energy*, **14** (5), 319–322.

72 Huot, J.Y., Trudeau, M.L., and Schulz, R. (1991) Low hydrogen overpotential nanocrystalline Ni-Mo cathodes for

alkaline water electrolysis. *J. Electrochem. Soc.*, **138** (5), 1316–1321.

73 Ma, H. *et al.* (2006) Study of ruthenium oxide catalyst for electrocatalytic performance in oxygen evolution. *J. Mol. Catal. A Chem.*, **247** (1–2), 7–13.

74 Marshall, A. *et al.* (2007) Hydrogen production by advanced proton exchange membrane (PEM) water electrolysers – Reduced energy consumption by improved electrocatalysis. *Energy*, **32** (4), 431–436.

75 Hamdani, M. *et al.* (2004) Physicochemical and electrocatalytic properties of Li-Co3O4 anodes prepared by chemical spray pyrolysis for application in alkaline water electrolysis. *Electrochim. Acta*, **49** (9–10), 1555–1563.

76 Zheng, Z. *et al.* (2013) Effects of CeO2 on the microstructure and hydrogen evolution property of Ni-Zn coatings. *J. Power Sources*, **222**, 88–91.

77 Zheng, Z. *et al.* (2012) Ni-CeO2 composite cathode material for hydrogen evolution reaction in alkaline electrolyte. *Int. J. Hydrogen Energy*, **37** (19), 13921–13932.

78 Zheng, Z. *et al.* (2013) Electrochemical synthesis of Ni-S/CeO2 composite electrodes for hydrogen evolution reaction. *J. Power Sources*, **230**, 10–14.

79 Lyons, M.E.G., Doyle, R.L., and Brandon, M.P. (2011) Redox switching and oxygen evolution at oxidized metal and metal oxide electrodes: iron in base. *Phys. Chem. Chem. Phys.*, **13** (48), 21530–21551.

80 Lyons, M.E.G. and Floquet, S. (2011) Mechanism of oxygen reactions at porous oxide electrodes. Part 2 – Oxygen evolution at RuO2, IrO2 and IrxRu1-xO2 electrodes in aqueous acid and alkaline solution. *Phys. Chem. Chem. Phys.*, **13** (12), 5314–5335.

81 Singh, R.N. *et al.* (2007) Novel electrocatalysts for generating oxygen from alkaline water electrolysis. *Electrochem. Commun.*, **9** (6), 1369–1373.

82 Kamat, P.V. (2007) Meeting the clean energy demand: nanostructure architectures for solar energy conversion. *J. Phys. Chem. C*, **111** (7), 2834–2860.

83 Giz, M.J., Bento, S.C., and Gonzalez, E.R. (2000) NiFeZn codeposit as a cathode material for the production of hydrogen by water electrolysis. *Int. J. Hydrogen Energy*, **25** (7), 621–626.

84 Song, S. *et al.* (2007) An improved catalyst-coated membrane structure for PEM water electrolyzer. *Electrochem. Solid-State Lett.*, **10** (8), B122–B125.

85 El-Deab, M.S. *et al.* (2007) Enhanced water electrolysis: electrocatalytic generation of oxygen gas at manganese oxide nanorods modified electrodes. *Electrochem. Commun.*, **9** (8), 2082–2087.

86 Wendt, H., Hofmann, H., and Plzak, V. (1989) Materials research and development of electrocatalysts for alkaline water electrolysis. *Mater. Chem. Phys.*, **22** (1–2), 27–49.

87 Crnkovic, F.C., Machado, S.A.S., and Avaca, L.A. (2004) Electrochemical and morphological studies of electrodeposited Ni-Fe-Mo-Zn alloys tailored for water electrolysis. *Int. J. Hydrogen Energy*, **29** (3), 249–254.

88 Han, Q. *et al.* (2003) Hydrogen evolution reaction on amorphous Ni-S-Co alloy in alkaline medium. *Int. J. Hydrogen Energy*, **28** (12), 1345–1352.

89 Sheela, G., Pushpavanam, M., and Pushpavanam, S. (2002) Zinc-nickel alloy electrodeposits for water electrolysis. *Int. J. Hydrogen Energy*, **27** (6), 627–633.

90 Hu, W. (2000) Electrocatalytic properties of new electrocatalysts for hydrogen evolution in alkaline water electrolysis. *Int. J. Hydrogen Energy*, **25** (2), 111–118.

91 Hu, W. and Lee, J.-Y. (1998) Electrocatalytic properties of Ti2Ni/Ni-Mo composite electrodes for hydrogen evolution reaction. *Int. J. Hydrogen Energy*, **23** (4), 253–257.

92 Los, P., Rami, A., and Lasia, A. (1993) Hydrogen evolution reaction on Ni-Al electrodes. *J. Appl. Electrochem.*, **23** (2), 135–140.

93 Raj, I.A. (1993) Nickel-based, binary-composite electrocatalysts for the cathodes in the energy-efficient industrial-production of hydrogen from alkaline-water electrolytic cells. *J. Mater. Sci.*, **28** (16), 4375–4382.

94 Stojic, D.L. et al. (2007) Electrocatalytic effects of Mo-Pt intermetallics singly and with ionic activators. Hydrogen production via electrolysis. *Int. J. Hydrogen Energy*, **32** (13), 2314–2319.

95 Marceta Kaninski, M.P. et al. (2004) Ionic activators in the electrolytic production of hydrogen -- cost reduction-analysis of the cathode. *J. Power Sources*, **131** (1–2), 107–111.

96 Endres, F. and Abedin, S.Z.El. (2006) Air and water stable ionic liquids in physical chemistry. *Phys. Chem. Chem. Phys.*, **8** (18), 2101–2116.

97 de Souza, R.F. et al. (2006) Dialkylimidazolium ionic liquids as electrolytes for hydrogen production from water electrolysis. *Electrochem. Commun.*, **8** (2), 211.

98 Jones, S.F., Evans, G.M., and Galvin, K.P. (1999) Bubble nucleation from gas cavities - a review. *Adv. Colloid Interface Sci.*, **80** (1), 27–50.

99 Vogt, H. (1989) The problem of the departure diameter of bubbles at gas evolving electrodes. *Electrochim. Acta*, **34** (10), 1429–1432.

100 Kiuchi, D. et al. (2006) Ohmic resistance measurement of bubble froth layer in water electrolysis under microgravity. *J. Electrochem. Soc.*, **153** (8), E138–E143.

101 Matsushima, H. et al. (2003) Water electrolysis under microgravity - Part 1. Experimental technique. *Electrochim. Acta*, **48** (28), 4119–4125.

102 Wei, Z.D. et al. (2007) Water electrolysis on carbon electrodes enhanced by surfactant. *Electrochim. Acta*, **52** (9), 3323–3329.

14
Dynamic Operation of Electrolyzers – Systems Design and Operating Strategies

Geert Tjarks, Jürgen Mergel, and Detlef Stolten

14.1
Introduction

Renewable energy sources are characterized by a fluctuated energy production. In particular, this applies for wind turbines and photovoltaic systems. The load profiles typically show a dynamic progression with peaks and shutdown periods. The increased grid integration of renewable energy sources leads to problems concerning the quality of supply [1]. Mainly this is expressed by gaps between the energy demand and the energy supply (Figure 14.1). Currently, conventional power plants can compensate for the occurring differences. However, further expansion of renewable energy sources would lead to temporary shut downs of entire power plants. In addition to an adverse influence on economic aspects, the energy supply will depend on fossil energy carriers for grid stabilization in the future. To support the necessary expansion of renewable energy sources, it is discussed presently that grid stabilization based on renewable energy sources could be realized by power to gas systems [2].

The power-to-gas technology uses surplus energy for hydrogen production. The produced hydrogen can be seen as a chemical energy carrier. Occurring gaps in the energy supply can be compensated by a reconversion of hydrogen in gas turbines for energy production. In addition, reconversion can be realized by fuel cells, for instance in mobile applications. In this manner, conventional fuels could be substituted by hydrogen for emission free transport applications [3]. Hydrogen offers further potentials as a chemical energy carrier [4,5]. This is mainly due to the high energy density of compressed hydrogen. In particular the automotive industry is showing a trend to hydrogen as energy carrier with the development of hydrogen fuel cell cars [6]. A subsequent chemical conversion into synthetic methane is also discussed in literature because of the easily handling of methane [7].

Emission free produced hydrogen can be achieved by water electrolysis. Contemporary technologies for the water electrolysis are alkaline electrolysis, polymer electrolyte membrane (PEM) electrolysis, and solid oxide electrolysis (SOEC) [8]. Solid oxide electrolysis can produce hydrogen with high efficiencies

Hydrogen Science and Engineering: Materials, Processes, Systems and Technology, First Edition.
Edited by Detlef Stolten and Bernd Emonts.
© 2016 Wiley-VCH Verlag GmbH & Co. KGaA. Published 2016 by Wiley-VCH Verlag GmbH & Co. KGaA.

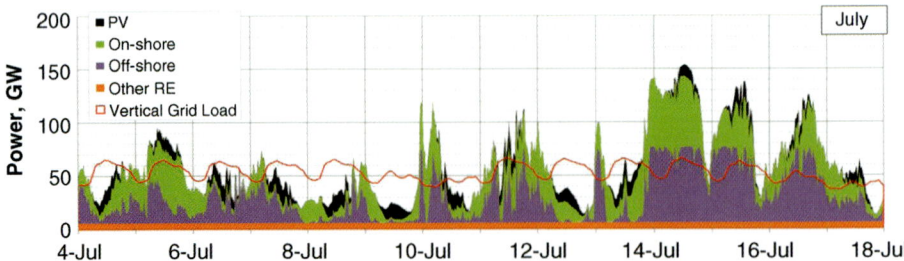

Figure 14.1 Simulated grid load and energy supply by renewable energy sources [11].

but development is still at the stage of basic research. The knowledge of entire systems either does not exist or is very limited [9]. For that reason this chapter is focused on alkaline and PEM electrolysis systems. Alkaline electrolysis systems have become very well established in recent decades. Several large scale systems are used for hydrogen production around the world. Thereby, power classes up to 150 MW are realized [10]. A major drawback of alkaline electrolysis systems is their low power densities. This is the reason for increased interest in PEM electrolysis systems in recent years. They offer high power densities, which leads to the potential for more compact megawatt systems. This chapter gives an overview of these technologies and their respective systems. The coupling with renewable energy sources and the resulting requirements is the main focus.

14.2
Process Steps and System Components

The production of hydrogen by water electrolysis implies several process steps independent of the employed electrolysis technology (Figure 14.2). This includes a conditioning of the power signal for the supply to the electrolyzer and a post-treatment of the produced gases. Conditioning of the power signal is needed since electrolyzers require direct current for supply. Alternate current from the grid or other energy sources has to be converted and fitted to the specification of the electrolyzer. The conditioning of the power signal is realized by power

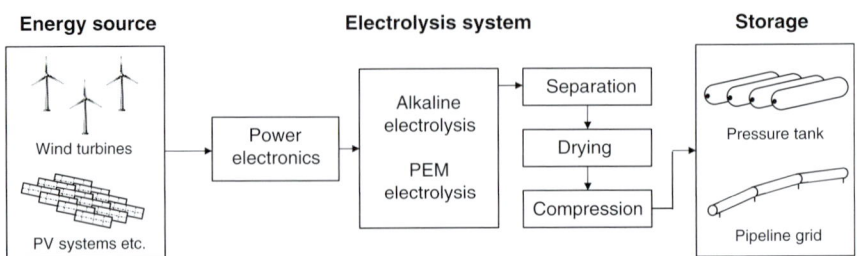

Figure 14.2 Process steps and system integration of water electrolysis system.

electronics between the primary energy source and the electrolyzer. The electrochemical splitting of water into hydrogen and oxygen is then executed in the electrolyzer. The produced gases are mixed with impurities like liquid and gaseous water when they leave the cells. In the post-treatment, these impurities have to be removed to guarantee application specified gas purities. Separation of the liquid phase is fulfilled in gas separators for both produced gases. The gases contain an amount of water vapor that has to be removed in a gas drying process. Usually, only hydrogen requires gas drying since high purity oxygen is not needed for most applications. For energy storage applications the oxygen is typically vented. After gas drying the hydrogen is compressed for storage applications. Since gas compression can be achieved during electrolysis, the compression process has to be considered for the system design. Still, high pressure storage applications up to 700 bar require additional mechanical compression.

The following sections describe the physical fundamentals of the process steps and gives an overview of the respective system components. First, alkaline and PEM electrolysis are introduced and compared. Then post electrolysis conditioning process and strategies are discussed.

14.2.1
Alkaline Electrolysis

For alkaline water electrolysis an aqueous KOH solution is used as electrolyte. Typical concentrations are at a level of 20–40% KOH. The produced gases are separated by a diaphragm that is placed between the two electrodes. The diaphragm must be conductive for hydroxide ions since they act as the charge carrier in alkaline electrolysis. Alkaline electrolysis systems are usually operated at temperatures of 80 °C and current densities in the range 0.2–0.4 A cm^{-2} [8]. If high production rates are needed the overall current can be increased by larger active cell areas or by an increase in cell number. Figure 14.3 illustrates the reaction equations at the anode and cathode.

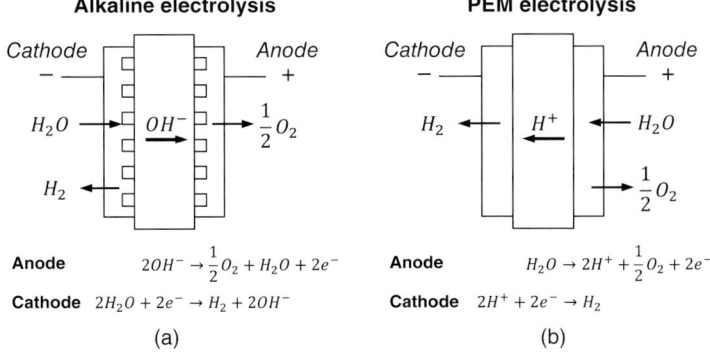

Figure 14.3 Operation principle of an (a) alkaline and (b) PEM electrolysis cell.

Alkaline water electrolysis systems have been produced commercially for several decades. Power classes up to 750 Nm^3 h^{-1} have been obtained for a single system [10]. For large power plants several single systems can be combined. The biggest power plant designed by this strategy was built at the Aswan dam in Egypt with an overall power of 156 MW [12]. Although the alkaline electrolysis technique was greatly improved in the 1980s and 1990s, market interest was very low. When interest in power-to-gas systems rose in the last few years, a trend to PEM water electrolysis was recognizable. This is because alkaline electrolysis has some drawbacks in comparison with PEM electrolysis. Namely, the lower power density, limited pressure operation, and lower part load range.

14.2.2
PEM Electrolysis

For PEM electrolysis the electrodes are separated by a solid polymer membrane. The membrane is characterized by its ionic conductivity and the isolating effect for electrons. These properties are achieved by sulfonation of the membrane. Owing to the ionic conductivity, the membrane is suitable as an electrolyte. In PEM electrolysis the positive loaded hydronium is used as load carrier. Beside the ionic conductivity, a water drag to the cathode side is recognizable. This leads to water on the cathode side, which has to be transported out of the cell.

Current PEM electrolysis systems are operated at about 50–80 °C. The current densities are up to 2 A cm^{-2} [13]. As by alkaline electrolysis, higher production rates can be achieved by enhancement of the cell area and the number of cells. Figure 14.3 shows the reaction equations that take place in PEM electrolysis.

The main advantages of PEM electrolysis are higher power densities in comparison with alkaline electrolysis. This is a major aspect for the design of large power plants. Alkaline electrolysis is typically operated at current densities up to 0.4 A cm^{-2} while present-day PEM electrolysis systems reach current densities up to 2.0 A cm^{-2}. This leads to higher production rates for a constant cell area. In addition, lower cell voltages are achieved for constant current densities in the PEM electrolysis. This increases the efficiency of the cell and the complete system. Figure 14.4 shows exemplary characteristic performances of alkaline and PEM electrolysis. Clearly, the performance of a present-day PEM electrolysis cell outperforms its alkaline electrolysis counterpart.

The development of PEM electrolysis started in the 1960s when General Electric presented the first PEM water electrolyzer. Recent research efforts have lead mainly to improved performances and reduced cost. However, current commercial systems are still reaching production rates under 50 Nm^3 h^{-1} at power levels up to 150 kW [10]. For construction of large power plants the stack technology has to achieve the MW range. Commercial manufactures like Proton OnSite and Hydrogenics have announced the demonstration of these technologies in the near future [14].

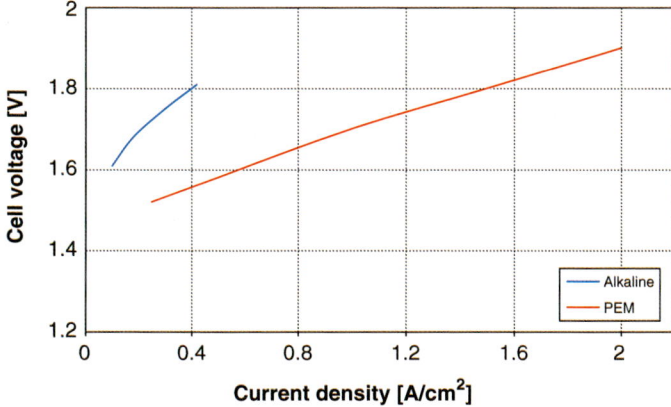

Figure 14.4 Characteristic cell performance for alkaline and PEM electrolysis from Hydrogenics Corp [15].

14.2.3
Gas Separation

The produced gases are either mixed with liquid water or KOH solution at the stack outlet depending on type of electrolysis. To purify the gas from the liquid phase, gravity gas separators are used. Thereby, the liquid phase is displaced to the bottom of a vessel. Owing to the water drag across the membrane this also applies to the cathode side of the PEM electrolyzer. The purified gas can be captured at the top of the vessel. The separated liquid phase can be reused for the stack supply.

In PEM electrolysis the separated water can be fed directly to the anode loop. An additional water loop can be realized on the cathode side, mainly for temperature regulation. Owing to higher investment costs for an extra pump, current systems typically contain a water loop only on the anode side. Irrespective of an additional water loop, the water level has to be regulated at the cathode side since water is not consumed at the cathode. The separated water has to be drained or recirculated to the anode side. This can lead to problems when hydrogen is produced at elevated pressures. At elevated pressures the amount of dissolved hydrogen in the liquid phase increases. This leads to a loss of produced hydrogen and impurities on the anode side. This has to be considered when selecting a system design and the operation strategy. If the anode side operated under pressure, the recirculation loop requires an addition pump. Some systems contain an expander to reduce the amount of dissolved hydrogen before re-using the water. Since water is consumed at the anode, the fresh water supply is provided on the anode side. Figure 14.5 illustrates the system architecture schematic.

In an alkaline electrolysis system the liquid phase is separated from the produced gases by gravity gas separators. The separation principle is similar to the gas separation in PEM electrolysis. However, regulation of the fluid level differs

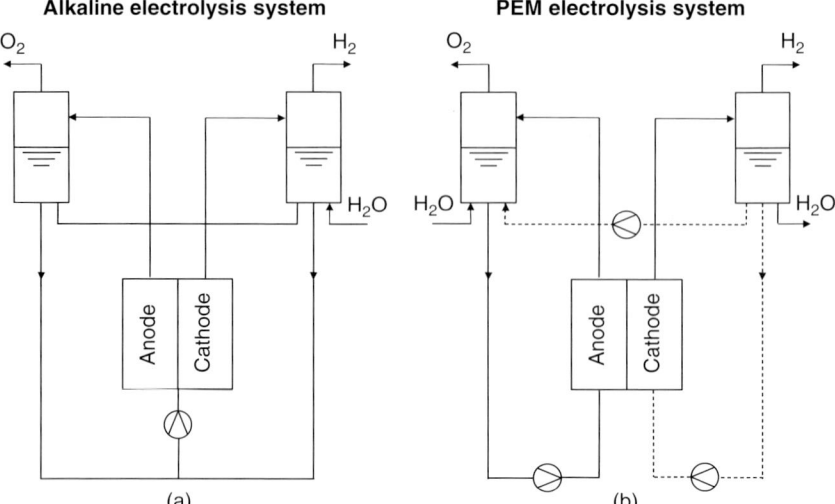

Figure 14.5 Schematic of (a) alkaline electrolysis and (b) PEM electrolysis.

from PEM electrolysis because the concentration of the KOH solution changes in the electrolysis process. The concentration increases on the cathode side when water is consumed for the electrolysis. On anode side the concentration decreases since water is produced on the anode side (Figure 14.3). For the stack supply the concentration has to be adjusted to a constant value by mixing the separated KOH solutions. This is realized in a mixing pipe in front of the stack. Since more water is consumed on the cathode than is produced on the anode side, the cathode side is fitted with a water supply. The KOH-solution level in the gas separation vessels is regulated by the pressure, the water supply, and a connection pipe between the two vessels. Figure 14.5a shows the schematic system architecture for gas separation in alkaline electrolysis systems.

14.2.4
Compression

For storage applications compressed hydrogen is required due to higher energy density of compressed gases. The level of compression depends on the respective application area. Fueling stations for mobile applications require pressures up to 700 bar, whereas gas pipelines are supplied with pressures up to 80 bar [16]. The compression and the required energy must be considered for the system design, since the electrolyzer is able to deliver hydrogen at elevated pressure.

Thermodynamics describe two ideal cases for compression. These cases are isothermal compression and adiabatic compression. For an isothermal compression it is assumed that all produced heat could be transported out of the system so that the temperature remains constant. This leads to much higher efficiencies

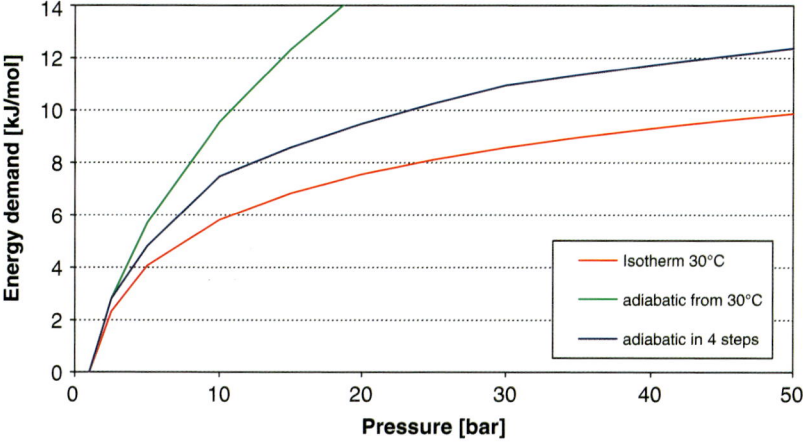

Figure 14.6 Energy demand per mol for isothermal, adiabatic, and multi-step adiabatic compression.

in comparison with adiabatic compression, where all produced heat remains in the system. Real compression processes are called polytropic. They are characterized by efficiencies between the two ideal cases. To increase the efficiency, the compression could be separated into single steps with gas cooling between the steps. An adiabatic compression process with pressure steps is more similar to an isothermal compression (Figure 14.6) and thus more efficient. A mechanical compression can take advantage of this effect by separating the compression process into several steps. A typical cost optimum for a compression up to 700 bar yields two compression steps.

The electrolyzer itself can execute a compression that is more isothermal and has therefor a higher efficiency than mechanical compression. That leads to the concept of high pressure electrolysis where the compression or at least a part of it is performed in the electrolyzer itself. Alkaline electrolysis is able to produce hydrogen with a pressure up to 30 bar. For PEM electrolysis pressure limits of 50 bar are commercially available [10]. However, higher pressure limits are reported in research projects for both technologies. Elevated pressure also leads to problems that have to be considered for the cell development of both electrolysis technologies. Permeation effects are supported in a pressure operation since the partial pressure gradients of the respective gases increase with an elevated overall pressure. This has a negative influence on the gas purities. Furthermore, the mechanical stress on the membrane of PEM electrolysis is increased at high pressure gradients. To lessen these drawbacks two different strategies for pressurization of PEM electrolyzers have been established.

Differential pressure operation is a mode of operation where the hydrogen side is operated at elevated pressures. It can be operated in special cases up to 350 bar [17]. This pressurization leads to different effects, for example, an enhanced gas crossover. To reduce the pressure gradient over the membrane, balanced pressure systems have evolved.

In balanced or equal pressure systems the oxygen side is operated at the same pressure level as the cathode side. This increases the system costs because the anode side has to be designed for elevated pressure. In addition compressed oxygen is difficult to handle due to the risk increased posed by gasket material and powder metal [18]. However, due to lower mechanical stress of the membrane, the lifetime of the cell could increase under balanced pressure operation.

Detailed analysis of the pros and cons of various pressure operation strategies for PEM electrolysis is still missing from the literature. A first approach has been performed by Schalenbach et al. [19]. In this work the energy demand for hydrogen compression with different pressure operations and compression strategies is calculated. It could be demonstrated that there is an optimum point at which compression in the electrolyzer is more efficient than complete mechanical compression. This optimum depends on the final pressure level desired. Similar results have been presented by Hanke-Rauschenbach et al. [20]. Both studies use a thermodynamic approach. For detailed research, further aspects, like technical feasibility and system efficiency, have to be considered.

14.2.5
Gas Drying

The gas is mixed with water vapor at the stack outlet. After the gaseous phase is separated from the liquid phase, the separated gas contains hydrogen and water vapor. This mixture is called humid hydrogen and can be described by the laws of an ideal humid gas. The water content (χ) is described by the relation of the mass portions. The mass portions are given by m_{wv} for water vapor and m_g for the gas. By Dalton's law the partial pressure of the water vapor p_{wv} and the total pressure p can be inserted as described in the following equation. The specific gas constants are given by R_g and R_w:

$$\chi = \frac{m_{wv}}{m_g} = \frac{R_g}{R_w} \frac{p_{wv}}{p - p_{wv}} \tag{14.1}$$

To remove the water vapor from the produced hydrogen, gas drying is required. For different storage applications a defined dew point of the hydrogen is required. If the dew point is too high, the water could condense from the hydrogen under changing environmental conditions, which could occur during the storage process. For example, dew points down to $-40\,°C$ are required for storage applications [21].

Usually, pressure swing adsorption dryers are used in current systems for gas drying. They contain two pillars that are filled with an adsorption material like silica gel. When the dryer is operated, just one pillar is used for gas drying by floating with the humid hydrogen. The silica gel adsorbs the water that is mixed with the hydrogen. The adsorption can precede until the water content in the silica gel becomes too high. To reactivate the adsorption material, the pillar can be floated with dry hydrogen which leads to a loss of the produced hydrogen.

Other gases for reactivation of the adsorption material would have a negative impact on the gas purity. When the gas dryer is in operation, one pillar is used for gas drying and the other pillar is reactivated. This process requires energy, so the total system efficiency decreases.

As Eq. (14.1) describes, the water content is lower at higher pressure. Consequently, for high pressure electrolysis the energy content required for gas drying is lower. This is also reported by Ayers *et al.* [22]. Since the energy demand for compression increases for high pressure, there is an optimum for the pressure level of the electrolyzer. Ayers *et al.* find this optimum at a pressure of 70 bar.

14.2.6
Power Supply

For the supply of water electrolysis, direct current is necessary. For coupling with the grid and renewable energy sources like wind turbines a power electronics system is needed that contains a commutation of alternating current to direct current. In addition, the power electronics regulate the current and voltages for the stack supply. This also applies for the supply from PV systems that deliver direct current. An additional commutation is not necessary for the supply from PV systems.

Ursúa *et al.* [23] report two different power supply technologies for electrolyzers. They are classified according to the semiconductor used in thyristor-based power supplies and the transistor-based power supplies. These two groups contain different designs for the power supply. Ursúa reports that thyristor-based power supplies are characterized by harmonic content of the power signal. Transistor-based power supplies deliver power signals without any harmonic content. The study investigated the influence of these two main groups on the efficiency of an alkaline electrolyzer. Therefore, the energy consumption was measured at constant hydrogen production rates, which can be correlated to an average stack current. Owing to the constant power signal the efficiency is higher for transistor-based power supplies. For a hydrogen production rate of $1\,\mathrm{Nm^3\,h^{-1}}$ the energy consumption for a thyristor-based power supply is measured as $5.4\,\mathrm{kWh\,Nm^{-3}}$. For the transistor-based power supply an energy of $4.9\,\mathrm{kWh\,Nm^{-3}}$ is required. The energy consumption and therefore the efficiency depend on the production rate. Ursúa reports up to 10% lower efficiencies for thyristor-based power supplies over the entire load range [23].

14.3
Dynamic Operation of Electrolyzers

In normal operation the electrolyzer is operated under fluctuating and dynamic conditions through the coupling to renewable energy sources. Therefore, the electrolyzer has to cover a wide load range that is fitted to the requirements of the power supply. Typically, an electrolyzer stack is characterized by its nominal

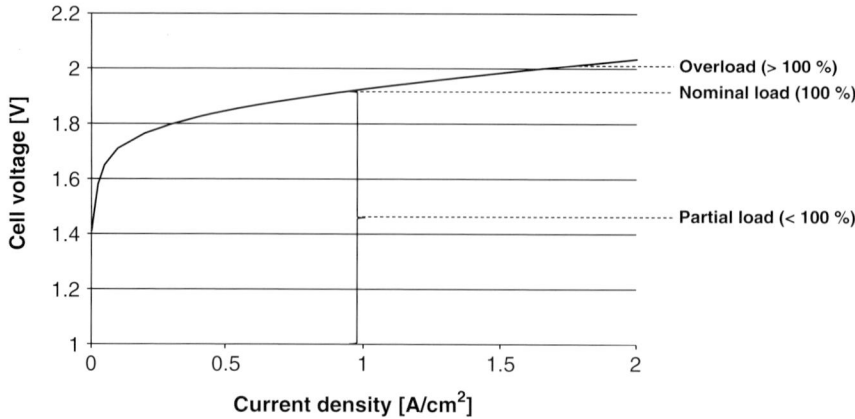

Figure 14.7 Exemplary load range of a PEM electrolyzer.

load. For higher loads than the nominal load, the stack is operated in an overload range. The partial load range is addressed for loads that are lower than the nominal load. A sensible definition of the nominal load point can be derived from an efficiency of 70% based on the lower heating value (LHV). Figure 14.7 illustrates the load range of a PEM electrolyzer. Alkaline electrolysis features the same nomenclature, just with a different operating range and nominal load point.

In addition to the load range, the dynamic behavior of the stack and the complete system is a relevant characteristic for the design and operation strategy. The dynamic behavior is given by the response time of different signals that occur in the operation of electrolyzer systems. Furthermore, the start- and shutdown behavior has to be investigated since power signals of renewable energy sources feature shutdown periods. This section reviews the known effects of the dynamic operation of electrolyzers.

14.3.1
Operation Range

The operation range of the electrolyzer based on the nominal load indicates its ability to operate under dynamic conditions. It is bounded downwards by the gas purities, while for the upper value of the operation range lifetime issues are the limiting aspects. This is due to the hazard conditions for the cell materials that are present at high power densities.

Different effects lead to a contamination of the produced hydrogen and the produced oxygen with respect to the other gas. Oxygen contamination on the cathode side lowers the gas purity of the produced hydrogen and with it the total efficiency since most of the oxygen combusts with hydrogen to regenerate water. The amount of reacted hydrogen is a loss to the hydrogen production but does not influence the dynamic operation behavior of the electrolyzer. This is not the case for the impurity of oxygen with hydrogen. A mixture of oxygen and hydrogen is spontaneously combustible over the concentration limit of 4% hydrogen in oxygen under atmospheric pressure and room temperature (lower explosion limit). The limit of 4% must not be exceeded in operation to avoid hazardous gas mixtures. For security reasons, in technical applications this limit is typically set to 2% hydrogen in oxygen. The amount of permitted gas strongly depends on the operation parameters like pressure and temperature. However, some effects that lead to gas crossover are independent of the current density. For higher production rates the contamination of oxygen with hydrogen decreases since the larger production rate of produced oxygen lowers the relative concentration of hydrogen. If the current density decreases and with it the production rate of oxygen, the concentration of hydrogen in oxygen increases because the crossover rate of hydrogen remains the same. Therefore, the low part of the load range of both electrolyzer technologies depends on the gas purity on the anode side.

For alkaline electrolysis there are mainly two effects concerning contamination of the product gases with the respective other component. The product gases can diffuse through the diaphragm that separates the two components. In addition, a small amount of the product gases is dissolved in the KOH solution. Since the separated solutions are mixed for a constant KOH-concentration, gas molecules cross over to other chamber. The purity of oxygen strongly depends on the cell temperature and the operation pressure. Janssen et al. have shown that an increase of 40 °C from 40 to 80 °C leads to an elevated gas impurity from 0.3% to 2.3% hydrogen in oxygen for a 200 bar operation strategy [24]. The gas contamination of oxygen increases linear with the pressure. Janssen reports a signal of 1.5% hydrogen in oxygen for 70 bar and 60 °C. For 130 bar and 60 °C it is 2.5%. For contemporary alkaline electrolysis systems the lower load range limit is typically specified to 20–40% of the entire load range [8,10].

In PEM electrolysis only gas permeation across the membrane has to be considered. However, dissolved hydrogen can be transported to the anode side by a recirculation of water from the cathode side of the cell. In present-day PEM electrolysis systems the water from the cathode side is clarified in an additional expander to reduce the impurity of the produced oxygen. Gas permeation across the membrane results from three physical effects: diffusion due to a concentration gradient, diffusion due to a pressure gradient, and a permeation due to water drag with dissolved hydrogen [19]. Figure 14.8 illustrates the effects that lead to a gas crossover in PEM electrolysis. The given effects are dependent on the temperature, pressure, and membrane thickness. Lower temperatures and thicker membranes can reduce the amount of permitted hydrogen. However, a lower temperature and a thicker membrane lead to an increase in the overvoltage. Simulation of the gas crossover by Schalenbach et al. at different operation conditions in PEM electrolysis shows a clear increase of permeated hydrogen for high pressure operations. For differential pressure operation, the concentration reaches much higher values in comparison to atmospheric pressure operation (Figure 14.8). This lowers the partial load range for low current densities. This can be improved by balanced pressure operation but the partial load range is still

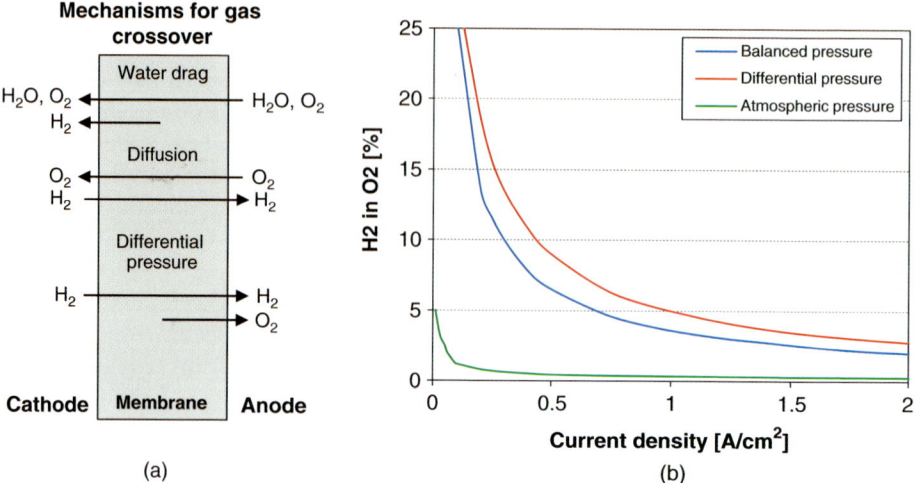

Figure 14.8 Permeation effects (a) and gas purity (b) for PEM electrolysis. Reprinted from Schalenbach et al. [19].

considerably smaller than for atmospheric pressure operation. Special treatment of the membrane is necessary to reduce the gas crossover or to recombine the permeated hydrogen to regenerate water. The membrane can be treated with a catalyst to ensure a recombination. In this case the permeated hydrogen is a loss but the recombination enables a higher operation range. However, electrolysis systems with balanced pressure strategies are characterized by higher investment costs since an addition pressure-proved design for the cathode side is required.

Both electrolyzer technologies can be operated in an overload range. The overload operation, however, lowers the lifetime of the cell components. For the operation strategy an optimum between the durability and the overload range has to be found. Alkaline electrolysis systems usually do not operate in an overload range since the operation range is more limited and therefore the nominal load is very high relative to the performance of an alkaline electrolyzer. PEM electrolysis offers a wide performance range, so that basic overload operation is possible, but sacrifices efficiency. However, the cell components have to resist the hazard conditions in overload operation. The cell and stack design should be fitted to these conditions. For the development of operation strategies a clear definition of the overload range is necessary. The overload range results from definition of the nominal load point, which can be derived from an efficiency of 70% based on the lower heating value.

14.3.2
Dynamic System Response

The response time of the electrolyzer system indicates how fast a specified signal can follow an input requirement from the energy source. The signal can come from the stack current and voltage, the stack temperature, the flow rates of the product gases, or the gas purity signals. The response time of the respective signals defines the capability of a dynamic operation.

A fast response of the stack current and voltage is important to use the entire energy supply. If the response of the stack current and voltage is low, peaks of the energy supply cannot be captured by the electrolyzer for hydrogen production. The response of the system strongly depends on the response time of the electrolyzer stack and the individual cells respectively. It has been reported that electrolysis technologies are characterized by short response times. Vaes *et al.* have investigated the response of alkaline electrolyzers for a pressure based control loop and a power based control loop [25]. For a pressure based control loop a response time of about 30 s was measured. Power based control loops enable response times of 2 s [25]. Ursúa *et al.* have investigated the coupling of a 1 $Nm^3 h^{-1}$ alkaline electrolyzer with a wind turbine and a photovoltaic system [26]. The current and voltage signals featured an adequate response time for the demands of renewable energy sources. A similar surveillance has been made by Gandía *et al.* [27]. For the response time of a PEM electrolyzer Miland has investigated the response time of the stack current with a 1.5 kW PEM electrolyzer. Response times under 0.2 s were measured with the PEM electrolysis system [28].

Highly dynamic operation of an electrolyzer stack presumes a respective dynamic of the system components, which must be designed to guarantee an adequate supply of the stack for occurring load steps. Time delays can occur due to delays of the power electronics or the pumps. As Vaes *et al.* have shown, the operation strategy of the system has a considerable influence on the system response time [25].

The response time is lower for signals that indicate concentration values like the concentration of hydrogen in oxygen. This influences the system design and operation strategy since a hazard situation cannot be compensated in an adequate way by the operation strategy. The response of concentration values has been researched by Kim *et al.* with a cell simulation of PEM electrolysis [29].

Another signal with a low response time is the stack temperature. Very slow values are reported by different studies [26,27]. Owing to cell overvoltages, like ohmic losses, energy is dissipated as heat. The stack temperature should be maintained within a definite range. Low temperatures decrease the efficiency while high temperatures can damage cell parts like the membrane and the catalyst layers and increase corrosion at the metal components. At high power densities the amount of dissipated energy is higher and therefore so is the amount of heat. Especially for large systems, heat management is required since the heat capacity increases with system size. For small systems with low heat capacities a cooling system is not necessary. In fact, small systems typically require an additional heating of the water supply to guarantee a required temperature of the stack. However, in large systems the temperature signal is the limiting factor for dynamic operation concerning the response times. Small heat capacities of the system should be preferred for fast dynamic response times.

14.3.3
Startup and Shutdown

The energy supply from renewable energy sources is characterized by gaps in the supply. In these periods the electrolyzer has to be operated in a standby mode or has to be shut down completely. A respective startup process has to be realized when the energy supply returns. The start period typically requires a warm-up time until the signals reaches a definite level. In particular, this applies for the pressure level and the stack temperature. In this startup period no gas flow exists since the produced gas is required to reach the pressure level. This can take a few minutes dependent on the pressure level and the volume of the gas separation vessels.

In a standby mode the production rate is reduced to a minimal value. If the production rate is lower than this limit, the concentration of hydrogen and oxygen can exceed the security limit of 2%. To prevent explosive gas mixtures the stack has to be supplied with energy for minimum production rate of oxygen. This lowers the efficiency of the system, so an optimum between a complete shut down and a standby mode has to be found in practice.

14.4
System Design Criterion

For system development different criteria must be considered. For technical aspects, the system efficiency is an important criterion. The system efficiency is influenced by nearly every system component and the operation strategy. From an economical point of view the costs are the major criterion. They can be divided into investment costs and hydrogen production costs. The costs, on the other hand, are influenced by different aspects like the system efficiency and the lifetime of the system components. The following section gives an overview of the criteria and their respective connections.

14.4.1
System Efficiency

The efficiency is generally given by the quotient of profit and effort. The profit in water electrolysis is the amount of produced hydrogen. Since the produced oxygen is typically vented for energy storage applications, only hydrogen is considered for the efficiency of the electrolyzer. The effort is the consumed electrical energy that is delivered by the energy source. A reasonable indication of the efficiency is therefore the energy demand per produced amount of hydrogen. The hydrogen amount can be indicated in kilograms or in standard cubic metres [10].

To evaluate the process in comparison to an ideal reversible process, a percentage indication is useful. For this purpose the contained energy of the produced hydrogen has to be considered. The stored energy of the hydrogen can be calculated by using the higher heating value (HHV) and the lower heating value (LHV). The HHV is mainly used when the system should be evaluated without consideration of the subsequent use of the produced hydrogen. For an evaluation of a process chain with a subsequent consumption of the hydrogen, the LHV is more reasonable since the consumption is typically related to the LHV [10]. The respective equations are given by the following equations with the amount of produced hydrogen (Δn_{H2}) and the energy consumption ($\Delta E_{electric}$):

$$\eta_{LHV} = \frac{\Delta n_{H2} LHV}{\Delta E_{electric}} \quad \eta_{HHV} = \frac{\Delta n_{H2} HHV}{\Delta E_{electric}} \tag{14.2}$$

Most of the electric energy from the system supply is needed for the stack. However, the portion depends on the operation strategy: typical values are between 60% and 75% [27]. For lower energy demand of the stack, the stack efficiency has to be reduced. The stack efficiency itself results mainly from the cell efficiency. The cell efficiency is given by the voltage efficiency and the faradaic efficiency. The voltage efficiency takes into account the overvoltages from activation energy, ohmic resistance, and mass transport. It is given by the ratio of the reversible cell voltage without losses to the actual cell voltage in operation. The faradaic efficiency describes the hydrogen loss due to gas crossover. Therefore, it is considered that a permitted oxygen molecule consumes two hydrogen molecules on the cathode side. The relationships are given in Eqs. (14.3). For the stack efficiency, additional ohmic losses and hydrogen losses have to be considered. They are very low in comparison to the cell losses:

$$\eta_{voltage} = \frac{E_{rev}}{E_{real}} \quad \eta_{faradaic} = \frac{\Delta n_{H2} - \Delta n_{H2}^{per} - 2\Delta n_{O2}^{per}}{\Delta n_{H2}} \tag{14.3}$$

The cell efficiency and therefore the stack efficiency are influenced by different operation parameters like the current density, temperature, and pressure level. For low current densities the cell efficiency is mainly influenced by the faradaic efficiency. For increasing current densities the overvoltages become more dominant so that the cell efficiency is mainly influenced by the voltage efficiency at high current densities. Figure 14.9 illustrates the efficiency of a PEM electrolyzer.

For both electrolysis technologies the efficiency increases for increasing temperatures. This is because the overvoltages are reduced for higher temperatures. This means lower activation energies are required and the ohmic losses of the membrane are reduced. A further parameter that influences the efficiency is the pressure level of the electrolyzer. The effect of different pressure strategies for the efficiency of a PEM electrolysis cell is presented by Schalenbach et al. [19]. The efficiencies of an atmospheric, balanced, and differential pressure strategy are illustrated in Figure 14.9. Figure 14.10 shows the effect of temperature and pressure on the performance of a ten-cell stack presented by Selamet et al. [30].

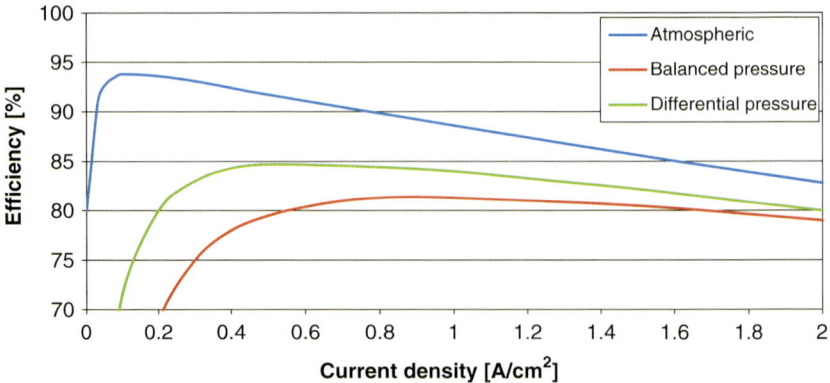

Figure 14.9 Efficiency of a PEM (polymer electrolyte membrane) electrolyzer. Reprinted from Schalenbach et al. [19].

The overall effect of elevated pressure is very low for alkaline electrolysis since two different effects compensate for each other when the pressure is elevated. On the on hand, more energy is needed for the electrolysis process since work against the pressurized hydrogen has to be done from a thermodynamic point of view. On the other hand, the increase in pressure leads to smaller gas bubbles in the electrolyte. The gas outlet is easier and the occupied area of catalyst is smaller. For alkaline electrolysis this effect has been examined in different studies [31]. Whether this can be transferred to PEM electrolysis has not been investigated.

Beside the efficiency of the stack, for the system efficiency, the energy demand of the respective system components has to be considered. This includes the power electronics, gas dryer, and heat management. Furthermore, electric energy

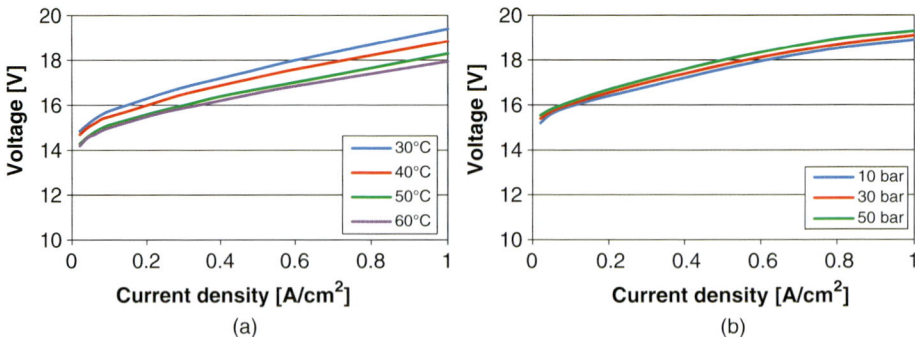

Figure 14.10 Effect of (a) temperature and (b) pressure on the performance of a ten-cell PEM stack. Reprinted from Selamet et al. [30].

is needed for control applications like valves and sensors. Depending on the focus of interest, the mechanical compression can be considered as well. For an adequate system design the system efficiency is an important technical evaluation criterion since it is influenced by every system component and their respective operation strategy. The respective portion of the energy demand of every system component strongly depends on the system size, system design, and operation strategy. For approximate values the energy demand of the power electronics is between 8% and 12%. Gas drying requires between 10% and 15% of the entire energy demand [32]. For large systems the efficiencies increase for both electrolysis technologies. This is because the system components can run in a more efficient way for large systems.

Typical system efficiencies for alkaline electrolysis systems operated under atmospheric conditions are in the range 4.1–4.5 kWh Nm^{-3}. This correlates to an efficiency of 85% related to the HHV. For pressurized alkaline electrolysis the efficiency decreases to 78% and 5.0 kWh Nm^{-3}. The efficiency of PEM electrolysis systems has been significantly improved in recent years. The energy demand for hydrogen production with a pressure of 30 bar is between 6.0 and 7.0 kWh Nm^{-3} for contemporary PEM electrolysis systems [10].

The system efficiency is influenced by nearly all operation parameters like the currents densities, pressure level, and temperature. Therefore, operation parameters can have opposite effects on the respective efficiencies of the system components. An increased pressure of the electrolyzer decreases the stack efficiency but reduces the energy demand for gas drying. In addition, the energy demand for mechanical compression is decreased. Another example is that a higher temperature improves the voltage efficiency of the cell but decreases the faradaic efficiency since the gas crossover rises with higher temperatures. A detailed study for the entire system and the dependence on operation parameters is still lacking.

14.4.2
Investment Cost and Hydrogen Production Costs

The relevant costs for water electrolysis are the investment costs and the hydrogen production costs. Investment costs depend on the electrolysis technology, pressure strategy, and system size. Typical values are given by the cost per installed power in kW.

The investment costs result from the stack costs and the costs of the system components. The stack costs are about 50% of the entire system cost for a PEM electrolyzer with an installed power of 60 kW [22]. For large power plants the investment cost per installed power becomes lower since the costs for system components can distributed over the large number of stacks. For an alkaline power plant with a production rate of 100 Nm^3 h^{-1} the investment costs are about €2400 kW^{-1}, whereas for a power plant with a production rate of 500 Nm^3 h^{-1} it is reduced to €1200 kW^{-1} [10]. A similar trend is assumed for

PEM electrolysis power plants. For present-day PEM electrolysis systems with a power of 1 MW, investment costs of about €1500 kW^{-1} have been given. It is assumed that these cost can be reduced to €500 kW^{-1} in the coming years [8].

The hydrogen production cost describes the costs per kilogram produced hydrogen. It contains the investment costs, system operation time over a year, system efficiency, and energy cost for the supply of the electrolyzer. Further costs are maintenance costs and other additional charges. To estimate these costs for future scenarios is very difficult. This leads to a wide range of the assumed hydrogen production cost in the literature. The example study by the Frauenhofer Institute for Solar Energy specified a hydrogen production cost of €3.17–3.85 kg^{-1} in the year 2050 [10]. However, a reliable estimation cannot be made due to insecure boundary conditions.

14.4.3
Lifetime

The lifetime of the electrolysis system mainly influences the investment costs and therefore the hydrogen production costs. The limiting aspect for the system lifetime is the stack. The lifetime of the stack is influenced by the materials used for the cell components, the quality of the water supply, and the operation conditions. Since in electrolyzers hazard conditions are present the used materials have to withstand these situations. Therefore, a lot of precious metals are used in electrolyzer technologies. The use of these materials leads to increased lifetimes of the stack but increases the cost simultaneously. Impurities in the supply water increase corrosion and can damage the cell components. Most early failures of electrolyzer stacks can be explained by poor water quality.

Since the condition of the cell depends on the operation strategy, this is a main route to improving stack lifetimes. A high voltage increases the corrosion and therefore leads to higher degradation. High voltages are reached in overload operation so that a reduced lifetime can be expected. The same is valid for high temperatures. They increase the degradation of the membrane and support the corrosion. The durability of the materials limits the temperature to about 80 °C for present-day electrolysis systems. How the lifetime is influenced by the dynamic operation has not been investigated in the literature. However, especially for differential pressure operation of PEM electrolyzers, a decreased lifetime can be expected since the membrane is exposed to mechanical stress with dynamic changes. For equal pressure this can be reduced due to lower stress on the membrane.

Alkaline electrolysis stacks, as a mature technology, can reach lifetimes of 90 000 h [13]. The lifetime is mainly influenced by the concentration of the KOH-solution and the temperature. PEM electrolysis stacks with lifetimes of 50 000 h have been presented [33]. However, notably, the lifetime strongly depends on the operation strategy and presently a standard process for measuring the stack lifetime is not available.

14.4.4
Safety

The system design and operation strategy have to fulfill the safety requirements. In operation, hazardous mixtures of hydrogen and oxygen have to be avoided. However, the use of ex-proof components is recommended. A spatial separation of the electronics and the hydrogen production is an option to prevent an ignition. Sensors for measuring the hydrogen concentration on the anode side and the surroundings of the electrolyzer are required. The system should provide an emergency shutdown to respond to a dangerous incident or situation.

14.5
Conclusion

The coupling of electrolysis systems to renewable energy sources with dynamic operation of the electrolyzer is possible from a technical perspective. This has been shown in various studies for alkaline and PEM electrolysis systems. Adequate dynamic response times under dynamic operation modes are reported. However, for energy storage systems for grid stabilization, systems are required that gain the multi MW to GW range. To obtain this advantage further research and development is necessary.

Alkaline electrolysis is a mature technology and capable of operation under dynamic conditions. For large-scale applications, though, the power density must be improved. This can be achieved by an improved cell performance. New electrocatalysts must be found that decrease the overpotentials and increase the cell efficiency. This applies to diaphragms with decreased ohmic losses. Furthermore, improved diaphragms with lower gas permeation characteristics would increase the operation range.

PEM electrolysis offers great potential for energy storage systems in the MW to GW range. This is due to higher current densities and better cell performance in comparison with alkaline electrolysis. A drawback of PEM electrolysis systems is the higher costs for both the stack and for the complete system. A cost reduction can be achieved through improved cell components like catalysts and bipolar plates. Significant progress towards improved cell components has been made in recent years and further potential is available. For large-scale applications large cell areas are required to gain the MW range for single stacks. The stacks have to offer adequate lifetimes for operation under dynamic conditions. Furthermore, improved membranes can reduce the gas crossover and therefore the operation range.

To achieve high system efficiencies the interaction of the system components and their respective operation strategies must be considered. Optimal operation points must be found with an analysis of the complete system. This especially applies for the pressure operation of the electrolyzer. An optimum pressure level for the pressurized electrolysis has to be found under consideration of the subsequent mechanical compression. However, detailed research of the influence of operation parameters on the system efficiency is still missing in the literature.

References

1. Pehnt, M. and Höpfner, U. (2009) Wasserstoff- und stromspeicher in einem energiesystem mit hohen anteilen erneuerbarer energien: analyse der kurz- und mittelfristigen perspektiven. Contract study for BMU by ifeu GmbH.
2. Gatzen, C. (2008) *The Economics of Power Storage. Theory and Empirical Analysis for Central Europe*, Oldenbourg Industrieverlag, Munich.
3. Winter, C. (2009) Hydrogen energy – abundant, efficient, clean: a debate over the energy-system-of-change. *Int. J. Hydrogen Energy*, **34**, 1–52.
4. Grube, T. and Stolten, D. (2010) Benefits of hydrogen use. *BWK: Das Energ.-Fachmagazin*, **62** (4), 16–17.
5. Grube, T., Fischedick, M., Pastowski, A., and Stolten, D. (2009) Perspectives for the expansion of hydrogen infrastructure on the example of NRW. *Chem.-Ing.-Tech.*, **81** (5), 591–598.
6. Rabbani, A., Rokni, M., Hosseinzadeh, E., and Mortensen, H. (2014) The start-up analysis of a PEM fuel cell system in vehicles. *Int. J. Green Energy*, **11**, 91–111.
7. Hiller, H. et al. (2000) Gas production, in *Ullmann's Encyclopedia of Industrial Chemistry*, Wiley-VCH Verlag GmbH, Weinheim.
8. Mergel, J., Carmo, M., and Fritz, D. (2013) Status on technologies for hydrogen production, in *Transition to Renewable Energy Systems* (eds. D. Stolten and V. Scherer), Wiley-VCH Verlag GmbH, Weinheim, pp. 423–450.
9. Laguna-Bercero, MA. (2012) Recent advances in high temperature electrolysis using solid oxide fuel cells: a review. *J. Power Sources*, **203**, 4–16.
10. Smolinka, T., Günther, M., and Garche, J. (2011) NOW-Studie: stand und entwicklungspotenzial der wasserelektrolyse zur herstellung von wasserstoff aus regenerativen energien. Technical report. Frauenhofer ISE.
11. Stolten, D., Grube, T., and Weber, M. (2013) Wasserstoff: Das Speichermerdium für erneuerbare Energie. Presented at the 12th Symposium Energieinnovation, 15–17 February 2012, University of Graz.
12. Zeng, K. and Zhang, D. (2010) Recent progress in alkaline water electrolysis for hydrogen production and applications. *Prog. Energy Combust.*, **36** (3), 307–326.
13. Carmo, M., Fritz, D., Mergel, J., and Stolten, D. (2013) A comprehensive review on PEM water electrolysis. *Int. J. Hydrogen Energy*, **38**, 4901–4934.
14. Proton Onsite (2013) Long life PEM water electrolysis stack experience and future directions. Technical Forum Presentation at the Hannover Fair 8–12 April 2013.
15. Hydrogenics Corp (2012) Electrolysis for grid balancing – Where are we? International Water Electrolysis Symposium, Copenhagen 10–11 May 2012.
16. US Department of Energy (2012) Hydrogen delivery (updated September 2012). Hydrogen, fuel cells and infrastructure technologies program. multi-year research, development and demonstration plan. Planned program activities for 2005–2015, http://www1.eere.energy.gov/hydrogenandfuelcells/mypp/pdfs/delivery.pdf (accessed 2 December 2013).
17. Proton Onsite (2012) Elimination of mechanical compressors using PEM-based electrochemical technology, 19th World Hydrogen Energy Conference 2012, 3–7 June Toronto, Canada.
18. Littman, F.E., Church, F.M., and Kinderman, E.M. (1961) A study of metal ignitions 1. The spontaneous ignition of titanium. *J. Less-Common Metals*, **3**, 367–378.
19. Schalenbach, M., Carmo, M., Fritz, D., Mergel, J., and Stolten, D. (2013)

Pressurized PEM water electrolysis: efficiency and gas crossover. *Int. J. Hydrogen Energy*, **38**, 14921–14933.

20 Bensmann, B., Hanke-Rauschenbach, R., Arias, P., and Sundmacher, K. (2013) Energetic evaluation of high pressure PEM electrolyzer systems for intermediate storage of renewable energies. *Electrochim. Acta*, **110**, 570–580.

21 Rothuizen, E., Mérida, W., Rokni, M., and Wistoft-Ibsen, M. (2013) Optimization of hydrogen vehicle refueling via dynamic simulation. *Int. J. Hydrogen Energy*, **38**, 4221–4231.

22 Ayers, K., Anderson, B., Capuano, C., and Blake, C. (2010) Research advances towards low cost, high efficiency PEM electrolysis. *ECS Trans.*, **33**, 3–15.

23 Ursúa, A., San Martín, I., Barrios, L., and Sanchis, P. (2013) Stand-alone operation of an alkaline water electrolyzer fed by wind and photovoltaic systems.
Int. J. Hydrogen Energy, **38**, 14952–14967.

24 Janssen, H., Bringmann, J.C., Emonts, B., and Schroeder, V. (2003) Safety-related studies on hydrogen production in high pressure electrolysers. *Int. J. Hydrogen Energy*, **29**, 759–770.

25 Vaes, J., Daneels, R., and Van Dingenen, D. (2013) Fast response alkaline water electrolysis for green energy storage. Presented at 4th European PEFC and H2 Forum, 2–5 July 2013, Lucerne.

26 Ursúa, A., San Martín, I., Barrios, E., and Sanchis, P. (2013) Stand-alone operation of an alkaline water electrolyser fed by wind and photovoltaic systems. *Int. J. Hydrogen Energy*, **38**, 14952–14967.

27 Gandía, L., Oroz, R., Ursúa, A., Sanchis, P., and Diéguez, P. (2007) Renewable hydrogen production: performance of an alkaline water electrolyzer working under emulated wind conditions. *Energy & Fuels*, **21**, 1699–1706.

28 Miland, H. (2005) Operational experience and control strategies for a stand-alone power system based on renewable energy and hydrogen. PhD Thesis, Norwegian University of Science and Technology, Trondheim, IMT-report 2004:66.

29 Kim, H., Park, M., and Soon-Lee, K. (2013) One-dimensional dynamic model of a high pressure water electrolysis system for hydrogen production. *Int. J. Hydrogen Energy*, **38**, 2596–2609.

30 Selamet, Ö., Acar, M., Mat, M., and Kaplan, Y. (2013) Effects of operating parameters on the performance of a high-pressure proton exchange membrane electrolyzer. *Int. J. Energy Res.*, **37**, 457–467.

31 Hausmann, E. (1976) Technical status of water electrolysis. *Chem. Eng. Technol.*, **48** (2), 100–103.

32 Briguglio, N., Brunaccini, G., Siracusano, S., and Randazzo, N. (2013) Design and testing of a compact PEM electrolyzer system. *Int. J. Hydrogen Energy*, **38**, 11519–11529.

33 Ayers, K., Dalton, L., and Anderson, E. (2012) Efficient generation of high energy density fuel from water. *ECS Trans.*, **41**, 27–38.

15
Stack Technology for PEM Electrolysis

Jürgen Mergel, David L. Fritz, and Marcelo Carmo

15.1
Introduction to Electrolysis

The production of hydrogen and oxygen from water using electrolysis is a technical process that has been well established throughout the world for over 100 years. However, only about 4% of hydrogen requirements worldwide are currently supplied by electrolysis [1]. One of the main reasons for this is the higher cost of producing electrolytic hydrogen compared to producing hydrogen from fossil fuels. About 600 billion Nm^3 of hydrogen are produced every year, primarily by steam reforming of natural gas, partial oxidation of mineral oil, and coal gasification.

Electrolyzers for hydrogen production are typically used in the glass, steel, and food industries, in power plants for generator cooling, in the manufacture of electronic components, and in the production of fertilizer via ammonia where cheap electrical energy is available from hydropower (Aswan Dam [2,3], 33 000 Nm^3-$H_2\,h^{-1}$).

The relevant techniques of water electrolysis are alkaline electrolysis using a liquid alkaline electrolyte, acidic electrolysis using a polymeric solid-state electrolyte (PEM electrolysis), and high-temperature electrolysis using a solid oxide electrolyte. The reactions at the cathode (hydrogen evolution reaction, HER) and at the anode (oxygen evolution reaction, OER) vary depending on the electrolysis technology used. Figure 15.1 provides an overview of the different operating principles [4].

Water is usually supplied at the cathode side in alkaline electrolysis and at the anode side in PEM electrolysis. In the case of high-temperature electrolysis, the required steam is supplied at the cathode side. Commercial electrolyzers are currently only available for alkaline electrolysis and PEM electrolysis. In the case of alkaline electrolysis, systems have been built commercially for nearly 100 years in various power classes up to approx. 750 $Nm^3\,h^{-1}$. Product development for PEM electrolysis only began 40 years ago; consequently, fewer commercial products (<65 $Nm^3\,h^{-1}$) are currently on the market. High-temperature electrolysis is not currently commercially available; however, it is being developed on the laboratory scale and there is some manufacturer interest. The two electrolysis

---

Hydrogen Science and Engineering: Materials, Processes, Systems and Technology, First Edition.
Edited by Detlef Stolten and Bernd Emonts.
© 2016 Wiley-VCH Verlag GmbH & Co. KGaA. Published 2016 by Wiley-VCH Verlag GmbH & Co. KGaA.

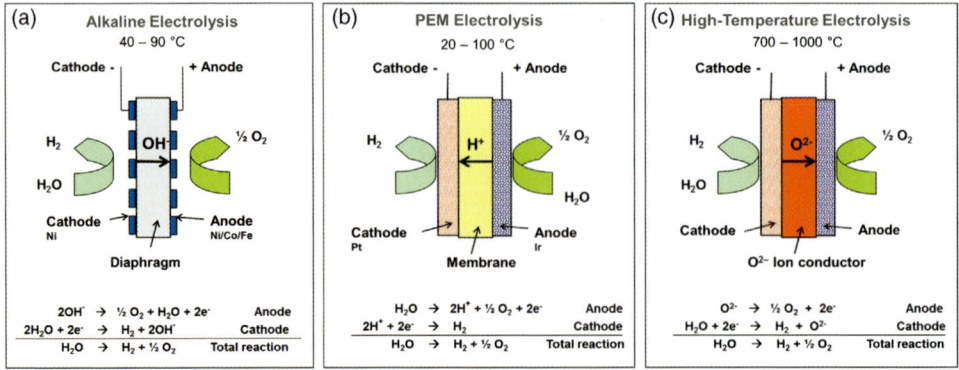

Figure 15.1 Operating principles of the different types of water electrolysis.

technologies already on the market both have advantages and disadvantages. For example, alkaline electrolysis does not require a catalyst made of platinum group metals, whereas PEM electrolysis does. However, this also means that achievable current densities are many times lower for alkaline water electrolysis than for PEM electrolysis (Figure 15.2). The use of polymer membranes makes the gas qualities and the partial load tolerance of PEM electrolyzers more suitable for intermittent operation and strongly fluctuating inputs than alkaline electrolyzers [4].

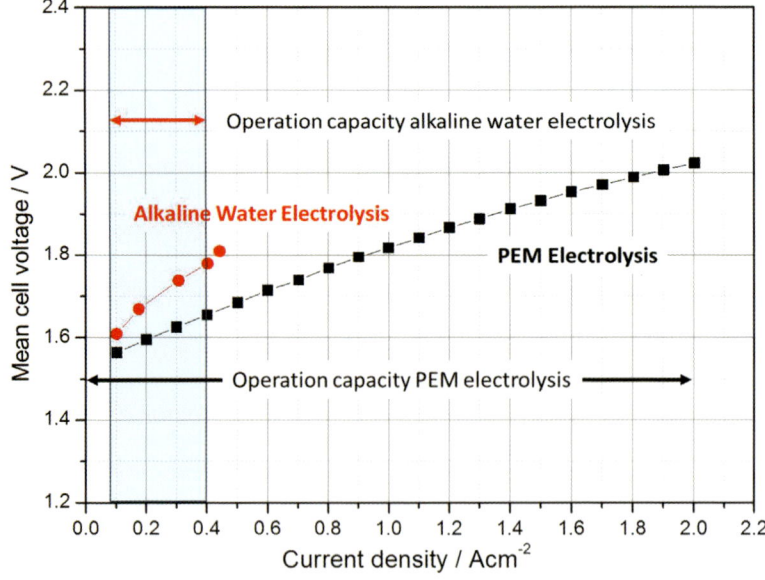

Figure 15.2 Typical current–voltage characteristics and operating ranges for alkaline and PEM electrolysis.

15.1.1
History of PEM Electrolysis

In the 1960s, the use of solid-state polymer electrolytes instead of the liquid electrolytes normally used in alkaline electrolysis led to the development of novel concepts for water electrolysis. The development of solid polymer electrolyte membranes enabled General Electric to realize solid polymer electrolyte water electrolysis (SPE electrolysis) for the first time. The two sides of the membrane were coated with catalysts that served as anode (iridium) and cathode (platinum) electrodes, respectively [5]. In the 1970s, General Electric developed this method from the laboratory scale with a single electrode surface of approx. 50 cm^2 up to 50 kW units with a single electrode surface of 2300 cm^2. They achieved current densities of almost 1.8 A cm^{-2} at a cell voltage of 2 V and an operating temperature of 80 °C using Nafion membranes that were about 250 µm thick [6].

At around the same time, from 1976 to 1989, ABB in Switzerland (formerly Brown Boveri Ltd.) developed PEM electrolysis up to a system size of 100 kW under the name of MEMBREL® [7,8]. In total, two 100 kW systems were installed. These systems were operated by Stellram SA in Nyon (Switzerland) from 1987 to 1990 and by Solar-Wasserstoff-Bayern GmbH (SWB) in Neuenburg vorm Wald (Germany) from 1990 to 1996. The commercial MEMBREL systems consisted of four modules with about 30 cells with an active surface of 20 × 20 cm^2 each, which were arranged in two vertical stacks or, in the system operated by SWB, in one vertical stack [8].

Figure 15.3 shows the components of a 400 cm^2 cell module and the 100 kW system with a hydrogen production rate of 20 Nm^3 h^{-1} at an operating pressure of 1–2 bar. Nafion 117 membranes were normally used, which were coated with

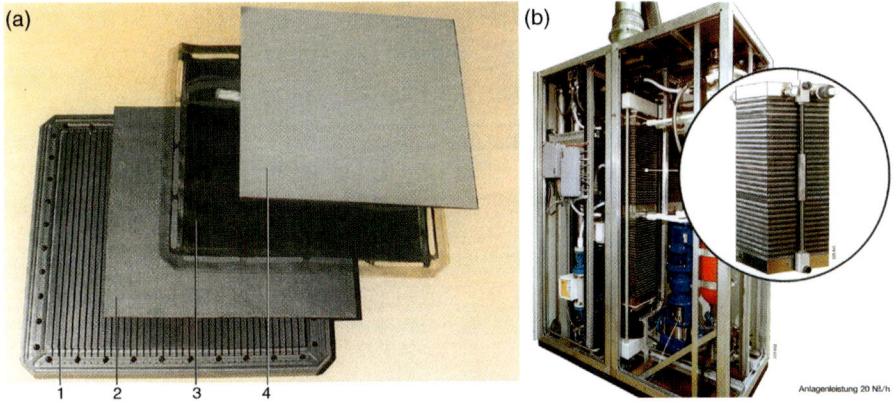

Figure 15.3 Components of a 400 cm^2 MEMBREL cell module (a): 1 = bipolar plate made of graphite/plastic, 2 = cathode current collectors made of graphite/plastic, 3 = membrane, and 4 = anode current collector made of sintered titanium; (b) 100 kW MEMBREL pilot plant for 20 Nm^3 h^{-1} at an operating pressure of 1–2 bar.

platinum on the cathode side and $Ru_xIr_{1-x}O_2$ mixed oxides on the anode side. Under nominal conditions (80 °C, 1 A cm^{-2}), cell voltages of 1.75 V were achieved [8].

In Japan, Fuji Electric Corporate Research and Development Ltd. have developed a PEM electrolyzer with a single-electrode surface of 2500 cm^2 as part of the World Energy Network (WE-NET) project funded by the Japanese Ministry of Economy, Trade and Industry (METI) from 1993 to 1998. With a five-cell short stack, cell voltages of 1.57 V were achieved at 80 °C and a current density of 1 A cm^{-2} under atmospheric conditions [9]. The objective of the WE-NET project was to develop PEM electrolyzers with a single cell area of 10 000 cm^2, current densities of 1–3 A cm^{-2}, and an efficiency of >90% relative to the higher heating value (HHV) of hydrogen. This corresponds to a cell voltage of <1.64 V.

The three projects presented here only generated a few commercial products for hydrogen production, because some of them were terminated prematurely or discontinued. This was undoubtedly because the demand for PEM electrolyzers for the production of hydrogen was insufficient at the time because the production costs were too high in comparison to hydrogen from fossil fuels. However, PEM electrolysis can also be used to generate oxygen at high current density, greatly reducing the systems overall size [5]. General Electric has, for example, used the PEM electrolysis technology as an oxygen generator, in US Navy and Royal Navy submarines.

15.1.2
Challenges Facing PEM Electrolysis

Energy technology is currently undergoing a radical transformation worldwide. The generally accepted factors driving this transition are climate change, supply security, industrial competitiveness, and local emissions. The expansion of renewable energy generation capacity creates new challenges associated with storing large amounts of energy, because increasing amounts of renewable energy lead to a rapid rise in intermittent energy from wind and solar power in the electricity grid. Apart from compressed air energy storage, and electrical and thermal storage, great significance is attached to storage in the form of renewable gases (power to gas) such as hydrogen or methane, with the hydrogen being produced by electrolysis from surplus electricity generated from renewable sources. This trend has led to increased interest in efficient, highly dynamic, cost effective electrolyzers that can be used as an operating reserve or reserve capacity module to stabilize the grid in an energy system based on renewables. The intermittent availability of renewable energies – for example in the form of wind energy – make particularly high demands on the stack and the process technology for the respective electrolysis techniques, especially if they are used to provide an operating reserve and are thus operated at strongly fluctuating outputs and frequent interruptions due to low input. If water electrolysis technology is to be widely and sustainably used for hydrogen production from surplus electricity generated from renewable sources by 2020, further steps are necessary to solve open technical issues. These include inadequate power density, insufficient stability, and excessive costs of the technologies currently employed. In the long term, PEM electrolysis may play a greater role due to its advantages over alkaline electrolysis. However, this requires the technology to be scaled up to, and even beyond, the MW range, and the costs of current PEM electrolyzers to be drastically reduced [4].

15.2 General Principles of PEM Electrolysis

15.2.1 State-of-the-Art

PEM electrolysis with solid polymer membranes (Figure 15.1) has been under development for over 40 years and only a few commercial products are presently available for industrial niche applications. In contrast to alkaline water electrolysis, PEM electrolysis uses platinum group metals for the electrodes, which in most cells are applied directly to the membrane. Noble metals or their oxides are necessary because of the high anode potential and the acidic proton-conducting ionomer used in PEM electrolysis. The high anodic overvoltage for the oxygen electrode is one reason for the increased energy requirements for water electrolysis. It is therefore important to identify an ideal catalyst for oxygen generation in order to minimize energy losses. For example, several studies have shown that oxides such as RuO_2 and IrO_2 are more suitable for oxygen electrodes than the pure elemental Ru, Ir, and other noble metals. This is why IrO_2 is often used as an anode catalyst in PEM electrolyzers, because in contrast to RuO_2 or mixed oxides containing Ru, IrO_2 has excellent electrochemical stability [10]. Carbon-supported platinum catalysts can be used as a cathode. In present-day commercial systems, around 2–4 mg cm^{-2} iridium or ruthenium are required for the anode, and around 2 mg cm^{-2} platinum for the cathode [11].

The structure of a PEM electrolysis stack is shown schematically in Figure 15.4. An electrolysis stack consists of several cells that are electrically connected in series and braced using tie rods and end plates. The catalyst layers on the membrane (CCM: catalyst-coated membrane) are contacted on both sides with thin, porous, and chemically resistant current collectors. Porous sintered compacts made of titanium are usually used on the anode side, while carbon-based material is an alternative for the cathode side. The individual cells are confined on both sides by a gas-tight, electroconductive plate referred to as a separator plate or bipolar plate. The separator plate is joined to the respective catalyst coated membranes and porous transport layers using a frame and appropriate seals. The cell frames and separator plates have the necessary inlets and outlets for water and gases.

Table 15.1 shows the specifications of state-of-the-art PEM electrolysis stacks. At the given operating conditions, these stacks run at voltages of about 2 V,

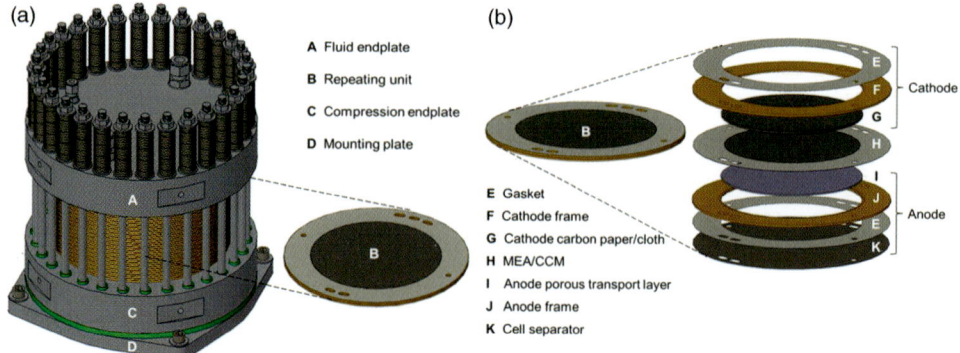

Figure 15.4 Components of a PEM electrolysis stack (a) and its individual components (b). Source; Giner, Inc.

A Fluid endplate
B Repeating unit
C Compression endplate
D Mounting plate

E Gasket
F Cathode frame
G Cathode carbon paper/cloth
H MEA/CCM
I Anode porous transport layer
J Anode frame
K Cell separator

current densities of approx. 2 A cm^{-2}, and operating pressures up to a maximum of 30 bar [12]. Although this corresponds to the same voltage efficiency, of approximately 67–82% (relative to the HHV), as alkaline water electrolysis, in comparison the current densities are much higher (0.6–2.0 A cm^{-2} instead of 0.1–0.4 A cm^{-2}). A comparison of the typical current–voltage characteristics for alkaline and PEM electrolysis with typical operating ranges is shown in Figure 15.2. In contrast to alkaline electrolysis, the lifetime of PEM electrolysis stacks is estimated at <20 000 h. However, Proton Onsite recently achieved a lifetime of more than 50 000 h for stacks used in the PEM electrolyzers of the HOGEN C series, for example [13].

Table 15.2 provides an overview of leading developers/manufacturers of commercially available PEM electrolysis stacks in the 1–300 kW class.

Table 15.1 Specifications of state-of-the-art PEM electrolyzer stacks as reported in the NOW-study [14].

Parameter	Value
Cell temperature	50–80 °C
Pressure	<30 bar
Current density	0.6–2.0 A cm^{-2}
Cell voltage	1.8–2.2 V
Cell voltage efficiency (HHV)	67–82%
Specific energy consumption per stack	4.2–5.6 kWh Nm^{-3}
Lower partial load range	0–10%
Cell area	≤300 cm^2
H$_2$ production rate per stack	Up to 10 Nm^3 h^{-1}
Lifetime of stack	<20 000 h

Table 15.2 Overview of leading developers/manufacturers of PEM electrolysis stacks.

Parameter (units)	Proton OnSite	Giner Inc.	Hydrogenics	Siemens	ITM Power	CETH2	H-Tec Systems
H_2 rate (Nm^3 h^{-1})	30–50	30	227 nominal; 320 overload	5	12	5, 10, and 15	0.3–1.8
Operating pressure (bar)	30	50	30	50	15–80	14 H_2/13 O_2	30
Operating temperature (°C)	35–60	80	60–80	20–90	55	60	60
Nominal current (A)	650–1300	725	3600	—	380	594	60
Number of cells	65–100	100	—	40	75	20, 40, and 60	72
Energy consumption (kWh Nm^{-3} H_2)	4.5–5.0	4.98	4.45	4.5–5.5	4.8	4.4	4.8

15.2.2
Stack Design

To design a PEM electrolysis stack many considerations need to be made. Each component is responsible for several functions and all components are in some way interdependent. This means that with operational or design changes to one component the performance of all other components risks being affected. In the stack design it is important to consider the individual parts, the stack as a whole, and how it will be operated to achieve the best possible performance with the desired stack lifetime. This section will first discuss the individual components as shown in Figure 15.4, then the advantages and disadvantages for various modes of system operation.

15.2.2.1 Catalyst Coated Membranes (CCMs)

Within the realm of water electrolysis, it is the employed electrolyte that differentiates each classification (Figure 15.1). PEM water electrolysis refers to polymer electrolyte membrane (PEM) or proton exchange membrane (PEM) where a proton conductive membrane is used as a solid electrolyte and, at the same time, it works to separate the produced gases at each electrode. The polymer electrolyte membrane macroscopically forms a single solid phase; however, at the microscopic level the membrane consists of two polymer phases chemically bonded to each other. The first chemical structure consists of a hydrophobic backbone resembling that of Teflon® connected to a second polymer structure with a hydrophilic moiety that, when hydrated, results in phase separation, forming continuous interconnected water-containing channels (Figure 15.5). The hydrophilic moieties are sulfonic acid groups that promote the proton transport from one side to the other. The most common polymer that provides good availability, low gas permeability, flexibility, strength, chemical stability, thermal stability, and high protonic conductivity is DuPont's Nafion® membrane.

Figure 15.5 shows the cross-section blueprint of a single cell, where each side of the membrane is in contact with each electrode (anode and cathode), forming the membrane electrode assembly (MEA). When the electrode layer is directly or indirectly coated on the membrane, this configuration is then called catalyst-coated membrane. The electrode layers can also be coated on the surface of the current collectors or porous transport layers and then built together in contact with the membrane, also forming the MEA. In general, for PEM electrolysis, the electrodes are preferably coated on the membranes, instead of the porous transport layers. When coated on the membrane, the electrodes generally present higher mechanical stability against the gas evolution (bubble release). In this case, the catalyst layer has a stronger adhesion to the membrane, and the gas bubbles are preferably released to the outermost side of the catalyst coated membrane and not between the membrane and catalyst layer. When the catalyst is coated on the porous transport layer, gas evolution between the membrane and catalyst layer would accelerate mechanical delamination of the catalyst layer, consequently decreasing the durability and performance (increase of ohmic losses inside the CCM).

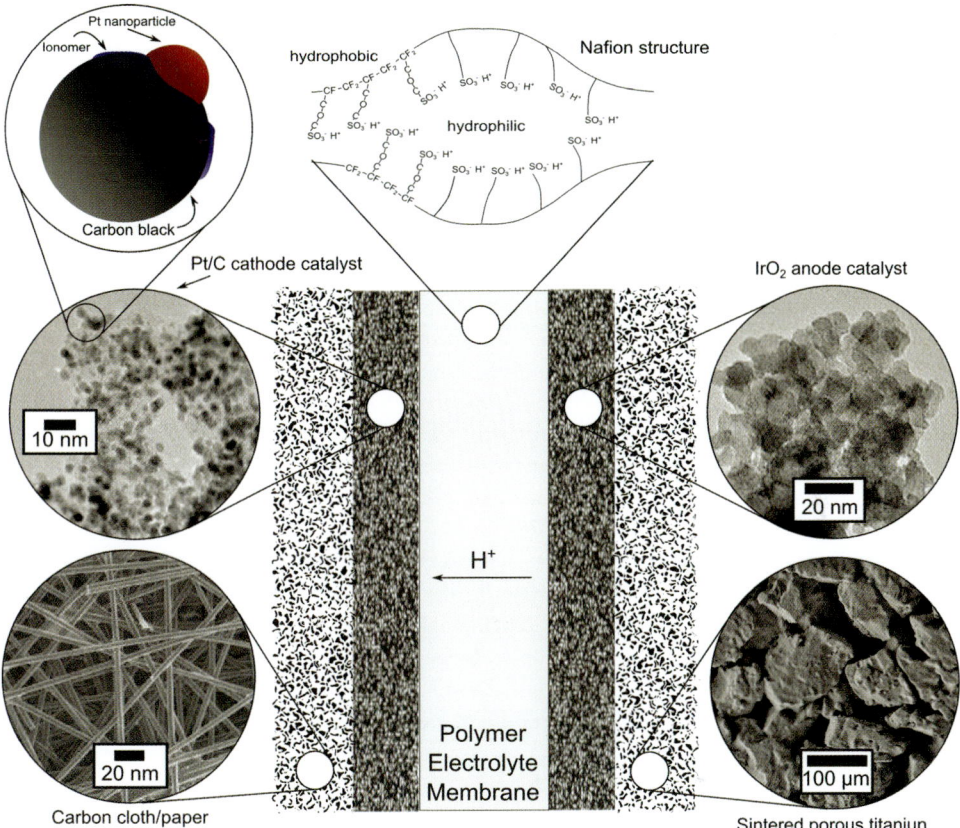

Figure 15.5 Cross-section schematic of a catalyst coated membrane (CCM) of a PEM water electrolyzer. Insets: micro/nano structure for each component obtained either by scanning electron microscopy (SEM) or by transmission electron microscopy (TEM).

Different methods can be used to fabricate the catalyst coated membranes, and each method will have a crucial influence on the catalyst coated membrane structure and will therefore directly affect the performance and durability of the catalyst coated membrane. Some of the available methods include hand brushing, spray coating, blade coating, screen printing, and ink-jet printing. These coating methods can be used to directly coat the polymer membrane with the catalyst layer, or also used to coat an inert layer such as a PTFE foil. The catalyst coated PTFE foil can then be hot-pressed with the membrane, transferring the catalyst layer to the membrane surface, which is the so-called decal method.

A common protocol used to fabricate catalyst coated membranes consists of mixing the catalyst powder (platinum nanoparticles supported on carbon black for the hydrogen evolution reaction and iridium oxide nanoparticles for the oxygen evolution reaction) with a solvent, Nafion ionomer, and additives. The powder catalyst and Nafion ionomer must be dispersed in the solvent, and the mixture must be homogenized and stabilized to a specific viscosity (the viscosity value is a specific requirement for each method). Then, the ink or paste mixture is coated on the membrane or over an inert foil, dried, and the final loading is usually gravimetrically controlled. The main critical steps to obtain high performance and durable catalyst coated membranes normally rely on the ink preparation procedure, coating, and drying step. The catalyst powder must be very well dispersed and homogenized with the Nafion ionomer so that the so-called triple phase boundary (nanoparticle–Nafion, nanoparticle–support, support–Nafion) is well constructed over the membrane. Figure 15.5 also shows the ideal micro- and nanostructure inside the catalyst layer. Since the electrochemical reactions occur at the triple-phase zones, it is desired that the reactants obtain full access to all elements inside the matrix, approaching full catalyst and electrolyte utilization.

Iridium or iridium oxide (IrO_2) is generally accepted as the state-of-the-art catalyst for the oxygen evolution reaction in PEM water electrolysis. Ruthenium oxide (RuO_2) is also highly active for the oxygen evolution reaction, but problems related to its chemical stability (corrosion) restrict its use. Owing to the highly corrosive conditions inside the catalyst layer during electrolysis, carbon based materials cannot be used to support or anchor the iridium oxide nanoparticles. The catalyst support is to maximize the catalyst utilization. Therefore, very high iridium loadings are generally used, ranging nowadays between 2 and 4 $mg_{Ir}\,cm^{-2}$.

The low overvoltage for the hydrogen evolution reaction permits the use of carbon based catalysts on the cathode side of the cell. Platinum supported on carbon black is presently accepted as the state-of-the-art catalyst for the hydrogen evolution reaction in PEM water electrolysis. Today, loadings for the cathode side range between 1 and 2 $mg_{Pt}\,cm^{-2}$, and further reductions will always be desired, aiming at the possibility of reaching values below 0.2 $mg_{Pt}\,cm^{-2}$.

Considering that the state-of-the-art catalyst materials for PEM electrolysis still present negative aspects that cannot be neglected, analogous but advanced catalyst materials and methods could be developed and employed for PEM electrolysis. The present-day technology is still essentially based on the technology developed in the last century. Well-grounded knowledge is already available in many publications related to PEM electrolysis, fuel cells, and electrochemical system catalysts. This knowledge could potentially be transferred to PEM water electrolysis as well in order to further develop this technology.

15.2.2.2 Porous Transport Layer (PTL)

The porous transport layer (PTL) and carbon paper/cloth are responsible for the transport of all the necessary components of the electrolysis reaction. The reactant water is supplied to the catalyst layer through the pores, the electrons are supplied to and from the catalyst layer through the rigid body, and the product oxygen in the gas phase is removed from the catalyst layer through the same pore space the liquid water is entering. These competing mass transport mechanisms can lead to unwanted increases in cell voltage and reduced overall performance.

Figure 15.6 illustrates the gas flows and developments from the carbon paper/cloth (cathode side) and porous transport layer (anode side) through the flow

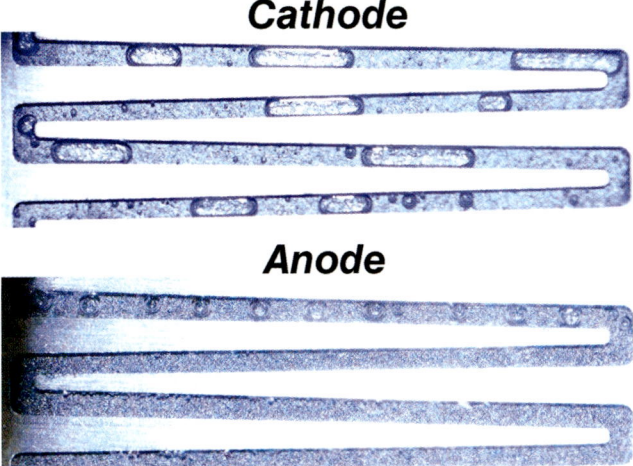

Figure 15.6 Gas flow from carbon paper/cloth (cathode) and porous transport layer (anode) through a serpentine flow channel supplied with water from the bottom.

channels in a 17.64 cm^2 cell operating at 0.6 A cm^{-2} equipped with transparent endplates. Owing to the nature of these flows it is important to select a porous transport layer with properly sized pores. If the pores are too small, transport losses affect the overall performance at lower current densities causing a loss of efficiency and limiting the operating range. If the pores are too large the porous transport layer may not support the delicate membrane adequately, increasing the chance of mechanical failure of the membrane. Figure 15.6 also shows that on the anode side the flow regime changes from a predominantly liquid flow to a bubbly flow, and at higher current densities or larger cells the flow transitions all the way to a slug flow like that seen on the cathode side.

The porous transport layer is also responsible for the conduction of electricity from the separator plates to the catalyst layers. The shape of the sintered particles or mesh used within the cell affects the conductivity of this layer. With a flatter contacting surface the porous transport layer offers higher conductivities and, thus, reduced electrical ohmic losses.

Figure 15.7 gives an example of the available powder based sintered porous transport layers, each of which have their own advantages and disadvantages. The spherically shaped particles demonstrate the best performance under high current densities due to the delayed onset of mass transport losses; however, electrical conductivity and material thickness tolerances of the pressed porous transport layers are typically the highest when comparing the sintered powder. Woven titanium meshes and expanded metals have been used as alternative flow fields and porous transport layers – examples of such materials can be seen in Figure 15.8. As for a substitute for flow fields, woven titanium cloths (Figure 15.8b) can be used in small cells; however, as the active area increases, more traditional flow fields are required, such as those found in the PEM fuel cells.

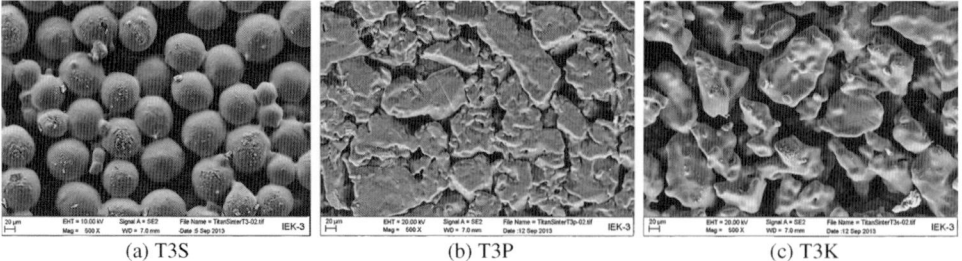

Figure 15.7 SEM micrographs of three different sintered powder porous transport layers with a similar particle size range of 125–200 μm: (a) T3S is an uncompressed, sintered titanium particle spherical in shape; (b) T3P is a pressed, sintered titanium particle of irregular shape; (c) T3K is an uncompressed, sintered titanium particle of irregular shape.

The porous transport layers exhibit a range of workability and good strength properties depending on the choice material. These properties are especially important to consider when operating a differential pressure as the membrane is pushed into the porous transport layer from the larger pressures on the cathode side. The sintered titanium porous transport layer behaves under mechanical compression much like solid titanium for the range of pressures foreseeable (0–350 bar) in a PEM electrolyzer [15]. In terms of workability, the sintered titanium porous transport layer is a bit more difficult than the expanded metal and wire mesh solutions as milling operations smear the pores closed and precision cutting requires the use of laser or water jet (Figure 15.9).

One of the biggest performance issues facing the uncoated porous transport layer is the ever-growing passivation layer that develops on the surface of even the titanium porous transport layers. This layer acts as an electrical insulator and continually reduces performance over the cells life. To avoid this phenomenon some manufactures coat even the porous transport layer in gold, improving

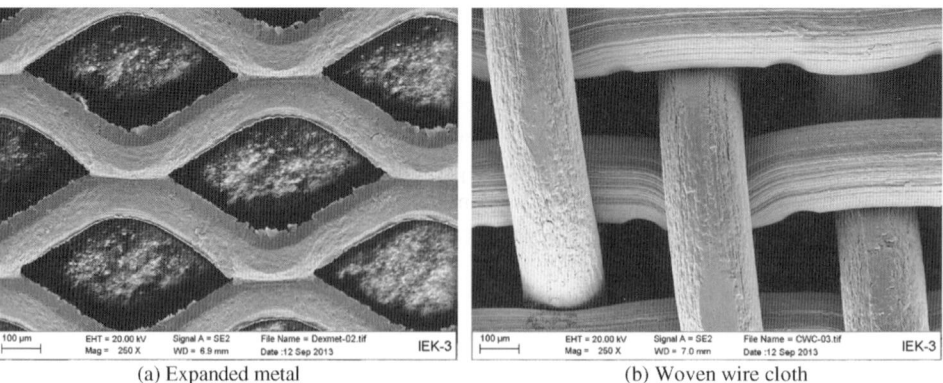

Figure 15.8 SEM micrographs of (a) expanded titanium foil and (b) woven titanium mesh.

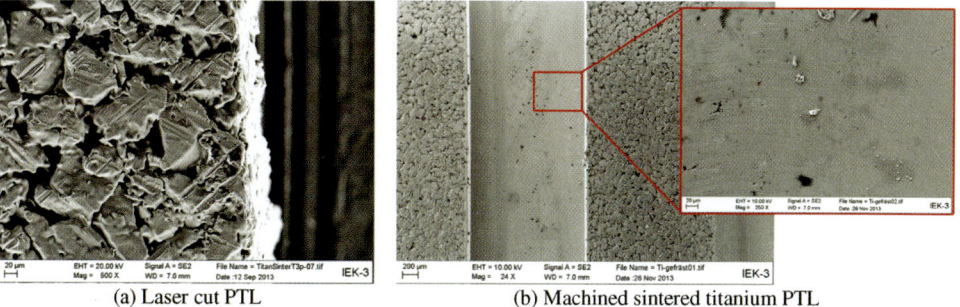

Figure 15.9 SEM micrographs of (a) laser cut edge of a sintered titanium porous transport layer and (b) machined channel in sintered titanium porous transport layer.

performance and lifetime but drastically increasing the cost of one of the cells already most expensive components [16].

15.2.2.3 Separator Plates

The separator plate (Figure 15.4) is responsible for separation of the multiple cells within a stack, and in most cases also for the even supply of water over the porous transport layer. As the active area of the cell increases this component is also helpful in reducing the pressure drop needed to circulate the reactant/cooling water.

Titanium offers the best combination of conductivity, corrosion resistance, and strength for the separator plate to avoid voltage induced corrosion on the oxygen-rich anode side. To avoid highly resistive passivation layers and embrittlement leading to higher overvoltages, titanium separator plates are coated with platinum on the anode side and gold on the cathode side. This combination causes the separator plates to be, along with the porous transport layer, one of the most expensive components of the PEM electrolysis stack [16].

To save costs and reduce ohmic losses the titanium separator plates are typically made as thin as possible with the minimum noble metal coating thicknesses possible. The lower limit to plate thickness depends on the channel/flow field design, assembly pressure, operating pressure, and manufacturing process employed. Less expensive coatings such as, titanium nitride, carbon, and epoxy with conductive gold points are also under consideration; however, the stabilities and conductivities of these coatings are not presently comparable to gold and platinum. Separator plate designs have used milled, stamped, and hydroformed techniques resulting in material thicknesses from as much as 5 mm in the milled plates to as thin as 0.1 mm in the hydroformed and stamped plates. Figure 15.10 illustrates several potential flow field designs for a square electrolysis cell. Of course, there is no limit to the number of potential designs available for flow fields; however, they all share similar properties to one of the flow fields shown in Figure 15.10. Many commercial cells in the past have been designed with a circular shape, which helps with high pressure operation but, however, increases

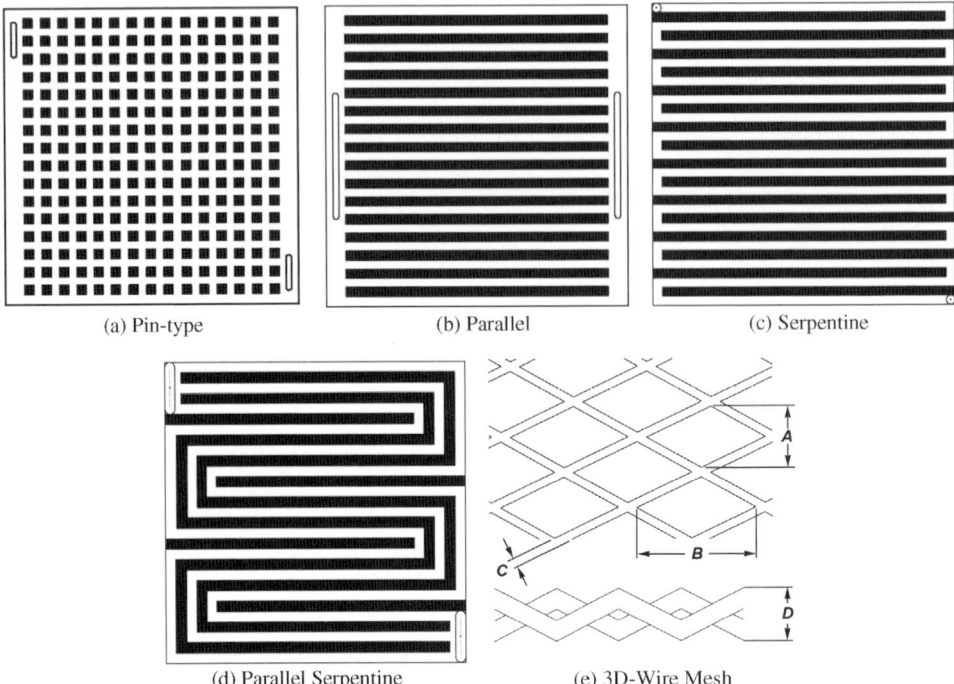

Figure 15.10 Examples of typical flow field designs: (a) pin-type flow field, (b) parallel flow fields, (c) serpentine flow field, (d) parallel serpentine, and (e) 3D-expanded titanium mesh (A–D are the controllable parameters).

waste of the individual components that must be cut such as the membrane, gas diffusion layer, porous transport layer, and even the separator plates. The design considerations between rectangular and round cells will be discussed in more detail in Section 15.2.3.

Separator plates with machined flow fields have the most flexibility in design. These flow fields have very few limitations in terms of channel size, land width, and flow path. The flow fields can even have different forms from the anode side to the cathode side. The major drawback to machining flow fields in the separator plate is the cost. Machining each individual part requires large processing times and incurs rather large material waste as well as increased material thicknesses, ultimately increasing ohmic losses. Alternatively, the stamped and hydroformed separator plates offer great material savings as well as speed of manufacture. In most cases the stamped flow field will have an inverse design on the cathode to that of the anode; this limits many design parameters and reduces the overall available contact area. Differing geometries on the cathode and anode can be obtained with stamped or hydroformed separator plates. This, however, then requires two layers and a welding process, which results in reduced conductivity and increased manufacture time.

15.2.2.4 Stack Operation

The PEM electrolysis system can be operated in many different modes: balanced pressure, differential pressure, and with or without water circulation on the cathode side. Each of the modes of operation has advantages and disadvantages and although a stack could be developed to handle all modes, with stack cost and performance in mind, they are typically designed specifically for only one mode of operation.

Balanced Pressure Operation

The balanced pressure mode of operation for a PEM electrolyzer means that the pressure on the anode and the cathode of the cell are equal. This equal pressure on both sides can either be atmospheric pressure or some elevated pressure.

In the balanced atmospheric pressure mode of operation the PEM electrolyzer stack has the simplest of designs and results in low mechanical stresses on the individual components, helping to reduce overall bulk of the system as well as any expensive strengthening strategies, such as operating a stack within a hyperbaric chamber. When operating at balanced atmospheric pressure the membrane is exposed to very low pressures, reducing the chance of possible tears or punctures that could lead to catastrophic failures [17]. The seals are also only responsible for pressures no higher than that of the hydraulic head of the water circulation.

The balance of plant in atmospheric operation can, however, become quite expensive. Since the hydrogen is usually stored at elevated pressures, ranging from 30 to 750 bar depending on its application, additional external compression is then required with an ambient system. Lower pressure hydrogen can also hold more water vapor, leading to additional system components, costs, and balance of plant efficiency losses to obtain the often necessary high purity hydrogen.

At elevated pressures the balanced pressure operation still offers good protection to the membrane. Since both anode and cathode have equal pressure the membrane is stressed in neither direction, while the high pressure within the cell can still be counter balanced by keeping the cell in a high-pressure chamber. This reduces the need for strengthening the individual components of the stack to be able to withstand the high pressures within the stack. Keeping the anode under pressure also delays the onset of mass transport losses, because the increased pressure reduces the volume of the product oxygen gas, keeping it from interfering with the delivery of water to the catalyst layer. Despite the low stresses on the membrane the safety aspects are not necessarily better as high-pressure oxygen in contact with the sintered titanium powder and the seal materials can be dangerous and cause the cell to degrade at higher rates.

Differential Pressure Operation

The differential pressure mode operation is when the cathode side of the PEM electrolyzer is operated at an elevated pressure. The anode side could be either operated at atmospheric or any pressure below that of the cathode. The differential pressure mode attempts to combine the advantages of both the balanced pressure modes.

As was the case with the balanced pressure mode the water vapor in the hydrogen gas is reduced at higher cathode pressures; thus, with a differential pressure operation the gas drying process can be reduced, decreasing the cost of the gas drying components and the balance of plant efficiency losses. The higher pressure does, however, decrease the Faraday efficiency from an increase in gas crossover, where the high-pressure hydrogen permeates through the membrane from the cathode to the anode side.

This gas crossover can create safety issues when the molar percent of hydrogen on the oxygen side reaches the lower explosive limit (LEL) of 4.0% hydrogen in oxygen. Because the crossover rate remains relatively constant, the low gas production at lower current densities will increase the overall percentage of hydrogen in oxygen. This means that the higher the differential pressure the smaller the part-load operational range is for the system [18].

In addition to the savings on the gas drying, the differential pressure mode reduces the need for additional compression. Proton OnSite has demonstrated a stack electrochemically compressing the product hydrogen to 350 bar with only a 3.1% reduction in stack performance when compared to a 13 bar system. The demonstrated stack was, however, limited in the part load range to above 0.6 A cm^{-2} to maintain a safe level of hydrogen in oxygen due to the gas crossover [15].

With high differential pressures it becomes difficult to manage the internal pressure with the normal components that are typically made of plastic, specifically the frame that holds the porous transport layer and sealing gaskets. To manage this a metal frame only on the cathode side is often chosen. Since the frame on the anode side does not see high pressure the plastic can remain, ensuring there are no short-circuits and the metal frame on the cathode side can then handle the high pressures without adding substantial amounts of material. This solution solves the pressure issue; however, it drastically increases the stack cost by adding a considerable amount of titanium to the stack.

Water Circulation

Controlling the water circulation through a stack is another operational parameter; a stack can run with the water circulation on only the anode side, or on both the anode and cathode side. In most cases, where there is no circulation on the cathode side the cathode is filled with liquid water and left without any circulation. This is possible because water is only required on the anode side of the reaction as shown in Figure 15.1b.

The flow rate of water through the stack on the anode side has little effect on the overall performance. Figure 15.11 illustrates a single cell PEM electrolyzer operating with varying water supply rates. At lower flow rates a reduction in performance is noticed due to the limited amount of water reaching the catalyst layer in comparison with the large amount of oxygen being developed. Faraday's law can explain the molar rate at which water is reacted in the PEM electrolyzer:

$$\dot{N}_{H_2O} = \frac{I}{2F}$$

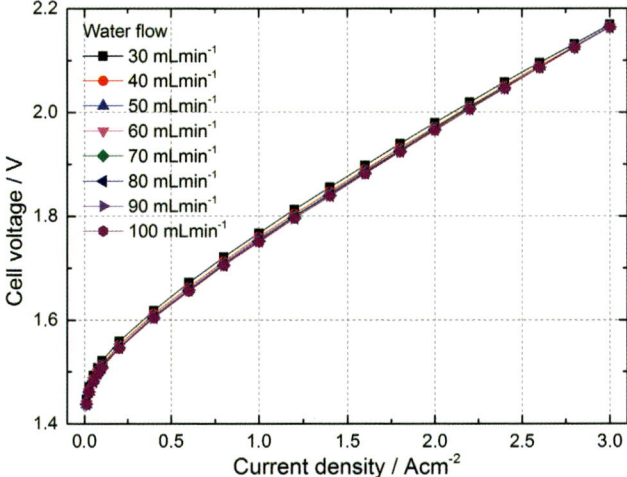

Figure 15.11 Performance of a 25 cm² PEM electrolysis cell under varying flows on the anode side with pin-type flow field.

where N_{H2O} is the rate of water produced (molar basis), I is the applied current, and F is the Faraday constant (96485.3 C mol⁻¹).

The water consumption rate calculated by Faraday's law is the amount of water that is consumed at a given current; the circulated water needs to be supplied at a much larger rate to ensure maximum catalyst utilization. For the cell shown in Figure 15.11 a flow rate of 20 mL min⁻¹ begins to demonstrate a loss due to insufficient water supply and in this case 35 times the required water for the reaction is being supplied to the cell at 4 A cm⁻². A typical system will supply around 50–300 times the water required at higher current densities [19,20]. This is to ensure enough water is reaching the catalyst and to control the cell temperature.

Operating a stack without circulation on the cathode side helps to reduce the system cost with little to no impact on cell performance (Figure 15.12). Some commercial systems will operate with the cathode side humidified, some with only stagnant water on the cathode and others with full circulation of water on the cathode side.

The systems operating with a full circulation on the cathode side bear the added cost of the additional peripheries and processes. The circulation itself will need a pump and the gas drying process will be more intensive than a system without water on the cathode side. Systems are chosen with circulation on the cathode side to ensure the membrane stays properly hydrated and the stack does not overheat. Systems with water on the cathode side, but no circulation still require additional work in the gas drying process; however, they save on the pumping losses and keep the membrane fully hydrated. Systems operating with a dry cathode side save on both gas drying and the pumping losses; however, they are subject to membrane dry out and the formation of hotspots that could

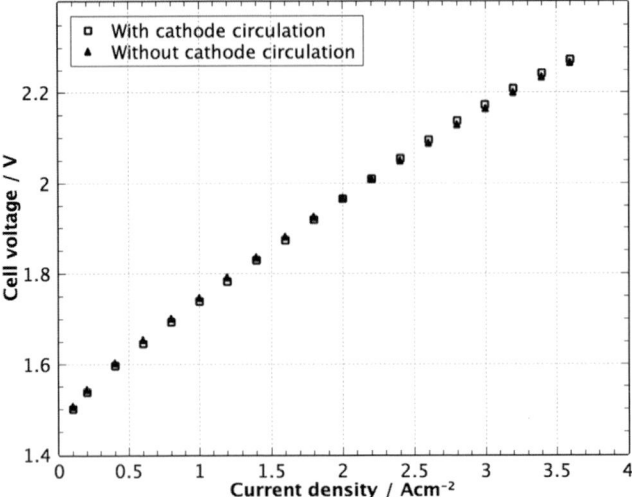

Figure 15.12 Performance of a 25 cm² PEM electrolysis cell with and without cathode water circulation.

eventually cause the stack to fail. With water on the cathode side the system will have a longer lifetime, but initial costs and operational costs will both be higher, so the optimum operational mode will depend on the system application.

Thermal Management
The control of temperature in PEM electrolysis is crucial to optimization of performance and stack lifetime. By simply increasing temperature the performance of a cell can be greatly improved. Figure 15.13 shows an example of a cell's performance at various temperatures at 1.85 V, a typical operational voltage to keep corrosion to a minimum; the current density nearly doubled from 30 to 80 °C.

Balancing the generated heat, removed heat, and supplied heat is an important balance to maintain a stable and efficient stack. The heat generated in the stack comes from the electrolysis reaction, and is a function of the rate of hydrogen production and system overvoltages (irreversibilities). The equation shown below calculates the heat production per cm² of active area of a single cell within the stack:

$$\dot{Q} = \left[\frac{(E_{cell} - E_{therm})2F}{\Delta H}\right]$$

where Q is the heat, E_{cell} is the cell voltage, E_{therm} is the thermal neutral voltage, F is the Faraday constant, and ΔH is the enthalpy change.

Heat is also generated through Joule heating; this heat generation originates from the resistance in the cell and, thus, is dependent on the materials, material

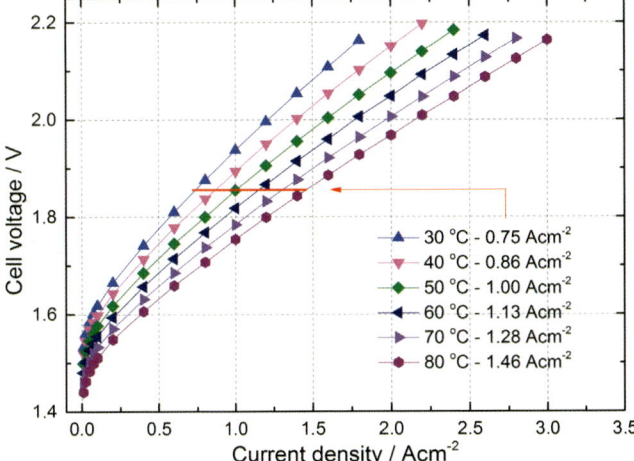

Figure 15.13 Performance of a 25 cm² PEM electrolysis cell operated at temperatures ranging from 30 to 80 °C.

thicknesses, and contact area of the cell components. The heat developed from Joule heating can be described by the following equation:

$$\dot{Q} = I^2 R_{cell}$$

where Q is the heat, I is the current, and R_{cell} is cell resistance.

These heat generation terms are then offset by the circulating feed water, which is temperature controlled to ensure the stack operates at the desired temperature. Owing to the heat capacity of liquid water the circulating feed water is capable of removing or adding large amounts of heat from the cell, thus allowing for stable temperature control even under very high loads. This attribute of the PEM electrolyzer removes nearly all need for any additional cooling flows, even when the cell is operated without water circulation on the cathode side.

15.2.3
Future Trends in PEM Electrolysis

As the need for renewable energy increases more and more, emphasis will be put on the development of an energy storage solution. This is mainly due to the intermittent nature of renewable energies like wind, solar, and even hydroelectric. PEM electrolysis offers an elegant solution to capture all available energy from the installed renewable resources and store it as hydrogen. Once stored as hydrogen, this energy could be either converted back into electricity through fuel cells and reinserted into the grid or injected into the transportation sector as a means of supplying fuel for zero-emission fuel cell vehicles.

Current commercial PEM electrolysis system manufactures have already demonstrated long lifetimes (>60 000 h^{-1}), high operational pressures (350 bar), high efficiency (80–90% relative to HHV, 65–75% relative to LHV), and Megawatt scale systems. Moving forward the biggest challenges remaining are to maintain this performance, increase the system load capacity, and reduce the cost of the system. Electrical grid loads reach well into the Gigawatt range and will likely only increase in the future, therefore electrolysis systems need to be scaled up to match these demands, all the while reducing the cost of produced hydrogen.

15.2.3.1 Cost Reduction

Despite PEM water electrolysis having been known for decades, a direct cost reduction approach on stack components and materials is found only sporadically in the literature. However, to achieve a competitive market position, especially when compared to alkaline electrolyzers, the costs for PEM systems must be reduced, focusing essentially on system design for manufacturability, scale up in capacity, and pressure capability [21].

Catalyst Coated Membranes (CCMs)

The cost breakdown for PEM electrolyzers stacks includes separator plates and flow fields, MEAs or CCMs, balance of stack, and balance of cell. It is well recognized that the flow fields and catalyst coated membranes drive the main costs of the stacks [16]. The catalyst coated membranes can be further divided into membrane and catalysts, and the high costs of the noble components and fabrications methods will be directly affected by their availability (abundance), manufacturability, efficiency, durability, and scale-up in capacity.

In terms of electrolyte materials (or membrane), PEM electrolyzer stacks mostly revolve around Dupont's Nafion® based membranes despite its high cost ($500–1000 m^{-1}) based technology. Other companies today also provide Nafion®-like membranes, and also hydrocarbon membranes, such as fumatech, 3M, Gore, and so on. Further development of electrolyte membranes can assist to reduce the costs of the membrane materials for PEM electrolysis. Various different approaches can be used and some of them can be listed as follows: (i) modifying PFSA (perfluorosulfonic acid) membranes or hydrocarbon based membranes to enhance their proton transport characteristics; (ii) decrease the gas crossover by crosslinking the polymer structure, or by forming composites with hygroscopic oxides, like MO_2 (where M = Zr, Si, Ti) or with inorganic proton conductors by designing inorganic organic composite membranes; (iii) increase the mechanical and chemical stability of the membrane at higher differential pressures and long-term operation.

PEM water electrolyzers have to overcome the barrier of high catalyst cost caused by the exclusive use of platinum and iridium based catalysts, in order to further reduce the costs of the catalyst coated membranes. The slow kinetics related to the electrochemical reactions on the anode side (oxygen evolution reaction) has concentrated R&D effort on developing new materials for the anode side for PEM water electrolysis. The large overpotential contribution coming from the anode side and the high corrosive conditions demand the use of unsupported noble iridium catalysts with high loading and poor catalyst utilization. Moreover, iridium has one particular limitation – it is one of the rarest elements in the Earth's crust, having an average mass fraction of 0.001 ppm in crustal rock. Conversely, gold and platinum are 40 times and 10 times more abundant, respectively. The major commercial sources of iridium are found in pyroxenite and the sulfide ore laurite in South Africa, as well as pentlandite from nickel $(Fe,Ni)_9S_8$ mining regions in Russia and Canada. Iridium demand has recently increased due to its use in crucibles employed to fabricate LEDs for smartphones, tablets, televisions, and automobiles. It is consequently expected that a high penetration of PEM electrolysis technology into the market would considerably affect the demand for iridium and consequently the price. Since the first studies on electrocatalysts for PEM electrolysis by General Electric in the 1960s, research groups have tried to overcome the high cost of iridium catalysts with numerous alternatives. The main focus relies on dissolving the IrO_2 particles or clusters in a second metal oxide matrix, forming a solid solution, to stabilize the catalyst structure against corrosion, improve iridium utilization for the oxygen evolution reaction, enhance efficiency, and reduce noble metal loading [10].

Porous Transport Layer (PTL)

Being one of the most expensive individual components in an electrolyzer (along with the separator plate) the porous transport layer has a lot of room for cost reduction [16]. The source of the PTLs high cost is the titanium powder and the thickness required by the typical sintering processes, gravity or pressure assisted sintering. By reducing component thickness the cost of the porous transport layer can be reduced; however, as cell areas expand the gravity and pressure assisted sintering processes require larger component thickness, thus leading to a necessary size optimization or selection of a different material/process.

Reducing the porous transport layer thickness by employing different sintering techniques such as metal film casting is a viable option. The gravity and pressure assisted sintering techniques offer a material thickness of as small as 1 mm while metal film casting can obtain material thicknesses down to 0.25 mm. This thickness reduction alone could reduce the raw material costs by a factor of four; however, when reducing the thickness the particle size must then be reduced. A reduced particle size from 75 to 45 μm increases the raw material costs by at least 10% and decreases the porosity and pore sizes in the porous transport layer. This can then result in mass transport losses and a reduction in the potential operating range. The metal film casting option also allows for potentially larger surface area porous transport layers due to the process allowing continuous operation in one direction (no limit to length of porous transport layer, only height and width). For the cost savings due to a reduction in thickness and an increase in active surface area to be realized the stack needs to be designed to accommodate the larger transport losses experienced in such a porous transport layer.

Substitution of the metal powder with a structured layer such as expanded metal or a woven metal cloth is another possible avenue for the reduction of material costs. These materials are easier to coat, leaving options open for different base materials, easier to cut (reducing manufacturing times), and are less expensive than the sintered powder option. The expanded metal and woven meshes do, however, pose a problem with minimum pore size as well as mechanical stability. With the smallest achievable pore sizes in the 100 μm diameter range the expanded and woven metal meshes cannot operate at larger differential pressure without endangering the membrane in terms of ripping or tearing. The flexibility of these layers also creates problems under differential pressure operation. This lack of structural integrity and membrane support, especially in larger cell areas, would result in a large deformation and stress in the membrane in the center of the cell.

Replacing titanium with a coated, less expensive base metal is another possible solution currently being explored. By replacing the titanium in the porous transport layer with a coated stainless steel the base metal cost could see an order of magnitude reduction and, depending on the coating, the passivation of the surface could be reduced, yielding even better performance over the lifetime of the stack. The success of this solution in reducing the cost of the cell depends on the cost of the coating and more importantly on the stability of the coating. If cracks and holes exist in the coating material the stainless steel base metal can corrode in the harsh anode atmosphere developing oxygen under the coating, leading to delamination of the protective coating. Even further cost reductions could be made by employing a combination of the aforementioned changes to the porous transport layer.

Separator Plates

The separator plate typically utilizes titanium as the substrate material like the porous transport layer, and has many of the same cost reduction possibilities. Material substitutions, coatings, thinner materials, and designs that reduce processing time such as the transfer of the flow field from the separator plate to the porous transport layer can all reduce the cost.

For the separator plates the coating process is easier due to the simplified geometry and absence of a complex pore structure. This allows for more coating options with less expensive processes than available to the porous transport layer. Despite the increase in possibilities the same corrosion/delamination problems remain when the applied coatings have imperfections. The coating materials can also be very expensive and add to the cell losses by introducing more ohmic losses; Table 15.3 shows some potential coating candidates and their properties. Graphene coatings are a potential substitute for the typically gold coated cathode, resulting in cost reduction; however, due to the large difference in thermal expansion coefficients with most substrate materials the lifetime in dynamic operation will be sacrificed. The PTFE coating, although not conductive, has been implemented with an array of gold dots that provide a conductive path to the substrate. This reduces the amount of gold needed to create a conductive separator plate and protects the substrate well against corrosion.

The use of thinner materials and designs reducing manufacturing time can present significant reductions in the overall stack cost. Owing to the need for flow fields to facilitate even water distribution in cells with larger active areas a

Table 15.3 Potential coatings to protect non-noble base metal separator plate substrates.

Parameter	Coating					
	Au	Ta	Nb	TiN	$C_{graphene}$	PTFE
Coating material	Metal	Metal	Metal	Ceramic	Non-metal	Polymer
Specific resistance [22,23] ($10^{-8}\,\Omega\,m$)	2.2	12.5	15.1	25	1.04 [24]	1.00×10^{6} – 1.00×10^{10}
Thermal expansion coefficient [22,23] (K^{-1})	14.4	6.5	7.1	9.4	−4.8 [25]	14–250
Commodity price ($€\cdot g^{-1}$)	39.10 [26]	0.07 [27]	0.01 [27]	1.70 [28]	—	8.80 [28]
Coating process	PVD/galvanic	PVD	PVD	PVD	CVD	Injection molding
Advantages	• Low ohmic resistance; • acid-resistant; • similar CTE to stainless steel[a]	• Corrosion-resistant; • acid-resistant	• Corrosion-resistant; • acid-resistant	• Corrosion-resistant; • acid-resistant	• High electric conductivity	• Highest corrosion resistance
Disadvantages	• Cost-intensive; • heavy price fluctuation	• Cost-intensive material and coating process; • deviant CTE compared to AISI 304[a]	• Expensive coating process; • deviant CTE compared to AISI 304[a]	• Expensive coating process; • deviant CTE compared to AISI 304[a]	• Expensive coating process; • not resistant under high potential	• Isolator

a) Coefficient of thermal expansion (CTE) (AISI 304)$_{20-100\,°C} \approx 16$ [29].

flow field must be incorporated. To include the flow field in the separator plate, much like the flow field of a PEM fuel cell, the plate thickness must be at least 2–3 mm while the stamped and hydroformed components can easily reach thicknesses as small as 0.1 mm. Avoiding the incorporation of the flow field in the separator plate can, however, drastically reduce the manufacturing costs, Figure 15.10e illustrates a 3D expanded metal mesh that can function as a flow field, thus removing the need for expensive machining or even stamping processes.

Cell Design
PEM electrolysis systems have many design considerations that influence the cost, many of which become tradeoffs between performance, lifetime, or manufacturability. As the active area of the cells increases these considerations become increasingly important. For small-scale electrolyzers the material waste, flow fields, mass transport, and sealing against higher internal pressures all play a roll in the overall stack cost.

The commercially sold electrolysis stacks have been historically predominately round. The round stack has the advantage of being easier to design than its rectangular counterpart. The round design is better equipped to deal with higher pressures and can more evenly distribute water than the rectangular design. The rectangular design can also handle such conditions; however, the design will need to be more carefully chosen to ensure even distribution of water to the entire catalyst layer, as well as additional support for the high pressures seen within a stack. As a round cell's active area expands, the amount of material waste (for separator plates, membrane, and sometimes porous transport layers) increases linearly. This is because most components that are made on the larger scales are rectangular; this means the typical round cell design will waste 1.27 times more material than a rectangular cell. With the price of the sintered titanium porous transport layers, membranes, and separator plates dominating the cost of the stack this waste plays a critical role in keeping stack costs to a minimum.

The ever-increasing demand for green energy has put great emphasis on increasing the active area of a PEM electrolysis cell. This is due to the potential cost savings on the periphery system equipment that are available when increasing the size of the individual stacks, in both number of cells and active area. This has caused commercial cell manufacturers to shift their next-generation PEM electrolyzers toward the rectangular design in order to reduce the amount of waste and effectively decrease the cost of the stack. The next generation of large-scale electrolyzers will migrate to the rectangular design solely on the basis of material cost and waste reduction.

Modular System Design
Despite the possible cost savings associated with assembling larger stacks with more cells, the modular design also has its benefits. With a modular design, instead of producing a single stack on the MW scale, this load could be split into multiple stacks each capable of a portion of the MW load. This allows the

components to be held at higher tolerances without greatly increasing the difficulty of manufacture. Splitting the MW load into multiple stacks also leaves the option of having different components in each stack, to help maintain high purity and a large operational load range with short dynamic operation response time.

Another strong argument for the modular system design is the component size. As the size of the components like the porous transport layer and the membrane reach a certain size the cost to produce increases drastically. This is mainly due to the present manufacturing sizes of the membrane and the porous transport layers. By simply increasing the number of cells and adding capacity to a system by adding additional electrolyzers the system then has the flexibility for an even larger operational range by toggling the various stacks on and off in accordance to the supplied load. This keeps gas crossover to a minimum and allows for an increased stack lifetime.

15.3 Summary

Advances in PEM stack technology will play a significant role in the future of the hydrogen energy economy. The historical feats and progressions should also not be overlooked as from the very start long lifetimes (20 000 h), low current densities (1.57 V @ 1 A cm^{-2}), and large cell areas (2500 cm^2) were achieved, with the drawback of high production cost. With the technological advances in the manufacturing of thinner supported membranes, thinner more easily manufactured separator plates, higher catalyst utilization catalyst layers, and larger size and lower cost porous transport layers, the system cost can be brought down to a commercially feasible level. With reduced investment cost the PEM electrolyzer exhibits an elegant solution to couple with intermittent green energies such as wind and solar to capture the off-peak energy and store it in large quantities as hydrogen.

The up-scaling and cost reduction of PEM electrolysis systems has become one of the greatest challenges delaying its introduction as an energy storage solution. The largest commercially available PEM electrolysis stacks are in the hundreds of kilowatt range. The present-day grid demand ranges near 100 GW and is continually growing, leaving a large gap to be filled (at least 40 GW of installed capacity assuming overload capability) in order for this technology to be used as a grid load balancing solution.

Present-day state-of-the-art the PEM electrolyzers show a lot of promise. The cell-level performance and gas purity levels are unrivaled by other electrolysis modes, the commercial systems have been increasing in size, efficiency, and lifetime, and from the advances in coating materials, manufacturing processes, and catalyst development a lot of room for improvement in stack cost still remains. Increasing the size of the individual cell while removing as much titanium, gold, platinum, and iridium as possible will be the key to reducing the system cost. However, due to manufacturing limited material widths for certain components (membrane, porous transport layer, carbon cloth/paper) the systems may need to consist of multiple smaller stacks to minimize cost. This is a balance that will always be changing with the currently available materials. The system controls in a modular electrolysis plant can be established to handle very fast and very large fluctuations in input load due to the rapid response time of the PEM electrolyzer from standby to operation so this poses no threat to overall system performance.

Intermittent renewable energy appears to be here to stay, and to maximize its potential an efficient expandable and dynamic energy storage system needs to be introduced. PEM water electrolysis is currently the best-suited technology for maximizing utilization and balancing grid loads with the installed intermittent renewable energy resources. As research and development brings larger and less expensive PEM electrolysis systems to the market a truly green hydrogen economy could soon be realized.

References

1 Wöhrle, D. (1991) Wasserstoff als energieträger – eine replik. *Nachr. Chem. Tech. Lab.*, **39** (11), 1256–1266.
2 Winter, C.J. (1989) *Wasserstoff Als Energieträger: Technik, Systeme, Wirtschaft*, 2nd revised and expanded edition, Springer, Berlin.
3 Kuron, D. and Graefen, H. (1985) Avoidance of corrosion damage at water electrolysis plants. *DECHEMA-Monogr.*, **98** (Tech. Elektrolysen), 329–339.
4 Mergel, J., Carmo, M., and Fritz, D.L. (2013) Status on technologies for hydrogen production by water electrolysis, in *Transition to Renewable Energy Systems* (eds D. Stolten and V. Scherer), Wiley-VCH Verlag GmbH, Weinheim, pp. 425–450.
5 Russell, J.H., Nuttall, L.J., and Fickett, A.P. (1973) Hydrogen generation by solid polymer electrolyte water electrolysis. *Am. Chem. Soc., Div. Fuel Chem., Prepr.*, **18** (3), 24–40.
6 Nuttall, L.J. and Russell, J.H. (1980) Solid polymer electrolyte water electrolysis - development status. *Int. J. Hydrogen Energy*, **5** (1), 75–84.
7 Oberlin, R. and Fischer, M. (1986) Status of the MEMBREL process for water electrolysis. *Hydrogen Energy Progress VI: Proceedings of the 6th World Hydrogen Energy Conference, Vienna, Austria, 20–24 July 1986* (eds T.N. Veziroğlu, N. Getoff, and P. Weinzierl), vol. **1**, Pergamon Press, pp. 333–340.
8 Stucki, S., Scherer, G.G., Schlagowski, S., and Fischer, E. (1998) PEM water electrolyzers: evidence for membrane failure in 100 KW demonstration plants. *J. Appl. Electrochem.*, **28** (10), 1041–1049.
9 Yamaguchi, M., Shinohara, T., Taniguchi, H., Nakanori, T., and Okisawa, K. (1998) Development of 2500 cm^2 solid polymer electrolyte water electrolyzer in We-Net. *Hydrogen Energy Prog.*, **1**, 747–756.
10 Carmo, M., Fritz, D.L., Mergel, J., and Stolten, D. (2013) A comprehensive review on PEM water electrolysis. *Int. J. Hydrogen Energy*, **38** (12), 4901–4934.
11 Sheridan, E., Thomassen, M., Mokkelbost, T., and Lind, A. (2010) The development of a supported iridium catalyst for oxygen evolution in PEM electrolyzers. Presented at the 61st Annual Meeting of the International Society of Electrochemistry, 26 September to 1 October 2010 Nice, France.
12 Smolinka, T., Rau, S., and Hebling, C. (2010) Polymer electrolyte membrane (PEM) water electrolysis, in *Hydrogen and Fuel Cells*, Wiley-VCH Verlag GmbH, Weinheim, pp. 271–289.
13 Ayers, K.E., Dalton, L.T., and Anderson, E.B. (2012) Efficient generation of high energy density fuel from water. *ECS Trans.*, **41** (33), 27–38.

14 Smolinka, T., Günther, M., and Garche, J. (2011) NOW-Studie: Stand und Entwicklungspotenzial der Wasserlektrolyse zur Herstellung von Wasserstoff aus Regenarativen Energien (NOW Studie), Fraunhofer ISE, FCBAT.

15 Anderson, E., Dalton, L., Carter, B., and Ayers, K. (2012) Elimination of mechanical compressors using PEM-based electrochemical technology. Paper ASR 14.5. Presented at the World Hydrogen Energy Conference 2012, 3–7 June 2012, Toronto, Canada.

16 Ayers, K.E., Anderson, E.B., Capuano, C.B., Carter, B.D., Dalton, L.T., Hanlon, G., Manco, J., and Niedzwiecki, M. (2010) Research advances towards low cost, high efficiency PEM electrolysis. *ECS Trans.*, **33** (1), 3–15.

17 Millet, P., Ranjbari, A., de Guglielmo, F., Grigoriev, S.A., and Aupretre, F. (2012) Cell failure mechanisms in PEM water electrolyzers. *Int. J. Hydrogen Energy*, **37**, 17478–17487.

18 Schalenbach, M., Carmo, M., Fritz, D.L., Mergel, J., and Stolten, D. (2013) Pressurized PEM water electrolysis: Efficiency and gas crossover. *Int. J. Hydrogen Energy*, **38** (35), 14921–14933.

19 Baglio, V., Siracusano, S., Monforte, G., Briguglio, N., Brunaccini, G., Di Blasi, A., Stassi, A., Ornelas, R., Trifoni, E., Antonucci, V., and Arico, A.S. (2011) Development and investigation of a 9-cell PEM stack for water electrolysis. Presented at the 220th ECS Meeting, 9–14 October 2011, Boston, USA.

20 Selamet, Ö.F., Caner Acar, M., Mat, M.D., and Kaplan, Y. (2013) Effects of operating parameters on the performance of a high-pressure proton exchange membrane electrolyzer. *Int. J. Energ. Res.*, **37**, 457–467.

21 Ayers, K.E., Capuano, C., and Anderson, E.B. (2012) Recent advances in cell cost and efficiency for PEM-based water electrolysis. *ECS Trans.*, **41** (10), 15–22.

22 ASM International (1990) *Properties and Selection: Nonferrous Alloys and Special-Purpose Materials*, ASM Handbook, vol. **2**, ASM International.

23 Lide, D.R. (ed.) (1999) *CRC Handbook of Chemistry and Physics*, 80th ed, CRC Press, Boca Raton, FL.

24 Novoselov, K.S. *et al.* (2004) Room temperature electric field effect and carrier-type inversion in graphene films. arXiv:cond-mat/0410631v1.

25 Yoon, D., Son, Y.W., and Cheong, H. (2011) Negative thermal expansion coefficient of graphene measured by Raman spectroscopy. *Nano Lett.*, **11**, 3227–3231.

26 USB Group (2013) Realtime indication www.finanzen.net (accessed 15 November 2013).

27 Kristof, K. and Hennicke, P. (2010) Materialeffizienz und Resourcenschonung, Wuppertal Institut für Klima, Umwelt, Energie GmbH.

28 Sigma-Aldrich Titanium nitride (2013) www.sigmaaldrich.com (accessed 15 November 2013).

29 Thyssen Krupp (2010) Material Data Sheet: X5CrNi18-10. http://www.edelstahl-service-center.de/tl_files/ThyssenKrupp/PDF/Datenblaetter/1.4301.pdf.

16
Reversible Solid Oxide Fuel Cell Technology for Hydrogen/Syngas and Power Production

Nguyen Q. Minh

16.1
Introduction

World energy consumption is projected to grow by about 56% from 2010 to 2040 (from 524 quadrillion to 820 quadrillion British thermal units (Btu)) with consumption increases from all fuel sources (fossil, nuclear, and renewable) and fossil fuels expected to continue supply much of the energy used worldwide (almost 80%) [1]. As use of fossil fuels is projected to increase, world energy-related CO_2 emissions will grow from 31 billion metric tons in 2010 to 45 billion metric tons in 2040, a 45% increase [1]. Thus, it is expected that future energy systems should be fuel-flexible (flexibility) (to facilitate integration with any type of energy sources) and environmentally compatible (compatibility) (to reduce CO_2 emissions). In addition, such systems should have the characteristics of capability (useful for different functions), adaptability (suitable for various applications and adaptable to local energy needs), and affordability (competitive in costs) (Figure 16.1). Reversible solid oxide fuel cells (RSOFCs), being developed for hydrogen/syngas production and power generation, potentially have all those desired characteristics and, therefore, can serve as a base technology for green, flexible, and cost-competitive future energy systems [2]. This chapter provides an overview of the RSOFC, discusses its hydrogen/syngas production and power generation operations, and reviews the status of the technology and its applications.

16.2
Reversible Solid Oxide Fuel Cell Overview

16.2.1
Operating Principles

A RSOFC is a device that can operate efficiently in both fuel cell mode for power generation and electrolysis mode for chemical production. Thus, in fuel cell

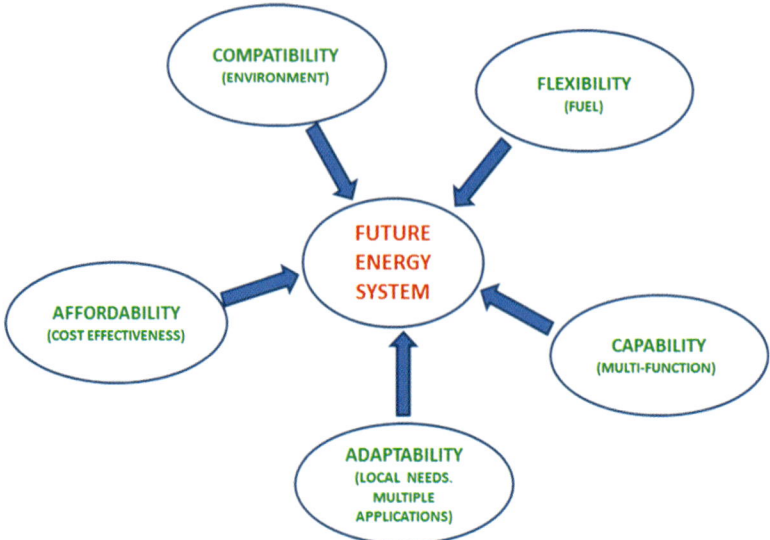

Figure 16.1 Features of future energy systems.

mode, a RSOFC functions as a SOFC, generating electricity by electrochemical combination of a fuel (hydrogen, hydrocarbons, alcohols, etc.) with air (oxygen in the air) across an oxide electrolyte. In electrolysis mode (reverse of fuel cell mode), a RSOFC functions as an electrolyzer (in this case, referred to as a solid oxide electrolysis cell (SOEC)), producing chemicals such as hydrogen from H_2O inputs when coupled with an energy source (fossil, nuclear, renewable). Figure 16.2 illustrates the operating principles of the RSOFC. RSOFCs can be based on oxygen ion conducting [3–5] or proton conducting [6,7] electrolytes. Notably, oxygen ion conducting electrolytes for RSOFCs permit power generation

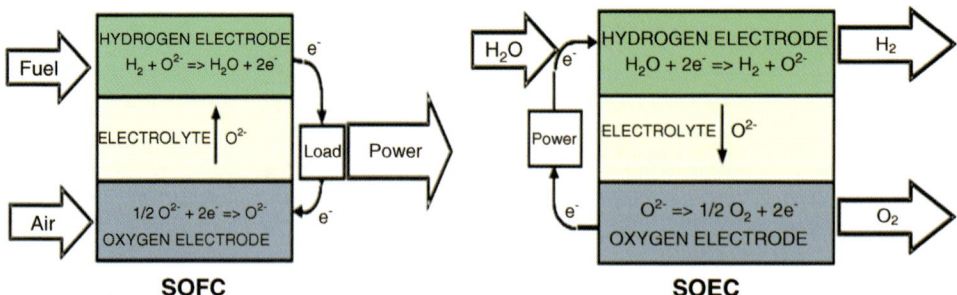

Figure 16.2 Reversible solid oxide fuel cell (RSOFC) operating principles (shown for oxygen ion conducting electrolyte, H_2 fuel in SOFC mode, and H_2O input in SOEC mode).

Table 16.1 Materials for RSOFCs/SOFCs/SOECs.

Component	Most common materials	Other materials
Electrolyte	Yttria-stabilized zirconia (YSZ)	Doped LaGaO$_3$, doped CeO$_2$, high-temperature proton conductors (e.g., doped SrCeO$_3$)
Cathode	Sr-doped LaMnO$_3$ (LSM) Sr-doped LaCo$_{0.2}$Fe$_{0.8}$O$_3$ (LSCF)	Several perovskite oxides (e.g., Sr-doped LaFeO$_3$ (LSF), Ba$_{0.5}$Sr$_{0.5}$Co$_{0.8}$Fe$_{0.2}$O$_{3-\delta}$ (BSCF)) and other oxides
Anode	Ni-YSZ	Cu-CeO$_2$, several conducting oxides (e.g., doped SrTiO$_3$)
Interconnect	Stainless steels Sr-doped LaCrO$_3$ (LSCr)	Other high-temperature alloys and doped oxides

with CO containing fuels and facilitate internal reforming and direct fuel utilization in fuel cell mode [8] and co-electrolysis of mixtures of $H_2O + CO_2$ to produce syngas in electrolysis mode [9,10].

The RSOFC is both the SOFC and SOEC incorporated in a single unit. Since the SOEC is the SOFC operated in reverse mode and traditionally derived from the SOFC, the RSOFC being developed is typically based on the more technologically advanced SOFC. Thus, materials for the RSOFC are those commonly used in the SOFC, for example, for single cells, yttria stabilized zirconia (YSZ) for the electrolyte, perovskites (such as lanthanum strontium manganese oxide (LSM), lanthanum strontium cobalt iron oxide (LSCF)) for the oxygen electrode, and nickel-YSZ cermet for the hydrogen electrode and for stacks, conductive oxides (such as lanthanum strontium chromium perovskite (LSCr)) or stainless steels for the interconnect (depending on the operating temperature) (Table 16.1) [11–13]. Like the SOFC, the RSOFC operates in the temperature range 600–1000 °C. The specific operating temperature depends on cell/stack designs and selected materials [13].

16.2.2
Features

The RSOFC is fundamentally and technologically based on SOFC technology. The RSOFC when used for power generation has all the advantages of a fuel cell system such as high energy conversion efficiency, reduced pollutant emission, modular construction, and cogeneration capability. The RSOFC possesses the distinctive and attractive features offered by the SOFC: cell and stack design flexibility, multiple fabrication options, operating temperature choices, and multi-fuel capability (for fuel cell mode). The RSOFC when used as a SOEC for hydrogen/chemical production has the characteristics of simple chemistry (in comparison with other processes), environmental compatibility (no CO_2 and

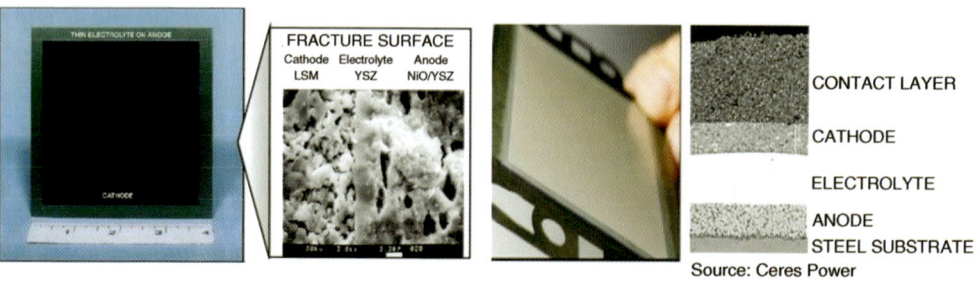

Figure 16.3 Examples of SOFC single cells and microstructures.

other pollutant emissions), and high conversion efficiency (electrical-to-hydrogen efficiency):

Cell and stack design flexibility: Various options are available for designing the RSOFC. Single cells can be configured using different components for structural support. Thus, cells can be self-supporting (electrolyte-supported, electrode-supported) or external supporting (interconnect-supported, substrate-supported). Figure 16.3 shows examples of an electrode (anode)-supported cell and a metal substrate-supported cell for SOFC operation. Stacks of cells can be built based on the tubular design, the segmented-cells-in-series design, the monolithic design, and the planar design (the planar stack design is currently the most common for the SOFC, and thus for the RSOFC). Figure 16.4 gives some examples. These design options permit flexibility to shape the RSOFC into a structure having the desired electrical and electrochemical performance along with required thermal management, mechanical integrity, and dimensional constraint (if any) to meet operating requirements of specified applications [14].

Multiple fabrication options: Numerous fabrication processes that have been developed for the SOFC can be used for making RSOFC cells. Selected processes depend on cell design and configuration of the cell in the stack. One of the prerequisites is the fabrication of dense electrolytes. Suitable processes range from conventional ceramic processing methods such as tape casting, tape calendering, screen printing to deposition techniques such as plasma spraying, chemical vapor deposition, and electrophoretic deposition. Figure 16.5 shows, as an example, the fabrication sequence based on tape calendering for making electrode-supported cells.

Operating temperature choices: In general, the conductivity requirement for the electrolyte determines the operating temperature of the RSOFC (SOFC/SOEC) [11,12,15]. The operating temperature of the RSOFC thus can be varied/reduced (for YSZ based cells) by modifying the electrolyte material and/or electrolyte thickness (Figure 16.6). Examples include operating temperatures of 900–1000 °C for thick (>50 μm) YSZ electrolytes [11–13], 700–800 °C for

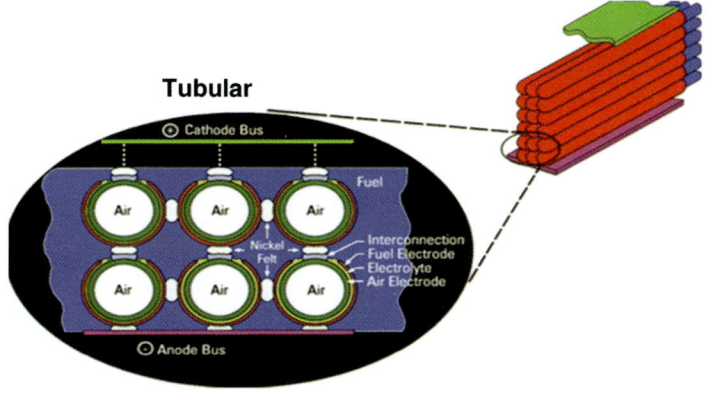
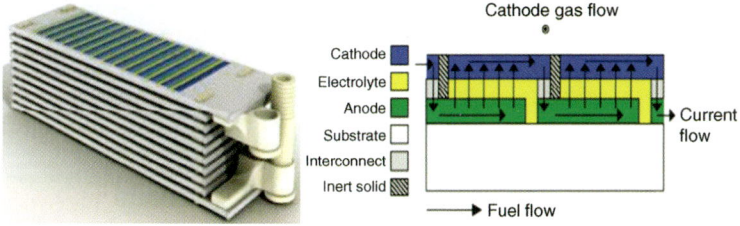

Figure 16.4 Examples of SOFC tubular and segmented-cells-in-series designs.

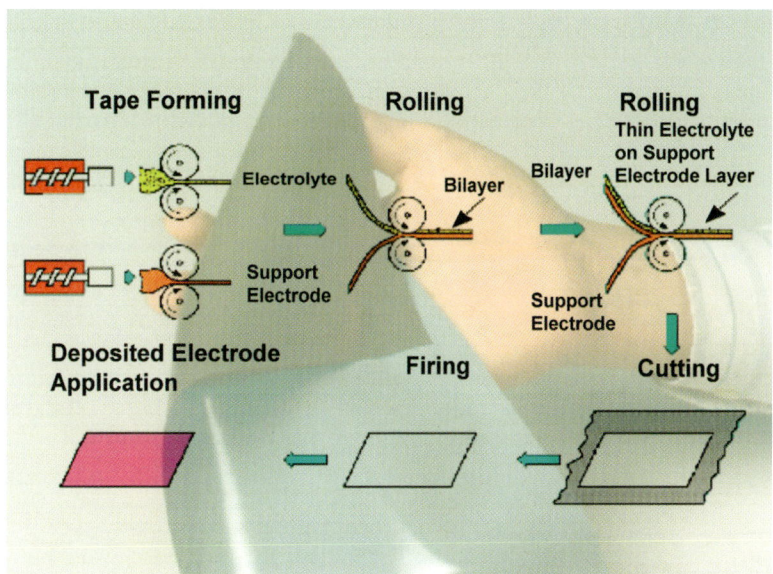

Figure 16.5 Fabrication sequence based on tape calendering for making electrode-supported cells.

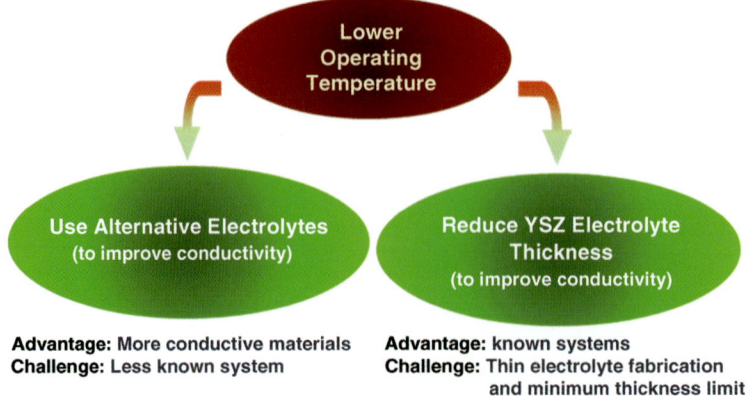

Figure 16.6 Approaches to reduce operating temperatures of YSZ based cells.

thin (<15 μm) YSZ electrolytes [11–13] or doped lanthanum gallate electrolytes [16,17], 500–600 °C for thin doped ceria electrolytes [18,19], and 400–500 °C for thin doped ceria/bismuth oxide bilayer electrolytes [20].

Multi-fuel capability in fuel cell mode: One of the key attributes of the RSOFC as a SOFC is its multi-fuel capability. For fuels other than pure hydrogen, the RSOFC can operate on reformates (via external reformation) or on hydrocarbons and other fuels (via internal reforming or direct fuel utilization) (Figure 16.7). (Further discussion on fuel utilization is given in Section 16.3.) An example of operation of a SOFC on reformate from a catalytic partial oxidation (CPOX) reformer is given in Figure 16.8.

Efficient production of hydrogen and chemicals in electrolysis mode: The high operating temperature of the RSOFC when operated as a SOEC provides the possibility of using heat from an external source (in addition to unavoidable

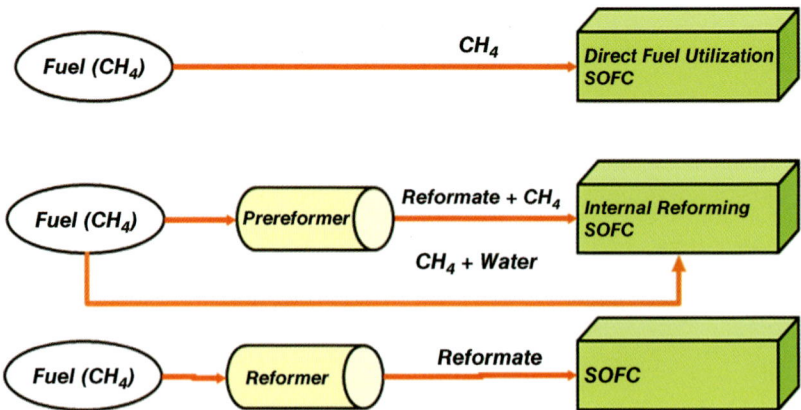

Figure 16.7 SOFC operation on fuel other than hydrogen (shown for methane).

16.2 Reversible Solid Oxide Fuel Cell Overview | 365

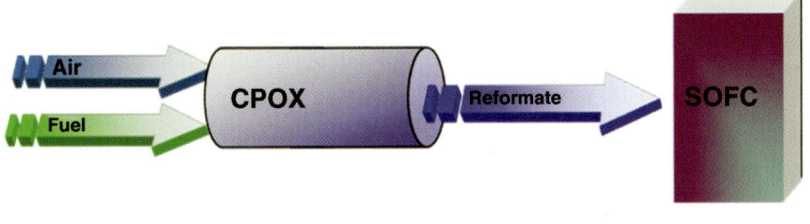

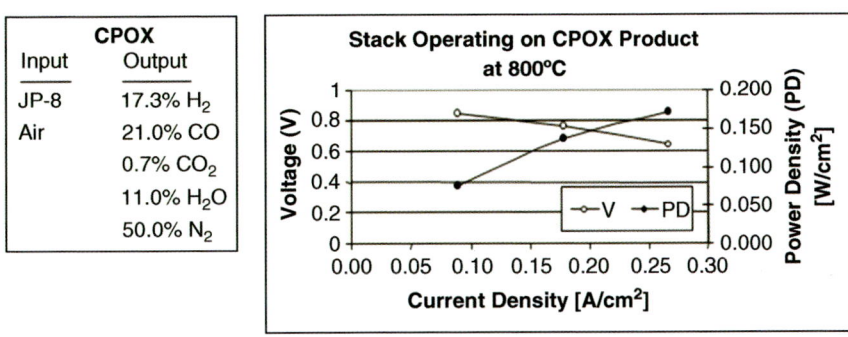

Figure 16.8 SOFC operation on reformate from a CPOX reformer.

Joule heat) to increase its electrical-to-hydrogen conversion efficiency. As seen in Figure 16.9 for steam electrolysis, the total energy demand is almost constant while the electrical energy demand decreases (and the heat demand increases) as temperature increases. Thus, a RSOFC operating in electrolysis mode can work under the so-called thermoneutral condition, that is, at the thermoneutral voltage (V_{tn}) the electricity input exactly matches the total energy demand of the electrolysis reaction. In this case, the electrical-to-hydrogen conversion efficiency is 100%. At cell operating voltages $<V_{tn}$, heat

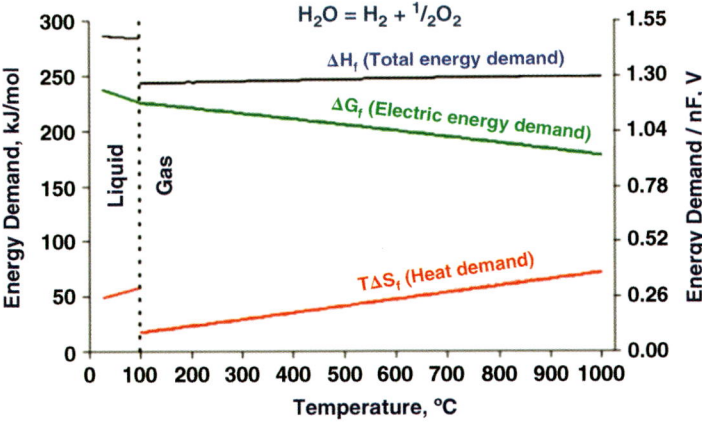

Figure 16.9 Energy demand for water electrolysis.

must be supplied to the system to maintain the temperature and the conversion efficiency is above 100%. At cell operating voltages $>V_{tn}$, heat must be removed from the system and the efficiency is below 100%. Another major advantage of the RSOFC/SOEC is the capability of CO_2 conversion. RSOFCs/SOECs based on oxygen ion conducting electrolytes can reduce CO_2 or $CO_2 + H_2O$ to CO or $CO + H_2$ (syngas). CO or syngas can be used as fuel for the RSOFC/SOFC or further processed to produce chemicals such as methane and methanol.

16.3
Solid Oxide Fuel Cell Technology

The RSOFC is a SOFC in fuel cell operating mode and fundamentally and technologically based on SOFC technology [8,11–13]. The SOFC has been considered for a broad spectrum of power generation applications with systems ranging from watt-size consumer portable devices to multi-megawatt central station power plants. Complete SOFC systems of power level up to several hundreds of kilowatts have been built and operated [21]:

Performance: Peak power densities of SOFC single cells (with pure hydrogen fuel and air oxidant, low fuel and air utilizations) can be as high as $2\,W\,cm^{-2}$ at temperatures as low as 650 °C, depending on electrolyte/electrode systems and cell configurations [20,22]. Figure 16.10 gives an example of SOFC single

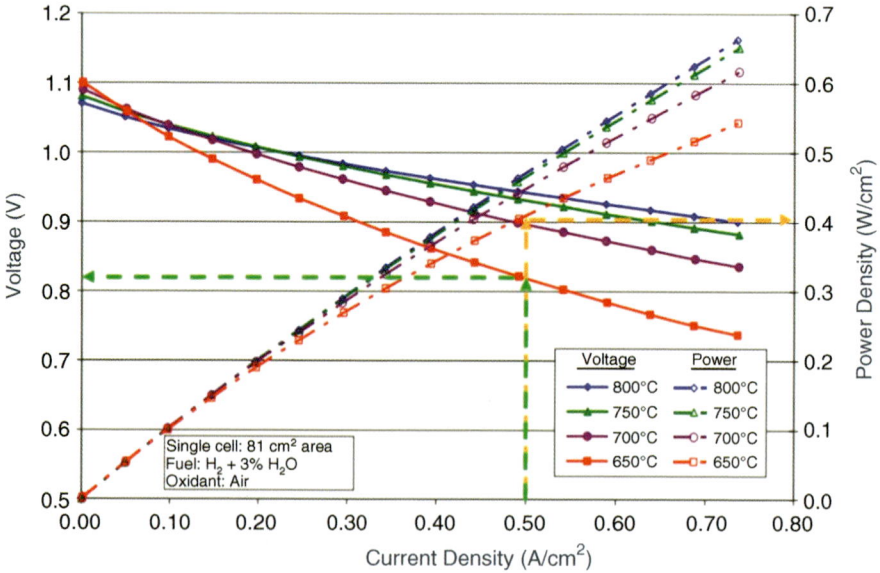

Figure 16.10 Example of SOFC single cell (anode-supported) performance [23].

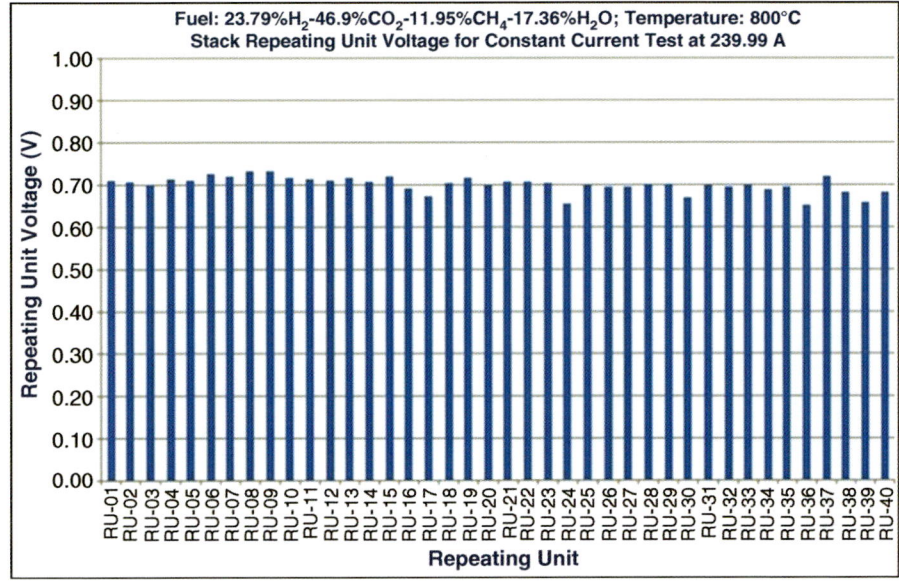

Figure 16.11 Example of cell-to-cell voltage variation in a 40-cell planar stack (anode-supported cells, metallic interconnects) [25].

cell performance [23]. A power density of about $0.4\,\text{W cm}^{-2}$ at $0.82\,\text{V}$ is achieved at $650\,°\text{C}$. SOFC stacks have demonstrated electrochemical performance under operating conditions appropriate for practical uses. For example, a 96-cell planar stack shows a power density of about $0.3\,\text{W cm}^{-2}$ (voltage of about $0.82\,\text{V}$ per cell at $0.364\,\text{A cm}^{-2}$), $715\,°\text{C}$ on air (air utilization (U_a) of 15%) and fuel containing 25.2% H_2–22.4% N_2–14.5% natural gas (NG)–37.8% H_2O (fuel utilization (U_f) of 68%) [24]. One important characteristic for stable and reliable stack performance is minimal cell-to-cell variation within the stack, which can be achieved via design and process control [25]. An example is given in Figure 16.11 [25].

For state-of-the-art SOFC single cells (having minimal ohmic resistance contributions from the components), cathode (oxygen electrode) polarization is generally the major contribution to cell performance losses. Cathode polarization can be reduced by improving catalytic activity of cathode materials and/or modifying electrode microstructures. For instance, use of mixed ionic electronic conducting materials for the cathode (e.g., LSCF) expands triple phase boundary sites for the oxygen reaction, thus reducing polarization and improving cathode performance. Infiltration of active components as dispersed particles or connected nanoparticulate networks to form nanostructures enhances cathode performance by modifying catalytic activities and/or conduction pathways of the electrode [26–28]. One example is infiltration of yttria-doped ceria (YDC) into LSM-YSZ cathode; this has been shown to increase cell peak power density from 208 to $519\,\text{mW cm}^{-2}$ at $700\,°\text{C}$ and

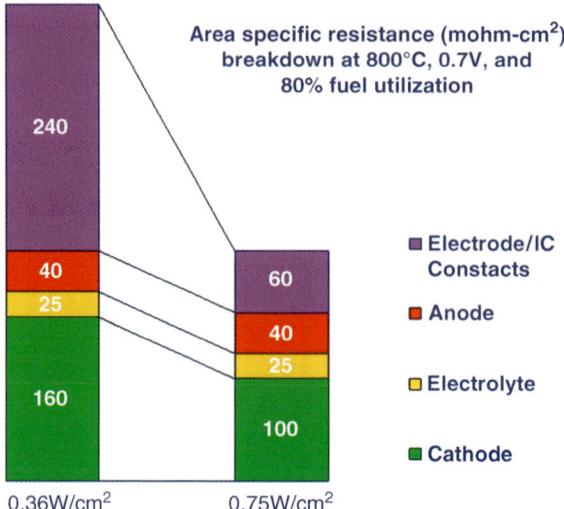

Figure 16.12 Area specific resistance (ASR) breakdown for a planar SOFC stack (anode supported cells, metallic interconnects).

power density at 0.7 V from 135 to 370 mW cm^{-2} [26]. The main issue is the stability of the nanostructure over extended periods of time at high operating temperatures. Operating the SOFC at reduced temperatures (e.g., <600 °C) or stabilizing the nanostructure are potential approaches to maintain sufficient long-term stability [26,29].

In SOFC stacks, especially planar stacks with metallic interconnects, contact resistance between the electrodes, especially the cathode, and the metallic interconnect is a major factor in stack performance losses (Figure 16.12) [30] and long-term performance degradation (discussed in the next section). Conductive contact pastes have been used in planar stacks to minimize contact resistance; however, the stability of such contact pastes over long duration is questionable.

Performance degradation and life: SOFC single cells, when properly prepared with conventional high-purity materials and operated on clean fuels and air, show minimal performance degradation for extended periods of time. For example, tubular cells were electrically tested for times as long as 8 years and showed satisfactory performance with less than 0.1% per 1000 h degradation [14]. SOFC cells, however, can experience significant performance degradation in realistic environments, depending on several factors such as gas input purities and component materials used in the stack/system [31,32]. On the anode side, sulfur is the most prevalent fuel impurity in many practical fuels and its poisoning effects on the Ni-YSZ anode are well known [11–13]. It has also been shown that silicon impurities present in the fuel (originating from, for example, stack glass sealants or silica containing insulations in the

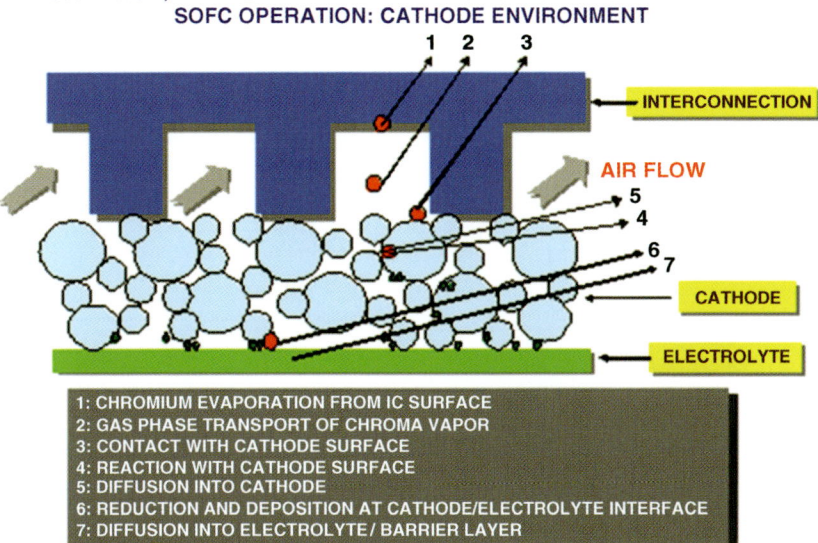

Figure 16.13 Potential steps in chromium poisoning of SOFC cathodes.

system) can also poison the Ni-YSZ anode [33]. On the cathode side, the presence of significant amounts of water or carbon dioxide in air can have deleterious effects on cell performance [34–36]. An important performance degradation mode is chromium poisoning of the cathode in long-term operation of planar stacks having metallic interconnects. Chromium present in the metallic interconnect can migrate to cathode reactive sites and interact with the cathode (Figure 16.13), poisoning the electrode and thereby increasing cathode polarization with time. At present, the most common mitigating approach is to use conductive coatings (e.g., Co-Mn spinel) on the metallic interconnect to minimize chromium transport and migration, thus reducing degradation rates [21]. Figure 16.14 shows an example of performance degradation of a three-cell stack with Co-Mn spinel coated interconnects.

As discussed in the previous section, contact resistance between the cathode and the metallic interconnect can be a major factor in long-term performance degradation of multi-cell stacks. The contact between the ceramic cathode and the metallic interconnect tends to change due to thermodynamic driving forces (cathode ceramic particles tend to sinter/agglomerate to minimize their surface free energy, thus moving away the metallic interconnect) and other operating characteristics such as temperature distribution and thermal expansion mismatch as operation proceeds These factors can lead to degradation in long-term operation. It is highly possible that during long-term operation chemical interaction develops and the electrical contact between the cathode

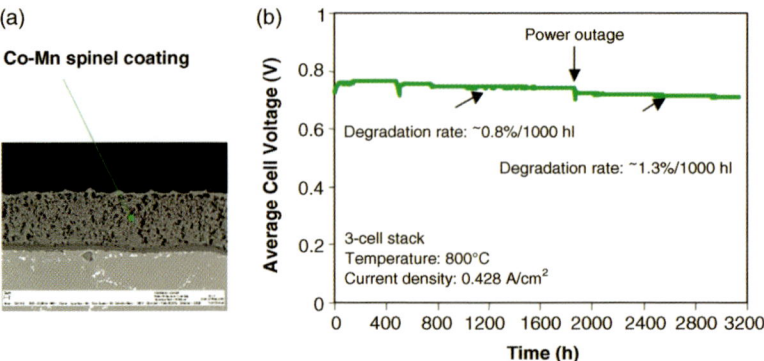

Figure 16.14 Micrograph of Co-Mn spinel coated interconnect (a) and performance degradation of a three-cell stack with a node-supported cells and coated interconnects (b).

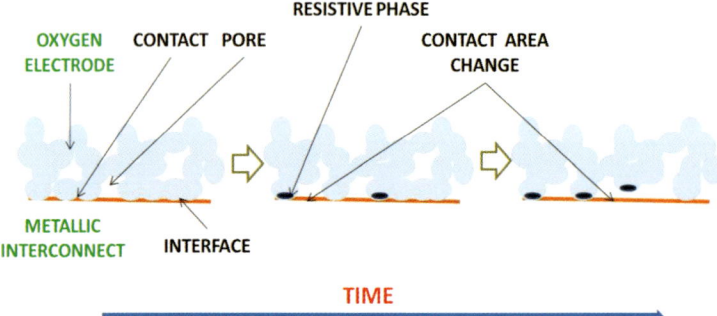

Figure 16.15 Cathode/metallic interconnect contact evolution.

and the interconnect evolves, ohmic resistance increases, and the contact area reduces, resulting in higher ohmic losses and thus degradation (Figure 16.15).

SOFC stacks and systems have been operated for tens of thousands of hours and durability has been demonstrated with low performance degradation rates under specified operating conditions. For example, a short planar SOFC stack (with uncoated metallic interconnects) has been in operation at 700 °C for more than 5 years with an overall voltage degradation of about 1% per 1000 h [37]. With coated metallic interconnects, a stack has been tested for more than 14 000 h with a reduced degradation rate of about 0.12% per 1000 h [37]. Figure 16.16 gives an example of the performance of a 96-cell stack showing 1.3% voltage degradation per 1000 h (with internal reforming) [24].

Fuel utilization: The SOFC can operate directly on fuels other than hydrogen (e.g., hydrocarbons, alcohols) via internal reforming (on fuel feeds with significant amounts of water) or direct utilization (on fuel feeds with no water). Internal reforming using on-anode reformation is well known and has been demonstrated for the SOFC. Instead of complete (100%) internal reforming, it

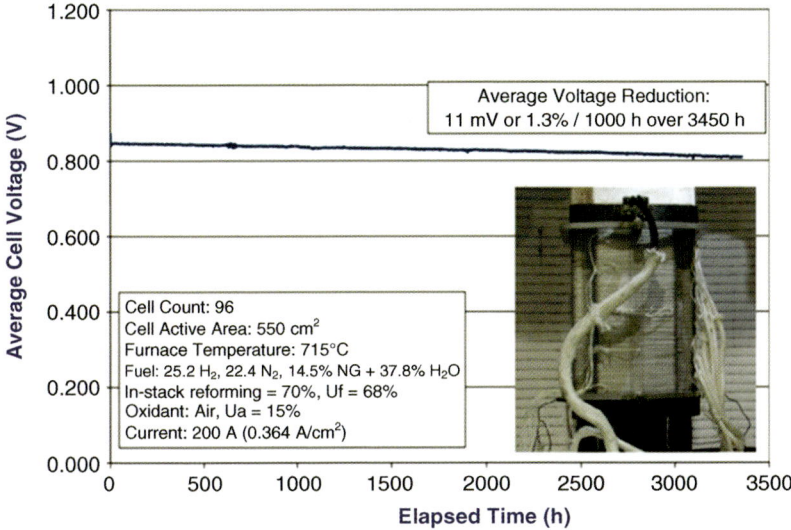

Figure 16.16 Long-term performance of 96-cell planar stack [24].

is possible to have a portion of the fuel reformed in an external reformer (referred to as a pre-reformer) and the resulting reformate plus the remaining fuel are fed to the SOFC where the fuel is internally reformed (via steam reforming) within the fuel cell. An example is the demonstration of operation of a 5 kW SOFC system with an autothermal reformation reformate containing about 7 vol.% methane slip [38]. This pre-reforming/internal reforming option has been employed to use the endothermic reforming reactions to reduce cooling requirements in thermal management of the SOFC.

The SOFC has been shown to have the capability for direct utilization of different types of fuel [39]. For direct fuel utilization operation, the anode material has been modified to address the carbon deposition issue associated with nickel commonly used in the anode composition (e.g., Cu-ceria instead of Ni-YSZ [40]). With modified anodes, high electrochemical performance can be achieved for direct SOFCs. For example, a peak power density of about 400 mW cm^{-2} at 800 °C was obtained with 7.3% ethanol balance He fuel and air oxidant for an anode-supported SOFC with a dual layer anode consisting of a Cu-CeO$_2$ impregnated Ni-YSZ support outer layer and a Ni-YSZ electroactive inner layer (Figure 16.17) [41]. Long-term performance stability of direct SOFCs without significant carbon deposition, however, remains to be demonstrated.

Applications: The SOFC has been considered for a broad spectrum of power generation applications and markets. Applications for the SOFC include power systems for various uses ranging from watt-sized consumer electronic devices to multi-megawatt central station power plants and potential markets cover portable, transportation, and stationary sectors (Table 16.2). Many of the

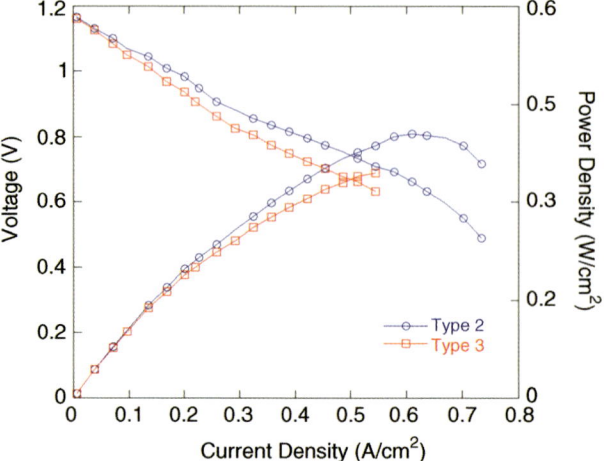

Figure 16.17 Performance at 800 °C with 7.3% ethanol–92.7% He fuel and air oxidant for an anode-supported SOFC (with a dual layer anode consisting of a Cu-CeO$_2$ impregnated Ni-YSZ support outer layer and a Ni-YSZ electroactive inner layer) (Type 2 and Type 3 indicate different thermal treatments of infiltrated anodes) [41].

Table 16.2 SOFC market sectors and applications.

Market	Example of application	Power size
Portable	Consumer electronic devices, mobile personal power	1–100 W
	Portable power, battery chargers	200–500 W
Transportation	Automobile and truck auxiliary power units (APUs)	5–50 kW
	Aircraft APUs	Up to 500 kW
Stationary	Residential/micro combined heat and power (CHP), uninterruptible power	1–10 kW
	CHP and distributed generation (DG)	100 kW–1 MW
	Base load centralized power plants	100–500 MW

applications for the SOFC have progressed to hardware demonstration and prototype/pre-commercial stages while several applications, especially those with large power outputs, are at the conceptual/design stage (Figure 16.18).

16.4
Solid Oxide Electrolysis Cell Technology

The RSOFC is a SOEC in electrolysis mode, that is, a SOFC in reverse operating mode. The SOEC has been considered for production of hydrogen from

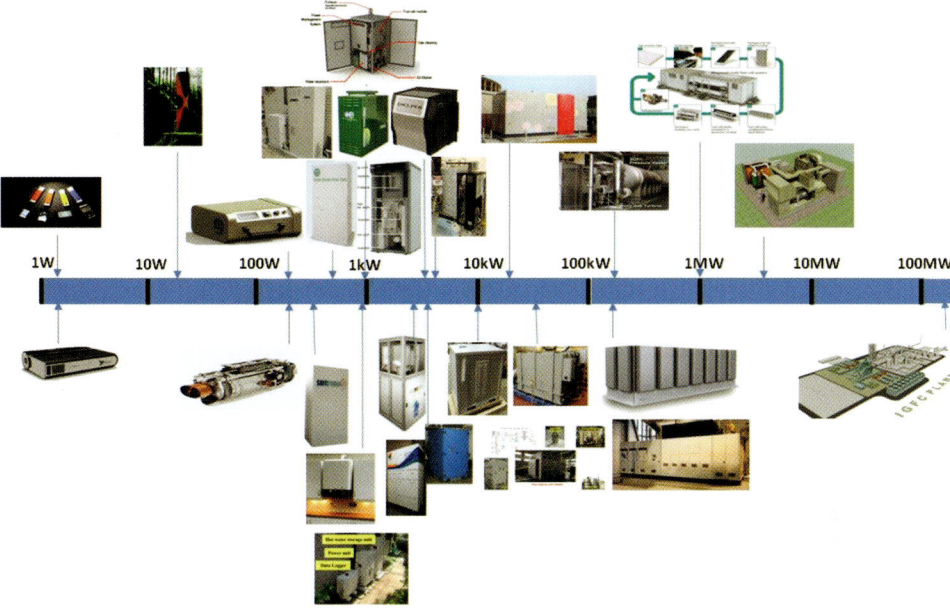

Figure 16.18 SOFC power systems (hardware demonstrators, prototypes, and pre-commercial systems up to 200 kW, concepts at 1 MW and above).

H_2O [42], syngas from mixtures of H_2O and CO_2 [43], and oxygen from CO_2 [44]. The SOEC is the only electrolysis cell having this multi-product capability and one of the cleanest methods for hydrogen production [45]:

Performance (for hydrogen production from steam): SOEC single cells have been shown to have the ability to perform well for hydrogen production from steam. For example, a cell voltage of 1.1 V (below V_{tn}) has been obtained for a Ni-YSZ electrode-supported SOEC cell of 45 cm² active area at a current density of about 1.4 A cm^{-2}, 900 °C, 93% H_2O balance H_2 [46]. Extraordinarily high current densities have also been reported for SOECs, for example, about 3 A cm^{-2} at 1.3 V (V_{tn}), 950 °C [43]. A typical performance metric for SOECs is the current density (A cm^{-2}) at V_{tn}. Table 16.3 summarizes selected data on current densities at thermoneutral voltage for YSZ-based SOECs. Current density variations for the same electrode material systems are due to starting material characteristics, processing, electrode microstructure, absolute humidity input, and flow rates. Figure 16.19 graphically shows data reported in the literature for current densities at V_{tn} between 500 and 900 °C along with the range of values for various types of SOEC [46–60]. The performance of SOEC single cells, like SOFCs, can be improved by using nanoparticles to modify electrode microstructures [61,62].

Operation of multi-cell (up to 60-cell) SOEC stacks for hydrogen production from steam with reasonable performance has been demonstrated. For

Table 16.3 Current densities for selected YSZ-based SOECs at thermoneutral voltage.

Materials	Cell support	Temperature (°C)	Voltage (V)[a]	Current density (A cm^{-2})
Ni-YSZ/YSZ/LSM [47]	YSZ electrolyte	1000	1.3	0.20
Ni-YSZ/YSZ/LSCr [47]	YSZ electrolyte	1000	1.3	0.08
Ni-YSZ/YSZ/LSM [46]	Ni-YSZ electrode	800	1.3	1.12
Ni-YSZ/YSZ/BSCF [48]	Ni-YSZ electrode	700, 750, 800, 850	1.3	0.15, 0.25, 0.45, 0.65
Ni-YSZ/YSZ/LSM-YSZ [48]	Ni-YSZ electrode	700, 750, 800, 850	1.3	0.10, 0.22, 0.30, 0.48
Ni-YSZ/YSZ/LSM-YSZ [49]	Ni-YSZ electrode	750	1.3	0.50
Ni-YSZ/YSZ/LSM-YSZ [50]	Ni-YSZ electrode	800, 900	1.3	0.40, 0.70
Ni-YSZ/YSZ/LSM-YSZ [51]	Ni-YSZ electrode	850, 900	1.3	0.20, 0.30

a) Between temperatures of 700 and 1000 °C, the thermoneutral voltage varies from 1.28 to 1.30 V.

example, ASRs of <0.6 Ω cm^2 have been obtained for five-cell planar stacks (100 cm^2 cell active area) at 800 °C [63]. Figure 16.20 shows a photograph of a ten-cell stack [64]. This stack is capable of producing up to 100 standard litres per hour (l h^{-1}) hydrogen [65]. In general, hydrogen production rates have been found to be in good agreement with dew point measurements and rate predictions based on current (Figure 16.21) [66].

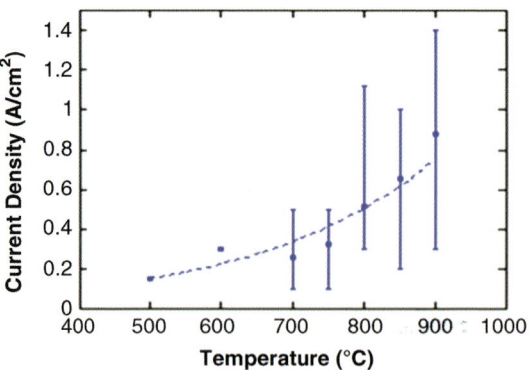

Figure 16.19 Current densities at thermoneutral voltage reported for various types of SOEC between 500 and 900 °C (the bars show the range of values) [46–60].

16.4 Solid Oxide Electrolysis Cell Technology | 375

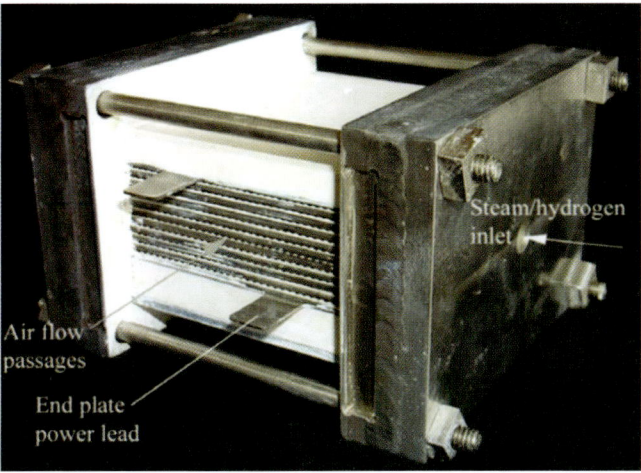

Figure 16.20 Photograph of a ten-cell SOEC stack [65].

Performance degradation and life: Current SOEC cells and stacks can have significant performance degradation in short-term and long-term operations. The level of degradation is very much dependent on electrode materials and microstructures and operating parameters. The presence of glassy phase impurities in Ni-YSZ based hydrogen electrodes (originating from starting materials or glass sealants) can cause cell performance to degrade [67].

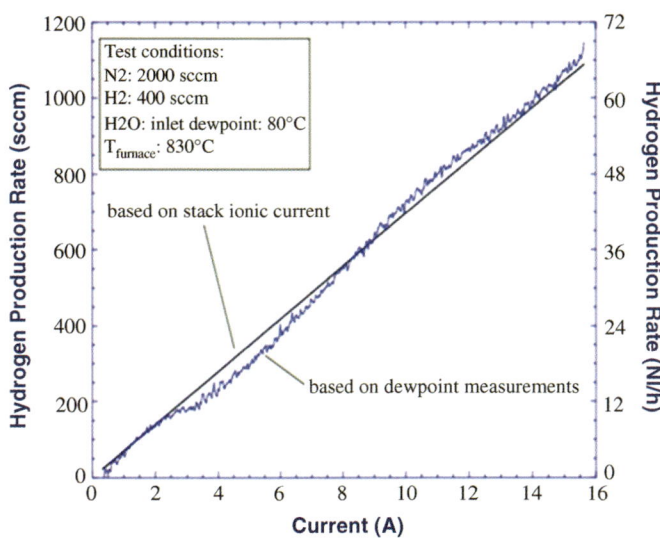

Figure 16.21 Hydrogen production rates (in standard cubic centimeter per minute or sccm and normal liter per hour (Nl h^{-1})) of ten-cell stack [66].

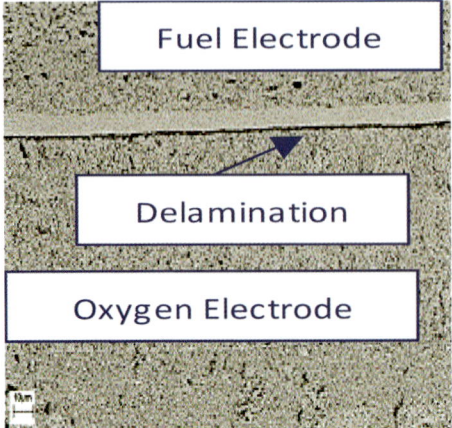

Figure 16.22 Micrograph showing oxygen electrode delamination [67].

An operating SOEC cell may experience rapid performance decay due to oxygen electrode delamination [68]. Figure 16.22 gives an example of oxygen electrode delamination. This type of degradation has been observed for oxygen electrodes based on predominantly electronic conducting oxides (e.g., LSM) when electrode microstructures are not designed to minimize oxygen pressure buildup (caused by oxygen evolution) at the oxygen electrode/electrolyte interface during operation. For oxygen electrodes based on mixed ionic electronic conducting oxides (e.g., LSCF) with similar microstructures, electrode delamination may occur, but at higher current densities (because of lower electrode overvoltage due to the spreading of triple phase boundary active sites on the mixed conducting surface). Short-term performance degradation can also occur in SOEC cells operating under conditions such as high current densities and high steam concentrations. High current densities can lead to cell ohmic resistance increase, which is attributed to oxygen formation in the electrolyte grain boundaries near the oxygen electrode/electrolyte interface [69]. High steam concentrations can accelerate the time constant for passivation of the SOEC caused by segregation of silica-containing impurities to the hydrogen electrode/electrolyte interface during electrolysis [70,71].

SOEC cells and stacks have been operated for up to 2500 h and hydrogen/syngas production (from $H_2O/H_2O + CO_2$) has been demonstrated on the laboratory scale. Performance degradation in long-term operation is typically of the order of 2–10% per 1000 h [57,72,73]. Figure 16.23 shows an example of performance degradation (in terms of ASR and hydrogen production rate) of a 25-cell stack [73]. Low performance degradation rates (<1% per 1000 h) for SOEC stacks have also been reported [74,75]. An example is given in Figure 16.24. In general, long-term degradation of SOEC cells is higher than that obtained for similar cells in fuel cell mode [76]. Root causes for this difference, however, are not fully understood.

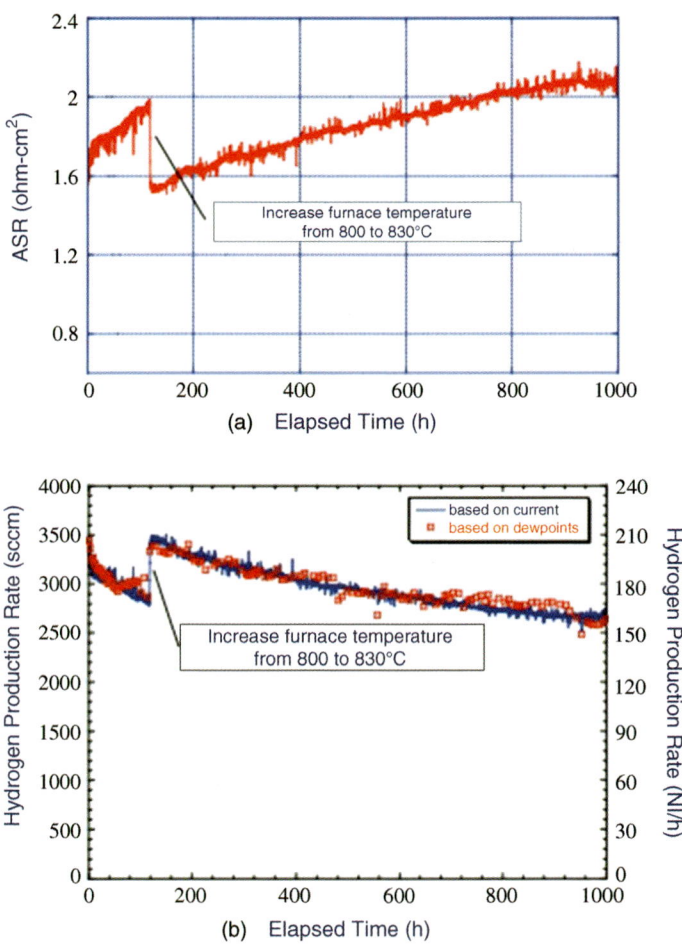

Figure 16.23 Performance degradation of 25-cell stack in terms of ASR (a) and hydrogen production rate (b) [73].

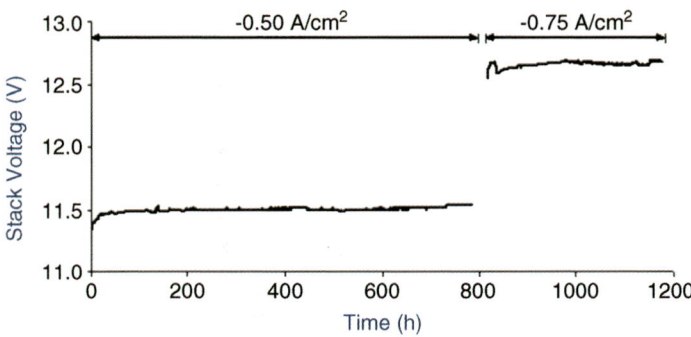

Figure 16.24 Voltage as a function of time for a 1 kW ten-cell stack of 12 cm × 12 cm footprint, temperature = 850 °C, and gas input = 10% H_2–45% H_2O–45% CO_2 [74].

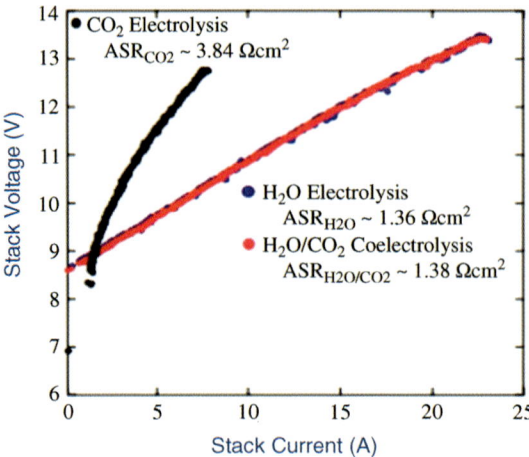

Figure 16.25 Performance of a ten-cell stack for electrolysis of H_2O, $H_2O + CO_2$, and CO_2 at 800 °C [84].

Co-electrolysis of H_2O and CO_2: In addition to steam electrolysis, the SOEC (based on oxygen ion conducting electrolytes) is also suitable for electrolyzing CO_2 and co-electrolyzing $H_2O + CO_2$. Electrolysis of CO_2 and $H_2O + CO_2$ has been considered to produce/recover oxygen for spacecraft life support and propulsion [77–80]. Co-electrolysis of $H_2O + CO_2$ can be used to produce syngas, an intermediate energy carrier that can be converted into various practical chemicals/fuels [81,82]. SOECs have been shown to be capable of syngas production by inputs of $H_2O + CO_2$ at similar current density ranges to steam electrolysis [71,83–85] although the ASR for electrolysis of CO_2 is generally higher than that of H_2O (Figure 16.25) [84]. Performance of the CO/CO_2 electrode, however, can be improved by modifying electrode materials and structures [86]. The mechanism for the CO_2 reaction is not well determined and may depend on electrode microstructures. In the case of $H_2O + CO_2$, it is possible that only H_2O is involved in the electrochemical reaction and the CO_2 in the mixture reacts with H_2 of the reaction products via a reverse water–gas shift reaction.

Applications: The SOEC has been considered for hydrogen production from steam [52,67] at various sizes, for example, distributed plant (e.g., 1500 kg H_2 per day) and central station (e.g., 150 000 kg H_2 per day) sizes [68], syngas production for industrial uses [81,87], and oxygen generation/recovery for space applications [88,89]. Integration of SOEC systems with nuclear [64,90–92] and renewable energy resources such as solar energy [82,93,94] and wind [95] has been envisioned. Figure 16.26 shows a concept of a SOEC based hydrogen production central system coupled with high-temperature gas-cooled nuclear reactors [63]. A concept combining SOECs with biomass based plant has been proposed [96]. In this concept, oxygen produced in the SOEC unit can be used for the gasification of biomass, and steam for the SOEC can be generated in

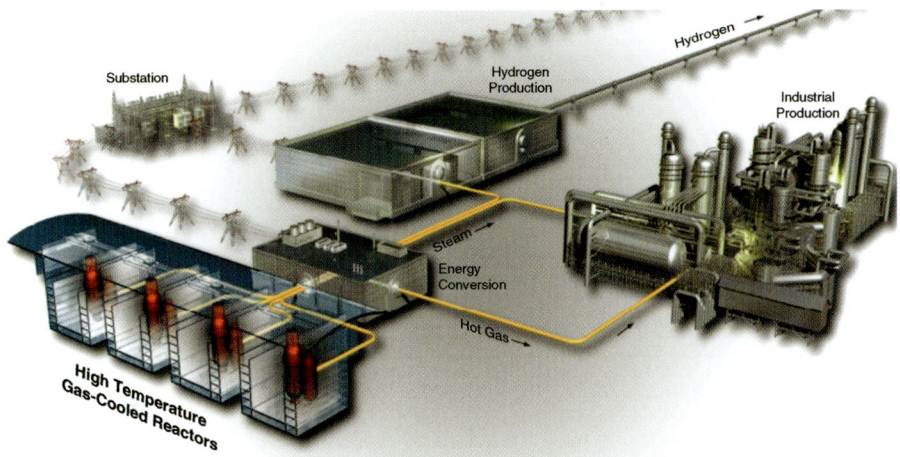

Figure 16.26 Concept for large-scale centralized nuclear hydrogen production [63].

the gasification plant. The SOEC generates hydrogen that is needed for conversion of all the carbon in the biomass into chemicals such as methanol, dimethyl ether, gasoline, and synthetic natural gas. This concept also includes a method of upgrading digested biogas using co-electrolysis of CO_2 (present) and steam (added) in biogas.

16.5
Reversible Solid Oxide Fuel Cell Technology

A RSOFC can function in both fuel cell and electrolysis modes. The RSOFC is thus suitable for the various applications being considered for the SOFC and the SOEC. A RSOFC integrated with a redox cycle forms a rechargeable battery [97–100]. An example is a YSZ based RSOFC to provide H_2/H_2O cycles for a Fe/FeO redox unit in Fe-air rechargeable batteries [97,99].

A RSOFC must operate efficiently in both SOFC and SOEC modes [2,101–103]. Thus, two key requirements for RSOFCs, in addition to those specified for SOFC/SOEC operation, are (i) acceptable electrode performance reversibility and stability and (ii) efficient and stable cyclic operation:

Electrode performance reversibility and stability for reversible operation: Electrode materials commonly used in SOFCs and SOECs can be used for the RSOFC. These materials, however, may be modified or microstructures engineered for stable reversible operation [104,105]. In general, the hydrogen electrode shows performance reversibility (symmetry) between fuel cell and electrolysis modes. The oxygen electrode, on the other hand, shows performance reversibility at low current densities (Figure 16.27) [68] but may exhibit irreversibility at higher current densities in electrolysis mode (Figure 16.28) [30]. This property, however, depends on several factors such as electrode

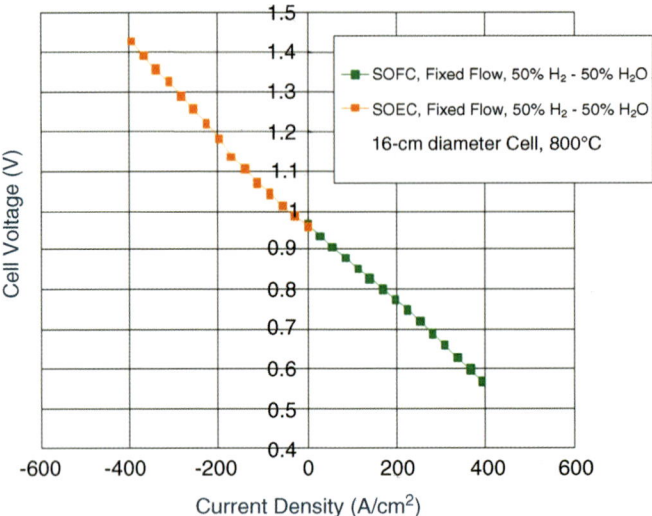

Figure 16.27 Performance curve of YSZ based RSOFC cell with LSM-YSZ oxygen electrode (Ni-YSZ hydrogen electrode) at current densities of <400 mA cm^{-2} [68].

microstructure, material, and operating parameters [68]. Performance stability of the electrodes is very much dependent on electrode microstructures. Oxygen electrode microstructures must be designed to circumvent oxygen pressure built-up at electrode/electrolyte interfaces during electrolysis. Hydrogen electrode microstructures must be engineered to facilitate both water and hydrogen transport to and from reaction sites.

Efficient and stable cyclic operation: RSOFC single cells and multicell stacks have been built and tested and their cyclic operation has been demonstrated.

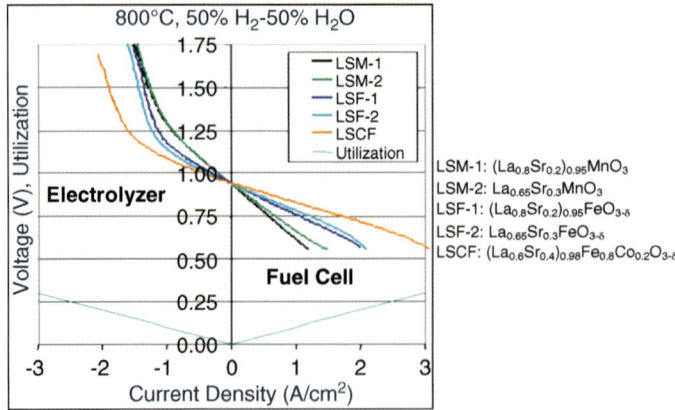

Figure 16.28 Voltage–current density curves of YSZ-based RSOFC cells with Ni-YSZ hydrogen electrodes and different oxygen electrodes showing irreversibility at high current densities [30].

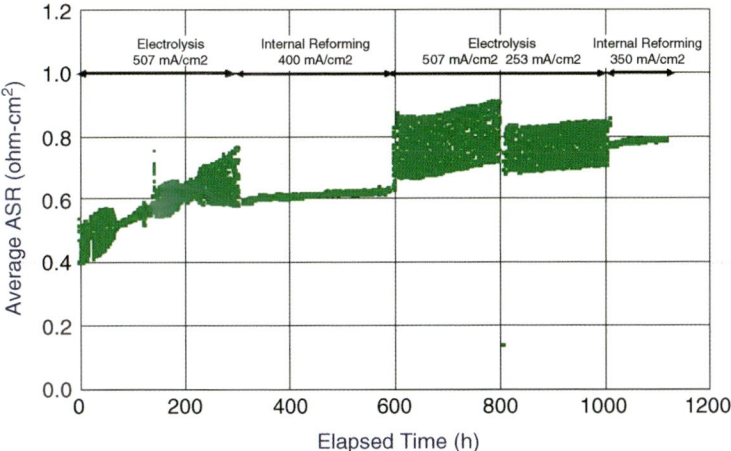

Figure 16.29 Photograph of a ten-cell stack (cell active area of 142 cm^2) and stack ASR in methane internal reforming fuel cell mode and steam electrolysis mode (fluctuations seen in steam electrolysis due to instability of steam generation and delivery) [30].

Cell and stack performance in electrolysis mode typically shows higher degradation rates than those in fuel cell mode [68,106–108]. Figure 16.29 shows an example of a ten-cell RSOFC stack and its performance (in terms of ASR) in operation for more than 1000 h (operating alternately between internal reforming fuel cell mode and steam electrolysis mode) [68]. This stack showed an initial power density of 480 mW cm^{-2} at 0.7 V, 800 °C with 64%H_2–35%N_2 as fuel and 80% fuel utilization in fuel cell mode and produced about 6.3 standard liters per minute of hydrogen at 1.26 V cell voltage, 0.62 A cm^{-2} with 30% H_2–70% H_2O and steam utilization of about 54%.

Design concepts have been proposed for RSOFC systems for power generation and hydrogen production. Figure 16.30 is an example of a schematic of a RSOFC system concept showing all the key components for fuel cell and electrolysis operation [2]. Sustainable energy systems based on the RSOFC for the future are feasible. An example of such a system is shown schematically in Figure 16.31. In this system, the RSOFC, operating in the electrolysis mode,

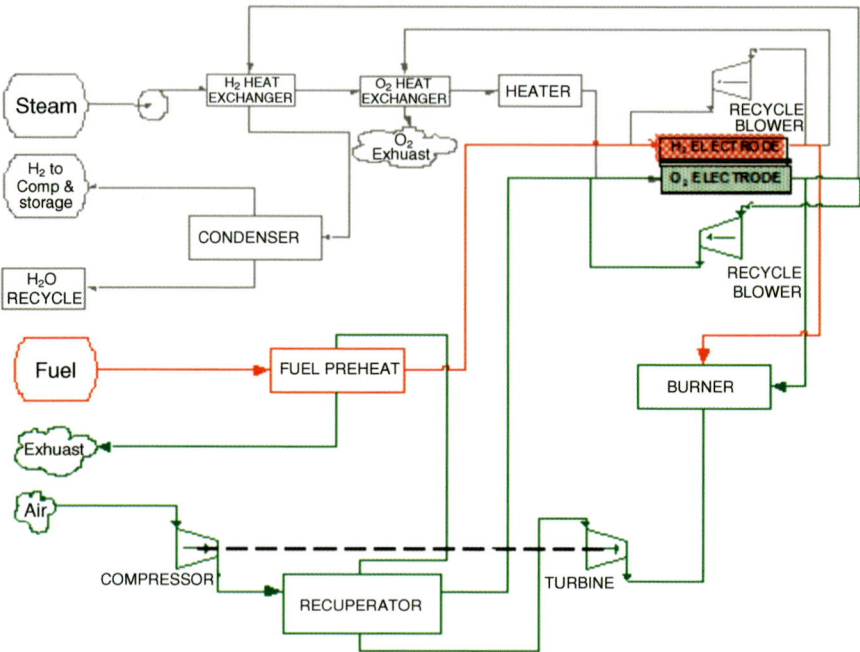

Figure 16.30 Simplified RSOFC system schematic (steam line for electrolysis mode operation and fuel and air lines for fuel cell mode operation) [2].

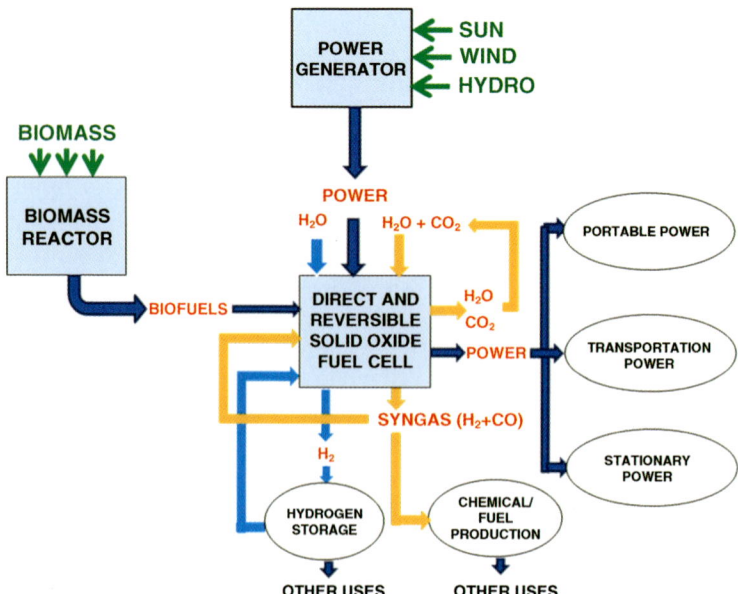

Figure 16.31 Example of a RSOFC based sustainable energy system.

uses a renewable energy supply (e.g., solar, wind, hydro) to produce hydrogen (from H_2O) or syngas ($H_2 + CO$) (from mixtures of H_2O and CO_2). The chemicals produced can be used to generate power by the same RSOFC operating in the fuel cell mode or can be stored or converted into other chemicals/fuels for subsequent use. Similarly, the RSOFC can generate power from biomass-derived fuels and the electricity generated can then be used for various power generation applications.

16.6
Summary

The RSOFC has several attractive features and several desired characteristics (demonstrated or potential) to serve as a base technology for future energy systems. A RSOFC can function as a SOFC to generate power from various fuels (fuel cell mode) and as a SOEC to produce hydrogen from steam or syngas from mixtures of steam and carbon dioxide when integrated with an energy source (electrolysis mode). The SOFC has been considered for a broad spectrum of power generation applications and significant progress has been made in demonstrating hardware/prototype/pre-commercial systems and improving several important operating properties such as performance, performance degradation/life, and fuel utilization. The SOEC has been considered for hydrogen and syngas production for commercial and other uses and oxygen generation/recovery for space applications. The SOEC has shown high performance under appropriate conditions; however, performance stability and life at present remain the key technical challenges. The RSOFC currently being developed derives from SOFC/SOEC technology. The feasibility of RSOFC operation has been demonstrated. Performance reversibility and stable reversible operation are the key requirements of this technology, in addition to those required for SOFCs/SOECs. The RSOFC is at its early stage of development and, to date, only limited work, mainly on a laboratory scale, has been conducted on this technology. Advancements in RSOFCs will continue to leverage progress in SOFC/SOEC technology.

References

1 U.S. Energy Information Administration (2013) International Energy Outlook 2013, Report DOE/EIA-0484.
2 Minh, N.Q. (2012) Direct and reversible solid oxide fuel cell energy systems. Presented at the Proceedings of 10th European SOFC Forum, Lucerne, Switzerland, Paper A1106.
3 Isenberg, A.O. (1981) Energy conversion via solid oxide electrolyte electrochemical cells at high temperatures. *Solid State Ionics*, **3–4**, 431–437.
4 Spacil, H.S. and Tedmon, C.S. Jr. (1969) Electrochemical dissociation of water vapor in solid oxide electrolyte cells: I. Thermodynamics and cell characteristics. *J. Electrochem. Soc.*, **116**, 1618–1626.
5 Spacil, H.S. and Tedmon, C.S. Jr. (1969) Electrochemical dissociation of water vapor in solid oxide electrolyte

cells: II. Materials, fabrication, and properties. *J. Electrochem. Soc.*, **116**, 1627–1633.

6 Iwahara, H., Uchida, H., and Yamasaki, I. (1987) High-temperature steam electrolysis using $SrCeO_3$-based proton conductive solid electrolyte. *Int. J. Hydrogen Energy*, **12**, 73–77.

7 Iwahara, H., Asakura, Y., Katahira, K., and Tanaka, M. (2004) Prospect of hydrogen technology using proton-conducting ceramics. *Solid State Ionics*, **168**, 299–310.

8 Minh, N.Q. (2012) System technology for solid oxide fuel cells, in *Fuel Cell Science and Engineering, Materials, Processes, Systems and Technology*, vol. **2** (eds D. Stolten and B. Emonts), Wiley-VCH Verlag GmbH, Weinheim, pp. 963–1010.

9 Weissbart, J. and Smart, W.H. (1967) Study of Electrolytic Dissociation of Carbon Dioxide-Water Using a Solid Oxide Electrolyte, NASA Report CR-680.

10 Weissbart, J., Smart, W.H., Inami, S.H., McCullough, C.M., and Ring, S.A. (1969) Development of a Carbon Dioxide-Water Solid Oxide Electrolyte Electrolysis System: First Annual Report, NASA Report CR-73358.

11 Minh, N.Q. (1993) Ceramic fuel cells. *J. Am. Ceram. Soc.*, **76**, 563–588.

12 Minh, N.Q. and Takahashi, T. (1995) *Science and Technology of Ceramic Fuel Cells*, Elsevier, Amsterdam.

13 Singhal, S.C. and Kendall, K. (eds) (2003) *High Temperature Sold Oxide Fuel Cells: Fundamentals, Design and Applications*, Elsevier, Oxford, UK.

14 Kendall, K., Minh, N.Q., and Singhal, S.C. (2003) Cell and stack designs, in *High Temperature Sold Oxide Fuel Cells: Fundamentals, Design and Applications* (eds S.C. Singhal and K. Kendall), Elsevier, Oxford, UK, pp. 197–228.

15 Fergus, J.W. (2006) Electrolytes for solid oxide fuel cells. *J. Power Sources*, **162**, 30–40.

16 Yamada, T., Chitose, N., Eto, H., Yamada, M., Hosoi, K., Komada, N., Inagaki, T., Nishikawi, F., Hashino, K., Yoshida, H., Kawano, M., Yamasaki, S., and Ishihara, T. (2008) Application of lanthanum gallate based oxide electrolyte in solid oxide fuel cell stack, in *Advances in Solid Oxide Fuel Cells III*, Ceramic and Engineering Science Proceeding, vol. **28** (eds N.P. Bansal, J. Salem, and D. Zhu), American Ceramic Society, Westerville, OH, pp. 79–89.

17 Akbay, T. (2009) Intermediate-temperature solid oxide fuel cells using $LaGaO_3$, in *Perovskite Oxide for Solid Oxide Fuel Cells* (ed. T. Ishihara), Springer, New York, pp. 183–203.

18 Mogensen, M., Sammes, N.M., and Tompsett, G.A. (2000) Physical, chemical and electrochemical of pure and doped ceria. *Solid State Ionics*, **129**, 63–94.

19 Tucker, M.C. (2010) Progress in metal supported solid oxide fuel cells: a review. *J. Power Sources*, **195**, 4570–4582.

20 Wachsman, E.D. and Lee, K.T. (2011) Lowering the temperature of solid oxide fuel cells. *Science*, **334**, 935–939.

21 Minh, N.Q., Singhal, S.C., and Williams, M.C. (2009) Solid oxide fuel cells: development activities, trends, and technological challenges, in *Fuel Cell Seminar 2008*, ECS Transactions, vol. **17**, No. (1) (eds M. Williams, K. Krist, and N. Garland), The Electrochemical Society, Pennington, NJ, pp. 211–219.

22 Kim, J.-.W., Virkar, V.A., Fung, K.-Z., Mehta, K., and Singhal, S.C. (1999) Polarization effects in intermediate temperature, anode-supported solid oxide fuel cells. *J. Electrochem. Soc.*, **146**, 69–78.

23 Ghezel-Ayagh, H. (2011) Progress in SECA coal-based program. Presented at the 12th Annual SECA Workshop. Available at http://seca.doe.gov/publications/proceedings/11/seca/index.html (accessed 8 December 2014).

24 Ghezel-Ayagh, H. (2012) Progress in coal-based SOFC technology development. Presented at the 13th Annual SECA Workshop. Available at http://www.alrc.doe.gov/publications/proceedings/12/seca/index.html (accessed 8 December 2014).

25 Narasimhamurthy, P. and Kerr, R. (2011) SOFC program review and update. Presented at the 12th Annual SECA Workshop. Available at http://seca.doe.gov/publications/proceedings/11/seca/index.html (accessed 8 December 2014).

26 Sholklapper, T.Z., Jacobson, C.P., Visco, S.J., and De Jonghe, L.C. (2008) Synthesis of dispersed and contiguous nanoparticles in solid oxide fuel cell electrodes. *Fuel Cells*, **5**, 303–312.

27 Vohs, J.M. and Gorte, R.J. (2009) High-performance SOFC cathodes prepared by infiltration. *Adv. Mater.*, **21**, 943–956.

28 National Energy Technology Laboratory (2013) Recent Solid Oxide Fuel Cell Cathode Studies, Report DOE/NETL-2013/1618.

29 Gong, Y., Palacio, D., Song, X., Patel, R.L., Liang, X., Zhao, X., Goodenough, J.B., and Huang, K. (2013) Stabilizing nanostructured solid oxide fuel cell cathode with atomic layer deposition. *Nano Lett.* **13** (9), 4340–4345.

30 Minh, N.Q. (2010) Solid oxide fuel cells for power generation and hydrogen production. *J. Korean Ceram. Soc.*, **47**, 1–7.

31 Yokokawa, H., Yamaji, K., Brito, M.E., Kishimoto, H., and Horita, T. (2011) General considerations on degradation of SOFC anodes and cathodes due to impurities in gas. *J. Power Sources*, **196**, 7070–7075.

32 Kiviaho, J. (2011) Overview of worldwide research on degradation issues in SOFC. Presented at the 2nd International Workshop on Degradation Issues of Fuel Cells, Thessaloniki, Greece.

33 Liu, Y.L., Primdahl, S., and Mogensen, M. (2003) Effects of impurities on microstructure in Ni/YSZ-YSZ Half-cells for SOFC. *Solid State Ionics*, **161**, 1–10.

34 Liu, R.R., Kim, S.H., Taniguchi, S., Oshima, T., Shiratori, Y., Ito, K., and Sasaki, K. (2011) Influence of water vapor on long-term performance and accelerated degradation of solid oxide fuel cell cathodes. *J. Power Sources*, **196**, 7090–7096.

35 Hagen, A., Neufeld, K., and Liu, Y.L. (2010) Effect of humidity in air on performance and long-term durability of SOFCs. *J. Electrochem. Soc.*, **157**, B1343–B1348.

36 Zhou, W., Liang, F., Shao, Z., and Zhu, Z. (2012) Hierarchival CO_2-protective shell for highly efficient oxygen reduction reaction. *Sci. Rep.*, **2**, 327. doi: 10.1038/srep00327

37 Blum, L., Packbier, U., Vinke, I.C., and de Haart, L.G.J. (2013) Long-term testing of SOFC stacks at Forschungszentrum Jülich. *Fuel Cells*, **13**, 646–653.

38 GE Hybrid Power Generation Systems (2006) Solid State Energy Conversion Alliance (SECA) Solid Oxide Fuel Cell Program, Final Report.

39 Park, S., Vohs, J.M., and Gorte, R.J. (2000) Direct oxidation of hydrocarbons in a solid-oxide fuel cell. *Nature*, **404**, 265–267.

40 McIntosh, S. and Gorte, R.J. (2004) Direct carbon solid oxide fuel cells. *Chem. Rev.*, **104**, 4845–4865.

41 Armstrong, E.N., Park, J.-W., and Minh, N.Q. (2012) High-performance direct ethanol solid oxide fuel cells. *Electrochem. Solid-State Lett.*, **15**, B75–B77.

42 Dönitz, W. and Erdle, E. (1985) High-temperature electrolysis of water vapor – status of development and perspectives of application. *Int. J. Hydrogen Energy*, **10**, 291–295.

43 Jensen, S.H., Larsen, P.H., and Mogensen, M. (2007) Hydrogen and synthetic fuel production from renewable energy source. *Int. J. Hydrogen Energy*, **32**, 3253–3257.

44 Guan, J., Doshi, D., Lear, G., Montgomery, K., Ong, E., and Minh, N. (2002) Ceramic oxygen generators with thin-film zirconia electrolytes. *J. Am. Ceram. Soc.*, **85**, 2651–2654.

45 Laguna-Bercero, M.A. (2012) Recent advances in high-temperature electrolysis using solid oxide fuel cells. *J. Power Sources*, **203**, 4–16.

46 Brisse, A., Schefold, J., and Zahid, M. (2008) High temperature water electrolysis in solid oxide cells. *Int. J. Hydrogen Energy*, **33**, 5375–5382.

47 Eguchi, K., Hatagishi, T., and Arai, H. (1996) Power generation and steam electrolysis characteristics of an electrochemical cell with a zirconia- or ceria-based electrolyte. *Solid State Ionics*, **86–8**, 1245–1249.

48 Kim-Lohsoontorn, P., Brett, D.J.L., Laosiripojana, N., Kim, Y.M., and Bae,

J.M. (2010) Performance of solid oxide electrolysis cells based on composite La0.8Sr0.2MnO3-delta - yttria stabilized zirconia and Ba0.5Sr0.5Co0.8Fe0.2O3-delta oxygen electrodes. *Int. J. Hydrogen Energy*, **35**, 3958–3966.

49 Jensen, S.H., Sun, X.F., Ebbesen, S.D., Knibbe, R., and Mogensen, M. (2010) Hydrogen and synthetic fuel production using pressurized solid oxide electrolysis cells. *Int. J. Hydrogen Energy*, **35**, 9544–9549.

50 Yang, C.H., Coffin, A., and Chen, F.L. (2010) High temperature solid oxide electrolysis cell employing porous structured (La0.75Sr0.25)(0.95)MnO3 with enhanced oxygen electrode performance. *Int. J. Hydrogen Energy*, **35**, 3221–3226.

51 Liang, M.D., Yu, B., Wen, M.F., Chen, J., Xu, J.M., and Zhai, Y.C. (2009) Preparation of LSM-YSZ composite powder for anode of solid oxide electrolysis cell and its activation mechanism. *J. Power Sources*, **190**, 341–345.

52 Yu, B., Zhang, W.Q., Chen, J., Xu, J.M., and Wang, S.R. (2008) Advance in highly efficient hydrogen production by high temperature steam electrolysis. *Sci. China Ser. B*, **51**, 289–304.

53 He, F., Song, D., Peng, R., Meng, G., and Yang, S. (2010) Electrode performance and analysis of reversible solid oxide fuel cells with proton conducting electrolyte of BaCe0.5Zr0.3Y0.2O3-[delta]. *J. Power Sources*, **195**, 3359–3364.

54 Osada, N., Uchida, H., and Watanabe, M. (2006) Polarization behavior of SDC cathode with highly dispersed Ni catalysts for solid oxide electrolysis cells. *J. Electrochem. Soc.*, **153**, A816–A820.

55 Wang, W.S., Huang, Y.Y., Jung, S.W., Vohs, J.M., and Gorte, R.J. (2006) A comparison of LSM, LSF, and LSCo for solid oxide electrolyzer anodes. *J. Electrochem. Soc.*, **153**, A2066–A2070.

56 Chauveau, F., Mougin, J., Bassat, J.M., Mauvy, F., and Grenier, J.C. (2010) A new anode material for solid oxide electrolyser: The neodymium nickelate Nd2NiO4+delta. *J. Power Sources*, **195**, 744–749.

57 Schiller, G., Ansar, A., Lang, M., and Patz, O. (2009) High temperature water electrolysis using metal supported solid oxide electrolyser cells (SOEC). *J. Appl. Electrochem.*, **39**, 293–301.

58 Ishihara, T., Jirathiwathanakul, N., and Zhong, H. (2010) Intermediate temperature solid oxide electrolysis cell using LaGaO3 based perovskite electrolyte. *Energy Environ. Sci.*, **3**, 665–672.

59 Elangovan, S., Hartvigsen, J.J., and Frost, L.J. (2007) Intermediate temperature reversible fuel cells. *Int. J. Appl. Ceram. Tec.*, **4**, 109–118.

60 Liu, Q.A., Yang, C.H., Dong, X.H., and Chen, F.L. (2010) Perovskite Sr2Fe1.5Mo0.5O6-delta as electrode materials for symmetrical solid oxide electrolysis cells. *Int. J. Hydrogen Energy*, **35**, 10039–10044.

61 Tsekouras, G., Neaqu, D., and Irvine, J.T.S. (2013) Step-change in high temperature electrolysis performance of perovskite oxide cathodes with exsolution of B-site dopants. *Energy Environ. Sci.*, **6**, 256–266.

62 Li, Y., Wang, Y., Doherty, W., Xie, K., and Wu, Y. (2013) Perovskite chromates cathodes with exsolved iron nanoparticles for direct high-temperature steam electrolysis. *ACS Appl. Mater. Interfaces*, **5**, 8553–8562.

63 O'Brien, J.E., Zhang, X., O'Brien, C.R., and Tao, G. (2011) High temperature steam electrolysis: demonstration of improved long term performance. Presented at Fuel Cell Seminar and Exposition, Orlando, FL, 31 October to 3 November 2011. Available at http://www.fuelcellseminar.com/media/9066/hrd34-5.pdf.

64 Stoots, C.M., O'Brien, J.E., Condie, K.G., and Hartvigsen, J.J. (2010) High-temperature electrolysis for large-scale hydrogen production from nuclear energy - experimental investigations. *Int. J. Hydrogen Energy*, **35**, 4861–4870.

65 Herring, S., Lessing, P.A., O'Brien, J.E., and Stoots, C.M. (2004) High Temperature Solid Electrolyzer System, 2004 DOE Hydrogen, Fuel Cells &

Infrastructure Technologies Program Review.
66. Herring, S. (2005) High Temperature Solid Electrolyzer System, 2005 DOE Hydrogen, Fuel Cells & Infrastructure Technologies Program Review.
67. Hauch, A., Ebbesen, S.D., Jensen, S.H., and Mogensen, M. (2008) Highly efficient high temperature electrolysis. *J. Mater. Chem.*, **18**, 2331–2340.
68. GE Hybrid Power Generation Systems (2008) High Performance Flexible Reversible Solid Oxide Fuel Cell, Final Technical Report.
69. Knibbe, R., Traulsen, M.L., Hauch, A., Ebbesen, S.D., and Mogensen, M. (2010) Solid oxide electrolysis cells: degradation at high current densities. *J. Electrochem. Soc.*, **157**, B1209–B1217.
70. Hauch, A., Jensen, S.H., Bilde-Sørensen, J.B., and Mogensen, M. (2007) Silica segregation in the Ni/YSZ electrode. *J. Electrochem. Soc.*, **154**, A619–A626.
71. Graves, C., Ebbesen, S.D., and Mogensen, M. (2011) Co-electrolysis of CO_2 and H_2O in solid oxide cells: performance and durability. *Solid State Ionics*, **192**, 398–403.
72. Hauch, A., Ebbesen, S.D., Jensen, S.H., and Mogensen, M. (2008) Solid oxide electrolysis cells: microstructure and degradation of the nickel/yttria-stabilized zirconia electrode. *J. Electrochem. Soc.*, **155**, B1184–B1193.
73. Sohal, M.S., O'Brien, J.E., Stoots, C.M., Sharma, V.I., Yildiz, B., and Virkar, A. (2010) Degradation Issues in Solid Oxide Cells During High Temperature Electrolysis, INL/CON-10-18163.
74. Ebbesen, S.D., Høgh, J., Nielsen, K.R., Nielsen, J.U., and Mogensen, M. (2011) Durable SOC stacks for production of hydrogen and synthesis gas by high temperature electrolysis. *Int. J. Hydrogen Energy*, **36**, 7363–7373.
75. Chen, M., Høgh, J.V.T., Nielsen, J.U., Benzen, J.J., Ebbesen, S.D., and Hendriksen, P.V. (2013) High temperature co-electrolysis of steam and CO_2 in an SOC stack: performance and durability. *Fuel Cells*, **13**, 638–645.
76. Hauch, A., Jensen, S.H., Ebbesen, S.D., and Mogensen, M. (2007) Durability of solid oxide electrolysis cells for hydrogen production. Presented at Energy Solutions for Sustainable Development Proceedings, Risø Energy International Conference 2007, Risø-R-1608 (EN)/ed. L.S. Petersen and H. Larsen, pp. 327–338. Available at http://orbit.dtu.dk/fedora/objects/orbit:53431/datastreams/file_479b4bff-4bfe-4cb4-a5c8-bebab1e3bcdb/content
77. Minh, N., Chung, B., Doshi, R., Montgomery, K., Ong, E., Reddig, M., MacKnight, A., and Fuhs, S. (1998) Zirconia Electrolysis Cells for Oxygen Generation from Carbon Dioxide for Mars In-Situ Resource Utilization, SAE Technical Paper 981655, Warrendale, PA SAE International.
78. Sridhar, K.R. and Vaniman, B.T. (1997) Oxygen production on Mars using solid oxide electrolysis. *Solid State Ionics*, **93**, 321–328.
79. Isenberg, A.O. and Verostko, C.E. (1989) Carbon Dioxide and Water Vapor High Temperature Electrolysis, SAE Technical Paper 891506, Warrendale, PA, SAE International.
80. McKellar, M.G., Stoots, C.M., Sohal, M.S., Mulloh, L.M., Luna, B., and Abney, M.B. (2010) The Concept and Analytical Investigation of Carbon Dioxide and Steam Co-Electrolysis for Resource Utilization in Space Exploration, AIAA Paper 2010–6273, Reston, VA, American Institute of Aeronautics and Astronautics.
81. Fu, Q. (2011) Role of electrolysis in regenerative syngas and synfuel production, in *Syngas: Production, Applications and Environmental Impact* (eds A. Indarto and J. Palgunadi), Nova Science Publishers, Hauppauge, NY, pp. 209–240.
82. Jensen, S.H., Larsen, P.H., and Mogensen, M. (2007) Hydrogen and synthetic fuel production from renewable energy sources. *Int. J. Hydrogen Energy*, **32**, 3253–3257.
83. O'Brien, J.E., Stoots, C.M., Herring, J.S., and Hartvigsen, J.J. (2007) High-Temperature Co-Electrolysis of Steam and Carbon Dioxide for Direct Production of Syngas; Equilibrium Model

and Single-Cell Tests, INL/CON-07-12241.
84 Stoots, C.M., O'Brien, J.E., Herring, J.S., Condie, K.G., and Hartvigsen, J.J. (2008) Idaho National Laboratory Experimental Research in High Temperature Electrolysis for Hydrogen and Syngas Production, INL/CON-08-14622.
85 Kim-Lohsoontorn, P. and Bae, J. (2011) Electrochemical performance of solid oxide electrolysis cell electrodes under high-temperature coelectrolysis of steam and carbon dioxide. *J. Power Sources*, **196**, 7161–7168.
86 Bidrawn, F., Kim, G., Corre, G., Irvine, J.T.S., Vohs, J.M., and Gorte, R.J. (2008) Efficient reduction of CO_2 in a solid oxide electrolyzer. *Electrochem. Solid-State Lett.*, **11**, B167–B170.
87 Stoots, C.M., Hartvigsen, J.J., O'Brien, J.E., and Herring, J.S. (2008) Syngas production via high-temperature coelectrolysis of steam and carbon dioxide. *J. Fuel Cell Sci. Technol.*, **6**, 011014.
88 Sridhar, K.R. and Foertner, R. (2000) Regenerative solid oxide fuel cell for mars. *J. Propulsion Power*, **16**, 1105–1111.
89 Weng, D. and Yates, S. (2010) Solid Oxide Electrolysis Stack for Closed Loop Oxygen Generation System, AIAA Paper 2010–8675, Reston, VA, American Institute of Aeronautics and Astronautics.
90 Rivera-Tinoco, R., Mansilla, C., Bouallou, C., and Werkoff, F.F. (2008) Hydrogen production by high temperature electrolysis coupled with an EPR, SFR or HTR: techno-economic study and coupling possibilities. *Int. J. Nuclear Hydrogen Prod. Appl.*, **1**, 249–266.
91 Utgikar, V. and Thiesen, T. (2006) Life cycle assessment of high temperature electrolysis for hydrogen production via nuclear energy. *Int. J. Hydrogen Energy*, **31**, 939–944.
92 Forsberg, C.W. and Kazimi, M.S. (2009) Nuclear Hydrogen Using High-Temperature Electrolysis and Light-Water Reactors for Peak Electricity Production, MIT-NES-TR-10.
93 Padin, J., Veziroglu, T.N., and Shahin, A. (2000) Hybrid solar high-temperature hydrogen production system. *Int. J. Hydrogen Energy*, **25**, 295–317.
94 Frost, L., Hartvigsen, J., and Elangovan, S. Formation of Synthesis Gas Using Solar Concentrator Photovoltaics (SCPV) and High Temperature Co-Electrolysis (HTCE) of CO_2 and H_2O. Available at http://www.ceramatec.com/documents/syngas-production/SOEC/Formation-of-Synthesis-Gas-Using-Solar-Concentrator-Photovoltaics-(SCPV)-and-High-Temperature-Co-electrolysis-(HTCE)-of-CO2-and-H2O.pdf.
95 Hartvigsen, J., Frost, L., and Elangovan, S. (2012) High Temperature Co-Electrolysis of Steam and Carbon Dioxide Using Wind Power, http://www.fuelcellseminar.com/media/51332/sta44.4.pdf.
96 Mogensen, M., Jensen, S.H., Ebbesen, S.D., Hauch, A., Graves, C., Høgh, J.V.T., Sun, X., Das, S., Hendriksen, P.V., Nielsen, J.U., Pedersen, A.H., Christiansen, N., and Hansen, J.B. (2012) Production of "green natural gas" using solid oxide electrolysis cells (SOEC): status of technology and costs. Presented at the 2012 World Gas Conference, Kuala Lumpur, Malaysia.
97 Xu, N., Li, X., Zhao, X., Goodenough, J.B., and Huang, K. (2011) A novel solid oxide redox flow battery for grid energy storage. *Energy Environ. Sci.*, **4**, 4942–4946.
98 Inoishi, A., Ida, S., Uratani, S., Okano, T., and Ishihara, T. (2012) High capacity of an Fe-air rechargeable air battery using $LaGaO_3$-based oxide ion conductor as an electrolyte. *Phys. Chem. Chem. Phys.*, **14**, 12818–12822.
99 Inoishi, A., Ju, Y.W., Ida, S., and Ishihara, T. (2013) Fe-air rechargeable battery using oxide ion conducting electrolyte of Y_2O_3 stabilized ZrO_2. *J. Power Sources*, **229**, 12–15.
100 Inoishi, A., Okamoto, Y., Ju, J.W., Ida, S., and Ishihara, T. (2013) Oxidation rate of Fe and electrochemical performance of Fe-air solid oxide rechargeable battery using $LaGaO_3$ based oxide ion conductor. *RSC Adv.*, **3**, 8820–8825.

101 Ruiz-Morales, J.C., Marrero-López, D., Canales-Vázpuez, J., and Irvine, J.T.S. (2011) Symmetric and reversible solid oxide fuel cells. *RSC Adv.*, **1**, 1403–1414.

102 Jung, G.B., Chen, J.Y., Lin, C.Y., and Chan, S.H. (2011) Development of reversible solid oxide fuel cell for power generation and hydrogen production, in *Energy Harvesting and Storage: Materials, Devices, and Applications II, 6 July 2011*, Proceedings SPIE, SPIE, vol. **8035**, doi: 10.1117/12.883654

103 Eguchi, K., Hatagishi, T., and Arai, H. (1996) Power generation and steam electrolysis characteristics of an electrochemical cell with a zirconia- or ceria-based electrolyte. *Solid State Ionics*, **86–88**, 1245–1249.

104 Laguna-Bercero, M.A., Kilner, J.A., and Skinner, S.J. (2011) Development of oxygen electrodes for reversible solid oxide fuel cells with scandia stabilized zirconia electrolytes. *Solid State Ionics*, **192**, 501–504.

105 Cheng, C.H., Su, A., and Chan, S.H. (2012) Computer aided design of the reversible solid oxide fuel cell air electrode. *Adv. Mater. Res.*, **485**, 532–535.

106 Tao, G. and Virkar, A. (2006) A Reversible Planar Solid Oxide Fuel-Fed Electrolysis Cell and Solid Oxide Fuel Cell for Hydrogen and Electricity Production Operating on Natural Gas/Biogas, 2006 DOE Hydrogen Program Annual Review. Available at http://www.hydrogen.energy.gov/pdfs/review06/pdp_33_tao.pdf.

107 Tao, G. and Virkar, A. (2011) Lessons Learned from SOFC/SOEC Development. Available at http://www1.eere.energy.gov/hydrogenandfuelcells/pdfs/rev_fc_wkshp_tao.pdf.

108 Brown, C. (2011) Progress on the Development of Reversible SOFC Stack Technology. Available at http://www1.eere.energy.gov/hydrogenandfuelcells/pdfs/rev_fc_wkshp_brown.pdf.

Part 2.4
H_2 from Biomass

17
Assessment of Selected Concepts for Hydrogen Production Based on Biomass

Franziska Müller-Langer, Konstantin Zech, Stefan Rönsch, Katja Oehmichen, Julia Michaelis, Simon Funke, and Elias Grasemann

17.1
Introduction

The multipurpose fuel hydrogen is applied as energy carrier or bulk chemical for many different applications. Global annual hydrogen production was about 53 mn t in 2010, mainly driven by the refinery and ammonia industry. Currently, merchant hydrogen accounts about 12% of the total on-purpose production, the rest is captive-produced [1]. Hydrogen is produced by numerous conversion pathways, today mainly based on natural gas, naphtha, and oil reforming, followed by coal gasification and to a small extent by electrolysis.

Hydrogen based on biomass can be provided by diverse conversion routes (Figure 17.1) which can be structured into (i) biomass production and supply, (ii) conversion into biohydrogen, (iii) infrastructure for hydrogen supply, and (iv) energetic application. Furthermore, biohydrogen can be produced via thermo- or biochemical conversion routes or via electrolysis from biomass-based electricity.

Thermochemical processes are characterized by the conversion of solid, liquid, or gaseous biofuels at defined reaction conditions (temperatures, pressure, catalysts, and/or other reactants). For biohydrogen, mainly gasification of solid biofuels or biomass based intermediates (e.g., via hydrothermal processing, torrefaction, or pyrolysis) followed by product gas treatment. However, gasification and gas cleaning have to be adjusted to generate a hydrogen-rich gas. Such concepts are currently in a development stage comparable to technology validation in the relevant environment. The other alternative is to apply reforming or partial oxidation of energy resources that have been produced previously via different physical- or biochemical and thermochemical routes. For instance, glycerin reforming is at the pilot stage. Like natural gas reforming, biomethane (biogas) can also be used [3].

Another option for biohydrogen is production via microorganisms, such as (i) biomass fermentation using heterotrophic or photoheterotrophic bacteria or (ii) direct or indirect water splitting by green algae or cyano bacteria [4–7]. Despite intensive research activities biochemically produced hydrogen is still at the laboratory stage of development [8].

Hydrogen Science and Engineering: Materials, Processes, Systems and Technology, First Edition.
Edited by Detlef Stolten and Bernd Emonts.
© 2016 Wiley-VCH Verlag GmbH & Co. KGaA. Published 2016 by Wiley-VCH Verlag GmbH & Co. KGaA.

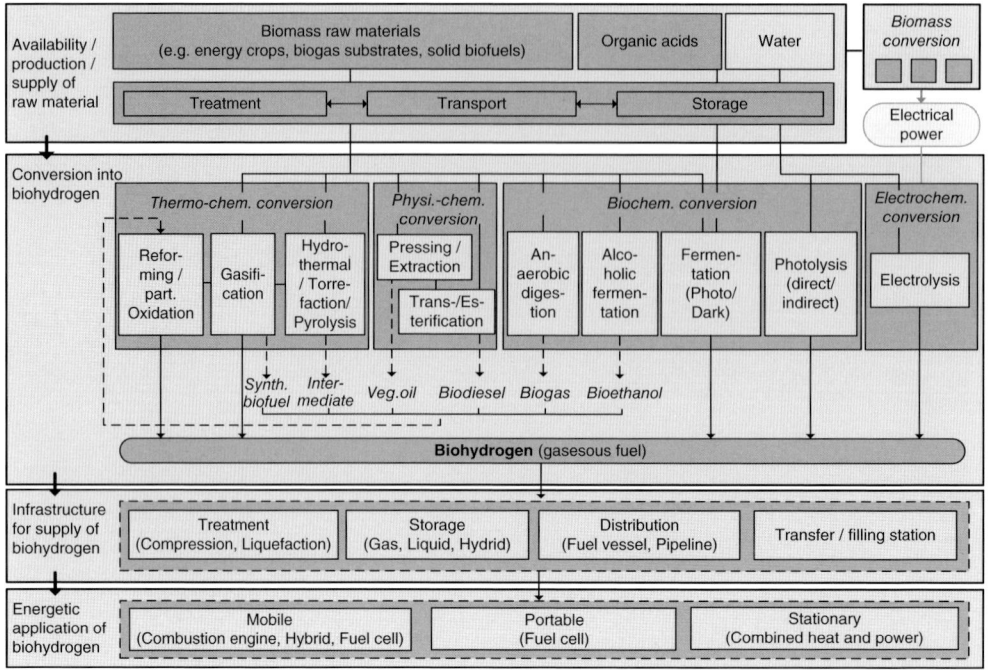

Figure 17.1 General overview of conversion routes for hydrogen based on biomass (based on Reference [2]).

Of course, biohydrogen can also be provided via different commercially available electrolysis processes using biomass-based electrical power.

Against this background, this chapter deals with a selection of biomass based hydrogen concepts, via thermochemical or in combination with biochemical production, that are close to demonstration and used as fuel for mobility. They are briefly characterized regarding their technical characteristics and analyzed regarding certain technical, economic, and environmental aspects. The focus is on the most important drivers for efficiency, greenhouse gas (GHG) emissions, and costs.

17.2
Characteristics of Selected Hydrogen Concepts

Here, an overview of certain biohydrogen concepts under development is given followed by a detailed description of exemplary concepts that will then be assessed.

17.2.1
Concepts under Discussion

Table 17.1 shows a brief summary of technical characteristics, state of technology and research, and industrial players for different hydrogen approaches,

Table 17.1 Technical characteristics of thermochemical and combined biohydrogen approaches (selection).

Biohydrogen approach[a]	Typical raw materials	Technical characteristics	State of development[b]	Player in research and industry[c]	Remarks on data availability[d], references
Chrisgas	Wood	Autothermal fluidized bed gasifier as CHP (combined heat and power) about 18 MW thermal input demonstrated, for hydrogen not implemented	Demo for CHP, technical lab for hydrogen, TRL 4	Foster Wheeler, Chrisgas project in Värnamo, Sweden	Moderate [9–12]
Renugas	Wood, residues from cellulose industry	Autothermal fluidized bed gasifier as CHP about 12 MW thermal input demonstrated, for hydrogen (here via membrane) not implemented	Demo for CHP, technical lab for hydrogen, TRL 4	Gas Technology Institute, USA	Moderate [13–16]
Fast internally circulating fluidized bed (FICFB)	Wood, partly straw	Allothermal twin bed fluidized bed gasifier as CHP in different scales (e.g., 4.5 to >8 MW thermal input), realized as demo for methanation to SNG (synthetic natural gas), adaption to hydrogen not implemented	Commercial for CHP, Demo plant for SNG, pilot for hydrogen TRL 5	Technical University of Vienna, Repotec, Austria	Good [9,17–20]
Absorption enhanced reforming (AER)	Wood, partly straw	Allothermal twin bed fluidized bed gasifier incl. chemical loop with focus on hydrogen production, 100 kW thermal input as pilot, tests in Güssing	Technical lab for hydrogen, test in demo plant in Güssing TRL 5	Center for solar energy and hydrogen research Baden-Württemberg (ZSW), Germany	Good [9,21–25]
Heat-pipe-reformer	Wood	Allothermal twin bed fluidized bed gasifier up to 5 MW thermal input, currently only CHP, for hydrogen not implemented	Technical lab for hydrogen, TRL 4	Technical University of Munich, Agnion, HSE, Germany	Moderate [26–30]

(continued)

Table 17.1 (Continued)

Biohydrogen approach[a]	Typical raw materials	Technical characteristics	State of development[b]	Player in research and industry[c]	Remarks on data availability[d], references
Milena	Wood	Allothermal twin bed fluidized bed gasifier up to 800 kW thermal input as pilot, demo plant for CHP, for hydrogen not implemented	Pilot for hydrogen TRL 4	Energy research Center of the Netherlands (ECN)	Moderate [31–36]
SilvaGas	Wood, agricultural residues, energy crops	Allothermal twin bed fluidized bed gasifier of 85 kW thermal input as CHP demo, other focus with adapted gasifier on BTL and hydrogen	Pilot for hydrogen TRL 4	Rentech Inc., USA	Poor [31,37–40]
BioLiq	Straw, wood, and the like	Pyrolysis to slurry combined with entrained flow gasifier (5 MW thermal input), focus on BTL production, adaption to hydrogen possible in general	Pilot for BTL, for hydrogen TRL 4	Karlsruhe Institute of Technology (KIT), Lurgi, CAC, Germany	Moderate [41–43]
Carbo-V	Wood	Multistep entrained flow gasifier realized as pilot in 1 MW thermal input and demo in 45 MW thermal input (not running), focus on BTL production, adaption to hydrogen possible in general	Demo for BTL, for hydrogen TRL 5	Linde Engineering (former owned by Choren), Germany	Poor [44–47]
Verena	For example, maize silage	Hydrothermal gasifier at technical scale for simultaneous hydrogen and methane production	Technical lab for hydrogen, TRL 4	Karlsruhe Institute of Technology (KIT), Germany	Moderate [48–50]

Concept	Feedstock	Description	Status	Examples	Data quality
Blue Tower	Grass cuttings	Staged reforming of pyrolysis gas as 1 MW thermal input pilot, demo planned, target polygeneration of hydrogen, CHP	Pilot for hydrogen TRL 5	Solar millennium	Poor [51,52]
Steam reforming of intermediates	Maize silage, manure, organic waste, straw	Anaerobic digestion to biogas followed by steam reforming realized as pilot. Glycerin pyro-reforming realized as pilot	Commercial as CHP and for grid injection of biomoethane, commercial components for hydrogen demo for hydrogen TRL 6	Various, for example, compact reformer by Haldor Topsoe, Denmark, H2Gen Inc., USA, Mahler AGS, Germany Pyro-reforming by Linde, Germany	Good [53–56]

a) Related to applied gasification concepts and/or overall concept.
b) Focus on hydrogen production concept, according to technology readiness level (TRL) of the European Commission, which outlines in detail the different research and deployment steps (1 – basic principles observed, 2 – technology concept formulated, 3 – experimental proof of concept, 4 – technology validation in laboratory, 5 – technology validation in relevant environment, 6 – demonstration in relevant environment, 7 – demonstration in operational environment, 8 – system completed, and qualified, 9 – successful mission operations) [57].
c) W/o entitlement of completeness.
d) Good – primary data available/published, moderate – comprehensive literature data, mainly based on manufacturer data, poor – in total only few data available.

mainly based on thermochemical biomass conversion. Moreover, the state of development and data quality are evaluated.

17.2.2
Concepts for Further Assessment

To emphasize the characteristics of different hydrogen production concepts, two exemplary concepts are analyzed. They have been selected with regard to short- and medium-term realization in practice, taking into account the (i) state of technology, (ii) demand for technical adaptions, and (iii) complexity of overall plant concept.

Following that, a thermochemical concept (concept A) is based on the FICFB (fast internally circulating fluidized bed) gasification technology using wood chips as fuel [58,59]. The combined concept (concept B) relies on anaerobic digestion technology using maize silage, manure, and organic waste as raw material followed by steam reforming [3,53]. To calculate the mass and energy flows of concept A and B a process simulation tool developed with the software MATLAB Simulink is used [60].

Moreover, a concept for hydrogen distribution is described related to the plant capacity of the production concepts.

17.2.2.1 Concept A – Wood Chip Gasification
Concept A is shown in Figure 17.2. In this concept wood chips are dried with preheated air. Subsequently, the wood chips are gasified in an allothermal fluidized bed gasifier with water steam as gasification agent. For the heat supply of the gasification reactions bed material circulating between the gasifier and a separate combustion chamber is used. The bed material is heated in the combustion chamber with tars, char, and recycle gas as fuel [58,59,61]. The raw gas leaving the gasifier is cooled in a heat exchanger, de-dusted in a precoated bag house filter, and fed into a sulfur resistant raw gas shift reactor to raise the hydrogen content of the gas. Tars are removed in an oil-based washing system, the gas is compressed, and carbon dioxide is removed in a pressurized water washing. The remaining methane from the gasification is separated from hydrogen by pressure swing adsorption. The methane is directed to a reformer unit and used as fuel for the combustion chamber of the gasification [62,63]. Process energy demand is generated within the process; the electricity demand is about 0.31 MW.

17.2.2.2 Concept B – Biogas Reforming
Figure 17.3 shows the combined bio- and thermochemical hydrogen production concept. In this concept, the raw material (i.e., maize silage, manure, and organic waste) is mashed with water. Subsequently, the biogas substrate is anaerobically digested, resulting in two product fractions, digestate and biogas. The digestate is mechanically dewatered and the water is predominantly used for the mashing process. The biogas is separated from sulfur compounds. The methane fraction of the biogas is then converted in an allothermal steam reforming process

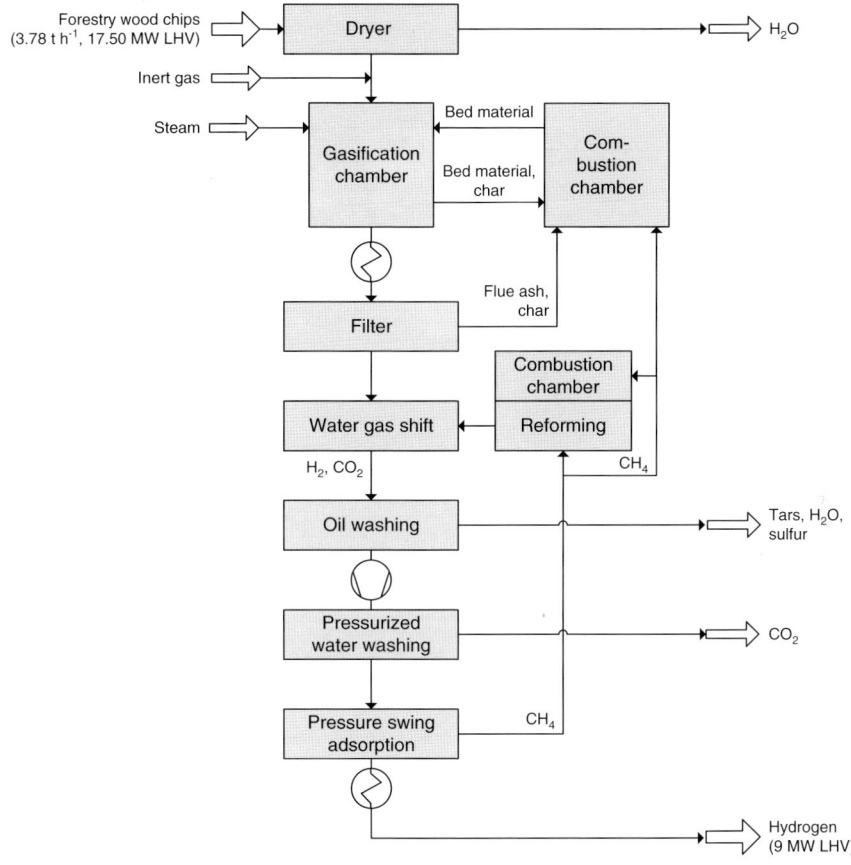

Figure 17.2 Process scheme of concept A.

followed by a water–gas-shift reactor to increase the hydrogen content of the gas. A fraction of the gas is used as fuel for the reforming process. Finally, the gas is compressed, dried, and purified by pressure swing adsorption. The total external process electricity demand is about 0.43 MW.

17.2.2.3 Concept for Hydrogen Distribution of Produced Hydrogen

Since the sites of hydrogen production and hydrogen use do not coincide, hydrogen must be transported to the consumers at, for example, filling stations. The choice of the means of transport depends on the volume and the transport distance. For large distances, the capital cost of a pipeline is high. Pipeline investment only makes sense if large quantities of hydrogen are to be transported [64]. With regard to plant capacity of the above-mentioned production concepts, it is assumed that transport is realized by trailers with pressure cylinders of 500 bar and a capacity of 1 ton of hydrogen per trailer [65].

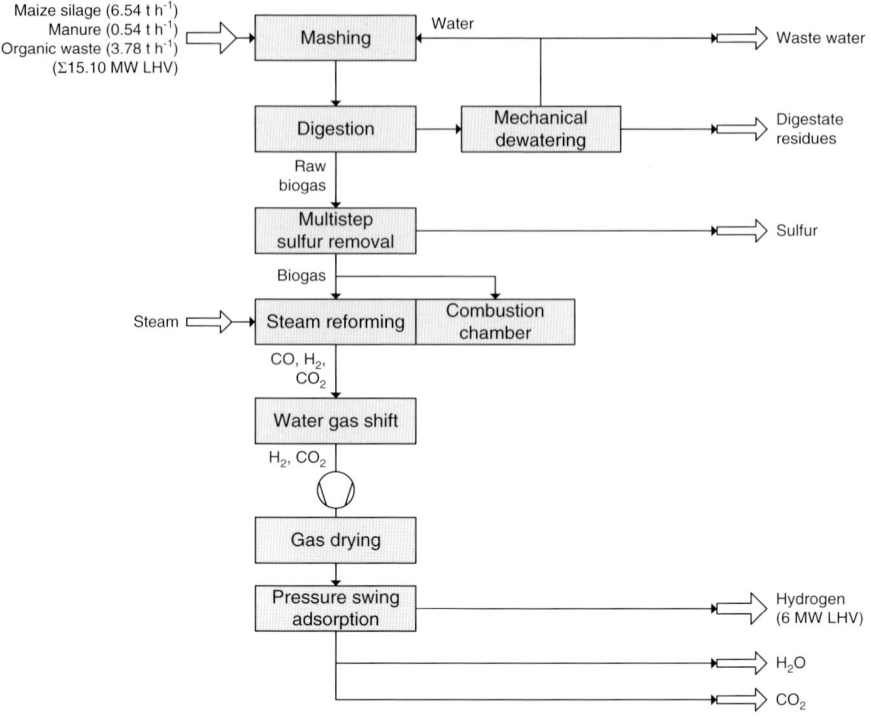

Figure 17.3 Process scheme of concept B.

The distribution concept includes compression and storage, transport, and distribution of hydrogen at the filling station. Close to or at the hydrogen production plant, hydrogen is compressed from 8 to 500 bar and filled into a storage tank. The storage tank is used directly as a trailer for hydrogen transport, avoiding additional reload. An additional trailer is included as a buffer tank. The power of a single compressor is 1.5 MW hydrogen. Depending on the plant size, six (concept A) or four (concept B) compressors are needed. Compression expenses are driven by the electricity consumption, which depends on the difference between inlet and outlet pressure level and on the necessity for a preliminary cooling. Compression in four stages including preliminary cooling consumes 0.081 kWh electricity per kWh of hydrogen.

The gaseous hydrogen is then transported in trailers to filling stations. As neither the production site nor the filling station is considered as a specific location, a mean transport distance of 150 km is assumed. At the filling station, the so-called "booster concept" is applied [65]. The delivered hydrogen is stored in stationary buffer storage of 300 bar. Before a vehicle (e.g., future fuel cell electric vehicles) is refueled, the hydrogen is compressed to a pressure level of 700 bar by a booster compressor.

The assumed annual volume of hydrogen fuel that will be sold at the filling station is 120 t of hydrogen.

17.3
Concept Assessment of Technical Aspects

In the following, concepts A and B are analyzed from a technical point of view. The methodical approach and related results are described below.

17.3.1
Methodical Approach

The energetic efficiency is the evaluation criterion for the technical assessment and is calculated based on a detailed mass and energy flow analysis. For hydrogen production it accounts for all in- and outgoing energy flows for each of concepts A and B and is defined as the ratio of outgoing to ingoing energy flows (Eq. (17.1)). For hydrogen distribution to filling station the energy demand (mainly electricity) is relevant:

$$\eta_{en,overall} = \frac{\dot{m}_{MP} H_{MP}}{\dot{m}_{RM} H_{RM} + \sum \dot{m}_{Aux} H_{Aux} + P_{ext}(+P_{dis})} \tag{17.1}$$

with:

$\eta_{en,overall}$	=	overall energetic efficiency for hydrogen production and supply;
$\dot{m}_i$	=	mass flow of main product (MP, hydrogen), raw material (RM, biomass free plant gate), and auxiliaries (Aux, including energy sources);
H_i	=	heating value of main product (MP, hydrogen), raw material (RM, biomass free plant gate), auxiliaries (Aux, including energy sources);
P_{ext}	=	demand process electricity external supply;
P_{dis}	=	demand electricity for hydrogen distribution.

17.3.2
Overall Efficiencies

Figure 17.4 shows the main results, summarizing the efficiency for hydrogen production and hydrogen distribution to a filling station.

Neglecting the energy demand of hydrogen distribution, plant concept A is characterized by an efficiency of about 50% whereas concept B reaches an efficiency of about 39%. The main reason for this efficiency contrast is the raw gas composition. While the biogas produced by concept B mainly contains carbon dioxide and methane, which has to be reformed, the raw gas produced by concept A additionally already contains hydrogen. Thus, a lower gas volume flow has to be reformed in concept A than in concept B.

Furthermore, the results show that there is a higher mass throughput for concept B than for concept A. To produce hydrogen with a power of 1 MW

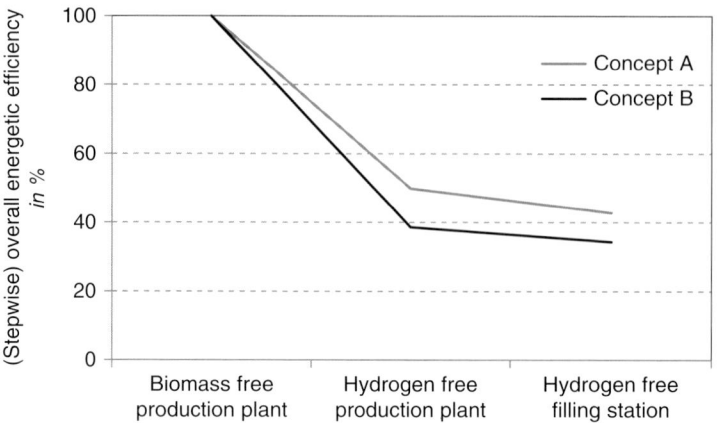

Figure 17.4 Overall energetic efficiency; "free" here means free at the factory gate.

concept B requires about $1.8\,t\,h^{-1}$ biomass (water content free plant gate), in concept A about $0.6\,t\,h^{-1}$ are used.

However, beside the efficiency benefits of concept A, the decision regarding investment in a hydrogen production plant will be influenced by several other factors. Other technical related aspects for example are the technological maturity and complexity. As biochemical conversion plants have been – in contrast to biomass gasification plants – commercially available for many years they, for example, profit from a high grade of technological maturity.

Compared to other international publications on similar thermochemical hydrogen production concepts the overall efficiency range (calculated by the same approach, cf. Eq. (17.1)) is 54–64% [66,67], which can be explained by larger hydrogen plant capacities and therefore lower heat losses of these concepts. Another reason for the lower efficiency of the thermochemical concept calculated in this work is the comparatively high raw gas methane content above 10% assumed for this calculation. Energetically, the higher raw gas methane content is a disadvantage as methane has to be reformed for hydrogen production.

Taking into account the energy demand related to the distribution of hydrogen at filling station as considered here the overall efficiency decreases to 42.8% for concept A and 34.3% for concept B.

17.4
Concept Assessment of Environmental Aspects

The increasing use of biofuels is mainly driven by concerns about the security of energy supplies and the intention to mitigate anthropogenic greenhouse gases (GHG). This is also relevant for other alternative fuels like biohydrogen. Here, both potential environmental impacts will be assessed for the considered biohydrogen concepts and compared to other international results.

17.4.1
Methodical Approach

The environmental performance of biofuels is often described by the LCA (life cycle assessment) methodology, which is defined within international standards [68,69]. Since the methodology of LCA aims to quantify potential environmental impacts throughout the production, use, and disposal of products, the system boundaries of this investigation are very broadly defined. In general, they consist of the overall product life cycle, starting with the extraction of raw materials ("from cradle") through to the utilization of the product and the disposal of all intermediate or waste products ("to grave"). The emissions linked to all production processes along the product life are considered.

According to the mentioned ISO standards, a LCA consists of four major iterative steps, which are summarized with regard to the assessment carried out here in Table 17.2.

17.4.2
GHG Emissions

Figure 17.5 summarizes the specific greenhouse gas (GHG) emissions of the considered concepts. The GHG emissions lie in a bandwidth of 4.08 kgCO$_2$-eq kg^{-1} for concept A to 5.28 kgCO$_2$-eq kg^{-1} for concept B.

The significant differences between both concepts result mainly from the different biomass raw materials used. By the exclusive use of residues the thermochemical concept is more advantageous regarding GHG emissions than the combined concept whose raw material provision causes higher GHG emissions due to a high share of maize silage in the substrate mix. The GHG emissions from maize cultivation are mainly influenced by the use of fossil diesel for machinery and the use of fertilizer. The share of climatic relevant emissions that come from nitrogen application is very large. GHG emissions arise on the one hand from fertilizer production, on the other hand as soil-derived field emissions from nitrogen application. The emissions from the woodchip provision for concept A are restricted to the processing of the forest residues to woodchips (due to residue definition upstream processes were not considered) and arise mainly at the burning of the fossil diesel in silvicultural machinery. Regarding GHG the process of transport/distribution is the most intensive one for both concepts, the thermochemical and the biochemical. The process of transport/distribution also consists of, beside other transport processes, the up- and downstream compression processes of hydrogen transport. The high GHG emissions of the high-input compression processes can be explained by the use of the German electricity generation mix. The combustion of fossil fuels, which makes up the largest part in the German electricity mix, is the cause for increased GHG emissions [71]. The use of electricity from the German grid is the defining parameter for the GHG emissions of the conversion processes, too.

Table 17.2 Methodology and assumptions for LCA.

Methodology step	Assumptions for biohydrogen concepts
Goal and scope definition	
Impact categories	GHG emissions, cumulated non-renewable energy demand
Functional unit	Emissions and demands related to 1 kg biohydrogen free filling station
System boundary for LCA	Well-to-tank-chain including biomass production for silo maize and (w/o direct or indirect land use changes), biomass collection of bioresidues as forestry wood residues, manure, and biowaste and their supply, biohydrogen production plant, biohydrogen treatment, and supply to filling station
	No consideration of infrastructure (i.e., buildup of plants, components, vehicles)
Consideration of by-products	Referring to European Renewable Energy Directive (2008/29/EC) allocation of by-products (i.e., here subdividing emissions and demands along the production chain between hydrogen and digestate residue) according to their energy content (lower heating value)
Inventory calculation	
Input/output analysis	Consideration of all relevant input and output streams (i.e., energy and raw material inputs, auxiliaries and utilities, products and by-products, waste, and emissions to air, water, and soil) within the system boundary
	Concepts based on process simulation, own data, and EcoInvent Version 2.1 [70]
	External electrical power based on German mix for 2010 [71], emissions according to Gemis [72]
Impact assessment	
Approach	Evaluation of data resulting from input/output analysis regarding potential environmental impacts by means of so-called characterizing factor aggregation with regard to one reference substance
GHG emissions	According 2008/29/EC, IPCC 2001 [73], w/o consideration of process-related biogenic CO_2 emissions
	CH_4 with 24 CO_2-eq; N_2O with 296 CO_2-eq
Cumulated non-renewable energy demand	MJ fossil and nuclear primary energy related to functional unit
Results interpretation	
	See text for discussion of results

Considering the fossil reference (here hydrogen based natural gas reforming [74]) the GHG mitigation potential is about 55% for concept A and 65% for concept B.

A comparison with other international publications is hard to realize since the production and distribution concepts and the methodology applied here (according 2008/29/EC) are different from those of other studies (e.g. Reference [67]).

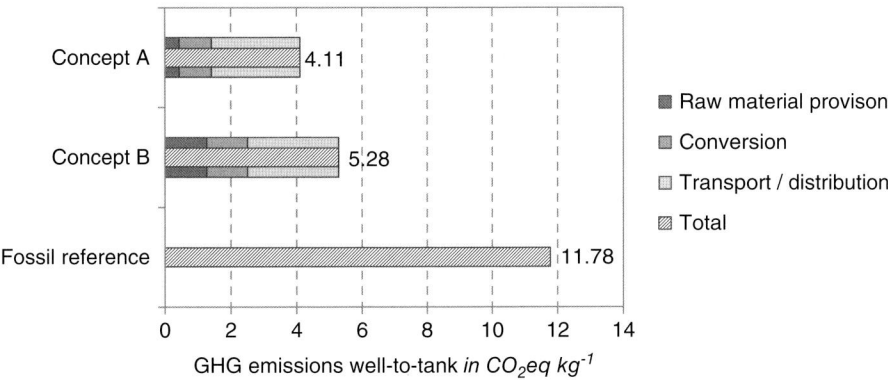

Figure 17.5 GHG emissions for hydrogen concepts.

17.4.3
Cumulated Non-renewable Energy Demand

The results for the cumulated non-renewable energy demand are summarized in comparison to the fossil reference [74] in Figure 17.6.

According to this, the cumulated non-renewable energy demand is mainly influenced by the use of fossil fuels for electricity provision. The values are similar for both concepts since the electricity demand for the processes conversion/processing and transportation/distribution as well as the expenditures for raw material provision are almost even. The energy demand of the woodchip provision for the gasification based concept A relates primarily to the diesel needed by silvicultural machines. For the raw material provision of the fermentation based concept B, in turn these are for the cultivation processes. The energy demand arises here, on the one hand, from the use of fossil diesel in agricultural machines and, on the other hand, from the production of the fertilizers.

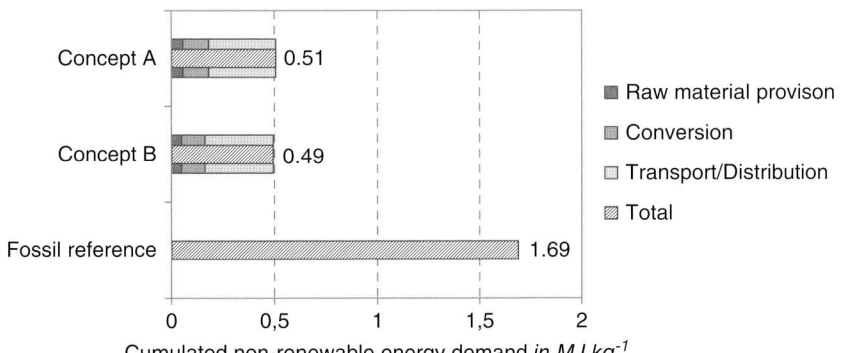

Figure 17.6 Cumulated non-renewable energy demand.

17.5
Concept Assessment of Economic Aspects

In addition to technical parameters and environmental impacts a concept assessment of hydrogen production and supply also requires consideration of economic criteria. Typically, analyzing economics is intended to evaluate different cost alternatives in order to identify relative advantages, to compare different options with regard to emissions, and to determine important influencing factors. In this respect, local conditions need to be taken into account (here, those in Germany).

17.5.1
Methodical Approach

The hydrogen production costs were calculated based on the approach defined in guideline VDI number 6025 by the association of German engineers [75]. Therein, different cost parameters at regional frame conditions and appropriate time horizons have to be taken into account: (i) capital expenditures (CAPEX; including total capital investments (TCIs), equity and leverage, interest rates, life time, maintenances), (ii) variable operational expenditures (OPEXs; raw material, auxiliaries, residues, annual full load), (iii) fixed OPEX (personnel, servicing, operation, insurances), and (iv) revenues (e.g., for by-products).

Regarding both financial risks and biofuel production costs, TCIs are of crucial importance. Depending on the state of technology development (Table 17.1), the calculation of TCI can be based on different approaches (e.g., rough, study, or approval estimations, all with different accuracies and therefore financial uncertainties). While for commercial plants approval estimations can often be applied (accuracy of $\pm 5-15\%$), for plants at pilot or demonstration stage TCI is based on study estimations (accuracy of $\pm 20-30\%$). According to this, plant equipment costs usually are determined by up- or downscaling (typical scale factor of about 0.6–0.7) of the known TCI for similar technology devices [76]. Additionally, device specific installation factors (for biofuels usually about 1.0–1.7) are taken into account [77].

The determination of distribution costs was carried out by calculation of the total costs of the different distribution stages. The necessary technical plants have to be purchased, installed, and maintained. The cost calculation is analogous to the calculation of the compression costs. Total costs consist of investment costs and operating costs. Investments were integrated in the cost calculation by using their annuity. Future investments were determined by a technical analysis followed by the use of learning curves to describe the development of future costs. Operating costs included essentially the costs for service and maintenance of the technical plants on the one hand and the costs of energy consumption (mainly for compression), that is, electricity costs, on the other hand. Owing to the high electricity consumption of the facilities, exemption from the German Renewable Energy Law apportionment and therefore a relatively low electricity price can be hypothesized [65]. Transport costs include investment and maintenance costs for the truck and trailer, fuel costs, and personnel costs for the truck driver.

Table 17.3 Assumptions for economic assessment.

Cost parameter	Assumption
General	
Base year	2011
Period under consideration	15 years
Composite interest rate (imputed)	8% a^{-1}
Hydrogen production	
Parameter CAPEX	
Repairs	2.5% of the investment pa
Parameter OPEX	
Maintenance and cleaning	2.5% of the investment pa
Insurance	2.5% of the investment pa
Full load hours	7500 h a^{-1}
Price of wood pellets (CFR)	EUR 85 t^{-1} dry mass
Price of maize silage (CFR)	EUR 35 t^{-1} fresh mass
Acceptance price of biological waste	EUR 35 t^{-1} fresh mass
Personnel cost per employee	EUR 50 000 a^{-1}
Number of employees	10 (concept A and B)
Hydrogen distribution	
Base year	2020
Full load hours	7500 h a^{-1}
Compression OPEX	
Maintenance and cleaning	1% of the investment pa
Storage OPEX	
Maintenance and cleaning	10% of the investment pa
Trailer truck	
Lifetime	5–7 a
Annual mileage	140 000 – 192 000 km
Fuel consumption	35 L diesel 100 km^{-1}
Diesel price	EUR 1.50 L^{-1}
Personnel cost (driver)	EUR 50 h^{-1}

Table 17.3 shows selected assumptions for the assessment of hydrogen production and distribution costs.

17.5.2
Total Capital Investments of Hydrogen Production

The total capital investments (TCI) of the two considered concepts A and B (Figure 17.7) are divided into a gas generating unit (including building, raw material storage, pretreatment, and exhaust gas treatment) and a gas upgrading unit. The TCIs were calculated based on offers from plant fabricators, expert

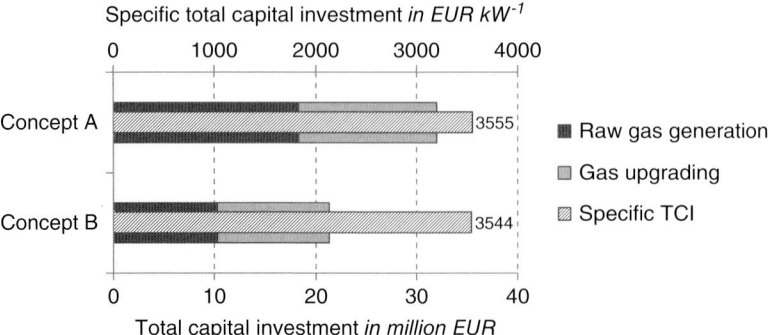

Figure 17.7 TCI (total capital investment) of selected hydrogen production concepts.

judgments, and references taken from recent publications that have been adapted to the reference year 2011 (e.g., References [78–82]).

Figure 17.7 shows that concept A has a significantly higher TCI, which can mostly be explained through the higher hydrogen capacity. The specific TCIs are nearly equal for both concepts. Therein, the higher complexity of concept is alleviated through higher economies of scale, while the lower efficiency of concept B and hence its relatively larger gas generating unit is alleviated though its lower complexity. Compared to other international publications on similar hydrogen production concepts and taking into account the Kölbel-Schulze price index for chemical plants until 2011 a range of EUR 1057–2466 kW^{-1} hydrogen output is given [67,82]. Notably, for the considered concepts the assumed oil price of US $50 bbl^{-1} is low and the assumptions seemed to have been optimistic at that time.

The specific TCI of distribution costs of the two concepts A and B are in the same range. Owing to the higher amount of hydrogen produced, total capital investments of concept A (EUR 2.4 million) exceed the investments of concept B (EUR 1.6 million). Investments were calculated on the basis of expert judgments and references taken from recent publications. Total investment costs for the trailer truck and the fuel station (each about EUR 0.45 million) are independent of the concept and have therefore been integrated as specific costs per kWh hydrogen assuming full capacity utilization.

17.5.3
Hydrogen Production Costs

The total hydrogen production costs split into the above-mentioned cost fractions are presented in Figure 17.8.

The hydrogen production costs are on a similar level for both concepts A and B, each with around 50% CAPEX (capital expenditure) and OPEX. Owing to its more complex equipment, concept A has a slightly higher share of CAPEX, while concept B has slightly higher OPEX due to its lower efficiency and hence higher specific raw material consumption. However, under the assumption of

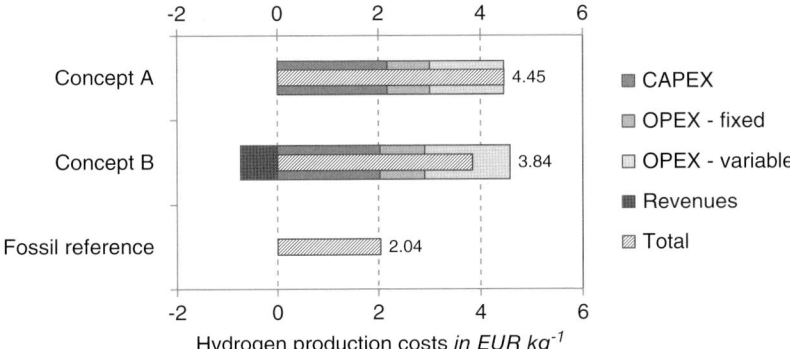

Figure 17.8 Hydrogen production costs.

using biological wastes as part of the raw materials, concept B has the advantage of receiving revenues for accepting them. At EUR 3.84 kg^{-1} the hydrogen production costs of concept B are roughly 14% below those of concept A (EUR 4.45 kg^{-1}). In both cases, the sourcing of raw materials accounts for about 55% of the OPEX. The depicted fossil reference shows the cost of producing hydrogen in large centralized facilities based on a price of natural gas of EUR 4ct kWh^{-1} and further assumptions based on References [67,74].

A sensitivity analysis shows that the specific hydrogen production costs are not influenced equally by all cost parameters. The most influential cost factors are the capital costs, the raw material price, the conversion efficiency, and the number of annual full load hours. The effects are very similar for both concepts A and B and hence only the sensitivity analysis for concept A is depicted in Figure 17.9. It shows that, for instance, a decrease in capital costs of 30% would lead to 19% lower specific hydrogen production costs.

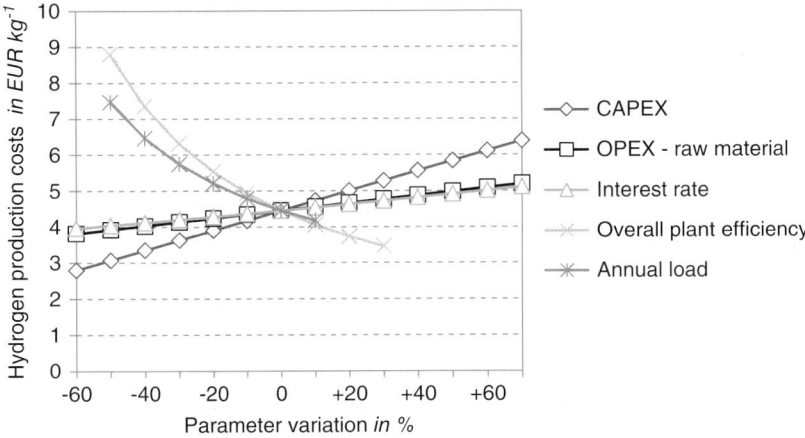

Figure 17.9 Sensitivity analysis for the specific hydrogen production costs – concept A.

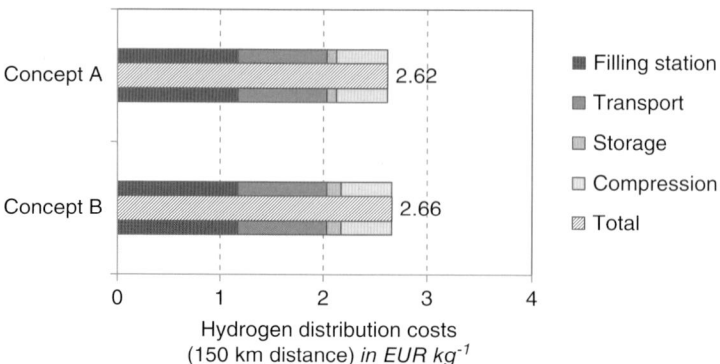

Figure 17.10 Distribution cost for a mean transport distance of 150 km.

International studies show a cost range for hydrogen production costs based on similar concepts of EUR 1.75–1.80 kg^{-1} [66,67]. These significantly lower costs can be caused by considerably higher plant capacities and thus often higher plant and cost efficiencies with lower TCI and biomass raw material costs at the same time [66]. This also assumes a considerably higher biomass conversion efficiency and much lower OPEX while Reference [67] is based on relatively old publications and a low oil price.

17.5.4
Hydrogen Distribution Costs

Total hydrogen distribution costs represent an essential part (up to 40%) of the overall hydrogen costs free filling station for the considered concepts (Figure 17.10).

Total costs are dominated by costs for transport and distribution at the filling station. These costs are the sum of the operational and capital costs of the different distribution parts, that is, the costs for compression and storage, transport, and distribution of hydrogen at the filling station. For the different production concepts the distribution costs only differ by the costs for the storage tanks as the same size is used for the different production capacities of the two concepts. Except for these storage tanks, the distribution facilities can be run with the same capacity for the different production concepts resulting in the same specific costs. In general, distribution costs for the different production concepts are in a similar range.

The critical parameter for transport and distribution costs as a whole is the transport distance. Following a sensitivity analysis with regard to transport distance, overall costs range from EUR 2.1 kg^{-1} for a transport distance of 50 km to EUR 3.75 kg^{-1} for a distance of 400 km (costs refer to concept A). Additionally, distribution costs at the filling station contain electricity costs that make up to 20% of these costs. The calculated cost ranges are comparable to those of studies that analyzed hydrogen production by electrolysis [83].

When cumulating hydrogen production and distribution costs, total costs at filling station are about EUR 7.07 kg^{-1} for concept A and EUR 6.50 kg^{-1} for concept B. The current price of hydrogen at existing filling stations in Germany is about EUR 8–10 kg^{-1}.

17.6 Summary

This chapter describes and assesses selected small-scale concepts for the provision of hydrogen based on biomass taking into account that these concepts may be realized in short- to medium-term due to their stage of technical development:

Technical aspects. For combined concepts (here biogas production combined with small-scale steam reforming) the technical assessment reveals drawbacks regarding the overall efficiency. Considering thermodynamic criteria, gasification based concepts seem to be more promising due to lower losses, which are especially relevant for larger plant capacities. When implementing such concepts to commercial demonstration, acceptance with regard to robustness of the overall concepts as well as the annual load are most important for a successful plant operation.

Environmental aspects. With regard to selected potential environmental impact categories, such as GHG and cumulated non-renewable energy demand, that have been calculated via life cycle analysis the results show that especially the hydrogen distribution from production plant to filling station by truck is crucial for the overall emissions and demands. The main influence came from the electrical power for the compression of hydrogen. For hydrogen production the differences of the considered concepts are small. However, for the raw materials there are significant differences; one is based on forestry wood residues and the other mainly on maize silage. Through a larger share of renewable energy in the future electricity mix, the GHG emissions of all concepts could be lowered substantially.

Economic aspects. In considering relevant costs for hydrogen production and distribution both concepts show similar specific hydrogen provision costs at the filling station. Generally, up to 40% of the total costs are caused by the distribution of the hydrogen. As for other renewable fuels technologies, high TCI increases the riskiness of investment. Hence, the key criteria for economic viability include ideal locations with appropriate infrastructure, a secure market for hydrogen as a fuel, and a long-term continuous raw material supply at adequate prices.

Acknowledgment

The work presented here was part of the collaborative project "Evaluation of processes and technologies for the provision of hydrogen based on biomass."

The project was funded by the National Organization Hydrogen and Fuel Cell Technology (NOW GmbH) in Germany.

References

1 Markets and Markets (2011) Hydrogen Generation Market – by Merchant & Captive Type, Distributed & Centralized Generation, Application & Technology – Trends & Global Forecasts (2011–2016). http://www.marketsandmarkets.com/Market-Reports/hydrogen-generation-market-494.html (accessed April 2013).

2 Müller-Langer, F. and Kaltschmitt, M. (2006) Wasserstofferzeugung aus biomasse in der gesamtschau - bio-chemische und thermochemische verfahren im vergleich, in *Gülzower Fachgespräche, Band 25*, Wasserstoff aus Biomasse, Gülzow, pp. S194–S216.

3 Müller-Langer, F., Tzimas, E., Kaltschmitt, M., and Peteves, S. (2007) Techno-economic assessment of hydrogen production processes for the hydrogen economy for the short and medium term. *Int. J. Hydrogen Energy*, **32** (16), 3797–3810.

4 Claassen, P.A.M. and De Vrije, T. (2006) Non-thermal production of pure hydrogen from biomass: HYVOLUTION. *Int. J. Hydrogen Energy*, **31**, 1416–1423.

5 Das, D. and Nejat Veziroglu, T. (2008) Advances in biological hydrogen production processes. *Int. J. Hydrogen Energy*, **33**, 6046–6057.

6 Keskin, T., Abo-Hashesh, M., and Hallenbeck, P.C. (2011) Photofermentative hydrogen production from wastes. *Bioresour. Technol.*, **102**, 8557–8568.

7 Modigell, M., Schumacher, M., and Claassen, P.A.M. (2007) Hyvolution-entwicklung eines zweistufigen bioprozesses zur produktion von wasserstoff aus biomasse. *Chem-Ing-Tech.*, **79**, 637–641.

8 Rechtenbach, D. (2009) *Fermentative Erzeugung von Biowasserstoff aus biogenen Roh-und Reststoffen*. Hamburger Berichte vol. **34**, Abfallwirtschaft, Technische Universität Hamburg-Harburg, Verlag Abfall aktuell, Stuttgart, ISBN 978-3-9812867-1-7.

9 Vogel, A. (2007) Dezentrale Strom- und Wärmeerzeugung aus biogenen Festbrennstoffen: Eine technische und ökonomische Bewertung der Vergasung im Vergleich zur Verbrennung. IE-Report 2/2007, Institut für Energetik und Umwelt gGmbH, Leipzig, ISSN 1862–8060.

10 Bengtsson, S. (2011) The CHRISGAS-project. *Biomass Bioenerg.*, **35**, S2–S7.

11 Bengtsson, S. (2011) VVBGC demonstration plant activities at Värnamo. *Biomass Bioenerg.*, **35**, S16–S20.

12 VVBGC AB (2011) VVBGC – Background 1991–2009. http://www.vvbgc.se/index.php/om-vvbgc/bakgrund-1991-2009 (accessed September 2011).

13 Roberts, M., Souleinmanova, R., Morreale, B., Davis, M., and Krueger, B. (2012) One Step Biomass Gas Reforming-Shift Separation Membrane Reactor. 2011 DOE Hydrogen Program Review, 17 May 2012.

14 Busch, V. (2008) Biomass gasification: experience and current projects. Presented at the International Seminar on Gasification – Production Technologies and Applications, 9 October 2008, Malmö, Sweden.

15 I/S Skive Fjernvarme (2009) First-of-its-kind at Skive. *Spectrum*, vol. 20. http://spectrum.andritz.com/index/iss_20/art_20_16.htm (accessed April 2013).

16 Gas Technology Institute (GTI) (2009) The GTI Gasification process. Available at http://www.netl.doe.gov/File%20Library/research/coal/energy%20systems/gasification/gasifipedia/GTIGasificationProcess9_18_07.pdf.)

17 REPOTEC - Renewable Power Technologies Umwelttechnik GmbH (2011) Biomasse-vergasung - wasserdampf-wirbelschicht. http://www.repotec.at/index.php/biomasse-vergasung-wasserdampf-wirbelschicht.html (accessed November 2011).

18 Hofbauer, H. (2008) Biomass steam gasification – Industrial experience and

future prospects. Presented at the International Seminar on Gasification – Production Technologies and Applications, 9 October 2008, Malmö, Sweden.

19 Bengtsson, K. (2007) Twin-Bed Gasification Concepts for Bio-SNG Production, Department of Chemical Engineering, Lund University, Sweden, 2007. Available at http://www.chemeng.lth.se/exjobb/E450.pdf.

20 Müller, S., Stidl, M., Pröll, T., Rauch, R., and Hofbauer, H. (2010) Hydrogen from biomass: Large scale hydrogen production based on a dual fluidized bed steam gasification system. Presented at the 10th International Conference on Polygeneration Strategies, 7–9 September 2010, Leipzig, Germany.

21 Zuberbühler, U. (2007) Neues thermochemisches biomasse-konversionsverfahren: erzeugung eines konditionierten gases zur herstellung von wasserstoff und synthetischen kraftstoffen. Zentrum für Sonnenenergie- und Wasserstoff-Forschung Baden-Württemberg (ZSW).

22 Zuberbühler, U., Specht, M., Marquard-Möllenstedt, T., Stürmer, B., Sichler, P., Dürrbeck, M., Rauch, R., Pfeiffer, C., Hofbauer, H., and Koch, M. (2007) AER-GAS II: Poly-Generation from solid biomass - Test of the AER-Process in the 8 MW gasification plant in Guessing. 15th European Biomass Conference & Exhibition, 7–11 May 2007, Berlin, Germany.

23 Koppatz, S., Pfeiffer, C., Rauch, R., Hofbauer, H., Marquard-Möllenstedt, T., and Specht, M. (2009) H_2 rich product gas by steam gasification of biomass with in situ CO_2 absorption in a dual fluidized bed system of 8 MW fuel input. *Fuel Process. Technol.*, **90**, 914–921.

24 Zentrum für Sonnenenergie- und Wasserstoff-Forschung Baden-Württemberg (ZSW) (2009) Thermochemische Biomassekonversion: Technische Daten zum AER-Prozess. Available at http://www.zsw-bw.de/fileadmin/ZSW_files/Themen/Brennstoffe_Wasserstoff/docs/20091202_AER_Verfahren_technische_Daten.pdf.

25 Sourkup, G. (2009) In situ CO_2 capture in a dual fluidized bed biomass steam gasifier – bed material and fuel variation. *Chem. Eng. Technol.*, **32** (3), 348–354.

26 Kienberger, T. (2010) *Methanisierung biogener Synthesegase mit Hinblick auf die direkte Umsetzung von höheren Kohlenwasserstoffen*, Fortschrittberichte VDI, Reihe 6, Energietechnik; Nr. 595, Verein Deutscher Ingenieure, ISBN 978-3-18-359506-8.

27 Karellas, S. and Karl, J. (2007) Analysis of the product gas from biomass gasification by means of laser spectroscopy. *Opt. Laser Eng.*, **45**, 935–946.

28 Karellas, S., Karl, J., and Kakaras, E. (2008) An innovative biomass gasification process and its coupling with microturbine and fuel cells systems. *Energy*, **33**, 284–291.

29 Karl, J. (2009) 10 years biomass peatpipe reformer. Presented at Second Semi-annual Task Meeting, IEA Bioenergy, Task 33: Thermal Gasification of Biomass, Breda, Netherlands.

30 Agnion Technologies GmbH (2011) Der heatpipe-reformer von agnion. agnion.de/index.php?id=31 (accessed November 2011).

31 Bengtsson, K. (2007) Twin-Bed Gasification Concepts for Bio-SNG Production, Department of Chemical Engineering, Lund University, Sweden.

32 Van der Meijden, C.M., Veringa, H.J., Vreugdenhil, B.J., and van der Drift, B. (2009) Bioenergy II: scale-up of the Milena biomass gasification process. *Int. J. Chem. Reactor Eng.*, **7** (1), article A53-1/21.

33 Van der Meijden, C.M. *Development of the MILENA gasification technology for the production of Bio-SNG*, TU Eindhoven, Netherlands.

34 Van der Meijden, C.M., Bergmann, P., van der Drift, A., and Vreugdenhil, B. (2010) Preparations for a 10 MWth Bio-CHP demonstration based on the MILENA gasification technology. Presented at the 18th European Biomass Conference and Exhibition, 3–7 May 2010, Lyon, France.

35 Van der Meijden, C.M. (2011) MILENA gasification process. http://www.ecn.nl/fileadmin/ecn/units/bio/Leaflets/b-08-021_MILENA.pdf.

36 Energy research center of the Netherlands (2011) MILENA gasification technology: status of development. http://www.milenatechnology.com/milena-gasification-technology/ (accessed November 2011).

37 Paisley, M.A., Irving, J.M., and Overend, R.P. (2001) A promising power option - the FERCO SILVAGAS biomass gasification process – operating experience at the Burlington gasifier. Presented at the ASME Turbo Expo Land, Sea & Air 2001, 4–7 June 2001, New Orleans, USA.

38 Pearson, J. (2012) Aviation biofuels: are you getting on board? Rentech, Presented at Biomass 2012: Confronting Challenges, Creating Opportunities Sustaining a Commitment to Bioenergy.

39 Rentech Inc. (2011) Rentech-SilvaGas Gasifier. http://www.rentechinc.com/silvaGas.php (accessed November 2011).

40 Niessen, W. and Dresser, C. (2007) Waste to energy (WTE) using combustion and conversion technologies. Presented at the Illinois Recycling & Solid Waste Management Conference and Trade Show.

41 Dahmen, N. *et al.* (2009) Synthesekraftstoffe aus biomasse, in *Erneuerbare Energie: Alternative Energiekonzepte für die Zukunft*, Wiley VCH Verlag GmbH, Weinheim.

42 Dahmen, N. and Dinjus, E. (2010) Synthetische chemieprodukte und kraftstoffe aus biomasse. *Chem-Ing-Tech.*, **82** (8), 1147–1152.

43 Karlsruher Institut für Technologie, Institut für Katalyseforschung und –Technologie (2011) Biomass to liquid - Der bioliq-Prozess. www.bioliq.de/55.php (accessed November 2011).

44 Wiese, L. (2008) *Energetische, exergetische und ökonomische Evaluation der thermochemischen Vergasung zur Stromproduktion aus Biomasse*, Technische Universität Hamburg-Harburg, ISBN: 978-3-18-35609-9.

45 Vogel, A., Thrän, D., Muth, J., Beiermann, D., Zuberbühler, U., Hervouet, V., Busch, O., and Biollaz, S. (2008) Comparative Assessment of Different Production Processes. Scientific report WP5.4. Technical Assessment, SES6-CT-2003-502705 RENEW – Renewable fuels for advanced powertrains, 2008.

46 CHOREN Industries GmbH i. I. (2011) Das Carbo-V(Verfahren. http://www.choren.com/carbo-v/ (accessed November 2011).

47 Linde Engineering (2012) Press release on Linde Engineering Dresden erwirbt CHORENs Carbo-V$^($-Technologie. http://www.the-linde-group.com/de/news_and_media/press_releases/news_120209.html (accessed April 2013).

48 Kruse, A. (2009) Hydrothermal biomass gasification. *J. Supercrit. Fluid.*, **47**, 391–399.

49 Boukis, N. (2007) Biomass gasification in supercritical water. Experimental progress achieved with the VERENA pilot plant. 15th European Biomass Conference & Exhibition, 7–11 May 2007, Berlin, Germany.

50 D'Jesús, P.M. (2007) *Die Vergasung von realer Biomasse in überkritischem Wasser: Untersuchung des Einflusses von Prozessvariablen und Edukteigenschaften*, Dissertation Universität Karlsruhe, KIT Scientific Publishing, ISBN: 978-3-86644-161-3.

51 Solar Millennium (2015) Blue Tower – functional description. http://www.solarmillennium.de/english/archives/technology/innovation/blue-tower/funktionsweise.html (accessed 19 January 2015).

52 Mühlen, H.J. (2011) The Blue Tower concept - a targeted development. Presented at SGC International Seminar on Gasification, 6–7 October 2011, Malmö, Sweden.

53 Kaltschmitt, M., Hartmann, H., and Hofbauer, H. (Hrsg.) (2009) *Energie aus Biomasse – Grundlagen, Techniken und Verfahren*, 2. Aufl., Springer Verlag, Berlin, Heidelberg, New York, ISBN: 978-3-540-85094-6.

54 Schmersahl, R., Ellner, J., and Scholz, V. (2008) Dampfreformierung von biogas für brennstoffzellenanwendungen. 6. Risaer Brennstoffzellenworkshop "Alternative und regenerative Brennstoffe- H2 für Brennstoffzellen", Glaubitz, Germany.

55 Scholz, V. and Schmersahl, R. (2006) Reformierung von biogas. Gülzower Fachgespräche, Band 25: Wasserstoff aus Biomasse, Gülzow, Germany, 2006.

56 Tamhankar, S. (2012) Green hydrogen by pyroreforming of glycerol. Linde Group, WHEC, Toronto, Canada, 6 June 2012.
57 European Commission (2011) *Key Enabling Technologies*, European Commission, Brussels.
58 Hofbauer, H., Rauch, R., Loeffler, G., Kaiser, S., Fercher, E., and Tremmel, H. (2002) Six year's experience with the FICFB-gasification process.Presented at the 12th European Conference and Technology Exhibition on Biomass for Energy, Industry and Climate Protection, 17–21 June 2002, Amsterdam.
59 Rauch, R. and Hofbauer, H. (2002) Zweibett-wirbelschichtvergasung in güssing (A) mit 2MWel/4,5MWth; konzept, betriebserfahrungen und wirtschaftlichkeit. 7. Holzenergie-Symposium – Luftreinhaltung und Explosionsschutz bei Holzfeuerungen und Stand der Technik der Holzvergasung; Zürich.
60 Rönsch, S. (2011) Optimierung und Bewertung von Anlagen zur Erzeugung von Methan, Strom und Wärme aus biogenen Festbrennstoffen. Dissertation, DBFZ Report Nr. 5; Leipzig.
61 Rönsch, S. and Kaltschmitt, M. (2012) Bio-SNG – concepts and their assessment. *Biomass Convers. Biorefin.*, **2**, 285–296.
62 Müller, S., Kotik, J., Pröll, T., and Hofbauer, H. (2011) Prozesssimulation als werkzeug für die entwicklung von innovativen und wirtschaftlichen anlagen für die industrie. Vortrag, Workshop Fließschemasimulationen in der Energietechnik, 23 June 2011, Leipzig.
63 Müller, S., Kotik, J., Pröll, T., and Hofbauer, H. (2011) Hydrogen from biomass for industry – biomass gasification for integration in refineries. Tagungsbandbeitrag. Presented at the International Conference on Polygeneration Strategies, 30 August to 1 September 2011, Vienna.
64 Ball, M., Weindorf, W., and Bünger, U. (2009) Hydrogen distribution, in *The Hydrogen Economy: Opportunities and Challenges*, Cambridge University Press, New York, pp. 322–347.
65 Bünger, U. and Weindorf, W. (2011) Well-to-Wheel-Analyse von Elektrofahrzeugen. Studie für das Büro für Technikfolgen-Abschätzung beim Deutschen Bundestag (TAB) im Rahmen des Innovationsreports 'Konzepte der Elektromobilität und deren Bedeutung für Wirtschaft, Gesellschaft und Umwelt'", Ludwig-Bölkow-Systemtechnik GmbH (LBST), München.
66 Müller, S., Stidl, M., Pröll, T., Rauch, R., and Hofbauer, H. (2011) Hydrogen from biomass: large-scale hydrogen production based on a dual fluidized bed steam gasification system. *Biomass Convers. Biorefin.*, **1**, 55–61. Doi 10.1007/s13399-011-0004-4.
67 EUCAR, CONCAWE & JRC/IES (2007) Well-to-Wheels Analysis Of Future Automotive Fuels and Powertrains in the European Context, Joint Research Centre.
68 DIN (2006) DIN EN ISO 14040, Environmental management – Life cycle assessment – Principles and framework, 2006.
69 DIN (2006) DIN EN ISO 14044, Environmental management – Life cycle assessment – Requirements and guidelines, 2006.
70 Swiss Centre for Life Cycle Inventories (2009) Ecoinvent v2.1 for umberto 5.5.
71 Thrän, D. et al. (Ed.) (2011) Methodenhandbuch Methoden zur Bestimmung von Technologiekennwerten, Gestehungskosten und Klimagaseffekten von Vorhaben im Rahmen des BMU-Förderprogramms. Version 2.1.
72 Öko-Institut (2011) Globales Emissions. Modell Integrierter Systeme (GEMIS), 2011.
73 IPCC (2001) Guidelines for National Greenhouse Gas Inventories.
74 Perimenis, A., Majer, S., Zech, K., Holland, M., and Müller-Langer, F. (2010) WP 4 Report, Lifecycle Assessment of Transportation Fuels, Technology Opportunities and Strategies towards Climate Friendly Transport (TOSCA).
75 Verein Deutscher Ingenieure e.V. (1996) Richtlinie VDI 6025 - Betriebswirtschaftliche Berechnungen für Investitionsgüter und Anlagen. VDI-Gesellschaft Technische Gebäudeausrüstung, Verein Deutscher Ingenieure, Düsseldorf, Beuth Verlag GmbH, Berlin.

76 Geldermann, J. (2006) *Mehrzielentscheidungen in der industriellen Produktion*, Kit Scientific Publishing, Karlsruhe.

77 Vogel, A., Brauer, S., Müller-Langer, F., and Thrän, D. (2008) Deliverable D 5.3.7 – Conversion Costs Calculation, Report SES6-CT-2003-502705 RENEW – Renewable fuels for advanced powertrains, 2008.

78 Hamelinck, C., Faaij, A., Den Uil, H., and Boerrigter, H. (2003) Production of FT transportation fuels from biomass; technical options, process analysis and optimisation, and development potential. *Energy*, **29**, 1743–1771.

79 Kreutz, T., Larson, E., Liu, G., and Williams, R. (2008) Fischer–Tropsch fuels from coal and biomass. Presented at the 25th Annual International Pittsburgh Coal Conference, 29 September to 2 October 2008, Pittsburgh, PA.

80 Urban, W., Zeidler-Fandrich, B. et al. (2009) Beseitigung technischer, rechtlicher und ökonomischer Hemmnisse bei der Einspeisung biogener Gase in das Erdgasnetz zur Reduzierung klimarelevanter Emissionen durch Aufbau und Anwendung einer georeferenzierten Datenbank. Abschlussbericht für das BMBF-Verbundprojekt "Biogaseinspeisung".

81 Zinoviev, S., Müller-Langer, F., Das, P., Bertero, N., Fornasiero, P., Kaltschmitt, M., Centi, G., and Miertus, S. (2010) Cover picture: Next-generation biofuels: Survey of emerging technologies and sustainability issues (ChemSusChem 10/2010). *ChemSusChem*, **3**, 1089. doi: 10.1002/cssc.201090039

82 Chemie Technik (2013) Kölbel-Schulze-Index. www.chemietechnik.de: CHEMIE TECHNIK exklusiv: Preisindex für Chemieanlagen (accessed: April 2013).

83 NOW GmbH (2013) Integration von Wind-Wasserstoff-Systemen in das Energiesystem. Nationale Organisation Wasserstoff- und Brennstoffzellentechnologie (NOW GmbH), Berlin, 2013.

18
Hydrogen from Biomass – Production Process via Fermentation

Balachandar G., Shantonu Roy, and Debabrata Das

18.1
Introduction

18.1.1
Current Energy Scenario

The present energy scenario is based mainly on fossil fuels, which have limited reserves. Worldwide demand for energy is growing at an alarming rate, rising from 82.2 to 86.7 MBD (million barrel per day) during the period 2004–2007. It has been strongly agreed that with current rate of consumption we will be running out of petroleum within the next 50 years, natural gas within 65 years, and coal in about 200 years [1]. Continuous and non-judicious use of fossil fuels for rapid industrialization and urbanization has threatened our environment and ecosystem by increasing carbon dioxide (CO_2) and other greenhouse gases in the atmosphere [2] (Table 18.1).

Fossil fuels used for electricity generation result in increased emissions of carbon dioxide with the reciprocal increase of power plant efficiency. The economy of an industrialized country is greatly dependent on fossil fuels. Most countries therefore, have already become involved in research seeking alternative sources of energy. A conscientious effort is required for clean energy alternatives to satisfy the growing demand for energy. Recently, due to the rising costs of fossil fuels and environmental problems, there has been growing interest in biohydrogen (Figure 18.1).

For long-term climate protection, the only alternative is exploration of energy resources that are carbon neutral and abundantly available from biomass. Gaseous energy generation from biomass through dark fermentation is one of the promising ways among the plethora of processes available in the current scenario. The use of biohydrogen from biomass might decrease greenhouse gas emissions (GH). Therefore, increased dependencies on biohydrogen would not only help to negate the challenges of global warming. Among all the available biofuels, hydrogen is considered as the fuel of the future by virtue of the fact that it is a carbon neutral renewable energy source [7–9].

Hydrogen Science and Engineering: Materials, Processes, Systems and Technology, First Edition.
Edited by Detlef Stolten and Bernd Emonts.
© 2016 Wiley-VCH Verlag GmbH & Co. KGaA. Published 2016 by Wiley-VCH Verlag GmbH & Co. KGaA.

Table 18.1 Different hydrogen production processes and their impact on the environment.

Process of hydrogen production	Raw material	Environmental impact	Reference
Steam reforming of methane	Fossil fuel	Produces carbon monoxide as impurity	[3]
Partial oxidation of hydrocarbon	Fossil fuel	Produces carbon monoxide as impurity	[4]
Coal gas	Coal	Produces carbon monoxide as impurity	[4]
Pyrolysis	Coal	Produces carbon monoxide as impurity	[4]
Electrolysis	Water	Very energy intensive process	[5]
Biological process	Biological waste/wastewater	Wastewater treatment with CO_2 sequestration from flue gas through biophotolysis	[6]

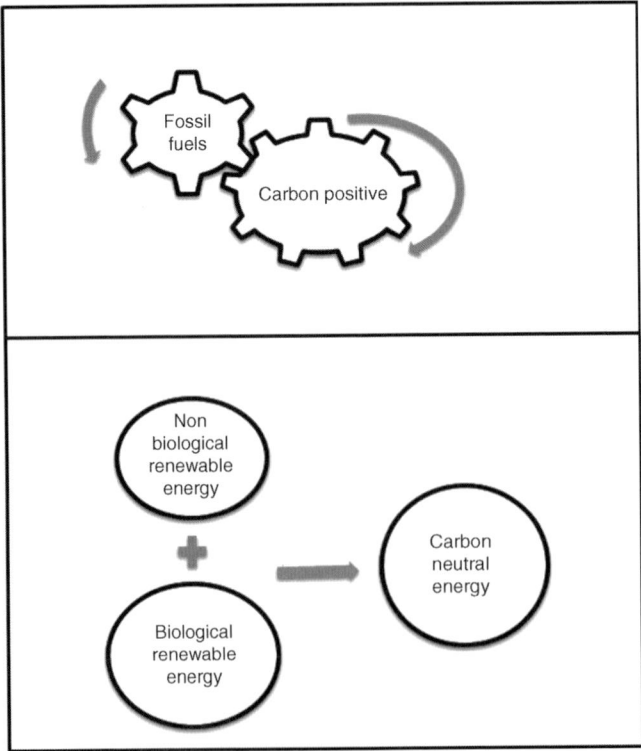

Figure 18.1 Carbon footprints of different energy sources.

18.1.2
Importance and Applications of Hydrogen as a Fuel

Hydrogen is an attractive alternative to carbon-based fossil fuels because it has the highest energy density (143 GJ ton^{-1}) and produces only water during its combustion [10]. On the other hand, burning of hydrogen does not cause any environmental hazards that may contribute to greenhouse gas emission, acid rain, or ozone depletion. Hydrogen can be utilized in high efficiency power generation systems, including fuel cells, for both vehicular transportation and distributed electricity generation [11,12]. Approximately 49% of hydrogen produced is used for the manufacture of ammonia, 37% for petroleum refineries, 8% for methanol production, and about 6% for miscellaneous smaller-volume applications. The future widespread use of hydrogen is likely to be in the transportation sector as well as electricity generation.

18.1.3
Conventional Hydrogen Production

At present, hydrogen is produced from fossil fuels, water, and biomass by various physicochemical processes. The production of hydrogen from nonrenewable energy resources like fossil fuels (coal, natural gas), or nuclear power causes the co-production of CO_2, which is largely responsible for the so-called greenhouse effect. Currently, the total annual world hydrogen production is near to 368 trillion cubic meters [13]. Out of the total global production of hydrogen, 48% is produced from steam methane reforming, about 30% from oil/naphtha reforming from refinery/chemical industrial off-gases, 18% from coal gasification, 3.9% from water electrolysis, and 0.1% from other sources [14]. These figures imply that globally 96% of hydrogen production comes from fossil fuels. However, such processes are incapable of reducing waste and require power derived from fossil energy or use fossil fuel as a raw material [15]. Hydrogen produced through a range of renewable primary energy sources such as wind, biomass, solar, hydro, or geothermal energy has gained importance in recent times [16]. The energy potential of biomass is considered to be the most promising among the renewable energy sources, due to its worldwide availability. Biomass and biomass-derived fuels can be used to produce hydrogen sustainably [17]. The use of biomass in place of fossil fuels for the production of hydrogen would reduce the net amount of CO_2 released to the atmosphere. In addition, biomass is mostly non-toxic, biodegradable, and free of sulfur and aromatic compounds as compared to gasoline and diesel fuels [18].

18.1.4
Biological Hydrogen Production

Biological hydrogen production processes are environmentally friendly and, moreover, renewable [19,20]. In addition, they can also use waste as substrate for hydrogen production. Thus, they may solve the dual problem of waste disposal and energy demand. Biological hydrogen production technologies provide a wide range of approaches to generate hydrogen, including direct and indirect biophotolysis mediated by cyanobacteria and green algae, photofermentative hydrogen production by photofermentative bacteria and dark hydrogen production by either mesophilic or thermophilic fermentative bacteria. However, each processes has its advantages and disadvantages. The major problem with biophotolysis of water and photofermentation is that they are light dependent and their hydrogen yield and rate of hydrogen production are too low for practical applications. Among the various other processes, dark fermentation appears to be more promising. The hydrogen yield and rate of hydrogen production from this process are more attractive than for other biological hydrogen production processes [21]. Dark fermentation at thermophilic temperatures favors the kinetics and stoichiometry of H_2 production and has a reduced risk of methanogenic contamination. At higher temperatures, metabolism becomes thermodynamically more favorable and less affected by the partial pressure of hydrogen (pH_2) in the liquid phase. Industrial wastewater such as distillery effluent discharges at higher temperatures. Cooling this wastewater to treat under mesophilic conditions is not cost effective and there is always a risk of losing biological activity while cooling [22]. Thermophilic hydrogen producing bacteria can be used for fermentation of hydrolysates obtained from hydrothermal pretreatment of lignocelluloses or high temperature industrial effluents where the temperature is already high.

18.1.4.1 Biochemistry behind Thermophilic Biohydrogen Production via Dark Fermentation

Thermophilic microorganisms have higher hydrogen yields with respect to mesophiles. High hydrogen yields in the range 314–473 ml g^{-1} sugars have been previously reported by using thermophiles such as *Clostridium thermocellum* and *Thermoanaerobacterium thermosaccharolyticum* and extreme thermophiles such as *Thermotoga elfii*, *Caldicellulosiruptor saccharolyticus*, and *Caldanaerobacter subterraneus*. Hydrogen production during carbohydrate fermentation is mediated by the hydrogenase enzyme which uses electrons from reduced ferredoxin (Fd_{red}) and NADH to reduce protons. The Fd_{red}, contains electrons from oxidative decarboxylation of pyruvate by pyruvate-ferredoxin oxidoreductase (PFOR) [23].

The electrons released during oxidation of substrate are transferred to carriers such as nicotinamide adenine dinucleotides (NAD) or ferredoxins (Fd). The reduced carriers (NADH, Fd) are reoxidized by transferring the electrons to protons or other cleavage products or their derivatives, which are thereby reduced and then excreted from the cells. The fermentation of carbohydrates and various others substances leads to the formation of the following products: ethanol, 2-propanol, 2,3-butanediol, *n*-butanol, formate, acetate (Eq. (18.1)), lactate, propanoate, butyrate (Eq. (18.2)), succinate, capronoate, acetone, carbon dioxide, and hydrogen either as sole products or as a mixture [24]:

$$C_6H_{12}O_6 + 2H_2O \rightarrow 2CH_3COOH + 2CO_2 + 4H_2 \quad (18.1)$$

$$C_6H_{12}O_6 + 2H_2O \rightarrow CH_3CH_2COOH + 2CO_2 + 2H_2 \quad (18.2)$$

This oxidation to acetyl-CoA requires a ferredoxin (Fd) reduction. Reduced Fd is oxidized by hydrogenase, which regenerates Fd(ox) and releases electrons as molecular H_2. The overall reaction of the processes can be described as follows:

$$\text{Pyruvate} + \text{CoA} + 2\text{Fd(ox)} \rightarrow \text{Acetyl-CoA} + 2\text{Fd(red)} + CO_2 \quad (18.3)$$

$$2H^+ + \text{Fd(red)} \rightarrow H_2 + \text{Fd(ox)} \quad (18.4)$$

Acetyl-CoA is further converted into acetoacetyl-CoA by the enzyme acetyl-coA acetyltransferase. This intermediate serves as electron acceptor for NADH and is converted into butyrate, n-butanol, acetone, 2-propanol, or capronoate. The synthesis of acetoacetyl CoA prevents the bacterium from regenerating ATP from acetyl-CoA and therefore results in a low biomass yield. Some fermentative bacteria contain an enzyme, NADH-ferredoxin oxido-reductase, which catalyzes the reaction (Eq. (18.5)):

$$\text{NADH} + 2\,\text{Fd(ox)} \rightarrow \text{NAD}^+ + 2\,\text{Fd (red)} \quad (18.5)$$

Despite having a higher evolution rate, the yield of H_2 from the fermentation process is lower than that of other chemical/electrochemical processes. The theoretical H_2 yield is $4\,\text{mol}\,\text{mol}^{-1}$ of glucose when the end product is acetic acid, while $2\,\text{mol}\,\text{mol}^{-1}$ of glucose will be obtained if the metabolic end product is butyric acid. In practice, the yields are low since the end products contain both acetate and butyrate. In addition, as yields increase the reaction becomes thermodynamically unstable [25].

18.1.4.2 Microbial Characteristics of Thermophilic Hydrogen Producing Bacteria

A few genera of thermophilic bacteria have been isolated from hot springs in Iceland. It was reported that 50 and 60 °C enrichments leads to the domination of a few genera like *Thermoanaerobacter*, *Thermoanaerobacterium*, and *Clostridium* species [26]. These genera have been reported to produce hydrogen. The 70 and 75 °C enrichments were found mostly to be *Caldicellulosiruptor* and *Thermotoga*, and in these genera the end product was directed more towards hydrogen than ethanol. *Thermoanaerobacterium* sp. were first described in 1993 when two xylan degrading bacteria were isolated from Frying Pan Springs in Yellowstone National Park [27]. The molecular characteristics used to classify the bacteria means they fall within clusters V, VI, and VII in phylogenetic interrelationships of *Clostridium* species [35]. These are straight rods, Gram negative, filamentous, motile, and peritrichous bacteria, capable of reducing thiosulfate to elemental sulfur. Their metabolic end products are diverse, like ethanol, acetate,

CO_2, H_2, and lactate. *Thermoanaerobacter* sp. were listed under the irregular, non-spore forming, Gram-positive rods [28]. The genera *Thermoanaerobacter* and *Thermoanaerobium* included the first thermophilic, anaerobic bacteria that produce hydrogen along with ethanol and lactate as sugar fermentation products. These are obligate anaerobes, often non-spore forming (exception: *Thermoanaerobacter finnii*) bacteria. Strains are saccharolytic, capable of degrading starch but cannot degrade cellulose. Their major metabolic end products are H_2, ethanol, lactate, acetate, and CO_2 [29]. *Clostridium* sp. are the most studied, because of their potential biofuel production characteristics. Diverse species among the genus have been found by molecular analysis. The genus *Clostridium* belongs to the family Clostridiaceae, order Clostridiales, class Clostridia and phylum Firmicutes. These species are Gram positive, motile rods, often spore forming, facultative anaerobic to obligate anaerobic. They contain a wide variety of cellulose degrading enzymes and are generally present in environments containing plant decaying material. *Caldicellulosiruptor* sp. belongs to thermophilic cellulolytic clostridia, which have been placed within the *Bacillus/Clostridium* subphylum on the basis of their physiological characteristics and phylogenetic position. These species are obligatory anaerobes, extremely thermophilic, and non-spore forming Gram-positive bacteria. They have a plethora of hydrolytic enzymes that help in solubilizing and utilizing a wide variety of substrates like cellulose, cellobiose, xylan, and xylose. These species were found to be superior hydrogen producers with reports of best hydrogen yields on lignocellulosic wastes. Their main by product channel is diverted towards the acetate pathway [30]. *Thermotoga* sp. have been isolated and characterized as a unique thermophilic bacteria from geothermally heated sea floors in Italy. These are extremophilic bacteria that grow at the highest reported temperature for hydrogen production (70 °C). They are rod shaped, obligate anaerobes containing an outer sheet-like structure called toga [31]. They are mainly isolated from extreme conditions like high temperature, pressure, and sulfur containing environments and are often reported to use elemental sulfur or thiosulfate or both as their electron source. The metabolic end products reported in this organism were mainly acetate, lactate, hydrogen, and CO_2, with ethanol in very trace amounts. The butyrate pathway was absent in this organism.

18.2
Hydrogen Production from Biomass as Feedstock

Biomass is a renewable energy resource derived from the carbonaceous waste of various human and natural activities. It is derived from numerous sources, including by-products from the timber industry, agricultural crops, raw material from the forest, and major parts of household waste and wood. Biomass has become an important source of energy, the most important fuel worldwide after coal, oil, and natural gas. Bioenergy, in the form of biogas, which is derived from biomass, is expected to become one of the key energy resources for global

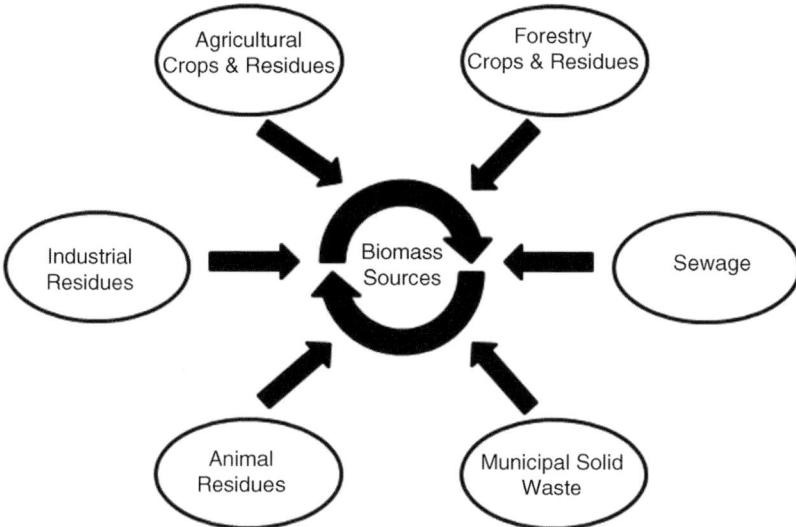

Figure 18.2 Different sources of biomass available for the generation of green energy.

sustainable development. The important criteria required for the selection of different biomasses as substrate for hydrogen production are based on their sustainability, availability, organic content, biodegradability, and cost [32]. Simple sugars such as glucose, sucrose, maltose, lactose, and so on are readily biodegradable and they are the most preferred substrates for hydrogen production [33]. However, they are expensive raw materials. Figure 18.2 shows that major biomass resources that can be used as substrate for hydrogen production by dark fermentation process include forestry residues, agricultural residues, animal residues, industrial residues, municipal solid waste, and sewage.

18.2.1
Agricultural Residues

Sugar and starch are suitable for biohydrogen production. In this context, biohydrogen production by dark fermentation from lignocellulosic biomass is an interesting alternative since lignocellulosic raw materials do not compete with food crops and are also less expensive than conventional agricultural feedstocks. Lignocellulosic materials are mostly composed of cellulose, hemicelluloses, and lignin. These are the products of photosynthesis that form the structural component of the plant cell wall [34]. Cellulose is the major constituent of plant biomass and is readily available in the agricultural residues and wastes. Agricultural residues are the major sources that generate waste rich in starch, protein, and/or cellulose. However, biodegradability of these residues may be affected due to their complex nature. As such, agricultural residues are rich in cellulose and thus may require some pretreatment process to convert them from complex form into simpler forms prior to use as substrate for hydrogen production by dark fermentation. There are different pretreatments methods, each with their own advantages and disadvantages. Among them, thermochemical and chemical are currently the most effective methods. Moreover, pretreatment process may significantly influence cost and performance of the dark fermentation process. Recently, a combination of different pretreatment methods has been considered to achieve higher yields. Although various literature is available on the effects of the different pretreatment processes on biomass composition and hydrogen yields, very limited information is available on the cost comparison of the pretreatment process [35]. On other hand, cellulose degrading bacteria are also present that can be used to directly degrade cellulose into simpler sugars. These simpler sugars may then be utilized as substrate for hydrogen production [36].

18.2.2
Municipal Solid Waste and Sewage

Sewage sludge can be considered and used as substrate for fermentative hydrogen production. With an increasing global population, the annual production of sewage sludge is rising significantly. It has several advantages over other wastes based on its availability and low cost. Despite these advantages, it has received little attention as source of seed or inoculums for hydrogen production [37]. It contains several macronutrients (carbohydrates, proteins, and lipids) and micronutrients (vitamins and minerals). Therefore, it may be considered as a suitable substrate for hydrogen fermentation. However, some of the organic components present in the sewage sludge make it unsuitable for utilization by microorganisms during fermentation [28]. Moreover, sludge as seed will contain methanogenic bacteria that will interfere with the hydrogen production process. Therefore, several studies suggest that, prior to use, pretreatment processes are required to make it suitable as substrate for dark fermentative hydrogen production [38]. In view of this, the possible utilization of sewage sludge to produce clean energy is highly promising and economically viable. In addition, it may be used for the bioremediation of waste.

18.2.3
Industrial Residues

Industrial effluents with high COD (chemical oxygen demand) can be used for many purposes, namely, biogas production, potash recovery by incineration, composting for producing green manure, or using the discharge for agricultural fields. Utilization of fermentable substrates present in these wastewaters for hydrogen production would serve the purpose of gaseous energy generation as well as waste management. Hydrogen production from industrial wastewaters depends upon the nature and composition of the industrial wastewater. Other influencing factors are biodegradability, bioreactor configuration, operation mode, operating pH, and substrate loading rate. Moreover, the characteristics of microorganisms used for fermentation also play an important role in hydrogen production. Hot industrial effluents may be used as substrate directly in a thermophilic dark fermentation process [39,40]. Thus, the high temperature wastewater generated from industry can directly be used in the bioreactor, without cooling it. The need for autoclaving can also be avoided in thermophilic dark fermentation because mesophilic contaminants would not grow at temperatures above 50 °C. It has been reported that the production of hydrogen from high temperature carbohydrate-rich wastewaters [41], sweet potato starch-manufacturing plant [42], brewery [43], and so on gives a higher hydrogen yield as compared to mesophiles. Major promising wastewaters that can be used as substrate for hydrogen production by dark fermentation process include domestic wastewater, paper mill wastewater, starch effluent, food processing wastewater, rice winery wastewater, molasses based wastewater, palm oil mill wastewater, glycerol based wastewater, chemical wastewater, cattle wastewater, dairy process wastewater, and designed synthetic wastewater (Table 18.2).

18.2.3.1 Distillery Industry Waste

Thermophilic dark fermentation is gaining in importance due to the high conversion rate of organic wastes into hydrogen without autoclaving the substrate. There are many industrial effluents like those from distillery and food processing industries that are often discharged at higher temperatures. Consequently, it would be suitable to treat these wastes in their native form, as cooling them down for mesophilic treatment is not cost-effective [50]. The suitability of the thermophilic dark fermentation process for hydrogen production from organic waste is shown in Table 18.2. It has been reported that distillery waste can produce hydrogen under thermophilic conditions (170 ml hydrogen per g of VSS at HRT of 4 days) (HRT, hydraulic retention time) [55].

18.2.3.2 Food Industry Waste

Food waste can be considered as one of the most promising substrates for hydrogen production, since it contains about 90% volatile suspended solids, favoring microbial degradation [46]. Generally, disposal of food waste is by land filling, resulting in environmental problems: foul smell and pollution of ground water. In view of this, utilization of the waste to produce clean energy appears highly promising and economically viable. Presently, researchers are focusing on thermophilic fermentation of food wastes for biohydrogen production. Institutional food wastes used for thermophilic hydrogen production gave 81 ml H_2 per g VSS as compared to 63 ml H_2 per g VSS by mesophilic dark fermentation [37,56]. Venkata Mohan *et al.* have been evaluating the suitability of dairy wastewater for hydrogen production. Hydrogen gas production gave a highest yield of 1.10 mmol-H_2 m^{-3} min^{-1} and 64.7% COD removal at an organic loading rate of 3.5 kg COD m^{-3} day^{-1} [45]. Yang *et al.* have examined a range of food to

Table 18.2 Biohydrogen production from organic residues by dark fermentation.

Substrate	Microorganism	Hydrogen yield	Reference
Food waste	Mixed culture	57 ml H_2 per g VS	[44]
Food waste	Mixed culture	0.39 l H_2 per g COD	[45]
Dairy waste	Mixed culture	1.10 mmol-H_2 m^{-3} min^{-1}	[46]
Cheese whey	Mixed culture	22 mmol per g COD	[47]
Cheese whey	*C. saccharoperbutylacetonicum* ATCC27 021	2.7 mol H_2 per mol lactose	[47]
Rice slurry	Mixed culture	346 ml H_2 per g carbohydrate	[48]
Starchy waste	*Thermoanaerobacterium* sp. mixed culture	92 ml H_2 per g starch	[50]
Distillery effluent	Co-culture of *Citrobacter freundii* 01, *Enterobacter aerogenes* E10, and *Rhodopseudomonas palustris* P2	2.76 mol H_2 per mol hexose	[51]
Domestic sewage	Mixed culture	6.01 mmol per g COD	[52]
Domestic sewage	Mixed culture	1.25 mmol per g COD	[46]
Cattle manure	Mixed culture	12.41 mmol per g COD	[53]
POME (palm oil mill effluent)	*Thermoanaerobacterium*-rich sludge	6.33 l H_2 per l-POME	[54]

microorganism (F/M) ratio of 1.0–1.5 to study its effect on hydrogen production [56]. Rice slurry is another major starch-based residue from industry as rice is the most common dietary food. Rice contains carbohydrate, protein, lipid, and water [49]. Fang et al. conducted an experiment on hydrogen production using rice slurry residue containing 5.5 g carbohydrate per litre by using *Clostridium* sp. rich seed sludge [48]. The hydrogen yield of 346 ml H_2 per g carbohydrate was obtained at low pH of 4.5, which was 62.6% of the theoretical yield at 553 ml hydrogen per gram of polysaccharides. Palm oil mill effluent (POME) is one of the many organic wastes released at higher temperatures [57]. It is highly enriched in organic content with a BOD of more than 20 g l^{-1} and nitrogen content of more than 0.5 g l^{-1} [58]. Thus, it can be considered a potential substrate for thermophilic dark fermentation processes.

18.3
Reactor Configurations and Scale-Up Challenges

18.3.1
Reactor Configurations

Biohydrogen production processes have been conducted with batch, semi-continuous, and continuous bioreactors in both mesophilic and thermophilic conditions. Usually, biohydrogen fermentation can be carried out in either batch or continuous mode. Moreover, the continuous mode of fermentation would be preferred by most industries due to higher productivity. Many studies have been conducted with continuous stirred tank reactors (CSTRs) because of simple construction, ease of operation, effective homogenous mixing, and the ability to keep the system at a certain hydraulic retention time controls the microbial growth rate. Besides the extensively studied CSTR, various types of bioreactor such as the anaerobic sequencing batch reactor (ASBR), carrier induced granular sludge bed reactor (CIGSB), agitated granular sludge bed reactor (AGSBR), membrane bioreactor (MBR), fixed-bed bioreactor, fluidized-bed bioreactor, and upflow anaerobic sludge blanket bioreactor (UASB) have been developed with high production yields. Table 18.3 gives various reactor configurations used for hydrogen production.

The choice of bioreactor relies on the nature of substrate, which could be converted into organic acids, alcohols, and biogas by microorganisms. It has been reported that higher hydrogen yields can be produced by physical retention of microbial biomass, including the use of naturally forming flocs or granules of self-immobilized microbes, microbial immobilization on inert materials, microbial-based biofilms, or retentive membranes. Very few studies have been conducted on a comparative analysis of the performance of these reactors [77]. Some specialized bioreactors need to be developed to improve biohydrogen production by dark fermentation under both mesophilic and thermophilic conditions.

Table 18.3 Biohydrogen production by dark fermentation using different bioreactors.

Reactor type	Volume (L)	Substrate	Microorganism	Hydrogen production rate ($l\,l^{-1}\,h^{-1}$)	Reference
UASB	—	Sucrose	*Clostridium* sp.	0.12	[59]
UASB	—	—	Mixed culture	0.28	[60]
CIGSB	1.3	Sucrose	*Clostridium* sp.	9.31	[61]
Granule-based CSTR	6.0	Glucose	Anaerobic bacteria	3.26	[62]
AFBR biofilm reactor	1.4	Glucose	Anaerobic bacteria	7.60	[63]
AFBR	—	—	Anaerobic bacteria	1.82	[64]
AFBR	—	—	Mixed culture	2.36	[65]
CIGSB	0.88	Sucrose	Mixed culture	7.66	[66]
MBR	1	Glucose	Mixed culture	1.48	[67]
		Sucrose		1.39	
		Fructose		1.36	
PBR	—	—	Anaerobic bacteria	7.41	[66]
PBR	—	Glucose	*E. cloacae* IIT-BT 08	1.69	[68]
PBR	0.48	Molasses	*E. cloacae* BL-21	2.17	[69]
GAC-AFBR	—	Glucose	Mixed culture	2.36	[70]
DTFBR	8	Sucrose	*Clostridium* sp.	2.27	[71]
TBFR	—	—	Mixed culture	1.07	[72]
CSTR	—	—	Mixed culture	15.09	[73]
CSTR	—	—	Mixed culture	3.20	[74]
TBFR	8	Glucose	Thermophilic mixed culture	0.20	[75]
FBR	5	Glucose	Thermophilic mixed culture	0.25	[75]
IBR	0.5	Glucose	Thermophilic mixed culture	1.8	[76]

IBR – immobilized bioreactor, CIGSB – carrier-induced granular sludge bed reactor, AFBR – anaerobic fluidized bed reactor, MBR – membrane bioreactor, GAC-AFBR – granular activated carbon anaerobic fluidized bed reactor, DTFBR – draft tube fluidized bed reactor, PBR – packed bed reactor, CSTR – continuous stirred tank reactor, UASB – upflow anaerobic sludge blanket reactor.

18.3.2
Current Status of Technologies Available on Scale Up

As has been already discussed, operation of fermentor at high temperature for a thermophilic dark fermentation process not only favors the reaction kinetics and stoichiometry but also makes the reactor less prone to contamination by methanogenic bacteria. In addition, operation at high temperature operation results in better mass transfer because of increased diffusion and reduction in viscosity. Despite such advantages, there are some technical problems associated with the operation of a fermentor at high temperature. The major problem is corrosion due to the high temperatures, salt concentrations, and presence of sulfide. However, it has been suggested that the use of bioreactors made of stainless steel with an added nickel, molybdenum, chromium, or Teflon coating could prevent corrosion [78]. However, this increases the cost of the fermentor several-fold. Another important problem associated with the thermophilic production of hydrogen is low cell density. It has been reported that the cell density of thermophiles is considerably lower compared to mesophiles. This problem of low cell density can be resolved by immobilization of whole cells. Coir has been reported as the potential matrix among different natural polymers for the immobilization *of Escherichia cloacae IIT-BT 08* [68]. Another way of improving cell density is by biomass retention such as the formation of a biofilm on the carrier molecules or the formation of microbial granules. A comparison between a trickle bed reactor packed and equipped with a heating coil and a fluidized bed reactor with nitrogen as stripping gas has been made for hydrogen production [75]. The above experimental reactor designs were used to increase the cell density and to reduce the partial pressure of nitrogen, which is found inhibitory. Both systems were reported to perform well, yielding high productivities [79]. A comparative study of hydrogen production by continuous stirred tank reactor (CSTR), up flow anaerobic sludge blanket reactor (UASB), and anaerobic filter (AF) reactor using the thermophilic organism *Caldicellulosiruptor owensensis* was reported by Peintner et al. [74]. Differences were observed in the reactor performances in terms of different solid retention times in the attached cell growth (UASB& AF), and suspended cell growth (CSTR). The thermophilic granular UASB system fed with glucose using mixed cultures gave a maximum hydrogen yield of $2.47 \pm 0.15\,\text{mol}\,\text{mol}^{-1}$ glucose [81]. The higher yield obtained on using the UASB reactor was attributed to the more favorable thermodynamic conditions at thermophilic temperatures. This indicated that UASB system could be a promising system for hydrogen production [82]. Hydrogen production using a CSTR has been reported (750 cm^3 working volume) using pig slurry as inoculum at 70 °C, with a hydraulic retention time (HRT) of 24 h and the organic loading rate of $24.9\,\text{g}\,\text{d}^{-1}$ of volatile solid (VS) [80]. The short HRT (24 h) and the thermophilic temperature applied in the reactor were enough to prevent methanogenesis. No pretreatment methods or other control methods for preventing methanogenesis were necessary.

Table 18.4 Properties of different types of insulation materials.

Type of insulation (thickness (mm))	Maximum heat flux (W m^{-2})	Thermal conductivity (W m^{-1} K^{-1})	Reference
Glass wool (25)	1000	0.04	[86]
Coconut coir (50)	1000	0.074	[86]
Polyurethane (30)	1140	0.040	[87]
Polystyrene (15)	880	0.030	[88]

The disadvantage with the CSTR system is the risk of cell washout. On the other hand, UASB and AF can be operated at higher dilution rates, approximately three times higher than CSTR due to higher solid retention times [83]. The factor that limits the design in scaling up of the process is the provision of maintaining temperature in the reactors as both reactors are non-ideal in comparison to CSTR [84]. A reactor is equipped with a perforated Teflon pipe for the distribution of hot gas showed better heat transfer [85]. A major problem associated with thermophilic dark fermentation is maintaining the high temperature in the reactor. Generally, in laboratory-scale reactors, temperatures can be maintained by using double jacketed vessels or using heating coils inside the reactor. There is no report on the use of jackets for temperature maintenance in a reactor more than 20 l in size because of the poor mixing and heat transfer characteristics at higher volumes. The main bottleneck of reactors operating at thermophilic temperature is heat loss. Heat loss can be minimized by insulating the reactor, which might lead to a decrease in energy input cost. Materials with better insulation properties that have been discussed in the literature are mainly the polymeric materials shown in Table 18.4. Glass wool is generally used for insulation of instruments related to temperature control. The high cost of glass wool makes it least feasible as an insulating material for scale up applications. Cheaper insulating materials like coconut coir and concrete can be used as insulating materials. It has been reported that heat loss in the thermophilic process is 57% higher than in the mesophilic process [89]. Thus, cheap and effective insulating materials can prevent this heat loss and can make the process more viable.

18.4
Economics and Barriers

Basic and applied research on biological hydrogen production has been carried out for several decades. The process has its own merits and demerits in terms of technology and productivity. Rigorous evaluation in terms of the cost for commercialization has not been achieved yet. Only a limited number of economic analyses of bio-hydrogenation processes are available. It has been reported that hydrogen yield is directly proportional to the operating costs while rate is directly proportional to the reactor costs or the installation costs. For dark fermentation the costs of operation can be reduced by using cheaper raw substances like sewage sludge, distillery waste, and so on. Reports on the cost of hydrogen production using locally available lignocellulosic feedstock suggest that the possibility of commercialization of the process is quite promising [90,91]. The efficient conversion of cheaper biomass into hydrogen is possible using potential microbial consortia. By utilizing municipal sewage/wastewater, the cost of hydrogen could be brought down to as low as $1.3 per MBTU, while the cost of natural gas and gasoline (2007) were $2–7 per MBTU and $23.5 per MBTU, respectively [92].

18.5
Future Prospects

Though biological systems can produce H_2, no commercial systems are yet available, and questions concerning the practical application of biohydrogen loom large. Several research groups have already made significant progress in biological H_2 production processes, but still the economy of the process is not attractive as compared to the conventional H_2 production processes. To make the process economically viable, the following points require immediate attention:

- use of cheaper raw materials to improve the H_2 yield;
- development of mixed microbial consortium or the metagenomic approach to develop an efficient microbial strain for better utilization of industrial wastewater;
- in two-stage processes, the major bottleneck lies in the scaling up of the photofermentation process; thermophilic MEC (microbial electrolysis cell) might become useful in the development of-two stage processes, thereby improving overall H_2 yield.

Thus, improvements in technology may reduce the overall cost of biohydrogen production technology significantly.

18.6
Conclusion

Dark fermentative hydrogen production provides the advantage of simultaneous waste stabilization and hydrogen generation and is relatively inexpensive, efficient, and requires low energy demands (no aeration, no light source, etc.). Dark fermentation processes can utilize the carbon available in the large-volume discharges of agro-industrial wastewater that often contains organic wastes, carbohydrates, or lignocellulosic biomass to recover available energy as well as purify the effluent. Thermophilic dark fermentation can use high temperature wastewater discharged from various industries such as palm oil mills, distilleries, and food processing plants. However, the establishment of biohydrogen as a source of energy is also very challenging with respect to environmental problems. Two major aspects need indispensable optimization, namely, suitable cheaper raw materials like renewable biomass or wastewater and ideal microbial consortia that can convert this biomass efficiently into H_2. Thus, the future of biological H_2 production depends not only on research advances, that is, improvement in efficiency through genetically engineered microorganisms and/or the development of bioreactors, but also on economic considerations (the cost of fossil fuels), social factors, and the development of H_2 energy systems.

Acknowledgment

The authors gratefully acknowledge financial assistance for biohydrogen research from the Department of Biotechnology, Council for Scientific and Industrial Research (CSIR), Ministry of New and Renewable Energy (MNRE), Government of India.

References

1 Soetaert, W. and Vandamme, E.J. (2009) *Biofuels* (eds W. Soetaert and E.J. Vandamme), John Wiley and Sons Ltd, Chichester, pp. 1–8.
2 Baxter, L.W. and Calandri, K. (1992) Global warming and electricity demand: a study of California. *Energy Policy*, **20**, 233–244.
3 Goswami, D.Y., Mirabal, S.T., Goel, N., and Ingley, H.A. (2003) A review of hydrogen production technologies, in *Proceeding of the First International Conference on Fuel Cell Science, Engineering and Technology, Rochester, New York, 21–23 April 2003* (eds R.K. Shah and S.G. Kandilkar), ASME, New York, pp. 61–74.
4 Abbas, H.F. and Wan Daud, W.M.A. (2010) Hydrogen production by methane decomposition: a review. *Int. J. Hydrogen Energy*, **35**, 1160–1190.
5 Holladay, J.D., Hu, J., King, D.L., and Wang, Y. (2009) An overview of hydrogen production technologies. *Catal. Today*, **139**, 244–260.
6 Kotay, S.M. and Das, D. (2008) Biohydrogen as a renewable energy resource-prospects and potentials. *Int. J. Hydrogen Energy*, **33**, 258–263.
7 Das, D. and Verziroglu, T.N. (2001) Hydrogen production by biological processes: a survey of literature. *Int. J. Hydrogen Energy*, **06**, 13–28.
8 Arent, D.J., Wise, A., and Gelman, R. (2011) The status and prospects of renewable energy for combating global warming. *Energy Econ.*, **33**, 584–593.
9 Suzuki, Y. (1982) On hydrogen as fuel gas. *Int. J. Hydrogen Energy*, **7**, 227–230.
10 Hallenbeck, P.C. (2005) Fundamentals of the fermentative production of hydrogen. *Water Sci. Technol.*, **52**, 21–29.
11 Lee, D.H. and Hung, C.P. (2012) Toward a clean energy economy: with discussion on role of hydrogen sectors. *Int. J. Hydrogen Energy*. doi: 10.1016/j.ijhydene.2012.02.064
12 Barreto, L., Makihira, A., and Riahi, K. (2003) The hydrogen economy in the 21st century: a sustainable development scenario. *Int. J. Hydrogen Energy*, **28**, 267–284.
13 Pandu, K. and Joseph, S. (2012) Comparisons and limitations of biohydrogen production processes: A

review. *Int. J. Adv. Eng. Technol.*, **02** (1), 342–356.
14 Baghchehsaree, B., Nakhla, G., Karamanev, D., and Argyrios, M. (2010) Fermentative hydrogen production by diverse Microflora. *Int. J. Hydrogen Energy*, **35**, 5021–5027.
15 Levin, D.B., Pitt, L., and Love, M. (2004) Biohydrogen production: prospects and limitations to practical application. *Int. J. Hydrogen Energy*, **29** (02), 173–185.
16 Khasnis, A.A. and Nettleman, M.D. (2005) Global warming and infectious disease. *Arch. Med. Res.*, **36**, 689–696.
17 Saxena, R.C., Adhikari, D.K., and Goyal, H.B. (2009) Biomass-based energy fuel through biochemical routes: a review. *Renew. Sustain. Energy Rev.*, **13**, 167–178.
18 Show, K.L., Lee, D.J., and Zhang, Z.P. (2011) Production of biohydrogen: current perspectives and future prospects, in *Biofuels: Alternative Feedstocks and Conversion Processes* (ed. A. Pandey *et al.*), Academic Press, Amsterdam, pp. 467–479.
19 Benemann, J.R. (1997) Feasibility analysis of photobiological hydrogen production. *Int. J. Hydrogen Energy*, **22**, 979–987.
20 Greenbaum, E. (1990) Hydrogen production by photosynthetic water splitting, in *Hydrogen Energy Progress VIII. Proceedings 8th WHEC, Hawaii* (eds T.N. Veziroglu and P.K. Takashashi), Pergamon Press, New York, pp. 743–754.
21 Das, D., Khanna, N., and Veziroglu, T.N. (2008) Recent developments in biological hydrogen production processes. *Chem. Ind. Chem. Eng. Q.*, **14**, 57–67.
22 Jo, J.H., Lee, D.S., Park, D., and Park, J.M. (2008) Biological hydrogen production by immobilized cells of Clostridium tyrobutyricum JM1 isolated from a food waste treatment process. *Bioresour. Technol.*, **99**, 6666–6672.
23 van Niel, E., Budde, M., de Haas, G., van der Wal, F., Claassen, P., and Stams, A. (2002) Distinctive properties of high hydrogen producing extreme thermophiles, *Caldicellulosiruptor saccharolyticus* and *Thermotoga elfii*. *Int. J. Hydrogen Energy*, **27**, 1391–1398.
24 Hallenbeck, P.C. (2009) Fermentative hydrogen production: principles, progress, and prognosis. *Int. J. Hydrogen Energy*, **34**, 7379–7389.
25 Argun, H. and Kargi, F. (2011) Bio-hydrogen production by different operational modes of dark and photo-fermentation: an overview. *Int. J. Hydrogen Energy*, **36**, 7443–7459.
26 Orlygsson, J., Sigurbjornsdottir, M.A., and Bakken, H.E. (2010) Bioprospecting thermophilic ethanol and hydrogen producing bacteria from hot springs in Iceland. *Icelandic Agr. Sci.*, **23**, 73–85.
27 Lee, Y.E., Jain, M.K., Lee, C.Y., Lowe, S.E., and Zeikus, J.G. (1993) Taxonomic distinction of saccharolytic thermophilic anaerobes - description of *Thermoanaerobacterium xylanolyticum* gen-nov, sp-nov, and *Thermoanaerobacterium saccharolyticum* gen-nov, sp-nov - reclassification of *Thermoanaerobium brockii*, *Clostridium thermosulfurogenes*, and *Clostridium thermohydrosulfuricum* E100-69 as *Thermoanaerobacter brockii* comb-nov, *Thermoanaerobacterium thermosulfurigenes* comb-nov, and *Thermoanaerobacter thermohydrosulfuricus* comb-nov, respectively - and transfer of *Clostridium thermohydrosulfuricum* 39E to *Thermoanaerobacter ethanolicus*. *Int. J. Syst. Bacteriol*, **43**, 41–51.
28 Wiegel, J.K.W. (2001) Genus Thermoanaerobacter in *Bergey's Manual of Systematic Bacteriology*, 2nd edn, The Williams & Wilkins Co., Baltimore, pp. 1379–1383.
29 Soboh, B., Linder, D., and Hedderich, R. (2004) A multi-subunit membrane-bound [NiFe] hydrogenase and an NADH-dependent Fe-only hydrogenase in the fermenting bacterium Thermoanaerobacter tengcongensis. *Microbiology*, **150**, 2451–2461.
30 Kadar, Z., de Vrijek, T., van Noorden, G.E., Budde, M.A.W., Szengyel, Z., and Reczey, K. (2004) Yields from glucose, xylose, and paper sludge hydrolysate during hydrogen production by the extreme thermophile *Caldicellulosiruptor saccharolyticus*. *Appl. Biochem. Biotechnol.*, **113**, 497–508.
31 Huber, R., Langworthy, T.A., Konig, H., Thomm, M., Woese, C.R., and Sleytr, U.B.

(1986) Thermotoga maritima sp. nov., represents a new genus of unique extremely thermophilic eubacteria growing up to 90°C. *Arch. Microbiol.*, **144**, 324–333.

32 Nath, K. and Das, D. (2003) Hydrogen from biomass. *Curr. Sci.*, **85**, 265–271.

33 Soloman, B.D., Barnes, J.R., and Halvorsen, K.E. (2007) Grain and cellulosic ethanol: history, economics and energy policy. *Biomass Bioenergy*, **31**, 416–425.

34 Hendriks, A.T.W.M. and Zeeman, G. (2009) Pretreatments to enhance the digestibility of lignocellulosic biomass. *Bioresour. Technol.*, **100**, 10–18.

35 de Vrije, T., De Haas, G.G., Tan, G.B., Keijsers, E.R.P., and Claassen, P.A.M. (2002) Pretreatment of Miscanthus for hydrogen production by Thermotoga elfii. *Int. J. Hydrogen Energy*, **27**, 1381–1390.

36 Kotay, S.M. and Das, D. (2007) Microbial hydrogen production with Bacillus coagulans IIT-BT S1 isolated from anaerobic sewage sludge. *Bioresour. Technol.*, **98**, 1183–1190.

37 Lay, J.J., Lee, Y.J., and Noike, T. (1999) Feasibility of biological hydrogen production from organic fraction of municipal solid waste. *Water Res.*, **33**, 2579–2586.

38 Kongjan, P. and Angelidaki, I. (2010) Extreme thermophilic biohydrogen production from wheat straw hydrolysate using mixed culture fermentation: effect of reactor configuration. *Bioresour. Technol.*, **101**, 7789–7796.

39 Hawkes, F., Hussy, I., Kyazze, G., Dinsdale, R., and Hawkes, D. (2007) Continuous dark fermentative hydrogen production by mesophilic microflora: principles and progress. *Int. J. Hydrogen Energy*, **32** (2), 172–184.

40 Ueno, Y., Otsuka, S., and Morimoto, M. (1996) Hydrogen production from industrial wastewater by anaerobic microflora in chemostat culture. *J. Ferment. Bioeng.*, **82**, 194–197.

41 Yokoi, H., Maki, R., Hirose, J., and Hayashi, S. (2002) Microbial production of hydrogen from starch manufacturing wastes. *Biomass Bioenerg.*, **22**, 389–395.

42 Shi, X.Y., Jin, D.W., Sun, Q.Y., and Li, W.W. (2010) Optimization of conditions for hydrogen production from brewery wastewater by anaerobic sludge using desirability function approach. *Renew. Energy*, **35**, 1493–1498.

43 Pan, J., Zhang, R., El-Mashad, H.M., Sun, H., and Ying, Y. (2008) Effect of food to microorganism ratio on biohydrogen production from food waste via anaerobic fermentation. *Int. J. Hydrogen Energy*, **33**, 6968–6975.

44 Han, S.K. and Shin, H.S. (2004) Biohydrogen production by anaerobic fermentation of food waste. *Int. J. Hydrogen Energy*, **29**, 569–577.

45 Venkata Mohan, S., Lalit Babu, V., and Sarma, P.N. (2007) Anaerobic biohydrogen production from diary wastewater treatment in sequencing batch reactor (AnSBR): effect of organic loading rate. *Enzyme Microb. Technol.*, **41**, 506–515.

46 Azbar, N., Cetinkaya Dokgoz, F.T., Keskin, T., Korkmaz, K.S., and Syed, H.M. (2009) Continuous fermentative hydrogen production from cheese whey wastewater under thermophilic anaerobic conditions. *Int. J. Hydrogen Energy*, **34**, 7441–7447.

47 Ferchichi, M., Crabbe, E., Gil, G.H., Hintz, W., and Almadidy, A. (2005) Influence of initial pH on hydrogen production from cheese whey. *J. Biotechnol.*, **120**, 402–409.

48 Fang, H.H.P., Li, C.L., and Zhang, T. (2006) Acidophilic biohydrogen production from rice slurry. *Int. J. Hydrogen Energy*, **31**, 683–692.

49 Zhang, T., Liu, H., and Fang, H.H.P. (2003) Biohydrogen production from starch in wastewater under thermophilic conditions. *J. Environ. Manag.*, **69**, 149–156.

50 Vatsala, T.M., Mohan Raj, S., and Manimaran, A. (2008) A pilot-scale study of biohydrogen production from distillery effluent using defined bacterial co-culture. *Int. J. Hydrogen Energy*, **33**, 5404–5415.

51 Fernandes, B.S., Peixoto, G., Albrecht, F.U., del Aguila, N.K.S., and Zaiat, M. (2010) Potential to produce biohydrogen from various wastewaters. *Energy Sustain. Dev.*, **14**, 143–148.

52 Tang, G.L., Huang, J., Sun, Z.J., Tang, Q.Q., Yan, C.H., and Liu, G.Q. (2008) Biohydrogen production from cattle wastewater by enriched anaerobic mixed consortia: influence of fermentation

temperature and pH. *J. Biosci. Bioeng.*, **106**, 80–87.

53 O-Thong, S., Prasertsan, P., Intrasungkha, N., Dhamwichukorn, S., and Birkeland, N.K. (2007) Improvement of biohydrogen production and treatment efficiency on palm oil mill effluent with nutrient supplementation at thermophilic condition using an anerobic sequencing batch reactor. *Enzyme Microb. Technol.*, **41**, 583–590.

54 O-Thong, S., Hniman, A., Prasertsan, P., and Imai, T. (2011) Biohydrogen production from cassava starch processing wastewater by thermophilic mixed cultures. *Int. J. Hydrogen Energy*, **36**, 3409–3416.

55 Chen, CC., Chuang, Y.S., Lin, C.Y., Lay, C.H., and Sen, B. (2012) Thermophilic dark fermentation of untreated rice straw using mixed cultures for hydrogen production. *Int. J. Hydrogen Energy*, **37** (20), 15540–15546.

56 Yang, P., Zhang, R., McGarvey, J.A., and Benemann, J.R. (2007) Biohydrogen production from cheese processing wastewater by anaerobic fermentation using mixed microbial communities. *Int. J. Hydrogen Energy*, **32**, 4761–4771.

57 O-Thong, S., Prasertsan, P., Intrasungkha, N., Dhamwichukorn, S., and Birkeland, N.K. (2007) Improvement of biohydrogen production and treatment efficiency on palm oil mill effluent with nutrient supplementation at thermophilic condition using an anerobic sequencing batch reactor. *Enzy. Microb. Technol.*, **41**, 583–590.

58 Zhao, B.H., Yue, Z.B., Zhao, Q.B., Mu, Y., Yu, H.Q., Harada, H., and Li, Y. (2008) Optimization of hydrogen production in a granule-based UASB reactor. *Int. J. Hydrogen Energy*, **33**, 2454–2461.

59 Chang, F.Y. and Lin, C.Y. (2004) Biohydrogen production using an up-flow anaerobic sludge blanket reactor. *Int. J. Hydrogen Energy*, **29** (01), 33–39.

60 Lee, K.S., Lin, P.J., and Chang, J.S. (2006) Temperature effects on biohydrogen production in a granular sludge bed induced by activated carbon carriers. *Int. J. Hydrogen Energy*, **31** (04), 465–472.

61 Show, K.Y., Zhang, Z.P., Tay, J.H., Liang, T.D., Lee, D.J., and Jiang, W.J. (2007) Production of hydrogen in a granular sludge-based anaerobic continuous stirred tank reactor. *Int. J. Hydrogen Energy*, **32** (18), 4744–4753.

62 Zhang, Z.P., Show, K.Y., Tay, J.H., Liang, D.T., and Lee, D.J. (2008) Biohydrogen production with anaerobic fluidized bed reactors – a comparison of biofilm-based and granule-based systems. *Int. J. Hydrogen Energy*, **33**, 1559–1564.

63 Lin, C.N., Wu, S.Y., Chang, J.S., and Chang, J.S. (2009) Biohydrogen production in a three phase fluidized bed bioreactor using sewage sludge immobilized by ethylene vinyl acetate copolymer. *Bioresour. Technol.*, **100** (13), 3298–3301.

64 Zhang, Z.P., Tay, J.H., Show, K.Y., Yan, R., Liang, D.T., Lee, D.J., and Jiang, W.J. (2007d) Biohydrogen production in a granular activated carbon anaerobic fluidized bed reactor. *Int. J. Hydrogen Energy*, **32** (2), 185–191.

65 Lee, K.S., Lo, Y.C., Lin, P.J., and Chang, J.S. (2006) Improving biohydrogen production in a carrier-induced granular sludge bed by altering physical configuration and agitation pattern of the bioreactor. *Int. J. Hydrogen Energy*, **31**, 1648–1657.

66 Lee, K.S., Lin, P.J., Fangchiang, K., and Chang, J.S. (2007) Continuous hydrogen production by anaerobic mixed microflora using a hollow-fiber microfiltration membrane bioreactor. *Int. J. Hydrogen Energy*, **32**, 950–957.

67 Kumar, N. and Das, D. (2001) Electron microscopy of hydrogen producing immobilized E. cloacae IIT-BT 08 on natural polymers. *Int. J. Hydrogen Energy*, **26** (11), 1155–1163.

68 Chittibabu, G., Nath, K., and Das, D. (2006) Feasibility studies on the fermentative hydrogen production by recombinant *Escherichia coli* BL-21. *Process Biochem.*, **41**, 682–688.

69 Zhang, Z.P., Tay, J.H., Show, K.Y., Yan, R., Liang, D.T., Lee, D.J., and Jiang, W.J. (2007) Biohydrogen production in a granular activated carbon anaerobic fluidized bed reactor. *Int. J. Hydrogen Energy*, **32**, 185–195.

70 Lin, C.Y. and Cheng, C.H. (2006) Fermentative hydrogen production from xylose using anaerobic mixed microflora. *Int. J. Hydrogen Energy*, **31**, 832–840.

71 Oh, Y.K., Kim, S.H., Kim, M.S., and Park, S. (2004a) Thermophilic biohydrogen production from glucose with trickling biofilter. *Biotechnol. Bioeng.*, **88** (06), 690–698.

72 Wu, S.Y., Hung, C.H., Lin, C.N., Chen, H. W., Lee, A.S., and Chang, J.S. (2006) Fermentative hydrogen production and bacterial community structure in high-rate anaerobic bioreactors containing silicone-immobilized and self-flocculated sludge. *Biotechnol. Bioeng.*, **93**, 934–946.

73 Zhang, Z.P., Show, K.Y., Tay, J.H., Liang, D.T., Lee, D.J., and Jiang, W.J. (2007) Rapid formation of hydrogen-producing granules in an anaerobic continuous stirred tank reactor induced by acid incubation. *Biotechnol. Bioeng.*, **96** (06), 1040–1050.

74 Peintner, C., Zeidan, A.A., and Schnitzhofer, W. (2010) Bioreactor systems for thermophilic fermentative hydrogen production: evaluation and comparison of appropriate systems. *J. Clean. Prod.*, **18**, 15–22.

75 Roy, S., Ghosh, S., and Das, D. (2012) Improvement of hydrogen production with thermophilic mixed culture from rice spent wash of distillery industry. *Int. J. Hydrogen Energy*, **37** (21), 15867–15874.

76 Murnleitner, E., Becker, T.M., and Delgado, A. (2001) State detection and control of overloads in the anaerobic wastewater treatment using fuzzy logic. *Water Res.*, **36**, 201–211.

77 Sharp, R.J. and Raven, N.D.H. (1997) Isolation and growth of hyperthermophiles, in *Applied Microbial Physiology* (eds. P.M. Rhodes and P.F. Stanbury), IRL Press, New York, pp. 23–52.

78 Hoist, O., Manelius, A., Krahe, M., Miirkl, H., Rawen, N.J., and Sharp, R. (1997) Thermophiles and fermentation technology. *Comp. Biochem. Physiol. A*, **118**, 415–422.

79 Akutsu, Y., Lee, D.Y., Chi, Y.Z., Li, Y.Y., Harada, H., and Yu, H.Q. (2009) Thermophilic fermentative hydrogen production from starch-wastewater with bio-granules. *Int. J. Hydrogen Energy*, **34** (12), 5061–5071.

80 Kotsopoulos, T.A., Fotidis, I.A., Tsolakis, N., and Martzopoulos, G.G. (2009) Biohydrogen production from pig slurry in a CSTR reactor system with mixed cultures under hyper-thermophilic temperature (70°C). *Biomass Bioenerg.*, **33**, 1168–1174.

81 Dong, F., Li, W.-W., Sheng, G.-P., Tang, Y., Yu, H.-Q., and Harad, H. (2011) An online-monitored thermophilic hydrogen production UASB reactor for long-term stable operation. *Int. J. Hydrogen Energy*, **36**, 13559–13565.

82 Kongjan, P., O-Thong, S., Kotay, M., Min, B., and Angelidaki, I. (2010) Biohydrogen production from wheat straw hydrolysate by dark fermentation using extreme thermophilic mixed culture. *Biotechnol. Bioeng.*, **105**, 899–908.

83 Kitamura, Y., Paquin, D., Gautz, L., and Liang, T. (2007) A rotational hot gas heating system for bioreactors. *Biosyst. Eng.*, **98** (02), 215–223.

84 Fang, H.H.P. and Liu, H. (2002) Effect of pH on hydrogen production from glucose by mixed culture. *Bioresour. Technol.*, **82**, 87–93.

85 Nahar, N.M. (2003) Year round performance and potential of a natural circulation type of solar water heater in India. *Energy Build.*, **35**, 239–247.

86 Chuawittayawuth, K. and Kumar, K. (2002) Experimental investigation of temperature and flow distribution in a thermosiphon solar water heating system. *Renew. Energy*, **26**, 431–448.

87 Abdullah, A.H. and Abou-Ziyan, H.Z. (2003) Thermal performance of flat plate solar collector using various arrangements of compound honeycomb. *Energy Conversion Manage.*, **44**, 3093–3112.

88 Andoh, H.Y., Gbaha, P., and Toure, S. (2007) Experimental study on the comparative thermal performance of a solar collector using coconut coir over the glass wool thermal insulation for water heating system. *J. Appl. Sci.*, **7** (21), 3187–3197.

89 de Vrije, T., Mars, A.E., Budde, M.A.W., Lai, M.H., Dijkema, C., and de Waard, P.

(2007) Glycolytic pathway and hydrogen yield studies of the extreme thermophile *Caldicellulosiruptor saccharolyticus*. *Appl. Microbiol. Biotechnol.*, **74** (06), 1358–1367.

90 Claassen, P.A.M., de Vrije, T., and Budde, M.A.W. (2004) Biological hydrogen production from sweet sorghum by thermophilic bacteria, in *Proceedings Of the 2nd World Conference on Biomass for Energy, Industry and Climate Protection, 10–14 May 2004, Rome, Italy* (eds W.P.M. van Swaaij, T. Fjällström, P. Helm, and A. Grassi), ETA-Florence and WIP-Munich, pp. 1522–1525.

91 Blencoe, G. (9 November 2009) Cost of hydrogen from different sources. http://www.h2carblog.com/?paged=12.

92 Das, D. (2009) Advances in biohydrogen production processes: an approach towards commercialization. *Int. J. Hydrogen Energy*, **34** (17), 7349–7357.

Part 2.5
Hydrogen from Solar Radiation and Algae

Hydrogen Science and Engineering: Materials, Processes, Systems and Technology, First Edition.
Edited by Detlef Stolten and Bernd Emonts.
© 2016 Wiley-VCH Verlag GmbH & Co. KGaA. Published 2016 by Wiley-VCH Verlag GmbH & Co. KGaA.

19
Photoelectrochemical Water Decomposition

Sebastian Fiechter

19.1
Introduction

In building a future energy supply based on renewable and sustainable energy sources the nearly infinite radiation energy of the sun opens the unique possibility to convert and to store sunlight in the form of chemical energy in large quantities. This converted energy can be used as a power source for all kinds of applications, especially for the time when no sunlight is available. The direct conversion of sunlight into chemical energy using an artificial membrane, which mimics the processes of photosynthesis, is a difficult and challenging task. Such a membrane can be compared with the thylakoid membrane of the chloroplasts in plants and green algae. Among biomimetic approaches using inorganic systems, one possible solution is the conversion of sunlight into chemical energy via photonic excitation of a thin film photovoltaic structure, directly connected to corrosion-stable front and back contacts of the light harvesting system, which is immersed in an aqueous electrolyte (Figure 19.1).

Hydrogen generated at the cathode–electrolyte interface can be stored as compressed gas, liquid-H_2, or in a metal hydride. In all cases, noble metal-free catalysts are needed to develop electrodes for a mass market.

Besides the components of such a device, capable of generating the required photovoltage and photocurrent in an electrochemical environment, catalysts are necessary to lower reaction barriers at the electrode–electrolyte interface in the process of light-induced hydrogen and oxygen evolution. To reduce costs, cheap semiconductor materials and non-precious catalysts are required to develop electrodes for a mass market; such materials should be highly stable in contact with the electrolyte under working conditions.

Hydrogen Science and Engineering: Materials, Processes, Systems and Technology, First Edition.
Edited by Detlef Stolten and Bernd Emonts.
© 2016 Wiley-VCH Verlag GmbH & Co. KGaA. Published 2016 by Wiley-VCH Verlag GmbH & Co. KGaA.

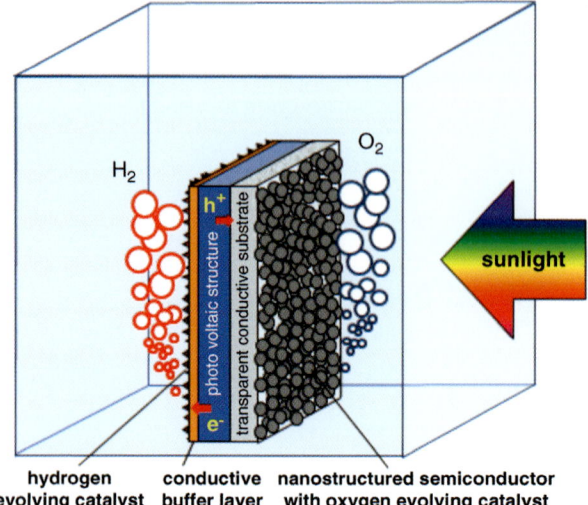

Figure 19.1 Idealized structure of a water splitting membrane immersed in an aqueous electrolyte where a transparent conductive substrate was used to deposit photoactive layers on both sides. Corrosion resistant buffer layers on top of the light absorbers act as tunneling barriers for light-excited charge carriers. At the electrode–electrolyte interfaces cheap and abundant electrocatalysts lower overvoltages of the electrochemical reaction.

19.2
Principles of Photoelectrochemical Water Splitting

Photoelectrochemical (PEC) cells, which consist of a vessel, two electrodes, and an electrolyte, are used two convert sunlight into electrical and chemical energy, respectively. Consequently, at least one of the electrodes has to be photoactive. To harvest electric energy a redox couple added to the electrolyte is necessary to shuttle charges between the electrodes. A prominent example is the redox couple iodine/iodide. In the case of chemical energy generation, only an adequate ion concentration in the electrolyte is necessary to lower the resistivity of the liquid. To split water by electrolysis a voltage of at least $U = 1.229$ V has to be applied between the electrodes, which are immersed in an aqueous solution (Figure 19.1). At room temperature, this potential is equivalent to the Gibbs free energy of the reaction:

$$H_2O(\text{liquid}) \Leftrightarrow H_2 + \tfrac{1}{2}O_2(\text{gaseous}) \quad \Delta G = 237 \text{ kJ mol}^{-1} \tag{19.1}$$

which can be calculated from the electromotive force $E_0 = -\Delta G^\circ/zF$, where F is the Faraday constant and z the number of electrons transferred to obtain 1 mol H_2.

In a water splitting photoelectrochemical device this voltage has to be generated by at least one photoactive semiconductor electrode. The lower edge of the conduction band (CB) of such a semiconductor has to be located above the

19.2 Principles of Photoelectrochemical Water Splitting

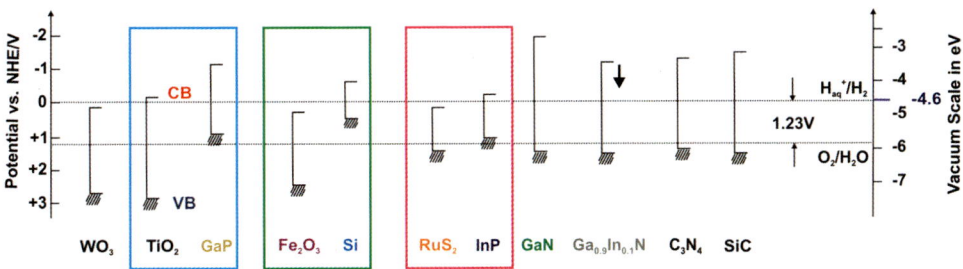

Figure 19.2 Position of conduction and valence band edges VB and CB for several semiconductors with respect to the electrochemical potentials H_2/H^+ and H_2O/O_2 at pH 0. The right-hand axis shows the position of the band edges of the semiconductors with respect to the vacuum scale, the left-hand side with respect to the electrochemical scale versus NHE (normal hydrogen electrode). While photoelectrodes made from TiO_2, GaN, $Ga_{1-x}In_xN$, C_3N_4, and SiC can in principle split water under illumination without any external bias the semiconductors in the green and red boxes can provide it only in a tandem configuration.

potential for hydrogen evolution (>-4.6 eV related to the vacuum scale and $U<0$ V related to the electrochemical scale; see Figure 19.2) so that the reduction reaction of electrons e^- with protons H^+ to evolve H_2 can proceed (Eq. (19.3)). On the other hand, the energetic position of the upper edge of the valence band (VB) has to be located below the redox potential for oxygen evolution H_2O/O_2 (at ≤ -5.9 eV related to the vacuum scale) at a voltage of >1.23 V (related to the electrochemical scale) to enable a four hole transfer for oxidation of water under O_2 evolution (Eq. (19.2)). The simplified electrochemical equations can be written as:

$$\text{anode side}: 2\,H_2O + 4\,h^+ \rightarrow O_2 + 4\,H^+ \tag{19.2}$$

$$\text{cathode side}: 4\,H^+ + 4\,e^- \rightarrow 2\,H_2 \tag{19.3}$$

Figure 19.2 shows band gap positions of transition metal oxides, metal pnictides, as well as Si, SiC, and C_3N_4 with respect to the electrochemical and vacuum scale.

19.2.1
Photoelectrochemical Cells with a Single Photoelectrode

Most of the semiconductors shown in Figure 19.2 are not able to split water under illumination as single electrodes, with the exception of TiO_2, GaN, $Ga_{x-}In_{1-x}N$, C_3N_4, and SiC. Fujishima and Honda [1] were the first to demonstrate water splitting using a rutile single crystal. Although the band gap of this TiO_2 modification is of adequate size and correct energy position the efficiency of the large band gap material as absorber for sunlight is naturally limited due to the low intensity of UV light in the solar spectrum at ground level.

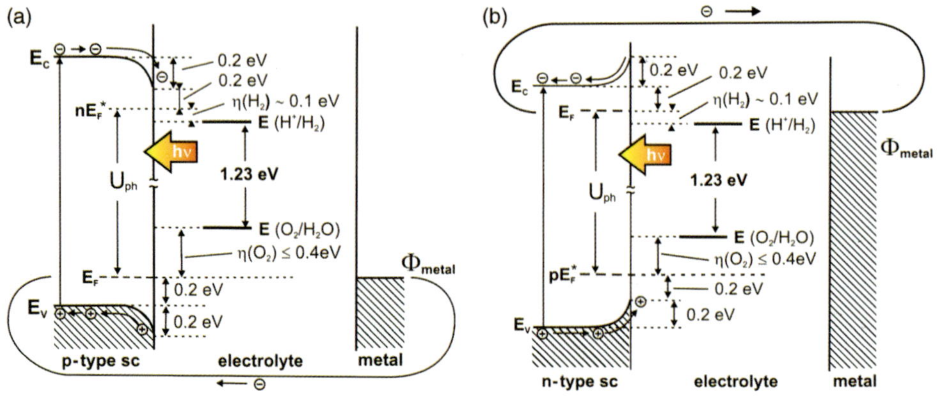

Figure 19.3 Position of valence and conduction band of a p-type (a) and n-type (b) photoactive semiconductor related to the redox potentials H_2/H_{aq}^+ and H_2O/O_2. Taking into account band bending, energetic distances of the Fermi level (E_F) and the quasi-Fermi level (E_F*) from the band edges as well as the overvoltages (η) needed for water splitting, a band gap of at least 2.2 eV is required for a single junction PEC. The photovoltage U_{ph} in (a) and (b) can theoretically achieve a value of 1.8 eV, which is large enough to split water into hydrogen and oxygen.

Other properties of the above-mentioned semiconductors also limit their applicability as light absorbers: C_3N_4 is highly resistive, n-type $Ga_xIn_{1-x}N$ is not stable against corrosion as photoanode [2]. But the main drawback is that all these semiconductors do not show photovoltages that are expected from the size of their band gaps. Therefore, even prominent materials such as n-type α-Fe_2O_3, n-type $BiVO_4$, and p-type Cu_2O need a bias voltage to split water. From a thermodynamic point of view the photovoltage of an appropriately doped semiconductor should amount to:

$$U_{photo} = E_g - 0.4 \text{ eV} \tag{19.4}$$

a value that has never been observed in semiconductors with band gaps $E_g > 1.5$ eV. Obviously, in all these compounds the recombination rates of excited charge carriers are high caused by impurities and intrinsic defects.

Figure 19.3 depicts the required position of valence and conduction band related to the electrochemical redox potentials for water oxidation and proton reduction. Assuming overvoltages of $\eta(O_2) < 0.4$ V and $\eta(H_2) \leq 0.1$ V for oxygen and hydrogen evolution, respectively, losses for band bending of ~ 0.2 eV and the distances of Fermi and quasi-Fermi level ($\sim 2 \times 0.2$ eV) from the band edges also have to be taken into account. The size of the band gap necessary to split water with a single photoelectrode can then calculated by the equations:

$$E_g = 1.23 \text{ V} + \left(E_C - nE_F^*\right) + (E_F - E_v) + iR + \eta(H_2) + \eta(O_2) \tag{19.5}$$

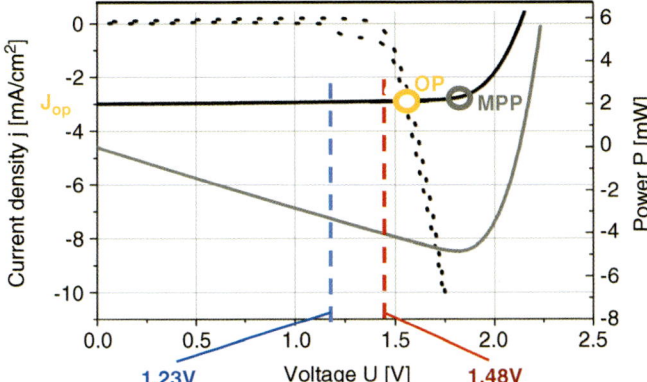

Figure 19.4 Characteristic of the photovoltaic component of a water splitting PEC (black line). From the point of intersection of the electrochemical curve (black dashed line) with the PV characteristic, the current density at the operation point (J_{op}) can be determined to calculate the solar-to-hydrogen efficiency using either ΔG (1.23 V) or ΔH (1.48 V) of the water splitting reaction. This point of intersection should be close to the maximum power point (MPP) of the PV component in the PEC Adapted from Reference [4].

of a n-type material and:

$$E_g = 1.23\,\text{V} + \left(pE_F^* - E_v\right) + (E_C - E_F) + iR + \eta(H_2) + \eta(O_2) \tag{19.6}$$

of a p-type material.

As a result of Eq. (19.5), a photoactive electrode should have a band gap of at least 2.2 eV, which limits the efficiency of the system to a value $\eta < 18\%$ due to the light intensity of the solar spectrum absorbed by the semiconductor of this band gap [3]. To determine the solar-to-hydrogen (STH) efficiency of a water splitting membrane the performance of the photovoltaic part of the electrochemical cell and the electrochemical behavior of the cell in the dark have to be determined (Figure 19.4; Reference [4]). The point of intersection of the electrochemical curve (dashed black line in Figure 19.4) with the J–U characteristic (solid line) assigns the current density J_{op} at the operation point of the water splitting cell, which has to be used to calculate STH in % according to the equation:

$$\text{STH}(\%) = [(E_{\Delta G} \times J_{op})/S] \times 100\% \tag{19.7}$$

where S denotes the total incident solar irradiance of 100 mW cm^{-2} (AM 1.5) and $E_{\Delta G}$ equals 1.23 V, equivalent to the Gibbs free energy ΔG of the reaction from Eq. (19.1). Keeping the entropy in mind, the STH can also be calculated using the potential 1.48 eV, the thermoneutral point, equivalent to $E_{\Delta H}$, which is related to the heat ΔH needed for splitting of water:

$$\text{STH}(\%) = [(E_{\Delta H} \times J_{op})/S] \times 100\% \tag{19.8}$$

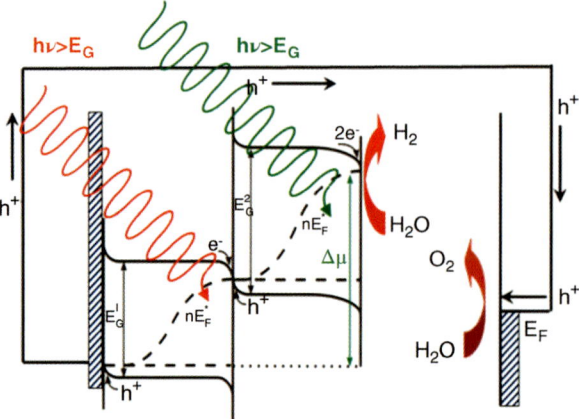

Figure 19.5 PEC for water a splitting device consisting of two p-type absorbers of different band gap and a counter-electrode; $\Delta\mu$ equals the photovoltage U_{ph}, E_{G1} and E_{G2} are the band gaps, and E_F and $nE_F{}^*$ represent Fermi and quasi-Fermi level, respectively, of the semiconductors under illumination [11].

19.2.2
Photoelectrochemical Cells with Two Photoelectrodes

To overcome the limitations of a single photoactive layer, two semiconductors with band gaps of $E_g < 2.2$ eV can be selected to realize a tandem PEC (Figure 19.5). In this case, the condition has again to be fulfilled that at least one band gap position of the first semiconductor has to be energetically located above and the second below the redox potentials for water splitting, respectively. In a two photon process, a photovoltage $U_{photo} > 1.23$ V can be obtained whose MPP should be close to the point of intersection of the electrochemical characteristic. The advantage of this PEC is that sunlight is collected by two photoelectrodes where semiconductors having high absorption coefficients and high mobilities can be used instead of a wide band gap material of modest light absorption and low mobility. In addition, the photovoltaic efficiency of such a system can accomplish values above 30% (Figure 19.6).

Figure 19.2 shows that this requirement is fulfilled by combining, for example, n-type RuS_2 with p-type InP, n-type α-Fe_2O_3 with n-Si [5], and p-type GaP with n-type TiO_2 [6]. From the absolute positions of valence and conduction band the pair RuS_2 and InP is in principal an excellent example because the band edges of this pair are close to the redox potentials $H_{aq}{}^+/H_2$ and H_2O/O_2. It has been demonstrated that p-type InP single crystal electrodes can be used as efficient photocathodes with efficiencies above 12% in a half cell [7,8].

The situation is more complex in the case of RuS_2: In a recent study nanostructured RuS_2 films have been prepared by reactive sputtering and investigated electrochemically with respect to their water oxidation ability. In contact with an electrolyte a thin amorphous RuO_x film is formed under anodic potential at the

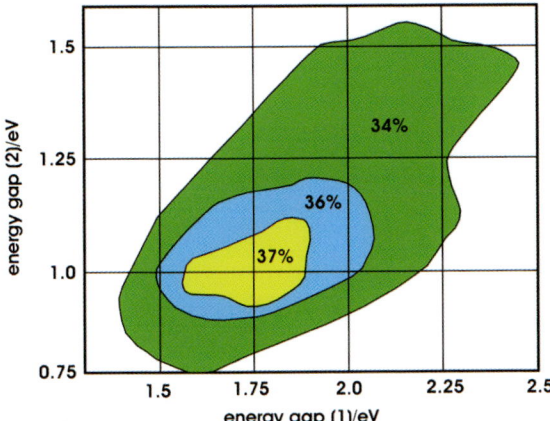

Figure 19.6 Expected efficiencies of PV tandem devices combining two semiconductors of different band gaps [11].

electrode–electrolyte interface, acting as efficient oxygen evolving catalyst [9]. Since RuS_2 is a d-band semiconductor with a band gap of 1.35 eV a photovoltage in the range of 0.9 V should be achievable. However, even with RuS_2 single crystals only small photovoltages ($\leq$0.2 V) could be found due to a high recombination rate of charge carriers in the bulk, at grain boundaries, and at the surface [10].

In pioneering work, Nozik [6] investigated n-GaP/p-GaP and n-TiO_2/p-GaP electrodes for light-induced water splitting. In the latter, he succeeded for the first time in light induced water splitting without any bias voltage. However, a high internal resistivity of the n-TiO_2 layer limited the efficiency of the cell.

Another issue concerns the surfaces of the electrodes, which have to be stable against corrosion when in contact with the electrolyte. Especially in the case of water oxidizing photoelectrodes, the risk of corrosion is high. Therefore, metal oxides and oxinitrides are of special interest as photoelectrodes although their optoelectronic properties are far from ideal. One way to protect the photoelectrodes is to deposit a transparent buffer layer of high electric conductivity (Figure 19.1).

For this reason, Grätzel et al. deposited a highly doped TiO_2 layer on top of a ZnO/Cu_2O photocathode [12]. Lewerenz et al. electrochemically grew an indium oxide layer on top of an epitaxially grown InP layer to protect the phosphide against corrosion [8]. In addition, the p-GaP electrode described by Nozik was protected by the deposition of the n-TiO_2 film. In 1976, Morisaki prepared n-Si/TiO_2 and n-Si/p^+-Si/TiO_2 hybrid electrodes for light-induced oxygen evolution. The protecting TiO_2 layers were deposited on silicon via a chemical vapor deposition technique [13].

In summary, semiconductors to be used for light induced water splitting in single and tandem solar cell configuration should exhibit the following properties:

- high optical absorption coefficients (α);
- a band gap > 1.8 eV for a single junction device;

- band gaps in the range 1–1.8 eV for a tandem configuration;
- high corrosion stabilities of the photoelectrodes in contact with the electrolyte under illumination and in the dark;
- high mobilities (μ) for electrons and holes;
- long life times (τ) for excited electrons and holes;
- space charge layers $d > 3/\alpha$ of the semiconductor;
- minority-carrier diffusion lengths $L_{Diff} > d$ for excited charge carriers.

All these properties are preconditions to obtain a high efficiency of the water splitting device. No semiconductor material is known that can accomplish all these properties. Therefore, combinations of materials have to be identified to approach the required properties for a specific water splitting device geometry.

19.2.3
Electrocatalysts and Overvoltage

As illustrated in Figure 19.1, on top of the electrodes catalysts should be deposited to lower kinetic barriers for proton reduction and water oxidation at the electrode–electrolyte interface, respectively. For InP, rhodium nanoparticles were successfully used as efficient hydrogen evolving catalyst (HEC) [8]. In an acid electrolyte prominent electrocatalysts are platinum as HEC and the noble-metal oxides $Ru_{1-x}Ir_xO_2$ as efficient oxygen evolving catalysts (OECs). Under alkaline and neutral conditions, catalysts such as CoO_x and $\alpha\text{-}Mn_2O_3$ can be used to replace expensive noble metal oxides [14,15]. Instead of platinum, metal alloys and transition metal compounds (e.g., MoNiZn [14], Co_xMo_y [16]; $NiFe_xO_y$ [16], MoS_2 [4]) have successfully been tested. The task of electrocatalysts is to lower the overvoltages. Typical operation points are associated with potentials in the range 1.6–1.7 V at a current density ranging from 5 to 20 mA cm^{-2}. The shape of the curve in Figure 19.7 indicates higher overvoltages at the anode side due to the fact that four charges have to be transferred at the anode–electrolyte interface to produce one O_2 molecule (Eq. (19.2)).

19.3
Design of Water Splitting Devices

In 1998, Khaselev and Turner reported on a PEC for water splitting using a tandem solar cell (p-GaAs/n-GaAs//p-GaInP$_2$) as photocathode and platinum as counter electrode [17]. The surface of the GaInP$_2$ electrode was coated with platinum as hydrogen evolving catalyst. An efficiency of 12.4% was reported. However, the electrodes corroded within a few hours in the acidic electrolyte (3M H$_2$SO$_4$). From Figure 19.8 it is obvious that a wired system was tested, in contrast to the geometry shown in Figure 19.1 where oxygen and hydrogen are generated at opposite sides of an integrated device. As discussed later, the disadvantage of this geometry is a long diffusion path of protons, generated at the

19.3 Design of Water Splitting Devices | 449

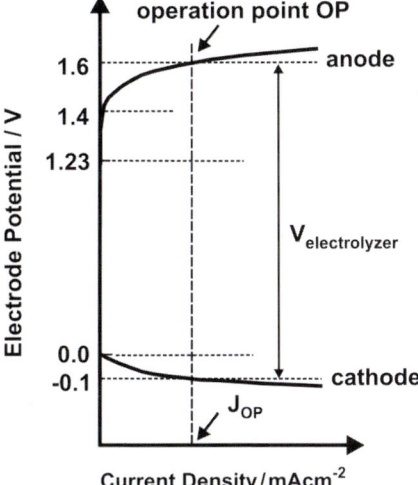

Figure 19.7 Exponential behavior of the electrode potential as a function of current density at the anode (oxygen evolution) and at the cathode (hydrogen evolution) side of an electrolyzer. Overvoltages are defined as electric energies that have to be applied in addition to the thermodynamic value of 1.23 V for oxygen evolution. Owing to a four-electron-transfer at the anode side in the process of water oxidation higher overvoltages are necessary as in the case of hydrogen evolution. Typical values for J_{OP} are in the range 5–20 mA cm^{-2}.

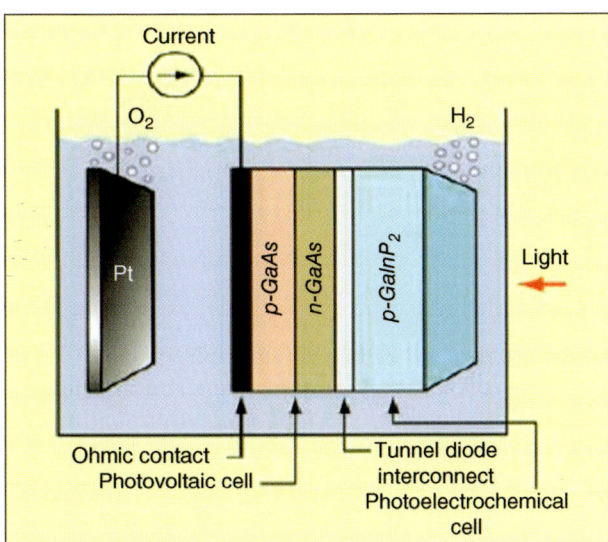

Figure 19.8 PEC with a tandem photocathode (p-GaAs/n-GaAs//p-GaInP$_2$) immersed in an acidic electrolyte connected to a wired anode (platinum) [17].

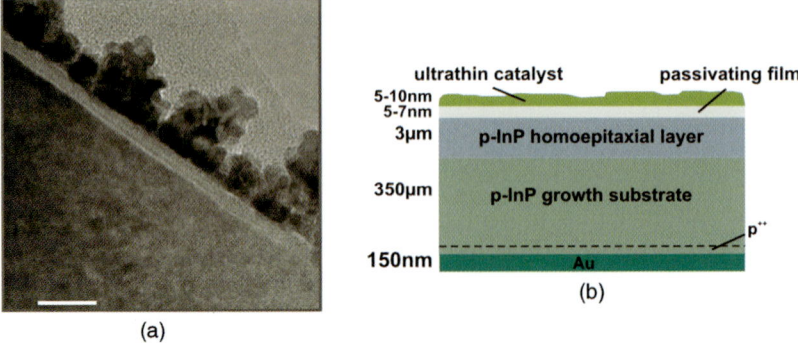

Figure 19.9 Transmission electrode microscopic (TEM) graph of a cross section (a) of a p-InP/p-InP+ photocathode provided with a conductive indium oxide layer and rhodium nanoparticles on top. (b) Schematic cross section of the photocathode [11].

anode side, in the process of water splitting (Eq. (19.2)), which have to diffuse to the cathode.

To improve the stability of the III-V semiconductor surface in the electrolyte Lewerenz et al. electrochemically grew a ~5 nm thick indium oxide layer on top of a p-InP photocathode (Figure 19.9). After deposition of rhodium nanoparticles as HEC, the InP photoelectrode with the conductive oxide layer in between was protected against corrosion. As an alternative procedure to protect the photocathode against corrosion, Deutsch et al. replaced phosphorus with nitrogen in the III-V absorber layer [18].

In 1998, Rocheleau, Miller, and Misra introduced an alternative PEC design as shown in Figure 19.10 and demonstrated an STH efficiency of ≤7.8% using a photocathode made from a commercial triple junction amorphous silicon solar (3j-a-Si) PV cell deposited on a glass substrate with the layer sequence glass/TCO/pin-pin-pin/metal [16]. While the size of the cell was of about $0.25\,\text{cm}^2$ the catalyst layers deposited on Ni sheets had an area of $1\,\text{cm}^2$. A sputtered $Co_{0.73}Mo_{0.27}$ layer served as hydrogen evolving catalyst (HEC), connected to the metal back contact of the PV cell employing an indium solder. The nickel counter-electrode, parallel-positioned at the opposite side of the PEC photoelectrode, was covered by a $NiFe_{0.19}O_{2.2}$ OEC layer, which was wired with the transparent conductive oxide film of the 3j-a-Si-PV cell. By this means, the distance between the electrodes could be precisely adjusted, minimizing the diffusion path of H^+ ions between anode and cathode. For this reason, both electrodes were integrated in a Teflon-sealed reactor using 1N KOH as electrolyte. Under illumination ($100\,\text{mW}\,\text{cm}^{-2}$) the working point of the device was at a current density of $6\,\text{mA}\,\text{cm}^{-2}$ at a voltage of 1.6 V. In this PEC geometry, the electrochemical part is separated from the PV part of the PEC (Figure 19.11), which simplifies water splitting device with respect to band alignment of the PV cell to metallic films. A further advantage is that solar light, illuminating the solar cell, is not shadowed by any catalyst films.

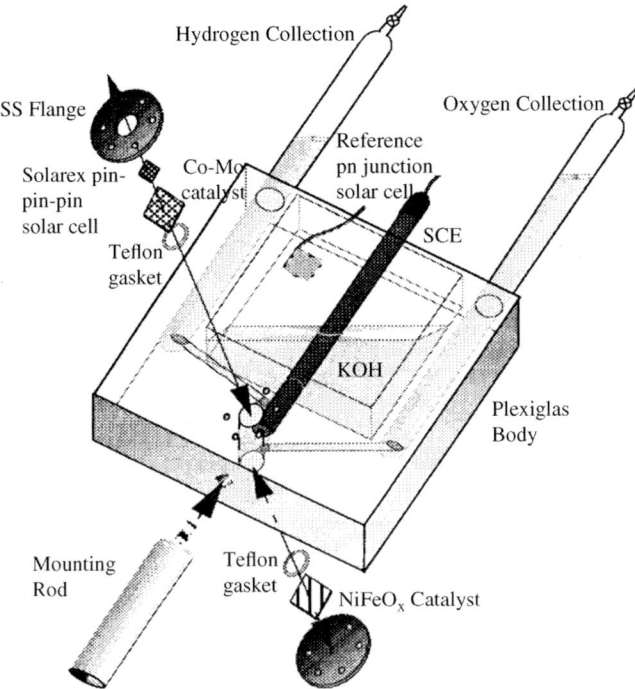

Figure 19.10 Test cell consisting of a Plexiglas body with a 1N KOH electrolyte reservoir connected to a PEC unit using a Co_xMo_y layer as hydrogen evolving catalyst mounted onto the back contact of a 3j-a-Si PV cell. The counter electrode, which is wired to the front contact of the PV cell, uses a $NiFe_vO_w$ catalyst for oxygen evolution [16].

In 2011, Nocera et al. reported on a thin film triple-junction silicon solar cell deposited on a steel foil that was – after modifying front and back contact by a CoO_x and a NiMoZn alloy catalyst – able to afford light-induced splitting of water into hydrogen and oxygen with an STH of 2.5% (in membrane geometry (Figure 19.1)) and 4.7% (in a wired system) [14]. Owing to the low chemical stability of the amorphous CoO_x catalytic layer the water splitting membrane had to be operated at pH 7. A self-healing effect was found on adding Co^{2+} ions to the electrolyte.

The photocurrents of triple-junction a-Si based solar cells are presently restricted to a current density $J < 11$ mA cm^{-2}. To further improve the SHE of a water splitting PEC other semiconductors and/or other junctions have to be tested that distinguish themselves by higher current densities at a photovoltage of <1.8 V at the MPP. This requirement is accomplished in the above-mentioned III-V based solar cell for a water splitting PEC as described by Khaselev and Turner (Figure 19.8). However, to avoid any corrosion phenomena of the III-V photocathode by the electrolyte, Dimroth et al. mounted a high efficient III-V

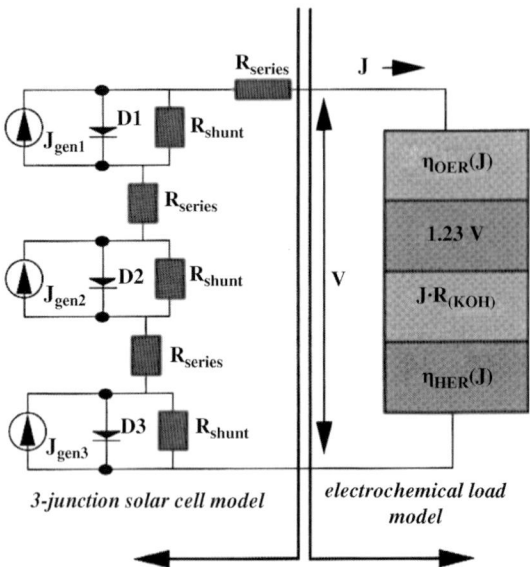

Figure 19.11 By wiring cathode and anode to the conductive front and back contact of the PV cell, the electrochemical system can be separated from the PV cell [16].

tandem solar cell on top of a graphite plate of a polymer electrolyte membrane electrolyzer (Figure 19.12). At a current density of 13 mA cm^{-2} (100 mW cm^{-2} radiation energy) an efficiency of 16% could be achieved, calculated for $U = 1.23$ V [19,20]. From the values given by the authors an efficiency $\eta = 1.23$ V/U_{op} for the electrolyzer of about 76% can be inferred. Although the features of this PEC are

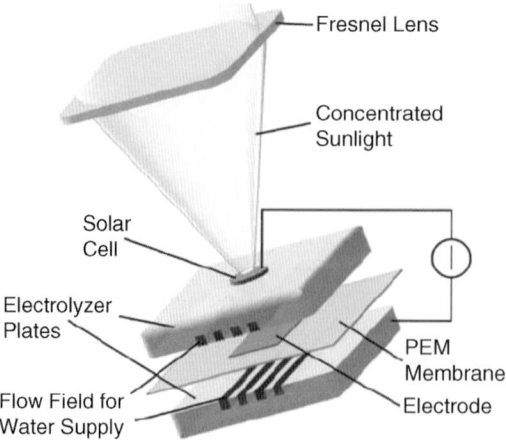

Figure 19.12 With a water splitting device wiring a $Ga_{0.35}In_{0.65}P/Ga_{0.83}In_{0.17}As$ tandem solar cell with a polymer electrolyte membrane electrolyzer, a STH of >16% can be realized [19,20].

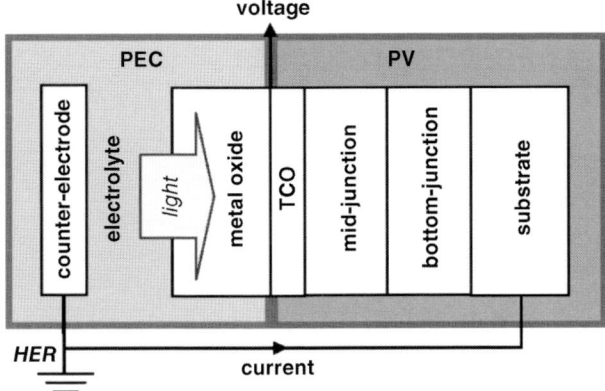

Figure 19.13 Monolithic water splitting device consisting of a tandem junction PV device and a photoactive n-type metal oxide layer both deposited on a glass substrate covered by a transparent conductive oxide [22].

impressive, showing that solar-to-hydrogen-efficiencies significantly higher than 10% are achievable, all components (Nafion membrane, graphite plates, noble metal catalysts (Pt and RuO_2), and the III–V tandem solar cell itself) are presently too expensive to achieve low costs for hydrogen production. The water used in the electrolyzer was demineralized to protect the RuO_2 catalyst from corrosion. The catalysts are directly deposited on the NAFION membrane to minimize the migration path of protons from the anode to the cathode side.

Having the cost issue in mind, the concept of Miller *et al.* has to be discussed [21]: in 2005 the authors described a hybrid photoelectrode as a monolithically stacked device operated in an acidic media (Figure 19.13). Here, the triple junction a-Si solar cell from Figure 19.10 is replaced by a tandem junction PV cell and the metal electrode by a photoactive oxide layer (Figure 19.13). As photoanode they used a WO_3 film to obtain a photovoltage high enough to split water. An efficiency of 0.6% could be demonstrated in this first approach. Using reactive sputtering at low temperatures (<300 °C) Miller *et al.* demonstrated photocurrents of 3.4 mA cm^{-2} under AM 1.5 illumination in a 0.33M H_3PO_4 at 1.23 V vs RHE. With this value an STH of 4.2% should be feasible [22]. In 2008, this cell concept was assimilated by Alexander, Augustynski *et al.* An STH of 3.6% was obtained by depositing a micro/nano-porous WO_3 layer on a fluorine-doped tin oxide layer by a sol–gel technique [23]. A bias voltage from the Si tandem structure of 1 V was needed. Since WO_3 has a band gap of 2.5 eV the visible part of sunlight was absorbed in this layer while the near-infrared part was absorbed in the silicon solar cell using a stacked system as shown in Figure 19.13.

Recently, Abdi, van de Krol *et al.* developed a water oxidizing device using the concept shown in Figure 19.13, but combining a silicon tandem PV cell with a $BiVO_4$ photoanode, obtaining an STH efficiency of 4.9% (Figure 19.14) [24]. The efficiency could be accomplished by introducing a gradient dopant concentration of tungsten in the $BiVO_4$ film of μm thickness creating a distribution n-n$^+$

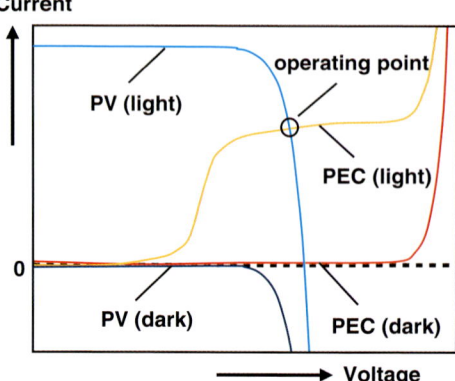

Figure 19.14 Characteristics of the tandem junction PV device and of the n-type metal oxide layer in contact with an electrolyte (PEC) in the dark and under illumination. The photovoltage necessary to split water is raised by the sum of photovoltage of the PEC and the PV part at the operating point. The current increase of the red curve indicates oxygen evolution at high voltages in the dark.

homojunction in the film. This is the highest STH efficiency ever reported for a stand-alone water splitting device based on a metal oxide photoanode. At the operation point of the light-driven electrolyzer a photocurrent of 4 mA cm^{-2} at a voltage of 1.1 V was found. Compared to WO_3, which has an indirect band gap of 2.5 eV, $BiVO_4$ is a direct semiconductor with $E_g = 2.4$ eV. The slightly lower band gap improves light absorption and therefore leads to an improved STH efficiency. In all these materials the proper control of defect chemistry and doping profiles is crucial to obtain an efficient photoelectrode.

Instead of a Si-based PV cell, Grätzel, Augustynski et al. combined a dye sensitized solar cell (DSSC) with nanostructured hematite (α-Fe_2O_3) and WO_3 photoanodes. Owing to the better optoelectronic properties of the WO_3 layer an STH efficiency of 3.1% could be realized [25].

Only in the concept described by Dimroth [19] was the issue addressed of how to collect and to separate the evolved gases, generated by light-driven water splitting. For safety reasons, the gases have to be kept in different compartments. Under operation conditions, an ion conducting membrane (e.g., NAFION) has then to be introduced in the light-driven water splitting device. An array of electrolyzers, consisting of thin film PV cells integrated in the outer face of the body housing of a water splitting device that is separated by an ion conducting membranes into two flat compartments, was proposed by Deng and Xu as a solution (Figure 19.15).

In summary, from an economic point of view a cheap PV cell (e.g., an amorphous silicon tandem solar cell or a DSSC) can be combined with a low-cost metal oxide photoanode layer in a monolithic device together with cheap, nontoxic, and abundant oxygen and hydrogen evolving catalysts, which are stable under alkaline or acidic conditions.

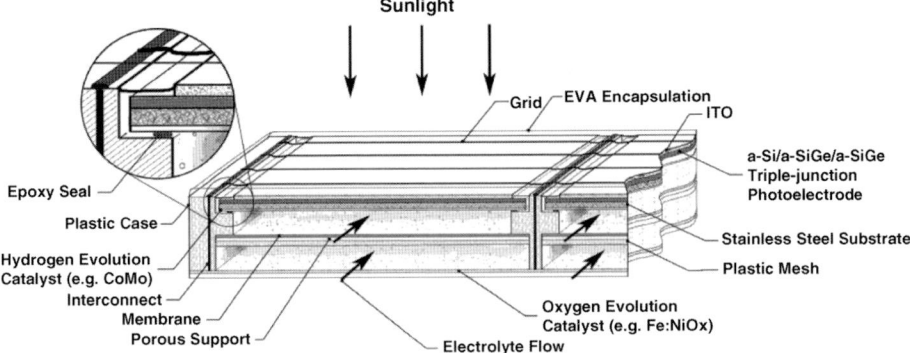

Figure 19.15 Array of flat and in-parallel oriented light-driven electrolyzer units. The photovoltage of every water splitting device is generated by triple junction Si-Ge-based PV cells. Front contact of the PV cell is wired to the anode side. The oxygen and hydrogen evolving compartments are separated by a proton conductive membrane. Reproduced with permission from X. Deng, Toledo.

19.4
Nano- and Microstructured Photoelectrodes

Although the above-mentioned oxides WO_3 and Fe_2O_3 are electrochemically stable and contain non-toxic elements the efficiencies of photoelectrodes made from these oxides remain so far limited due to low current densities caused by small mobilities for minority carriers, short carrier diffusion lengths, and short charge carrier life times. One possibility to overcome these drawbacks is to nano- or microstructure the surface of the photoelectrodes as an alternative to working with ultrathin films deposited on a fractal substrate. Indeed, micro- as well as nano-structuring of the layers applying sophisticated deposition protocols for absorbers and catalysts has been successfully applied to improve current densities in oxide and oxinitride anode films significantly. For a highly porous WO_3 layer about 2.5 μm thick, light absorption could be enhanced due to the thickness of the layer. However, the thickness of the individual platelet-like particles was in the nanometer range, allowing efficient minority charge carrier transport in the stacked and intergrown WO_3 platelets of <1000 nm lateral extension. A current density of 3 mA cm^{-2} under AM 1.5 conditions could be obtained [23].

A challenging, but promising approach has been proposed Lewis and co-workers who introduced the idea of a regular array of nano- or micro-wires leading to interfacial features that increase light trapping, but decrease the semiconductor thickness normally required in a non-structured absorber layer by a factor of $4n^2$, where n is the refraction index (Figure 19.16) [26]. Since photogenerated minority charge carriers only have a short distance to travel to the reactive interface the requirements in terms of the purity of the semiconductor material are not as high as for a bulk absorber layer. As a prototype, microwire arrays are under development that decouple light absorption from carrier

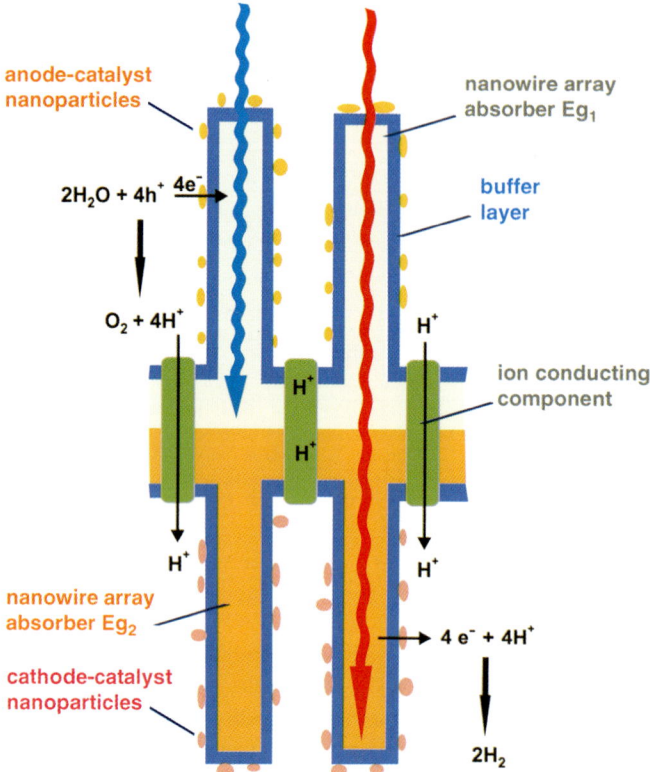

Figure 19.16 Schematic of a water splitting membrane where two arrays of micro/nanostructured semiconducting wires allow an efficient absorption of photons in the visible and near-infrared range compared with a flat configuration. By introducing proton conducting segments between the arrays of different band gap, generated protons only have a short distance to travel from the anode to the cathode side of the electrolyzer [26].

collection. In addition, the increased surface area of the structured absorber, decorated by nanosized catalytic particles, also enhances the catalytic activity at the electrode–electrolyte interface and lowers overvoltages in the process of water splitting. However, recombination centers on the surface have to be passivated by growing a thin conformal, but highly conductive and transparent, layer on top of the structured electrode surface, which can be deposited by an atomic layer deposition (ALD) technique. A schematic drawing of such a water splitting membrane is shown in Figure 19.16.

As an example of a nanowired cathode structure, p-type (In,Ga)N nanowires were grown by plasma-assisted molecular beam epitaxy on a highly doped silicon wafer. Under illumination p-(In,Ga)N nanowires showed a cathodic photocurrent that could be correlated with the evolution of H_2. After photodeposition of Pt on the nanowires, the photocurrent density could be increased significantly to $5\,\text{mA}\,\text{cm}^{-2}$ at a potential of $-0.5\,\text{V}$ (vs NHE). Although an incident photon-to-current conversion efficiency of around 40% was found at −0.45 V (vs NHE) the overpotential for hydrogen evolution has to be improved [2].

Recently, Lewis *et al.* published a report on a nanowired anode structure according to the concept shown in the upper part of Figure 19.16. Starting from a core–shell n–p$^+$-Si nanostructure, the silicon wires were first covered by a thin In-doped SnO$_2$ (ITO) film and afterwards by a thicker WO$_3$ layer. The tandem device exhibited open-circuit potentials of $E_{oc} = 1.21$ V versus $E°(O_2/H_2O)$, demonstrating the additive effect of photovoltages from Si and WO$_3$. Using concentrated light unassisted H$_2$ was found at the platinum counter-electrode; however, the efficiency was still very low [27].

19.5
Economic Aspects

Instead of a water splitting membrane, which is exclusively driven by sunlight, water electrolysis can also be realized by combining conventional electrolyzing systems with established PV cells as power supply. However, in the case of such a water splitting system, separate stands, frames, wiring, and electronic control for the electrolyzer and the solar cells are necessary while the above-discussed photoelectrochemical systems are characterized by a monolithic design and can be operated not only under bright sunlight but also under weak, diffuse, and even rapidly fluctuating light conditions without degradation of the electrodes.

In a recent publication, Zhai *et al.* estimated the net primary energy balance of a solar-driven PEC water splitting device [28]. Obviously, when introducing a new renewable energy generation technology, the device should produce more energy during its lifetime than required to manufacture it. Considering the STH efficiency of different PEC systems in the range 3–10% and their longevity ranging from 5 to 30 years, the authors concluded that at an STH efficiency of 10% the life time of the PEC system should be ten years. Concerning the materials needed for the PEC devices highest primary energy requirements are needed for allocation of the gas separation membrane and the housing of the electrolyzing system. In the fabrication process most energy is consumed for photoelectrode preparation (50%), membrane fabrication (15%), and environmental control. Since the technologies used to fabricate PEC systems for light-induced water splitting are still at an early stage significant uncertainties in the assessment of STH efficiency and longevity remain [28].

19.6
Concluding Remarks

Hybrid PECs consisting of a low cost PV cell of high durability and a corrosion stable photoanode arranged in a monolithic device are presently the most promising systems in terms of technical and economic feasibility. The photoanode materials and their preparation techniques, which are presently under investigation, have to be further improved to optimize the optoelectronic properties of the absorber materials and their interfaces as well as the activity of the electrocatalysts. A crucial role is played by the recombination of minority charge carriers at the absorber/window and the photoanode/electrolyte interface. Consequently, detailed knowledge of the defect chemistry of the absorbers as a function of preparation conditions is a prerequisite to tailor the catalytic properties of the material's surface in contact with an electrolyte. On the other hand, alternative routes using molecular approaches should also be kept in mind. In all cases stability, availability, environmental sustainability, and efficiency are crucial basic conditions for a useful application to convert sunlight into chemical energy.

References

1 Fujishima, A. and Honda, K. (1972) Electrochemical photolysis of water at a semiconductor electrode. *Nature*, **238**, 37–38.

2 Kamimura, J., Bogdanoff, P., Lähnemann, J., Hauswald, C., Geelhaar, L., Fiechter, S., and Riechert, H. (2013) Photoelectrochemical properties of (In, Ga)N nanowires for water splitting investigated by in situ electrochemical mass spectroscopy. *J. Am. Chem. Soc.*, **135**, 10242–10245.

3 Fiechter, S.P., Bogdanoff, P., Bak, T., and Nowotny, J. (2012) Basic concepts of photoelectrochemical solar energy conversion systems. *Adv. Appl. Ceram.*, **111**, 39–43.

4 Stellmach, D., Bogdanoff, P., Gabriel, O., Stannowski, B., Schlatmann, R., van de Krol, R., and Fiechter, S. (2013) Nanostructured MoS2 particles as a novel hydrogen evolving catalyst integrated in a PV-hybrid electrolyzer, in *Materials and Processes for Energy: Communicating Current Research and Technological Developments* (ed. A. Méndez-Vilas), Formatex Research Center, Badajoz, Spain, pp. 880–886.

5 van de Krol, R. and Liang, Y. (2013) An n-Si/n-Fe$_2$O$_3$ hetrojunction tandem photoanode for solar water splitting. *CHIMIA*, **67**, 168–171.

6 Nozik, A.J. (1976) p-n Photoelectrolysis cells. *Appl. Phys. Lett.*, **29**, 150–153.

7 Heller, A. (1984) Hydrogen-evolving solar cells. *Science*, **223**, 1141–1148.

8 Lewerenz, H.J., Heine, C., Skorupska, K., Szabo, N., Hannappel, T., Vo-Dinh, T., Campbell, S.A., Klemm, H.W., and Munoz, A.G. (2010) Photoelectrocatalysis: principles, nanoemitter applications and routes to bio-inspired systems. *Energy Environ. Sci.*, **03**, 748–761.

9 Bogdanoff, P., Zachäus, C., Brunken, S., Ellmer, K., Martinez Moreno, E., Kratzig, A., and Fiechter, S. (2013) Ruthenium sulphide thin layers as a catalyst for the electrooxidation of water. *Phys. Chem. Chem. Phys.*, **15**, 1452–1459.

10 Kühne, H.-M. and Tributsch, H. (1983) Visible light photo-oxidation of water with single-crystal RuS$_2$ electrodes. *J. Electrochem. Soc.*, **130**, 1448–1450.

11 Kaiser, B., Jaegermann, W., Fiechter, S., and Lewerenz, H.J. (2011) Direct photoelectrochemical conversion of sun light into hydrogen for chemical energy storage. *Bunsen-Mag.*, **13**, 104–111.

12 Paracchino, A., Laporte, V., Sivula, K., Grätzel, M., and Thimsen, E. (2011) Highly active oxide photocathode for photoelectrochemical water reduction. *Nat. Mater.*, **10**, 456–461.

13 Morisaki, H., Watanabe, T., Iwase, M., and Yazawa, K. (1976) Photoelectrolysis of water with TiO$_2$ covered solar-cell electrodes. *Appl. Phys. Lett.*, **29**, 338–340.

14 Reece, S.Y., Hamel, J.A., Sung, K., Jarvi, T.D., Esswein, A.J., Pijpers, J.J.H., and Nocera, D.G. (2011) Wireless solar water splitting using silicon-based semiconductors. *Science*, **334**, 645–648.

15 Gorlin, Y. and Jaramillo, T.F. (2010) A bifunctional nonprecious metal catalyst for oxygen reduction and water oxidation. *J. Am. Chem. Soc.*, **132**, 13612–13614.

16 Rocheleau, R.E., Miller, E.L., and Misra, A. (1998) High-efficiency photoelectrochemical hydrogen production using multijunction amorphous silicon photoelectrodes. *Energy Fuels*, **12**, 3–10.

17 Khaselev, O. and Turner, J.A. (1998) A monolithic photovoltaic-photoelectrochemical device for hydrogen production via water splitting. *Science*, **280**, 425–427.

18 Deutsch, T.G., Koval, C.A., and Turner, J.A. (2006) III-V nitride epilayers for photoelectrochemical water splitting: GaPN and GaAsPN. *J. Phys. Chem. B*, **110**, 25297–25307.

19 Dimroth, F., Peharz, G., Wittstadt, U., Hacker, B., and Bett, A.W. (2006) Hydrogen production in a PV concentrator using III-V multi-junction solar cells, in: *IEEE 4th World Conference on Photovoltaic Energy Conversion, 7–12 May 2006, Waikoloa, Hawaii*, IEEE, Piscataway, vol. **1**, pp. 640–643.

20 Peharz, G., Dimroth, F., and Wittstadt, U. (2007) Solar hydrogen production by water splitting with a conversion efficiency of 18%. *Int. J. Hydrogen Energy*, **32**, 3248–3252.

21 Miller, E.L., Marsen, B., Paluselli, D., and Rocheleau, R.E. (2005) Optimization of hybrid photoelectrodes for solar water-splitting. *Electrochem. Solid-State Lett.*, **08**, A247–A249.

22 Miller, E.L., Marsen, B., Cole, B., and Lum, M. (2006) Low-temperature reactively sputtered tungsten oxide films for solar-powered water splitting applications. *Electrochem. Solid-State Lett.*, **09**, G248–G250.

23 Alexander, B.D., Kulesza, P.J., Rutkowska, I., Solarska, R., and Augustynski, J. (2008) Metal oxide photoanodes for solar hydrogen production. *J. Mater. Chem.*, **18**, 2298–2303.

24 Fatwa, F.A., Han, L., Smets, A.H.M., Zeman, M., Dam, B., and van de Krol, R. (2013) Efficient solar water splitting by enhanced charge separation in a bismuth vanadate-silicon tandem photoelectrode. *Nat. Commun.* **4**, 2195–2201.

25 Brillet, J., Yum, J.-H., Cornuz, M., Hisatomi, T., Solarska, R., Augustynski, J., Graetzel, M., and Sivula, K. (2012) Highly efficient water splitting by a dual-absorber tandem cell. *Nat. Photon.*, **06**, 824–828.

26 McKone, J. and Lewis, N. (2013) Structured materials for photoelectrochemical water splitting, in *Photoelectrochemical Water Splitting – Materials, Processes and Architectures* (eds H.-J. Lewerenz and L. Peter), RSC Publishing, Cambridge, pp. 52–82.

27 Shaner, M.R., Fountaine, K.T., Ardo, S., Coridan, R.H., Atwater, H.A., and Lewis, N.S. (2013) Photoelectrochemistry of core–shell tandem junction n–p$^+$-Si/n-WO$_3$ microwire array photoelectrodes. *Energy Environ. Sci.* **7**, 779–790.

28 Zhai, P., Haussener, S., Ager, J., Sathre, R., Walczak, K., Greenblatt, J., and McKone, T. (2013) Net primary energy balance of a solar-driven photoelectrochemical water-splitting device. *Energy Environ. Sci.*, **06**, 2380–2389.

20
Current Insights to Enhance Hydrogen Production by Photosynthetic Organisms

Roshan Sharma Poudyal, Indira Tiwari, Mohammad Mahdi Najafpour, Dmitry A. Los, Robert Carpentier, Jian-Ren Shen, and Suleyman I. Allakhverdiev

20.1 Introduction

Hydrogen is a tasteless, colorless, odorless, non-toxic, and highly combustible diatomic gas with the molecular formula H_2. Hydrogen is abundant in several resources: natural gas, fossil fuels, coal, and living organisms (animals, plants, and microorganisms). This gas constitutes an intriguing avenue by which to store solar energy. It is a feedstock in petroleum refining, which can be used in the chemical industry as a new generation fuel for the future [1]. Hydrogen gas is highly flammable in air with a high enthalpy of combustion:

$$2H_2(g) + O_2(g) \rightarrow 2H_2O(l) + 572 \text{ kJ mol}^{-1}$$

The above properties of H_2 make its usage attractive as a basic energy source. There are several ways to produce H_2. The water splitting mechanism from H_2O to hydrogen (H_2) and oxygen (O_2) remains the basic principle to produce H_2. Water splitting by solar energy can store the energy in the form of fuel (H_2):

$$2H_2O \rightarrow 4H^+ + O_2 + 4e^- \tag{20.1}$$

$$4H^+ + 4e^- \rightarrow 2H_2 \tag{20.2}$$

$$2H_2O \rightarrow 2H_2 + O_2 \tag{20.3}$$

However, water oxidation (Eq. (20.1)) involving multi-electron transfer is difficult to perform due to thermodynamic and kinetic limitations [2]. Thus, to produce hydrogen in a sustainable manner, a *"super catalyst"* for water oxidation is needed [2]. Significant attention has been devoted to the development of novel water oxidation catalysts that can be used in electrolyzers, or as a component of photoelectrochemical devices [2].

For this purpose the water oxidizing complex in photosystem II (PSII) can be considered as a natural model system for water splitting (Figure 20.1).

Hydrogen Science and Engineering: Materials, Processes, Systems and Technology, First Edition.
Edited by Detlef Stolten and Bernd Emonts.
© 2016 Wiley-VCH Verlag GmbH & Co. KGaA. Published 2016 by Wiley-VCH Verlag GmbH & Co. KGaA.

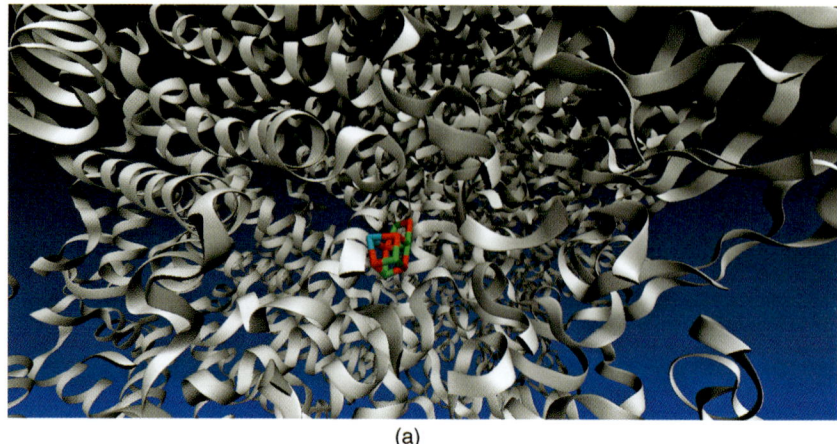

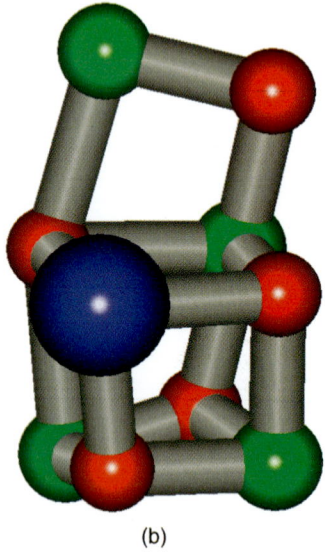

Figure 20.1 CaMn$_4$O$_5$ cluster and the surrounding amino acids in photosystem II (PSII) (a). The CaMn$_4$O$_5$ cluster in PSII has dimensions of about $\sim 0.5 \times 0.25 \times 0.25$ nm^3 (oxygen: red, Mn: green, and calcium: blue) (b). These images were produced with VMD and are owned by the Theoretical and Computational Biophysics Group, NIH Resource for Macromolecular Modeling and Bioinformatics, at the Beckman Institute, University of Illinois at Urbana-Champaign. The original data is from Reference [3] (PDB: 3ARC).

Besides the water splitting mechanism of PSII other methods are:

1) steam reforming: convert methane into hydrogen and carbon monoxide;
2) electrolysis: splitting water into hydrogen and oxygen by electric power;
3) steam electrolysis: splitting of water into hydrogen and oxygen by heating;

4) photochemical splitting of water: splitting of water into hydrogen and oxygen by semiconductor materials;
5) photobiological splitting of water: splitting of water into hydrogen and oxygen by an organisms in the presence of sunlight;
6) thermal splitting of water: splitting of water into hydrogen and oxygen by intense temperature (more than 1000 °C);
7) gasification: to break biomass and coal into a H_2.

Thus, H_2 can be produced by heat (including that provided by nuclear power plants) from gasified biomass and fossil fuels, and by water electrolysis. However, in the long term, photosynthetic H_2 production may be considered as an effective way to produce H_2 directly from solar energy.

Many biologically active organisms may serve as a potent source for the production of H_2 in nature. In the presence of sunlight, several microorganisms are able to produce H_2. Our task includes: identification of such organisms, study of the mechanisms of H_2 production, and the ways of optimizing such biological systems for more efficient production of H_2.

Biological H_2 production, especially by photosynthetic microorganisms, has several advantages because it requires rather simple techniques and low cost energy (natural sunlight) as compared to electrochemical H_2 production based on water splitting.

Owing to the unlimited source of solar energy, lowered rate of pollution, relatively easy storage and transportation, photosynthetic H_2 production is considered as a sustainable source of renewable energy for human and animals [4]. However, the production of H_2 by photosynthetic organisms is affected by many factors and biochemical pathways [5–7]. Several microorganisms, such as green algae, cyanobacteria, and photosynthetic bacteria generate H_2 under light by the activity of their own catalytic metalloprotein, hydrogenase (Figures 20.2 and 20.3). This enzyme plays a crucial role in the reversible oxidation of H_2 [8].

20.2
Biological H_2 Production

Sunlight, water, and biomass are plentiful resources in nature by which to produce H_2. The utilization of these resources for biological H_2 production is important for bio-products and chemical industries. Therefore, biological H_2 production has many advantages as compared to others methods as this gas can be generated by photosynthesis, fermentation processes, and by microbial electrolytes [9].

Several higher plants, algae, and cyanobacteria can produce biological hydrogen. Some legume plants have a symbiotic association with bacteria (*Rhizobium*) and fix atmospheric nitrogen into ammonia [10] with H_2 as the byproduct of the activity of nitrogenase enzymes. Such H_2 production is a high energy consuming process. During nitrogen fixation, the H_2 produced is oxidized and recycled [11].

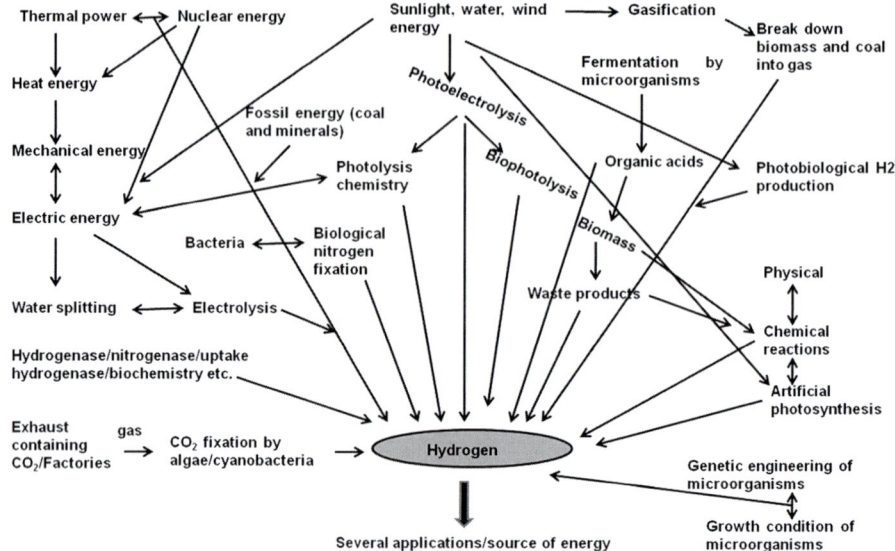

Figure 20.2 Diverse pathway and sources of H₂ production in nature; H$_2$ can be produced by different physical, chemical, and biological process.

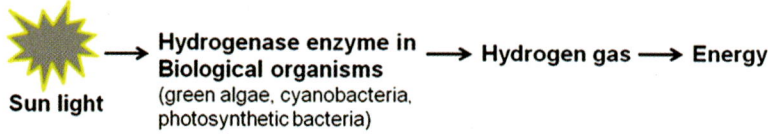

Figure 20.3 Schematic flow diagram showing the linear route to hydrogen production by microorganisms in the presence of sunlight.

Some symbiotic bacteria (*Rhizobium*) also possess an uptake hydrogenase (HUP) to evolve hydrogen. Das and Veziroğlu (2001) [12] classified the possibilities of biological hydrogen production as follows:

1) biophotolysis of water by algae and cyanobacteria;
2) photodecomposition of organism compounds by photosynthetic bacteria;
3) hydrogen production by using organic compounds during fermentation;
4) hydrogen production by a hybrid system of photosynthetic and fermentative bacteria.

Hydrogen can be also be produced from some waste products, though with low yield. Such processes require huge amounts of waste products and a substantial area for the bioreactors [13]. Thus, H$_2$ production by algae from water and sunlight remains one of the most efficient, economic, and sustainable methods.

20.3
Physiology and Biochemistry of Algae and Cyanobacteria for H_2 Production

In algae and cyanobacteria, H_2 is produced by a combination of different physiological and biochemical processes. Hydrogen production in green algae by hydrogenase and in cyanobacteria by nitrogenase enzymes was studied several decades ago [14]. In cyanobacteria, hydrogenase and nitrogenase are two enzymes that generate H_2. Nitrogenase enzyme is present in the heterocyst of filamentous cyanobacteria. When they grow in nitrogen limiting conditions they catalyze the production of H_2. Similarly, uptake hydrogenases (HUPs) catalyze the consumption of H_2 produced by nitrogenase. In addition, a bidirectional hydrogenase can also produce and take up H_2.

More recently, different pathways for H_2 production in algae and cyanobacteria has been reported [5,15,16]. In oxygenic or anoxygenic photosynthesis, sunlight, absorbed by the light harvesting complexes of photosystems generates electrons that pass through an electron transport chain producing ATP and NADPH as energy intermediate molecules. These molecules are consumed during carbon fixation in the dark reactions of photosynthesis [17]. The first pathway of H_2 production refers to the electron transport chain in PSI and PSII that is able to reduce the oxidized ferredoxin and/or nicotinamide adenine dinucleotide phosphate ($NADP^+$). The reducing equivalents (ferredoxin) can be used by hydrogenase to reduce protons for evolution of molecular hydrogen (under special conditions). In green algae, ferredoxin (PetF) is essential for electron transfer between PSI and HydA1 (the latter encodes the H_2 evolving group of [Fe-Fe] hydrogenases, which catalyze H_2 production in anaerobic conditions) [18]. The second pathway refers to photofermentative processes that occur at two different places, that is, mitochondrial respiration and the PQ pool of the thylakoid membrane of chloroplast. The third pathway refers to photosynthetically derived electrons and protons catalyzed by nitrogenase to produce H_2 in cyanobacteria; a detail explanation of this pathway has been provided [4,15,16,19].

The physiology of cyanobacteria is also important for H_2 production, since its efficiency varies in different strains [20]. Hydrogen production by hydrogenase has several biochemical problems in cyanobacteria because the hydrogenase enzyme activity is inhibited by molecular oxygen (O_2) [21]. It is important to overcome these problems by enhancing H_2 production efficiency with biotechnological and genetic approaches [18,22–26] and to engineer new strains for high H_2 production [20].

Nitrogen fixing and non-nitrogen fixing cyanobacteria produce H_2 through the activity of nitrogenase and hydrogenase under certain conditions [27,28] and many strains of cyanobacteria can produce H_2 directly [29]. Previous findings have shown that facultative photoautotrophic or photoheterotrophic microalgae are good candidates for H_2 production [30,31]. In addition, green microalgae and cyanobacteria can produce H_2 indirectly by dark fermentation [32,33]. It was shown recently that cyanobacteria produces hydrogen by the activity of both nitrogenase and hydrogenases enzymes [32] and the quantification of H_2 production under different conditions has been carried out [34]. The crucial role of these two enzymes is the catalysis of $2H^+ + 2X_{(reduced)} = 6H_2 + 2X_{(oxidized)}$ by hydrogenase and $N_2 + 8H^+ + 8e^- + 16ATP \rightarrow 2NH_3 + H_2 + 16ADP + 16Pi$ by nitrogenase.

Factors that may influence biological H_2 production are as follows:

1) photosystem I activity [35],
2) redox potential of PetF [36],
3) electron transport interaction between HydA1 and PetF [18,37],
4) interaction of PetF–ferrodoxin–NADPH oxidoreductase [36],
5) redox balance of glycolysis under anaerobic condition [38],
6) increase in amounts of fatty acids, lipids, and starch [39–41],
7) acclimation to sulfur deprivation [26,42].

Recently, Oey et al. (2013) [43] reached higher efficiencies of photosynthetic H_2 production by knock-out of three light harvesting complex proteins (LHCMB1, 2, and 3) in the *Chlamydomonas reinhardtii* mutant *Stm6Glc4*. Therefore, physiological and biochemical phenomena are most important factors in algae and cyanobacteria for efficient H_2 production.

20.4
Hydrogenase and Nitrogenase for H_2 Production

Hydrogenases and nitrogenases enzymes are structurally and functionally diverse groups of catalytic enzymes discovered early on in green algae [44] and photosynthetic bacteria [45–47]. Since then, practical applications of these enzymes have been under consideration and discussion.

In green algae and in some cyanobacteria, hydrogenase is inhibited in the presence of oxygen therefore an anaerobiosis condition is required to activate hydrogenase. In algae, [FeFe]-hydrogenase activation is required under anaerobiosis conditions but in most cyanobacteria [NiFe]-hydrogenase is activated in the presence of oxygen [48–50]. Therefore, based on their activity there are several (FeFe) hydrogenases [51] and (NiFe) hydrogenases, which have been summarized recently by Marshall et al. [52].

20.4.1
Hydrogenase and H_2 Production

As mentioned above, hydrogen is being considered as a major fuel for the future that can be produced by the catalytic activity of hydrogenase enzymes in algae and cyanobacteria. Therefore, more understanding and research on hydrogenase is required. Three classes of hydrogenases have been identified: [NiFe], [FeFe], and [Fe] hydrogenases [8]. [NiFe] hydrogenases are the most abundant

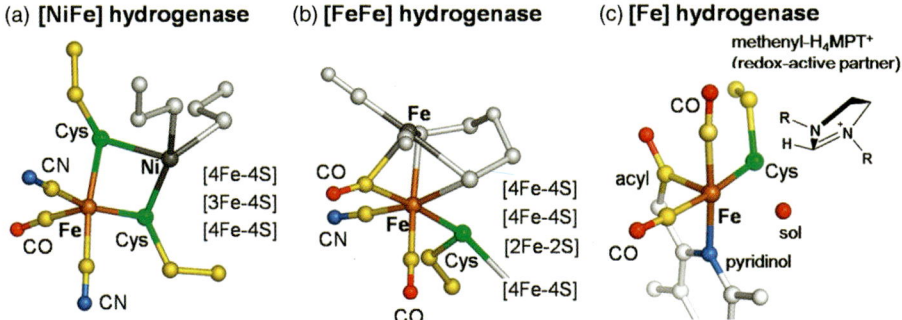

Figure 20.4 Active-site structures of three microbial hydrogenases. (a) [NiFe]-hydrogenase from *Desulfovibrio gigas*; (b) [FeFe]-hydrogenase from *Clostridium pasteurianum* and *Desulfovibrio desulfuricans*; (c) [Fe]-hydrogenase from *Methanocaldococcus jannaschii*. Figures courtesy of Ulrich Ermler, modified from Shima et al., 2008 [58].

hydrogenases to uptake H_2 in photosynthetic bacteria [53]. A nickel atom, sulfur-coordinated by four cysteine residues, is the active site of [NiFe] hydrogenases (Figure 20.4). These [NiFe] hydrogenases uptake H_2 molecules and ensure H_2 evolution. When the bridge ligand of [Ni-Fe] is removed the enzyme is activated and molecular hydrogen is produced. [NiFe] hydrogenases are bidirectional because they couple the reversible cleavage of hydrogen to the redox conversion of NADH [54]. In [FeFe] hydrogenases, [4Fe-4S] and [2Fe-2S] moieties are bound by four cysteine residues [55–57]. Similarly, in [Fe] hydrogenases, the mononuclear iron is coordinated by the sulfur of cysteine-176, two carbon monoxide molecules, and the sp^2-hybridized nitrogen of 2-pyridinol compounds with back bonding properties similar to those of cyanide [58].

20.4.2
Uptake Hydrogenase and Hydrogen Production

The hydrogen uptake hydrogenase was identified several decades ago in the nitrogen fixing bacteria *Bradyrhizobium japonicum* and the hyperthermophilic archaebacterium *Pyrodictium brockii* [59]. Many microorganisms contain an uptake hydrogenase (HUP). Some cyanobacteria, like *Nostoc* species, possess all three enzymes involved in H_2 production (Figure 20.5):

1) nitrogenase, which catalyzes nitrogen fixation and produces H_2 as a byproduct;
2) HUP, which reoxidizes H_2 as a byproduct of nitrogen fixation;
3) bidirectional hydrogenase, which produces and takes up H_2 under different environmental conditions [60–62].

Cyanobacterial HUP consists of two different subunits, HupS and HupL, whereas other eubacteria carry a third subunit of HUP, HupC or HoxZ [63].

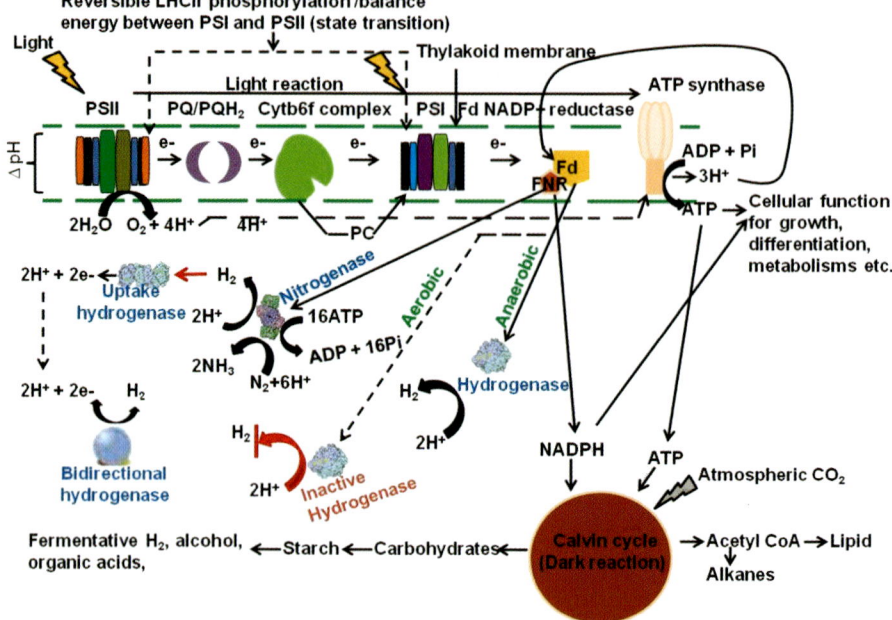

Figure 20.5 Photosynthetic electron transport pathway with detailed possible route for H_2 and biofuel production in photosynthetic microorganisms. A typical Z-scheme in green algae, cyanobacteria, and photosynthetic organisms shows the light reaction and H_2 production by hydrogenase and nitrogenase. NADPH and ATP are consumed in dark reactions of the Calvin cycle for starch synthesis and other cellular functions. Starch and acetyl CoA are the precursor molecules for biofuel production. The red arrow shows the inhibitory pathway for H_2 production, smooth arrows show the direct regulatory pathway, and dashed arrows indicate the alternative pathway, which depends on the physiological and biochemical condition.

However, genes that affect hydrogenases pleitropically (*hyp*) and genes encoding the hydrogenase specific endopeptidases (*hupW* and *hoxW*) have been characterized. Apart from this, NtcA and LexA are involved in the transcription of uptake hydrogenase and bidirectional hydrogenase in cyanobacteria [64]. Although HUPs consume hydrogen in nitrogen fixing cyanobacteria, high levels of H_2 production were observed in *Cyanothece* even in the presence of a HUP [65]. Filamentous cyanobacteria possess the uptake hydrogenase HupSL in heterocysts. The role of HUP is the oxidation of hydrogen in the small subunit of HupS [66]. However, the detailed properties of HupSL are still unknown. Knock out of the corresponding gene enhanced H_2 production [67–69], stressing the importance of detailed studies of different HUP subunits for understanding the mechanism of hydrogen production [67].

20.4.3
Nitrogenase and H_2 Production

Biological nitrogen fixation in purple photosynthetic bacteria might become a major source of H_2 production because the nitrogenase reduces atmospheric N_2 into ammonia with concomitant obligate production of molecular hydrogen [70]. The FeMo cofactor, which is a [Fe-S] cluster, is the active site of nitrogenase. It has been studied intensively and the crystal structure analyzed of nitrogenase FeMo-protein [71].

However, nitrogenase is not very efficient for the production of hydrogen during nitrogen fixation because it is a highly endergonic reaction [63] and most of the energy consumed by nitrogenase is utilized to generate ammonia ($N_2 + 8H^+ + 8e^- + 16ATP \rightarrow 2NH_3 + H_2 + 16ADP + 16P_i$). Nevertheless, nitrogenase is the key enzyme for production of hydrogen in cyanobacteria. Mo-nitrogenase, V-nitrogenase, and Fe-nitrogenase are three different nitrogenases involved in nitrogen fixation [48]. Nitrogen fixation and hydrogen production in cyanobacteria represent a significant part of their metabolism [72–74]. Nitrogen fixation by nitrogenase is observed in both unicellular and heterocystous cyanobacteria; therefore, they can be modified for mass production and for potential application of hydrogen production.

20.5
Photosystems and H_2 Production

In photosynthetic green algae and cyanobacteria, PSII and PSI exist as two multiprotein complexes. Owing to the presence of light harvesting complexes in PSII and PSI, both protein complexes have the ability to absorb sunlight for the light reactions of photosynthesis. However, an excess of light may also damage PSII, therefore, its repair cycle is most important to maintain the functional PSII in higher plants [75,76]. Similarly, in damaged PSII of green algae and cyanobacteria, the efficiency of electron transport is low, which indirectly affects H_2 generation. On the other hand, the light harvesting protein complex, known as "LHCII" protein, is also important to balance the electrons between PSII and PSI because LHCII becomes phosphorylated in light, and this phosphorylation causes migration of the light harvesting protein complex from PSII to PSI (Figure 20.5). In green algae such as *Chlamydomonas*, this phenomenon facilitates the distribution of excitation energy between the two photosystems (PSII and PSI) by the so-called state transition [77,78]. During the light reactions of photosynthesis in algae and cyanobacteria, H_2O is oxidized into O_2 and electrons are used for ferredoxin and $NADP^+$ reduction. When hydrogenase enzymes activate, the reduced NADPH and ferredoxin are used for hydrogen production [9]. Hydrogen production was observed at greater rates during activation of PSI and deactivation of PSII in cyanobacteria [79]. There are two pathways for H_2

production associated with photosystems. One is the direct pathway, in which PSII supplies electrons to hydrogenase via the electron transfer chain. Another pathway operates in the absence of PSII via PSI and cyclic electron flow around PSI [80] by non-photochemical reduction of plastoquinones [81]. Starch catabolism has been proposed to play its roles in both pathways. In the direct pathway, starch catabolism sustains mitochondrial respiration and allows the anaerobic conditions. In the indirect pathway, starch catabolism supplies electrons to hydrogenase via PSI-dependent processes [81,82]. Thus, we can have the following idea about the hydrogen production by photosystems: after water splitting the electrons pass from PSII to cytochrome b6f into PSI. When electrons are transported in PSI, protons (H^+), which are outside the thylakoid membrane, pass into the inner thylakoid space. Alternatively, when light is absorbed by PSI, it excites and facilitates electron transport into the electron acceptor outside the thylakoid membrane toward ferredoxin. Under some conditions, ferredoxin carries electrons to $NADP^+$, but under anaerobic conditions the electrons are passed to hydrogenase, which reduces protons to molecular hydrogen (Figure 20.5). The general equation is given as: $4H^+ + \text{ferredoxin}_{(oxidized)} = \text{ferredoxin}_{(reduced)} + 2H_2$.

20.6
Factors Affecting Hydrogen Production

Several physiological, biochemical, environmental, and metabolic factors play important roles in biological H_2 production by algae and cyanobacteria. Some such factors such are growth conditions, light intensity, carbon source, nitrogen source, sulfur, salinity, oxygen, and temperature. There are diverse groups of cyanobacteria that are useful to produce H_2. The efficiency of H_2 production also depends on the environmental growth conditions and the organism's physiology. Importantly, light intensity is required to produce H_2 in cyanobacteria and the effect depends on the species [83]. For example, *Arthrospira platensis* produce H_2 during day and night, under anaerobic conditions. However, *Spirulina platensis* and *Synechococcus elongatus* PCC7942 produces H_2 only in the dark, under anaerobic conditions [84], but some species (*Azospirillum brasilense* and *Azospirillum lipoferum*) produce H_2 only in the presence of light [85]. Thus, one should consider the role of photoperiod and light intensity for H_2 production in cyanobacteria. In general, temperature values around 30–40 °C are optimal for cyanobacterial H_2 production. This certainly depends on the biology of the species: some produce more H_2 at 22 °C and others at 30 or 40 °C [86–88].

There are other factors that directly or indirectly control H_2 production in algae and cyanobacteria. Oxidative stress can affect growth of cyanobacteria [89] and influence the activity of hydrogenases and the rate of nitrogen fixation. Improvement of H_2 production can be obtained by protection from oxygen [57]. Excessive H_2 production was observed in cyanobacteria cultured under low

(0.3%) CO_2, (2%) N_2, at pH 7.5 with a 16/8 h dark–light photoperiod at 27 °C. Inhibition of photosynthetic electron flow with DCMU or supplementation of the culture medium with various macro- and micro-nutrients also enhanced H_2 production in cyanobacteria [90].

In addition, some specific culture conditions of cyanobacteria were shown to enhance H_2 production [91]. In some cyanobacteria (*Cyanothece* 51142) kept under dark and nitrogen fixing conditions, the *hupSL* genes of uptake hydrogenase were strongly induced [92]. Green algae produce photosynthetic H_2 under sulfur-deprived conditions [93–95] when the reversible hydrogenase is induced to evolve H_2 in light [96]. Photosynthetic H_2 production in *Chlamydomonas reinhardtii* was induced under nitrogen starvation [23].

To improve the biological production of H_2 some strategies have been proposed [97]. Algae, cyanobacteria, and other microorganisms required different growth conditions to generate H_2. On the basis of their nutritional consumption for better growth, organisms are classified into: (i) autotrophs – organisms that use inorganic CO_2 as a sole carbon source, which are subdivided into photoautotrophs, like cyanobacteria, and chemoautotrophs that derive energy from oxidation of inorganic compounds; (ii) heterotrophs – organisms that use the organic carbon as a carbon source, which are subdivided into photoheterotrophs that utilize organic carbon and sunlight to produce ATP and chemoheterotrophs that utilize only organic compounds, for example, sugars, acetate, or glycerol.

20.7 Designing the Photosynthetic H_2 Production

In algal cells, like *Chlamydomonas*, H_2 and O_2 are evolved in one single cell. Hydrogenase is sensitive to inactivation by O_2, and this is a serious problem for generation of H_2. An effective way for the production of H_2 is to design genetic and biochemical modifications of the proton channel for photolytic H_2 production [98]. The proton channel in the thylakoid membrane has many advantages for H_2 generation: (i) electron transport from water to ferredoxin/hydrogenase is prevented by a proton gradient; (ii) competitive inhibition of photosynthetic H_2 production by the Calvin cycle is eliminated.

Three other strategies can be applied to produce H_2:

1) optimization of biological and genetically modified organisms for photobiological H_2 production;
2) semi-artificial systems of biological catalysts isolated from microorganisms and immobilized on electrodes;
3) biomimetics obtained by coupling of an artificial photosynthetic water splitting system with an artificial H_2 production system [99].

Our knowledge of the mechanisms by which cyanobacteria can produce H_2 by oxygenic photosynthesis in photosynthetic cells is far from optimal [100].

One major problem is that water oxidation by photosystem II during photosynthesis evolves oxygen that inhibits the activity of hydrogenase enzyme that is essential to catalyze H_2 evolution [101]. To overcome this problem, genetic modification of the active site of [FeFe] hydrogenase and a comparative study with [NiFe] hydrogenase has been explored in green algae [102]. Finding possible mechanisms for inhibition of oxygen evolution to protect hydrogenase activity for photosynthetic H_2 production constitutes another potential avenue.

20.8
Leaf and Solar H_2 Production

Many attempts have been made to harvest the energy and fuel directly from sunlight. Modern semiconductor photoelectrochemical materials may force water to dissociate directly into hydrogen and oxygen. Some progress has been made in artificial photosynthesis to generate hydrogen [103,104]. As an example, an electrocatalyst has been applied to link light-driven charge separation for utilization of solar energy [105].

In the presence of sunlight, PSII of plants, algae, and cyanobacteria can split water [5,6,16,106]. In algae ferredoxin, which is an electron mediator, is reduced by PSI to generate hydrogen by [FeFe] hydrogenase [107]. Nowadays, the conception of an "*artificial leaf*" has been proposed, which can be employed as a source of clean fuel. An artificial leaf can be developed using a silicon coated sheet, which split water into hydrogen and oxygen [1]. Silicon became an attractive material to design an artificial leaf; silicone is cheap, abundant and such cells can capture and store the energy. Although, the development of an artificial leaf is still a scientific challenge, some concepts have been developed to overcome technical difficulties:

1) An anode has been made that uses sunlight to oxidize water and collect the electrons.
2) A cathode has been proposed that interacts with a conductive substrate. When it contacts with a membrane, the latter becomes permeable to protons that generate H_2 [108].
3) A membrane was developed that enables rapid transport of electrons. In such way we can further develop a novel photoelectrochemical cell (PEC) and other light harvesting systems to produce solar fuel at low cost [1,109]. A new artificial leaf is based on a triple-junction silicon layer used as a photovoltaic device for hydrogen and oxygen production after water splitting [1], and on mesoporous TiO_2 electrodes with *in situ* deposited PbS/CdS/QDSs heterodimers (quasi-artificial leaf) for light harvesting of visible and near-infrared light. This cell generated H_2 at a rate of 60 mL hydrogen cm^{-1} per day [110].

20.9
Biofuel and Hydrogen Production by Other Organisms

Biofuel can be produced by several species of algae and such biofuel has highest significance as an alternative source of energy. As we know, biodiesel, triglycerides, fatty acids, lipids, carbohydrates, ethanol, alcohols cellulose or the biomass of organisms are major biofuels and these can be produced by several species of algae, bacteria, or yeast. Therefore, identification of desired strains of algae and modification of those desired strains to obtain the biofuel is one of the challenging operations to generate biofuel from algae. Based on the current situation, several researchers are concerned about the production of biofuel in an efficient way by using different strains of microorganisms. Algae are able to fulfill the photosynthesis and to produce hydrogen needed for bioenergy, although the engineering development for massive culture and harvesting of genetically modified algae in outdoor ponds is challenging [111]. There are several species of algae capable of high biomass production (in terms of carbohydrates, proteins, lipids) that can be used as an alternative source of bioenergy. *Spirulina maxima* gives 60–71% proteins, *Porphyidium cruentum* 40–57% carbohydrates, and *Schizochytrium* species 50–77% lipids. Hence, microalgae can be a good source of biomass under certain conditions [112]. In addition, microalgae can produce higher amounts of biodiesel than cotton or palm plants [113]. For example, genetic modification of a keto-acid decarboxylase gene in the cyanobacteria *Synechococcus elongates* PCC 7942 can improve the production of isobutyraldehyde and butanol [114]. Some species of green algae such as *Botryococcus braunii* and *Chlorella protothecoides* contain high levels of terpenoid hydrocarbons and glyceryl lipid that can be converted into shorter hydrocarbons as major crude oil [115]. Hence, algae have great potential for producing petroleum products such as bioethanol, triterpenic hydrocarbons, isobutyraldehyde, and isobutanol [112].

Many species of bacteria such as *Escherichia coli* and *Bacillus subtilis* with their genetic engineering yielded higher amounts of biofuel such as alcohols, isoprenoids, and fatty acid derivatives and therefore they have unique properties for the production of biofuel. Some species of bacteria such as *Clostridium acetobutylicum* and *Clostridium leijerinckii* have been used for the production of biofuel by acetone–butanol–ethanol fermentation [116]. Bacteria not only produce biofuel but also some species of *Bacillus* and *Escherichia coli* produce some chemicals such as lactic acid and glutamic acids [117]. Several species of bacteria are able to produce ethanol. However, genetic studies in *Caldicellulosiruptor, Thermococcus, Pyrococcus,* and *Thermotoga* species of bacteria indicated the production of higher amounts of hydrogen and low amounts of ethanol production, which mean, that the role of each gene for the regulation of biochemical pathways in the bacterial cell is different for the production of each compound [118]. Apart from this, recently it has been shown that a co-culture of *Thermoanaerobactor* species with cellulolytic organisms is effective for the production of biofuel (ethanol) [119]. As another example of a microorganism *Saccharomyces cerevisiae* is a model organism for the efficient production of ethanol and lipid by fermentative processes [120–122].

20.10
Available Methods to Enhance Photosynthetic Hydrogen Production

Hydrogen can be produced by various means but water electrolysis and thermocatalytic reformation of hydrogen-rich compounds are the main methods used. However, these methods require high energy input and cost. Therefore, biological hydrogen production by microorganisms constitutes an effective approach to generate H_2.

There are several methods available to enhance biological photosynthetic H_2 production. Among them the best and effective methods are summarized as: (i) photolytic H_2 production by microorganisms, (ii) photosynthetic bacterial H_2 production, (iii) dark fermentative H_2 production, (iv) microbial-aided electrolysis for H_2 production, (v) genetic engineering of microorganisms, (vi) thermochemical H_2 by coal and biomass gasification, (vii) distributing natural gas reforming, (viii) heat driven chemical reaction to split water, and (ix) water electrolysis by electricity. Here we only focus on the methods of biological H_2 production that use microorganisms, as briefly summarized below.

20.10.1
Photolytic H_2 Production by Microorganisms

As mentioned above, green algae and cyanobacteria are able to produce hydrogen after splitting water in the presence of sunlight. Water is split into oxygen, electrons, and protons. These microorganisms utilize and consume water to produce hydrogen as a product of their natural metabolic process. This biophotolysis process involves both hydrogenase and nitrogenase [123]. However, hydrogenase is sensitive to oxygen, hence the yield of H_2 production is low. To overcome this problem, new developments to improve H_2 production by increasing the supply of H^+ and e^- to the hydrogenases in the state transition mutant (*stm6*) of *Chlamydomonas* have been proposed for high production rates of H_2 [124]. Recently, improved performance for photobiological processes, focusing on existing photobioreactors, to enhance hydrogen production by photosynthetic and non-photosynthetic microorganisms for future purposes has been discussed [32,125]. Similarly, the main biological pathway of biohydrogen production in microorganisms has been discussed in detail [126]. However, the efficiency of natural photosynthesis for solar energy conversion for biofuel generation with present technologies is relatively low [127]. Hence two major approaches have been proposed to improve the efficiency of photosynthesis for efficient biofuel production: (i) expansion of solar spectrum utilized in photosynthesis and reduction of antenna system that couple to the photochemical reaction center [128] and (ii) detailed cultivation system of microalgae, biomass harvesting methods, and biofuel production methods [129].

20.10.2
Photosynthetic Bacterial H_2 Production

Hydrogen production by some species of photosynthetic bacterium was identified several decades ago [130]. Some photosynthetic bacteria, for example, purple non-sulfur bacteria, utilize sunlight and break down their organic compounds to produce hydrogen. These bacteria grow as photoheterotrophs, photoautotrophs, and chemoheterotrophs. As they do not produce oxygen, the hydrogenase activity is not inhibited, which is essential for hydrogen production. Hence this type of photosynthetic bacteria are most promising microorganisms for biohydrogen production because they lack oxygen evolving activity, use a wide range of light wavelength, and are capable of using organic compounds [123]. The addition of a metal ion such as of molybdenum, iron, and manganese enhanced the production of hydrogen in these bacteria. Even light driven processes can be incorporated into non-photosynthetic bacteria to increase the metabolic biosynthetic pathway, as also proposed [131]. In addition, the photosynthetic bacteria *Rhodospirillum rubrum* grown in cheese whey in the presence of light also produced hydrogen [132]. During the process of photobiological hydrogen production by bacteria, some photosynthetic purple sulfur and purple non-sulfur bacteria also produced H_2 in the presence of light [125]. Even during the co-cultivation of green algae, photofermentative bacteria also produce hydrogen [32]. Besides this, some photosynthetic bacteria, that is, *Rhodopseudomonas palustris* CQK 01, produce hydrogen when immobilized on the surface of a thin glass slide [133]. Hence, different approaches have led to H_2 production in these photosynthetic bacteria.

20.10.3
Dark Fermentative H_2 Production

Biohydrogen can also be produced either by photofermentation or by dark fermentation and these methods are also useful as alternative methods to produce biohydrogen [134]. Some bacteria have the capacity to decompose organic compounds to produce hydrogen in the absence of sunlight. In this process, those bacteria, which grow in dark with high amounts of carbohydrates, directly ferment an organic material into hydrogen [135]. During dark fermentation, it has been estimated that the yield of hydrogen was 4 mol-H_2 per mol-hexose sugar converted into other substrates, acetate, and butyrate [136]. However, acetate and butyrate are toxic and hence to solve this problem the combination of dark fermentation with photofermentation has been applied to improve the production of hydrogen [137]. Woodward et al. [138] studied the role of [Fe] hydrogenase and glucose dehydrogenase for hydrogen production in fermentative processes. In dark fermentation, [NiFe]-hydrogenases and [FeFe]-hydrogenases are utilized by microorganisms; neither produced nor consumed oxygen. It can be considered that dark fermentation methods to generate H_2 are less expensive than direct photolysis and photofermentation [139]. Several strategies to enhance biohydrogen production by fermentative methods using bacteria have been summarized [126]. However, several factors such as pH, hydrogen partial pressure, metal ions, and so on also affect biohydrogen production from biomass by dark fermentation method [140]. There are several species of bacteria that produce hydrogen by mean of dark fermentative processes in which they can convert the carbohydrates into biohydrogen and other biofuels [141,142]. Recently, Cheng et al. [143] observed high yields of hydrogen evolution (2.53 mol-H_2 per mol hexose sugar) in dark fermentation with mixed anaerobic bacteria (*Clostridium* sp.).

20.10.4
Genetic Engineering to Enhance Hydrogen Production

Currently, the main challenging issue is the improvement of biohydrogen production by using microorganisms. To solve this challenge several researchers are concentrating on the genetic study of photosynthetic microorganisms. Genetic manipulation of green algae, cyanobacteria, and bacteria is considered to be a powerful approach to enhance biohydrogen production. Some research has focused on the mutation of *Rhodobactor sphaeroides* RV by UV irradiation and mutation of *pucBA* depleting B800-850 light harvesting complex to increase the hydrogen production rate [144,145]. Similarly, mutation of *cyt cbb3 oxidase* in *Rhodobactor* species enhanced hydrogen production because the deletion of *cyt cbb3* oxidase induced nitrogenase activity and inhibited uptake hydrogenase [146]. Recently, deletion of light harvesting antenna complexes in *Rhodobactor sphaeroides* and the deletion of uptake hydrogenase showed higher hydrogen production rates [147,148]. Similarly, the composition of gene encoding pathways and genetic biomarkers have been studied to identify potential methods of optimization of hydrogen production and metabolic engineering for biofuel production in bacteria [118,149].

Green algae and cyanobacteria have good ability to produce hydrogen under anaerobic conditions. In cyanobacterial *Synechocystis* sp. PCC6803, genetic engineering was employed to obtain transformed strains to generate renewable biofuel [150]. In *Chlamydomonas reinhardtii*, the deletion of *ble* gene, which encodes the fan enzyme with zeomycin resistant properties, also produced hydrogen [151]. Besides the above, several genetic engineering procedures have been conducted to generate higher hydrogen production in cyanobacteria [17,152]. This includes mutation of some genes and other biochemical modifications [18,22–26], as well as producing new strains of cyanobacteria to enhance H_2 production [20].

20.11
Application of Biohydrogen

Owing to the increasing population on earth, people are looking for alternative sources of renewable energy to sustain a better life. There are several applications and approaches for biohydrogen production. Biohydrogen can be one of the most economic and eco-friendly gas for the future. Some of advantages of biohydrogen are (i) biohydrogen is eco-friendly and its production can be controlled according to demand; (ii) utility of waste materials for the biohydrogen production, hence waste materials, is recycled or consumed; (iii) biohydrogen production cost is less than fossil fuel; (iv) biohydrogen is more easily stored than fossil fuel, coal, and gasoline, for example, 1 kg (2.2 pounds) of hydrogen has the same energy as 1 US gallon of gasoline (6.2 pounds), (v) hydrogen can be used for metal processing, manufacture of glasses, refining the petroleum product; (vi) energy carriers move energy in a usable form from one place to another; (vii) hydrogen allows us to store energy from many sources and bring it to where we need it; (viii) replacement by H_2 production of petroleum refining, coal gasification, and thermochemical, which are hazardous processes; (ix) the design of hydrogen fuel cells, with many times the energy content of gasoline, thus leading to more versatile energy careers.

20.12
Conclusion and Future Prospectus

It is known that H_2 can be produced by several microorganisms such as cyanobacteria, green algae, and bacteria, which constitute the best source of hydrogen generation. The detailed pathway is given in Figure 20.5. Some higher legume plants that have symbiotic association with some bacterium (*Rhizobium*) are also able to produce hydrogen as a byproduct of nitrogen fixation catalyzed by nitrogenase. Genetic modification of some specific genes by using genetic engineering in cyanobacterial and algal cells has become a powerful technique to enhance the production of hydrogen by the activity of hydrogenase. Enzymes responsible for uptake of H_2 and thus inhibiting hydrogen generation can also be modified by using several molecular and genetic techniques. Growth conditions and environmental factors for the microorganisms can be modified to liberate more H_2. In addition, the study of kinase enzymes, whose key role is to phosphorylate different thylakoid membrane proteins for efficient PSII repair cycle in photosynthetic microorganisms, can be also considered for further research.

Fermentative processes are also useful to generate H_2. However, they have some limitations as the yield is less with a high demand of biochemical oxygen. Hydrogen generation by oxygenic photosynthesis in several microorganisms presents an important molecular problem due to the inactivation of hydrogenase by oxygen. In addition, achievement of higher quantities of hydrogen production by microbial electrolysis cells also has some challenges, which implies the loss of energy.

Therefore, it is still difficult to meet the needs for hydrogen production in large quantities. However, owing to the improvement of molecular techniques, methods, skills, and finding better culture conditions for mass cultivation of cyanobacteria, green algae, as well as bacteria, there should be important improvement for the large-scale production of hydrogen to meet our needs and demands for alternative sources of renewable energy in the future.

Acknowledgments

This work was supported by grants from the Russian Foundation for Basic Research and Molecular and Cell Biology Programs of the Russian Academy of Sciences to SIA.

MMN is grateful to the Institute for Advanced Studies in Basic Sciences for financial support.

Abbreviations

ADP	adenosine diphosphate
ATP	adenosine triphosphate
CO_2	carbon dioxide
DCMU	3-(3,4-dichlorophenyl)-1,1-dimethylurea
e^-	electron
FNR	ferredoxin-$NADP^+$-reductase
H^+	hydrogen ion (proton)
H_2	hydrogen gas
LHCII	light harvesting complex of PSII
NADH	nicotinamide adenine dinucleotide
NADP	nicotinamide adenine dinucleotide phosphate
N_2	nitrogen gas
O_2	oxygen gas
PEC	photoelectrochemical cells
P_i	inorganic phosphate
PQ	plastoquinone (oxidized)
PQH_2	plastoquinone (reduced)
PSI	photosystem I
PSII	photosystem II

References

1 Reece, S.Y., Hamel, J.A., Sung, K., Jarvi, T.D., Esswein, A.J., Pijpers, J.J., and Nocera, D.G. (2011) Wireless solar water splitting using silicon-based semiconductors and earth-abundant catalysts. *Science*, **334** (6056), 645–648.

2 Bockris, O.M. (1977) *Energy – The Solar Hydrogen Alternative*, John Wiley & Sons, Inc., New York.

3 Umena, Y., Kawakami, K., Shen, J.R., and Kamiya, N. (2011) Crystal structure of oxygen-evolving photosystem II

at a resolution of 1.9 Å. *Nature*, **473**, 55–60.

4 Sakurai, H., Masukawa, H., Kitashima, M., and Inoue, K. (2013) Photobiological hydrogen production: bioenergetics and challenges for its application. *J. Photochem. Photobiol. C: Photochem. Rev.*, **17**, 1–25.

5 Allakhverdiev, S.I., Thavasi, V., Kreslavski, V.D., Zharmukhamedov, S.K., Klimon, V.V., Ramakrishna, S., Los, D.A., Mimuro, M., Nishihara, H., and Carpentier, R. (2010) Photosynthetic hydrogen production. *J. Photochem. Photobiol. C: Photochem. Rev.*, **11** (2–3), 101–113.

6 Allakhverdiev, S.I., Kreslavski, V.D., Thavasi, V., Zharmukhamedov, S.K., Klimov, V.V., Nishihara, H., Ramakrishna, S., Mimuro, M., Carpentier, R., and Nagata, T. (2010) Photosynthetic energy conversion: hydrogen photoproduction by natural and biomimetic systems, in *Biomimetics, Learning from Nature*, (ed. A. Mukherjee), IN-TECH, Vukovar, Croatia, pp. 49–76.

7 Antal, T.K., Krendeleva, T.E., Pashchenko, V.Z., Rubin, A.B., Stensjo, K., Tyystjärvi, E., Ramakrishna, S., Los, D.A., Carpentier, R., Nishihara, H., and Allakhverdiev, S.I. (2012) Photosynthetic hydrogen production: mechanisms and approaches, in *State of the Art and Progress in Production of Biohydrogen* (eds N. Azbar, and D. Levin), Bentham Science Publishers, Canada, pp. 25–53.

8 Vignais, P.M., Billoud, B., and Meyer, J. (2001) Classification and phylogeny of hydrogenases. *FEMS Microbiol. Rev.*, **25** (4), 455–501.

9 Lee, H.-S., Vermaas, W.F.J., and Tittmann, B.E. (2010) Biological hydrogen production: prospects and challenges. *Trends Biotechnol.*, **28** (5), 262–271.

10 Hungria, M. and Bohrer, T.R.J. (2002) Variability of nodulation and dinitrogen fixation capacity among soybeans cultivars. *Biol. Fertility Soils*, **31** (1), 45–52.

11 Dong, Z. and Layzell, D.B. (2001) H_2 oxidation, O_2 uptake and CO_2 fixation in hydrogen treated soils. *Plant Soil*, **229** (1), 1–12.

12 Das, D. and Veziroğlu, T.N. (2001) Hydrogen production by biological processes: a survey of literature. *Int. J. Hydrogen Energy*, **26** (1), 13–28.

13 Kapdan, I.K. and Kargi, F. (2006) Bio-hydrogen production from waste materials. *Enzyme Microb. Technol.*, **38** (5), 569–582.

14 Miura, Y., Yokovama, H., Kanaoka, K., Saito, S., Iwasa, K., Okaxami, M., and Komemushi, S. (1980) Hydrogen evolution by a thermophelic blue-green alga Mastigocladus laminosus. *Plant Cell Physiol.*, **21** (1), 149–156.

15 Allakhverdiev, S.I., Kreslavski, V.D., Thavasi, V., Zharmukhamedov, S.K., Klimov, V.V., Nagata, T., Nishihara, H., and Ramakrishna, S. (2009) Hydrogen production by photosynthetic organisms and biomimetic systems. *Photochem. Photobiol. Sci.*, **8** (2), 148–156.

16 Allakhverdiev, S.I. (2012) Photosynthetic and biomimetic hydrogen production. *Int. J. Hydrogen Energy*, **37** (10), 8744–8752.

17 Sriranjan, K., Pyne, M.E., and Chou, C.P. (2011) Biochemical and genetic engineering strategies to enhance hydrogen production in photosynthetic algae and cyanobacteria. *Bioresour. Technol.*, **102** (18), 8589–8604.

18 Winkler, M., Kuhlgert, S., Hippler, M., and Happe, T. (2009) Characterization of key step for light-driven hydrogen evolution in green algae. *J. Biol. Chem.*, **284** (52), 36620–36627.

19 Najafpour, M.M. and Allakhverdiev, S.I. (2012) Manganese compounds as water oxidizing catalysts for hydrogen production via water splitting: from manganese complexes to nano-sized manganese oxides. *Int. J. Hydrogen Energy*, **37** (10), 8753–8764.

20 Tiwari, A. and Pandey, A. (2012) Cyanobacterial hydrogen production – A step towards clean environment. *Int. J. Hydrogen Energy*, **37** (1), 139–150.

21 Govindjee and Shevela, D. (2011) Adventures with cyanobacteria: a personal perspective. *Front. Plant Sci.*, **2**, Article 28, 1–17. doi: 10.3389/fpls.2011.00028

22 Papazi, A., Andronis, E., Loannidis, N.E., Chaniotakis, N., and Kotzabasis, K.

(2012) High yields of hydrogen production induced by meta-substituted dichlorophenols biodegradation from the green alga Scenedesmus obliquus. *PLoS ONE*, **07** (11), e49037. doi: 10.1371/journal.pone.0049037

23 Philipps, G., Happe, T., and Hemschemeir, A. (2012) Nitrogen deprivation results in photosynthetic hydrogen production in Chlamydomonas reinhardtii. *Planta*, **235** (04), 729–745.

24 Bonente, G., Formighieri, C., Mantelli, M., Catalanotti, C., Giuliano, G., Morosinotto, T., and Bassi, R. (2011) Mutagenesis and phenotypic selection as a strategy toward domestication of Chlamydomonas reinhardtii strains for improved performance in photobioreactors. *Photosynth. Res.*, **108** (2–3), 107–120.

25 Nguyen, A.V., Toepel, J., Burgess, S., Uhmeyer, A., Blifernez, O., Doebbe, A., Hankamer, B., Nixon, P., Wobbe, L., and Kruse, O. (2011) Time-course global expression profiles of Chlamydomonas reinhardtii during photo-biological H_2 production. *PLoS ONE*, **06** (12), e29364. doi: 10.1371/journal.pone.0029364

26 Ghysels, B. and Franck, F. (2010) Hydrogen photo-evolution upon S deprivation stepwise: an illustration of microalgal photosynthetic and metabolic flexibility and a step stone for future biotechnological methods of renewable H_2 production. *Photosynth. Res.*, **106** (1–2), 145–154.

27 Lindbald, P., Christensson, K., Lindberg, P., Fedorov, A., pinti, F., and Tysgankov, A. (2002) Photoproduction of H_2 by wildtype Anabaena PCC7120 and a hydrogen uptake deficient mutant: from laboratory experiments to outdoor culture. *Int. J. Hydrogen Energy*, **27** (11–12), 1271–1281.

28 Howarth, D.C. and Codd, G.A. (1985) The uptake and production of molecular hydrogen by unicellular cyanobacteria. *J. Gen. Microbiol.*, **131** (7), 1561–1569.

29 Pinto, F.A.L., Troshima, O., and Lindbald, P. (2002) A brief look at three decades of research on cyanobacterial hydrogen evolution. *Int. J. Hydrogen Energy*, **27** (11–12), 1209–1215.

30 Griesbeck, C., Kobi, I., and Heilter, M. (2006) Chlamydomonas reinhardtii: a protein expression system for pharmaceutical and biotechnological proteins. *Mol. Biotechnol.*, **34** (2), 213–223.

31 White, A.L. and Melis, A. (2006) Biochemistry of hydrogen metabolism in Chlamydomonas reinhardtii wild type and a Rubisco-less mutant. *Int. J. Hydrogen Energy*, **31** (4), 455–464.

32 Eroglu, E. and Melis, A. (2011) Photobiological hydrogen production: recent advances and state of the art. *Bioresour. Technol.*, **102** (18), 8403–8413.

33 Dauvillee, D., Chochois, V., Steup, M., Haebel, S., Eckermann, N., Ritte, G., Ral, J.P., Colleoni, C., Hicks, G., Wattebled, F., Deschamps, P., Dhulst, C., Lienard, L., Cournac, L., Putax, L., Dupeyre, D., and Ball, S.G. (2006) Plastidal phosphorylase is required for normal starch synthesis in Chlamydomonas reinhardtii. *Plant J.*, **48** (2), 274–285.

34 Anjana, K., Kaushik, A., Kiran, B., and Nisha, R. (2007) Biosorption of Cr(VI) by immobilized biomass of two indigenous strains of cyanobacteria isolated from metal contaminated soil. *J. Hazard. Mater.*, **148** (1–2), 383–386.

35 Melis, A. and Happe, T. (2001) Hydrogen production. Green algae as a source of energy. *Plant Physiol.*, **127** (3), 740–748.

36 Hurley, J.K., Weber-Main, A.M., Stankovich, M.T., Benning, M.M., Thoden, J.B., Vanhooke, J.L., Holden, H.M., Chae, Y.K., Xia, B., Cheng, H., markley, J.L., Martinez-Júlvez, M., Gómez-Moreno, C., Schmeits, J.L., and Tollin, G. (1997) Structure-function relationship in Anabaena ferredoxin: correlations between X-ray crystal structures, reduction potentials, and rate constants of electron transfer to ferredoxin:NADP+ reductase for site-specific ferredoxin mutants. *Biochemistry*, **36** (37), 11100–11117.

37 Lo Conte, L., Chothia, C., and Janin, J. (1999) The atomic structure of protein-protein recognition sites. *J. Mol. Biol.*, **285** (5), 2177–2198.

38 Posewitz, M.C., Dubini, A., Meuser, J.E., Seibert, M., and Ghirardi, M.L. (2009)

The Chlamydomonas Sourcebook. Organellar and Metabolic Processes, vol. **2**, 2nd edn (eds D.B. Stern and E.H. Harris), Academic Press, San Diego, pp. 217–256.

39 Matthew, T., Zhou, W., Rupprecht, J., Lim, L., Thomas-Hall, S.R., Doebbe, A., Kruse, O., Hanamer, B., Marx, U.C., Smith, S.M., and Schenk, P.M. (2009) The metabolome of Chlamydomonas reinhardtii following induction of anaerobic H2 production by sulfur depletion. *J. Biol. Chem.*, **284** (35), 23415–23425.

40 Zhang, L. and Melis, A. (2002) Probing green algal hydrogen production. *Philos. Trans. R. Soc. London B Biol. Sci.*, **357** (1426), 1499–1509, discussion 1507–1511.

41 Tsygankov, A., Kosourov, S., Seibert, M., and Ghirardi, M.L. (2002) Hydrogen photoproduction under continuous illumination by sulfur-deprived, synchronous Chlamydomonas reinhardtii cultures. *Int. J. Hydrogen Energy*, **27** (11), 1239–1244.

42 Antal, T.K., Krendeleva, T.E., and Rubin, A.B. (2011) Acclimation of green algae to sulfur deficiency: underlying mechanisms and application for hydrogen production. *Appl. Microbiol. Biotechnol.*, **89** (1), 3–15.

43 Oey, M., Ross, I.L., Stephens, E., Steinbeck, J., Wolf, J., Radzun, K.A., Kügler, J., Ringsmuth, A.K., Kruse, O., and Hankamer, B. (2013) RNAi knockdown of LHCBM1, 2 and 3 increases photosynthetic H_2 production efficiency of the green alga Chlamydomonas reinhardtii. *PLoS ONE*, **8** (4), e61375. doi: 10.1371/journal.pone.0061375

44 Gaffron, H. and Rubin, J. (1942) Fermentative and photochemical production of hydrogen in algae. *J. Gen. Physiol.*, **26** (2), 219–240.

45 Zirngibl, C., Van Dongen, W., Schwörer, B., Von Bünau, R., Richter, M., Klein, A., and Thauer, R.K. (1992) H_2-forming methylenetetrahydromethanopterin dehydrogenase, a novel type of hydrogenase without iron-sulfur clusters in methanogenic archaea. *Eur. J. Biochem.*, **208** (2), 511–520.

46 Thauer, R.K. (1998) Biochemistry of methanogenesis: a tribute to Marjory Stephenson. *Microbiology*, **144** (9), 2377–2406.

47 Burke, D.H., Hearst, J.E., and Sidowt, A. (1993) Early evolution of photosynthesis: clues from nitrogenase and chlorophyll iron proteins. *Proc. Natl. Acad. Sci. U.S. A.*, **90** (15), 7131–7138.

48 Bothe, H., Schmitz, O., Yates, M.G., and Newton, W.E. (2010) Nitrogen fixation and hydrogen metabolism in cyanobacteria. *Microbiol. Mol. Biol. Rev.*, **74** (4), 529–551.

49 Tsygankov, A. (2007) Nitrogen-fixing cyanobacteria. A review. *Appl. Biochem. Microbiol.*, **43** (3), 250–259.

50 Ghirardi, M.L., Posewitz, M.C., Maness, P.C., Dubini, A., Yu, J., and Seibert, M. (2007) Hydrogenase and hydrogen photoproduction in oxygenic photosynthetic organisms. *Annu. Rev. Plant Biol.*, **58**, 71–91.

51 Stripp, S.T. and Happe, T. (2009) How algae produce hydrogen-news from the photosynthetic hydrogenase. *Dalton Trans.*, (45), 9960–9969.

52 Marshall, I.P.G., Berggren, D.R.V., Azizian, M.F., Burow, L.C., Semprini, L., and Spormann, A.M. (2012) The hydrogenase chip: a tiling oligonucleotide DNA microarray technique for characterizing hydrogen-producing and -consuming microbes in microbial communities. *ISME J.*, **6** (4), 814–826.

53 Vignais, P.M. and Billoud, B. (2007) Occurrence, classification and biological function of hydrogenase: an overview. *Chem. Rev.*, **107** (10), 4206–4272.

54 Horch, M., Lauterbach, L., Lenz, O., Hildebrandt, P., and Zebger, I. (2012) NAD(H)-coupled hydrogen cycling-structure-function relationships of bidirectional [NiFe] hydrogenase. *FEBS Lett.*, **586** (5), 545–556.

55 Vo Abendroth, G., Stripp, S., Silakov, A., Croux, C., Soucaille, P., Girbal, L., and Happe, T. (2008) Optimized over expression of [FeFe] hydrogenase with high specific activity in Clostridium acetobutylicum. *Int. J. Hydrogen Energy*, **33** (21), 6076–6081.

56 Krassen, H., Stripp, S., von Abendroth, G., Atak, K., Happe, T., and Heberle, J. (2009) Immobilization of the [FeFe]-hydrogenase CrHydA1 on a gold electrode: design of a catalytic surface for the production of molecular hydrogen. *J. Biotechnol.*, **142** (1), 3–9.

57 Stripp, S.T., Goldet, G., Brandmayr, C., Sanganas, O., Vincent, K.A., Haumann, M., Armstrong, F.A., and Happe, T. (2009) How oxygen attacks [FeFe] hydrogenases from photosynthetic organisms. *Proc. Natl. Acad. Sci. U.S.A*, **106** (41), 17331–17336.

58 Shima, S., Pilak, O., Vogt, S., Schick, M., Stagni, M.S., Meyer-Klaucke, W., Warkentin, E., Thauer, R.K., and Ermler, U. (2008) The crystal structure of [Fe]-hydrogenase reveals the geometry of the active site. *Science*, **321** (5888), 572–575.

59 Pihl, T.D., Schichot, R.N., Kellyt, R.M., and Maier, R.J. (1989) Characterization of hydrogen-uptake activity in the hyperthermophile Pyrodictium brockii. *Proc. Natl. Acad. Sci. U.S.A*, **86** (1), 138–141.

60 Lindberg, P., Devine, E., Stensjö, K., and Lindbald, P. (2011) HupW protease specifically required for processing of the catalytic subunit of the uptake hydrogenase in the cyanobacterium Nostoc sp. *Appl. Environ. Microbiol.*, **78** (1), 273–276.

61 Boison, G., Schmitz, O., Mikheeva, L., Shestakov, S., and Bothe, H. (1996) Cloning, molecular analysis and insertional mutagenesis of the bidirectional hydrogenase genes from the cyanobacterium Anacystis nidulans. *FEBS Lett.*, **394** (2), 153–158.

62 Peschek, G.A. (1979) Anaerobic hydrogenase activity in Anacystis nidulans H_2- dependent photoreduction and related reactions. *Biochim. Biophys. Acta (BBA)-Bioenergetics*, **548** (2), 187–202.

63 Tamagnini, P., Axelsson, R., Lindberg, P., Oxelfelt, F., Wünschiers, R., and Lindbland, P. (2002) Hydrogenases and hydrogen metabolism of cyanobacteria. *Microbiol. Mol. Biol. Rev.*, **66** (01), 1–20.

64 Tamagnini, P., Leitão, E., Oliveria, P., Ferreira, D., Pinto, F., Harris, D.J., Heidorn, T., and Lindbald, P. (2007) Cyanobacterial hydrogenase: diversity, regulation and applications. *FEMS Microbiol. Rev.*, **31** (6), 692–720.

65 Min, H. and Sherman, L.A. (2010) Genetic transformation and mutagenesis via single-stranded DNA in the unicellular, diazotrophic cyanobacteria of the genus Cyanothece. *Appl. Environ. Microbiol.*, **76** (22), 7641–7645.

66 Raleiras, P., Kellers, P., Lindbald, P., Styring, S., and Magnuson, A. (2013) Isolation and characterization of the small submit of the uptake hydrogenase from the cyanobacterium Nostoc punctiforme. *J. Biol. Chem.*, **288** (25), 18345–18352.

67 Yonemoto, I.Y., Matteri, C.W., Nguyen, T.A., Smith, H.O., and Weyman, P.D. (2013) Dual organism design cycle reveals small subunit substitutions that improve [NiFe] hydrogenase hydrogen evolution. *J. Biol. Eng.*, **7** (17), 1–11.

68 Khetkorn, W., Lindbald, P., and Incharoensakdi, A. (2012) Inactivation of uptake hydrogenase leads to enhanced and sustained hydrogen production with high nitrogenase activity under high light exposure in the cyanobacterium Anabaena siamensis TISTR 8012. *J. Biol. Eng.*, **6**, 19.

69 Vardar-Schara, G., Maeda, T., and Wood, T.K. (2008) Metabolically engineered bacteria for producing hydrogen via fermentation. *Microb. Biotechnol.*, **1** (2), 107–125.

70 Rey, F.E., Heiniger, E.K., and Harwood, C.S. (2007) Redirection of metabolism for biological hydrogen production. *Appl. Environ. Microbiol.*, **73** (5), 1665–1671.

71 Einsle, O., Tezcan, F.A., Andrade, S.L., Schmid, B., Yoshida, M., Howard, J.B., and Rees, D.C. (2002) Nitrogenase MoFe-protein at 1.16 A resolution: a central ligand in the FeMo-cofactor. *Science*, **297** (5587), 1696–1700.

72 Newton, W.E. (2007) Physiology, biochemistry and molecular biology of nitrogen fixation, in *Biology of Nitrogen Cycle* (eds H. Bothe, S.J. Ferguson, and W.E. Newton), Elsevier, Amsterdam, pp. 109–129.

73 Rees, D.C., Akif tezcan, F., Haynes, C.A., Walton, M.Y., Andrade, S., Einsle, O., and Howard, J.B. (2005) Structural basis of biological nitrogen fixation. *Phil. Trans R. Soc. A*, **363** (1829), 971–984, discussion 1035–1040.

74 Vignais, P.M. and Colbeau, A. (2004) Molecular biology of microbial hydrogenases. *Curr. Issue Mol. Biol.*, **6** (2), 159–188.

75 Kirchhoff, H., Hall, C., Wood, M., Herbstová, M., Tsabari, O., Nevo, R., Charuvi, D., Shimoni, E., and Reich, Z. (2011) Dynamic control of protein diffusion within the granal thylakoid lumen. *Proc. Natl. Acad. Sci. U.S.A.*, **108** (50), 20248–20253.

76 Tikkanen, M., Nurmi, M., Kangasjärvi, S., and Aro, E.-M. (2008) Core protein phosphorylation facilitates the repair of photodamaged photosystem II at high light. *Biochim. Biophys. Acta (BBA)-Bioenerg.*, **1777** (11), 1432–1437.

77 Depège, N., Bellafiore, S., and Rochaiz, J.-D. (2003) Role of chloroplast protein kinase Stt7 in LHCII phosphorylation and state transition in Chlamydomonas. *Science*, **299** (5612), 1572–1575.

78 Lemeille, S., Turkina, M.V., Vener, A.V., and Rochaix, J.-D. (2010) Stt7-dependent phosphorylation during state transitions in the green alga Chlamydomonas reinhardtii. *Mol. Cell. Proteomics*, **9** (6), 1281–1295.

79 Hoshino, T., Johnson, D.J., and Cuello, J.L. (2012) Design of strategy for green algal photo-hydrogen production: spectral-selective photosystem I activation and photosystem II deactivation. *Bioresour. Technol.*, **120**, 233–240.

80 Chochois, V., Dauvillée, D., beyly, A., Tolleter, D., Cuiné, S., Timpano, H., Ball, S., Cournac, L., and Peltier, G. (2009) Hydrogen production in Chlamydomonas: photosystem II-dependent and -independent pathways differ in their requirement for starch metabolism. *Plant Physiol.*, **151** (2), 631–640.

81 Melis, A. (2007) Photosynthetic H_2 metabolism in Chlamydomonas reinhardtii (unicellular green algae). *Planta*, **226** (5), 1075–1086.

82 Fouchard, S., Hemschemeir, A., Caruana, A., Pruvost, J., Legrand, J., Happe, T., Peltier, G., and Cournac, L. (2005) Autotrophic and mixotrophic hydrogen photoproduction in sulfur-derived Chlamydomonas cells. *Appl. Environ. Microbiol.*, **71** (10), 6199–6205.

83 Dutta, D., De, D., Chaudhari, S., and Bhattacharya, S.K. (2005) Hydrogen production by cyanobacteria. *Microb. Cell Fact.*, **4**, 36.

84 Asada, Y. and Miyake, J. (1999) Photobiological hydrogen production. *J. Biosci. Bioeng.*, **88** (1), 1–6.

85 Stal, L.J. and Krumbein, W.E. (1985) Oxygen protection of nitrogenase the aerobically nitrogen fixing, non-heterocystous cyanobacterium Oscillatoria sp. *Arch. Microbiol.*, **143** (1), 72–76.

86 Ernst, A., Kerfin, W., Spiller, H., and Boger, P. (1979) External factors influencing light-induced H_2 evolution by the blue-green algae, Nostoc muscorum. *Z. Naturforsch.*, **34**, 820–825.

87 Datta, M., Nikki, G., and Shah, V. (2000) Cyanobacterial hydrogen production. *World J. Microbiol. Biotechnol.*, **16** (8–9), 757–767.

88 Serebryakova, L.T., Shremetieva, M.E., and Lindbald, P. (2000) H_2-uptake and evolution in the unicellular cyanobacterium Chroococcidiopsis thermalis CALU 758. *Plant Physiol. Biochem.*, **38** (6), 525–530.

89 Sundaram, S. and Soumya, K.K. (2011) Study of physiological and biochemical alternations in cyanobacterium under organic stress. *Am. J. Plant Physiol.*, **06** (1), 1–16.

90 Prabina, B.J. and Kumar, K. (2010) Studies on the optimization of cultural conditions for maximum hydrogen production by selected cyanobacteria. *ARPN J. Agric. Sci. Biol. Sci.*, **5** (5), 22–31.

91 Aryal, U.A., Callister, S.J., Mishra, S., Zhang, X., Shutthanandan, J.I., Angel, T.E., Shukla, A.K., Monroe, M.E., Moore, R.J., Koppenaal, D.W., Smith, R.D., and Sherman, L. (2013) Proteome analysis of

strains ATCC 51142 and PCC 7822 of the diazotrophic cyanobacterium Cyanothece sp. Under culture conditions resulting in enhanced H_2 production. *Appl. Environ. Microbiol.*, **79** (4), 1070–1077.

92 Toepel, J., McDermott, J.E., Summerfield, T.C., and Sherman, L.A. (2009) Transcriptional analysis of the unicellular, diazotrophic cyanobacterium Cyanothece sp. ATC 51142 grown under short day/night cycles. *J. Phycol.*, **45** (3), 610–620.

93 Jo, J.H., Lee, J.D., and Park, J.M. (2006) Modeling and optimization of photosynthetic H2 production by green alga Chlamydomonas reinhardtii in sulfur-deprived circumstance. *Biotechnol. Prog.*, **22** (2), 431–437.

94 Kosourov, S., Makarova, V., Fedorov, A.S., Tsygankov, A., Seibert, M., and Ghirardi, M.L. (2005) The effect of sulfur re-addition on H_2 photoproduction by sulfur-deprived green algae. *Photosynth. Res.*, **85** (3), 295–305.

95 Kosourov, S., Patrusheva, E., Ghirardi, M.L., Seibert, M., and Tsygankov, A. (2007) A comparison of hydrogen photoproduction by sulfur-deprived Chlamydomonas reinhardtii. *J. Biotechnol.*, **128** (4), 776–787.

96 Kosourov, S., Tsygankov, A., Seibert, M., and Ghirardi, M.L. (2002) Sustained hydrogen photoproduction by Chlamydomonas reinhardtii: effect of culture parameters. *Biotechnol. Bioeng.*, **78** (7), 731–740.

97 Hallenbeck, P.C., Abo-Hashesh, M., and Ghosh, D. (2012) Strategies for improving biological hydrogen production. *Bioresour. Technol.*, **110**, 1–9.

98 Lee, J.W. and Greenbaum, E. (2003) A new oxygen sensitivity and its potential application in photosynthetic H_2 production. *Appl. Biochem. Biotechnol.*, **105–108**, 303–313.

99 Waschewski, N., Bernát, G., and Rögner, M. (2010) Engineering photosynthesis for H_2 production from H_2O: cyanobacteria as design organisms, in *Biomass to Biofuels: Strategies for Global Industries* (eds A. Vertes, N. Qureshi, H. Blaschek, and H. Yakuwa), John Wiley & Sons, Ltd, Chichester, pp. 387–401.

100 Ghirardi, M.L. and Mohanthy, P. (2010) Oxygenic hydrogen photoproduction current status of the technology. *Curr. Sci.*, **98** (4), 499–507.

101 Melnicki, M.R., Pinchuk, G.E., Hill, E.A., Kucek, L.A., Fredrickson, J.K., Konopka, A., and Beliaev, A.S. (2012) Sustained H_2 production driven by photosynthetic water splitting in a unicellular cyanobacterium. *mBio*, **3** (4), article number e00197-12. doi: 10.1128/mBio.0097-12.

102 Ghirardi, M.L., Dubini, A., Yu, J.P., and Maness, P.-C. (2009) Photobiological hydrogen-producing systems. *Chem. Soc. Rev.*, **38** (1), 52–61.

103 Lewis, N.S. and Nocera, D.G. (2006) Powering the planet: chemical challenges in solar energy utilization. *Proc. Natl. Acad. Sci. U.S.A*, **103** (43), 15729–15735.

104 Nocera, D.G. (2009) Personalized energy: The home as a solar power station. *ChemSusChem*, **2** (5), 387–390.

105 Tran, P.D., Won, L.H., Barber, J., and Loo, J.S.C. (2012) Recent advantages in hybrid photocatalysts for solar fuel production. *Energy Environ. Sci.*, **5** (3), 5902–5918.

106 Barber, J. (2009) Photosynthetic energy conversion: natural and artificial. *Chem. Soc. Rev.*, **38** (1), 185–196.

107 Yacoby, I., Pochekailov, S., Toporik, H., Ghirardi, M.L., King, P.W., and Zhang, S. (2011) Photosynthetic electron partitioning between [FeFe]-hydrogenase and ferredoxin:NADP+-oxidoreductase (FNR) enzymes in vitro. *Proc. Natl. Acad. Sci. U.S.A.*, **108** (23), 9396–9401.

108 Centi, G., Perathoner, R., Passalacqua, R., and Ampelli, C. (2011) *Carbon-Neutral Fuels and Energy Carriers* (eds N.Z. Myradov, and T.N. Veziroğlu), CRS Press (Tayler & Francis Group), Boca Raton, FL, pp. 291–323.

109 Grätzel, M. (2001) Review article photoelectrochemical cells. *Nature*, **414** (6861), 338–344.

110 Trevisan, R., Rodenas, P., Gonzalez-Pedro, V., Sima, C., Sanchez, R.S., Barea, E.M., Mora-Sero, I., Fabregat-Santiago, F., and Gimenez, S. (2013) Harnessing

infrared photons for photoelectrochemical hydrogen generation. A PbS quantum dot based "quasi-artificial leaf". *J. Phys. Chem. Lett.*, **04** (1), 141–146.

111 Gimpel, J.A., Specht, E.A., Georgianna, D.R., and Mayfield, S.P. (2013) Advances in microalgae engineering and synthetic biology applications for biofuel production. *Curr. Opin. Chem. Biol.*, **17** (3), 489–495.

112 Razaghifard, R. (2013) Algal biofuels. *Photosynth. Res.*, **117** (1–3), 207–219.

113 Singh, A., Nigam, P.S., and Murphy, J.D. (2011) Renewable fuels from algae: an answer to debatable and based fuels. *Bioresour. Technol.*, **102** (1), 10–16.

114 Atsumi, S., Higashide, W., and Liao, J.C. (2009) Direct photosynthetic recycling of carbon dioxide to isobutyraldehyde. *Nat. Biotechnol.*, **27** (12), 1177–1180.

115 Tran, N.H., Bartlett, J.R., Kannangara, G.S.K., Milev, A.S., Volk, H., and Wilson, M.A. (2010) Catalytic upgrading of biorefinery oil from micro-algae. *Fuel*, **189** (2), 265–274.

116 Gronenberg, L.S., Marcheschi, R.J., and Liao, J.C. (2013) Next generation biofuel engineering in prokaryotes. *Curr. Opin. Chem. Biol.*, **17** (3), 462–471.

117 Hasunuma, T., Okazaki, F., Okai, N., Hara, K.Y., Ishii, J., and Kondo, A. (2013) A review of enzymes and microbes for lignocellulosic biorefinery and the possibility of their application to consolidated bioprocessing technology. *Bioresour. Technol.*, **135**, 513–522.

118 Carere, C.R., Rydzak, T., Verbeke, T.J., Cicek, N., Levin, D.B., and Sparling, R. (2012) Linking genome content biofuel production yields: a meta-analysis a major catabolic pathways among select H_2 and ethanol-producing bacteria. *BMC Microbiol.*, **12**, 295.

119 Verbeke, T.J., Zhang, X., Henrissat, B., Spicer, V., Rydzak, T., Krokhin, O.V., Fristensky, B., Levin, D.B., and Sparling, R. (2013) Genetic evaluation of Thermoanaerobactor spp. For the construction of designer co-cultures to improve lignocellulosic biofuel production. *PLoS ONE*, **8** (3), e59362. doi: 10.1371/journal.pone.0059362

120 Ilmén, M., den Hann, R., Brevnova, E., Mcbride, J., Wiswall, E., Froehlich, A., Koivula, A., Voutilainen, S.P., Siika-Aho, M., la Grange, D.C., Thorngren, N., Ahlgren, S., Mellon, M., Deleault, K., Rajgarhia, V., van Zyl, W.H., and Penttilä, M. (2011) High level secretion of cellobiohydrolases by Saccharomyces cerevisiae. *Biotechnol. Biofuels*, **4**, 30.

121 Tai, M. and Stephanopoulos, G. (2013) Engineering the push and pull of lipid biosynthesis in oleaginous yeast Yarrowia lipolytica for biofuel production. *Metab. Eng.*, **15**, 1–9.

122 Buijs, N.A., Siewers, V., and Nielsen, J. (2013) Advanced biofuel production by the yeast Saccharomyces cerevisiae. *Curr. Opin. Chem. Biol.*, **17** (3), 480–488.

123 Basak, N. and Das, D. (2007) The prospect of purple non-sulfur (PNS) photosynthetic bacteria for hydrogen production: the present state of art. *World J. Microbiol. Biotechnol.*, **23** (1), 31–42.

124 Kruse, O., Rupprecht, J., Bader, K.-P., Thoman-Hall, S., Schenk, P.M., Finazzi, G., and Hankamer, B. (2005) Improved photobiological H_2 production in engineered green algal cells. *J. Biol. Chem.*, **280** (40), 34170–34177.

125 Dasgupta, C.N., Gilbert, J.J., Lindbald, P., Heidorn, T., Borgvang, S.A., Skjanes, K., and Das, D. (2010) Recent trends on the development of photobiological processes and photobioreactors for the improvements of hydrogen production. *Int. J. Hydrogen Energy*, **35** (19), 10218–10238.

126 Mathews, J. and Wang, G. (2009) Metabolic pathway engineering for enhanced biohydrogen production. *Int. J. Hydrogen Energy*, **34** (17), 7404–7416.

127 Blankenship, R.E., Tiede, D.M., Barber, J., Brudvig, G.W., Flemming, G., Ghirardi, M., Gunner, M.R., Junge, W., Kramer, D.M., Melis, A., Moore, T.A., Moser, C.C., Nocera, D.G., Nozik, A.J., Ort, D.R., Parson, W.W., Prince, R.C., and Sayre, R.T. (2011) Comparing photosynthetic and photovoltaic efficiencies and recognizing the potential for improvement. *Science*, **332** (6031), 805–809.

128 Blankenship, R.E. and Chen, M. (2013) Spectral expansion and antenna reduction can enhance photosynthesis for energy production. *Curr. Opin. Chem. Biol.*, **17** (03), 457–461.

129 Chen, C.-Y., Yeh, K.-L., Aisyah, R., Lee, D.-J., and Chang, J.-S. (2011) Cultivation, photobioreactor design and harvesting of microalgae for biodiesel production: a critical review. *Bioresour. Technol.*, **102** (1), 71–81.

130 Zürrer, H. and Bachofen, R. (1979) Hydrogen production by the photosynthetic bacterium Rhodospirillum rubrum. *Appl. Environ. Microbiol.*, **37** (5), 789–793.

131 Johnson, E.T. and Schmidt-Dannert, C. (2008) Light-energy conversion in engineered microorganisms. *Trends Biotechnol.*, **26** (12), 682–689.

132 Salih, F.M. and Maleek, M.I. (2010) Influence of metal ions on hydrogen production by photosynthetic bacteria grown in Escherichia coli pre-fermented cheese whey. *J. Environ. Prot.*, **1** (4), 426–430.

133 Wang, Y.J., Liao, Q., Wang, Y.-Z., Zhu, X., and Li, J. (2011) Effect of flow rate and substrate concentration on the formation and H_2 production of photosynthetic bacterial biofilms. *Bioresour. Technol.*, **102** (13), 6902–6908.

134 Levin, D.B., Pitt, L., and Love, M. (2004) Biohydrogen production: prospects and limitations to practical application. *Int. J. Hydrogen Energy*, **29** (2), 173–185.

135 Martínez-Pérez, N., Cherryman, S.J., Premier, G.C., Dinsdale, R.M., Hawkes, D.L., Hawkes, F.R., Kyazze, G., and Guwy, A.J. (2007) The potential for hydrogen-enriched biogas production from crop: scenarios in the UK. *Biomass Bioenerg.*, **31** (2–3), 95–104.

136 Hawkes, F.R., Hussy, I., Kyazze, G., Dinsdale, R., and Hawkes, D.L. (2007) Continuous dark fermentative hydrogen production by mesophilic microflora: principles and progress. *Int. J. Hydrogen Energy*, **32** (2), 172–184.

137 Su, H., Cheng, J., Zhou, J., Song, W., and Cen, K. (2009) Improving hydrogen production from cassava starch by combination of dark and photo fermentation. *Int. J. Hydrogen Energy*, **34** (4), 1780–1786.

138 Woodward, J., Orr, M., Cordray, K., and Greenbaum, E. (2000) Enzymatic production of biohydrogen. *Nature*, **405** (6790), 1014–1015.

139 Pinto, F.A.L., Troshima, O., and Lindbald, P. (2002) A brief look at three decades of research on cyanobacterial hydrogen evolution. *Int. J. Hydrogen Energy*, **27** (11–12), 1209–1215.

140 Chong, M.L., Sabaratnam, V., Shirai, Y., and Hassan, M.A. (2009) Biohydrogen production from biomass and industrial wastes by dark fermentation. *Int. J. Hydrogen Energy*, **34** (8), 3277–3287.

141 Guwy, A.J., Dinsdale, R.M., Kim, J.R., Massanet-Nicolau, J., and Premier, G. (2011) Fermentative biohydrogen production systems integration. *Bioresour. Technol.*, **102** (18), 8534–8542.

142 Nath, K. and Das, D. (2011) Modeling and optimization of fermentative hydrogen production. *Bioresour. Technol.*, **102** (18), 8569–8581.

143 Cheng, J., Su, H., Zhou, J., Song, W., and Cen, K. (2011) Hydrogen production by mixed bacteria through dark and photo fermentation. *Int. J. Hydrogen Energy*, **36** (1), 450–457.

144 Kondo, T., Arakawa, M., Hirai, T., Wakayama, T., Hara, M., and Miyake, J. (2002) Enhancement of hydrogen production by a photosynthetic bacterium mutant with reduced pigments. *J. Biosci. Bioeng.*, **93** (2), 145–150.

145 Kim, E.-J., Kim, J.-S., Kim, M.-S., and Lee, J.K. (2006) Effect of changes in the level of light harvesting complexes of Rhodobactor sphaeroides on the photoheterotrophic production of hydrogen. *Int. J. Hydrogen Energy*, **31** (4), 5311–5538.

146 Ozturk, Y., Yücel, M., Daldal, F., Mandaci, S., Gündüz, U., Türker, L., and Eroğlu, I. (2006) Hydrogen production by using Rhodobactor caspsulatus mutants with genetically modified electron transfer chains. *Int. J. Hydrogen Energy*, **31** (11), 1545–1552.

147 Eltsova, Z.A., Vasilieva, L.G., and Tsygankov, A.A. (2010) Hydrogen

production by recombinant strains of Rhodobactor sphaeroides using a modified photosynthetic apparatus. *Appl. Biochem. Microbiol.*, **46** (5), 487–497.

148 Liang, Y., Wu, X., Gan, L., Xu, H., Hu, Z., and Long, M. (2009) Increased biological hydrogen production by deletion of hydrogen-uptake system in photosynthetic bacteria. *Microbiol. Res.*, **164** (6), 674–679.

149 Cha, M., Chung, D., Elkins, J.G., Guss, A.M., and Westpheling, J. (2013) Metabolic engineering of Caldicellulosiruptor bescii yields increased hydrogen production from lignocellulosic biomass. *BMC Microbiol.*, **6**, 85.

150 Lindberg, P., Park, S., and Melis, A. (2010) Engineering a platform for photosynthetic isoprene production in cyanobacteria, using Synechocystis as the model organism. *Metab. Eng.*, **12** (1), 70–79.

151 Wu, S., Xu, L., Wang, R., Liu, X., and Wang, Q. (2011) A high yield of Chlamydomonas reinhardtii for photoproduction of hydrogen. *Int. J. Hydrogen Energy*, **36** (21), 14134–14140.

152 Kars, G. and Gündüz, U. (2010) Towards a super H_2 producer: improvements in photofermentative biohydrogen production by genetic manipulations. *Int. J. Hydrogen Energy*, **35** (13), 6646–6656.

Part 2.6
Gas Clean-up Technologies

21
PSA Technology for H$_2$ Separation
Carlos A. Grande

21.1
Introduction

Hydrogen is an indispensable compound in chemical, fuel, and energy industries. Nowadays, most hydrogen is produced and used in refineries and to produce ammonia and methanol [1,2]. However, the amount of hydrogen produced to be used as energy vector in the automotive industry has been increasing over the years [3,4].

The most common H$_2$ production route is by steam reforming of hydrocarbons, like natural gas, refinery gas, LPG, and light naphtha [2,5,6]. The major product in the stream is H$_2$, but it is contaminated with carbon oxides (CO + CO$_2$), some unreacted CH$_4$ (and possibly some C$_{2+}$), and H$_2$O [7–9]. Other contaminants that might be present in the system are N$_2$, O$_2$, Ar, H$_2$S, and so on [10]. The reforming reaction takes place at very high temperatures (~1000 K). The gas exiting the reformer is cooled down (passing through a shift conversion to produce more hydrogen and convert most CO into CO$_2$). After the shift reactor, the gas is cooled (condensing water) to low temperature (~300 K) for H$_2$ purification in a given separation process.

The required purity of hydrogen used in most applications is >98%, but purities up to 99.999% are not unusual. This means that a purification process is required. Since the 1980s, the traditional separation process used to produce high purity hydrogen has been pressure swing adsorption (PSA) [10].

The inlet molar composition for the separation process is around 65–80% H$_2$, 15–25% CO$_2$, 2–6% CH$_4$, and 1–4% CO while other possible contaminants represent less than 1%, like N$_2$, O$_2$, Ar, and so on. The stream is saturated with humidity at ambient temperature, which is around 3%. Although for some applications it would be of interest to have a higher temperature operation, the normal feed temperature used is around 300–310 K. The inlet pressure varies, depending on the source and reforming technology used, but the normal pressure is between 20 and 40 bar. The pressure of the stream can be much higher when hydrogen is produced from biomass conversion [11]. Typical blowdown

Hydrogen Science and Engineering: Materials, Processes, Systems and Technology, First Edition.
Edited by Detlef Stolten and Bernd Emonts.
© 2016 Wiley-VCH Verlag GmbH & Co. KGaA. Published 2016 by Wiley-VCH Verlag GmbH & Co. KGaA.

pressure is 1–3 bar. The pressure drop across one column filled with particles of adsorbent of 1–4 mm diameter can be almost 1 bar [2,12,13].

A PSA operates at a temperature close to ambient and satisfies even very stringent requirements in H_2 purity. In the initial years of utilization of PSA, the hydrogen recovery was around 75% but nowadays, after several process optimizations, hydrogen recovery is approaching 90%. Furthermore, the scale of the units has been gradually increasing, extending the flexibility of design from around 30 up to 400 000 $Nm^3 h^{-1}$, ranging from single-column units to PSAs with up to 16 columns in parallel. Moreover, PSA technology is not limited only to steam reformers but can also handle most refinery off-gases including coke oven off gas, which has numerous impurities. The inlet pressure of refinery off-gases is normally less than 10 bar.

The operating costs of a PSA for H_2 purification are minor once the feed is already pressurized, the unit operation is highly automatized and the adsorbent is known to last for several years. The capital costs of the PSA unit are around 10% of the total cost of the H_2 production plant [14].

This chapter aims to describe the operation of PSA units for H_2 purification and to provide an idea of R&D trends to improve them and extend their future utilization.

21.2
Basics of PSA Technology

Adsorption processes are based on the selectivity that some solids have to attract specific molecules to their surface [15–17]. Thus, when a multicomponent gas phase is put into contact with a certain surface of a solid, the molecules that are less attracted, or adsorbed, can be separated from molecules that are more strongly adsorbed. The adsorbent is normally porous to enhance the surface available per unit volume to retain gases. The interactions between the solid surface and the gas molecules can be diverse [18], but the ones used for separation by adsorption processes are the type of bonding that correspond to physical adsorption, which have adsorption energies lower than 50 kJ mol^{-1}. The physical adsorption is spontaneous and thus exothermal, releasing energy. This also means that to desorb a molecule, energy has to be provided to the system and this normally corresponds to the energy expenditure of the adsorption processes.

The amount adsorbed (also known as loading) increases with pressure and decreases with temperature. The adsorption behavior of a porous material is normally assessed by measuring the amount adsorbed at equilibrium state (thermodynamic data) at different pressures at constant temperature, the so-called adsorption equilibrium isotherms. The "shape" of these isotherms varies depending on the type of binding that the gas molecules has to the surface. Isotherms Type I (Figure 21.1) according to IUPAC nomenclature are the most common. For utilization in PSA units, the more linear the isotherms, the better. When isotherms are strongly nonlinear, a large portion of the loading is achieved at

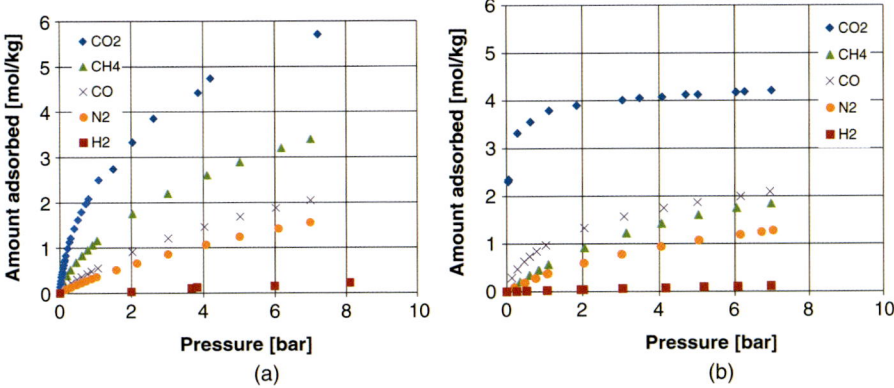

Figure 21.1 Adsorption equilibrium of CO_2, CH_4, CO, N_2, and H_2 on activated carbon (a) and zeolite (b) at 303 K [19].

low pressures and then – to regenerate the adsorbent and desorb the loaded gases – vacuum has to be employed, increasing the energy consumption of the process and, thus, reducing the economic viability of the technique.

Once the adsorbent has reached equilibrium with the surrounding gas phase, it loses its separation ability. For this reason, to use the same adsorbent material, it has to be "regenerated." The technology implemented for the regeneration is determinant in defining the apparatus specifications that will be used. In fact, the regeneration method is typically used to name the separation process. When the regeneration method is the reduction of pressure, the process is called pressure swing adsorption, which is the technique focused on here. Changes in pressure can be accomplished quite rapidly and thus a PSA processes can be employed in the removal of bulk components. The duration of a PSA cycle is around 10 min, but can be decreased greatly to less than 1 min. Alternatively, for removal of small amounts of contaminants, the temperature swing adsorption (TSA) process can be used, where desorption is accomplished by increasing the temperature of the adsorbent. TSA cycles last for several hours because the energy storage of a column filled with adsorbent is high.

The design of a PSA process is a quite complex task. It involves the interplay between material science and process/control engineering; a selective material has to be identified and used in a specific (sometimes tailored) manner.

Regarding the adsorbent material, the most common porous materials used in H_2 PSA are activated carbon and zeolites. Examples of isotherms of the normal gases present in steam–methane reforming off gases (CO_2, CH_4, CO, and H_2) at 303 K on zeolite and activated carbon are shown in Figure 21.1 [19]. In both materials, the adsorption of hydrogen is quite small. Although not shown in the figure, water is the most adsorbed compound in both adsorbents. The gases are removed from the feed stream according to their affinity to the adsorbent. In an off-gas from steam–methane reformers (SMRs) the order of removal is: H_2O, CO_2, CH_4, CO, and N_2/Ar.

A normal strategy to deal with the adsorption of so many gases is to employ different layers of adsorbents per column [20–27]. This is done to maximize the adsorption capacity of the bed to all components. Three major adsorbents are used:

1) Alumina: since the stream is saturated with water, a first layer of adsorbent is specially located in the bottom of the column to retain water. Silica gel can also be employed. Porous alumina can adsorb as much as 30 mol of water per kilogram of adsorbent when the stream is saturated [28–30]. Adsorption isotherms are normally Type V (IUPAC nomenclature), corresponding to materials having an initial plateau after the formation of a monolayer of water, followed by multilayer adsorption until capillary filling of the pores. Since the initial steepness of the water isotherm is not very high, regeneration is possible by changes of pressure in the system.

2) Activated carbon: this adsorbent can also retain water so the utilization of alumina is not mandatory [31,32]. However, the loading of water is lower than for aluminas (around 20 mol kg^{-1} has been reported) and also the adsorbent density is smaller, which is why volumetric loadings are considerably reduced. Water isotherms on activated carbons follow Type IV isotherms (IUPAC nomenclature) that are shown by materials presenting repulsive forces at low partial pressure of water followed by multilayer adsorption and capillary filling. Activated carbon is a generic and very flexible adsorbent that can remove selectively almost all contaminants from the gas feed, while also presenting a low adsorption capacity towards hydrogen [33–35]. This adsorbent is also quite cheap and withstands the presence of several gases without significantly changing its adsorption properties. The order of removal of the main contaminants is: $H_2O > CO_2 > CH_4 > CO$ [36–39].

3) Zeolites: since the purity requested for H_2 is very high, intensive polishing of the gas is necessary. To remove CO to ppm level (normally below 10 ppm), a top layer of zeolite is employed in most H_2 PSA units. In this adsorbent, the loading of CO (eventually N_2, O_2, and Ar) is also higher (around twice that of the activated carbon at low partial pressures), which is it is more efficient for polishing purposes [19,40–47]. However, the utilization of zeolite as a top-layer imposes an extra constraint on the system design. The zeolite layer cannot receive either water or CO_2: their adsorption is very strong and they cannot be desorbed, thereby reducing the available loading for other gases [48–51].

The adsorbents used in the PSA process have to withstand the continuous presence of these compounds (H_2O, CO_2, CH_4, and CO), maintaining their stability and performance. The adsorbent life of adsorbents used in industrial PSA units is around 30 years.

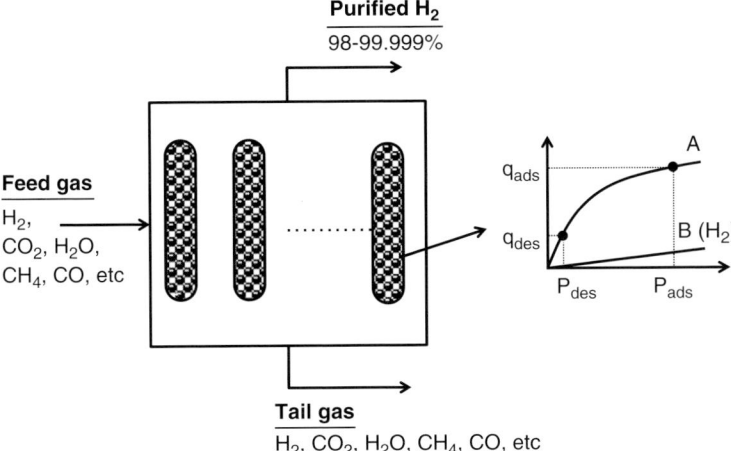

Figure 21.2 Basic black-box scheme of a PSA process for H_2 purification.

Although the operation of PSA might seem complex since it is essentially an unsteady process, an initial understanding of the process can be carried out by considering a traditional separation process. Figure 21.2 shows the basic "black-box" scheme of the PSA process.

Components are removed from the feed gas to produce high-purity hydrogen with a purity that ranges between 98 and 99.999%, depending on the final application. The components removed together with some hydrogen lost in the process are obtained as "tail gas." The tail gas is routed to the "tail gas drum," which is the large tank that is normally found together with the set of parallel columns. The tail gas is a low-grade fuel that is used as local fuel in the burners of a SMR, for example. The existence of the tail gas drum is to allow the delivery of a constant flow (that can be controlled) to the burners.

When the cyclic steady state (CSS) is reached, the mass that enters the feed stream has to be recovered in the two outlet streams (product and tail gas). The typical process performance parameters are [52,53]: H_2 purity, H_2 recovery, process or adsorbent productivity, and energy consumption. The purity is imposed by the final application and the process has to be adjusted to deliver it with a high H_2 recovery (that is around 90% in modern large PSA units).

The productivity and energy consumption cannot be assessed via black-box design and so insights into the process have to be available. In H_2-PSA, the energy consumption is not very important. Since H_2 is produced at high pressure and the tail gas is normally produced at sufficient pressure to combust, the only direct energetic consumption of the process (when operating) is for valves and for control of the unit.

The fundamental operation of a PSA process can be understood by observing the isotherms of a contaminant (A) and hydrogen, designated as gas B in

Figure 21.2. In the adsorbent used in the example of Figure 21.2, B is much less adsorbed than A.

At the feed conditions, when a gas mixture A + B enters into contact with the adsorbent packed in a fixed-bed column, A is selectively removed from the feed gas while B is recovered at the other end of the column since it is adsorbed to much lesser extent. The partial pressure of component A is P_{ads} and it is loaded in the adsorbent until q_{ads}. Since the loading of component B (H_2) is much lower, this product can be recovered at high pressure. Once the column is saturated, the adsorbent has to be regenerated by lowering the pressure to P_{des} and lowering the loading of component A to q_{des}. After the high pressure is restored, the adsorbent can adsorb to q_{ads} again and thus the cyclic loading (also termed cyclic capacity) of the adsorbent towards component A is ($q_{ads} - q_{des}$).

Before the cyclic steady state is achieved, the cyclic loading is slightly higher since part of the adsorbent is initially clean (the initial loading is from 0 to q_{ads}) and thus the mass exiting the two outlets of the PSA unit is smaller (there is accumulation of mass in the adsorbent).

The productivity of the PSA units is a very important parameter since it determines its final size and thus the capital cost of the H_2 separation. The productivity is calculated as the amount of hydrogen produced per unit time divided by the amount of adsorbent used, with units of mass of H_2/(adsorbent weight × time). The factors influencing the calculation of the productivity are diverse and will be discussed in the sections below.

As referred to above, the layers of adsorbents placed in the PSA columns undergo a cyclic adsorption–desorption operation by swinging the pressure between low and high pressure. The adsorption is spontaneous and is thus controlled by the adsorption equilibrium properties of the adsorbents used, taking into account some diffusional resistances. On the other hand, desorption is not spontaneous (promoted) and this is where the real engineering efforts should be made. Desorption is normally accomplished by a series of steps with different purposes. The sum of all the adsorption and desorption steps makes up a PSA cycle that is repeated over time, turning a PSA into a cyclic process [54].

The PSA process that stands in a refinery has to operate in continuous mode. For this reason, a PSA should have an array of columns that operate in parallel. The simplest unit that can be idealized is a two-column PSA: when one column is removing the contaminants from the feed, the other column is being regenerated.

The standard and reference cycle of a PSA process bears the name of its inventor: Charles Skarstrom [55]. The Skarstrom cycle is a sequence of four steps:

1) Adsorption: the contaminants of the stream (H_2O, CO_2, CH_4, CO) are removed from the feed stream and retained in the different layers of adsorbent in the packed columns. In this step, hydrogen is produced at high pressure. This step operates at the highest pressure of the system.
2) Counter-current blowdown: immediately before breakthrough of the contaminants, the feed is interrupted and the pressure in the column is

reduced, partially removing the contaminants. The pressure in this step is variable and the step stops when the pressure reaches the desired lowest pressure of the cycle. Since the flow exiting the column is pressure-driven, caution has to be taken to avoid a large gas velocity from the blowdown that can smash the adsorbent in the column. This is the first step of the regeneration process. This step is also known as "depressurization" or "dump".

3) Counter-current purge: after initial desorption of the contaminants promoted in the blowdown step, the column is "washed" with hydrogen. For this reason, this step is also known as "light-recycle". This hydrogen recycle reduces the partial pressure of all contaminants, promoting further desorption. In addition, the hydrogen recycled to the column is used to displace the gas molecules of the contaminants in the gas phase, preparing the column for the next cycle. The purge is also carried out at the lower pressure of the system to minimize consumption of H_2 and have a maximum effect in reduction of the partial pressure of all contaminants.

4) Pressurization: this is the last step of the cycle and aims to recondition the column in order to start the next cycle. The major aim of this step is to increase the pressure from the low pressure to the highest pressure of the system. In PSA for H_2 purification, the pressurization step is performed counter-currently to the feed direction with recycled enriched gas (purified hydrogen). The counter-current recycle of H_2 also helps in producing a "cleaner" top of the bed in the next feed step.

Figure 21.3 shows an example of the pressure levels of the different steps for a PSA unit with two beds following the Skarstrom cycle. This figure also illustrates how the streams are connected. The mass balance of the PSA as a black-box process is performed knowing the amount of gas and the composition of all streams crossing the dashed line.

Note that in using the Skarstrom cycle, when the first desorption step starts (blowdown), there is a large amount of hydrogen in the gas phase of the column since the column was at the highest pressure of the cycle. This hydrogen is thus going to be part of the tail gas, reducing the overall hydrogen recovery of a two-column PSA unit.

To improve the H_2 recovery, the amount of hydrogen that goes to the tail gas has to be minimized. In the basic Skarstrom cycle, most of the H_2 losses occur because, in the initial moments of the blowdown, H_2 is removed from the column at high pressure. Therefore, a strategy to reduce the amount of hydrogen in the column before regeneration has to be realized. There are at least two different procedures to reduce the amount of hydrogen present in the column before the blowdown step.

One of the alternatives to reducing the amount of H_2 before blowdown is to sequentially reduce the pressure of the column before blowdown in a "pressure equalization" step [56–58]. A pressure equalization is a step where two columns are put in contact, with one being at higher pressure than the other, and time is

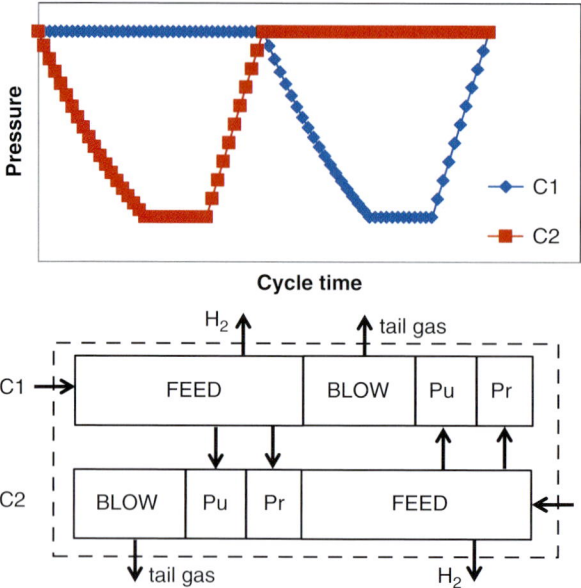

Figure 21.3 Pressure history in a two-column PSA process and cycle steps showing inflow and outflow streams. The streams crossing the dashed line should be used to make a mass balance of the unit. Pu denotes the "purge" step and PR denotes the "pressurization" step.

allowed for gas transfer until pressure equilibration. As both columns are connected from the top, the flow from the high-pressure column is co-current to feed while for the low-pressure column the flow is counter-current.

Since there is a pressure reduction in the pressure equalization steps, the feed step has to be stopped before the maximum loading is achieved; when the pressure is reduced, there will be desorption of the contaminants from the last portion of the adsorbent of the column receiving the gas. Note that when two columns are undergoing a pressure equalization step, another column should be processing the feed stream, so a minimum of three columns is required to have a continuous process with one pressure equalization. If more pressure equalizations are introduced in the cycle, more columns are required.

The increase in recovery obtained by adding pressure equalizations is counterbalanced by increasing the number of columns in the system, thus decreasing the overall unit productivity. It has been shown that after three pressure equalizations the increase in recovery is not significant.

Another possibility for reducing the amount of hydrogen from the column before the blowdown step and "recycling" it to the system is in a step called "provide purge" [59,60]. Since the purge step is carried out at the lowest pressure of the system, the source of hydrogen can be taken from any other step in the cycle that will be at a higher pressure. Some cycles use a combination of pressure equalization steps and also a provide-purge step.

21.3
Selective Adsorbents; Commercial and New Materials

As mentioned above, the mixture that is processed in the PSA unit for H_2 purification requires removal of several gases. Since it is very difficult to obtain a material that has a high selectivity towards all the components, utilization of different layers of adsorbents are common.

As the columns used in large commercial units can be several meters high the mechanical properties of the adsorbents are an issue. All the materials need to have good mechanical stability. Moreover, the flows of the different steps (particularly the ones driven by pressure differences) have to be well regulated to avoid particle movement and crushing.

The inlet stream of the PSA for H_2 purification is saturated in water. Water is the first component to be removed. To eliminate water from the stream, a layer of alumina or silica gel is used. The water loading of a basic gamma alumina can reach up to 30 mol kg^{-1} with a Type V isotherm according to IUPAC nomenclature: favorable monolayer followed by capillary condensation in the pores. Alumina also has excellent mechanical properties and a density of around 900 kg m^{-3}. The amount of water adsorbed in silica gel is lower (around 15 mol kg^{-1}); the density of large-pore silica gel is also around 900 kg m^{-3} [61].

Alternatively, activated carbon can be used as water selective adsorbent [32,62]. Depending on the structure of the carbon, the loading can reach more than 15 mol kg^{-1} of water adsorbed. The density of activated carbons is also lower but can reach up to 900 kg m^{-3}. The isotherms of water on most activated carbons are Type IV, which is unfavorable adsorption followed by capillary condensation. Although the "water concentration front" will penetrate deeper into the bed, desorption of water at low partial pressures is favorable and thus it is feasible.

The utilization of activated carbon as water-selective adsorbent has only been reported in small PSA units. Large commercial processes normally prefer the utilization of alumina since it has a higher volumetric adsorption equilibrium.

Some metal–organic frameworks (MOFs) can adsorb a significant amount of water. In HKUST-1 adsorbent (also known as CuBTC), the loading can be as high as 40 mol kg^{-1} while in CPO-27-Ni, the loading can be up to 30 mol kg^{-1}, being particularly high at low water concentrations [63]. Even when the water loadings obtained are interesting, the lower density of MOFs (and thus lower volumetric adsorption equilibrium) can be a competitive problem. Furthermore, their high hypothetical price (such an adsorbent is not on the market in large scale) makes it difficult to compete with alumina.

In the initial moments of desorption, a major amount of water is released from the adsorbent (due to bulk desorption of water from the pores) and the concentration of water in the gas phase will exceed the vapor–liquid equilibrium and thus liquid water is also obtained in the blowdown step line.

Once water is removed, the next components to be removed are the acid gases and heavy hydrocarbons. For steam-reforming off-gases, the bulk removal of

CO_2 takes place together with a large portion of hydrocarbons and also H_2S, if present.

Since the stream is always available at high pressures, the preferred adsorbent for this task is activated carbon. The adsorption of CO_2 on activated carbons at high pressures can be quite high (Figure 21.1), reaching up to 8 mol kg^{-1} for some activated carbons [36–38,64–66]. Furthermore, the isotherms are noticeably less nonlinear than those of zeolites, allowing regeneration at pressures around atmospheric.

The layer of activated carbon also serves to selectively adsorb most of the present CH_4 in the feed gas [10,21–27,65]. The preferential order of adsorption in activated carbons is H_2S followed by hydrocarbons (C_{2+}), CO_2, and finally CH_4. For the steam–methane reforming off gases, normally there is no H_2S (removed before reforming) and CO_2 is initially adsorbed best, followed by methane and CO as shown in Figure 21.1.

To evaluate the efficiency of the activated carbon for CO_2 removal from the stream, it has to be kept in mind that the adsorption loading has to be evaluated on a volumetric basis (mol m^{-3}), so that when higher mass loadings are reported (mol kg^{-1}) the density of the adsorbent also has to be known, and it is their product (volumetric loading) that should be used as final criteria.

As a general trend, the volumetric loading of CO_2 of commercial activated carbon samples is quite similar at the partial pressures of H_2 purification. If the activated carbon has larger pores, its density and mechanical resistances are smaller but the CO_2 diffusion can be faster. If pores are narrow, density can be higher and the carbon can have good mechanical properties, but CO_2 diffusion (as well as CH_4) can be an issue. The development of activated carbons for this application has been delimited by such a trade-off.

The number of publications in materials science for porous solids able to remove CO_2 from various streams is quite large. In particular, new metal–organic frameworks (MOFs) have reported with quite high loadings of CO_2 [67–71].

For H_2 purification, the performance of the adsorbent has to be measured by the cyclic CO_2 capacity between 1–5 bar of CO_2 (the specific interval depends on the particular CO_2 partial pressure used in the feed and the blow-down pressure). The adsorption equilibrium isotherms of CO_2 on activated carbons are only slightly nonlinear when compared with other materials, meaning that regeneration of activated carbons is simpler. In that sense, to claim that a new material has an opportunity to be used in this market the adsorbent has to have a higher volumetric loading than activated carbon and its regenerability should also be at least similar. The price of a new material is also an important issue since activated carbons are known by their low prices, in the range US$2000–4000 ton^{-1}.

In addition to the specifications in H_2 purity, there is a further specification in the amount of CO that the H_2 should have, which should normally lower than 30 or 10 ppm. The loading of activated carbons towards CO is quite low at the partial pressures of CO in the feed mixture. If activated carbons are used to remove all the contaminants, the last portion should be dedicated to CO. Since the loading is low and the polishing should be quite intensive, the length of the PSA columns would increase.

To improve the volumetric loading of CO in the bed, a zeolite layer is used after the activated carbon layer. As shown in Figure 21.1, the amount adsorbed of CO in zeolites in the low partial pressure range is higher than for activated carbons. Furthermore, zeolites have a higher density than activated carbons. Consequently, the overall volumetric loading of CO can be easily doubled. The cost of zeolites is strongly influenced by the materials used for its preparation and lies in the range US$5000–8000 ton^{-1} (common synthetic zeolites) to as high as US$50 000 ton^{-1} (for specialized materials).

Zeolites with small pores (eight-membered ring) can present diffusion problems and are thus avoided; zeolite 5A or 13X are the ones used most often. An important restriction imposed by the utilization of a layer of zeolites is that "all" the CO_2 has to be contained in the activated carbon layer in all the steps of the cycle, including the pressure equalizations. The same restriction is applied to water, particularly if only activated carbon is used in the unit. If H_2O or CO_2 enters the zeolite layer, its adsorption is so strong that regeneration is not possible unless vacuum or higher temperature is applied. The process should not need vacuum for normal operation and thus utilization of a vacuum will generate unnecessary and inefficient power consumption.

New adsorbent materials like MOFs can be an alternative for this separation, although very few studies report its utilization for selective adsorption of CO from H_2 [72]. Most of the literature is concerned with CO_2/H_2 separation. It has to be remembered that the volumetric loading of CO should be the main focus, and then only MOFs that can reach a high volumetric loading with a relatively linear isotherm are required.

21.4
Improving the PSA Cycle

In adsorption processes, improving the performance of the material is only one of ways to enhance the performance of the process. Since each of the columns operates cyclically running in unsteady state, improving the steps of the cycle and also adjusting the total time of the cycle are other alternatives to enhance the performance of H_2 PSA [73,74].

Such interplay between the adsorbent properties and the cycle efficiency can be observed in the calculation of the unit productivity, given by:

$$\text{Productivity} = \frac{\text{Feed (mass } H_2 \text{ s}^{-1}) \times \text{feed time} \times H_2 \text{ Recovery}}{H_2 \text{ Purity} \times \text{cycle time} \times w_{ads}} \quad (21.1)$$

The properties of the adsorbent define the amount of time that it can be on feed mode for a given amount of adsorbent per column. However, the H_2 purity and recovery, as well as the cycle time, depend on the regeneration strategies and thus on the engineering of the PSA cycle.

Numerical simulation has greatly enhanced the possibilities of improving PSA cycles [75–77]. Unfortunately, there is not yet a theoretical solution by which to calculate the optimized cycle sequence and duration; the cycle is optimized by experience and is quite a complex procedure. Details of the different possibilities of improvement of the H_2-PSA unit by engineering design and optimization are discussed in this section.

As explained previously, regeneration of the H_2-PSA cycle consists of an arrangement of different pressure equalizations, a provide-purge step, a blowdown and purge steps, and a re-pressurization step. Generically, the more hydrogen that can be removed from the column before the blowdown step, the higher the H_2 recovery will be. However, when more gas is removed from the column, the concentration front of the contaminants also goes deeper inside the column and thus the feed step has to be interrupted earlier. The two most common approaches are:

1) Use several pressure equalizations between different columns. Up to four pressure equalization steps are used in large-scale units [13,78]. A balance between the number of equalization steps and the time consumed for each of them also plays a role that should be considered; the total cycle time increases and so the productivity decreases.
2) Use less pressure equalizations and use a provide-purge step [75,79]. Utilization of this step reduces the consumption of H_2 from the feed gas and also allows better control of gas flow exiting the unit. The stream used for the provide-purge step is the one after the pressure equalization steps.

The arrangement of the different cycle steps in a multi-column PSA has to take into consideration that the feed has to be processed continuously and also that the times of the pressure equalization and provide-purge/purge steps match between the columns. In some cases, other alternatives were explored, like tanks used to store the purge gas that will be used at a different time when the source for the purge is not matched with the purge step time [80].

Another engineering approach to handle large flows of gas is to split the feed step into different columns [80–82]. Splitting the flow into different columns reduces the velocity of the gas inside each column, reducing the size of the PSA and thus the installation and capital costs. Furthermore, since these units have over ten columns, a large number of pressure equalizations and provide-purge steps can be performed, further increasing the H_2 recovery of the process. Figure 21.4 shows an example of a PSA cycle with 12 columns in feed mode [81]. Since such units also have dedicated lines for purge gas, the purge step and the provide-purge step from one column to another does also not need to match and the stream can be obtained from different columns.

Another interesting approach to improve PSA units for H_2 purification concerns the impact of the total cycle time [83–89]. If the total cycle time is reduced, the productivity increases (Eq. (13.1)) and thus smaller columns can be realized to perform the separation. The cycle times of PSA units are around

	FEED			1	2	3	4	PP	PP/BD	BD	Purge			4'	3'	2'	1'	R								
1'	R	FEED			1	2	3	4	PP	PP/BD	BD	Purge			4'	3'	2'									
3'	2'	1'	R	FEED			1	2	3	4	PP	PP/BD	BD	Purge			4'									
Pu	4'	3'	2'	1'	R	FEED			1	2	3	4	PP	PP/BD	BD	Purge										
Purge		4'	3'	2'	1'	R	FEED			1	2	3	4	PP	PP/BD	BD	Pu									
BD	Purge			4'	3'	2'	1'	R	FEED			1	2	3	4	PP	PP/BD									
PP	PP/BD	BD	Purge			4'	3'	2'	1'	R	FEED			1	2	3	4	PP								
PP		PP/BD	BD	Purge			4'	3'	2'	1'	R	FEED			1	2	3	4								
3	4	PP		PP/BD	BD	Purge			4'	3'	2'	1'	R	FEED			1	2								
1	2	3	4	PP		PP/BD	BD	Purge			4'	3'	2'	1'	R	FEED										
FEED	1	2	3	4	PP		PP/BD	BD	Purge			4'	3'	2'	1'	R	FEED									
FEED			1	2	3	4	PP		PP/BD	BD	Purge			4'	3'	2'	1'	R	FEED							

Figure 21.4 Twelve-column PSA cycle with feed divided into various columns [81]. Steps with numbers 1–4 correspond to the columns delivering gas for pressure equalization steps and with numbers 1'–4' the receiving gas. PP refers to "provide purge" and BD to blowdown.

10 min (up to 30 min for longer cycles). The utilization of faster cycles, to around 1 min, can improve the productivity by an order of magnitude. It has been reported that by using a cycle of 8 min a productivity of ~100 mol-H_2 kg_{ads}^{-1} day^{-1} can be obtained [2,83]. If the total cycle time is reduced to around 1 min, the productivity can increase up to ~600 mol-H_2 kg_{ads}^{-1} day^{-1} [83], which represent a significant saving in materials and thus investment costs. The traditional configuration of a PSA can be improved by reducing the total cycle time, but the decrease of cycle time can produce some attrition of the adsorbents and more importantly can result in gas diffusion limitations. In addition, the operational life of the valves can be a problem.

Rotary valves can be used to develop rapid pressure swing adsorption (RPSA) units [84]. These units have been commercialized for H_2 purification on small and intermediate scales (30–12 000 $Nm^3 h^{-1}$). The first such unit has been operated in an ExxonMobil refinery since 2007. The RPSA technology relies not only on the change of valves, but the adsorbent also needs to be different. Instead of classical pellets, laminated or structured adsorbents are required to avoid attrition and also to reduce the pressure drop [85–87].

21.5
Summary

The utilization of pressure swing adsorption (PSA) for H_2 purification is standard in modern refineries. This technology can produce hydrogen with a purity from 98% up to 99.999%. Hydrogen recovery in large units is close to 90%.

Process design from single small columns up to 16 multiple large columns in parallel are available, treating streams from 30 to 400 000 Nm^3 h^{-1}.

H$_2$ PSA can also be used to treat different streams, the most common ones being the steam–methane reforming off-gas and the refinery off-gas. Different types of contaminants are removed such as H$_2$O, H$_2$S, CO$_2$, C$_{2+}$, CH$_4$, CO, N$_2$, and so on.

Intensive work is under way to improve the performance of this process, either in terms of H$_2$ recovery or/and in unit size (productivity). Innovative materials can provide enhanced cyclic capacity that will reduce the size of the units. The engineering of such units is also of great importance in defining the final performance. Diverse types of cycle sequences are being developed to efficiently use the adsorbent loading. Moreover, the total cycle time has been severely reduced in rapid pressure swing adsorption (RPSA) units that operate with rotary valves and structured adsorbents.

The future of H$_2$ PSA is very promising since the market for hydrogen is continuously increasing and the process has demonstrated unique flexibility in operating under different conditions.

Acknowledgments

The author acknowledges the financial support of the Research Council of Norway through the RENERGI program by the project Novel Materials for Utilization of Natural Gas and Hydrogen (grant 190 980/S60).

References

1 Damle, A. (2009) *Hydrogen Fuel: Production, Transport and Storage* (ed. R.B. Gupta), CRC Press, Boca Raton.
2 Sircar, S. and Golden, T.C. (2010) *Hydrogen and Syngas Production and Purification Technologies* (eds K. Liu, C. Song, and V. Subranami), John Wiley & Sons, Inc., Hoboken NJ.
3 PATH Associations (2011) Annual Report on World Progress in Hydrogen. Available at http://www.hpath.org/ReportBook.pdf.
4 IEA Technology Essentials (2007) Hydrogen Production & Distribution. Available at http://www.iea.org/techno/essentials5.pdf.
5 Uehara, I. (2008) Separation and purification of hydrogen, in *Energy Carriers and Conversion Systems with Emphasis on Hydrogen*, vol. **1** (eds T. Ohta and T. Nejat), UNESCO, Eolss Publisher, Paris, p. 268.
6 Rostrup-Nielsen, J.R. and Rostrup-Nielsen, T. (2002) Large-scale hydrogen production. *Cattech*, **6**, 150–159.
7 The Linde Group (2012) Hydrogen. Available at http://www.linde-engineering.com/internet.global.lindeengineering.global/en/images/H2_1_1_e_12_150dpi19_4258.pdf.
8 UOP (2010) Refinery-Wide H$_2$ Optimization for Improved Profitability. Available at http://www.uop.com.php53-7.ord1-1.websitetestlink.com/?document=uop-hydrogen-management-solutions-brochure&download=1.
9 Air Products (2013) Steam Methane Reformer Overview. Available at http://www.airproducts.com/~/media/Files/PDF/

industries/energy-hydrogen-steam-methane-reformer-datasheet.pdf.
10 Stöcker, J., Whysall, M., and Miller, G.Q. (1998) *30 Years of PSA Technology for Hydrogen Purification*, UOP LLC, Des Plaines, IL.
11 Ni, M., Leung, D.Y.C., Leung, M.K.H., and Sumathy, K. (2006) An overview of hydrogen production from biomass. *Fuel. Process. Technol.*, **87**, 461–472.
12 Biswas, P., Agrawal, S., and Sinha, S. (2010) Modeling and simulation for pressure swing adsorption system for hydrogen purification. *Chem. Biochem. Eng. Q.*, **24**, 409–414.
13 Waldron, W.E. and Sircar, S. (2000) Parametric study of a pressure swing adsorption process. *Adsorption*, **6**, 179–188.
14 Simbeck, D.R. and Chang, E. (2002) Hydrogen Supply: Cost Estimate for Hydrogen Pathways – Scoping Analysis, NREL/SR-540-32525, November 2002.
15 Ruthven, D.M. (1984) *Principles of Adsorption and Adsorption Processes*, John Wiley & Sons, Inc., New York.
16 Yang, R.T. (1986) *Gas Separation by Adsorption Processes*, Butterworths, Boston.
17 Wankat, P.C. (2007) *Separation Process Engineering*, 2nd edn, Prentice Hall, London.
18 Humphrey, J.L. and Keller, G.E. II (1997) *Separation Process Technology*, McGraw-Hill, New York.
19 Lopes, F.V.S., Grande, C.A., Ribeiro, A.M., Loureiro, J.M., and Rodrigues, A.E. (2009) Adsorption of H_2, CO_2, CH_4, CO and N_2 in activated carbon and zeolite for hydrogen production. *Sep. Sci. Technol.*, **44**, 1045–1073.
20 Lee, C.-H., Yang, J., and Ahn, H. (1999) Effects of carbon-to-zeolite ratio on layered bed H_2 PSA for coke oven gas. *AIChE J.*, **45**, 535–545.
21 Malek, A. and Farooq, S. (1998) Hydrogen purification from refinery fuel gas by pressure swing adsorption. *AIChE J.*, **44**, 1985–1992.
22 Ahn, H., Lee, C.-H., Seo, B., Yang, J., and Baek, K. (1999) Backfill cycle of a layered bed H_2 PSA process. *Adsorption*, **5**, 419–433.
23 Sircar, S. and Golden, T.C. (2000) Purification of hydrogen by pressure swing adsorption. *Sep. Sci. Technol.*, **35**, 667–687.
24 Park, J.-H., Kim, J.-N., and Cho, S.-H. (2000) Performance analysis of a four-bed H_2 PSA process using layered beds. *AIChE J.*, **46**, 790–802.
25 Yang, J. and Lee, C.-H. (1998) Adsorption dynamics of a layered bed PSA for H_2 recovery from coke oven gas. *AIChE J.*, **44**, 1325–1334.
26 Park, J.-H., Kim, J.-N., Cho, S.-H., Kim, J.-D., and Yang, R.T. (1998) Adsorber dynamics and optimal design of layered beds for multicomponent gas adsorption. *Chem. Eng. Sci.*, **53**, 3951–3963.
27 Nikolic, D., Kikkinides, E.S., and Georgiadis, M.C. (2009) Optimization of multibed pressure swing adsorption processes. *Ind. Eng. Chem. Res.*, **48**, 5388–5398.
28 Komarneni, S., Pidugu, R., and Menon, V.C. (1996) Water adsorption and desorption isotherms of silica and alumina mesoporous molecular sieves. *J. Porous Mater.*, **3**, 99–106.
29 Serbezov, A. (2003) Adsorption equilibrium of water vapor on F-200 activated alumina. *J. Chem. Eng. Data*, **48**, 421–425.
30 Ribeiro, A.M.P., Sauer, T., Grande, C.A., Moreira, R.F.P.M., and Rodrigues, A.E. (2008) Adsorption equilibrium and kinetics of water vapor on different adsorbents. *Ind. Eng. Chem. Res.*, **47**, 7019–7026.
31 Mahle, J.J. and Friday, D.K. (1989) Water adsorption equilibria on microporous carbons correlated using a modification to the Sincar isotherm. *Carbon*, **27**, 835–843.
32 Ribeiro, A.M., Grande, C.A., Lopes, F.V.S., Loureiro, J.M., and Rodrigues, A.E. (2009) Four beds pressure swing adsorption for hydrogen purification: case of humid feed and activated carbon beds. *AIChE J.*, **55**, 2292–2302.
33 Bansal, R.C., Donnet, J.-B., and Stoeckli, F. (1998) *Active Carbon*, Marcel Dekker, Inc, New York.
34 Rodríguez-Reinoso, F., Molina-Sabio, M., and González, M.T. (1995) The use of steam and CO_2 as activating agents in the

preparation of activated carbons. *Carbon*, **33**, 15–23.
35 Py, X., Guillot, A., and Cagnon, B. (2004) Nanomorphology of activated carbon porosity: geometrical models confronted to experimental facts. *Carbon*, **42**, 1743–1754.
36 Ritter, J.A. and Yang, R.T. (1987) Equilibrium adsorption of multicomponent gas mixtures at elevated pressures. *Ind. Eng. Chem. Res.*, **26**, 1679–1686.
37 Jee, J.-G., Kim, M.-B., and Lee, C.-H. (2001) Adsorption characteristics of hydrogen mixtures in a layered bed: binary, ternary and five-component mixtures. *Ind. Eng. Chem., Res.*, **40**, 868–878.
38 Grande, C.A, Lopes, F.V.S., Ribeiro, A.M., Loureiro, J.M., and Rodrigues, A.E. (2008) Adsorption of off-gases from steam methane reforming (H_2, CO_2, CH_4, CO and N_2) on activated carbon. *Sep. Sci. Technol.*, **43**, 1338–1364.
39 Siriwardane, R.V., Shen, M.-S., Fisher, E.P., Poston, J.A., and Shamsi, A. (2001) Adsorption and desorption of CO_2 on solid sorbents. Presented at the First National Conference on Carbon Sequestration, 14–17 May 2001, Washington, DC. Session 3B. Capture & Sequestration III: Adsorption Studies, NETL. Available at http://www.alrc.doe.gov/publications/proceedings/01/carbon_seq/3b3.pdf.
40 Reiss, G. (1984) Molecular sieve zeolite for producing hydrogen by pressure variation adsorption technique, U.S. Patent 4,477,267.
41 Bec, R.L. (2005) Method for purifying hydrogen-based gas mixtures using calcium X-zeolite, U.S. Patent 6,849,106.
42 Kärger, J. and Ruthven, D.M. (1992) *Diffusion in Zeolites and Other Microporous Solids*, John Wiley & Sons, Inc., New York.
43 Coe, C.G., Kirner, J.F., Pierantozzi, R., and White, T.R. (1992) Nitrogen adsorption with Ca and or Sr exchanged lithium X-zeolites, U.S. Patent 5,152,813.
44 Hutson, N.D., Rege, S.U., and Yang, R.T. (1999) Mixed cation zeolites: LixAgy-X as a superior adsorbent for air separation. *AIChE J.*, **45**, 724–734.
45 Gorbach, A., Stegmaier, M., and Eigenberger, G. (2004) Measurement and modeling of water vapor adsorption on zeolite 4A – equilibria and kinetics. *Adsorption*, **10**, 29–46.
46 Lopes, F.V.S., Grande, C.A., Ribeiro, A.M., Loureiro, J.M., and Rodrigues, A.E. (2010) Effect of ion-exchange on the adsorption of SMR off-gases on zeolite 13X. *J. Chem. Eng. Data*, **55**, 184–195.
47 Wang, Y. and LeVan, M.D. (2009) Adsorption equilibrium of carbon dioxide and water vapor on zeolites 5A and 13X and silica gel: pure components. *Ind. Eng. Chem. Res.*, **54**, 2839–2844.
48 Lu, Y., Doong, S.-J., and Bulow, M. (2004) Pressure-swing adsorption using layered adsorbent beds with different adsorption properties: II – experimental investigation. *Adsorption*, **10**, 267–275.
49 Khelifa, A., Derriche, Z., and Bengueddach, A. (1999) Sorption of carbon dioxide by zeolite X exchanged with Zn^{2+} and Cu^{2+}. *Microporous Mesoporous Mater.*, **32**, 199–209.
50 Chlendi, M. and Tondeur, D. (1995) Dynamic behavior of layered columns in pressure swing adsorption. *Gas Sep. Pur.*, **9**, 231–242.
51 Cavenati, S., Grande, C.A., and Rodrigues, A.E. (2006) Separation of $CH_4/CO_2/N_2$ mixtures by layered pressure swing adsorption for upgrade of natural gas. *Chem. Eng. Sci.*, **61**, 3893–3906.
52 Rota, R. and Wankat, P.C. (1990) Intensification of pressure swing adsorption processes. *AIChE J.*, **36**, 1299–1312.
53 Da Silva, F.A., Silva, J.A., and Rodrigues, A.E. (1999) A general package for the simulation of cyclic adsorption processes. *Adsorption*, **5**, 229–244.
54 Ruthven, D.M., Farooq, S., and Knaebel, K.S. (1994) *Pressure Swing Adsorption*, VCH Publishers, New York.
55 Skarstrom, C.W. (1960) Method and apparatus for fractionating gas mixtures by adsorption. U.S. Patent 2,944,627.
56 Marsh, W.D., Pramuk, F.S., Hoke, R.C., and Skarstrom, C.W. (1964) Pressure

equalization depressurising in heatless adsorption. U.S. Patent 3,142,547.

57 Warmuziński, K. and Tańczyk, M. (1997) Multicomponent pressure swing adsorption Part I. Modelling of large-scale PSA installations. *Chem. Eng. Proc.*, **36**, 89–99.

58 Delgado, J.A. and Rodrigues, A.E. (2008) Analysis of the boundary conditions for the simulation of the pressure equalization step in PSA cycles. *Ind. Eng. Chem. Res.*, **63**, 4452–4463.

59 Fuderer, A. (1985) Pressure swing adsorption with intermediate product recovery, U.S. Patent 4,512,780.

60 Yang, S.I., Choi, D.-Y., Jang, S.-C., Kim, S.-H., and Choi, D.-K. (2008) Hydrogen separation by multi-bed pressure swing adsorption of synthesis gas. *Adsorption*, **14**, 583–590.

61 Grande, C.A. and Rodrigues, A.E. (2001) Adsorption equilibria and kinetics of propane and propylene in silica gel. *Ind. Eng. Chem. Res.*, **40**, 1686–1693.

62 Golden, T.C. and Weist, E.L. (2003) Activated carbon as sole absorbent in rapid cycle hydrogen PSA. U.S. Patent 6,660,064.

63 Liu, J., Wang, Y., Benin, A.I., Jakubczak, P., Willis, R.R., and LeVan, M.D. (2010) CO_2/H_2O adsorption equilibrium and rates on metal-organic frameworks: HKUST-1 and Ni/DOBDC. *Langmuir*, **26**, 14301–14307.

64 Lopes, F.V.S., Grande, C.A., Ribeiro, A.M., Loureiro, J.M., and Rodrigues, A.E. (2009) Enhancing capacity of activated carbons for hydrogen purification. *Ind. Eng. Chem. Res.*, **48**, 3978–3990.

65 Ribeiro, A.M., Grande, C.A., Lopes, F.V.S., Loureiro, J.M., and Rodrigues, A.E. (2008) A parametric study of layered bed PSA for hydrogen purification. *Chem. Eng. Sci.*, **63**, 5258–5273.

66 Grande, C.A., Blom, R., Möller, A., and Möllmer, J. (2013) High-pressure separation of CH_4-CO_2 using activated carbon. *Chem. Eng. Sci.*, **89**, 10–20.

67 Schell, J., Casas, N., Blom, R., Spjelkavik, A.I., Andersen, A., Hafizovic Cavka, J., and Mazzotti, M. (2012) MCM-41, MOF and UiO-67/MCM-41 adsorbents for pre-combustion CO_2 capture by PSA: adsorption equilibria. *Adsorption*, **18**, 213–227.

68 Casas, N., Schell, J., Blom, R., and Mazzotti, M. (2013) MOF and UiO-67/MCM-41 adsorbents for pre-combustion CO_2 capture by PSA: breakthrough experiments and process design. *Sep. Purif. Technol.*, **112**, 34–48.

69 Herm, Z.R., Swisher, J.A., Smit, B., Krishna, R., and Long., J.R. (1022) Metal-organic frameworks for hydrogen purification and precombustion carbon dioxide capture. *J. Am. Chem. Soc.*, **133**, 5664–5667.

70 Moellmer, J., Lange, M., Moeller, A., Patzschke, C., Stein, K., Laessig, D., Lincke, J., Glaeser, R., Krautscheid, H., and Staudt, R. (2012) Pure and mixed gas adsorption of CH_4 and N_2 on the metal-organic framework Basolite® A100 and a novel copper-based 1,2,4-triazolylisophthalate MOF. *J. Mater. Chem.*, **22**, 10274–10286.

71 Jasra, J.R. and Walton, K.S. (2008) Effect of open metal sites on adsorption of polar and nonpolar molecules in metal-organic framework Cu-BTC *Langmuir*, **24**, 8620–8626.

72 Banu, A.-M., Friedrich, D., Brandani, S., and Düren, T. (2013) A multiscale study of MOFs as adsorbents in H_2 PSA purification. *Ind. Eng. Chem. Res.*, **52**, 9946–9957.

73 Voss, C. (2005) Applications of pressure swing adsorption technology. *Adsorption*, **11**, 527–529.

74 Von Gemmingen, U. (2011) Technical adsorption – modern processes. *Chem.-Ing.-Tech*, **83**, 36–43.

75 Kumar, R. (1994) Pressure swing adsorption process: performance optimum and adsorbent selection. *Ind. Eng. Chem. Res.*, **33**, 1600–1605.

76 Agarwal, A., Biegler, L.T., and Zitney, S.E. (2009) Simulation and optimization of pressure swing adsorption systems using reduced-order modeling. *Ind. Eng. Chem. Res.*, **48**, 2327–2343.

77 Khajuria, H. and Pistikopoulos, E.N. (2011) Dynamic modelling and explicit multi-parametric MPC control of pressure swing adsorption systems. *J. Proc. Control*, **21**, 151–163.

78 Casas, N., Schell, J., and Mazzotti, M. (2010) Pre-combustion CO_2 capture by PSA for IGCC plants. Presented at the 10th International Conference on Fundamentals of Adsorption, 23–28 May 2010, Awaji, Hyogo, Japan.

79 Whysall, M. (1984) Pressure swing adsorption process. U.S. Patent 4,482,361.

80 Zhou, L., Lü, C.-Z., Bian, S.-J., and Zhou, Y.-P. (2002) Pure hydrogen from dry gas of refineries via a novel pressure swing adsorption process. *Ind. Eng. Chem. Res.*, **41**, 5290–5297.

81 Xu, J., Rarig, D.L., Cook, T.A., Hsu, K.-K., Schoonover, M., and Agrawal, R. (2003) Pressure swing adsorption process with reduced pressure equalization time. U.S. Patent 6,565,628.

82 Baksh, M.S.A. and Simo, M. (2012) Ten bed pressure swing adsorption process operating in normal and turndown modes. U.S. Patent Appl. 2012/0174775.

83 Lopes, F.V.S., Grande, C.A., and Rodrigues, A.E. (2012) Fast-cycling VPSA for hydrogen purification. *Fuel*, **93**, 510–523.

84 Connor, D.J., Doman, D.G., Jeziorowski, L., Keefer, B.G., Larisch, B., Mclean, C., and Shaw, I. (2002) Rotary pressure swing adsorption apparatus. U.S. Patent 6,406,523.

85 Golden, T.C., Weist, E.L., Jr, and Novosat, P.A. (2008) Adsorbents for rapid cycle pressure swing adsorption processes. U.S. Patent 7,404,846.

86 Keefer, B.G., Carel, A., Sellars, B., Shaw, I., and Larisch, B. (2004) Adsorbent laminate structures. U.S. Patent 6,692,626.

87 Kopaygorodsky, E.M., Guliants, V.V., and Krantz, W.B. (2004) Predictive dynamic model of single-stage ultra-rapid pressure swing adsorption. *AIChE J.*, **50**, 953–962.

88 Todd, R.S. and Webley, P.A. (2006) Mass-transfer models for rapid pressure swing adsorption simulation. *AIChE J.*, **52**, 3126–3145.

89 Alizadeh-Khiavi, S., Sawada, J.A., Gibbs, A.C., and Alvaji, J. (2007) Rapid cycle syngas pressure swing adsorption system. U.S. Patent Application 2007/0125228.

22
Hydrogen Separation with Polymeric Membranes
Torsten Brinkmann and Sergey Shishatskiy

22.1
History

The transport of molecules through polymeric materials can be traced back to 1748 when Abbè Nolet studied permeation of water through a pig's bladder and introduced the term "osmosis" [1]. In the middle of nineteenth century Thomas Graham carried out experiments on "diffusion" of water through a bladder and formulated solution-diffusion theory [2,3]. The first patent on the application of thin rubber membranes for separation of hydrocarbons was issued for F. E. Frey in 1939 [4]. The first large-scale application of gas separation by means of membranes was carried out from 1943 to 1945 as separation of the uranium isotopes as part of the Manhattan project. Microporous metallic membranes with Knudsen [5] diffusion as the separation mechanism were used. A process of such low efficiency as the separation of ^{235}U and ^{238}U by the difference in molecular weight could be carried out only under the funding umbrella of a governmental military budget and was not realized in commercial application [6].

Only after development of the phase inversion process by Loeb and Sourirajan [7], it became possible to produce defect-free layers of polymers on top of porous supports thin enough to provide reasonable membrane permeance. The required selective membrane thickness for gas separation applications was in the range of several hundred nanometers and the probability of defect formation which would compromise membrane quality by providing viscous and not selective gas flow was high. The problem was solved by Henis and Tripodi, who developed the "resistance model" for membranes [8]. They showed that coverage of a selective layer containing defects with a highly permeable but low selective polymer would block defects and increase the selectivity of the membrane to values of the intrinsic selectivity of the polymer of the selective layer.

The first commercial membrane system implemented for gas separation was built by Monsanto in 1980 [9]. Further development of membrane separation systems have led to the application of polymeric membranes in adjustment of the synthesis gas ratio and separation of hydrogen from refinery gases as well as

Hydrogen Science and Engineering: Materials, Processes, Systems and Technology, First Edition.
Edited by Detlef Stolten and Bernd Emonts.
© 2016 Wiley-VCH Verlag GmbH & Co. KGaA. Published 2016 by Wiley-VCH Verlag GmbH & Co. KGaA.

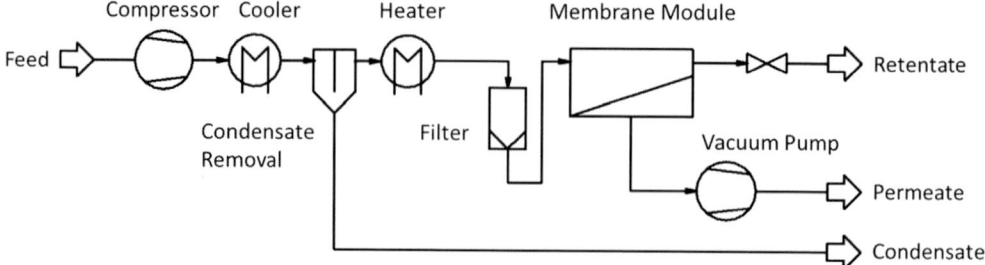

Figure 22.1 Main elements of a gas separation membrane process.

the consideration of this technology for usage in the production of renewable energies and industrial feedstocks [10–12].

22.2
Basics of Membrane Gas Separation

Gas separation by membrane technology is nowadays an established unit operation. Figure 22.1 shows the general features of a gas permeation process [13]. The driving force of the process is the difference in chemical potential, realized by the pressure difference. Either the process feed gas mixture is already pressurized or a compressor has to be employed. Some membrane materials are adversely affected by the formation of condensates. Hence, cooling the gas below its dew point and condensate removal might be required. Subsequently, the gas has to be heated up again. Even if no condensate removal is required, adjusting the temperature is often necessary since the separation performance of membrane materials is highly temperature dependent. When the gas stream contains particulates or droplets, a filter/coalescer should be employed. The apparatus in which the membrane material is installed is called the membrane module. The material retained on the high pressure side of the membrane is called the retentate, whilst the material passing through the membrane is called the permeate. Reducing the permeate pressure to levels below atmospheric pressure in order to increase the pressure ratio across the membrane and apply additional driving force can be advantageous.

Four different basic types of membranes can be employed to separate gaseous mixtures [14]:

- polymeric membranes
- ceramic membranes
- carbonous membranes
- metal alloy membranes.

Ceramic and carbonous membranes effect the separation of hydrogen due to the defined microporous structure which allows hydrogen to pass through the

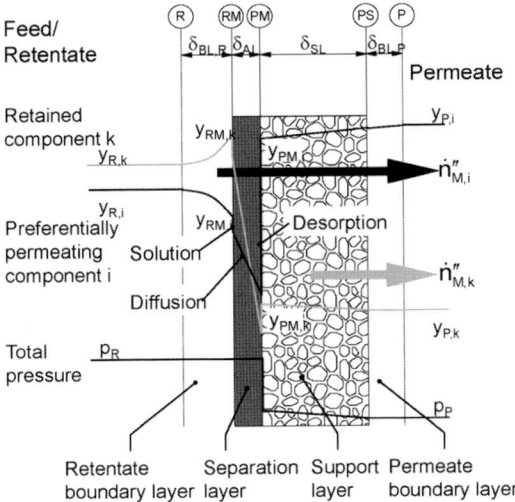

Figure 22.2 Transport resistances in gas permeation.

membranes whilst other, bulkier components are retained on the feed side. Hence the separation mechanism is due to molecular sieving [15,16].

Metal and metal alloy membranes for hydrogen separation consist of palladium or its alloys where the major alloying metal is silver. Palladium readily adsorbs hydrogen and once a driving force is applied across the membrane hydrogen will start diffusing through the metal lattice. To prevent embrittlement on temperature cycling, palladium has to be alloyed with other metals such as silver [17–19].

Polymeric membranes used for gas separation are asymmetric, that is, they consist of a thin, dense separation layer supported by a porous structure with pores gradually decreasing in diameter towards the separation layer. There are two basic types. For integral asymmetric membranes the same polymer is used for the porous support and the dense separation layer. Thin film composite membranes also consist of a porous support structure made of a glassy or rubbery semicrystalline polymer PVDF is rubbery ($Tg = -40°C$) but semicrystalline polymer. The dense separation layer is applied in an additional fabrication step on top of the porous support by a coating technique [20,21].

Gas transport through dense polymer films can be described by the solution-diffusion mechanism [6,13]. This mechanism consists of three steps (Figure 22.2, positions RM to PM):

- solution of the gas molecule into the dense polymer film;
- diffusion through the polymer matrix;
- desorption of the molecule into the gaseous state.

The mechanism requires the application of a driving force, that is, the difference in chemical potential between the two membrane sides. This can be simplified to the difference in fugacities or, when the permeating gases can be assumed to be in ideal gas state, to the difference in partial pressures [13]. The solution and desorption steps can be described by a thermodynamic equilibrium relationship. The linear, Henry's law approach can often be employed:

$$c_{M,i} = \underbrace{S_i}_{\text{sorption coefficient}} \cdot f_i = S_i \cdot \varphi_i \cdot y_i \cdot p \stackrel{\text{ideal gas}}{=} S_i \cdot y_i \cdot p \quad (22.1)$$

For higher permeant concentration the dual mode sorption model [22,23] is used for glassy polymers since it can describe adsorption in the non-relaxed free volume of the polymer structure:

$$c_{M,i} = S_i \cdot f_i + \frac{a_i \cdot k_i \cdot f_i}{1 + k_i \cdot f_i} \quad (22.2)$$

For rubbery polymers, the description according to Henry's law is sufficient for most instances. However, at high concentrations of permeants capable of initiating swelling of the polymer the employment of an appropriate model such as, for example, the Flory–Huggins model becomes necessary [24,25].

Mass transport through the polymer matrix can be described by an applicable mass transfer model for diffusion processes such as Fick's law [26]:

$$\dot{n}''_{M,i} = -c_{M,\text{tot}} \cdot D_{M,i} \cdot \frac{dy_{M,i}}{dz} \quad (22.3)$$

Although the derivation using Fick's law is common, more evolved concepts employing general thermodynamic descriptions are possible [13,26,27]. Integration of Fick's law along the thickness of the active separation layer from location RM to PM according to Figure 22.2 results in:

$$\dot{n}''_{M,i} = \frac{D_{M,i}}{\delta_{AL}} \cdot (c_{RM,i} - c_{PM,i}) \quad (22.4)$$

Assuming linear solution behavior, the concentration at the membrane interfaces can be calculated using Henry's law, resulting in:

$$\dot{n}''_{M,i} = \frac{D_{M,i} \cdot S_i}{\delta_{AL}} \cdot (f_{RM,i} - f_{PM,i}) \stackrel{\text{ideal gas}}{=} \frac{D_{M,i} \cdot S_i}{\delta_{AL}} \cdot (p_R \cdot y_{RM,i} - p_{PM} \cdot y_{PM,i}) \quad (22.5)$$

Equation (22.5) shows that the transmembrane flux of a component i depends on the solution of the permeant into the polymer matrix, the diffusion in dissolved state in the polymer, the thickness of the separation layer, and the driving force applied. In the most general case the flux is not only dependent on the permeant–polymer interaction but also on the interaction of permeating molecules of different species among each other or the interaction of molecules of one species with the polymer influencing the permeation of molecules of other species, for example, due to swelling of the polymer structure. Scenarios like this are more common for applications where rubbery membranes are employed [28,29].

Equation (22.5) also shows that the permeation mechanism can be diffusion controlled, sorption controlled, or influenced by both phenomena. It is apparent that a technical membrane should be as thin as possible to realize a high flux. The permeation in glassy polymers is often diffusion controlled, that is, the high diffusion coefficient of small molecules like hydrogen is exploited to affect separation. Conversely, for separation of condensable gases from permanent gases rubbery polymers are commonly employed, where separation is mainly due to preferred sorption of the condensable gases in the polymer. These influences can be expressed by defining a selectivity for a gas pair and a given membrane material. It is defined as:

$$\alpha_{ik} = \frac{D_{M,i} \cdot S_i}{D_{M,k} \cdot S_k} = \alpha_{i,k}^D \cdot \alpha_{i,k}^S \tag{22.6}$$

that is, the overall membrane selectivity is the product of diffusion selectivity and sorption selectivity. The product of diffusion and sorption coefficient is defined as the permeability:

$$P_i = D_{M,i} \cdot S_i \tag{22.7}$$

Dividing the permeability by the thickness of the active separation layer results in the permeance:

$$L_i = \frac{P_i}{\delta_{AL}} = \frac{D_{M,i} \cdot S_i}{\delta_{AL}} = \frac{\dot{n}''_{M,i}}{f_{RM,i} - f_{PM,i}} \tag{22.8}$$

The single gas permeance especially is easily accessible by permeation experiments. For glassy polymers the temperature dependency can often be described using an Arrhenius type relationship, that is:

$$L_i = L_i^\infty \cdot \exp\left(\frac{-E}{RT}\right) \tag{22.9}$$

Furthermore no pressure (i.e., concentration) dependency is observed for a lot of components of technical interest in glassy polymers. Should such a behavior be observed, the pressure dependency has to be accounted for along with its influence on the transport of other permeating components through the polymer [28,29].

Applying Eqs (22.5) and (22.6) together with the definition of the pressure ratio:

$$\phi = \frac{p_R}{p_{PM}} \tag{22.10}$$

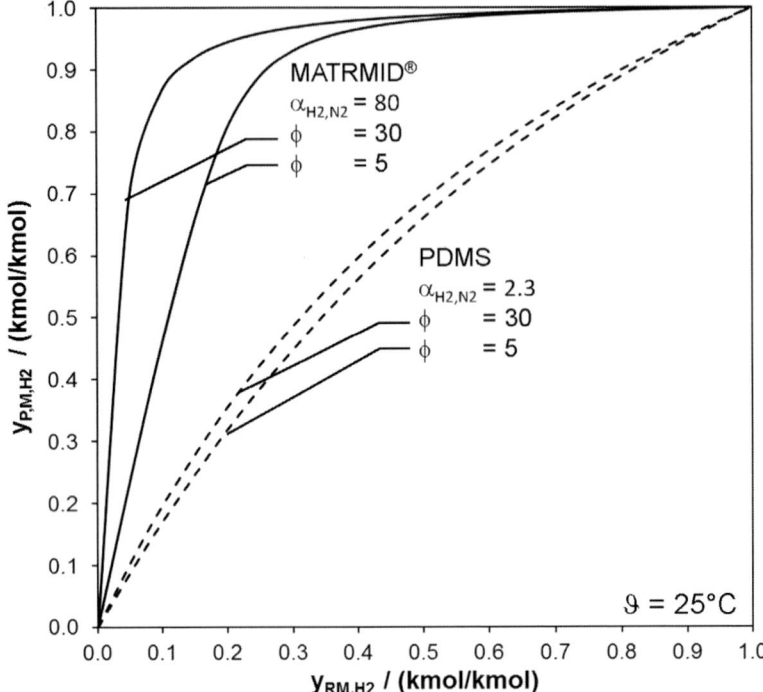

Figure 22.3 Permeate mole fraction as function of retentate mole fraction for two different membranes and pressure ratios.

to a binary mixture of ideal gases permeating through a polymer film results in an expression for the permeate composition in dependence of the feed composition:

$$y_{PM,i} = \frac{1}{2}\left[1 + \phi \cdot \left(y_{RM,i} + \frac{1}{\alpha - 1}\right)\right] \\ - \sqrt{\left[\frac{1}{2}\left[1 + \phi \cdot \left(y_{RM,i} + \frac{1}{\alpha - 1}\right)\right]\right]^2 - \frac{\alpha \cdot \phi \cdot y_{RM,i}}{\alpha - 1}} \quad (22.11)$$

Equation (22.11) allows us to evaluate the influences of the different parameters that affect the permeation process. This is illustrated in Figure 22.3 for the separation of hydrogen from nitrogen using two membrane types and two pressure ratios for each membrane. It is apparent that a highly selective membrane such as, for example, made from the polyimide Matrimid® 5218 (Huntsman Advanced Materials, http://www.huntsman.com/advanced_materials/a/Home, retrieved February 11, 2014) allows for higher hydrogen permeate compositions than those of limited selectivity, for example, employing polydimethylsiloxane (PDMS). However, next to selecting the correct membrane material, also designing the process such that the process conditions allow for the exploitation of the membrane material's selectivity is important. This is illustrated by the different pressure ratios: a low pressure ratio will not allow the selectivity of the material to be realized in the process, especially at low feed concentrations of the component to be enriched on the permeate side.

Along with the desired, selective mass transfer resistance of the active separation layer, the other layers making up the technical membrane as well as fluid dynamical boundary layers exert mass transfer resistances detrimental to separation. These are also depicted in Figure 22.2. These effects are:

- the concentration boundary layer between the feed side bulk and the feed side membrane surface, that is, concentration polarization;
- the pressure drop in the porous support layer of the membrane;
- the occurrence of concentration gradients in the porous support layer;
- analogously to the feed side, the concentration boundary layer on the permeate side.

In gas permeation, typically the feed side concentration polarization and the pressure drop in the porous support can be important. An additional phenomenon that can influence the operating performance significantly is the Joule–Thomson effect. This effect is due to the isenthalpic throttling of permeate on passing the membrane. It is especially pronounced for real, condensable gases and not that strong for hydrogen [13,30]. To describe the permeation correctly, real gas behavior should also be considered in the driving force term of Eq. (22.8), that is, fugacities should be employed. These can be easily calculated employing equations of state [31]. Mathematical treatise of these effects can be found elsewhere [13,28,30,32]. The described effects result in a system of equations that can be employed to calculate the material and heat transfer through a membrane, provided the feed and permeate side of the membrane are ideally mixed, that is, feed composition, pressure, and temperature are constant and the permeate side pressure is known. However, membranes are installed in separation apparatuses, that is, membrane modules. In these modules concentration, pressure, and temperature profiles will build up from feed to retentate side along the membrane surface. Analogously, profiles will build up on the permeate side as well. However, often an unhindered permeate withdrawal can be assumed, meaning that the local permeate can be assumed to be withdrawn orthogonally from the membrane backside and subsequently ideally mixed to form the permeate stream of the module. Figure 22.4 illustrates the concept of unhindered permeate withdrawal [13,30]. The composition and velocity profiles shown result from the solution of the differential material balance:

$$\frac{d\dot{n}_{R,i}}{dA} + \dot{n}''_{M,i} = 0; \ i = 1, \ldots, nc \qquad (22.12)$$

The transmembrane flux $\dot{n}''_{M,i}$ is calculated according to Eq. (22.8) with the additional consideration of the influence of the concentration boundary and support layers, if required. The boundary conditions are defined by the feed and the

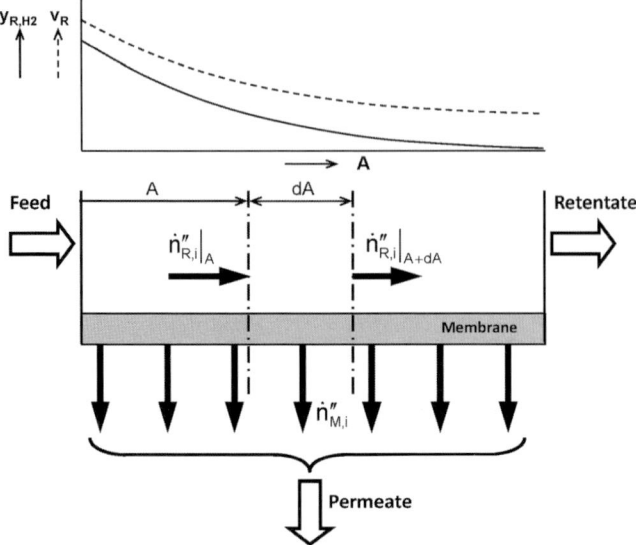

Figure 22.4 Concept of a model for membrane modules. Unhindered permeate withdrawal is assumed.

applied permeate pressure:

$$\dot{n}_{R,i}\big|_{A=0} = \dot{n}_F \cdot y_{F,i} \tag{22.13}$$

This is the simplest model allowing for the calculation of distributed variables along a membrane surface. The determination of the velocity requires additional thermodynamical calculations to estimate the density and information on the cross-sectional area. If temperature and pressure profiles are to be determined, energy and momentum balances have to be solved, respectively. More complex flow patterns, for example, consideration of evolving variable profiles on the permeate side require the setup of additional sets of equations depending on the geometry of the membrane module. References [30,32–37] give information on how such models are developed and solved. The models can be implemented in commercial, equation oriented process simulators such as Aspen Custom Modeler (AspenTech www.aspentech.com/ (accessed 11 February 2014) and gProms (Process Systems Enterprise http://www.psenterprise.com/gproms.html, retrieved 12 November 2015) or using programming languages such as Fortran, C, C++, Java or Visual Basic for Applications. Interfacing with steady state process simulation tools is generally possible and hence the performance of a gas separation membrane stages can be investigated as part of the overall process.

22.3
Hydrogen Separation and Fractionation by Gas Permeation

Hydrogen is employed in numerous production and energy generation processes and forms one of the most important reactants in the chemical and petrochemical industries. Within the changing raw material and energy supply scenario caused by ever more depleting sources of fossil fuels hydrogen is expected to become even more important. On the one hand, it can be used directly for generating energy. On the other hand, it can be employed as a reactant in conversions involving carbon dioxide as another reactant for the production of intermediates used in the chemical industry. Converting carbon dioxide would not be possible without using a reactant with a high energy level, such as, for example, hydrogen. Processes involving hydrogen separation by membrane technology will be outlined in the following section. Especially, the established, large-scale industrial applications are part of the product portfolio of internationally operating companies such as Air Products (http://www.airproducts.com/products/Gases/supply-options/prism-membranes/prism-membrane-products.aspx, retrieved 12 November 2015), Air Liquide (http://www.medal.airliquide.com/en/hydrogen-membrane-gas-separation.html, retrieved 12 November 2015), Generon (http://igs-global.com/products/, retrieved 12 November 2015), Honeywell-UOP (http://www.uop.com/equipment/hydrogen-separation/#membrane-systems, retrieved 12 November 2015), Ube (https://www.ube.com/content.php?pageid=16, retrieved 12 November 2015) Evonik Industries (http://www.sepuran.com/product/sepuran/en/Pages/default.aspx, retrieved 12 November 2015), and Borsig Membrane Technology (Borsig Membrane Technology GmbH, http://mt.borsig.de/en/home.html, retrieved 12 November 2015).

22.3.1
Hydrogen/Nitrogen Separation

The application of gas permeation in ammonia production was among its first industrial applications. The synthesis gas used in ammonia production is a nitrogen/hydrogen mixture in the ratio of 1:3. The reaction is equilibrium restricted. Hence it is important to recover hydrogen from the product gas and recycle it to feed. Furthermore, additional components such as methane and argon are present and should not be recycled to the reactor to prevent a build-up of their concentrations, that is, they should be separated from the product gas as far as possible (http://www.airproducts.com/products/Gases/supply-options/prism-membranes/prism-membrane-products.aspx, retrieved 12 November 2015, http://www.medal.airliquide.com/en/hydrogen-membrane-gas-separation.html, retrieved 12 November 2015, http://www.uop.com/equipment/hydrogen-separation/#membrane-systems, retrieved 12 November 2015, https://www.ube.com/content.php?pageid=16, retrieved 12 November 2015, http://www.sepuran.com/product/sepuran/en/Pages/default.aspx, retrieved 12 November 2015, Borsig Membrane Technology GmbH, http://mt.borsig.de/en/home.html, retrieved 12 November 2015) [38]).

22.3.2
Hydrogen/Carbon Monoxide Separation

A similar application is the recovery and separation of carbon monoxide and hydrogen from the purge gas of methanol production processes. In addition, the adjustment of the composition of the hydrogen/carbon monoxide synthesis gas in order to meet the stoichiometric requirements of oxo-alcohol synthesis is addressed by membrane technology. Membrane systems are also employed to purify and dry carbon monoxide streams (http://www.airproducts.com/products/Gases/supply-options/prism-membranes/prism-membrane-products.aspx, retrieved 12 November 2015, http://www.medal.airliquide.com/en/hydrogen-membrane-gas-separation.html, retrieved 12 November 2015, http://www.sepuran.com/product/sepuran/en/Pages/default.aspx, retrieved 12 November 2015, Borsig Membrane Technology GmbH, http://mt.borsig.de/en/home.html, retrieved 12 November 2015)).

22.3.3
Hydrogen/Hydrocarbon Separation

In refineries hydrogen is employed for numerous purposes (http://www.airproducts.com/products/Gases/supply-options/prism-membranes/prism-membrane-products.aspx, retrieved 12 November 2015, http://www.medal.airliquide.com/en/hydrogen-membrane-gas-separation.html, retrieved 12 November 2015, https://www.ube.com/content.php?pageid=16, retrieved 12 November 2015, http://www.sepuran.com/product/sepuran/en/Pages/default.aspx, retrieved 12 November 2015, Borsig Membrane Technology GmbH, http://mt.borsig.de/en/home.html, retrieved 12 November 2015, http://igs-global.com/products/, retrieved 12 November 2015). Examples are hydrocracker and hydrotreater product gases as well as fluidized catalytic cracker and catalytic reformer off gases. Hence several separations by gas permeation are possible in this industry segment. The process streams consist of light hydrocarbons and hydrogen at different concentration levels. The resulting hydrogen streams typically have purities in the range 95–99.9 vol.% (http://www.airproducts.com/products/Gases/supply-options/prism-membranes/prism-membrane-products.aspx, retrieved 12 November 2015, http://www.medal.airliquide.com/en/hydrogen-membrane-gas-separation.html, retrieved 12 November 2015). In cases were higher purities are required, pressure swing adsorption (PSA) processes are employed (http://www.airproducts.com/products/Gases/supply-options/prism-membranes/prism-membrane-products.aspx, retrieved 12 November 2015, http://igs-global.com/products/, retrieved 12 November 2015). However, these processes can also benefit from membrane gas separation since this unit operation can be employed to treat the purge gas of the PSA and recover hydrogen otherwise used as boiler fuel [6] (http://www.airproducts.com/products/Gases/supply-options/prism-membranes/prism-membrane-products.aspx, retrieved 12 November 2015). Baker [6] also describes a process where hydrogen is separated from the light hydrocarbons produced by cracking prior to cold train distillation. This process allows for significantly increased temperature of the demethanizer column of the cold train and hence reduces investment and operating costs.

22.3.4
Hydrogen Separation and Fractionation Applications in Renewable Hydrogen Production

Renewable energy technology and advanced energy generation processes require separation technologies for hydrogen. Examples are the treatment of product streams from biomass gasification, separation of hydrogen from Fischer–Tropsch reaction product gases, and separation of hydrogen from electrolysis gases [39–41]. In addition, advanced power plant concepts such as integrated gasification combined cycle power plants employing coal gasification with subsequent water–gas shift reaction to produce hydrogen allow for the use of hydrogen selective membranes to recover hydrogen [41,42]. Along with hydrogen selective membranes, membranes preferentially allowing higher hydrocarbons and carbon dioxide to permeate can also be beneficial in these processes since they allow hydrogen to remain on the high pressure, feed side of the membrane module and hence do not require recompression of the hydrogen stream. An example is the adjustment of reactant ratio for Fischer–Tropsch synthesis feed gases [40].

22.4
Membrane Materials and Modules

22.4.1
Membrane Classification

Hydrogen separation with polymeric membranes is based on the interaction of gas molecules penetrating through the bulk of polymer and the polymer itself [43,44].

Hydrogen is the second smallest molecule and has a weak affinity to all known polymers resulting in high diffusion and low solubility coefficients [45]. Membranes for gas separation utilize the ability of polymers to discriminate molecules permeating through the bulk of the material by size exclusion or solubility exclusion mechanisms. Size exclusion or diffusion selectivity is dominant in the case of glassy polymers where segmental movement of the macromolecular chain is restricted and only movement in the local range close to the size of the monomeric unit is allowed. Solubility selectivity, determined by the affinity of the gas to polymer, is realized in rubbery polymers where large ensembles of monomeric units called Kuhn segments are allowed to move [28]. This opens up the possibility for penetrants as small as molecules of gases or vapors to move through the large openings in the bulk of polymers caused by segmental mobility. The mobility of the macromolecular chains and the glassy or rubbery state of the polymer is determined by the temperature of experiment or application. The temperature of transition between the glassy and the rubbery state is called glass transition temperature (T_g) [46]. This parameter is crucial for polymers for membrane application because below T_g the polymer is much more rigid than above T_g and can form self-supporting structures able to withstand significant mechanical stress. At the same time the state of polymer determines the quantity and quality of the "free volume" [47], or the quantity and organization of voids within the polymer bulk caused by imperfections in polymer chain packing and mobility of polymeric chains.

Mechanical and thermal properties of the polymer predetermine the structure and geometry of the gas separation membranes, which can be classified according to the cross-sectional morphology or geometry [21].

22.4.1.1 Morphological Classification

The main task of membrane development is achieving the formation of a defect free thin layer of polymer since permeance of the membrane linearly depends on the thickness of the selective layer (Eq. (22.8)). The absence of defects on the surface of the membrane prevents the occurrence of viscous Poiseuille or Knudsen flow though the membrane. These flow regimes would cause a low selectivity. The size of the macromolecular coil in a solid, solvent free polymer bulk can be estimated from molecular weight and density values. For polymers having a weight average molecular weight $M_w = 100\,\text{kg}\,\text{mol}^{-1}$ and a density of $1\,\text{g}\,\text{cm}^{-3}$, the characteristic size of the polymer molecule, considered as a sphere, is 6.8 nm. This value determines the minimal possible thickness of the polymer layer in the case it will be built as monolayer. In practice the thinnest selective layers achieved to date are in the range 30–40 nm [48] and the usual thickness is 70–90 nm [49] meaning that about ten layers of macromolecules are forming a continuous layer on top of the membrane.

Layers as thin as 30 or 90 nm cannot be handled without a porous supporting structure providing the mechanical strength to the membrane and drainage of the substances permeating through the selective layer. Gas separation membranes are divided to integral asymmetric (Figure 22.5) and thin film composite (Figure 22.6) according to the nature of their support structure [9,50]. It has to be mentioned that morphological classification deals with the selective layer and support responsible for this layer's stability. Additional layers of the membranes are not considered as vital for the classification.

Integral asymmetric membranes use the same polymeric material for the selective layer and the porous support structure. They are usually produced by the phase inversion process according to Loeb and Sourirajan [7], where a polymer solution in the a form of a flat sheet or a hollow fiber is immersed into a non-solvent. In contact with the non-solvent the polymer solution loses stability and separates into two phases: polymer rich and polymer depleted. The polymer rich phase forms the pore walls of the support. To obtain the selective layer, a portion of the volatile solvent is evaporated before the polymer solution is immersed into the non-solvent [51]. The partial evaporation of the solvent causes a significant increase of the polymer concentration in the layer being in contact with air and induces continuous defect free selective layer formation. The obtained asymmetric structure is subjected to the solvent exchange procedures in order to extract any residual solvent which can cause plasticization of pores walls and hence cause the collapse of the porous structure.

Figure 22.5 Cross-section of the supported asymmetric gas separation membrane: (a) overview with non-woven fabric; (b) morphology of porous structure; (c) magnified region of gas tight selective layer with porous substructure. Membrane is coated with PDMS for defect curing.

Integral asymmetric membranes can contain several layers of support, as in the case of double layer hollow fiber or non-woven supported flat-sheet membranes. The selective layer can be coated with an additional defect blocking layer [52].

According to the morphology of the integral asymmetric membranes one can conclude that the polymer of the selective layer and the porous substructure should be selective and permeable toward the gases to be separated and also possess a high mechanical stability. Hence only glassy polymers can be employed for membrane formation of integral asymmetric membranes.

The class of thin film composite membranes opens up the possibility of using glassy and rubbery polymers as materials of the selective layer [16,53]. The TFC membrane usually consists of several layers, of which two are important to determine its position in this classification. The porous substructure can be prepared from virtually any material (including inorganic) that would provide mechanical support for the selective layer and efficient drainage of the permeating molecules. The selective layer is responsible only for the molecular transport through its bulk and can be formed of virtually any polymer or combination of materials able to form continuous defect free film on top of the porous support [54].

As discussed above, the thickness of the selective layer is crucial for membrane separation process efficiency. It influences directly investment costs (required

Figure 22.6 Cross-section (a) and surface porosity (b) of polyacrylonitrile asymmetric porous membrane and Polyactive™ TFC membrane on PAN porous support (c). The selective layer of TFC membrane consists of two layers: PDMS gutter layer and Polyactive™. Effective thickness of Polyactive™ layer, calculated using resistance model is 120 nm of total selective layer thickness 235 nm.

membrane surface) and has to be as thin as possible. Hence the requirements are high for the quality of the porous support with respect to pore size distribution, surface porosity and roughness, affinity to the material of the selective layer, and solvents used for the layer formation.

The pore size distribution and surface porosity of the support are often in relation to each other: the smaller the average pore size, the smaller is the porosity of the support. For example, the HZG developed polyacrylonitrile porous support widely used for TFC membrane formation has an average pore size of 20 nm and a surface porosity of 15.5%, meaning that the rest (84.5%) of the membrane surface is barely available for the gas transport [55].

To overcome this porosity limitation the concept of a multilayer TFC membrane was developed. The multilayer TFC consists of several layers, each having characteristic thickness in the range of 100 nm. Starting from the feed side of the membrane layers are: (i) protective layer of fast and non-selective polymer; (ii) selective layer of target polymer; (iii) "gutter" layer of fast, non- or low-selective, adhesive polymer (this layer usually is the thickest of all and provides lateral distribution of penetrant); (iv) porous support; and (v) non-woven fabric providing mechanical strength to the whole structure.

Some additional layers can be introduced, for example, a compatibilizer between selective and gutter layers.

Non-woven fabric is widely used for membrane formation due to its good and uniform mechanical properties, high surface quality, and possibility to process the membrane in further manufacturing steps.

As a conclusion for this classification it can be stated that two types of membranes are present in industrial separation processes. Each of them has benefits and drawbacks. Integral asymmetric membranes can be and usually are manufactured in continuous technological processes while TFC membranes are generally subject to a discontinuous process where different treatment steps are not necessarily following each other immediately.

22.4.1.2 Geometrical Classification

Two major problems are to be solved during the development of a membrane and the separation process based on it: (i) the highest possible packing density, (m^2 of membrane)/(m^3 of module), of the membrane in module to reduce the volume of the module housing, often working at high pressures, and (ii) efficient handling of streams within the module (feed to retentate, permeate, sweep gas). Inefficient membrane packing causes an increased number of modules required for the separation process and thus leads to higher investment costs. Ill designed fluid flow distribution, undesired pressure drops, and inefficient membrane surface usage leads to reduced process selectivity or inability to realize the potential of the membrane in a technological process and thus increases exploitation costs [56].

Two major shapes of membranes have been developed and are in competition with each other [9,57].

Hollow fiber membranes are spun from polymer solutions and have a total diameter down to 100 μm and less, selective layer thickness down to 30 nm, and wall thickness of app. 20 μm. All industrially produced hollow fiber membranes for hydrogen separation are Loeb–Sourirajan integral asymmetric membranes, meaning that the selective layer and supporting porous structure are obtained in one technological step and are formed of the same polymer (http://igs-global.com/products/, retrieved 12 November 2015) [58,59].

Recent development in fiber spinning allows for the formation of two layers of two different polymers: one for the porous substructure and one for the selective layer itself with a thin porous layer contacting the first porous layer. Such an arrangement is believed to reduce consumption of the polymer of the selective layer and to improve mechanical properties [60].

The process of fiber spinning is continuous and can be done using multiple spinnerets at one time. After several treatment stages ensuring complete solvent removal the fibers are collected in bundles and assembled in membrane modules of various configurations.

The selective layer can be formed on both sides of the hollow fiber and can be exposed to the pressure of the feed stream from bore or shell side of the fiber.

Hollow fibers can be coated with a thin defect curing and protection layer which significantly increases the selectivity of the membrane as was shown by Henis and

Tripodi and realized in Monsanto's and later Air Products' PRISM membranes [8] (http://www.airproducts.com/products/Gases/supply-options/prism-membranes/prism-membrane-products.aspx, retrieved 12 November 2015).

Flat sheet membranes can be formed as integral asymmetric membranes with or without additional support or as thin film composite membranes [50].

An important feature of flat sheet membranes is the possibility of membrane quality control during membrane production at each technological step from phase inversion to module formation.

22.4.2
Membrane Defect Curing

Formation of 100 nm thin selective layers on a scale of thousands of square meters is not possible without defects. Any defect representing a hole in the membrane surface, often having size of a few nanometers, causes non-selective or nearly non-selective flow of the feed gas stream through the membrane. Gas flows according to Knudsen or Poiseuille mechanisms are much faster than solution-diffusion flow and severely deteriorate membrane selectivity.

To solve the problem of non-selective defects the resistance model of membranes was developed [8,59]. The model considers different parts of the membrane as resistances for the gas flow and presents the membrane in analogy to an electrical circuit. It was shown that the application of highly permeable and low selective layers of polymer on top of the membrane's selective layer blocks viscous flow defects converting them into slow solution-diffusion mechanism membranes with low selectivity. Since the fraction of the membrane covered by the desired selective layer is much bigger than the fraction of defects plugged with non-selective polymer, the total selectivity of the membrane approaches the value of the intrinsic properties of the polymer of the selective layer (Figure 22.7):

$$\frac{1}{L} = R = R_1 + \frac{R_2 \cdot R_3}{R_2 + R_3} + R_4 \tag{22.14}$$

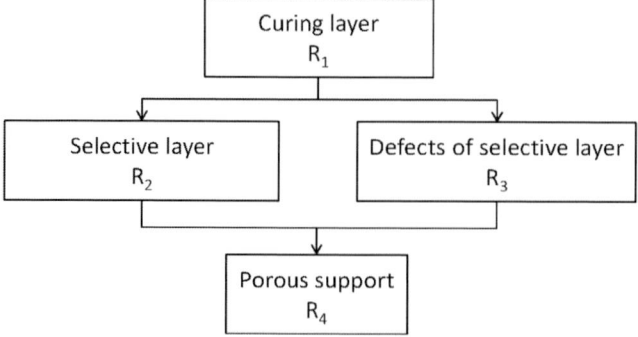

Figure 22.7 Resistance model of TFC membrane with defect curing layer.

22.4.3
Membrane Module Classification

Membrane module types have to be designed according to the geometry of the membrane, the separation characteristics of the membrane, and the operating conditions of the process where the modules are to be applied. The aim is to transfer the intrinsic membrane material's properties into an application with minimized losses, for example, due to pressure drop and concentration polarization, and a maximized usage of the available driving force and membrane area.

22.4.3.1 Hollow Fiber Modules

In most industrial applications for hydrogen separation the hollow fiber module types are employed (http://www.airproducts.com/products/Gases/supply-options/prism-membranes/prism-membrane-products.aspx, retrieved 12 November 2015, http://www.medal.airliquide.com/en/hydrogen-membrane-gas-separation.html, retrieved 12 November 2015, http://www.uop.com/equipment/hydrogen-separation/#membrane-systems, retrieved 12 November 2015, https://www.ube.com/content.php?pageid=16, retrieved 12 November 2015, http://www.sepuran.com/product/sepuran/en/Pages/default.aspx, retrieved 12 November 2015, Borsig Membrane Technology GmbH, http://mt.borsig.de/en/home.html, retrieved 12 November 2015, http://igs-global.com/products/, retrieved 12 November 2015). Hollow fibers having an external diameter of 80–800 μm and an internal diameter of 40–500 μm [6,13] are assembled as bundles, the ends of which are fixed in a polymeric resin [6,13]. These pottings are mounted into a cylindrical pressure vessel. Due to the high pressures usually applied, the feed is supplied to the outside of the fiber bundle with permeate drawn from the bore side. An exception is the Evonik Industries design: the information currently available implies a bore side feed with both ends of the fiber left open (http://www.sepuran.com/product/sepuran/en/Pages/default.aspx, retrieved 12 November 2015). However, the typical configuration is that the bore side is sealed at one end and open at the other, that is, permeate is withdrawn from only on side of the fiber bundle. This geometry allows for a counter-current flow pattern which in theory allows for the best application of the available driving force [13,32,61]. Figure 22.8 shows the flow patterns schematically. Furthermore, high packing densities of up to $10\,000\,m^2$ membrane area per m^3 module volume can be realized [13]. However, it can be expected that the shell side flow is not sufficiently evenly distributed to allow the assumption of an ideal plug flow in parallel to the membrane surface. The high packing density can be expected to cause splitting and recombination of the flow as well as convective transport orthogonal to the desired flow direction. Hence some degree of axial dispersion is probable for hollow fiber type membrane modules. The feed is either supplied in a longitudinal configuration as, for example, in the Air Production PRISM and in Ube and Evonik Industries Sepuran® designs (http://www.airproducts.com/products/Gases/supply-options/prism-membranes/prism-membrane-products.aspx, retrieved 12 November 2015, https://www.ube.com/

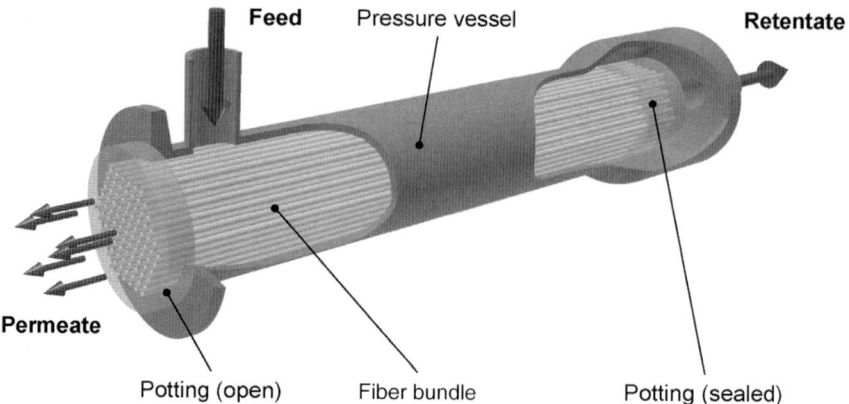

Figure 22.8 Schematic representation of a hollow fiber module.

content.php?pageid=16, retrieved 12 November 2015, http://www.sepuran.com/product/sepuran/en/Pages/default.aspx, retrieved 12 November 2015) or radially as, for example, in Air Liquide's Medal or UOPs Polysep modules (http://www.medal.airliquide.com/en/hydrogen-membrane-gas-separation.html, retrieved 12 November 2015, http://www.uop.com/equipment/hydrogen-separation/#membrane-systems, retrieved 12 November 2015).

22.4.3.2 Flat Sheet Modules

Although most hydrogen selective membranes are produced in a hollow fiber configuration, some developments such as the HZG PEI and Matrimid® membranes have flat sheet geometries [51]. UOP also offers a version of their Polysep membranes as flat sheets (Cnop, T., Dortmundt, D., Schott, M. (2007). *UOP Continued Development of Gas Separation Membranes*. Retrieved February 11, 2014, from http://www.uop.com). Furthermore are membranes selective for carbon dioxide or hydrocarbons, that is, selective layers made of poly(ethylene oxide) containing block copolymers or silicon polymers, respectively, are generally of flat sheet geometry [53]. The most common module type for these membranes is the spiral wound module, allowing feed and permeate to flow perpendicularly and hence realizing a cross flow pattern (Figure 22.9) [6,13,32]. To manufacture a spiral wound element, two rectangular sheets of membrane material are glued together at three edges with the active separation layers on the outside and spacer material on the inside of the thus produced envelope. The open edges of several envelopes are attached to a perforated permeate collection pipe. The membrane envelopes are wrapped around the permeate pipe with layers of spacer material separating the active layers. Spiral wound elements used for gas separation are produced that are up to 203 mm in diameter and up to 1016 mm long, containing up to 40 m² of membrane area [9].

Another flat sheet membrane module is the envelope type module (Figure 22.10). This module is also successfully applied to the separation of, for example, hydrocarbons or carbon dioxide from gas streams [28,30,62] (*BORSIG Membrane Technology GmbH – Innovative Membrane Processes*. Retrieved February 11, 2014, from

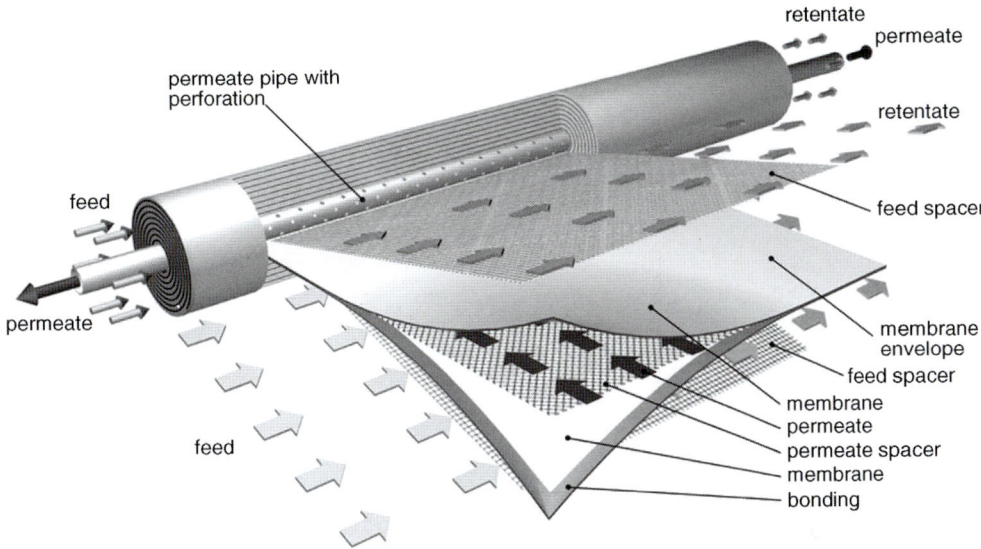

Figure 22.9 Spiral wound membrane element.

http://mt.borsig.de/en/home.html), (Sterlingsihi Advanced Materials. (n.d.) Retrieved February 11, 2014, from http://www.sterlingsihi.com/cms/en/home/products-services/engineeredsystems/membrane-systems.html). The membrane material is configured as envelopes which are thermally welded at the outer circumference with the active separation layer facing to the outside. A spacer layer on the inside allows for a radial withdrawal of permeate towards a hole in the center. These holes allow the envelopes to be stacked on a perforated permeate pipe which is mounted into a pressure vessel. The module can be segmented internally by means of baffle plates, hence it allows for the reduction of the cross sectional area on the pathway from feed to retentate. Thus the flow velocity can be kept nearly constant although part of the feed flow is withdrawn as permeate. This configuration hence features a minimized influence of concentration polarization and short permeate pathways resulting in low permeate side pressure drops.

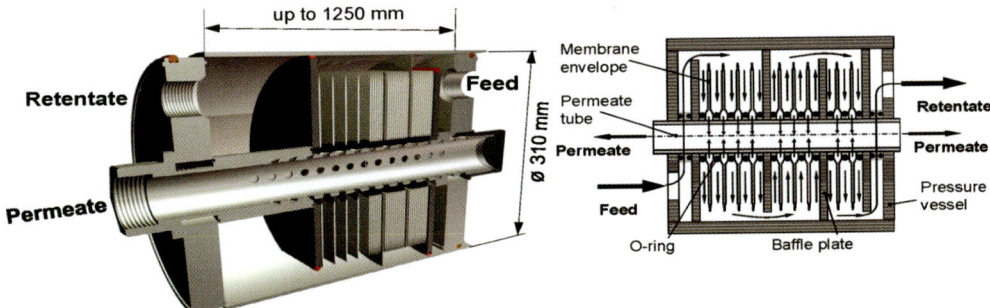

Figure 22.10 Envelope type membrane module.

Although only packing densities up to $1000\,m^2\,m^{-3}$ can be realized using flat sheet membrane modules, such modules are superior in utilization of available membrane area compared to hollow fiber modules due to the more advantageous flow distribution. Furthermore do they allow for straightforward application of modern multilayer composite membranes with high fluxes.

22.4.4
Membrane Material Classification

22.4.4.1 Membranes Based on Rubbery Polymers

As discussed above, in rubbery polymers diffusion of penetrating molecules is much less restricted then in glassy polymers and the molecular sieving effect of free volume voids can hardly be used for separation. The permeability selectivity of rubbers depends on the properties of the gas or vapor itself, namely on the critical point and on the affinity of the gas towards the polymer (Figure 22.11).

Poly(ethylene glycol) (PEG) based polymers offer high CO_2/H_2 selectivity and can be processed into TFC membranes [63–66]. Examples of PEG containing polymers studied for gas separation and implemented as selective materials of membranes are Pebax® 1657 (Archema) and Polyactive™ (PolyVation). The structure of the Polyactive™ 1500 is shown in Figure 22.12. The polymer consists of blocks of PEG having an optimum for gas separation application molecular weight of $1500\,g\,mol^{-1}$, and a hard polybutylene terephthalate block responsible for the

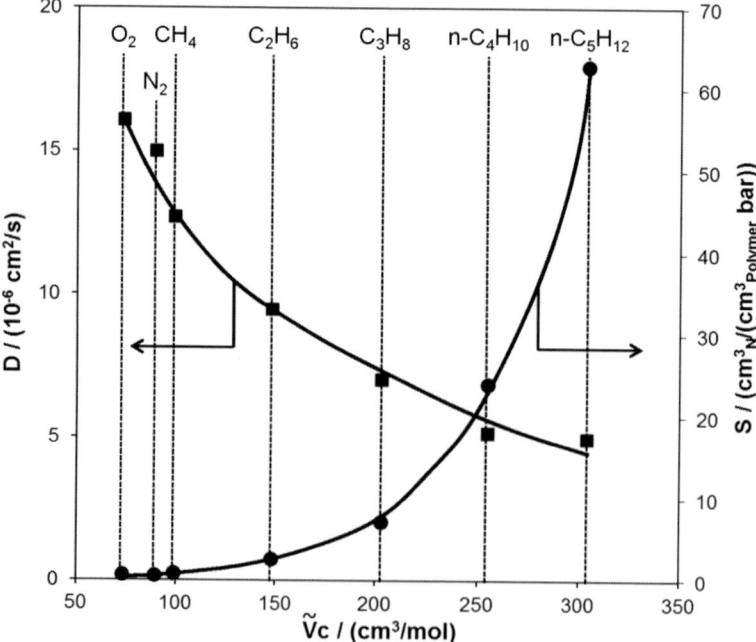

Figure 22.11 Diffusion coefficients and solubilities of different gases in silicon rubber [28].

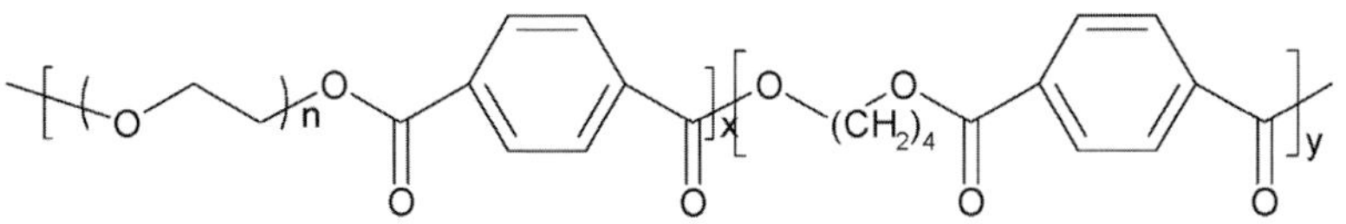

Figure 22.12 Chemical structure of Polyactive™ 1500 (Polyactive™1500PEGT77PBT23) (PolyVation, USA).

mechanical stability of the polymer. Gas transport properties of the Polyactive™ (Table 22.1) show two possible applications: separation of CO_2/N_2 and CO_2/H_2. Membranes based on PEG containing polymers are under pilot tests for the separation of CO_2 from hydrogen and nitrogen [67,68]. Another rubbery polymer widely used in membranes for gas separation is polydimethylsiloxane

Table 22.1 Gas properties important for transport in polymers in comparison with gas transport properties of selected polymers.

	Properties of gases influencing transport in polymers						
	H_2	CO_2	N_2	CH_4	H_2/CO_2	H_2/N_2	H_2/CH_4
T_c (K)[a]	33	304	126	191	0.11	0.26	0.17
d_k (nm)[b]	0.289	0.33	0.364	0.38	0.88	0.79	0.76

		Gas transport properties of polymers						
		H_2	CO_2	N_2	CH_4	H_2/CO_2	H_2/N_2	H_2/CH_4
Polymer	T_g (K)[c]	Permability (Barrer)[d]				Permeability selectivity		
Matrimid	588	24	8.9	0.25	0.22	2.70	96	109
Polyactive 1500	224	18	181	3.39	10	0.099	5.3	1.8
PDMS	144	820	3500	390	1100	0.23	2.1	0.75
		Diffusion × 10^7 (cm^2 s^{-1})				Diffusion selectivity		
Matrimid		11	0.029	0.024	0.0057	379	458	1930
Polyactive 1500		70	9.1	15.5	2.7	7.7	4.5	26
PDMS		650	158	180	140	4.1	3.6	4.6
		Solubility × 10^3 (cm^3(STP) (cm^3 cmHg)$^{-1}$)				Solubility selectivity		
Matrimid		2.2	310	11	39	0.0071	0.20	0.056
Polyactive 1500		0.25	20	0.22	3.7	0.013	1.1	0.068
PDMS		1.3	22	2.2	7.8	0.059	0.59	0.17

a) Critical temperature.
b) Kinetic diameter [43].
c) Glass transition temperature.
d) Permeability coefficient in Barrer (10^{-10} cm^3(STP) cm (cm^2 s cmHg)$^{-1}$).

Table 22.2 Permeability coefficients and permeability selectivity of polymer used in industrial membrane gas separation. Data compilation from Reference [11].

Polymer	Permeability (Barrer)				Selectivity		
	H_2	CO_2	N_2	CH_4	H_2/CO_2	H_2/N_2	H_2/CH_4
Cellulose acetate	12	4.8	0.15	0.15	2.5	80	80
Polysulfone	14	5.6	0.25	0.25	2.5	56	56
Poly(phenylene oxide)[a]	80	27	2.6	2.7	3	31	30
Aramid	24,5	—	—	0.1	—	—	245

a) Data for poly(2,6-dimethylphenylene oxide) from Reference [72].

(PDMS) [69–71]. The selectivity of the PDMS for permanent gas pairs is relatively low but it is accompanied by high permeability coefficients which provoke its use as a "gutter" layer or as a pin-hole defects blocking agent.

22.4.4.2 Glassy Polymer Based Membranes

Gas separation membranes based on glassy polymers where the first and are the most widely used in hydrogen separation systems. The usage of polymers in the glassy state predetermines separation performance based on diffusion selectivity.

Industrial membrane gas separation is based on polymers presented in Table 22.2. As it can be seen the permeability coefficient of hydrogen for all of these polymers is relatively low and the combination of the permeability and selectivity place them relatively far from the "upper bound" of the Robeson plot [73]. This observation reveals the importance of the various polymer properties on their success in the membrane separation market: it is important to have good gas transport properties but it is also important to be available on the market, have stable molecular characteristics independent of batch, and be processable according to the standard membrane formation routines.

Several new polymer classes such as polyacetylenes [74], polymers of intrinsic microporosity (PIM) [75], or thermally rearranged polymers [76,77,78] have been discovered during recent decades. Some of the polymers demonstrate outstanding gas transport properties [77] and most probably will find their way into practical membrane separation processes.

Among all polymers tested for gas separation properties, polyimides are the most rapidly growing family [79]. Synthesis of polyimides is carried out by a polycondensation reaction between and dianhydride and diamine (in rare cases diisocyanate) mixed in equimolar quantity (Figure 22.13). In some cases additional monomers can be added to alter the composition of the main chain. A broad variety of monomers and relative simplicity of the synthesis has allowed the accumulation of data on properties of more than a thousand polyimides. Some of them have an outstanding combination of gas transport properties and are close to the upper bound of the Robeson plot [73] setting a mark for further membrane related polymer synthesis. In the last few years several polyimides able to undergo thermal rearrangement in a solid state when under the influence of high temperature have been synthesized [80]. Some moieties afford the

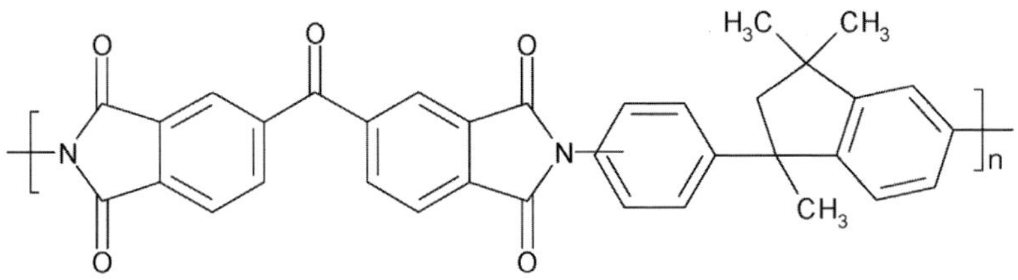

Figure 22.13 Chemical structure of Matrimid® 5218 (Huntsman, USA).

possibility of reacting with each other to form a volatile compound leaving the polymer while the main chain of the polymer undergoes significant rearrangement of its chemical structure (Scheme 22.1) to become significantly higher in T_g than the precursor polymer, losing solubility and in some cases obtaining gas transport properties superior to most polymers of similar nature [80].

22.5
Process Examples

This section presents some examples of the application of hydrogen separation by means of membrane technology. The examples are selected based on the application of the majority of realized systems and with respect to possible future applications.

22.5.1
Hydrogen Separation from Purge Streams in Ammonia Production

Figure 22.14 shows the integration of gas permeation membrane stages into an ammonia production process [38]. The purge gas has to be withdrawn from the process in order to prevent the buildup of inert components in the reaction loop. The membrane modules are operated at high pressures on the feed and on the permeate sides, that is, at limited pressure differences and ratios in order to minimize compression costs and to enable recycling of permeate to the appropriate compressor stages. The limited purities of the gases are sufficient to ensure a hydrogen recovery of 87.7%. Membrane modules employed are typically the hollow fiber modules of the commercial suppliers as discussed in Section 22.4.3.1.

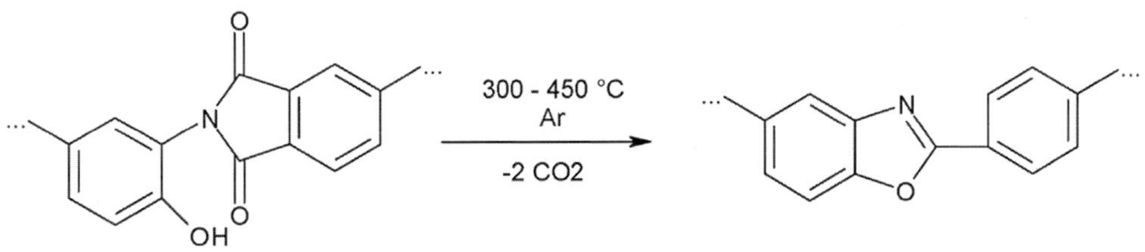

Scheme 22.1 Scheme of polyimide conversion into thermally rearranged polymer.

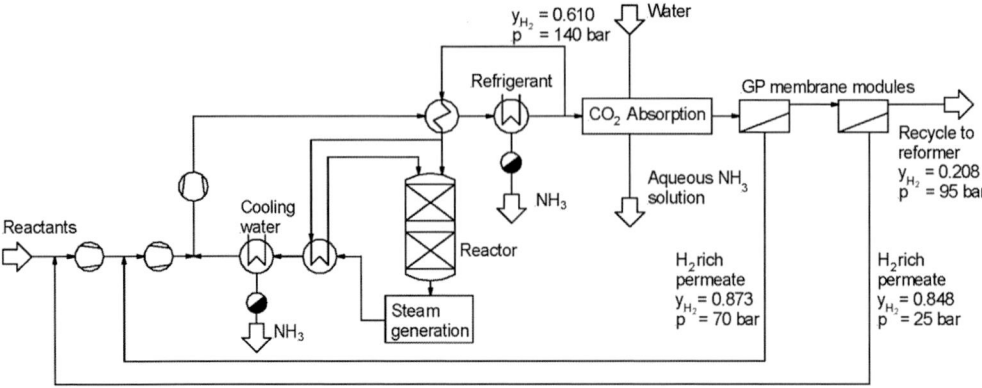

Figure 22.14 Integration of gas permeation modules for purge gas treatment in ammonia production [38].

22.5.2
Hydrogen Separation from Hydrocracker Flash Gas

One application in refineries is the recovery of hydrogen from flash gas generated when treating the product stream of a hydrocracking unit. These units are employed to generate different fuel grades from vacuum distillate [81]. Figure 22.15

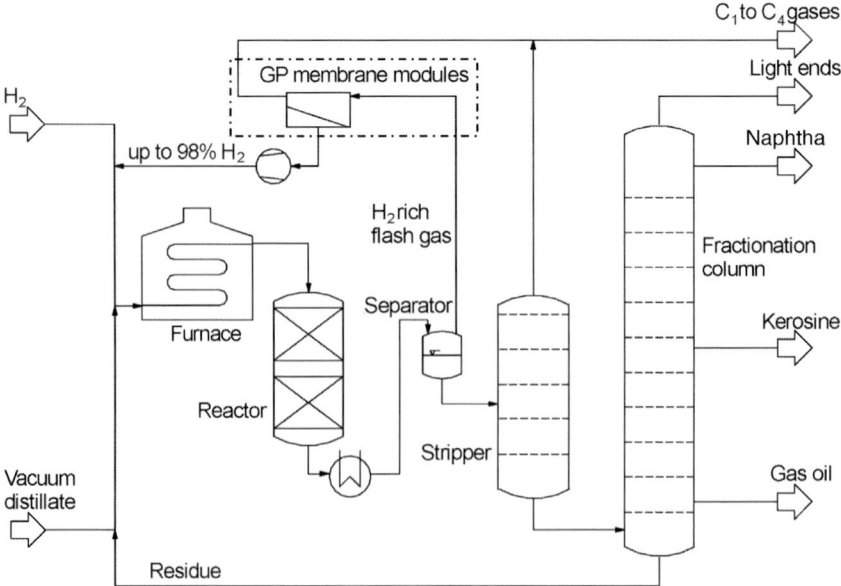

Figure 22.15 Integration of a gas permeation unit for the separation of hydrogen from hydrocracker flash gas.

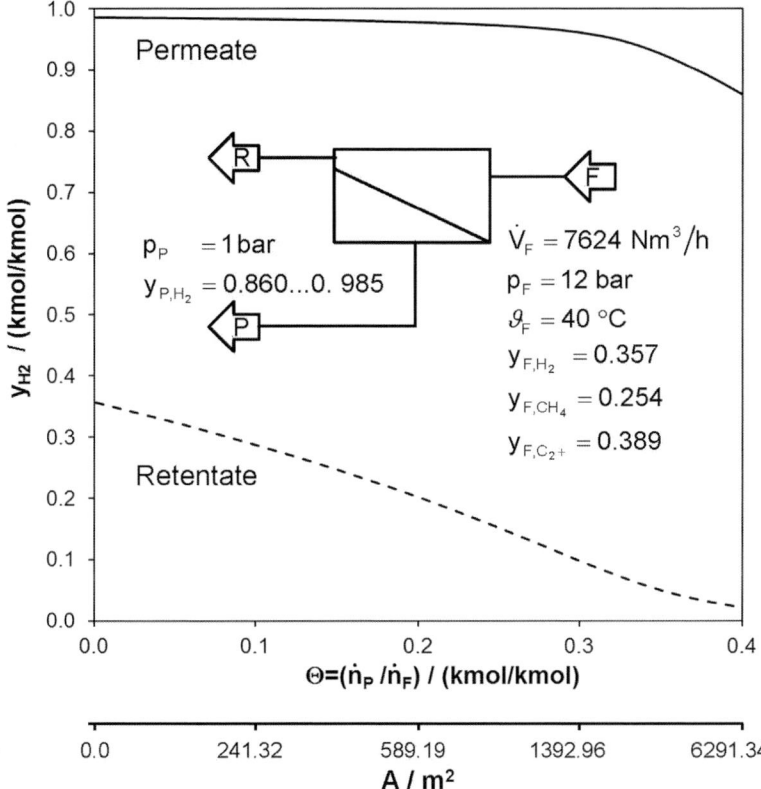

Figure 22.16 Hydrogen content in retentate and permeate for the separation of hydrogen from hydrocracker flash gas using a Matrimid membrane as function of the stage cut; operating data from [82].

shows the possible integration of a gas permeation stage into such a process where it separates a large amount of the hydrogen from flash gas. The hydrogen rich permeate is recompressed and recycled to the feed hydrogen stream of the hydrocracker. These processes are also commercially addressed by the hollow fiber modules described in Section 22.4.3.1. Figure 22.16 shows how retentate and permeate hydrogen contents develop as functions of the stage cut (i.e., the ratio permeate to feed flow-rate) and employed membrane area for the gas permeation module section of the process (dash-pointed box in Figure 22.15). The operating data is taken from Reference [82] and the permeances are that of Matrimid [51]. The model used for the simulations was described in Section 22.2. It is apparent that with increasing stage cut the hydrogen recovery is increasing (i.e., up to 99%) at the expense of permeate purity. Furthermore, the membrane area is increasing exponentially due to the decreasing driving force across the membrane caused by the almost full removal of hydrogen from the retentate side.

22.5.3
Carbon Dioxide Removal from Biomass Gasification Product Gas

The gasification of biomass is intensively discussed for the generation of synthesis gas for further conversion into intermediates required by the chemical industry or into motor fuels [39,83]. Next to carbon monoxide and hydrogen, the gases produced typically contain large amounts of carbon dioxide. Separating carbon dioxide is often necessary in order to adjust the concentration ratio of the gasification product for further usage as a feed for subsequent reactions. Polyethylene containing block copolymer membranes such as those employing PolyActive^{TM} or Pebax$^®$ for the active separation layer can be used for this separation since they allow for the preferential permeation of carbon dioxide and water, which is also contained in large amounts in the product gas. Figure 22.17 shows a simple one-stage design of such a process. The membrane employed is a multilayer thin film composite membrane with PolyActive^{TM} as sepa-ration layer. The membrane and its permeation characteristics are described elsewhere [32,84,85]. The feed conditions are taken from Reference [83].

A large advantage of this process is that hydrogen remains on the high pressure side whilst a disadvantage is the limited carbon dioxide selectivity. This can be somewhat rectified by using the temperature dependence of the selectivity, that is, by cooling the feed gas at the expense of an increasing membrane area. The process selectivity can be further increased by applying vacuum to the permeate side of the process. Figure 22.18 shows calculations for the process conditions depicted in Figure 22.17, that is, for a fixed feed composition and for a product hydrogen mole fraction of 0.6. The varied parameters were the permeate pressure and the temperature directly upstream of the module, indicated as ϑ_1 in Figure 22.18. It is apparent that the hydrogen recovery can be increased by cooling the feed and applying a vacuum to the permeate side. The disadvantage of this strategy is the increased demand of membrane area and even more importantly the necessity of employing a vacuum pump, implying additional investment and operating costs. However, which strategy is the economically and ecologically most favorable can only be decided on a case by case basis.

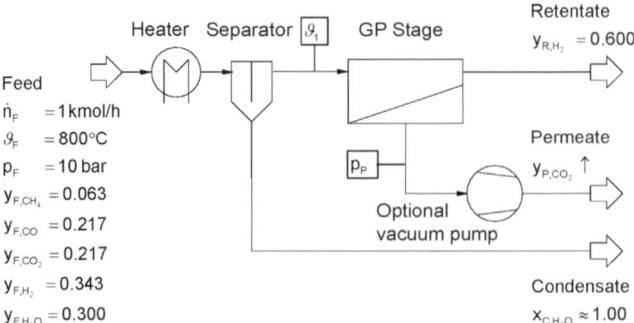

Figure 22.17 Separation of carbon dioxide from biomass gasification product gas.

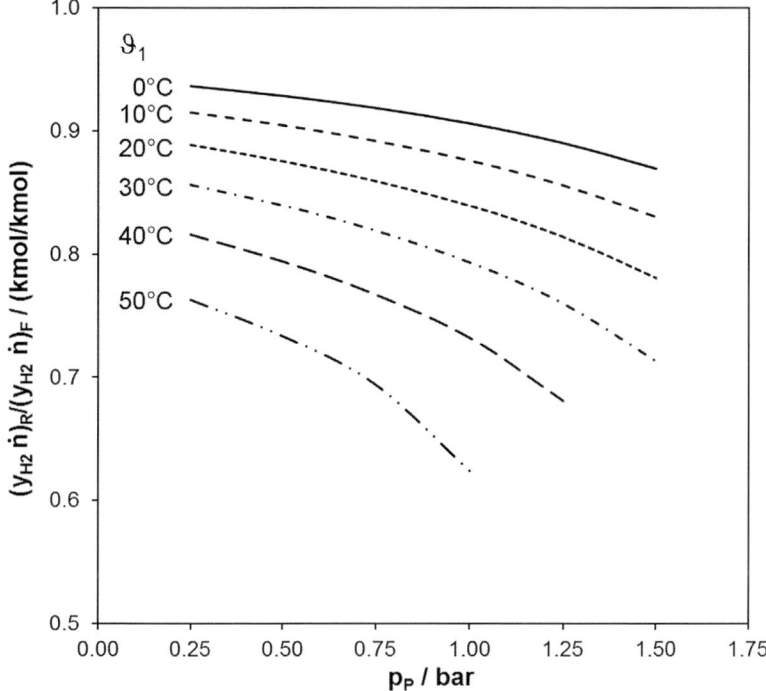

Figure 22.18 Separation performance for the removal of carbon dioxide from biomass gasification product gas using PolyActive™ membranes as a function of operating temperature and pressure.

22.6 Conclusions

Hydrogen separation by means of polymeric membranes is an established unit operation in ammonia production, synthesis gas applications, and refineries. It was among the first gas permeation processes realized industrially. Nowadays various new applications can be addressed in the context of hydrogen technology as one important means to tackle the depletion of fossil fuels. Furthermore additional processes are possible that address the changing energy and raw material supply of today's societies.

Currently only a few polymers such as polysulfone and polyimides are used for the preparation of hydrogen selective membranes. Hollow fiber modules operated in the dominant outside–in module geometry are well proven in the mentioned industrial applications.

Thin film composite flat sheet membranes possess the capability to employ novel, high performance polymers with superior permeation characteristics in a way that is easy to scale up to technical scale. In combination with flat sheet membrane modules that allow for better fluid dynamics and a better utilization of the available driving force, especially for vacuum assisted operation, than hollow fiber modules, this type of membrane is especially well suited for emerging, small scale applications and for scenarios where hydrogen is supposed to remain on the feed side of the membrane module, that is, components such as carbon dioxide are to be separated. Potentially, these materials can lead to an efficiency increase for established technologies and address novel hydrogen separation tasks.

Nomenclature

A	Coefficient in dual sorption model	kmol m^{-3}
A	Membrane area	m^2
C	Concentration	kmol m^{-3}
D	Diameter	nm
D	Diffusion coefficient	m^2 s^{-1}
E	Activation energy	kJ kmol^{-1}
f	Fugacity	bar
k	Coefficient in dual sorption model	bar^{-1}
L	Permeance	m^3_N (m^2 h bar)$^{-1}$ kmol (m^2 s bar)$^{-1}$
M	Molecular weight	kmol kg^{-1}
$\dot{n}$	Molar flow-rate	kmol h^{-1}, kmol s^{-1}
$\dot{n}''$	Molar flux	kmol (m^2 h)$^{-1}$, kmol (m^2 s)$^{-1}$
p	Pressure	bar
P	Permeability	kmol (s m bar)$^{-1}$, Barrer = 10^{-10} cm^3_N cm (cm^2 s cmHg)$^{-1}$
R	Universal gas constant	8.31 433 kJ kmol^{-1} K^{-1}
R	Resistance	m^2 h bar m$^{-3}_N$, m^2 s bar kmol^{-1}
S	Solution coefficient	kmol (m^3 bar)$^{-1}$
T	Temperature	K
v	Velocity	m s^{-1}
$\dot{V}$	Volumetric flow rate	m^3_N h^{-1}
$\tilde{V}$	Molar volume	cm^3 mol^{-1}
x	Mole fraction liquid phase	kmol kmol^{-1}
y	Mole fraction gas phase	kmol kmol^{-1}
z	Spatial co-ordinate	m

Greek

α	Selectivity	—
δ	Layer thickness	m
ϕ	Pressure ratio	—

φ	Fugacity coefficient	—
ϑ	Temperature	°C
Θ	Stage cut	kmol kmol^{-1}

Superscripts

D	Diffusion
S	Solution or sorption
∞	For $T \to \infty$

Subscripts

AL	Active separation layer
BL	Boundary layer
c	Critical
C	Condensate
F	Feed
g	Glass transition
k	Kinetic
M	Membrane
nc	Number of components
P	Permeate
PS	Interface support - permeate side
PM	Membrane surface, permeate side
R	Retentate
RM	Membrane surface, retentate side
SL	Support layer
tot	Total

References

1 Nollet, J.A. (1995) Investigations on the causes for the ebullition of liquids. *J. Membr. Sci.*, **100** (1), 1–3.
2 Graham, T. (1995) On the law of the diffusion of gases. *J. Membr. Sci.*, **100** (1), 17–21.
3 Böddeker, K.W. (1995) Commentary: tracing membrane science. *J. Membr. Sci.*, **100** (1), 65–68.
4 Frey, F.E. (1939) Process for concentrating hydrocarbons. US Patent 2159434A.
5 Knudsen, M. (1995) The laws of molecular flow and of inner friction flow of gases through tubes. *J. Membr. Sci.*, **100** (1), 23–25.
6 Baker, R. (2007) *Membrane Technology and Applications*, 2nd edn, John Wiley & Sons, Ltd., Chichester.
7 Loeb, S. and Sourirajan, S. (1962) High-flow semipermeable membranes for separation of water from saline solutions. *Adv. Chem. Ser.*, **38**, 117–132.
8 Henis, J.M.S. and Tripodi, M.K. (1980) A novel approach to gas separations using composite hollow fiber membranes. *Sep. Sci. Technol.*, **15** (4), 1059–1068.
9 Baker, R.W. (2002) Membrane technology, in *Encyclopedia of Polymer Science and Technology*, John Wiley & Sons, Inc., Hoboken.

10 Perry, J.D., Nagai, K., and Koros, W.J. (2006) Polymer membranes for hydrogen separations. *MRS Bull.*, **31** (10), 745–749.

11 Sanders, D.F., Smith, Z.P., Guo, R., Robeson, L.M., McGrath, J.E., Paul, D.R., and Freeman, B.D. (2013) Energy-efficient polymeric gas separation membranes for a sustainable future: A review. *Polymer*, **54** (18), 4729–4761.

12 Bernardo, P., Drioli, E., and Golemme, G. (2009) Membrane gas separation: a review/state of the art. *Ind. Eng. Chem. Res.*, **48** (10), 4638–4663.

13 Rautenbach, R. (1997) *Membranverfahren*, Springer Verlag, Berlin, Heidelberg.

14 Ma, Y.H. (2008) *Advanced Membrane Technology and Applications* (eds N. Li, A. G. Fane, W.S. Ho, and T. Matsuura), John Wiley and Sons, Inc., New York.

15 Iwahara, H. (2004) Prospect of hydrogen technology using proton-conducting ceramics. *Solid State Ionics*, **168** (3–4), 299–310.

16 Adhikari, S., Fernando, S., Engineering, B., and State, M. (2006) Hydrogen membrane separation techniques. *Ind. Eng. Chem. Res.*, **45** (3), 875–881.

17 Koros, W.J. (2001) A more aggressive approach is needed for gas separation membrane development. *МАКЦ*, **31** (3), 157–160.

18 Huang, T.-C., Wei, M.-C., and Chen, H.-I. (2001) Permeation of hydrogen through palladium/alumina composite membranes. *Sep. Sci. Technol.*, **36** (2), 199–222.

19 Keizer, K. and Nijmeijer, A. (2006) *Encyclopedia of Surface and Colloid Science* (ed. P. Somasundaran), Taylor & Francis Group LLC, Boca Raton.

20 Mulder, M. and Mulder, J. (1996) *Basic Principles of Membrane Technology*, Kluwer Academic Publishers, Dordrecht.

21 Baker, R., Cussler, E., Eykamp, W., and Koros, W. (1990) Membrane Separation Systems – A Research and Development Needs Assessment. DOE/ER/30133--H1-Vol.2 DE 90 011770, 2.

22 Petropoulos, J.H. (1970) Quantitative analysis of gaseous diffusion in glassy polymers. *J. Polym. Sci. Polym. Phys.*, **8** (10), 1797–1801.

23 Vrentas, J.S. and Duda, J.L. (1976) Diffusion of small molecules in amorphous polymers. *Macromolecules*, **9** (5), 785–790.

24 Paterson, R., Yampol'Skii, Y.P., Fogg, P.G., Bokarev, A., Bondar, V., Ilinich, O., and Shishatskii, S. (1999) IUPAC-NIST solubility data series 70. The solubility of gases in glassy polymers. *J. Phys. Chem. Ref. Data*, **28** (5), 1255–1452.

25 Favre, E., Nguyen, Q.T., Schaetzel, P., Clement, R., and Neel, J. (1993) Sorption of organic solvents into dense silicone membranes. Part 1. – Validity and limitations of Flory-Huggins and related theories. *J. Chem. Soc. Faraday Trans.*, **89** (24), 4339–4346.

26 Bird, R.B., Stewart, W.E., and Lightfoot, E. N. (1960) *Transport Phenomena*, John Wiley and Sons, Inc., New York.

27 Taylor, R. and Krishna, R. (1993) *Multicomponent Mass Transfer*, John Wiley and Sons, Inc., New York.

28 Ohlrogge, K., Wind, J., and Brinkmann, T. (2010) *Comprehensive Membrane Science and Engineering* (eds E. Drioli and L. Giorno), Academic Press, Oxford, pp. 213–242.

29 Fang, S.-M., Stern, S.A., and Frisch, H.L. (1975) A "free volume" model of permeation of gas and liquid mixtures through polymeric membranes. *Chem. Eng. Sci.*, **30** (8), 773–780.

30 Brinkmann, T. (2006) *Membranen* (eds K. Ohlrogge and K. Ebert), Wiley-VCH Verlag GmbH, Weinheim.

31 Gmehling, J. and Kolbe, B. (1989) *Thermodynamik*, Georg Thieme Verlag, Stuttgart.

32 Brinkmann, T., Pohlmann, J., Withalm, U., Wind, J., and Wolff, T. (2013) Theoretical and experimental investigations of flat sheet membrane module types for high capacity gas separation applications. *Chem-Ing-Tech.*, **85** (8), 1210–1220.

33 Marriott, J., Sørensen, E., and Bogle, I.D. (2001) Detailed mathematical modelling of membrane modules. *Comput. Chem. Eng.*, **25** (4–6), 693–700.

34 Marriott, J. and Sørensen, E. (2003) A general approach to modelling membrane modules. *Chem. Eng. Sci.*, **58** (22), 4975–4990.

35 Scholz, M., Harlacher, T., Melin, T., and Wessling, M. (2012) Modeling gas permeation by linking nonideal effects. *Ind. Eng. Chem. Res.*, **52** (3), 1079–1088.

36 Coker, D.T., Freeman, B.D., and Fleming, G.K. (1998) Modeling multicomponent gas separation using hollow-fiber membrane contactors. *AICHE J.*, **44** (6), 1289–1302.

37 Coker, D.T., Allen, T., Freeman, B.D., and Fleming, G.K. (1999) Nonisothermal model for gas separation hollow-fiber membranes. *AICHE J.*, **45** (7), 1451–1468.

38 Rautenbach, R. (1991) *Vorlesungsumdruck Membranverfahren*, RWTH Aachen, Aachen.

39 Makaruk, A., Miltner, M., and Harasek, M. (2012) Membrane gas permeation in the upgrading of renewable hydrogen from biomass steam gasification gases. *Appl. Therm. Eng.*, **43**, 134–140.

40 Brinkmann, T. (2012) Fischer–Tropsch-Verfahren zur Herstellung von Kohlenwasserstoffen aus Biogas. EU Patent Application EP12191124.

41 Grätz, M., Shishatskiy, S., and Brinkmann, T. (2013) Membrane processes for H2 recovery in fossil fuelled power plants and related systems, in *Engineering with Membranes, Towards a Sustainable Future: Proceedings: EWM2013*, Saint-Pierre D'Oléron (France), 3–7 September 2013, publisher: As. C. Ingeniería con Membranas, 2013.

42 Franz, J. and Scherer, V. (2010) An evaluation of CO2 and H2 selective polymeric membranes for CO2 separation in IGCC processes. *J. Membr. Sci.*, **359** (1–2), 173–183.

43 Yampolskii, Y.P., Freeman, B., and Pinnau, I. (eds) (2006) *Materials Science of Membranes for Gas and vapor Separation*, John Wiley & Sons Ltd., Chichester.

44 Yampolskii, Y. (2012) Polymeric gas separation membranes. *Macromolecules*, **45** (8), 3298–3311.

45 Van Krevelen, D.W. and Te Nijenhuis, K. (2009) Properties determining mass transfer in polymeric systems, in *Properties of Polymers*, Fourth Edition, Elsevier, Amsterdam, ch. 18, pp. 655–702.

46 Bower, D.I. (2002) *An Introduction to Polymer Physics*, Cambridge University Press, Cambridge, UK.

47 Vrentas, J.S. and Duda, J.L. (1978) A free-volume interpretation of the influence of the glass transition on diffusion in amorphous polymers. *J. Appl. Polym. Sci.*, **22** (8), 2325–2339.

48 Filiz, V., Shishatskiy, S., Rangou, S., Gacal, B.N., and Abetz, V. (2011) Synthesis of poly(vinyl trimethylsilane) and block-copolymers of vinyl trimethylsilane with isoprene, their functionalization and gas transport properties, in *Proceedings of the ICOM 2011 Conference*, Amsterdam, The Netherlands, July 24–29, 2011.

49 Yave, W., Huth, H., Car, A., and Schick, C. (2011) Peculiarity of a CO2-philic block copolymer confined in thin films with constrained thickness: "a super membrane for CO2-capture.". *Energy Environ. Sci.*, **4** (11), 4656–4661.

50 Shishatskiy, S., Nistor, C., Popa, M., Nunes, S.P., and Peinemann, K.-V. (2006) Comparison of asymmetric and thin-film composite membranes having Matrimid 5218 selective layer. *Desalination*, **199** (1), 193–194.

51 Shishatskiy, S., Nistor, C., Popa, M., Nunes, S.P., and Peinemann, K.V. (2006) Polyimide asymmetric membranes for hydrogen separation: influence of formation conditions on gas transport properties. *Adv. Eng. Mater.*, **8** (5), 390–397.

52 Le, N.L., Tang, Y.P., and Chung, T.-S. (2013) The development of high-performance 6FDA-NDA/DABA/POSS/Ultem® dual-layer hollow fibers for ethanol dehydration via pervaporation. *J. Membr. Sci.*, **447**, 163–176.

53 Car, A., Stropnik, C., Yave, W., and Peinemann, K.-V. (2008) Pebax®/polyethylene glycol blend thin film composite membranes for CO2 separation: Performance with mixed gases. *Sep. Purif. Technol.*, **62** (1), 110–117.

54 Hägg, M.-B. (2009) *Handbook of Membrane Separation* (eds A.K. Pabby, S.S.H. Rizvi, and A.M. Sastre Requena), Taylor & Francis Group LLC, Boca Raton, FL.

55 Abetz, V., Brinkmann, T., Dijkstra, M., Ebert, K., Fritsch, D., Ohlrogge, K., Paul, D., Peinemann, K.-V., Pereira-Nunes, S., Scharnagl, N., and Schossig, M. (2006) Developments in membrane research: from material via process design to industrial application. *Adv. Eng. Mater.*, **8** (5), 328–358.

56 Low, B.T., Zhao, L., Merkel, T.C., Weber, M., and Stolten, D. (2013) A parametric study of the impact of membrane materials and process operating conditions on carbon capture from humidified flue gas. *J. Membr. Sci.*, **431**, 139–155.

57 Sandru, M., Haukebø, S.H., and Hägg, M.-B. (2010) Composite hollow fiber membranes for CO2 capture. *J. Membr. Sci.*, **346** (1), 172–186.

58 Favvas, E.P., Papageorgiou, S.K., Nolan, J.W., Stefanopoulos, K.L., and Mitropoulos, A.C. (2013) Effect of air gap on gas permeance/selectivity performance of BTDA-TDI/MDI copolyimide hollow fiber membranes. *J. Appl. Polym. Sci.*, **130** (6), 4490–4499.

59 Henis, J.M.S. and Tripodi, M.K. (1981) Composite hollow fiber membranes for gas separation: the resistance model approach. *J. Membr. Sci.*, **8** (3), 233–246.

60 Widjojo, N., Chung, T.-S., and Kulprathipanja, S. (2008) The fabrication of hollow fiber membranes with double-layer mixed-matrix materials for gas separation. *J. Membr. Sci.*, **325** (1), 326–335.

61 Scholz, M., Harlacher, T., Melin, T., and Wessling, M. (2012) Modeling gas permeation by linking nonideal effects. *Ind. Eng. Chem. Res.*, **52** (3), 1079–1088.

62 Ohlrogge, K., Wind, J., Scholles, C., and Brinkmann, T. (2005) Membranverfahren zur abtrennung organischer dämpfe in der chemischen und petrochemischen industrie. *Chem-Ing-Tech.*, **77** (5), 527–537.

63 Barillas, M.K., Enick, R.M., O'Brien, M., Perry, R., Luebke, D.R., and Morreale, B.D. (2011) The CO2 permeability and mixed gas CO_2/H_2 selectivity of membranes composed of CO_2-philic polymers. *J. Membr. Sci.*, **372** (1–2), 29–39.

64 Car, A., Yave, W., Peinemann, K.-V., and Stropnik, C. (2010) Tailoring polymeric membrane based on segmented block copolymers for CO2 separation, in *Membrane Gas Separation*, John Wiley & Sons, Ltd., Chichester, pp. 227–253.

65 Reijerkerk, S.R., Wessling, M., and Nijmeijer, K. (2011) Pushing the limits of block copolymer membranes for CO_2 separation. *J. Membr. Sci.*, **378** (1–2), 479–484.

66 Lin, H. and Freeman, B.D. (2004) Gas solubility, diffusivity and permeability in poly(ethylene oxide). *J. Membr. Sci.*, **239** (1), 105–117.

67 Lin, H., He, Z., Sun, Z., Vu, J., Ng, A., Mohammad, M., Kniep, J., Merkel, T.C., Wu, T., and Lambrecht, R.C. (2014) CO_2-selective membranes for hydrogen production and CO_2 capture Part I: Membrane development. *J. Membr. Sci.*, J. Memb. Sci. **457** (1) 149–161.

68 Merkel, T.C. (2013) New Project in Alberta Will Use Membranes to Capture CO2 from Syngas, Enhance Methanol and Hydrogen Production (July 2013), Membrane Technology and Research. http://www.mtrinc.com/news_archive.html (accessed 20 August 2013).

69 Merkel, T.C., Bondar, V.I., Nagai, K., Freeman, B.D., and Pinnau, I. (2000) Gas sorption, diffusion, and permeation in poly (dimethylsiloxane). *J. Polym. Sci. Pol. Phys.*, **38** (3), 415–434.

70 Henis, J. and Tripodi, M. (1983) The developing technology of gas separating membranes. *Science*, **220**, 11–17.

71 Yave, W., Car, A., Wind, J., and Peinemann, K.-V. (2010) Nanometric thin film membranes manufactured on square meter scale: ultra-thin films for CO_2 capture. *Nanotechnology*, **21** (39), 395301.

72 Alentiev, A., Drioli, E., Gokzhaev, M., Golemme, G., Ilinich, O., Lapkin, A., Volkov, V., and Yampolskii, Y. (1998) Gas permeation properties of phenylene oxide polymers. *J. Membr. Sci.*, **138** (1), 99–107.

73 Robeson, L.M. (2008) The upper bound revisited. *J. Membr. Sci.*, **320** (1–2), 390–400.

74 Ichiraku, Y., Stern, S.A., and Nakagawa, T. (1987) An investigation of the high gas permeability of poly (1-trimethylsilyl-1-propyne). *J. Membr. Sci.*, **34** (1), 5–18.

75 McKeown, N.B., Budd, P.M., Msayib, K.J., Ghanem, B.S., Kingston, H.J., Tattershall, C.E., Makhseed, S., Reynolds, K.J., and Fritsch, D. (2005) Polymers of intrinsic microporosity (PIMs): bridging the void between microporous and polymeric materials. *Chem-Eur. J.*, **11** (9), 2610–2620.

76 Park, H.B., Jung, C.H., Lee, Y.M., Hill, A.J., Pas, S.J., Mudie, S.T., van Wagner, E.,

Freeman, B.D., and Cookson, D.J. (2007) Polymers with cavities tuned for fast selective transport of small molecules and ions. *Science*, **318** (5848), 254–258.

77 Choi, J.I., Jung, C.H., Han, S.H., Park, H.B., and Lee, Y.M. (2010) Thermally rearranged (TR) poly(benzoxazole-co-pyrrolone) membranes tuned for high gas permeability and selectivity. *J. Membr. Sci.*, **349** (1–2), 358–368.

78 Do, Y.S., Seong, J.G., Kim, S., Lee, J.G., and Lee, Y.M. (2013) Thermally rearranged (TR) poly(benzoxazole-co-amide) membranes for hydrogen separation derived from 3,3′-dihydroxy-4,4′-diaminobiphenyl (HAB), 4,4′-oxydianiline (ODA) and isophthaloyl chloride (IPCl). *J. Membr. Sci.*, **446**, 294–302.

79 Kase, Y. (2008) *Gas separation by polyimide membranes. in: Advanced Membrane Technology and Applications*, John Wiley & Sons, Inc., Chichester, pp. 581–598.

80 Han, S.H., Misdan, N., Kim, S., Doherty, C.M., Hill, A.J., and Lee, Y.M. (2010) Thermally rearranged (TR) polybenzoxazole: effects of diverse imidization routes on physical properties and gas transport behaviors. *Macromolecules*, **43**, 7657–7667.

81 Onken, U. and Behr, A. (1996) *Chemische Prozesskunde - Lehrbuch der Technischen Chemie Band 3*, Georg Thieme Verlag, Stuttgart.

82 Kaldis, S., Kapantaidakis, G., and Sakellaropoulos, G. (2000) Simulation of multicomponent gas separation in a hollow fiber membrane by orthogonal collocation — hydrogen recovery from refinery gases. *J. Membr. Sci.*, **173** (1), 61–71.

83 Tijmensen, M. (2002) Exploration of the possibilities for production of Fischer Tropsch liquids and power via biomass gasification. *Biomass Bioenerg.*, **23** (2), 129–152.

84 Brinkmann, T., Hoting, B., Wind, J., and Wolff, T. (2010) Separation of CO2 from biogas by gas permeation, in *Proceedings of the 13th Aachen Membrane Colloquium, 27–28 October 2010*, RWTH Aachen University, Aachen.

85 Car, A., Stropnik, C., Yave, W., and Peinemann, K.-V. (2008) Tailor-made polymeric membranes based on segmented block copolymers for CO_2 separation. *Adv. Funct. Mater.*, **18** (18), 2815–2823.

23
Gas Clean-up for Fuel Cell Systems – Requirements & Technologies

Matthias Gaderer, Stephan Herrmann, and Sebastian Fendt

23.1
Introduction

Fuel cells are considered promising devices for future energy conversion systems. Owing to their nature as electrochemical devices they promise higher efficiencies than conventional systems in the field of power generation, especially at small- to medium-scale [1,2]. However, there are still major drawbacks with regards to versatility of fuel cells for most applications. Besides high costs, which are mainly attributable to expensive materials and lack of mass production, the required fuel gas quality is an especially important issue. As an example, the combination of thermochemical gasification systems and fuel cells could lead to high system efficiencies, of up to 55%, while opening up the whole range of solid fuels for utilization in fuel cells requiring gaseous fuels [2]. However, at present, without extensive gas cleaning measures, fuel cells show severe degradation rates when operated on syngas from gasification systems.

To overcome this problem, different technologies have been and currently still are under investigation. The following section provides an overview of cleaning requirements and promising technologies for fulfilling them in order to make fuel cell systems powered by a wide range of fuels available for the future.

23.2
Background

There is a wide range of potential sources of fuel gases available for fuel cells. Conventional sources include natural gas, shale gas, other type of fracking, and reforming residues as well as synthesis gas from gasification of coal or bitumen, to name just a few. Equally, from renewable sources one can find biomass-derived gases from anaerobic digestion, gasification, or similar such as biogas, landfill gas, synthesis gas derived from sewage, agro waste, municipal/industrial organic waste fraction, wood residues, and others [3]. Furthermore, other

Hydrogen Science and Engineering: Materials, Processes, Systems and Technology, First Edition.
Edited by Detlef Stolten and Bernd Emonts.
© 2016 Wiley-VCH Verlag GmbH & Co. KGaA. Published 2016 by Wiley-VCH Verlag GmbH & Co. KGaA.

sources like product gases retrieved from electrolysis based on excess electricity from renewable and/or conventional sources are going to gain in significance in the coming decades.

Depending on the type of fuel cell, different types of gaseous substances are in principle feasible for utilization as a fuel. In the case of low temperature fuel cells, such as polymer electrolyte membrane fuel cells (PEMFCs, also called proton exchange membrane fuel cells), mainly pure hydrogen is used as a fuel. To date, this hydrogen is mostly produced by reformation of natural gas [4], as is described in more details in other chapters of this book.

This is inherently due to the ion species transported through the electrolyte, which are, for example, H^+ ions in the case of PEMFC. If only H^+ ions can penetrate the electrolyte, consequently only hydrogen can be oxidized on the air side electrode.

The part of a fuel cell that is most sensitive to degradation due to contaminants is the catalyst material. Since it is the chemically most active part of the cell, it is also the most affected. The standard catalyst material for low temperature fuel cells is platinum due to its unique activity in catalyzing the water formation reaction from H_2 and O_2; in rarer cases ruthenium or mixed platinum/ruthenium are also used [5,6]. In the literature there are studies aimed at utilizing more inexpensive materials such as copper or iron; however, to achieve conversion rates appropriate for fuel cell application, their electronic configuration must be strongly chemically modified by integration in more complex compounds. To date, there are no inexpensive catalysts that exhibit comparable performance and stability to platinum [6–8].

For high temperature fuel cells, such as solid oxide fuel cells (SOFCs) or molten carbonate fuel cells (MCFC), a large variety of gaseous substances can be applied. The transported ion species are O^{2-} in the case of SOFC and CO_3^{2-} for MCFC, which effectively also results in transporting oxygen ions by splitting into CO_2 and O^{2-} at the anode. If O^{2-} ions are transported through the electrolyte, basically any substance capable of reacting with oxygen, which is present on the fuel side, can act as a fuel. Typical examples are, besides H_2, carbon monoxide (CO), methane (CH_4), higher hydrocarbons, and even ammonia (NH_3) [9]. Nevertheless, there are some limitations also for the high temperature fuel cells with regard to fuel composition. These result mainly from side effects of the fuel conversion, such as, for example, carbon deposition on the catalyst material of the anodes [10–13]. Consequently, for various substances the amount makes the potential fuel a poison and vice versa. In this respect many of the substances described below can act as a fuel for high temperature fuel cells at low concentrations, but become a contaminant at higher concentrations.

Comparted to low temperature fuel cells, high temperature fuel cells, due to the high operating temperature, can use more inexpensive catalyst. The standard catalyst material for SOFCs and MCFCs is nickel or, more rarely, one can find researchers applying iron, copper, rhodium, and others. However, for high temperature fuel cells not only the catalyst material is important. In addition, the bulk material and microstructure can be affected and contribute to degradation phenomena or, on the other hand, it can add to the degradation resilience of a cell [14,15]. In the case of SOFCs, for example, the standard anode (fuel side) material composition today is NiYSZ (Ni-yttrium stabilized zirconia) [1]. However, several studies in the literature indicate that, for example, NiGDC (Ni-gadolinium doped ceria, also called NiCGO, for nickel-ceria-gadolinium-oxide) anodes are more resilient to degradation by sulfur compounds, which are among the most notorious poisons for high temperature fuel cells [16]. The same applies for lanthanum-strontium-vanadium-oxide (LSV) based anodes and others [17,18]. However, similar to the low temperature fuel cells the performance of the "standard" material NiYSZ is, presently, not matched by the other material combinations.

As well as the fuel side, degradation might also occur at the air side of the fuel cell. This is due to air pollutants, like NO_x, SO_x, and metallic ions from stack material [5] in the case of low temperature fuel cells and effects such as poisoning of high temperature fuel cell cathodes by chromium from metallic piping and stack components [19]. However, since these effects do not result from contaminants present in the fuel they are not considered further in this chapter.

There is a wide variety of potential contaminants present in fuel gases. The relevant contaminant species and amounts are dependent on the type of fuel/gas and upstream-processes and treatments. The most prominent representatives are discussed briefly in the following.

23.3
Fuel and Pollutants

23.3.1
Main Gas Components

Natural gas is the most commonly used gaseous fuel today. Its main component is methane with varying shares of ethane, propane, and other hydrocarbons. Furthermore, significant amounts of nitrogen, carbon dioxide, and other inert gases can be contained in natural gas, as well as hydrogen sulfide, mercaptans, carbonyl sulfide, and others, depending on the source [20]. Biogas mainly consists of methane and carbon dioxide at different ratios [21]. Both natural gas and biogas need to be reformed before utilization in fuel cells. Reforming is usually carried out by either steam reforming, autothermal reforming using air or oxygen, or catalytic reforming [22,23].

Gasification synthesis gas is again a more complex mixture and strongly dependent on the chosen gasification process. Especially, the gasification medium is important, as well as the bed material and many other parameters. Some examples are given in Table 23.1. For more detailed information on the different gasification processes the reader is referred to the relevant literature [24–35].

Table 23.1 Gas composition of untreated product gas from different gasifiers. Steam contents can vary between 20% and 50%. Heating values are between 3.7 and 10 MJ Nm$_{dry}^{-3}$ [23–35].

Gasifier	Gasification medium	Composition (vol.% dry base)				
		H$_2$	CO	CH$_4$	CO$_2$	N$_2$
Fixed bed	Air	10–23	15–25	2–6	10–15	40–55
Stationary fluidized bed	Air, 20 bar	12	18	17	6	50
Stationary fluidized bed IGCC	Steam, O$_2$, 12 bar	25–29	19	13.5	38–46	0.1
AER fluidized bed	Steam, CaO, allothermal	68–78	4	12	10	4
HPR fluidized bed	Steam, allothermal	40–50	15–25	5–10	15–25	5–10
Fast internally circulating fluidized bed	Steam allothermal	40	25	10	20	4
Entrained flow	Air	7–15	20	0	11	46
Entrained flow	O$_2$	27	50	0	14	4

23.3.1.1 Hydrogen

Hydrogen is the most important fuel for fuel cell applications, as discussed previously.

23.3.1.2 Carbon Monoxide

Carbon monoxide and carbon dioxide are unproblematic for high temperature fuel cells; CO even acts as a fuel. However, for low temperature fuel cells they are among the most critical contaminants. Carbon monoxide will bond irreversibly to platinum catalyst material and deactivate it. Thus, to avoid carbon monoxide poisoning, very low concentrations at the level of a few ppm must be achieved [36].

23.3.1.3 Methane

Methane and higher hydrocarbons and oils and similar substances can be the result of, or leftover from, fuel reforming processes, lubricants for turbo machinery, and others. In general for high temperature fuel cells they can act as a fuel up to certain concentrations since they are cracked and/or reformed to H$_2$ and CO at high temperatures in the presence of steam or CO$_2$ [14]. However, especially chemically active substances with C=C double bonds, such as ethylene, show great tendency towards carbon deposition effects at least on nickel catalysts [14,37]. In the case of low temperature fuel cells condensation of higher hydrocarbons can occur. Furthermore, they are susceptible to catalytic reforming as well, forming carbon monoxide and thus can act indirectly as catalyst poison [5,38].

23.3.1.4 Carbon Dioxide

Carbon dioxide itself is not problematic. Nevertheless, in low temperature fuel cells some of it can react to give carbon monoxide by shift reactions catalyzed by the platinum and thus also lead to poisoning of the electrode material. From thermodynamic equilibrium calculations de Bruijn *et al.* [38] found that 20–100 ppm carbon monoxide can be generated by the reverse water gas shift reaction from 20% CO_2 in the gas stream even at temperatures as low as 40 °C when catalytically active materials such as platinum are involved.

23.3.1.5 Nitrogen

Nitrogen on the air side electrode of the fuel cell is regarded as mostly neutral due to its low reactivity. Even for high temperature fuel cells reactions involving nitrogen are negligible. However, a source of significant NO_x emissions can be nitrogen compounds resulting from nitrogen present in the fuel, as discussed later. High amounts of nitrogen in the fuel gas can result in a significant performance decrease of fuel cells because of dilution of the fuel.

23.3.1.6 Steam

Steam in the fuel gas results from water content of the feedstock, process steam for reforming, gasification, and similar. Steam is not harmful to fuel cells; however, at very low fuel concentrations in high temperature fuel cells it can lead to re-oxidation of anode catalyst material and thus damage the cells. This effect actually limits the applicable fuel utilization. Furthermore, high amounts of steam lead to dilution and performance loss, similar to other neutral substances like nitrogen and carbon dioxide [16]. However, a certain amount of steam is necessary for most fuel cell types in order to operate properly, for example, in the case of PEMFC, where the electrolyte has to be humidified for conducting H^+ ions.

23.3.2
Trace Gas Components

The concentration of contaminants in the fuel gas stream mainly depends on the origin of the gas. There are numerous processes for generating gas from renewable or conventional sources. Table 23.2 shows typical ranges for the most prominent contaminants present in biomass gasification synthesis gas; there is a large variety of contaminants due to the versatility of elements present in different biomass feedstock and the large number of possible reactions occurring at the typically high gasification temperatures.

23.3.2.1 Tars

Tars are by definition mostly cyclic hydrocarbons/aromatics larger than benzene [15]. They are a prominent contaminant in gasification synthesis gases (Table 23.3). The main effects of tars in potential fuel gases are: On the one hand, tars can condense at temperatures below their specific dew point,

Table 23.2 Typical contaminants present in biomass gasification synthesis gas [39].

Contaminant	Range
Tars	1–100 g Nm^{-3}
Particles	2–20 g Nm^{-3}
Alkali metals	0.5–5 ppm
H_2S	50–100 ppm
COS	2–10 ppm
NH_3	200–2000 ppm
Halogens	0–300 ppm

Table 23.3 Tar content in untreated product gas from different gasifiers [44–51].

Gasifier	Tar content (g Nm^{-3})	Favored tar species
Downdraft	0.05–5	Tertiary
Updraft	10–160	Primary and secondary
Stationary fluidized bed	1–23	Secondary and tertiary
Circulating fluidized bed	1–30	Secondary and tertiary
Entrained flow	<0.05	Tars are completely cracked

depending on the concentration in the fuel gas, which results in blocking of piping, filters, and so on. On the other hand, tars can lead to carbon deposition on catalyst material at elevated temperatures, which means the formation of solid carbon of different possible structures (filaments, graphite, others) [13,14]. In low temperature fuel cells hydrocarbon compounds are a catalyst poison since they can be partially reformed by the catalyst, resulting in the formation of catalyst poisons like carbon monoxide.

Similar to other hydrocarbon compounds, tars can be a fuel for high temperature fuel cells, depending on their concentration. Many studies have been performed on the effect of tars from biomass gasification on SOFC, mostly utilizing model tars [10,40–42]. Owing to their high operating temperatures SOFC are considered as most suitable for operating on fuels containing carbon compounds. Small amounts of tars can even increase the cell power since they can be reformed at the anode similarly to other hydrocarbons like, for example, methane [13]. However, available studies indicate that larger tar compounds, for example, naphthalene and pyrene can hinder the reformation of other hydrocarbons. Addition of naphthalene, for example, increases the optimal temperature for reformation of methane and toluene by around 50 K. At high concentrations of tars, high temperature fuel cells also fail completely [43].

However, today almost all available studies are conducted as short-term experiments of only several hours; long-term results are still rare [11].

Table 23.4 Typical particle content in untreated product gas from different gasifiers [33,34].

Gasifier	Particle content (g Nm^{-3})
Downdraft	0.1–8
Updraft	0.1–3
Fluidized bed	20–100

23.3.2.2 Particulate Matter (Particles)

Particulate matter/particles are mainly ash, and unconverted carbon particles. They result from combustion and/or gasification processes of solid fuels (Table 23.4). Their size, shape, and composition can vary over a large range, which makes it difficult to clean them completely from fuel gases. The main effects of particles are: on the one hand, particles can cause blocking of feeding pipes and of pores of fuel cell anodes, and, on the other hand, they can release further volatile gaseous substances directly to the sensitive components, once they have arrived on the surface of the anode material.

23.3.2.3 Alkali Compounds

Alkali compounds lead to deposition and corrosion of stack components and cell material for high temperature fuel cells due to their reactivity [11]. Furthermore, this corrosion can also free other substances that can have negative effects on the fuel cells. For low temperature fuel cells alkali compounds are not very problematic, since they are not present in the gas phase at low temperatures.

23.3.2.4 Sulfur Compounds

Typical sulfur compounds present in fuel gases are H_2S, COS, SO_x, and others. Sulfur compounds at varying concentrations are present in almost all available fuel gas sources, except for clean gases available from electrolysis. In biomass derived gases they result from the inherent sulfur content. For natural gas, sulfur compounds are added artificially as an odorant to provide a scent detectable for humans [52]. Sulfur contamination causes performance loss by reversible (deactivation) or irreversible (degradation) reactions of the contaminant materials and the fuel cell catalyst. The effect of sulfur compounds has been widely studied in the literature for high temperature fuel cells [53–56]. Concerning the degradation mechanism, at least for SOFC, with H_2S, there is an indication that the coverage of catalyst material with contaminants might be a decisive parameter. This theory has been described in detail by Hansen *et al.*. For further information the reader is referred to the relevant article [57]. Similar mechanisms might be possible for other contaminants, but further investigation is needed for more fundamental insight.

In the case of low temperature fuel cells the degradation mechanism is supposed to be similar. Reversible and irreversible bonding of the sulfur to catalyst particles leads to deactivation and/or degradation of the latter [5].

23.3.2.5 Nitrogen Compounds

Nitrogen compounds (NH_3, HCN, NO_x, and others) can in general be a fuel for high temperature fuel cells, or mostly neutral in the case of NO_x. NO_x in large quantities could in theory lead to re-oxidation of nickel catalysts, although to the knowledge of the authors this effect has not been observed yet. Overall, the effect of nitrogen compounds on high temperature fuel cells has not, to date, been studied enough. For low temperature fuel cells nitrogen compounds cause degradation due to irreversible bonding to catalyst material [5]. Thus, very low concentrations are necessary for safe operation.

23.3.2.6 Halogen Compounds

Halogen compounds such as HCl, HF, C_2Cl_4, and others are considered to act similarly to sulfur compounds. Mainly they induce the formation of halides of catalyst metals. This leads to deactivation of the catalyst and, at higher amounts, to irreversible degradation. Furthermore, of importance for high temperature fuel cells is the fact that halides of, for example, nickel can have a lower melting or sublimation point than the operating temperature [56,58]. This precipitates loss of the anode material and thus leads to irreversible degradation. However, overall the effect of halogen compounds seems to be an order of magnitude smaller than that of sulfur compounds [11,56].

23.3.2.7 Siloxanes

Siloxanes can lead to deposition of silicon compounds on the catalyst material of high temperature fuel cells [56,58]. Initial results show a huge effect of even very low concentrations (few ppm to sub-ppm). However, the relevant effects have to be studied in more detail. Siloxanes are mainly present in biogas derived from wastewater treatment at levels up to several ppm [3]. Their origin is anticipated to be silica compounds containing shampoos and other cosmetics, which finally end up in wastewater treatment plants. For low temperature fuel cells, to the knowledge of the authors the effect of siloxanes is as yet unknown.

23.3.2.8 Other Potential Contaminants

Other potential contaminants like phosphorus, metal–organic compounds, and so on can be present in specific fuel sources. However, to date their effects have not been so well investigated and thus they are not discussed in further detail here.

23.4
Pollutant Level Requirements

Table 23.5 summarizes the main impurities along with typical limits considered for safe operation of low and high temperature fuel cells. The values given represent the results of experiments performed at very different operation conditions. This is the reason why the ranges given are quite broad; depending on cell type

Table 23.5 Contaminant limits suggested for safe operation of fuel cells [11,59–64].

Contaminant	Typical limit for low temperature fuel cells	Typical limit for high temperature fuel cells
Particles	Unknown	1 ppm
Sulfur compounds	0.1–1 ppm	1 ppm
Alkali metals	Unknown	1–10 ppm
Halogen compounds	1 ppm	10–100 ppm
Siloxanes	Unknown	<1 ppm
Tars	Unknown	<100 mg Nm^{-3}
CO	5–50 ppm	—
CO_2	1%	—
Nitrogen compounds	<10 ppm	—

and material, for different operating temperatures, pressures, current density, and so on the impact of certain contaminants changes tremendously. However, the lower value given usually represents a safe value for most operating conditions and cell types.

In general in the literature an observable trend is that the contaminant resistivity of low temperature fuel cells decreases with increasing current density, whereas for high temperature fuel cells the opposite is true. This is probably because in low temperature fuel cells, such as PEMFC, the fuel is transported through the electrolyte and the contaminants remain at the electrode, while for high temperature fuel cells the oxidant is transported and an increased amount of (clean) oxidant is able to clean away a larger amount of contaminant species.

For low temperature fuel cells the effect of high temperature contaminants has not been studied in detail and is thus often unknown, since at low temperatures they are not in the gaseous phase, like alkali metals and tars, or are fairly easily cleaned from the fuel gas like particles.

23.5
Technologies to Remove Pollutants

This section briefly describes state of the art technologies for gas cleaning. Starting from requirements defined before, proper selection of the technological gas clean-up for fuel cell systems has to be selected individually depending on various factors. Available technologies are summarized and classified.

Gas clean-up methodology can be separated into hot gas clean-up (HGC) and cold gas clean-up (CGC) systems, either at moderate temperatures (mostly below 80 °C) or high temperatures (up to 1000 °C). While some technologies are capable of removing several or all pollutions in one step (e.g., liquid scrubbing technologies), others are designed for specific removal of a single pollutant. All technologies described in the following consider downstream removal and purification after gas production, for example, from reforming, meaning that the focus is on secondary downstream removal technologies while *in situ* or so-called primary measures within the gas supply systems may very well be at least equally important.

23.5.1
Cold Gas Clean-Up

Cold gas clean-up technology is mainly based on liquid scrubbing technology and adsorption processes with high process maturity and significant amounts of industrial know-how. However, due to low temperatures, approaching ambient temperature, significant cooling and subsequent reheating of the gas stream for downstream synthesis are often required, lowering overall efficiencies and exergy losses, respectively. Specifically, considering high temperature fuel cell applications, cold gas clean-up might not be an option due to the high efficiency losses. The main focus in this work is given to the hot gas clean-up process and the single reduction and conversion steps within.

23.5.2
Hot Gas Clean-Up

The need for hot gas clean-up originates from the goal of minimizing downstream maintenance of syngas process equipment by removal of, mainly, particles and tar. With increasing utilization of syngas towards chemical synthesis and conversion processes and away from simple combustion systems the desire for highly efficient and robust hot gas clean-up has increased again recently. With temperatures well above 200 °C (as compared to cold gas clean-up) many process configurations benefit thermochemically, for example, by reduced waste streams and increased efficiencies. However, many hot gas clean-up steps are still at the research stage, with only a limited number of industrial-scale applications.

23.5.3
Particulate Matter

High temperature removal of particles is the initial clean-up step, with significant improvements in commercial applications over the last 30 years [65]. Most technologies are based on the physical principles of inertial separation (mass and acceleration), barrier filtration, and electrostatic interaction. Figure 23.1 shows the possible technologies.

Cyclones are the most commonly employed initial separation devices, using the density difference and centripetal acceleration principle to separate the heavier solids from the lighter gases. Impact separators and dust agglomerates make up only a small minority of the techniques. Cyclones exist in various designs

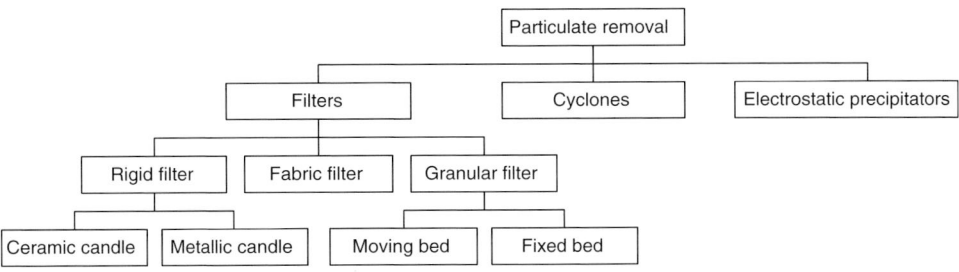

Figure 23.1 Overview of technologies for particulate removal.

depending, for example, on the desired cleaning quality and separation efficiency. Separation efficiencies have been reported recently of up to 99.6%, which is already comparable to low temperature devices [65]. The main design feature of all cyclones is their simplicity, with no moving parts, which enables high temperature operation that is limited merely by the mechanical strength of the construction material. Fouling and corrosion mainly from condensation of water and other compounds is usually avoided by operating at sufficiently high temperatures. Particle sizes down to the μm-range can be removed.

Filtration, mainly carried out by barrier filters is one of the most common technologies for particulate removal in both hot and cold gas clean-up configurations. Particles are separated while passing through a solid barrier of, for example, granules, fibers, or a porous monolith based on a combination of the four basic mechanisms diffusion, inertial impaction, direct interception, and gravitational settling [65]. In addition, particles are also separated by simple restriction of their size due to a specific pore size. This effect can be enhanced by the building of an additional barrier, the filter cake, which in turn hinders particles from passing through. This leads to an increased pressure drop over the filter with a maximum value, after which the filter cake has to be removed, for example, by a reverse jet impulse.

Common filter types are fabric and rigid filters as well as fixed and moving bed granular filters differing in range of maximum operating temperature and particle removal efficiency. Sometimes, pre-coating of the gas stream and the particles with, for example, CaO is also applied to reduce particle penetration into the filter matrix and thus to improve subsequent filtration efficiency. Removal to levels of $1\,\mu g\,m^{-3}$ for particles larger than $1\,\mu m$ is possible with filtration. Further information and detailed description of particulate removal is available in the literature [66,67].

Most fabric filters are limited in their maximum operating temperature to around 250 °C, which is why rigid filters, especially ceramic and metallic candle filters are mostly used for hot gas clean-up. After a rough cleaning step with a cyclone, most hot gas clean-up configurations use either arrays of porous ceramic or sinter metal candle filters with the dirty gas passing from the outside of the long, closed-end tube, depositing the particles as filter cake at the surface. The construction and material of the candle filter elements determine the material's porosity (granules or fibers), strength, and even catalytic activity, which can also be influenced by additional layers of, for example, metals [66]. There are also some variations in filter design, for example cross-flow tube filters with rectangular gas channels instead of filter candles. As for all components of hot gas clean-up technology, filter life time and operational availability are main factors and critical issues for commercial success. The lifetime is decreased mainly by thermal stress and shock due to the reverse jet pulse removal of the filter cake, the aggressive corrosion of gas phase alkali components, and high operating temperatures [66]. Recent developments aim to reduce the filter down times, for example, by using glass ceramic tube filters [68] or by an innovative design of the jet removal system that tries to maintain a constant filter cake layer at the surface [69].

To address the low thermal stress resistance and corrosion problems, sintered metal filters have been developed. Depending on the alloys, operating temperatures of up to $1000\,°C$ can be reached with very high filtration efficiencies. Advantages are: simple construction, high resistance to thermal shock, which simplifies reverse jet pulse cleaning, and high corrosion resistance to, for example, alkali components [66,70].

However, ceramic as well as sintered metal filters can be constructed out of fibers or powders. While fiber-based filters show lower flow resistance at similar strength, which translates to smaller pressure drops, powder-based filters seem to cause fewer problems with plugging and thus show higher reliability.

Fixed or moving bed granular filters use granular material, for example, limestone or sand to clean the particle-laden gas stream. With particle sizes typically in the range of several hundred micrometres, filtration efficiencies can be reached similar to those of bag house fabric filters while the temperature limitations can be exceeded by far (550–600 $°C$ and above). Depending on the design (moving or fixed bed), cleaning or renewal of the surface is achieved by reverse pulses or by replacing the particles at the surface. High availability can be reached with continuous operating moving bed filters, keeping particulate build-up and pressure drop constant while replacing the laden material with clean bed media. Various technological designs have been developed over the years, for example, in combination with electrostatic forces or by using potential adsorbents and catalysts and temperatures up to 870 $°C$ are reported with efficiencies above 99.9% [71].

In general, incorporation of catalytic materials into various filter constructions has been a focus of recent research and development efforts. For example, the addition of catalytic materials for tar reduction, mostly active Ni-based components, is a promising area of research [72,73]. Reduction of complexity, elimination of secondary cleaning measures, and cost savings are the main drivers for such efforts struggling with remaining challenges like small contact times, complex change of catalyst charge, system integration, availability, and high temperature life time.

While state of the art for many flue gas cleaning systems, removing fly ash, for example, from coal-fired power plants, electrostatic precipitators (ESPs) are not

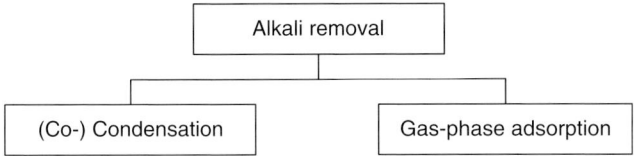

Figure 23.2 Overview of alkali removal options.

commonly used as part of a hot gas clean-up system. However, on using electrostatic forces for particulate removal, high separation efficiencies can be achieved. The accumulating particles are collected on the electrode and removed periodically by a mechanical rapping device. Electrostatic devices have been used at temperatures up to 400 °C and even for some special applications, for example, oil vapor separation, at up to 450 °C. Beyond that, research has been conducted at temperatures up to 1000 °C, indicating, though, lower separation efficiencies [66]. However, high temperatures cause problems like reduced separation efficiencies due to reduced operation range, and plugging problems due to low melting regions of biomass ash. Further research is required here to maintain high availability and improve temperature ranges.

23.5.4
Alkali Components

For hot gas clean-up basically two alkali removal principles are possible: co-removal with other contaminants by condensation mechanism (e.g., on particles) and adsorption of the gaseous alkali component on getter materials (Figure 23.2). If not separated from the gas stream, alkali components cause severe corrosion problems downstream on surfaces below the condensation temperature, like heat exchanger areas, gas turbine blades, or also catalysts.

Depending on the condensation temperature, alkali components nucleate and agglomerate on particulate matter in the gas stream. Around 600–650 °C and below, most alkali vapors can be effectively removed be condensation. Above the condensation temperature, so-called getter materials have to be applied that feature high temperature stability, high adsorption rate, and high loading capacity. Materials such as natural minerals, bauxite, kaolinite, and others have to be selected for specific applications due to process requirements, depending on sorbent life time, temperature, and regeneration ability. For high temperature applications (up to about 800 °C) materials like activated Al_2O_3 and silica SiO_2 may even simultaneously remove chlorine components.

23.5.5
Tars and Higher Hydrocarbons

In general there are different approaches to clean a gas stream of tar contamination. The most obvious is the preventive destruction of the tar *in situ* within the

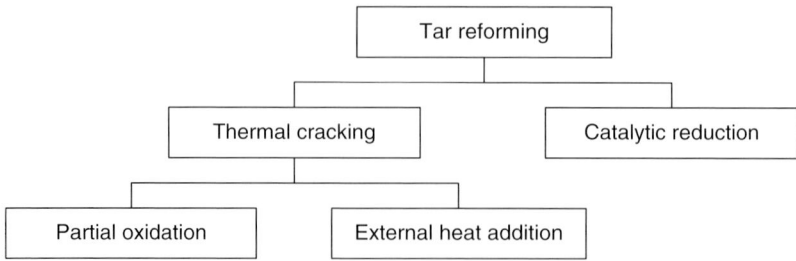

Figure 23.3 Hot gas tar reforming.

gas production process, for example, through using a catalytic active bed material in a gasifier. Downstream processes can be distinguished mainly by the operating temperature and the utilization of a catalyst.

Tars are usually separated by gas scrubbers at low temperatures. Operating at temperatures of up to about 60–80 °C depending on the scrubbing liquid, single- and multi-stage water scrubbers, RME (rapeseed methyl ether) scrubber, or biodiesel scrubbers are used most widely. Through recycling of the loaded scrubbing liquid and thermal utilization of the burnable liquid in the combustion chamber of the gasifier, elaborate cleaning of the liquid (as in case of water as scrubbing liquid) is not necessary. However, a significant cooling of the gas stream is necessary, which negatively affects the cold gas efficiency of the process – more so if higher temperatures are necessary again in any downstream process steps.

At high temperatures tar can be converted and reformed, respectively, by partial oxidation (steam reforming) and thermal or catalytic reforming/cracking (Figure 23.3). Depending on the gasifier technology upstream, the tar content and compositions favor different technologies. All tar conversion technologies have in common that the converted tar is not separated from the gas stream and thus the chemical energy content does not leave the system, but of course the conversion also influences the main gas composition, which may be favorable or not.

23.5.5.1 Thermal Cracking

Thermal cracking uses only the high temperatures for thermal cracking of the tar compounds into smaller gas components. Temperatures starting from 1100 °C have to be applied to break the chemical bonds within the larger molecules. To decrease required residence times even higher temperatures of up to 1300 °C are typically applied. These temperatures can be generated either by external heat exchangers and hot surfaces, respectively, or by partial oxidation of the gas stream. Injection of a small amount of pure oxygen or air can increase the temperature to that required for cracking. While external heating has the disadvantage of high energy input, partial oxidation can lead to dilution of the synthesis gas with lowering of the calorific value. However, thermal cracking can only reduce the tar concentration to a certain amount, for example, down to 15 mg m^{-3} at around 1300 °C [74]. While this may be sufficient for internal combustion engines, it is not sufficient for fuel cells. In addition, the quite simple design is not that simple to implement and is also expensive. Furthermore, there are technological problems like soot formation.

23.5.5.2 Catalytic Reduction

Introduction of a catalyst lowers the temperature required for cracking of the tars, by reduction of the activation energy, to around 850–950 °C. This reduces problems with supplying high temperature heat and high contact times but leads to additional problems associated with catalysis like poisoning, attrition, and carbon deposition.

Various classifications for tar cracking catalysts are discussed in literature, distinguishing, for example, between chemical mechanism, between temperature levels, and complex design parameters. In the following, catalysts are classified simply according to their origin into mineral-based and synthetic catalysts:

Mineral-based catalysts are chemical-untreated homogeneous solids with variable compositions, such as olivine, dolomites, clay minerals, and ferrous metal oxides. Mineral-based catalysts are usually abundant and can thus be supplied with lower costs. Olivine is a silicate-based mineral with high Fe and Mg content, which enhance the catalytic activity. Compared to dolomite, olivine shows slightly lower activity but higher attrition resistance, which in turn favors *in situ* utilization or fluidized bed application, for example, in a gasifier. Clay minerals, with the active components Al_2O_3 and SiO_2, show even lower activity and promote coking [75]. In general, a high iron content in minerals seems to enhance catalytic activity.

Synthetic catalysts require chemical treatment and include alkali metal carbonates, zeolites, transition metal-based catalysts (nickel and iron), and activated alumina or char. However, nickel-based catalysts have proven to be the best choice in many cases and are used widely. Activities are up to ten-times higher compared to dolomite [76]. Nickel-based catalysts are also used for various applications like naphtha and methane reforming, they show water-gas-shift capability, and are also tested for ammonia decomposition. However, Ni-based catalysts have the disadvantage of significant sulfur poisoning (mainly H_2S) and coking tendency especially at lower temperatures of <900 °C. Transition and precious metal-based catalysts consisting of elements like Pt, Rh, Ru, Zr, and Rh show even higher activities (e.g., $Rh/CeO_2/SiO_2$) and better coking resistance but are inherently more expensive [77]. In addition, aluminium-based catalysts (e.g., in the form of zeolites from FCC (fluid catalytic cracking) in the petroleum industry) may be applied for tar conversion, with better sulfur tolerance and good stability but again showing coking problems besides other poisoning. Even char (e.g., from thermochemical conversion of carbon-rich feedstock) can show significant tar reduction potential depending on the widespread properties and types such as charcoal, activated carbon, and char from different woody biomasses.

A further minor method may be the utilization of non-thermal plasma for tar destruction. Pulsed corona, RF plasma, or microwave plasma can be applied

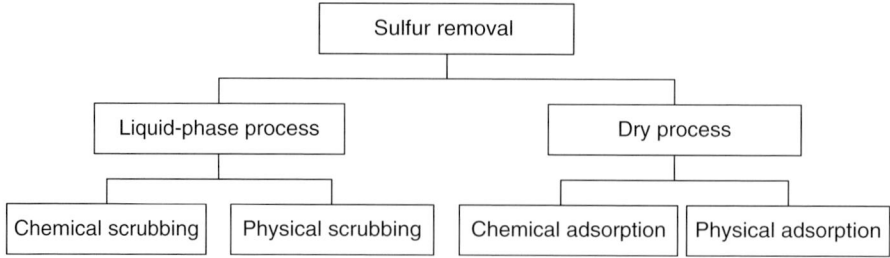

Figure 23.4 Sulfur removal options.

effectively even at elevated temperatures. However, cost, complexity, lifetime, and high required auxiliary operation energy limit the commercial potential of such systems.

23.5.6
Sulfur Components

Removal of sulfur components basically focusses on SO_2 and H_2S as most common and relevant contaminants. Clean-up can be performed either through wet processes or dry removal (Figure 23.4). Many syngas applications require very low sulfur concentrations, for example, H_2S levels of <1 ppm for subsequent catalysis. While SO_2 is mainly removed because of environmental standards for emissions, H_2S accounts for most downstream process failures and lifetime limitations. Furthermore, sulfur removal can be used in large-scale applications (oil, petroleum, and natural gas industry) for elemental sulfur recovery (e.g., Claus process), accounting for a significant share in today's overall sulfur production worldwide. Chemical scrubbing technologies using, for example, amines like diethanolamine (DEA) and methyl-diethanolamine (MDEA) but also sodium hydroxide (NaOH) solutions, as well as physical scrubbing technologies, for example, methanol- or poly(ethylene glycol)-based scrubbing, favor high volume streams and large-scale applications, respectively, but are limited to low temperatures.

For hot gas clean-up of sulfur components, adsorption plays a major role. Technologies can use either a physical or a chemical adsorption mechanism, determining the binding character and thus the binding strength and reversibility of the contaminant sulfur species to the adsorbent. Whereas chemical adsorption involves strong molecular binding, physical adsorption is based on weaker intermolecular interaction allowing, however, multiple layers of adsorbed sulfur species and easier desorption in return. Materials for high temperature adsorption like metal oxides based on Zn, Fe, Cu, Mn, Mo, Co, and V are selected due to desulphurization capability. The most common high temperature sorbents and their capacities are listed in Reference [78]. Combinations of metal oxides with support material based on Al, Si, Ti, and additives are supposed to enhance specific characteristics like thermal durability, capacity, or tolerance towards contaminations. For example, Mn mixed with V or Cu is reported to enable high sulfur removal capacities at temperatures above 600 °C [79]. In the range 400–650 °C and above, Cu_2O, Fe_3O_4, CaO, and more so ZnO as well as combinations thereof can be applied. Coking, for example, on iron- and zinc-based sorbents, as well as sorbent sintering at high temperatures, co-adsorption, and poisoning by other contaminants (e.g., HCl) and, in general, attrition are the main problems to be addressed for each specific technical application.

References

1 Ormerod, R.M. (2003) Solid oxide fuel cells. *Chem. Soc. Rev.*, **32**, 17–28.
2 Wongchanapai, S., Iwai, H., Saito, M., and Yoshida, H. (2012) Performance evaluation of an integrated small-scale SOFC-biomass gasification power generation system. *J. Power Sources*, **216**, 314–322.
3 Allegue, L.B. and Hinge, J. (2012) Report on Biogas and Bio-Syngas Upgrading, Danish Technological Institute, Aarhus.
4 Dicks, A.L. (1996) Hydrogen generation from natural gas for the fuel cell systems of tomorrow. *J. Power Sources*, **61** (1–2), 113–124.
5 Cheng, X., Shi, Z., Glass, N., Zhang, L., Zhang, J., Song, D., Liu, Z.-S., Wang, H., and Shen, J. (2007) A review of PEM hydrogen fuel cell contamination: Impacts, mechanisms, and mitigation. *J. Power Sources* 165, 739–756.
6 Liu, H., Song, C., Zhang, L., Zhang, J., Wang, H., and Wilkinson, D.P. (2006) A review of anode catalysis in the direct methanol fuel cell. *J. Power Sources*, **155** (2) 95–110.
7 Zhang, L., Zhang, J., Wilkinson, D.P., and Wang, H. (2006) Progress in preparation of non-noble electrocatalysts for PEM fuel cell reactions. *J. Power Sources*, **156** (2), 171–182.
8 Liu, S.-H., Wu, J.-R., Pan, C.-J., and Hwang, B.-J. (2014) Synthesis and characterization of carbon incorporated Fe–N/carbons for methanol-tolerant oxygen reduction reaction of polymer electrolyte fuel cells. *J. Power Sources*, **250**, 279–285.
9 Fuerte, A., Valenzuela, R.X., Escudero, M.J., and Daza, L. (2009) Ammonia as efficient fuel for SOFC. *J. Power Sources*, **192** (1), 170–174.
10 Liu, M., van der Kleij, A., Verkooijen, A.H.M., and Aravind, P.V. (2013) An experimental study of the interaction between tar and SOFCs with Ni/GDC anodes. *Appl. Energy*, **108**, 149–157.
11 Aravind, P.V. and de Jong, W. (2012) Evaluation of high temperature gas cleaning options for biomass gasification product gas for solid oxide fuel cells. *Prog. Energy Combust.*, **38**, 737–764.
12 Song, C. (2002) Fuel processing for low-temperature and high-temperature fuel cells: challenges, and opportunities for sustainable development in the 21st century. *Catal. Today*, 77 (1–2) 17–49.
13 Frank, N. (2009) Teerumwandlung in SOFC-Brennstoffzellen, Dissertation, TU München.
14 Takeguchi, T., Kani, Y., Yano, T., Kikuchi, R., Eguchi, K., Tsujimoto, K., Uchida, Y., Ueno, A., Omoshiki, K., Aizawa, M. (2002) Study on steam reforming of CH4 and C2 hydrocarbons and carbon deposition on Ni-YSZ cermets. *J. Power Sources*, **112** (2), 588–595.
15 Rabou, L.P.L.M., Zwart, R.W.R., Vreugdenhil, B.J., and Bos, L. (2009) Tar in biomass producer gas, the Energy research Centre of The Netherlands (ECN) experience: an enduring challenge. *Energy Fuels*, **23**, 6189–6198.
16 Mermelstein, J., Millan, M., and Brandon, N. (2010) The impact of steam and current density on carbon formation from biomass gasification tar on Ni/YSZ, and Ni/CGO solid oxide fuel cell anodes. *J. Power Sources*, **195** (6) 1657–1666.

17 Gong, M., Liu, X., Trembly, J., and Johnson, C. (2007) Sulfur-tolerant anode materials for solid oxide fuel cell applications. *J. Power Sources*, **168**, 289–298.

18 Mukundan, R., Brosha, E.L., and Garzon, F.H. (2004) Sulfur tolerant anodes for SOFCs. *Electrochem. Solid-State Lett.*, 7, A5–A7.

19 Kornely, M., Neumann, A., Menzler, N.H., Weber, A., and Ivers-Tiffée, E. (2011) Degradation of solid oxide fuel cell performance by Cr-poisoning. *ECS Trans.*, **35** (1), 2009–2017.

20 Mokhatab, S. and Poe, W.A. (2012) *Handbook of Natural Gas Transmission and Processing*, Gulf Professional Publishing, ISBN 0123868145.

21 Weiland, P. (2010) Biogas production: current state and perspectives. *Appl. Microbiol. Biotechnol.*, **85**, 849–860.

22 Heinzel, A., Vogel, B., and Hübner, P. (2002) Reforming of natural gas - hydrogen generation for small scale stationary fuel cell systems. *J. Power Sources*, **105** (2), 202–207.

23 Ersoz, A., Olgun, H., and Ozdogan, S. (2006) Reforming options for hydrogen production from fossil fuels for PEM fuel cells. *J. Power Sources*, **154** (1), 67–73.

24 Spliethoff, H. (2010) *Power Generation from Solid Fuels*, Springer, Berlin, Heidelberg, ISBN 978-3-642-02855-7.

25 Stahl, K. and Neergard, M. (1998) Experiences from the biomass fuelled IGCC Plant at Vernamo, in *Proceeding of the 10th International Conference Biomass for Energy, 8–11 June 1998, Wurzburg, Germany* (H. Kopetz, T. Weber, W. Pals, P. Chartier, and G.L. Ferrero), C.A.R.M.E. N. Publishers, Rimpar, Germany, pp. 291–294.

26 Basu, P. (2006) *Combustion and Gasification in Fluidized Beds*, CRC Press Taylor & Francis, Boca Raton, FL, ISBN 978-0-8493-3396-5.

27 Stemmler, M. (2010) Chemische Heißgasreinigung bei Biomassevergasungsprozessen, Dissertation, Hochschule Aachen.

28 Ståhl, K. (2005) CHRISGAS project- clean hydrogen-rich gas through biomass gasification an hot gas upgrading. Presented at 14th European Biomass Conference and Exhibition, 17–21 October 2005, Paris.

29 Ståhl, K. (2004) Biomass IGCC at Värnamo, Sweden – past and future. Presented at GCEP Energy Workshop 27 April 2004, Stanford University, USA.

30 Bengtsson, S. (2009) CHRISGAS syngas Project and the Värnamo Biomass gasification plant. Presented at Training Course and Summer School on Next Generation Biofuels: Development of sustainable chemical processes for production of biofuels and bio-based chemicals from agricultural waste and non-food biomass, 14–18 September 2009, Bologna, Italy.

31 Karl, M. (2009) Wirbelschichtfeuerungen für Biomasse und Ersatzbrennstoffe, Diplomarbeit, TUM/LES.

32 Paisley, M.A. and Overend, R.P. (2002) The Silvagas Process from future energy resources – a commercialisation success. Presented at 12th European Conference and Technology Exhibition on Biomass for Energy, Industry and Climate Protection, 17–21 June 2002, Amsterdam.

33 Hofbauer, H., Veronik, G., Fleck, T., Rauch, R., Mackinger, H., and Fercher, E. (1997) The FICFB – gasification process, in *Developments in Thermochemical Biomass Conversion* (eds A.V. Bridgwater and D.G.B. Boocock), vol. **2**, Blackie Academic & Professional, London, pp. 1016–1025.

34 Marquard-Möllenstedt, T., Stürmer, B., Zuberbühler, U., and Specht, M. (2009) Fuels hydrogen product-absorption enhanced reforming in *Encyclopedia of Electrochemical Power Sources* (editor in chief J. Garche), Elsevier, pp. 249–258.

35 Karl, J. (2004) *Dezentrale Energiesysteme: neue Technologien im liberalisierten Energiemarkt* Oldenbourg Verlag GmbH, ISBN 3-486-27505-4.

36 Ghenciu, A.F. (2002) Review of fuel processing catalysts for hydrogen production in PEM fuel cell systems. *Curr. Opin. Solid State Mater. Sci.*, **6** (5) 389–399.

37 Laosiripojana, N. and Assabumrungrat, S. (2007) Catalytic steam reforming of methane, methanol, and ethanol over

Ni/YSZ: the possible use of these fuels in internal reforming SOFC. *J. Power Sources*, **163** (2), 943–951.

38 de Bruijn, F.A., Papageorgopoulos, D.C., Sitters, E.F., and Janssen, G.J.M. (2002) The influence of carbon dioxide on PEM fuel cell anodes. *J. Power Sources*, **110** (1), 117–124.

39 Bandi, A. (2003) Verfahrensübersicht Gasreinigungsverfahren, in: FVS Fachtagung.

40 Namioka, T., Naruse, T., and Yamane, R. (2011) Behavior and mechanisms of Ni/ScSZ cermet anode deterioration by trace tar in wood gas in a solid oxide fuel cell. *Int. J. Hydrogen Energy*, **36**, 5581–5588.

41 Namioka, T., Nagai, Y., Yoshikawa, K., and Min, T. (2012) A tolerance criterion for tar concentration in a model wood gas for a nickel/scandia-stabilized zirconia cermet anode in a solid oxide fuel cell. *J. Hydrogen Energy*, **36**, 17245–17252.

42 Gorte, R.J., Kim, H., and Vohs, J.M. (2002) Novel SOFC anodes for the direct electrochemical oxidation of hydrocarbon. *J. Power Sources*, **106**, 10–15.

43 Nadine, F. (2010) Umsetzung von Kohlenwasserstoffen in SOFCs, Dissertation, Chair for Energy Systems, Technical University of Munich.

44 Kübel, M. (2007) Teerbildung und Teerkonversion bei der Biomassevergasung - Anwendung der nasschemischen Teerbestimmung nach CEN-Standard, Dissertation Universität Stuttgart.

45 Kaltschmitt, M., Hartmann, H., and Hofbauer, H. (eds) (2009) *Energie aus Biomasse: Grundlagen, Techniken und Verfahren, 2nd* edn, Springer, ISBN 978-3-540-85094-6.

46 Nussbaumer, T. (2002) Luftreinhaltung und explosionsschutz bei holzfeuerungen und stand der technik der holzvergasung, 7 Holzenergie Symposium, 18 October 2002, ETH Zürich.

47 Higman, C. and van der Burgt, M. (2003) *Gasification*, Elsevier, ISBN 978-0-7506-8528-3.

48 Evans, R.J. and Milne, T.A. (1997) Chemistry of tar formation and maturation in the thermochemical conversion of biomass" in *Developments in Thermochemical Biomass Conversion*, vol. **2** (eds A.V. Bridgwater and D.G.B. Boocock), Blackie Academic & Professional, London, pp. 803–816.

49 Volz, F. (2012) Bundesmessprogramm zur Weiterentwicklung der kleintechnischen Biomassevergasung, Verbundprojekt BMU, Förderkennzeichen 03KB017F, ZAE Bayern.

50 Milne, T.A. and Evans, R.J. (1998) Biomass Gasifier "Tars": Their Nature, Formation and Conversion, NREL, Task No BP811010.

51 Schmitz, W. (2001) *Konversion Biogener Brennstoffe für die Nutzung in Gasturbinen*, Fortschr. -Ber. VDI Reihe 6 Nr. 459, VDI-Verlag, Düsseldorf, ISBN 3-18-345906-X.

52 Katikaneni, S., Yuh, C., Abens, S., and Farooque, M. (2002) The direct carbonate fuel cell technology: advances in multi-fuel processing and internal reforming. *Catal. Today*, **77** (1–2), 99–106.

53 Matsuzaki, Y. and Yasuda, I. (2000) The poisoning effect of sulphur-containing impurity gas on a SOFC anode: Part I. Dependence on temperature, time, and impurity concentration. *Solid State Ionics*, **132** (3–4), 261–269.

54 Xia, S.J. and Birss, V.I. (2005) *Solid Oxide Fuel Cells IX* (eds S.C. Singhal and J. Mizusaki), The Electrochemical Society, Pennington, NY, p. 1275.

55 Sasaki, K., Susuki, K., Iyoshi, A., Uchimura, M., Imamura, M., Kusaba, H., and Teraoka, Y. (2006) H_2S poisoning of solid oxide fuel cells. *J. Electrochem. Soc.*, **153** (11), A2023–A2029.

56 Sasaki, K., Haga, K., Yoshizumi, T., Minematsu, D., Yuki, E., Liu, R.R., Uryu, C., Oshima, T., Ogura, T., Shiratori, Y., Ito, K., Koyama, M., and Yokomoto, K. (2011) Chemical durability of solid oxide fuel cells: influence of impurities on long-term performance. *J. Power Sources*, **196** (22), 9130–9140.

57 Hansen, J.B. (2008) Correlating sulphur poisoning of SOFC nickel anodes by a Temkin isotherm. *Electrochem. Solid State Lett.*, **11** (10), B178–B180.

58 Haga, K., Adachi, S., Shiratori, Y., Itoh, K., and Sasaki, K. (2008) Poisoning of SOFC

59 Dayton, D.C., Ratcliff, M., and Bain, R. (2001) Fuel Cell Integration – A study of the impacts of gas quality and impurities, Milestone Completion Report. National Renewable Energy Laboratory (NREL), NREL-MP-510-30298.

58 anodes by various fuel impurities. *Solid State Ionics*, **179** (27–32) 1427–1431.

60 EG&Technical Services, Inc. (2004) Fuel Cell Handbook seventh edition, Contract No. DE-AM26-99FT40575. U.S. Department of Energy, Office of Fossil Energy, National Energy Technology Laboratory.

61 Lymberopoulos, N. (2005) Fuel Cells and their Application in Bio-energy, Centre for Renewable Energy Sources (C.R.E.S).

62 Firor, R.L. (2002) Chemical Analysis in Fuel Cell Systems: Application of the Agilent 5000a Real-time Gas Analyser for Monitoring Low-level Sulfur, Agilent Technologies.

63 McPhail, S.J., Aarva, A., Devianto, H., Bove, R., and Moreno, A. (2011) SOFC and MCFC: Commonalities and opportunities for integrated research. *Int. J. Hydrogen Energy*, **36**, 1033–10345.

64 Lampe, S. (2006) Assessment of Fuel Gas Cleanup Systems for Waste Gas Fueled Power Generation, EPRI, Palo Alto, CA 1012763.

65 Woolcock, P.J. and Brown, R.C. (2013) A review of cleaning technologies for biomass-derived syngas. *Biomass Bioenergy*, **52**, 54–84.

66 Seville, J.P.K. (ed.) (1997) *Gas Cleaning in Demanding Applications*, vol. **xv**, 1st edn, Blackie Academic & Professional, New York, London.

67 Lee, K.W. and Liu, B.Y.H. (1982) Theoretical study of aerosol filtration by fibrous filters, *Aerosol Sci. Technol.*, **1** (2), 147–161.

68 Sharma, S.D., Dolan, M., Park, D., Morpeth, L., Ilyushechkin, A., McLennan, K. *et al.* (2008) A critical review of syngas cleaning technologies - fundamental limitations and practical problems. *Powder Technol.*, **180**, 115–121.

69 Sharma, S.D., Dolan, M., Ilyushechkin, A.Y., McLennan, K.G., Nguyen, T., and Chase, D. (2010) Recent developments in dry hot syngas cleaning processes. *Fuel*, **89** (4), 817–826.

70 Cummer, K.R. and Brown, R.C. (2002) Ancillary equipment for biomass gasification. *Biomass Bioenerg.*, **23** (2), 113–128.

71 Smid, J., Hsiau, S.S., Peng, C.Y., and Lee, H.T. (2005) Hot gas cleanup: new designs for moving bed filters. *Filtr. Separat.*, **42** (10), 36–39.

72 Ma, L. and Baron, G.V. (2008) Mixed zirconia-alumina supports for Ni/MgO based catalytic filters for biomass fuel gas cleaning. *Powder Technol.*, **180** (1–2), 21–29.

73 Rapagna, S., Gallucci, K., Di Marcello, M., Foscolo, P.U., Nacken, M., and Heidenreich, S. (2009) In situ catalytic ceramic candle filtration for tar reforming and particulate abatement in a fluidized-bed biomass gasifier. *Energy Fuel*, **23** (7), 3804–3809.

74 Brandt, P. and Henriksen, U. (2001) Decomposition of tar in gas from updraft gasifier by thermal cracking, in *1st World Conference on Biomass for Energy and Industry, 5–9 June 2000, Sevilla, Spain* (eds S. Kyritsis, A.A.C.M. Beenackers, P. Helm, A. Grassi, and D. Chiaramonti), James & James Ltd., London, pp. 1756–1758.

75 Simell, P.A. and Bredenberg, J.B.S. (1990) Catalytic purification of tarry fuel gas. *Fuel*, **69** (10), 1219–1225.

76 Abu El-Rub, Z., Bramer, E.A., and Brem, G. (2004) Review of catalysts for tar elimination in biomass gasification processes. *Ind. Eng. Chem. Res.*, **43** (22), 6911–6919.

77 Han, J. and Kim, H. (2008) The reduction and control technology of tar during biomass gasification/pyrolysis: an overview. *Renew. Sustain Energy Rev.*, **12** (2), 397–416.

78 Vamvuka, D., Arvanitidis, C., and Zachariadis, D. (2004) Flue gas desulfurization at high temperatures: a review. *Environ. Eng. Sci.*, **21** (4), 525–547.

79 Torres, W., Pansare, S.S., and Goodwin, J.G. (2007) Hot gas removal of tars, ammonia, and hydrogen sulfide from biomass gasification gas. *Catal. Rev. Sci. Eng.*, **49** (4), 407–456.

Edited by Detlef Stolten and Bernd Emonts

Hydrogen Science and Engineering

Materials, Processes, Systems and Technology

Volume 2

Verlag GmbH & Co. KGaA

Editors

Prof. Dr. Detlef Stolten
Forschungszentrum Jülich GmbH
IEK-3: Fuel Cells
Leo-Brandt-Straße
52425 Jülich
Germany

Dr. Bernd Emonts
Forschungszentrum Jülich GmbH
IEK-3: Fuel Cells
Leo-Brandt-Straße
52425 Jülich
Germany

All books published by **Wiley-VCH** are carefully produced. Nevertheless, authors, editors, and publisher do not warrant the information contained in these books, including this book, to be free of errors. Readers are advised to keep in mind that statements, data, illustrations, procedural details or other items may inadvertently be inaccurate.

Library of Congress Card No.: applied for

British Library Cataloguing-in-Publication Data
A catalogue record for this book is available from the British Library.

Bibliographic information published by the Deutsche Nationalbibliothek
The Deutsche Nationalbibliothek lists this publication in the Deutsche Nationalbibliografie; detailed bibliographic data are available on the Internet at http://dnb.d-nb.de.

© 2016 Wiley-VCH Verlag GmbH & Co. KGaA, Boschstr. 12, 69469 Weinheim, Germany

All rights reserved (including those of translation into other languages). No part of this book may be reproduced in any form – by photoprinting, microfilm, or any other means – nor transmitted or translated into a machine language without written permission from the publishers. Registered names, trademarks, etc. used in this book, even when not specifically marked as such, are not to be considered unprotected by law

Print ISBN: 978-3-527-33238-0
ePDF ISBN: 978-3-527-67429-9
ePub ISBN: 978-3-527-67428-2
Mobi ISBN: 978-3-527-67427-5
oBook ISBN: 978-3-527-67426-8

Cover Design   Formgeber, Mannheim
Typesetting   Thomson Digital, Noida, India
Printing and Binding   Markono Print Media Pte Ltd, Singapore

Printed on acid-free paper

Contents

List of Contributors *xxxi*

Volume 1

Part 1 Sol–Gel Chemistry and Methods *1*

1 **Hydrogen in Refineries** *3*
   James G. Speight
1.1 Introduction *3*
1.2 Hydroprocesses *4*
1.2.1 Hydrotreating Processes *6*
1.2.2 Hydrocracking Processes *8*
1.2.3 Slurry Hydrocracking *10*
1.2.4 Process Comparison *10*
1.3 Refining Heavy Feedstocks *11*
1.4 Hydrogen Production *12*
1.5 Hydrogen Management *14*
   References *17*

2 **Hydrogen in the Chemical Industry** *19*
   Florian Ausfelder and Alexis Bazzanella
2.1 Introduction *19*
2.2 Sources of Hydrogen in the Chemical Industry *22*
2.2.1 Synthesis Gas-Based Processes *22*
2.2.2 Steam Reforming *23*
2.2.3 Process Variations *25*
2.2.3.1 Partial Oxidation *25*
2.2.3.2 Autothermal Reforming *25*
2.2.3.3 Pre-reforming *25*
2.2.3.4 Water–Gas Shift Conversion *26*
2.2.3.5 Gasification *26*
2.2.3.6 Other Waste and Coupled Streams *26*
2.2.4 Electrolytic Processes *26*
2.2.4.1 Alkaline Electrolysis *27*

2.2.4.2	PEM Electrolysis	*27*
2.2.4.3	High-Temperature Electrolysis	*27*
2.2.5	Hydrogen Production Steam Reforming versus Electrolysis	*28*
2.2.6	Hydrogen as Coupled Stream in the Electrolytic Production of Chlorine	*28*
2.2.6.1	Membrane Cell Process	*29*
2.2.6.2	Mercury Cell Process	*30*
2.2.6.3	Daphragm Cell Process	*31*
2.2.6.4	New Developments	*31*
2.3	Utilization of Hydrogen in the Chemical Industry	*32*
2.3.1	Ammonia	*32*
2.3.2	Methanol	*34*
2.3.3	Other Uses and Applications of Hydrogen	*36*
2.3.4	Current Developments and Outlook	*37*
	References *38*	

3 Chlorine–Alkaline Electrolysis – Technology and Use and Economy *41*
Alessandro Delfrate

3.1	Introduction	*41*
3.2	Production Technologies	*42*
3.2.1	Electrochemistry of Chlorine Production	*42*
3.2.2	Mercury Electrolyzer Technology	*43*
3.2.3	Diaphragm Electrolyzers	*45*
3.2.4	Ion Exchange Membrane Electrolyzers	*46*
3.2.5	Research	*49*
3.2.6	Breakthrough Technologies: Chlorine-Alkali Production with Oxygen Depolarized Cathode (ODC)	*49*
3.3	Use of Chlorine and Sodium Hydroxide	*52*
3.3.1	Chlorine	*52*
3.3.2	Sodium Hydroxide	*53*
3.3.3	Economy of Chlorine and Caustic Soda	*53*
3.3.4	Energy Savings	*55*
	References *56*	

Part 2 Hydrogen as an Energy Carrier *57*

Part 2.1 Introduction and National Strategies *57*

4 Hydrogen Research, Development, Demonstration, and Market Deployment Activities *59*
Jochen Linssen and Jürgen-Friedrich Hake

4.1	Introduction	*59*
4.2	Germany	*60*

4.2.1	Energy Framework and Relevant Policies	*60*
4.2.2	Hydrogen Related Energy Policy Strategies	*60*
4.2.3	Hydrogen Research, Development, Demonstration, and Deployment Activities	*61*
4.2.3.1	Transportation	*61*
4.2.3.2	Hydrogen Production	*63*
4.2.3.3	Stationary and Residential Applications	*63*
4.2.3.4	Special Markets	*64*
4.2.3.5	Industry Activity	*64*
4.3	Norway	*65*
4.3.1	Energy Framework and Relevant Policies	*65*
4.3.2	Hydrogen Related Energy Policy Strategies	*65*
4.3.3	Hydrogen Research, Development, Demonstration, and Deployment Activities	*66*
4.3.3.1	HyNor Project	*67*
4.3.3.2	ZEG Power	*67*
4.3.3.3	Transnova Hydrogen Projects	*68*
4.4	European Union	*68*
4.4.1	Energy Framework and Relevant Policies	*68*
4.4.2	Hydrogen Related Energy Policy Strategies	*69*
4.4.3	Hydrogen Research, Development, Demonstration, and Deployment Activities	*69*
4.5	Canada	*70*
4.5.1	Energy Framework and Relevant Policies	*70*
4.5.2	Hydrogen Related Energy Policy Strategies	*74*
4.5.3	Hydrogen Research, Development, Demonstration, and Deployment Activities	*74*
4.6	United States of America	*76*
4.6.1	Energy Framework and Relevant Policies	*76*
4.6.2	Hydrogen Related Energy Policy Strategies	*76*
4.6.3	Hydrogen Research, Development, Demonstration, and Deployment Activities	*77*
4.7	Japan	*78*
4.7.1	Energy Framework and Relevant Policies	*78*
4.7.2	Hydrogen Related Energy Policy Strategies	*79*
4.7.3	Hydrogen Research, Development, Demonstration, and Deployment Activities	*80*
4.8	International Networks	*80*
4.8.1	Hydrogen Implementing Agreement of the IEA	*81*
4.8.2	IPHE	*81*
4.8.3	EHA	*81*
4.8.4	International Association for Hydrogen Energy	*82*
	Acknowledgment	*82*
	References	*82*

Part 2.2 Thermochemical Hydrogen Production 85

5 Thermochemical Hydrogen Production – Solar Thermal Water Decomposition 87
Christian Sattler, Nathalie Monnerie, Martin Roeb, and Matthias Lange
5.1   Introduction 87
5.2   Historical Development 88
5.3   Present State of Work 89
5.3.1   Metal/Metal Oxide Thermochemical Cycles 89
5.3.1.1 FeO/Fe_3O_4 90
5.3.1.2 Ferrites 91
5.3.1.3 Hercynite Cycle 93
5.3.1.4 Manganese Ferrite plus Activated Sodium Carbonate 94
5.3.1.5 Zn/ZnO 94
5.3.1.6 CeO_2/Ce_2O_3 95
5.3.2   Sulfur Cycles 96
5.3.2.1 Hybrid Sulfur Cycle (Westinghouse, Ispra Mark 11) 96
5.3.2.2 Mark 13 V2 98
5.3.2.3 Mark 13A 98
5.3.2.4 Sulfur–Iodine or General Atomics Process (ISPRA Mark 16) 99
5.3.3   Other Cycles 100
5.3.3.1 UT3 (Ca/Fe/Br Cycle) 101
5.3.3.2 Hybrid Copper–Chlorine Cycle 101
5.3.3.3 Uranium–Europium Cycle 102
5.4   Conclusion and Outlook 102
        Nomenclature 103
        References 103

6 Supercritical Water Gasification for Biomass-Based Hydrogen Production 109
Andrea Kruse
6.1   Introduction 109
6.1.1   Hydrothermal Biomass Conversions 109
6.1.2   Properties of Water 111
6.2   Model Compounds 113
6.2.1   Glucose 113
6.2.2   Cellulose 115
6.2.3   Amino Acids 115
6.2.4   Phenols 115
6.2.5   Others 116
6.3   Biomass 116
6.3.1   Influence of Salts 118
6.3.2   Influence of Proteins 118
6.3.3   Influence of Lignin 118
6.4   Catalysts 119

6.5	Challenges	*119*
6.5.1	Heating-Up	*119*
6.5.2	Heat Recovery	*120*
6.5.3	Yields	*120*
6.5.4	Salt Deposition	*121*
6.5.5	Material Choice	*121*
6.5.6	Catalyst Stability	*122*
6.6	Scale-Up and Technical Application	*122*
6.7	New Developments	*122*
6.8	Conclusion	*123*
	References	*123*
7	**Thermochemical Hydrogen Production – Plasma-Based Production of Hydrogen from Hydrocarbons**	***131***
	Abdullah Aitani, Shakeel Ahmed, and Fahad Al-Muhaish	
7.1	Introduction	*131*
7.2	Non-thermal Plasma	*132*
7.2.1	Gliding-Arc Plasma	*132*
7.2.2	Microwave Plasma	*136*
7.2.3	Dielectric Barrier Discharge (DBD) Plasma	*139*
7.2.4	Corona Discharge	*142*
7.2.5	Spark and Pulsed Plasmas	*142*
7.3	Thermal Plasma	*144*
7.3.1	DC Torch Plasma	*144*
7.3.2	Three-Phase AC Plasma	*145*
7.3.3	DC-RF Plasma	*145*
7.4	Concluding Remarks	*146*
	Acknowledgment	*147*
	References	*147*
8	**Solar Thermal Reforming**	***151***
	Christos Agrafiotis, Henrik von Storch, Martin Roeb, and Christian Sattler	
8.1	Introduction	*151*
8.2	Hydrogen Production via Methane Reforming	*152*
8.2.1	Thermochemistry and Thermodynamics of Reforming	*152*
8.2.2	Current Industrial Status	*153*
8.3	Solar-Aided Methane Reforming	*154*
8.3.1	Solar Concentration Systems	*154*
8.3.2	Coupling Reforming with Solar Energy: Solar Receiver–Reactor Concepts	*154*
8.3.3	Worldwide Research into Solar Thermal Reforming of Methane	*157*
8.3.3.1	Indirectly Heated Reactors	*159*
8.3.3.2	Directly Irradiated Reactors	*163*
8.4	Current Development Status and Future Prospects	*167*
	References	*169*

9	**Fuel Processing for Utilization in Fuel Cells** *173*
	Ralf Peters
9.1	Introduction *173*
9.2	Scope of the Work and Methodical Approach *174*
9.3	Chemical Engineering Thermodynamics *175*
9.3.1	Thermodynamic Property Relations *175*
9.3.2	Chemical Equilibrium *178*
9.4	Unit Operations *180*
9.4.1	Catalytic Reactors *180*
9.4.1.1	Hydrogen Generation *181*
9.4.1.2	Water–Gas Shift Reaction *184*
9.4.1.3	Preferential Oxidation *186*
9.4.1.4	Selective CO Methanation *187*
9.4.1.5	Catalytic Combustion *188*
9.4.2	Separation Devices *189*
9.4.2.1	Adsorption Process *189*
9.4.2.2	Membrane Process *189*
9.4.3	Balance-of-Plant Components *191*
9.4.3.1	Pumps and Compressors *191*
9.4.3.2	Heat Exchangers and Evaporators *192*
9.5	Subsystems of Fuel Processing *192*
9.5.1	Optimization of the Subsystem Reforming *193*
9.5.1.1	Process Selection and Optimization *193*
9.5.1.2	Catalyst Development *200*
9.5.1.3	Current Challenges in Fuel Processing *201*
9.5.1.4	Technical Outlook *204*
9.5.2	Design of Complete Gas Cleaning System *205*
9.6	Conclusion *208*
	Acknowledgments *209*
	References *209*
10	**Small-Scale Reforming for On-Site Hydrogen Supply** *217*
	Ingrid Schjølberg, Christian Hulteberg, and Dick Lieftink
10.1	Introduction *217*
10.2	Definition *218*
10.3	Reforming Technologies *219*
10.3.1	Steam Methane Reforming (SMR) *219*
10.3.2	Partial Oxidation *220*
10.3.3	Hydrogen Purity *221*
10.4	Feedstock Options *223*
10.4.1	Upgraded Biogas *223*
10.4.2	Biomethanol and bioDME *224*
10.4.3	Biodiesel *224*
10.4.4	Biopropane *225*
10.5	Suppliers and Products *225*

10.5.1	Cost Trends *225*
10.5.1.1	Central Plant Production *225*
10.5.1.2	On-Site SR *226*
10.5.1.3	Economics: On-Site versus Central Plant Production *227*
10.6	Emerging Technologies *228*
10.6.1	Material Development *229*
10.6.2	Reactor Improvements and Design Aspects *230*
10.6.3	Clean-Up Technology *230*
10.6.4	Small-Scale CO_2 Capture *231*
10.7	Process Control *232*
10.7.1	Condition Monitoring *232*
10.7.2	Control Structures *233*
10.8	Safety *234*
10.9	Conclusion *235*
	References *235*

11   Industrial Hydrogen Production from Hydrocarbon Fuels and Biomass *237*
Andreas Jess and Peter Wasserscheid

11.1	Options to Produce Hydrogen from Fuels–An Overview *237*
11.2	Hydrogen Production from Solid Fuels (Coal, Biomass) *242*
11.2.1	Basic Principles and Reactions of Syngas Production from Solid Fuels *242*
11.2.2	Hydrogen Production by Gasification of Solid Fuels *242*
11.3	Syngas by Partial Oxidation of Heavy Oils *244*
11.4	Syngas by Steam Reforming of Natural Gas *246*
11.5	Conclusions *249*
	References *251*

Part 2.3 H_2 from Electricity *253*

12   Electrolysis Systems for Grid Relieving *255*
Filip Smeets and Jan Vaes

12.1	Introduction *255*
12.2	Energy Policies around the Globe Drive Demand for Energy Storage *256*
12.3	The Options for Integration of Intermittent Renewable Energy Sources *261*
12.4	The Evolution of the Demand for Energy Storage *268*
12.5	The Role of Electrolyzers in the Energy Transition *270*
12.6	The Overall Business Case and Outlook *274*
12.6.1	De-carbonization of all Energy Sectors and the Chemical Industry *275*
12.6.2	Combination of Low-Cost Power Generation with Low-Cost Gas Infrastructure *276*

12.6.3	Grid-Scale Renewable Energy Storage 276
12.6.4	Optimization of Power System Operation with Intermittent Generators 277
12.7	Conclusions 278
	References 279

13   Status and Prospects of Alkaline Electrolysis 283
Dongke Zhang and Kai Zeng

13.1	Introduction 283
13.2	Thermodynamic Consideration 285
13.2.1	Theoretical Cell Voltages 285
13.3	Electrode Kinetics 287
13.3.1	Hydrogen Generation Overpotential 288
13.3.2	Oxygen Generation Overpotential 290
13.3.3	Cell Overpotential 291
13.4	Electrical and Transport Resistances 292
13.4.1	Electrical Resistances 292
13.4.2	Bubble Phenomena 293
13.4.2.1	Bubble Departure Diameter Predictions 293
13.4.2.2	Comparison of Model Predictions with Experimental Observations 296
13.5	Research Trends 297
13.5.1	Electrodes 297
13.5.2	Electro-Catalysts 298
13.5.3	Electrolyte and Additives 302
13.5.4	Bubble Management 302
13.6	Summary 303
	References 304

14   Dynamic Operation of Electrolyzers – Systems Design and Operating Strategies 309
Geert Tjarks, Jürgen Mergel, and Detlef Stolten

14.1	Introduction 309
14.2	Process Steps and System Components 310
14.2.1	Alkaline Electrolysis 311
14.2.2	PEM Electrolysis 312
14.2.3	Gas Separation 313
14.2.4	Compression 314
14.2.5	Gas Drying 316
14.2.6	Power Supply 317
14.3	Dynamic Operation of Electrolyzers 317
14.3.1	Operation Range 318
14.3.2	Dynamic System Response 320
14.3.3	Startup and Shutdown 322
14.4	System Design Criterion 322

14.4.1	System Efficiency	*322*
14.4.2	Investment Cost and Hydrogen Production Costs	*325*
14.4.3	Lifetime	*326*
14.4.4	Safety	*327*
14.5	Conclusion	*327*
	References	*328*

15 Stack Technology for PEM Electrolysis *331*
Jürgen Mergel, David L. Fritz, and Marcelo Carmo
15.1 Introduction to Electrolysis *331*
15.1.1 History of PEM Electrolysis *333*
15.1.2 Challenges Facing PEM Electrolysis *334*
15.2 General Principles of PEM Electrolysis *335*
15.2.1 State-of-the-Art *335*
15.2.2 Stack Design *338*
15.2.2.1 Catalyst Coated Membranes (CCMs) *338*
15.2.2.2 Porous Transport Layer (PTL) *340*
15.2.2.3 Separator Plates *343*
15.2.2.4 Stack Operation *345*
15.2.3 Future Trends in PEM Electrolysis *349*
15.2.3.1 Cost Reduction *350*
15.3 Summary *355*
References *356*

16 Reversible Solid Oxide Fuel Cell Technology for Hydrogen/Syngas and Power Production *359*
Nguyen Q. Minh
16.1 Introduction *359*
16.2 Reversible Solid Oxide Fuel Cell Overview *359*
16.2.1 Operating Principles *359*
16.2.2 Features *361*
16.3 Solid Oxide Fuel Cell Technology *366*
16.4 Solid Oxide Electrolysis Cell Technology *372*
16.5 Reversible Solid Oxide Fuel Cell Technology *379*
16.6 Summary *383*
References *383*

Part 2.4 H$_2$ from Biomass *391*

17 Assessment of Selected Concepts for Hydrogen Production Based on Biomass *393*
Franziska Müller-Langer, Konstantin Zech, Stefan Rönsch, Katja Oehmichen, Julia Michaelis, Simon Funke, and Elias Grasemann
17.1 Introduction *393*
17.2 Characteristics of Selected Hydrogen Concepts *394*

17.2.1 Concepts under Discussion *394*
17.2.2 Concepts for Further Assessment *398*
17.2.2.1 Concept A – Wood Chip Gasification *398*
17.2.2.2 Concept B – Biogas Reforming *398*
17.2.2.3 Concept for Hydrogen Distribution of Produced Hydrogen *399*
17.3 Concept Assessment of Technical Aspects *401*
17.3.1 Methodical Approach *401*
17.3.2 Overall Efficiencies *401*
17.4 Concept Assessment of Environmental Aspects *402*
17.4.1 Methodical Approach *403*
17.4.2 GHG Emissions *403*
17.4.3 Cumulated Non-renewable Energy Demand *405*
17.5 Concept Assessment of Economic Aspects *406*
17.5.1 Methodical Approach *406*
17.5.2 Total Capital Investments of Hydrogen Production *407*
17.5.3 Hydrogen Production Costs *408*
17.5.4 Hydrogen Distribution Costs *410*
17.6 Summary *411*
Acknowledgment *411*
References *412*

18 Hydrogen from Biomass – Production Process via Fermentation *417*
Balachandar G., Shantonu Roy, and Debabrata Das
18.1 Introduction *417*
18.1.1 Current Energy Scenario *417*
18.1.2 Importance and Applications of Hydrogen as a Fuel *419*
18.1.3 Conventional Hydrogen Production *419*
18.1.4 Biological Hydrogen Production *419*
18.1.4.1 Biochemistry behind Thermophilic Biohydrogen Production via Dark Fermentation *420*
18.1.4.2 Microbial Characteristics of Thermophilic Hydrogen Producing Bacteria *421*
18.2 Hydrogen Production from Biomass as Feedstock *422*
18.2.1 Agricultural Residues *423*
18.2.2 Municipal Solid Waste and Sewage *424*
18.2.3 Industrial Residues *424*
18.2.3.1 Distillery Industry Waste *425*
18.2.3.2 Food Industry Waste *425*
18.3 Reactor Configurations and Scale-Up Challenges *427*
18.3.1 Reactor Configurations *427*
18.3.2 Current Status of Technologies Available on Scale Up *429*
18.4 Economics and Barriers *430*
18.5 Future Prospects *431*
18.6 Conclusion *431*
Acknowledgment *432*
References *432*

Part 2.5 Hydrogen from Solar Radiation and Algae *439*

19 Photoelectrochemical Water Decomposition *441*
Sebastian Fiechter
19.1 Introduction *441*
19.2 Principles of Photoelectrochemical Water Splitting *442*
19.2.1 Photoelectrochemical Cells with a Single Photoelectrode *443*
19.2.2 Photoelectrochemical Cells with Two Photoelectrodes *446*
19.2.3 Electrocatalysts and Overvoltage *448*
19.3 Design of Water Splitting Devices *448*
19.4 Nano- and Microstructured Photoelectrodes *455*
19.5 Economic Aspects *457*
19.6 Concluding Remarks *457*
References *458*

20 Current Insights to Enhance Hydrogen Production by Photosynthetic Organisms *461*
Roshan Sharma Poudyal, Indira Tiwari, Mohammad Mahdi Najafpour, Dmitry A. Los, Robert Carpentier, Jian-Ren Shen, and Suleyman I. Allakhverdiev
20.1 Introduction *461*
20.2 Biological H_2 Production *463*
20.3 Physiology and Biochemistry of Algae and Cyanobacteria for H_2 Production *465*
20.4 Hydrogenase and Nitrogenase for H_2 Production *466*
20.4.1 Hydrogenase and H_2 Production *466*
20.4.2 Uptake Hydrogenase and Hydrogen Production *467*
20.4.3 Nitrogenase and H_2 Production *469*
20.5 Photosystems and H_2 Production *469*
20.6 Factors Affecting Hydrogen Production *470*
20.7 Designing the Photosynthetic H_2 Production *471*
20.8 Leaf and Solar H_2 Production *472*
20.9 Biofuel and Hydrogen Production by Other Organisms *473*
20.10 Available Methods to Enhance Photosynthetic Hydrogen Production *474*
20.10.1 Photolytic H_2 Production by Microorganisms *474*
20.10.2 Photosynthetic Bacterial H_2 Production *475*
20.10.3 Dark Fermentative H_2 Production *475*
20.10.4 Genetic Engineering to Enhance Hydrogen Production *476*
20.11 Application of Biohydrogen *477*
20.12 Conclusion and Future Prospectus *477*
Acknowledgments *478*
Abbreviations *478*
References *478*

Part 2.6 Gas Clean-up Technologies *489*

21 PSA Technology for H_2 Separation *491*
Carlos A. Grande
21.1 Introduction *491*
21.2 Basics of PSA Technology *492*
21.3 Selective Adsorbents; Commercial and New Materials *499*
21.4 Improving the PSA Cycle *501*
21.5 Summary *503*
Acknowledgments *504*
References *504*

22 Hydrogen Separation with Polymeric Membranes *509*
Torsten Brinkmann and Sergey Shishatskiy
22.1 History *509*
22.2 Basics of Membrane Gas Separation *510*
22.3 Hydrogen Separation and Fractionation by Gas Permeation *516*
22.3.1 Hydrogen/Nitrogen Separation *517*
22.3.2 Hydrogen/Carbon Monoxide Separation *517*
22.3.3 Hydrogen/Hydrocarbon Separation *518*
22.3.4 Hydrogen Separation and Fractionation Applications in Renewable Hydrogen Production *519*
22.4 Membrane Materials and Modules *519*
22.4.1 Membrane Classification *519*
22.4.1.1 Morphological Classification *520*
22.4.1.2 Geometrical Classification *523*
22.4.2 Membrane Defect Curing *524*
22.4.3 Membrane Module Classification *525*
22.4.3.1 Hollow Fiber Modules *525*
22.4.3.2 Flat Sheet Modules *526*
22.4.4 Membrane Material Classification *528*
22.4.4.1 Membranes Based on Rubbery Polymers *528*
22.4.4.2 Glassy Polymer Based Membranes *530*
22.5 Process Examples *531*
22.5.1 Hydrogen Separation from Purge Streams in Ammonia Production *531*
22.5.2 Hydrogen Separation from Hydrocracker Flash Gas *532*
22.5.3 Carbon Dioxide Removal from Biomass Gasification Product Gas *534*
22.6 Conclusions *535*
Nomenclature *536*
References *537*

Contents | XVII

23 **Gas Clean-up for Fuel Cell Systems – Requirements & Technologies** *543*
Matthias Gaderer, Stephan Herrmann, and Sebastian Fendt
23.1 Introduction *543*
23.2 Background *543*
23.3 Fuel and Pollutants *545*
23.3.1 Main Gas Components *545*
23.3.1.1 Hydrogen *546*
23.3.1.2 Carbon Monoxide *546*
23.3.1.3 Methane *546*
23.3.1.4 Carbon Dioxide *547*
23.3.1.5 Nitrogen *547*
23.3.1.6 Steam *547*
23.3.2 Trace Gas Components *547*
23.3.2.1 Tars *547*
23.3.2.2 Particulate Matter (Particles) *549*
23.3.2.3 Alkali Compounds *549*
23.3.2.4 Sulfur Compounds *549*
23.3.2.5 Nitrogen Compounds *550*
23.3.2.6 Halogen Compounds *550*
23.3.2.7 Siloxanes *550*
23.3.2.8 Other Potential Contaminants *550*
23.4 Pollutant Level Requirements *550*
23.5 Technologies to Remove Pollutants *551*
23.5.1 Cold Gas Clean-Up *552*
23.5.2 Hot Gas Clean-Up *552*
23.5.3 Particulate Matter *552*
23.5.4 Alkali Components *555*
23.5.5 Tars and Higher Hydrocarbons *555*
23.5.5.1 Thermal Cracking *556*
23.5.5.2 Catalytic Reduction *557*
23.5.6 Sulfur Components *558*
References *559*

Volume 2

Part 3 **Hydrogen for Storage of Renewable Energy** *563*

24 **Physics of Hydrogen** *565*
Carsten Korte, Tabea Mandt, and Timm Bergholz
24.1 Introduction *565*
24.2 Molecular Hydrogen *565*
24.2.1 The H_2 Molecule *565*
24.2.1.1 Covalent Bonding and Molecular Orbitals *565*
24.2.1.2 Natural Isotopes *566*
24.2.1.3 Nuclear Spin-States, Ortho- and Para-H_2 *566*

24.2.2 Thermodynamic Properties 569
24.2.2.1 Pressure–Temperature Phase Diagram 569
24.2.2.2 Pressure–Volume Phase Diagram 571
24.2.2.3 Joule–Thomson Effect 572
24.2.3 Reaction Kinetics 574
24.2.3.1 Reaction with O_2 – Thermodynamics 574
24.2.3.2 Reaction with O_2 – Microscopic Mechanisms 574
24.2.3.3 Reaction with O_2 – Explosion Limits 577
24.2.4 Transport Kinetics 578
24.2.4.1 Thermal Conductivity 578
24.2.4.2 Diffusion in Gasses 579
24.2.4.3 Permeation and Diffusion in Polymers 580
24.2.4.4 Permeation and Diffusion in Metals 583
24.3 Hydrides 588
24.3.1 Classification and Properties of Hydrides 588
24.3.1.1 Ionic Hydrides 588
24.3.1.2 Covalent Hydrides 590
24.3.1.3 Complex Hydrides (Hydrido Complexes) 591
24.3.1.4 Interstitial (Metallic) Hydrides 592
24.3.2 Formation of Hydrides 593
24.3.2.1 Ionic and Covalent Hydrides 593
24.3.2.2 Interstitial (Metallic) Hydrides 594
24.3.3 Clathrates 597
References 598

25 Thermodynamics of Pressurized Gas Storage 601
Vanessa Tietze and Detlef Stolten
25.1 Introduction 601
25.2 Calculation of Thermodynamic State Variables 602
25.2.1 Ideal Gases 602
25.2.2 Real Gases 603
25.3 Comparison of Thermodynamic Properties 606
25.3.1 Compressibility Factor 606
25.3.2 Joule–Thomson Coefficient 607
25.3.3 Isentropic Exponent 609
25.4 Thermodynamic Analysis of Compression and Expansion Processes 610
25.4.1 Isothermal and Isentropic Compression 611
25.4.2 Isenthalpic and Isentropic Expansion 615
25.5 Thermodynamic Modeling of the Storage Process 617
25.5.1 Governing Equations 617
25.5.2 Heat Transfer Equations 619
25.6 Application Examples 620
25.6.1 Refueling of a Vehicle Storage Tank 620
25.6.2 Salt Cavern 622
25.7 Conclusion 624
References 625

26	**Geologic Storage of Hydrogen – Fundamentals, Processing, and Projects** *629*	

Axel Liebscher, Jürgen Wackerl, and Martin Streibel

- 26.1 Introduction *629*
- 26.2 Fundamental Aspects of Geological Hydrogen Storage *631*
- 26.2.1 Physicochemical Properties of Hydrogen and Hydrogen Mixtures *631*
- 26.2.2 Interaction between Hydrogen and Microbial Inventories *634*
- 26.2.3 General Types of Geological Storage Option *635*
- 26.2.4 Hydrogen Storage in Porous Rocks *636*
- 26.2.5 Hydrogen Storage in Caverns *640*
- 26.3 Process Engineering *642*
- 26.3.1 Gas Transport: Pipelines and Tubing *642*
- 26.3.2 Compressors *643*
- 26.3.3 Metering *644*
- 26.3.4 Controlling and Safety Components *644*
- 26.3.5 Geologic Storages: Survey *645*
- 26.3.6 Geologic Storages: Construction *646*
- 26.3.7 Geologic Storages: Initial Testing *647*
- 26.3.8 Geologic Storages: Operation *648*
- 26.4 Experiences from Storage Projects *649*
- 26.4.1 Hydrogen Storage Projects *650*
- 26.4.2 Pure Hydrogen Storage Projects *650*
- 26.4.2.1 Clemens Dome, Texas, USA *650*
- 26.4.2.2 Moss Bluff, Texas, USA *650*
- 26.4.2.3 Teesside Project, Yorkshire, UK *651*
- 26.4.3 Town Gas Storages *651*
- 26.4.3.1 Town Gas Storage at Ketzin, Germany *652*
- 26.4.3.2 Town Gas Storage at Lobodice, Czech Republic *652*
- 26.4.3.3 Town Gas Storage at Beynes, Ile de France, France *654*
- 26.5 Concluding Remarks *654*
- References *655*

27 Bulk Storage Vessels for Compressed and Liquid Hydrogen *659*

Vanessa Tietze, Sebastian Luhr, and Detlef Stolten

- 27.1 Introduction *659*
- 27.2 Stationary Application Areas and Requirements *660*
- 27.3 Storage Parameters *661*
- 27.4 Compressed Hydrogen Storage *662*
- 27.4.1 Hydrogen Compression *662*
- 27.4.2 Hydrogen Pressure Vessels *663*
- 27.4.2.1 Conventional Small-Scale Bulk Storage *663*
- 27.4.2.2 Current and Potential Small- to Medium-Scale Bulk Storage *665*
- 27.4.2.3 New Ideas for Medium-Scale Bulk Storage *669*
- 27.5 Cryogenic Liquid Hydrogen Storage *670*
- 27.5.1 Hydrogen Liquefaction *670*

27.5.2	Liquid Hydrogen Storage Tanks	*671*
27.6	Cost Estimates and Economic Targets	*675*
27.7	Technical Assessment	*678*
27.8	Conclusion	*683*
	References	*684*

28 Hydrogen Storage in Vehicles *691*
Jens Franzen, Steffen Maus, and Peter Potzel

- 28.1 Introduction: Requirements for Hydrogen Storage in Vehicles *691*
- 28.2 Advantages of Pressurized Storage over Other Storage Methods *693*
- 28.3 Design of a Tank System *695*
- 28.3.1 Flow Diagram and Description of the Components *695*
- 28.3.2 Container *697*
- 28.4 Specific Requirements for Compressed Gas Systems for Vehicles *699*
- 28.4.1 Legal and Normative Requirements *699*
- 28.4.2 Refueling *700*
- 28.5 Special Forms of Compressed Gas Storage *704*
- 28.5.1 Parallel Hydride and Compressed Gas Storage *704*
- 28.5.2 Metal Hydride-Filled Pressure Containers *705*
- 28.5.3 Conformable Containers *705*
- 28.6 Conclusion *707*
- References *707*

29 Cryo-compressed Hydrogen Storage *711*
Tobias Brunner and Oliver Kircher

- 29.1 Motivation for Cryo-compressed Hydrogen Vehicle Storage *711*
- 29.2 Thermodynamic Opportunities *714*
- 29.3 Refueling and Infrastructure Perspectives *717*
- 29.4 Design and Operating Principles *719*
- 29.4.1 System Design *720*
- 29.4.2 Operating Principles *722*
- 29.5 Validation Challenges of Cryo-compressed Hydrogen Vehicle Storage *725*
- 29.5.1 Validation Procedure *726*
- 29.5.2 Validation Challenges and Opportunities *729*
- 29.5.3 Hydrogen Safety Validation *730*
- 29.6 Summary *731*
- References *731*

30 Hydrogen Liquefaction *733*
Alexander Alekseev

- 30.1 Introduction *733*
- 30.2 History of Hydrogen Liquefaction *734*
- 30.3 Hydrogen Properties at Low Temperature *735*
- 30.3.1 Thermodynamic Properties *735*
- 30.3.2 Ortho and Para Modifications of Hydrogen *735*

30.3.2.1 Underlying Physics *735*
30.3.2.2 Ortho-to-Para Conversion and Liquefaction of Hydrogen *737*
30.3.2.3 Available Data *738*
30.3.2.4 Some Useful Thermodynamics *738*
30.4 Principles of Hydrogen Liquefaction *739*
30.4.1 Power Requirements *739*
30.4.2 General Principle *740*
30.4.3 Simple Joule–Thomson Process with Nitrogen Precooling *742*
30.4.3.1 Basic Process *742*
30.4.3.2 Integration of Ortho-to-Para Conversion *744*
30.4.4 Evolution of the Hydrogen Liquefaction Processes *745*
30.4.5 Process Design of the Precooling Part *745*
30.4.6 Precooling by Nitrogen Brayton Cycle *746*
30.4.6.1 Process Description *746*
30.4.6.2 Evaluation *748*
30.4.7 Mixed Gas Refrigeration for Precooling Purposes *748*
30.4.8 Final Cooling *749*
30.5 Key Hardware Components *751*
30.5.1 Compression *752*
30.5.1.1 Impact of the Isentropic Exponent *752*
30.5.1.2 Low Density *753*
30.5.1.3 Screw Compressor *753*
30.5.1.4 Reciprocating (Piston) Compressors *755*
30.5.2 Expansion Turbine (or Expander or Turbine) *755*
30.5.2.1 Oil Bearing *756*
30.5.2.2 Gas Bearing *757*
30.5.3 Heat Exchangers *758*
30.6 Outlook *760*
References *761*

31 Hydrogen Storage by Reversible Metal Hydride Formation *763*
Ping Chen, Etsuo Akiba, Shin-ichi Orimo, Andreas Zuettel, and Louis Schlapbach
31.1 Introduction *763*
31.2 Summary of Energy Relevant Properties of Hydrogen and its Isotopes *764*
31.3 Hydrogen Interaction with Metals, Alloys and Other Inorganic Solids *764*
31.4 Hydrogen Storage in Intermetallic Compounds *767*
31.4.1 AB_5 Type Compounds *771*
31.4.2 AB_2 Type Hydrogen Absorbing Alloys *771*
31.4.3 AB Type Alloy TiFe *772*
31.4.4 Intermetallic Hydrides *773*
31.5 Hydrogen Storage in Complex Hydrides *773*
31.5.1 Alanates (tetrahydroaluminate) *774*

31.5.2 Amides *775*
31.5.3 Borohydrides (Tetrahydroborate) *779*
31.6 Physisorption and High Open-Porosity Structures for Molecular Hydrogen Storage *781*
31.7 Other Energy Relevant Applications of Hydrogen Interacting Materials *784*
31.8 Conclusions and Outlook *785*
References *786*

32 Implementing Hydrogen Storage Based on Metal Hydrides *791*
R.K. Ahluwalia, J.-K. Peng, and T.Q. Hua
32.1 Introduction *791*
32.2 Material Requirements *792*
32.2.1 Operating Temperatures *792*
32.2.2 Material Thermodynamics *794*
32.2.3 Containment Tank *794*
32.2.4 Buffer Hydrogen *796*
32.2.5 Desorption Kinetics *797*
32.2.6 Sorption Kinetics *798*
32.2.7 Material Compaction and Heat Transfer *799*
32.3 Reverse Engineering: A Case Study *800*
32.3.1 MH Refueling: Temperature Profile and Conversion *802*
32.3.2 System Analysis Model *803*
32.3.3 Reference Targets *804*
32.3.4 Sensitivity Study *805*
32.4 Summary and Conclusions *807*
Acknowledgments *808*
References *808*

33 Transport and Storage of Hydrogen via Liquid Organic Hydrogen Carrier (LOHC) Systems *811*
Daniel Teichmann, Wolfgang Arlt, Eberhard Schlücker, and Peter Wasserscheid
33.1 Hydrogen Storage and Transport for Managing Unsteady Renewable Energy Production *811*
33.2 Liquid Organic Hydrogen Carrier (LOHC) Systems *814*
33.3 Development of LOHC-Based Energy Storage Systems *819*
33.4 Applications of LOHC-Based Energy Storage Systems *822*
33.4.1 LOHC Systems for the Storage of Renewable Energy Equivalents, in Particular for Decentralized Storage in Heat–Storage Coupling *823*
33.4.2 Energy Transport over Long Distances with LOHC *824*
33.4.2.1 Import of Solar Energy from Northern Africa *825*
33.4.2.2 Import of Renewable Energy from Iceland *826*
33.4.3 Mobile Applications *827*
33.5 Conclusions *828*
References *829*

Contents | XXIII

Part 4 Traded Hydrogen *831*

34 Economics of Hydrogen for Transportation *833*
Akiteru Maruta
34.1 Introduction *833*
34.2 Hydrogen Transportation System *833*
34.2.1 FCEVs *833*
34.2.2 Hydrogen Infrastructure *835*
34.3 Economics of Hydrogen for Transportation *836*
34.3.1 Hydrogen Cost *836*
34.3.1.1 Two Approaches for Hydrogen Cost Calculation *836*
34.3.1.2 Bottom-Up Approach for Hydrogen Cost *836*
34.3.1.3 Top-Down Approach for Hydrogen Cost *839*
34.3.1.4 Total Cost of Ownership (TCO) Approach *841*
34.3.2 Economics of Social Cost and Benefits *841*
34.3.2.1 Social Costs *842*
34.3.2.2 Social Benefits *843*
34.4 Conclusion *845*
References *846*

35 Challenges and Opportunities of Hydrogen Delivery via Pipeline, Tube-Trailer, LIQUID Tanker and Methanation-Natural Gas Grid *849*
Krishna Reddi, Marianne Mintz, Amgad Elgowainy, and Erika Sutherland
35.1 Introduction *849*
35.2 Variation in Demand for Hydrogen *850*
35.3 Refueling Station Components and Layout *852*
35.3.1 Gaseous Hydrogen Refueling Station *852*
35.3.2 Cryo-Compressed and Gaseous Refueling Station with Liquid Delivery *854*
35.3.3 Refueling Station Challenges *854*
35.4 Distributed Production of Hydrogen *856*
35.5 Central or Semi-central Production of Hydrogen *857*
35.5.1 Gaseous Hydrogen Delivery *857*
35.5.1.1 Pipeline Delivery Pathway *857*
35.5.1.2 Tube-Trailer Delivery Pathway *860*
35.5.1.3 Challenges *863*
35.5.2 Liquid Hydrogen Delivery Pathway *864*
35.5.2.1 Components Layout *864*
35.5.2.2 Cost Estimates *864*
35.5.2.3 Challenges *865*
35.6 Power-to-Gas Mass Energy Solution (Methanation) *866*
35.6.1 Hydrogen Methanation Process *866*
35.6.2 Current Applications *867*
35.6.3 Challenges and Opportunities *869*
35.7 Outlook and Summary *870*
Note *871*
References *872*

36	**Pipelines for Hydrogen Distribution** *875*
	Sabine Sievers and Dennis Krieg

- 36.1 Introduction *875*
- 36.2 Overview *875*
- 36.2.1 Pipelines in Comparison to Other Transportation Possibilities *875*
- 36.2.2 An Overview of Existing Hydrogen Pipelines *876*
- 36.2.3 Some Material Concerns *877*
- 36.3 Brief Summary of Pipeline Construction *879*
- 36.3.1 Planning and Approval *879*
- 36.3.2 Assessment of Preferred Pipeline Routes and Alternative Routes *880*
- 36.3.3 Project Execution, Construction and Commissioning *883*
- 36.4 Operation of an H_2 Pipeline *886*
- 36.4.1 The Control Center *886*
- 36.4.2 Operations *887*
- 36.5 Decommissioning/Dismantling/Reclassification *888*
- 36.6 Conclusion *888*
- References *889*

37	**Refueling Station Layout** *891*
	Patrick Schnell

- 37.1 Introduction *891*
- 37.2 Basic Requirements for a Hydrogen Refueling Station *892*
- 37.2.1 Car Refueling *892*
- 37.2.2 Bus Refueling *894*
- 37.3 Technical Concepts for Hydrogen Filling Stations *895*
- 37.3.1 Hydrogen Production *895*
- 37.3.1.1 Hydrogen Production from Biomass *896*
- 37.3.1.2 Electrolysis *897*
- 37.3.1.3 Steam Reforming *900*
- 37.3.1.4 Byproduct Hydrogen *900*
- 37.3.1.5 Biological Hydrogen Production *900*
- 37.3.2 Hydrogen Delivery *900*
- 37.3.2.1 CGH_2 Delivery *901*
- 37.3.2.2 LH_2 Delivery *901*
- 37.3.2.3 Pipeline Delivery *901*
- 37.3.3 Major Components of Hydrogen Refueling Stations *902*
- 37.3.3.1 Production *902*
- 37.3.3.2 Hydrogen Storage *902*
- 37.3.3.3 Compressors *903*
- 37.3.3.4 Pre-cooling *904*
- 37.3.3.5 Dispenser *905*
- 37.3.3.6 Controls *905*
- 37.3.4 Integration of Hydrogen Refueling Stations *906*
- 37.3.5 Facility Size/Space Requirements *907*
- 37.4 Challenges *907*

37.4.1   Standardization   *907*
37.4.2   Costs   *909*
37.4.3   Reliability   *910*
37.4.4   Approval Processes   *910*
37.4.5   Gauged H_2-Metering   *911*
37.4.6   Refueling According to Technical Guidelines   *911*
37.4.7   Hydrogen Quality   *912*
37.5   Conclusion   *913*
        References   *914*

Part 5   Handling of Hydrogen   *917*

38   Regulations and Codes and Standards for the Approval of Hydrogen Refueling Stations   *919*
        Reinhold Wurster
38.1   Introduction   *919*
38.1.1   Explanation of the term "Regulations, Codes and Standards (RCS)"   *919*
38.1.1.1 Regulations   *919*
38.1.1.2 Codes of Practice   *920*
38.1.1.3 Standards   *921*
38.1.1.4 Referencing of standards   *922*
38.1.1.5 Why Globally Harmonized Standards are Needed   *923*
38.1.2   General Requirements for the Approval of Hydrogen Refueling Stations (HRSs)   *923*
38.2   European Union and Germany   *924*
38.2.1   Europe   *924*
38.2.2   Germany   *928*
        References   *930*

39   Safe Handling of Hydrogen   *933*
        William Hoagland
39.1   Introduction   *933*
39.2   Hydrogen Safety and the Elements of Risk   *934*
39.2.1   Assessing Risk   *935*
39.2.2   Current Shortcomings of QRA for Hydrogen Safety   *935*
39.3   The Unique, Safety-Related Properties of Hydrogen   *937*
39.4   General Considerations for the Safe Handling of Hydrogen   *938*
39.4.1   Gaseous Hydrogen   *939*
39.4.2   Liquid Hydrogen   *939*
39.4.3   Handling Emergencies   *940*
39.5   Regulations, Codes, and Standards   *940*
39.6   International Collaborations to Prioritize Hydrogen Safety Research   *942*

39.6.1　Survey of Hydrogen Risk Assessment Methods, 2008　*942*
39.6.2　Knowledge Gaps White Paper, 2008　*942*
39.6.3　Comparative Risk Assessment Studies of Hydrogen and Hydrocarbon Fueling Stations, 2008　*943*
39.7　Current Directions in Hydrogen Safety Research [6]　*943*
39.7.1　Research on the Physical Behavior of Leaked or Leaking Hydrogen　*943*
39.7.2　Hydrogen Storage Systems Safety Research　*944*
39.7.3　Research that Supports Early Market Applications　*945*
39.7.4　Research on Risk Mitigation Measures　*945*
39.7.5　Simplified Tools to Assess and Mitigate Risk　*947*
39.8　Summary　*947*
　　　References　*948*
　　　Bibliography　*948*

Part 6　Existing and Emerging Systems　*949*

40　Hydrogen in Space Applications　*951*
Jérôme Lacapere
40.1　Liquid Hydrogen for Access to Space　*951*
40.1.1　Liquid Storage　*951*
40.1.2　Constraints Due to Liquid Hydrogen Use　*952*
40.1.3　Insulation　*953*
40.2　To Go Beyond GTO　*954*
40.2.1　Coasting Phase　*954*
40.2.2　Re-ignition Preparation Phase　*955*
40.3　Relevant Tests in Low Gravity Environment　*958*
40.4　In-Space Propulsion　*960*
40.5　Conclusion　*961*
　　　References　*963*

41　Transportation/Propulsion/Demonstration/Buses: The Design of the Fuel Cell Powertrain for Urban Transportation Applications (Daimler)　*965*
Wolfram Fleck
41.1　Introduction　*965*
41.2　Operational Environment　*966*
41.3　Requirements　*967*
41.3.1　Propulsion Power to Drive　*967*
41.3.2　Auxiliary Power Demand　*970*
41.3.3　Heating and Air Condition Power Demand　*971*
41.4　Design Solutions　*973*
41.4.1　NEBUS　*973*
41.4.1.1　Design Solution　*973*
41.4.2　CUTE and HyFleet: CUTE Program　*975*
41.4.3　Citaro FuelCELL-Hybrid　*978*

41.5   Test and Field Experience   *982*
41.5.1   NeBus   *982*
41.5.2   CTA, CMBC   *982*
41.5.3   CUTE and HyFleet:CUTE Program   *984*
41.5.4   NaBuz DEMO and CHIC   *984*
41.6   Future Outlook   *986*
41.6.1   Transit Bus Applications   *986*
41.6.2   Truck Applications   *987*
41.6.3   Life Time and Product Costs   *989*
41.6.4   Summary   *989*
         References   *990*

**42   Hydrogen and Fuel Cells in Submarines   *991*
*Stefan Krummrich and Albert Hammerschmidt***
42.1   Background   *991*
42.1.1   When it All Began . . .   *992*
42.2   The HDW Fuel Cell AIP System   *992*
42.3   PEM Fuel Cells for Submarines   *993*
42.3.1   Introduction   *993*
42.3.2   The Oxygen/Hydrogen Cell Design   *994*
42.3.2.1 Constructive Features/Cell Design of Siemens PEM Fuel Cell   *994*
42.3.2.2 Results from Fuel Cell Operation   *995*
42.3.3   Constructive Feature of Fuel Cell Module for Submarine Use   *995*
42.3.3.1 Preconditions   *995*
42.3.3.2 Cascaded Fuel Cell Stacks [2]   *996*
42.3.3.3 Pressure Cushion for Uniform Current Distribution [4]   *998*
42.3.3.4 Fuel Cell Module   *999*
42.3.3.5 Results from Fuel Cell Module Operation   *1000*
42.3.3.6 Safety Features of Submarine Fuel Cell Modules   *1001*
42.4   Hydrogen Storage   *1002*
42.5   The Usage of Pure Oxygen   *1004*
42.6   System Technology – Differences Between HDW Class 212A and Class 214 Submarines   *1005*
42.7   Safety Concept   *1006*
42.8   Developments for the Future – Methanol Reformer for Submarines   *1006*
42.8.1   System Configuration   *1006*
42.8.2   Challenges of the Methanol Reformer Development   *1007*
42.8.3   Hydrogen Purification Membranes   *1008*
42.8.4   High Pressure Catalytic Oxidation   *1008*
42.8.5   Integration on Board a Submarine   *1008*
42.9   Conclusion   *1009*
         References   *1010*

43 Gas Turbines and Hydrogen *1011*
Peter Griebel
43.1 Introduction *1011*
43.2 Combustion Fundamentals of Hydrogen relevant for Gas Turbines *1012*
43.2.1 Ignition Delay *1014*
43.2.2 Flame Speed *1016*
43.2.3 Flame Temperature, Stability, and Emissions *1018*
43.3 State-of-the-art Gas Turbine Technology for Hydrogen *1019*
43.4 Research and Development Status, New Combustion Technologies *1022*
43.4.1 New Combustion Technologies *1024*
43.4.1.1 MILD Combustion or Flameless Oxidation *1024*
43.4.1.2 Lean Direct Injection, Multi-injection, and Micro-mixing *1026*
43.5 Concluding Remarks *1028*
References *1028*

44 Hydrogen Hybrid Power Plant in Prenzlau, Brandenburg *1033*
Ulrich R. Fischer, Hans-Joachim Krautz, Michael Wenske, Daniel Tannert, Perco Krüger, and Christian Ziems
44.1 Introduction *1033*
44.2 Description of the Concept of the Hybrid Power Plant at Prenzlau *1035*
44.2.1 Overview *1035*
44.2.2 Alkaline Electrolyzer *1036*
44.2.3 Safety Engineering *1039*
44.3 Operating Modes of the Hybrid Power Plant *1042*
44.3.1 Hydrogen Production Mode *1042*
44.3.2 Base Load Mode *1043*
44.3.3 Forecast Mode *1044*
44.3.4 EEX Mode *1045*
44.4 Operational Management and Experiences *1045*
44.4.1 Dynamic Load Operation *1045*
44.4.2 Temperature Influence on Stack Voltage *1047*
44.4.3 Influence of Activated Electrodes *1049*
44.4.4 Voltage Efficiency of the Electrolyzer *1050*
44.5 Outlook *1050*
References *1051*

45 Wind Energy and Hydrogen Integration Projects in Spain *1053*
Luis Correas, Jesús Simón, and Milagros Rey
45.1 Introduction *1053*
45.2 The Role of Hydrogen in Wind Electricity Generation *1055*
45.2.1 Mini Grids *1056*
45.2.2 Electricity Storage *1056*

45.2.3	Fuel Production *1058*
45.2.4	Comparison of the Three Configurations *1059*
45.3	Description of Wind–Hydrogen Projects *1059*
45.3.1	RES2H2 Project *1060*
45.3.2	Hidrólica Project *1061*
45.3.3	ITHER Project *1063*
45.3.4	Sotavento Project *1064*
45.4	Operation Strategies Tested in the Sotavento Project *1066*
45.4.1	Peaking Plant Strategy *1067*
45.4.2	Strategy of Deviation Correction *1069*
45.4.3	Strategy for Increasing the Capacity Factor of the Wind Farm *1070*
45.4.4	Load Leveling Enabling Distributed Generation or Island *1071*
45.5	Conclusions *1071*
	References *1072*

46 Hydrogen Islands – Utilization of Renewable Energy for an Autonomous Power Supply *1075*
Frano Barbir

46.1	Introduction *1075*
46.2	Existing Hydrogen Projects on Islands *1077*
46.3	System Design/Configuration *1082*
46.4	Key Technologies *1083*
46.4.1	Electrolyzer *1083*
46.4.2	Hydrogen Storage *1085*
46.4.3	Fuel Cell *1086*
46.5	System Issues *1087*
46.5.1	Capacity Factor *1087*
46.5.2	Coupling Efficiency *1087*
46.5.3	Intermittent Operation *1088*
46.5.4	Water Consumption *1088*
46.5.5	Performance Degradation with Time *1088*
46.6	Sizing *1088*
46.7	Energy Management *1090*
46.8	Other Uses/System Configurations *1092*
46.8.1	Demand Side Management *1092*
46.8.2	Seawater Desalination *1092*
46.8.3	Oxygen Use *1093*
46.8.4	Fuel for Vehicles *1093*
46.9	Conclusions *1093*
	References *1094*

Index *1097*

List of Contributors

Christos Agrafiotis
Deutsches Zentrum für Luft- und
Raumfahrt e.V.
Linder Höhe
51147 Köln
Germany

Rajesh Ahluwalia
Argonne National Laboratory
Nuclear Engineering Division
9700 S. Cass Avenue
Lemont, IL 60439
USA

Abdullah M. Aitani
King Fahd University of Petroleum
& Minerals (KFUPM)
Center of Research Excellence in
Petroleum Refining &
Petrochemicals
University Blvd. 4597
34463 Dhahran
Saudi Arabia

Etsuo Akiba
Kyushu University
International Institute for
Carbon-Neutral Energy Research
(WPI-I2CNER)/Department of
Mechanical Engineering
Fukuoka City, Kyushu
Japan

Alexander Alekseev
Linde AG
Dr.-Carl-von-Linde-Str. 6–14
82049 Höllriegelskreuth
Germany

Suleyman I. Allakhverdiev
Institute of Plant Physiology
Russian Academy of Sciences
Botanicheskaya St. 35
Moscow 127276
Russia

and

Institute of Basic Biological
Problems
Russian Academy of Sciences
Pushchino
Moscow Region 142290
Russia

and

Department of Plant Physiology
Faculty of Biology
M.V. Lomonosov Moscow State
University
Leninskie Gory 1-12
Moscow 119991
Russia

List of Contributors

Wolfgang Arlt
Lehrstuhl für Thermische
Verfahrenstechnik der Friedrich-
Alexander-Universität (FAU)
Erlangen-Nürnberg
Egerlandstrasse 3
91058 Erlangen
Germany

Florian Ausfelder
DECHEMA Gesellschaft für
Chemische Technik und
Biotechnologie e.V.
Theodor-Heuss-Allee 25
60486 Frankfurt am Main
Germany

Frano Barbir
University of Split
Faculty of Electrical Engineering
Mechanical Engineering and Naval
Architecture (FESB)
R. Boskovica 32
21000 Split
Croatia

Alexis Bazzanella
DECHEMA Gesellschaft für
Chemische Technik und
Biotechnologie e.V
Theodor-Heuss-Allee 25
60486 Frankfurt
Germany

Timm Bergholz
Forschungszentrum Jülich GmbH
Institut für Energie- und
Klimaforschung
Elektrochemische
Verfahrenstechnik (IEK-3)
Wilhelm-Johnen-Straße
52428 Jülich
Germany

Torsten Brinkmann
Helmholtz-Zentrum Geesthacht,
Centre for Materials and Coastal
Research
Insitute of Polymer Research
Verfahrenstechnik
Max-Planck-Straße 1
21502 Geesthacht
Germany

Tobias Brunner
BMW Group
Knorrstr.147
80788 München
Germany

Marcelo Carmo
Forschungszentrum Jülich GmbH
Institute of Energy and Climate
Research
IEK-3: Electrochemical Process
Engineering
Wilhelm-Johnen-Straße
52425 Jülich
Germany

Robert Carpentier
Groupe de Recherche en Biologie
Végétale, Université du Québec à
Trois-Rivières
C.P.500, Trois-Rivières
QC G9A 5H7
Canada

Ping Chen
Chinese Academy of Sciences
Dalian Institute of Chemical Physics
457 Zhongshan Road
Dalian 116023
China

Luis Correas
Fundación para el Desarrollo de las
Nuevas Tecnologías del Hidrógeno
en Aragón
Parque tecnológico Walqa
22197 Huesca
Spain

Debabrata Das
Indian Institute of Technology
Kharagpur
Department of Biotechnology
Kharagpur 721302
India

Alessandro Delfrate
ThyssenKrupp Uhde Chlorine
Engineers (Italia) S.r.l.
Via L. Bistolfi 35
20134 Milan
Italy

Amgad Elgowainy
Argonne National Laboratory
9700 S. Cass Avenue
Lemont, IL 60439
USA

Sebastian Fendt
Technical University Munich
Institute Energy Systems
Boltzmannstrasse 15
85748 Garching
Germany

Sebastian Fiechter
Helmholtz-Zentrum Berlin für
Materialien und Energie GmbH
Institute for Solar Fuels
Hahn-Meitner-Platz 1
14109 Berlin
Germany

Ulrich R. Fischer
BTU Cottbus
H2-Forschungszentrum
03013 Cottbus
Germany

Wolfram Fleck
Daimler AG
73230 Kirchheim / Teck- Nabern
Germany

Jens Franzen
Daimler AG
HPC NAB
73230 Kirchheim unter Teck
Germany

David L. Fritz
Forschungszentrum Jülich GmbH
Institute of Energy and Climate
Research
IEK-3: Electrochemical Process
Engineering
Wilhelm-Johnen-Straße
52425 Jülich
Germany

Simon Funke
Fraunhofer Institute for Systems
and Innovation Research ISI
Breslauer Str. 48
76139 Karlsruhe
Germany

Balachandar G.
Indian Institute of Technology
Kharagpur
Department of Biotechnology
Kharagpur 721302
India

Matthias Gaderer
Technical University Munich
Institute Regenerative Energy Systems
Schulgasse 16
94315 Straubing
Germany

Carlos A. Grande
Senior Research Scientist
SINTEF Materials and Chemistry
Forskningsveien 1
0373 – Oslo
Norway

Elias Grasemann
Deutsches Biomasseforschungszentrum Gemeinnützige GmbH
Department Biorefineries
Torgauer Str. 116
04347 Leipzig
Germany

Peter Griebel
German Aerospace Center (DLR)
Institute of Combustion Technology
Pfaffenwaldring 38–40
70569 Stuttgart
Germany

Jürgen-Friedrich Hake
Forschungszentrum Jülich
Institute of Energy and Climate Research – Systems Analysis and Technology Evaluation (IEK-STE)
Wilhelm-Johnen-Straße
52425 Jülich
Germany

Albert Hammerschmidt
Siemens AG
Coburger Str. 47a
91056 Erlangen
Germany

Stephan Herrmann
Technical University Munich
Institute Energy Systems
Boltzmannstrasse 15
85748 Garching
Germany

William Hoagland
Element One, Inc.
7253 Siena Way
Boulder, CO 80301
USA

Thanh Hua
Argonne National Laboratory
Nuclear Engineering Division
9700 S. Cass Avenue
Lemont, IL 60439
USA

Christian Hulteberg
Lund University
Chemical Engineering
Box 124
22100 Lund
Sweden

Andreas Jess
Lehrstuhl für Chemische Verfahrenstechnik der Universität Bayreuth
Universitätsstrasse 30, 95440 Bayreuth
Germany

Oliver Kircher
BMW Group
Knorrstr.147
80788 München
Germany

Carsten Korte
Forschungszentrum Jülich GmbH
Institut für Energie- und
Klimaforschung
Elektrochemische
Verfahrenstechnik (IEK-3)
Wilhelm-Johnen-Straße
52428 Jülich
Germany

Hans-Joachim Krautz
BTU Cottbus
H2-Forschungszentrum
03013 Cottbus
Germany

Dennis Krieg
AIR LIQUIDE Advanced
Technologies GmbH
Hans-Güther-Sohl-Str. 5
40235 Düsseldorf
Germany

Perco Krüger
BTU Cottbus
H2-Forschungszentrum
03013 Cottbus
Germany

Stefan Krummrich
ThyssenKrupp Marine Systems
GmbH
Werftstraße 112–114
24143 Kiel
Germany

Andrea Kruse
Universität Hohenheim
Institute of Agricultural
Engineering (44U1)
Garbenstraße 9
70599 Stuttgart
Germany

Jérôme Lacapere
Absolut System
2 Rue des Murailles
38170 Seyssinet Pariset
France

Matthias Lange
Deutsches Zentrum für Luft- und
Raumfahrt e.V.
Linder Höhe
51147 Cologne
Germany

Axel Liebscher
GFZ German Research Centre for
Geosciences
Centre for Geological Storage
Telegrafenberg
14473 Potsdam
Germany

Dick Lieftink
HyGear
P.O. Box 5280
6802 EG Arnhem
Norway

Jochen Linssen
Forschungszentrum Jülich
Institute of Energy and Climate
Research – Systems Analysis and
Technology Evaluation (IEK-STE)
Wilhelm-Johnen-Straße
52425 Jülich
Germany

Dmitry A. Los
Institute of Plant Physiology
Russian Academy of Sciences
Botanicheskaya St. 35
Moscow 127276
Russia

Sebastian Luhr
Forschungszentrum Jülich GmbH
Institute für Elektrochemische
Verfahren
Wilhelm-Johnen-Straße
52425 Jülich
Germany

Tabea Mandt
Forschungszentrum Jülich GmbH
Institut für Energie- und
Klimaforschung
Elektrochemische
Verfahrenstechnik (IEK-3)
Wilhelm-Johnen-Straße
52428 Jülich
Germany

Akiteru Maruta
Technova Inc.
13th Floor Imperial Hotel Tower
Uchisaiwai-cho, Chiyoda-ku
Tokyo 100-0011
Japan

Steffen Maus
Daimler AG
HPC T332
70546 Stuttgart
Germany

Jürgen Mergel
Forschungszentrum Jülich GmbH
Institute of Energy and Climate
Research
IEK-3: Electrochemical Process
Engineering
Wilhelm-Johnen-Straße
52425 Jülich
Germany

Julia Michaelis
Fraunhofer Institute for Systems
and Innovation Research ISI
Breslauer Str. 48
76139 Karlsruhe
Germany

Nguyen Q. Minh
University of California, San Diego
Center for Energy Research
9500 Gilman Drive
La Jolla, CA 92093-0417
USA

Marianne Mintz
Argonne National Laboratory
9700 S. Cass Avenue
Lemont, IL 60439
USA

Nathalie Monnerie
Deutsches Zentrum für Luft- und
Raumfahrt e.V.
Linder Höhe
51147 Cologne
Germany

Franziska Müller-Langer
Deutsches
Biomasseforschungszentrum
Gemeinnützige GmbH
Department Biorefineries
Torgauer Str. 116
04347 Leipzig
Germany

Mohammad Mahdi Najafpour
Department of Chemistry
Institute for Advanced Studies in
Basic Sciences (IASBS)
Zanjan, 45137-66731
Iran

and

Center of Climate Change and
Global Warming
Institute for Advanced Studies in
Basic Sciences (IASBS)
Zanjan, 45137-66731
Iran

Katja Oehmichen
Deutsches
Biomasseforschungszentrum
Gemeinnützige GmbH
Department Bioenergy Systems
Torgauer Str. 116
04347 Leipzig
Germany

Shin-ichi Orimo
Tohoku University
WPI Advanced Institute for
Materials Research (WPI-AIMR)/
Institute for Materials Research
(IMR)
Sendai, Miyagi
Japan

Jui-Kun Peng
Argonne National Laboratory
Nuclear Engineering Division
9700 S. Cass Avenue
Lemont, IL 60439
USA

Ralf Peters
Forschungszentrum Jülich GmbH
Institute of Energy and Climate
Research
IEK-3: Energy Process Engineering
Wilhelm-Johnen-Straße
52428 Jülich
Germany

Peter Potzel
Daimler AG
HPC X456
71059 Sindelfingen
Germany

Roshan Sharma Poudyal
Department of Molecular Biology
Pusan National University
Busan 609-735
Republic of Korea

and

Central Department of Botany
Tribhuvan University
Kirtipir, Kathmandu
Nepal

Krishna Reddi
Argonne National Laboratory
9700 S. Cass Avenue
Lemont, IL 60439
USA

Milagros Rey
Gas Natural, S.A.
Plaza del Gas No. 1
08003 Barcelona
Spain

Martin Roeb
Deutsches Zentrum für Luft- und
Raumfahrt e.V.
Linder Höhe
51147 Cologne
Germany

Stefan Rönsch
Deutsches Biomasseforschungszentrum Gemeinnützige GmbH
Department Biorefineries
Torgauer Str. 116
04347 Leipzig
Germany

Shantonu Roy
Indian Institute of Technology Kharagpur
Department of Biotechnology
Kharagpur 721302
India

Christian Sattler
Deutsches Zentrum für Luft- und Raumfahrt e.V.
Linder Höhe
51147 Cologne
Germany

Ingrid Schjølberg
Norwegian University of Science and Technology (NTNU)
IVT-IMT
7491 Trondheim
Norway

Louis Schlapbach
NIMS National Institute for Materials Science
Tsukuba, Ibaraki
Japan

and

Kyushu University
WPI I2CNER
Fukuoka City, Kyushu
Japan

Eberhard Schlücker
Lehrstuhl für Prozessmaschinen und Anlagenbau, Friedrich-Alexander-Universität (FAU) Erlangen-Nürnberg
Cauerstrasse 4
91058 Erlangen
Germany

Patrick Schnell
TOTAL Deutschland GmbH
Jean-Monnet-Straße 2
10557 Berlin
Deutschland
Germany

Jian-Ren Shen
Photosynthesis Research Center
Graduate School of Natural Science and Technology/Faculty of Science
Okayama University
Okayama 700-8530
Japan

Sergey Shishatskiy
Helmholtz-Zentrum Geesthacht, Zentrum für Material- und Küstenforschung
Verfahrenstechnik
Max-Planck-Straße 1
21502 Geesthacht
Germany

Sabine Sievers
AIR LIQUIDE Deutschland GmbH
H2 Energy Deutschland
Hans-Güther-Sohl-Str. 5
40235 Düsseldorf
Germany

Jesús Simón
Fundación para el Desarrollo de las
Nuevas Tecnologías del Hidrógeno
en Aragón
Parque tecnológico Walqa
22197 Huesca
Spain

Filip Smeets
Hydrogenics Europe NV
Nijverhiedstraat 48c
2260 Oevel
Belgium

James G. Speight
CD&W Inc.
2476 Overland Road
Laramie, WY 82070
USA

Detlef Stolten
Forschungszentrum Jülich GmbH
Institute of Energy and Climate
Research
IEK-3: Electrochemical Process
Engineering
Wilhelm-Johnen-Straße
52425 Jülich
Germany

Martin Streibel
GFZ German Research Centre for
Geosciences
Centre for Geological Storage
Telegrafenberg
14473 Potsdam
Germany

Erika Sutherland
US Department of Energy
Fuel Cell Technologies Office
1000 Independence Avenue, SW
Washington, DC 20585
USA

Daniel Tannert
BTU Cottbus
H2-Forschungszentrum
03013 Cottbus
Germany

Daniel Teichmann
Hydrogenious Technologies GmbH
Weidenweg 13
91058 Erlangen
Germany

Vanessa Tietze
Forschungszentrum Jülich GmbH
Institute für Elektrochemische
Verfahren
Wilhelm-Johnen-Straße
52425 Jülich
Germany

Indira Tiwari
Department of Microbiology
Pusan National University
Busan 609-735
Republic of Korea

Geert Tjarks
Forschungszentrum Jülich GmbH
Institute of Energy and Climate
Research
IEK-3: Electrochemical Process
Engineering
Wilhelm-Johnen-Straße
52425 Jülich
Germany

Jan Vaes
Hydrogenics Europe NV
Nijverhiedstraat 48c
2260 Oevel
Belgium

Henrik von Storch
Deutsches Zentrum für Luft- und Raumfahrt e.V.
Professor-Rehm-Straße 1
52428 Jülich
Germany

Jürgen Wackerl
Research Centre Jülich
Institute of Energy and Climate Research (IEK)
52425 Jülich
Germany

Peter Wasserscheid
Lehrstuhl für Chemische Reaktionstechnik der Friedrich-Alexander-Universität (FAU) Erlangen-Nürnberg, Egerlandstrasse 3, 91058 Erlangen
Germany

and

Forschungszentrum Jülich, „Helmholtz-Institut Erlangen-Nürnberg" (IEK-11), Nägelsbachstrasse 49b, 91052 Erlangen
Germany

Michael Wenske
BTU Cottbus
H2-Forschungszentrum
03013 Cottbus
Germany

Reinhold Wurster
Ludwig-Bölkow-Systemtechnik GmbH
Daimlerstr. 15
85521 Ottobrunn
Germany

Konstantin Zech
Deutsches Biomasseforschungszentrum Gemeinnützige GmbH
Department Biorefineries
Torgauer Str. 116
04347 Leipzig
Germany

Kai Zeng
The University of Western Australia
UWA Centre for Energy
35 Stirling Highway
Crawley, WA 6009
Australia

Dongke Zhang
The University of Western Australia
UWA Centre for Energy
35 Stirling Highway
Crawley, WA 6009
Australia

Christian Ziems
BTU Cottbus
H2-Forschungszentrum
03013 Cottbus
Germany

Andreas Zuettel
Empa, Swiss Federal Institute for Material Science & Technology
Überlandstrasse 129
8600 Dübendorf
Switzerland

Part 3
Hydrogen for Storage of Renewable Energy

Hydrogen Science and Engineering: Materials, Processes, Systems and Technology, First Edition.
Edited by Detlef Stolten and Bernd Emonts.
© 2016 Wiley-VCH Verlag GmbH & Co. KGaA. Published 2016 by Wiley-VCH Verlag GmbH & Co. KGaA.

24
Physics of Hydrogen

Carsten Korte, Tabea Mandt, and Timm Bergholz

24.1
Introduction

Accounting for about 75% of the mass of all normal matter, hydrogen is the most abundant element in the universe. On the Earth's surface, hydrogen is the third most abundant element, predominantly bound in water. A minor amount is bound in other compounds, mainly hydrocarbons. There is only a very small concentration of molecular hydrogen in the Earth's atmosphere, of the order of 1 ppm. Most of the atmospheric hydrogen from the formation of the solar system and the Earth, as well as from biologic or geologic sources, dissipated over geological time periods. Its small atomic weight of only 2.016 g mol^{-1} allows it to escape from the Earth's gravity more easily than other much heavier atmospheric gasses [1,2].

24.2
Molecular Hydrogen

24.2.1
The H$_2$ Molecule

24.2.1.1 Covalent Bonding and Molecular Orbitals

Hydrogen forms diatomic molecules over a wide temperature and pressure range. A single bond is formed by an overlapping of two 1s atomic orbitals, each occupied with an electron. As a result of the superposition of the two 1s-wave functions a bonding molecular orbital (MO), with a decreased energy compared to the orbitals of the non-interacting atoms, and an antibonding MO with an increased energy are formed (Figure 24.1).

The bonding MO is occupied and the antibonding MO is unoccupied, resulting in a more stable state than the unbonded atoms. The bond in molecular H$_2$ has the highest dissociation enthalpy of any single bonded homonuclear molecule:

$$H_2 \rightleftarrows 2H \quad \Delta H^0_{\text{Diss.}} = 435 \text{ kJ mol}^{-1} \quad (24.1)$$

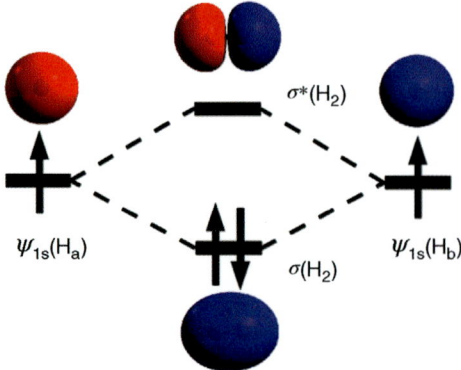

Figure 24.1 Bonding and antibonding molecular orbitals (MO) of a H_2 molecule and their energy levels.

Only single bonds in heteronuclear molecules with fluorine as a partner reach higher values. The bond length of an H_2 molecule in the ground state is 74.166 pm [1,2].

24.2.1.2 Natural Isotopes

Hydrogen from natural sources consists of three isotopes. The vast majority are hydrogen atoms with only a single proton as nucleus, that is, ^1H. This isotope, referred as "protium," has an atomic mass of $1.00\,782\,504\,\text{g mol}^{-1}$ and a natural abundance of 99.9855%. The second natural isotope consists of atoms with an additional neutron, that is, ^2H. It is also referred as "deuterium" or "heavy hydrogen" and has a natural abundance of 0.0145%. The atomic mass amounts to $2.01\,410\,178\,\text{g mol}^{-1}$. The natural abundances and atomic masses of ^1H and ^2H yield the well-known atomic mass of the natural isotope mixture of $1.0797\,\text{g mol}^{-1}$.

Atoms with two additional neutrons can only be found in very small traces with an abundance of the order of 10^{-15}%. The isotope ^3H is a beta emitter with a half-life of 12.32 years. It is also known as "tritium." A natural source of tritium is the interaction of atmospheric nitrogen with fast neutrons from cosmic rays [1,2].

Because of the high mass difference between ^1H and ^2H, chemical reactions involving compounds with deuterium atoms are considerably slower compared to reactions involving compounds only with protium atoms. During water electrolysis deuterated water (^2H_2O or $^1H^2HO$) is accumulated because of its slower redox kinetics. Depending on the electrode material, the ratio between ^1H and ^2H is up to 16 times higher (Pt- and Au-electrodes) in the gas produced than in the liquid [1,2].

24.2.1.3 Nuclear Spin-States, Ortho- and Para-H_2

Fermions, like electrons and protons, have a half-integer spin. In the H_2-molecule the spins of the electrons, occupying the bonding orbital, are always

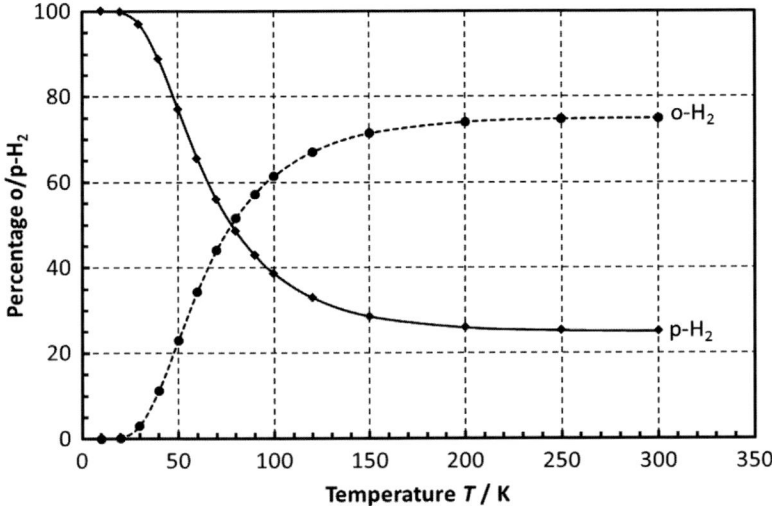

Figure 24.2 Equilibrium composition of (normal) n-H_2. Percentage of o-H_2 and p-H_2 as a function of temperature [3].

antiparallel. However, two nuclear spins states are allowed. The form with parallel nuclear spins is called ortho-H_2 (o-H_2), the form with antiparallel nuclear spins para-H_2 (p-H_2). The energetically more stable form is p-H_2. Thus, the composition of an o/p-mixture in equilibrium is temperature dependent [2]. At 0 K only p-H_2 exists. At high temperatures (room temperature) a limiting composition of ¾ o-H_2 and ¼ p-H_2 is reached (Figure 24.2). In the following, this 3 : 1 mixture is denoted as "normal" H_2 (n-H_2), that is, the equilibrium composition at room temperature.

The conversion between both forms is quite slow, especially at low temperatures. At room temperature there is no noticeable conversion on a timescale of weeks, unless catalytically active contaminations are present. For this reason, p-H_2 and o-H_2 can be separated. The equilibration of both forms is catalyzed by activated charcoal or paramagnetic materials.

The majority of the physico-chemical properties of p-H_2 and o-H_2 are only slightly different. The melting and boiling points differ only by 0.1 K [2]. However, there are considerable differences in heat capacity and related properties, such as thermal conductivity. The molar heat capacity of a gas can be measured under constant pressure ($c_{p,m}$) or constant volume ($c_{V,m}$). For a quasi-ideal gas like hydrogen, the following applies:

$$R = c_{p,m} - c_{V,m} \qquad (24.2)$$

where R is the molar gas constant. The molar heat capacity of a gas consists of translational, rotational, and vibrational parts:

$$c_{V,m} = c_{V,m}^{Trans} + c_{V,m}^{Rot} + c_{V,m}^{Vib} \qquad (24.3)$$

In the case of hydrogen with diatomic molecules there is no appreciable occupation of higher vibrational states for temperatures below 1000 K ($T_{vib} = 6320$ K). Thus, the vibrational part ($c_{V,m}^{Vib}$) can be neglected [4]. The translational part ($c_{V,m}^{Trans}$) of a (quasi) ideal gas is equal to $\frac{3}{2}R$ (3 translational degrees of freedom $\times \frac{1}{2}R = 12.5$ J mol^{-1} K^{-1}). The rotational part ($c_{V,m}^{Rot}$) will reach the limiting value R at room temperature (T_{rot} = 88 K, 2 rotational degrees of freedom $\times \frac{1}{2}R$ = 8.3 J mol^{-1} K^{-1}). Thus, at 300 K we get $c_{V,m} = \frac{5}{2}R = 20.5$ J mol^{-1} K^{-1} and $c_{p,m} = \frac{7}{2}R = 28.8$ J mol^{-1} K^{-1} (c_p = 14.3kJ kg^{-1} K^{-1}) [5]. In the range between room temperature and 1000 K the heat capacity of hydrogen does not change very much.

Due to quantum effects at low temperatures, the rotational part ($c_{V,m}^{Rot}$) is considerably different for o-H$_2$ and p-H$_2$. The symmetry of the total wave function of all electrons and protons in the case of p-H$_2$ only allows the occupation of rotational states with even quantum numbers (J = 0, 2, 4, . . . with even symmetry). The odd numbered rotational states (J = 1, 3, 5, . . . with odd symmetry) are restricted to o-H$_2$ [6].

Only p-H$_2$ can populate the ground state (J = 0). The energy levels of the molecule rotation have a quadratic dependence on the quantum number, which results in different partition sums and thus in a different temperature dependence for the thermodynamic state functions. In the range between 150 and 200 K the heat capacity of p-H$_2$ exceeds the value of n-H$_2$ (3 : 1 mixture) by a factor of 1.4 (Figure 24.3) [3,7].

Due to its dependence on the heat capacity (c_V) the thermal conductivity (κ) of p-H$_2$ is also higher than for o-H$_2$ (see more detailed information in Section 24.2.4 (Transport Kinetics)). In the range between 150 and 200 K the thermal conductivity of p-H$_2$ is higher than n-H$_2$ by up to a factor of 1.3 (Figure 24.4) [7–9].

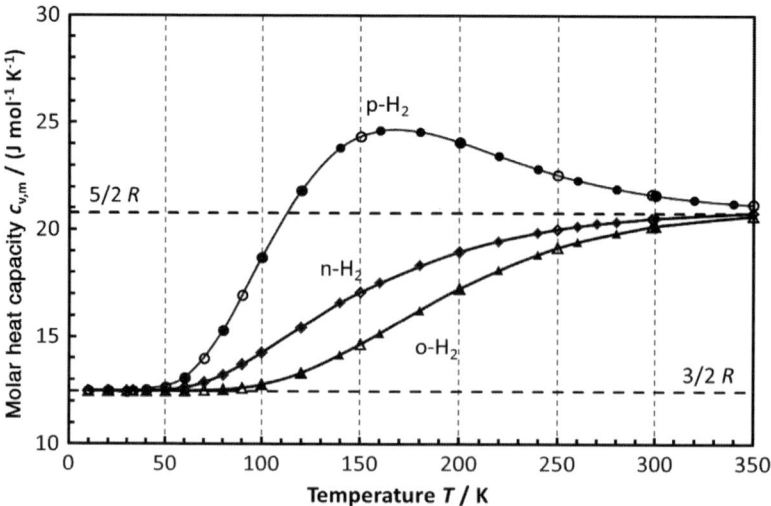

Figure 24.3 Molar heat capacity ($c_{V,m}$) at constant volume of p-H$_2$, o-H$_2$, and n-H$_2$ (3 : 1 mixture) as a function of the temperature (T) [3,7].

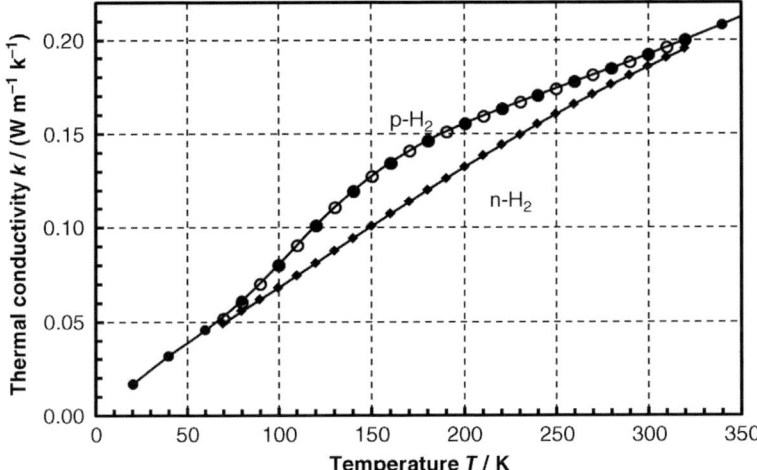

Figure 24.4 Thermal conductivity (κ) of p-H_2 and n-H_2 (3:1 mixture) as a function of temperature (T) [7–9].

The enthalpy of the conversion of pure o-H_2 into p-H_2 is equal to the energetic difference between the occupied rotational states of these spin isomers, taking the temperature dependent population into consideration. Thus, at low temperatures, when mainly the lowest possible states are populated, the conversion enthalpy of gaseous hydrogen can be estimated as the energetic difference of the rotational states with $J = 1$ (o-H_2) and $J = 0$ (p-H_2):

$$\text{o-}H_2 \rightleftarrows \text{p-}H_2 \quad -1.063 \text{ kJ mol}^{-1} \text{ (at the boiling point 21.15 K)} \quad (24.4)$$

This exceeds the (absolute) value of the evaporation enthalpy ($\Delta_{\text{evap}}H^0$) of +0.904 kJ mol^{-1}. Without any special precautions, liquefied hydrogen will always contain a non-equilibrium concentration of o-H_2 due to the slow conversion kinetics. The heat generation in liquefied hydrogen, when o-H_2 converts subsequently into p-H_2, would result in a considerably increased evaporation. The o-H_2 into p-H_2 conversion causes a loss of up to 18% of the liquid hydrogen at 20 K within 24 h after liquefaction [10]. To avoid these complications, hydrogen is usually converted catalytically into the para state at low temperatures before liquefaction.

24.2.2
Thermodynamic Properties

24.2.2.1 Pressure–Temperature Phase Diagram

Hydrogen exists in the molecular form H_2 over a wide temperature and pressure range. At standard pressure (1013.25 hPa) and temperature (298.15 K) hydrogen is a molecular gas with a density of 0.08 g l^{-1} [2]. Figure 24.5 shows the p–T phase diagram of hydrogen over a wide temperature and pressure range. The

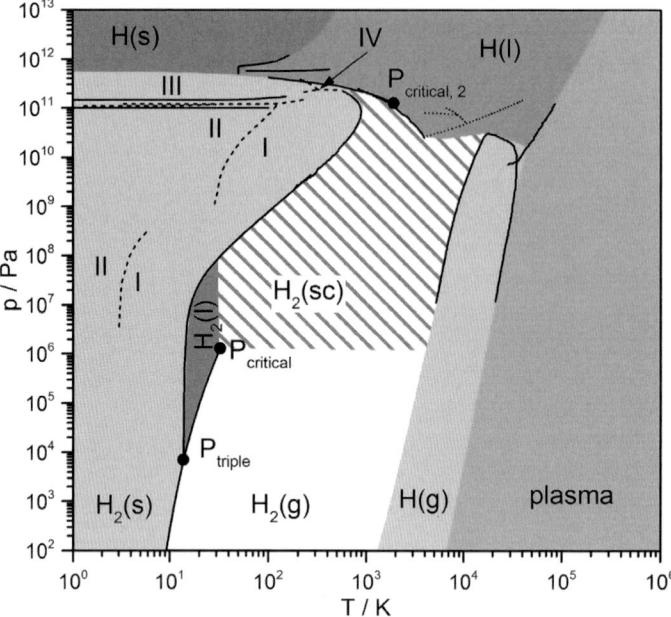

Figure 24.5 Complete hydrogen phase diagram. Lines and data points are from the literature [11–23]. Solid (s), liquid (l), gaseous (g), and supercritical (sc) phase regions for molecular and atomic hydrogen are estimated. Solid lines correspond to phase transitions. Dashed lines correspond to phase transitions between the solid H_2 phases I–IV. Phase transitions shown as dotted lines are from Reference [18] and do not correspond to the other literature data.

lines correspond to phase transitions and the shaded areas are the estimated phase regions from literature data [11–23]. Unless explicitly specified all data in the following are given for n-H_2 (3 : 1 mixture of o- and p-H_2). At standard pressure, the boiling point of liquid hydrogen $H_2(l)$ is 20.38 K (p-H_2: 20.27 K) [15].

The triple point, where gaseous $H_2(g)$, liquid $H_2(l)$, and solid $H_2(s)$ coexist, is at a temperature of 13.96 K and a pressure of 7.36 kPa (p-H_2: 13.80 K, 7.04 kPa) [15]. There are no density anomalies between gaseous, liquid, and solid hydrogen. Thus, the equilibrium lines between solid and gaseous [17], solid and liquid [13], and liquid and gaseous phases [15] all have a positive slope. The equilibrium line between the liquid and gaseous phase ends at the critical point, namely, at a temperature T_{crit} of 33.15 K and a pressure p_{crit} of 1.30 MPa (p-H_2: 32.94 K, 1.29 MPa) [15]. At temperatures and pressures above the critical point the gaseous and the liquid H_2 phase are indistinguishable and only a supercritical fluid exists, $H_2(sc)$ [19].

At standard pressure the melting point of solid hydrogen $H_2(s)$ is 14.2 K. According to the available literature, the phase transitions of solid molecular hydrogen are well known for pressures up to 10^{11} Pa. From Raman and IR measurements, it is supposed that solid hydrogen exists in up to four phases. In phase I, freely rotating H_2 molecules form a (rotational disorder) hexagonal close-packed lattice (hcp). Depending on the occupation of the rotational states, the o-H_2/p-H_2 composition, cubic closed-packed structures (fcc) are also reported. In phase II, the H_2 molecules are still in the same lattice positions as in phase I, but the mean molecular orientations of the H–H bond axes are ordered (broken symmetry). A lower symmetry lattice structure derived from hcp is formed. In phase III, the rotational degree of freedom of the H–H bond axes is even more restricted. In the literature, an orthorhombic and a monoclinic structure are reported [19,22,23]. There is evidence of a fourth phase around 10^{11} Pa and 200 K [23,24].

Notably, at very high pressures, hydrogen undergoes a phase transition to metallic phases consisting of H atoms with metallic bonds and a high electronic conductivity. At temperatures below $\sim 10^2$ to 10^3 K and pressures above 8×10^{11} Pa there is a transformation into a metallic phase, $H_{metal}(s)$, consisting of a lattice of H atoms [16,21]. At higher temperatures and pressures above 3×10^{11} Pa a liquid metallic phase, $H_{metal}(l)$, occurs [19]. (Liquid) metallic hydrogen might be present in the center of gas giant planets like Jupiter and Saturn, where temperatures of 30.000–40.000 K and pressures of 10^{15} Pa occur.

At very high temperatures above 10^4–10^5 K (depending on pressure) an (thermal) atomic plasma is formed [2,14]. High-pressure thermal plasmas of hydrogen can be found in stars like the Sun.

24.2.2.2 Pressure–Volume Phase Diagram

Figure 24.6 shows the p–V_m phase diagram of hydrogen, compiled from literature data [3,15,17,25,26]. The isotherms for temperatures of 16 to 1000 K are indicated with black lines. The phase boundary lines, separating single phase from two-phase fields, are indicated with red lines. The single phase fields for solid, liquid, gaseous, and supercritical hydrogen are marked with $H_2(s)$, $H_2(l)$, $H_2(g)$, and $H_2(sc)$, respectively.

For pressures above the critical point (p_{crit}), liquid and gaseous phases are indistinguishable. At the critical point the density of the supercritical liquid/gas phase ($H_2(sc)$) is 31.26 g l^{-1} (p-H_2: 31.32 g l^{-1}) [15]. In the subsequent pressure range up to $\sim 4 \times 10^7$ Pa, solid hydrogen ($H_2(s)$, light gray area) and liquid hydrogen ($H_2(l)$, dark gray are) coexist in a two-phase-field, $H_2(s)$/$H_2(l)$ (red hatched area). Above this pressure solid hydrogen, $H_2(s)$, and a supercritical liquid/gas ($H_2(sc)$, black hatched area) with a density of 80–150 g l^{-1} coexist. In the supercritical field, $H_2(sc)$, the isotherms deviate considerably from the behavior of an ideal gas (dashed black lines).

For pressures below the critical point (p_{crit}) the (subcritical) gaseous, $H_2(g)$, and the liquid phases, $H_2(l)$, coexist, as do the $H_2(l)$ and the $H_2(s)$ phases. Solid, liquid, and gaseous hydrogen coexist at the triple line at 7.36 kPa (not included in Figure 24.6). The triple line separates the two-phase-field between liquid and gaseous hydrogen, $H_2(l)$/$H_2(g)$, and solid and liquid hydrogen, $H_2(s)$/$H_2(l)$, from the two-phase-field between solid and gaseous hydrogen, $H_2(s)$/$H_2(g)$. The densities (g l^{-1}) of the coexisting phases are 86.02 $H_2(s)$, 76.60 $H_2(l)$, and 0.13 $H_2(g)$ (p-H_2: 85.82, 76.41, and 0.12, respectively) [26]. At standard pressure

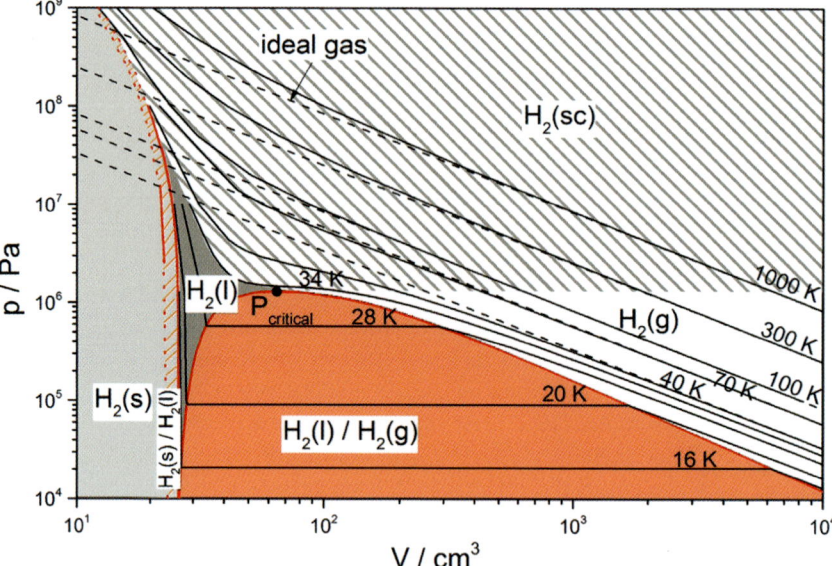

Figure 24.6 p–V_m Phase diagram of hydrogen compiled from literature data [3,15,17,25,26]. The phase fields of solid (s), liquid (l), gaseous (g), and supercritical (sc) H_2 are labeled. The black solid lines are the isotherms for the temperature range 16–1000 K. The course of the isotherms in the case of ideal behavior is indicated by black dashed lines. The phase equilibrium lines are marked in solid red. The dotted lines are extrapolated.

(101.325 kPa) the densities (g l^{-1}) of the coexisting gas, H_2(g), and liquid phase, H_2(l), are 1.33 and 70.85 (p-H_2: 1.34 and 70.83, respectively) [15].

In the subcritical pressure range the isotherms can be described in good approximation with the state equation of ideal gasses:

$$pV_m = RT \tag{24.5}$$

In the log-log plot used in Figure 24.6 this results in straight isotherms with a slope of −1. At higher pressures the molar volume is larger than for an ideal gas because the particle volume cannot be neglected. At (very) low temperatures and intermediate pressures, a deviation to lower volumes can be observed due to increasing intramolecular attraction. For example, at 300 K hydrogen behaves like an ideal gas up to pressures of 6×10^6 Pa. At 40 K, it deviates to smaller molar volumes compared to an ideal gas even for pressures higher than 10^5 Pa.

24.2.2.3 Joule–Thomson Effect

In contrast to an ideal gas, the temperature of a real gas can be changed in an isenthalpic expansion/compression experiment. The particles in a gas interact attractively or repulsively. In the case of attractive interactions, the potential energy increases with increasing volume. Thus, to preserve the total internal energy (U), the kinetic energy of the particles is reduced, leading to a decrease in the temperature. If the gas particles interact repulsively, the potential energy decreases with increasing volume. Thus, during expansion, the kinetic energy and the temperature increase. The isenthalpic change in temperature with pressure is described by the Joule–Thomson coefficient (μ_{JT}):

$$\mu_{JT}(T,p) = \left(\frac{\partial T}{\partial p}\right)_H = \frac{V_m}{c_{v,m}}(T\alpha - 1) \tag{24.6}$$

where $\alpha = 1/V \, (\partial V/\partial T)_p$ is the thermal expansion coefficient, V_m is the molar volume, and $c_{v,m}$ is the molar heat capacity. For an ideal gas μ_{JT} is zero ($\alpha = 1/T$). If μ_{JT} is positive, the gas cools down, $\Delta T < 0$, in the case of isenthalpic expansion, $\Delta p < 0$. If μ_{JT} is negative, the gas heats up, $\Delta T > 0$. The value of μ_{JT} is temperature dependent and decreases with increasing temperature. The temperature at which it changes its sign is denoted as the inversion temperature.

Figure 24.7 shows the Joule–Thomson coefficients (μ_{JT}) of p-H_2, n-H_2, and nitrogen (N_2). Due to the dependence on the heat capacity, the value of μ_{JT} is different for o-H_2, p-H_2, and n-H_2. The inversion temperature of hydrogen at 202 K is considerably lower than that of other gasses [27]. Nitrogen and oxygen have inversion temperatures at 621 and 764 K, respectively. Thus, at room temperature the μ_{JT} of hydrogen is negative. It has a value of -0.30 K MPa^{-1} (n-H_2). In industrial H_2-liquefaction, hydrogen must first be cooled to below the inversion temperature, before the Joule–Thomson effect can be used to decrease the temperature further.

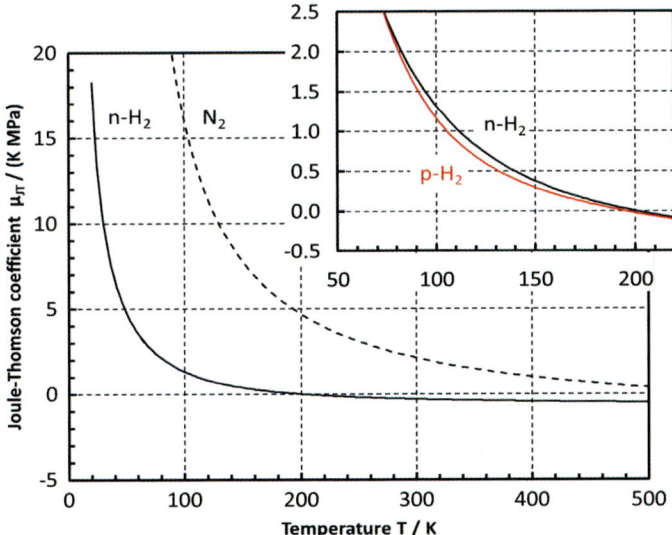

Figure 24.7 Joule–Thomson coefficient of hydrogen (n-H_2: solid black line, p-H_2: solid red line) and nitrogen (dashed line). Data taken from the National institute of Standards and Technology (NIST).

24.2.3
Reaction Kinetics

24.2.3.1 Reaction with O_2 – Thermodynamics

(Liquid) water has the highest formation enthalpy ($\Delta_f H^0$) of all known binary hydrides. When comparing water vapor with other gaseous hydrides, the high enthalpy of condensation ($\Delta_{cond} H^0$) is only exceeded by that of hydrogen fluoride HF[1] [28]:

$$H_2 + \tfrac{1}{2} O_2 \rightleftarrows H_2O\,(g)$$
$$\Delta_f H^0_{H_2O(g)} = -241.98 \text{ kJ mol}^{-1} \text{ (at 298.15 K)} \tag{24.7}$$

$$H_2 + \tfrac{1}{2} O_2 \rightleftarrows H_2O\,(l)$$
$$\Delta_f H^0_{H_2O(l)} = \Delta_f H^0_{H_2O(g)} + \Delta_{cond} H^0_{H_2O} = -286.02 \text{ kJ mol}^{-1} \text{ (at 298.15K)} \tag{24.8}$$

To evaluate the temperature dependent state of equilibrium of hydrogen, oxygen, and water (vapor), the free formation enthalpy ($\Delta_f G^0_{H_2O(g)}$) is needed. Using the partial pressures of the reactants and the mass action law, the equilibrium constant (K) is:

$$K = e^{-\frac{\Delta_f G^0_{H_2O(g)}}{RT}} = \frac{p_{H_2} p_{O_2}^{\frac{1}{2}}}{p_{H_2O}} \left(p^0\right)^{-\frac{1}{2}} \tag{24.9}$$

The $\Delta_f G^0_{H_2O(g)}$ of gaseous water at room temperature is $-228.58 \text{ kJ mol}^{-1}$ and decreases with increasing temperature (Figure 24.8) [7]. At nearly 4000 K the free formation enthalpy becomes zero and the equilibrium constant (K) reaches a value of 1. Thus, a considerable thermal dissociation of water vapor can be observed only at temperatures above 3000 K [29,30].

24.2.3.2 Reaction with O_2 – Microscopic Mechanisms

The reaction of H_2 and O_2 proceeds by a chain-type radical mechanism. The elementary reactions involved in the mechanism as well as their kinetic parameters are still under discussion. According to the literature, about 20–30 elementary reactions can be considered [2,31–37].

In the following, only the most important and widely established reaction paths are shown. The first step involves the (thermally activated) formation of hydrogen radicals H•:

$$H_2 + M \rightarrow 2H^\bullet + M \quad \Delta H^0_{Diss.} = 435 \text{ kJ mol}^{-1},\ E_a = 440 \text{ kJ mol}^{-1} \tag{24.10}$$

where M is another particle or the wall of the surrounding vessel, which delivers energy during the collision. Due to the high activation energy for the formation of initial radicals, the reaction rate in a H_2/O_2 mixture below 400 °C is

---

1) Formation enthalpy of gaseous hydrogen fluoride at 298.15 K: $\Delta_f H^0_{HF} = 271 \text{ kJ mol}^{-1}$.

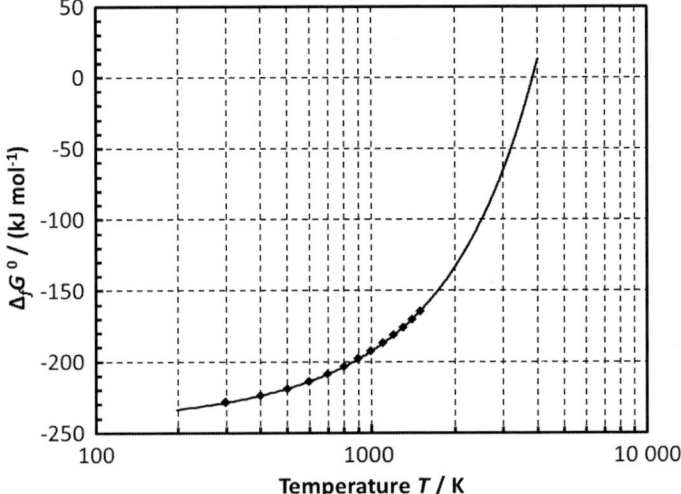

Figure 24.8 Free formation enthalpy ($\Delta_f G^0$) of H_2O vapor as a function of temperature (T) [7].

undetectable. In the temperature range 400–560 °C, the reaction rate and the prevailing mechanism are controlled by the pressure and temperature. The following intermediate radical species were observed: hydrogen (H•), oxygen (O••), hydroxyl (HO•), and hydroperoxyl (HOO•) as well as hydrogen peroxide (H_2O_2). The generation rate and lifetime of the intermediate radicals depend on the probability of two- and three-body collisions and collisions with the walls of the reaction vessel.

Depending on pressure there are two reaction paths for H• radicals when reacting with O_2 molecules. In a low pressure range it forms a hydroxyl radical (HO•) and an oxygen radical (O••) in a two-body collision (Figure 24.9, light gray area). Subsequently, both radicals react with H_2, forming new H• radicals. This results in a branching chain reaction, an exponentially increasing reaction rate, and thus an isothermal explosion. Below a critical pressure the probability that the H•, HO•, and O•• radicals will be destroyed by a collision with the walls of the vessel is higher than a reaction with H_2 or O_2 molecules propagating the chain reaction (1st explosion limit). This results in a slow, non-explosive reaction.

The reaction chains can also be terminated by the formation of a hydroperoxyl radical (HOO•) in a three-body collision between H• radicals, O_2 molecules, and a third molecule (M), which absorbs excessive energy. The HOO• radicals are relatively inert compared to the other radical species. They react only slowly with H_2 molecules and do not maintain the branched chain reaction. Due to their relatively long lifetime, they can diffuse preferentially to the walls of the vessel or recombine by forming hydrogen peroxide. The probability of three-body collisions increases with increasing pressure. Above a second critical pressure the formation of HOO• radicals dominates (2nd explosion limit). This

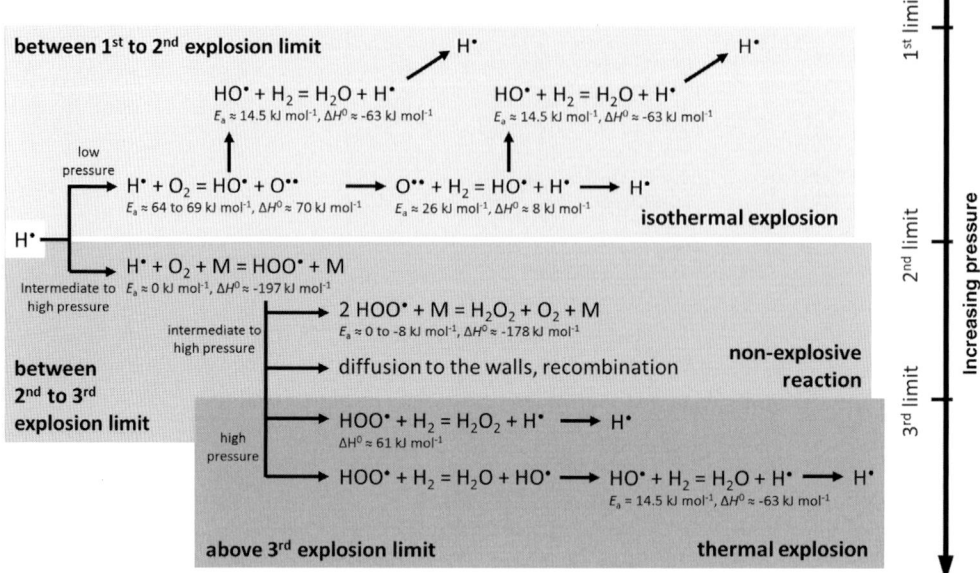

Figure 24.9 Prevailing elementary steps in the reaction between hydrogen and oxygen. Depending on pressure, the reaction alternates between an isothermal explosion, non-explosive water formation, and a thermal explosion.

results in an intermediate pressure range with a slow, non-explosive reaction (Figure 24.9, medium gray area).

When the pressure increases above a third critical value, a self-accelerating explosive reaction occurs again (3rd explosion limit). Above this critical pressure the probability of HOO• radicals reacting with H_2 to form new H• or HO• radicals is high enough to maintain an (unbranched) chain reaction. A thermal explosion occurs due to the high reaction enthalpy released and the increasing temperature (Figure 24.9, dark gray area).

As outlined in Figure 24.9, hydrogen peroxide is formed in addition to water. At the high reaction temperatures in H_2/O_2 flames, only traces of hydrogen peroxide exist. As an intermediate it decays by forming hydroxyl radicals (HO•):

$$H_2O_2 + M \rightarrow 2HO^\bullet + M \tag{24.11}$$

or it is consumed by the reaction with H•, O••, or HO• radicals:

$$H_2O_2 + H^\bullet \rightarrow H_2O + HO^\bullet \tag{24.12}$$

$$H_2O_2 + H^\bullet \rightarrow HOO^\bullet + H_2 \tag{24.13}$$

$$H_2O_2 + O^{\bullet\bullet} \rightarrow HOO^\bullet + HO^\bullet \tag{24.14}$$

$$H_2O_2 + HO^\bullet \rightarrow H_2O + HOO^\bullet \tag{24.15}$$

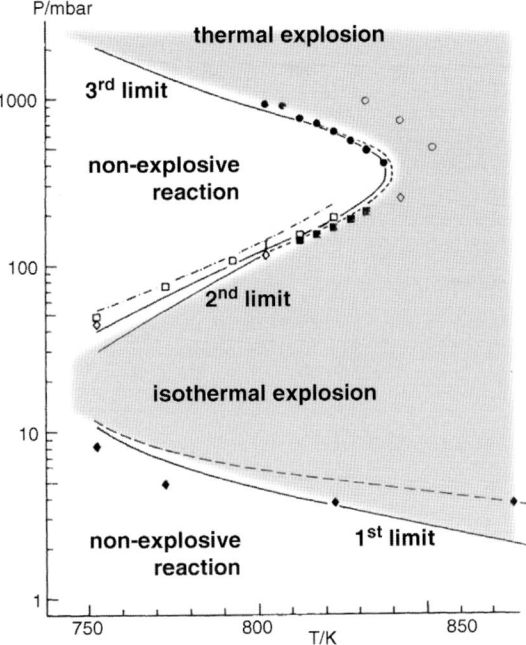

Figure 24.10 Explosion limits for 2:1 mixtures of H_2 and O_2 as a function of total pressure (p) and temperature (T). Solid, dashed, and dot-and-dashed lines are calculated [34]. Experimental data using KCl-coated vessels: ■, ○, ● and ◇ [93,94]. Experimental data using Pyrex vessels: | [94]. Experimental data using fused silica vessels: ◆ [96]. Experimental data using B_2O_3-coated vessel: □ [95].

By quenching the combustion gas of a H_2/O_2 flame at a cooled surface, hydrogen peroxide[2] can be isolated.

24.2.3.3 Reaction with O_2 – Explosion Limits

Despite the high free formation enthalpy ($\Delta_f G^0$), mixtures of H_2 and O_2 are metastable. As described in the preceding section, there is no detectable reaction below 400 °C. In the temperature range 400–570 °C, whether the reaction is explosive or non-explosive depends on the composition of the H_2/O_2 mixture, the total pressure, the temperature, and the dimensions of the vessel (autoignition temperature, T_{ig}) [28].

Figure 24.10 shows the explosion limits for a 2:1 mixture of H_2 and O_2 as a function of total pressure (p) and temperature (T). Isothermal explosion occurs between the 1st and 2nd explosion limit and thermal explosion above the 3rd limit. Between the 2nd and the 3rd explosion limit, a steady, non-explosive reaction can be seen. Above the 3rd limit a thermal explosion occurs. At a total

---

2) The formation of H_2O_2 can be easily demonstrated by touching ice with a H_2-flame. When titanyl sulfate ($TiOSO_4$) solution is added to the melting water, orange colored peroxotitanyl cations (TiO_2^{2+}) are formed.

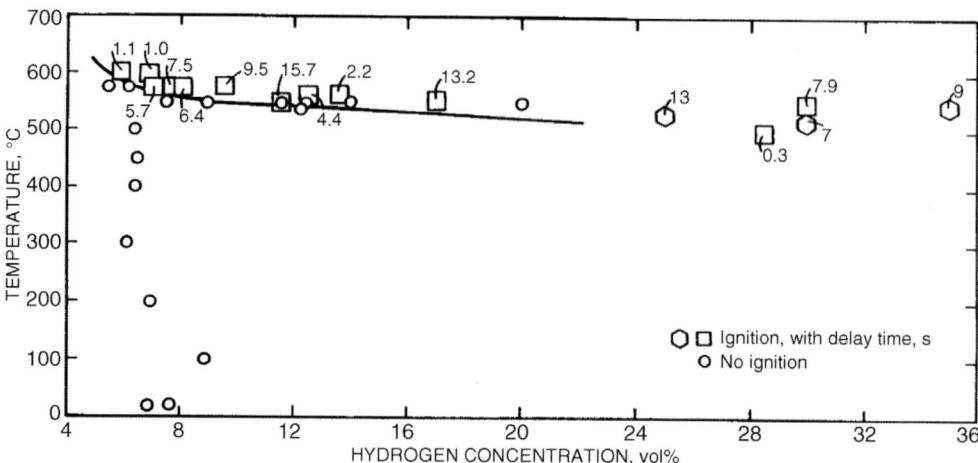

Figure 24.11 Autoignition temperature (T_{ig}) of H_2/air-mixtures at a total pressure of 1 bar as a function of composition. Illustration taken from Reference [28]. Journal of Fire Sciences, SAGE Publications.

pressure of 1 bar, this explosion limit is reached at a temperature of ~520 °C (~790 K). The highest autoignition temperature (T_{ig}) of 570 °C (840 K) can be found at a lower total pressure of ~0.4 bar.

Figure 24.11 shows the T_{ig} for H_2/air mixtures as a function of hydrogen content (vol.%) at a constant total pressure of 1 bar. At a composition of 28.6 vol.% H_2 a 2:1 mixture of H_2 and O_2 is present (+57.1 vol.% inert nitrogen). At this composition autoignition occurs at ~520 °C. The T_{ig} increases with decreasing hydrogen content. The T_{ig} of hydrogen is notably high compared with common, liquid hydrocarbon fuels like gasoline or diesel with $T_{ig} \approx 250-300$ °C.

24.2.4
Transport Kinetics

24.2.4.1 Thermal Conductivity

The thermal conductivity (κ) is defined as the proportionality constant connecting a heat flux ($\vec{j}_Q$) with the driving temperature gradient (∇T) (Fourier's law):

$$\vec{j}_Q = -\kappa \nabla T \tag{24.16}$$

Over a wide temperature and pressure range hydrogen can be assumed as a (quasi-) ideal gas. For an ideal gas, κ depends on the mean velocity ($\bar{v}$), mean free (Maxwell) path length (λ_M), and molar heat capacity (c_V):

$$\kappa = \tfrac{1}{2} c_V \frac{p}{RT} \bar{v} \lambda_M = \sqrt{\frac{kT}{\pi m}} \frac{c_V}{\sigma N_A} \quad \text{with} \quad \lambda_M = \frac{kT}{\sqrt{2} p \sigma} \quad \text{and} \quad \bar{v} = \sqrt{\frac{8kT}{\pi m}} \tag{24.17}$$

The mean free path length (λ_M) depends on the collisional cross-section, σ. The kinetic diameter (d) of hydrogen molecules can be (experimentally) determined as 0.289 nm ($\sigma = \pi d^2$). As long as λ_M is smaller than the dimensions of the vessel, κ is independent of pressure (p). At standard pressure (1013.25 hPa) and standard temperature (298 K), the value of λ_M for pure hydrogen is about 0.11 µm.

The value of κ for an (ideal) gas will increase with increasing temperature (T) and decreasing molecular mass (m). Thus, hydrogen with the lowest m and the smallest collisional cross-section (σ) has the highest value of κ of all gases. Due to the dependence on the heat capacity (c_V), the thermal conductivity of hydrogen depends on the state of the o-H_2/p-H_2 equilibrium. As depicted in Figure 24.4, the thermal conductivity of p-H_2 is higher than that of n-H_2 [7–9]. At room temperature n-H_2 reaches a value of 0.1805 W m^{-1} K^{-1}.

24.2.4.2 Diffusion in Gasses

Generally, the diffusion coefficient (D_i) of species i is defined as the proportionality constant connecting its mass flux density ($\vec{j}_i$, in mol m^{-2} s^{-1}) with the driving concentration gradient, ∇c_i:

$$\vec{j}_i = -D_i \nabla c_i \tag{24.18}$$

As stated above, hydrogen can be assumed as an (quasi) ideal gas. The self-diffusion coefficient (D) for an ideal gas can be derived from the kinetic theory of gasses. For an ideal gas, D depends on the mean velocity ($\bar{v}$) and the mean free (Maxwell) path length (λ_M) of the molecules:

$$D = \tfrac{1}{3}\bar{v}\lambda_M = \sqrt{\frac{(kT)^3}{6\pi m}} \frac{1}{p\sigma} \tag{24.19}$$

In the case of a mixture of gasses with different atomic masses, m_i, there are different velocity distributions for each component i, and thus different mean velocities ($\bar{v}_i$) and mean free path length ($\lambda_{M,i}$). Using a simple approach, the interdiffusion coefficient D_{12} ($= D_{21}$) of two gasses, 1 and 2, can be calculated as follows:

$$D_{12} = D_{21} = \sqrt{\frac{(kT)^3}{6\pi\mu_{12}}} \frac{1}{p_{tot}\sigma_{12}} \quad \text{with} \quad \frac{1}{\mu_{12}} = \frac{1}{m_1} + \frac{1}{m_2}, \tag{24.20}$$

$$p_{tot} = p_1 + p_2 \quad \text{and} \quad \sigma_{12} = \pi(r_1 + r_2)$$

where μ_{12} is the reduced mass (the relative velocities have to be taken into account), p_{tot} is the total pressure, r_1 and r_2 are the kinetic radii of the particles, and σ_{12} is the (mutual) collisional cross section. Due to the temperature dependence of $\bar{v}_i$ and of $\lambda_{M,i}$, the D_{12} (and the self-diffusion coefficient, D) is found to have a $T^{3/2}$ temperature dependence and a reciprocal pressure dependence.

As hydrogen has the lowest molecular mass and smallest collisional cross section, the value of its self-diffusion coefficient (D_{H_2}) exceeds the values of most other gasses by an order of magnitude (Figure 24.12). It is only comparable with He. At 298 K and 1013.25 hPa (mbar) a value of $1.2 \times 10^{-4}\,\mathrm{m^2\,s^{-1}}$ can be measured. In a 1 : 1 mixture with another gas with a considerably higher molar mass, a D_{12} of about 0.6–$0.8 \times 10^{-4}\,\mathrm{m^2\,s^{-1}}$ can found (298 K and 1013.25 hPa).

24.2.4.3 Permeation and Diffusion in Polymers

When performing a gas permeation experiment on a membrane, a pressure difference, Δp, is applied. A solubility equilibrium is established between the gas phase and the membrane surface. Hydrogen is dissolved in the form of undissociated H_2 molecules in the polymer. Thus, the activity (a_{H_2}) of H_2 dissolved in the polymer is proportional to the fugacity (f_{H_2}). At the limit of an ideal gas, the concentration (c_{H_2}) and the (partial) pressure (p_{H_2}) can be used (law of mass action):

$$\frac{a_{H_2,\text{Polymer}}}{f_{H_2,\text{Gas}}} = k_{H_2,\text{sol}} \quad \text{for ideal behaviour} \quad \frac{c_{H_2}}{p_{H_2}} \approx \frac{c^0}{p^0} k_{H_2,\text{sol}} \approx K_{H_2} \qquad (24.21)$$

The concentration of H_2 in the surface layer on both sides of the membrane, that is, the applied concentration difference, Δc_{H_2}, can be calculated using the solubility, K_{H_2}, which is identical to the reciprocal value of Henry's constant. The temperature dependence of the equilibrium constant, $k_{H_2,\text{sol}}$, and of the solubility, K_{H_2}, can be described by the solution enthalpy ($\Delta_{\text{sol}} H_{H_2}$):

$$K_{H_2} = K_{H_2}^0 e^{-\frac{\Delta_{\text{sol}} H_{H_2}}{RT}} \qquad (24.22)$$

The H_2 molecules are located in voids between the polymer chains. The diffusional movement of the H_2 molecules from one void to a neighboring one is a thermally activated process. The diffusion coefficient (D_{H_2}) has an Arrhenius-type temperature dependence:

$$D_{H_2} = D_{H_2}^0 e^{-\frac{\Delta_{\text{mig}} H_{H_2}^{\#}}{RT}} \qquad (24.23)$$

where $D_{H_2}^0$ is the pre-exponential factor and $\Delta_{\text{mig}} H_{H_2}^{\#}$ is the activation enthalpy for the migration of H_2 molecules. Using Fick's first law in Eq. (24.18) and the solubility equilibrium in Eq. (24.21), the flux density (j_{H_2}) through a polymer membrane with the thickness Δx is proportional to the applied pressure difference, Δp_{H_2}:

$$j_{H_2} = -D_{H_2} \frac{\Delta c_{H_2}}{\Delta x} = -\Phi_{H_2} \frac{\Delta p_{H_2}}{\Delta x} \quad \text{with} \quad \Phi_{H_2} = K_{H_2} D_{H_2} \qquad (24.24)$$

The product of the diffusion coefficient (D_{H_2}) and the solubility (K_{H_2}) of molecular H_2 is introduced as the permeation coefficient (Φ_{H_2}) of the polymer material.[3] The temperature dependence of Φ_{H_2} is controlled by the permeation

---

3) When using volume at standard pressure ($p^0 = 101\,325$ Pa) and standard temperature ($T^0 = 273.15$ K) (cm^3 STP) instead of mole as a measure for the amount of H_2, the permeation coefficient (Φ_{H_2}) and the flux density (j_{H_2}) of hydrogen can be converted by multiplying with RT^0/p^0 (assuming an ideal gas).

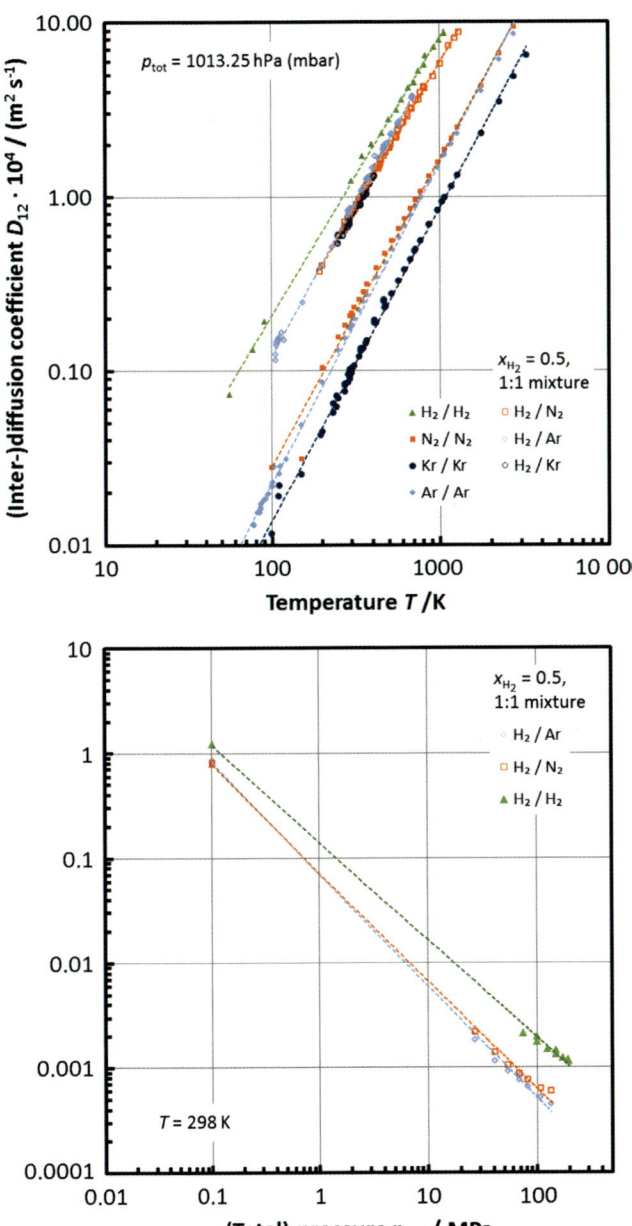

Figure 24.12 Self-diffusion coefficients (D) of pure H_2, N_2, Ar, and Kr (filled symbols) and interdiffusion coefficients (D_{12}) of 1:1 H_2/N_2-, H_2/Ar-, and H_2/Kr-mixtures (empty symbols) as a function of temperature at a total pressure of 1013.25 hPa (top) and as a function of the total pressure at a temperature of about 298 K (bottom) [38].

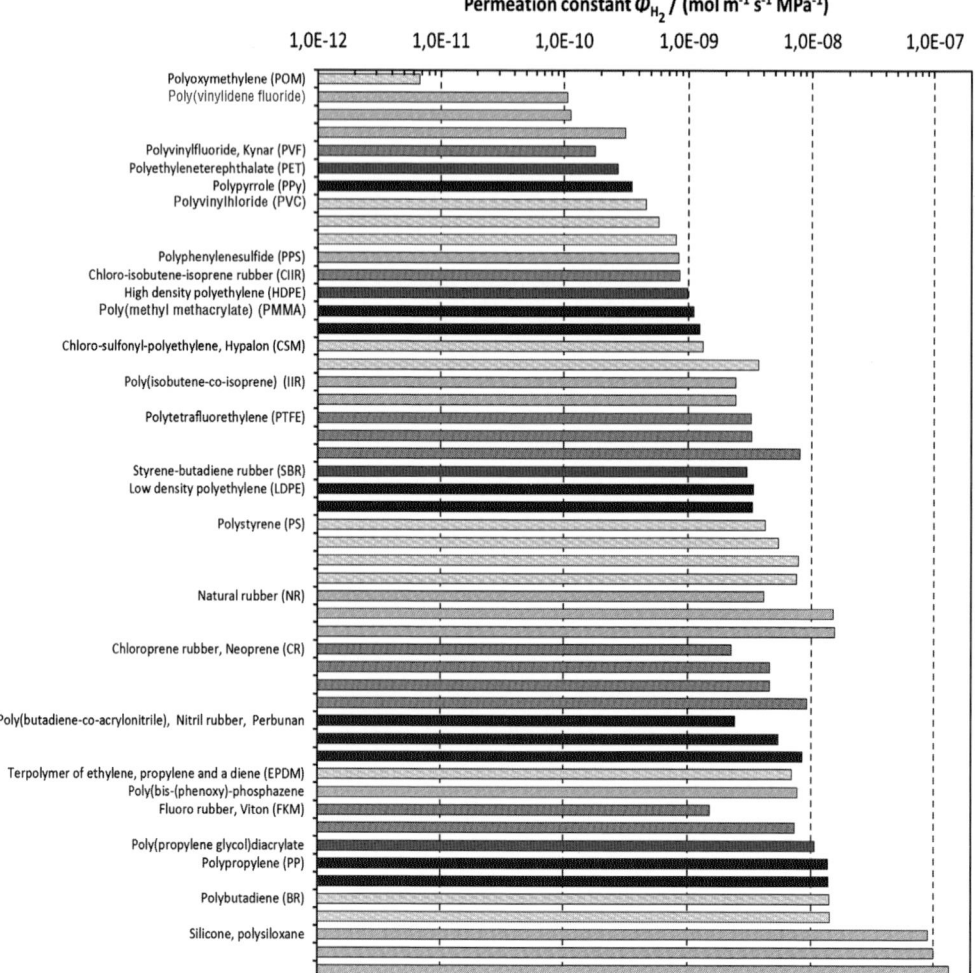

Figure 24.13 Permeation coefficients (Φ_{H_2}) of hydrogen for polymer and elastomer materials [39,40].

enthalpy ($\Delta_{perm}H^0_{H_2}$), which is the sum of the activation enthalpy ($\Delta_{mig}H^{\#}_{H_2}$) and the solution enthalpy ($\Delta_{sol}H^0_{H_2}$):

$$\Phi_{H_2} = \Phi^0_{H_2} e^{-\frac{\Delta H^0_{H_2,perm}}{RT}} = \Phi^0_{H_2} e^{-\frac{\Delta_{mig}H^{\#}_{H_2} + \Delta_{sol}H^0_{H_2}}{RT}} \quad (24.25)$$

According to Eq. (24.24), a Φ_{H_2} of, for example, 10^{-9} mol-H_2 m^{-1} s^{-1} MPa^{-1} results in a permeation flux density (j_{H_2}) of 10^{-6} mol-H_2 m^{-2} s^{-1} or 89.2 sccm-H_2 m^{-2} h^{-1}, when applying a (partial) pressure difference of 1 bar (0.1 MPa) H_2 to a 100 μm thick membrane.

Figure 24.13 summarizes the permeation coefficients (Φ_{H_2}) of hydrogen for various polymer materials at standard temperature (25 °C). The absolute value of

Φ_{H_2} varies in a range of about four orders of magnitude, that is, 10^{-11} to 10^{-7} mol-H_2 m^{-1} s^{-1} MPa^{-1}. As shown exemplarily for some selected polymer materials in Table 24.1, the corresponding permeation enthalpies ($\Delta_{perm}H^0_{H_2}$) are in the range 20–40 kJ mol^{-1} (0.2–0.4 eV).

A qualitative analysis of the data in Figure 24.13 reveals that the permeation coefficients (Φ_{H_2}) of polymers with polar groups and strong dipole moments are mainly in the lower range of 10^{-10}–10^{-9} mol-H_2 m^{-1} s^{-1} MPa^{-1}. In nonpolar polymers, values that are about one order of magnitude higher, in the range of 10^{-9}–10^{-8} mol-H_2 m^{-1} s^{-1} MPa^{-1}, can be observed. The highest values for Φ_{H_2} were found for polysiloxane based elastomers, which had the lowest migration enthalpies, $\Delta_{perm}H^0_{H_2}$. The value of Φ_{H_2} reaches about 10^{-7} mol-H_2 m^{-1} s^{-1} MPa^{-1}.

The diffusion coefficients (D_{H_2}) of molecular hydrogen at room temperature are in the range 10^{-11}–10^{-9} m^2 s^{-1} and the solubilities (K_{H_2}) in the range 10^1–10^2 mol m^{-3} MPa^{-1}. The D_{H_2} at room temperature is about five orders of magnitude lower than transport in the gas phase. The activation enthalpies ($\Delta_{mig}H^{\#}_{H_2}$) for the migration of H_2 are in a similar range as the permeation enthalpies. The solution enthalpies ($\Delta_{sol}H^0_{H_2}$) are usually small. Values of only about a few kJ mol^{-1} can be measured.

24.2.4.4 Permeation and Diffusion in Metals

The mechanism of H_2 transport in transition metals differs significantly from that in polymers. Molecular H_2 dissociates prior to dissolution in the metal. In the case of only a small amount of hydrogen, the hydrogen atoms (H) are incorporated in interstitial positions of the original metal lattice. For higher hydrogen contents, for example, defective (and distorted) rock salt, wurzite (MeH$_{1-\delta}$), fluorite (MeH$_{2-\delta}$), or AlCu$_2$Mn-type (MeH$_{3-\delta}$) structures are formed. There is metallic bonding, that is, the conduction and valence band is mainly formed from the d-orbitals of the transition metal and the 1s orbitals of the hydrogen atoms.

The activity (a_H) of the dissolved H atoms is proportional to the square-root of the fugacity, f_{H_2}. Again, at the limits of an ideal gas, the concentration (c_H) and the (partial) pressure (p_{H_2}) can be used (law of mass action, Sievert's law):

$$\frac{a^2_{H,Metal}}{f_{H_2,Gas}} = k_{H,sol} \quad \text{for ideal behaviour} \quad \frac{c_H}{p^{\frac{1}{2}}_{H_2}} \approx \frac{c^0}{(p^0)^{\frac{1}{2}}} k^{\frac{1}{2}}_{H,sol} \approx K_H \quad (24.26)$$

The relation between the concentration (c_H) of hydrogen atoms and the applied pressure (p_{H_2}) is given by the solubility, K_H. The flux density (j_{H_2}) of molecular H_2 through a metallic membrane is equal to that of atomic H in the membrane. Using Fick's first law in Eq. (24.18) and Sievert's law in Eq. (24.26), the j_{H_2} of H_2 in mole-H_2 per area and time through a membrane with a thickness of Δx is proportional to the difference of the square roots of the applied pressures ($\Delta p^{\frac{1}{2}}_{H_2}$):

$$j_{H_2} = \tfrac{1}{2} j_H = -\tfrac{1}{2} D_H \frac{\Delta c_H}{\Delta x} = -\Phi_{H_2} \frac{\Delta p^{\frac{1}{2}}_{H_2}}{\Delta x} \quad \text{with} \quad \Phi_{H_2} = \tfrac{1}{2} K_H D_H \quad (24.27)$$

Table 24.1 Permeation coefficients (Φ_{H_2}), diffusion coefficients (D_{H_2}), and solubilities (K_{H_2}) for 298 and 308 K (italic entries) and the corresponding Arrhenius parameters of some selected polymer (and elastomer) membrane materials (permeation enthalpy, activation enthalpy for H_2 migration, solution enthalpy of H_2, and the pre-exponential factors).

Material [Reference]	T (K)	$\Phi_{H_2} \times 10^9$ (mol m^{-1} s^{-1} MPa^{-1})		$D_{H_2} \times 10^{12}$ (m^2 s^{-1})		K_{H_2} (mol m^{-3} MPa^{-1})	
		$\Phi^0_{H_2} \times 10^3$ (mol m^{-1} s^{-1} MPa^{-1})	$\Delta_{perm}H^0_{H_2}$ (kJ mol^{-1})	$D^0_{H_2} \times 10^6$ /(m^2 s^{-1})	$\Delta_{mig}H^{\#}_{H_2}$ /(kJ mol^{-1})	$K^0_{H_2}$ (mol m^{-3} MPa^{-1})	$\Delta_{sol}H^0_{H_2}$ (kJ mol^{-1})
PVF (Kynar) [39]	308	*0.180*		*33.6*		*5.36*	
PET	298	*~0.13*		*~20*		*~7.4*	
[40,41]	298–313	0.000 115	22.5	2.12	28.6	0.546	−6.46
PVC (unplasticized) [39]	298	*~0.5*		*~50*		*~10*	
[40,41]	298–353	0.651	34.5	55.7	34.5	11.7	~0
[40,41]	298–313	0.000 173	20.4	0.697	23.6	2.48	−3.20
HDPE (0.955 g cm^{-3}) [39]	298	*~1*		*~350*		*~3.2*	
[40,41]	298–313	0.49	32.5	2.83	22.2	172	9.8
PTFE [39]	298	*3.23*		*14.7*		*220*	
LDPE (0.914 g cm^{-3}) [39]	293–403	0.0185	21.4	47.4		70.5	
PS [39]	298	*3.3*					
[40,41]	298	*~4.2*		*~320*		*~13*	
Neoprene G (CR) [39]	298–313	0.000 534	12.0	0.335	17.3	1.59	−5.2
Poly(butadiene-co-acrylonitrile), Perbunan (NBR) [39]	298	*4.55*		*380*		*12.8*	
[39]	288–323	3.93	33.9	26.2	27.6	162	6.3
PP (0.907 g cm^{-3}) [39]	298	*8.43*		*643*		*13.2*	
	298–323	1.57	30.1	23.2	26	69.1	4.1
Poly(butadiene) (BR) [39]	293	*13.8*		*210*		*65.9*	
	293–343	99.9	38.5	960		14.3	
Polysiloxane [39]	298	*14.1*					
	298–323	0.959	27.6	5.20	21.3	185	6.3
[39]	298	*~100*	6–18				

The data are taken from References [39–41].

In the case of metallic membranes, the permeation coefficient (Φ_{H_2}) is introduced as the product of half of the solubility (K_H) of atomic H and the diffusion coefficient (D_H) of H atoms. Using the apparent concentration (c_{H_2}) of "molecular hydrogen H_2" in the membrane, Φ_{H_2} can be defined in an identical form as for polymer membranes, see Eq. (24.24):

$$\Phi_{H_2} = K_{H_2} D_{H_2} \quad \text{with} \quad \tfrac{1}{2}\Delta c_H \equiv \Delta c_{H_2}, D_H \equiv D_{H_2} \quad \text{and} \quad \tfrac{1}{2} K_H \equiv K_{H_2} \tag{24.28}$$

The temperature dependences of K_H and D_H are given as follows (see Eqs (24.22) and (24.23), respectively, for polymers):

$$K_H = K_H^0 e^{-\frac{\Delta_{sol} H_H^0}{RT}} \quad \text{and} \quad D_H = D_H^0 e^{-\frac{\Delta_{mig} H_H^\#}{RT}} \tag{24.29}$$

The temperature dependence of the Φ_{H_2} is again controlled by the sum of the activation enthalpy of migration ($\Delta_{mig} H_H^\#$) and the solution enthalpy ($\Delta_{sol} H_H^0$):

$$\Phi_{H_2} = \Phi_{H_2}^0 e^{-\frac{\Delta_{perm} H_{H_2}^0}{RT}} = \Phi_{H_2}^0 e^{-\frac{\Delta_{mig} H_H^\# + \Delta_{sol} H_H^0}{RT}} \tag{24.30}$$

Table 24.2 summarizes the permeation coefficients (Φ_{H_2}), diffusion coefficients (D_{H_2}), and the solubilities (K_{H_2}) of hydrogen for some transition metals (calculated) for 500 K (227 °C) as well as the corresponding Arrhenius parameters. The absolute value of the permeation coefficient (Φ_{H_2}) varies in a range of more than eight orders of magnitude, that is, $<10^{-14}$ to 10^{-6} mol-H_2 m^{-1} s^{-1} MPa$^{-\frac{1}{2}}$. The permeation enthalpies ($\Delta_{perm} H_{H_2}^0$) are in the range 15–150 kJ mol^{-1} (0.16–1.6 eV). Apart from palladium and α-iron with values of only 15 and 34 kJ mol^{-1}, this is on average up to twice that of polymer materials.

As shown in Table 24.2, palladium and α-iron exhibit the Φ_{H_2} of all metals with values of about 10^{-8} and 5×10^{-6} mol-H_2 m^{-1} s^{-1} MPa$^{-\frac{1}{2}}$, respectively, at 500 °C. At room temperature the permeation coefficients (Φ_{H_2}) of palladium and α-iron decrease, respectively, by two and one order of magnitude to 4×10^{-11} and 5×10^{-7} mol-H_2 m^{-1} s^{-1} MPa$^{-\frac{1}{2}}$.

The diffusion coefficient (D_{H_2}) for molecular hydrogen at 500 K is in the range 10^{-11}–10^{-8} m^2 s^{-1} and the solubility (K_{H_2}) is in the range 10^{-4}–10^3 mol m^{-3} MPa$^{-\frac{1}{2}}$. The values of D_{H_2} at 500 K are comparable to those of polymer materials, measured at room temperature. Due to the necessary dissociation of the H_2 molecules and the interaction with the electron band structure of the metal, the solution enthalpies ($\Delta_{sol} H_H^0$) are usually significantly higher than for the solution process in polymer materials.

Especially in the case of aluminum, the solubility (K_{H_2}) is extraordinarily small. Despite the fact that the D_{H_2} of aluminum is comparable to, for example, that of nickel, this results in a very small Φ_{H_2} of less than 10^{-14} mol-H_2 m^{-1} s^{-1} MPa$^{-\frac{1}{2}}$ at 500 K and (presumably) less than 10^{-17} mol-H_2 m^{-1} s^{-1} MPa$^{-\frac{1}{2}}$ at room temperature [39].

Table 24.2 Permeation coefficient (Φ_{H_2}), diffusion coefficient (D_{H_2}), and the solubility (K_{H_2}) for 500 K (italic entries) and the corresponding Arrhenius parameters for some selected transition metal membrane materials as well as for two (common) stainless steels (permeation enthalpy, activation enthalpy for H_2 migration, solution enthalpy of H_2, and the pre-exponential factors).

Material	T (K)	$\Phi_{H_2} \times 10^9$ (mol m^{-1} s^{-1} MPa$^{-1/2}$)	$\Phi^0_{H_2} \times 10^3$ (mol m^{-1} s^{-1} MPa$^{-1/2}$)	$\Delta_{perm} H^0_{H_2}$ (kJ mol^{-1})	$D_{H_2} \times 10^{12}$ (m^2 s^{-1})	$D^0_{H_2} \times 10^6$ (m^2 s^{-1})	$\Delta_{mig} H^\#_H$ (kJ mol^{-1})	K_{H_2} (mol m^{-3} MPa$^{-1/2}$)	$K^0_{H_2}$ (mol m^{-3} MPa$^{-1/2}$)	$\Delta_{sol} H^0_H$ (kJ mol^{-1})
Tungsten [42,43]	1100–2400		<0.00001	132						
	500	*0.76*			*~20–600*			*~0.001–0.0001*		
Aluminum[a)]										
[42]	420–520		58	123						
[42,44,45]	633–873		3.8	157		11	40.98		690	58.2
[45,46]	573–673					26	58.7		720 000	96.7
[47]	753–913					0.5	33.3			
	500	*~0.001*								
Molybdenum										
[42,48]	500–1700		0.233	80.7						
	500	*~0.01*								
Copper										
[49]	573–1073		0.0366	60.5		0.226	29.3			31.2
[42,50]	470–710		0.84	76.6					137	
[42,51]	620–970		0.586	75.7						
	500	*~0.02*			*~196*			*~0.075*		
γ-Iron										
[49]	573–1073		0.634	71.6		0.663	44.9		180.1	23.66
	500	*~0.1*			*~5*			*~20*		
X5CrNi18-10 (1.4301, 304 SS)										
[49]	573–1073		0.625	65.4		0.825	49.7		758	15.7
[52]	373–873		1.04	64.0		2.72	54.4		384	9.6

X5CrNiMo17-12-2 (1.4401, 316 SS)							
[49]	500	~0.1	61.7	~6	47.8	~15	13.9
	573–1073	0.270		0.632		427	
Nickel	500	~1		~40		~25	
[49]	573–1073	0.708	54.8	0.743	44.1	953	10.7
[53]	373–623	0.335	54.24	0.712	40.64	471	13.6
[54]	500–1000	0.337	54.6	0.703	39.48	479	15.1
[42,55]	300–550	0.39	55.4				
[56]	371–500	0.580	53.8				
[57]	575–770	0.238	52.21				
α-Iron	500	~10		~25 000		~0.4	
[63]	342–619	0.048	33.9	0.101	6.68	474	27.2
[52]	373–873	0.0359	34.3	0.233	6.68	154	27.6
[58]	298–353			0.0793	7.34		
[42,59]	>375	0.041	34.9				
Palladium	500	~5000		~1500		~3500	
[42,60]	300–709	0.220	15.7				
[61,62]	322–449	0.140	13.65	0.290	22.02	483	−8.37

Data is taken from References [39,42–62].

a) Values are strongly scattered. High experimental errors due to a very small solubility of hydrogen in aluminum. The solubility is also sensitive to the pretreatment of the metal surface (etching, grinding, sputtering).

24.3
Hydrides

24.3.1
Classification and Properties of Hydrides

Hydrogen forms stable binary chemical compounds of the general formula EH_n with many elements (except noble gasses, Pm, Mo, W, Mn, Tc, Re, Fe, Ru, Os, Co, Rh, Ir, Pt, Ag, Fr, and Ra). In the following, in accordance with the literature, not only compounds with negatively charged (e.g., CsH, BaH_2) or polarized hydrogen (e.g., AlH_3, CuH) are referred to as "hydrides," but also compounds with positively polarized hydrogen in covalent compounds (e.g., HCl, H_2O, NH_3) or hydrogen in metallic phases (e.g., $CrH_{1-\delta}$, $PdH_{2-\delta}$) [2,64].

In comparison to only physisorbed hydrogen on a surface, hydrogen in binary hydrides has a much higher bonding energy (0.1 versus 2–3 eV). As indicated in Figure 24.14, binary hydrides can basically be divided into three categories according to their prevailing bonding types: ionic, interstitial (metallic), and covalent hydrides. The specific properties of the hydrides, such as bonding type, formation enthalpy $\Delta_f H^0$, melting T_{melt} and evaporation point T_{evap}, acidity, thermal decomposition temperature T_{decomp} and tendency for hydrolytic decomposition, vary in a systematic way along the periods and groups of the elements in the periodic table.

24.3.1.1 Ionic Hydrides
Elements with low electronegativity EN on the left side of the periodic table, such as alkali and alkaline earth metals (except Be), will have a large positive ΔEN relative to hydrogen. In these cases, hydrides with a predominant ionic character are formed. These compounds are salt-like crystalline solids with M^{n+} cations and H^- hydride anions. Hydride anions act as very strong reducing agents, very strong nucleophiles and as very strong bases, which are also able to deprotonate very weak acids. In contact with water, for example, as an amphoteric compound, ionic hydrides are easily decomposed under the release of hydrogen and the formation of the corresponding hydroxide:

$$EH_n + nH_2O \rightarrow E^{n+} + nOH^- + \tfrac{n}{2}H_2 \quad pK_B(H^-) \sim -21 \quad (24.31)$$

Typical salt-like hydrides have very high thermal decomposition temperatures T_{decomp} of several hundred degrees Celsius up to 1000 °C, due to their high bonding energies and high lattice energies (strong negative formation enthalpies $\Delta_f H^0$, see Figure 24.14). All salt-like hydrides react with water or even weak Brønsted acids (such as CH_4) by formation of H_2 and are used in organic synthesis as strong bases [65]. Dehydrogenation can thus be accomplished by hydrolysis, which is normally irreversible [66].

With decreasing ΔEN, for example, from BaH_2 to MgH_2, the ionicity of the binary hydrides decreases and the covalency increases, as the electron density of the hydride anion shifts more and more towards the cation. This results in a

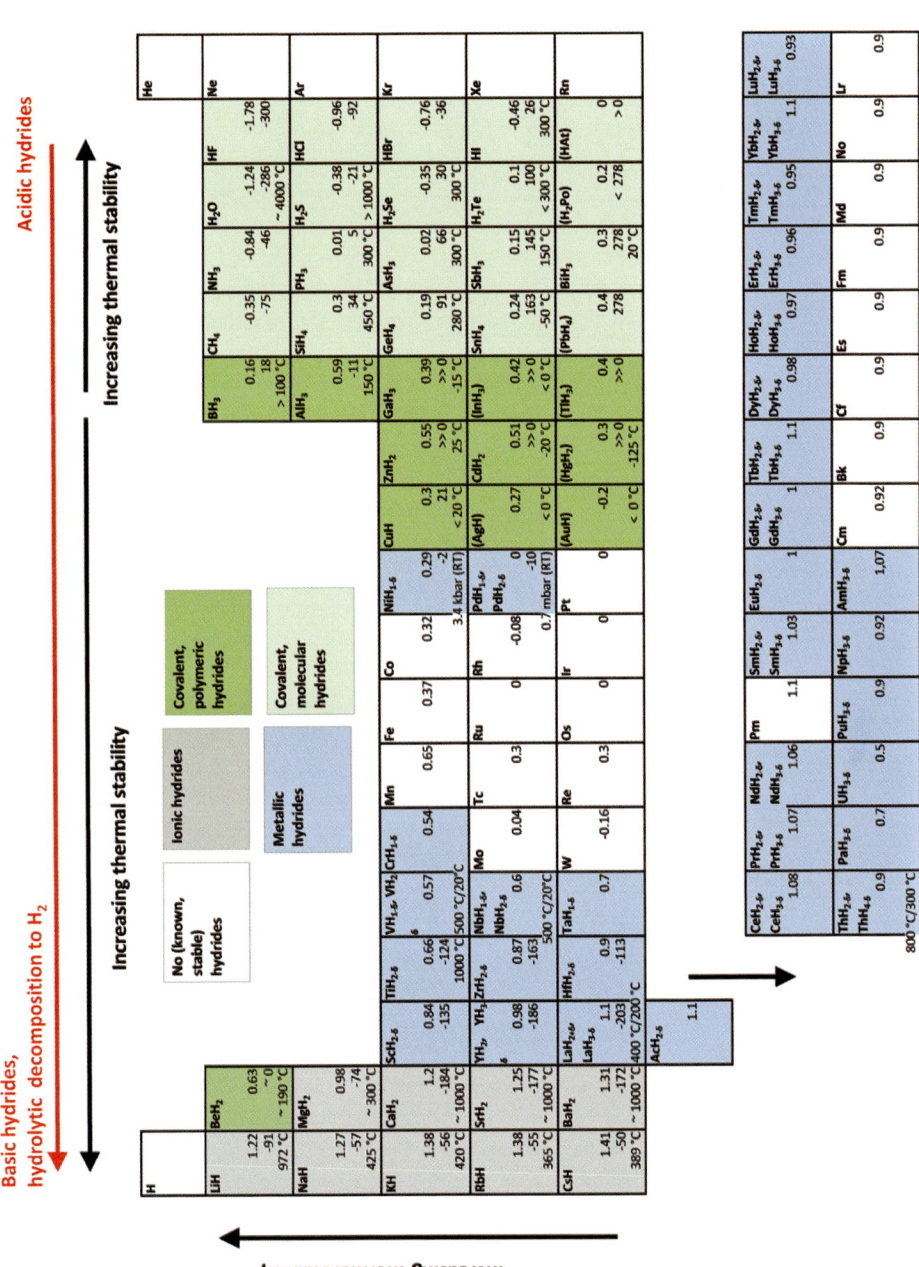

Figure 24.14 Overview of the (stable) binary hydrides (EH_n) of all elements. In addition to the stoichiometry, the electronegativity difference (ΔEN), the formation enthalpy ($\Delta_f H^0$ in kJ mol^{-1}), and the decomposition temperature(s) (T_{decomp}) for the hydride(s) of each element are given. The data is taken from References [1,2].

decreasing formation enthalpy $\Delta_f H^0$, a decreasing thermal stability, a decreasing basicity, decreased reactivity towards electrophiles.

In case of the earth-alkali elements, the transition from a hydride with covalent character to a hydride with ionic character can be observed from MgH_2 to BeH_2. Whereas MgH_2 is a crystalline ionic solid with a rutile-type structure [67], BeH_2 is a covalent compound with $(BeH_2)_n$ polymer chains [68]. The radius of H^- in ionic hydrides depends strongly on the countercation and the lattice structure due to its high polarisability [69]. The radius varies between 139 pm in CsH and 114 pm in LiH, which corresponds to those of F^- (133 pm) and Cl^- (181 pm). The hydride-fluoride analogy results in fluorite-type (SrH_2, CaH_2) or NaCl-type (LiH, NaH) crystal structures for the ionic hydrides [70].

Ionic hydrides can be formed thermally or chemically. Electrochemical formation is difficult to accomplish, because of the hydrides' low hydrogen diffusion coefficients (e.g., MgH_2 at 20 °C, $D_H = 5.8 \times 10^{-14}$ cm^2 s^{-1}) [71] and the formation of oxide or hydroxide surface layers [72]. The adsorption and desorption kinetics of macroscopic particles without the presence of a catalysts are usually slow. In addition, the formation enthalpies $\Delta_f H^0$ of ionic hydrides are the highest of all hydrides. Therefore, the equilibrium temperatures for thermal decomposition/formation are comparatively high for a pressure $p_{H2,eq}$ of 1 bar (e.g., MgH_2, $T_{1\,bar} = 300$ °C). Accelerated reaction kinetics can be realised by reducing the particle size and adding catalysts. As an example, pure and unmilled MgH_2 shows irreversible desorption of 7 wt% H_2 after about 50 min at a temperature of 350 °C and a pressure of 0.15 bar. The addition of 1 mol.% Cr_2O_3 and ballmilling reduces the temperature (300 °C) and the desorption time (10–35 min) and enables a reversible H_2 uptake of 6.4 wt% for 1000 cycles [73]. The optimized process conditions can be explained by a reduction in the particle size and the formation of defects and amorphisation during the ball milling procedure. This results in a decreased transport length for hydrogen diffusion and a reduced expansion enthalpy during hydride phase formation. The catalyst accelerates hydrogen dissociation and provides fast adsorption and desorption kinetics [74]. The most interesting binary ionic hydrides for H_2-storage are LiH and MgH_2 due to their high H_2-capacities (12.5 and 7.6 wt%). By optimizing of those materials using ball milling, destablization of hydrides by alloying with other elements (e.g., Ni) or adding of catalysts, the thermodynamic properties illustrated in Figure 24.14 can be achieved [74,75].

24.3.1.2 Covalent Hydrides

As shown in Figure 24.14, non-metals and semi-metals on the right side of the periodic table with a higher electronegativity form covalent hydrides. In the case of semi-metals or metals in the 3rd to 5th main group, the covalent bond is weakly polar with a negatively polarized $H^{\delta-}$ (e.g., B_2H_6, AlH_3, SiH_4). In compounds of the high electronegative elements on the very right side of the periodic table, highly polar bonds with positively polarized $H^{\delta+}$ can be found (e.g., HF, H_2O, HCl). The intermediate main-group elements form less polar to nonpolar bonds (e.g., CH_4, PH_3).

Covalent hydrides with negatively polarized hydrogen are decomposed in contact with acidic or amphoteric agents like water, because of the nucleophilic character of the negatively polarized hydrogen (B_2H_6, SiH_4, AlH_3):

$$EH_n + nH_2O \rightarrow E(OH)_4 + \tfrac{n}{2}H_2 \tag{24.32}$$

Accompanied by the nucleophilic character, these hydrides are still strong reducing agents. Their reactivity increases with an increasing negative partial charge on the hydrogen atoms.

Covalent hydrides with positively polarized hydrogen act as acids in contact with basic or amphoteric agents like water, because of the electrophilic character of the positively polarized hydrogen (H_2S, HF):

$$EH_n + H_2O \rightarrow EH_{n-1}^- + H_3O^+ \tag{24.33}$$

In contrast to free hydride anions or hydrides with negatively polarized hydrogen, hydrides with positively polarized hydrogen can act as oxidizing agents.

Covalent hydrides with unsaturated electronic shells from the 2nd and 3rd main group like AlH_3, BH_3 or BeH_2 form three-center-two-electron bonds, which result in molecules of high-molecular weight or solid polymers, for example, $(AlH_3)_n$, $(BeH_2)_n$, B_2H_6 or B_4H_{12}. These compounds contain corner or edge-bridged tetrahedrons (BeH_4, BH_4) or octahedrons. Covalent hydrides with saturated electronic shells like CH_4, SiH_4, PH_3 or H_2S exist as molecules with only weak London forces and are usually volatile at standard pressure and temperature. Only the highly polar covalent hydrides in the second period, such as NH_3, H_2O, and HF are able to form so-called hydrogen bonds. Due to the increased interactions between the molecules by dynamic formation of hydrogen bonds, the boiling and melting points of these hydrides are remarkably higher than those of the homolog hydrides from the other periods.

The formation enthalpy and thus the thermal stability of the covalent hydrides increase with increasing electronegativity of the elements (towards the upper right side of the periodic table). The stability against electrophilic or nucleophilic agents and oxygen increases with increasing coordinative saturation (toward the upper rows of the periodic table).

24.3.1.3 Complex Hydrides (Hydrido Complexes)

In complex hydrides hydrogen atoms are (partially) covalently bound in a polyatomic anion. In a ternary compound the complex hydride anion is ionically bound to an alkali or earth alkali cation, forming a salt-like crystal structure [76]. The most important compounds are hydrido alanates (AlH_4^-), hydrido borates (BH_4^-), amides (NH_2^-) and imides (NH^{2-}). Depending on the size and charge of the cation and the anion, different crystallographic structures are realised. Whereas the hydrido borates of alkali metals crystallize in a NaCl-type structure (except for $LiBH_4$ with an orthorhombic structure), the hydrido borates of divalent earth alkali metals have more complex structures [77].

The formation enthalpies (per mole hydrogen) are comparable to the binary ionic hydrides. This results in similar high thermal decomposition temperatures, for example, $NaBH_4$: $T_{decomp} = 406\,°C$. As they contain negatively polarized hydrogen, hydrido alanates and hydrido borates are strong bases and they react with weak acids and amphoteric compounds like water by hydrogen evolution:

$$Na[AlH_4] + 4H_2O = Na[Al(OH)_4] + 4H_2 \tag{24.34}$$

At a high pH, hydrido borates decompose only very slowly. In addition to the ionic hydrides, they are strong reducing agents (Brønsted and Lewis bases).

24.3.1.4 Interstitial (Metallic) Hydrides

Hydrogen can be dissolved as a solid solution in transition metals from the ppm range up to several at.% [97]. It is incorporated in an atomic form in the original metal structure and occupies tetrahedral and/or octahedral interstitial sites [78].

At higher hydrogen contents, binary compounds are formed with a lattice structure different from the original metal lattice structure. The structure of binary interstitial hydrides varies according to the stoichiometry of the compounds. The occupation of all octahedral interstitial sites in an fcc or hcp lattice results in a $1:1$ stoichiometry ($NiH_{1-\delta}$: distorted NaCl-type, $CrH_{1-\delta}$: NiAs-type). Occupying all tetrahedral sites in an fcc lattice yields a $1:2$ stoichiometry ($TiH_{2-\delta}$, $VH_{2-\delta}$, $PdH_{2-\delta}$: fluorite-type). If all interstitial sites in an fcc lattice are occupied, a $1:3$ stoichiometry results ($LaH_{3-\delta}$). Often a substoichiometry δ of several percent can be observed. The transition metal hydrides are crystalline solids with a metallic or semi-metallic character and high thermal and electronic conductivities. There is a metallic bonding of the hydrogen atoms with a comparably less ionic character. Often, remarkable high diffusivities ($D_H = 10^{-6}\text{–}10^{-14}$ $cm^2 s^{-1}$ at $20\,°C$) for hydrogen atoms can be observed [72].

The formation enthalpy $\Delta_f H^0$ and thus the thermal stability decreases with increasing electronegativity of the transition metal (towards the right side of the transition metal block in the periodic system). Hydrides such as $TiH_{2-\delta}$ and $LaH_{3-\delta}$ have high thermal decomposition temperatures of about 1000 and $200\,°C$, respectively, whereas $NiH_{1-\delta}$ and $PdH_{2-\delta}$ decompose below room temperature (Figure 24.14).

The hydrides from metals of the lower left-hand side of the transition metal block, with more positively polarized hydrogen atoms, decompose on contact with water, such as ionic hydrides ($LaH_{3-\delta}$, $CeH_{3-\delta}$). With decreasing negative polarity and less ionic character of the hydrogen atoms, the metallic hydrides react only with acids and are less or not affected by water ($TiH_{2-\delta}$, $NiH_{1-\delta}$).

Ternary Interstitial Hydrides Ternary interstitial hydrides can be classified as AB_5, AB_2, AB_3, AB, and A_2B type compounds with a metal A, which forms a stable metal hydride, and another transition metal B (additional classes are A_2B_4- and A_2B_7-type compounds and V-based solid solution alloys). As indicated in Table 24.3, the compound types have different structures as well as varying H-capacities and formation enthalpies.

Table 24.3 Classification and properties of ternary interstitial hydrides.

Type	Example	Structure	m_H (wt%)	ΔH (kJ mol^{-1} H$_2$)	Note
AB$_5$	LaNi$_5$	CaCu$_5$-type, hexagonal	1.4	−30	Low m_H
AB$_2$	TiCr$_2$	Laves phase, hexagonal (C14) or cubic (C15)	2.3	−23	—
AB$_3$	CeNi$_3$	Rhombohedral	1.3	−44	Poor cyclic stability
AB	TiFe	CsCl-structure, cubic	1.75–1.9	−34	Poor kinetics, high ΔV
A$_2$B	Mg$_2$Ni	AlB$_2$-structure, cubic	3.8	−60	High ΔH

24.3.2
Formation of Hydrides

24.3.2.1 Ionic and Covalent Hydrides

Direct Reaction of the Elements, Thermally Activated Reaction Ionic and covalent hydrides with high (negative) formation enthalpies ($\Delta_f H$) can be easily formed by direct reaction of the elements. The stoichiometric compounds are formed with (nearly) 100% yield, as shown exemplarily for the formation of LiH, BaH$_2$, CaH$_2$, H$_2$O, HF, or HCl:

$$\text{Li} + {}^1\!/_2 \text{H}_2 \rightarrow \text{LiH} \tag{24.35}$$

$$\text{Cl}_2 + \text{H}_2 \rightarrow 2\text{HCl} \tag{24.36}$$

In the case of the metals Li, Ba, Ca, Sr, and Na, the exothermic reaction with hydrogen starts above 300 °C. Oxygen, chlorine, and fluorine react at room temperature after ignition or spontaneously. Nitrogen reacts only at elevated temperatures in the presence of iron as a catalyst (Haber–Bosch process).

Conversion with Other Hydrides Ionic and covalent hydrides with only moderate (negative) or positive formation enthalpies ($\Delta_f H$) cannot be readily prepared from the elements or only with a poor yield. These hydrides can be prepared by conversion of for example, the metal halides with LiH:

$$\text{MgCl}_2 + 2\text{LiH} \rightarrow \text{MgH}_2 + \text{LiCl} \tag{24.37}$$

In the same way complex hydride compounds can be prepared:

$$\text{AlCl}_3 + 4\text{LiH} \rightarrow \text{LiAlH}_4 + 3\text{LiCl} \tag{24.38}$$

$$\text{B(OCH}_3)_3 + 4\text{NaH} \rightarrow \text{NaBH}_4 + 3\text{NaOCH}_3 \tag{24.39}$$

Conversion with Complex Hydrides Another method is conversion with a complex hydride or hydrogenation agent, like LiAlH$_4$ or NaBH$_4$, as shown, for example, for the generation of disilane and arsine:

$$2Si_2Cl_6 + 3LiAlH_4 \rightarrow 2Si_2H_6 + 3LiCl + 3AlCl_3 \quad (24.40)$$

$$4AsCl_3 + 3NaBH_4 \rightarrow 4AsH_3 + 3NaCl + 3BCl_3 \quad (24.41)$$

Hydrolysis or Acidolysis Non-metal hydrides can be gained by the hydrolysis or acidolysis of a binary anionic compound, for example, silicides, phosphides, sulfides, or iodides:

$$Mg_2Si + 4H_2O \rightarrow SiH_4 + 2Mg(OH)_2 \quad (24.42)$$

$$Ca_3P_2 + 6H_2O \rightarrow 2PH_3 + 3Ca(OH)_2 \quad (24.43)$$

$$FeS + 2HCl \rightarrow FeCl_2 + H_2S \quad (24.44)$$

$$3KI + H_3PO_4 \rightarrow 3HI + K_3PO_4 \quad (24.45)$$

24.3.2.2 Interstitial (Metallic) Hydrides

Direct Reaction of the Elements, Thermally Activated Reaction Transition metals can react directly with hydrogen gas (thermal formation). Depending on temperature and hydrogen partial pressure a solid solution of hydrogen and hydride phases is formed. At a low partial pressure of hydrogen and small hydrogen concentrations in the metal, a solid solution of atomic hydrogen can be observed in the original metal lattice (α-phase). The host lattice expands with increasing hydrogen concentration by approximately $3\,\text{Å}^3$ per H atom [79]. For small hydrogen concentrations in the metal, this is described by Sievert's law, see Eq. (24.26).

At a higher partial pressure and higher hydrogen concentrations in the metal, the H atoms, which are initially randomly distributed on interstitial sites, form a hydride phase with a higher hydrogen content and their own structure [80]. Accompanied by the transition from the solid-solution to the hydride phase, a volume expansion of the metal lattice of 8–30% can occur [81]. This can lead to pulverization of the material [82].

In Figure 24.15, the equilibrium between the solid-solution (α-phase) and the hydride phase (β-phase) at different temperatures is shown exemplarily for $LaNi_5H_{6-\delta}$ [79]. The dependence of the phase equilibrium on the hydrogen pressure (p_{H_2}) can be treated by the law of mass action:

$$\tfrac{x}{2}H_2 + M(s) \rightleftarrows MH_x(s) \quad (24.46)$$

$$\frac{a_{MH_x}}{f_{H_2,Gas}^{x/2} a_M} = k_{MH_x} \quad \text{for ideal behaviour} \quad \frac{a_{MH_x}}{a_M} \approx \left(\frac{p_{H_2}}{p^0}\right)^{x/2} k_{MH_x} \quad (24.47)$$

At the equilibrium hydrogen partial pressure, $p_{H_2,eq}$, α- and β-phase coexist, that is, $a_M = a_{MH_x} = 1$ but with different hydrogen concentrations, $c_{H,M} \neq c_{H,MH_x}$. As shown in Figure 24.15, a plateau with a constant equilibrium hydrogen partial pressure ($p_{H_2,eq}$) is reached.

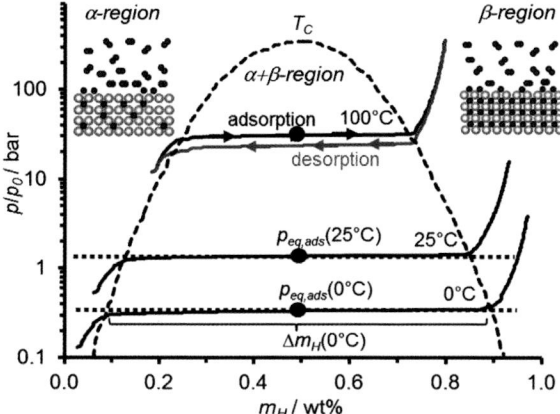

Figure 24.15 Pressure isotherms for the hydrogen incorporation of a typical interstitial transition metal hydride (LaNi$_5$H$_{6-\delta}$) at different temperatures, and the different phase-regions [79].

The value of $p_{H_2,eq}$ depends on the free formation enthalpy ($\Delta_f G^o_{MH_x}$) of the hydride and thus on temperature T:

$$\left(\frac{p_{H_2,eq}}{p^0}\right)^{\frac{x}{2}} = \frac{1}{k_{MH_x}} = e^{\frac{\Delta_f G^o_{MH_x}}{RT}} \quad \text{and} \quad \left(\frac{\partial \ln k_{MH_x}}{\partial T}\right)_p = \frac{\Delta_f H^o_{MH_x}}{RT^2} \quad (24.48)$$

The hydride formation can be endothermic or exothermic. In the case of a negative formation enthalpy ($\Delta_f H^o_{MH_x}$) the equilibrium constant (k_{MH_x}) decreases with increasing temperature, resulting in an increased equilibrium pressure, $p_{H_2,eq}$, as found for LaNi$_5$H$_{6-\delta}$. At a hydrogen partial pressure (p_{H_2}) above the equilibrium value ($p_{H_2,eq}$) the α-phase is completely transferred to the β-phase and additional hydrogen is dissolved as a solid solution in the β-phase. As the temperature increases, so too does the solubility of hydrogen in the α- and the β-phase. This results in a narrowing of the equilibrium plateaus in Figure 24.15. In the case of LaNi$_5$H$_{6-\delta}$, the two-phase region disappears above a critical temperature T_c [72].

It was found experimentally that the formation entropy ($\Delta_f S^o_{MH_x}$) of transition metal hydrides is mainly provided by the entropy change in gaseous hydrogen, $\Delta S^o_{H_2}(25\,°C) = -130.9\,\text{J}\,\text{mol}^{-1}\,\text{K}^{-1}$ [83]. Hence, the formation enthalpy ($\Delta_f H^o_{MH_x}$) of a hydride can be estimated from the equilibrium temperature (T_{eq}) for a pressure of 1 bar:

$$\Delta_f H^0_{MH_x}(1\,\text{bar}) = \Delta_f S^0_{MH_x} \cdot T_{eq}(1\,\text{bar}) = -130.9\,\left(\text{J}\,\text{mol}^{-1}\,\text{K}^{-1}\right) T_{eq}(1\,\text{bar}) \quad (24.49)$$

Electrochemical Formation Hydrogen can also be incorporated in transition metals by electrochemical (cathodic) reduction in an aqueous alkaline solution. Important prerequisites are a sufficiently high diffusion coefficient for atomic hydrogen at room temperature and stability of the solid solution formed, that is, of the hydride against water. The total cathodic reaction yields:

$$\tfrac{x}{2}H_2O + M(s) + \tfrac{x}{2}e^- \rightarrow MH_x(s) + \tfrac{x}{2}OH^- \quad (24.50)$$

In a first step, the atomic hydrogen adsorbed on the metal surface is generated by a charge transfer to a water molecule (Volmer reaction). In a second step the adsorbed H-atoms are incorporated into the metal and diffuse into the bulk metal:

$$H_2O + e^- \rightarrow H_{ad} + \tfrac{x}{2}OH^- \quad (24.51)$$

$$H_{ad} \rightarrow H_{solv} \quad (24.52)$$

The recombination of adsorbed H (Tafel reaction) or the dehydrogenation of water (Heyrovski reaction) leads to the formation of molecular hydrogen [84]. This is regarded as a side reaction:

$$2H_{ad} \rightleftarrows H_2 \quad (24.53)$$

$$H_{ad} + H_2O + e^- \rightarrow H_2 + OH^- \quad (24.54)$$

The electrode reaction can be described by Butler–Volmer kinetics [85]. The exchange current density reflects the capability of the utilized material to adsorb and incorporate hydrogen, as well as the kinetics of the side reactions [72].

When using a standard hydrogen electrode (SHE) as a reference the equilibrium potential (U_{eq}) of the electrode depends on the hydrogen activity (a_{H_2}) in the solid solution and in the hydride (Nernst equation):

$$U_{eq} = -\frac{RT}{2F}\ln a_{H_2} = -\frac{RT}{2F}\ln\frac{p_{H_2,eq}}{p^0} \quad (24.55)$$

The hydrogen activity (a_{H_2}) in the electrode material corresponds to the equilibrium pressure ($p_{H_2,eq}$) of hydrogen.

In Figure 24.16, the change in the electrode potential (U) versus the degree of hydrogenation (x) is plotted for a PdH$_x$ electrode. A Pd electrode is

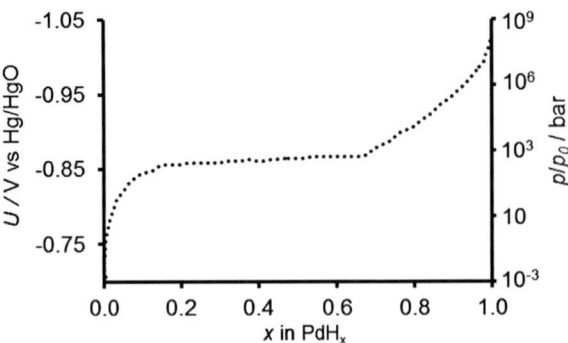

Figure 24.16 Equilibrium potential and derived equilibrium hydrogen pressure for the electrochemical hydrogenation of a 200 nm PdH$_x$ electrode at 25 °C [85].

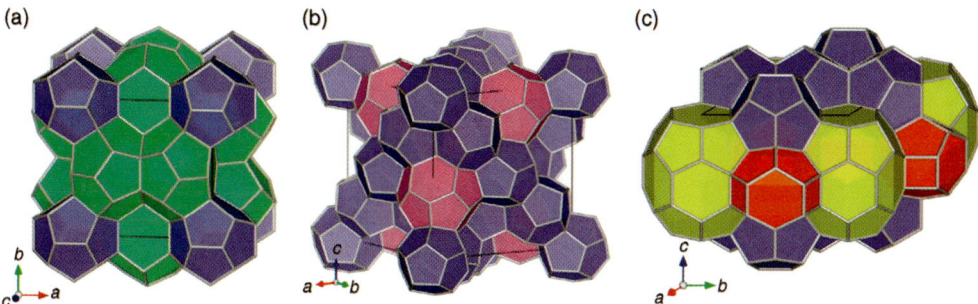

Figure 24.17 Structure of the cubic clathrates sI and sII (a and b) and the hexagonal clathrate sH (c). The oxygen atoms of the water molecules are situated on the corners of the polyhedron. Crystallographically identical polyhedrons are depicted in the same color. The guest molecules are situated within the polyhedrons. Illustration taken from Reference [88]. Nature publishing group.

electrochemically hydrogenated under galvanostatic conditions at 25 °C [85]. Qualitatively, one obtains the same course for the hydrogen activity isotherms as for the hydrogen pressure isotherms in the case of a direct (thermally activated) reaction of the elements (Figure 24.15) [72].

The mean squared displacement, $(\Delta x)^2 = Dt$, can be used to estimate the equilibration time when changing the hydrogen concentration (electrochemically) on the surface. According to the diffusion coefficient (D_H) of atomic hydrogen in Pd at 25 °C of $1.5 \times 10^{-9}\,\mathrm{m^2\,s^{-1}}$ a 0.2 μm thick film will equilibrate nearly instantaneously within a timescale of $\sim 10^{-5}$ s.

24.3.3
Clathrates

Another option for the physical storage of hydrogen in clathrates was first described in 1993 for a mixture of H_2 in H_2O at pressures of 3.1 GPa [86]. Clathrates consist of a lattice of a molecule matrix that traps another molecule type. Without the guest molecule, the clathrate structure is thermodynamically unstable. There are three different types of polyhedral cages, normally formed by hydrogen-bonded water molecules: cubic sI-hydrate, cubic sII-hydrate, and hexagonal sH-hydrate (Figure 24.17) [87,88]. The size and the number of the trapped molecules (N_2, H_2, THF, THF + H_2, CH_4, CO_2 + H_2, TBAF + H_2, HQ + H_2, MTBE + H_2) determine the structure type as well as the temperature and pressure of clathrate formation [89]. Whereas sII-type hydrates with THF (tetrahydrofuran) and H_2 [90] or TBAF (tetra-n-butylammoniumfluoride) and H_2 [91] as guest molecules have formation pressures of around 100–140 bar at temperatures between 0 and 30 °C, the pressure rises to 2500 bar (0 °C) for H_2 trapped in a sII-hydrate [92]. The hydrogen capacity of the materials varies between 3.8 wt% (31 g l^{-1}) for the sII-hydrate fully loaded with H_2 and 0.37 wt% (3.8 g l^{-1}) for a sI-hydrate with ethylene oxide and H_2 as guest molecules [89].

References

1. Greenwood, N.N. and Earnshaw, A. (1990) *Chemie der Elemente*, 1st edn, VCH, Weinheim.
2. Holleman, A. and Wiberg, N. (2007) *Lehrbuch der Anorganischen Chemie*, de Gruyter, Berlin.
3. Woolley, H.W., Scott, R.B., and Brickwedde, F.G. (1948) *J. Res. NBS*, **41**, 379–475.
4. Moesta, H. (1979) *Chemische Statistik*, Springer Verlag, Berlin - Heidelberg.
5. Wedler, G. (1997) *Lehrbuch der Physikalischen Chemie*, Wiley-VCH Verlag GmbH, Weinheim.
6. Bonhoeffer, K.F. and Harteck, P. (1929) *Z. Phys. Chem. B*, **4**, 113–141.
7. Haynes, W.M. (2012) *CRC Handbook of Chemistry and Physics*, 93rd edn, Taylor & Francis Ltd., Boca Raton, FL.
8. Roder, H.M. (1984) *Int. J. Thermophys.*, **5**, 323–350.
9. Saxena, S.C. and Saxena, V.K. (1970) *J. Phys. A*, **3**, 309.
10. Weitzel, D.H., Loebenstein, W.V., Draper, J.W., and Park, O.E. (1958) *J. Res. NBS.*, **60**, 221–227.
11. Bonev, S.A., Schwegler, E., Ogitsu, T., and Galli, G. (2004) *Nature*, **431**, 669–672.
12. Chen, J., Li, X.-Z., Zhang, Q., Probert, M.I.J., Pickard, C.J., Needs, R.J., Michaelides, A., and Wang, E. (2013) *Nat. Commun.*, **4**, 1–5.
13. Diatschenko, V., Chu, C.W., Liebenberg, D.H., Young, D.A., Ross, M., and Mills, R.L. (1985) *Phys. Rev. B*, **32**, 381–389.
14. Fortney, J.J., Glenzer, S.H., Koenig, M., Militzer, B., Saumon, D., and Valencia, D. (2009) *Phys. Plasmas*, **16**, 041003.
15. Leachman, J.W., Jacobsen, R.T., Penoncello, S.G., and Lemmon, E.W. (2009) *J. Phys. Chem. Ref. Data*, **38**, 721–748.
16. Leung, W.B., March, N.H., and Motz, H. (1976) *Phys. Lett. A*, **56**, 425–426.
17. McCarty, R.D. (1975) *Hydrogen Technological Survey - Thermophysical Properties*, National Aeronautics and Space Administration, Washington, D.C.
18. Magro, W.R., Ceperley, D.M., Pierleoni, C., and Bernu, B. (1996) *Phys. Rev. Lett.*, **76**, 1240–1243.
19. Silvera, I. (2010) *Proc. Natl. Acad. Sci. U.S. A.*, **107**, 12743–12744.
20. Tamblyn, I. and Bonev, S.A. (2010) *Phys. Rev. Lett.*, **104**, 065702.
21. Young, D.A. (1991) *Phase Diagrams of the Elements*, University of California Press, Berkley and Los Angeles.
22. March, N.H. (2000) *Int. J. Quantum Chem.*, **80**, 692–700.
23. Goncharov, A.F., Hemley, R.J., and Mao, H.-K. (2011) *J. Chem. Phys.*, **134**, 174501.
24. Howie, R.T., Guillaume, C.L., Scheler, T., Goncharov, A.F., and Gregoryanz, E. (2012) *Phys. Rev. Lett.*, **108**, 125501.
25. Hemmes, H., Driessen, A., and Griessen, R. (1986) *J. Phys. C Solid State*, **19**, 3571.
26. McCarty, R.D., Roder, H.M., and Hord, J. (1981) *Selected Properties of Hydrogen (Engineering Design Data)*, National Bureau of Standards Monograph 168, NBS, Washington, DC.
27. Züttel, A. (2004) *Naturwissenschaften*, **91**, 157–172.
28. Conti, R.S. and Hertzberg, M. (1988) *J. Fire Sci.*, **6**, 348.
29. Jellinek, H.H.G. (1986) *J. Chem. Educ.*, **63**, 1029–1037.
30. Lédé, J., Lapicque, F., and Villermaux, J. (1983) *Int. J. Hydrogen Energy*, **8**, 675–679.
31. Hong, Z., Davidson, D.F., and Hanson, R.K. (2011) *Combust. Flame*, **158**, 633–644.
32. Konnov, A.A. (2008) *Combust. Flame*, **152**, 507–528.
33. Li, J., Zhao, Z., Kazakov, A., and Dryer, F.L. (2004) *Int. J. Chem. Kinet.*, **36**, 566–575.
34. Maas, U. (1988) *Combust. Flame*, **74**, 53.
35. O'Conaire, M., Curran, H.J., Simmie, J.M., Pitz, W.J., and Westbrook, C.K. (2004) *Int. J. Chem. Kinet.*, **36**, 603–622.
36. Willbourn, A.H. and Hinshelwood, C.N. (1946) *Proc. R. Soc. A*, **185**, 353–369.
37. Willbourn, A.H. and Hinshelwood, C.N. (1946) *Proc. R. Soc. A*, **185**, 369–376.
38. Winkelmann, J. (2007) *Landolt-Börnstein – Numerical Data and Functional Relationships in Science and Technology*, Springer Verlag.
39. Marchi, C.S. and Somerday, B.P. (2012) Technical Reference for Hydrogen Compatibility of Materials, Sandia Report

SAND2012-7321, Sandia National Laboratories, Albuquerque, USA.

40 Maxwell, A.S. and Roberts, S.J. (2008) Review of Data on Gas Migration through Polymer Encapsulants, Technical & Assurance Services Serco, Harwell, GB.

41 Toi, K., Takeuchi, K., and Tokuda, T. (1980) *J. Polym. Sci.*, **18**, 189–198.

42 Steward, S.A. (1983) Review of Hydrogen Isotope Permeability Through Materials, Laurence Livermore National Laboratory, Livermore.

43 Frauenfelder, R. (1968) *J. Chem. Phys.*, **48**, 3955–3965.

44 Eichenauer, W., Hattenbach, K., and Pebler, A. (1961) *Z. Metall.*, **52**, 682–684.

45 Scully, J.R., Young, G.A. Jr., and Smith, S.W. (2000) *Mater. Sci. Forum*, **331–336**, 1583–1600.

46 Hashimoto, H. and Kino, T. (1983) *J. Phys. F: Metal. Phys.*, **13**, 1157–1165.

47 Feichtinger, H.K., Locarnini, J.-M., and Marincek, B. (1978) *Mater. Technik*, **2**, 67–70.

48 Chandler, W.T. and Walter, R.J. (1968) in: *Refractory Metal Alloys* (eds I. Machlin, R.T. Begley, and E.D. Weisert), Plenum, New York, pp. 197–249.

49 Tanabe, T., Yamanishi, Y., Sawada, K., and Imoto, S. (1984) *J. Nucl. Mater.*, **122–123**, 1568–1572.

50 Perkins, W.G. and Begeal, D.R. (1972) Permeation and Diffusion of Hydrogen in Ceramvar, Copper, Ceramvar-Copper Laminates, Report SC-DC-714493, Sandia National Laboratories, Albuquerque, USA.

51 Gorman, J.K. and Nardella, W.R. (1962) *Vacuum*, **12**, 19–24.

52 Nelson, H.G. and Stein, J.E. (1973) Gas-Phase Hydrogen Permeation Through Alpha Iron, 4130 Steel and 304 Stainless Steel from Less than 100 °C to Near 600 °C, NASA TN D-7265, National Aeronautics and Space Administration, Washington D.C., USA.

53 Altunoglu, A.K., Blackburn, D.A., and Braithwaite, N.S.J. (1991) *J. Less Common Met. B*, **172–174**, 718–726.

54 Katz, L., Guinan, M., and Borg, R.J. (1971) *Phys. Rev. B*, **4**, 330–341.

55 Louthan, M.R., Donovan, J.A., and Caskey, G.R. (1975) *Acta Metall. Mater.*, **23**, 745–749.

56 Meunier, G. and Manaud, J.P. (1992) *Int. J. Hydrogen Energy*, **17**, 599–602.

57 Gala, H.B., Kara, A.M., Sung, S., Chiang, S.H., and Klinzing, G.E. (1981) *AlChE J.*, **27**, 159.

58 Zheng, Y.-P. and Zhang, T.-Y. (1998) *Acta Mater.*, **46**, 5035–5043.

59 Gonzales, O.D. (1967) *Trans. Metal. Soc. AIME*, **239**, 929–930.

60 Koffler, S.A., Hudson, J.B., and Ansell, G.S. (1969) *Trans. Metal. Soc. AIME*, **245**, 1735.

61 Holleck, G.L. (1970) *J. Phys. Chem.*, **74**, 503–511.

62 Swansiger, W.A., Swisher, J.H., Darginis, J.P., and Schoenfelder, C.W. (1976) *J. Chem. Phys.*, **80**, 308–312.

63 Miller, R.F., Hudson, J.B., and Ansell, G.S. (1975) *Metall. Mater. Trans. A*, **6**, 118.

64 Dalebrook, A.F., Gan, W., Grasemann, M., Moret, S., and Laurenczy, G. (2013) *Chem. Commun.*, **49**, 8735–8751.

65 Clayden, J., Greeves, N., Warren, S., and Wothers, P. (2009) *Organic Chemistry*, Oxford University Press, Oxford.

66 Kong, V.C.Y., Foulkes, F.R., Kirk, D.W., and Hinatsu, J.T. (1999) *Int. J. Hydrogen Energy*, **24**, 665–675.

67 Noritake, T., Towata, S., Aoki, M., Seno, Y., Hirose, Y., Nishibori, E., Takata, M., and Sakata, M. (2003) *J. Alloys Compd*, **356–357**, 84–86.

68 Grochala, W. and Edwards, P.P. (2004) *Chem. Rev.*, **104**, 1283–1316.

69 Bronger, W. (1996) *Z. Anorg. Allg. Chem.*, **622**, 9–16.

70 Bouamrane, A., Laval, J.P., Soulie, J.P., and Bastide, J.P. (2000) *Mater. Res. Bull.*, **35**, 545–549.

71 Čermák, J. and Král, L. (2008) *Acta Mater.*, **56**, 2677–2686.

72 Tliha, M., Khaldi, C., Boussami, S., Fenineche, N., El-Kedim, O., Mathlouthi, H., and Lamloumi, J. (2014) *J. Solid State Electrochem.*, **18**(3), 577–593. DOI: 10.1007/s10008-013-2300-3.

73 Dehouche, Z., Klassen, T., Oelerich, W., Goyette, J., Bose, T.K., and Schulz, R. (2002) *J. Alloy. Compd*, **347**, 319–323.

74 Sakintuna, B., Lamari-Darkrim, F., and Hirscher, M. (2007) *Int. J. Hydrogen Energy*, **32**, 1121–1140.

75 Varin, R.A., Czujko, T., and Wronski, Z.S. (2009) *Nanomaterials for Solid State*

Hydrogen Storage, Springer-Verlag US, Boston, MA.
76 Jain, I.P., Jain, P., and Jain, A. (2010) *J. Alloys Compd*, **503**, 303–339.
77 Černý, R., Filinchuk, Y., Hagemann, H., and Yvon, K. (2007) *Angew. Chem. Int. Ed.*, **46**, 5765–5767.
78 Smithson, H., Marianetti, C.A., Morgan, D., Van der Ven, A., Predith, A., and Ceder, G. (2002) *Phys. Rev. B*, **66**, 144107.
79 Züttel, A. (2003) *Mater. Today*, **6**, 24–33.
80 Schlapbach, L. and Zuttel, A. (2001) *Nature*, **414**, 353–358.
81 Joubert, J.-M., Latroche, M., and Percheron-Guégan, A. (2002) *MRS Bull.*, **27**, 694–698.
82 Westlake, D.G. (1983) *J. Less Common Met.*, **90**, 251–273.
83 Andreasen, A. (2004) *Predicting Formation Enthalpies of Metal Hydrides*, RISØ National Laboratory, Roskilde, DK.
84 Machida, K., Enyo, M., Adachi, G., and Shiokawa, J. (1984) *Electrochim. Acta*, **29**, 807–815.
85 Ledovskikh, A., Danilov, D., Vermeulen, P., and Notten, P.H.L. (2010) *J. Electrochem. Soc.*, **157**, A861–A869.
86 Vos, W.L., Finger, L.W., Hemley, R.J., and Mao, H.-K. (1993) *Phys. Rev. Lett.*, **71**, 3150–3153.
87 Hu, Y.H. and Ruckenstein, E. (2006) *Angew. Chem. Int. Ed.*, **45**, 2011–2013.
88 Momma, K., Ikeda, T., Nishikubo, K., Takahashi, N., Honma, C., Takada, M., Furukawa, Y., Nagase, T., and Kudoh, Y. (2011) *Nat. Commun.*, **2**, 1–7.
89 Strobel, T.A., Hester, K.C., Koh, C.A., Sum, A.K., and Sloan, Jr., E.D. (2009) *Chem. Phys. Lett.*, **478**, 97–109.
90 Florusse, L.J., Peters, C.J., Schoonman, J., Hester, K.C., Koh, C.A., Dec, S.F., Marsh, K.N., and Sloan, E.D. (2004) *Science*, **306**, 469–471.
91 Sakamoto, J., Hashimoto, S., Tsuda, T., Sugahara, T., Inoue, Y., and Ohgaki, K. (2008) *Chem. Eng. Sci.*, **63**, 5789–5794.
92 Dyadin, Y.A., Larionov, E.G., Manakov, A.Y., Zhurko, F.V., Aladko, E.Y., Mikina, T.V., and Komarov, V.Y. (1999) *Mendeleev Commun.*, **9**, 209–210.
93 Heiple, H.R. and Lewis, B. (1941) The Reaction Between Hydrogen and Oxygen: Kinetics of the Third Explosion Limit, *J. Chem. Phys.*, **9**, 584.
94 von Elbe, G. and Lewis, B. (1942) Mechanism of the Thermal Reaction Between Hydrogen and Oxygen, *J. Chem. Phys.*, **10**, 366.
95 Egerton, A. and Warren, D.R. (1951) Warren, Kinetics of the Hydrogen/Oxygen Reaction. I. The Explosion Region in Boric Acid-Coated Vessels, *Proc. R. Soc. Lond. A*, **204**, 465–476.
96 Hinshelwood, C.N. and Moelwyn-Hughes, E.A. (1932) Moelwyn-Hughes, The Lower Pressure Limit in the Chain Reaction between Hydrogen and Oxygen, *Proc. R. Soc. Lond. A*, **138**, 311–317.
97 Wipf, H. (2001) Solubility and diffusion of hydrogen in pure metals and alloys, *Phys. Scr.*, **T94**, 43–51. DOI: 10.1238/Physica.Topical.094a00043.

25
Thermodynamics of Pressurized Gas Storage

Vanessa Tietze and Detlef Stolten

25.1
Introduction

Hydrogen from renewable resources has the potential to replace common fossil energy carriers and to serve as a long-term environmentally friendly alternative in a sustainable energy system. Especially, the use of hydrogen as a future carbon-free road transport fuel represents a promising option, since it offers both ecological and economic benefits compared to alternatives [1]. Efficient hydrogen storage on-board the vehicle is currently the subject of research and development with various technologies under examination [2,3]. This work essentially concentrates on storing hydrogen as compressed gas, cryogenic liquid, and chemically bonded in solids with high-pressure storage being the technology most applied in vehicles [4]. Correspondingly, most fueling stations are equipped exclusively for refueling with compressed hydrogen [4,5]. Therefore, this chapter concentrates on the thermodynamics of pressurized gas storage. Further information on the thermodynamics of liquid hydrogen storage can be found in References [6–8].

For nationwide application of compressed gaseous hydrogen as a road transport fuel in Germany the usage of a delivery infrastructure that employs pipelines for transportation and distribution and large-scale underground storage in salt caverns to even out daily and seasonal fluctuations is most suitable [1,9]. Within this framework analysis of the technical requirements is of great relevance. Since structural analogies and technical similarities exist between this future hydrogen supply system and the existing one for natural gas, a comparison of the behavior of hydrogen and natural gas with respect to typical thermodynamic processes in terms of storage is useful. Natural gas consists of 70–99% methane [10]. For simplicity, methane alone is analyzed as representative of natural gas in the following.

First, the equations of state for real and ideal gases used for the modeling of hydrogen and methane within this thermodynamic analysis are specified. Subsequent important thermodynamic properties of the two gases are then presented and compared. Based on this thermodynamic difference, borderline processes for expansion and compression are investigated by means of an indirect and direct calculation methods. After this, a theoretical model based on mass and energy balance, thermodynamic laws, and simple equations to consider heat transfer is presented. This model is then used for the computation of two different examples to demonstrate its applicability. In the first example the refueling of a vehicle storage tank is examined, with particular emphasis on the influence of the Joule–Thomson effect. In the second example the pressure and temperature change within a salt cavern for periods of injection, withdrawal, and stand-by is analyzed. All computations within this chapter were performed using the numerical software MATLAB as the simulation environment.

25.2
Calculation of Thermodynamic State Variables

The equations of state, which comprise the mathematical relationship between different thermodynamic state variables, represent the fundamental basis for describing gas behavior. For the sake of completeness, the thermal and caloric equations of state for both ideal and real gases are therefore specified briefly in the following.

25.2.1
Ideal Gases

The relationship of the thermal state variables of an ideal gas is defined by the well-known classical ideal gas law:

$$pV = mR_i T = nR_u T \quad (25.1)$$

where p is the pressure, V the volume, T the temperature, m the mass, n the molar mass, and R_i and R_u the specific and universal gas constants, respectively. The isobaric heat capacity (c_p^0) of an ideal gas is only a function of the temperature (see Eq. (25.2)). Here, a simple polynomial given in Reference [11] was used for its calculation. The isochoric heat capacity (c_v^0) and the isentropic exponent (κ^0) of ideal gases can then be determined according to Eqs (25.3) and (25.4), respectively:

$$c_p^0 = f(T) \quad (25.2)$$

$$c_v^0 = c_p^0 - R_i \quad (25.3)$$

$$\kappa^0 = \frac{c_p^0}{c_v^0} \quad (25.4)$$

The enthalpy (h^0) and the internal energy (u^0) of an ideal gas at a certain temperature can be obtained by integrating the isobaric and isochoric heat capacity as specified in Eqs (25.5) and (25.6), respectively. The reference temperature (T_{Ref}) can be arbitrary chosen; here the standard temperature of 273.15 K was selected:

$$h^0(T) = \int_{T_{\text{Ref}}}^{T} c_p^0(T) \cdot dT + h_{\text{Ref}}, \quad h_{\text{Ref}} = 0 \tag{25.5}$$

$$u^0(T) = \int_{T_{\text{Ref}}}^{T} c_v^0(T) \cdot dT + u_{\text{Ref}}, \quad u_{\text{Ref}} = -R_i \cdot T_{\text{Ref}} \tag{25.6}$$

25.2.2
Real Gases

The idea of the ideal gas is a theoretical concept based on the assumption that molecules can be modeled as point particles of negligible volume and without forces of attraction or repulsion. For many applications the simplified equation of state of ideal gases can predict the behavior of a gas with a sufficient degree of accuracy. However, especially at high pressures and low temperatures, these simplifications can no longer be considered valid and lead to errors. Deviations between ideal and real gas behavior increase as the critical point is approached. To consider these divergences it is a convenient approach to include a dimensionless property termed the compressibility factor (Z) in the thermal equation of state for real gases (Eq. (25.7)):

$$pV = mZR_iT = nZR_uT \tag{25.7}$$

Several different models for the prediction of real gas behavior exist. Here, equations of state for hydrogen and methane in the form of a fundamental equation explicit in the Helmholtz free energy are used. The Helmholtz free energy (a) can be expressed according to Eq. (25.8) as a dimensionless quantity, which is a function of the reciprocal reduced temperature (τ) and reduced density (δ):

$$\frac{a(T,p)}{RT} = \alpha(\delta, \tau), \quad \delta = \frac{\rho}{\rho_c}, \quad \tau = \frac{T_c}{T} \tag{25.8}$$

This dimensionless quantity is termed the reduced Helmholtz free energy (α) and is composed of the ideal gas contribution (α^0) and the residual contribution (α^r) (Eq. (25.9)). With knowledge of the reduced Helmholtz free energy other thermodynamic properties can be calculated with the equations given in Table 25.1:

$$\alpha(\delta, \tau) = \alpha^0(\delta, \tau) + \alpha^r(\delta, \tau) \tag{25.9}$$

To calculate the ideal gas part and the residual contribution of the reduced Helmholtz free energy and their derivatives, expressions given in References [12,13] were employed for hydrogen and methane, respectively. Both are

Table 25.1 Definitions of some thermodynamic properties and their expression in terms of the reduced Helmholtz free energy (α) [15–17].

Property and definition	Relation to α^o and α^r and their derivatives
Compressibility factor: $Z(T,\rho) = \dfrac{p}{\rho RT}$	$Z(\tau,\delta) = 1 + \delta\alpha^r_\delta$ \quad (25.10)
Pressure: $p(T,p) = -\left(\dfrac{\partial a}{\partial v}\right)_T$	$p(\tau,\delta) = \dfrac{\rho R_u T}{M}\left[1 + \delta\alpha^r_\delta\right]$ \quad (25.11)
Isobaric heat capacity: $c_p(T,p) = \left(\dfrac{\partial h}{\partial T}\right)_p$	$c_p(\tau,\delta) = \dfrac{R_u}{M}\left[-\tau^2\left(\alpha^0_{\tau\tau} + \alpha^r_{\tau\tau}\right) + \dfrac{\left(1 + \delta\alpha^r_\delta - \delta\tau\alpha^r_{\delta\tau}\right)^2}{\left(1 + 2\delta\alpha^r_\delta + \delta^2\alpha^r_{\delta\delta}\right)}\right]$ \quad (25.12)
Isochoric heat capacity: $c_v(T,p) = \left(\dfrac{\partial u}{\partial T}\right)_v$	$c_v(\tau,\delta) = \dfrac{R_u}{M}\left(-\tau^2\left[\alpha^0_{\tau\tau} + \alpha^r_{\tau\tau}\right]\right)$ \quad (25.13)
Enthalpy: $h(T,p) = u + pv$	$h = \dfrac{R_u T}{M}\left[\tau\left(\alpha^0_\tau + \alpha^r_\tau\right) + \delta\alpha^r_\delta + 1\right]$ \quad (25.14)
Internal energy: $u(T,p) = a + Ts$	$u = \dfrac{R_u T}{M}\tau\left[\alpha^0_\tau + \alpha^r_\tau\right]$ \quad (25.15)

Entropy:
$$s(T,p) = -\left(\frac{\partial a}{\partial T}\right)_v$$
$$s = \frac{R_u}{M}\left[\tau(\alpha_\tau^0 + \alpha_\tau^r) - \alpha^0 - \alpha^r\right] \quad (25.16)$$

Joule–Thomson coefficient:
$$\mu(T,p) = \left(\frac{\partial T}{\partial p}\right)_h$$
$$\mu(\tau,\delta) = \frac{-(\delta\alpha_\delta^r + \delta^2\alpha_{\delta\delta}^r + \delta\tau\alpha_{\delta\tau}^r)M}{\left[(1+\delta\alpha_\delta^r - \delta\tau\alpha_{\delta\tau}^r)^2 - \tau^2(\alpha_{\tau\tau}^0 + \alpha_{\tau\tau}^r)(1+2\delta\alpha_\delta^r + \delta^2\alpha_{\delta\delta}^r)\right]R_u\rho} \quad (25.17)$$

Real isentropic exponent:
$$\kappa_{p,v}(T,p) = -\left(\frac{v}{p}\right)\left(\frac{\partial p}{\partial v}\right)_s$$
$$\kappa_{p,v} = \frac{1+2\delta\alpha_\delta^r + \delta^2\alpha_{\delta\delta}^r}{1+\delta\alpha_\delta^r}\left[1 - \frac{(1+\delta\alpha_\delta^r - \delta\tau\alpha_{\delta\tau}^r)^2}{\tau^2(\alpha_{\tau\tau}^0 + \alpha_{\tau\tau}^r)(1+2\delta\alpha_\delta^r + \delta^2\alpha_{\delta\delta}^r)}\right] \quad (25.18)$$

Derivatives of α^0 and α^r with respect to δ and τ (for detailed expressions see Reference [18] for hydrogen and Reference [13] for methane):

$$\alpha_\delta^r = \left(\frac{\partial\alpha^r}{\partial\delta}\right)_\tau, \alpha_{\delta\delta}^r = \left(\frac{\partial^2\alpha^r}{\partial\delta^2}\right)_\tau, \alpha_{\tau\tau}^r = \left(\frac{\partial^2\alpha^r}{\partial\tau^2}\right)_\delta, \alpha_{\delta\tau}^r = \left(\frac{\partial^2\alpha^r}{\partial\delta\partial\tau}\right), \alpha_\tau^0 = \left(\frac{\partial\alpha^0}{\partial\tau}\right)_\delta, \alpha_{\tau\tau}^0 = \left(\frac{\partial^2\alpha^0}{\partial\tau^2}\right)_\delta$$

implemented in the NIST Chemistry WebBook [14] and are currently considered to be the most accurate equation of state available. In the case of hydrogen the parameters and coefficients for normal hydrogen, which is an equilibrium mixture of 75% ortho-hydrogen and 25% parahydrogen at normal conditions, were used. The equation of state for normal hydrogen is valid for temperatures from the triple-point of 13.957 K up to 1000 K and for pressures up to 2000 MPa [12]. The equation of state for methane is valid from the triple point (90.6941 K, 0.011696 MPa) up to temperatures of 625 K and pressures of 1000 MPa [13].

25.3
Comparison of Thermodynamic Properties

The pressure and temperature dependency of the compressibility factor, the Joule–Thomson coefficient, and the real isentropic exponent of hydrogen and methane are examined and compared. These thermodynamic properties are especially important for pressurized gas storage, which is explained in the following.

25.3.1
Compressibility Factor

The compressibility factor is also known as the compression factor and serves as a measure of the divergences from ideal gas behavior. Therefore, the closer to the critical point, the stronger the rate of change in the compressibility factor [19]. The critical point of normal hydrogen lies at 33.1 K and 1.3 MPa [12]. This is much lower than the critical point of methane with a temperature of 190.6 K and a pressure of 4.6 MPa [13]. The compressibility factor can be expressed, for instance, as a function of temperature and pressure. Figure 25.1

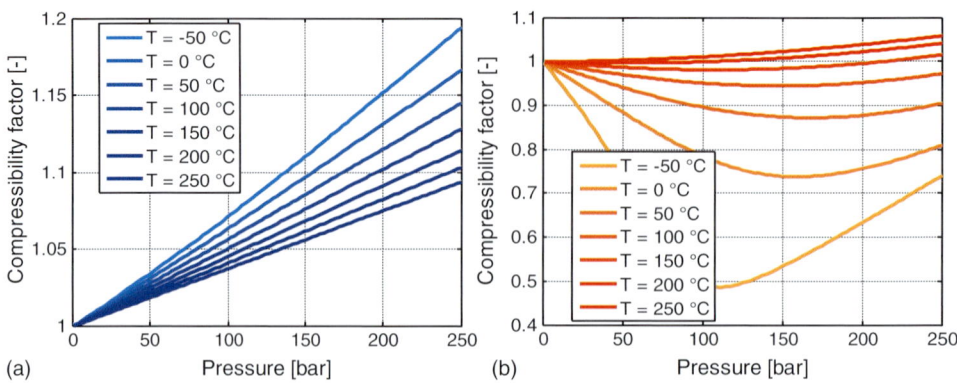

Figure 25.1 Compressibility factor of hydrogen (a) and methane (b).

shows the rate of change of the compressibility factor of hydrogen and methane for a temperature and pressure range of −50 to 250 °C and of 1 to 250 bar, respectively. For hydrogen the compressibility factor rises nearly linearly with pressure, with the temperature dependency being more pronounced at higher pressures. For methane no linear behavior can be observed in the chosen temperature and pressure range. At colder temperatures the rate of change clearly increases, as expected due to proximity to the critical point.

The compressibility factor of hydrogen is above, and that of methane below, 1 at ambient temperature conditions and pressures below 250 bar. The meaning of this can be best explained with the help of a sample calculation. A storage tank with a geometric volume of 1 m³, temperature of 20 °C, and a pressure of 250 bar contains 18 kg hydrogen or 194 kg methane when the compressibility factor is taken into account. However, under ideal gas assumption it contains 21 kg hydrogen and 165 kg of methane. Consequently, the mass stored in the tank is underestimated by 18% in the case of methane and overestimated by 14% in the case of hydrogen when using the ideal gas law.

25.3.2
Joule–Thomson Coefficient

The adiabatic throttling of a fluid at constant enthalpy through an insulated porous plug, valve, or similar device is often referred to as Joule–Thomson expansion [20]. Depending on the conditions of state the temperature of the fluid can decrease, increase, and, seldomly, even stay constant. This temperature change is called the Joule–Thomson effect, concerning which the differential and integral Joule–Thomson effect are discerned. The first describes the differential temperature change with reference to a differential pressure change, which is expressed by the Joule–Thomson coefficient:

$$\mu(T, p) = \left(\frac{\partial T}{\partial p}\right)_h \tag{25.19}$$

The integral Joule–Thomson effect is the temperature change for a pressure drop from p_1 to p_2, which is defined according to Eq. (25.20) as the integral over the Joule–Thomson coefficient from p_1 to p_2. The relationship of temperature and pressure change can be expressed by a proportionality factor, which is termed the integral Joule–Thomson coefficient [21,22]:

$$\Delta T_h = T_2(h, p_2) - T_1(h, p_1) = \int_{p_1}^{p_2} \mu(T, p) \mathrm{d}p = \mu_{\text{int}}(p_2 - p_1) \tag{25.20}$$

For both the differential and integral Joule–Thomson effect, inversion curves can be determined [21,23], but the latter is not as common and is therefore neglected here. The differential inversion curve is derived from the

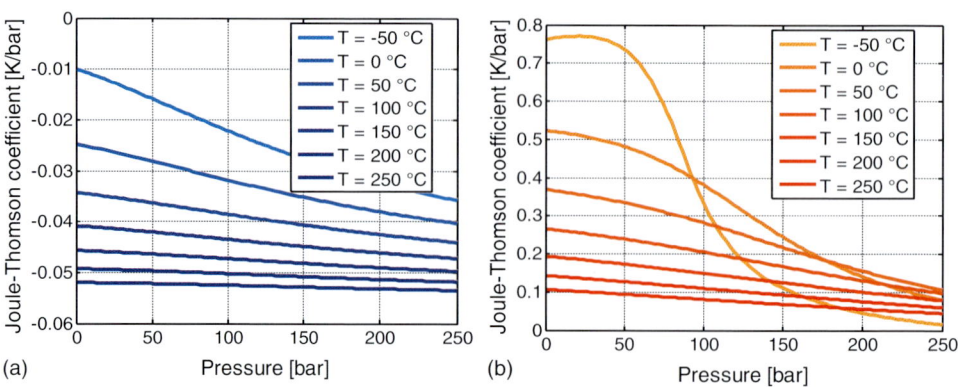

Figure 25.2 Joule–Thomson coefficient of hydrogen (a) and methane (b).

thermodynamic states, which satisfy Eq. (25.21) in the p,h-diagram. For an ideal gas, Eq. (25.21) is satisfied for all states, and, thus, the isenthalpic expansion does not cause a temperature change:

$$\mu(T,p) = \left(\frac{\partial T}{\partial p}\right)_h = 0 \tag{25.21}$$

Consequently, the inversion temperature is a function of pressure and enthalpy. With the help of the differential inversion curve two areas of thermodynamic states can be distinguished. One with a negative Joule–Thomson coefficient, where a reduction in pressure leads to an increase of temperature and heating of the gas, and one with a positive Joule–Thomson coefficient, were a reduction in pressure leads to a decrease in temperature and cooling of the gas [20]. Reference [24] gives the differential inversion curves of hydrogen and methane. Above a temperature of 200 K the Joule–Thomson coefficient of hydrogen is always negative. Conversely, methane possesses, for temperatures between 300 and 650 K and pressure below 40 MPa, a positive Joule–Thomson coefficient.

Figure 25.2 shows the Joule–Thomson coefficient for hydrogen and methane as a function of pressure for different temperatures. For cold temperatures and low pressures the Joule–Thomson coefficient of hydrogen is close to zero and the steepness of the curve increases, since it is closer to the inversion temperature. With increasing temperature, the negative value increases, whereas the pressure dependency of it vanishes. In the range considered this means an increase of the temperature change at higher temperatures. For methane, a very strong nonlinear influence of pressure can be observed for a temperature of −50 °C, which can be attributed to the fact that with increasing pressure the differential inversion curve is approached. In contrast, similar to hydrogen, the pressure dependency for higher temperatures adopts a linear, slightly decreasing character.

Table 25.2 Isentropic relations for ideal and real gases.

Independent variables	Isentropic relation for ideal gases	Isentropic relation for real gases
p, v	$pv^\kappa = \text{const.}$ (25.22)	$pv^{\kappa_{p,v}} = \text{const.}$ (25.23)
T, p	$T^\kappa p^{(1-\kappa)} = \text{const.}$ (25.24)	$T^{\kappa_{T,p}} p^{(1-\kappa_{T,p})} = \text{const.}$ (25.25)
T, v	$Tv^{(\kappa-1)} = \text{const.}$ (25.26)	$Tv^{(\kappa_{T,v}-1)} = \text{const.}$ (25.27)

25.3.3
Isentropic Exponent

The isentropic exponent allows the use of simple relations (Table 25.2) to describe the change of the thermal variables of state during an isentropic process, which means the entropy of the system is assumed to remain constant. Consequently, it is an important and often used material characteristic for the calculation of compression and expansion processes based on thermal equations of state (Section 25.4).

For ideal gases the isentropic exponent κ_0 is defined according to Eq. (25.4) as the ratio of specific heats and depends only on temperature or is a constant. For real gases three different isentropic exponents ($\kappa_{p,v}$, $\kappa_{T,p}$, and $\kappa_{T,v}$) can be defined depending on the pair of independent variables used [25,26]. These three isentropic exponents of real gases can be connected by a relation. Depending on the type of gas the calculated values for the three exponents $\kappa_{p,v}$, $\kappa_{T,p}$, and $\kappa_{T,v}$ vary differently with temperature and pressure. Nevertheless, to describe isentropic processes they are equivalent and only one of the three isentropic relations for real gases in Table 25.2 is needed. However, the correct relationships of p, v, and T according to the thermal equation of state for real gases have to be inserted, which leads to different expressions. Here, p and v are selected as independent variables. The corresponding isentropic relation is given in Eq. (25.23) and the definition of $\kappa_{p,v}$ is given in Eq. (25.18) (Table 25.1). The temperature after an isentropic change can be calculated according to Eq. (25.28), which is derived by substituting Eq. (25.7) in Eq. (25.23):

$$T_2 = T_1 \frac{Z_1}{Z_2} \left(\frac{p_2}{p_1}\right)^{\frac{\kappa_{p,v}-1}{\kappa_{p,v}}} \quad (25.28)$$

However, Eq. (25.28) assumes a constant isentropic exponent during the state change, which actually is a function of temperature and pressure. Assuming a linear change of the isentropic exponent, a more precise approach can be derived by utilizing its arithmetic mean. However, the more its behavior varies from the assumed linear character, the higher the result deviates from the actual value. Therefore, this estimation method should only be used for small or

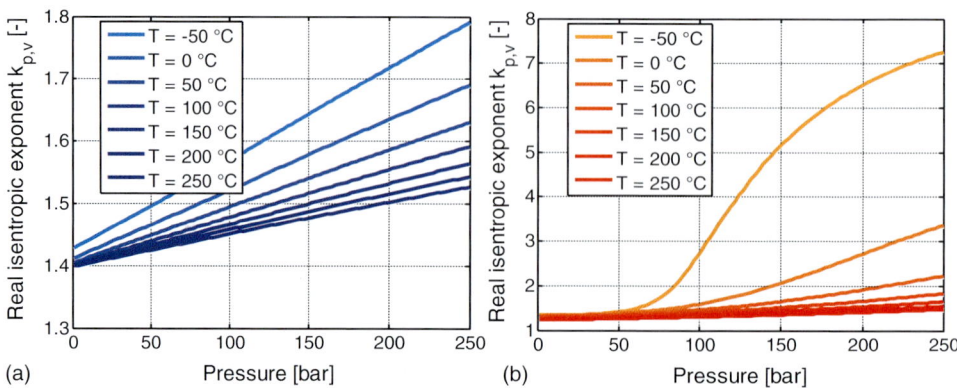

Figure 25.3 Real isentropic exponent of hydrogen (a) and methane (b).

differential changes in state when there is a strong nonlinear temperature and pressure dependency of the isentropic exponent between the considered states.

Figure 25.3 shows the real isentropic exponent $\kappa_{p,v}$ of hydrogen and methane as a function of pressure for different temperatures. The real isentropic exponent of hydrogen increases nearly linear with rising pressure, whereby the temperature dependency is stronger at higher pressures. The real isentropic exponent of methane shows no linear behavior with rising pressure for low temperatures. With increasing temperature, the change of the isentropic exponent decreases and the curve approaches a straight line with a significantly lower gradient. For both methane and hydrogen, the value of the isentropic exponent decreases with increasing temperature. However, the change of the isentropic exponent of hydrogen with temperature is not as distinctive as the one of methane. Especially at low temperatures the isentropic exponent of methane is much higher in comparison to hydrogen. However, at high temperatures and pressures the deviations between the two gases become smaller.

25.4
Thermodynamic Analysis of Compression and Expansion Processes

For loading and unloading of pressurized gas storage facilities, compression and expansion processes are used. From a technical point of view multiple options exist for their implementation, which differ in terms of their thermodynamic process. As a consequence, the final state of the gas can differ significantly, depending on the utilized process of compression or expansion. In this section, the thermodynamic processes that present borderline cases for compression and expansion are introduced briefly. Different calculation methods are then presented, demonstrated, and analyzed by applying them to hydrogen and methane.

25.4.1
Isothermal and Isentropic Compression

The actual change of state during compression can only be determined with great effort. Therefore, it is common to use a thermodynamic reference process to which the actual process can be referred to. The thermodynamic borderline cases for reversible compression are constituted by the isothermal process resulting in the minimum required work and the isentropic process resulting in the maximum required work. During an isothermal process the temperature is assumed to stay constant. Therefore, it is generally selected as reference process for single- and multistage compressors with cooling [19]. Since the isentropic process describes an adiabatic change of condition, which means no heat exchange with the environment, the temperature rises during compression. It is generally used as reference process for uncooled single- and multistage turbocompressors of moderate compression ratio and single-stage displacement compressors with jacket cooling [19]. As mentioned above, both processes are reversible, which means that losses are not considered, and consequently they are idealized interpretations of the compression process. Deviations exist between the thermodynamic reference process and the real process, because various losses occur. Their influence is usually considered by applying a corresponding efficiency as proportional factor. In this manner, the higher temperatures and compression work of the real process can be taken into account. For completeness, it should be mentioned that beside the isothermal and isentropic reference process a polytropic reference process also exists. However, strictly speaking, it does not describe a reversible change of condition and will not be discussed further here.

Principally, two different options for calculation of the compression work exist. The first option is derived by solving the integral given in Eq. (25.29) and allows calculation of the compression work using only thermal variables of state. As an advantage, the caloric equations of state are not required and the computing effort is generally lower:

$$w_{t,12} = \int_1^2 v \, dp \tag{25.29}$$

The second option is based on the caloric equations of state, which are required to calculate the compression work directly from the first law of thermodynamics for an open system. They arise according to Eq. (25.30) from the enthalpies before and after compression and the heat energy removed from the process. This method is known to be more precise, since the caloric values are determined directly instead of utilizing the thermal values [27]. Here, following Reference [27] the first option is called an indirect method and the second option a direct method. In the following, the respective equations for isothermal and isentropic compression are presented, starting with the indirect method and finishing with the direct method:

$$w_{t,12} = (h_2 - h_1) - q_{12} \tag{25.30}$$

The equation used to calculate the specific compression work for an isothermal process of a real gas according to the indirect method can be obtained by inserting Eq. (25.7) into Eq. (25.29), which results in Eq. (25.31):

$$w_{t,12} = R_i T \int_1^2 \frac{Z}{p} dp \approx R_i T \overline{Z} \ln\left(\frac{p_2}{p_1}\right), \overline{Z} = \frac{Z(p_1, T) + Z(p_2, T)}{2}, T = \text{const.}$$

(25.31)

The compressibility factor is a function of the pressure, which must be considered when solving the integral. For simplicity, the compressibility factor (Z) can be assumed to be constant, whereby a suitable mean value should be used [19]. Here, the arithmetic mean value of the state before and after compression is applied. Since the temperature is constant and the exit pressure is given, the state after compression is known and the compression work can be determined by directly solving Eq. (25.31).

The equation used to calculate the compression work for an isentropic process of a real gas according to the indirect method is given in Eq. (25.32), which is obtained by integrating the isentropic relation Eq. (25.23) into Eq. (25.29):

$$w_{t,12} = R_i T_1 Z_1 p_1^{\frac{1-\overline{\kappa}_{p,v}}{\overline{\kappa}_{p,v}}} \int_1^2 p^{\frac{-1}{\overline{\kappa}_{p,v}}} dp \approx \frac{\overline{\kappa}_{p,v}}{\overline{\kappa}_{p,v} - 1} R_i T_1 Z_1 \left[\left(\frac{p_2}{p_1}\right)^{\frac{\overline{\kappa}_{p,v}-1}{\overline{\kappa}_{p,v}}} - 1 \right], \overline{\kappa}_{p,v}$$

$$= \frac{\kappa_{p,v}(p_1, T_1) + \kappa_{p,v}(p_2, T_2)}{2}$$

(25.32)

Similar to the isothermal process, solution-finding can be simplified by neglecting the pressure and temperature dependency of the isentropic exponent and by assuming a constant value. As before, a suitable mean value has to be used [19]. Here, the arithmetic mean value of the state before and after compression is again applied. However, this time the state after compression is not known. Therefore, Eq. (25.32) has to be solved together with Eq. (25.28) iteratively.

Principally, the indirect method is based on formulas of an ideal gas that were adjusted with the compressibility factor and the real isentropic exponent to correspond to the real gas behavior. In addition, the simplifications made for solution-finding lead to errors when larger changes in state are considered or when there is a noticeable nonlinear temperature and pressure dependency of the compressibility factor and/or of the real isentropic exponent. Therefore, this simplified calculation should only be used within certain limits, which are given in Reference [19]. The direct method is a more precise alternative. It requires,

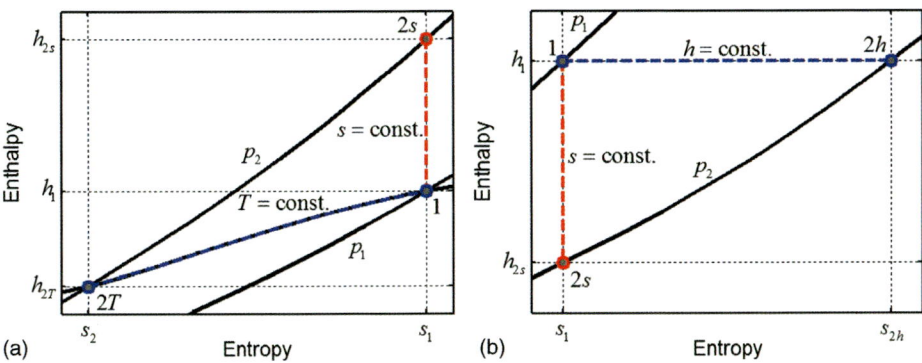

Figure 25.4 Schematic principle of the isothermal and isentropic compression (a) and the isenthalpic and isentropic expansion (b) shown in the (h,s)-diagram.

however, the implementation of caloric equations of state of real gases, which can require great effort. With the help of a suitable computer based calculation program, at least, these equations can be quickly solved.

The equation used to calculate the specific compression work for an isothermal process of a real gas according to the direct method is specified in Eq. (25.33), which is based on Eq. (25.30). In the isothermal case, the heat removed from the process to keep the temperature constant also has to be considered:

$$w_{t,12} = (h_{2T} - h_1) - T(s_2 - s_1)$$
$$= \left[h(T,p_2) - h(T,p_1)\right] - T\left[s(T,p_2) - s(T,p_1)\right] \quad (25.33)$$

During the isentropic process no heat is removed and Eq. (25.30) reduces to Eq. (25.34). Notably, for calculations of the enthalpies in Eqs (25.33) and (25.40) different thermodynamic variables are used. The schematic principle of the isothermal and isentropic compression in the (h,s)-diagram is illustrated in Figure 25.4a. The temperature after isentropic compression can be obtained from the corresponding enthalpy and pressure by solving Eq. (25.14) iteratively:

$$w_{t,12} = (h_{2s} - h_1) = h(s_1,p_2) - h(s_1,p_1) \quad (25.34)$$

For a comparison the compression work and the isentropic temperature of hydrogen and methane were computed with the equations given above and the equations of state for real gases in Section 25.2.2. To clarify the difference between ideal and real gas behavior, they were computed according to the indirect method by using the equations of state for ideal gases in Section 25.2.1 as well. The temperature and pressure at the suction side has been set to 10 °C and 50 bar, respectively. The compression ratio p_2/p_1 was increased until a final pressure of 250 bar was reached.

The results with respect to the compression work are shown in Figure 25.5 and with respect to the isentropic temperature in Figure 25.6. From Figure 25.5 it can be seen that the compression of hydrogen requires more energy on a

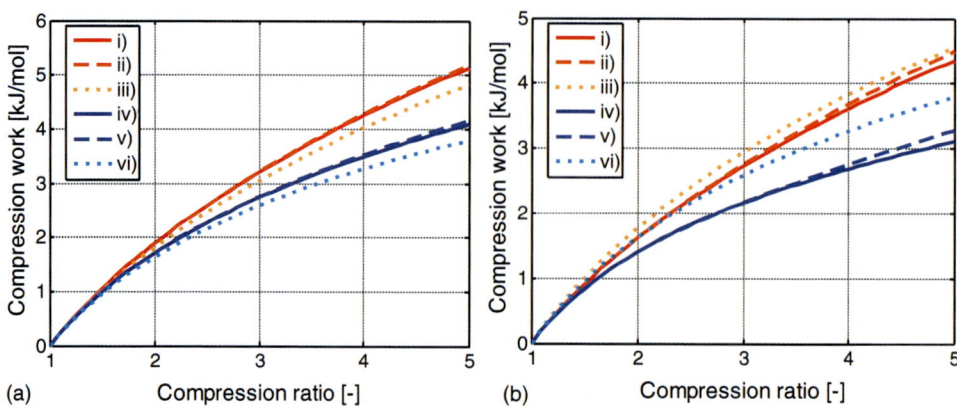

Figure 25.5 Compression work of hydrogen (a) and methane (b) as a function of the compression ratio for different processes and calculation methods: (i) isentropic process, direct method, real gas; (ii) isentropic process, indirect method, real gas; (iii) isentropic process, indirect method, ideal gas; (iv) isothermal process, direct method, real gas; (v) isothermal process, indirect method, real gas; (vi) isothermal process, indirect method, ideal gas.

molar basis than methane. Since hydrogen is considerably lighter than methane, with reference to the mass also a significantly higher value is obtained, for example, 2545 kJ kg^{-1} for hydrogen compared to 271 kJ kg^{-1} for methane under the same conditions (isentropic process, compression ratio of 5). It is also interesting to refer the compression work to the lower heating value (LHV) of the gas. For a compression ratio of 5, the isentropic and isothermal compression of hydrogen requires 2.12% and 1.69% of its LHV, respectively. Under the same conditions methane needs 0.54% und 0.39% of its LHV.

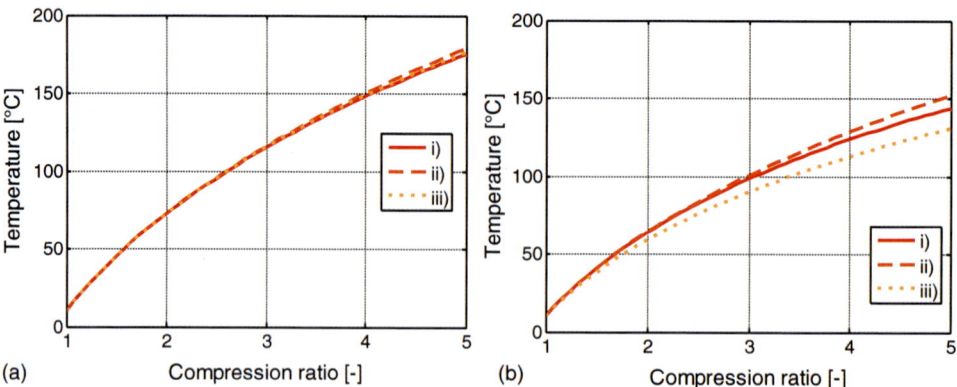

Figure 25.6 Temperature of hydrogen (a) and methane (b) after isentropic compression as a function of the compression ratio for different calculation methods: (i) direct method, real gas, (ii) indirect method, real gas, and (iii) indirect method, ideal gas.

Comparing the results of the indirect and direct method for real gases, it can be seen that they are rather close with the deviations increasing at higher compression ratios. These deviations are greater for methane, which can be traced back to the stronger temperature and pressure dependency of the compressibility factor and the real isentropic exponent. For the determination of the work input of pipeline compressors with a compression ratio below 1.6, an inaccurateness of the compressibility factor of 0.5% and of the real isentropic exponent of 2% is tolerable according to [28]. However, for the determination of the isentropic compression efficiency or the final temperature, a high accuracy of both properties is needed and the error should be below 0.5% [28].

Comparing the ideal and real gas behavior, as expected, the discrepancies are more pronounced at higher compression ratios. They are especially obvious in the case of the isothermal compression of methane, which is due to the influence of the compressibility factor. Generally, the ideal gas law under predicts the compression work in the case of hydrogen and over predicts it in the case of hydrogen. With respect to the temperature the opposite can be observed in Figure 25.6. Furthermore, it is obvious, that the final temperature of hydrogen is generally higher than the one of methane.

25.4.2
Isenthalpic and Isentropic Expansion

Similar to the compression process, for the expansion process two different thermodynamic borderline cases can be distinguished. The first is the isenthalpic or Joule–Thomson expansion (Section 25.3), which describes the expansion in a perfect throttle valve. The second is the isentropic expansion, which describes the expansion in a perfect gas turbine (100% efficiency) [20,29].

These two different thermodynamic borderline cases also can be calculated by means of a direct and an indirect method. Since the isentropic expansion is the opposite of the isentropic compression (Section 25.4.1), the same relationships used before in Eqs (25.28), (25.32), (25.34) can be used. However, it has to be considered that the final pressure p_2 is smaller than p_1, which then results in a pressure ratio p_2/p_1 below one. For this reason, it is only necessary to specify the equations of the direct and indirect method for an isenthalpic expansion.

The indirect method uses the Joule–Thomson coefficient for the calculation and thus avoids the need to implement the caloric equations of a real gas. Of course, this requires the knowledge of the Joule–Thomson coefficient in relation to the temperature and the pressure. In the natural gas business it is usual to determine the temperature change for an isenthalpic expansion according to Eq. (25.20) [22], which is the isenthalpic integral Joule–Thomson effect. However, in doing so the temperature dependency of the Joule–Thomson coefficient is neglected and a mistake of varying degree is created [30]. Therefore, a step-by-step approximation method is applied in [30], whereat the temperature change is calculated according to Eq. (25.35) over small pressure intervals. Here, Eq. (25.35) is used for the indirect method as well, but for the sake of simplicity and as a contrast to the direct method the step-by-step calculation over small pressure intervals is not applied and instead the initial and final state is inserted directly:

$$T_2 = T_1 + \left[\frac{\mu(T_1,p_1) + \mu(T_2,p_2)}{2}\right](p_2 - p_1) \tag{25.35}$$

The direct method uses the fact that the enthalpy has to stay constant (see Eq. (25.35)). The temperature after isenthalpic expansion can then be calculated from the constant enthalpy and the known final pressure by solving Eq. (25.14) iteratively. The schematic principle of the isenthalpic and isentropic compression in the (h,s)-diagram is illustrated in Figure 25.4b:

$$h(T_1,p_1) = h(T_2,p_2) \rightarrow T_2 = f(h,p_2) \tag{25.36}$$

The different calculation methods specified above were used to compute the temperature change during expansion of hydrogen and methane. As before, the computation was carried out for ideal and real gases. In addition, two different initial states and pressure drops were considered. In the first case the pressure drops from 100 to 50 bar and in the second from 400 to 200 bar. Both expansion processes have a pressure ratio (expansion ration) p_2/p_1 of 0.5 and an initial temperature of 10 °C.

The results in Figure 25.7 clearly illustrate the different behavior of the two gases. As expected from the Joule–Thomson coefficient, the temperature of hydrogen increases, while the temperature of methane decreases during an isenthalpic expansion. In the case of the first expansion process hydrogen heats up only by 1.6 K, while for the second expansion process the temperature difference

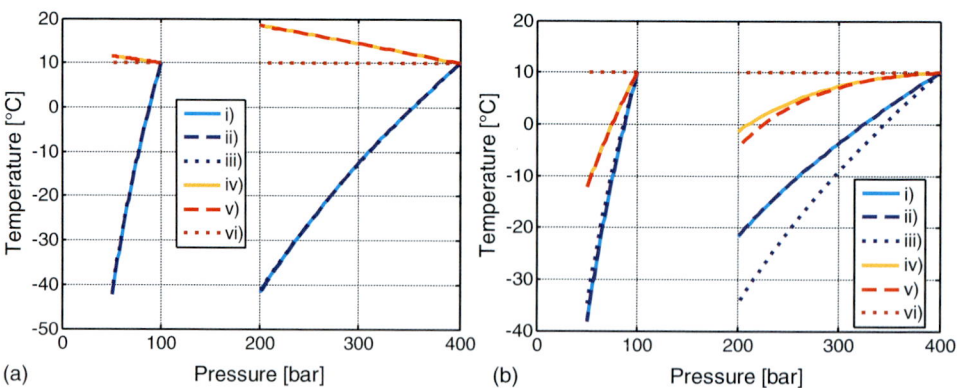

Figure 25.7 Temperature change during expansion of hydrogen (a) and methane (b) as a function of the pressure for different processes and calculation methods: (i) isentropic process, direct method, real gas; (ii) isentropic process, indirect method, real gas; (iii) isentropic process, indirect method, ideal gas; (iv) isenthalpic process, direct method, real gas; (v) isenthalpic process, indirect method, real gas; (vi) isenthalpic process, indirect method, ideal gas.

amounts to 8.5 K. The temperature of methane drops by 22.3 and 11.4 K, respectively. This can be traced back to the fact that for an isenthalpic expansion it is not the pressure ratio but the pressure difference that is crucial. Furthermore, the influence of temperature and pressure on the Joule–Thomson coefficient depends on the condition of state. This becomes apparent when looking at the discrepancies of the direct and indirect method for real gases at the second expansion process of methane. For the rest, the results of the two methods correlate very well, even though the arithmetic mean value over the entire process was used for the indirect method. Of course, the deviations with respect to the ideal gas are great, since its temperature stays constant.

Since the pressure ratio is decisive for the isentropic expansion, both expansion processes result, for an ideal gas, in the same temperature irrespective of the initial state, which is 41.5 °C for hydrogen and −34.71 °C for methane. For a real gas different final states are found. However, for hydrogen the deviation between ideal and real gas behavior is not very pronounced in the range considered. In addition, the pressure ratio is not high. Therefore, the different calculation methods and the two expansion processes show only minor disagreements. For methane the direct and indirect methods for real gas deliver similar results, but with respect to the ideal gas large discrepancies exist. Furthermore, the final states of the two expansion processes clearly differ. This can be traced back to the strong pressure dependency at low temperatures of the compressibility factor and the isentropic exponent.

25.5
Thermodynamic Modeling of the Storage Process

The fundamental model of a storage system is based on mass and energy conservation equations. Depending on the required precision, the thermodynamic properties can be calculated by using equations of state for either ideal or real gas. In addition the storage system can be assumed to be adiabatic or heat transfer equations can be included. For simplicity, for the latter, modeling equations based on a steady state overall heat transfer coefficient are presented here.

25.5.1
Governing Equations

The temporal change of mass inside the storage system can be described by using the well-known law of mass conservation:

$$\frac{dm_{Storage}}{dt} = V_{Storage} \frac{d\rho_{Storage}}{dt} = a \frac{dm_{in}}{dt} - b \frac{dm_{out}}{dt} \qquad (25.37)$$

where t is the time, m_{in}/dt is the mass flow rate entering the system, m_{out}/dt is the mass flow rate leaving the system, $V_{Storage}$ is the geometric storage volume,

$m_{Storage}/dt$ and $\rho_{Storage}/dt$ are the rate of change of mass and density inside the storage, respectively. The parameters a and b are used here to distinguish the different operating modes with $a=1$, $b=0$ for injection and $a=0$, $b=1$ for withdrawal.

The energy balance of the storage system can be determined from the first law of thermodynamics for an open system with no work added. Since the kinetic and potential energy changes are often insignificant, they are neglected here. Considering heat exchange, the general energy balance of the storage system is:

$$\frac{dU_{Storage}}{dt} = a\left(h_{in}\frac{dm_{in}}{dt}\right) - b\left(h_{out}\frac{dm_{out}}{dt}\right) + c\left(\frac{dQ}{dt}\right) \quad (25.38)$$

where $U_{Storage}/dt$ is the rate of change of internal energy inside the storage system, $dQ/dt = \dot{Q}$ is the rate of heat exchanged with the surroundings, and h_{in} and h_{out} are the specific enthalpy entering and leaving the system, respectively. The parameter c is used here to distinguish whether heat exchange is considered or neglected, with $c=1$ for a diathermal process and $c=0$ for an adiabatic process. The rate of change of internal energy can be transformed by using the products rule when differentiating with respect to time:

$$\frac{dU_{Storage}}{dt} = \frac{d(m_{Storage}u_{Storage})}{dt} = m_{Storage}\frac{du_{Storage}}{dt} + u_{Storage}\frac{dm_{Storage}}{dt} \quad (25.39)$$

To solve this system of ordinary differential equations, the time dependent change of one thermodynamic variable has to be known, which then becomes the controlling equation of the system. Principally, two different cases can be distinguished here. In the first the mass or volume flow entering and leaving the storage vessel as a function of time is given and in the second the change of one of the thermodynamic variables inside the storage vessel is known. The latter can be achieved, for example, by controlling the pressure increase in the storage as was done in Reference [31]. However, the first case is more general, thus it is selected here and a constant mass flow rate is assumed in the following. With this assumption Eq. (25.37) can be solved and results in:

$$m_{Storage}(t) = V_{Storage}\rho_{Storage}(t) = m_{Storage,0} + a\dot{m}_{in}t - b\dot{m}_{out}t \quad (25.40)$$

where $m_{storage,0}$ is the mass in the storage at time zero, which means prior to injection or withdrawal, $\dot{m}_{in}$ and $\dot{m}_{out}$ are the constant mass flow entering and leaving the system, respectively. Inserting Eqs. (25.39) and (25.40) into Eq. (25.38) results in Eq. (25.41), which can be numerically solved with the help of an adequate simulation algorithm:

$$\frac{du_{Storage}}{dt} = \frac{[h_{in}(t)a\dot{m}_{in} - h_{out}(t)b\dot{m}_{out}] - u_{Storage}(a\dot{m}_{in} - b\dot{m}_{out}) + c\dot{Q}(t)}{(m_{Storage,0} + a\dot{m}_{in}t - b\dot{m}_{out}t)}$$

$$(25.41)$$

Notably, the specific enthalpy entering and leaving the system and the heat exchange are functions of thermal state variables, which in turn can be time dependent. However, an analytical solution of Eq. (25.41) can be derived for the injection process if the enthalpy entering the system is maintained constant and no heat exchange takes place. Given these assumptions, Eq. (25.41) can be transformed into Eq. (25.42):

$$\frac{du_{Storage}}{dt} = \frac{h_{in}\dot{m}_{in} - u_{Storage}\dot{m}_{in}}{m_{Storage,0} + \dot{m}_{in}t} \tag{25.42}$$

Integrating Eq. (25.42) over time results in:

$$u_{Storage}(t) = \frac{h_{in} \cdot \dot{m}_{in} \cdot t + m_{Storage,0} \cdot u_{Storage,0}}{m_{Storage,0} + \dot{m}_{in} \cdot t} \tag{25.43}$$

where $u_{Storage,0}$ is the initial specific internal energy of hydrogen in the storage system. The storage temperature in turn can be determined iteratively via Eq. (25.15) from the known time dependent density and specific internal energy of the storage system.

25.5.2
Heat Transfer Equations

The heat exchange with the surroundings is considered by using a steady state overall heat transfer coefficient. With this the rate of heat transfer across the walls of the storage system can be conveniently expressed by:

$$\dot{Q} = U_{HT}A_{HT}(T_\infty - T_{Storage}) \tag{25.44}$$

where U_{HT} is the overall heat transfer coefficient, A_{HT} is the heat transfer area of the storage system, and T_∞ and $T_{Storage}$ are the temperature of the surroundings and inside the storage system, respectively. Since the temperature inside the storage system changes with time, the heat flow is also unsteady. Storage systems are often cylindrical. The overall thermal resistance R_{HT} can then be calculated by using:

$$R_{HT} = \frac{1}{U_{HT}A_{HT}} = \frac{1}{\pi L}\left[\frac{1}{\alpha_i d_i} + \frac{\ln\left(\frac{d_o}{d_i}\right)}{2\lambda} + \frac{1}{\alpha_o d_o}\right] \tag{25.45}$$

where L is the cylinder height or length, λ is thermal conductivity, d_i and d_o are the inner and outer diameter, and α_i and α_o are the heat transfer coefficients of the gas inside and outside the storage system, respectively.

25.6
Application Examples

To demonstrate the applicability of the model presented in Section 25.5 two different examples were selected. In the first the refueling of a vehicle storage tank is examined and in the second the pressure and temperature change within a salt cavern is analyzed.

25.6.1
Refueling of a Vehicle Storage Tank

Different approaches for the modeling and analysis of the filling process of a vehicle storage tank with hydrogen can be found in the literature [5,18,31–36]. For compressed natural gas different studies are available as well [10,37,38]. The focus here is to give a basic understanding of the involved thermodynamic processes and their influence. Therefore, a very simple model was chosen, which is shown in Figure 25.8. The vehicle storage tank is refueled completely from the single high pressure reservoir tank at the fueling station. This storage reservoir is assumed to be sufficiently large, so that its temperature and pressure remain constant during the refueling process. For the reservoir temperature an ambient temperature of 20 °C is assumed. The reservoir pressure has to be at least 1.2 times the target pressure of the vehicle storage tank [4], for which a 350 bar Type III cylinder was selected [33]. Its internal volume is 74 L and at 15 °C and 350 bar a total of 1.79 kg of hydrogen can be stored [33]. The initial tank pressure prior to refueling is assumed to be 5 bar, since this is the current minimum delivery pressure for the fuel cell [39]. The initial tank temperature generally is not equal to the ambient temperature, because it can cool down during driving or heat up when parked in the sun [4]. According to Reference [39] it has to lie in the range −30 and 50 °C and therefore an arbitrary initial tank temperature of 15 °C was assumed. The refueling time should not exceed 3 min [4]. On this basis and the values given above a constant standard volume flow of 0.1078 m³$_{(STP)}$ s^{-1} was calculated for the injection. Simplistically, the process can be assumed to be adiabatic due to the rapid refueling. Detailed studies that include thermodynamic modeling of the heat transfer of a vehicle storage tank have been published [31,40–43].

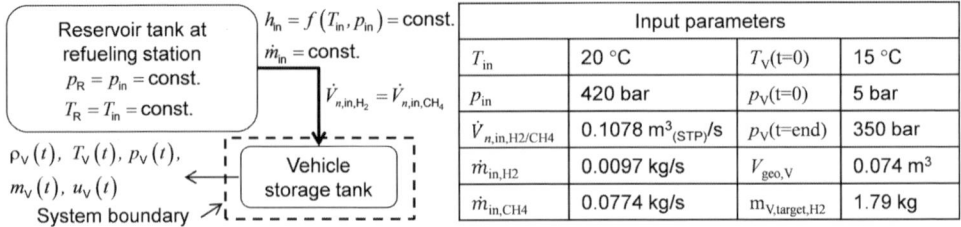

Figure 25.8 Schematic diagram and input parameters of the refueling process example.

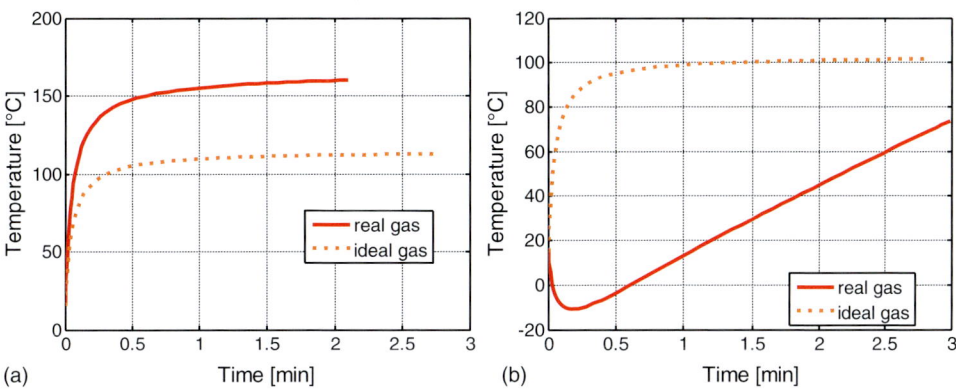

Figure 25.9 Temporal variation of the storage tank temperature of hydrogen (a) and methane (b).

Although the selected input parameters of the model were appropriate for hydrogen, the calculation was performed for methane as well, since it is also interesting to observe the behavior of another gas under the same conditions. However, notably, filling stations for compressed natural gas typically operate at lower pressure ranges [38]. To ensure comparability an equal feed rate on a volume basis of the two gases was assumed.

The computation was carried out for both ideal and real gas to clarify the influence of the Joule–Thomson effect. The program stopped when the target pressure of 350 bar was reached.

Figures 25.9 and 25.10 show the results with respect to the storage tank temperature and pressure, respectively. The deviation between real and ideal gas is clearly visible, which can be traced back mainly to the influence of the Joule–Thomson effect. It is especially strong at the early part of filling when the gas expands into the vehicle storage tank and results in heating for hydrogen and cooling for methane. As the tank pressure increases this effect diminishes and

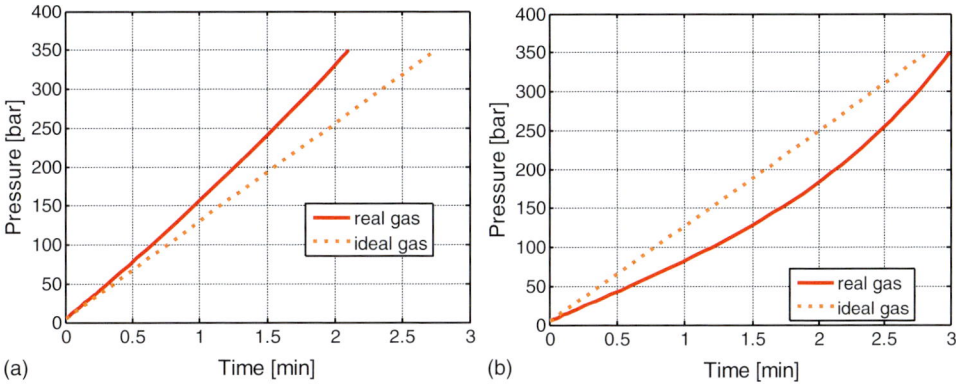

Figure 25.10 Temporal variation of the storage tank pressure of hydrogen (a) and methane (b).

heating occurs due to the compression, which can be clearly seen in the case of methane. The higher temperature of hydrogen also ensures that the target pressure is reached earlier compared to methane. However, this high hydrogen temperature also has an unwelcome side effect regarding the fill level of the vehicle storage tank, which amounts only to 70% referred to target. In addition, the temperature inside the vehicle storage tank should remain within the range −40 to 85 °C [39]. Therefore, depending on the conditions, precooling is advisable [4,31,44].

25.6.2
Salt Cavern

The debate about balancing fluctuations of renewable energies led to increased interest in compressed air energy storage (CAES) and hydrogen storage, with several pilot projects announced [45,46]. They both prefer the use of underground storage in salt rock formations, which are characterized by their tightness and flexibility [47,48]. In addition, in the natural gas sector a further expansion of salt cavern storage is also planned [49]. Accurate predictions of the varying storage pressure and temperature during the cyclical injection and withdrawal are essential to assure operation within the facilities safety limits. This can be achieved with the help of thermodynamic modeling, which has been the subject of different approaches reported in the literature [50–55]. Most of them, however, concentrate on compressed air and natural gas storage.

The intention here is to examine the real gas behavior of hydrogen and methane under the same operating conditions by using the model presented in Section 25.5. The design of the salt cavern is adapted to parameters given in Reference [56]. It has a geometric cavern volume of 500 000 m^3, which corresponds to a diameter of 46 m by assuming a cylindrical shape and a height of 300 m. The last cemented casing string lies at 1000 m, which results in 175 and 58 bar for the maximum and minimum cavern pressure, respectively The maximum permissible change in pressure within the cavern is given as 10 bar day^{-1} [56]. To calculate the undisturbed salt temperature surrounding the cavern an average thermal gradient of 3 K per 100 m starting from a surface temperature of 10 °C [50,57] is assumed and referred to the middle of the cavern height, which then gives a value of 44 °C. In addition, the cavern is assumed to be in perfect thermal equilibrium with the surrounding salt rock prior to injection and the initial pressure was chosen to be 65 bar. Since the maximum pressure gradient should not be exceeded, a constant standard volume flow of 35 m$^3_{(STP)}$ s^{-1} was selected, which is equal for injection and withdrawal of both hydrogen and methane. The injection and withdrawal time was arbitrarily chosen to be 8 days. It is assumed that injection and withdrawal are separated by a standby time, which was selected to be 24 days to ensure that thermal equilibrium with the surroundings is achieved. By using standard equations given in Reference [11], a natural convective film coefficient of 133 and 120 W m^{-2} K^{-1} could be determined for hydrogen and methane, respectively. The thermal

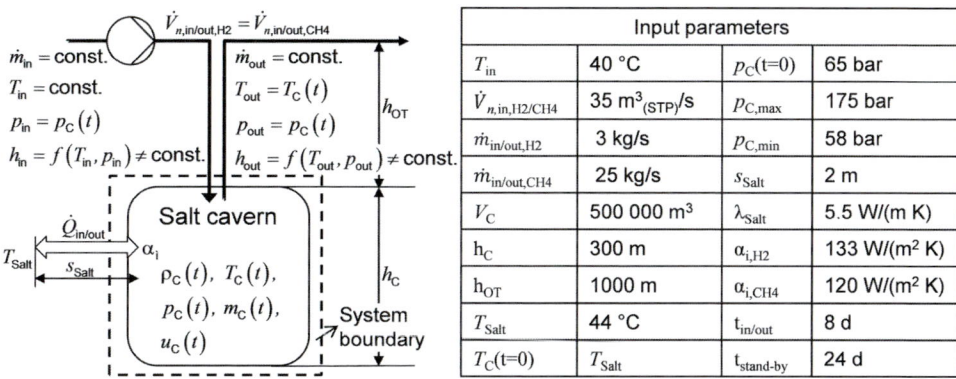

Figure 25.11 Schematic diagram and input parameters of the salt cavern example.

conductivity of salt was assumed to be $5.5\,\text{W}\,\text{m}^{-2}\,\text{K}^{-1}$ on the basis of data reported in Reference [52]. All input parameters and assumptions are summarized in Figure 25.11, together with a schematic diagram. During injection, compression according to counter pressure and subsequent cooling to a constant inlet temperature of 40 °C is assumed. At the system boundary the conditions of the gas stream leaving the cavern are equal to the conditions inside the cavern. Therefore, the enthalpy of the incoming and outgoing gas stream varies with time.

Figures 25.12 and 25.13 present the results of the computation. There are only minor differences in temperature of the two gases, but hydrogen reaches thermal equilibrium sooner. The deviations become more obvious with regard to the pressure, where hydrogen achieves higher values during injection. However, it has to be mentioned that the simplified heat transfer equations used here lack

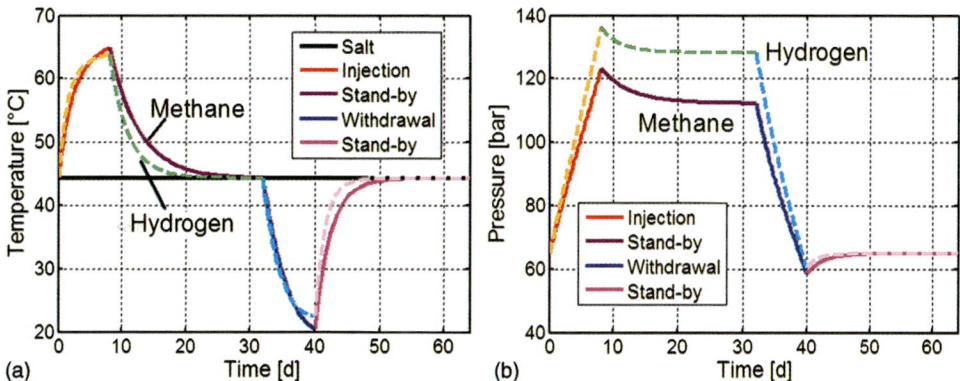

Figure 25.12 Temporal variation of the cavern temperature (a) and pressure (b) of hydrogen and methane.

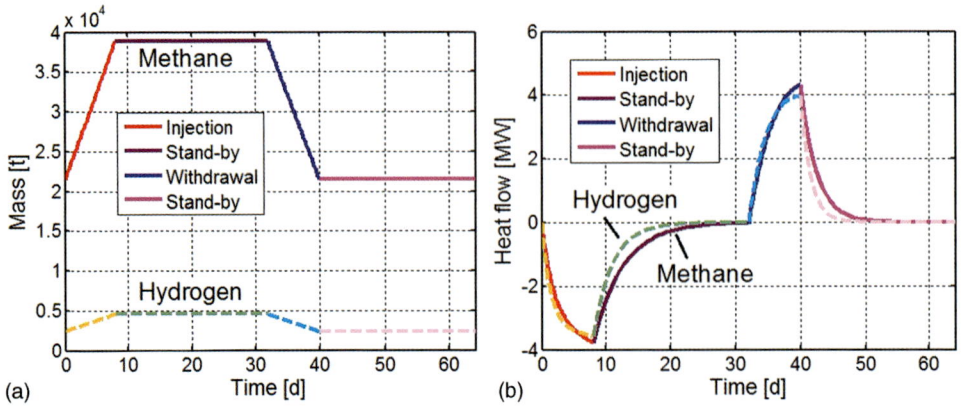

Figure 25.13 Temporal variation of the stored mass (a) and the heat flow (b) of hydrogen and methane.

accuracy since the thermal coefficient is not constant and also the thermal storage capacity of the rock salt has an influence.

25.7
Conclusion

Pressurized gas storage is already a common technology for storing hydrogen and in view of current developments it can be expected that its field of application will be broadened. Thermodynamic simulations are a convenient way towards understanding the processes associated with pressurized gas storage. Different theoretical models together with equations of state to describe the behavior of ideal and real gases have been presented within this chapter. These models have been used to examine the behavior of hydrogen as a real and ideal gas in different compression and expansion processes as well as in selected application examples. Besides hydrogen, methane has also been included in the analysis for reasons of comparison.

Some fundamental differences between the two gases have been made visible in a comparison of the compressibility factor, the Joule–Thomson coefficient, and the real isentropic exponent, which are important thermodynamic properties with respect to pressurized gas storage. Within the pressure range considered and for ambient temperature conditions, the compressibility factor of hydrogen is greater than 1 and that of methane is less than 1. As a consequence, for example, mass stored at a temperature of 20 °C and a pressure of 250 bar is underestimated by 18% in the case of methane and overestimated by 14% in the case of hydrogen when using the ideal gas law. In addition, it could be observed that the Joule–Thomson coefficient of hydrogen is negative while that of methane is positive at conditions relevant for pressurized gas storage applications.

An isenthalpic expansion then results in heating in the case of hydrogen and in cooling in the case of methane. Generally, it became apparent that with decreasing temperatures the rate of change in the thermodynamic properties increases. This is much more pronounced for methane, since its critical point is higher.

With respect to compression, the theoretical work required to compress hydrogen in an isothermal and isentropic process was found to be higher than the corresponding value for methane. The temperature after isentropic compression is also much higher for hydrogen than for methane. Very big differences in gas behavior exist when comparing an isenthalpic and isentropic expansion process. In this context, it should also be remembered that in an isenthalpic expansion the decisive quantity is the pressure difference, while in an isentropic expansion the pressure ratio is crucial. In a sample calculation of an isenthalpic expansion, the different Joule–Thomson effects could be clearly observed with the temperature of hydrogen increasing and that of methane decreasing. Conversely, for the isentropic expansion the temperature of hydrogen was below that of methane.

Finally, two different application examples have been selected. In the first example the refueling of a vehicle storage tank was examined. It was revealed that the Joule–Thomson effect has a detrimental influence and that the ideal gas assumption leads to large errors. In the second example, the pressure and temperature change within a salt cavern for periods of injection, withdrawal, and stand-by have been analyzed by taking the heat exchange with the surroundings into account. Generally, the behavior of both gases was quite similar. Only with respect to the pressure increase during injection did hydrogen achieve noticeably higher values than methane. All this indicates that from the point of view of thermodynamics no obstacles exist to storing hydrogen in salt caverns, as is common today for methane.

References

1 Stolten, D., Grube, T., Mergel, J. (2012) Beitrag elektrochemischer Energietechnik zur Energiewende. *Innovative Fahrzeugantriebe 2012, VDI-Berichte 2183, 8. VDI-Tagung*, 199–215.

2 Satyapal, S., Petrovic, J., Read, C. et al. (2007) The U.S. department of energy's national hydrogen storage project: progress towards meeting hydrogen-powered vehicle requirements. *Catal. Today*, **120** (3–4), 246–256.

3 Jensen, J.O., Vestbø, A.P., Li, Q. et al. (2007) The energy efficiency of onboard hydrogen storage. *J. Alloys Compd.*, **446–447**, 723–728.

4 Maus, S., Hapke, J., Ranong, C.N. et al. (2008) Filling procedure for vehicles with compressed hydrogen tanks. *Int. J. Hydrogen Energy*, **33** (17), 4612–4621.

5 Farzaneh-Gord, M., Deymi-Dashtebayaz, M., Rahbari, H.R. et al. (2012) Effects of storage types and conditions on compressed hydrogen fuelling stations performance. *Int. J. Hydrogen Energy*, **37** (4), 3500–3509.

6 Klell, M. (2010) Storage of hydrogen in the pure form, in *Handbook of Hydrogen Storage: New Materials for Future Energy Storage/* (ed. M. Hirscher), Wiley-VCH Verlag GmbH, Weinheim, 1–38.

7 Eichlseder, H. and Klell, M. (2010) *Wasserstoff in der Fahrzeugtechnik Erzeugung, Speicherung, Anwendung*, Vieweg+Teubner Verlag, Wiesbaden.

8 Klell, M., Kindermann, H., and Jogl, C. (2007) Thermodynamics of gaseous and liquid hydrogen storage. Presented at International Hydrogen Energy Congress and Exhibition IHEC 2007, 13–15 July 2007, Istanbul, Turkey.

9 Krieg, D. (2012) *Konzept und Kosten eines Pipelinesystems zur Versorgung des deutschen Strassenverkehrs mit Wasserstoff*. PhD thesis, Schriften des Forschungszentrums Jülich. Reihe Energie und Umwelt/energy and environment, Band/Volume 144, Jülich, 1–288.

10 Deymi-Dashtebayaz, M., Gord, M.F., and Rahbari, H.R. (2012) Studying transmission of fuel storage bank to NGV cylinder in CNG fast filling station. *J. Brazil. Soc. Mech. Sci. Eng.*, **34**, 429–435.

11 Verein Deutscher Ingenieure VDI-Gesellschaft Verfahrenstechnik und Chemieingenieurwesen (GVC) (eds) (2006) *VDI-Wärmeatlas*, Springer, Berlin and Heidelberg.

12 Leachman, J.W., Jacobsen, R.T., Penoncello, S.G. et al. (2009) Fundamental equations of state for parahydrogen, normal hydrogen, and orthohydrogen. *J. Phys. Chem. Ref. Data*, **38** (3), 721–748.

13 Setzmann, U. and Wagner, W. (1991) A new equation of state and tables of thermodynamic properties for methane covering the range from the melting line to 625 K at pressures up to 100 MPa. *J. Phys. Chem. Ref. Data*, **20** (6), 1061–1155.

14 National Institute of Standards and Technology (2011) NIST Chemistry WebBook. http://webbook.nist.gov/chemistry/fluid/ (accessed 20 December 2012).

15 Lemmon, E.W. and Jacobsen, R.T. (2005) A new functional form and new fitting techniques for equations of state with application to pentafluoroethane (HFC125). *J. Phys. Chem. Ref. Data*, **34** (1), 69–108.

16 Span, R., Lemmon, E.W., Jacobsen, R.T. et al. (2000) A reference equation of state for the thermodynamic properties of nitrogen for temperatures from 63.151 to 1000 K and pressures to 2200 MPa. *J. Phys. Chem. Ref. Data*, **29** (6), 1361–1433.

17 Kunz, O. (2006) A new equation of state for natural gases and other mixtures for the gas and liquid regions and the phase equilibrium. PhD thesis, University of Bochum.

18 Yang, J.C. (2009) A thermodynamic analysis of refueling of a hydrogen tank. *Int. J. Hydrogen Energy*, **34** (16), 6712–6721.

19 VDI (August 1993) Acceptance and performance test on turbo compressors and displacement compressors. Theory and Examples, VDI 2045.

20 Yang, J.C. and Huber, M.L. (2008) Analysis of thermodynamic processes involving hydrogen. *Int. J. Hydrogen Energy*, **33** (16), 4413–4418.

21 Doering, E., Schedwill, H., Dehli, M. (2008) *Grundlagen der Technischen Thermodynamik*. Vieweg + Teubner Verlag, Wiesbaden.

22 Mischner, J., Fasold, H.G., and Kadner, K. (eds) (2011) *Gas2energy.net: Systemplanung in der Gasversorgung; gaswirtschaftliche Grundlagen*, Oldenbourg Industrieverla, Munich.

23 Maytal, B.Z. and Pfotenhauer, J.M. (2012) *Miniature joule–thomson cryocooling: principles and practice*, Springer.

24 Andreas, T. (2011) Thermodynamische Zustandsänderung realer Gase. Cahier Scientifique: Revue Technique Luxembourgeoise 1.

25 Kouremenos, D.A. and Kakatsios, X.K. (1985) The three isentropic exponents of dry steam. *Forschung Ingenieurwesen*, **51** (4), 117–122.

26 Scholz, R. (1974) Verschiedene Definitionen und Verwendungen des Isentropenexponenten bei realen Gasen. *Gas Wärme Int.*, **23** (12), 486–488.

27 Fasold, H.-G. (2011) Ein thermodynamisches Modell zur Berechnung der Verdichtungsarbeit von Turboverdichtern, in gas2energy.net: Systemplanerische Grundsatzfragen der Gasversorgung (eds J. Mischner, H.G. Fasold, and K. Kadner), Oldenbourg Industrieverlag, Munich, 652–668.

28 Fasold, H.G. (2011) Berücksichtigung des Realgasverhaltens bei der Planung von

Erdgasversorgungssystemen, in gas2energy.net *Systemplanung in der Gasversorgung; Gaswirtschaftliche Grundlagen* (eds J. Mischner, H.G. Fasold, and K. Kadner), Oldenbourg Industrieverla, Munich, 159–185.

29 Fasold, H.G. (2011) Gasentspannung in Expansionsmaschinen unter Berücksichtigung des Realgasverhaltens, in gas2energy.net *Systemplanung in der Gasversorgung; Gaswirtschaftliche Grundlagen* (eds J. Mischner, H.G. Fasold, and K. Kadner), Oldenbourg Industrieverla, Munich, 307–326.

30 Wolowski, E. and Hartmann, H. (1964) Über den Joule-Thomson Effekt von Gasen. *Ruhrgas-Forschungsheft*, **14**, 13–22.

31 Rothuizen, E., Mérida, W., Rokni, M. et al. (2013) Optimization of hydrogen vehicle refueling via dynamic simulation. *Int. J. Hydrogen Energy*, **38** (11), 4221–4231.

32 Hosseini, M., Dincer, I., Naterer, G.F. et al. (2012) Thermodynamic analysis of filling compressed gaseous hydrogen storage tanks. *Int. J. Hydrogen Energy*, **37** (6), 5063–5071.

33 Dicken, C.J.B. and Mérida, W. (2007) Measured effects of filling time and initial mass on the temperature distribution within a hydrogen cylinder during refuelling. *J. Power Sources*, **165** (1), 324–336.

34 Galassi, M.C., Baraldi, D., AcostaIborra, B. et al. (2012) CFD analysis of fast filling scenarios for 70 MPa hydrogen type IV tanks. *Int. J. Hydrogen Energy*, **37** (8), 6886–6892.

35 Galassi, M.C., Papanikolaou, E., Heitsch, M., Baraldi, D., Iborra, B.A., and Moretto, P. (2014) Assessment of CFD models for hydrogen fast filling simulations. *Int. J. Hydrogen Energy*, **39** (11), 6252–6260.

36 Heitsch, M., Baraldi, D., and Moretto, P. (2011) Numerical investigations on the fast filling of hydrogen tanks. *Int. J. Hydrogen Energy*, **36** (3), 2606–2612.

37 Farzaneh-Gord, M., Deymi-Dashtebayaz, M., and Rahbari, H.R. (2011) Studying effects of storage types on performance of CNG filling stations. *J. Nat. Gas Sci. Eng.*, **3** (1), 334–340.

38 Farzaneh-Gord, M., Reza Rahbari, H., and Deymi-Dashtebayaz, M. (2014) Effects of natural gas compositions on CNG fast filling process for buffer storage system. *Oil & Gas Sci. Technol.–Rev. d'IFP Energies Nouv.*, **69** (2), 319–330.

39 US Department of Energy Office of Energy Efficiency and Renewable Energy and The FreedomCAR and Fuel Partnership (September 2009) Targets for onboard hydrogen storage systems for light-duty vehicles. Available at http://energy.gov/sites/prod/files/2014/03/f11/targets_onboard_hydro_storage_explanation.pdf (last accessed 3 February 2015).

40 Daney, D.E. (1976) Turbulent natural convection of liquid deuterium, hydrogen and nitrogen within enclosed vessels. *Int. J. Heat Mass Transfer*, **19** (4), 431–441.

41 Monde, M., Mitsutake, Y., Woodfield, P.L. et al. (2007) Characteristics of heat transfer and temperature rise of hydrogen during rapid hydrogen filling at high pressure. *Heat Transfer—Asian Res.*, **36** (1), 13–27.

42 Monde, M., Woodfield, P., Takano, T. et al. (2012) Estimation of temperature change in practical hydrogen pressure tanks being filled at high pressures of 35 and 70 MPa. *Int. J. Hydrogen Energy*, **37** (7), 5723–5734.

43 Woodfield, P., Monde, M., and Mitsutake, Y. (2007) Measurement of averaged heat transfer coefficients in high-pressure vessel during charging with hydrogen, nitrogen or argon gas. *J. Thermal Sci. Technol.*, **2** (2), 180–191.

44 Schneider, J., Klugman, J., Boyd, B., and Ward, J. (2010) The implementation of SAE J2601: hydrogen fuelling protocol guideline for demonstration projects, in *Proceedings 18th World Hydrogen Energy Conference; 2010 –WHEC 2010, 16–21 May 2010, Essen* (ed. D. Stolten and T. Grube), Energy and Environment, vol. **78**, Forschungszentrum Jülich, Book 1, 419–422.

45 Hypos (2013) Hypos-Projekt bietet Ostdeutschland vielfältige Chancen. http://www.hypos-eastgermany.de/hypos-%E2%80%93-die-idee-im-idealen-umfeld (accessed 1 December 2013).

46 IVG Caverns (2012) IVG Caverns plant Pilotanlage zur Speicherung von Windwasserstoff. http://www.kaverneninformationszentrum-etzel.de/detail-aktuelles/ivg-pilotanlage.html (accessed 20 November 2013).

47 Crotogino, F. and Huebner, S. (2008) Energy storage in salt caverns. Developments and concrete projects for adiabatic compressed air and for hydrogen storage. Presented at the SMRI Spring Meeting 2008, 27–30 April 2008, Porto, Portugal.

48 Crotogino, F., Donadei, S., Bünger, U. et al. (2010) Large-scale hydrogen underground storage for securing future energy supplies, in *Proceedings 18th World Hydrogen Energy Conference; 2010 –WHEC 2010, 16–21 May 2010, Essen* (ed. D. Stolten and T. Grube), Energy and Environment, vol. **78**, Forschungszentrum Jülich, Book 4, 37–45.

49 Landesamt für Bergbau, Energie und Geologie (LBEG) (2013) Untertage-Gasspeicherung in Deutschland. *Erdöl Erdgas Kohle*, **129** (11), 378–388.

50 Wirths, A. (2006) Thermische Energiespeicherung in Druckluftspeicherkraftwerken – Theorie der Speicherung im fluiden Einphasengebiet diploma thesis, FH Gießen-Friedberg, Dresden.

51 Tek, M.R. (1996) *Natural gas underground storage: inventory and deliverability*, PennWell Pub, Tulsa, Oklahoma.

52 Kushnir, R., Dayan, A., and Ullmann, A. (2012) Temperature and pressure variations within compressed air energy storage caverns. *Int. J. Heat Mass Trans.*, **55** (21–22), 5616–5630.

53 Raju, M. and Kumar Khaitan, S. (2012) Modeling and simulation of compressed air storage in caverns: A case study of the Huntorf plant. *Appl. Energy*, **89** (1), 474–481.

54 Raju, M. and Khaitan, S.K. (2012) System simulation of compressed hydrogen storage based residential wind hybrid power systems. *J. Power Sources*, **210**, 303–320.

55 Steinberger, A., Civan, F., Hughes, R.G. et al. (2002) Phenomenological inventory analysis of underground gas storage in salt caverns. Presented at the SPE Annual Technical Conference and Exhibition, 29 September to 2 October 2002, San Antonio, Texas.

56 Acht, A. and Donadei, S. (2012) Hydrogen storage in salt caverns. State of the art, new developments and R&D projects. Presented at the SMRI Fall Conference, 30 September to 2 October 2012, Bremen, Germany.

57 Voigt, H.D. (2011) *Lagerstättentechnik: Berechnungsmethoden für das Reservoir Engineering*. Springer, Heidelberg.

26
Geologic Storage of Hydrogen – Fundamentals, Processing, and Projects

Axel Liebscher, Jürgen Wackerl, and Martin Streibel

26.1
Introduction

The use and burning of fossil fuels emit large quantities of the greenhouse gas carbon dioxide (CO_2). These man-made greenhouse gas emissions are the main driving force for the observed global warming [1]. To reduce CO_2 emissions and to mitigate global warming and its negative impacts, the use of fossil fuels needs to be reduced and energy systems have to switch over to renewable, CO_2 free or neutral, energy sources. These may include wind, solar, and hydro power or geothermal energy. However, energy supply by these renewable energy sources fluctuates highly and is only partly predictable and may not meet an energy demand that is rather static and predictable. A way to handle the mismatch between the supply and demand of mainly electrical energy is the intermediate storage of energy surplus. Several options for this are shown in Figure 26.1.

On a short time scale or small energy levels in the range of MWh, batteries, compressed air storages, and pumped storage hydro power stations can be used. On larger time scales or large energy surplus levels in the range of GWh or more, the conversion of electrical energy into a chemical intermediate is needed. This can be achieved using power-to-gas concepts [2] as indicated in Figure 26.2.

The most convenient way is the conversion of the excess electrical energy into hydrogen gas from water using electrolyzers [3]. The produced hydrogen is intermediately stored and later either converted into natural substitute gas or used and burned directly, dependent on the energy demand. Therefore, the storage of hydrogen gas will play a central role in a future, especially purely hydrogen based, energy economy. Large storage capacities for longer-term grid stability and seasonal balancing with required storage capacities above about 5 GWh and discharge times above several days are needed and can only be provided by geological storage options [4].

So far, explicit geological storage of hydrogen is restricted to only a few examples worldwide and present knowledge and understanding of geological storage of hydrogen comes from transfer of the scientific and operational knowledge gained during the geological storage of natural gas and town gas, which has

Hydrogen Science and Engineering: Materials, Processes, Systems and Technology, First Edition.
Edited by Detlef Stolten and Bernd Emonts.
© 2016 Wiley-VCH Verlag GmbH & Co. KGaA. Published 2016 by Wiley-VCH Verlag GmbH & Co. KGaA.

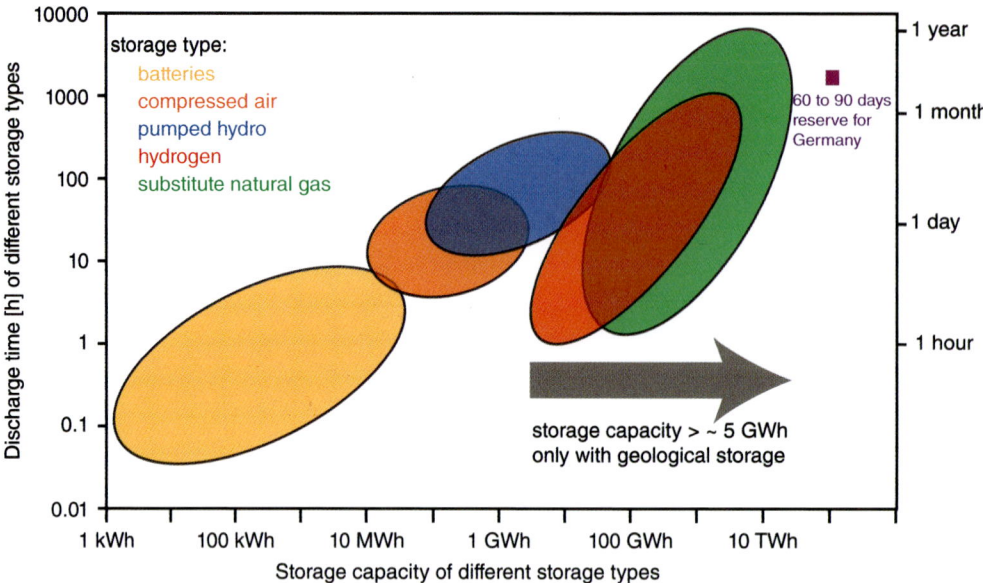

Figure 26.1 Different storage options according to storage capacity and discharge time. Modified after Reference [2].

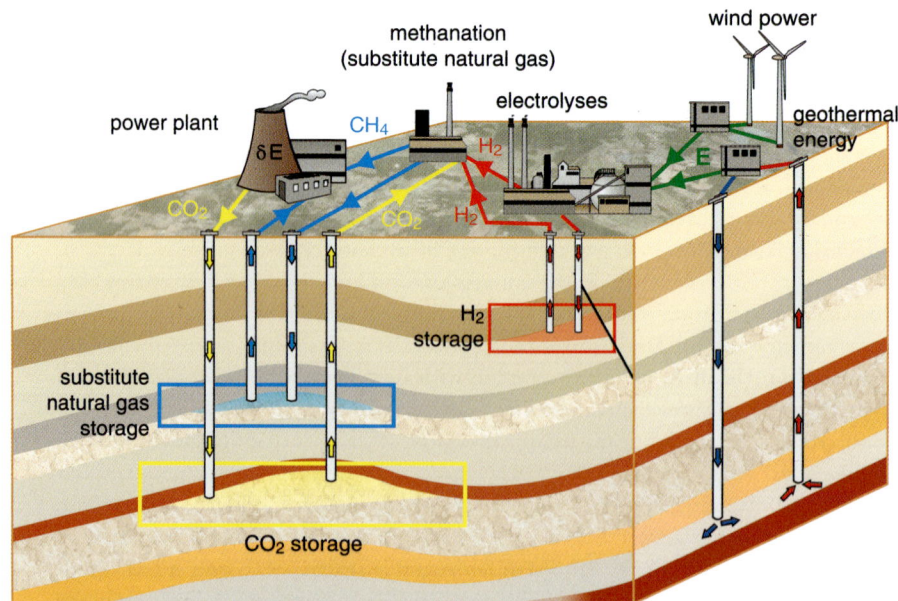

Figure 26.2 Schematic drawing of an integrated power-to-gas concept with the different storage operations needed. Modified after Reference [3].

been a proven technology for several decades. Despite there being few operating geological hydrogen storage operations, the conceptual, scientific, and technological aspects of geological hydrogen storage have been addressed for a few decades. Overviews and seminal key publications on the different aspects of geological storage of hydrogen are given by References [5–11].

This chapter provides (i) an overview of fundamental aspects of geological hydrogen storage, focusing on thermodynamic properties and potential organic and inorganic geochemical interactions between injected hydrogen and the geological reservoir, and the different geological storage options; (ii) an overview of the different process engineering aspects relevant to the geological storage of hydrogen; and (iii) a compilation of the few operating geological hydrogen storage operations as well as the operational experiences from the storage of hydrogen rich town gas. Given the diverse and interdisciplinary aspects relevant for geological storage of hydrogen ranging from geoscientific questions over technological and material questions to operational and engineering issues, such a chapter cannot provide a comprehensive presentation of all the different topics. For further information the reader is referred to the referenced publications.

26.2
Fundamental Aspects of Geological Hydrogen Storage

26.2.1
Physicochemical Properties of Hydrogen and Hydrogen Mixtures

Physicochemical properties like compressibility, pressure–volume–temperature (PVT) data, and phase relations in hydrogen and hydrogen bearing gas mixtures have been long and extensively studied (see, for example, References [12–20]). The solubility of hydrogen in water, which is important for underground storages, has been determined [21–24]. The most accurate equation of state (EOS) for gas/fluid mixtures including hydrogen is the GERG 2004 XT08 EOS [25]. It allows the calculation of thermodynamic properties for gas/fluid mixtures with up to 21 components (H_2, N_2, O_2, He, Ar, H_2O, H_2S, CO_2, CO, CH_4 to $C_{10}H_{22}$). This EOS was developed to meet the needs of the hydrocarbon industry and their typical compositional ranges. Whether the EOS also applies with the requested precision to hydrogen storage with its respective high concentrations of hydrogen is not clear and currently under investigation. A notable shortcoming of the GERG 2004 XT08 EOS is the lack of data for salts or other ionic species. This is especially important for hydrogen storage in porous reservoirs (see below) where highly saline formation fluids may be present and interact with the stored hydrogen. However, the effect of salt on the hydrogen solubility in water has already been studied [23,24].

The mutual solubility of hydrogen and water is rather low (Figure 26.3). For the pressure–temperature range of <30 MPa and <200 °C to be expected for hydrogen storage, the water content in the hydrogen gas will generally be less

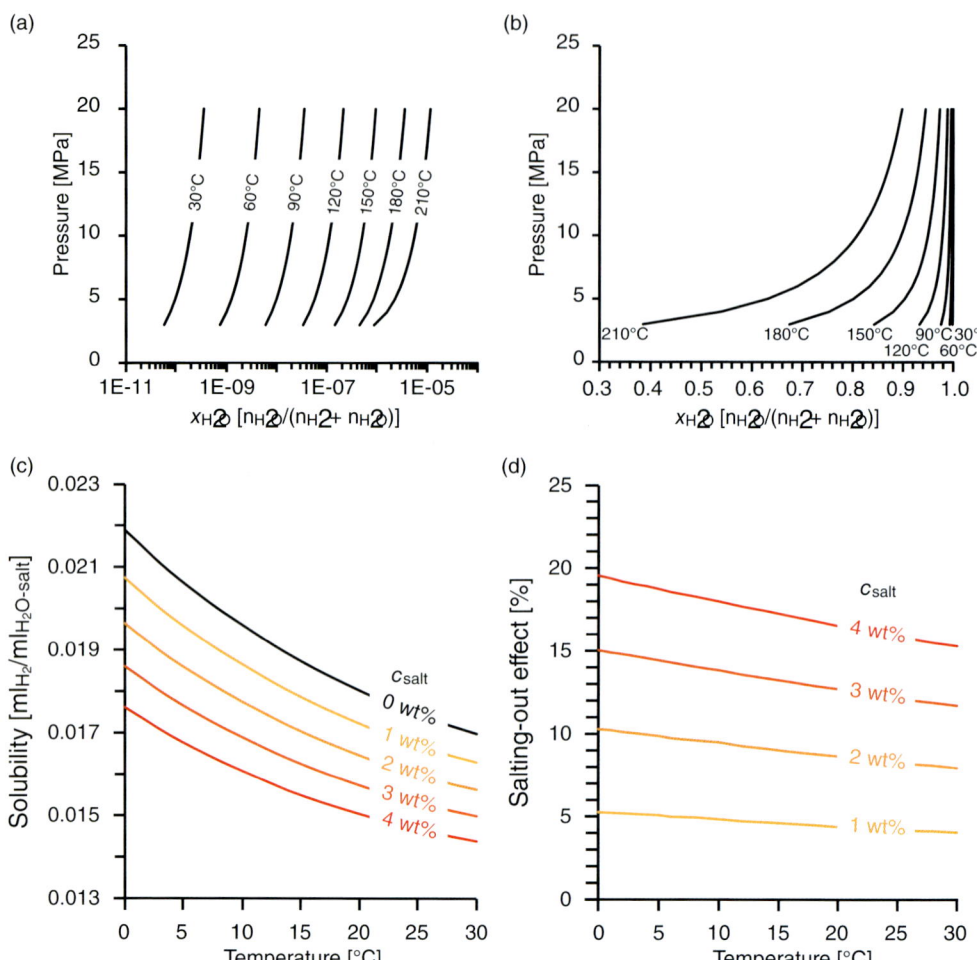

Figure 26.3 Mutual solubility of hydrogen and water (a, b) as calculated with the GERG 2004 XT08 EOS [25] and the effect of salt on the hydrogen solubility in brine (c, d) according to the data by Reference [23].

than $x_{(H2O)} = 10^{-4}$ or below about 1000 ppm (Figure 26.3a). While these water contents are probably negligible from the reservoir and cap rock perspective, they may become relevant for technical aspects of the storage operation like corrosion of technical installations. Hydrogen solubility in coexisting water only becomes notable for temperature above 90 °C and is enhanced by low pressure conditions (Figure 26.3b). For temperatures below 90 °C, the hydrogen concentration in coexisting water is typically $x_{(H2)} < 0.01$ or <1100 ppm. As is the case for other H_2O–nonpolar gas–salt systems like H_2O–methane–NaCl and H_2O–methane–$CaCl_2$ [26], the presence of salt further decreases hydrogen solubility in the brine (Figure 26.3c and d). This salting-out effect may account for up to a

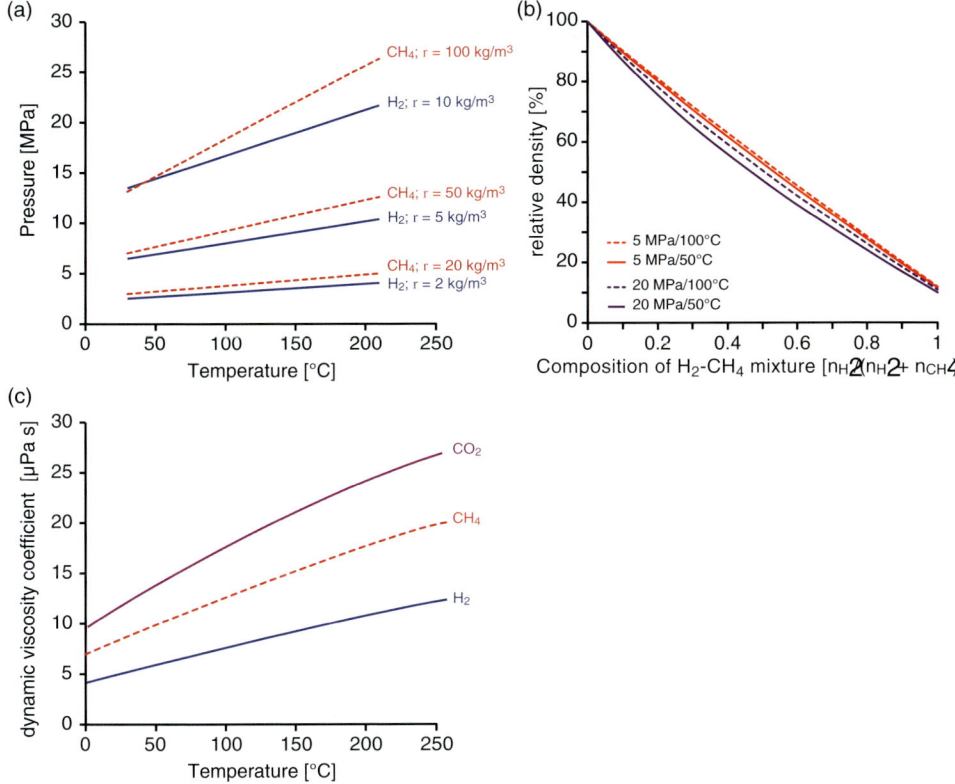

Figure 26.4 Comparison of the density of hydrogen and methane (a) and density of hydrogen–methane gas mixtures (b) as calculated with the GERG 2004 XT08 EOS [25]. (c) Temperature dependency of the coefficient of dynamic viscosity of H_2, CH_2 and CO_2 (for comparison) at atmospheric pressure, modified after Reference [27].

20% reduction in hydrogen solubility for a brine with 4 wt.% NaCl [23]. Hydrogen loss by dissolution into the formation brine in case of saline aquifer storage (see below) therefore will only become relevant under very specific circumstances. Hydrogen losses will be enhanced by low pressure and high temperature conditions, and large contact area between hydrogen and formation brine. Additionally an enhanced brine circulation replacing hydrogen saturated brine at the brine–gas contact by fresh, hydrogen unsaturated brine are reasons for hydrogen losses.

Even though H_2 has more than double the heat value compared to natural gas when related to mass, the volumetric heat value is just about a third of natural gas. This because, at storage conditions, hydrogen has a notably lower density than methane, which is taken as a proxy for natural gas (Figure 26.4). Depending on pressure and temperature, the density of hydrogen is about ten times lower than that of methane at identical conditions. The data also show that, within realistic pressure and temperature conditions, changes in pressure and temperature do not change this density difference significantly (Figure 26.4b). This difference in density means that for a given storage site or pore-volume only about 10% by mass can be stored in hydrogen storage operation when compared to natural gas storage operation. Therefore, the storage sites for hydrogen need significantly bigger dimensions than existing natural gas storage sites for the same energy content. However, newer excavation technologies are available nowadays allowing the construction of larger storage sites. Besides, the hydrogen storages are only one part of a longer energy conversion chain in which immediate usage and the gas to liquids technology reduce the actual amount needed to store as hydrogen gas. As well as density the dynamic viscosity of hydrogen is also lower than that of methane and shows a smaller temperature dependency (Figure 26.4c) [27]. This indeed gives H_2 the advantage of enabling a faster filling or draining of storage sites compared to natural gas since the H_2 gas can permeate faster and also pressure gradients are leveled faster.

26.2.2
Interaction between Hydrogen and Microbial Inventories

Recent studies on recovered material from deep drilling operations indicate an active deep biosphere consisting of different groups of microorganisms even in the absence of light. These microorganisms may play an important role in transformation processes within the Earth crust [28–30] and also within potential porous reservoirs (see below) for hydrogen storage. The influence of high hydrogen concentrations on these microbiological processes as they occur during geological storage of hydrogen are not yet sufficiently studied. While some studies expect a predominantly inert behavior of hydrogen (e.g., Reference [6]) new studies provide evidence for measurable interactions between reservoir rocks, microbial community, and injected hydrogen (e.g., References [31,32]). As an electron donor, hydrogen can serve as energy source for many microorganisms and, by this, hydrogen injection may accelerate the microbial metabolism processes in the reservoir that normally are very slow due to the lack of suitable energy sources. Microorganisms that may use hydrogen as energy source include methanogenic archaea as well as sulfate reducing microorganisms [33]. Typical, simplified reactions for the microbiological triggered transformation of hydrogen in the presence of CO_2 or CO include [31]:

$$2CO_2 + 4H_2 \rightarrow CH_3COOH + 2H_2O$$

$$CO_2 + 4H_2 \rightarrow CH_4 + 2H_2O$$

$$CO + 3H_2 \rightarrow CH_4 + H_2O$$

In the presence of sulfate compounds:

$$5H_2 + 2SO_2^- + 2H^+ \rightarrow 2H_2S + 4H_2O$$

The microbial triggered oxidation of hydrogen with concomitant reduction of sulfate, carbon dioxide, and carbon monoxide produces methane and hydrogen sulfite. In the presence of iron oxides in the reservoir oxidation of hydrogen may also result in formation of reduced iron which can act as catalyst for H_2 splitting. While the generation of methane may only influence the energy balance of hydrogen storage, the formation of hydrogen sulfite may cause corrosion of the technical installations and may also trigger precipitation of sulfites and by this may negatively influence permeability and performance of the storage reservoir. In addition, a noticeable formation of hydrogen sulfite has significant negative impacts on the energy balance of hydrogen storage. It depends on the actual storage type as to which of the above-mentioned processes will become relevant for a particular hydrogen storage site. While reduction of carbon dioxide and carbon monoxide may be most relevant for storage in depleted gas/oil reservoirs or in converted former natural or town gas storages due to the presence of these species in the residual hydrocarbons or gas, the reduction of sulfate is probably more important in hydrogen storage in saline aquifers, where sulfate is available in the formation brine. Sulfate reduction may also be relevant in cases where the contacting rock is anhydrite bearing or even dominant.

In addition to these direct consequences of hydrogen storage on microbial activities, there may be also indirect or secondary effects. Injection of hydrogen will result in changes in concentrations of specific species as well as in temperature and pressure, pH, and redox conditions and will also induce different fluid mobilizations. These effects may result in mobilization of organic and inorganic substances that in addition to the hydrogen itself may serve as new or additional energy source for microorganisms and may affect the microbial metabolism. This metabolism may precipitate inorganic (bio-mineralization) or organic substances that may reduce the reservoir's permeability and performance but may also act as a self-healing process for small-scale fractures and fissures. On the other hand, formation of organic acids decreases the pH and may trigger pH-dependent dissolution of specific minerals and by this enhance the reservoir's porosity and permeability. However, if this dissolution occurs within the cap rock it creates new fluid pathways and negatively affects the cap rock's integrity. Knowledge of the direct and secondary effects of hydrogen injection on the inorganic/organic interactions between injected hydrogen, formation fluids, reservoir and cap rocks, and microbial inventories is therefore essential to address potential consequences of these interactions on the injection operation, reservoir performance, cap rock integrity, and technical installations.

26.2.3
General Types of Geological Storage Option

Future geological storage of hydrogen will not differ in its basic storage options from geological storage of other liquids and gases as it is performed today for storage of natural gas, liquids, and also CO_2. Two principally different geologic storage options can be distinguished: (i) hydrogen storage in the pore space of sedimentary rocks and (ii) hydrogen storage in caverns. Storage options in porous rocks include saline aquifers and depleted gas and oil fields whereas storage options in caverns include artificially constructed salt caverns and mined rock caverns as well as

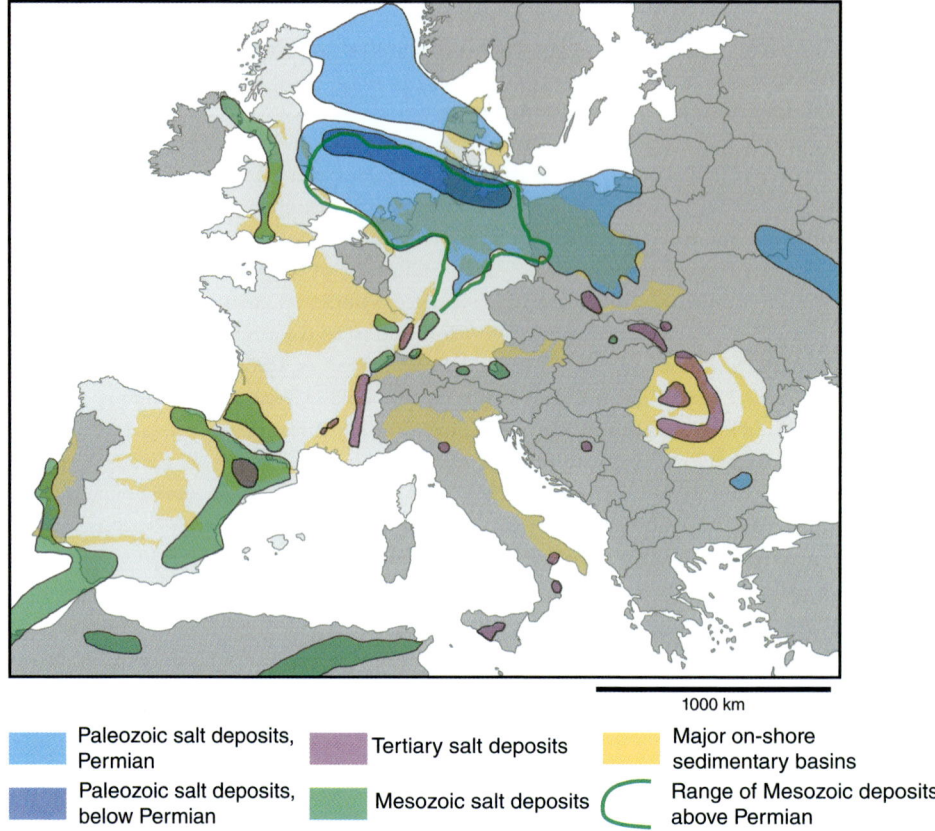

Figure 26.5 Geographic distribution of major on-shore sedimentary basins and salt deposits in Europe, that potentially provide geologic settings for hydrogen storage in saline aquifers, depleted gas/oil reservoirs, and salt caverns. Modified and compiled after Reference [34].

abandoned conventional mines. Based on decade long experiences from the storage of natural and town gas, and due to their favorable intrinsic geologic properties, technological advantages, and widespread occurrence of the required geologic formations (Figure 26.5), hydrogen storage in saline aquifers, depleted gas fields, and artificially constructed salt caverns will be the most important storage options for hydrogen [34]. The other options will only become important for specific niche operations or in areas where the geologic setting does not allow for use of saline aquifers, depleted gas/oil reservoirs, or salt caverns.

26.2.4
Hydrogen Storage in Porous Rocks

Hydrogen storage in porous rocks, either in saline aquifers or in depleted gas and/or oil fields, requires a porous and permeable reservoir rock that provides

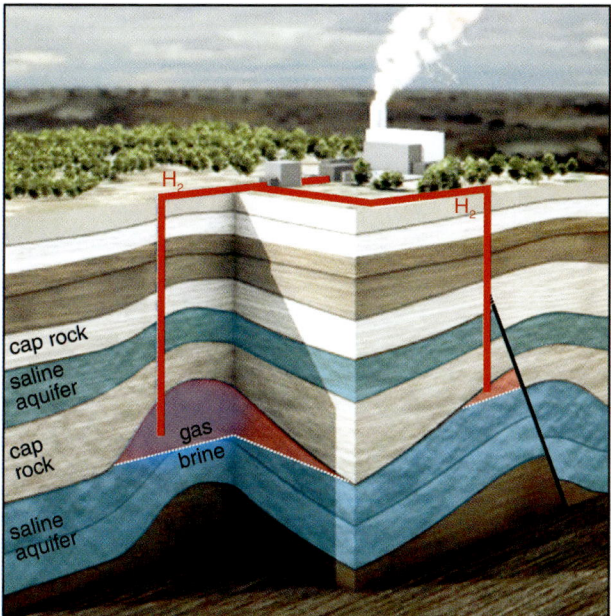

Figure 26.6 Schematic drawing of geological storage options for hydrogen in porous rocks. Principal features apply to storage in saline aquifers (as shown in figure) and depleted gas/oil fields. Storage in porous rocks requires the porous reservoir rock itself (typically sandstone or carbonate), an impermeable cap rock that prevents the hydrogen from upward migration (typically shale, salt, or evaporates), and a geologic trap that defines the lateral extent of the storage complex. These traps may be anticlines (left part in figure), faults (right part in figure) or stratigraphic/lithological traps due to facies changes or sedimentary unconformities (not shown in figure).

the actual storage volume, an impermeable cap rock overlying the reservoir rock that prevents the stored hydrogen from buoyancy driven upward migration out of the storage complex, and finally a three-dimensional structure, the so-called trap, that localizes the stored hydrogen in a predefined area (Figure 26.6). Typical reservoir rocks are sandstones and other clastic sediments or carbonates. These sedimentary rocks typically provide a certain volume of porosity, which can be used for gas storage. The amount of porosity varies between the different sedimentary rock types and also depends on the diagenetic history, for example, its chemical and physical transformations, of the respective rock and may account for more than 25 vol.% of the rock. While sandstones have an initial intergranular porosity that may be diagenetically reduced due to compaction and cementation, carbonates often also show secondary porosity due to dissolution processes during burial diagenesis. In saline aquifers the pore space of the reservoir rocks is filled with fresh water or – more commonly – brines of variable salinity and composition. In contrast, in depleted gas and oil fields the pore space of the reservoir rocks has been initially filled with formation fluid and the extracted hydrocarbons. The stored hydrogen will have a lower density than the surrounding rocks and the formation fluids initially present in the reservoir rocks. It therefore has a strong tendency for buoyancy driven upward migration. To prevent the hydrogen from this upward migration, the reservoir rocks must be overlain by impermeable cap rocks. These cap rocks need to have a very low gas permeability and high capillary entry pressures for hydrogen. Shales, salts, and evaporates like anhydrite are examples of typical suitable cap rocks. Due to its buoyancy, hydrogen not only tends to vertical migration but also to lateral up-dip migration. The presence of a hydrogen tight cap rock is therefore not sufficient to keep the stored hydrogen in place and to prevent it from migration out of the storage complex. Lateral migration of the stored hydrogen must be controlled by a geological trap that defines the lateral extension of the storage complex. Such geological traps may be formed either structurally by anticlines or faults (Figure 26.6) or stratigraphically by, for example, sedimentary unconformities or discordances. In any case, the lateral extension of the storage complex is defined by the trap's spill point, which is the deepest point at which the trap is closed for the stored hydrogen.

While the main geologic requirements and characteristics are the same for porous rock storage in saline aquifers and depleted gas/oil reservoirs, the pressure conditions are notably different. In saline aquifers the pore space is filled with saline formation brine and can reasonably well be assumed to represent more or less hydrostatic pressure conditions. Injection of hydrogen (as injection of any other gas) therefore has to first displace the formation brine to create available pore space. This brine displacement results in overall lateral outward directed flow of brine and an instantaneous increase in reservoir pressure above hydrostatic with onset of injection. In depleted gas/oil reservoirs, the pore space is already partly available due to former extraction of hydrocarbons and the reservoir is typically below-hydrostatic pressure. Due to the below-hydrostatic pressure, overall fluid movement is inward directed. Injection of hydrogen (as injection of any other gas) therefore first has to re-raise the reservoir pressure to a level at which economic injection and withdrawal is possible.

Depending on the type of geologic trap, storage reservoirs may be further classified as open, confined, or semi-confined (Figure 26.7). Open reservoirs are not limited in their lateral extension by any low- to impermeable barriers. Lateral movement of the injected hydrogen or displaced formation fluids is therefore not hindered and the lateral extent of the reservoir is only defined by the spillpoint. In the case of saline aquifers, the pressure build-up in the reservoir due to injection is lower compared to closed or semi-closed reservoirs due to the principal possibility of lateral fluid movement and displacement. In confined reservoirs, geologic features such as faults or lithological unconformities act as lateral boundaries to the reservoir in all directions. A lateral movement of either injected hydrogen or displaced formation brine across these boundaries is not possible. For saline aquifers, confined reservoirs respond with a notably stronger pressure increase due to gas injection than open reservoirs. Due to this stronger pressure increase, the storage capacity in confined reservoirs is generally smaller than in open reservoirs as the pre-defined fracture or threshold pressure limit is

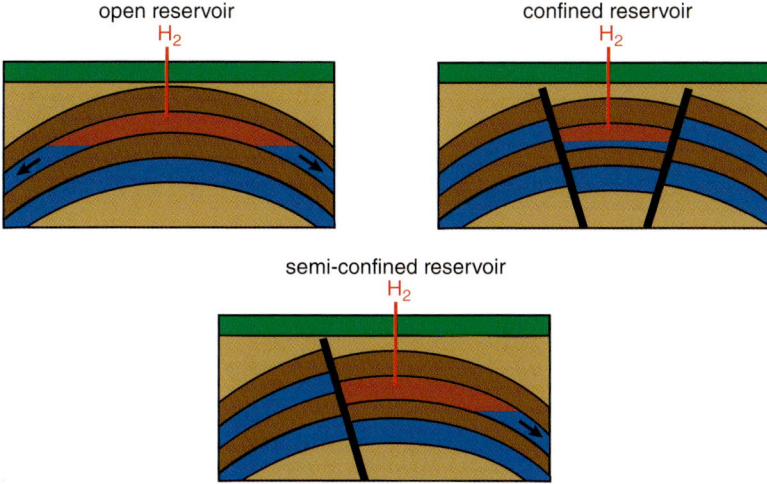

Figure 26.7 Schematic drawings of open, confined, and semi-confined porous reservoirs. Arrows indicate possible movement of injected gas and displacement of formation brine. Geological boundaries may be faults as shown in the figure or lithological boundaries.

reached after smaller amounts of injected hydrogen. For confined depleted gas/oil reservoirs, the lateral boundaries prohibit back-flow of formations brine and pressure increase due to H_2 injection is only moderate. In reality, reservoirs will in most cases neither be perfectly open nor perfectly confined.

In addition to the type of geologic trap the pressure built up in the reservoir also depends on the size of the reservoir itself, the reservoir's permeability for both the displaced formation brine (in the case of saline aquifers) and the injected hydrogen and also on the compressibility of the reservoir's rocks and formation fluids. Larger reservoir size and/or higher permeability generally result in lower pressure built up and larger lateral spread of the injected hydrogen. In general, porous rock reservoirs can store large volumes of gas. However, flow velocities in porous reservoirs are always low when compared to cavern storage. They are therefore rather inflexible with respect to injection and withdrawal rates when compared to hydrogen storage in, for example, salt caverns. Due to this inflexibility and low flow rates at concomitantly high storage capacities, porous reservoirs are typically used for seasonal gas storage within the frame of natural gas storage. It is reasonable to assume that this will also hold for future hydrogen storage in porous rocks. Typical injection and withdrawal rates for natural gas storage in saline aquifers are around 2×10^6 kg per day or 2.5×10^6 m^3(st) per day [35]. Compared to typical working gas capacities of saline aquifer storage of the order of 10^8 m^3 to 10^9 m^3(st) [35], these injection and withdrawal rates are rather small and typically only allow for one gas turn over per year. Due to its lower viscosity, comparable volumetric flow rates may be expected for hydrogen storage in saline aquifers. Given its lower density, volumetric flow rates of 2.5×10^6 m^3(st) per day transform into hydrogen mass flow rates during injection and withdrawal of about 0.24×10^6 kg per day [35]. Experiences from natural gas storage in depleted gas/oil reservoirs indicate injection and withdrawal rates of about 2×106 to 2×107 kg per day or 2.4×106 m3(st) to 2.4×107 m3(st) for natural gas. Comparable volumetric flow rates can be assumed for hydrogen storage in depleted gas/oil reservoirs resulting in mass flow rates during injection and withdrawal of about a tenth of the aquifer ones [35].

26.2.5
Hydrogen Storage in Caverns

Caverns are artificially created cavities in the subsurface. For storage of hydrogen, these cavities may be salt caverns that were built for storage purposes in thick salt deposits, mined rock caverns that were excavated by conventional mining techniques for the purpose of storing hydrogen, or conventional, abandoned mines that were originally constructed for the extraction of natural resources. Salt caverns are by far the most important cavern storage options for natural gas and it is reasonable to assume that this also holds for the future storage of hydrogen [34]. During certain geologic periods, for example, Permian, Triassic, Tertiary in Europe, thick salt deposits accumulated (see Figure 26.5). These may be present as so-called bedded or stratiform salt deposits reflecting the original bedding of the rocks or as salt structures (Figure 26.8). Typical depth ranges for bedded salt deposits in Europe are <2000 m for Permian salt deposits and up to about 1000 m for Triassic and Tertiary salt deposits. At a certain overburden salt behaves plastically due to increased pressure and temperature and starts to flow within the frame of salt tectonics. Due to its lower density compared to the surrounding rocks, the mobilized salt tends to migrate upwards and to accumulate in large salt structures, bending or even fracturing overlying strata during upward migration. Depending on shape, these structures may be salt domes, salt diapirs or salt pillows, which may be some thousand meters thick with up to tens of kilometres lateral extension. Salt deposits are mainly composed of halite (NaCl) but may also contain different types of lower soluble claystones and sulfates (gypsum, anhydrite) as well as highly-soluble salts like sylvine (KCl). In bedded salt deposits these compounds follow a characteristic succession from low-soluble compounds, that is, sulfates, at the bottom through halite, forming the major part of the salt deposit, to the most soluble compounds like sylvine at the top. In salt structures, however, the different compounds may be complexly folded due to the migration and salt tectonic process. While this folding in general has no impact on the overall tightness and integrity of the salt structure it may cause technical problems for cavern excavation.

Due to its specific properties, salt is gas tight and no additional gap rock is needed in the case of salt cavern storage. In addition to its gas tightness, salt is inert to reactions with the injected hydrogen and therefore no degeneration of the injected hydrogen is to be expected. Nevertheless, due to the excavation process a certain amount of brine will always remain at the bottom of the cavern and will increase the moisture of the stored hydrogen and subsequent drying after withdrawal may be necessary depending on the foreseen use of the hydrogen.

Figure 26.8 Schematic drawing of geological storage options for hydrogen in salt caverns. Salt caverns may be constructed in large salt structures like salt diapirs (left part in figure) or within bedded salt (right part in figure). Caverns constructed in salt structures are typically vertical whereas caverns constructed in bedded salt may be horizontal due to limited thickness of bedded salt.

Contrary to salt caverns, the two other potential cavern storage options, namely, mined rock caverns and abandoned mines, will probably only play a subordinate role for hydrogen storage if any. Mined rock caverns have been constructed for the purpose of storing gas or liquids. They require geologic formations that provide the necessary lateral and vertical extension and homogeneity and rocks intrinsic stability and competence that allow for the construction of large, long-term stable caverns. Typical formations include magmatic bodies such as granites but also crystalline, metamorphic rocks. Although these rocks typically have a very low porosity and permeability they may be fractured during construction, providing fracture permeability, and are therefore not intrinsically gas tight. The tightness may be intensified by engineering techniques like water introduction or technical lining installation. A conceptual design for compressed hydrogen storage in mined caverns is given in Lindblom [36]. The storage option using abandoned mine utilizes already existing mines. Because these mines have not been constructed for the purpose of storing gas or liquids but to extract different natural resources, the type, quality, and suitability of this storage option are very heterogeneous. Tightness and integrity of abandoned mines is generally questionable. Tightness may be increased as in mined rock caverns; however, as abandoned mines were not constructed initially for gas storage, their complex shapes and geometries due to their extraction history will in most cases prohibit technical lining installation such as is possible in mined rock caverns.

26.3
Process Engineering

As for all larger technological projects, geological storages cannot be simply constructed and operated out of the box. They are a complex part of an even bigger, complicated conglomerate of different subsystems. Process engineering is used not only to design and operate but also monitor, control, and optimize the storage sites. Obviously, in this context, this includes not only the storage itself but also its peripheral parts. Aspects for hydrogen storage to be considered include the:

- gas connection,
- gas compression/expansion,
- monitoring,
- storage properties.

This list is of course incomplete and certainly misses many additional components. However, due to space constraints it is not possible to cover all aspects in detail in this chapter. Nevertheless, the above given list represents the major components and problems found for one item appears often with the others. Therefore, a carefully chosen material and device mix is needed to operate a storage site in a proper and safe manner. Especially, the use of hydrogen as process gas implies some challenges compared to natural gas.

26.3.1
Gas Transport: Pipelines and Tubing

At first glance, the gas transport in tubes seems to be a trivial point. However, here more is involved than for a normal pipelining as found in homes or laboratories. First of all, the dimensions are different. The volumes to be transported are in the range of several thousands of standard cubic meters (scms) per day. Typical diameters are in the range 22–32 inches (56–81 cm). These diameters are industry standards and are needed to achieve laminar flow in the tubes even at high mass flows. Typical pressure levels for the transmission grid, where the geologic storages are attached to, are in the range 1.5–8.0 MPa [37]. Lower pressure levels would require less material since the tubes are less mechanically stressed. However, higher pressure levels are needed to bridge long distances since a pressure drop appears due to friction of the gas at the tube walls but also as compression waves even for these large diameters. This can be calculated using the Darcy–Weisbach equation. Height differences due to the landscape also have an influence on the local pressure in the tubes. Therefore, the pipelines are routed preferentially at constant height although their length might therefore increase.

A common issue is electrochemical corrosion of the pipelines [38] since the tubing is made of steel for economic reasons. Routing of the pipeline parallel to train tracks or electrical power lines, especially those operated with high currents, is avoided and wherever needed a 90° angle crossing is applied. This reduces the effect of induced eddy currents in the tube mantle due to changing magnetic fields. These eddy currents can locally heat the tubes and also produce a voltage drop along the tube due to the finite electrical resistance of the metal. If the voltage drop gets too high, corrosion occurs due to local element formation with the surrounding soil. In some cases, special polymer coatings are applied to electrically insulate the tube from the surrounding. Since this imposes additional costs this measure is only applied at specific parts of the line. Besides manmade influences natural ones can also produce corrosion. The deep soil itself provides an effective iso-potential reference to the metal of the tube; the surface soil besides acts as a local battery towards the tube. Different pH values and chemical compositions of the soil along the pipeline can produce high currents due to electrical potential differences. Therefore, the pipelines are equipped as standard with a cathodic protection. This is accomplished by electrically connecting magnesium or zinc anodes buried in the soil to the pipeline at critical points. The anodes are used up over time and have therefore to be replaced occasionally. As an alternative, electrical sources are connected to the pipeline to electrochemically "lift" the metal parts of the pipeline to an electrical potential suppressing corrosion.

Hydrogen specific properties, however, limit the material choice for the tubing materials [39–41] compared to natural gas. Since hydrogen is a very small molecule it tends to diffuse in metals more easily and faster than other gas components. When the same steels are used as for the natural gas transport, embrittlement including the formation of fissures is a common reason for failure [42]. This can be practically suppressed by using austenitic steels with 18% chromium and 8% nickel (Duplex stainless steel, 300 series) or chromium-molybdenum blended stainless steels [43]. Pipelines made of this material have already been in operation for several decades now without any significant failure conditions. An example is the Rheine-Ruhr H_2 pipeline in Germany, which was introduced in 1938 to distribute H_2 gas to the chemical industries. Additional polymer coatings further prevent interactions with water acting as catalyst for corrosion. Alternatively the tubing can also be embedded in concrete tubes. This is especially applied where the pipeline crosses swampy ground since it not only blocks water and electrically isolates but is also an effective countermeasure for buoyancy and keeps the line in the ground due to the weight [44].

26.3.2
Compressors

Compressors are an essential part in the gas distribution infrastructure. Albeit building up certain pressure in the line, in the distribution grid they have to provide a high mass flow with a high efficiency. The normal pressure for the grid is

1.5 to 8.0 MPa. The lower the pressure differences to bridge, the more efficient these machines can operate. Depending on the pressure difference, different kinds of compressors are used. However, the most common type is the encapsulated centrifugal compressor. A major common issue in compressor technology operated with hydrogen gas is the bearings. The traditional lubing of bearings involving hydrocarbon products like mineral oils cannot be used for hydrogen gas since these lubes would react with H_2 and in most cases decompose rapidly. For small mass flows where piston type compressors can be employed, ionic liquids as lubricants can be used [45,46]. For application at a high mass flow in the range of 5×10^5 kg per day H_2 where encapsulated centrifugal compressors are normally used [47], currently only a few commercial products exist. Most of the available ones for hydrogen can only bridge small pressure differences [48]. Compressors operating from 1.5 MPa input pressure levels directly to 8.0 MPa output pressure are still under development [49].

26.3.3
Metering

The most common gas meters are based on measuring the real volumetric flux. These well-known meters have a revolving bucket and are used mainly as home counters. However, this concept is limited to small mass flows. Fortunately there are many different metering methods [50,51], but only a few are feasible in providing high accuracy at high mass flows. The most common used especially for natural gas are the turbine based meters measuring the flux also on a volume flux base. Although they involve bearings and rotation parts, they are also used for hydrogen. An alternative design used for hydrogen gas is the orifice meter, which measures the flux by pressure difference. The main advantage of this type is that no movable parts are involved. Another option uses ultrasound. Even though the exact gas composition, temperature, and pressure have to be known, this method allows measurement of the flow from the outside of the pipeline. Therefore it is used not only as inspection tool but also for permanent monitoring of hydrogen gas flow since no direct contact with the gas and no electric parts inside the gas are needed.

26.3.4
Controlling and Safety Components

Besides the above-mentioned components, safety and controlling components are one of the most important parts of a pipeline system. One of the most obvious components involves valves. They are used to control the gas flow either in operation, hence to limit the flux, or as generic or emergency shutdown units. For all cases, as with compressors, there are movable parts involved. The materials used within the valves are again mostly of Duplex stainless steel type, especially of the 300 series, to handle not only the chemical but also mechanical stresses [52] which arise mainly from the gas stream itself. In addition, the design of the valves compared to natural gas is different, since sealing issues have to be addressed. Especially for hydrogen transport, metal-on-metal sealing is used where a soft elastic material like graphite or compounds thereof presses on a stiffer one. This is an established method and ensures a gas tight operation [53]. A special variety of valves are the blocking ones used for emergency shutdowns. Since they are operated mainly only once in a lifetime but need a fast response time and fast switching, they are normally equipped with explosives to insert a blocking bolt in the gas stream. Since this is not feasible for hydrogen gas pipelines, an alternative using hydraulic pistons is applied.

A safety part mostly invisible from the outside is the flame arrestor. It prevents the backdraft of a flame in a tube system in the case of an accident [54]. The flame arrestor itself consists in most cases of fine meshes either of metal or ceramic in the case of hydrogen gas. The mesh prevents the inflowing gas from ignition by physical means. Consequently, in the case of a serious incident, the gas is burned in a controlled manner at the arrestor and propagation of the flame front back to the gas source is prevented. The flame arrestors are located at a defined distance to critical points of failure like valves or maintenance openings.

26.3.5
Geologic Storages: Survey

Except for planned hydrogen storage in depleted gas/oil fields, converted natural gas storages or abandoned mines, for all of which comprehensive geological site information are available, extensive exploration surveys are needed for any possible hydrogen storage site. Several techniques are used in combination to investigate a possible geological storage site. The procedure varies only a little for hydrogen gas storages compared to the established natural gas storage site survey. A first overview can be made using two mainly airborne techniques. The first measures the magnetic field strength and orientation of the earth magnetic field. Geological structures down to several 100 m lateral resolution can be detected. The second measures the gravimetric properties, for example, the gravimetric acceleration. This value indicates the average density of a local point and gives a general overview about the possible nature of the rock formations underground. The smaller the gravimetric acceleration on an on-ground spot, the lower the average mass of the rocks underground and therefore the lower their densities. From such data, conclusions about the rock formations underground are drawn. The two state-of-the-art techniques can be applied also on-ground to enhance the lateral resolution, but with that the in-depth information also decreases.

If a prospective area is found, active 2D and 3D seismic measurements are applied and investigated to unveil the local structure of rock layers. Since natural seismic waves are rare and uncontrolled, controlled detonations or vibrators are therefore used. For the detonations, small bore holes are drilled and filled with small amounts of explosives. The hole is then closed with concrete, and after a short settling time the detonation can be triggered. Acoustic and seismic sensors are located over a wide radius around the detonation spot and the response to the detonation is recorded. Since each rock type has its own specific wave speed, reflections of the seismic waves at the interface between rocks can appear due to a more or less expressed acoustic impedance mismatch. The rock formations can be identified by the time delay and the attenuation of the detonation wave appearing at the sensors. The location is performed by triangulation using several sensors. Where detonations should be avoided, the so-called vibrators are applied. In principle these are huge and heavy plates mounted on heavy trucks which can be lifted and released. The impact of the plate on the ground induces a small seismic wave, which is recorded and investigated. The vibrators have the additional advantage that they can be moved and multiple waves in short intervals can be generated while the detonation is more or less a one-shot measurement. Consequently, the vibrators help to reduce the survey time significantly although the seismic waves created are usually smaller and therefore the depth information is reduced due to the damping of the rocks.

Finally, at selected spots, exploration wells are drilled and core samples are taken to investigate the actual rocks. This gives the most precise information. However, the drilling cores are only the proof of the other measurements since they are very specific to the location taken and represent the overall situation only in rare cases. The number of drills is kept low to minimize the impact on the possible storage site. In best case the exploration drill hole can be further exploited as an operating drill hole later.

26.3.6
Geologic Storages: Construction

If a suitable site for a geological storage is found and all formal prerequisites are met, the construction phase starts with preparation of the ground for drilling and equipment with infrastructure. The feed/drain hole "compound" to the site consists mostly of a 28 inch (approx. 71 cm) diameter conductor pipe to protect the inner tubes from lose rocks and water. Within that an approximate 18.5 inch (about 47 cm) diameter surface casing tube is inserted. It mechanically supports the inner cemented casing with a diameter to about 13.5 inch (about 35 cm) and provides an additional barrier for water from the outside but also catches gas pressure from the inside. In addition, the installations on the bottom and the top of the tube are mounted on that casing. The inner casing provides the gas tightness and acts as support for the final tubing. This part is the only one penetrating through the cap rock and ends several meters below it.

In the case of hydrogen storage in saline aquifers, typically several vertical wells have to be drilled through the cap rock into the reservoir. In specific circumstances, a smaller number of horizontal wells may be drilled. Drilling of several wells that are spread over the storage structure not only allow for more efficient injection and withdrawal but also allow for monitoring of the gas behavior in the reservoir and the movement of the gas–water contact. For hydrogen storage in depleted gas/oil reservoirs, former productions wells and additional surface infrastructure are typically in place and may be used for hydrogen storage. However, the former production wells are not necessarily designed and well-suited for storage operations. Therefore new wells for the storage operation have to be drilled in most cases.

In the case of hydrogen storage in salt caverns, the excavation process by solution mining is applied through wellbore. Water is injected through an outer leaching string and dissolves the salt. This salt bearing brine is recovered through an inner leaching string. A blanket with a lower density than the used water or brine is injected to prevent the solution of salt above the end of the cemented casing. This blanket could be gas (typically N_2) or oil. The dissolution process starts slowly since first the brine has to be concentrated to achieve a reasonably fast dissolution of the salt. This can take up to several months until the nominal dissolution speed is reached. During the excavation, progress is monitored constantly using sonar to control the shape of the later storage site. Once the cavern has reached its planned size the leaching process is stopped. After the end of the leaching process, the cavern is filled with brine from the leaching process. This brine has to be back-produced during the first fill of the cavern with gas. During the whole leaching process and the first fill, the back produced brine has to be disposed of. This may be achieved by selling the brine to a chemical engineering process, by re-injection into the subsurface, or by discharging it to the sea in cases where cavern construction is carried out in a near-shore environment. Overall, the construction of a salt cavern can take up to five years.

26.3.7
Geologic Storages: Initial Testing

Once the construction phase is over, the final testing of the storage site is carried out. The drilled wells will be pressure tested to ensure proper well completion and cementation. In the case of porous rocks storage, these tests are typically done with fluids as pressure media, so the tests provide only information about the well tightness for fluids. In addition, pressure test between wells at different depths in the reservoir or within different compartments of the reservoir and also between wells in the reservoir and above the cap rock are performed to test for hydraulic communication between different horizons within the reservoir and also across the cap rock or faults. For hydrogen storage in salt caverns, the site is normally filled using an inert gas like nitrogen. However, not the complete site is filled with gas, but only the top with the inner casing tube since this is the most crucial part for leakage problems – the rest is filled with the liquid used for excavation. Both maximum pressure and potential leakage are tested. This can already be investigated upon filling, since the amount of gas inserted is monitored and the approximate volume of the top volume is already known from the construction phase. If significantly more gas is needed to induce a certain pressure change, a severe leakage of the storage is likely. In addition, the maximum pressure level has to be tested to ensure low leakage even at this extreme operating condition. This normally takes several weeks since the storage site has first to be filled with its nominal pressure and the site itself needs to be settled and thermally equilibrated after excavation. Shape changes are normal during that testing period and should not affect the safety.

26.3.8
Geologic Storages: Operation

After successful testing and approval of the gas tightness of the site and acquiring the legal operating permission, the storage site is filled with the operating gas. Injection and withdrawal rates in saline aquifer storage are controlled by the reservoir's permeability and size, the type of trap, and the maximum allowable reservoir pressure, which is defined to ensure that any damaging or fracturing of the reservoir and its cap rock is avoided. Because pressure is the most critical parameter, monitoring thereof is essential for gas storage operations. For saline aquifer storage, the driving force for withdrawal is the pressure difference between the reservoir and the production wells. This pressure difference forces the gas to flow back to the wells with concomitant back-flow of the displaced formation fluid. However, a certain amount of the injected gas is trapped in the pore space by capillary forces (residual gas saturation) and cannot be back-produced. This is in contrast to depleted gas/oil reservoir storage where this pore space is already filled by the residual gas. Flow of the stored gas through the pore space of the reservoir depends on the reservoir's permeability, the viscosity of the stored gas at reservoir pressure and temperature and also on the relative permeability of gas and brine. In general, flow velocities and therefore injection and withdrawal rates are much lower for porous rock storage than for gas storage in caverns. This is the main reason why porous reservoirs are typically used for seasonal gas storage.

In the case of a dry salt cavern storage site, first the brine has to be removed. However, up to 30% of the total storage volume may be occupied permanently by the brine since it may contain insoluble solids too big to be removed during the excavation. Therefore, the stored hydrogen gas is always humidified. This leads to a cooling effect upon filling the storage with dry gas, but also causes problems on draining, especially when fed in the pipeline system. Corrosion of the steels is enhanced in the vicinity of water vapor; therefore, dehumidifying of the drained hydrogen gas is mandatory. Salt cavern operation is typically carried out by compression and decompression and therefore variable pressure regimes between a pre-defined maximum and minimum pressure. The initially injected gas volume therefore splits into the working gas, which can be injected and withdrawn without falling below the pre-defined minimum pressure threshold, and the cushion gas, which must remain in the cavern to ensure minimum pressure conditions and mechanical stability of the cavern. Calculations and experiences from gas storage activities indicate that only about two-thirds of the total gas volume can be used as working gas. Besides these pre-defined minimum and maximum pressure threshold values, operation of salt caverns is also governed and restricted by the maximum flow velocities inside the injection/withdrawal wells and the maximum possible injection and withdrawal rates that still guarantee mechanical integrity and stability of the cavern. However, in specific circumstances the cavern can also be operated at constant pressure conditions. In these cases, the gas is replaced during withdrawal by co-injected brine from a surface brine pond, keeping the cavern pressure constant. During gas injection, the brine is again displaced and stored in the surface brine pond. This technique allows reduction of the required cushion gas to almost zero. However, it requires large surface installations for the brine ponds.

If the hydrogen storage is of the type "depleted gas/oil reservoir" even a filtering to remove hydrocarbons is needed to first maintain the original hydrogen gas purity level but second to avoid coking due to cracking of the hydrocarbons.

Another thermal effect arises from the rather low Joule–Thomson inversion temperature of hydrogen. This means that at typical operating conditions hydrogen heats up upon expansion, which is in contrast to other gases. Upon fast filling, care has to be taken to reduce the heating effect since a thermal equilibration may take over up to several weeks and heated gas reduces the effective (working) gas capacity [55]. This can be overcome by using special designed in- and outlet tubes inside the storage to reduce the effect.

In addition, the structural integrity of the geological storage has to be monitored at least from time to time. This is because each filling cycle, for example, pressure release and build-up or exchange of hydrogen gas with brine, can induce fissures and break-offs at the wall of the storage, for example, the shape of the storage is not fixed and therefore changes with time. This can be monitored using sonar and radar techniques from inside the storage, for example, the drain/feed lines are used to "inject" the probes. Another issue to address while operating a geological storage is the convergence, for example, the shrinking of the total open space volume of the storage due to surrounding rock pressure. This not only reduces the capacity, but also imposes depressions on the ground above. Down leveling of the ground above and in a certain range around the storage site are known and can cause severe damage to buildings and infrastructure in this area. However, this effect can be addressed and controlled by adjusting the operation conditions like maximum drain rate or average pressure level. Shape changes in rare cases also open leak paths. This is monitored by estimating of the leakage rate from the actual operation conditions, for example, feed and drain volume, pressure, and temperature.

26.4
Experiences from Storage Projects

Geological hydrogen storage as an option for large-scale energy storage has not yet been applied and implemented on an industrial scale. Therefore, only limited experiences from operational hydrogen storage projects exist. So far, hydrogen storage is almost exclusively related to the chemical industry, which stores hydrogen to meet its operational hydrogen demand, and storage is exclusively done in salt caverns. Compared to future underground hydrogen storage in the frame of energy storage, these projects are comparatively small. Projects include the Clemens Dome and Moss Bluff Projects in Texas, United States, and the Teesside Project in Great Britain, and also hydrogen storage operations in Russia [27]. Additional information on the subsurface behavior of stored hydrogen comes from storage of so-called town gas, which is produced by gasification of coal and contains variable amounts of hydrogen, carbon monoxide, methane, and other gases such as carbon dioxide and nitrogen. Storage and use of town-gas was mostly applied in former Eastern Europe to guarantee a safe and secure energy supply but was also applied, for example, in former West Germany and France. General information and experiences on geological storage of gas come from the storage of natural gas, which has been performed successfully for several decades.

26.4.1
Hydrogen Storage Projects

Worldwide, few storage sites for hydrogen gas exist. Furthermore, these sites can be grouped in two fractions: the pure hydrogen sites, which are more interesting for future use, and the mixed hydrogen–natural gas (town gas) ones, which were of interest in the past.

26.4.2
Pure Hydrogen Storage Projects

26.4.2.1 Clemens Dome, Texas, USA
At Clemens Dome hydrogen is stored in a salt cavern at depths from 850 to 1150 m with a volume of about 580 000 m^3. The operation pressure ranges from 7.0 to 13.5 MPa. The cushion gas amounts to about 2.98×10^6 kg H_2 or 34.9×10^6 m^3(st.) H_2 and the working gas to about 2.56×10^6 kg H_2 or $30.1 \cdot 10^6$ m^3(st.) H_2. The energy storage capacity is about 100 GWh for the working gas compared to about 117 GWh for the cushion gas. The underground storage facility is operated by ConocoPhillips and has been in operation (commissioned) since 1983 [35].

26.4.2.2 Moss Bluff, Texas, USA
Storage at Moss Bluff is done in a salt cavern that is 580 m high and about 60 m in diameter, located at a depth of from 820 to 1400 m. The operation pressure ranges from 5.5 to 15.2 MPa. The cushion gas amounts to about 2.30×10^6 kg H_2 or 27.0×10^6 m^3(st.) H_2 and the working gas to about 3.72×10^6 kg H_2 or 43.7×10^6 m^3(st.) H_2. The energy storage capacity is about 147 GWh for the working gas compared to about 91 GWh for the cushion gas. The underground storage facility is operated by Praxair and has been in operation (commissioned) since 2007.

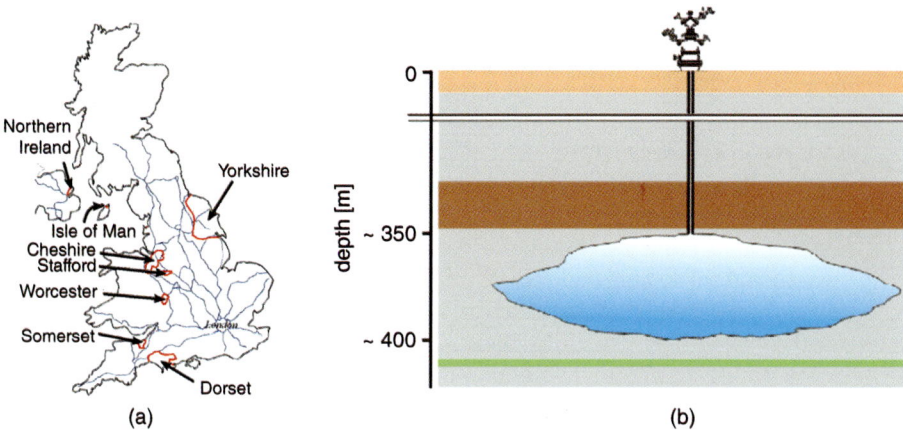

Figure 26.9 Location of potential salt cavern storage locations in UK (a) and schematic drawing of the elliptic salt caverns at Teesside, constructed in bedded Permian salt deposits (b). Modified and compiled after Reference [58].

26.4.2.3 Teesside Project, Yorkshire, UK

Hydrogen storage at Teesside consists of three quite shallow, elliptic salt caverns, located at depths of between 350 and 400 m in Upper Permian bedded salt deposits (Figure 26.9). The stored gas consists of about 95% H_2 and 3–4% CO_2. Each cavern has a volume of about 70 000 m^2 resulting in an overall storage volume of about 210 000 m^3. The caverns at Teesside are operated at a constant pressure of 4.6 MPa by alternating injection and withdrawal of hydrogen and brine. Due to the operation at constant pressure no cushion gas is needed and the working gas equals the total hydrogen gas volume and amounts to about 0.76×10^6 kg H_2 or 8.9×10^6 m^3(st.) H_2. The energy storage capacity is about 30 GWh for the working gas. The underground storage facility is operated by Sabic Petro and is has been in operation (commissioned) since about 1972.

26.4.3
Town Gas Storages

Due to the only small number of pure hydrogen storage facilities and projects, important information on storage and subsurface behavior of hydrogen comes from the underground storage of so-called town gas. Town gas is produced by gasification of coal and contains notable amounts of hydrogen and other components. Compositional ranges are 10–33% CH_4, 25–60% H_2, 12–20% $CO_2 + CO$, up to 30% N_2, and minor amounts of hydrocarbons and O_2. Experiences include town gas storage operations in Germany (Kiel, Bad Lauchstädt, Kirchheilingen, Ketzin, Burggraf-Bernsdorf), Czech Republic (Lobodice), and France (Beynes, Ile de France).

At Kiel, Germany, town gas with up to 60% H_2 has been stored in a salt cavern with a volume of 32 000 m^3. Storage pressure ranged between 8.0 and 10.0 MPa.

Extensive town gas storage has been performed by the VNG in the former GDR [31]. Storage operations included storage in salt caverns at Bad Lauchstädt, in a depleted reservoir at Kirchheilingen, in a saline aquifer at Ketzin, and in an abandoned salt mine at Burggraf-Bernsdorf.

26.4.3.1 Town Gas Storage at Ketzin, Germany

At Ketzin, gas losses of the order of $2 \times 10^8 \, m^3$ were detected between 1964 and 1985 at a working gas volume of about $1.30 \times 10^8 \, m^3$. Detailed investigations proved tightness and integrity of cap rock and wells. Chemical and microbiological processes in the reservoir have been determined as the main reason for the observed gas losses. By adjusting the injection and withdrawal regime gas losses could be reduced. In addition to the observed gas losses, corrosion of the technical underground installations, changes in reservoir permeability, and changes in gas composition between injection and withdrawal periods have been observed (Figure 26.10). The available data indicate a general loss of CO and gain of CO_2 between injection and withdrawal periods. The overall picture for H_2 and CH_4 is not that clear but suggests an overall gain of H_2 and CH_4. The observed changes in composition differ from those observed at Lobodice, Czech Republic (see below) and cannot be explained by simple microbial degradation of the stored hydrogen with concomitant reduction of carbon dioxide and carbon monoxide and formation of methane. The exact reasons for the observed changes in gas composition at Ketzin are not fully resolved.

26.4.3.2 Town Gas Storage at Lobodice, Czech Republic

At Lobodice, Czech Republic, town gas with about 54% H_2 has been stored in a saline aquifer at a depth of about 400–500 m. Reservoir rocks were heterogeneous Miocene sands, gravels, and sandstones. Between injection and withdrawal after about seven months of storage the composition of the town gas showed notable losses of H_2 (from 54% to 37%) and CO (from 9.0% to 3.3%) and gains of CH_4 (from 22% to 40%) and N_2 (from 2.5% to 8.6%) [32]. This change in chemical composition was accompanied by a slight pressure decrease in the reservoir that was below the value calculated and predicted by material balance. Formation fluid samples from the reservoir were extracted and experimentally treated and proved the presence of methanogenic microorganisms in the reservoir, suggesting that microbial activities are responsible for the observed compositional changes. Microbial origin of the additional methane is also supported by $\delta^{13}C$ carbon isotope analyses [32] that showed a notable shift from $\delta^{13}C \sim -34.5‰$ for the injected town gas to $\delta^{13}C \sim -80‰$, which is typical for biogenic methane, for the withdrawn town gas. Detailed carbon isotope studies also allowed distinction between microbial processes and hydrogen losses by migration, a distinction hardly possibly by purely chemical analyses [56]. The applicability of carbon isotopes to distinguish between injected gas and residual formation fluids and to unravel reservoir processes like mixing or migration has also been shown in Reference [57] for three natural gas storages in depleted gas reservoirs in the Czech Republic.

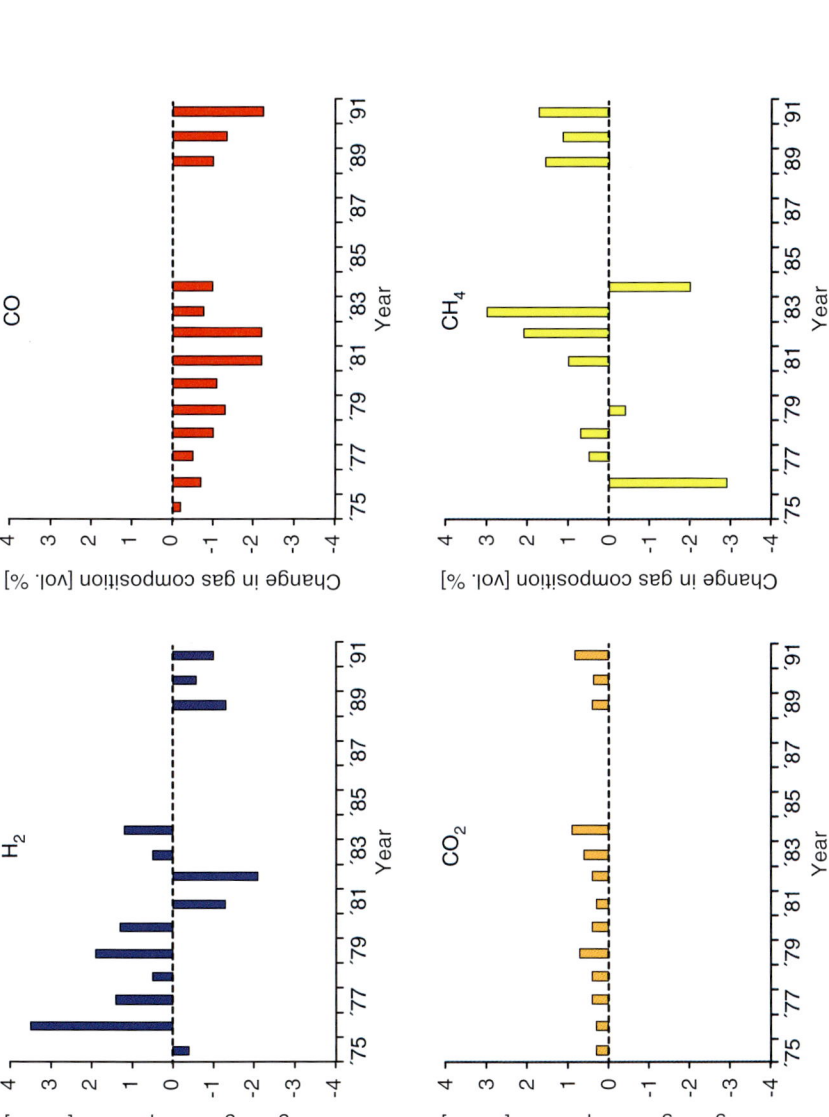

Figure 26.10 Changes in composition of stored town gas between injection and withdrawal periods at the town gas storage site Ketzin for the period 1975–1991. Drawn and modified after Reference [31].

26.4.3.3 Town Gas Storage at Beynes, Ile de France, France

The town gas storage at Beynes was operated by Gas de France from 1956 to 1974, when the storage facility was converted into natural gas storage. The stored town gas contained 50% to 16% H_2. Storage was done in a saline aquifer with a storage volume of $3.3 \times 10^8 \, m^3(st)$. No gas losses or changes in gas composition or any safety issues are reported.

26.5
Concluding Remarks

Geological storage of gas is a proven and mature technology going back several decades. Most geological and technological aspects relevant to the geological storage of natural gas and town gas will also apply to geological storage of hydrogen. Implementation of geological storage of hydrogen can therefore built on long-lasting technological and scientific experience. The most important differences between storage of natural and town gas and storage of hydrogen relate to the stored medium itself. Hydrogen has a lower density, lower viscosity, and lower volumetric heating value than natural gas. While the latter only affects the energy storage capacity, the first two also influence the subsurface behavior of hydrogen and by this the performance of the reservoir. This has consequences for storage capacity, injection and withdrawal rates, and also within reservoir processes like fingering. Besides this, flow properties like relative permeability and capillary entry pressure for hydrogen in the reservoir–cap rock system are largely unknown. In addition, the organic and inorganic chemical processes during hydrogen storage will differ from those occurring during natural gas storage operations.

Process engineering of a geological storage involves much more complexity and infrastructure compared to a normal on-ground facility. It is not limited to the storage itself, but also to the connected pipeline system. Especially the safety measures involve more components. Hydrogen as storage gas imposes additional challenges since embrittlement and fissures can lead to fatal incidents more easily compared to hydrocarbons. However, there are already existing components from the hydrogen industry that can be and are already used. Only the market for compressors is small, especially for the high volume flux applications needed for the huge geological storages.

The experiences and operational data gained from geological storage of town gas provide a very promising knowledge base for addressing and studying the various aspects of hydrogen storage. For hydrogen storage in saline aquifers and depleted gas/oil fields, experiences from town gas storages can provide "real-life" insights into organic and inorganic hydrogen–reservoir fluid(s)–rock interactions. A thorough and careful re-inspection of the available data from the different town gas storage operations is an important step towards understanding geological hydrogen storage and will provide a very fruitful field of future research.

References

1. IPCC (2005) *IPCC Special Report on Carbon Dioxide Capture and Storage*. Prepared by Working Group III of the Intergovernmental Panel on Climate Change (eds Metz, Davidson, de Coninck, Loos, and Meyer), Cambridge University Press, Cambridge.
2. DVGW (2012) *Mit Gas-Innovationen in die Zukunft*, 2nd edn, DVGW Deutscher Verein des Gas- und Wasserfaches e.V., Bonn, 36 pp.
3. Kühn, M., Nakaten, N., Streibel, M., and Kempka, T. (2013) Klimaneutrale flexibilisierung regenerativer überschussenergie mit untergrundspeichern. *Erdöl Erdgas Kohle*, **129**, 348–352.
4. VDE (2009) *Energiespeicher in Stromversorgungssystemen mit hohem Anteil erneuerbarer Energieträger – Bedeutung, Stand der Technik, Handlungsbedarf*, Energietechnische Gesellschaft im VDE, Frankfurt,.
5. Panfilov, M. (2010) Underground storage of hydrogen: in situ self-organisation and methane generation. *Transport Porous Med.*, **85**, 841–865.
6. Walters, A.B. (1976) Technical and environmental aspects of underground hydrogen storage, in *Proceedings 1st World Hydrogen Energy Conference Miami Beach, 1–3 March 1976* (ed. T.N. Veziroglu) vol. **2**, Pergamon Press, New York, pp. 65–79.
7. Carden, P.O. and Paterson, L. (1979) Physical, chemical and energy aspects of underground hydrogen storage. *Int. J. Hydrogen Energy*, **4**, 559–569.
8. Foh, S., Novil, M., Rockar, E., and Randolph, P. (1979) Underground Hydrogen Storage. Final Report. Brookhaven National Laboratory, Upton, New York. 145 pp.
9. Taylor, J.B., Alderson, J.E.A., Kalyanam, K.M., Lyle, A.B., and Phillips, L.A. (1986) Technical and economic assessment of methods for the storage of large quantities of hydrogen. *Int. J. Hydrogen Energy*, **11**, 5–22.
10. Lord, A.S. (2008) Overview of Geological Storage of Natural Gas with Emphasis on Assessing the Feasibility of Storing Hydrogen. Sandia Report, 28 pp.
11. Crotogino, F., Donadei, S., Bünger, U., and Landinger, H. (2012) Large-scale hydrogen underground storage for securing future energy supplies, in *18th World Hydrogen Energy Conference 2010, Essen Germany, 16–21 April 2010 – WHEC 2010 Proceedings: Speeches and Plenary Talks* (eds T. Grube, and D. Stolten) Schriften des Forschungszentrums Jülich/Energy & Environment, No. 78, Forschungszentrum, Zentralbibliothek Jülich, pp. 37–45.
12. Bartlett, E.P., Cupples, H.L., and Tremearne, T.H. (1928) The compressibility isotherms of hydrogen, nitrogen and a 3:1 mixture of these gases at temperatures between 0 and 400° and at pressures to 1000 atmospheres. *J. Am. Chem. Soc.*, **50**, 1275–1288.
13. Kritschewsky, I.R. and Markov, V.P. (1940) The compressibility of gas mixtures. I. The P-V-T data for binary and ternary mixtures of hydrogen, nitrogen and carbon dioxide. *Acta Physicochim. URSS*, **12**, 59–66.
14. Kritschewsky, I.R. and Levchenko, G.T. (1941) The compressibility of gas mixtures. II. The P-V-T data for binary and ternary mixtures of methane, nitrogen and hydrogen. *Acta Physicochim. URSS*, **14**, 271–278.
15. Michels, A. and Goudeket, M. (1941) Compressibilities of hydrogen between 0 °C and 150 °C up to 3000 atmospheres. *Physica*, **8**, 347–352.
16. Klink, A.E., Chen, H.Y., and Amick, E.H. (Jr.) (1975) The vapor-liquid equilibrium of the hydrogen–n-butane system at elevated pressures. *AIChE J.*, **21**, 1142–1148.
17. Mihara, S., Sagara, H., Arai, Y., and Saito, S. (1977) The compressibility factors of hydrogen–methane, hydrogen–ethane and hydrogen–propane gaseous mixtures. *J. Chem. Eng. Jpn.*, **10**, 395–399.
18. Seward, T.M. and Franck, E.U. (1981) The system hydrogen-water up to 440 °C and 2500 bar pressure. *Ber. Bunsen- Ges.*, **85**, 2–7.
19. Magee, J.W., Pollin, A.G., Martin, R.J., and Kobayashi, R. (1985) Burnett-isochoric P-V-T measurements of a nominal 20 mol%

hydrogen–80 mol% methane mixture at elevated temperatures and pressures. *Fluid Phase Equilib.*, **22**, 155–173.

20 Seward, T.M., Suleimenov, O.M., and Franck, E.U. (2000) pVT data for binary H2-H2O mixtures in the homogeneous region up to 450 °C and 2500 bar, in *Steam, Water, and Hydrothermal Systems: Physics and Chemistry Meeting the Needs of Industry, Proceedings of the 13th International Conference on the Properties of Water and Steam, Toronto, ON, Canada, 12–16 September 1999* (ed. P. Tremaine, P. Hill, D.E. Irish, and P.V. Balakrishnan), NRC Research Press, pp. 104–109.

21 Wiebe, R. and Caddy, V.L. (1934) The solubility of hydrogen in water at 0, 50, 75 and 100 °C from 25 to 1000 atmospheres. *J. Am. Chem. Soc.*, **56**, 76–79.

22 Pray, H.A., Schweickert, C.E., and Minnich, B.H. (1952) Solubility of hydrogen, oxygen, nitrogen, and helium in water. *Ind. Eng. Chem.*, **44**, 1146–1151.

23 Crozier, T.E. and Yamamoto, S. (1974) Solubility of hydrogen in water, seawater, and NaCl solutions. *J. Chem. Eng. Data*, **19**, 242–244.

24 Lassin, A., Dymitrowska, M., and Azaroual, M. (2011) Hydrogen solubility in pore water of partially saturated argillites: application to Callovo-Oxfordian clayrock in the context of a nuclear waste geological disposal. *Phys. Chem. Earth*, **36**, 1721–1728.

25 Kunz, O., Klimeck, R., Wagner, W., and Jaeschke, M. (2007) The GERG-2004 Wide-Range Equation of State for Natural Gases and Other Mixtures. GERG Technical Monograph 15. Fortschritts-Berichte VDI, Reihe 6, Nr. 557, VDI Verlag, Düsseldorf, 2007. Available at http://www.gerg.eu/public/uploads/files/publications/technical_monographs/tm15_04.pdf.

26 Krader, T. and Franck, E.U. (1987) The ternary systems water-methane-sodium chloride and water-methane-calcium chloride to 800 K and 250 MPa. *Ber. Bunsen- Ges.*, **91**, 627–634.

27 Basniev, K.S., Omelchenko, R.J., and Adzynova, F.A. (2010) Underground hydrogen storage problems in Russia, in *18th World Hydrogen Energy Conference 2010, Essen Germany, 16–21 April 2010 – WHEC 2010 Proceedings: Speeches and Plenary Talks* (eds T. Grube, and D. Stolten) Schriften des Forschungszentrums Jülich/Energy & Environment, No. 78, Forschungszentrum, Zentralbibliothek Jülich, pp. 47–53.

28 Parkes, R.J. and Wellsbury, P. (2004) Deep biospheres, in *Microbial Diversity and Bioprospecting* (ed. A.T. Bull), ASM Press, pp. 120–129.

29 D'Hondt, S., Jorgensen, B.B., Miller, D.J. et al. (2004) Distributions of microbial activities in deep subseafloor sediments. *Science*, **306**, 2216–2221.

30 Sahl, J.W., Schmidt, R., Swanner, E.D., Mandernack, K.W., Templeton, A.S., Kieft, T.L., Smith, R.L., Sanford, W.E., Callaghan, R.L., Mitton, J.B., and Spear, J.R. (2008) Subsurface microbial diversity in deep-granitic-fracture water in Colorado. *Appl. Environ. Microb.*, **74**, 143–152.

31 Höcher, T. (2013) Erfahrungen mit wasserstoff in stadtgas-untergrundspeichern. Presentation at the Innovation Forum Power to Gas to Power, Leipzig, 25 April 2013.

32 Smigan, P., Greksak, M., Kozankova, J., Buzek, F., Onderka, V., and Wolf, I. (1990) Methanogenic bacteria as a key factor involved in changes of town gas stored in an underground reservoir. *FEMS Microbiol. Ecol.*, **73**, 221–224.

33 Parkes, R.J., Linnane, C.D., Webster, G., Sass, H., Weightman, A.J., Hornibrook, E.R.C., and Horsfield, B. (2011) Prokaryotes stimulate mineral H2 formation for the deep biosphere and subsequent thermogenic activity. *Geology*, **39**, 219–222.

34 Kruck, O. and Crotogino, F. (2013) Benchmarking of Selected Storage Options. Public Deliverable No. 3.3 of Project HyUnder "Assessment of the potential, the actors and relevant business cases for large scale and seasonal storage of renewable electricity by hydrogen underground storage in Europe", 32 pp.

35 Kruck, O., Crotogino, F., Prelicz, R., and Rudolph, T. (2013) Overview on all Known Underground Storage Technologies for Hydrogen. Public

Deliverable No. 3.1 of Project HyUnder "Assessment of the potential, the actors and relevant business cases for large scale and seasonal storage of renewable electricity by hydrogen underground storage in Europe", 94 pp.

36 Lindblom, U.E. (1985) A conceptual design for compressed hydrogen storage in mined caverns. *Int. J. Hydrogen Energy*, **10**, 667–675.

37 Schmura, E. and Klingenberg, M., (2005) Existing Natural Gas Pipeline Materials and Associated Operational Characteristics, DOE Hydrogen Program FY 2005 Progress Report, 8 pp. http://www.hydrogen.energy.gov/pdfs/progress05/v_g_1_schmura.pdf (accessed 18 July 2013).

38 U.S. Department of Transportation - Pipeline and Hazardous Materials Safety Administration (PHMSA) (2000) Operator Qualification Training – Chapter VIII: Corrosion Control, 27 pp. http://www.phmsa.dot.gov/staticfiles/PHMSA/DownloadableFiles/smalllpgas-chapt8.pdf (accessed 17 July 2013).

39 Dodds, P.E. and Demoullin, S. (2013) Conversion of the UK gas system to transport hydrogen. *Int. J. Hydrogen Energy*, **38**, 7189–7200.

40 Briottet, L., Moro, I., and Lemoine, P. (2012) Quantifying the hydrogen embrittlement of pipeline steels for safety considerations. *Int. J. Hydrogen Energy*, **37**, 17616–17623.

41 Nanninga, N.E., Levy, Y.S., Drexler, E.S., Condon, R.T., Stevenson, A.E., and Slifka, A.J. (2012) Comparison of hydrogen embrittlement in three pipeline steels in high pressure gaseous hydrogen environments. *Corros. Sci.*, **59**, 1–9.

42 Fassina, P., Brunella, M.F., Lazzari, L., Re, G., Vergani, L., and Sciuccati, A. (2013) Effect of hydrogen and low temperature on fatigue crack growth of pipeline steels. *Eng. Fract. Mech.*, **103**, 10–25.

43 European Industrial Gases Association (EIGA) (2004) Hydrogen Transportation Pipelines, 83 pp. http://h2bestpractices.org/docs/Doc121_04%20H2TransportationPipelines.pdf (accessed 17 July 2013).

44 Gilette, J.L. and Kolpa, R.L. (2007) Overview of Interstate Hydrogen Pipeline Systems. (Argonne National Laboratory), 52 pp. http://corridoreis.anl.gov/documents/docs/technical/APT_61012_EVS_TM_08_2.pdf (accessed 17 July 2013).

45 Linde Industrial Gases (2013) Linde Ionic Compressors. http://www.linde-gas.com/en/innovations/hydrogen_energy/fuelling_technologies/ionic_compressor.html (accessed 17 July 2013).

46 Stubinitzky, A. (2012) paper presented at the World Hydrogen Energy Conference 2012 (WHEC2012), Toronto Centre Sheraton, Toronto, ON, Canada, 3–7 June 2012.

47 Di Bella, F.A. (2012) Development of a Centrifugal Hydrogen Pipeline Gas Compressor, DOE Hydrogen and Fuel Cells Program FY 2012 Annual Progress Report. (Department of Energy), 5 pp. http://www.hydrogen.energy.gov/pdfs/progress12/iii_7_dibella_2012.pdf (accessed 17 July 2013).

48 Atlas Copco Group (2011) Air & Industrial Gas Compressors, 40 pp. http://www.atlascopco.com.au/Images/Air%2026%20industrial%20gas%20compressors_2935%200598%2011_tcm795-2928482.pdf (accessed 18 August 2013).

49 Heshmat, H. (2012) Oil-Free Centrifugal Hydrogen Compression Technology Demonstration, 28 pp. http://www.hydrogen.energy.gov/pdfs/review12/pd016_heshmat_2012_o.pdf (accessed 18 August 2013).

50 ABB (2011) Industrial flow measurement - Basics and practice, 290 pp. http://search.abb.com/library/Download.aspx?DocumentID=D184B075U02&LanguageCode=en&DocumentPartId=&Action=Launch (accessed 18 August 2013).

51 Hoffer Flow Controls (2013) Premier Gas Series Turbine Flowmeters. http://www.hofferflow.com/datasheets/premgas.pdf (accessed 17 July 2013).

52 Yu, W., Dianbo, X., Jianmei, F., and Xueyuan, P. (2010) Research on sealing performance and self-acting valve reliability in high-pressure oil-free hydrogen compressors for hydrogen

refueling stations. *Int. J. Hydrogen Energy*, **35**, 8063–8070.

53 Sequeira, T. (2012) Update on valves used in hydrogen service. *Hydrocarb. Process.*, **91** (6), V83, V85–V88.

54 Elmac Technologies Limited (2009) Why use a flame arrester? A concise guide, 12 pp. http://www.elmactechnologies.com/datacentre-pdf/ETL_flame_arrester_guide.pdf (accessed 17 July 2013).

55 Zheng, J., Liu, X., Xu, P., Liu, P., Zhao, Y., and Yang, J. (2012) Development of high pressure gaseous hydrogen storage technologies. *Int. J. Hydrogen Energy*, **37**, 1048–1057.

56 Buzek, F., Onderka, V., Vancura, P., and Wolf, I. (1994) Carbon isotope study of methane production in a town gas storage reservoir. *Fuel*, **73**, 747–752.

57 Buzek, F. (1992) Carbon isotope study of gas migration in underground gas storage reservoirs, Czechoslovakia. *Appl. Geochem.*, **7**, 471–480.

58 Schmitz, S. (2013) Innovationsforum PGP – überblick zu wasserstoff-untergrundgasspeichern – H2-UGS. Presentation at the Innovations Forum Power to Gas to Power, Leipzig, 25 April 2013.

27
Bulk Storage Vessels for Compressed and Liquid Hydrogen

Vanessa Tietze, Sebastian Luhr, and Detlef Stolten

27.1
Introduction

[1][1)] In future sustainable energy systems hydrogen could be employed in addition to its current chemical usage as a versatile energy carrier. This requires the buildup of a hydrogen infrastructure with facilities for stationary large- to small-scale storage. One option for large-scale storage is to use underground storage. Compressed hydrogen storage in salt caverns is especially suitable for very large quantities [2,3], but good geological conditions are limited to a few sites. For example, in Germany they appear mainly in northern areas. For this reason, alternatives for the storage of large quantities of hydrogen could be of high importance in future.

The most common methods for stationary hydrogen storage are as compressed gaseous hydrogen, cryogenic liquid hydrogen, and in metal hydrides [4]. However, the latter has only recently become commercially available and has been put to practical use only to a limited extent [5]. Furthermore, they are usually applied for the storage of small amounts of hydrogen [4]. In addition, employing metal hydrides for stationary storage in the range of tons would be too expensive compared to other options (Section 27.6).

In addition to the common technologies, there are a few novel approaches for stationary hydrogen storage currently under investigation. These approaches include representatives of the reversible solid-state hydrogen storage group such as clathrate hydrates [6–8] and cryoadsorbing storage systems [3] as well as some irreversible chemical hydrogen storage concepts such as the methyl–cyclohexane–toluene–hydrogen cycle [9,10], benzene–water systems [11], and aqueous hydrogen peroxide [12]. This shows that very different storage methods are conceivable for the purpose of bulk storage. However, these methods still require further improvement before they can be put into practice [5,6,9].

---

1) Parts of this chapter are identical to Reference [1].

Hydrogen Science and Engineering: Materials, Processes, Systems and Technology, First Edition.
Edited by Detlef Stolten and Bernd Emonts.
© 2016 Wiley-VCH Verlag GmbH & Co. KGaA. Published 2016 by Wiley-VCH Verlag GmbH & Co. KGaA.

The suitability of a storage technology is directly linked to the requirements and boundary conditions of the intended application purpose. Therefore, first a general overview of the field of application for stationary hydrogen storage is given, which is then limited to hydrogen use in road transport for further consideration. For the sake of clarity, some important storage parameters are explained and defined in the next section. The main part of this chapter concentrates on the description of different storage vessel types that can be located above ground or buried a few meters below the earth's surface. As well as technical data some cost estimates and economic targets are also presented. In addition, finally, a technical assessment of the storage efficiency and capacity of selected storage facilities with respect to the daily hydrogen demand of a fueling station is given.

27.2
Stationary Application Areas and Requirements

Nearly all hydrogen produced is currently used in the industrial sector. It is common practice for major hydrogen consumers to possess their own captive hydrogen production facilities. Up to 2008 this captive hydrogen accounted for 95% of the global production with the remainder constituting so-called merchant or traded hydrogen, which is for sale or use by an entity other than the producer at the location [13]. Consequently, storage options at industrial sites are only required where onsite production and immediate consumption are not possible or if it is important to prevent the interruption of operation in the event of production system failure. Therefore, presently only a few companies possess large-scale hydrogen storage facilities, which have mainly been realized in the form of underground storage facilities such as those of Sabic Petrochemicals in Teesside, UK, and the ones of Praxair and ConocoPhillips in Texas, USA [14,15]. Other larger storage facilities currently realized belong to special niche applications such as the space flight sector. This shows that, to date, very few stationary large-scale hydrogen storage facilities exist, which are realized either as gaseous underground or liquid storage [16].

In future, however, it is conceivable that hydrogen will be employed, in addition to its current chemical usage, as a versatile energy carrier in the transport and/or the energy utility sector including power generation and heating. Both hydrogen usage as transportation fuel and reconversion into electricity require the buildup of an adequate delivery infrastructure, although to different extents. Delivery describes the process of transporting and distributing hydrogen from the production site to the final utilization application [17]. Stationary storage facilities are a fundamental part of every hydrogen delivery system. Therefore, it is probable that the demand for stationary bulk hydrogen storage facilities will rise in future. Hydrogen use as a transportation fuel is in economic and environmental terms with respect to CO_2 reduction the more efficient option compared to alternative uses [18]. Therefore, this is chosen as the main final utilization application within this chapter.

Widespread introduction of hydrogen as a fuel for the transport sector requires, besides the question of the best automotive on-board storage solution, the question of a feasible and cost-effective way for its delivery to be addressed. Possible options are gaseous, liquid, and carrier-based delivery pathways, with the latter storing hydrogen in any chemical state other than free molecules [19]. Presently, the state-of-the-art for onboard hydrogen storage are compressed gaseous tanks with pressures of 350 and 700 bar and cryogenic liquid tanks [20] with high-pressure storage being the technology most applied in vehicles [21]. An analysis in Reference [22] concludes that using hydrogen carriers in a pathway that discharges hydrogen at the station and supplies compressed hydrogen to vehicles will hardly, or not at all, reduce the costs for fueling stations because there is still a need for compressed hydrogen equipment at the fueling station. Therefore, this chapter is limited to consideration of a delivery infrastructure designed for the transport of hydrogen in its molecular form.

For nationwide application of compressed gaseous hydrogen as a road transport fuel it is most suitable to build up a delivery infrastructure incorporating pipelines for transportation and distribution as well as storage facilities to compensate fluctuations [18,23]. This future hydrogen supply system has structural analogies and technical similarities to the existing one for natural gas, which has comparable storage requirements. In this context, a minimum and maximum demand with respect to the storage period can be defined: a short-term capacity to compensate the hourly fluctuations in refueling station demand and a long-term capacity to compensate the seasonal fluctuations in refueling station demand and production plant outages [24]. In the latter, reserves to overcome intermittencies in renewable energy production also need to be included. Consequently, storage facilities of different sizes are required in strategically favorable locations. Small-scale storage facilities are typically utilized at the distribution or final-user level, while large-scale storage facilities are typically applied at the production or transport level including the production sites, the terminals of pipelines, and other transportation paths [3,25,26]. Terms such as small-scale and large-scale are often used in the literature without any specific magnitude of size. For clarity, here, the following definitions are introduced: small-scale bulk storage is storage below 5 tons of hydrogen, medium-scale bulk storage reaches up to 500 tons of hydrogen, and large-scale bulk storage covers everything above 500 tons.

27.3
Storage Parameters

The suitability of different storage methods for a certain application field is best assessed by using appropriate storage parameters. It is common usage to compare storage technologies on the basis of their gravimetric and volumetric capacity. Gravimetric and volumetric capacity are defined as the amount of hydrogen stored per unit weight and volume, respectively [27–29]. However, these terms can be misleading since the amount of hydrogen stored can be given in terms of thermal energy, mass, or volume. To avoid ambiguity the terms gravimetric and volumetric capacity are applied here only with respect to the mass of hydrogen stored per unit weight or volume. The terms specific energy and energy density are used here to define the amount of thermal energy per unit weight or volume, respectively. To express the amount of hydrogen stored in terms of thermal energy, mass, or volume the terms energy, mass, or volume capacity are chosen. Thereby, a difference is made between the total and usable amount of hydrogen stored, which are defined here as gross and net storage capacity, respectively.

Furthermore, care has to be given to which basis the specific energy, energy density, gravimetric, and volumetric capacity are referred to. This can be the storage material solely or can include the vessel as well as the total storage system with all the necessary components. The material-only value is a measure of the absolute capacity while the total-system based values represent an upper limit to the amount of useable hydrogen [27]. In the following the term physical is used when reference is made solely to the storage material. The term technical is used when reference is made to the storage vessel or the total system.

When hydrogen is stored in molecular form, the corresponding thermal energy can be calculated on basis of the lower heating value (LHV). The physical specific energy is $33.33\,\mathrm{kWh\,kg^{-1}}$ ($120\,\mathrm{MJ\,kg^{-1}}$) and the energy density is $2.99\,\mathrm{kWh\,m^{-3}}$ ($10.78\,\mathrm{MJ\,m^{-3}}$) at standard temperature and pressure (STP) conditions. The first value stays constant, while only the second one changes with density. Consequently, in this case only the physical energy density is a crucial storage parameter.

27.4
Compressed Hydrogen Storage

The technique of storing hydrogen under pressure has been applied for many years, and gaseous pressure vessels are currently the most common means of storing hydrogen [19]. After a short overview of the thermodynamics and methods for hydrogen compression, different containments that can be used to store pressurized hydrogen are described.

27.4.1
Hydrogen Compression

To reduce the required storage volume, the volumetric energy density of gaseous hydrogen must be increased by means of compression. For example, at an ambient temperature of 20 °C hydrogen has a density of $0.084\,\mathrm{kg\,m^{-3}}$ at a pressure of 1.013 bar, whereas this value increases to $7.797\,\mathrm{kg\,m^{-3}}$ for a pressure of 100 bar. Continuing the compression even further to 1000 bar leads to a volumetric energy density of $49.938\,\mathrm{kg\,m^{-3}}$, corresponding to the non-ideal characteristics of hydrogen at elevated pressures.

The required compression work differs depending on the type of compression and the applied cooling. Isothermal compression poses the most energy-efficient method and leads to a theoretical minimum work for the compression of hydrogen to pressures of 253 and 1013 bar (250 and 1000 atm) of 1.5 and 2.0 kWh kg^{-1}, respectively [30]. Since in practice there are also losses such as parasitic compression losses and the general energy consumption of the compressor station have to be considered [31], the power demand rises to 2.5 and 4.0 kWh kg^{-1} [30]. This corresponds to 7.5% and 12% of the LHV of hydrogen. Other sources state a consumption of approximately 5% and 20% of the LHV when compressing up to 350 and 700 bar, respectively [32].

Hydrogen compressors are employed satisfactorily in relatively large numbers in industrial gas handling as well as the chemical and oil industry. They can be regarded today as state of the art [3]. Different compressor types such as reciprocating and rotary displacement compressors as well as centrifugal compressors are generally used for the compression of gases. However, because of the special characteristics of hydrogen only reciprocating displacement compressors are of practical use. In this category, piston, diaphragm, and ionic compressors can be discerned. State-of-the-art for high-volume applications are reciprocating pistons, while for small-volume applications pistons or diaphragms are used [19]. Another interesting approach is the use of non-mechanical hydrogen compressors such as high pressure electrolyzers, metal hydrides, and electrochemical compressors [31].

27.4.2
Hydrogen Pressure Vessels

To meet the demand of different application areas, various containments with different sizes, shapes, and pressure levels have been developed to store pressurized gas. For clarity, the following is divided into three categories. First, the main features of different pressure tube types are explained, which are categorized as conventional small-scale bulk storage. Pressure vessels typically used for the storage of natural gas or town gas are also supposed to be suitable for hydrogen storage [3,33]. By employing them, hydrogen could be stored up to the low medium-scale range. Finally, some new ideas for medium-scale bulk hydrogen storage are described.

27.4.2.1 Conventional Small-Scale Bulk Storage
Typical bulk gas storage systems commercially available today consist of a number of pressure tubes that can be manifolded together to extend storage volumes and are often referred to as ground storage assemblies. The systems are modular in design and hence can be sized according to customer use rates [34]. Depending on the purpose the tubes can be erected horizontally or vertically. Generally, four different types of high pressure tubes can be discerned by their structural element and their permeation barrier according to the classification of the European Integrated Hydrogen Project [19,35]. These types have different main

Table 27.1 Classification and main features of hydrogen pressure vessel types in 2006, after References [19,35].

Type I	Type II	Type III	Type IV
All-metal cylinder	Load-bearing metal liner hoop wrapped with resin-impregnated continuous filament	Non-load-bearing metal liner axial and hoop wrapped with resin-impregnated continuous filament	Non-load-bearing, non-metal liner axial and hoop wrapped with resin impregnated continuous filament
Technology mature: ++ Pressure limited to 300 bar ($\rightarrow$ density: $-$)	Technology mature: + Pressure not limited ($\rightarrow$ density: +)	Technology mature for $p \leq 350$ bar; 700 bar under development	Technology mature for $p \leq 350$ bar; 700 bar under development
Cost performance: ++	Cost performance: +	Cost performance: $-$	Cost performance: $-$
Weight performance: $-$	Weight performance: 0	Weight performance: +	Weight performance: +

features, which are briefly presented in Table 27.1. For stationary gaseous storage, commonly pressure vessels of Type I are applied, which have the highest weight but are the cheapest [19,35,36]. For stationary applications with higher pressures Type II is preferred. Types III and IV are much more expensive, but their lower weight is a benefit with respect to transport [35]. However, research and development is underway to make Type III and IV more cost-effective for high pressure stationary storage [19].

Diameter, length, and pressure rating of each tube determine the quantity of hydrogen stored. A high pressure level enables the storage of more hydrogen per unit volume, but this requires thicker walls and the cost increases. Therefore, larger storage capacities are commonly realized on a modular basis. Typical industrial pressure tubes of Type I have a volume of 54 l and contain approximately 0.61 kg of hydrogen at a pressure of 156 bar at 21 °C. Modules are available in configurations of 3 to 18 tubes and can reach a hydrogen storage capacity of about 700 kg at 165 bar [19]. Large cylindrical storage vessels for industrial applications typically are in the range 50–70 bar [37,38] with a water volume of approx. 100–150 m^3 [39]. One example in this category is the hydrogen storage tank built next to the hydrogen application center in Herten, Germany (Figure 27.1). It is 22 m high and consists of an unalloyed structural steel. At the maximum operating pressure of 50 bar it has a volume capacity of 5300 m$^3_{(STP)}$ and a mass capacity of 470 kg hydrogen (K. Klug, Westfälisches Energieinstitut, Gelsenkirchen, personal communication, 26 November 2013) [40]. For fast-fill refueling of high-pressure automobile tanks fueling stations require very high pressures tubes that can reach up to 1000 bar [19,36]. One example here is the Type II vessels developed by FIBA Technologies, which are certified to withstand

Figure 27.1 Hydrogen storage tank in Herten, Germany (a) [43] and hydrogen gas holder of Infraserv in Höchst/Frankfurt, Germany (b) [44].

pressures of up to 15 000 psi (1034 bar) and range from 6′ to 40′ with diameters up to 24 inch [41,42].

27.4.2.2 Current and Potential Small- to Medium-Scale Bulk Storage

Storage vessels typically used in the natural gas utility sector are gas holders, spherical pressure vessels, and pipe storage facilities. Gas holders store gas at a low pressure of only a few mbar above atmospheric pressure level. They were built between 1900 and 1960 [45] with many different ways of construction. Generally, they can be divided according to the mechanism of sealing in water-sealed gas holders and dry type gas holders. At operating pressures of up to 1.5 bar a gas volume of 600 000 to 700 000 m^3 can be stored in some types [46]. However, the space requirements are huge due to the low pressure (Figure 27.1).

Gas holders were widely applied for town gas at the beginning of the twentieth century. However, some were even used for hydrogen, as for example a gas holder with a capacity of 56 000 m^3 in Neustadt/Coburg, which was built to fuel airships in 1914 [47]. Although this technique is now considered obsolete, there are still a few hydrogen gas holders in operation today (Table 27.2). One recent example is a gasholder, 26 m in diameter and 30 m high (L. Küchler, F.A. Neuman Anlagentechnik GmbH, Eschweiler, personal communications 14 December 2012) [1] at the Höchst Industrial Park (Figure 27.1). It collects hydrogen, which is a chemical byproduct, for use at a multi-fuel filling station [48].

Spherical pressure vessels (Figure 27.2) were erected from 1930 to 1990 [45] and allow relatively large amounts of gas to be stored in a small space. In comparison to low pressure gas holders they offer the benefit of reduced procurement and operating costs [51]. Spherical pressure vessels can store gas volumes of up to 300 000 m^3 and can be operated at pressures of up to 20 bar [52]. The first spherical pressure vessels was built in 1923 by the Chicago Bridge & Iron Company N.V [53]. As with gas holders, they were usually employed to store

Table 27.2 Technical data of hydrogen gas holders.

		Operator		
		Infraserv Höchst	**BASF**	**CABB GmbH**
Site/year built		Höchst/1997[a]	Ludwigshafen/not available[a]	Gersthofen/1929[b]
Type/material		Dry type gas holder/S235 JRG2[a]	Water-sealed gas holder/not available[a]	Water-sealed gas holder/black steel[b]
Parameter	Unit			
Pressure	bar	1.07[c]	Not available	1.025[b]
Net volume	m^3	10 000[a),c]	30 000[a]	1000[b]

a) From L. Küchler, F.A. Neuman Anlagentechnik GmbH, Eschweiler, personal communications 14 December 2012 [1].
b) From T. Borucinski, CABB GmbH, Gersthofen, personal communication, 20 February 2013 [1].
c) From Reference [48].

town gas; Table 27.3 gives some technical data. Today, spherical pressure vessels are a widespread technology. In the literature [14,54–58] the existence of a spherical pressure vessel for hydrogen with a volume of about 15 000 m^3 and an operating pressure between 12 and 16 bar is stated. Despite intensive efforts no proof for the actual existence of this storage tank was found. Instead, tracing back the cited literature lead to Reference [3], published in 1988 with the statement that the application range of typical low-pressure spherical containers (>15 000 m^3/12–16 bar) in use by the gas business might be similar for hydrogen storage. Since, in addition, no other evidence for hydrogen containing spherical pressure vessel could be found, Table 27.4 uses technical data from natural gas spherical pressure vessels to calculate hydrogen storage parameters under the assumption that the vessels would be the same. Consequently, these values can only be considered as an estimate.

Figure 27.2 Spherical pressure vessel of Wuppertaler Stadtwerke in Möbeck/Sonnborn, Germany (a) [49] and pipe storage facility of Erdgas Zürich AG in Urdorf, Switzerland (b) [50].

Table 27.3 Technical data of spherical pressure vessels used for town gas (L. Küchler, F.A. Neuman Anlagentechnik GmbH, Eschweiler, personal communications 14 December 2012) [1,51].

		Operator			
		STAWAG	GASAG	Public utility	Public utility
Site/year built		Aachen	Berlin	Lahr	Stockholm
Form/material		Spherical/STE 29	Spherical/STE 51	Spherical/STE 47	Spherical/FG 51
Parameter	Unit				
Diameter	M	27.16	39.5	26.74	22.4
Pressure	bar	5.5	10.4	10	17
Water volume	m^3	10 490	32 057	10 000	5885
Wall thickness	mm	21	34	25.3	32–34

Pipe storage facilities have been installed by gas utility companies since 1980 until today [45]. This kind of storage vessel is constructed from standard tubes commonly used in trade with a high nominal diameter of about DN 1400 and is able to withstand high operating pressures of 64–100 bar [62]. Table 27.5 summarizes the technical data of different storage facilities for natural gas operated by public utilities and gas supply companies. One of the most recent projects is the construction of a pipe storage facility for natural gas in Urdorf, Switzerland, by Erdgas Zürich Transport AG, which will be one of the largest in Europe (Figure 27.2). The investment costs amount to CHF 21 million (~€17 million) and work was completed at the end of 2013. The entire construction pit has a dimension of $210 \times 50 \, m^2$. A total of 260 single tubes, each 14 t in weight, are welded together, resulting in 20 pipe strings and a total length of 4140 m. The tube diameter is 1422 mm, with a wall thickness of 25.5 mm. The total geometric volume is $6112 \, m^3$ and offers a net volume capacity of $720\,000 \, m^3$ between an upper and lower operating pressure of 100 and 7 bar, respectively [50,63–65]. Transferring this net volume capacity to hydrogen results in a net mass of 64 714 kg and a thermal energy equivalent of 2156 MWh stored. This demonstrates that, with this type of facility, storage at higher pressures in the GWh range is possible.

In comparison to spherical pressure vessels buried pipe storage facilities offer technical and economic advantages [63,66,67]. The geometric shape of the ball permits minimal surface areas and low wall thicknesses leading to reduced material expenditure per net gas volume [62]. However, the maximum pressure is limited since the wall thickness has to increase with higher stress levels. In contrast, pipe storage facilities can be operated at much higher pressure ranges, require comparatively small storage volumes, and have an unproblematic

Table 27.4 Calculated hydrogen storage parameters based on technical data of spherical pressure vessels for natural gas of different public utility companies.

Parameter	Unit	Site/year built			
		Heilbronn am Neckar/1964[a]	Reutlingen/ 1965[b]	Gießen/ 1961[c]	Wuppertal/ 1956[d]
Diameter[e]	m	34[a]	34.11[b]	16.88[c]	47.3[d]
Wall thickness	mm	30[a]	30[b]	20[c]	30[d]
Pressure[f]	bar	7[a]	9.3[b]	8[c]	5.05[d]
Weight[g]	t	759[a]	965[h]	175[c]	1944[d]
Calculated hydrogen storage parameters[i]					
Geometric volume[j]	m^3	20 471	20 670	2 500	55 199
Gross volume capacity[k]	$m^3_{(STP)}$	141 422	189 721	19 742	275 108
Net volume capacity[l]	$m^3_{(STP)}$	101 015	148 921	14 807	166 154
Gross mass capacity[k]	kg	12 711	17 052	1 774	24 727
Net mass capacity[l]	kg	9 079	13 385	1 331	14 934
Gross energy capacity[k]	MWh	424	568	59	824
Net energy capacity[l]	MWh	303	446	44	498
Gross specific energy[k,m]	kWh t^{-1}	524.39	578.60	334.48	418.52
Net specific energy[l,m]	kWh t^{-1}	374.57	454.17	250.86	252.77
Gross energy density[k,m]	kWh m^{-3}	20.58	27.34	23.48	14.87
Net energy density[l,m]	kWh m^{-3}	14.70	21.46	17.61	8.98

a) From References [59,60].
b) From K. Leibfritz, FairEnergie GmbH, Reutlingen, personal communication, 30 January 2013 [1].
c) From D. Althaus, Stadtwerke Gießen AG, Gießen, personal communication, 31 January 2013 [1].
d) From B. Seipenbusch, WSW Energie & Wasser AG, Wuppertal, personal communication, 31 January 2013 [1].
e) Assumed to be the external diameter.
f) Assumed to be the maximum operating pressure.
g) Assumed to be the weight of empty vessel.
h) From Reference [61].
i) Data in italics are our own calculations.
j) Calculated from the inner radius (external radius minus wall thickness).
k) Gross capacity is calculated at maximum operating pressure.
l) The net capacity is calculated assuming a minimum operating pressure of 2 bar.
m) Technical value (on the basis of the storage vessel including hydrogen).

Table 27.5 Technical data of pipe storage facilities for natural gas, after Reference [62].

Operator	Length (m)	Diameter (DN)	Pressure (bar)	Water volume (m³)	Net volume (m³)
SBL, Linz	1720	1600	5–22	5000	85 000
SW Bietigheim-Bissingen	852	1400	22	1330	25 000
EV Hildesheim	6000	1400	75	9200	700 000
Gas- und E-Werk Singen	1200	1400	16–80	1800	144 000
Erdgas Zürich	5500	1500	7–70	9540	714 000

technical construction, which in total results in low investment cost [62]. Furthermore, pipe storage facilities are typically buried a few meters below ground. This has the benefit of protection against adverse weather conditions and external mechanical impacts [66]. In addition, the land area can still be used for other purposes. For example, at the pipe storage facility in Urdorf the land will be made agriculturally usable again [63].

Notably, hydrogen has a different impact on materials than natural gas and consequently requires special attention. Vessels have to be constructed with materials resistant to hydrogen embrittlement and fatigue. In addition, they have to maintain structural integrity under high-pressure cycling environments [19]. Since hydrogen storage adds significantly to the cost of a delivery infrastructure, research into new cost-effective materials, coatings, and fiber or other composite structures is needed [17,19].

27.4.2.3 New Ideas for Medium-Scale Bulk Storage

In Reference [68] an initial feasibility study was performed on the use of prestressed concrete pressure vessel (PCPV) structures for on-shore storage of hydrogen gas at pressures up to 100 bar. These structures could be spherical, hemispherical, or silo shaped. To prevent hydrogen leakage an impermeable liner has to be installed. Metallic liners are commonly used in hydrogen storage. However, for high pressures (700 bar) alternative liners have been developed using high molecular weight carbon composites and polymers. Preliminary estimates indicate that the cost for a structure with 5 GWh of energy storage capacity may be in the range of £1.0–3.0 (~€1.2–€3.6) per kWh.

Another new approach [69] is to utilize wind turbine towers for the storage of pressurized hydrogen. The optimum pressure levels are identified to be in the range 10–15 bar. The additional costs for an 84 m tall tower of a 1.5 MW wind turbine are estimated to be $83 000 (~€63 000), with a capability of storing 940 kg hydrogen at 11 bar. Hence, the specific storage cost would amount to 88 $ kg^{-1} (~67 € kg^{-1}). However, design issues around fatigue, burst, corrosion, and hydrogen-induced cracking have to be further investigated to assess future potential for this storage option [69,70].

27.5
Cryogenic Liquid Hydrogen Storage

Liquefaction of hydrogen is a well-established technology, which is especially applied in space exploration and science. However, liquefaction is an energy intensive process. First, the reasons for this are explained and the energy demand is quantified. Then, the different types of containment for storing liquefied hydrogen are described.

27.5.1
Hydrogen Liquefaction

Cryogenic liquid hydrogen has a higher physical energy density than compressed gaseous hydrogen. The volume reduction of liquid hydrogen makes it especially attractive for bulk delivery by truck, train, or ship since the required number of runs for transporting the same energy content reduces drastically. The reason for this is the difference in densities. Liquid hydrogen has a density of 70.8 kg m^{-3} at the normal boiling point (Table 27.6). In contrast, the volumetric density of gaseous hydrogen is only 49.938 kg m^{-3} at an ambient temperature of 20 °C and at a pressure of 1000 bar.

However, to liquefy hydrogen a considerable amount of energy is required. Several parameters influence the minimum theoretical liquefaction power of hydrogen [71]. Beside process parameters like the feed pressure and temperature, one important feature is the existence of two different forms of the hydrogen molecule, namely, ortho- and para-hydrogen. Their difference is caused by the orientation of the nuclear spin, which results in slightly different properties. Para-hydrogen, whose electrons rotate in the opposing direction, has a lower energy level than ortho-hydrogen, whose electrons rotate in the same direction. At normal conditions, molecular hydrogen consists of about 75% ortho- and 25% para-hydrogen, which is commonly defined as normal hydrogen [72,73]. The equilibrium between ortho- and para-hydrogen shifts towards a higher fraction

Table 27.6 Physical and chemical properties relevant for hydrogen liquefaction.

Properties	Value	Units	Conditions	Reference
Normal boiling temperature	20.268	K	$p_{(NTP)} = p_{(STP)} =$ 1.01 325 bar	[74]
Density of liquid hydrogen	70.78	kg m^{-3}	At normal boiling point	[74]
Density of gaseous hydrogen	1.338	kg m^{-3}	At normal boiling point	[74]
Physical energy density	8.49	MJ l^{-1}	$p_{(NTP)} = p_{(STP)} = 1.01$ 325 bar	[37]
Heat of ortho-to-para conversion	555	kJ kg^{-1}	At normal boiling point	[75]
Heat of vaporization	445.6	kJ kg^{-1}	At normal boiling point	[74]

of para-hydrogen with decreasing temperature. For example, at 19 K an equilibrium sample of hydrogen contains 99.75% para-hydrogen [73].

If liquid hydrogen that has not yet attained its equilibrium concentration is stored over a longer period, the ortho-hydrogen present will slowly transform into para-hydrogen. This exothermic conversion produces a non-negligible amount of heat leading to vaporization and loss of liquid stored hydrogen in the long term even though the storage vessel is excellently insulated. Consequently, the pressure inside the vessel increases, which is regulated via a pressure relief system, which contributes to hydrogen losses of the storage tank called boil-off.

Therefore, normal hydrogen has to be converted into nearly pure para-hydrogen with the help of catalyzed reactions during the liquefaction process [47–49]. Commercial liquefaction plants normally operate with a product para fraction of at least 95%. This conversion makes up a significant portion of the liquid hydrogen exergy content. For example, the minimum specific power amounts to 3.94 kWh kg^{-1} for the transformation of normal hydrogen at 300 K into saturated liquid hydrogen with an ortho-para composition in equilibrium and feed and product stream at a pressure of 1 bar. The conversion of ortho- into para-hydrogen thereby makes up 0.59 kWh kg^{-1}, which is about 15% of the total reversible work [71,76].

Several liquefaction processes for hydrogen exist, but only a few are actually applied at liquefaction plants. For example, Linde Kryotechnik uses a helium pre-cooled Joule–Thomson process for small capacity liquefaction plants up to 1000 l h^{-1} and a liquid nitrogen pre-cooled Claude process for higher capacities [39,77]. Common to all hydrogen liquefaction processes is the need for upstream purification so that all impurities have a concentration below 1 ppm. Otherwise these materials would cause clogging in the liquefier because they are solid at 20 K [31,78].

The typical energy demand of existing hydrogen liquefaction plants is in the range 36–54 MJ kg^{-1} [31], which corresponds to 30–45% of the LHV of hydrogen. The cost and the energy requirements per kilogram of liquefied hydrogen decrease as the plant capacity increases [25]. Currently, only a few plants exist worldwide, with more than nine located in the USA, four in Europe, and eleven in Asia, with production capacities ranging from 5–34, 5–10, and 0.3–11.3 t d^{-1}, respectively. The exergy efficiencies of these plants are 20–30%, but there are some proposed conceptual plants with efficiencies of 40–50% [79]. For future large-scale hydrogen production it is especially important to increase the liquefaction capacity and to lower the specific power requirements by process optimization and new techniques [33].

27.5.2
Liquid Hydrogen Storage Tanks

Storage and transportation of industrial and medical gases has been accomplished with cryogenic tanks, also called dewars, for more than 40 years [35].

The storage of liquid hydrogen is similar to that of liquid helium and is state-of-the-art today, especially due to intensive applications in space flight [33].

Liquid storage tanks can have either a spherical or cylindrical form. The latter is available in a horizontal and vertical configuration. Larger tanks for long-term storage usually have a spherical form to ensure the smallest surface to volume ratio and hence to decrease evaporative losses [39]. Besides stationary application these vessels can also be applied in the bulk transportation sector. The spherical vessel is more advantageous for ships, while the cylindrical form is better for transportation via railway or truck.

A typical installation with subsequent gaseous hydrogen use normally consists of a cryogenic storage tank, ambient air vaporizer, and controls. Generally, bulk liquid storage systems can be selected based on customer volume, desired pressure, purity level, flow rate, and operating pattern [34]. Common tank capacities range from 1500 l or 100 kg [14] to 95 000 l or 6650 kg [19] of hydrogen. The maximum pressure in the tank is 20–30 bar [80].

There are only a few manufacturers for cryogenic vessels worldwide. Well-known companies and their countries of origin are, for example, Linde in Germany, Air Liquide in France, Air Products and Praxair in the USA, British Oxygen Company in the UK, Kobe Steel in Japan, and JSC Cryogenmash in Russia [39]. Table 27.7 gives exemplarily technical data for cylindrical storage tanks in horizontal and vertical constructions and various sizes commercially available today. As in the case of bulk ground storage modules for compressed

Table 27.7 Technical data of horizontal and vertical tanks for liquid hydrogen storage, after Reference [81].[a]

Parameter	Unit	Type specification					
		TLH-1500	VLH-4500	TLH-4500	TLH-9000	VLH-15000	TLH-18150
Diameter	ft-inch	7'-3"	8'	8'	8'-8"	10'-8"	10'-8"
Height	ft-inch	17'-5"	23'-5"	23'-1/8"	38'-8"	41'-4"	43'-3"
Weight (empty vessel)	t	3.86	9.25	9.07	21.05	28.89	34.84
Gross volume capacity	m^3(STP)	5.7	16.3	16.4	33.3	53.8	64.9
Gross volume capacity	m^3(STP)	*4517*	*12 813*	*13 000*	*26 377*	*42 119*	*51 404*
Gross mass capacity	kg	*408*	*1179*	*1247*	*2363*	*3765*	*4604*
Gross specific energy[b]	MWh kg^{-1}	*13.6*	*39.28*	*41.54*	*78.70*	*125.38*	*153.32*

a) Data in italics are our own calculations.
b) Technical value (on basis of storage vessel including hydrogen).

Table 27.8 Technical data of liquid hydrogen storage tanks depending on their size, after Reference [39].

Parameter	Unit	"Local"	Large/"today"	Large/"long-term"
Water volume	m^3	60	3000	100 000
Net mass capacity	t	3.8	191	6371
Net energy capacity	MWh	127	6370	212 349
Technical system lifetime	a	30	30	30
Utilization by time[a]	h/a	8400	8400	NA
Utilization by volume	%	90	90	90
Loss rate due to boil-off	%	0.4	0.07	0.01

a) According to Reference [57].

gaseous hydrogen, liquid storage tanks can be used in modular fashion as well. Table 27.8 contains technical data provided by Reference [39] assuming that local storage ("Local") at filling stations or other consumer sites requires storage volumes of approximately 60 m^3 and that future storage tanks (Large/"long-term") can be built that are larger than the largest tanks of today (Large/"today").

Stationary large-scale tanks are required for space applications, since the space shuttles have to be loaded with huge amounts of liquid hydrogen and oxygen just a few hours before take-off. Presently, NASA has the largest spherical storage tank (Figure 27.3) [14,33,39]. The tank is located at the northeast corner of Launch Pad 39A at Cape Canaveral in Florida and stores about 3400 m^3 (850 000 gallons) of liquid hydrogen at −253 °C (−423 °F) [82]. The cited data for its diameter and for its mass capacity range from 20 m [14,19,33] to 22 m [39] and from 230 t [39] to 270 t [14,33], respectively.

(a)   (b)

Figure 27.3 Spherical storage tank of NASA (a) [82] and horizontal storage tank at the Linde Hydrogen Center in Unterschleißheim, Germany (b) [87].

Besides launch sites for space flights, liquid hydrogen storage tanks are also applied at hydrogen fueling stations. One example is the Linde Hydrogen Center near Munich, Germany, which started operation in 2007. The fueling station has a vertical aboveground liquid hydrogen storage tank, which is super-insulated and provides a storage volume of 17 600 l. The fueling process for both liquid and compressed gaseous hydrogen can be accomplished in a few minutes, whereby the liquid hydrogen is kept at a cryogenic temperature of −253 °C and the compressed gaseous hydrogen is provided at 350 bar through corresponding dispenser systems. The boil-off is utilized as supply for compressed hydrogen [84]. A similar fueling station was erected in Berlin, Germany, in 2006 [85].

As in the case of compressed gaseous storage vessels, liquid hydrogen tanks can be placed above and below ground. Although the costs for underground liquid hydrogen storage would probably be higher than for traditional aboveground pressurized hydrogen system, this approach could combine advantages from both underground and liquid storage. For an underground storage tank the area aboveground is saved for other purposes. This is especially advantageous where space is limited, such as at filling stations in cities. Furthermore, because of the higher density of liquid hydrogen less space is needed compared with storage of gaseous hydrogen. In addition, this concept makes a fueling station inherently safer [19]. Currently, there are only a few fueling stations with a underground liquid hydrogen tank; they are in London, Munich, and Washington DC [14,54].

Special attention has to be given to the selection of materials for the construction of liquid hydrogen storage tanks. Hydrogen embrittlement can be neglected at the boiling point because the hydrogen solubility is low [31,35]. For example, in the case of unstable austenitic stainless steels the hydrogen embrittlement effect is maximum at −100 °C, but negligible for temperatures below −150 °C. However, the brittleness of metals at cryogenic temperatures limits the choice of materials. Therefore, nickel ferritic steels, which can be applied down to −200 °C, or stabilized austenitic stainless steels and aluminum alloys, which are adaptable down to the absolute zero, are commonly used for cryogenic tanks [35].

To avoid undesired evaporation of the stored liquid hydrogen the vessel must have excellent insulation to minimize conductive, convective, and radiative heat transfer into the inside. Therefore, cryogenic vessels are composed of an inner pressure vessel and an external protective container with an insulating vacuum layer between them. There are different options for insulation. Multilayer insulation, which consists of several layers of aluminum foil alternating with glass fiber matting to avoid heat radiation, or perlite vacuum insulation are often applied [14,33,35,39,72].

Despite excellent vessel insulation, evaporation occurs to a certain extent, which causes a pressure increase over longer storage periods [72]. Therefore, a liquid hydrogen storage tank has to be designed as an open system and equipped with a pressure relief system leading to hydrogen losses, which are commonly called boil-off [72]. The released gaseous hydrogen can either be directly utilized or stored in an auxiliary system or it can be returned to the liquefaction plant when the storage tank is situated next to it [31,39].

27.6 Cost Estimates and Economic Targets

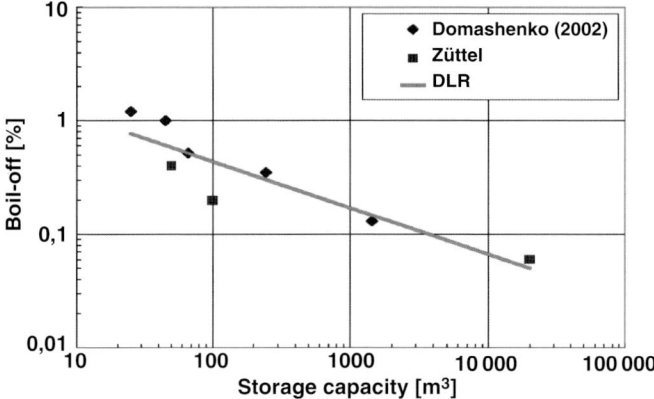

Figure 27.4 Boil-off rate of large liquid hydrogen storage tanks. Reproduced with permission from Reference [39].

Heat transfer from the environment to the inside of the storage tank depends on the thermal insulation, tank size, and shape [31,39]. As explained above, the ortho-to-para ratio of the liquid hydrogen also influences the boil-off rate. The spherical form is advantageous because it offers a minimal surface area to volume ratio. With increasing storage capacity or volume, respectively, this ratio increases so that larger vessels have a lower evaporation rate than smaller vessels assuming equal insulation (Figure 27.4) [39]. Typical values for boil-off rates are below 0.03% per day for large storage spherical tanks with perlite vacuum insulation such as the NASA tank, 0.4% per day for vacuum-superinsulated tanks, and 1–2% per day for large tanks with vacuum powder insulation depending on their geometry [33].

27.6
Cost Estimates and Economic Targets

Costs for bulk stationary hydrogen storage are a decisive factor [6,19]. This section presents economic data for different storage systems together with cost targets of the DOE [17] and the European Commission (EC) [88]. A literature search found that some detailed assessments and review articles [3,26,56,57,87–90] were published before 2000, while only two sources [4,91] with newer data published in 2005 and 2004, respectively, could be obtained. In Reference [39] investment costs for bulk liquid and gaseous hydrogen storage cited from References [57,87] were adjusted to $€_{2000}$, but when comparing this data with newer data from References [4,91] no substantial matching could be found. Consequently, the economic data given before 2000 is ranked as outdated and is not included here.

Figure 27.5 depicts the investment costs of different hydrogen storage options depending on the gross storage volume capacity. The economic data in Figure 27.5, which can be obtained from [91], is an estimate on the basis of data

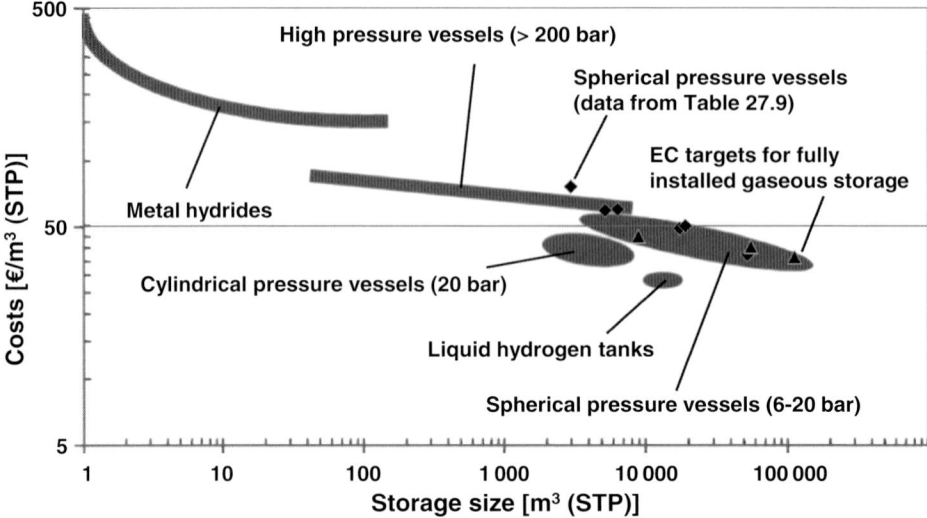

Figure 27.5 Investment cost estimates of different hydrogen storage options depending on the storage size and targets of the EC for fully installed gaseous storage. Reproduced with permission from Reference [91]. Copyright European Communities, 2004.

requested from the industry (M. Altmann, Ludwig-Bölkow-Systemtechnik GmbH, personal communications 17 December 2012) [1]. In addition economic data listed in Table 27.9 and targets of the EC listed below in Table 27.11 are included in Figure 27.5 for comparative purposes. Notably, the EC targets are on a fully installed basis, while the specific investment given in References [4,91] are not specified in detail. The different cost basis also has to be considered with respect to Table 27.10. Consequently, direct comparison and assessment of these data is not possible.

In Figure 27.5 and Table 27.11 the high cost level of solid state storage is clearly visible. It also becomes clear that gaseous storage costs are influenced by the size, the pressure range, and the shape of the vessel. In Figure 27.5 it can be

Table 27.9 Technical and economic data of spherical pressure vessels, after Reference [4].

Volume (m³)	Max. pressure (MPa)	Min. pressure (MPa)	Volume capacity (m³(STP))	Investment (€)	Spec. investment (€ m⁻³(STP))
300	1.2	0.1	3000	230 000	76
300	2	0.1	5200	307 000	59
1000	0.8	0.1	6400	383 000	60
1000	2	0.1	17 400	844 000	49
3000	0.8	0.1	19 300	971 000	50
3000	2	0.1	52 100	1 917 000	37

Table 27.10 Economic data and targets of the DOE for stationary gaseous hydrogen storage, after Reference [17].[a]

Category	Unit[b]	2005 Status	FY 2011 status	FY 2015 target	FY 2020 target
Low pressure (160 bar) purchased capital cost	$ kg^{-1}	1000	1000	850	700
	€ m^{-3}(STP)	68	68	58	48
Moderate pressure (430 bar) purchased capital cost	$ kg^{-1}	1100	1100	900	750
	€ m^{-3}(STP)	75	75	61	51
High pressure (860 bar) purchased capital cost	$ kg^{-1}	Not available	1450	1200	1000
	€ m^{-3}(STP)	Not available	99	82	68

a) Data in italics are our own calculations.
b) It is assumed that $1 is equivalent to €0.76.

recognized that liquid storage tanks are cheaper than gaseous storage vessels. However, in an overall analysis the costly and energy-intensive liquefaction process also has an influence. Therefore, an economic assessment of the different storage alternatives is only reasonable on the basis of a detailed application

Table 27.11 Economic data and targets of the EC, after Reference [86].[a]

Category	Unit	Current 2010	Target 2015	Target 2020
Distributed storage of gaseous hydrogen				
Capacity per site (high end of size range)	t	0.8	5	10
	m^3(STP)	8899	55 617	111 235
Capital expenditure (fully installed) per capacity	M€ t^{-1}	0.5	0.45	0.4
	€ m^{-3}(STP)	45	40	36
Storage of hydrogen in solid materials				
Capacity per site (high end of size range)	t	3	5	10
	m^3(STP)	33 370	55 617	111 235
Capital expenditure (fully installed) per capacity	M€ t^{-1}	5	1.5	0.83 [b]/0.85 [c]
	€ m^{-3}(STP)	450	135	75/76

a) Data in italics are our own calculations.
b) Value given in annex on page 37 of Reference [86].
c) Value given on page 7 of Reference [86].

scenario including all relevant aspects such as the charging–discharging schedule [87,89].

27.7
Technical Assessment

Dependent on the application area different solutions can be beneficial since each storage methods has its own advantages and disadvantages and consequently can fulfill distinct requirements. However, for evaluation of the suitability of a technology the whole storage system including charging, storing, and release of the hydrogen has to be considered with respect to the application purpose and the boundary conditions. In this context, aspects concerning the transport of hydrogen to and from the storage system are also important.

The technical assessment presented in the following is based on two different cases. Both assume that gaseous hydrogen is delivered to the storage system via pipeline at 30 bar. According to References [17,23] this value represents the lower operating pressure of a transmission pipeline. In the first case, no special application purpose or means of transport is considered for the hydrogen after storage. Instead, the lower storage pressure, which is set to 1.5 bar for all storage systems, is taken as the reference point. This leaves room for other delivery options such as transport by truck in either liquid or gaseous state. In the second case it is assumed that the hydrogen has to be transported again by pipeline, for which also a pressure of 30 bar was chosen. This value lies in the middle of the pressure range of a hydrogen distribution pipeline [17]. Furthermore, this assumption allows calculation of the storage efficiency from the same reference point before and after storage.

The storage efficiency describes the ratio of useful energy output to the required energy input of the storage system. Thereby, the power demand for loading and unloading is also considered. The difference between the two cases, explained above, is that for the first one no energy is required for unloading. For clarity, the terms storage efficiency at 1.5 bar and storage efficiency at 30 bar are used in the following. In addition to this another performance indicator was defined here, which is termed supply equivalent. It describes the number of days a fueling stations with a hydrogen consumption of $1500\,\mathrm{kg\,d^{-1}}$ [23] could be supplied and thus gives a realistic impression of the net mass capacity of the storage system. To ensure traceability the assessment is kept deliberately simple, which means that the dynamic behavior of the storage systems is neglected in large parts and the ideal gas equation is used. In addition, losses caused by leakage and boil-off are not considered.

Selected storage types of this assessment are gas holders, spherical pressure vessels, pipe storage facilities, and spherical dewars (Table 27.12). For the calculation the different containments were designed in such a way that they represent the upper range in terms of size. Since large spherical pressure vessels and pipe storage facilities for hydrogen have yet not been built, the design of the

Table 27.12 Technical assessment of different bulk hydrogen storage technologies.[a]

Storage technology		Gas holder	Spherical pressure vessel	Pipe storage facility	Spherical dewar
Parameter	Unit				
Average operating temperature	°C	10	10	10	−253
Min. operating pressure	bar	1.5[b]	1.5	1.5	1.5
Max. operating pressure	bar	1.5[b]	20[c]	100[d]	1.5
Inner diameter	m	23.2	31.1	1.4[e]	19.4
Height/length	m	100	—	6496	—
Footprint	m²	420	760	15 391[f]	295
Geometric volume/water volume	m³	42 014	15 755	10 000[e]	3814
Gross volume capacity	m³ (STP)	60 000[g]	300 000[c]	952 068	3 004 005
Net volume capacity	m³ (STP)	54 000	277 500	937 787	2 703 605
Gross mass capacity	kg	5393	26 964	85 572	270 000[h]
Net mass capacity	kg	4854	24 942	84 288	243 000
Gross energy capacity	MWh	180	899	2851	8997
Net energy capacity	MWh	162	831	2809	8097
Utilization by volume	%	90	93	99	90[j]
Supply equivalent[i]	d	3	17	56	162
Specific power demand for loading	kWh kg⁻¹	—	—	<0.7[k),l]	9.2[m]

(continued)

Table 27.12 (Continued)

Storage technology		Gas holder	Spherical pressure vessel	Pipe storage facility	Spherical dewar
Parameter	Unit				
Total power demand for loading	MWh	—	—	<40 [k]	2236
Storage efficiency at 1.5 bar[n]	%	100	100	>98.6 [k]	78.4
Specific power demand for unloading	kWh kg^{-1}	1.8 [o]	<1.8 [k,o]	<1.8 [k,o]	1.8 [o]
Total power demand for unloading	MWh	9	<45 [k]	<44 [k]	435
Storage efficiency at 30 bar[p]	%	94.9	>94.9 [k]	>97.1 [k]	75.2

a) Data in bold are the design value/assumptions; data in italic are our own calculation.
b) According to Reference [46].
c) According to Reference [52].
d) According to Reference [65].
e) Selected according to Table 27.5.
f) A length of 200 m for the pipe stings and space of 1 m between them is assumed.
g) According to Reference [47].
h) According to References [14,33].
i) According to Reference [39].
j) Represents the number of days a fueling station with a hydrogen consumption of 1500 kg d^{-1} could be supplied.
k) In reality less power is required for compression and the storage efficiency is higher, because of the dynamic nature of the storage system.
l) Compression from 30 to 100 bar with isentropic efficiency of 70%, motor efficiency of 94%, two stages with intercooling to 30 °C and calculation according to ideal gas law.
m) According to References [71,95] considering 30 bar feed pressure.
n) Reference point is 1.5 bar and only energy required for loading has to be considered.
o) Compression from 1.5 to 30 bar with isentropic efficiency of 70%, motor efficiency of 94%, three stages with intercooling to 30 °C and calculation according to ideal gas law.
p) Reference point is 30 bar and energy required for loading and unloading as to be considered.

vessel is aligned to existing ones for natural gas and values found in the literature and thus describe potential future technologies. The storage capacity of the gas holder and the spherical dewar are chosen according to the gas holder that existed in Neustadt/Coburg to fuel airships and the spherical dewar of the NASA, respectively. For calculation of the latter a tank temperature near to the normal boiling point is assumed. According to References [92–94] the annual average temperature of the air and the soil do not show much variation and therefore, for simplicity, a temperature of 10 °C is chosen and it is assumed that the containments are always in thermal equilibrium with their surroundings.

The operating pressure of gas holders remains approximately constant, because the geometric volume changes with storage capacity. Therefore, here a constant operating pressure of 1.5 bar was selected according to Reference [46]. Since the minimal and maximal operating pressure are the same, for the calculation of the net volume capacity a utilization by volume of 90% is assumed, which represents the geometric dead volume. With this the net volume capacity amounts to 54 000 m^3$_{(STP)}$ and the net mass capacity to 4854 kg. This is enough for a supply equivalent of 3 d, which means that one fueling station with a hydrogen consumption of 1500 kg d^{-1} can be supplied for 3 days. During loading no energy is required because the hydrogen is delivered at a pressure of 30 bar and filled to the gas holder after expansion. Consequently, the storage efficiency at 1.5 bar is 100%. However, in the second case the hydrogen has to be fed into the distribution grid at a pressure of 30 bar. With a total power demand of 9 MWh for compression of the net mass capacity, which is taken from the storage system, the storage efficiency at 30 bar is 94.9%.

According to Reference [52] the gross volume capacity and the maximum operating pressure of the spherical pressure vessel are set to 300 000 m^3$_{(STP)}$ and 20 bar, respectively. The minimum operating pressure is for all storage types fixed to 1.5 bar for comparability reasons. Consequently, the utilization by volume amounts to 93% with a net volume capacity of 277 500 m^3$_{(STP)}$. This equals a net mass capacity of 24 942 kg or a supply equivalent of 17 d. As in the case of the gas holder the storage efficiency at 1.5 bar is 100%. For calculation of the storage efficiency at 30 bar it was assumed that the total net mass capacity has to be compressed from a constant outlet pressure of 1.5 bar to the distribution pipeline pressure of 30 bar. However, in reality the outlet pressure depends on the amount of hydrogen stored and drops from 20 to 1.5 bar during unloading. Therefore, in practice the total power demand for compression is less than 45 MWh and accordingly the storage efficiency at 30 bar is higher than 94.9%.

Based on Table 27.5 a water volume of 10 000 m^3 and an inner tube diameter of 1.4 m were selected for the design of the pipe storage facility. The maximum operating pressure was assumed to be 100 bar, according to Reference [65]. Since the minimal operating pressure is, at 1.5 bar, quite low, the storage range is, with a utilization by volume of 99%, nearly fully exploited. Accordingly, the net volume capacity, the net mass capacity, and the supply equivalent are, at 937 787 m^3$_{(STP)}$, 84 288 kg, and 56 d, respectively, the highest of all storage types for gaseous hydrogen. To calculate the power demand for loading it is assumed that no energy is required to fill the pipe storage facility up to a pressure of 30 bar. Then the power demand is calculated as if the remaining mass of 69 900 kg had to be compressed from 30 to 100 bar. Again, this is not the case in reality, because during loading the pressure rises and a specific power demand of 0.7 kWh kg^{-1} is only necessary at the end. Therefore, the storage efficiency at 1.5 bar is higher than 98.6%. Similarly, to calculate the power demand for unloading, it is assumed that no energy is required until an equal vessel pressure of 30 bar is reached. Afterwards, the total power demand was calculated as was done for the spherical pressure vessel. Consequently, the storage efficiency at 30 bar also has to be higher than 97.1% in practice. A change of the compression efficiency shows only minor deviations. For example, with an isentropic efficiency of 56% and 85% and a motor efficiency of 92% and 96%, selected according to References [24] and [71,76], the storage efficiency at 30 bar amounts to 96.3% and 97.7%, respectively.

For the spherical dewar a gross mass capacity of 270 t corresponding to literature data given in References [14,33] for the NASA tank was taken. Assuming a utilization by volume of 90%, the net volume capacity and the net mass capacity amounts to 2 703 605 m^3$_{(STP)}$ and 243 000 kg, respectively. This is nearly three times the net capacity of the pipe storage facility and equals a supply equivalent of 162 d. With respect to the stored amount of hydrogen the spherical dewar exceeds the other storage types. However, the power demand is also highest. With an exergy efficiency of 30%, which represents good state-of-the-art liquefaction processes [79], and an inlet feed pressure of 30 bar the specific power demand for liquefaction is approximately 9.2 kWh kg^{-1} according to References [71,95]. In the calculation of storage efficiency, only the energy demand that is required to load and unload the same amount of hydrogen is considered. This means that although the entire spherical dewar has to be filled with liquid hydrogen the first time, only the energy demand for liquefaction of the net mass capacity, which is 2236 MWh, has to be used here. With this the storage efficiency at 1.5 bar amounts to 78.4%. To compress the entire usable mass of 243 000 kg a power demand of 435 MWh is needed and the specific efficiency at 30 bar drops to 75.2%. Improving the exergy efficiency of the liquefaction process to 50%, which is postulated in some theoretical concepts [79], results in a much higher storage efficiency at 1.5 and 30 bar of 84.3% and 81.8%, respectively.

In the following the results of the technical assessment are summarized briefly. With respect to the usable mass of hydrogen, the supply equivalent, and footprint the spherical dewar performs best. However, the power demand for loading and unloading are quite high under the chosen boundary conditions and, consequently, the storage efficiency for the two reference points set at 1.5 and 30 bar are also the lowest. In this regard the gas holder and the spherical pressure vessel achieve the highest result in the first case and the pipe storage facility in the second case. However, it has to be considered that, finally, at a fueling station a very high pressure is required to load the compressed gaseous on-board storage tank. Since the compression work depends on the pressure ratio, especially a drop near to atmospheric pressure should be avoided.

27.8
Conclusion

Presently, very few near-surface bulk hydrogen storage facilities exist. They store hydrogen either in gaseous or liquid state mainly for further utilization in the chemical industry or in the space flight sector. In future, however, when hydrogen is employed in addition to its current chemical usage as a versatile energy carrier, a greater number of storage facilities are needed. One promising application field is to employ hydrogen as a carbon free fuel, which also allows integrating renewable energy into the transport sector.

One challenge for future pipeline grids will be to balance fluctuations resulting from irregular injection from renewable energies and inconsistent withdrawal at fueling stations. Pipeline grids are able to provide some built-in buffering storage, but only to a small extent. Therefore, further management measures for compensating fluctuations might be required. It is conceivable that for this purpose pressure vessels similar to the ones already used for the storage of natural gas or town gas will be applied. Vessel types in this category are gas holders, spherical pressure vessels, and pipe storage facilities.

Gas holders are able to store large gas volumes but only at pressures slightly above atmospheric pressure level. Since electrolyzers produce hydrogen at elevated pressure, which will even increase in future, and also a hydrogen grid would be operated with higher pressure, gas holders are rather unsuitable in this case. However, they could be an interesting option for other applications, where hydrogen is produced at low pressure. The characteristics of spherical pressure vessels are medium operating pressures of up to 20 bar and gas volumes of up to 300 000 m^3. The advantages of the spherical shape are a minimal surface area and a low wall thickness leading to reduced material expenditure per net gas volume. Furthermore, at the same pressure it permits storage of the biggest gas volume at the smallest footprint. Pipe storage facilities can achieve net volume capacities above 700 000 m^3 and operating pressures up to 100 bar, which makes storage in the GWh range possible. They are typically buried below ground, where they are protected against adverse weather conditions and external mechanical impacts. In addition, this permits use of the land area for other purposes. Beyond the existing gas storages technologies, new storage concepts have been investigated for medium-scale bulk hydrogen storage. Among these ideas are the utilization of prestressed concrete pressure vessel structures and wind turbine towers.

Storage capacities exceeding those for gaseous pressure storage can be achieved when hydrogen is stored in liquid form. Presently, the largest existent spherical dewar is that of NASA, which has a mass storage capacity between 230 and 270 t hydrogen and a boil-off rate of 0.03% per day. However, the disadvantages of this technology are the energy and cost intensive liquefaction process and the losses due to boil-off.

Investment cost estimates for gaseous and liquid hydrogen storage vessels where shown here as well. The costs for gaseous storage vessels clearly depend on the size, pressure range, and shape of the vessel. For the same storage size, liquid storage tanks were estimated to be cheaper than gaseous storage vessels. However, only considering the storage vessels can be misleading, since this neglects other important aspects such as, for example, the cost and energy requirements for charging and discharging. In addition, other questions such as safety aspects or the utilization of hydrogen after storage are of relevance here. Therefore, the different storage technologies have to be assessed in close context to their application purpose within an overall analysis.

References

1 Tietze, V. and Luhr, S. (2013) Near-surface bulk storage of hydrogen, in *Transition to Renewable Energy Systems* (eds D. Stolten and V. Scherer), Wiley-VCH Verlag GmbH, Weinheim, pp. 659–690.
2 Crotogino, F., Donadei, S., Bünger, U. et al. (2010) Large-scale hydrogen underground storage for securing future energy supplies, in *18th World Hydrogen Energy Conference 2010 – WHEC 2010: Parallel Sessions Book 4: Storage Systems/ Policy Perspectives, Initiatives and Cooperations* (eds D. Stolten and T. Grube), Forschungszentrum Jülich GmbH, Essen, pp. 37–45.
3 Carpetis, C. (1988) Storage, transport and distribution of hydrogen, in *Hydrogen as an Energy Carrier: Technologies, Systems, Economy* (eds C.J. Winter and J. Nitsch), Springer Verlag, pp. 249–290.
4 Blandow, V., Schmidt, P., Weindorf, W. et al. (2005) Earth & Space-Based Power Generation Systems – A Comparison Study. A Study for ESA Advanced Concepts Team, Final Report. LBST.
5 Takahashi, K. (2009) Hydrogen storage, in *Energy Carriers and Conversion Systems* (ed. T. Ohta), Eolss Publishers, Oxford, UK.
6 Shibata, T., Yamachi, H., Ohmura, R. et al. (2012) Engineering investigation of hydrogen storage in the form of a clathrate hydrate: conceptual designs of underground hydrate-storage silos. *Int. J. Hydrogen Energy*, **37**, 7612–7623.
7 Strobel, T.A., Hester, K.C., Koh, C.A. et al. (2009) Properties of the clathrates of hydrogen and developments in their applicability for hydrogen storage. *Chem. Phys. Lett.*, **478** (4–6), 97–109.
8 Nakayama, T., Tomura, S., Ozaki, M. et al. (2010) Engineering investigation of hydrogen storage in the form of clathrate hydrates: conceptual design of hydrate production plants. *Energy Fuel.*, **24** (4), 2576–2588.
9 Alhumaidan, F., Cresswell, D., and Garforth, A. (2011) Hydrogen storage in liquid organic hydride: producing hydrogen catalytically from methylcyclohexane. *Energy Fuel.*, **25** (10), 4217–4234.
10 Newson, E., Haueter, T., Hottinger, P. et al. (1998) Seasonal storage of hydrogen in stationary systems with liquid organic hydrides. *Int. J. Hydrogen Energy*, **23**, 905–909.
11 Huang, H., Yu, Y., and Chung, K.H. (2012) Seasonal storage of electricity by hydrogen in benzene–water system. *Int. J. Hydrogen Energy*, **37** (17), 12798–12804.
12 Disselkamp, R.S. (2008) Energy storage using aqueous hydrogen peroxide. *Energy Fuel.*, **22** (4), 2771–2774.
13 Horwitz, J. (2009) *Developments in Hydrogen Storage*, Pira International, Leatherhead.
14 Barbier, F. (2010) Hydrogen distribution infrastructure for an energy system: present status and perspectives of technologies, in *Hydrogen and Fuel Cells: Fundamentals, Technologies and Applications* (ed. D. Stolten), Wiley-VCH Verlag GmbH, Weinheim, pp. 121–148.
15 Ajayi-Oyakhire, O. (2012) Hydrogen – Untapped Energy?, Institution of Gas Engineers and Managers. Available at

http://www.igem.org.uk/media/232929/Hydrogen-Report-Complete-web.pdf (accessed 12 March 2013).

16 Larsen, H., Feidenhans'l, R., and Petersen, L.S. (eds) (November 2004) Risø Energy Report 3 – Hydrogen and its Competitors, Risø National Laboratory. Available at http://www.cphcleantech.com/media/149597/ris%C3%B8%20energy%20report%203%20-%20hydrogen%20and%20its%20competitors_2004.pdf (accessed 19 March 2012).

17 U.S. Department of Energy - Energy Efficiency and Renewable Energy (2012) Hydrogen Delivery (updated September 2012). Hydrogen, Fuel Cells & Infrastructure Technologies Program. Multi-Year Research, Development and Demonstration Plan. Planned Program Activities for 2005–2015. https://www1.eere.energy.gov/hydrogenandfuelcells/mypp/pdfs/delivery.pdf (accessed 6 December 2012).

18 Stolten, D., Grube, T., and Mergel, J. (2012) Beitrag elektrochemischer energietechnik zur energiewende. Innovative Fahrzeugantriebe 2012, VDI-Berichte 2183, 8. VDI-Tagung.

19 U.S. Department of Energy – Energy Efficiency and Renewable Energy (2005) Hydrogen Delivery Technology Roadmap http://www1.eere.energy.gov/vehiclesandfuels/pdfs/program/delivery_tech_team_roadmap.pdf (accessed 10 January 2013).

20 U.S. Department of Energy – Energy Efficiency and Renewable Energy (2009) Hydrogen Storage (updated September 2011), in Hydrogen, Fuel Cells & Infrastructure Technologies Program. Multi-Year Research, Development and Demonstration Plan. Planned Program Activities for 2005–2015. https://www1.eere.energy.gov/hydrogenandfuelcells/mypp/pdfs/storage.pdf (accessed 6 December 2012).

21 Maus, S., Hapke, J., Ranong, C.N. et al. (2008) Filling procedure for vehicles with compressed hydrogen tanks. Int. J. Hydrogen Energy, 33 (17), 4612–4621.

22 Nexant Inc. (2010) Hydrogen Delivery Infrastructure Options Analysis. Final Report, DOE Award Number: DE-FG36-05GO15032, http://www1.eere.energy.gov/hydrogenandfuelcells/pdfs/delivery_infrastructure_analysis.pdf (accessed 16 February 2013).

23 Krieg, D. (2012) Konzept und Kosten eines Pipelinesystems zur Versorgung des deutschen Strassenverkehrs mit Wasserstoff. PhD thesis, Schriften des Forschungszentrums Jülich. Reihe Energie und Umwelt/energy and environment, Band/Volume 144, Jülich, 1–288.

24 Nexant Inc. et al. (May 2008) H2A Hydrogen Delivery Infrastructure Analysis Models and Conventional Pathway Options Analysis Results. Interim Report, http://www1.eere.energy.gov/hydrogenandfuelcells/pdfs/nexant_h2a.pdf.

25 Kevin, M. and Warren, V. (2012) Hydrogen infrastructure: production, storage, and transportation, in Hydrogen Energy and Vehicle Systems (ed. S.E. Grasman), Green Chemistry and Chemical Engineering, CRC Press, Boca Raton, FL, pp. 23–44.

26 Carpetis, C. (1994) Technology and cost of hydrogen storage. TERI Information Digest Energy Environ., 4 (1), 1–13.

27 McWhorter, S., Read, C., Ordaz, G. et al. (2011) Materials-based hydrogen storage: attributes for near-term, early market PEM fuel cells. Curr. Opin. Solid State Mater. Sci., 15 (2), 29–38.

28 Broom, D.P. (2011) Hydrogen Storage Materials: The Characterisation of Their Storage Properties, Springer, London.

29 Marrero-Alfonso, E.Y., Beaird, A.M., Davis, T.A. et al. (2009) Hydrogen generation from chemical hydrides. Ind. Eng. Chem. Res., 48 (8), 3703–3712.

30 Berry, G.D., Martinez-Frias, J., Espinosa-Loza, F. et al. (2004) Hydrogen storage and transportation, in Encyclopedia of Energy (ed. J.C. Cutler), Elsevier, New York, pp. 267–281.

31 Godula-Jopek, A., Jehle, W., and Wellnitz, J. (2012) Hydrogen Storage Technologies: New Materials, Transport and Infrastructure, Wiley-VCH Verlag GmbH, Weinheim.

32 Zheng, J., Liu, X., Xu, P. et al. (2012) Development of high pressure gaseous hydrogen storage technologies. Int. J. Hydrogen Energy, 37 (1), 1048–1057.

33 Hydrogen Strategy Group of the Federal Ministry of Economics and Labour (January 2005) Strategy Report on Research Needs in the Field of Hydrogen Energy Technology. http://www.wiba.de/download/SKH2_english_01022005.pdf.

34 Air Products (2012) Typical Bulk Liquid Storage Systems. http://www.airproducts.com/products/Gases/supply-options/bulk-deliveries-and-storage-systems/typical-bulk-liquid-storage-systems.aspx (accessed 22 December 2012).

35 Barthélémy, H. (2012) Hydrogen storage – industrial prospectives. *Int. J. Hydrogen Energy*, **37** (22), 17364–17372.

36 ASME (2005) *Hydrogen Standardization Interim Report for Tanks, Piping, and Pipelines*, ASME Standards Technology, New York.

37 Wolf, J. (2003) Die neuen Entwicklungen der Technik – Wasserstoff-Infrastruktur: von der Herstellung zum Tank, in Linde Technology 2/2003 (ed. A. Belloni), A.G. Linde, Wiesbaden, pp. 20–25.

38 Reitzle, W. (2003) Eine Vision für die Wirtschaft 4 – Potenziale und Marktchancen von Wasserstoff, in Linde Technology 2/2003 (ed. A. Belloni), A.G. Linde, Wiesbaden, pp. 4–10.

39 Krewitt, W. and Schmid, S. (September 2005) CASCADE Mints, WP 1.5 Common Information Database, D 1.1 Fuel Cell Technologies and Hydrogen Production/Distribution Options. Final Report. DLR, Stuttgart. http://www.dlr.de/fk/Portaldata/40/Resources/dokumente/publikationen/2005-09-02_CASCADE_D1.1_fin.pdf (accessed 16 February 2013).

40 Huss, A. (March 2013) Wind Power and Hydrogen: Complementary Energy Sources for Sustainable Energy Supply. The Town of Herten Leads the Field with its Innovative Wind Power Electrolysis Plant. White Paper, Anwenderzentrum h2herten GmbH.

41 Felbaum, J. (2012) An Overview of CNG Storage Options, July 2012 http://blog.fibatech.com/cng/cng-storage/ (accessed 19 February 2013).

42 FIBA Technologies, Inc. (2013) ASME Ground Storage. http://www.fibatech.com/~fibatech/sites/default/files/clients/ACME_Co/asme_tubes_brochure.pdf (accessed 3 July 2012).

43 Anwenderzentrum h2herten GmbH (2000) Pressebild, Anwenderzentrum mit Wasserstofftank. http://www.infraserv.com/de/aktuelles/pressefotos/index.html (accessed 15 December 2013).

44 Infraserv GmbH & Co. Höchst KG (2000) Anlagen im Industriepark Höchst, Wasserstoffgasometer. http://www.infraserv.com/de/aktuelles/pressefotos/index.html (accessed 20 December 2012).

45 Langner, T., Küster, M., and Müller-Kirchenbauer, J. (2013) Lokale Erdgasspeicheranlagen in Deutschland - Teil 1. *Energ. Wasser-Praxis*, **2** (64), 54–58.

46 Rummich, E. (2011) *Energiespeicher: Grundlagen- Komponenten- Systeme und Anwendungen*, Expert-Verlag GmbH, Renningen.

47 Sven Bardua (2012) Gasbehälter und Gaswerke in Deutschland. industrie-kultur. Magazin für Denkmalpflege, Landschaft, Sozial-, Umwelt- und Technikgeschichte. Available at www.industrie-kultur.de/index.php?module=html01pages&func=display&pid=90 (accessed 20 November 2012).

48 Industriepark Höchst (2012) Innovation. Visionary Ideas from Industriepark Höchst. http://www.industriepark-hoechst.com/en/index/industriepark/innovation.htm (accessed 17 December 2012).

49 Wuppertaler Stadtwerke (2012) Pressebilder, Gaskugel in Sonnborn. http://www.wsw-online.de/unternehmen/presse/Fotoarchiv/Energie_Wasser/Startseite_Energie_Wasser (accessed 20 December 2012).

50 Erdgas Zürich AG (2012) Röhrenspeicher Urdorf, Tag der offenen Baustelle, Medienmitteilungen. http://www.erdgaszuerich.ch/de/netzbau-planauskunft/urdorf/tag-der-offenen-baustelle.html (accessed 20 December 2012).

51 Ackermann, W., Gutsme, H., Jerusalem, E.G. *et al.* (1975) Bau von Hochdruckkugelgasbehältern in Berlin Charlottenburg. Der Stahlbau, Jahrgang 44 (Heft 11), pp. 321–330.

52 Arenz, B. (2008) *Netzmeister: Technisches Grundwissen Gas Wasser Fernwärme Bereich Umwelt*, Oldenbourg Industrieverlag, München.

53 Chicago Bridge & Iron Company N.V. (2012) History. http://www.cbi.com/about-cbi/history/ (accessed 23 November 2012).

54 Weber, M. and Perrin, J. (2008) Hydrogen transport and distribution, in *Hydrogen Technology* (ed. A. Léon), Springer, Berlin, Heidelberg, pp. 129–149.

55 Häussinger, P., Lohmüller, R., and Watson, A.M. (2000) Hydrogen, in *Ullmann's Encyclopedia of Industrial Chemistry*, Wiley-VCH Verlag GmbH, Weinheim.

56 Padro, C.E.G. and Putsche, V. (September 1999) Survey of the Economics of Hydrogen Technologies. NREL/TP-570-27079. National Renewable Energy Laboratory. Available at http://www1.eere.energy.gov/hydrogenandfuelcells/pdfs/27079.pdf.

57 Amos, W.A. (1998) Costs of Storing and Transporting Hydrogen. NREL/TP-570-25106. National Renewable Energy Laboratory. Available at http://www.nrel.gov/docs/fy99osti/25106.pdf.

58 Hart, D. (1997) *Hydrogen Power: The Commercial Future of the Ultimate Fuel*, Financial Times Energy Publishing, London.

59 Baars-Werner, I. (2010) Heilbronner Gaskugel wird im November demontiert. http://www.stimme.de/heilbronn/nachrichten/stadt/sonstige-Heilbronner-Gaskugel-wird-im-November-demontiert;art1925,1948974 (accessed 30 January 2013).

60 Anon. (2011) Heilbronner Gaskugel vor Abriss im Frühjahr? http://www.stimme.de/heilbronn/nachrichten/stadt/Heilbronner-Gaskugel-vor-Abriss-im-Fruehjahr;art1925,2026437 (accessed 30 January 2013).

61 Frühschütz, O. (2012) Allgemeines über die Gasbehälter, Gasometer oder Gaskessel. www.gaswerk-augsburg.de/gasbehaelter.html (accessed 20 December 2012).

62 Möller, A. and Niehörster, C. (2003) Optimierung des Gasbezugs durch Röhrenspeicher. ET Energiewirtschaftliche Tagesfragen 6/2003. Büro for Energiewirtschaft und technische Planung GmbH, Aachen. Available at http://www.bet-aachen.de/fileadmin/redaktion/PDF/Veroeffentlichungen/2003/BET-Artikel_Gasbezug_0306.pdf.

63 Fuoli, F. (2012) Für 21 Millionen entsteht ein riesiger Erdgasspeicher. Limmattaler Zeitung, 15 August 2012. Available at http://www.limmattalerzeitung.ch/limmattal/region-limmattal/fuer-21-millionen-entsteht-ein-riesiger-erdgasspeicher-125016057.

64 Erdgas Zürich AG (2012) Baustart des Erdgas-Röhrenspeichers in Urdorf, Medienmitteilung. http://www.erdgaszuerich.ch/medien/archiv/mm-1142012-baustart-des-erdgas-roehrenspeichers-in-urdorf.html (accessed 19 September 2012).

65 Erdgas Zürich AG (2012) Erdgas Röhrenspeicher in Urdorf. Technische Daten http://www.erdgaszuerich.ch/fileadmin/media/ueber_uns/publikationen/produktebroschueren/Roehrenspeicher-Urdorf_Technische-daten.pdf (accessed 19 September 2012).

66 Kühn, U. (2008) Safe and Cost-Effective Pipe Storage Facility in Bocholt (Germany). Available at http://www.tuev-sued.de/uploads/images/1207896537119176350047/Pipe_storage_facility_Bocholt.pdf.

67 Erdgas Zürich AG, Erdgas erfolgreich speichern, Medienmitteilung. http://www.erdgaszuerich-transport.ch/de/medien/mm-1482012-erdgas-erfolgreich-speichern.html (accessed 16 February 2013).

68 Thistlethwaite, C., Jones, R., and Holmes, N. (June 2012) Potential for Grid-Scale Storage of Hydrogen in on and Offshore Concrete Structures, Scottish Hydrogen and Fuel Cell Association, 30 pp.

69 Kottenstette, R. and Cotrell, J. (2004) Hydrogen storage in wind turbine towers. *Int. J. Hydrogen Energy*, **29** (12), 1277–1288.

70 Gao, M. and Krishnamurthy, R. (2009) Hydrogen transmission in pipelines and storage in pressurized and cryogenic tanks, in *Hydrogen Fuel: Production, Transport, and Storage*

(ed. R.B. Gupta), CRC Press, Boca Raton, FL, pp. 341–380.
71 Berstad, D.O., Stang, J.H., and Nekså, P. (2009) Comparison criteria for large-scale hydrogen liquefaction processes. *Int. J. Hydrogen Energy*, **34** (3), 1560–1568.
72 Klell, M. (2010) Storage of hydrogen in the pure form, in *Handbook of Hydrogen Storage: New Materials for Future Energy Storage* (ed. M. Hirscher), Wiley-VCH Verlag GmbH, Weinheim, pp. 1–38.
73 Leachman, J.W., Jacobsen, R.T., Penoncello, S.G. et al. (2009) Fundamental equations of state for parahydrogen, normal hydrogen, and orthohydrogen. *J. Phys. Chem. Ref. Data*, **38** (3), 721–748.
74 HySafe (June 2007) Biennal Report on Hydrogen Safety, Chapter 1: Hydrogen Fundamentals. Available at http://www.hysafe.org/download/1196/BRHS_Chap1_V1p2.pdf.
75 Leachman, J. (2007) Fundamental Equations of State for Parahydrogen, Normal Hydrogen, and Orthohydrogen, MSc Thesis, College of Graduate Studies, University of Idaho.
76 Quack, H. (2001) Conceptual design of a high efficiency large capacity hydrogen liquefier. *Adv. Cryogenic Eng.*, **47**, 255–263.
77 Linde Kryotechnik AG (2005) Hydrogen Liquefiers. http://www.linde-kryotechnik.ch/1259/1260/1308/1309.asp (accessed 28 December 2012).
78 Léon, A. (2008) Hydrogen storage, in *Hydrogen Technology: Mobile and Portable Applications* (ed. A. Léon), Springer, Berlin, pp. 82–128.
79 Krasae-in, S., Stang, J.H., and Neksa, P. (2010) Development of large-scale hydrogen liquefaction processes from 1898 to 2009. *Int. J. Hydrogen Energy*, **35** (10), 4524–4533.
80 Rigas, F. and Amyotte, P. (2012) *Hydrogen Safety*, Taylor & Francis, Boca Raton, FL.
81 Praxair (2012) Bulk Supply Systems for Nitrogen, Oxygen, Argon and Hydrogen. http://www.praxair.com/praxair.nsf/0/6c0316bc10aa86518525654b00583771/$FILE/BulkSupply.pdf (accessed 20 December 2012).
82 National Aeronautics and Space Administration (2009) Kennedy Media Gallery. mediaarchive.ksc.nasa.gov/detail.cfm?mediaid=48266 (accessed 12 December 2012).
83 The Linde Group (2007) Linde Hydrogen Center. http://www.the-linde-group.com/de/news_and_media/image_library/index.html?imageLibraryCategory=Linde+Gas (accessed 20 December 2012).
84 Linde AG (2007) Linde Hydrogen Center in Munich, Germany. Liquid and gaseous hydrogen fuelling station. http://www.linde-gas.com/internet.global.lindegas.global/en/images/Linde%20Hydrogen%20Center%20in%20Munich,%20Germany17_15303.pdf (accessed 28 December 2012).
85 Linde AG (2006) CEP/TOTAL at Heerstrasse, Berlin, Germany. Liquid and gaseous hydrogen fuelling station. http://www.linde-gas.com/internet.global.lindegas.global/en/images/CEP-TOTAL%20hydrogen%20fuelling%20stations%20at%20Heerstrasse,%20Berlin,%20Germany.17_15299.pdf (accessed 18 December 2012).
86 European Commission (November 2011) Fuel Cells and Hydrogen Joint Undertaking (FCH-JU). Multi-Annual Implementation Plan 2008–2013. Document FCH JU 2011 D708. http://www.fch-ju.eu/sites/default/files/MAIP%20FCH-JU%20revision%202011%20final.pdf.
87 Taylor, J.B., Alderson, J.E.A., Kalyanam, K.M. et al. (1986) Technical and economic assessment of methods for the storage of large quantities of hydrogen. *Int. J. Hydrogen Energy*, **11** (1), 5–22.
88 Carpetis, C. (1980) A system consideration of alternative hydrogen storage facilities for estimation of storage costs. *Int. J. Hydrogen Energy*, **5** (4), 423–437.
89 Carpetis, C. (1982) Estimation of storage cost for large hydrogen storage facilities. *Int. J. Hydrogen Energy*, **7** (2), 191–203.
90 Venter, R.D. and Pucher, G. (1997) Modelling of stationary bulk hydrogen storage systems. *Int. J. Hydrogen Energy*, **22** (8), 791–798.
91 Altmann, M., Schmidt, P., Wurster, R. et al. (March 2004) Potential for Hydrogen as a Fuel for Transport in the Long Term (2020 - 2030). Full

Background Report, EUR 21090 EN. European Commission, http://www.europarl.europa.eu/stoa/webdav/shared/3_activities/energy_technology/hydrogen/ipts_fuel_en.pdf (accessed 4 February 2015).

92 Umwelt Bundesamt (2012) Trends der Lufttemperatur. http://www.umweltbundesamt-daten-zur-umwelt.de/umweltdaten/public/theme.do?nodeIdent=2355 (accessed 16 February 2013).

93 Umwelt Bundesamt (2013) Klimadaten und Bodentemperatur. http://www.umweltbundesamt.de/boden-und-altlasten/altlast/web1/berichte/langzeit/langzeit-3.5.html (accessed 14 January 2012).

94 ENREGIS (2012) Temperatur des Bodens. http://www.enregis.de/pop_popupshowimg.htm?img=auftritt/daten/bilder/popup/Waerme_Temperaturverteilung_gr.jpg&x=698&y=750&print=0 (accessed 14 January 2012).

95 Berstad, D., Stang, J., and Nekså, P. (2009) A Future Energy Chain Based on Liquefied Hydrogen. SINTEF Energy Research, http://www.sintef.no/Publikasjonssok/Publikasjon/Download/?pubid=SINTEF+S13047 (accessed 4 February 2015).

28
Hydrogen Storage in Vehicles
Jens Franzen, Steffen Maus, and Peter Potzel

28.1
Introduction: Requirements for Hydrogen Storage in Vehicles

The use of hydrogen in a fuel cell drive train of a vehicle combines the possibility of emissions-free driving with a high range potential while simultaneously permitting quick refueling. This combination makes the development of corresponding vehicles an even greater priority. The tank system plays a crucial role in this and has a significant influence on the range and refueling time. Hydrogen is predominantly stored in a compressed form in passenger vehicles.

Users of modern vehicles have become accustomed to ranges of several hundred kilometers, which particularly have an effect on long-distance driving. A vehicle with an alternative drive train must have equivalent ranges for many customers to find it attractive. More importantly, it will take some time before the filling station network has the same density as the current network of conventional filling stations.

From the customer's point of view, the required refueling time is based on what they are used to with conventional vehicles. A refueling time that is several factors greater than the time they currently spend at a filling station can be a reason for the customer to reject a vehicle with an alternative drive train. Therefore, there is a refueling standard for tank systems that are operated with compressed hydrogen at a working pressure of 70 MPa that enables a refueling within 3 min (Section 28.4).

A vehicle with a compressed hydrogen tank system must be able to operate under various climatic conditions. The Mercedes Benz B-Class F-Cell has already been sold in markets such as California and Norway, which can experience extreme temperatures. The diversity of markets for these vehicles should increase, meaning that the design temperatures for a tank system that lie between −40 and 85 °C may occur more frequently during operation of the vehicle. Notably, the exhaust of a fuel cell system, when operated on dark asphalt in hot countries in summer, can result in these high temperatures in the tank system. It must also be possible to operate at very low outside temperatures. This range of temperatures must be taken into account in the design and validation of

Hydrogen Science and Engineering: Materials, Processes, Systems and Technology, First Edition.
Edited by Detlef Stolten and Bernd Emonts.
© 2016 Wiley-VCH Verlag GmbH & Co. KGaA. Published 2016 by Wiley-VCH Verlag GmbH & Co. KGaA.

the tank system. A decisive aspect of this is the tightness of elastomer seals at high pressures and simultaneously low temperatures as well as the permeation of hydrogen through plastics at higher temperatures.

The leakage and permeation behavior determines the tightness of the system in the different operating states. One practicable method for verifying the tightness in the vehicle is to use leak detection spray to check that all junctions of the system are free of bubbles. This method is the reason that the tank system is required to be bubble-free. This is checked regularly, both during production and in the workshop. In production, however, a check using leak detection spray is only practicable in very small production runs. If tank systems are produced in larger production runs, then it will be a challenge to develop a reliable, quick leak detection. One possibility for this is to use concentration measurements. Since it is often possible for there to be several leakage points in the region being tested, the sum of all leaks is determined. The permissible leakage quantity of the sum of all leaks is naturally greater than the leakage quantity permitted for a single leak. There is a danger here of overlooking small leaks. In addition it is often then necessary to also determine their location. The total leakage quantity in the vehicle during operation is determined through concentration measurement by means of H2 sensors.

Operation of the vehicle subjects the tank system to vibration and impacts. But these must not be allowed to influence the tightness or the function of the tank system in any way. This must be verified through corresponding mechanical tests during the development process of the tank system. The leakage quantity is determined in the course of this; in addition, it is not permissible for tank system valves to unintentionally close or for similar malfunctions to occur.

A tank system for high-pressure hydrogen will undergo extensive safety testing. This begins at the subcomponent level. The containers are designed so that their bursting pressure is at least 157.5 MPa (2.25 × 70 MPa) [1,2]; the other components have even higher bursting pressures. As part of their certification process, they are also cycled with a multiple of the number of expected refueling operations. An important development goal with the containers is the so-called "leak before break" behavior, that is, if the container is damaged, it is permissible for it to leak, but under no circumstances can it be permitted to burst and thus abruptly release the stored mechanical energy. This behavior is checked, among other things, with a ballistics test. This simulates the penetration by the point of a sharp vehicle component in a serious crash. Here, too, it is permissible for the hydrogen to leak after the ballistics test, but the container must not under any circumstances be permitted to burst. Particular attention must be paid to fire safety. Without additional safety equipment, the container would be mechanically weakened in the event of a fire. At the same time, the heat would increase the gas pressure. This would cause the container to burst. For this reason, a thermally controlled pressure relief valve is provided, which opens at a certain excess temperature and releases the hydrogen via an additional conduit (see also Section 28.4.1). Finally, a vehicle with a high-pressure hydrogen tank system must pass the same crash tests as a conventional vehicle. No tank system leakage is permitted.

The customer who drives such a vehicle should have by and large the same driving experience as with a conventional vehicle. For the tank system, in addition to meeting the range requirements and providing a simple, quick refueling, this also means avoiding irritating noises such as flow noise when refueling and during operation. In addition, valves must operate silently.

The service life of a tank system for high-pressure hydrogen can be up to 20 years. It is certified for this value, for example, in the European Union [1,2]. This requires regular testing, but legally mandated testing methods have not yet been finalized.

Last but not least, there is one more important criterion for customer acceptance of such a vehicle. The price for the drive train must be comparable to that of a conventional vehicle.

28.2
Advantages of Pressurized Storage over Other Storage Methods

Hydrogen can be stored in various ways. When stored in compressed form, hydrogen for use in vehicles is typically compressed to 35 or 70 MPa and stored in cylindrical high-pressure containers. The storage density of pure hydrogen is then approx. 25 or approx. $40\,\text{kg}\,\text{m}^{-3}$.

Another method is storage in liquid form. In this case, the hydrogen has a temperature of approx. 20 K and a density of approx. $70\,\text{kg}\,\text{m}^{-3}$. The low temperature requires an expensive insulation of the container since, otherwise, heat input would cause the pressure in the container to rise to impermissible levels due to the evaporation of hydrogen. Since heat input into the container cannot be entirely prevented, hydrogen must be discharged from the container after a certain amount of time (boil-off). Thereby, the pressure is reduced and the stored hydrogen is cooled due to evaporation. The container of such a tank system is likewise cylindrical since this achieves a favorable volume/surface area ratio, which is convenient for minimizing heat input. The storage in liquid form can also be combined with high-pressure storage as a hybrid system in order to achieve greater storage densities [3].

Hydrogen can also be stored in a chemical form. In the past, metal hydrides have been tested, which can absorb hydrogen even at room temperature. The main disadvantage lies in the low material storage density of 1–2 mass% [4]. For this reason, efforts in past years were focused on developing materials that enable significantly higher storage densities. But these require a higher temperature level for an optimum reaction speed and have higher reaction enthalpies. On the one hand, the reaction enthalpy leads to heat release during the filling process while, on the other hand, the hydride must be heated to release hydrogen. As a result, the temperature level of the fuel cell thermal discharge is not high enough to permit the endothermic hydrogen release in the container and the chemical reaction comes to a standstill again. The level of the reaction enthalpy is of decisive importance in refueling. It is exothermic, that is, the reaction heat must be dissipated during the refueling. On the one hand, this results in the need for expensive heat exchanger structures inside the tank system. These must be capable of dissipating several 100 kW of thermal output from the tank system if the system is to be refueled in a matter of minutes. On the other hand, the heat at the filling station can then be processed further and/or dissipated into the environment. The focus in storage material development is therefore on reducing the reaction enthalpy and temperature of materials that have an intrinsically high storage density. When determining the storage density of a metal hydride storage device, it is also necessary to take into account the complex tank system geometry (including the heat exchanger). This reduces the system storage density by a factor of 1.5–2 in comparison to the pure material [5].

The above-mentioned characteristics of the different storage methods have various effects on the operation of such a storage system in a vehicle. The operation of the tank system within its specified temperature range does not play a significant role in hydrogen removal in compressed hydrogen and liquid hydrogen storage systems. In compressed hydrogen systems, it is important to note that during the removal the hydrogen cools as it expands, that is, during operation at the lower limit of the specified temperature the container and the valve can cool further.

The operation of a metal hydride-based hydrogen tank system at high ambient temperatures requires a high performance of the heat management system in the storage device. An excessively high temperature in the metal hydride during refueling results in excessively long refueling times. Overcooling of the metal hydride during the hydrogen removal decreases the possible removal rate.

For a liquid hydrogen container the hydrogen release when the vehicle is parked for longer periods is significant due to the heat input into the container (boil-off). In small tank systems with an unfavorable volume/surface area ratio, a boil-off occurs after 3–4 days. The hydrogen loss in this case is approx. 2–3 vol.% per day [6].

The service life of a compressed hydrogen or liquid hydrogen tank system can be verified by testing, provided that the subcomponents are correctly designed. In metal hydride storage systems, the cycle resistance of the materials is also important. The material properties of many hydrides degrade with frequent hydration; this reduces the storage capacity of the material and reduces the reaction speed, thus reducing the hydrogen absorption in the tank system.

The design for high internal pressures results in a considerable wall thickness of the container in high-pressure storage systems. This results in a high mechanical stability relative to external loads, which has positive effects on crash safety. The "leak before break" behavior of such a container results in a hydrogen release without bursting. The hydrogen can disperse very quickly in air. The container of a liquid hydrogen system is not as stable and is therefore more sensitive to external loads. In the metal hydride storage device, the hydrogen is chemically bonded. Depending on the temperature level of the metal hydride, a hydrogen release can take place if the container is destroyed, but comes to a halt for lack of heat input, provided that it does not burn. The basic prerequisite for

Table 28.1 Storage densities of previously implemented tank systems for passenger vehicles. Taken from Reference [5].

	Approximate gravimetric storage density (mass%)	Volumetric storage density (kg l^{-1})
Compressed hydrogen tank system (3 containers, 70 MPa)	4	0.025
Liquid hydrogen tank system (1 container)	8	0.05
Metal hydride	1–2	0.03

this, however, is that the metal hydride, which is in finely powdered form, does not ignite in air.

The storage density potential is highest in liquid hydrogen-based storage devices (Table 28.1) [5]. This is especially true of large storage systems. For small hydrogen tank systems of the kind particularly used in passenger vehicles, this advantage decreases as compared to compressed hydrogen systems. This is particularly the case when the vehicle geometry makes it necessary to provide a multi-container system. At present, metal hydride storage devices have the lowest storage density potential. In this case, the gravimetric storage density is a criterion for excluding use in vehicles.

28.3
Design of a Tank System

28.3.1
Flow Diagram and Description of the Components

A tank system for compressed hydrogen (Figure 28.1) must perform numerous secondary functions in order to achieve its primary intended use as described in the preceding sections. As a result, in addition to the main component – the actual container (component 1), whose design is described in greater detail in Section 28.3.2 – various other components are needed. A central function is performed by the tank valve unit (component 2), which is typically screwed directly into the container or containers. It integrates an array of safety-related functions. First, this is the actual opening of the container by means of an electrical signal and the closing of it as soon as the vehicle is switched off or as soon as the vehicle's power train does not require any gas. This ensures that in case of a leak while the vehicle is at a standstill, the gas remains confined in the containers; in addition, when a malfunction is detected during operation, the containers can be closed and an escape of hydrogen from the containers can be prevented. As an additional fall-back arrangement in the event of a failure of the electrical valve function, mechanical excess flow valves (EFVs) are provided, which largely shut

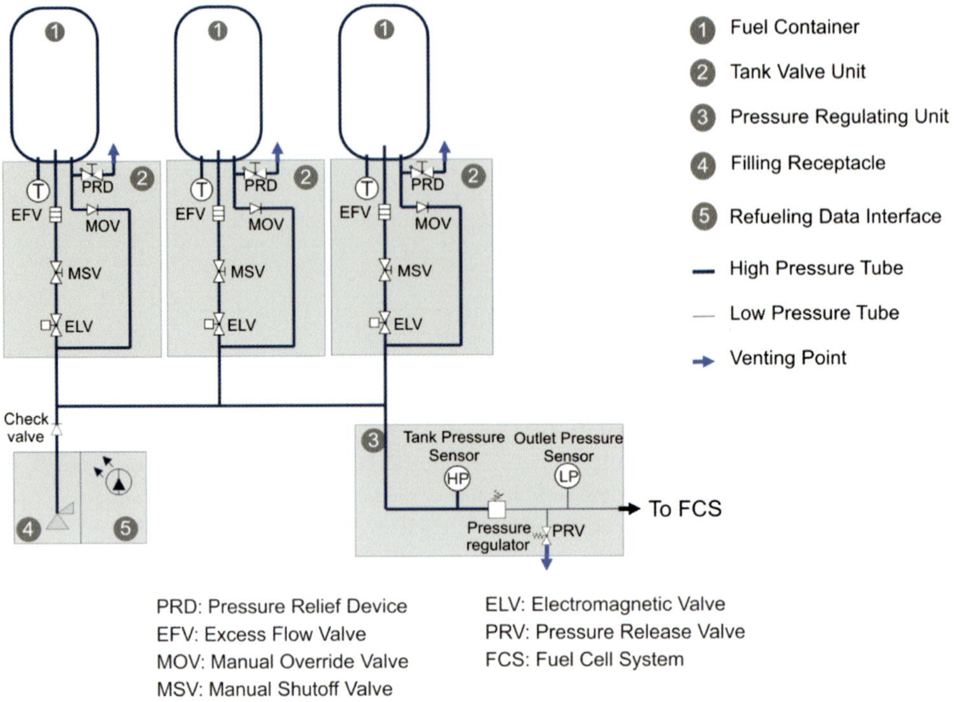

Figure 28.1 Flow diagram of a tank system for compressed hydrogen.

PRD: Pressure Relief Device
EFV: Excess Flow Valve
MOV: Manual Override Valve
MSV: Manual Shutoff Valve
ELV: Electromagnetic Valve
PRV: Pressure Release Valve
FCS: Fuel Cell System

off the container in the event of an excessively rapid gas removal, for example, a line rupture. By law (Section 28.4.1), each container must be equipped with a pressure relief device (PRD), which in the event of a fire empties the container and thus prevents it from bursting. This is usually implemented by means of a liquid-filled piston of the kind known from sprinkler systems. At a defined temperature, the liquid vaporizes and the piston bursts, thus releasing a mechanism that empties the container. During operation of the vehicle, typically all of the containers are opened simultaneously and gas is removed from all of the containers uniformly so that their pressures are each at the same level. This prevents gas from flowing back and forth due to pressure differences between the containers. But it is not possible to completely rule out the existence of different container pressures: due to its expansion, the gas in the container cools as it is defueled. In addition, the temperature change depends on its shape and the volume/surface area ratio. Furthermore, the containers can experience external temperature influences of different intensities, for example, through the dissipation of heat from the power train. All of these effects together result in the fact that the gas temperatures in the containers can drift apart from one another during operation of the vehicle. If the vehicle is switched off in this state of thermal imbalance and the tank valves are closed, then the temperatures equalize with one another while the vehicle is off, resulting in different pressures. The gas flow between containers that occurs upon reopening must be taken into account in the system design. The tank valve unit also includes the temperature sensor, which among other things, can be used to control the refueling (Section 28.4.2).

The third central component of the tank system is the pressure regulating unit (component 3). First of all, it regulates the tank system pressure, that is, between 1.5 and 87.5 MPa, depending on the fill level of the system, down to a constant value for supplying the power train. The design should take into account the fact that at typical temperatures hydrogen has an inverse Joule–Thompson effect, that is, the gas heats up as it is throttled by the pressure regulator. This effect results in a temperature increase of the gas of up to approx. 40 K. The pressure regulating unit also integrates sensor technology for measuring the tank pressure to determine the fill level and possibly also for controlling refueling. The gas side, which is not designed to withstand the high pressure and is located on the other side of the pressure regulator, is protected by a pressure-relief valve. It is thus possible, among other things, to limit a creeping pressure increase over a long parking period caused by a slight leakage flow through the regulating unit.

28.3.2
Container

There are basically four types of containers for storing hydrogen in the form of compressed gas in vehicles (Table 28.2); for mobile applications, only types III and IV are used for weight reasons [7]. In these container types, the actual gas container, the so-called liner, has a resin-impregnated, endless fiber wound around it, which has a low density, but high strength properties. This fiber matrix gives the container the mechanical properties required for high pressures, with a simultaneously low wall thickness and weight. In the type III container, the liner is usually made of aluminum, in the type IV container it is usually made of a polymer. Figure 28.2 shows the schematic design of the fully-wound container types.

Although the pressure container technology for storing hydrogen is in fact based on the technology of natural gas containers, because of the significantly lower energy density of hydrogen in comparison to natural gas, a nominal maximum storage pressure of 35 or 70 MPa is required to achieve an acceptable

Table 28.2 Container types for compressed gas storage [1].

Container type	Description
Type I	Seamless metallic container
Type II	Hoop wrapped container with a seamless metallic liner
Type III	Fully wrapped container with a seamless or welded metallic liner
Type IV	Fully wrapped container with a non-metallic liner

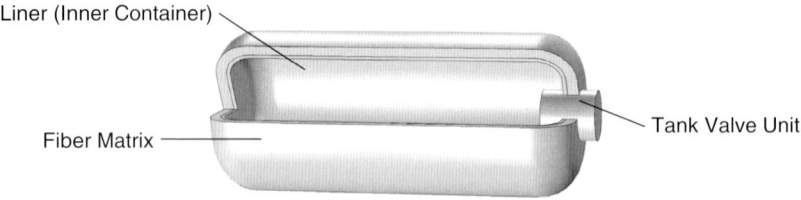

Figure 28.2 Schematic design of a fully wound type III or IV container.

range, as compared to 20 or 25 MPa with natural gas [8]. This places high demands not only on the container, but also on all other system components (Section 28.3.1).

The liner, as the actual gas container that is in direct contact with the hydrogen gas, has a decisive influence on container functionality. It guarantees gas-tightness and forms the interface for accommodating the tank valve unit. At the same time, the mechanical properties of the liner must permit expansions that occur during refueling and emptying cycles, without permitting leaks to occur (see also Section 28.4.1). Metallic liners that are used in type III containers are usually made of aluminum [9]. As compared to polymer liners, metallic liners generally have the advantage that the gas permeation is negligible [10]. In addition, a metallic liner absorbs a part of the load exerted by the internal pressure. Although aluminum is not susceptible to hydrogen embrittlement, at high pressures it is nevertheless not automatically cycle resistant [8]. Failure occurs through the growth of production-related microcracks. To increase finite-life fatigue strength, an autofrettage process is carried out. With this process, hydraulic internal pressure loads the liner material to a level greater than its elastic limit, resulting in a local flow and, after the pressure is relieved, in residual compressive stresses that limit the crack growth during operation [11]. Nevertheless, there are no type III containers currently available on the market that meet the legally mandated requirements [2] for an approval of vehicle service life with regard to cycle stability at a nominal operating pressure of 70 MPa. Polymer liners on the other hand, which are generally made of a thermoplastic, have the best mechanical properties for pressure containers, which is why only type IV containers currently meet all of the legal requirements, including cycle stability, and therefore only this container type is used in current vehicles with a storage pressure of 70 MPa. By contrast with type III containers, because of the non-negligible hydrogen permeation through plastics, certification requires performance of a permeation test, whose result must not exceed a defined limit value (e.g., <6 standard cm^3 per hour per liter water volume as per Reference [2]). Polymer liners are usually made of high-density polyethylene (HDPE) or a polyamide (PA).

For the next step of the winding, which takes place after liner production is completed, the liner, in addition to serving as a gas container, also serves as a stable core around which the winding takes place [12]. The fiber, which has been previously impregnated with resin, is laid onto the liner, which is rotated around its cylinder axis. Usually, epoxy resin is used for this because of its good mechanical properties and stability [13]. The movement of the fiber laying unit parallel to the container axis according to a specific sequence, with a simultaneous rotation of the container, winds both crosswise layers from pole to pole for absorbing the axial load and circumference layers for absorbing the circumferential forces in the cylindrical region [14]. Despite a high material price, carbon fibers are preferably used in hydrogen pressure containers because of their high rigidity for limiting expansion [14], high strength, long service life, and creep resistance [15] as compared to other fiber types such as aramid or glass fibers.

28.4
Specific Requirements for Compressed Gas Systems for Vehicles

In addition to the above-mentioned technical requirements, there are legal and normative requirements that a tank system must fulfill. Other requirements arise from the refueling process, which must meet worldwide standards.

28.4.1
Legal and Normative Requirements

There are significant differences between countries and economic regions when it comes to the approval of a high-pressure hydrogen storage system for use in a vehicle. Within the European Union and Japan, for example, there are requirements for the design and testing of the components and the system. A technical agency (e.g., the TÜV in Germany) monitors compliance with these requirements and national authorities recognize its testing protocols in the approval process. In the North American Free Trade Area (NAFTA), which includes the countries USA, Canada, and Mexico, with the USA and Canada being significant markets for hydrogen applications in vehicles, the manufacturer is responsible for having the product conform to the current state of the art. No testing by a technical agency or approving authority takes place.

In the European Union since 2012, the approval of hydrogen-powered vehicles has been contingent on obligatory compliance with regulation (EC) No 79/2009 on type-approval of hydrogen-powered motor vehicles [1] and regulation (EU) No 406/2010 for implementation of regulation (EC) No 79/2009 [2]. These regulations define, among other things, materials testing for verifying corrosion or hydrogen resistance, durability tests for verifying pressure cycle resistance, burst pressure requirements for carbon fiber-wound containers, and safe emptying of the tank system in case of fire or leak testing. Similar tests are stipulated for Japan, for example in References [16,17]. On the normative level, for tank system components and compressed gas storage systems, standards, References [18] and [19], are mentioned, for example. Chinese regulations for hydrogen compressed gas systems are currently in the process of being drafted and, up to this point, have approved only 35 MPa as the maximum storage pressure [20].

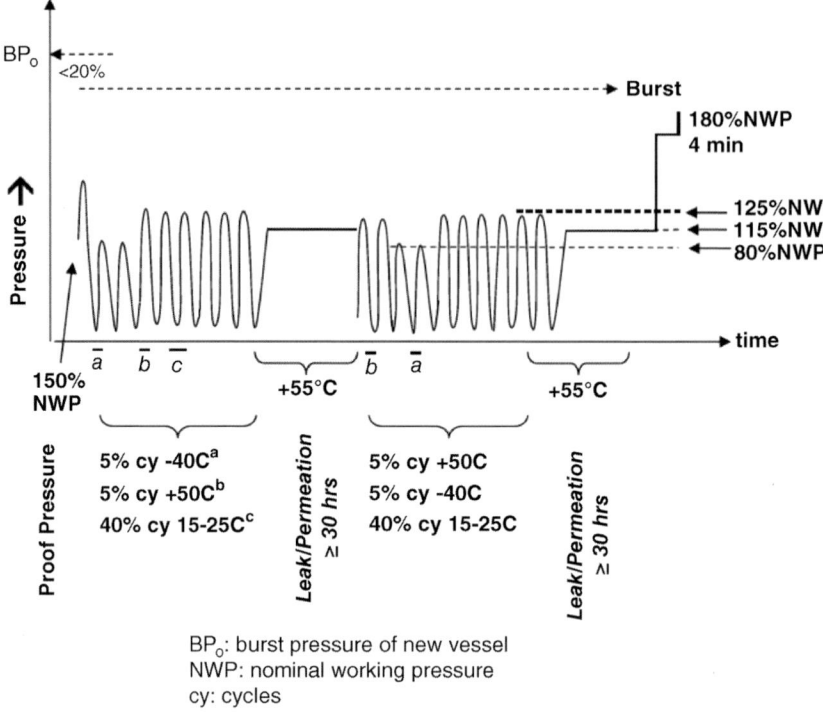

Figure 28.3 Sequence of the pneumatic gas cycle test of the Global Technical Regulation on Hydrogen and Fuel Cell Vehicles [21].

Different regulations currently exist in different countries and economic regions. Automobile manufacturers have been seeking a global harmonization of pressure storage systems for the use of hydrogen in vehicles. These efforts resulted in the ratification in 2013 at the United Nations of the Global Technical Regulation on Hydrogen and Fuel Cell Vehicles [21], whose design requirements and testing procedures are applicable for all countries participating in the convention. Figure 28.3 shows an example of a gas cycle test of the kind that must be carried out with the combination of the container and the shutoff valve to demonstrate the life cycle of the vehicle. In this case, a total of 500 pressure cycles are carried out at different ambient temperatures as a function of the nominal working pressure (NWP) and, in addition, leak and permeation inspections are carried out along with a final burst pressure test.

28.4.2
Refueling

The filling procedure in hydrogen-powered vehicles should offer the user the same convenience as current refueling with liquid fuels. There are thus requirements to be met both with regard to the sequence of the procedure, which

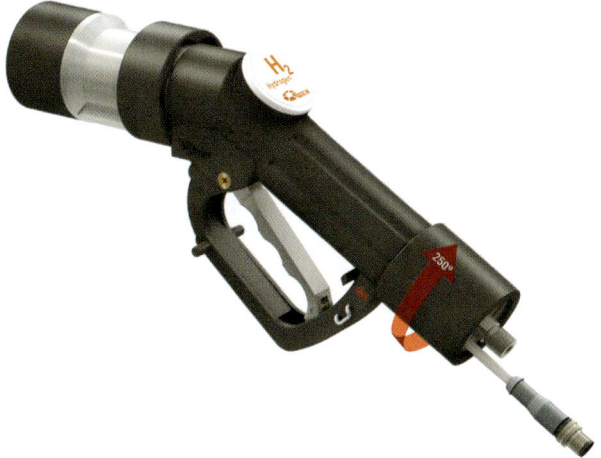

Figure 28.4 Filling receptacle for gaseous hydrogen at 700 bar. Figure from Weh GmbH.

should mirror the one that the customer is already accustomed to, and particularly with regard to the amount of time required for refueling. To achieve the first criterion, manufacturers are developing filling station components that resemble the design known from conventional filling stations (Figure 28.4).

The second requirement was fulfilled by a standardized filling procedure, which specifies a filling time of less than 3 min [22], which corresponds to those of current conventional filling stations. In addition, California law [23] specifies, among other things, a maximum filling time of 10 min for the highest category of zero-emissions vehicles.

Since the hydrogen is filled and stored in a compressed gaseous phase, a different method for determining the fill level is required than with liquid fuels. The standardized measurement quantity for compressed hydrogen tank systems is still the gas density, which can be calculated based on the pressure and temperature of the stored medium. By comparison with using only the gas pressure, this has the advantage that the indicated fill level no longer changes with the temperature; this deviation would otherwise be about 20% during normal operation, as evidenced by the comparison with CNG vehicles [24].

The compression of the gas during filling heats it in the container, thus requiring a compensation of this temperature effect [25]. The standard fill level of the system then relates to a temperature of 15 °C so that a 70 MPa tank system achieves a maximum storage density of 40 kg m^{-3}. If the gas is now heated during filling, then the target filling pressure is correspondingly adapted upward to up to 87.5 MPa at the maximum permissible temperature of 85 °C.

The sequence of a filling procedure has been largely standardized. First, a high-pressure tank is briefly opened to the vehicle until the pressure in the filling line has equalized. It is thus possible to determine the residual pressure in the vehicle's tank and to establish an estimate for its volume [26]. Based on these

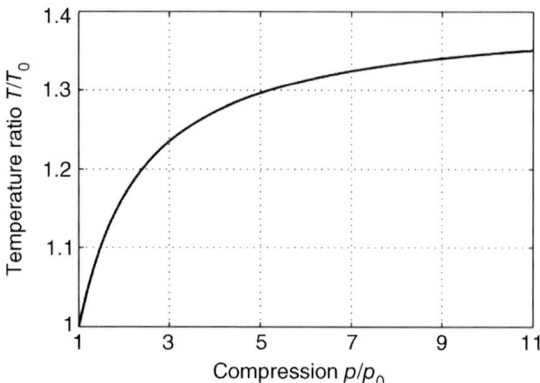

Figure 28.5 Temperature curve when filling a vehicle tank. Figure taken from Reference [28].

data, the temperature evolution of the gas is calculated and this curve is used to calculate the necessary target pressure for a complete filling [27]. The vehicle is then filled up to this target pressure.

During filling, particularly when this is correspondingly carried out in less than 3 min [22], the compression of the gas generates a large heat input. The energy release through convection and through heat conduction from the gas to the container occurs slowly in comparison to this so that one can almost assume that an adiabatic compression takes place. This results in a heating of the gas, which with a starting temperature of 20 °C, for example, results in a maximum value of approx. 140 °C [28] (Figure 28.5).

To respect the upper temperature limit of 85 °C under all ambient conditions, it is necessary to sufficiently precool the gas at the filling station; a value of −40 °C has been established, which can be used without expanding the temperature limits for tank components.

The heating of the gas in the vehicle's tank, however, depends on parameters that are not immediately accessible to the filling station. In particular, these are the initial temperature of the gas in the vehicle's tank – which does not generally have to be the same as the ambient temperature, the heat transmission properties of the container or containers, and their ratios of volume to surface area [29]. Despite the temperature compensation, the filling therefore cannot be carried out with pinpoint accuracy; to avoid an overfilling, a targeted fill level of less than 100% must be selected. Under conditions close to those of actual use, tolerances in the filling precision of up to 10% are ascertained [30]. As a consequence, tanks are on average underfilled by this 10%, which requires a tank that is this much larger in order to achieve a given range. From the technical and economic standpoint, it is therefore a good idea to counteract the temperature-related underfilling. This is achieved by transmitting to the filling station the vehicle data required for controlling the filling procedure, that is, the temperature and pressure of the gas in the vehicle's tank. With these data, the filling station can exactly determine the shut-off time and the precision of the filling

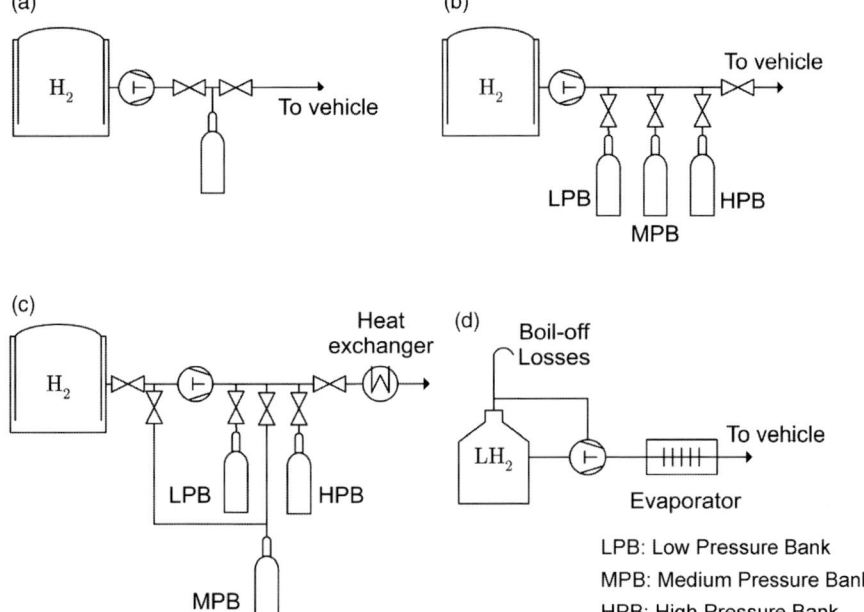

Figure 28.6 Types of common filling stations: (a) system with one bank of pressure tanks; (b) system with three banks of pressure tanks; (c) system with three banks of pressure tanks and a booster compressor; (d) liquid compressor for supercooled hydrogen. Figure taken from Reference [28].

then depends only on the tolerance of the vehicle's sensor system. This typically permits the filling precision to be reduced to less than 2%. The transmission of the data is standardized in References [22,31], making it possible to ensure functionality across various manufacturers.

The centerpiece of each filling station, however, is the technology for preparing the gas at the appropriate pressure. In this regard, different approaches have been established depending on the size of the filling station, that is, on the maximum number of vehicles that can be serviced per hour (Figure 28.6).

The technically simplest method is to store the gas in a stationary pressure tank (Figure 28.6a). Up to now, this method has only become common primarily in Japan. In this case, the vehicle is filled through a simple overflow. The disadvantages of this are the large storage volumes required and the comparatively large amount of energy required for compressing the gas. These disadvantages are counteracted by a storage at different pressure levels from which the filling is carried out in succession with increasing pressure (Figure 28.6b). When refilling the storage banks, it is thus only necessary to compress a small portion to the required maximum pressure of more than 87.5 MPa. The storage at the maximum pressure can be completely eliminated by using a booster compressor (Figure 28.6c). The booster is then designed so that during filling it draws gas from an intermediate pressure tank, for example, at 30 MPa, which then only has to be additionally compressed by a small amount. It is thus possible to achieve a high output with moderate compressor sizes. The filling station design that is the most energy efficient at least locally and is the most compact can be implemented by storing supercooled liquid hydrogen (Figure 28.6d). The liquid is then brought to the necessary target pressure and is only evaporated after compression in a heat exchanger. The downside to this technology, however, is the very high energy cost required to liquefy the hydrogen.

28.5
Special Forms of Compressed Gas Storage

Beyond the conventional hydrogen storage in cylindrical containers, several approaches have been made to improve the storage density or to adapt the geometry of the containers or the concept of the system to the needs in vehicle applications.

28.5.1
Parallel Hydride and Compressed Gas Storage

Owing to the thermal behavior of metal hydride storage devices, thermal management is required when charging and discharging them. If a metal hydride is being charged, then energy in the form of heat must be dissipated. During discharging, on the other hand, heat must be supplied to avoid inhibiting the reaction kinetics (Section 28.2). This is achieved structurally by means of a heat exchanger with large ribbed surface, which is embedded in the hydride powder. In developing vehicle concepts, designers have tried to make advantageous use of this effect for the thermal management of the vehicle by placing a small-dimension metal hydride storage device parallel to the actual fuel storage for driving the vehicle, the compressed gas container (Figure 28.7) [32]. For example, at high outside temperatures, the discharging of the metal hydride storage device can be used to assist the cooling of the fuel cell system and at low outside temperatures the charging can be used to improve the cold starting properties.

Despite the positive properties from a thermal management standpoint, a vehicle concept with parallel compressed gas and metal hydride storage devices

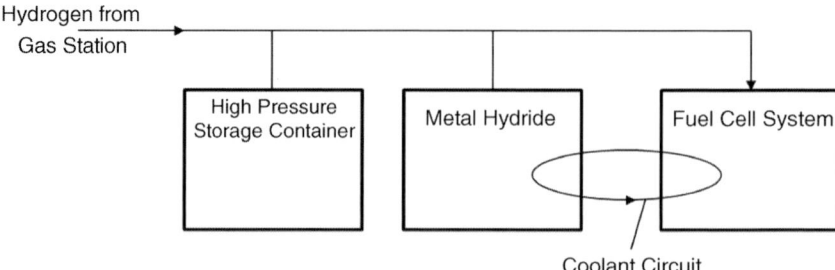

Figure 28.7 Schematic depiction of a vehicle concept with parallel compressed gas and metal hydride storage devices. Figure taken from Reference [32].

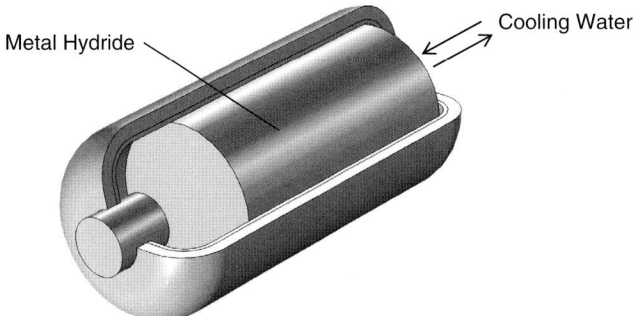

Figure 28.8 Schematic depiction of a metal hydride-filled pressure container.

is unlikely to be used in a production vehicle. Integration of the cooling circuit of the metal hydride storage device into the conventional vehicle cooling system increases the design and control-related complexity, which makes the overall system even more expensive than the use of two different hydrogen storage systems does already.

28.5.2
Metal Hydride-Filled Pressure Containers

With the goal of increasing the volumetric and gravimetric storage density of pressure tank systems, research is being carried out on hybrid storage devices, in which a metal hydride is placed in a pressure container (Figure 28.8). But having gravimetric advantages in addition to volumetric advantages as compared to pure pressure storage requires a high hydride-specific storage density [33,34]. In the known metal hydride materials, the formation enthalpy is higher the higher the storage density is [35]. Since high energy quantities in the form of heat must be dissipated in a short time in order to achieve a rapid filling, a complex heat exchanger is required, which must be integrated into the metal hydride storage device inside the pressure container and must be integrated into both the structure and control of the cooling circuits of the overall vehicle [36]. The goal, therefore, is to find a hydride material with a high storage capacity, but a low reaction enthalpy [35]. A metal hydride material with these properties is currently not known, but is the crucial element for the potential use in production vehicles, in order to increase the range with a simultaneously justifiable equipment engineering cost.

28.5.3
Conformable Containers

Because of the high mechanical load due to the internal gas pressure on the container, cylindrical containers with hemispherical ends are used in modern vehicles. In this form, the internal pressure can be absorbed by a fiber winding in the

Figure 28.9 Cylindrical container with flat ends. Figure from BaltiCo GmbH.

circumference and longitudinal directions. For use in vehicles, the limitation to cylindrical containers is disadvantageous because the small installation space cannot always be optimally used. For this reason, it would be desirable to have a more flexible design of the container. In each design, notably, the container must absorb the forces exerted on it by the gas pressure, which places strict limits on the shape.

One possible promising approach is the development of containers with flat ends [37] in which the forces acting in the radial direction are absorbed in the conventional way by a carbon fiber-reinforced cylinder, but the forces acting in the axial direction are absorbed by carbon fibers stretched taut by the container (Figure 28.9). It is thus possible to reduce the size of the region around the container ends that is difficult to use in vehicle construction. In addition, there is a very good utilization of the fibers of the container because the force vectors for circumferential and longitudinal fibers always extend in the direction of the fibers. In classic containers, however, the longitudinal fibers must be guided in a somewhat inclined fashion around the container ends, as a result of which a portion of the force extends transversely to the fiber. This technology is currently in the development stage.

A similar approach is being taken in research into conformable containers with the so-called "ribbed design" (Figure 28.10b), in which the radial forces in a flattened container are absorbed by carbon fiber ribs [38]. The issue of how to attach the ribs to the container has currently not yet been finally resolved.

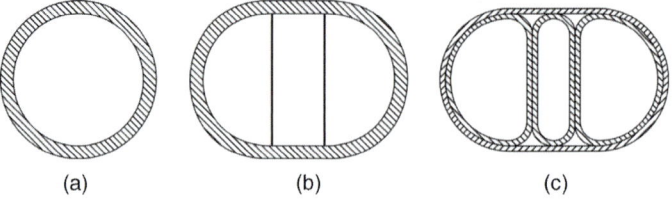

Figure 28.10 Cross-sections through different container shapes: (a) classical cylindrical container; (b) ribbed design; (c) bucked design.

One solution can be to embody the container using the "bucked design" (Figure 28.10c) in which several narrow containers are combined in a shared carbon fiber winding [39]. Both designs, however, share the fact that with regard to the storable hydrogen quantity more carbon fibers are required, accompanied by the corresponding disadvantages in weight and potential manufacturing costs.

28.6
Conclusion

Storage of compressed hydrogen aboard a vehicle is a mature technology. Compared to other storage methods such as liquid or chemical in metal hydrides it is most advantageous with regards to short refueling time, hermetically sealed system without boil-off, system complexity, and gas handling. Regulations for vehicle approval are already in place worldwide. Refueling technology is available; station network is currently being built up in many countries and regions.

References

1 European Parliament (2009) Regulation (EC) No 79/2009 of the European Parliament and of the Council on type-approval of hydrogen-powered motor vehicles, and amending Directive 2007/46/EC.

2 European Union (2010) Commission Regulation (EU) No 406/2010 implementing Regulation (EC) No 79/2009 of the European Parliament and of the Council on type-approval of hydrogen-powered motor vehicles.

3 Aceves, S.M., Espinosa-Loza, F., Ledesma-Orozco, E., Ross, T.O., Weisberg, A.H., Brunner, T.C., and Kircher, O. (2010) High-density automotive hydrogen storage with cryogenic capable pressure vessels. *Int. J. Hydrogen Energy*, **35**, 1219–1226.

4 Verlag TÜV Rheinland (1989) Alternative Energien für den Straßenverkehr - Wasserstoffantrieb in der Erprobung. Projektbegleitung Kraftfahrzeuge und Straßenverkehr, TÜV Rheinland e.V. by authority of the Federal Minister for Research and Technology (ed.).

5 Franzen, J. (2009) *Modellierung und Simulation eines Wasserstoffspeichers auf der Basis von Natriumalanat*, VDI Verlag, Düsseldorf.

6 Reijerkerk, C.J.J. (2004) Potential of cryogenic hydrogen storage in vehicles. Presented at the 15th World Hydrogen Energy Conference, 27 June to 2 July 2004, Yokohama, Japan.

7 Krainz, G., Bartlok, G., Meinert, M., and Novak, P. (2008) Anwendungsorientierte Entwicklung von Wasserstoff-Speichersystemen für den automotiven Einsatz. Presented at the 4th German Hydrogen Congress, 20–21 February 2008, Essen, Germany.

8 Tiller, D., Newhouse, N., and Eihusen, J. (2002) Development of an all-composite tank for high pressure hydrogen storage. 14th World Hydrogen Energy Conference, 9–14 June 2002, Montreal, Canada.

9 Inada, T., Chou, I., Eguchi, H., Shigegaki, Y., and Tanaka, T. (2004) Challenges to the ultra high pressure hydrogen tank for fuel cell vehicles. 15th World Hydrogen Energy Conference, 27 June to 2 July 2004, Yokohama, Japan.

10 Adams, P., Bengaouer, A., Cariteau, B., Molkov, V., and Venetsanos, A.G. (2011) Allowable hydrogen permeation rate from road vehicles. *Int. J. Hydrogen Energy*, **36**, 2742–2749.

11 Hu, J. and Chandrashekhara, K. (2009) Fracture analysis of hydrogen storage composite cylinders with liner crack accounting for autofrettage effect. *Int. J. Hydrogen Energy*, **34**, 3425–3435.

12 Tiller, D.B., Newhouse, N.L., and Eihusen, J.A. (2000) High pressure hydrogen storage for fuel cell vehicles. 33rd International Symposium on Automotive Technology and Automation, 25–29 September 2000, Dublin, Ireland.

13 Barthélémy, H. (2012) Hydrogen storage - industrial prospectives. *Int. J. Hydrogen Energy*, **37**, 17364–17372.

14 Krieger, J. and Multhoff, J.B. (2008) Auslegung von composite-hochdruckbehältern für H2-speicherung bei 700bar (im H2 700 NRW projekt). 4th German Hydrogen Congress, 20–21 February 2008, Essen, Germany.

15 Léon, A. (ed.) (2008) *Hydrogen Technology: Mobile and Portable Applications*, Springer, Berlin.

16 The High Pressure Gas Safety Institute of Japan KHK (Ed.) (2011) Technical Standard for 70MPa Containers of Compressed Hydrogen Fueled Vehicle. KHKS 0128.

17 Japan Automobile Research Institute (Ed.) (2004) Technical Standard for Accessories to Containers for Compressed Hydrogen Vehicle Fuel Devices. JARI S002.

18 SAE (2013) Norm SAE J2579:2013: Standard for Fuel Systems in Fuel Cell and Other Hydrogen Vehicles.

19 CSA (2013) Norm CSA HGV 3.1–2013 - Fuel system components for compressed hydrogen gas powered vehicles.

20 Standardization Administration of China (Ed.) (2011) Fuel Cell Electric Vehicles – Onboard Hydrogen System – Specifications. GB/T 26990.

21 United Nations Economic Commission for Europe (Ed.) (2013) Global technical regulation (GTR) on hydrogen and fuel cell vehicles. ECE/TRANS/180/Add. 13.

22 SAE (2010) Norm SAE TIR J2601:2010: Fueling Protocols for Light Duty Gaseous Hydrogen Surface Vehicles.

23 Cross, R. (2004) Type III Zero-Emission Vehicle (ZEV) Requirements State of California – Air Resources Board.

24 Newhouse, N. and Liss, W. (1999) Fast Filling of NGV Fuel Containers SAE Technical Paper 1999-01-3739.

25 Martinez, J.R. (2000) Compressed hydrogen fuelling station for fuel cell vehicles, Master's thesis. Hamburg, Advanced Technical College, Technology and Management.

26 Kountz, K.J. et al. (1998) Computer readable medium having a computer program for dispensing compressed natural gas. United States Patent No. 5771947.

27 Leon, A. (2005) Air Liquide hydrogen experience & State of the art of H2 filling stations. Presented at the 1st World Hydrogen Technologies Convention, 3–5 October 2005, Singapore.

28 Maus, S. (2007) *Modellierung und Simulation der Betankung von Fahrzeugbehältern mit komprimiertem Wasserstoff*, VDI Verlag, Düsseldorf.

29 Liss, W. et al. (2004) Development and validation testing of hydrogen fast-fill fueling algorithms. 5th National Hydrogen Association Annual Conference, 26–30 April 2004, Los Angeles, on CD ROM.

30 Kirsch, M. (2005) Charakterisierung und optimierung des betankungsprozesses von brennstoffzellenfahrzeugen. Degree dissertation. Ulm, Advanced Technical College, Department of Production Engineering and Production Management.

31 SAE (2007) Norm SAE IR J2799:2007: 70MPa Compressed Hydrogen Surface Vehicle Fueling Connection Device and Optional Vehicle to Station Communications.

32 Wenger, D. (2009) *Metallhydridspeicher zur Wasserstoffversorgung und Kühlung von Brennstoffzellenfahrzeugen*, VDI Verlag, Düsseldorf.

33 Takeichi, N., Senoh, H., Yokota, T., Tsuruta, H., Hamada, K., Takeshita, H.T., Tanaka, H., Kiyobayashi, T., Takano, T., and Nobuhiro, K. (2003) "Hybrid hydrogen storage vessel", a novel high-pressure hydrogen storage vessel combined with hydrogen storage material. *Int. J. Hydrogen Energy*, **28**, 1121–1129.

34 Corgnale, C., Hardy, B.J., and Anton, D.L. (2012) Structural analysis of metal hydride-based hybrid hydrogen storage systems. *Int. J. Hydrogen Energy*, **37**, 14223–14233.

35 Mori, D. and Hirose, K. (2009) Recent challenges of hydrogen storage technologies for fuel cell vehicles. *Int. J. Hydrogen Energy*, **34**, 4569–4574.

36 Mori, D., Haraikawa, N., Kobayashi, N., Shinozawa, T., Matsunaga, T., Kubo, H., Toh, K., and Tsuzuki, M. (2005) High-pressure metal hydride tank for fuel cell vehicles. Presented at the IPHE International Hydrogen Storage Technology Conference, 19–22 June 2005, Lucca, Italy.

37 Büchler, D. (2004) Neuartiger leichter Hochdruckbehälter (energiegeladen - Denkwerkstatt zu Entwicklungspotenzialen bei Brennstoffzellensystemen und gemeinsamen Projekten Berlin).

38 Aceves, S., Weisberg, A., Espinosa-Loza, F., Perfect, S., and Berry, G. (2005) Advanced concepts for containment of hydrogen and hydrogen storage materials. DOE Hydrogen Program – 2005 Progress Report.

39 Aceves, S., Perfect, S., and Weisberg, A. (2004) Optimum utilization of available space in a vehicle through conformable hydrogen vessels. DOE Hydrogen Program – 2004 Progress Report.

29
Cryo-compressed Hydrogen Storage
Tobias Brunner and Oliver Kircher

29.1
Motivation for Cryo-compressed Hydrogen Vehicle Storage

Compact onboard energy storage with a quick recharge option is a key challenge for future zero emission mobility. Although the energy density and capacity of vehicle batteries are expected to increase continuously over the coming decades, quick recharging capability remains subject to fundamental advances in battery and charging technology. In particular, replacing conventional gasoline refueling stations by a network of quick charging stations would bring technical and economic challenges, as can be seen from Figure 29.1: while one single conventional gasoline dispenser can fuel more than 100 000 km range per day to several hundred hybrid electric passenger vehicles, electric charging reaches 1/16 of equivalent range, even when assuming a 120 kW quick charge option. Power requirement, footprint, and cost of hundreds of such quick charging units required to replace all dispensers of an average gasoline station are prohibitive. Hydrogen refueling of a fuel cell electric vehicle (FCEV) with (cryo-)compressed gas according to currently developing and upcoming standards is still falling short behind gasoline in terms of refueling rate and time and, thus, produced range. However, it already is on the order of its magnitude with close to 100 000 kilometers range produced per day. Assuming that in the longer term hydrogen can be produced cost-effectively in large quantities from fossil resources, for example, by steam methane reforming of natural gas, as well as from renewable sources, for example, by large central and small decentralized electrolysis using excess electricity from wind or solar power plants, the main challenges to overcome are efficient and economically viable delivery, station storage, refueling, and vehicle onboard storage.

Vehicle onboard storage options cover solid and physical storage methods. All currently investigated solid storage options as visualized in Figure 29.2 are rated as at the stage of fundamental research and are not capable of simultaneously fulfilling the most relevant automotive requirements, for example, as summarized in the US Department of Energy's "Targets for Onboard Hydrogen Storage Systems for Light-Duty Vehicles" [1]. Among the physical storage options, liquid

Hydrogen Science and Engineering: Materials, Processes, Systems and Technology, First Edition.
Edited by Detlef Stolten and Bernd Emonts.
© 2016 Wiley-VCH Verlag GmbH & Co. KGaA. Published 2016 by Wiley-VCH Verlag GmbH & Co. KGaA.

29 Cryo-compressed Hydrogen Storage

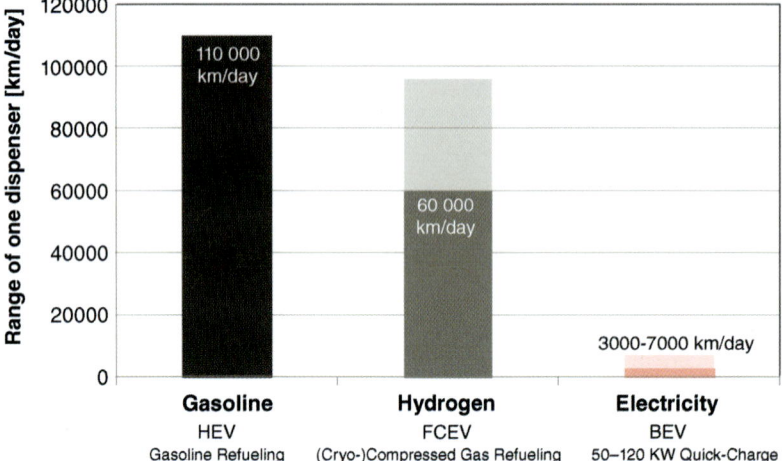

Figure 29.1 Estimated range of one dispenser or quick charging unit per day assuming an E segment vehicle with different state-of-the-art power train and refueling technologies: hybrid electric vehicle (HEV) with gasoline refueling, fuel cell electric vehicle (FCEV) with hydrogen (cryo-) compressed hydrogen gas refueling, and battery electric vehicle (BEV) with electricity quick charging option; 50% dispenser utilization rate per day assumed.

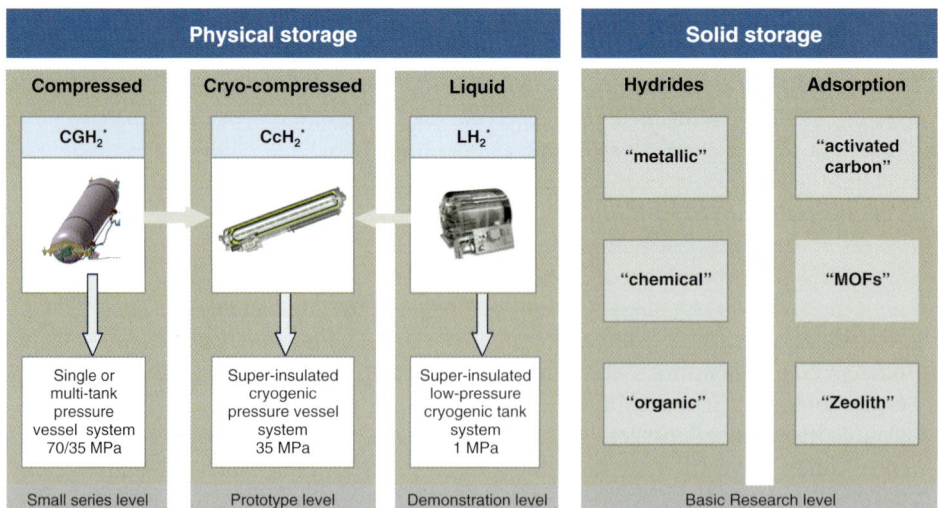

*) CGH_2 := Compressed Gaseous Hydrogen (70/35 MPa) CcH_2 := Cryo-compressed Hydrogen (35 MPa) LH_2 := Liquid/Liquefied Hydrogen (1 MPa)

Figure 29.2 Hydrogen storage options – automotive hydrogen storage is currently limited to physical storage, current industry mainstream is compressed gaseous hydrogen storage (CGH_2).

hydrogen (LH$_2$) storage has demonstrated feasibility in, for example, BMWs small series vehicle Hydrogen 7 [2], whereas high-pressure compressed gaseous hydrogen (CGH$_2$) storage up to 70 MPa has become the industry mainstream and is currently expected to see first relevant commercial application in FCEVs between 2015 and 2017. The initial concept of liquid hydrogen storage in super-insulated pressure vessels dates back to Aceves *et al.* [3]. The first ideas of a cryo-compressed hydrogen (CcH$_2$) vehicle storage linked to a cryogenic compressed hydrogen gas refueling by use of a station side LH$_2$ compressor were developed by the authors in 2005 based on the experience with BMW LH$_2$ vehicle tank and refueling technology. The thermodynamic analysis summarized in Section 29.2 as well as a fundamental validation of performance and safety discussed in Section 29.5 could prove the initial assumptions of the benefits of super-insulated carbon-fiber overwrapped metallic pressure vessels (Type 3) and their refueling with a cryogenic compressed gas, as described in Section 29.3:

- High volumetric and gravimetric storage system densities due to a high physical density of cryogenic compressed hydrogen gas (Section 29.2).
- Significantly higher heat receptivity compared to a low-pressure LH$_2$ tank and, thus, higher dormancy (loss-free duration from start of parking to start of venting at maximum operating pressure) as well as lower performance requirements on the insulation (Section 29.2).
- Capability to control pressure and temperature in the storage vessel by the use of internal and external heat exchangers (Section 29.4). Such control option allows for reliable performance at all climate and operating conditions, for example, high load operation at low ambient temperature.
- Storage heat receptivity in cold operation: The storage can work as a heat sink for vehicle waste heat, for example, from the PEM fuel cell and, thus, can help in overcoming the fuel cell drive train cooling challenge.
- Self-sustaining cooling: vessel and insulation are cooled-down by expansion cooling during discharge of stored hydrogen; no need for active cooling in typical automotive use profiles.
- Quick and efficient refueling capability with no need for transient flow rate control or communication between refueling station and vehicle (Section 29.3).
- A very high intrinsic safety level due to double-layer design with a protective outer vacuum enclosure, redundant safety release devices, and low adiabatic expansion energy (Section 29.5).
- Competitive system cost improving with system size due to lower material cost (mainly carbon fiber), which helps compensate for higher system complexity compared to CGH$_2$ storage.

With rising storage size and capacity the high physical density of cryogenic compressed hydrogen gas outbalances the volumetric effort for auxiliary components such as insulation material, valves, pressure release devices, internal and

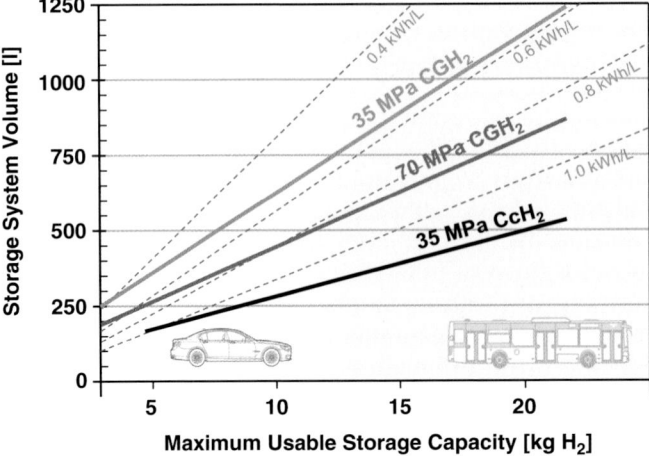

Figure 29.3 Volumetric efficiency of (cryo-)compressed hydrogen storage options; storage system volume is dependent on usable storage capacity.

external heat exchangers, and so on. The scalability over usable storage capacity in FCEV application of a CcH_2 storage system compared to 35 MPa and 70 MPa CGH_2 storage systems is illustrated in Figure 29.3. Evidently, the advantage in size, and consequently cost and weight, becomes more prominent when applied to vehicles with a high energy demand such as long distance large passenger vehicles, trucks, light duty commercial vehicles, city and transit busses, as well as heavy duty vehicles.

29.2
Thermodynamic Opportunities

Regarding physical hydrogen storage methods the density–temperature diagram of hydrogen shown in Figure 29.4 is framed by a region with hydrogen in liquid phase at temperatures below 33 K (LH_2) and a region with pressurized gas between −40 and +85 °C (CGH_2). The latter typically constrains the operating range of Type 4 automotive 70 MPa pressure vessels due to either sealing material limitations at low temperature or liner material limitations at high temperature. The region above 33 up to 233 K signifies a cold or cryogenic gas region, which above ambient pressure represents the cryo-compressed hydrogen (CcH_2) gas region. The authors' starting point for the concept of cryo-compressed hydrogen vehicle storage is based on liquid hydrogen delivered to and stored in a fueling station's LH_2 storage tank, from which it can be filled into LH_2 vehicle tanks by the use of a low-pressure transmission pump. Such fill process starting from LH_2 at ambient pressure (point "L" in Figure 29.4) into a LH_2 vehicle tank with a minimum operating pressure of, for example, 0.4 MPa ("T") at thermodynamic equilibrium goes along with a decrease in density from 71 to 63 g l^{-1}. If

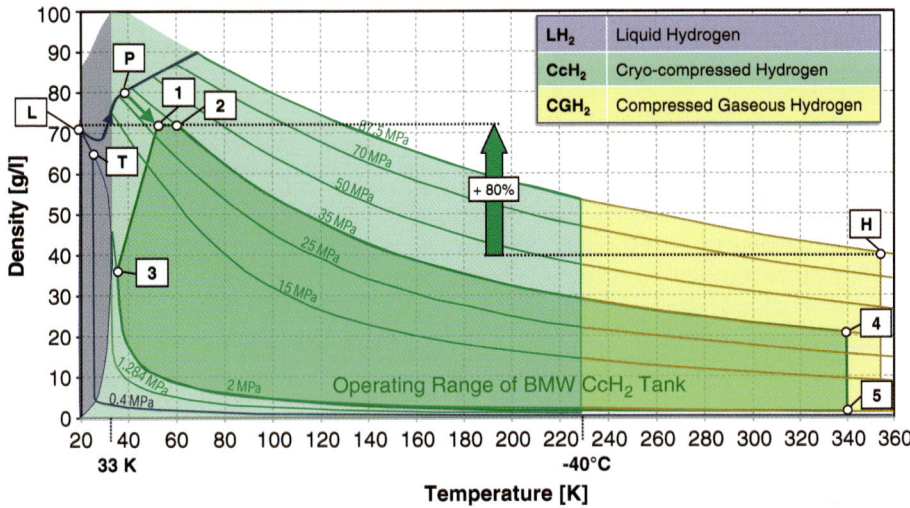

Figure 29.4 Operating range of cryo-compressed hydrogen storage in a density–temperature diagram of hydrogen.

due to technical constraints a minimum volume of gas needs to cover the liquid reservoir (e.g., to prevent liquid "boil-off" during sloshing or to enable capacity measurement via a liquid level sensor) the maximum hydrogen fuel density of such automotive LH_2 storage can shrink down to $50\,g\,l^{-1}$. In 2006 the authors claimed that by using and improving existing cryogenic compression technology a cold pressurized gas with a density significantly above that of liquid hydrogen could be produced and used to fill an insulated pressure vessel up to a density above $71\,g\,l^{-1}$. In 2010, BMW and Linde proved that an optimized Linde cryogenic high-pressure two-stage piston pump (schematically shown below in Figure 29.6) can pressurize liquid hydrogen up to a pressure of 30 MPa while keeping the temperature increase moderate to 38 K, resulting in a physical density of $80\,g\,l^{-1}$ at the outlet of the high-pressure cryo pump ("P"). Owing to compression heating during the fill of CcH_2 gas into a cold CcH_2 storage system a moderate density decrease to $72\,g\,l^{-1}$ could be expected and experimentally verified in 2010 at a BMW proving ground. The maximum storage density of $72\,g\,l^{-1}$ after CcH_2 refueling exceeds the density of 70 MPa CGH_2 by 80%. This result given, the authors and their teams between 2008 and 2013 developed two generations of cryo-compressed hydrogen vehicle storage system prototypes, the latter of which is shown below in Figure 29.8.

The hydrogen containing pressure vessels were designed for a maximum operating pressure of 35 MPa in order to guarantee a minimum pressure stroke between fill pressure at 30 MPa and vent release pressure at 35 MPa. The process of self-pressurization is a result of heat leaking through the passive insulation into the cryogenic pressure vessel ($1 \rightarrow 2$ in Figure 29.4). The minimum loss-free dormancy time representing the time after parking with a full cold CcH_2 storage

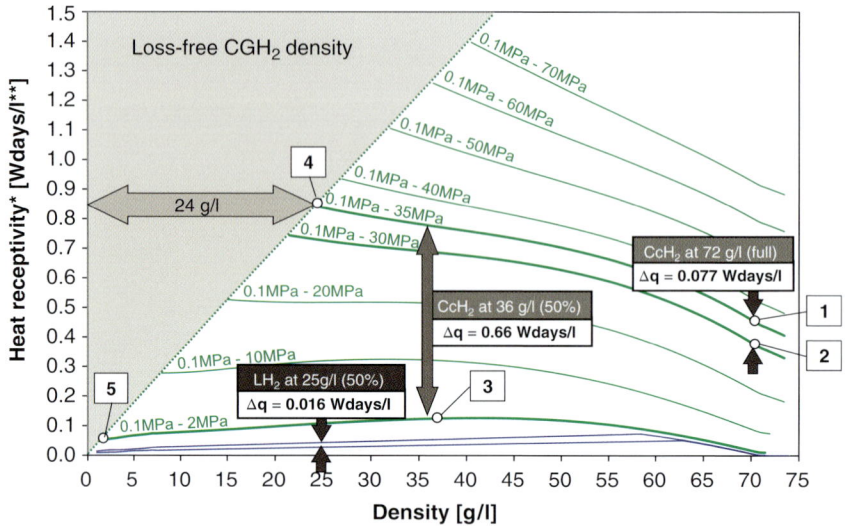

Figure 29.5 Heat receptivity of cryo-compressed and liquid hydrogen storage. Iso-lines representing delta value of volumetrically normalized internal energy over volumetric density between ambient (0.1 MPa) and nominal iso-pressure level.

until the heat leak induced self-pressurization leads to vent release can be derived from the heat receptivity diagram illustrated in Figure 29.5. With a given net storage volume of, for example, 115 l and an average heat leak of 5 W the loss-free dormancy time can easily be calculated to $0.077 \times 115/5$ days $= 1.7$ days, more than three times longer than in a LH_2 storage of the same size, but with lower heat leak of 3.5 W due to a more sophisticated, but less cost-effective insulation. Once venting has started (2 in Figure 29.4) it takes up to two months of additional parking time to reach the state of an ambient compressed hydrogen storage at 35 MPa (2 → 4). The remaining loss-free density equals that of ambient 35 MPa hydrogen gas and reaches $24\,\mathrm{g\,l^{-1}}$ at 288 K (15 °C) (4) or $21\,\mathrm{g\,l^{-1}}$ at 340 K. Given a typical use profile of a passenger vehicle, however, it is more likely that the storage pressure sees a decrease after refueling due to discharge from the storage during a consequent driving operation. If such driving operation starting from a full storage is continuous (e.g., long-distance driving) temperature and pressure can decrease to their respective minima of $\sim$35 K and 2 MPa (1 → 3). The decrease in temperature goes along with a cool down of storage container, insulation, and connection lines, which in typical automotive driving pattern with concurring driving and parking events can result in a self-sustaining balance of discharge cool down and idle heat up with self-pressurization. Once the minimum storage pressure (3) required to appropriately supply the drive train with hydrogen is reached, the pressure can be kept by re-feeding heat into the storage volume until the storage reaches its minimum density (3 → 5). A pure compressed gas operation can start from warm full and lead to warm empty state by continuous discharge (4 → 5), when either the storage has heated up during parking at hot ambient temperature conditions or when the CcH_2 storage is filled with compressed gas at ambient temperature. The described operating states (1–5 in Figure 29.4) and their interlinks define the thermodynamic boundaries of the operating range of BMWs 2010–2014 CcH_2 storage system prototypes.

29.3
Refueling and Infrastructure Perspectives

A major advantage of LH_2 compression compared to gaseous hydrogen compression is related to the significantly higher physical density of liquid hydrogen. The density of LH_2 of $63\,g\,l^{-1}$ at 0.4 MPa up to $71\,g\,l^{-1}$ at ambient pressure is about 40 times higher than the density of hydrogen gas at 0.2 MPa and ambient temperature and still doubles the density of pressurized warm gas at 50 MPa. As a consequence of such thermodynamic property disparity, compression of liquid hydrogen can by an order of magnitude be less energy consuming than compression of gas at ambient temperature. Moreover, quick CGH_2 refueling at 70 MPa in 3 min or less requires compression up to 87.5 MPa and, to avoid tank overheating during the fill process, pre-cooling of the pressurized gas down to −40 °C before it can enter the vehicle tank system through station nozzle and vehicle receptacle. Hence, onsite cryogenic compression of LH_2 stored at the station in amounts of up to several thousand kilograms in an insulated LH_2 storage tank combined with a cryogenic compression system from, for example, Linde as illustrated in Figure 29.6 [4], can keep the energy demand of a future refueling station including hydrogen fuel option at the order of magnitude of today's energy consumption, even with 70 MPa CGH_2 refueling. Furthermore, CcH_2 allows for quick refueling of hydrogen at highest purity for several hundreds of vehicles per day with a frequency of one LH_2 station tank refill per several days, similar to today's filling stations. The intrinsic disadvantage of onsite LH_2 storage, its boil-off production and venting that particularly occurs in idle times, becomes less relevant with rising hydrogen throughput, the result of an increasing hydrogen vehicle population to be expected beyond 2020. In the ramp-up phase of a hydrogen mobility the higher investment required for LH_2 based stations with cryogenic onsite compression up to 70 MPa that is intrinsically designed for large amounts of hydrogen throughput remains a major hurdle compared with competing smaller station technology based on gaseous delivery and conditioning. Thus, in the early phase of an implementing hydrogen infrastructure LH_2 based hydrogen stations with cryogenic compression could most likely be expected at locations with significant throughput, for example, linked to larger fleets of FCEVs (material handling vehicles, busses, light-duty commercial or passenger vehicles), at key locations of major cities and along central city-connecting highways. In a longer term perspective, given a successful FCEV market roll-out, scalability and cost-efficiency of large LH_2 based hydrogen filling stations with cryo-compression technology should open the door to reach a competitive market position.

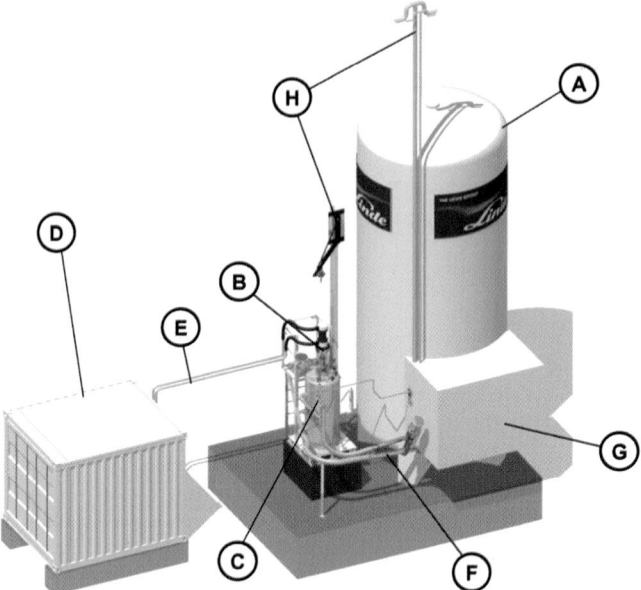

Figure 29.6 Cryo-compressed refueling unit with cryogenic high-pressure pump (B) driven by a hydraulic drive system (D) connected via hydraulic lines (E), immersed in a LH_2 side reservoir (C) connected to a LH_2 station storage (A) by insulated lines (F) through control unit (G), secured by safety and release funnels (H) [4].

Figure 29.7 summarizes the well-to-tank pathway of a liquid hydrogen filling station with cryogenic onsite compression and a dual dispensing option: mainstream 70 MPa CGH_2 refueling requires gas heat up to the minimum required temperature (−40 °C, according to SAE J2601 [5]). Direct filling via cryo pump and heat exchanger may be complemented by adequately adding high-pressure gas from station buffers at up to 87.5 MPa in order to match the required flow rates and pressure ramps according to J2601 refueling protocol or to re-use captured boil-off gas from the LH_2 station storage tank. CcH_2 refueling enables direct use of the cryo pump output and to fill via an insulated CcH_2 nozzle (recently developed and validated by Linde and BMW) into a CcH_2 vehicle storage system via a CcH_2 vehicle receptacle and insulated fueling lines (as explained in Section 29.4). Hence, there are evident synergies between 70 MPa CGH_2 and 30 MPa CcH_2 refueling processes, once the LH_2 based station design as of Figure 29.7 starts becoming a wide spread option. There are several opportunities resulting from CcH_2 refueling that need further investigation and validation:

- Simple, safe, and reliable CcH_2 refueling process: no need for communication between vehicle and station via, for example, the refueling coupling (as currently required by 70 MPa CGH_2 fast refueling); high safety level due to leak check after coupling.

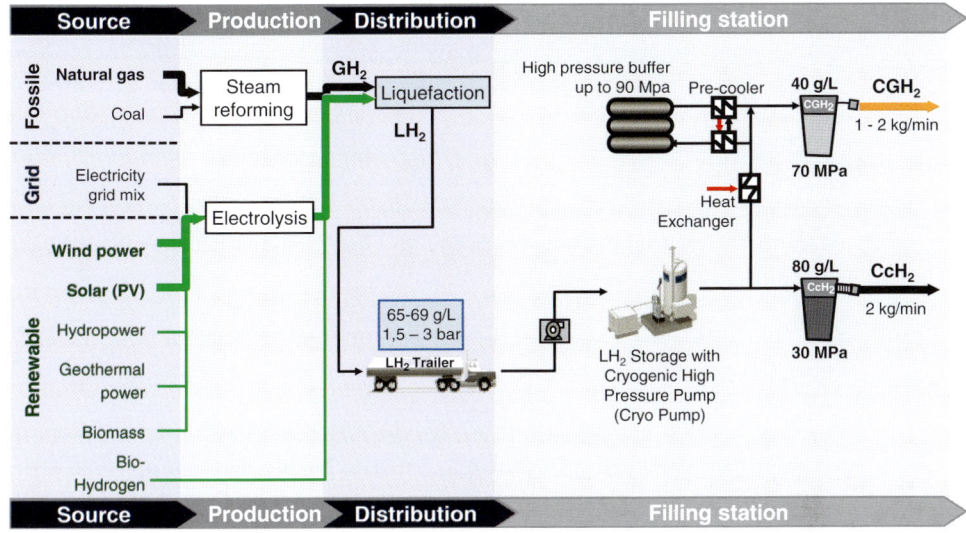

Figure 29.7 Schematic well-to-tank pathway of CcH$_2$ refueling from a LH$_2$ supplied station with cryogenic compression via cryo pump and combined CcH$_2$ and CGH$_2$ dispensing.

- Lower energy consumption related to compression resulting from a moderate maximum output pressure of 30 MPa CcH$_2$ (compared with up to 87.5 MPa for 70 MPa CGH$_2$ refueling) as well as related to further conditioning (e.g., cooling or heating to −40 °C for 70 MPa CGH$_2$ refueling). However, such advantage compares to the energy demand for liquefaction (currently 11–12 kWh kg^{-1}, future perspective 6–7 kWh kg^{-1}).
- Investment and operating cost savings, when designing the entire station to 35 MPa operating pressure: in the long term with rising efficiency of FCEVs and more purpose designed vehicle concepts CcH$_2$ gas refueling of high density could replace 70 MPa CGH$_2$ as an option for long distance and/or large vehicles, whereas an additional 35 MPa CGH$_2$ dispensing option could be used for short distance and/or smaller vehicles; as a result station and refueling complexity could be reduced: for example, no further need for communication and less effort for pre-conditioning, significantly lower boil-off gas production during the compression process.

29.4
Design and Operating Principles

Given the advantages and challenges of CcH$_2$ thermodynamics as described in Section 29.2, the BMW team around the authors tried to find an appropriate system and control design for a first automotive CcH$_2$ storage system, which allows us to take utmost advantage of its benefits such as high volumetric and gravimetric capacity, quick refueling, and passive discharge cooling leading to sufficient loss-free dormancy time. At the same time the main challenges had to be minimized, for example, potential hydrogen loss through venting after longer dormancy periods or variable density after refueling due to a storage temperature dependent on previous discharge and parking events. Hence, system design including operating conditions, vehicle compliant storage system dimensions with an optimum volume to surface ratio as well as appropriate insulation, mounting, component, and control concepts were developed to allow an optimum between high storage capacity and loss-free operation for all relevant vehicle use profiles.

29.4.1
System Design

The CcH$_2$ storage system shown in Figures 29.8 and 29.9 is designed to store cryogenic and ambient compressed gas between 30 and 340 K at up to 35 MPa. The storage vessel ("a") is a carbon fiber reinforced aluminum liner (Type 3). The carbon fiber composite reinforcement, in particular the choice of the resin–fiber combination, is optimized for the operating temperature range between cryogenic and ambient temperatures. The storage vessel is mounted into the vacuum enclosure ("c") by thermally optimized suspensions ("e") in order to minimize heat transport into the storage vessel. The vacuum super-insulation ("b") consists of a vacuum space carrying a pre-fabricated set of multi-layer reflecting aluminum foils embedded in glass fiber spacer material that have the aim to minimize radiation heat transfer to the cryogenic pressure vessel ("a"). Vacuum quality is controlled by a getter concept that allows maintaining a

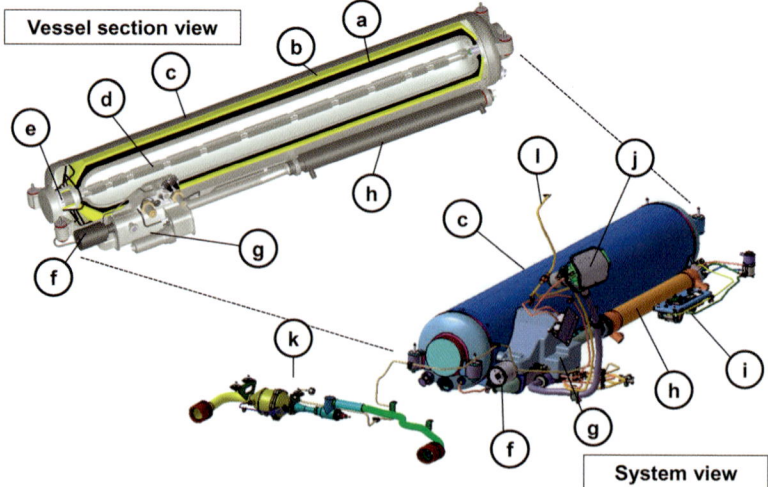

Figure 29.8 Layout of BMW cryo-compressed hydrogen storage prototype 2013. Left: storage system; right: vehicle fuel supply system for integration in the vehicle central tunnel.

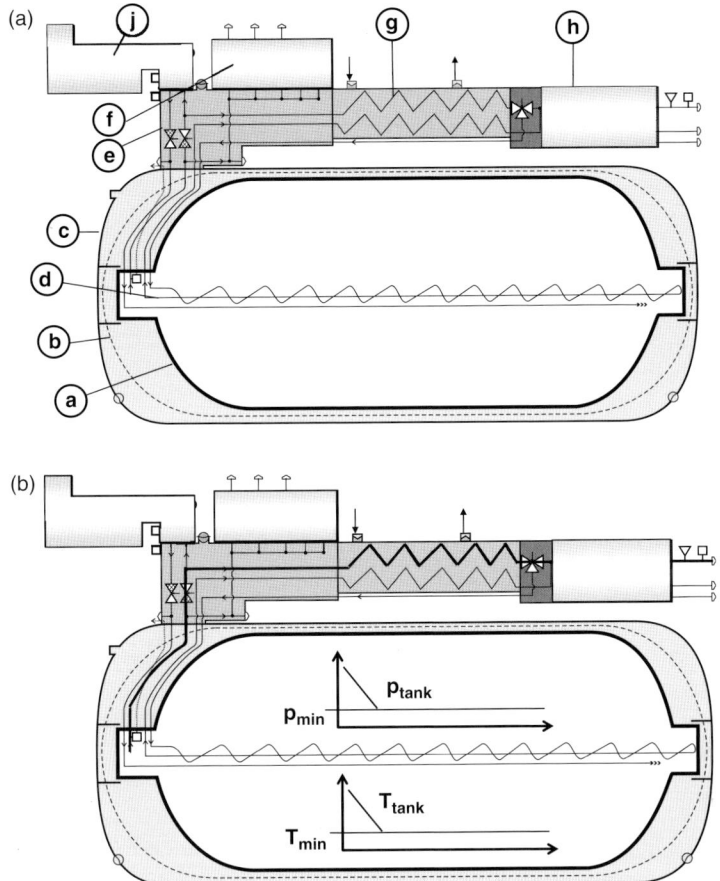

Figure 29.9 Schematic layout (a) and operating principle of pressure and temperature control in a cryo-compressed hydrogen storage system (b, c). While discharging, cryo-compressed hydrogen gas pressure and temperature decrease (b). When reaching a given pressure minimum, once extracted and heated, hydrogen recirculates into the tank heat exchanger in order to dissipate heat to the storage volume and, thus, control storage pressure (c).

sufficiently low vacuum pressure over lifetime (not depicted in Figure 29.8). A fail-safe tank shut-off valve ("f") located in a secondary vacuum module ("g") outside of the insulated cryogenic pressure vessel keeps the pressure vessel tight when not operating or in case of a safety hazard, for example, a vehicle crash. In discharge operation, hydrogen leaves the cryogenic pressure vessel ("a") through lines passing through the secondary vacuum module ("g"). It heats up into a coolant heat exchanger ("h") that is connected to the coolant circuit of the drive train (e.g., a low temperature PEM fuel cell) in order to re-use waste heat and, thus, improve vehicle drive train cooling. The pressure is regulated down to inlet pressure of the drive train aggregate by a pressure control unit ("i") carrying a

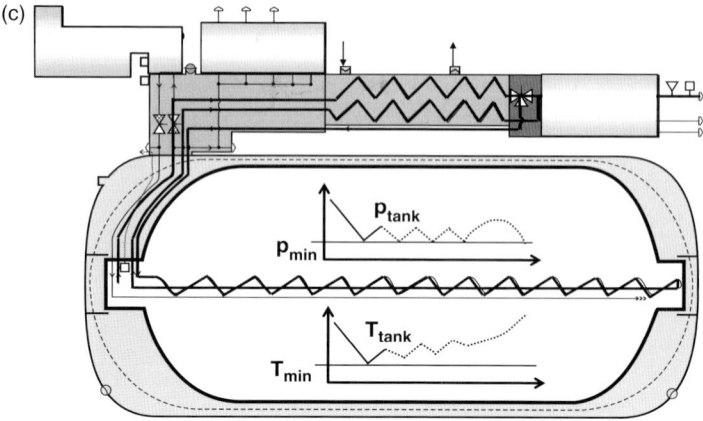

Figure 29.9 (Continued)

pressure regulator as well as a control valve. The control valve can be used to recirculate the heated hydrogen gas into the internal tank heat exchanger ("d"), where it dissipates heat before leaving the pressure vessel and passing through the coolant heat exchanger ("h") in a separated channel, again. Here, it heats up before it passes through the control unit ("i") and enters into the feed line of the drive train aggregate. A refueling line leads to the vehicle receptacle ("j") that, during a refill, is latched to the station dispensing nozzle. In case of longer dormancy with self-pressurization a blow-off valve located in the secondary vacuum module ("g") opens when the pressure in the vessel exceeds the maximum operating pressure. The hydrogen gas release is transformed into water vapor in a catalytic converter located in the blow-off management system ("k"). In case of malfunction or failure of the vacuum leading to sudden pressurization a pressure relief device, which is located in the secondary vacuum module ("g"), opens and safely releases the hydrogen through a safety vent line ("l"). Protection against overpressure in the vacuum enclosure as a consequence of hydrogen leakage into the vacuum space is realized via a burst disc integrated in the vacuum enclosure. In case of fire accident a hidden thermal pressure relief device opens and releases the hydrogen before structural damage of the cryogenic pressure vessel can occur.

29.4.2
Operating Principles

A CcH_2 storage system as designed by the authors and their teams is capable of storing cryogenic compressed hydrogen gas as well as ambient compressed hydrogen gas. The main purpose of such CcH_2 storage system is to store high-density cryogenic gas and supply gas at ambient temperature at a required inlet pressure to the fuel cell drive train aggregate of a FCEV (typically between 0.5 and 1.5 MPa). Owing to its ability to control temperature and pressure via an internal heat exchanger ("d") it could potentially supply a significantly higher pressure, for example, to a charged hydrogen internal combustion engine (e.g., 5–10 MPa) until the storage volume reaches ambient temperature and cannot longer hold the required supply pressure. Since onboard capacity calculation for an accurate state of charge indication relies on pressure and temperature measurements, only, the stored hydrogen should be kept in gaseous phase and not fall into the two-phase region below the critical temperature of 33 K, where temperature and pressure are not independent and, thus, not suited for deriving the storage density. Whereas in typical park and drive cycles the stored hydrogen gas never comes close to the critical temperature, a continuous high mass flow discharge, for example, during a longer highway drive, can bring down the temperature close to the critical temperature. To keep a distance from the two-phase region and to guarantee sufficient pressure supply to the PEM fuel cell via the pressure regulator, the pressure control mechanism is activated when a pressure level of ~2 MPa is reached (3 in Figure 29.4 and 29.9b). To avoid falling below the minimum pressure level once extracted and heated hydrogen recirculates into the internal heat exchanger ("d") realized as a closed pipe with fin heat exchange promoters that is mounted at the longitudinal axis of the storage pressure vessel ("a") and led forth and back to the exit port. The internal heat exchanger pipe is designed to cool the passing hydrogen down to storage temperature and, thus, maximize heat dissipation into the storage volume. After leaving the storage vessel through the internal heat exchanger exit and passing through the external coolant heat exchanger ("h") in a second recirculation channel separated from the primary flow line the re-conditioned hydrogen gas is led to the pressure control unit ("i"), where the pressure is regulated down to the input level of the fuel cell drive system. Through the described closed-loop recirculation method with heat transfer into the storage volume the pressure increases instantaneously and only ends when the recirculation is stopped or the storage volume is close to ambient temperature. The recirculation could be re-iterated continuously to keep the storage pressure close to the minimum operating pressure. However, such a control procedure would lead to a high number of switching operations of the control valve as well as to additional thermal cycles of internal heat exchanger and connection pipes. To minimize such additional load to the storage components a saw tooth control concept has been realized with a pressure stroke of several hundred kPa (Figure 29.9c). In case the CcH_2 storage reaches a temperature level close to ambient conditions the control is set to permanent recirculation in order to maximize usable capacity. The empty state of the CcH_2 storage system is reached at around $2\,g\,l^{-1}$ remaining hydrogen density. That coincides with the moment when the temperature gradient between storage volume and coolant is too low for dissipating enough heat into the storage required to keep the pressure above its minimum level. When the recirculation is applied for a longer duration a significant amount of heat dissipates into the cold storage until the storage pressure increases close to the vent pressure level (35 MPa). In such continuous recirculation mode the heat flux into the CcH_2 storage via the external coolant heat exchanger ("g") can almost double; the storage works in a heat sink mode. Given that cooling of a low temperature PEM fuel cell in a FCEV becomes difficult under high load and high ambient temperature levels (e.g., uphill highway driving), such heat sink mode can temporarily improve drive train cooling and, thus, vehicle performance and efficiency.

During longer dormancy, for example, vehicle parking events, a CcH_2 storage system behaves similarly to a LH_2 storage system. Self-pressurization due to heat transfer through the insulation can increase the storage pressure up to the vent release pressure. In contrast to LH_2 storage systems, in CcH_2 storage systems venting occurs significantly less frequent and vent losses are limited:

- Vent rates are lower than in a LH_2 storage system due to higher heat receptivity as illustrated in Figure 29.5. Vent rates decrease over time since the storage volume heats up while slowly approaching ambient temperature.
- Below around 30–35% of the maximum state of charge a CcH_2 storage system reaches the CGH_2 mode at ambient temperature ("4" in Figure 29.5). Vent loss through the mechanical blow-off valve stops.
- The average dormancy time depends on the pressure and temperature of the storage volume at the start of the dormancy event. With a state of charge of 50% ($36\,g\,l^{-1}$) and an initial pressure around the minimum pressure level that would occur after a longer driving event, the heat receptivity of the CcH_2 storage system reaches 0.66 Wdays per liter (Figure 29.5). With an assumed 115 l net storage volume and average heat leak of 5 W the loss-free dormancy time would reach 15 days. Infrequent driving would lead to a reduced average storage density since the average temperature of the CcH_2 storage would be too high to refuel to the maximum density of $72\,g\,l^{-1}$. A frequent driving profile with a lower average storage temperature and a higher average density reduces dormancy periods by its use profile. Hence, simulation results with a typical drive and park cycle pattern show that average losses of more than 2% per year are unlikely.

The state of charge and, thus, the maximum density after CcH_2 refueling without pre-cooling of storage or fill lines depends on the initial storage temperature and pressure. Hence, it is of significant relevance to gather realistic data of storage behavior in real world drive and park cycles and, in particular, the capability to switch between warm CGH_2 and cold CcH_2 operating modes. Figure 29.10 shows first experimental data on consecutive single-flow cold refueling and discharge events starting from a CcH_2 storage at ambient temperature. After three refueling and discharge events, the latter simulating continuous driving after refueling, the storage reaches 88% of its maximum state of charge. Evidently, a CcH_2 storage system at initially ambient temperature cools down by consecutive CcH_2 refueling and discharge events. Hence, even infrequent driving (<8000 km per year) that would make a CcH_2 storage work in a low density range of around $45\,g\,l^{-1}$ can reach a maximum state of charge during occasional longer driving events. However, the optimum use pattern in terms of maximum density

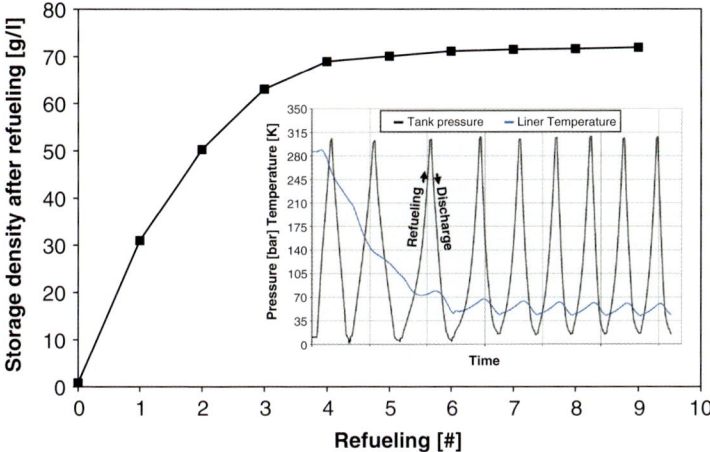

Figure 29.10 Storage density change during sequential single-flow cold refueling and discharge events, starting from an empty tank at ambient temperature; experimental results with BMWs CcH_2 storage and refueling system prototypes.

relates to a more frequent vehicle use, since a frequent discharge and refueling leaves the average density above $55\,\mathrm{g\,l^{-1}}$, enabling the potential of a CcH_2 passenger vehicle storage system. As a next important step a real world vehicle validation is required to optimize the CcH_2 storage control concept.

29.5
Validation Challenges of Cryo-compressed Hydrogen Vehicle Storage

For mainstream 70 MPa CGH_2 storage systems as well as LH_2 storage systems first standards and regulations already exist, for example, the regulation of the European Commission on type-approval of hydrogen-powered motor vehicles [6] or the SAE Standard for fuel systems in fuel cell and other hydrogen vehicles [7]. Several car OEMs already demonstrate performance and reliability of automotive CGH_2 storage systems under real world conditions. CcH_2 storage as a new automotive hydrogen storage concept still needs to prove safety, performance, and reliability over vehicle lifetime. Given hydrogen as fuel, a major validation focus is directed to safety and durability under load cycles equivalent to 15 years of worst case vehicle use. Generally, mechanical stress loads by pressure cycles are less demanding than in a 70 MPa CGH_2 storage system due to its operating pressure limited to 35 MPa. However, due to the expansion of the operating temperature range down to 30 K, additional thermo-mechanical loads and superimposed stresses arise in particular for the metallic liner and fiber composite materials of the storage vessel as well as for all components that are in contact with varying cold and warm hydrogen flow.

The critical real world load conditions of an automotive CcH_2 storage system that need to be reflected in a validation cycle are summarized in Figure 29.11 [9].

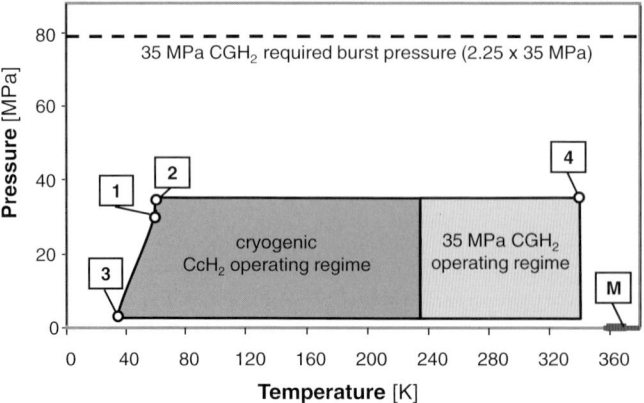

Figure 29.11 Operating range of cryo-compressed hydrogen vehicle storage; pressure–temperature diagram. Load points 1 to 4 according to Figure 29.4; "M" signifies manufacturing related soak conditions to de-humidify carbon fiber composite and insulation during vacuum generation.

Out of load points 1–5 explained in Section 29.2, the most relevant loads for a Type 3 cryogenic pressure vessel result from points 2, 3, and 4. Load point 3 is related to lowest operating pressure and temperature after a long continuous discharge following a refueling event. Such condition puts high compressive strain on the liner and, thus, maximizes the delaminating force between liner and carbon fiber composite overwrap (criticality of delamination is still under investigation). Load point 2 marks the start of venting from a cold storage at maximum operating pressure (35 MPa), which is linked to highest tensile stress in the metallic liner induced by superimposed mechanical and thermal load. Load point 4 represents the storage condition after a warm CGH_2 fill up to 35 MPa, which can be related to thermally induced transient stresses if a warm refueling follows a previous cold operation. Whereas points 1–4 are repeating events, "M" is related to the singular vacuum generation process during manufacturing, when a high thermal load is applied to the insulation as well as to the pressure vessel (340–390 K for at least several hours).

29.5.1
Validation Procedure

In a validation procedure of a hydrogen storage system, whether it is a CGH_2 or CcH_2 storage system, appropriate durability and safety test procedures have to be designed in a way to minimize the risk of one of the following incidents during real world operation:

- burst of pressure vessel or components that are exposed to high static or dynamic pressure loads, including severe accidents or vehicle fires;
- leakage that could lead to flammable or explosive hydrogen air mixture;

- malfunction or misuse of the storage system leading to hydrogen leakage or overpressure, for which the storage system might not be designed;
- any severe condition of the storage system due to misuse by the costumer.

It has been decided to follow a performance oriented verification procedure (see, for example, Reference [8]). The expected loads from refueling and discharge combined with vehicle vibration loads and additional loads arising from extreme weather conditions are taken into account. Expected service loads are multiplied by a safety factor in order to count for scattering of customer use as well as for unexpected divergence of component performance [8]. A "leak before burst" guideline has been pursued during component and system design to guarantee a maximized safety level, which minimizes the risk of fatal failure in early validation stages. The validation procedure of a 70 MPa CGH_2 hydrogen storage system typically consists of the following tests:

- burst tests of several virgin pressure vessels and auxiliary components;
- repeating burst tests after pressure cycling in order to determine the safety factor and the statistical spread at the beginning and the end of life of the CGH_2 storage system;
- hydraulic pressure cycling tests of the pressure vessel and the auxiliary components until failure or until a predefined number of cycles has been reached without failure;
- gas pressure cycling test of the pressure vessel and the auxiliary components in order to prove save operation during an expected service life of the CGH_2 storage system;
- bonfire and localized fire tests to prove that the vessel does not burst and hydrogen is discharged in a controlled way in case of a vehicle fire;
- penetration test;
- several tests on misuse scenarios that may occur during production or vehicle operation.

Most of the validation tests, which are currently designed for CGH_2 storage systems, can also be applied to a CcH_2 storage system. Owing to conceptual difference, however, the following tests will have to be performed in addition, partly resulting from required adaptations of existing tests:

- verification of safe and reliable function of pressure limiting devices integrated in the blow-off management system;
- verification of safe release of hydrogen due to overpressure caused by insulation damage or failure;
- cycle life test of the CcH_2 storage system at cryogenic temperature;
- verification of insulation stability.

Figure 29.12 schematically illustrates the proposed validation test procedure for a CcH_2 storage system. Prior to a combined pressure–temperature cycling

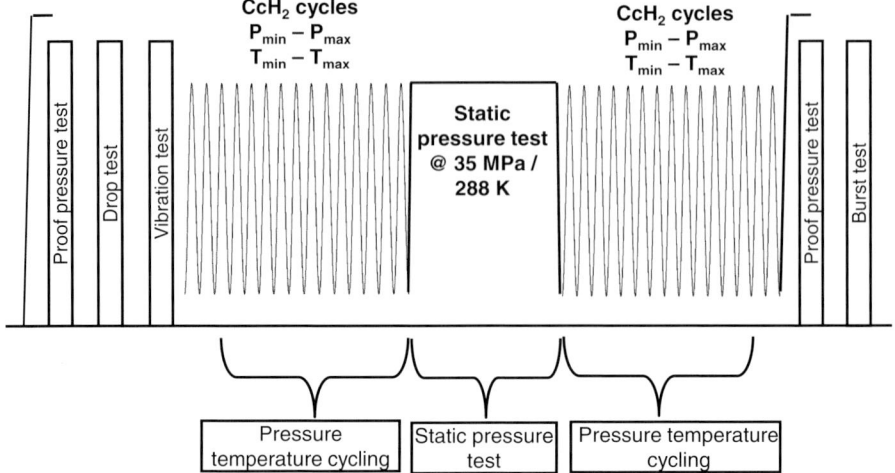

Figure 29.12 Basic validation procedure for a CcH$_2$ storage system.

test the pressure vessel and key auxiliary components undergo a proof test (typically at 1.5 times the maximum operating pressure), a drop test, and a vibration test reflecting mechanical loads in a vehicle use profile. Following these tests the storage system undergoes a cycle test with hydrogen gas exerting both, pressure and temperature loads to the storage vessel and related auxiliary components (e.g., insulation, valves, internal heat exchanger). The proposed combined pressure–temperature validation cycle is schematically shown in Figure 29.13. The cycle has been applied during first validation tests of BMWs CcH$_2$ storage prototypes. It consists of three subsequent cold single-flow refueling and discharge events starting with a storage vessel at ambient temperature. After the third refueling the internal heat exchanger is operated over five recirculation cycles before a warm gas refueling is performed in two steps, which simultaneously

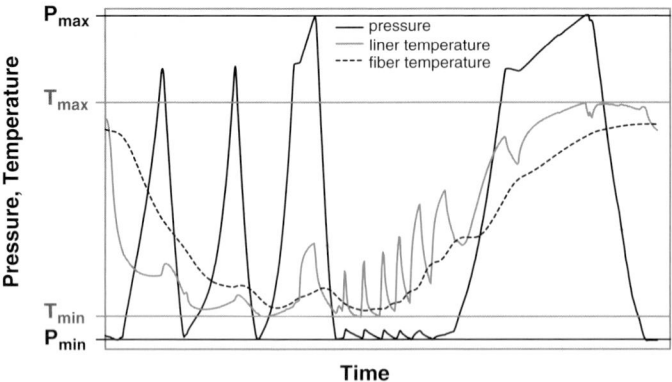

Figure 29.13 Schematic hydrogen durability test cycle.

heats up the storage vessel to its maximum in-cycle temperature. The pressure–temperature cycling is interrupted by a static pressure test to validate fatigue strength against high static loads at maximum operating pressure. After the cycle test a proof test with 150% of the maximum operating pressure is carried out followed by a final burst test to check the potential degradation of the storage vessel. The test is successful when a defined minimum burst pressure well above 2.25 times the maximum operating pressure of 35 MPa can be achieved. The adequate quantity of pressure and temperature cycles is determined by a probabilistic analysis as, for example, discussed by Kircher et al. [8].

29.5.2
Validation Challenges and Opportunities

In addition to durability, major challenges for the validation of an automotive CcH_2 storage system are linked to the verification of vacuum stability and reliability of the blow-off management system over lifetime. Since vacuum generation can be a lengthy process that might be prohibitive for mass production, a key focus in CcH_2 storage system design is set on materials with low disposition to outgassing in the operating temperature range of a CcH_2 storage system. That mainly concerns the pressure vessel's composite resin, but is also true for the multi-layer insulation materials. Furthermore, appropriate getter and adsorbent concepts are to be applied to maintain vacuum quality over lifetime and potentially improve the vacuum generation process. The blow-off management system that converts released hydrogen into water vapor needs to prove highest reliability over lifetime. The chosen solely passive catalytic combustion process has already proven its potential to fulfill the reliability requirement. Besides, these known challenges that require additional validation efforts compared to 70 MPa CGH_2 storage systems, CcH_2 also offers validation opportunities:

- In contrast to 70 MPa CGH_2 storage systems CcH_2 offers the opportunity of quick refueling without communication between vehicle and station dispenser and no further temperature conditioning required after compression. Thus, validation of refueling components and process will be less demanding than for 70 MPa CGH_2.
- By its super-insulation and choice of materials a CcH_2 storage is quasi-insensitive to ambient as well as thermodynamically induced temperature variations. Rapid discharge cooling of hydrogen does not lead to material challenges as in a 70 MPa CGH_2 storage system, where a minimum temperature of currently $-40\,°C$ needs to be guaranteed for Type 4 pressure vessels and sealing systems. Hence, performance validation under extreme weather conditions requires less effort.
- Cycling of a CcH_2 storage system is quick due to the thermal tolerance and temperature control means of a CcH_2 storage system (around 10 min for one refueling and discharge cycle compared to between 1 and 1.5 h in a 70 MPa CGH_2 storage).

29.5.3
Hydrogen Safety Validation

A CcH$_2$ storage system offers several intrinsic safety opportunities that guarantee a very high level of safety under all operating conditions and failure modes:

- The vacuum enclosure protects the storage pressure vessel against overheating in case of a fire as well as against mechanical and chemical intrusion. Even a localized fire, which may lead to a delayed response of the thermal pressure relief devices is not critical for a CcH$_2$ storage system, whereas this issue is one of the major safety challenges of 70 MPa CGH$_2$ storage systems.
- The vacuum super insulation enables very sensitive leak detection by simple means, for example, by measuring the pressure increase rate that significantly changes when the vacuum is broken by external or internal leaks. The vacuum furthermore guarantees a non-humid atmosphere over the lifetime of the pressure vessel and, thus, prevents carbon fiber composite aging due to external influence.
- The impact of a sudden failure of the pressure vessel that should be excluded as good as possible would be significantly lower for a CcH$_2$ storage system in cold operation than in a 70 MPa CGH$_2$ storage system. This is a consequence of the low adiabatic expansion energy (a measure for the energy release by sudden mechanical failure of a loaded pressure vessel) of cold hydrogen compared to warm hydrogen gas (Figure 29.14).

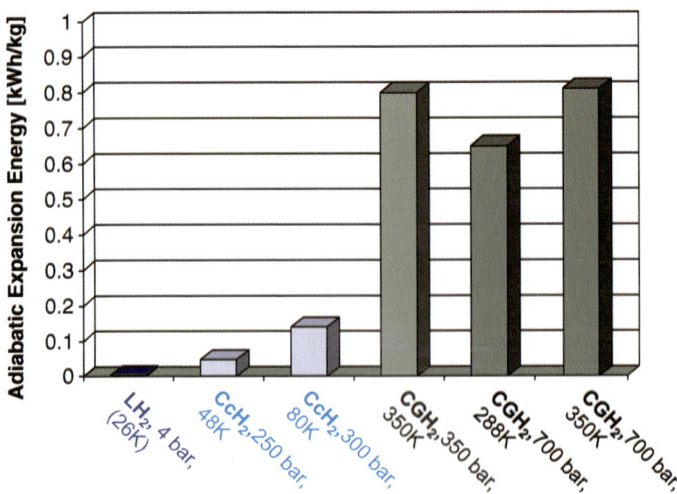

Figure 29.14 Comparison of adiabatic expansion energies per kg hydrogen for different physical hydrogen storage systems.

The remaining safety challenges of a CcH_2 storage system are related to a potential vacuum failure and the consecutive need for a quick hydrogen release. It, thus, is a hard design and validation requirement to prove the durability and crash resistance of the vacuum super insulation. This given, combined with a double independent pressure-triggered mechanical safety release system, a high overall hydrogen safety level can be achieved.

29.6
Summary

With the suggested concept of an automotive cryo-compressed hydrogen storage system (CcH_2) and its validation up to the date of publication, the authors, their teams, and co-workers offer a complementary hydrogen vehicle storage option, particularly suitable for larger vehicles with frequent long distance driving requirement. The thermodynamic opportunities of the proposed CcH_2 storage system seem to be a good compromise between maximized system energy density, refueling time and efficiency as well as high safety level and additional performance opportunities related to improved vehicle cooling and integration. Its bi-fuel capability that enables cryogenic CcH_2 gas refueling as well as ambient CGH_2 gas refueling up to 35 MPa outlines a direction to a long-term hydrogen infrastructure with a moderate maximum pressure level that evidently brings technical and economic advantages for both vehicle and refueling infrastructure. However, before CcH_2 vehicle storage systems can start into the mainstream competition, the remaining challenges related to validation of performance, durability, and safety under real world conditions as well as open questions on mass production capability and, thus, production cost must be addressed and overcome.

References

1 US Department of Energy (2012) Targets for Onboard Hydrogen Storage Systems for Light-Duty Vehicles. Available at http://energy.gov/sites/prod/files/2015/01/f19/fcto_myrdd_table_onboard_h2_storage_systems_doe_targets_ldv.pdf (accessed 15 February 2015).
2 Müller, C., Fürst, S., von Klitzing, W., and Hagler, T. (2007) Hydrogen safety: new challenges based on BMW hydrogen 7, Proceedings of the Second International Conference on Hydrogen Safety, 11–13 September 2007, San Sebastian, on CD ROM.
3 Aceves, S.M., Berry, G.D., and Rambach, G.D. (1998) Insulated pressure vessels for hydrogen storage on vehicles. *Int. J. Hydrogen Energy*, **23** (7), 583–591.
4 The Linde Group (2013) Linde Cryo Pump. http://www.linde-gas.com/en/innovations/hydrogen_energy/fuelling_technologies/cryo_pump.html (accessed 15 February 2015).
5 SAE (2014) Norm SAE TIR J2601: Fueling Protocols for Light Duty Gaseous Hydrogen Surface Vehicles.
6 European Parliament (2009) Regulation (EC) No 79/2009 of the European Parliament and of the Council on Type-approval of Hydrogen-powered Motor Vehicles, and amending Directive 2007/46/EC, 14 January 2009.

7 SAE (2013) J2579 – Standard for Fuel Systems in Fuel Cell and Other Hydrogen Vehicles 2013.
8 Kircher, O., Greim, G., Burtscher, J., and Brunner, T. (2011) Validation of cryo-compressed hydrogen storage (CcH2) – a probabilistic approach. Presented at the Fourth International Conference on Hydrogen Safety, 12–14 September, 2011, San Francisco.
9 Espinosa-Loza, F., Aceves, S.M., Ledesma-Orozco, E., Ross, T.O., Weisberg, A.H., Brunner, T.C., and Kircher, O. (2010) High density automotive hydrogen storage with cryogenic capable pressure vessels. *Int. J. Hydrogen Energy*, **35**, 1219–1226.

30
Hydrogen Liquefaction
Alexander Alekseev

30.1
Introduction

The use of liquefied hydrogen (LH2) is discussed in some future scenarios, mainly for transport purposes or for mobile applications, because of some important and advantageous properties like high energy density, light weight, and low environmental impact by using as fuel. The economics of hydrogen liquefaction and storage will be crucial for the attractiveness and competitiveness of this kind of transport chain. In addition, due to "economies of scale" a large system has a better chance to become a reality. This chapter gives some advice for the development of such a large future hydrogen liquefier.

On surfing the internet one will be surprised by the huge number of publications about the process design of hydrogen liquefiers in comparison to technologically very similar processes for nitrogen, natural gas, or helium liquefaction. It is a little in contradiction to today's reality: the total hydrogen liquefaction capacity worldwide, at less than 400 TPD (tons per day), is very small in comparison with installed nitrogen or natural gas liquefaction capacities (ca. 800 000 TPD for natural gas).

Another observation is that most of modern investigations [1–3] are focused more or less on "pure" process simulation. But the process design includes more than process simulation in a process simulator: design of main hardware components (compressors, turbines, heat exchangers, pumps, adsorbers, columns, etc.) is an essential part of the whole design procedure. The total efficiency of the plant (and therefore OPEX (operating expenditure)) can be estimated for specified hardware components only. Design/specification of main hardware components provides input for cost calculation (and therefore defines CAPEX (capital expenditure)), and also enables estimation for reliability of the plant and maintenance expenses. All these numbers are necessary to calculate the economics of an application. A description of hardware components for hydrogen liquefaction is therefore an important part of this publication.

Hydrogen Science and Engineering: Materials, Processes, Systems and Technology, First Edition.
Edited by Detlef Stolten and Bernd Emonts.
© 2016 Wiley-VCH Verlag GmbH & Co. KGaA. Published 2016 by Wiley-VCH Verlag GmbH & Co. KGaA.

30.2
History of Hydrogen Liquefaction

Hydrogen was liquefied first by James Dewar in 1897–1898:

> "He cooled hydrogen by means of liquid air at −200 °C and 200 atmospheres pressure and forced it through a fine nozzle. He obtained a jet of gas mixed with a liquid that he could not collect. The temperature of the hydrogen jet was very low, and by spraying it on liquid air or oxygen he transformed them into solids. Dewar was convinced that he could reach still lower temperatures, and he spent a year making a large liquid-air machine. In 1898 his endeavor ended in success. Cooled, compressed hydrogen liquefied on escaping from a nozzle into a vacuum vessel." [4].

For a relatively long period of time (40–45 years) hydrogen liquefaction was a kind of "esoteric" science–research at such low temperatures could be carried out only at a few universities and laboratories throughout the world before the Second World War (WW II).

The next important milestone was connected to development of first nuclear and later thermonuclear weapons by the USA, starting in the 1940s. These activities lead to development of a (still relative small) hydrogen liquefier with a capacity of $350 \, l \, h^{-1}$ in 1952 by the National Bureau of Standards (NBS) in Boulder, CO [5]. The most important fact from the present author's perspective is that not only was the first notable liquefaction capacity installed, but the appropriate infrastructure and dedicated equipment (storage capabilities, pumps, instrumentation, pipelines, safety procedures) were developed and installed at this time, and valuable practical experience and knowledge in handling of liquid hydrogen were also gained.

Attention to liquid hydrogen increased dramatically in 1957 in the USA with the launch of the world's first earth satellite, Sputnik I, by the Soviet Union:

> "almost overnight many of real and imagined impediments to using hydrogen as a rocket fuel evaporated . . . The first tonnage hydrogen liquefaction plant (capacity ≈ 1TPD) was designed, constructed and put into operation in 1957 by Air Products Inc. . . . During the 1960s, plant capacity and annual production again increased substantially with entry of Linde[1] into the field and continued growth in the number of Air Products plants"
>
> *(Reference [5] pp 381–382).*

---

1) "Linde" means here the company "Linde Air Products" – a former US subsidiary of German Linde AG, disappropriated by US administration after the WW II, and later taken over and integrated into the company Union Carbide.

These hydrogen activities in the USA were unique and unprecedented. The main competitors – Soviet scientists and engineers – started to work on a first LH2-rocket engine in the late-1960s. The first Soviet spacecraft with an LH2 engine was a combination of the spaceship "Buran" with the rocket "Energia" – a project started in 1976 and canceled in the late 1980s, unfortunately without any successor [6].

The most important "lessons learned" from this history are:

- The industry was able to develop considerable hydrogen liquefaction capacities in a very short time, starting from a level close to zero.
- It gives us hope that today's industry with:
  - essentially better tools for process design,
  - comprehensive experience in low temperature engineering,
  - considerably higher capabilities/capacities,
  - better materials and advanced technologies available will, if required, be able to develop an appropriate large liquid hydrogen infrastructure with a scale- up factor of 100–1000 in the future.

30.3
Hydrogen Properties at Low Temperature

30.3.1
Thermodynamic Properties

Table 30.1 [7] summarizes some important thermodynamic properties of hydrogen. The data in the table are given for two different molecular modifications of hydrogen, as described in the next section. The normal boiling point is very low and amounts to approx. 20 K (ca. −253 °C). The specific mass evaporation heat (in kJ kg^{-1}) seems to be relatively high, but the volumetric evaporation heat (in kJ l^{-1}) is essentially lower because of the very low hydrogen density.

An important issue from a safety perspective is the flammability of hydrogen in air: ignition temperature is approx. 580 °C for a very wide molar concentration range of 5–75%.

30.3.2
Ortho and Para Modifications of Hydrogen

30.3.2.1 Underlying Physics
Different relative orientations of the two nuclear spins in the diatomic molecule H_2 give rise to the molecular modifications designated by the prefixes "ortho" (nuclear spins in the same direction) and "para" (nuclear spins in opposite directions), known as o-H_2 and p-H_2 (Figure 30.1).

The equilibrium ortho–para composition (abbreviation: e-H_2) is temperature dependent (see Figure 30.3 below). At room temperature it corresponds

Table 30.1 Properties of hydrogen.

		Molecular weight (g mol^{-1})	Triple point			Normal boiling point (1.013 bar)			Critical point			Normal gaseous state (0 °C, 1.013 bar)		
			Temperature (K)	Pressure (bar)	Temperature (K)	Liquid density (kg m^{-3})	Evaporation heat (kJ l^{-1})	Evaporation heat (kJ kg^{-1})	Temperature (K)	Pressure (bar)	Density (kg m^{-3})	Density (kg m^{-3})	Kappa	Cp (kJ kg^{-1} K^{-1})
Normal hydrogen	n-H$_2$	2016	13.96	0072	20.36	70.9	31.76	448	33.26	13	30	0090	1.42	14.1
Para hydrogen	p-H$_2$	2016	13.86	0070	20.26	70.8	31.58	446	32.96	12.9	31.4	0090	1.38	15.1

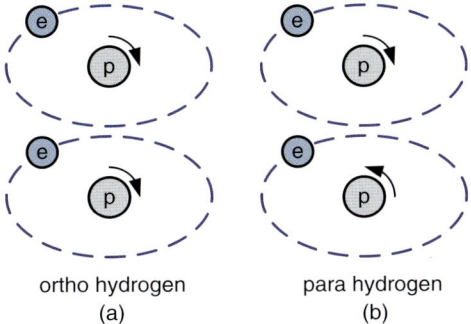

Figure 30.1 Ortho-hydrogen (a) and para-hydrogen (b); p – proton, e – electron, the arrows correspond to spin direction of the nucleus.

to approx. 75% ortho-hydrogen and 25% para-hydrogen and is referred to as so-called "normal hydrogen" (abbreviation: n-H_2). At the boiling point of liquid hydrogen the equilibrium ortho and para concentrations significantly shift to para-hydrogen, amounting to 0.21% o-H_2 and 99.79% p-H_2.

Most of the physical properties of hydrogen, such as vapor pressure, liquid density, triple point temperature, and pressures, depend slightly upon the ortho–para composition.

The interaction between the atomic nucleus and the electrons is extremely weak. Therefore, the natural, non-catalyzed ortho-to-para conversion is a very slow process and can take several days and weeks. However, the conversion reaction can be accelerated considerably by use of catalysts based on metal oxides.

30.3.2.2 Ortho-to-Para Conversion and Liquefaction of Hydrogen

From the liquefaction perspective the most important fact is that the energy content of ortho-hydrogen is higher than that of para-hydrogen. This means that the conversion from ortho into para hydrogen leads to heat generation.

Hydrogen can be cooled down and liquefied very quickly (in a few seconds or maximum minutes). In this case, the composition of the LH2 will correspond to the normal composition (composition at room temperature), because the natural ortho-to-para conversion hardly occurs in such a short period of time. Unfortunately, liquid hydrogen produced this way is not suitable for long-term storage: the natural ortho-to-para conversion will continue slowly. As the corresponding heat generation exceeds the heat required for evaporation of the LH2, this conversion will eventually lead to complete evaporation of LH2.

Consequently, liquefaction and conversion have to be carried out simultaneously, such that the ortho–para content of the liquid will correspond to the equilibrium mixture at the liquefaction temperature, mainly consisting of para-hydrogen, and hence be suitable for long-term storage.

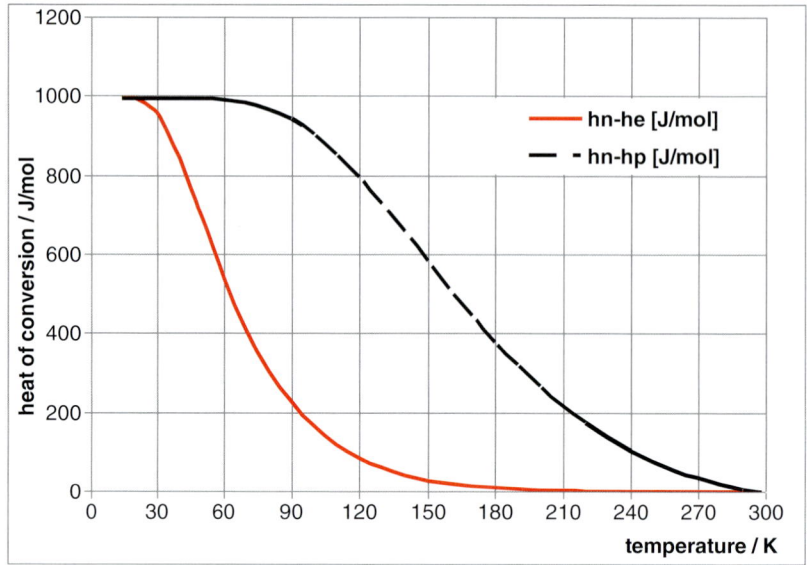

Figure 30.2 Heat of conversion, hn–he: from normal hydrogen to equilibrium hydrogen; hn–hp: from normal hydrogen to para hydrogen.

30.3.2.3 Available Data

Owing to the considerable heat of conversion, calculation of the ortho–para conversion becomes an important aspect in hydrogen liquefaction. Therefore, the quality of these data defines the accuracy of the whole process design.

Although collected and published more than 50 years ago, the data available from the National Institute of Standards (NIST, former NBS) are still unique; process design engineers worldwide use these data as main reference. The publication *Selected Properties of Hydrogen (Engineering Design Data)* [8] can be downloaded from the official NIST site. The following two figures are based on numbers from this publication.

According to the curve in Figure 30.2 the heat of ortho-to-para conversion at temperatures below 70 K is more or less temperature independent and amounts to approx. 1000 J mol^{-1}; at temperatures above 80 K this heat becomes lower and lower. The ortho-to-para conversion itself mostly occurs between 20 and 150 K (see below); therefore, the heat production by conversion is concentrated at this temperature range as well, with the maximum at 50–70 K.

30.3.2.4 Some Useful Thermodynamics

A good estimation for the ortho–para equilibrium is given by following equation using a Boltzmann distribution [9]:

$$\frac{y_{para}}{y_{ortho}} = \frac{1}{3} \cdot \frac{1 + 5 \cdot \exp\left(-6\frac{B}{T}\right)}{3 \cdot \exp\left(-2\frac{B}{T}\right) + 7 \cdot \exp\left(-12\frac{B}{T}\right)} \tag{30.1}$$

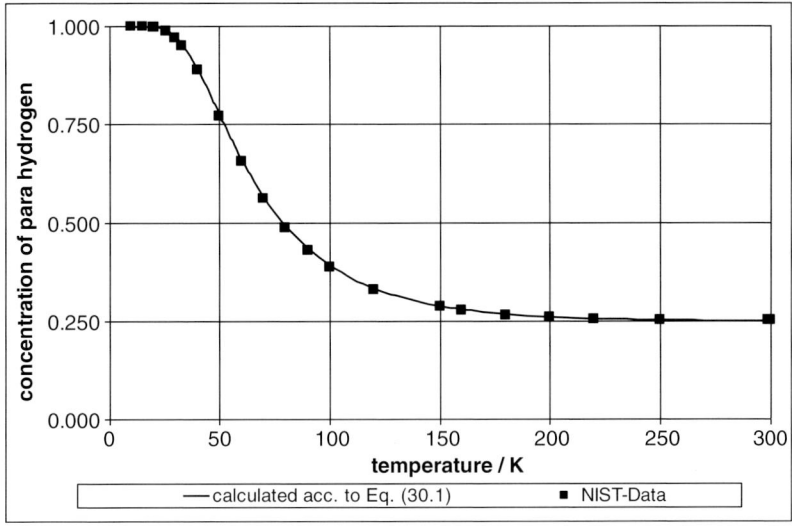

Figure 30.3 Equilibrium hydrogen composition depending on temperature.

where:

$B = \frac{h_P^2}{8\pi^2 I_T k_B} = 86.2 \text{ K}$

$h_P = 6.626 \times 10^{-34}$ J s

$k_B = 1.3806 \times 10^{-23}$ J K^{-1}

$I_T = 4.67 \times 10^{-48}$ kg m^2

T = temperature (K)

y = concentration (mol mol^{-1}).

Figure 30.3 shows the calculated concentration profile according to Eq. (30.1).

30.4
Principles of Hydrogen Liquefaction

30.4.1
Power Requirements

The theoretical minimum demand for hydrogen liquefaction amounts to 3.92 kWh (kg LH$_2$)$^{-1}$ according to Reference [12]. With 10–13 kWh kg^{-1}, today's hydrogen liquefaction plants need considerably more energy, corresponding to an overall thermodynamic efficiency of 30–40%. For comparison: the most efficient thermodynamic devices converting heat into power known today are combined cycle power plants with efficiencies of max. 61%. Applying a similar level of plant complexity to a hydrogen liquefier, and hence assuming a comparably high efficiency, a future hydrogen liquefier may reach power consumption values

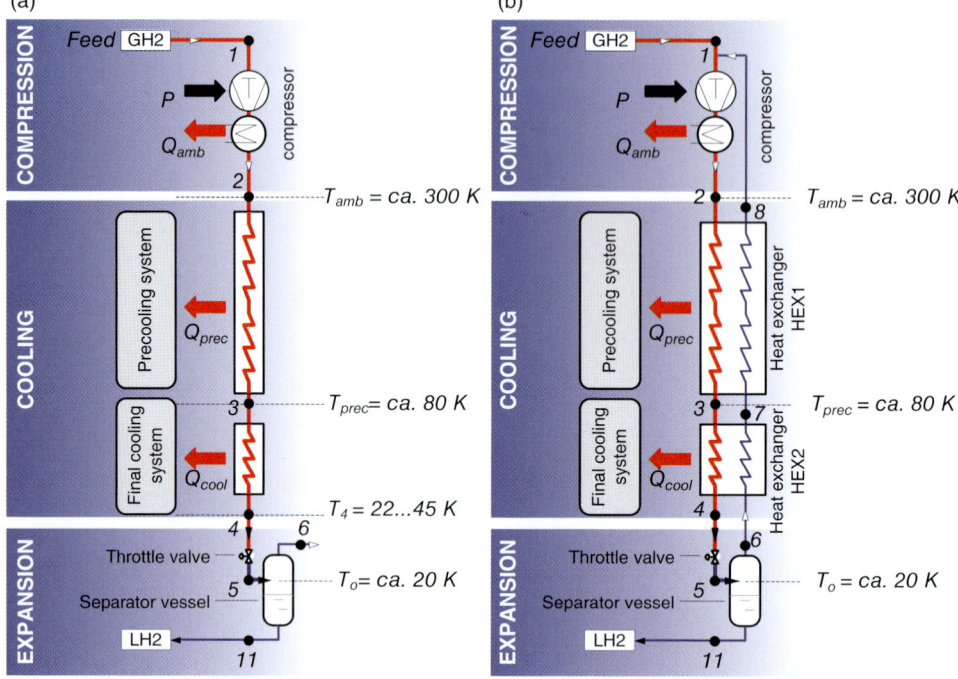

Figure 30.4 Hydrogen liquefaction: (a) general principle and (b) utilization of the flash gas cold.

of 6.5 kWh $(kg\ H_2)^{-1}$. This value may seem to be a realistic target for the long-term (20 years) development.

30.4.2
General Principle

The conventional way to liquefy hydrogen consists of the following three process steps (Figure 30.4):

1) **Compression** (process $1 \rightarrow 2$ in Figure 30.4) of pure and dry hydrogen (sometimes called "feed stream" or simply "feed") in a compressor to a high pressure p_{HP}, at least higher than the critical pressure (p_c): $p_{HP} > p_c$.
2) **Cooling** (process $2 \rightarrow 4$ in Figure 30.4) to low temperatures in two steps with the objective of cooling the gas to as low a temperature as possible:
   – precooling step (process $2 \rightarrow 3$): cooling hydrogen from ambient temperature $T_{amb} \approx 300$ K to temperatures close to $T_{prec} = 80 \pm 15$ K, the so-called liquid nitrogen temperature level – the heat flow Q_{prec} is removed from the hydrogen feed during this procedure;
   – cooling step or final cooling step (process $3 \rightarrow 4$): cooling to temperature $T_4 = 22$–45 K – the heat flow Q_{cool} is removed from the hydrogen feed in this phase.

Notably, ortho–para conversion and the corresponding removal of the conversion heat is an integrated part of the described cooling procedure.

3) **Expansion** (process 4 → 5 in Figure 30.4) of the hydrogen from high pressure p_{HP} to low pressure p_{LP} (a pressure slightly above ambient pressure), usually by means of a simple valve, in this case denominated as a throttle valve.

From a thermodynamic point of view this expansion is an adiabatic and an isenthalpic process, as neither heat nor power are applied or removed from the hydrogen. Hence the enthalpy (energy of the stream) remains unchanged during the throttling procedure.

During this expansion process the cooled pressurized hydrogen becomes colder and partially liquefies. The liquid hydrogen is separated from the remaining vapor in a simple vessel (referred to as phase separator); the liquid is collected in this vessel or can be removed from the liquefier.

Further, the vapor fraction (the so-called flash gas) is heated up to temperatures close to ambient temperature in heat exchangers, whereas the valuable cold of this cold vapor stream is utilized to cool the hydrogen feed stream. It is then injected into the feed stream prior to compression as shown in Figure 30.4b.

Part of the temperature–entropy diagram for hydrogen shown in Figure 30.5 helps us to visualize the throttling process. The green curves represent isobars – lines of equal pressure, given here for 1, 25, 30, 40, and 110 atm; all other (blue, red, and pink) curves are isenthalpic lines – lines of equal specific enthalpy of hydrogen.

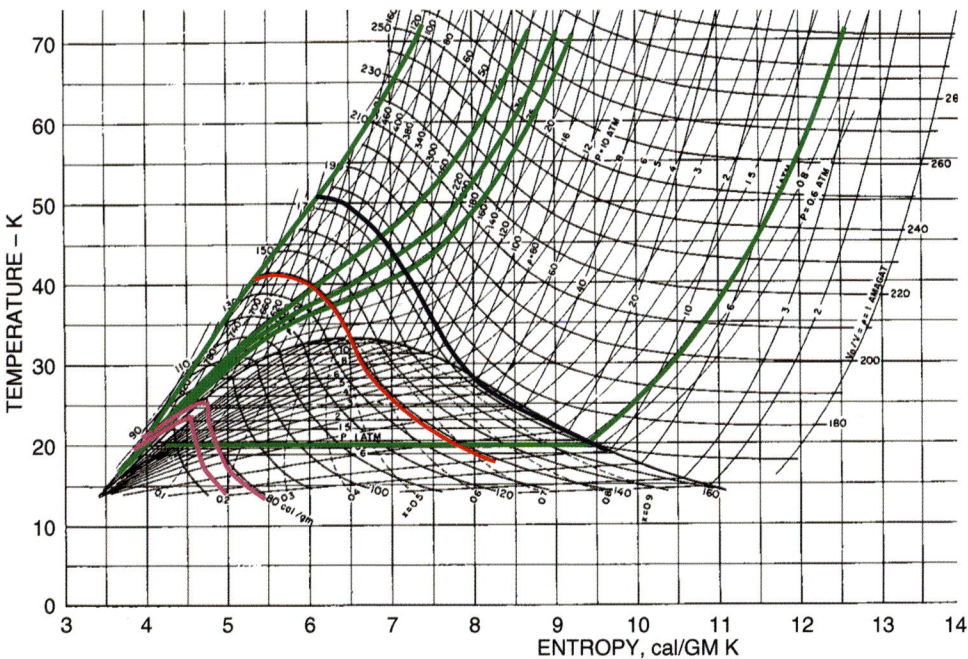

Figure 30.5 Cooling and throttling of the pressurized hydrogen in the T–s diagram.

Depending on the temperature before the throttle valve, three typical behaviors can be described:

1) The blue line describes an isenthalpic expansion from temperatures above 52 K. This scenario will not result in any liquefaction and is hence not applicable for a liquefaction plant.
2) The pink line outlines an expansion from a temperature level of 22–25 K to a pressure of 1 atm, resulting in very high liquefaction rate: The liquid fraction at the outlet of the throttle valve amounts to more than 85%, almost irrespective of the pressure level before the throttling device.
3) The entropy increase during the expansion and hence the thermodynamic losses are rather low.
4) One consequence of high liquefaction rate is that the amount of produced flash gas (vapor) is relatively low, such that the cold of the flash gas is not sufficient for cooling of the feed before throttling as shown in Figure 30.4b. Therefore, this kind of process can be realized by means of external cooling ($Q_{cool} > 0$) only.
5) An expansion from temperatures around 40 K, as represented by the red line, is a mixture of the scenarios described above: The liquid fraction at the valve outlet is less than 50%, which indicates a significant entropy increase and hence relatively high thermodynamic losses. On the other hand, a considerable amount of cold flash gas is produced during the throttling. After separation in phase separator this gas flows through the heat exchanger HEX2 and cools the feed hydrogen stream. Under certain conditions (usually if the feed gas pressure is higher than 50 bar[2] the amount of the cold flash gas is sufficient for cooling of the feed stream to temperatures below 45 K, and produces some liquid after throttling. This way the whole process below 80 K becomes "auto-refrigerated;" this means it does not need any external cooling below the precooling temperature level ($Q_{cool} = 0$). This kind of process is often referred to as a Joule–Thomson (JT) process.

30.4.3
Simple Joule–Thomson Process with Nitrogen Precooling

30.4.3.1 Basic Process
Figure 30.6 shows such a JT process with a so-called liquid nitrogen precooling in a very simplified form. Here the liquid nitrogen is used as a cold source for precooling of the hydrogen feed stream to liquid nitrogen temperature. The precooling part consists of two devices:

1) liquid nitrogen bath – a vessel with integrated heat exchanger (the simplest design is a spiral wound pipe) filled with liquid nitrogen;
2) heat exchanger HEX1 where the cold nitrogen vapor and cold flash gas is used for cooling of the warm pressurized feed stream.

---

2) This value depends on the pre-cooling temperature.

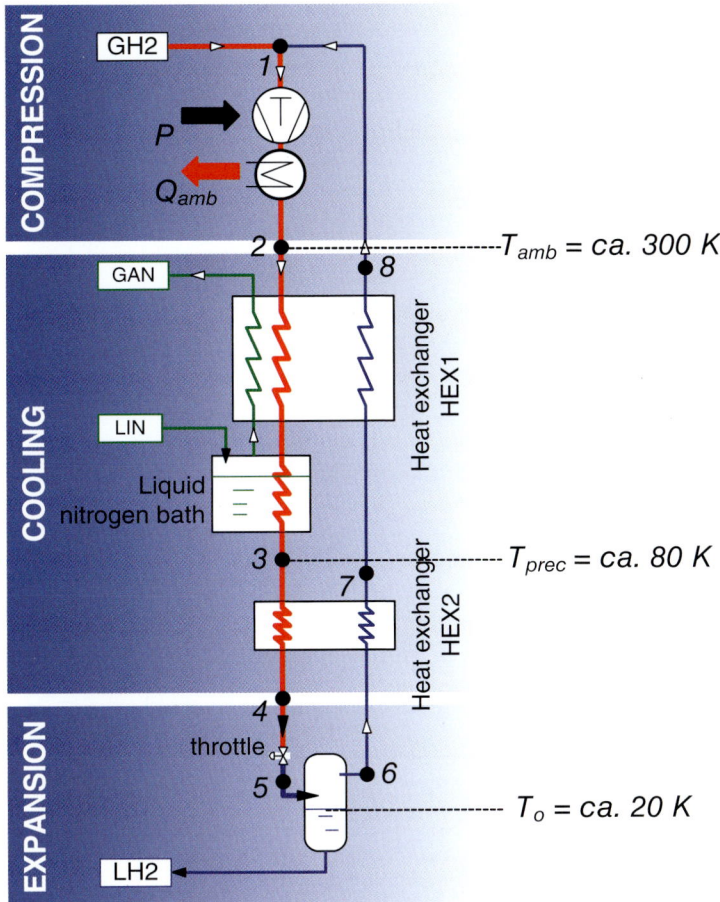

Figure 30.6 Joule–Thomson process with liquid nitrogen precooling.

At the beginning of the cool-down process the warm compressed (very high pressures close to 100 bar are required for this process) hydrogen feed flows through the liquid nitrogen bath and therefore becomes colder. The liquid nitrogen evaporates, this vapor leaves the bath, flows through the heat exchanger HEX1 to the ambient and helps to cool the hydrogen feed. This process is sometimes supported by a vacuum pump installed at the warm end of heat exchanger HEX1.

For final cooling, the Joule–Thomson process (described in the previous section) is used; the hydrogen feed is cooled with the heat exchanger HEX2 by the cold flash gas to a temperature of approx. 40 K (3 → 4) and is eventually expanded in a throttle valve to low pressures of approx. 1.2–1.4 bar (4 → 5). Through the throttling process the hydrogen feed becomes considerably colder, and partially liquefies. The produced liquid is separated from the cold vapor in a phase separator vessel, the cold vapor flows back through the heat exchanger HEX2 (6 → 7) and cools the inlet feed stream.

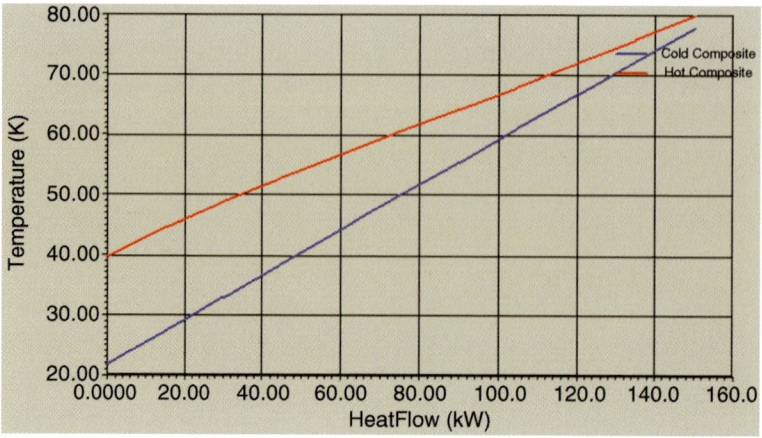

Figure 30.7 Temperature profile in final cooling stage, for 10 000 nm^3 h^{-1} hydrogen feed.

Figure 30.7 shows the corresponding temperature profile in the final cooling part (heat exchanger HEX2) for a process with p_{HP} = ca 120 bar (here without o–p conversion). A relatively high temperature difference of 18–20 K at the cold end of this heat exchanger leads to high irreversibility and as a consequence to a relatively low efficiency of the whole process.

30.4.3.2 Integration of Ortho-to-Para Conversion

Figure 30.8 shows two JT-processes with integrated ortho-to-para conversion in a very simplified form. The difference between these processes is the integration of the ortho-to-para conversion:

- The process shown in Figure 30.8a corresponds to very early design of hydrogen liquefiers. It consists of three separate o–p reactors (vessels filled with catalysts), placed on three temperature levels: liquid nitrogen level as well as before and after the throttle valve. Generally, the number of ortho-to-para conversion reactors and temperature levels can differ from the configuration shown here; the point is that the conversion reaction is focused on few locations. It causes a relatively high local heat generation (and temperature increase in the gaseous phase) inside of reactors and therefore additional thermodynamic losses.
- Figure 30.8b shows a more sophisticated and modern process design with catalyst placed inside the heat exchangers. The idea behind it is to distribute the ortho-to-para conversion homogeneously through the whole temperature range from room temperature to 20 K. Conversion hence takes place at the highest temperatures possible and sudden temperature changes with a respective high increase in entropy are avoided. As a consequence, the process becomes more efficient and fewer pieces of equipment are needed in the cold box design.

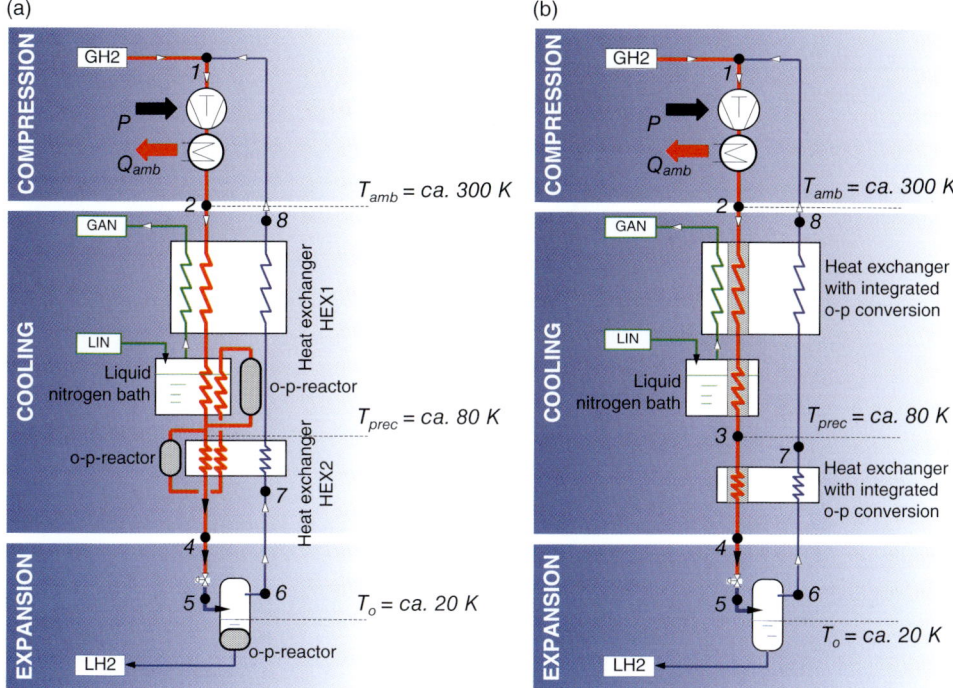

Figure 30.8 Ortho–para conversion in a Joule–Thomson process with liquid nitrogen precooling: (a) with separate ortho-to-para-reactors and (b) conversion is integrated into heat exchangers.

30.4.4
Evolution of the Hydrogen Liquefaction Processes

Table 30.2 shows the evolution of the hydrogen liquefaction processes from the beginning to recent times with respect to the changing requirements.

30.4.5
Process Design of the Precooling Part

The engineering phase for the whole system usually starts with the design of the cooling system (including precooling and final cooling), because this part of the process predefines the parameters of other subsystems (compression and consequently expansion).

In principle the cooling system and precooling system can be designed independently from each other, especially if the precooling temperature T_{prec} is fixed, for example, by use of nitrogen precooling. Alternatively, an optimal T_{prec} can be found as result of optimization for the whole process including compression and expansion.

Table 30.2 Evolution of hydrogen liquefaction processes.

Time period	General background and requirements	Process design
1898–1960	Main targets: • simple process design, • development/adaptation of suitable hardware components for hydrogen liquefaction, • development of hydrogen infrastructure including collection of operation experience	• Compression to very high pressures of around 100 bar, • cooling to temperature of 65–80 K by liquid nitrogen and flash gas, • cooling to 40–45 K by flash gas only (excluding external cooling applied below 65 K), • expansion in a throttle valve to low pressure.
1960–2000	Process and cost efficiency become more and more important, therefore: • main object of investigations is re-design of the process part below 80 K; • main goal is to reduce high thermodynamic losses in this part of system	New: • compression of hydrogen feed to moderate pressures of 25–40 bar, • cooling to 22–30 K by an external cold stream (and flash gas), for example by using a helium or hydrogen Brayton process
From 2000s	Process and cost efficiency become essential for a potential future hydrogen infrastructure, therefore: • main subject of investigations is further optimization of the process part below 80 K, • re-design of the process part above 80 K also becomes increasingly important.	New: • precooling to 65–80 K by an external cold stream (and flash gas), for example by using a nitrogen Brayton process or a mixed gas refrigerator, • multiple expansion turbines in low temperature part, • investigating JT-turbine instead of JT-valve, • further development of hardware components (compressors/turbines/ortho-to-para conversion devices)

30.4.6
Precooling by Nitrogen Brayton Cycle

30.4.6.1 Process Description

The most common option employed to cool the hydrogen feed to liquid nitrogen temperature level (precooling) is the so-called nitrogen Brayton process – a technology developed after WWII and widely applied in the air separation industry for cold generation and liquefaction of air gases like nitrogen, oxygen, and argon. In principle it is a low temperature version of the famous Brayton process

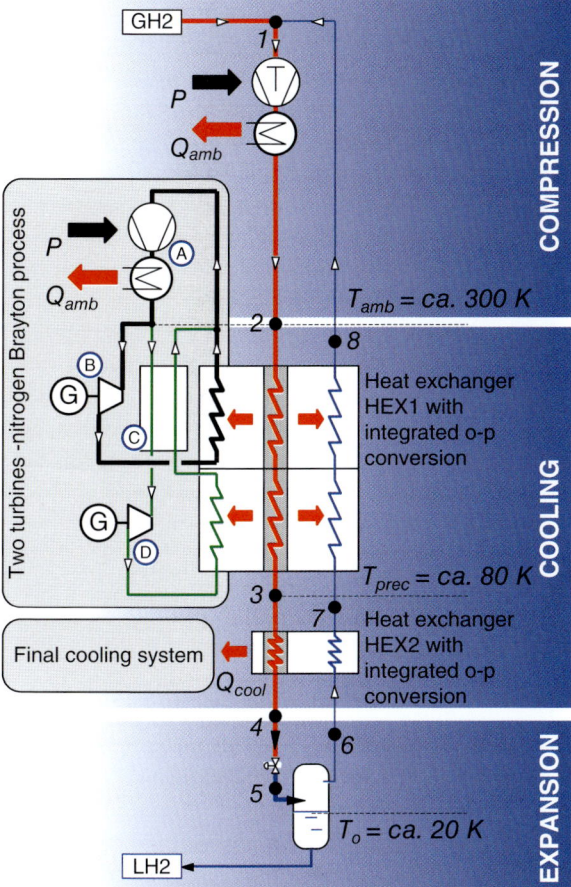

Figure 30.9 Precooling by two-stage nitrogen Brayton cycle.

known from power generation industry and realized in the form of a device called a gas turbine.

Figure 30.9 shows the so-called two-stage version of the Brayton process (in simplified form), which consists of two Brayton cycles switched in parallel and driven by the same compressor. The warm cycle is marked by a black line, the cold cycle by a green line. Similar to the gas turbine process, the working fluid (air in the case of gas turbines or nitrogen in low temperature processes) is first compressed in a compressor ("A" in Figure 30.9) from low pressure (usually 1–6 bar) to high pressure (8–80 bar). After compression the gas is cooled to ambient temperature by means of cooling water or cooling air and then expanded to low pressure in a gas expander ("B" in Figure 30.9) producing mechanical power and driving a generator device.

A considerable amount of mechanical power is produced during the expansion of the high pressure gas in this expander, which means that the enthalpy of the gas at the expander outlet is lower than at the inlet. This leads to a significant temperature reduction (60–100 K depending on pressure ratio). The temperature of the gas at the expander outlet reaches −40 to −80 °C ($\approx$190–230 K) or even colder. This cold stream flows through the heat exchanger HEX1 back to the compressor inlet and cools the hydrogen feed to the corresponding temperatures.

To produce the cold at temperatures below −80 °C another gas expander ("D" in Figure 30.9) in combination with heat exchanger ("C" in Figure 30.9) is used. A fraction of pressurized nitrogen after compression is cooled here in the heat exchanger to temperatures below −80 °C and after that expanded to low pressure in expander D resulting in a remarkable temperature reduction. The targeted temperature range for the gas after expander D is in the range of −150 and −190 °C (80–120 K), depending on the overall process design. The hydrogen feed gas can be cooled to the corresponding temperature by the cold stream from outlet of expander D. It first flows through the lower part of the heat exchanger HEX1 and subsequently through the heat exchanger C and cools the nitrogen before expansion in expander D.

30.4.6.2 Evaluation

The systems based on such a two-stage nitrogen Brayton process are well established; all required hardware components are available.

Leakage between channels of the heat exchanger HEX1 can be considered as a potential problem. Especially, the diffusion of nitrogen into the hydrogen feed has to be avoided, because it leads to solidification of nitrogen at temperatures below 60 K in the final cooling part and blockage of the hydrogen feed stream. The risk of uncontrolled leakage is partially mitigated by the fact that the pressure of the hydrogen feed is higher than the pressure of nitrogen on the cold side of the Brayton cycle. Nevertheless the tightness of process heat exchangers is an important issue for this kind of system.

Figure 30.10 shows a typical temperature profile in heat exchanger HEX1. Very tiny temperature differences indicate only small thermodynamic loss in this heat exchanger.

Another large source of thermodynamic losses in a Brayton process relates to gas expanders. The total cold capacity (power produced by turbines) for a hydrogen feed of 10 000 Nm^3 h^{-1} amounts to approx. 0.65 MW. With a relatively high isentropic efficiency of 85% this means nevertheless a loss of circa 0.11 MW on expanders and corresponds to approx. 0.22 MW additional power required for compression of the working fluid. This additional power consumption naturally becomes more relevant with increasing size of the plant.

30.4.7
Mixed Gas Refrigeration for Precooling Purposes

As a result of the above, another process idea has been discussed intensively in more recent times – the use of a so-called mixed gas refrigerator for precooling

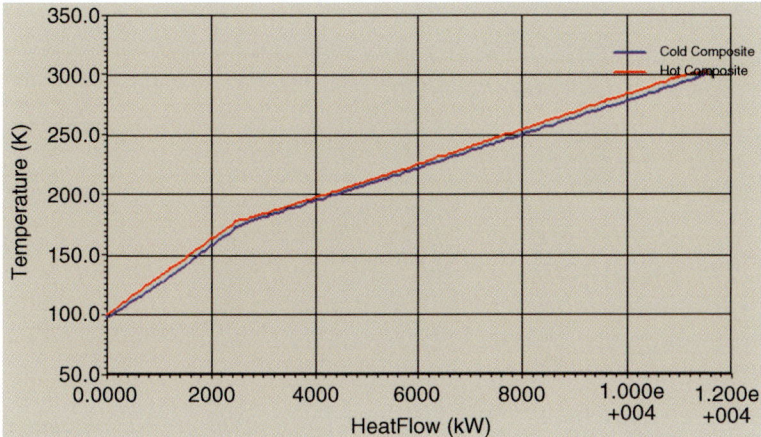

Figure 30.10 Temperature profile in heat exchanger HEX1, hydrogen feed 26 bar, here calculated without o–p conversion.

purposes [3]. This kind of cooling system does not include expanders at all (it consists mainly of heat exchangers); therefore, the corresponding expansion losses can potentially be saved. Of course this potential cannot be fully realized in a real system, but it opens up an opportunity for process design engineers to improve the overall efficiency of the system.

This advantage in energy efficiency is more or less balanced by several disadvantages:

- the lowest precooling temperature is only ca 90 K, because of solidification of almost all hydrocarbon based mixture components;
- extremely difficult design of the heat exchanger causes high costs;
- leakage: leakage inside the heat exchanger (from high pressure mixed gas to hydrogen feed) is a potential problem for reliable operation of the system and leakages of the mixed gas to the ambient is an environmental as well as safety issue;
- if the mixed gas contains flammable components, special safety measures are necessary, resulting in considerable additional costs;
- maintenance of the mixed gas composition is necessary to manage and control the operation especially with regard to temperature levels.

Therefore, a precooling stage based on a mixed gas refrigerator does not seem to be a viable solution today. But it may potentially become an interesting option for larger plants in the future.

30.4.8
Final Cooling

In Section 4.3 the use of the Joule–Thomson process for the final cooling was discussed. A further possible solution is the Brayton process – a technology

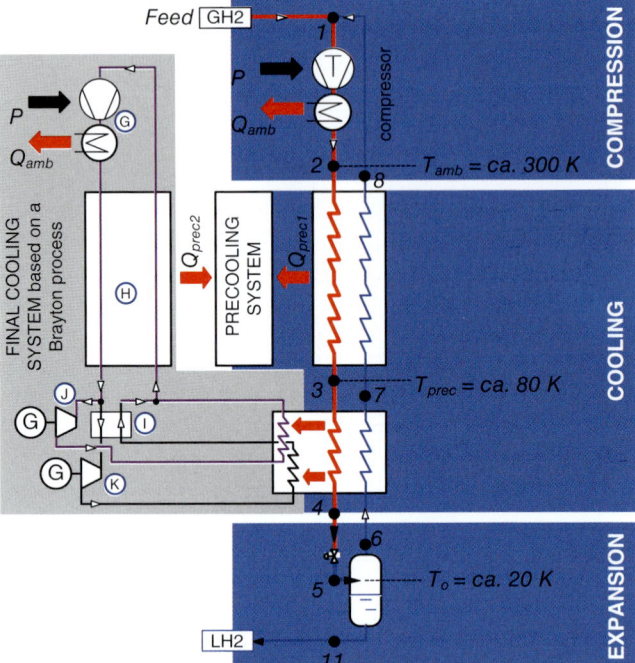

Figure 30.11 System with final cooling system based on a two-stage Brayton cycle.

discussed in the previous section for precooling purposes. An adapted cycle can also be used for final cooling, affording temperatures of 20–80 K (Figure 30.11). Appropriate pure working fluids for this temperature range are helium and hydrogen; some samples of the corresponding process design are described for example in Reference [10].

The process shown in Figure 30.12 corresponds to the Brayton process with helium as working fluid. The final cooling system is designed as an external cooling system: it is separated from the hydrogen feed loop, and it is driven by its own separate compressor ("G"), such that at least two compressors are needed – the hydrogen feed compressor and he helium Brayton cycle compressor (recycle compressor).

The hydrogen version of the low temperature Brayton process would look very similar to the process shown in Figure 30.12. However, in this case the whole process could theoretically be realized with a single compressor only, which would pressurize both streams: the hydrogen feed as well as the hydrogen for the low temperature Brayton stage(s).

However, in real engineering practice this advantage cannot be realized. For a helium Brayton cycle a low cost oil-lubricated compressor can be used, whereas for the flammable hydrogen Brayton cycle expensive piston type compressors are required which additionally are much more cost-intensive in terms of maintenance. Furthermore, most of the hydrogen feed gases come from steam reformer applications and are thereby available at sufficiently high pressure for processing

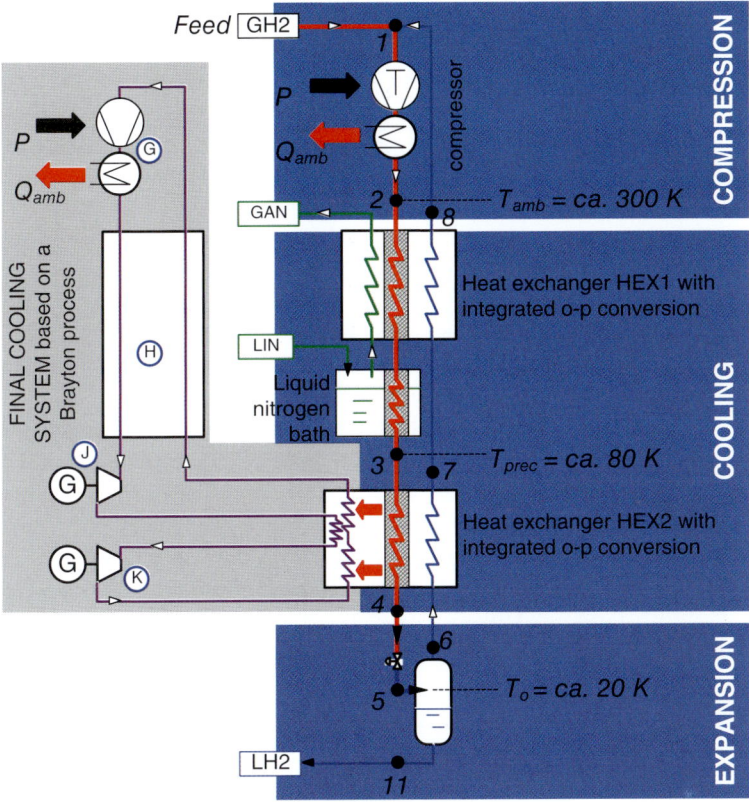

Figure 30.12 System with final cooling system based on a single-stage Brayton cycle with two separate impellers and intercooling in-between.

in a liquefaction plant. Hence hydrogen feed compression is often not required. As a result, helium Brayton cycles are the better choice for small capacities due to lower CAPEX, whereas for higher capacities the hydrogen Brayton cycles are applied with a higher CAPEX but savings of operational costs due to higher efficiency and hence lower power consumption.

30.5
Key Hardware Components

The process design is a scientific art on the border between thermodynamics and engineering: it is about how to design the most energy-efficient and cost-efficient system based on available hardware components: on one hand the availability of hardware components limits the creativity of engineers, but on the other hand the thermodynamic simulations performed for conceptually new processes set the new targets for developing new hardware components or for improving existing hardware.

Several key hardware components were developed in recent decades to fulfill some special requirements of helium refrigeration, such as very low operation temperature and related special properties of cryogenic fluids. The main components are compressors, expanders, and heat exchangers.

30.5.1
Compression

The compression of helium or hydrogen gas is a very special process, because of some special properties of these gases:

- helium and hydrogen are a very light substances (molecular weight 4 and 2, respectively) and have very low density'
- helium is a single-atom molecule, which leads to a relatively high isentropic exponent[3] $\kappa = 1.7$, while hydrogen has a relatively convenient κ of 1.4 because of the two atoms in its molecular structure.

30.5.1.1 Impact of the Isentropic Exponent

The following equations describe an ideal isentropic compression process. The outlet temperature (T_{out}) depends on inlet temperature (T_{in}), pressure ratio (p_{out}/p_{in}), and isentropic exponent:

$$T_{out} = T_{in} \left(\frac{p_{out}}{p_{in}}\right)^{\frac{\kappa-1}{\kappa}} \quad \text{Power} = \dot{M} R T_{in} \frac{\kappa}{\kappa-1} \left[\left(\frac{p_{out}}{p_{in}}\right)^{\frac{\kappa-1}{\kappa}} - 1\right]$$

The impact of the isentropic exponent is illustrated in Table 30.3. An ideal isentropic single-stage compression from ambient pressure (1.013 bar) to 2.2 bar is calculated here for a gas flow of $M = 100 \text{ mol s}^{-1}$, the inlet temperature

Table 30.3 Ideal isentropic compression of helium in comparison to compression of hydrogen.[a]

Parameter	H$_2$	He
κ	1.4	1.7
p_{out}/p_{in}	2.2	2.2
T_{out} (K)	374	413
T_{out} (°C)	101	140
Power (kW)	217	228

a) $R = 83\,135 \text{ J mol}^{-1} \text{ K}^{-1}$; $M = 100 \text{ mol s}^{-1}$; $T_{in} = 300 \text{ K}$; $p_{in} = 1013 \text{ bar}$; $p_{out} = 2.2 \text{ bar}$.

3) The isentropic exponent (κ, kappa) describes the connection between pressure and temperature of a gas during an isentropic process. A relatively high value of 1.7 means that a small pressure change (pressure ratio) leads to a big temperature change. For an ideal gas the isentropic exponent is equal to the "heat capacity ratio" $\kappa = \gamma = c_p/c_v$ – that is, it is the ratio of the heat capacity at constant pressure (c_p) to the heat capacity at constant volume, c_v.

is given by $T_{in} = 300$ K. The calculation is made for two different substances – hydrogen and helium, that is, for two fluids with different isentropic exponent values ($\kappa = 1.4$ for hydrogen).

This difference means that the outlet temperature of 140 °C is extremely high for the helium case – approx. 40 °C higher in comparison to the hydrogen case ($T_{out} = 101$ °C). The required power of approx. 228 kW for helium is approximately 5% higher than the compression power for hydrogen (ca. 217 kW).

This very simple calculation indicates very high compression temperatures and makes it obvious that an efficient cooling procedure is required during the helium compression.

30.5.1.2 Low Density

The low density of helium/hydrogen is the main reason and explanation for the fact that highly efficient turbo compressors are not applied for compression of these gases although this kind of equipment is always used in chemical engineering for compression of gaseous fluids.

A turbo compressor consists of several compressor stages, and every compressor stage includes three elements: guide vane, impeller (wheel), and diffusor (volute camber). The gas is accelerated firstly in the guide vane, but mainly in the impeller (the mechanical energy is transformed here into the kinetic energy of the gas with a final velocity close to 300 m s^{-1}), and after the acceleration the gas is decelerated in the diffusor (flow channel with widened cross-section). The kinetic energy of the gas is transferred into potential energy (pressure). The simplified Bernoulli equation describes this process:

$$p_1 + \frac{1}{2}\rho_1 v_1^2 = p_2 + \frac{1}{2}\rho_2 v_2^2$$

where $p_1, p_2, \rho_1, \rho_2, v_1, v_2$ are, respectively, the pressures, densities, and velocities at the inlet (1) and outlet (2) of the diffusor.

The achieved pressure difference in a single compression stage depends therefore on the velocity of the impeller as well as on the gas density. This means that a relatively high pressure difference (and therefore pressure ratio) can be achieved by compressing a heavy gas (like argon, krypton, and xenon). However, the pressure ratio for light gases like helium/hydrogen is limited to 1.05–1.2. Consequently, numerous compression stages are necessary instead of a single stage for heavy gases. Taking into account a corresponding number of intermediate coolers (necessary after every stage because of the high temperature after compression stage) along with pressure losses and so on, the turbo machine for helium/hydrogen would become very complex and expensive. As a result, today only piston and screw compressors are used for hydrogen and helium compression, respectively. Figure 30.13 shows the limitations for the three types of compressors.

30.5.1.3 Screw Compressor

The twin-screw type compressor is a positive displacement type device that operates by pushing the working fluid through a pair of meshing

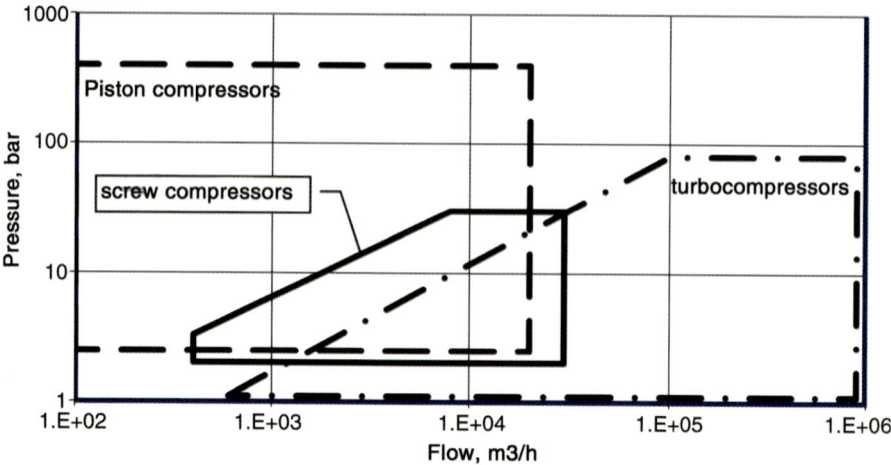

Figure 30.13 Application ranges for different compressors.

close-tolerance screws similar to a set of worm gears. Each rotor is radially symmetrical, but laterally asymmetrical (Figure 30.14). The working area is the inter-lobe volume between the male and female rotors. It is larger at the intake end, and decreases along the length of the rotors until the exhaust port. This change in volume results in the compression. [11]

A very common compression device is an oil lubricated screw compressor. The compressor oil is injected here into the compression cavities, it bridges the space between the rotors, both providing a hydraulic seal and transferring mechanical energy between the driving and driven rotor. Another important

Figure 30.14 Lysholm screw rotors. Reproduced from Reference [11].

function of the compressor oil is to provide an efficient cooling sink for the hot gas and hence to approach isothermal compression as much as possible.

On the other hand, the compressor oil causes problems in the low temperature part of the systems because it freezes out and consequently clogs valves and heat exchangers. Therefore, the oil has to be separated from the discharge stream before the coldbox inlet.

A screw compressor is a low cost device, because this kind of compressors belongs to bulk commodity: they are used to supply compressed air for general industrial applications or in conventional refrigeration. These air screw compressors can be adapted for helium compression, because of the similar pressures (1 bar at the inlet) and pressure ratio (usually 8–10) in a typical helium Brayton cycle.

Theoretically, the screw compressors can be applied for hydrogen compression. However, the use of mineral compressor oil for this purpose is practically impossible because of hydrogenation – a chemical reaction between molecular hydrogen and some organic compounds contained in the mineral oil, usually in the presence of a catalyst, but non-catalytic hydrogenation takes place at the high temperatures typical for compression, too.

30.5.1.4 Reciprocating (Piston) Compressors

Another kind of compressor widely used in helium/hydrogen refrigeration is the conventional reciprocating compressor. It is used in helium refrigeration for small scale liquefaction ($<100\,\mathrm{l\,h^{-1}}$) or for compression of helium/hydrogen to high pressure (>30 bar), if required.

For hydrogen compression, reciprocating compressors, especially the so-called "dry" compressors operating without lubricating oil are considered as equipment commonly used.

30.5.2
Expansion Turbine (or Expander or Turbine)

The heart of a helium/hydrogen liquefier is the expansion turbine removing the energy from the gas and hence enabling and supporting efficient cool-down. Consequently, an essential part of the know-how is related to building an efficient, reliable, and service friendly expansion turbine.

During the expansion from high pressure to low pressure the gas produces mechanical power. The produced mechanical power theoretically can be integrated into the refrigeration process. At temperatures below 80 K this option is still relatively uncommon as the additional complexity of the process would not justify the rather comparably low amount of energy of 0.3–1% which can be recovered. Engineers hence prefer to simply dissipate this energy.

The next figure shows two different turbines in terms of bearing supports. The common features are the following:

- The impeller is connected to the shaft, which is supported by at least two bearings.

- The operation temperature of the impeller is relatively low, whereas the bearings work at temperatures close to ambient or higher and, therefore, the distance between these two parts has to be as long as possible to avoid heat leakage by conductivity.
- Both areas (bearing and impeller) are separated spatially to minimize interchange of gas between the process section and the gas or medium inside the bearing chamber.
- The impeller of the turbines with gas bearings is usually located below the bearing chamber; in contrast, the bearing part of the turbine with an oil bearing is located below the impeller. It is necessary to prevent contamination of the impeller by the oil.

30.5.2.1 Oil Bearing

The oil bearing option (Figure 30.15a) was developed in the early 1960s: the oil film guarantees the load bearing capacity for both axial and radial loads. The oil is pumped to high pressure by a separate oil pump and injected into the bearing gap. It then flows down to the bottom of the bearing chamber and further to a small vessel placed below. This vessel is required for separation of the gas and oil. Simultaneously it works as a feed vessel for the oil pump. The gas in the bearing space comes from the main process stream through the gap between the impeller space and bearing chamber and is required to avoid any oil leakage

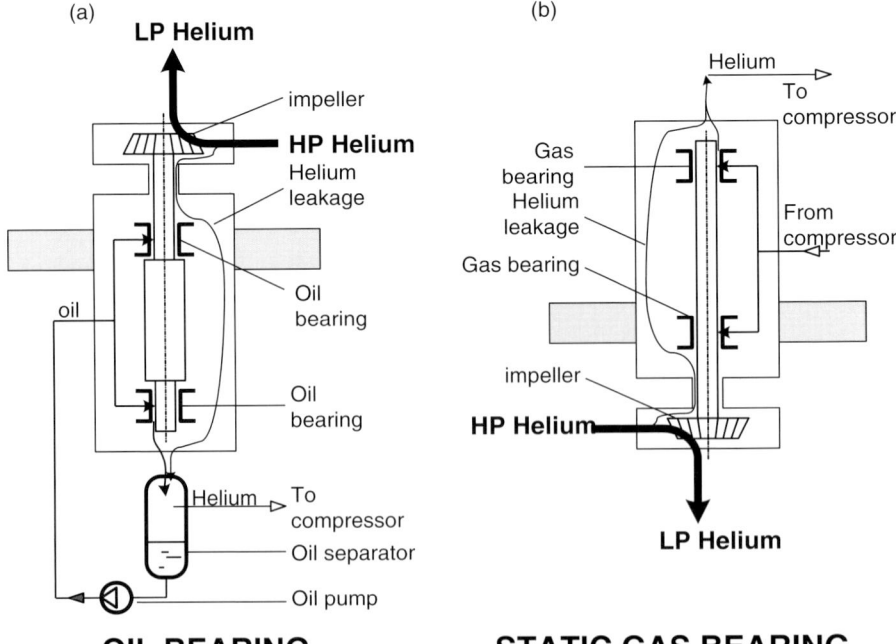

Figure 30.15 Typical bearing systems.

from the bearing location to the process section. This gas is then routed to the suction line of the main helium compressor. It is a kind a parasitic by-pass stream that reduces the efficiency of the whole cycle.

30.5.2.2 Gas Bearing

Later, another solution was found using a gas film instead of an oil film in the bearing.

Initially, the so-called static gas bearing (Figure 30.15b) was introduced to the market: the compressed gas from an external source (usually main recycle gas compressor) is injected into the tiny gap between shaft and support. The problem of potential contamination of gas by oil was elegantly solved this way.

Afterwards, the so-called dynamic bearing was invented. The external high pressure gas source and corresponding by-pass stream is completely eliminated in this solution, based on an appropriate design of the bearing, which allows build-up to the required pressure inside of the bearing chamber internally. The efficiency of the gas expansion turbines with dynamic bearings is therefore higher than for turbines with oil bearings and static gas bearings.

Figure 30.16 shows a modern turbine expander with dynamic gas bearings. The high pressure feed gas expands in the impeller and leaves the unit downwards axially. An impeller of a small turbo compressor is mounted on the opposite end of the turbine shaft. Consequently, the turbine impeller drives this turbo compressor, which is part of a secondary closed cycle (the so-called braking cycle): the gas is compressed, subsequently cooled against cooling water, and finally throttled again in a valve prior to being routed to the compressor inlet. This way the mechanical power produced by the turbine is eventually dissipated to ambient via the cooling water.

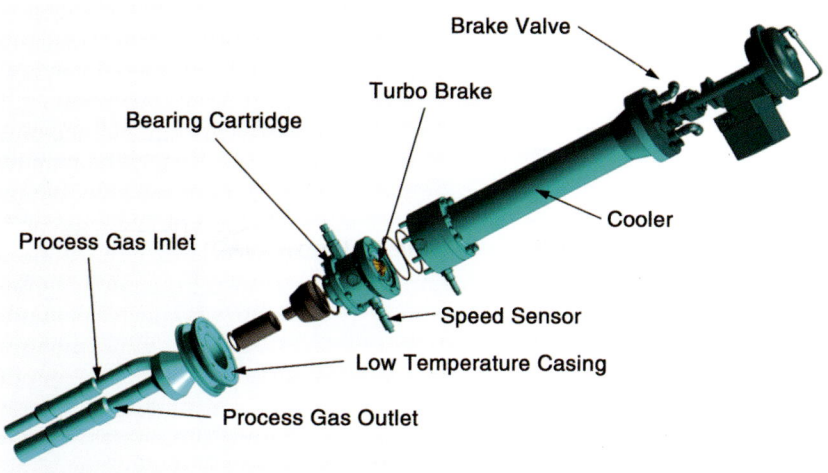

Figure 30.16 Gas expander with dynamic bearings (Linde).

Figure 30.17 Inside of a plate-fin-heat exchanger.

Unfortunately, expanders based on gas bearing for high power loads (MW range) are currently not available. Only the option based on oil bearing is discussed for large-scale hydrogen liquefaction at present.

30.5.3
Heat Exchangers

The heat exchangers in the cryogenic section are almost exclusively of the aluminum plate fin type.

Figures 30.17 and 30.18 show the structure of a plate fin heat exchanger module: the process streams are led through passages. Up to 200 of these passages are stapled one on top of the other. The large number of passages makes it possible to bring several streams into thermal contact within one unit (Table 30.4).

Table 30.4 Typical parameters of aluminum plate fin heat exchangers.

Parameter	Value
Size:	Up to $1.8 \times 1.5 \times 8.0$ m
Specific surface:	500–1800 $m^2 m^{-3}$
Fin:	(Plain perforated)
	(serrated)
	fin height: 4–10 mm
	fin thickness: 0.1–0.6 mm
Temperature:	−269 to +65 °C
Pressure:	< 115 bar
Materials:	ASTM 3003 (DIN AlMnCu), ASTM 5083, 6061 (DIN AlMg4.5 Mn)

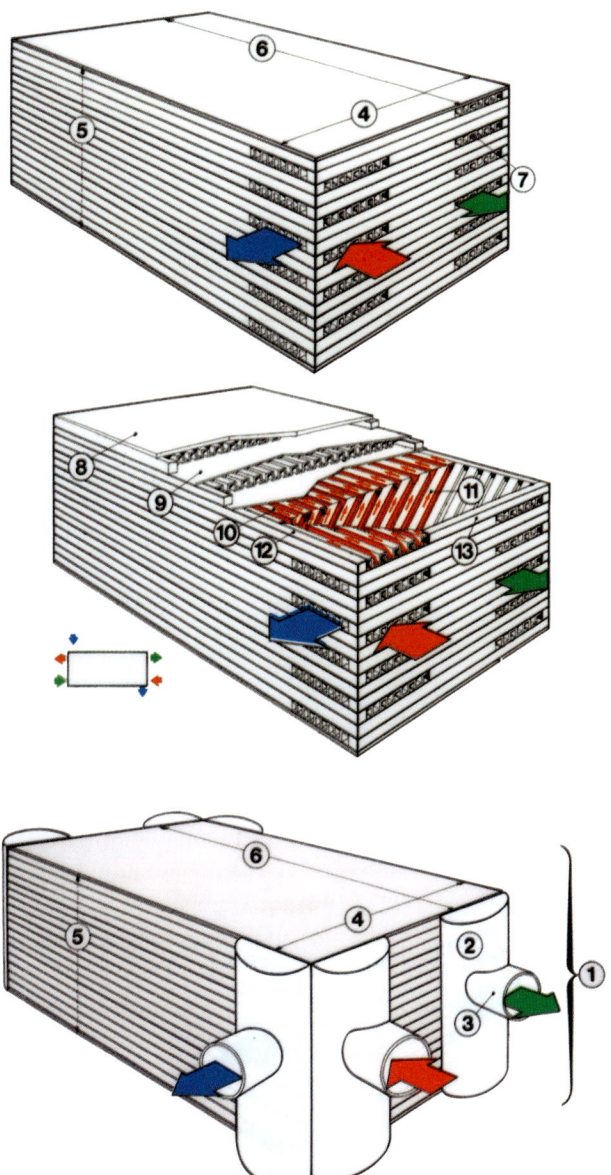

Figure 30.18 Aluminum plate fin heat exchanger: 1 – core, 2 – header, 3 – nozzle, 4 – width, 5 – height, 6 – length, 7 – passage outlet, 8 – cover sheet, 9 – parting sheet, 10 – heat transfer fins, 11 – distribution fins, 12 – side bar, and 13 – front bar.

The outlet frame is formed by 10–25 mm side bars, which are only interrupted for passage inlets and outlets. A fluid enters the passage via nozzles and headers. Beginning from here the flow is distributed with special fins over the entire cross-section of the passage and passed to the main section with heat transfer fins. The arrangement of the passages as well as fin types can be selected by process design engineers according to the process requirements.

The advantages of plate fin heat exchangers are:

- high flexibility concerning the number of process streams: several of them can be passed through a single block, hence allowing a complex and sophisticated process design;
- high specific heat exchange surface: this helps to realize a very efficient process because of small temperature differences and pressure losses;
- low pressure losses;
- low specific costs.

30.6
Outlook

The main purpose of this chapter is to give an overview on available hydrogen liquefaction technology, including process design and corresponding hardware components. A positive effect of this kind of process design analysis is that some routes and directions for technology development/improvement become increasingly visible. They are summarized as follows:

A) The whole life cycle liquefier costs has to be taken into account for fair comparison of different systems, and both portions – CAPEX and OPEX – can already be evaluated with high accuracy today.
B) The general impression is that the potential for improvement is well understood today. Therefore, more or less conventional engineering routes concerning process optimization have to be pursued in the next decades (short- and middle-term development):
  – adaptation of a two-stage nitrogen Brayton process for the precooling purposes;
  – further optimization of the low temperature Brayton process – optimization parameters are:
    • number of Brayton stages,
    • number of impellers per Brayton stage,
    • number of intermediate cooling/heating stages between expanders,
    • temperature at impeller outlet,
    • pressure ratio per impeller (if more than 1);
  – minimization of temperature difference and pressure losses in heat exchangers;
  – optimization of pressure levels;

- choice of best working fluid (hydrogen or helium or neon–helium–hydrogen mixture as suggested in Reference [3]), especially in combination with choice of the best suitable hardware components;
- further cost reduction.

C) From the long-term perspective (for the time after 2020) further development of hardware components is necessary. This primarily concerns rotating equipment like:
- compressors (biggest potential for improvement);
- hydrogen and helium gas expanders for higher capacity;
- Joule–Thomson expanders instead of Joule–Thomson throttling valve.

D) We see some energy saving potential not only in the liquefaction process, but in optimization of the whole LH_2-logistic chain: utilization of the cold and para-to-ortho-conversion-cold during the evaporation and warm-up of the LH2 could help to minimize the total energy requirements further.

References

1 Krasae-in, S. (2013) Efficient hydrogen liquefaction processes. PhD thesis, Norwegian University of Science and Technology (NTNU).
2 Staats, W.L. Jr (2008) Analysis of supercritical hydrogen liquefaction cycle. Ph.D. thesis, Massachusetts Institute of Technology (MIT).
3 Quack, H., Essler, J., Haberstroh, Ch., Walnum, H.T., Berstad, D., Drescher, M., and Nekså, P. (2012) Search for the best processes to liquefy hydrogen in very large plants, in *12th IIR International Conference on Cryogenics, 2012 (CRYOGENICS 2012): Proceeding of a Meeting Held 11–14 September 2012, Dresden Germany*, International Institute of Refrigeration, Paris, pp. 281–286.
4 Costa, A.B. (2008) James Dewar, in *Complete Dictionary of Scientific Biography*, Charles Scribner's Sons, Detroit.
5 Scurlock, R.G. (1993) *History and Origins of Cryogenics*, Clarendon Press, Oxford. ISBN 10: 0198548141.
6 Gubanov, B.I. (1998) *The Triumph and Tragedy of Energiya: The Reflections of a Chief Designer*, vol. **3**, NIER, Nizhniy Novgorod.
7 Hausen, H. and Linde, H. (1985) *Tieftemperaturtechnik*, Springer Verlag.
8 McCarty, R.D., Hord, J., and Roder, H.M. (1981) *Selected Properties of Hydrogen (Engineering Design Data)*, Monograph 168, National Bureau of Standards, Washington. Available at http://www.boulder.nist.gov/div838/Hydrogen/PDFs/McCarty.Hord.Weber.1981.monograph168.pdf.
9 Kinard, G.E. (1998) The commercial use of liquid hydrogen over the last 40 years, in *Proceedings of the 17th International Cryogenic Engineering Conference, Bournemouth, UK, 14–17 July 1998* (eds D. Dew-Hughes, R.G. Scurlock, and J.H.P. Watson), Institute of Physics, Bristol, pp. S39–44.
10 Ohlig, K. and Decker, L. (2013) Hydrogen, 04. Liquefaction, in *Ullmann's Encyclopedia of Industrial Chemistry*, Wiley-VCH Verlag GmbH, Weinheim. DOI: 10.1002/14356007.o13_o05.pub2.
11 Wikipedia (2015) Rotary screw compressors. http://en.wikipedia.org/wiki/Rotary_screw_compressor.
12 Peschka, W. (1992) *Liquid Hydrogen – Fuel of the Future*, Springer-Verlag, Wien.

31
Hydrogen Storage by Reversible Metal Hydride Formation

Ping Chen, Etsuo Akiba, Shin-ichi Orimo, Andreas Zuettel, and Louis Schlapbach

31.1
Introduction

Hydrogen, the simplest element and a carbon free synthetic fuel, reacts with most solids when temperature and pressure are appropriate. The storage challenge is to find or design solids and reversible hydrogen reactions at moderate temperature and pressure, which enable high density, safe, sustainable, and economic short- and medium-term stationary or mobile storage and handling of that precious element, which is a molecular gas at ambient temperature and pressure. Since the discovery of hydrogen as a component of water by Lavoisier (1783) and the first observation of "occlusion" of hydrogen in palladium metal by T. Graham (1866), work on hydrogen and metal interactions has seen various motivations in science and technology with many ups and downs, dreams and reality. Room temperature superconductivity of metallic hydrogen, failure of structures due to embrittlement, efficient catalysis, super-permanent magnet fabrication, aircraft sensors, and hybrid vehicle battery electrodes are just a few examples.

In a use-inspired classification of rather well-studied solid state hydrogen bonding materials we distinguish:

- hydrides of elemental metals, alloys and intermetallic compounds, so-called metal hydrides, generally with atomic hydrogen in bulk interstitial lattice sites and metallic bonding, some of them reversible around room temperature with rapid kinetics, volumetric density comparable with that of liquid hydrogen, gravimetric density so far limited to 1–3 wt% [1–8];
- complex hydrides of metals or metal–nonmetal compounds, partly charged hydrogen in bulk lattice site positions, covalent or ionic bonding, reversible at temperatures of 200–400 °C with limited kinetics, gravimetric density 5–10 wt% or higher [1–3,8–13];
- high porosity or nanosized cage structures of high specific surface area, inorganic or organic, with weakly adsorbed (physisorbed) molecular hydrogen, so far operating at cryogenic temperatures only [14–20].

Hydrogen Science and Engineering: Materials, Processes, Systems and Technology, First Edition.
Edited by Detlef Stolten and Bernd Emonts.
© 2016 Wiley-VCH Verlag GmbH & Co. KGaA. Published 2016 by Wiley-VCH Verlag GmbH & Co. KGaA.

In this chapter, we summarize extensive research results on the more classical hydrides of elemental metals and intermetallic compounds with references to available reviews, and focus more on the ongoing research on the formation of complex hydrides from solids; we also provide an overview of results elaborated for hydrogen storage by physisorption.

A sustainable low-carbon society does not base its energy supply on further exploitation and consumption on fossil forms of carbon. The great potential of carbon-free renewable energy is strongly linked with storage. Technically feasible forms of storable energy are gravitational, electric, heat, and chemical and synthetic fuel storage. Among these various types the production of hydrogen and its reversible storage in metal hydrides as well as the use of suitable metal hydrides for heat storage and functionalization of various materials for energy technologies have high promising potential [3,6–9,19,21].

31.2
Summary of Energy Relevant Properties of Hydrogen and its Isotopes

Physics distinguishes four types of interactions or forces: gravitation, Coulomb, strong and weak nuclear. All chemical reactions and their related energy changes are based on the Coulomb interaction. Hydrogen is the best chemical energy carrier in the sense that the H atom is made of one proton and one electron, no neutrons to be carried on, with the strong electron binding energy of 13.5 eV. Hydrogen and its isotopes deuterium (D, one proton, one electron, one neutron; stable) and tritium (T, one proton, one electron, two neutrons; radioactive, 12.5 year half-life) form diatomic molecules. As the H_2–H_2 interaction is small, hydrogen is a gas at ambient temperature and pressure. Liquefaction occurs below 21 K (pressure dependent); at even lower temperatures (14 K) an insulating transparent solid of hydrogen molecules is formed, which transforms under very high pressure ($>2 \times 10^{11}$ Pa) into a metallic solid of H atoms, very likely a superconducting simple metal [22].

The higher heating value of hydrogen gas amounts to 142×10^3 kJ kg^{-1} (methane 55×10^3 kJ kg^{-1}, gasoline 48×10^3 kJ kg^{-1}). The dissociation/recombination reaction $H_2 \leftrightarrow 2H$ is strongly endo/exothermic involving 432 kJ mol^{-1} at room temperature. With respect to safety the broad flammability limits of 4–75 vol.% and low minimum energy for ignition of 0.02 mJ in air are, to some extent, compensated by the high diffusion coefficient (0.61 cm^2 s^{-1} as compared to 0.16 and 0.05 cm^2 s^{-1} for methane and gasoline, respectively).

31.3
Hydrogen Interaction with Metals, Alloys and Other Inorganic Solids

All metals and many other inorganic solids dissolve at appropriate temperature and pressure conditions at least small quantities of oxygen, carbon, nitrogen, and of course also of hydrogen. In view of energy storage the solution and reversible sorption of hydrogen in the bulk of solids offers attractive phenomena. Many metals dissolve atomic, that is, non-molecular, hydrogen and form concentrated metal–hydrogen phases, named metal hydrides, under mild conditions. The source of hydrogen can be molecular H_2 gas, a proton providing electrolyte or acid, and in rare cases just H_2O. For hydrogen storage purposes, storage density (volumetric, gravimetric) and reversibility at around ambient conditions are the most important physicochemical properties; in addition, for economic and sustainability reasons abundant materials and low cost processing are needed. Gas phase charging and electrochemical charging are applicable. For the reversible storage of hydrogen the bulk properties of the solid dominate storage capacity or hydrogen density, thermodynamic properties, and diffusion, the surface and interface are crucial for the dissociation/recombination reaction $H_2 \leftrightarrow 2H$ and the related reaction kinetics (Figure 31.1).

The thermodynamics of metal hydride formation are best described [1–3,5] by pressure–composition isotherms, also known as *pcT* curves, and corresponding Van't Hoff plots (Figure 31.2).

The host metal dissolves a few at.% hydrogen as a solid solution. As the hydrogen gas pressure or the equivalent electrochemical potential together with the concentration of H in the metal is increased, interactions among hydrogen atoms become locally important leading to nucleation and growth of the hydride phase. While the two phases coexist, the isotherms show a flat plateau, the length of which determines how much hydrogen can be stored reversibly with small pressure variations. At higher hydrogen pressure, further plateaus and further hydride

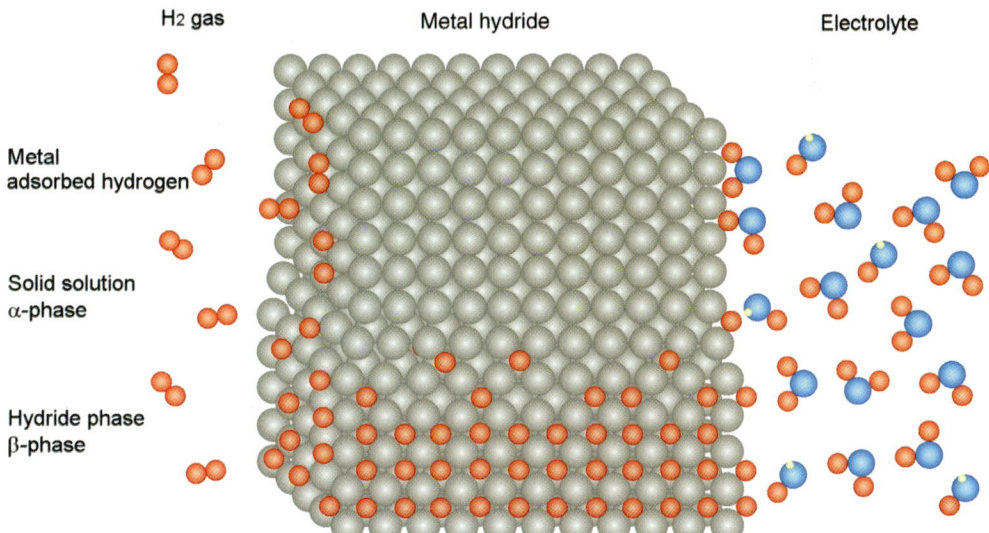

Figure 31.1 Schematic hydride formation from the gas phase (left) and from the electrolyte (right) [2].

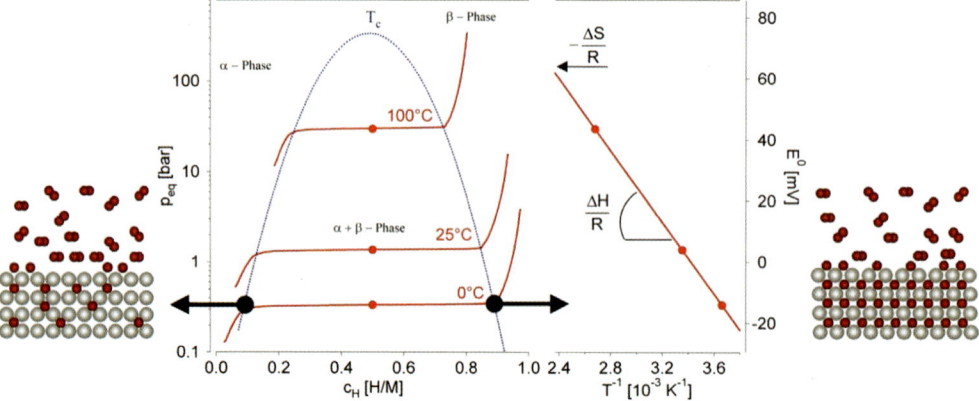

Figure 31.2 Pressure–composition isotherms for the hydrogen absorption in a typical metal on the left-hand side. Shown are the solid solution (α-phase), the hydride phase (β-phase), and the region of the coexistence of the two phases. The coexistence region is characterized by the flat plateau and ends at the critical temperature (T_c). The construction of the Van't Hoff plot is shown on the right-hand side. The slope of the line is equal to the enthalpy of formation divided by the gas constant and the intercept with the axis is equal to the entropy of formation divided by the gas constant [3].

phases may be formed. The two-phase region ends in a critical point, above which the phase transition is continuous. The plateau or equilibrium pressure depends strongly on temperature and is related to the changes ΔH and ΔS (of enthalpy and entropy, respectively). As the entropy change corresponds mostly to the change from molecular hydrogen gas to dissolved hydrogen, it is roughly 130 J K mol^{-1} for all metal–hydrogen systems under consideration. The enthalpy term characterizes the stability of the metal–hydrogen bond. To reach an equilibrium pressure of 1 bar at 300 K, ΔH should amount to 19.6 kJ mol^{-1}.

The phase diagram together with the Van't Hoff plot of Figure 31.2 directly illustrate not only the use for hydrogen storage but also for battery electrode applications (electrochemical hydride formation and decomposition) and thermal machines like heat pumps or heat storage systems. For thermal energy storage – an issue with renewables – the large amount of heat of exothermic formation of selected hydrides offers technical solutions much superior to those of molten salt systems [5,9,21].

Hydrogen atom to host atom ratios can be as high as 3.75 (Th$_4$H$_{15}$) for binary hydrides and 4.5 in BaReH$_9$ [11,12].

The solution of hydrogen and the formation of hydride phases changes crystal lattice properties as well as the behavior of electrons and phonons and with that also cohesion. Anisotropic lattice expansion and loss of symmetry are often observed as well as order–disorder phenomena. Linked with them, the electronic structure changes; metal to nonmetal and semiconductor transitions with changes of optical and magnetic behavior and even superconductivity occur and are used for various applications including sensing.

Especially in view of storage of energy the dynamics and kinetics of hydride formation and decomposition are as important as the storage capacity. The hydrogen atoms and their isotopes vibrate about their equilibrium positions in the lattice and perform local motions and long-range diffusion, all temperature dependent. By using small sized powder rather than bulk material the diffusion path is shortened so that diffusion is not the rate-limiting step. In technical storage systems, the limited thermal conductivity requires engineering support. After exposure of hydride forming materials to air or poisonous impurities (S, CO) of the hydrogen gas, the hydrogen dissociation/recombination $H_2 \leftrightarrow 2H$ limits kinetics; reactivation mostly by elevated temperature is needed.

The metals Pd, Ti, Ni, and Mg are the best studied hydride forming elemental metals; FeTi, LaNi$_5$, and TiMn$_2$/ZrMn$_2$ the prototypes of hydride forming intermetallics [1988, 1992 Schlapbach, H in Intermetallics] and among complex hydride formers Mg$_2$Ni, 2Mg-Fe, Na-Al, Li-B, and Li-N offer promising bonding with hydrogen at high gravimetric density [10–12].

The reaction with hydrogen is able to stabilize or to destabilize a compound, which is illustrated by the following example: Fe$_2$Mg does not exist; however, starting with MgH$_2$ and Fe the stable hydride Fe$_2$MgH$_6$ can be formed. Several rare earth-3d metal compounds react with hydrogen and disintegrate into rare earth hydrides and a residual compound. In that sense, many ternary hydrides are thermodynamically metastable compounds; however, the kinetics of decomposition around room temperature is so slow that storage applications are not affected even after cycling many thousand times.

What we positively see as an energy storage material of great potential future, that is, the solution of hydrogen and formation of hydride phases, is a serious problem with respect to the hydrogen embrittlement of some metallic structural materials and fatigue phenomena under a hydrogen atmosphere. Fortunately, the knowledge gained on hydride formation was extremely useful in learning how to prevent embrittlement and fatigue [23].

31.4
Hydrogen Storage in Intermetallic Compounds

Many intermetallic compounds with transition metal components react with hydrogen and form reversibly metal hydrides at pressures in the range 1–10 bar and temperature range of 300–400 K. Typically, hydrogen to metal atom ratios between 0.8 and 1.3 are reached and hydrogen volumetric densities are close to that of 20 K liquid hydrogen. Together with rapid kinetics and excellent cycling behavior they present attractive hydrogen storage materials. However, as the transition metal based metal host atoms have rather high mass, the gravimetric hydrogen densities range between 1.5 and 3 mass% only, which is considered to be too low for automotive applications of hydrogen as a fuel. Conversely, transition metal based metal hydride electrodes exhibit excellent performance in powering hybrid cars electrically.

Table 31.1 The most important families of hydride forming compounds including the prototype. A is an element with a high affinity to hydrogen, and B is an element with a low affinity to hydrogen.

Compound	Prototype	H/M	H/M (wt%)
AB_5	$LaNi_5H_6$	1	1.2
AB_2	$ZrV_2H_{5.5}$	2	1.9
AB_3	$CeNi_3H_4$	1	1.2
A_2B_7	$Y_2Ni_7H_3$	0.3	0.5
A_6B_{23}	$Ho_6Fe_{23}H_{12}$	0.3	0.5
AB	$TiFeH_2$, $ZrNiH_2$	1	1.9
A_2B	Mg_2NiH_4, Ti_2NiH_4	1.3	3.6
Alanate	$Na[AlH_4]$	2	7.4
Borohydride	$Li[BH_4]$	2	18
Amide	$LiNH_2$	1	9
BN compounds	BH_4NH_4	4	24.3

For large stationary and small portable use, hydrides of intermetallics remain an interesting option. Table 31.1 lists the most studied hydride forming compounds.

Detailed results on synthesis, crystallographic structure, thermodynamic, electronic and magnetic properties, diffusion and kinetics, as well as on surface properties and activation behavior can be found in Reference [5] and soon in Reference [1]. Many of the AB_5 and AB_2 bulk compounds in the form of brittle molten lumps react readily with hydrogen gas without any specific activation treatment and disintegrate into powder due to brittleness and anisotropic and large lattice expansion. Some AB compounds (TiFe) need specific activation processing such as elevated temperatures, partial substitutions, or mechanical pretreatment. This was successfully explained by the formation of catalytically active precipitates of the 3d component which act as dissociation/recombination catalyst in the hydride formation and decomposition process [24,25].

Several empirical models allow the estimation of the stability and the concentration of hydrogen in an intermetallic hydride. The maximum amount of hydrogen in the hydride phase is given by the number and size of interstitial sites in the intermetallic host and the energy involved for the adaptation of bonding and structure. According to the geometrical Westlake criterion the distance between two hydrogen atoms on interstitial sites [26] is at least 2.1 Å and the radius of the largest sphere on an interstitial site touching all the neighboring metallic atoms is at least 0.37 Å. The theoretical maximum volumetric density of hydrogen in a metal hydride, assuming a closed packing of the hydrogen, is therefore 254 kg m^{-3}, which is 3.6 times the density of liquid hydrogen.

More sophisticated is Miedema's rule of reversed stability [27]: The more stable an intermetallic compound the less stable the corresponding hydride and the

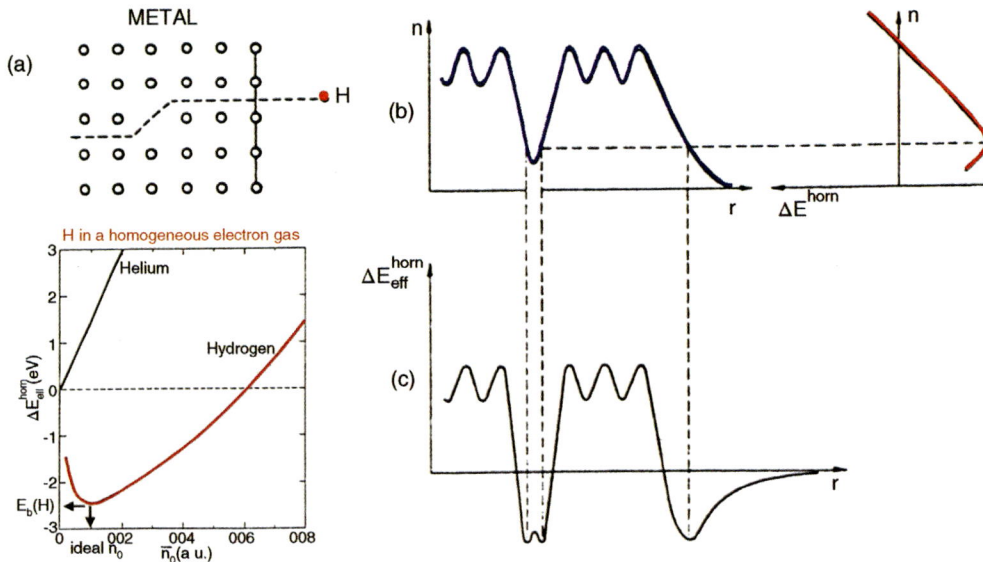

Figure 31.3 Hydrogen absorption in accordance with the effective medium theory: (a) hydrogen atom on a path to the surface of a metal and through the lattice; (b) electron density along the path of the hydrogen atom; (c) potential energy of the hydrogen atom along the path [28].

other way round. This model is based on the fact that hydrogen can only participate in a bond with a neighboring metal atom if the bonds between the metal atoms are at least partially broken.

The binding energy of a hydrogen atom as a function of the electron density of the environment can be calculated by means of the *effective medium theory* [28]. The binding energy of the hydrogen atoms shows a maximum at low electron densities and decreases with increasing electron density (Figure 31.3).

Hydrogen absorption is electronically an incorporation of electrons and protons into the *electronic structure of the host lattice*. The electrons have to fill empty states at the Fermi energy (E_F) while the protons lead to the hydrogen induced s-band approximately 4 eV below the E_F. The heat of formation of binary hydrides MH_x is related linearly to the characteristic band energy parameter $\Delta E = E_F - E_s$, where E_F is the Fermi energy and E_s the center of the host metal electronic band with a strong s character at the interstitial sites occupied by hydrogen. For most metals, E_s can be taken as the energy that corresponds to one electron per atom on the integrated density-of-states curve.

The semi-empirical models mentioned above allow an estimation of the stability of binary hydrides as long as a rigid band model makes sense. However, the interaction of hydrogen with the electronic structure of the host metal in some binary hydrides and especially in the ternary and quaternary hydrides is often more complicated. In many cases, the crystal structure of the host metal and, therefore, also the electronic structure changes upon the phase transition and

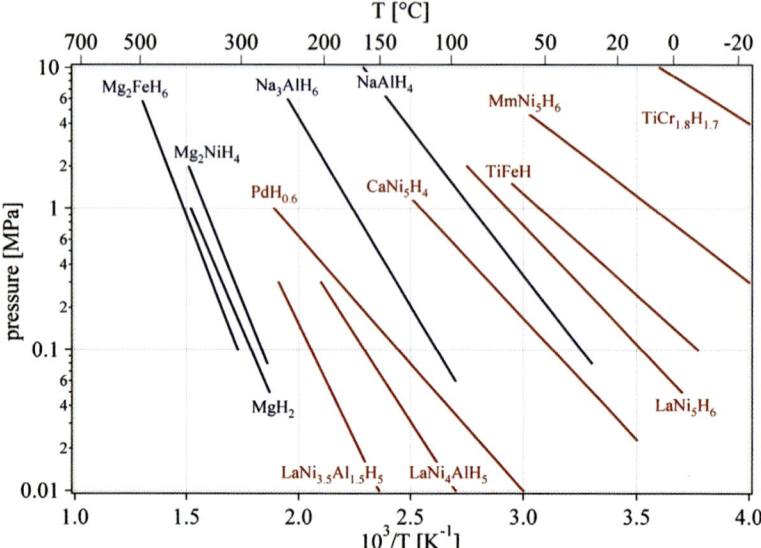

Figure 31.4 Van't Hoff plots of some selected hydrides. Stabilization of the hydride of LaNi$_5$ by the partial substitution of nickel with aluminum in LaNi$_5$ is shown as well as the substitution of lanthanum with mischmetal (e.g., 51% La, 33% Ce, 12% Nd, 4% Pr) [3].

the theoretical calculation of the stability of the hydride becomes very challenging [29,30].

The stability of metal hydrides is usually presented in the form of Van't Hoff plots (Figure 31.4). The most stable binary hydrides have enthalpies of formation of $\Delta H_f = -226$ kJ mol^{-1} H$_2$, for example, HoH$_2$. The least stable hydrides are FeH$_{0.5}$, NiH$_{0.5}$, and MoH$_{0.5}$ with enthalpies of formation of $\Delta H_f = +20, +20,$ and $+92$ kJ mol^{-1} H$_2$, respectively [4].

Due to the phase transition upon hydrogen absorption, metal hydrides have the very useful property of absorbing large amounts of hydrogen at a constant pressure, that is, the pressure does not increase with the amount of hydrogen absorbed as long as the phase transition takes place. The characteristics of the hydrogen absorption and desorption can be tailored by partial substitution of the constituent elements in the host lattice. Some metal hydrides absorb and desorb hydrogen at ambient temperature and close to atmospheric pressure.

One of the most interesting features of the metallic hydrides is the extremely high volumetric density of the hydrogen atoms present in the host lattice. The highest *volumetric hydrogen density* known today is 150 kg m^{-3} found in Mg$_2$FeH$_6$ and Al(BH$_4$)$_3$. Both hydrides belong to the complex hydrides and will be discussed in the next section. Metallic hydrides reach a volumetric hydrogen density of 115 kg m^{-3}, for example, LaNi$_5$. Most metallic hydrides absorb hydrogen up to a hydrogen to metal ratio of H/M = 2.

Greater ratios up to H/M = 4.5, for example, BaReH$_9$, have been found [11,12]; however, all hydrides with a hydrogen to metal ratio above 2 are ionic or covalent compounds and belong to the complex hydrides. Accordingly, as a function of hydrogen content metal–semiconductor and metal–insulator transitions are observed. Well-studied examples are the transitions from metallic rare earth dihydrides to nonmetallic trihydrides [5,30].

31.4.1
AB_5 Type Compounds

AB_5 type compounds (A: rare earth metal, B: d-transition metal), especially $LaNi_5$ forming $LaNi_5H_6$ and substitutes are the best investigated hydrides of intermetallics; crystallographic, thermodynamic, surface, electronic and magnetic properties, and hydrogen diffusion and kinetics were studied in detail in the 1980s [5], a new comprehensive scientific update will appear soon in the Landolt-Börnstein Series [1]. The discovery of its attractive hydrogen storage properties was a by-product of R&D work on $SmCo_5$ permanent magnets [27,31]. $LaNi_5$ melts congruently and thus is easily produced from laboratory scale up to industry scale by melting. It reacts with hydrogen gas and room temperature and slight overpressure without any activation treatment. Due to strong anisotropic lattice expansion, it disintegrates into powder. Reversible hydrogen sorption cycling up to $LaNi_5H_{6.7}$ and several tens-of-thousand cycles were proven. Upon cycling at elevated temperatures, disintegration into a stable rare earth hydride and a nickel rich alloy starts. The easy activation of AB_5 compounds was successfully explained by spontaneous surface segregation with selective oxidation of the rare earth and formation of metallic Ni-precipitates, which catalyze the hydrogen dissociation–recombination reaction. Studies of electronic and magnetic properties including band-structure calculations reveal strong hydrogen bonding due to Ni-d-states below the Fermi level and relatively high RE-d character at the Fermi level [30,32].

The relatively low reversible storage capacity (<2 wt%) and high price finally prevented a broader market entry of AB_5 type hydrides with one exception, a very successful one: The use of $LaNi_5$ hydride based electrodes in so-called nickel-metal hydride (NiMH) rechargeable batteries was developed in the 1970s together with many other applications (www.nilar.com) [33], and entered the market of consumer type small batteries in the 1990s prior to the Li-based batteries and reached the important market of hybrid automobiles. Thanks to superior safety and cyclic stability at high power and energy density it retains that position; AB_5 alloys of compositions around $La_{0.6}Ce_{0.8}Pr_{0.1}Nd_{0.2}Ni_{5.9}Co_{1.2}Mn_{0.7}Al_{0.5}$ are used as cathodes with an alkaline 6M KOH electrolyte and nickel hydroxide cathode [9,33,34].

31.4.2
AB_2 Type Hydrogen Absorbing Alloys

AB_2 type hydrogen absorbing alloys, which are usually called *Laves phase* alloys, from their crystal structures were first reported by Shaltiel *et al.* [35]. Laves phase themselves are one of the most popular alloys among the AB_2 compounds. There are three major types of crystal structures, hexagonal C14, cubic C15, and hexagonal C36, but their difference is the structure (sacking) of layers [36]. If the size of elements consisted of Laves phase, AB_2 is in an ideal ratio of $A:B = 1.225 : 1$, the structure is topologically close-packed. In general, Laves phase alloys have higher hydrogen capacity than AB_5. One of the reasons is that the hydrogen to metal ratio (H/M) is sometimes larger than 1. Especially alloys such as ZrV_2 with a C15 structure contain more hydrogen than $H/M = 1.5$.

Because Laves phase alloys can be used below 25 °C and have a slightly higher capacity than AB_5, they are favorably applied for heat pumps and refrigerators. For the same reason, application for Ni-hydrogen batteries has been tried but a Laves phase has not been commercially applied. The fact that the Laves phase lattice expands isotropically upon hydrogen absorption, that is, without discontinuous phase transition, results in good cycling behavior.

Another type of AB_2 alloys for hydrogen storage is the Zintl phase characterized by electron transfer between the components of different electronegativities. $SrAl_2H_2$ is the first example of a Zintl phase hydride that has two-dimensional infinite $[AlH]^-$ anion with a Sr^{2+} cation [37].

If the A atom is a rare earth element, there are phases between AB_5 and AB_2. Their crystal structure is a combination of AB_5 and AB_2 units [38]. Hydrogen absorption occurs in these alloys but substitution of Mg for the rare earth gives a dramatic improvement of hydrogen storage performance [39]. This type of alloy is appropriate for the use as a negative electrode of Ni/metal hydride (Ni/MH) battery [34]. Today, the most advanced Ni-hydrogen battery utilizes rare earth-Mg-Ni-M alloys with a combination structure of AB_5 and AB.

31.4.3
AB Type Alloy TiFe

Hydrogen storage using the *AB type alloy TiFe* was reported for the first time in 1974 [40]. The rather economic alloy TiFe reaches, after activation, storage capacities close to 2 wt% under ambient conditions and remains an attractive material for stationary storage with excellent cyclic lifetime. High purity crystalline TiFe requires an activation treatment at 400 °C under 30 bar of hydrogen to initialize hydrogen absorption. In the 1980s, many studies were devoted to the activation issues of TiFe. The convincing explanation is the selective oxidation of Ti and formation of metallic Fe or Fe based small aggregates, in analogy to the surprisingly easy activation of $LaNi_5$ at room temperature. The aggregates are active for the dissociation–recombination of hydrogen. The main difference is that TiFe oxidizes selectively at elevated temperatures only [25].

Recently, especially after the big earthquake in Japan in 2011, stationary energy storage using hydrogen has attracted much attention again. TiFe is one of the ideal materials for energy storage because of low cost and ambient working temperature. Research on activation of TiFe has been re-started. Edalati *et al.* [41]

reported that TiFe absorbs hydrogen without heat pretreatment when morphologically modified using a high-pressure torsion (HPT) technique. Other approaches to improve activation properties are based on off-stoichiometric composition (Ti-rich) [42], additions of Mn, Ni, and Pd [43], as well as nanostructuring by ball milling [44,45].

31.4.4
Intermetallic Hydrides

The hydrides *Mg_2NiH_4, Mg_2CoH_5, and Mg_2FeH_6,* discovered by Reilly and Wiswall [46] and Yvon *et al.* [11,12] are characterized by very high hydrogen content and strong hydrogen–metal bonding with nonmetallic appearance. Consequently, hydrogen desorption temperatures are too high for typical hydrogen storage applications. Attempts to destabilize the hydrogen bonding are described elsewhere [11]. However, the potential of these Mg-compounds for heat storage purposes is attractive [9,21].

Mg_2FeH_6 has the highest volumetric storage density ($150\,g\,l^{-1}$, 5.5 wt%) is made of inexpensive and abundant metals, and is a hydrogen stabilized compound, that is, the intermetallic compound Mg_2Fe does not exist. Structurally it contains Mg^{2+} cations and octahedral complex $[FeH_6]^{4-}$ anions. The enthalpy of formation is high; temperatures of around 320 °C are needed for the release of hydrogen gas at ambient pressure [11].

31.5
Hydrogen Storage in Complex Hydrides

Most research activities in materials development for hydrogen storage were mainly on metals or metal alloys in the last century. Compounds containing nonmetallic elements, such as N, B, or O, had not been widely considered as hydrogen absorbers before, despite the abundant chemistry knowledge of H_2 and these compounds. Group I and II metals with complex anions of $[AlH_4]^-$, $[BH_4]^-$, and $[NH_2]^-$, namely, alanate (tetrahydroaluminate), borohydride (tetrahydroborate), and amide, have high hydrogen gravimetric densities and are, in some cases, commercially available. Thus, a priori, they would seem to be viable candidates for application as practical, on-board mobile hydrogen-storage materials. Most of these materials are plagued by high kinetics barriers to dehydrogenation and/or re-hydrogenation in the solid-state and rather high temperatures are needed for desorbing hydrogen at around atmospheric pressure. Traditionally, only stable complex hydrides were synthesized and known and no catalyst for the sorption reaction was found. However, this situation changed after the pioneering studies of Bogdanovic and Schwickardi [47,48] demonstrating that, upon doping with selected titanium compounds, the dehydrogenation of $NaAlH_4$ was kinetically enhanced allowing reversibility under moderate conditions. This breakthrough has led to a worldwide effort to develop doped alanates hydrides as practical hydrogen storage materials; the effort quickly expanded to include amides and borohydrides [10,49].

31.5.1
Alanates (tetrahydroaluminate)

The dehydrogenation of NaAlH$_4$ was extensively studied by Ashby and Kobetz [50]. They found that controlled heating at 210–220 °C for 3 h evolved 3.7 mass% hydrogen to give Na$_3$AlH$_6$. This work established that the first step of the dehydrogenation proceeds according to Eq. (31.1) and that further elimination of hydrogen to give aluminum and NaH occurs through a separate reaction that takes place at ~250 °C, seen in Eq. (31.2):

$$3\text{NaAlH}_4 \rightarrow \text{Na}_3\text{AlH}_6 + 2\text{Al} + 3\text{H}_2 \qquad (31.1)$$

$$\text{Na}_3\text{AlH}_6 \rightarrow 3\text{NaH} + \text{Al} + \tfrac{3}{2}\text{H}_2 \qquad (31.2)$$

The original Bogdanovic materials were prepared by evaporation of suspensions of NaAlH$_4$ in diethyl ether solutions of the soluble titanium compounds, titanium tetra-n-butoxide, Ti(OBu^n)$_4$, and beta-titanium trichloride, β-TiCl$_3$. It was subsequently found that catalytic enhancement of NaAlH$_4$ also occurs upon mechanically mixing the titanium catalyst precursors with the aluminum hydride host. The materials resulting from this method of preparation have kinetic and cycling properties that are much closer to those required for a practical hydrogen storage medium (Figure 31.5).

The kinetic enhancement upon Ti-doping can be extended to the reversible dehydrogenation of LiNa$_2$AlH$_6$ to LiH, 2NaH, and Al. Bogdanovic and Schwickardi also found [48] the mixed alkali metal alanate to have a significantly lower plateau pressure at 211 °C and thus a higher value of ΔH of dehydriding reaction than Na$_3$AlH$_6$. Fossdal et al. [51] conducted hydrogen pressure–composition (p–c) isotherm measurements over the range 170–250 °C and, from the Van't Hoff plot of their data, determined ΔH to be 56 kJ mol^{-1} H$_2$. Graetz and Reilly [52] observed

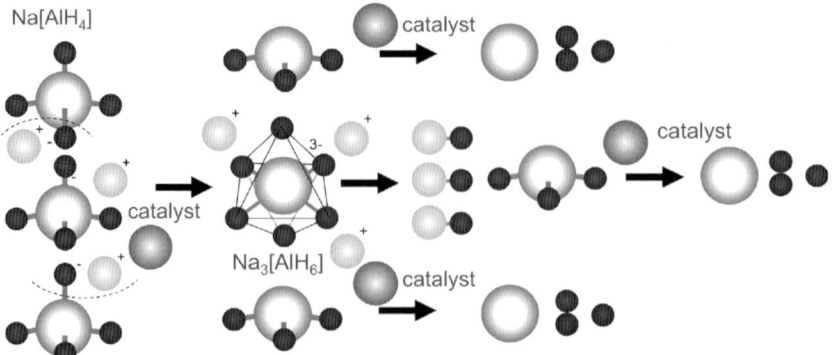

Figure 31.5 Schematic mechanism of hydrogen desorption from sodium alanate (NaAlH$_4$).

that the undoped mixed alanate will also undergo reversible dehydrogenation. It was found that while doping 4 mol.% $TiCl_3$ results in a highly pronounced kinetic enhancement it introduces no effect on the thermodynamic properties of the hydride [53].

31.5.2
Amides

Metal nitrides (M_3N_x), imides ($M_2(NH)_x$), and amides ($M(NH_2)_x$), where x is the valence of M, are examples showing continuous replacement of M by H in the metal–N–H system. The first experimental result showing the interaction between H_2 and Li_3N, which was reported by Dafert and Miklauz, dates back to the early twentieth century. Modern interest in using Li_3N as hydrogen storage material was ignited by Chen's work in 2002 [11,54]. It was observed that Li_3N absorbed H_2 following the reaction (31.3) giving rise to Li_2NH-LiH and $LiNH_2$-2LiH in a stepwise manner. The resulting material, that is, the composite of $LiNH_2$ and LiH, could also desorb H_2, giving a reversible hydrogen storage system. In total, ~10.5 wt% H_2 can be desorbed/absorbed in the composite:

$$Li_3N + 2H_2 \leftrightarrow Li_2NH + LiH + H_2 \leftrightarrow LiNH_2 + 2LiH \tag{31.3}$$

LiH, a highly stable hydride with a formation enthalpy of -91 kJ mol^{-1}, decomposes to H_2 at temperatures higher than 600 °C. The thermal decomposition of $LiNH_2$, on the other hand, takes place above 300 °C, giving off NH_3 rather than H_2. Upon forming the composite, $LiNH_2$ interacts with LiH and releases H_2 at relatively low temperatures (starting at <180 °C). The key point is the formation of stable product, that is, Li_2NH or Li_3N depending on the molar ratio of $LiNH_2$ to LiH and the condition applied, which alters the reaction pathway and, thus, tunes the thermodynamic parameters of the corresponding reaction.

Compared with those hydrogen storage materials reported previously, the $LiNH_2$-LiH composite has three distinct features:

1) it contains nonmetallic elemental N;
2) it is a composite; H_2 generation relies on the chemical interaction of the corresponding components;
3) both protic hydrogen ($H^{\delta+}$ in $LiNH_2$) and hydridic hydrogen ($H^{\delta-}$ in LiH) co-exist; the high chemical potential of the combination of $H^{\delta+}$ and $H^{\delta-}$ to molecular hydrogen (Eq. (31.4)) can be regarded as one of the driving forces for the dehydrogenation:

$$H^+ + H^- \leftrightarrow H_2 \qquad \Delta H = -17.37 \text{ eV} \tag{31.4}$$

The understanding of the $LiNH_2$-LiH composite system enables new materials design for adapted performance by probing the interactions between various amides and hydrides to explore potential hydrogen storage materials. Most alkali and alkaline earth metals have amide forms, as do several mixed metal amides.

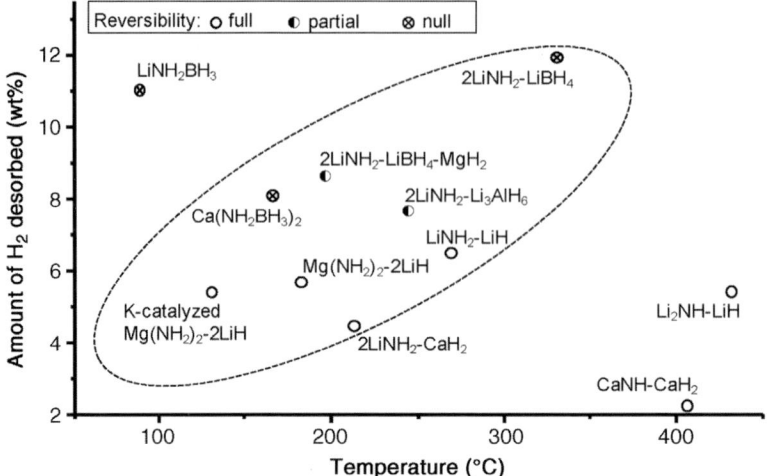

Figure 31.6 Hydrogen storage properties of a few representative amide–hydride composites/complexes. The x-axis shows the temperature of maximum rate of dehydrogenation. The value may change slightly with pretreatment of material and testing condition applied. In most cases, high hydrogen content systems release hydrogen at higher temperatures as indicated in the dotted cycle.

The choice of hydrides can be either elemental hydrides, such as LiH and MgH_2, or complex hydrides, such as $LiAlH_4$ or $LiBH_4$. A number of composites and complexes containing both $H^{\delta+}$ and $H^{\delta-}$ components have been developed since 2002. Figure 31.6 shows a few representative systems, in which the hydrogen content, temperature for dehydrogenation, and reversibility are indicated. The hydrogen desorption enthalpies of these systems vary from highly endothermic (e.g., Li_2NH-LiH [11]) to exothermic (e.g., $LiNH_2BH_3$) [54] showing an effective approach to tune the thermodynamics of dehydrogenation to compositional change. The hydrogen content varies from about 2 wt% for CaNH-CaH_2 [11] to about 11.9 wt% for $LiBH_4$-$2LiNH_2$ [55]. Most of the amide-elemental metal hydride composites desorb hydrogen endothermically and exhibit full or partial reversibility in hydrogen absorption and desorption. In particular, the $Mg(NH_2)_2$-2LiH composite possesses both relatively high reversible hydrogen capacity of 5.6 wt% and suitable thermodynamic properties ($\Delta H_{des} = 39$ kJ mol^{-1} H_2, $\Delta S_{des} = 112$ J K^{-1} mol^{-1}-H_2) [56–59], which illustrates the attractiveness of the system for onboard hydrogen storage. $Mg(NH_2)_2$ can interact with LiH in molar ratios of 1:2, 3:8, or 1:4 giving rise to different solid products, respectively. For the most studied 1:2 composite, hydrogen desorption is also a two-step process forming $Li_2Mg_2(NH)_3$ and $LiNH_2$ in the first step and $Li_2Mg(NH)_2$ in the second step, respectively (see reaction Eq. (31.5)). Hydrogen desorption from composites or complexes containing Al or B, however, are generally exothermic or mildly endothermic in nature, which is, to a large extent, due to the formation of stable Al–N or B–N bonds, and, thus, suffer from irreversibility under normal conditions [55,60,61]:

$$2Mg(NH_2)_2 + 4LiH \leftrightarrow Li_2Mg_2(NH)_3 + LiNH_2 + LiH + 3H_2$$
$$\leftrightarrow 2Li_2Mg(NH)_2 + 4H_2 \quad (31.5)$$

Making composites or complexes to improve hydrogen storage properties can be effectively applied to other systems. A few successful examples, such as LiH-Si and LiBH$_4$-MgH$_2$ [62], evidenced that elements or compounds can also be mixed with hydrides and alter the hydriding/dehydriding thermodynamics by forming more stable products or reactants.

Hydrogen desorption from the amide–hydride composite materials usually occurs by multistep reactions, and possibly co-exists with competing reaction pathways. Interestingly, most of the amide–elemental hydride composites, such as LiNH$_2$-LiH, CaNH-CaH$_2$, Mg(NH$_2$)$_2$-LiH, and Ca(NH$_2$)$_2$-LiH, have notably good reversibility in desorbing and absorbing hydrogen, which is likely due to the significant structural similarities between amides and their corresponding dehydrogenated products (i.e., imides). As examples, structural similarities were found between LiNH$_2$ and Li$_2$NH [63], Mg(NH$_2$)$_2$ and MgNH, CaNH and Ca$_2$NH, Mg(NH$_2$)$_2$ and Li$_2$Mg(NH)$_2$, [64] Ca(NH$_2$)$_2$ and Li$_2$Ca(NH)$_2$, and Ca(NH$_2$)$_2$ and MgCa(NH)$_2$. As shown in Figure 31.7 NH$_2^-$ and NH^{2-} anions in the structures of Mg(NH$_2$)$_2$ and Li$_2$Mg(NH)$_2$ are arranged in a close-packed configuration. The metal cations (Li$^+$ and/or Mg^{2+}) are tetrahedrally coordinated by the anions. The difference between the amides and imides, from a structural viewpoint, can be regarded as different extents of cation occupation in the tetrahedral interstices in the nitrogen framework (i.e., more interstices are occupied by cations when in the imide form). The apparent structural similarity between the reactant and product implies that the transport and exchange of small cations of Li$^+$ and H$^+$ in the N framework of amide/imide phases may play important roles in the hydrogenation/dehydrogenation of LiNH$_2$-LiH and Mg(NH$_2$)$_2$-2LiH systems [65].

There are controversial interpretations of hydrogen *desorption* from the amide–hydride composite system, among which the direct solid–solid reaction mechanism proposed by Chen *et al.* [66] and the ammonia-mediated mechanism suggested by Hu [67] and Ichikawa *et al.* [68] have been widely discussed. The experimental evidence supporting the first interpretation is the pronounced

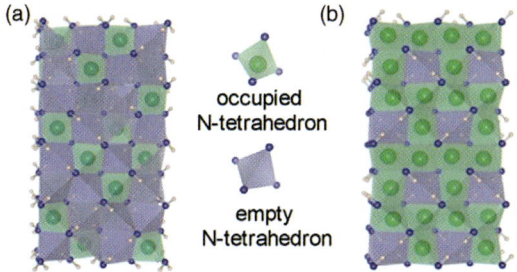

Figure 31.7 Structures of (a) Mg(NH$_2$)$_2$ (I4$_1$/abc) and (b) Li$_2$Mg(NH)$_2$ (Iba2, 8 cells).

lower hydrogen desorption temperatures of the various amide–hydride composites as compared to the decomposition temperatures of amides alone [66]. *In situ* TEM characterization showed that the particles of $LiNH_2$ kept essentially unchanged while LiH particles shrank upon hydrogen desorption, evidencing the immobile of N in the amide phase [69]. The ammonia-mediated mechanism, on the other hand, consists of two steps: (i) the self-decomposition of $LiNH_2$ gives rise to Li_2NH and NH_3 and (ii) the emitted NH_3 reacts with LiH to form H_2 and $LiNH_2$. Such a mechanism is supported by the facts that NH_3 reacts with LiH in an ultra-fast manner [67] and NH_3 by-product accompanies the evolution of H_2. It is also proposed that the Frenkel defect pair that is associated with the migration of both Li^+ and H^+ in amide phase may induce the formation of NH_3 with a series of non-stoichiometric intermediates [70].

Recently, experimental results have revealed that the main route for hydrogen desorption depends on the conditions applied. Under the following conditions where (i) the ion migration is substantially constrained as in the case of Li_2NH_2Br-LiH composite, or (ii) amide and hydride particles are not in good contact, or (iii) the reaction temperature is high enough to allow NH_3 formation at sufficient rate (Figure 31.8), hydrogen desorption is likely to follow the NH_3 mediated pathway. On the other hand, the reaction will follow a more energy favorable path via ion migration if amide and hydride particles are sufficiently contacted. Under such a condition, H^+ from $LiNH_2$ and Li^+ from LiH exchange at the interface of $LiNH_2$ and LiH leading to the formation of H_2 and the Li-rich layer in the amide phase. Li^+ will move into the amide phase and H^+ will move out to minimize the concentration gradient leading to the formation of the non-stoichiometric amide phase detected by David [70].

Hydrogen desorption from the amide–hydride composite materials generally encounters severe *kinetic* barriers that may originate from interface reactions, nucleation/nuclei growth, or/and diffusion processes. The $Mg(NH_2)_2$-2LiH

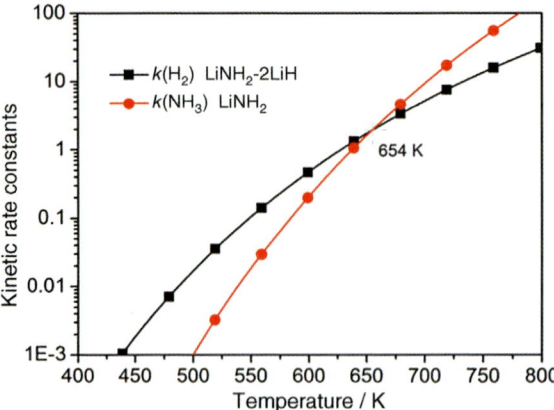

Figure 31.8 Comparison of rate constants of hydrogen desorption from the thoroughly mixed $LiNH_2$-2LiH ($k_{(H2)}$) and self-decomposition of $LiNH_2$ ($k_{(NH3)}$) samples.

system, as an example, releases H$_2$ at an appropriate rate only at temperatures above 180 °C although from the thermodynamic point of view the temperature for hydrogen release at 1 bar pressure is below 90 °C [71]. Various strategies including adequate mixing, size control, and an introduction of catalytic additives have proven to be effective in improving the hydriding/dehydriding kinetics.

Hydrogen desorption from composite materials occurs via interaction of at least two phases; therefore, thorough mixing of reactants is indispensable. Mechanical ball milling can efficiently reduce the particle size of reactants and increase the homogenous distribution as well as contact of amide and hydride particles in the resulting mixture, and thus lead to a significantly decreased kinetic barrier of dehydrogenation. However, the aggregation of small particles during the subsequent hydriding/dehydriding cycles is unavoidable. Introducing proper dispersants [72] can suppress the aggregation/growth of reactant particles to some degree.

Titanium-based additives were found to have positive effects in H$_2$ desorption from LiNH$_2$-LiH and Mg(NH$_2$)$_2$-2LiH systems [73,74]. Efforts to improve the Li$^+$ conductivity of materials by introducing LiBH$_4$ [75], and halides [76], yielded significant kinetic enhancement of hydrogen sorption as well as suppression of ammonia generation. A notable kinetic enhancement was achieved by introducing ~3 mol.% KH to the Mg(NH$_2$)$_2$-2LiH system. A mechanistic investigation showed that KH interacts with Mg(NH$_2$)$_2$ and the intermediate LiNH$_2$ forming K$_2$Mg(NH$_2$)$_4$ and Li$_3$K(NH$_2$)$_4$, and, thus, creates a different reaction pathway leading to a reduced kinetic barrier [77].

31.5.3
Borohydrides (Tetrahydroborate)

The study of M(BH$_4$)$_n$ started with LiBH$_4$. It releases 13.8 mass% of hydrogen by decomposition into LiH and B (Figure 31.9), according to the following reaction:

$$\text{LiBH}_4 \leftrightarrow \text{LiH} + \text{B} + \tfrac{3}{2}\text{H}_2 \tag{31.6}$$

The thermodynamic stabilities for the series of metal borohydrides M(BH$_4$)$_n$ (M = Li, Na, K, Mg, Sc, Cu, Zn, Zr, and Hf; $n = 1$–4) have been systematically

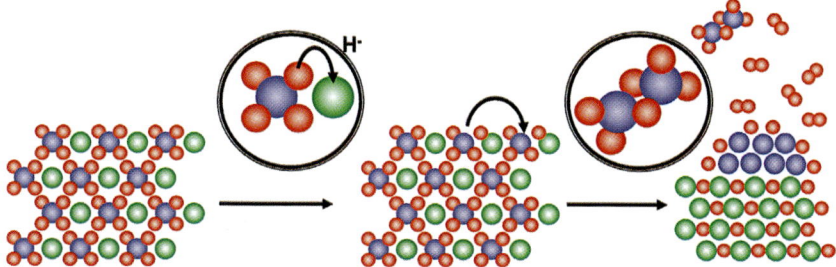

Figure 31.9 Hydrogen desorption mechanism in borohydrides. First H$^-$ is transferred from BH$_4^-$ to the cation and subsequently the BH$_3$ decomposes to give B and H$_2$ or two BH$_3$ form diborane (B$_2$H$_6$).

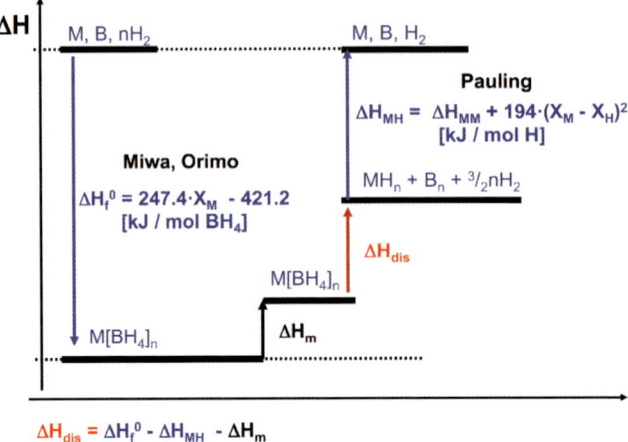

Figure 31.10 Schematic energy diagram for borohydrides.

investigated by first-principles calculations [78]. The heat of formation of the M(BH$_4$)$_n$ depends linearly on the Pauling electronegativity of M (Figure 31.10). Thus, the dehydrogenation temperatures of M(BH$_4$)$_n$ decrease with increasing electronegativity of M.

Dehydrogenation of LiBH$_4$ is reversible because the end products, LiH and B, react with hydrogen at 873 K at a hydrogen pressure of 35 MPa for 12 h or at 1000 K under a hydrogen pressure of 15 MPa for over 10 h to form LiBH$_4$ [79,80]. The high rehydrogenation temperature of above 873 K might be due to the sluggish kinetics of rehydrogenation, which requires recombination of the dehydrogenated LiH and B. Dehydrogenation of the LiBH$_4$/MgH$_2$ system proceeds as follows:

$$2\text{LiBH}_4 + \text{MgH}_2 = 2\text{LiH} + \text{MgB}_2 + 4\text{H}_2 \tag{31.7}$$

This approach not only reduces the enthalpy of the dehydrogenation, which decreases dehydrogenation temperature, but also kinetically enhances the rehydrogenation reaction.

By incorporation into nanoporous carbon scaffolds with a 13-nm pore size, the dehydrogenation rates of LiBH$_4$ were found to be up to 50 times faster than those in the bulk materials measured at 573 K. In addition, the capacity loss over three cycles was reduced from 72% for bulk LiBH$_4$ to ~40% for nanoconfined LiBH$_4$. Furthermore, a recent report confirmed the synergetic effects of nanoconfinement and Ni addition on the dehydrogenation and rehydrogenation properties of LiBH$_4$. The nanoconfinement of the mixture of LiBH$_4$ and Ni addition in a nanoporous carbon scaffold shows a higher rehydrogenation rate and larger rehydrogenation amount than those of the samples without Ni addition at 593 K under a hydrogen pressure of 4 MPa. This suggests that the combination of nanoconfinement and additives would be a valid way to improve the hydrogen storage properties of M(BH$_4$)$_n$ [81].

31.6
Physisorption and High Open-Porosity Structures for Molecular Hydrogen Storage

Today hydrogen storage by physisorption at large specific surface areas including open porous structures requires low temperature operation: bonding by physisorption is too weak. Better understanding of physisorption and chemisorption mechanisms hopefully will lead to intermediate bonding.

The origin of physisorption of gas molecules at a flat surface of a solid are resonant fluctuations of the charge distributions leading to Van der Waals interactions. A gas molecule interacts with several surface atoms of the solid. The interaction is composed of two terms: an attractive term, which diminishes with the distance of the molecule from surface to the power of −6, and a repulsive term, which diminishes with the distance to the power of −12. Therefore, the potential energy of the molecule shows a minimum at a distance of approximately one molecular radius of the adsorbate. The energy minimum is of the order of 0.01 to 0.1 eV (1–10 kJ mol^{-1}) [82]. Due to the weak interaction, significant physisorption is only observed at low, even cryogenic, temperatures.

Once a monolayer of adsorbate molecules is formed, the gaseous molecule interacts with the surface of the liquid or solid adsorbate. Therefore, the binding energy of the second layer of adsorbate molecules can be approximated by the latent heat of sublimation or vaporization (ΔH_V) of the adsorbate. Consequently, adsorption of the adsorbate at a temperature greater than the boiling point at a given pressure leads to the adsorption of a single monolayer [83].

The Brunauer–Emmett–Teller (BET) model assumes that the enthalpy of adsorption for the first monolayer of molecules is ΔH_{ads} and for all additional layers ΔH_V (Figure 31.11). Furthermore, it assumes that all layers are

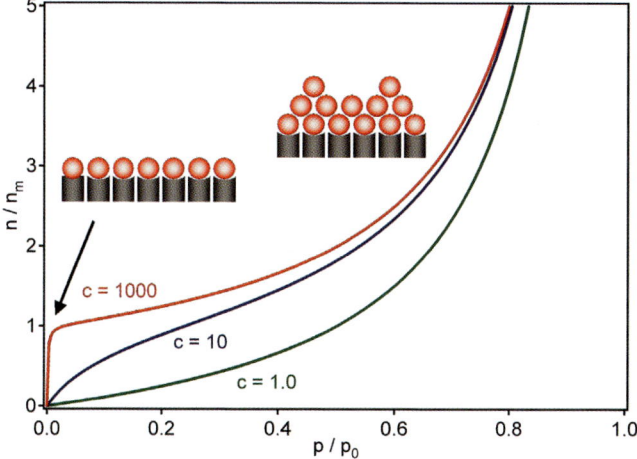

Figure 31.11 Brunauer–Emmett–Teller (BET) model for gas adsorption on surfaces [83].

in equilibrium. With the following definitions:

$$c = e^{\frac{\Delta H_{ads} - \Delta H_V}{kT}} \quad \text{and} \quad \beta = F \cdot \frac{k_a^0}{k_d^0} e^{\frac{\Delta H_V}{kT}} = \frac{p}{p_0}$$

the number of adsorbed molecules is given by:

$$\frac{n}{n_m} = \frac{c\beta}{(1-\beta)[1+(c-1)\beta]}$$

To estimate the quantity of adsorbate in the monolayer, the density of the liquid adsorbate and the volume of the molecule must be used. If the liquid is assumed to consist of a closed packed fcc structure, the amount of adsorbate in a monolayer on a substrate with a specific surface area (S_{Sub}) can be calculated from the density of the liquid (ρ_{liq}) and the molecular mass of the adsorbate (M_{ads}). The amount of adsorbate (m_{ads}) on a substrate material with a specific surface area (S_{Sub}) is then given by:

$$\frac{m_{ads}}{m_{Sub}} = \frac{2 S_{Sub} M_{ads}}{\sqrt{3} \cdot \left(\sqrt{2 N_A} \cdot \frac{M_{ads}}{\rho_{liq}} \right)^{\frac{2}{3}}}$$

where N_A stands for the Avogadro constant ($N_A = 6.022 \times 10^{23}$ mol^{-1}). In the case of carbon as the substrate and hydrogen as the adsorbate, the maximum specific surface area of carbon is $S_{spec} = 1315$ m^2 g^{-1} (single side graphene sheet) and the maximum amount of adsorbed hydrogen is $m_{ads} = 3.0$ mass%. From this theoretical approximation, we may conclude that the amount of adsorbed hydrogen is proportional to the specific surface area of the adsorbent with $m_{ads}/S_{Sub} = 2.27 \times 10^{-3}$ mass%·m^{-2} g and can only be observed at very low temperatures.

Hydrogen interacts with all forms of carbon. The main difference between carbon nanotubes and high surface area graphite is the curvature of the graphene sheets and the cavity inside the tube (Figure 31.12). In microporous solids with capillaries that have a width not exceeding a few molecular diameters, the potential fields from opposite walls will overlap so that the attractive forces which act upon adsorbate molecules will be increased as compared with that on a flat carbon surface [84]. This phenomenon is the main motivation for the investigation of hydrogen interaction with carbon nanotubes.

Rzepka et al. [15] used a grand canonical ensemble Monte Carlo program to calculate the amount of absorbed hydrogen for a slit pore and a tubular geometry. The amount of absorbed hydrogen depends on the surface area of the sample, the maximum is at 0.6 mass% ($p = 6$ MPa, $T = 300$ K) (Figure 31.13). The calculation was experimentally verified with excellent agreement. At a temperature of 77 K, the amount of absorbed hydrogen is about one order of magnitude higher compared to 300 K.

Measurement of the latent heat of condensation of nitrogen on carbon black [85] showed that the heat for the adsorption of one monolayer is between 11 and 12 kJ·mol^{-1} (0.11–0.12 eV) and drops for subsequent layers to the latent heat of condensation for nitrogen, which is 5.56 kJ mol^{-1} (0.058 eV). If we

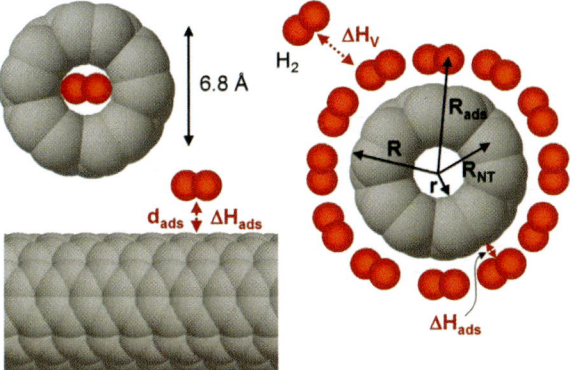

Figure 31.12 Hydrogen molecule on the surface of a carbon single-wall nanotube (SWNT) (5,5). The Van der Waals interaction between H$_2$ and the curved surface of the nanotube is weaker on the outside surface and stronger on the inner surface. Furthermore, the specific surface area for the adsorbate layer (R_{ads}) is larger than the specific surface area of the nanotube (R_{NT}).

assume that hydrogen behaves in a similar manner to nitrogen, then hydrogen would only form one monolayer of liquid at the surface of carbon at temperatures above the boiling point. Geometrical considerations of the nanotubes lead to the specific surface area and, therefore, to the maximum amount of condensed hydrogen in a surface monolayer [18]. The maximum amount of

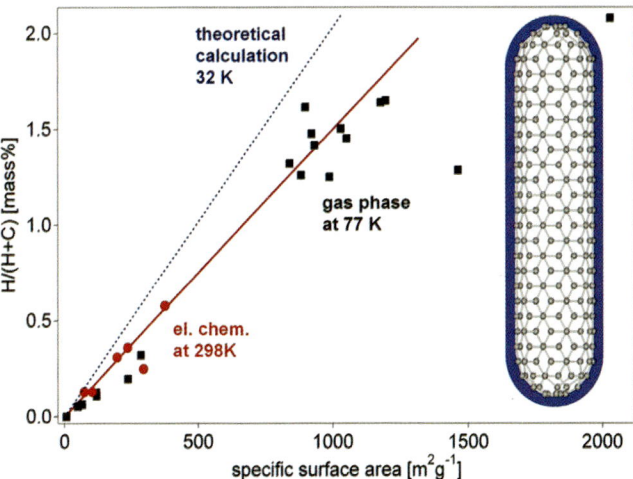

Figure 31.13 Reversible amount of hydrogen (electrochemical measurement at 298 K) versus the BET surface area (round symbols) of a few carbon nanotube samples including two measurements on high surface area graphite (HSAG) samples together with the fitted line. Hydrogen gas adsorption measurements at 77 K from Nijkamp et al. [16] (square symbols) are included. The dotted line represents the calculated amount of hydrogen in a monolayer at the surface of the substrate.

adsorbed hydrogen is 2.0 mass% for carbon single-wall nanotubes (SWNTs) with a specific surface area of $1315\,\mathrm{m^2\,g^{-1}}$ at 77 K.

As well as carbon nanostructures, other nanoporous materials have been investigated for hydrogen absorption. The hydrogen absorption of zeolites of different pore architecture and composition, for example, A, X, Y, was analyzed in the temperature range 293–573 K and pressure range 2.5–10 MPa [17]. Hydrogen was absorbed at the desired temperature and pressure. The sample was then cooled to room temperature and evacuated. Subsequently, hydrogen release upon heating of the sample to the absorption temperature was detected. The absorbed amount of hydrogen increased with increasing temperature and increasing absorption pressure. The maximum amount of desorbed hydrogen was found to be 0.08 mass% for a sample loaded at a temperature of 573 K and a pressure of 10 MPa. The adsorption behavior indicates that the absorption is due to a chemical reaction rather than physisorption. At liquid nitrogen temperature (77 K) the zeolites physisorb hydrogen in proportion to the specific surface area of the material. A maximum of 1.8 mass% of adsorbed hydrogen was found [20] for a zeolite (NaY) with a specific surface area of $725\,\mathrm{m^2 \cdot g^{-1}}$.

The low temperature physisorption (type I isotherm) of hydrogen in zeolites is in good agreement with the adsorption model mentioned above for nanostructured carbon. The desorption isotherm followed the same path as the adsorption, which indicates that no pore condensation occurred. The hydrogen adsorption in zeolites depends linearly on the specific surface areas of the materials and is in very good agreement with the results on carbon nanostructures [86].

A microporous metal–organic framework of the composition $Zn_4O(1,4$-benzenedicarboxylate$)_3$ has been proposed as hydrogen storage material [14]. The material absorbs hydrogen at a temperature of 298 K in proportion to the applied pressure. The slope of the linear relationship between the gravimetric hydrogen density and the hydrogen pressure was found to be $0.05\,\mathrm{mass\% \cdot bar^{-1}}$. No saturation of the hydrogen absorption was found, which is very unlikely for any kind of a hydrogen absorption process. At 77 K the amount of adsorbed hydrogen was detected to be 3.7 mass% already at very low hydrogen pressure and gave a slight almost linear increase with increasing pressure. This behavior is not a type I isotherm as the authors claimed and the results should be viewed with care.

The big advantages of the physisorption for hydrogen storage are the low operating pressure, the relatively low cost of the materials involved, and the simple design of the storage system. The rather small amount of adsorbed hydrogen on carbon together with the low temperatures necessary are significant drawbacks of hydrogen storage based on physisorption.

31.7
Other Energy Relevant Applications of Hydrogen Interacting Materials

Application of hydrogen storage materials can be classified into two groups: energy storage and chemical functions. Knowledge elaborated in studies of hydrogen–solid interactions generally and in metal hydride and complex hydride research are used in numerous applications apart from gas storage. Examples are:

- Ni-MH battery electrodes [9,33,87];
- preparation of highest purity hydrogen gas for the semiconductor industry (hydrogen desorbed from metal hydrides has typically one order of magnitude lower impurity contents) [33];
- the surfaces of metal hydrides are sources of atomic hydrogen, catalysis;
- CO_2 hydrogenation over Mg_2NiH_4, formation of synthetic hydrocarbons [88];
- storage of thermal energy, for example, solar thermal power, heat pumps [9,33];
- storage of hydrogen isotopes, isotope separation [33];
- various sensors, for example, Pd thin film based MOS-FET sensor [89];
- H-saturated a-Si for photovoltaics;
- growth of carbon nanostructures and synthetic diamond;
- control of optical properties, optical switching, smart windows [90];
- superconductivity [91].

A rather special use of complex hydrides concerns catalytic hydrolysis: exothermic hydrogen generation from water using sodium borohydride ($NaBH_4$) and a catalyst (e.g., Mg_2NiH_4) in following the reaction:

$$NaBH_4 + H_2O \rightarrow 4H_2 + NaBO_2 \qquad (31.8)$$

was developed for controlled hydrogen delivery in systems adaptable to cars [13].

31.8
Conclusions and Outlook

Hydrogen storage by hydride forming "classical" intermetallics remains an attractive solution for safe, high cyclic life-time stationary storage as well as for small portable storage devices. AB_5-type metal hydride electrodes are reliable battery components in hybrid cars on the market. For the storage of hydrogen gas as a synthetic fuel, they are too heavy, and some are also to expensive. Extensive research and development of complex hydrides (alanates, amides, borides, and others) over the past decade have considerably expanded the material scope of hydrogen storage. However, pursuing high performance sustainable hydrogen storage materials with both high hydrogen capacity and suitable thermodynamics is still challenging. The rich metal–nitrogen chemistry offers considerable room for further materials design and development.

Figure 31.14 gives an overview of different volumetric and gravimetric storage densities.

There is a wide gap between bonding energies for chemisorption and physisorption of hydrogen; an intense search for intermediate bonding is needed.

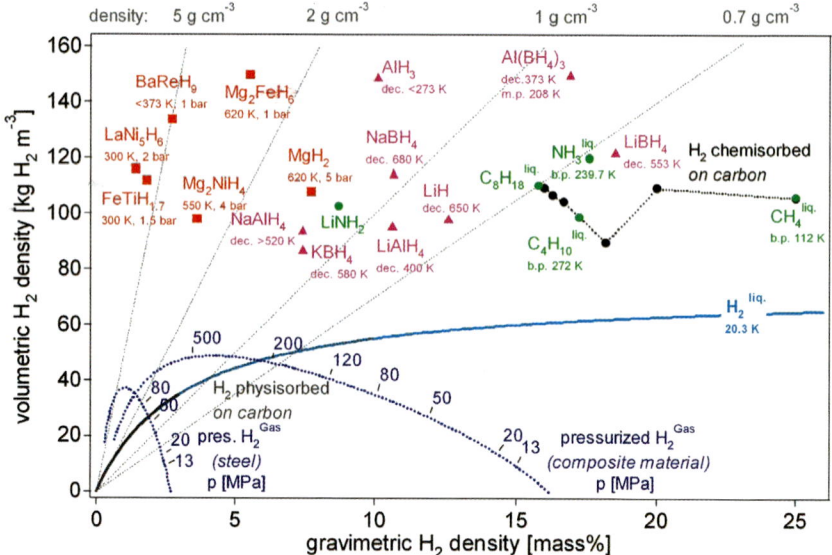

Figure 31.14 Comparison of gravimetric and volumetric hydrogen densities for various hydrogen storage methods [3].

Nanosized cages with functionalized inner surfaces possibly offer adsorption mechanisms different from those of flat large surfaces, and different phase diagrams.

Efficient use of energy remains a main issue for sustainable energy technologies.

References

1 Burzo, E. (ed.) (2014 or later) Hydrogen storage Landolt-Börnstein, Numerical data and functional relationships, New series, Volume VIII/8, in preparation, including chapter, in *Metal Hydrides, AB_5 Hydrides* (eds J.-M. Joubert, A. Percheron-Guégan et al.), Springer.

2 Schlapbach, L. and Züttel, A. (2001) Hydrogen-storage materials for mobile applications. *Nature*, **414**, 353–358.

3 Züttel, A. (2008) Metal hydrides, in *Hydrogen as a Future Energy Carrier* (eds A. Züttel, A. Borgschulte, and L. Schlapbach), Wiley-VCH Verlag GmbH, Weinheim, ch. 6.3; Züttel, A. (September 2003) Materials for hydrogen storage. *Mater. Today*, 18–27.

4 Rittmeyer, P. and Wietelmann, U. (2000) Hydrides, in *High-Performance Fibers to Imidazole and Derivatives*, Ullmann's Encyclopedia of Industrial Chemistry (5th edn), vol. **A13**, Wiley-VCH Verlag GmbH, Weinheim, pp. 199–226.

5 Schlapbach, L. (ed.) (1988 & 1992) *Hydrogen in Intermetallic Compounds I & II*, Springer Series Applied Physics, vol. **64 & 67**, Springer, Berlin.

6 Léon, A. (ed.) (2008) *Hydrogen Technology*, Springer Series Green Energy and Technology, Springer, Berlin Heidelberg; Fichtner, M. (2005) Nanotechnological aspects in materials for hydrogen storage: invited review. *Adv. Eng. Mater.*, **6**, 443–455.

7 Walker, G. (ed.) (2008) *Solid-State Hydrogen Storage*, Woodhead Publishing/CRC, Cambridge/Boston, 580 p.
8 Hirscher, M. (ed.) (2009) *Handbook of Hydrogen Storage*, Wiley-VCH Verlag GmbH, Weinheim.
9 Rönnebro, C.E. and Majzoub, E.H. (guest ed.) (2013) Recent advances in metal hydrides for clean energy applications. *MRS Bull.*, **38** (6), 452–508.
10 Chen, P., Xiong, Z.T., Luo, J.Z., Lin, J.Y., and Tan, K.L. (2002) Interaction of hydrogen with metal nitrides and imides. *Nature*, **420**, 302–304.
11 Yvon, K. and Renaudin, G. (2005) Hydrides: solid state transition metal complexes, in *Encyclopedia of Inorganic Chemistry*, 2nd edn, vol. **III** (editor in chief R.B. King), John Wiley & Sons Ltd., Chichester, pp. 1814–1846. ISBN 0-470-86078-2.
12 Yvon, K. (2005) Complex transition metal hydrides, in *Hydrogen as a Future Energy Carrier* (eds A. Züttel, A. Borgschulte, and L. Schlapbach), Wiley-VCH Verlag GmbH, Weinheim, ch. 6.4; Yvon, K. (1998) Complex transition metal hydrides. *Chimia*, **52** (10), 613–619.
13 Suda, S. and Kelly, M.T. (2008) Indirect hydrogen storage via metals and complexes using exhaust water, in *Hydrogen as a Future Energy Carrier* (eds A. Züttel, A. Borgschulte, and L. Schlapbach), Wiley-VCH Verlag GmbH, Weinheim, ch. 6.8.
14 Rosi, N.L., Eckert, J., Eddaoudi, M., Vodak, D.T., Kim, J., O'Keeffe, M., and Yaghi, O.M. (2003) Hydrogen storage in microporous metal-organic frameworks. *Science*, **300** (5622), 1127–1129.
15 Rzepka, M., Lamp, P., and de la Casa-Lillo, M.A. (1998) Physisorption of hydrogen on microporous carbon and carbon nanotubes. *J. Phys. Chem. B*, **102**, 10849.
16 Nijkamp, M.G. *et al.* (2001) Hydrogen storage using physisorption - materials demands. *Appl. Phys. A*, **72**, 619.
17 Weitkamp, J., Fritz, M., and Ernst, S. (1995) Zeolites as media for hydrogen storage. *Int. J. Hydrogen Energy*, **20** (12), 967.
18 Züttel, A. *et al.* (2002) Hydrogen storage in carbon nanostructures. *Int. J. Hydrogen Energy*, **27**, 203.
19 Schlapbach, L. (2009) Hydrogen fueled vehicles. *Nature*, **460**, 809–811.
20 Langmi, H.W. *et al.* (2003) Hydrogen adsorption in zeolites A, X, Y and RHO. *J. Alloy. Compd.*, **356–357**, 710.
21 Fellet, M. (2013) Research on metal hydrides revived for next-generation solutions to renewable energy storage. *MRS Bull.*, **38**, Energy Quarterly, 1012.
22 Nellis, W.J., Louis, A.A., and Ashcroft, N.W. (1998) Metallization of fluid hydrogen. *Phil. Trans. R. Soc. London, Ser. A*, **356**, 119–135.
23 Murakami, Y., Kanezaki, T., and Sofronis, P. (2013) Hydrogen embrittlement of high strength steels: determination of the threshold stress intensity for small cracks nucleating at nonmetallic inclusions. *Eng. Fract. Mech.*, **97**, 227–243; Yamabe, J., Matsumoto, T., Matsuoka, S., and Murakami, Y. (2012) A new mechanism in hydrogen-enhanced fatigue crack growth behavior of a 1900-MPa-class high-strength steel. *Int. J. Fract.*, **177**, 141–162.
24 Schlapbach, L. (1992) Surface properties and applications, in *Hydrogen in Intermetallic Compounds II*, (ed. L. Schlapbach), Springer Series Applied Physics, vol. **67**, Springer, ch. 2.
25 Schlapbach, L. and Riesterer, T. (1983) The activation of FeTi for hydrogen absorption. *Appl. Phys. A*, **32**, 169–182.
26 Westlake, D.J. (1983) A geometric model for the stoichiometry and interstitial site occupancy in hydrides (deuterides) of $LaNi_5$, $LaNi_4Al$ and $LaNi_4Mn$. *J. Less-Common Met.*, **91**, 275–292.
27 Van Mal, H.H., Buschow, K.H.J., and Miedema, A.R. (1974) Hydrogen absorption in $LaNi_5$ and related compounds: experimental observations and their explanation. *J. Less-Common. Met.*, **35**, 65.
28 Nordlander, P., Norskov, J.K., and Besenbacher, F. (1986) *J. Phys. F. Metal Phys.*, **16** (9), 1161–1171; Norskov, J.K. and Besenbacher, F. (1987) Theory of hydrogen interaction with metals. *J. Less-Common Met.*, **130**, 475–490.

29 Griessen, R. and Riesterer, T. (1988) Heat of formation models, in *Hydrogen in Intermetallic Compounds I* (ed. L. Schlapbach), Springer Series Applied Physics, vol. **64**, Springe; Griessen, R. (1988) Heats of solution and lattice-expansion and trapping energies of hydrogen in transition metals. *Phys. Rev. B*, **38**, 3690–3698.

30 Gupta, M. and Schlapbach, L. (1988) Electronic properties, in *Hydrogen in Intermetallic Compounds I* (ed. L. Schlapbach), Springer Series Applied Physics, vol. **64**, Springer.

31 Van Vucht, J.H.N. Kuijpers, F.A., and Bruning, H. (1970) Reversible room-temperature absorption of large quantities of hydrogen by intermetallic compounds. *Philips Res. Rept.*, **25**, 133.

32 Orgaz, E. and Gupta, M. (2002) *J. Alloy. Compd*, **330–332**, 323–327; Singh, D.J., Gupta, M., and Gupta, R. (2007) *Phys. Rev. B*, **75**, 35103.

33 Sandrock, G., Suda, S., and Schlapbach, L. (1992) Applications, in *Hydrogen in Intermetallic Compounds II* (ed. L. Schlapbach), vol. **67**, Springer; Ye, Z., Noreus, D., and Howlett, J.R. (2013) Metal hydrides for high power batteries. *MRS Bull.*, **38**, 504.

34 Kohno, T. *et al.* (2000) Hydrogen storage properties of new ternary system alloys: La_2MgNi_9, $La_5Mg_2Ni_{23}$, La_3MgNi_{14}. *J. Alloy. Compd*, **311**, L5.

35 Shaltiel, D. *et al.* (1977) Hydrogen absorption and desorption properties of AB_2 laves-phase pseudobinary compounds. *J. Less-Common Met.*, **53**, 117.

36 Komura, Y. *et al.* (1967) The crystal structure of the laves phase in Mg-Zn-Ag-system I. *J. Phys. Soc. Jpn.*, **23** (2), 398.

37 Gingl, F., Vogt, T., and Akiba, E. (2000) Trigonal $SrAl_2H_2$: the first Zintl phase hydride. *J. Alloys Compd*, **306**, 127–132.

38 Yamamoto, T. *et al.* (1997) Microstructures and hydrogen absorption/desorption properties of La-Ni alloys in the composition range of La-77.8–83.2 at. %Ni. *Acta Mater.*, **45**, 5213–5221.

39 Kadir, K. *et al.* (1997) Synthesis and structure determination of a new series of hydrogen storage alloys; RMg_2Ni_9 (R=La, Ce, Pr, Nd, Sm and Gd) built from $MgNi_2$ Laves-type layers alternating with AB_5 layers. *J. Alloy. Compd*, **257**, 115.

40 Reilly, J.J. and Wiswall, R.H. (1974) Formation and properties of iron titanium hydride. *Inorg. Chem.*, **13**, 218–222.

41 Edalati, K. *et al.* (2013) High-pressure torsion of TiFe intermetallics for activation of hydrogen storage at room temperature with heterogeneous nanostructure. *Int. J. Hydrogen Energy*, **38**, 4622; Edalati, K. *et al.* (2013) Mechanism of activation of TiFe intermetallics for hydrogen storage by severe plastic deformation using high-pressure torsion. *Appl. Phys. Lett.*, **103**, 143902.

42 Mizuno, T. and Morozumi, T. (1982) Titanium concentration in FeTi, ($1 \leq X \leq 2$) alloys and its effect on hydrogen storage properties. *J. Less-Common Met.*, **84**, 237–244.

43 Chung, H.S. and Lee, J.Y. (1985) Hydriding and dehydriding reaction rate of FeTi intermetallic compound. *Int. J. Hydrogen Energy*, **10**, 537–542; Bratanich, T.I., Solonin, S.M., and Skorokhod, V.V. (1995) Mechanical activation of hydrogen sorption with intermetallic compounds $LaNi_5$ and TiFe in powder systems. *Int. J. Hydrogen Energy*, **20**, 353–355; Kulshresshtha, S.K., Jayakuma, O.D., and Bhatt, K.B. (1993) Hydriding characteristics of palladium and platinum alloyed FeTi. *J. Mater. Sci.*, **28**, 4229–4223.

44 Trudeau, M.L., Dignard-Bailey, L., Schulz, R., Tessier, P., Zaluski, L., and Ryan, D.H. (1992) The oxidation of nanocrystalline FeTi hydrogen storage compounds. *Nanostruct. Mater.*, **1**, 457–464.

45 Haraki, T., Oishi, K., Uchida, H., Miyamoto, Y., Abe, M., and Kokaji, T. (2008) Properties of hydrogen absorption by nano-structured FeTi alloys. *Int. J. Mater. Res.*, **99**, 507–512.

46 Reilly, J.J. and Wiswall, R.H. (1968) The reaction of hydrogen with alloys of magnesium and nickel and the formation of Mg_2NiH_4. *Inorg. Chem.*, **7**, 2254.

47 Bogdanovic, B. and Schwickardi, M. (1997) Ti-doped alkali metal aluminium hydrides as potential novel reversible hydrogen storage materials. *J. Alloy. Compd*, **253–254**, 1–9.

48 Bogdanovic, B. *et al.* (2007) Complex aluminum hydrides. *Scripta Mater.*, **56**, 813–816.

49 Orimo, S. *et al.* (2007) Complex hydrides for hydrogen storage. *Chem. Rev.*, **107**, 4111–4132.

50 Ashby, E.C. and Kobetz, P. (1966) The direct synthesis of Na_3AlH_6. *Inorg. Chem.*, **5**, 1615–1617.

51 Fossdal, A., Brinks, H.W., Fonneløp, J.E., and Hauback, B.C. (2005) Pressure-composition isotherms and thermodynamic properties of TiF_3-enhancedNa_2LiAlH_6. *J. Alloy. Compd*, **397**, 135–139.

52 Graetz, J. and Reilly, J. (2007) Kinetically stabilized hydrogen storage materials. *Scripta Mater.*, **56**, 835–839.

53 Eberle, U. *et al.* (2009) Chemical and physical solutions for hydrogen storage. *Angew. Chem. Int. Ed.*, **48**, 6608–6630.

54 Xiong, Z.T. *et al.* (2008) High-capacity hydrogen storage in lithium and sodium amidoboranes. *Nat. Mater.*, **7** (2), 138–141.

55 Pinkerton, F.E. *et al.* (2005) Hydrogen Desorption exceeding ten weight percent from the new quaternary hydride $Li_3BN_2H_8$. *J. Phys. Chem. B*, **109** (1), 6–8.

56 Xiong, Z.T. *et al.* (2004) Ternary imides for hydrogen storage. *Adv. Mater.*, **16**, 1522.

57 Orimo, S. *et al.* (2004) Destabilization and enhanced dehydriding reaction of $LiNH_2$: an electronic structure viewpoint. *Appl. Phys. A-Mater.*, **79**, 1765–1767.

58 Leng, H.Y. *et al.* (2004) New metal-N-H system composed of $Mg(NH_2)(2)$ and LiH for hydrogen storage. *J. Phys. Chem. B*, **108**, 8763 and 12628.

59 Luo, W.F. (2004) $LiNH_2$-MgH_2: a viable hydrogen storage system. *J. Alloy. Compd*, **381**, 284–287.

60 Kojima, Y. *et al.* (2006) Hydrogen absorption and desorption by the Li-Al-N-H system. *J. Phys. Chem. B*, **110** (19), 9632–9636.

61 Lu, J., Fang, Z.Z., and Sohn, H.Y. (2006) A new Li-Al-N-H system for reversible hydrogen storage. *J. Phys. Chem. B*, **110** (29), 14236–14239.

62 Vajo, J.J. *et al.* (2004) Altering hydrogen storage properties by hydride destabilization through alloy formation: LiH and MgH_2 destabilized with Si. *J. Phys. Chem.*, **108** (37), 13977–13983; Vajo, J.J., Skeith, S.L., and Mertens, F. (2005) Reversible storage of hydrogen in destabilized $LiBH_4$. *J. Phys. Chem. B*, **109** (9), 3719–3722.

63 Ohoyama, K. *et al.* (2005) Revised crystal structure model of Li_2NH by neutron powder diffraction. *J. Phys. Soc. Jpn*, **74** (1), 483–487.

64 Wang, Y. and Chou, M.Y. (2007) First-principles study of cation and hydrogen arrangements in the Li-Mg-N-H hydrogen storage system. *Phys. Rev. B*, **76** (1), 14116.

65 Nakamori, Y. and Orimo, S. (2004) Li–N based hydrogen storage materials. *Mat. Sci. Eng. A-Struct.*, **108**, 48–50; Nakamori, Y. *et al.* (2004) Synthesis and dehydriding studies of Mg–N–H systems. *J. Power Sources*, **138**, 309–312.

66 Chen, P. *et al.* (2003) Interaction between lithium amide and lithium hydride. *J. Phys. Chem. B*, **107** (39), 10967–10970.

67 Hu, Y.H. and Ruckenstein, E. (2003) Ultrafast reaction between LiH and NH_3 during H-2 storage in Li_3N. *J. Phys. Chem. A*, **107** (46), 9737–9739.

68 Ichikawa, T. *et al.* (2004) Mechanism of novel reaction from $LiNH_2$ and LiH to Li_2NH and H-2 as a promising hydrogen storage system. *J. Phys. Chem. B*, **108**, 7887–7892.

69 Zhang, T.F. *et al.* (2013) A solid-solid reaction enhanced by an inhomogeneous catalyst in the (de)hydrogenation of a lithium-hydrogen-nitrogen system. *Royal Soc. Chem. Adv.*, **3** (18), 6311–6314.

70 David, W.I.F. *et al.* (2007) A mechanism for non-stoichiometry in the lithium amide/lithium imide hydrogen storage reaction. *J. Am. Chem. Soc.*, **129** (6), 1594–1601.

71 Xiong, Z.T. *et al.* (2005) Thermodynamic and kinetic investigations of the hydrogen storage in the Li-Mg-N-H system. *J. Alloy. Compd*, **398** (1–2), 235–239.

72 Wang, J.H. *et al.* (2009) Effects of triphenyl phosphate on the hydrogen storage performance of the $Mg(NH_2)(2)$-2LiH system. *J. Mater. Chem.*, **19** (15), 2141–2146.

73 Isobe, S. et al. (2005) Effect of Ti catalyst with different chemical form on Li-N-H hydrogen storage properties. *J. Alloy. Compd*, **404**, 439–442.

74 Lu, J. et al. (2007) Potential of binary lithium magnesium nitride for hydrogen storage applications. *J. Phys. Chem. C*, **111** (32), 12129–12134.

75 Hu, J.J. et al. (2008) Improvement of hydrogen storage properties of the Li-Mg-N-H system by addition of LiBH$_4$. *Chem. Mater.*, **20** (13), 4398–4402.

76 Anderson, P.A. et al. (2011) Hydrogen storage and ionic mobility in amide-halide systems. *Faraday Discuss.*, **151**, 271–284.

77 Wang, J.H. et al. (2009) Potassium-modified Mg(NH$_2$)(2)/2 LiH system for hydrogen storage. *Angew. Chem. Int. Ed.*, **48** (32), 5828–5832.

78 Nakamori, Y. et al. (2006) Correlation between thermodynamic stabilities of metal borohydrides and cation electronegativities: first-principles calculations and experiments. *Phys. Rev. B*, **74**, 045126.

79 Züttel, A. (2003) et al. Hydrogen storage properties of LiBH$_4$. *J. Alloy. Compd*, **356–357**, 515; Züttel, A. et al. (2007) tetrahydroborates as new hydrogen storage materials. *Scripta Mater.*, **56**, 823.

80 Orimo, S. et al. (2005) Dehydriding and rehydriding reactions of LiBH4. *J. Alloy. Compd*, **404–406**, 427.

81 Li, H. et al. (2011) Recent progress in metal borohydrides for hydrogen storage. *Energies*, **4**, 185.

82 London, F. (1930) Zur theorie und systematik der molekularkräfte. *Z. Phys.*, **63**, 245; London, F. (1930) Properties and applications of molecular forces. *Z. Phys. Chem.*, **11**, 222.

83 Brunauer, S., Emmett, P.H., and Teller, E. (1938) Adsorption of gases in multimolecular layers. *J. Am. Chem. Soc.*, **60**, 309.

84 Gregg, S.J. and Sing, K.S.W. (1967) *Adsorption, Surface Area and Porosity*, Academic Press, London and New York.

85 Beebe, R.A. et al. (1947) Heats of adsorption on carbon black. *J. Am. Chem. Soc.*, **69**, 95.

86 Harris, R. et al. (2004) Hydrogen storage: the grand challenge. *Fuel Cell Rev.*, **1** (1), 17.

87 Kohno, T. et al. (2000) Hydrogen storage properties of new ternary system alloys: La$_2$MgNi$_9$, La$_5$Mg$_2$Ni$_{23}$, La$_3$MgNi$_{14}$. *J. Alloy. Compd*, **311**, L5.

88 Peterson, A.A. and Nørskov, J.K. (2012) Activity descriptors for CO$_2$ electroreduction to methane on transition-metal catalysts. *J. Phys. Chem. Lett.*, **3**, 251–258.

89 Slaman, M. et al. (2007) Fiber optic hydrogen detectors containing Mg-based metal hydrides. *Sens. Actuators B Chem.*, **123**, 538–545; Yoshimura, K. et al. (2009) New hydrogen sensor based on sputtered Mg–Ni alloy thin film. *Vacuum*, **83**, 699–702.

90 Huiberts, J.N. et al. (1996) Yttrium and lanthanum hydride films with switchable optical properties. *Nature*, **380**, 231–234; Griessen, R. et al. (2008) Optical properties of metal hydrides: switchable mirrors in *Hydrogen as a Future Energy Carrier* (eds A. Züttel, A. Borgschulte, and L. Schlapbach), Wiley-VCH Verlag GmbH, Weinheim, ch. 7.2, p. 275.

91 Eremets, M.I. et al. (2008) Superconductivity in hydrogen dominant materials: silane. *Science*, **319**, 1506–1509.

32
Implementing Hydrogen Storage Based on Metal Hydrides

R.K. Ahluwalia, J.-K. Peng, and T.Q. Hua

32.1
Introduction

Hydrogen storage in metal hydrides (MH) continues to be regarded as a promising material based solution [1,2] that with continued development may replace gaseous hydrogen storage in high-pressure tanks for automotive fuel cell systems (FCSs). Many types of hydriding alloys and complexes have been investigated including intermetallic compounds (AB_5, AB_2, AB, A_2B, etc.), complexes of borohydrides and alanates, and metastable hydrides [3]. A recent multi-year, multi-organization study evaluated four classes of developmental materials [4]: destabilized metal hydrides, complex anionic materials, amide/imide materials, and off-board regenerable materials (AlH_3 and $LiAlH_4$). The study concluded that a metal hydride suitable for automotive hydrogen storage does not as yet exist and recommended that future work should focus on $LiBH_4/MgH_2$, $LiBH_4/Mg_2NiH_4$, $Mg(BH_4)_2$, $2LiNH_2/MgH_2$, and $LiNH_2/MgH_2$. The purpose of this work is to establish specific material targets for guiding such future efforts.

The scope of this chapter is limited to on-board reversible metal hydrides. Off-board regenerable materials, such as alane [5] and ammonia borane [6,7], have also attracted considerable attention recently but transporting these materials, loading and off-loading, and off-board regeneration cost and efficiencies remain as major obstacles [8,9].

The scope of this chapter is further limited to low-temperature metal hydride (LTMH) systems, shown in Figure 32.1a, in which hydrogen desorbs at a temperature below the temperature at which the FCS coolant leaves the stack. In an LTMH system, the waste heat generated in the stack can be used to liberate hydrogen from the metal hydride medium. It is distinguished from a medium-temperature metal hydride (MTMH) system, shown in Figure 32.1b, in which hydrogen desorbs at a temperature higher than the temperature at which the coolant leaves the stack. In an MTMH system, some hydrogen must be burned to heat an intermediate fluid that is circulated through the MH bed to supply the enthalpy of desorption. System simulations indicate a 18–25% penalty in net system efficiency if an MTMH configuration is used with catalyzed sodium alanates [10].

Hydrogen Science and Engineering: Materials, Processes, Systems and Technology, First Edition.
Edited by Detlef Stolten and Bernd Emonts.
© 2016 Wiley-VCH Verlag GmbH & Co. KGaA. Published 2016 by Wiley-VCH Verlag GmbH & Co. KGaA.

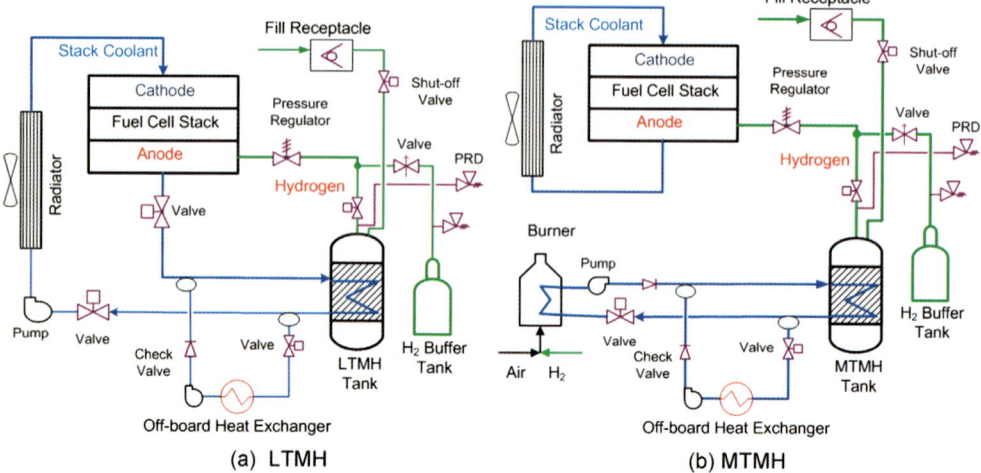

Figure 32.1 Thermal integration of metal hydride storage systems with low-temperature fuel cell systems.

32.2
Material Requirements

Table 32.1 lists some of the near-term and ultimate technical system targets for on-board hydrogen storage for light-duty fuel cell vehicles [11] that are the material covered here. There are additional targets for storage system cost, fuel cost, on-board efficiency (important for MTMH), and fuel quality (i.e., purity of H_2 from storage system) but they are not specifically addressed here. The purpose of this section is to lay the rationale and foundation for translating these system level targets to material level requirements for metal hydrides.

32.2.1
Operating Temperatures

Figure 32.2a shows a conceptual 80-kW_e, automotive, pressurized, proton exchange fuel cell (PEFC) system that has been extensively analyzed for performance and cost in many studies [12,13]. We wish to analyze the boundary conditions for a perspective metal-hydride hydrogen storage system that may potentially replace the 700-bar hydrogen storage tank in this FCS. Figure 32.2b presents the FCS efficiency and the ethylene glycol coolant temperature at stack exit that vary between 47.5% and 85 °C at rated power and 59.3% and 62 °C under idling conditions. For a thermally integrated metal hydride system in which the enthalpy of desorption (ΔH) comes from the waste heat produced in the stack and transferred to the stack coolant, the idling condition limits the maximum available ΔH to 100 kJ mol^{-1} and the desorption temperature to about 60 °C.

Table 32.1 Technical system targets for on-board hydrogen storage for light-duty fuel cell vehicles.

Storage parameter	Units	2017	Ultimate
System gravimetric capacity: Usable, specific-energy from H_2 (net useful energy/max system mass)	kWh kg^{-1} (kg H_2 per kg system)	1.8 (0.055)	2.5 (0.075)
System volumetric capacity: Usable energy density from H_2 (net useful energy/max system volume)	kWh l^{-1} (kg H_2 per l system)	1.3 (0.040)	2.3 (0.070)
Durability/operability:			
Operating ambient temperature	°C	−40/60 (sun)	−40/60 (sun)
Min/max delivery temperature	°C	−40/85	−40/85
Operational cycle life (¼ tank to full)	Cycles	1500	1500
Min delivery pressure from storage system	bar (abs)	5	3
Max delivery pressure from storage system	bar (abs)	12	12
Onboard efficiency	%	90	90
Charging/discharging rates:			
System fill time (5 kg)	min (kg H_2 per min)	3.3 (1.5)	2.5 (2.0)
Minimum full flow rate	g s^{-1} kW^{-1}	0.02	0.02
Start time to full flow (20 °C)	s	5	5
Start time to full flow (−20 °C)	s	15	15
Transient response at operating temperature 10–90% and 90–0%	s	0.75	0.75

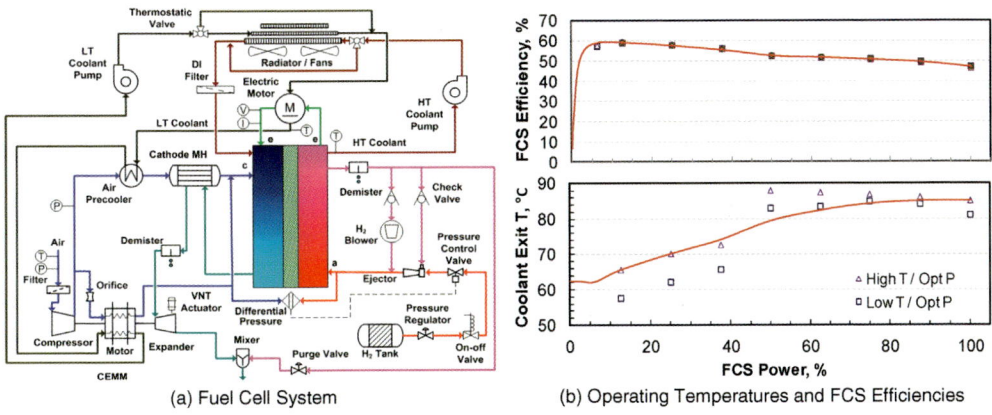

Figure 32.2 Conceptual automotive pressurized fuel cell system with compressed hydrogen storage tank.

32.2.2
Material Thermodynamics

The maximum ΔH is further restricted by the requirement that, during discharge, the equilibrium pressure (P_{eq}) at the minimum FCS coolant temperature (60 °C) should be higher than the target minimum delivery pressure (5 bar). Using the van't Hoff equation for plateau pressure [14]:

$$\ln(P_{eq}) = -\frac{\Delta H}{RT} + \frac{\Delta S}{R} \tag{32.1}$$

Figure 32.3a shows the relationship between ΔH_{max} and ΔS. There is also a lower limit on ΔH since, during refueling, the equilibrium pressure at the maximum allowable bed temperature should be lower than the H_2 supply pressure. Figure 32.3a includes the relationship between ΔH_{min} and ΔS for 100 °C maximum temperature and 50–200 atm refueling pressures. An acceptable material should have ΔH between ΔH_{min} and ΔH_{max}. Figure 32.3b shows that some of the known materials in the MH database have an enthalpy–entropy relationship that falls within this thermodynamic acceptability band [15] although they do not satisfy one or more other requirements discussed below.

32.2.3
Containment Tank

Type 3 tanks are generally preferred for containing metal hydrides if temperatures above 80 °C are encountered during refueling [16]. These consist of an inner layer of aluminum liner that is impervious to H_2 and an outer layer of high strength carbon fiber (CF) composite designed to withstand 225% of the nominal storage pressure. Table 32.2 summarizes the mechanical properties of the most-common in use high-strength and high-modulus T700S fiber and the fiber-resin composite with 60% T700S fiber by volume [17]. Prior to formal

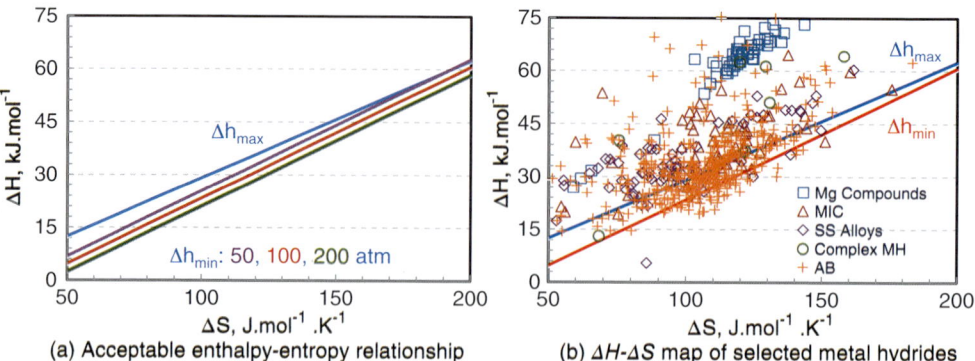

Figure 32.3 Thermodynamic requirement of metal hydrides suitable for on-board hydrogen storage.

Table 32.2 Properties of T700S carbon fiber and composite (60% fiber volume).

Property	Unit	Fiber	Composite
Tensile strength	MPa	4900	2550
Tensile modulus	GPa	230	135
Tensile strain	%	2.1	1.7
Density	kg m^{-3}	1800	137.5% NWP
Compressive strength	MPa		1470
Flexural strength	MPa		1670
Flexural modulus	GPa		120

finite-element analysis using codes such as ABAQUS [18,19], the tank geometry, optimal dome shape, winding angle, and preliminary helical and hoop layer thicknesses of the CF overwrap are generally determined by a netting analysis. A geodesic winding design for the dome, in which all fibers are uniformly stressed and display no shearing or bending stiffness, has a minimum mass of the CF composite for a given cylinder diameter and storage pressure [20–22]. Typically, the helical thickness is about half the hoop thickness in the cylinder section, and increases in the dome. Type 3 pressure vessels that are subjected to fluctuating pressure are auto-frettaged to generate residual compressive stresses in the liner, shift tensile loads to the CF composite, and extend the fatigue life of the metal liners [23]. SAE J2579 requires a fatigue life of 5500 pressure cycles at 125% of the nominal working pressure [24].

Figure 32.4 shows the weight and volume of 97-l Type-3 containment tanks as a function of storage pressure for designs in which the CF overwrap bears 85–90% of the pressure load with the balance being absorbed by the liner. The amount of CF needed is directly proportional to the storage pressure. For storage pressures under 100 atm, the liner thickness is held at a minimum but

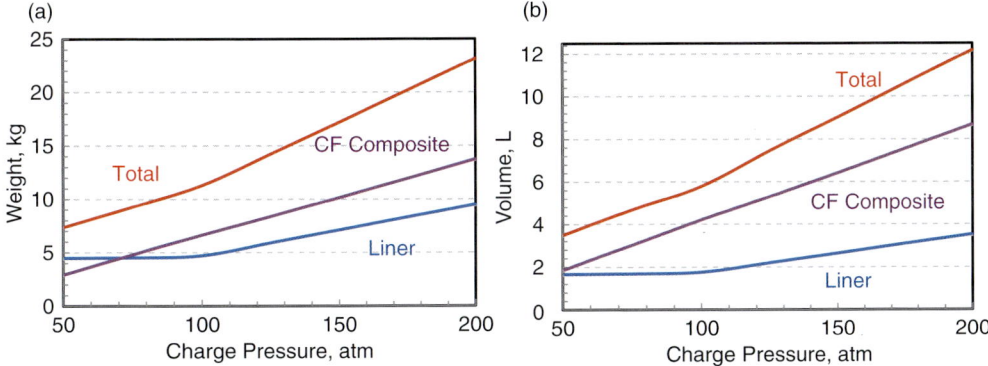

Figure 32.4 (a) Weight and (b) volume of CF composite and liner in Type-3 containment tank with ~95 l internal volume.

increases with pressure increasing above 100 atm. At the highest storage pressure of 200 atm, the containment tank accounts for ~23% of the on-board storage system weight. For comparison, the containment tank for a 700 bar compressed hydrogen storage system accounts for ~70% of the total system weight [16].

32.2.4
Buffer Hydrogen

While starting from cold (ambient temperature T_a), a buffer tank may be needed to provide H_2 for the duration of time that the bed takes to reach temperature (T_s) at which the equilibrium pressure exceeds the minimum delivery pressure and the discharge kinetics is sufficiently fast to provide H_2 at the target minimum full flow rate. The buffer capacity ($w_{H_2}^s$) is determined by the amount of H_2 consumed in the PEFC stack while the stack and the MH bed heat to reach the startup temperature. Recognizing that in a storage system without a burner, the bed (subscript b) is heated indirectly by using the waste heat produced in the stack (subscript FC), the required buffer capacity may be estimated from the following equation:

$$w_{H_2}^s = \frac{(w_{FC} C_{FC} + w_b C_b)(T_s - T_a)}{\Delta H_{H_2}(1 - \eta_{FC})} - w_{H_2}^p \tag{32.2}$$

where w denotes the mass, C is the heat capacity, ΔH_{H_2} is the lower heating value of hydrogen, η_{FC} is the stack efficiency during startup, and $w_{H_2}^p$ is the usable amount of gaseous H_2 in the void space of the MH bed (i.e., amount of H_2 available above 5 atm).

Figure 32.5a presents an illustrative method of coupling the buffer H_2 tank with the MH tank. The MH tank is charged with H_2 at 100 atm and may reach a peak temperature of 16 °C during refueling if the material has a ΔH of 28 kJ mol^{-1} and ΔS of 110 J mol^{-1} K^{-1}. The buffer is discharged and supplies H_2 to the PEFC stack during startup while the bed temperature is below 20 °C.

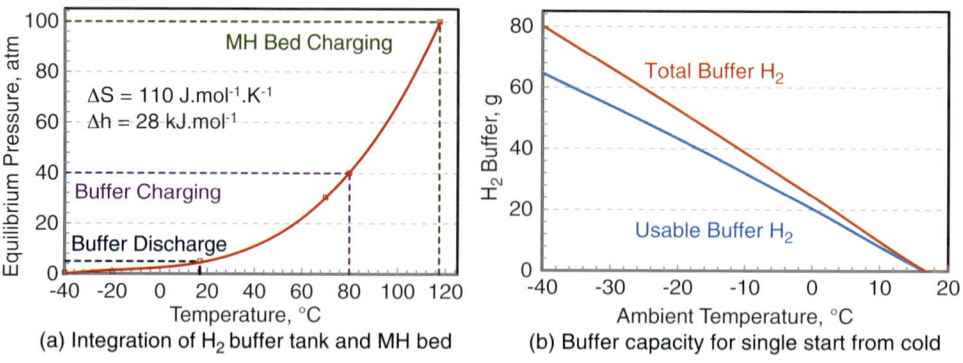

Figure 32.5 (a) Integration of buffer H_2 tank and MH bed and (b) buffer capacity needed for cold start.

Between refuelings, the buffer is replenished with H_2 released from the MH bed during normal operation when the stack coolant is at near the peak temperature. The buffer pressure is limited to 40 bar, which is the equilibrium pressure at 80 °C for this material.

Figure 32.5b shows the minimum total and usable (above 5 atm delivery pressure) amounts of buffer H_2 needed for a single start from the indicated ambient temperatures. The various parameters for the results in Figure 32.5 are: $w_{FC}C_{FC} = 20$ kJK^{-1}, $w_b C_b = 70$ kJ K^{-1}, $\Delta H_{H_2} = 120$ MJ kg^{-1}, $\eta_{FC} = 35\%$, $T_s = 16.8$ °C, and $w_{H_2}^p = 0$.

32.2.5
Desorption Kinetics

Not all the intrinsic H-capacity of a storage material may be accessible. For a simple material whose pressure–composition–temperature (PCT) isotherm can be represented by the van't Hoff equation, there is a lower limit of the state-of-charge (X) below which the equilibrium pressure declines rapidly with decrease in X. Hydrogen stored below this lower limit is not accessible for release at reasonable back pressures. Similarly, there is an upper limit of the state-of-charge (SOC) above which the equilibrium pressure rises rapidly with increase in X. The metal hydride cannot be charged above this upper limit at reasonable refueling pressures. The accessible lower and upper limits of SOC are functions of temperature, and the difference between the two limits for simple metal hydrides becomes smaller at higher temperatures [14].

Chemical kinetics may further limit the usable portion of the accessible intrinsic H-capacity (β). Let X_{min} and X_{max} denote the usable minimum and maximum state of charge, respectively. The minimum SOC (also called the maximum state of discharge) is determined by the kinetic requirement that the MH bed should be able to supply the minimum full flow of H_2 at the lowest operating temperature [10]. Consider a simple representation of the desorption kinetics in which the kinetic rate ($\dot{r}_d$) is first order in X and ($P_{eq} - P$):

$$\beta \frac{dX}{dt} = -k_d X (P_{eq} - P) = -\dot{r}_d \tag{32.3}$$

For isothermal discharge at constant back pressure (P), Eq. (32.3) can be integrated to obtain the following equation for discharge time (τ_d) between X_{min} and X_{max}:

$$\tau_d = \left[\frac{\beta}{k_d (P_{eq} - P)} \right] \ln \left(\frac{X_{max}}{X_{min}} \right) \tag{32.4}$$

Knowing τ_d, Eqs. (32.3) and (32.4) can be used to determine X_{min} at which the H_2 discharge rate ($\dot{w}_{H_2}$) from a bed containing w_{MH} amount of MH equals 1.6 g s^{-1} at the minimum coolant temperature:

$$X_{min} \ln \left(\frac{X_{max}}{X_{min}} \right) = \frac{\dot{w}_{H_2} \tau_d}{\beta w_{MH}} \tag{32.5}$$

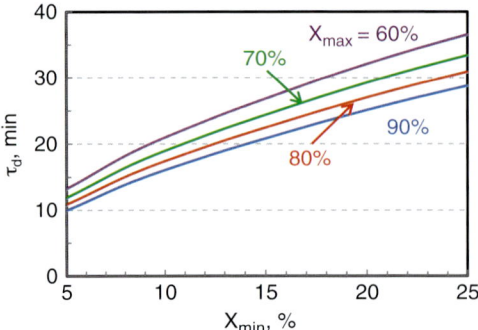

Figure 32.6 Desorption kinetic requirements defined in terms of time to discharge MH from X_{min} to X_{max} at 60 °C, 5 atm back pressure.

Figure 32.6 shows the relationship between τ_d and X_{min} for a bed that contains 5.6 kg of usable H_2 (w_{H_2}):

$$\beta(X_{max} - X_{min})w_{MH} = w_{H_2} \tag{32.6}$$

Naturally, for given X_{max}, the faster the desorption kinetics, the smaller is τ_d, and the lower is the allowable X_{min}.

32.2.6
Sorption Kinetics

Just as the desorption kinetics determines the minimum SOC, so the sorption kinetics determines the maximum SOC [10]. The relationship between X_{max} and the target 3.7 min refueling time is best illustrated by considering, analogous to Eq. (32.3), a simple representation of the sorption kinetics in which the kinetic rate ($\dot{r}_c$) is first order in $(1-X)$ and $(P - P_{eq})$, and the rate constant k_c has Arrhenius dependence on temperature:

$$\beta \frac{dX}{dt} = k_c(1-X)(P - P_{eq}) = \dot{r}_c \tag{32.7}$$

$$k_c = k_{c0}e^{-E_c/RT} \tag{32.8}$$

Because k_c increases and $(P - P_{eq})$ decreases with increase in temperature, there is an optimum temperature at which $\dot{r}_c$ is fastest. At this optimum refueling temperature (T_{op}), the equilibrium pressure is related to the refueling pressure by the following equation:

$$P_{eq}(T_{op}) = \left(\frac{E_c/\Delta H}{1 + E_c/\Delta H}\right)P \tag{32.9}$$

It follows that the off-board coolant that removes the heat of sorption from the MH tank should always be at a temperature below T_{op}. Figure 32.7a

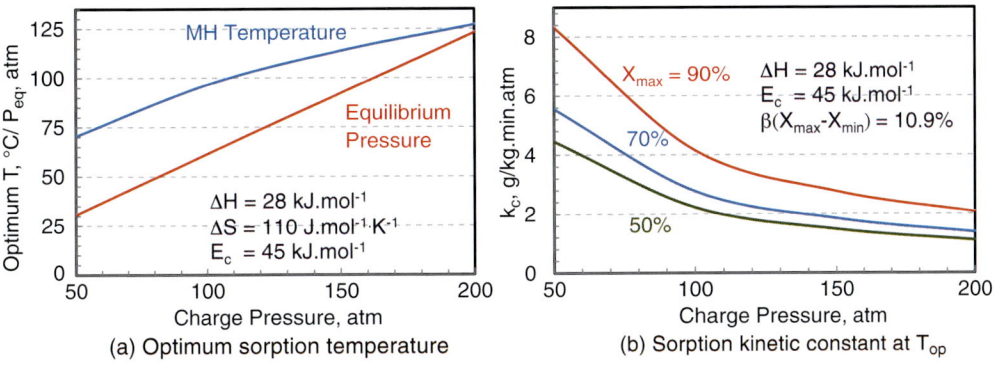

Figure 32.7 Sorption kinetic requirements defined by considering isothermal refueling at constant pressure.

presents T_{op} and $P_{eq}(T_{op})$ as a function of refueling pressure for a set of ΔH, ΔS, and E_c.

For isothermal refueling at constant pressure, Eq. (32.7) can be integrated to determine the kinetic refueling time (τ_r). The actual refueling time will be greater than τ_r due to heat transfer limitations:

$$\beta \ln\left(\frac{1-X_{min}}{1-X_{max}}\right) = k_c(P - P_{eq})\tau_r \tag{32.10}$$

Taking τ_r as 50% of the target refueling time, Figure 32.7b presents the relationship between k_c at T_{op} and refueling pressure for different X_{max}. It quantifies the levels of $k_c(T_{op})$ needed to meet the refueling time target and the enhancement in kinetic rates needed to widen the operating SOC window.

32.2.7
Material Compaction and Heat Transfer

In powder form, metal hydrides and adsorbents have low bulk density and poor thermal conductivity. Meeting the stringent volumetric target requires that the powders be mechanically compacted and consolidated to form pellets and disks. Attention must be paid to allow for the volumetric changes that occur as the material expands upon hydriding and the contact resistance between the compact and heat transfer surfaces. The compacts also must possess sufficient permeability for acceptably uniform hydrogen flow during rapid charging so that the MH sorption kinetics are not greatly impaired.

Many methods have been investigated to augment heat transfer in metal hydride beds including use of metal foams, wires, and fins [25,26]. Densified metal hydride compacts with high conductivity materials, such as aluminum powder and expanded natural graphite (EG), as additives appear particularly promising. Sanchez et al. [26] have shown that the measured effective thermal

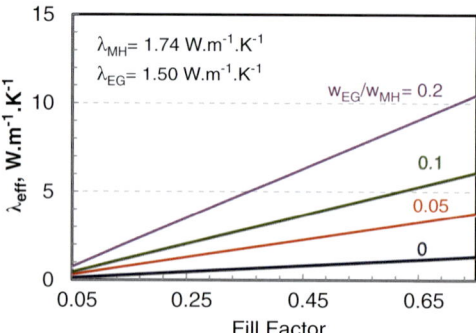

Figure 32.8 Effective thermal conductivity of composites as function of fill factor and EG to MH weight ratio.

conductivity (λ_{eff}) of these compacts can be correlated by a simple model with parallel thermal resistances of metal hydride and heat transfer additive:

$$\lambda_{eff} = \varepsilon_{MH}\lambda_{MH} + \varepsilon_{EG}\lambda_{EG} \qquad (32.11)$$

In the above equation, ε is the material fill factor and it can be represented in terms of the mass fractions (Y), material densities, and the bulk density (ρ_b) of the bed composed of these compacts:

$$\varepsilon_{MH} = Y_{MH}\left(\frac{\rho_b}{\rho_{MH}}\right) \qquad (32.12)$$

$$\varepsilon_{EG} = Y_{EG}\left(\frac{\rho_b}{\rho_{EG}}\right) \qquad (32.13)$$

$$\varepsilon = \varepsilon_{MH} + \varepsilon_{EG} \qquad (32.14)$$

Figure 32.8 presents the λ_{eff} of a compact with $\lambda_{MH} = 1.74\,\text{W m}^{-1}\,\text{K}^{-1}$ and $\lambda_{EG} = 150\,\text{Wm}^{-1}\,\text{K}^{-1}$. It shows that the effective thermal conductivity can be enhanced by increasing the fill factor (densification) and by increasing the EG weight fraction.

The existing data for metal hydrides suggests that there may be a correlation between the material density of metal hydrides and their H-capacity [27,28]. The data in Figure 32.9 indicates that ρ_{MH} is smaller for metal hydrides with higher β.

32.3
Reverse Engineering: A Case Study

The concepts presented in Section 32.2 can be employed to determine the envelope of material properties needed to satisfy the system level targets in Table 32.1. The case study discussed in this section pertains to a low-temperature metal-hydride hydrogen storage system that is thermally integrated

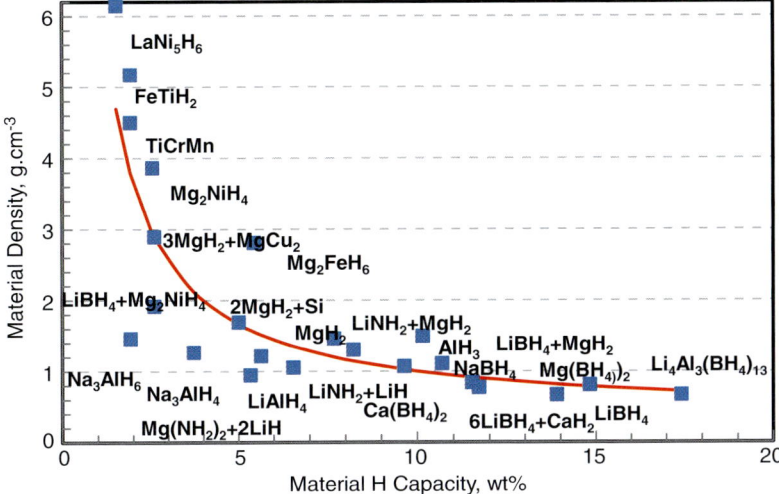

Figure 32.9 Correlation between MH material density and intrinsic H-capacity, data from Van Hessel et al. [27].

with the FCS as in Figure 32.1a. The storage system uses the FCS coolant to supply heat during discharge and an off-board coolant to remove heat liberated during refueling. There is no H_2 burner in the system so that the on-board storage system efficiency is 100%. Consistent with the discussion in the previous section, the scope of this study is limited to a Type 3 containment tank, buffer H_2 tank for FCS startup from −40 °C, and expanded natural graphite for conductivity enhancement. Table 32.3 lists the reference values of coolant

Table 32.3 Reverse engineering study parameters.

Parameter	Units	Reference values	Range of values	Comments
FCS coolant temperature	°C	62–85		Reference FCS [12]
Off-board coolant temperature	°C	80	60–100	
MH thermodynamics	ΔS, J mol^{-1} K^{-1}	110	TBD	$\Delta H_{min} = 26.8$ kJ mol^{-1}
	ΔH, kJ mol^{-1}	28	TBD	$\Delta H_{max} = 32.2$ kJ mol^{-1}
Sorption activation energy	E_c, kJ mol^{-1}	45	TBD	NaAlH$_4$ data
Minimum delivery pressure	atm	5		DOE target [1]
Minimum full flow rate of H_2	g s^{-1}	1.6		DOE target [1]
Refueling pressure	atm	100	50–200	
EG/MH weight ratio		0.1	0.05–0.25	IJHE (2003) 515–527 [26]
System gravimetric capacity	wt%	5.5		DOE target [1]
System volumetric capacity	g l^{-1}	40		DOE target [1]

temperatures, MH ΔH, ΔS, and E_c, and Y_{EG} considered in this study. Table 32.3 also includes the specifications for minimum delivery pressure and refueling pressure and the targets for system gravimetric and volumetric capacities.

32.3.1
MH Refueling: Temperature Profile and Conversion

An important aspect of designing metal hydride tanks is heat removal during fast refueling with hydrogen. A dynamic model is needed to track the temperature profile and hydrogen uptake which change continuously with time. Consider a unit annular cell of outer radius r_2 that consists of a representative heat transfer tube of radius r_1 surrounded by MH-EG compact. Assuming that the gas and solid are in local equilibrium, the following equations describe the dynamic changes in temperature and conversion (state-of-charge):

$$\rho_b C_b \frac{\partial T}{\partial t} = \frac{\lambda_{\text{eff}}}{r} \frac{\partial}{\partial r}\left(r \frac{\partial T}{\partial r}\right) + \rho_{\text{MH}} r_c \left(\frac{\Delta H}{M_{H_2}}\right) \tag{32.15}$$

$$\beta \frac{dX}{dt} = k_c(1-X)(P - P_{\text{eq}}) = r_c \tag{32.7}$$

$$-\lambda_{\text{eff}} \frac{\partial T}{\partial r} = h(T_1 - T) \quad \text{at } r = r_1 \tag{32.16}$$

$$\frac{\partial T}{\partial r} = 0 \quad \text{at } r = r_2 \tag{32.17}$$

where r is the radial coordinate, t is the time, ρ_{MH} is the bulk density of MH, M is the molecular weight, h is the convective heat transfer coefficient, and T_1 is the temperature of the coolant. Equations (32.15–32.17) can be solved numerically using an implicit finite difference scheme with the spatial derivative (heat conduction term) discretized by central differencing.

Figure 32.10 summarizes the results from simulations run to investigate the effect of off-board coolant temperature on refueling at constant pressure ($P = 100$ atm), starting from $T = T_1$ and $X = X_{\min} = 0.1$ everywhere. The size of the unit cell has been selected to achieve complete refueling, defined as average $X = 0.9$, in 3.7 min (t_r). The results indicate that even with EG additive λ_{eff} is low, and the refueling time is short (i.e., sorption kinetics is fast and the heat release rate is high), so that the cell periphery heats up rapidly, overshoots the optimum temperature well within $t/t_r = 0.2$, and approaches the equilibrium temperature corresponding to 100 atm refueling pressure. At longer times, the cell actually cools because of the heat removed by the coolant even though the absorption reaction is exothermic. The bed temperature profile evolves continuously during the refueling process, that is, there is no steady-state temperature profile.

Figure 32.10 shows that the conversion at cell periphery (X_2) is the slowest and is limited more by heat transfer rather than sorption kinetics. At 60 °C

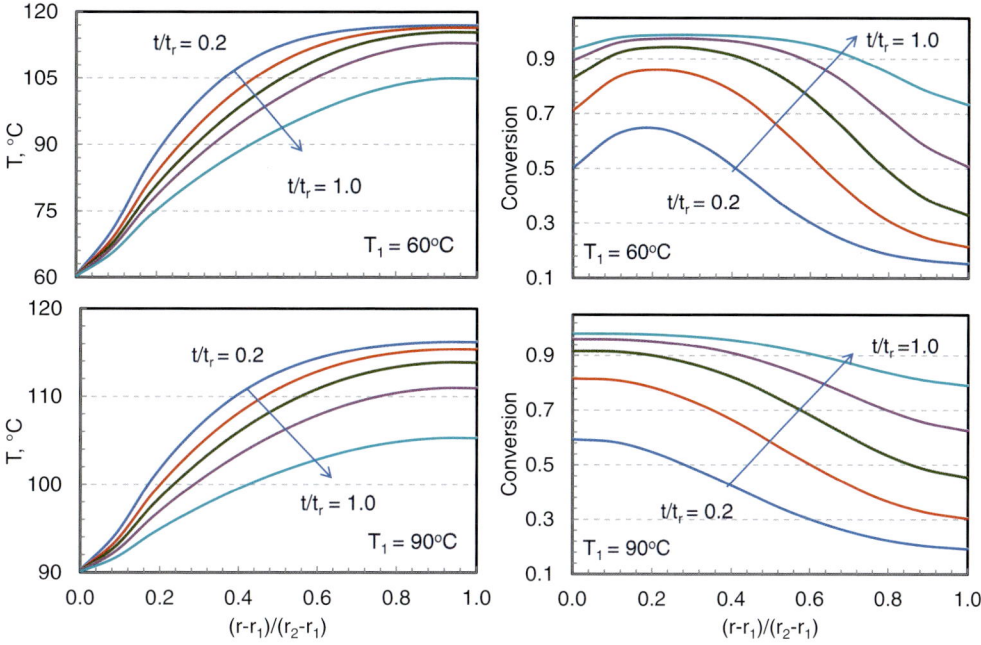

Figure 32.10 Temperature profile and conversion during refueling from X_{min} to X_{max} at constant pressure (100 atm).

refueling temperature, the maximum conversion occurs at a non-dimensional distance of 0.15 at $t/t_r = 0.2$ and slowly moves farther from the heat exchanger tube. At 90 °C refueling temperature, the maximum conversion occurs at the surface of the heat exchanger tube and does not move with time.

32.3.2
System Analysis Model

As discussed in an earlier publication [10], a system model has been developed to analyze the performance of low-temperature metal hydride systems.

As indicated in Tables 32.3 and 32.4, the following are the assumed specified variables: material thermodynamic properties (ΔH, ΔS, E_c); minimum and maximum limits of SOC (X_{min}, X_{max}); on-board and off-board coolant temperatures; refueling pressure; heat transfer tube parameters (aluminum, tube diameter, tube thickness); EG weight fraction (Y_{EG}); weight and volume of BOP components; and H_2 collection and distribution system (five stainless steel sintered metal tubes).

The following are the target performance parameters to be determined: charge kinetic rate; discharge kinetic rate; bulk density of MH/EG composites; number of heat transfer tubes; liner thickness; CF composite thickness; vessel diameter, vessel length; and amount of buffer H_2 and weight/volume of buffer tank.

Table 32.4 Reference metal hydride targets.

Variables	Related variables	Reference values	Constraints
MH intrinsic capacity		13.6% H capacity	5.5 wt% gravimetric
Fill ratio	Bulk density	24.7% bed porosity	40 g l^{-1} volumetric
	Thermal conductivity	589 kg m^{-3} MH bulk density	
		8.4 W m^{-1} K^{-1} bed conductivity	
Desorption kinetics	$X_{min} = 10\%$	$\tau_d = 16.8$ min	1.6 g s^{-1} min full flow
Sorption kinetics	$X_{max} = 90\%$	$k_c = 4.2$ g kg-MH^{-1} min^{-1} atm^{-1}	X_{min} to X_{max} in 1.85 min
HX tube spacing	Number of HX tubes	$r_2/r_1 = 3.1$	1.5 kg min^{-1} refueling
		85 U tubes	
Mass of MH	Mass of expanded Natural graphite	51.2 kg MH	5.6 kg usable H$_2$
		5.1 kg EG	
Buffer tank capacity	Weight of Al tank	11.1 kg buffer tank weight	Start up from $-40\,°C$
		33.7 l buffer tank volume	

The following are the constraints to be satisfied: amount of usable H$_2$; system gravimetric capacity; system volumetric capacity; minimum full flow rate of H$_2$; refueling time; startup from cold; SAE code for liner fatigue; SAE code for burst safety factor; and vessel aspect ratio (study parameter).

There is not a one-to-one correspondence between the performance parameters and the constraints. For example, bulk density affects both the volumetric capacity and the thermal conductivity and hence the refueling time. A multi-dimensional nonlinear equation solver is needed and one based on steepest descent method with quasi-Newton update has been found to be particularly suitable. The challenge is to embed the finite-difference algorithm discussed in Section 32.3.1 for determining the temperature profile, conversion, and refueling time.

32.3.3
Reference Targets

Table 32.4 lists the reference material properties needed to satisfy the system targets as determined using the system model outlined in Section 32.3.2. The system model indicates that a perspective MH candidate should have greater than 13.5% accessible intrinsic H-capacity (β) for the system to have 5.5 wt% gravimetric capacity. When mixed with 10% expanded natural graphite, the material

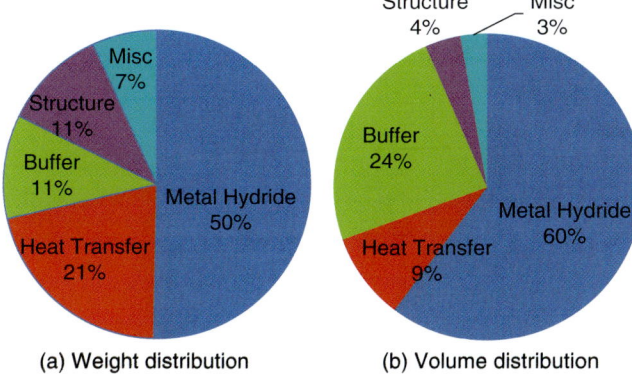

Figure 32.11 Weight and volume distribution for a H_2 storage solution using a metal hydride with material properties that combine to satisfy all the system-level requirements listed in Table 32.1.

should be capable of being compacted to $589 \, \text{kg m}^{-3}$ for the system to reach $40 \, \text{g l}^{-1}$ volumetric capacity. The resulting MH/EG compact has 24.7% bed porosity and $8.4 \, \text{W m}^{-1} \text{K}^{-1}$ thermal conductivity. The 3.7 min refueling time target can be met if the rate constant for charge kinetics is greater than $4.2 \, \text{g (kg-MH)}^{-1} \text{min}^{-1} \text{atm}^{-1}$ for isothermal refueling at the optimum temperature corresponding to 100 atm back pressure. The $1.6 \, \text{g s}^{-1}$ full flow rate requirement can be met if the metal hydride can be discharged from an SOC of 0.9 to 0.1 in 16.8 min at 60 °C and 5 atm back pressure.

Table 32.4 indicates that the 85 U tubes are needed for circulating the off-board coolant at 80 °C and charging the MH tank at $1.5 \, \text{kg min}^{-1}$ refueling rate. The mean spacing between the tubes is 3.1 times the tube diameter. The tank needs to be loaded with 51.2 kg of MH (and 5.1 kg of EG) for 5.6 kg usable H_2 between X_{min} and X_{max}.

A 33.7-l buffer tank capable of storing 80 g of H_2 at 80 °C and 40 atm is needed for a single start from −40 °C. The tank weighs 11.1 kg if it is made of Al 6061-T6 alloy.

Figure 32.11 presents the weight and volume distribution of a MH storage system that meets all system requirements. It shows that the MH accounts for 50% of the overall weight and that 21% of the weight is due to the heat transfer subsystem that includes EG and in-bed heat exchanger tubes. On a volume basis, MH and the buffer H_2 tank are the largest components accounting for 60% and 24% of the overall system volume, respectively.

32.3.4 Sensitivity Study

The reference material targets in the above section were based on certain assumptions regarding the off-board coolant temperature, refueling pressure,

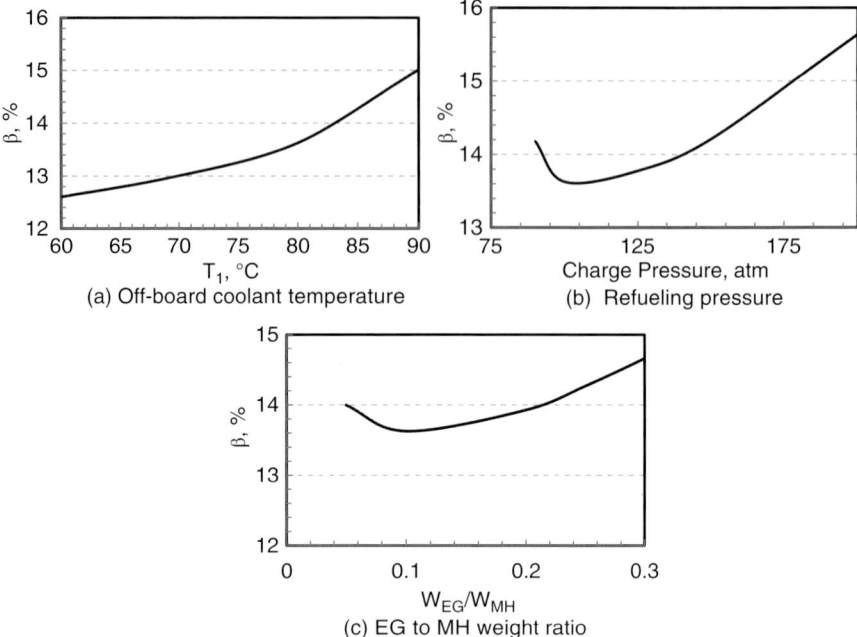

Figure 32.12 Single-parameter sensitivity study to examine the effect of (a) off-board coolant temperature, (b) refueling pressure, and (c) EG to MH weight ratio on intrinsic material capacity target.

EG to MH weight ratio, and minimum and maximum state-of-charge. This section briefly discusses the effect of relaxing these assumptions.

Figure 32.12a presents the effect of off-board coolant temperature on the target accessible H-capacity of the MH. It shows that for a MH with ΔH, ΔS, and E_c listed in Table 32.4, the target β is smaller if the coolant temperature is lowered to 60 from 80 °C. This is due to the higher heat rejection rate at 60 °C and the resulting saving in weight as fewer heat exchanger tubes are needed.

Figure 32.12b shows that β is lowest at 100–110 atm refueling pressure. At higher refueling pressures, β has to be larger to compensate for the heavier liner and CF composite. At lower refueling pressures, T_{op} is lower (see Eq. (32.9)), and more heat transfer tubes are needed to maintain the bed at a lower temperature during refueling.

Figure 32.12c indicates that an EG to MH weight ratio of 0.1 is about the optimum. The effective thermal conductivity increases at higher EG to MH ratio but the added weight of EG is greater than the resulting saving in weight of the heat transfer tubes.

Figure 32.13a shows the effect of desorption kinetics, k_d, on material targets. Doubling k_d (same as halving τ_d) from the reference value lowers X_{min} to 0.05 from 0.1 and results in a small decrease (beneficial) in β; this decrease is

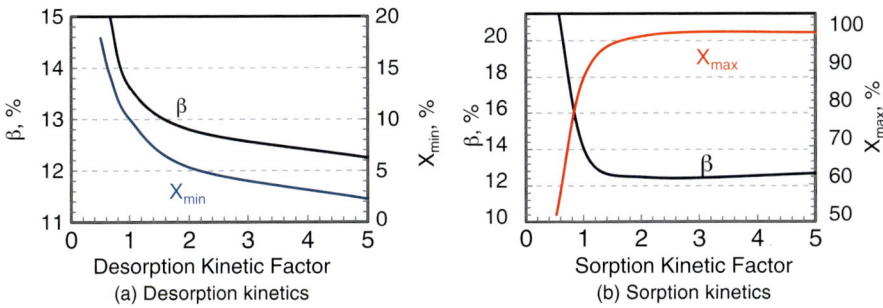

Figure 32.13 Single-parameter sensitivity study to examine the effect of (a) desorption and (b) sorption kinetics on intrinsic material capacity target.

proportional to $(X_{max} - X_{min})$. However, a 50% reduction in k_d results in a larger increase (penalty) in β.

Finally, Figure 32.13b shows the effect of sorption kinetics, $k_c(T_{op})$, on the material targets. Doubling k_c from the reference value raises X_{max} to ~1 from 0.9 while lowering X_{min} to 0.06 from 0.1, and results in an 11% decrease in β. However, a 50% decrease in k_c lowers X_{max} to 0.52 (with an accompanying small increase in X_{min}) and results in >63% increase (penalty) in β.

32.4
Summary and Conclusions

Automotive duty cycles impose some unique and stringent requirements on on-board hydrogen storage systems. These requirements can be used to derive certain minimum material properties that metal hydrides must have to be viable candidates for automotive application. The following is a summary of the minimum thermodynamic, physical, kinetic, and thermal requirements for prospective metal hydrides:

1) The equilibrium pressure should be greater than 5 atm at 60 °C and less than 100 atm at 100–150 °C.
2) The intrinsic hydrogen capacity (β), accessible between 5 to 100 atm at 60–150 °C, should be higher than 13.5 wt% (135 g-H_2 kg^{-1}).
3) The material should discharge from 90% to 10% of accessible intrinsic capacity in <17 min at 60 °C and 5 atm back pressure.
4) The material should be capable of being charged isothermally from 10% to 100% of accessible intrinsic capacity within 1.85 min at constant 100 atm. The temperature can be between 60 and 150 °C.
5) All properties should be measured on samples prepared by mixing the metal hydride with a high conductivity material (such as expanded natural graphite) and after compacting the sample to 600 kg m^{-3} bulk density. The compact should have a minimum effective thermal conductivity

of $8.4\,\mathrm{W\,m^{-1}\,K^{-1}}$. The heat transfer additive to metal hydride weight ratio preferably should be less than 0.1.

Acknowledgments

This work was supported by the Fuel Cell Technologies Office of the US Department of Energy's (DOE) Office of Energy Efficiency and Renewable Energy. Ms. Grace Ordaz was the DOE technology development manager for this work. Argonne is a DOE, Office of Science Laboratory operated under Contract No. DE-AC02-06CH11357 by UChicago, Argonne, LLC.

References

1 Klebanoff, L. and Keller, J. (2012) Final Report for the DOE Metal Hydride Center of Excellence. Sandia Report SAND2012-0786.
2 Bowman, B. and Klebanoff, L. (2012) Historical perspectives on hydrogen, its storage, and its applications, in *Hydrogen Storage Technology: Material and Applications* (ed. L. Klebanoff), CRC Press, Boca Raton, ch. 3, pp 65–91. ISBN 13: 978-1439841075.
3 Sandrock, G. (1999) A panoramic overview of hydrogen storage alloys from a gas reaction point of view. *J. Alloy. Compd*, **293–295**, 877.
4 Klebanoff, L. and Keller, J. (2013) 5 Years of hydrogen research in the U.S. DOE Metal Hydride Center of Excellence (MHCoE). *Int. J. Hydrogen Energy*, **38**, 4533–4576.
5 Ahluwalia, R., Hua, T., and Peng, J. (2009) Automotive storage of hydrogen in alane. *Int. J. Hydrogen Energy*, **34**, 7731–7740.
6 Ahluwalia, R., Peng, J., and Hua, T. (2011) Hydrogen release from ammonia borane dissolved in an ionic liquid. *Int. J. Hydrogen Energy*, **36**, 15698–15697.
7 Ott, K. (2010) Final Report for the DOE Chemical Hydrogen Storage Center of Excellence. Los Alamos Report LA-UR-20074.
8 Hua, T. and Ahluwalia, R. (2011) Alane hydrogen storage for automotive fuel cells – off-board regeneration processes and efficiencies. *Int. J. Hydrogen Energy*, **36**, 15259–15265.
9 Hua, T. and Ahluwalia, R. (2012) Off-board regeneration of ammonia borane for use as a hydrogen carrier for automotive fuel cells. *Int. J. Hydrogen Energy*, **37**, 14382–14392.
10 Ahluwalia, R. (2007) Sodium alanate hydrogen storage system for automotive fuel cells. *Int. J. Hydrogen Energy*, **32**, 1251–1261.
11 US DOE (2012) DOE targets for on-board hydrogen storage systems for light-duty vehicles, 2012 published on DOE/FCT web site: http://energy.gov/sites/prod/files/2015/01/f19/fcto_myrdd_table_onboard_h2_storage_systems_doe_targets_ldv.pdf.
12 Ahluwalia, R., Wang, X., Kwon, J., Rousseau, A., Kalinoski, J., James, B., and Marcinkoski, J. (2011) Performance and cost of automotive fuel cell systems with ultra-low platinum loadings. *J. Power Sources*, **196**, 4619–4630.
13 Ahluwalia, R. and Wang, X. (2008) Fuel cell systems for transportation: status and trends. *J. Power Sources*, **177**, 167–176.
14 Schlapbach, L. and Zuttel, A. (2011) Hydrogen storage materials for mobile application. *Nature*, **414**, 353–358.
15 Ragheb, M. (2011) Hydrides Alloys for Hydrogen Storage. Available at http://mragheb.com/NPRE%20498ES%20Energy%20Storage%20Systems/Metal%20Hydrides%20Alloys%20for%20Hydrogen%20Storage.pdf.
16 Hua, T., Ahluwalia, R., Peng, J., Kromer, M., Lasher, S., McKenney, K., Law, K., and Sinha, J. (2011) Technical assessment of

compressed hydrogen storage tank systems for automotive applications. *Int. J. Hydrogen Energy*, **36**, 3037–3049.
17 *T700S* DATA SHEET—*Toray* Carbon Fibers America. www.toraycfa.com/pdfs/t700sdatasheet.pdf, 2013.
18 Simulia Inc. ABAQUS analysis user's manual; 2011.
19 Roh, H., Hua, T., and Ahluwalia, R. (2013) Optimization of carbon fiber usage in Type 4 hydrogen storage tanks for fuel cell automobiles. *Int. J. Hydrogen Energy*, **38**, 12705–12802.
20 Zickel, J. (1962) Isotensoid pressure vessels. *ARS J.*, **32**, 950–951.
21 Vasiliev, V. and Morozov, E. (2001) *Mechanics and Analysis of Composite Materials*, Elsevier, New York.
22 Peter, S., Humphrey, W., and Foral, R. (1991) *Filament Winding: Composite Structure Fabrication*, SAMPE, Covina, California.
23 Ahluwalia, R., Hua, T., and Peng, J. (2012) On-board and off-board performance of hydrogen storage options for light-duty vehicles. *Int. J. Hydrogen Energy*, **37**, 2891–2910.
24 SAE J2579. *Technical Information Report for Fuel Systems in Fuel Cell and Other Hydrogen Vehicles*. SAE International, 2013.
25 Ranong, C. *et al.* (2009) Concept, design and manufacture of a prototype hydrogen storage tank based on sodium alanate. *Chem. Eng. Tech.*, **32** (8), 1154–1163.
26 Sanchez, A., Klein, H., and Groll, M. (2003) Expanded graphite as heat transfer matrix in metal hydride beds. *Int. J. Hydrogen Energy*, **28**, 515–517.
27 Van Hassel, B., Mosher, D., Pasini, J., Gorbounov, M., Holowczak, J., Tang, X., Brown, R., Laube, B., and Pryor, L. (2012) Engineering improvement of $NaAlH_4$ system. *Int. J. Hydrogen Energy*, **37**, 2756–2766.
28 Pasini, J., Corgnale, C., Van Hassel, B., Motyka, T., Kumar, S., and Simmons, K. (2013) Metal hydride material requirements for automotive hydrogen storage system. *Int. J. Hydrogen Energy*, **38**, 9755–9765.

33
Transport and Storage of Hydrogen via Liquid Organic Hydrogen Carrier (LOHC) Systems

Daniel Teichmann, Wolfgang Arlt, Eberhard Schlücker, and Peter Wasserscheid

33.1
Hydrogen Storage and Transport for Managing Unsteady Renewable Energy Production

Today's energy supply system is mainly based on the large-scale consumption of carbon-based fossil fuels. The rapid depletion of natural storages is not sustainable for various reasons. First of all the timespan that was needed for the formation and accumulation of these fuels – over hundreds of millions of years – is several magnitudes larger than the depletion rate. Further arguments for the lack of sustainability are the limited natural availability of fossil fuels, their unequal geographic distribution, and the consequences of their increasing use in form of anthropogenic climate change. In the context of this contribution, renewable energy is described as an energy source that provides energy obtained from the sun in either a direct (e.g., photovoltaics) or indirect way (e.g., wind or biomass). No resources are being depleted that have been accumulated over a much longer period than the period of their consumption.

It is expected that the key for a future sustainable energy system lies in the implementation and usage of renewable energy sources. Many countries worldwide have decided on ambitious targets to raise the share of solar, wind, geothermal, and biomass energy substantially in the upcoming years. Besides the substitution of fossil-fuel-based electricity generation, the mobility sector as well as heating are large contributors to overall energy consumption and are therefore to be thoroughly considered as well. The European Union has set its goal for renewable energy in gross final energy consumption to be 20% in 2020, which sets the share in electricity production at about 30–40%.

A key hurdle for the massive integration of renewable energy into our energy system is their intermittent character. The production output of, for example, wind turbines and solar PV units is primarily determined by meteorological parameters. There exist strong seasonal, day-time, and weather dependencies that render a production-to-demand-approach impossible. Especially, countries with a large share of wind and solar energy already experience energy over-production on very windy or sunny days while realizing significant energy shortages

Hydrogen Science and Engineering: Materials, Processes, Systems and Technology, First Edition.
Edited by Detlef Stolten and Bernd Emonts.
© 2016 Wiley-VCH Verlag GmbH & Co. KGaA. Published 2016 by Wiley-VCH Verlag GmbH & Co. KGaA.

during unfavorable weather conditions. The consequences of such behavior result in the rising need for extension of electrical grids, negative wholesale electricity prices, and net instabilities. A stable and reliable energy supply by renewable energies can consequently only be realized by the integration of storage technologies that enable balancing of energy over- and underproductions. Due to the magnitude of installed renewable power and the necessary timespan for seasonal or day-time buffering, storage technologies have to be developed and implemented that allow for the long-term (>5 h) storage of large amounts of energy, a task that can merely be accomplished by mechanical and electrochemical methods alone.

Due to its high gravimetric storage density, hydrogen is being considered a suitable energy carrier. It can be produced from renewable energy via electrolysis without any emissions or any consumption of limited fuels. In electrolysis water is split into hydrogen and oxygen by electrical energy. Predominant concepts that are used today are alkaline electrolysis and polymer electrolyte membrane (PEM) electrolysis. For high power ratios a multitude of electrolysis cells can be coupled. Due to its high gravimetric energy content (120 MJ kg^{-1}) hydrogen is a potentially important secondary energy vector and may form the basis of a future energy system. If hydrogen is being produced from water by renewable energy, the cycle can be completely closed as the generation of electricity, heat, or mechanical power from hydrogen and oxygen produces only water – the starting point of the conversion process.

Despite its high gravimetric energy density the low physical density of gaseous and liquid hydrogen lowers the amount of energy that can be stored per volume dramatically. Consequently, the volumetric storage density of hydrogen is much lower than that of conventional fossil fuels. But even in the form of compressed gaseous hydrogen (CGH2, typically at 700 bar pressure) or in the form of liquefied cryogenic hydrogen (LH2, typically at −253 °C) the volumetric storage density of elemental hydrogen is still quite low (Figure 33.1). Note also that hydrogen compression and cooling lead to additional energy consumption and require special infrastructure such as compressor stations, pipelines, and cryogenic tank systems.

Additional complexity is induced by the physical properties of hydrogen in regard to safety issues. Hydrogen has been used in the chemical industry for a long time without any great hazards, but the chemical industry is a professional, protected environment exclusively operated by experts. For the broad and untrained public the handling of a gas at very high pressures or very low temperatures is in contrast quite unfamiliar and requires safety mechanisms, training, and some public awareness.

As a consequence of the described challenges of storing and handling molecular hydrogen, researchers around the world are working on alternative hydrogen storage technologies. A very promising idea that avoids new infrastructure and builds on existing storage facilities is the reversible chemical conversion of hydrogen into hydrogen-rich chemicals. Figure 33.2 shows the general concept of binding hydrogen from renewable sources to

33.1 Hydrogen Storage and Transport for Managing Unsteady Renewable Energy Production

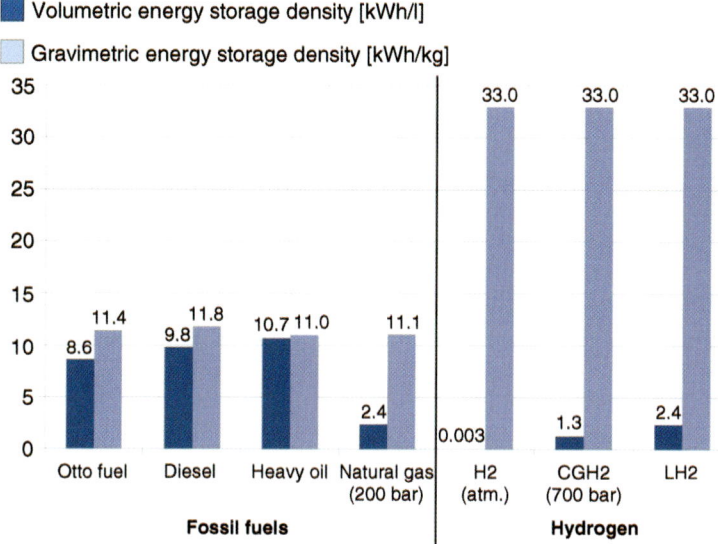

Figure 33.1 Comparison of the volumetric and gravimetric energy storage density of elemental hydrogen compared to traditional fossil fuels.

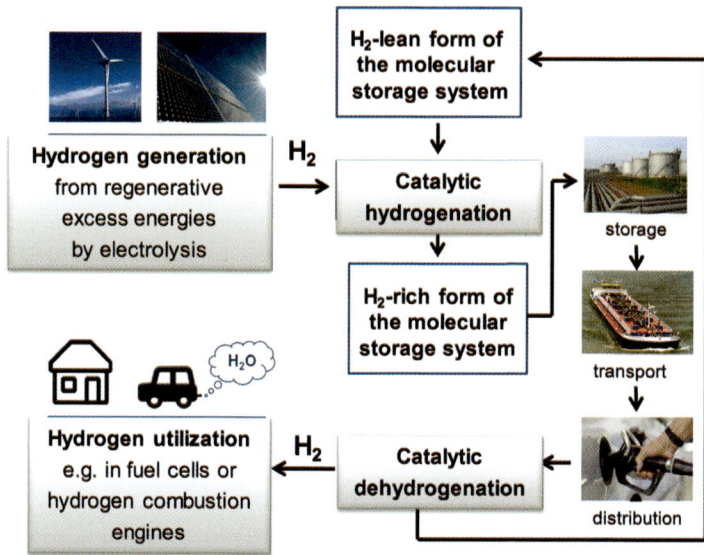

Figure 33.2 Storage and transport of renewable energy equivalents in the form of hydrogen-rich liquid chemicals based on reversible hydrogenation/dehydrogenation cycles.

a hydrogen-lean molecule that can be stored and transported in existing infrastructure.

The concept aims at making hydrogen from energy overproduction available at energy-lean times and energy-lean places. Based on the hydrogen-lean molecule nitrogen, the corresponding hydrogen-rich form is ammonia [1]. Based on the hydrogen-lean molecule CO_2, the corresponding hydrogen-rich forms are formic acid [2], methane [3], methanol [4], or Fischer–Tropsch products [5], respectively, depending on the applied catalyst and the process parameters during the hydrogenation reaction. It is particularly interesting, however, to use liquids as hydrogen-lean molecules as this allows for a closed material cycle without binding or releasing other substances from the atmosphere apart from water and oxygen. In contrast, systems that use gaseous substances as energy-lean molecules (e.g., N_2 or CO_2) release mixtures of hydrogen and the hydrogen-lean gas during dehydrogenation. This difference is important for many applications and has led to the definition of "liquid organic hydrogen carriers" (LOHCs), systems that allow hydrogen storage in reversible hydrogenation/dehydrogenation cycles. The conceptual rationale behind LOHC energy storage is that energy is stored during energy-rich times (high solar intensity, strong wind, cheap energy) in the form of hydrogen-rich, liquid LOHC compounds (large quantities, no losses over time, easy handling) and is later released at energy-lean times (low solar intensity, still air, high energy prices). The LOHC material itself can be stored and distributed using the existing infrastructure for mineral oil based fuels. There is no need to build up a completely new infrastructure which would cost billions of dollars.

In particular, reversible binding of hydrogen to liquid organic molecules is ideally suited for decentralized energy storage scenarios. Decentralization of energy storage processes in the context of renewable energy production is an intuitive choice given the fact that most installations to produce energy from wind or sun are relatively small and decentralized energy storage allows for the utilization of unavoidable conversion of electrical energy into heat during the storage process in an effective storage/heat coupling. Typical efficiencies of electrolyzers are 70%, typical efficiencies of fuel cells are 55%. Detailed energetic and economic calculations for such a local storage/heat combination have been recently published and confirm the high potential of this approach [6].

33.2
Liquid Organic Hydrogen Carrier (LOHC) Systems

LOHCs are typically liquid aromatic or heteroaromatic compounds that are converted during a catalytic hydrogenation reaction under hydrogen pressure into the corresponding alicyclic or heteroalicyclic compounds. The following list gives some of the most important requirements for suitable LOHC systems:

- The hydrogen capacity of the applied LOHC system is an important property as it defines the amount of material that has to be treated to store a certain amount of hydrogen as, respectively, energy. The LOHC-concept is somewhat limited to a hydrogen capacity of 7.2 mass% as this is the achievable limit of aromatic/alicyclic storage systems; a hydrogen capacity of 7.2 mass% corresponds to a releasable energy content of the hydrogen-rich molecule of 2.4 kWh$_{therm}$ kg^{-1}.
- No by-product should be formed during the hydrogen loading step. Note that water formation in the hydrogenation of, for example, CO_2 leads to an unavoidable loss in energy storage efficiency as some part of the hydrogen produced from water during the electrolysis forms water again during the synthesis step.
- The hydrogenation step should not be limited by thermodynamics as such a limitation (e.g., as is the case in methanol or ammonia synthesis) would lead to partial conversion of the hydrogen-lean compound and would require tedious recycling of unconverted material in the hydrogen-loading process.
- The hydrogen storage system should be liquid in the fully dehydrogenated, fully hydrogenated, and all partially hydrogenated stages for easy handling in tanks and easy transport by pumping or within liquid fuel logistics.
- The boiling points of the fully dehydrogenated, fully hydrogenated, and all partially hydrogenated stages of the LOHC system should be as high as possible to allow easy isolation of very pure hydrogen from the dehydrogenation process.
- The viscosity of the fully dehydrogenated, fully hydrogenated, and all partially hydrogenated stages of the LOHC system should be as low as possible to allow easy heat transfer from (exothermic hydrogenation) and to (endothermic dehydrogenation) the LOHC system.
- Thermal stability of all forms of the LOHC system should be as high as possible. Formation of thermal decomposition products under process conditions would make the concept of reversible hydrogenation/dehydrogenation cycles of the LOHC material impractical. Furthermore, it is to be expected that many decomposition products will act as catalyst poisons over time.
- Toxicological and ecotoxicological properties of the LOHC system should be as benign as possible, at least comparable to today's liquid fuels, such as Otto fuel and diesel.
- At least one of the involved LOHC structures should be widely industrially available at reasonable price. It is anticipated that LOHC systems suitable for larger scale applications should have a market price below €10 kg^{-1}.

Some readers may wonder why we do not refer to the hydrogenation/dehydrogenation enthalpy as an important prerequisite for suitable LOHC systems. The enthalpy of reaction is, via the van't Hoff equation, the important factor in changing the free enthalpy of reaction to the desired side and, thus, to making the reversible reaction thermodynamically possible. While it is true that high enthalpies of hydrogenation/dehydrogenation result in a high proportional

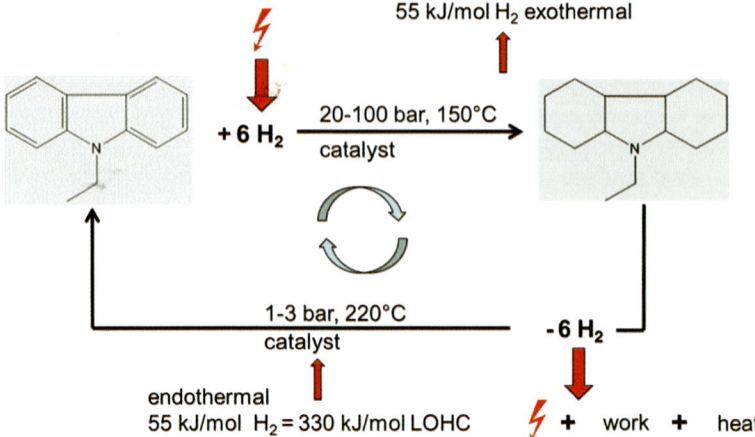

Figure 33.3 Schematic view of the relevant hydrogenation and dehydrogenation reactions involved with hydrogen storage in the system NEC/H12-NEC.

energetic investment into the hydrogen release, a more detailed energetic balance of complete energy storage systems reveals a surprisingly low impact of this particular property. Because the endothermic hydrogen release is in most cases directly combined with the strongly exothermic energetic use of the released hydrogen (typically with efficiencies below 60%) heat integration between hydrogen combustion in either high temperature fuel cells or combustion engines can provide the dehydrogenation heat. For cases where the temperature level of the hydrogen combustion is too low (e.g., in a PEM fuel cell) there is still the possibility to apply an external heat source or heat storage to generate the dehydrogenation heat instead of hydrogen combustion.

The system N-ethyl-carbazole (NEC)/dodecahydro-N-ethylcarbazole (H12-NEC) is the best investigated and most prominent LOHC system so far (Figure 33.3). It was first suggested by the company Air Products and Chemicals [7] and its thermodynamic suitability has been recently confirmed by Arlt and coworkers [8]. Several papers have been published highlighting, for example, the choice of catalyst and the reaction conditions of the relevant hydrogenation and dehydrogenation reactions [9]. It has been pointed out that Ru on alumina is among the best catalyst systems for NEC hydrogenation while Pd on alumina and Pt on alumina are both suitable for H12-NEC dehydrogenation.

Importantly, however, the NEC/H12-NEC system has some fundamental shortcomings that have recently motivated the search for better alternatives:

- The commercial availability of NEC is very limited. There is an annual production of ca 10 000 tons for dye manufacturing which is covered by coal tar distillation. Apart from the relatively small quantity, sulfur- and nitrogen-containing compounds are frequently found in commercial NEC samples.

These are catalyst poisons so that re-crystallization or distillation of NEC is required prior to the use as LOHC material.
- Fully dehydrogenated, pure NEC is a solid at room temperature (mp: 68 °C), which complicates practical dehydrogenation processes. It is possible to remain with a liquid dehydrogenated mixture by limiting the dehydrogenation degree to less than 90%; however, this comes with a loss in overall capacity.
- The thermal stability of NEC is limited by the dealkylation of the compound to carbazole, a reaction that becomes relevant at temperatures above 270 °C in the presence of typical dehydrogenation catalysts [10] and is clearly detectable in surface science experiments at model surfaces already at much lower temperatures [11,12].
- NEC is a registered industrial compound but only as a process intermediate. This means that no detailed toxicological studies of the compound are available at this point in time. Such studies would be required to distribute large quantities of the material into the hands of non-expert users.

Recently, dibenzyltoluene (MSH)/perhydro-dibenzyltoluene (H18-MSH) as well as benzyltoluene (MLH)/perhydro-benzyltoluene (H12-MLH) have been suggested as alternative LOHC systems [10]. These pairs have the great advantage of being based on commercial heating oils (Marlotherm LH© and Marlotherm SH©, respectively) which have many practical advantages like known toxicology, high thermal stability, and well-known and very favorable physico-chemical properties (Table 33.1).

Compared to the NEC system, the isomeric mixtures of benzyl- and dibenzyltoluenes provide a 7% higher absolute hydrogen capacity (6.2 wt% versus 5.8 wt%). The technical challenge of solidification of the pure unloaded hydrogen carrier that limits the practicability of the H12-NEC/NEC system is effectively solved for these eutectic mixtures (mp (NEC): 68 °C vs. mp (MSH): −34 °C). It has been demonstrated that MSH and MLH can be readily and fully hydrogenated using a commercial Ru on aluminum oxide catalyst [10].

The dehydrogenation of H12-MLH and H18-MSH requires the use of Pt on carbon or Pt on alumina catalysts. For thermodynamic reasons, the complete dehydrogenation of H12-MLH and H18-MSH requires a higher temperature than for the dehydrogenation of H12-NEC. However, the same rate of hydrogen production can be reached from H18-MSH compared to H12-NEC if the reaction temperature is set 80 °C higher in the first case. Hydrogen release from H12-MLH is consistently faster than from H18-MSH, with high hydrogen production dynamics from H12-MLH being observed already at 270 °C [10].

Note that the properties of H12-MLH/MLH and H18-MSH/MSH are even closer to diesel than H12-NEC/NEC. Heat capacities, heat conductivities, densities, viscosities, surface tensions, boiling points, flammabilities, and material compatibility aspects of MLH [14] and MSH [15] have been extensively documented due to the wide industrial use of these liquids as heat transfer oils in a

Table 33.1 Relevant properties of important LOHC systems.

LOHC system	NEC/H12-NEC	MSH/H18-MSH	MLH/H12-MLH
H$_2$-lean form			
H$_2$-rich form			
Mp (H$_2$-lean form) (°C)	68	−34	−30
Bp (H$_2$-lean form) (°C)	270	390 (diesel: 170–390)	280
H$_2$-capacity (wt%)	5.8	6.2	6.2
Energy content (kWh kg^{-1})	1.91	2.05 (Li ion battery= ca. 0.15)	2.05
Heat of hydrogenation (kJ mol^{-1} H$_2$)	55	65	65
Cost (€ kg^{-1}) (1 ton scale)	About 40	About 4	About 4
Hazard symbols H$_2$-lean form	Xn	— (Diesel: Xn, Otto-fuel: toxic)	Xn

Data from References [13–15].

wide range of industries. From the very similar physicochemical properties of MLH and MSH with typical hydrocarbon fuels, it can be expected that today's infrastructure for fuel distribution (tank ships, pipelines, storage tanks) can be made fully available for these LOHC systems [16]. Remarkably, documented toxicology and ecotoxicology properties of MSH are not only more favorable than NEC but are also clearly less problematic than common diesel. While MLH and MSH have some structural and mechanistic similarities to the earlier reported hydrogen storage systems benzene/cyclohexane [17], toluene/methylcyclohexane [18], or naphthalene/decalin [19], they are characterized by a significantly lower vapor pressure than the earlier reported systems and thus allow much more easily the liberation of very pure hydrogen.

From our understanding, we consider MLH/H12-MLH and MSH/H18-MSH as excellent LOHC systems, in particular for storing large amounts of energy for long periods of time.

33.3
Development of LOHC-Based Energy Storage Systems

An important next step on the way to commercial, LOHC-based, energy storage applications is the development of system components that represent building blocks of the final energy storage system. Our research, both in the academic (WA, ES and PW) as well as in the industrial context (DT), focusses very much on this aspect. Our analysis shows that some of the required components are well known and can be directly adopted from other technologies (e.g., pumps, coolers, heat exchangers), some others have to be modified or down-scaled (e.g., the trickle-bed reactor for LOHC hydrogenation), while again others need completely new development (e.g., tank filling indicator, dehydrogenation reactor). It is our near-time goal that components of all three categories will work together in a robust energy storage demonstrator that will be operated to collect real life performance data, for example, with respect to efficiencies, catalyst stabilities, or process dynamics.

Our first demonstration unit has been operating since early 2014. The LOHC hydrogenation part successfully accomplished 700 h of dynamic operation in 2014 without any relevant loss in performance. During this time, more than 800 kg of H18-Marlotherm have been produced to feed our own catalytic dehydrogenation units. The applied dead-end trickle bed reactor system with jacketed reactor is shown in Figure 33.4. The reactor is constructed to allow for an hourly hydrogenation capacity of up to 6 kg of Marlotherm SH.

Following the hydrogen charging process and storage or transport of the hydrogen-rich LOHC system, the catalytic dehydrogenation for hydrogen release requires special attention. The reaction is endothermic so that heat has to be introduced to the reactor to promote the reaction. At the same time, the enormous amount of hydrogen gas produced in the reactor (ca. 700 ml hydrogen per ml of LOHC) is challenging and may restrict heat transfer, liquid residence time in the reactor, and liquid contact at the catalyst. Therefore, a dehydrogenation reactor with high volumetric efficiency should allow effective heat transfer to the LOHC liquid, efficient removal of hydrogen gas from the catalytic surface, and quick separation of hydrogen gas and LOHC liquid.

Figure 33.5 shows an electrically heated dehydrogenation unit (V(HRU) = 500 ml; 200 g catalyst; 1 g precious metal = about €30 precious metal cost) constructed for a hydrogen release capacity of 6 kW$_{therm}$ or 3 KW$_{electric}$ (at a subsequent fuel cell). While electric heating is reasonable for kinetic studies and catalyst optimization experiments in development laboratories it is certainly impractical from the point of view of energy efficiency. Given the fact that the efficiency of the fuel cell is around 55% and 28% of the thermal energy content

Figure 33.4 Dead-end trickle bed reactor for the hydrogenation of dibenzyltoluene isomeric mixtures – capacity 6 kg h^{-1}; typical hydrogenation conditions: 30 bar hydrogen, 150 °C, Ru on alumina catalyst. The reactor is the vertical tube in the middle, right is the tank for hydrogen-lean LOHC, left is the tank for hydrogen-rich LOHC, pumps and controllers are below the tanks.

of the released hydrogen is required to promote full dehydrogenation, more than 50% of the produced electricity has to be invested in reactor heating in such a scenario. In real life application, reactor heating will therefore be realized with hot heat transfer fluids. The latter will be heated by a hydrogen oxidation process (e.g., SOFC (solid oxide fuel cell), hydrogen engine), by an independent heat source (e.g., high pressure steam), or by a suitable heat storage system.

On looking more closely at the energy efficiencies of the overall energy-storage process it is obvious that the major losses are in the electrolysis and in the hydrogen oxidation step. While the hydrogenation unit is heat producing and requires only a small energy contribution for liquid compression (note that hydrogen under pressure is directly provided by the electrolyzer), the temperature level of the energetic hydrogen use is crucial for the energetic efficiency of the LOHC dehydrogenation. Hydrogen oxidation processes operating at a high temperature level, such as, for example, SOFCs, produce their waste heat at

Figure 33.5 Continuous dehydrogenation unit with a capacity of 6 kW$_{therm}$. Heating is realized here via electrical heating, which is impractical from the point of view of energy efficiency – in all real-life demonstrators heating will be provided by hot heat transfer fluids heated, for example, by hydrogen oxidation; typical dehydrogenation conditions: 250–330 °C; 1.2–5 bar pressure.

temperatures above 400 °C and allow full heat integration with the LOHC dehydrogenation step. In contrast, conversion of hydrogen into electricity using a PEM fuel cell operates at temperatures of 80 °C and waste heat at this level is useless in promoting the LOHC dehydrogenation step. Consequently, the reaction heat for the dehydrogenation step has to be provided from other sources, in a stand-alone process, for example, by burning a part of the liberated hydrogen for this purpose only. This leads to the interesting fact that the given efficiency differences in the hydrogen oxidation process between a hydrogen combustion engine and PEM fuel cell do not lead to significant gains in the overall energy storage process (Table 33.2).

For a fair evaluation of the total storage efficiencies in Table 33.2 it is important to note that the heat losses of the different processes are available at suitable temperature levels (150–180 °C from exothermic hydrogenation; 80–250 °C from the different hydrogen oxidation processes after heat integration) for heating or cooling purposes so that a fully heat integrated system (e.g., a building using the LOHC-system as electricity storage and heating system) can easily realize much higher overall efficiencies. It is also fair to note that no electricity storage system that builds on hydrogen as secondary vector will ever reach electricity-to-electricity efficiencies above 38% if the efficiency of electrolysis and fuel cell technology presently remain in the indicated range. Compared to other options of chemical energy storage, all LOHC-based storage systems are highly efficient as the formation of water as by-product is avoided. The latter is encountered in

Table 33.2 Electricity-to-electricity efficiencies (%) of LOHC-based energy storage using different options for the energetic utilization of the hydrogen released from the storage system.

	H_2 combustion engine	PEM-FC	SOFC
Electrolysis	70	70	70
Hydrogenation	98	98	98
Dehydrogenation	100	$80^{a)}/72^{b)}/100^{c)}$	100
H_2 to electricity	42	55	50
Total efficiency	29	$30^{a)}/27^{b)}/38^{c)}$	34

a) LOHC system is NEC/H12-NEC.
b) LOHC system is MSH/H18-MSH.
c) Dehydrogenation heat is provided by a solar-heated external heat storage.

most storage options based on CO_2 (e.g., formation of methane, methanol, fuels by Fischer–Tropsch synthesis) and leads to an inevitable efficiency loss as hydrogen produced by electrolysis from water is re-converted into water during the storage process.

33.4
Applications of LOHC-Based Energy Storage Systems

LOHC systems are considered a promising energy storage technology for various applications. In techno-economic studies the usage of LOHC for different scenarios has been evaluated in detail. The most important applications will be presented shortly in the following, namely, decentralized storage of electricity and heat in LOHCs, the transport of energy over long distances, and the usage of LOHC as alternative fuel for mobile applications (Figure 33.6).

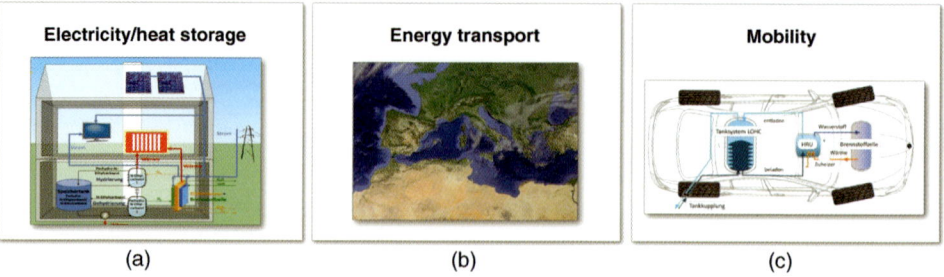

Figure 33.6 Applications of LOHC systems as energy storage technology: (a) decentralized electricity/heat storage, (b) energy transport over long distances, and (c) mobile applications.

33.4.1
LOHC Systems for the Storage of Renewable Energy Equivalents, in Particular for Decentralized Storage in Heat–Storage Coupling

Residential and commercial buildings account for a large part of overall energy consumption. Usually, around 20–33% of the total energy in buildings is electricity demand, but the major part is thermal energy used for heating.

Due to their high storage density and good manageability, LOHC substances are highly applicable for the local storage of excess energy in buildings. Following the approach of a CHP system ("combined heat and power" or in this case "combined heat and storage" CHS system), thermal losses from the storage processes can be used for heating or cooling purposes to increase the overall efficiency of the system. As a detailed evaluation of the economic feasibility shows, the co-usage of exhaust heat in decentralized storage systems significantly improves the profitability by providing a considerable financial contribution that is usually not exploitable for centralized storage units.

A LOHC storage system for buildings consists of the following five elements: electrolysis, LOHC hydrogenation, storage tank, LOHC dehydrogenation, and fuel cell (also gas turbines or combustion engines are conceivable). The building interacts with the outside world by electric and thermal connection. As energy input, local PV units or small wind turbines can be used as well as electricity that is purchased from the grid in times of low energy prices (usually in energy-rich times). Figure 33.7 shows the concept of LOHC electricity/heat storage in buildings.

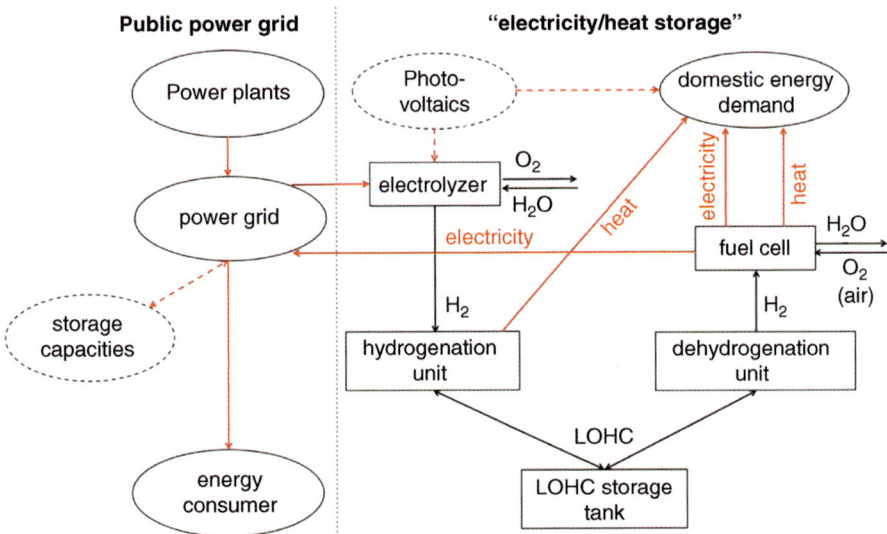

Figure 33.7 Illustration of an electricity/heat storage unit used for energy supply of domestically, commercially, or industrially used buildings. Adapted from Reference [6].

The electrical efficiency of any storage system based on hydrogen is usually limited to around 33–38%. This results from the efficiencies of power-to-hydrogen-conversion (electrolysis, ~70%) and hydrogen-to-power-conversion (fuel cell, ~50–55%). As the "energy losses" are heat, the overall efficiency of the system can be increased significantly if the waste heat is used, for example, to heat the building. The ability of LOHC systems to be implemented in decentralized scenarios facilitates the sensible usage of heat considerably. The electricity-to-electricity efficiencies of LOHC storage systems depicted in Table 33.2 show that the ratio of electrical energy to thermal energy fits the actual consumption ratio in buildings quite well.

The economic feasibility of the described LOHC electricity/heat-storage system has been evaluated in detail in Reference [6]. In this study different business models to create income from storage services have been compared:

- Provision of control power: There is an existing market for "energy balancing" services. Due to the increase in renewable generation an increasing need exists for shiftable generating capacities and consumers that can react within very short time intervals to balance short-term discrepancies between energy production and load. "Positive control power" refers to the provision of energy by generating capacities while "negative" control power, on the other hand, implies that the operator of a storage unit or a controllable load agrees to act as a consumer of electrical energy when requested. This willingness is reimbursed and generates income for the owner of storage systems.
- Price arbitrage: A storage system can also act as a trader of energy. It buys electricity from the grid whenever it is cheap (energy-rich times) and sells electricity at high prices (energy-lean times). Income is generated by the spread between purchase and sales prices.
- Heat: the usage of heat from LOHC storage processes considerably diminishes the consumption of fossil fuels for heating. As can be seen in detailed analyses, the value of the heat obtained from the system significantly promotes the economic feasibility of storage units.

Figure 33.8 shows results of the evaluation [6]. Three scenarios regarding price spread between purchase and sales price and capacity factor of the system have been assumed: case A – 4 Ct./kWh/22%. Case B – 3 Ct./kWh/34%. Case C – 12 Ct./kWh/29%. The figure shows that the income that is achievable through the mentioned business models and savings for heating can very well exceed the costs.

33.4.2
Energy Transport over Long Distances with LOHC

In the future, a significant share of domestic energy consumption might be satisfied by the import of energy that has been produced in regions with high

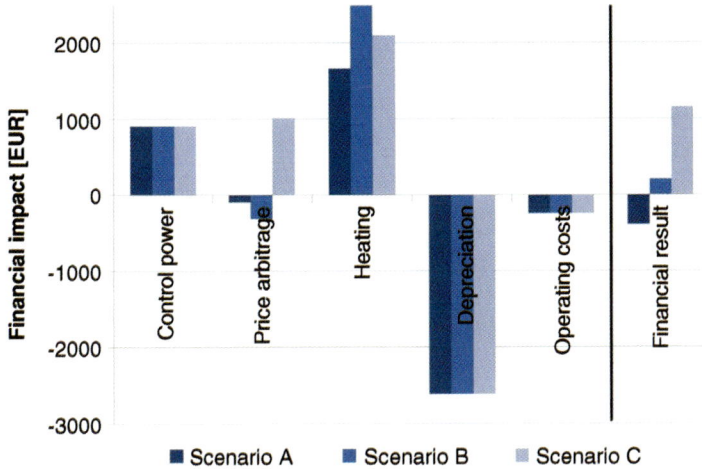

Figure 33.8 Composition of income and costs for operation of a LOHC electricity/heat storage system. Adapted from Reference [6].

potential for renewables. For example, the import of solar power from Northern Africa to Europe has been presented and evaluated in detail. In many discussions, electrical transmission is the predominant option for energy transport, but it comes along with several heavy drawbacks, such as, for example, huge investment, low acceptance by the public, no storage functionality, and potentially unstable states as transit countries.

LOHC systems as liquid, diesel-like substances can be transported and stored in the same infrastructure as today's mineral oil based fossil fuels. Due to their high storage density and benign safety profile, the storage of energy in the form of LOHCs can be considered as a very promising option to transport energy even over long distances. Due to the usage of an existing infrastructure (e.g., oil tankers, tank trucks, pipelines, etc.), the costs of building up this sort of "new" energy distribution system are comparably low.

A detailed evaluation of LOHC energy transport as well as a comparison with other forms of energy transmission has been conducted [16]. In the following, two interesting scenarios of energy transport via LOHC are presented as well as their financial impact (Figure 33.9).

33.4.2.1 Import of Solar Energy from Northern Africa

The annual energy demand of the whole world could theoretically be satisfied by the solar irradiation on a comparably small fraction of the Sahara desert in Northern Africa [16]. The "Desertec" concept has envisaged the installment of huge CSP (concentrated solar power) plants. Part of the produced energy would be exported to Europe via HVDC (high voltage direct current) transmission; the use of LOHC systems for the same function would

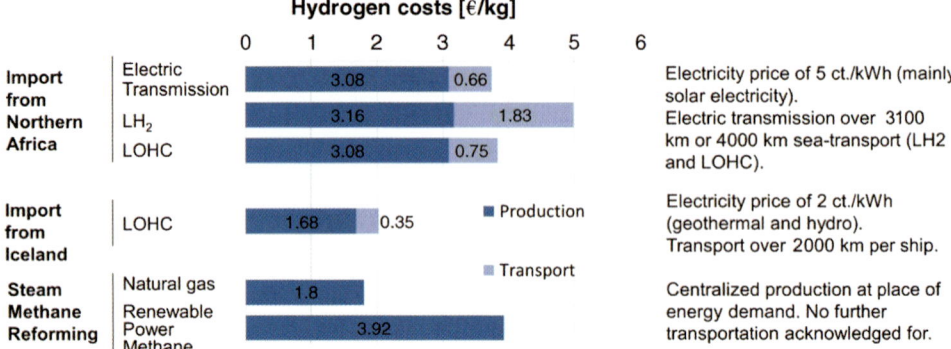

Figure 33.9 Hydrogen procurement costs for various production and transport scenarios. Adapted from Reference [16].

in addition provide a storage functionality. The harvested energy could be transported to any place with energy demand and could be released whenever energy demand is highest.

33.4.2.2 Import of Renewable Energy from Iceland

Iceland is geologically positioned on the tectonically active Mid-Atlantic ridge. Iceland is able to cover almost all of its energy consumption from renewable sources, mainly from hydroelectric and further from geothermal sources. Due to its comparably low electricity prices, Iceland has attracted energy-intensive industries like aluminum production over the years. While considerable additional potential for the production of energy from renewable sources exists, any export of these energy surpluses to other countries can only be realized through reliable, cheap, and efficient technologies for energy transportation, such as liquid transport of hydrogen-rich LOHC systems by tank ships to the industrial centers of the world.

The alternatives for energy transport, especially the alternatives of electrical transmission versus the transformation into a chemical carrier like hydrogen, have been discussed in various publications. For the latter the transport of liquefied hydrogen has been proposed. As the evaluations (Figure 33.8) show, the usage of LOHC for the storage and transport of solar energy might be highly cost-competitive. Furthermore, there exist substantial additional advantages for chemical energy transport in the form of LOHC systems:

- Required capital investment: as large parts of a LOHC distribution infrastructure already exist, capital investment would be much lower than for HVDC lines (in addition, incremental step-by-step-introduction is possible).
- Vulnerability of the energy system: When relying on electrical energy transmission one breakdown of a line or transformer has the potential to shut

down the whole energy supply. This endangerment becomes especially relevant for very long lines that run through politically unstable countries. The fragmented delivery of energy via ship transports of a chemical carrier enables – in contrast – high reliability.
- Storage capability: Pure electrical transmission of energy does not at all address the challenge of balancing intermittently produced renewable energy. Most solar energy production from Northern Africa would occur at the same time that the sun peaks in Europe. The transport of energy in LOHCs would intrinsically solve the storage problem, because the carrier can be stored as long as needed until the energy demand occurs.

33.4.3
Mobile Applications

Finally, we see a long-term perspective to use LOHC systems to make temporary over-production of energy from renewable sources available for mobile applications (Figure 33.10). Cruising ranges for cars operating on hydrogen-rich LOHCs could be close to those of gasoline-based cars. Range-extender concepts – already well established in today's battery-powered vehicles – operating on fuel cells with hydrogen supply from LOHC storage systems will circumvent to a very large extent dynamics requirements. From an infrastructural point of view the here suggested concept of using LOHC systems for the mobile sector is very attractive given the fact that only slight modifications of the existing fuel distribution infrastructure will be necessary to provide hydrogen-rich LOHC at every filling station [20].

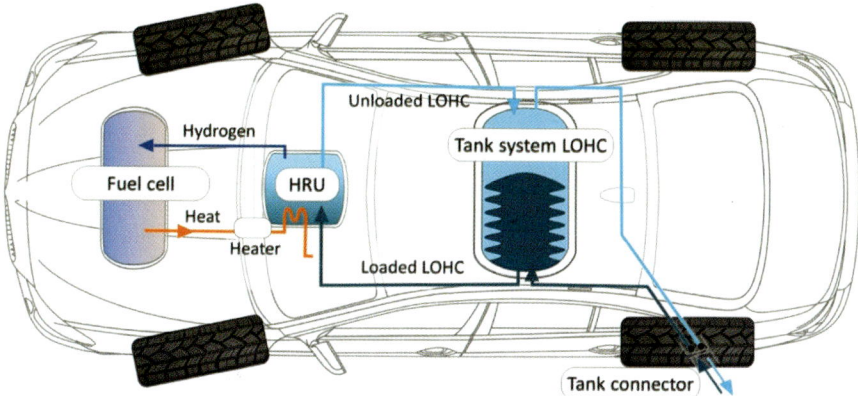

Figure 33.10 Concept of a fuel cell vehicle, hydrogen supplied by LOHC system. The HRU (hydrogen release unit) generates hydrogen on-board from LOHC. The unloaded LOHC material is exchanged for fresh (loaded) material at a gasoline station. Adapted from Reference [20].

33.5
Conclusions

The transformation of our current carbon-based energy system into a more sustainable, renewable-based one is timely, highly desirable, and already on-going. The share of energy production from wind and solar is increasing steadily in many parts of the world and – although total contributions to the overall energy system are still rather low – the intermittent character of these renewable sources creates several relevant challenges. There is no doubt that any future energy system that will build to a larger extent on sun and wind power will require technologies to store large amounts of electric energy safely, cheaply, and without loss over longer periods of time. Such storage technologies are simply needed to link electricity production and demand with respect to time and place. In this context, chemical energy storage is a very interesting option as fuels are characterized by high storage densities. Infrastructure for their production, storage and transport is available, and public confidence in this way of energy handling is very high.

Liquid organic hydrogen carrier (LOHC) systems have a very high potential to become a leading concept within the different options for chemical energy storage. Hydrogen loaded on liquid organic molecules in a reversible manner avoids the tedious provision of a second gaseous reaction partner in sufficient quantity and quality, prevents the formation of by-products, and allows very safe handling of large amounts of hydrogen in a future hydrogen-based energy system. The LOHC concept offers the potential to gradually substitute liquid fossil fuels over time. Most important, however, is the fact that the use of LOHC systems for energy storage is ideally suited for decentralized applications. Such decentralized storage is a natural choice given the decentralized character of electricity production from wind and sun and allows operation at very high efficiency in electricity storage–heat production couplings.

The development of LOHC technologies is still at an early stage but shows high dynamics. With Hydrogenious Technologies GmbH (www.hydrogenious.net), a first engineering company specialized in the LOHC technology has been founded and offers many services around the technology (hydrogenated LOHC materials, reactor concepts, dedicated storage solutions). Many other companies and academic groups have started to show interest in this highly promising field. Obviously, the many different application options for LOHC systems attract stakeholders from very different backgrounds, for example, public utility companies, owners of wind farms and solar parks, chemical industry, as well as power suppliers, net operators, or liquid logistics companies. This broad interest is very encouraging as it creates the required, substantial, joint effort to bring the LOHC technology into broad industrial and commercial application.

References

1 Hua, T.Q. and Ahluwalia, R.K. (2012) Off-board regeneration of ammonia borane for use as a hydrogen carrier for automotive fuel cells. *Int. J. Hydrogen Energy*, **37**, 14382–14392.

2 Loges, B., Boddien, A., Gärtner, F., Junge, H., and Beller, M. (2010) Catalytic generation of hydrogen from formic acid and its derivatives: useful hydrogen storage materials. *Top. Catal.*, **53**, 902–914.

3 Saxena, S., Kumar, S., and Drozd, V. (2011) A modified steam-methane-reformation reaction for hydrogen production. *Int. J. Hydrogen Energy*, **36**, 4366–4369.

4 Kobayashi, T. and Takahashi, H. (2004) Novel CO_2 electrochemical reduction to methanol for H_2 storage. *Energy Fuel*, **18**, 285–286.

5 Kaiser, P., Unde, R.B., Kern, C., and Jess, A. (2013) Production of liquid hydrocarbons with CO_2 as carbon source based on reverse water-gas shift and Fischer–Tropsch synthesis. *Chem-Ing-Tech.*, **85** (4), 489–549.

6 Teichmann, D., Stark, K., Müller, K., Zöttl, G., Wasserscheid, P., and Arlt, W. (2012) Energy storage in residential and commercial buildings via liquid organic hydrogen carriers (LOHC). *Energy Environ. Sci.*, **5** (10), 9044–9054.

7 Pez, G.P., Scott, A.R., Cooper, A.C., and Cheng, H. (2006) Air Products and Chemicals Inc., US Patent US7101530 B2.

8 Müller, K., Völkl, J., and Arlt, W. (2013) Thermodynamic evaluation of potential organic hydrogen carriers. *Energy Technol.*, **1**, 20–24.

9 (a) Teichmann, D., Arlt, W., Wasserscheid, P., and Freymann, R. (2011) A future energy supply based on liquid organic hydrogen carriers (LOHC). *Energy Environ. Sci.*, **4**, 2767; (b) Eblagon, K.M., Tam, K., and Tsang, S.C.E. (2012) Comparison of catalytic performance of supported ruthenium and rhodium for hydrogenation of 9-ethylcarbazole for hydrogen storage applications. *Energy Environ. Sci.*, **5**, 8621–8630; (c) Sotoodeh, F., Huber, B.J.M., and Smith, K.J. (2012) The effect of the N atom on the dehydrogenation of heterocycles used for hydrogen storage. *Int. J. Hydrogen Energy*, **37**, 2715–2722.

10 Brückner, N., Obesser, K., Bösmann, A., Teichmann, D., Arlt, W., Dungs, J., and Wasserscheid, P. (2014) Evaluation of industrially applied heat-transfer fluids as liquid organic hydrogen carrier systems. *ChemSusChem*, **7** (1), 229–235.

11 Sobota, M., Nikiforidis, I., Amende, M., Zanon, B.S., Staudt, T., Höfert, O., Lykhach, Y., Papp, C., Hieringer, W., Laurin, M., Assenbaum, D., Wasserscheid, P., Steinrück, H.-P., Görling, A., and Libuda, J. (2011) Dehydrogenation of dodecahydro-N-ethylcarbazole on Pd/Al_2O_3 model catalysts. *Chem-Eur. J.*, **17** (41), 11542–11552.

12 Gleichweit, C., Amende, M., Schernich, S., Zhao, W., Lorent, M.P.A., Höfert, O., Brückner, N., Wasserscheid, P., Libuda, J., and Steinrück, H.-P. (2013) Dehydrogenation of dodecahydro-N-ethylcarbazole on Pt(111). *ChemSusChem*, **6** (6), 974–977.

13 Clariant Produkte GmbH (Frankfurt) (2015) Safety data sheet "N-ethylcarbazol" (CAS-Number: 86-28-2; EG Number: 201-660-4).

14 Global Heat Transfer (2015) MSDS – Marlotherm LH: Material Safety Data Sheet. Available at http://formosa.msdssoftware.com/imagedir/i011B4D0.pdf.

15 Global Heat Transfer (2015) Material Safety Data Sheet: Marlotherm® SH. Available at http://www.julabo.com/sites/default/files/Downloads/Sicherheitsdatenblaetter/Julabo_Thermal_H350_de.pdf.

16 Teichmann, D., Arlt, W., and Wasserscheid, P. (2012) Liquid organic hydrogen carriers as an efficient vector for the transport and storage of renewable energy. *Int. J. Hydrogen Energy*, **37** (23), 18118–18132.

17 Tsuji, T., Shinya, Y., Hiaki, T., and Itoh, N. (2005) Hydrogen solubility in a chemical hydrogen storage medium, aromatic hydrocarbon, cyclic hydrocarbon, and their mixture for fuel cell systems. *Fluid Phase Equilib.*, **228**, 499–503.

18 Wang, B., Goodman, D.W., and Froment, G.F. (2008) Kinetic modeling of pure hydrogen production from decalin. *J. Catal.*, **253**, 229–238.

19 Lázaro, M.P., García-Bordejé, E., Sebastián, D., Lázaro, M.J., and Moliner, R. (2008) In situ hydrogen generation from cycloalkanes using a Pt/CNF catalyst. *Catal. Today*, **138**, 203–209.

20 Zenner, M., Teichmann, D., Di Pierro, M., and Dungs, J. (2012) Liquid hydrogen carriers as potential passenger car fuel. *Automobiltechnische Z. (ATZ)*, **114**, 940–947.

Part 4
Traded Hydrogen

34
Economics of Hydrogen for Transportation
Akiteru Maruta

34.1
Introduction

Clearly, hydrogen has environmental benefits, with no emission at end-use, especially if it is produced from renewables. Even when made from fossil fuels like natural gas, hydrogen still has certain advantages over fossil fuels, because of the high efficiency of such fuel cell systems. Therefore, it is useful to review hydrogen's economic benefits to fully understand hydrogen's advantages for society. This chapter summarizes the economic benefits owing to a hydrogen transportation system.

In this chapter, hydrogen transportation means fuel cell electric vehicles (FCEVs), which are powered by fuel cell systems fueled by pure hydrogen. Other hydrogen transportation technologies, such as hydrogen ICE vehicles and fuel cell systems with other kinds of reformates, such as methanol and gasoline, are not considered.

34.2
Hydrogen Transportation System

34.2.1
FCEVs

Leading automakers are planning to introduce FCEVs onto the market in 2014–2020 timeframe. Toyota, Honda, and Hyundai announced their intention to commercialize FCEVs in 2014–2015. Daimler, Nissan, and Ford formed an alliance to produce FCEVs in 2017. Now, FCEV is not a dream, but a reality. Figure 34.1 shows Toyota's new FCEV at the Tokyo Motor Show in November 2013.

The main components of FCEV are summarized in Table 34.1. FCEVs are considered as electric vehicles, powered by an electric motor. They are usually

Hydrogen Science and Engineering: Materials, Processes, Systems and Technology, First Edition.
Edited by Detlef Stolten and Bernd Emonts.
© 2016 Wiley-VCH Verlag GmbH & Co. KGaA. Published 2016 by Wiley-VCH Verlag GmbH & Co. KGaA.

Figure 34.1 Toyota's new FCEV at the Tokyo Motor Show in November 2013. Photograph Technova Inc.

hybridized with a secondary battery system for better energy efficiency by storing and releasing energy during breaking and accelerating.

With 4–6 kg of hydrogen onboard, FCEVs have a driving range of over 500 km, just like conventional vehicles. In general, 1 kg of hydrogen allows an FCEV to run for around 100 km. (The highest fuel economy recorded was 141.0 km per kg-H_2 by the Japanese JHFC project [1], and the US DOE's Technology Validation Project recorded a fuel economy of 58 miles kg^{-1} (= 93 km kg^{-1}) [2].)

Table 34.1 Main components of FCEVs.

Component	Functions
Hydrogen tank	• Stores 4–6 kg of hydrogen at 70 MPa; • type III tank: carbon fiber-winding on metal liner; • type IV tank: carbon fiber-winding on plastic liner.
Fuel cell stack	• Generating electricity from hydrogen (from hydrogen tank) and oxygen (from air); • typically PEM technology is adapted; • typical output power is 100 kW.
Secondary battery	• Stores regenerative electricity upon breaking; • provides power upon accelerating; • typically Li-ion battery technology is adapted; • capacity depends on design concept; • capacity size depends on design concept (balance between power output and cost).
Motor	• Typical output power is around 100 kW; • works also as electricity generator upon breaking for regeneration.
Power control unit	• Converting DC current into AC current to power electric motor, and DC into DC to change the voltage; • controls overall input and output of all components.

Source: Technova Inc.

34.2.2
Hydrogen Infrastructure

In preparation for the FCEV launch, major countries are working on the deployment of hydrogen supply infrastructure – the network of hydrogen stations. In general, there are two types of hydrogen stations, (i) on-site hydrogen stations and (ii) off-site hydrogen stations (Figure 34.2).

On-site hydrogen stations are equipped with hydrogen production systems, usually using either reforming technology or electrolyzer technology. This hydrogen infrastructure system is sometimes called a "distributed hydrogen production" system.

Off-site hydrogen stations do not have hydrogen production systems at site, and hydrogen is supplied either by hydrogen tube trailer or by pipeline. Since hydrogen is produced at large facilities like refineries and delivered to such stations, this infrastructure system is also called a "centralized hydrogen production" system.

For economic analysis, hydrogen cost is the key factor, and hydrogen cost can be divided into production cost and refueling cost (compression, storage, and dispensing cost), and delivery cost if necessary.

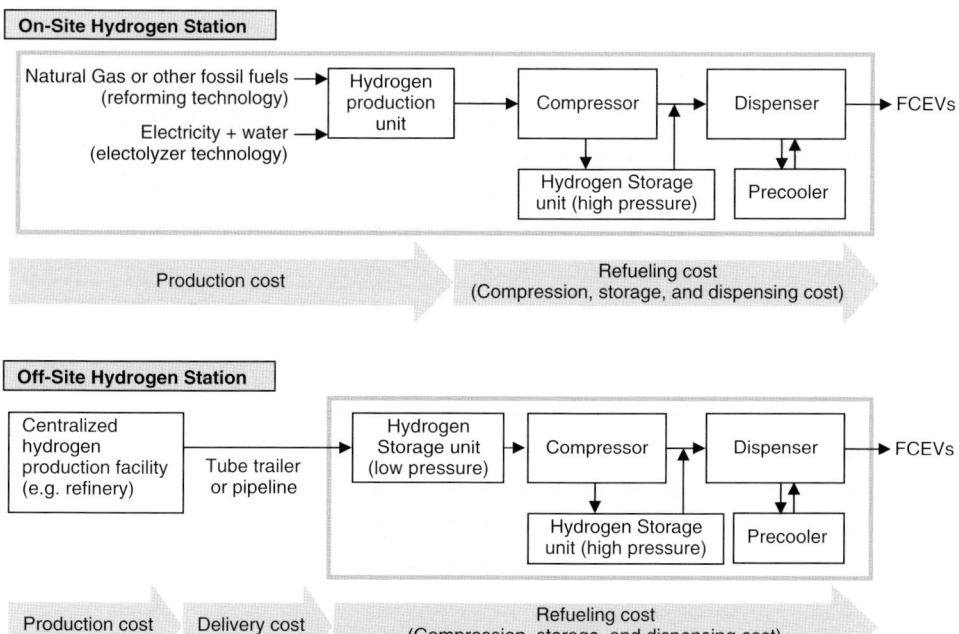

Figure 34.2 Hydrogen station systems. Source: Technova Inc.

34.3
Economics of Hydrogen for Transportation

34.3.1
Hydrogen Cost

34.3.1.1 Two Approaches for Hydrogen Cost Calculation

Hydrogen cost can be set or calculated by a (i) bottom-up approach, (ii) top-down approach, and (iii) total cost of ownership (TCO) approach.

Hydrogen cost based on the bottom-up approach is the realistic cost for today and the near future, whereas a hydrogen cost based on the top-down approach is rather an ideal cost for a future target. The TCO approach is totally different, since it focuses on the total cost over the FCEV lifetime, including FCEV purchase cost and running cost including hydrogen cost.

34.3.1.2 Bottom-Up Approach for Hydrogen Cost

In the bottom-up approach, hydrogen cost is calculated based on the realistic assumption of supply infrastructure.

According to the pathway analysis by the National Renewable Energy Laboratory (NERL) [3], distributed natural gas reforming technology produces the most cost competitive hydrogen today in the USA, which is $4.5 kg^{-1}, including production, compression, storage, and dispensing (Figure 34.3). This cost is also proved by US DOEs Program Record [4], which indicates that hydrogen cost at distributed facilities at high-volume refueling stations using current technology

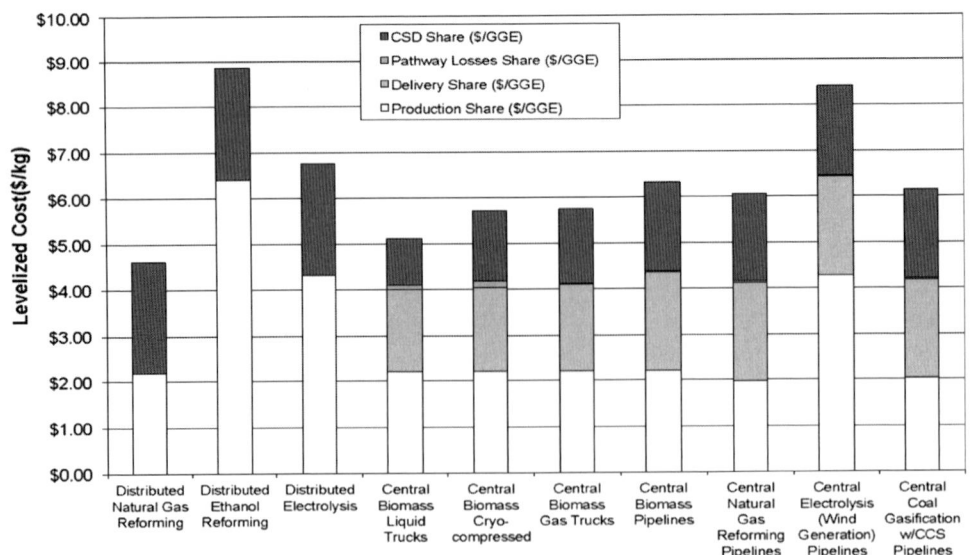

Figure 34.3 Today's hydrogen cost in the USA. Source: Reference [3].

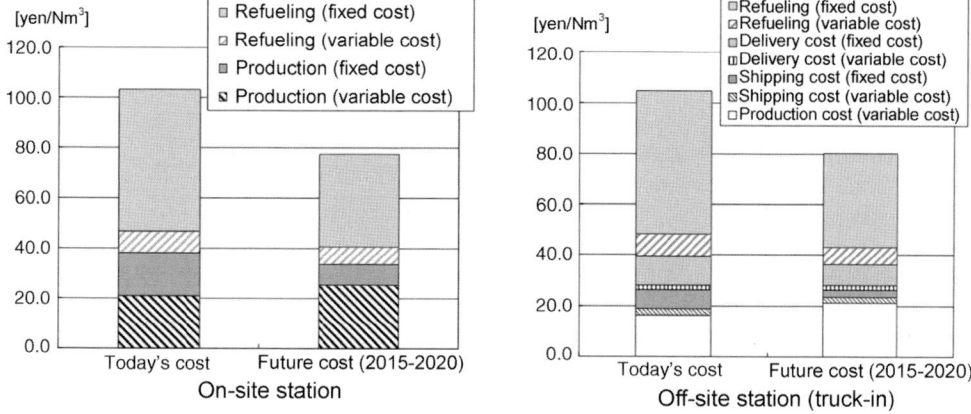

Figure 34.4 Today's hydrogen cost in Japan (70 MPa, 300 Nm³ h⁻¹, compressor-fill). Source: Reference [5].

can be $4.2–4.5 gge^{-1} ($4.2–4.5 kg^{-1}), depending on natural gas price for industry use.

Under the Japanese "JHFC project," a government-supported FCV demonstration project, hydrogen costs were calculated based on the bottom-up approach (Figure 34.4) [5]. For both on-site and off-site stations, today's hydrogen cost is estimated to be around ¥100 Nm^{-3} ($11 kg^{-1}), and the future cost in 2015–2020 will be ¥80 Nm^{-3} ($9 kg^{-1}).

For this calculation, the basic formula is given by Eq. (34.1), which indicates that the annual station cost (especially fixed cost) largely depends on hydrogen station cost, and the annual hydrogen volume depends on the utilization of stations. In addition, the annual hydrogen production cost will increase in line with the increase of natural gas cost. Therefore, future hydrogen cost is the product of decreasing hydrogen station cost (¥630 million per station today to ¥360 million in 2015–2020) and increasing production variable cost [5]:

$$\text{Hydrogen cost} = \frac{\text{Annual station (refueling) cost [fixed cost \& variable cost]} + \text{Annual hyrogen production cost [fixed cost \& variable cost]}}{\text{Annual hydrogen volume}}$$

(34.1)

The New Energy and Industrial Technology Development Organization (NEDO), a Japanese funding agency, announced hydrogen roadmap in 2010, which indicates that 2010 hydrogen cost was ¥120 Nm^{-3} ($13 kg^{-1}) based on an on-site station of ¥500 million, and the 2015 hydrogen cost is ¥90 Nm^{-3} ($10 kg^{-1}), based on an on-site station of ¥300 million. These numbers are comparable with JHFC project's cost analysis.

According to McKinsey & Company's analysis [7], the today's hydrogen cost was around €16.6 kg^{-1} ($23 kg^{-1}) in 2010 (Figure 34.5), which quickly came down to the €10 kg^{-1} ($13 kg^{-1}) level by 2015. The analysis also indicates that

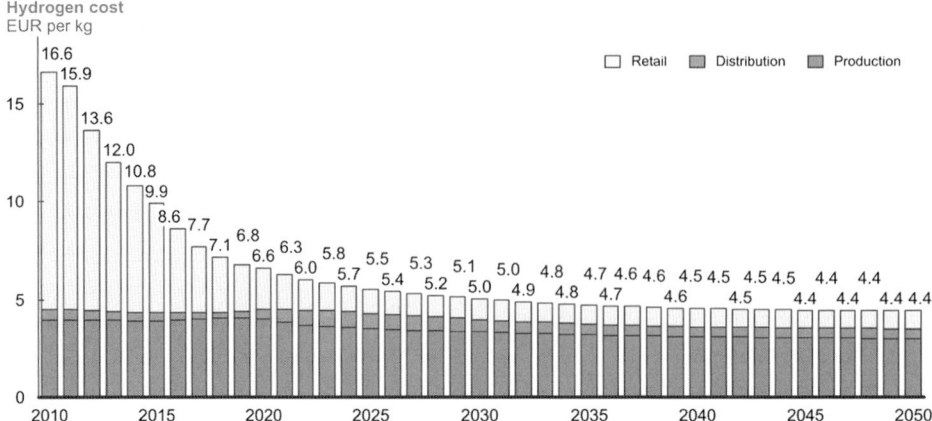

Figure 34.5 Today's hydrogen cost and future expectation in Europe. Source: Reference [7].

hydrogen retail cost (dispensing cost) will decrease significantly by 2020, whereas hydrogen production cost and distribution cost (delivery cost) will remain at the same levels toward 2050.

Both US analysis and JHFC/NEDO analysis are assuming high-volume refueling stations with current technology, so their hydrogen costs are somewhat imaginary even using the bottom-up approach. On the other hand, McKinney's analysis is based on industry's inputs, so that its hydrogen cost can be the realistic industry price today.

GermanHy [8], Germany's hydrogen roadmap issued in 2008, presents future hydrogen cost based on the bottom-up approach (Figure 34.6). According to this

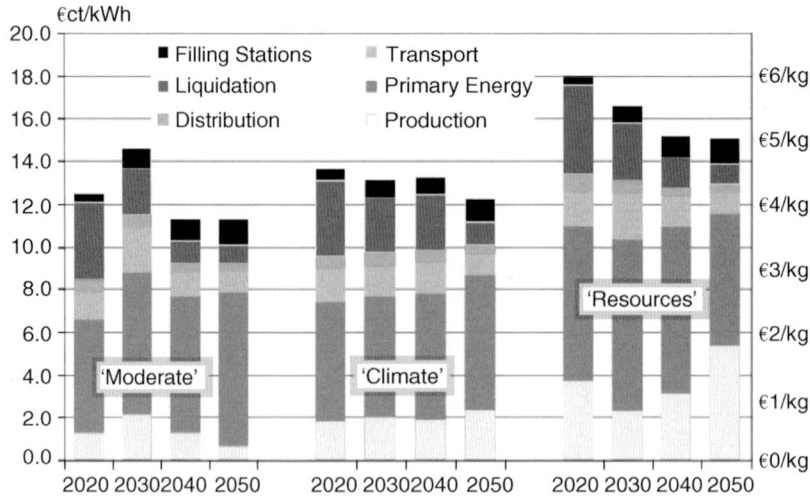

Figure 34.6 Future hydrogen cost in Germany. Source: Reference [8].

roadmap, the 2020 hydrogen cost will be around €4 kg^{-1} ($5.6 kg^{-1}) given a moderate development case (oil price: $54 per bbl) and €6 kg^{-1} ($8.4 kg^{-1}) given shortage-of-resources case (oil price: $248 per bbl).

Therefore, it can be said that today's hydrogen cost is in the range $10–20 kg^{-1} in Europe and Japan, depending on assumptions and hydrogen volume. The US hydrogen cost is very low at around $4.5 kg^{-1} due to the low natural gas price in the US market.

34.3.1.3 Top-Down Approach for Hydrogen Cost

In this approach, the hydrogen cost can be calculated as "would-be" cost, which is usually set to be competitive with conventional vehicles and fuels, for example, ICEs or hybrids with gasoline.

At this calculated cost, FCEV drivers may pay the same fuel cost per km-driven as other conventional vehicle drivers do. This cost is an ideal cost for the future, usually much cheaper than the current cost (or the cost calculated by bottom-up approach).

Since it is an ideal cost to be competitive with conventional fuel, it really depends on the energy efficiency (fuel mileage) of FCEVs and conventional vehicles, and the conventional fuel cost (in short, gasoline price, and then, oil price). Therefore, this cost also varies country by country.

The US DOE has set the target hydrogen cost (in 2020) using following equations [9]:

$$\text{Hydrogen/FCV-related cost per mile } (\$\text{mile}^{-1}) \\ = \text{HEV-related cost per mile } (\$\text{mile}^{-1}) \quad (34.2)$$

$$\frac{\text{Hydrogen cost } (\$\text{gee}^{-1})}{\text{FCEV fuel economy } (\text{mile gge}^{-1})} + \text{FCEV's incremental cost } (\$\text{mile}^{-1})$$

$$= \frac{\text{Projected gasoline cost } (\$\text{gal}^{-1})}{\text{HEV fuel economy } (\text{mile gal}^{-1})} \quad (34.3)$$

$$\text{Hydrogen cost } (\$\text{gee}^{-1}) = \left[\frac{\text{Projected gasoline cost } (\$\text{gal}^{-1})}{\text{HEV fuel economy } (\text{mile gal}^{-1})} \right. \\ \left. \text{FCEV's incremental cost}(\$\text{mile}^{-1}) \right] \quad (34.4) \\ \times \text{FCEV fuel economy } (\text{mile gge}^{-1})$$

The idea of Eq. (34.2) is that a hydrogen-FCEV owner should pay the same cost as a HEV owner on a mileage-driven basis. The DOE translates Eq. (34.2) into Eqs (34.3) and (34.4), where the FCEVs incremental cost is the difference in non-fuel ownership costs between the FCEV and the HEV, and includes vehicle depreciation, financing, maintenance, tires, repairs, insurance, and registration costs as well as taxes, fees, and tax credits with all terms converted into a

Table 34.2 DOEs target of hydrogen cost: production ($ kg^{-1}).

Production technology	2011 Status	2015 Target	2020 Target	2020 (Ultimate target)
Electrolysis from grid electricity	4.20	3.90	2.30	1–2
Bio-derived liquids (based on ethanol reforming case)	6.60	5.90	2.30	
Electrolysis from renewable electricity	4.10	3.00	2.00	
Biomass gasification	2.20	2.10	2.00	
Solar thermochemical	NA	14.80	3.70	
Photoelectrochemical	NA	17.30	5.70	
Biological	NA	NA	9.20	

Source: US DOE Annual Merit Review, June 2013.

$ mile^{-1} basis. The introduction of the incremental cost is intentional, because the hydrogen cost has to be low to compensate for the incremental cost.

In its 2011 calculation the DOE uses $3.13 gallon^{-1} for projected gasoline cost, 41.8 mile gallon^{-1} for HEV fuel economy, 57.5 mile gge^{-1} for FCEV fuel economy, and $0–0.04 mile^{-1} for incremental cost. The target hydrogen cost becomes $4.31 gge^{-1} (at an incremental cost of $0 mile^{-1}) to $2.01 gge^{-1} (at an incremental cost of $0.04 mile^{-1}). Therefore, US DOE sets the hydrogen cost target for 2020 at $2–4 gge^{-1} ($2–4 kg^{-1}), which is divided into a hydrogen production cost of $1–2 gge^{-1} ($1–2 kg^{-1}) and a hydrogen delivery cost of $1–2 gge^{-1} ($1–2 kg^{-1}). Tables 34.2 and 34.3 show detailed targets for hydrogen cost (production cost and delivering & dispensing cost, respectively) [10,11].

The Japanese hydrogen roadmap (NEDO's hydrogen technology roadmap [6]) also uses the same rationale, but in much simpler form:

$$\text{Hydrogen cost } (\$\,\text{gee}^{-1}) = \text{Projected gasoline cost } (\$\,\text{gal}^{-1}) \times \left[\frac{\text{FCEV fuel economy } (\text{mile gee}^{-1})}{\text{HEV fuel economy } (\text{mile gal}^{-1})}\right] \quad (34.5)$$

Table 34.3 DOEs target of hydrogen cost: delivering and dispensing ($ kg^{-1}).

	Delivering and dispensing cost	
	2015	2020 (Ultimate target)
Off-site station	Around 3.0	1.0–2.0
transport and distribution of hydrogen from central production plant	<1.40	<1.30
cost of compression, storage, and dispensing	<1.60	<0.70
On-site station	<2.15	<1.70

Source: US DOE Annual Merit Review, June 2013.

The item in square brackets in Eq. (34.5) refers to the FCEV's fuel economy advantage over HEV. If we apply the projected gasoline cost of ¥100 l^{-1} (in 2020) and FCEVs fuel economy advantage of 1.7–2.0, the target hydrogen cost becomes ¥170–200 l^{-1}, or ¥55–65 Nm^{-3}. The roadmap took the middle number, which is ¥60 Nm^{-3} for 2020. Therefore, the target cost is around $7 kg^{-1} (US$1 = ¥100).

According to the Fuel Cells and Hydrogen Joint Undertaking (FCH JU), a European public–private partnership, the hydrogen cost should be competitive with fossil fuel solutions in 2020 (excluding taxes). Its "Annual implementation plan (AIP) 2013" [12] indicates that the target hydrogen cost is <€10 kg^{-1} ($14 kg^{-1}) for the short term and <€5 kg^{-1} ($7 kg^{-1}) for the long term.

HyWays [13], a European hydrogen roadmap issued in 2008, also indicates a long-term hydrogen cost of €4 kg^{-1} ($5.6 kg^{-1}) in 2020, and €3 kg^{-1} ($4 kg^{-1}) in 2030, to be equivalent to diesel price.

Even today, the German Clean Energy Partnership (CEP) program intentionally sets the hydrogen selling price at around €9 kg^{-1}, which is competitive with ICE fuel economy.

Therefore, by this top-down approach, the hydrogen cost target is around $5–9 kg^{-1}, with the exception of the DOE's target of $2–4 kg^{-1}, which reflects the low gasoline price in the US market.

34.3.1.4 Total Cost of Ownership (TCO) Approach

This approach focuses on the TCO, which is calculated as the sum of FCEV purchase cost and running costs including fuel cost and maintenance costs over vehicle lifetime, for example, 15 years. This means that the calculated cost is no longer independent of vehicle cost, and the FCEV cost greatly affects the hydrogen cost.

A typical TCO study for future transportation powertrains was carried out by McKinsey & Company in 2010 [7]. Its TCO analysis for European C/D segment vehicles is shown in Figure 34.7. This analysis indicates that TCOs of future powertrains converge by 2025–2030, due to the improvement of infrastructure utilization as well as a rapid decrease for the fuel cell system.

Notably, TCO varies depending on assumptions made on vehicle type and cost, fuel type and cost, as well as vehicle usage and lifetime. Since the purchase price of a FCEV will be higher than conventional vehicles at the initial deployment, the running costs, a large part of which is fuel costs, should be lower than conventional costs, in order to be competitive with ICEs. In general, the TCO is more like a purpose-oriented number, rather than a fact-based number.

34.3.2 Economics of Social Cost and Benefits

Another aspect of the economics of hydrogen transportation is the social costs and benefits, which should be widely shared by society. If the social benefits are reasonably larger than social costs, the introduction and deployment of such a new powertrain can be justified socially and nationally.

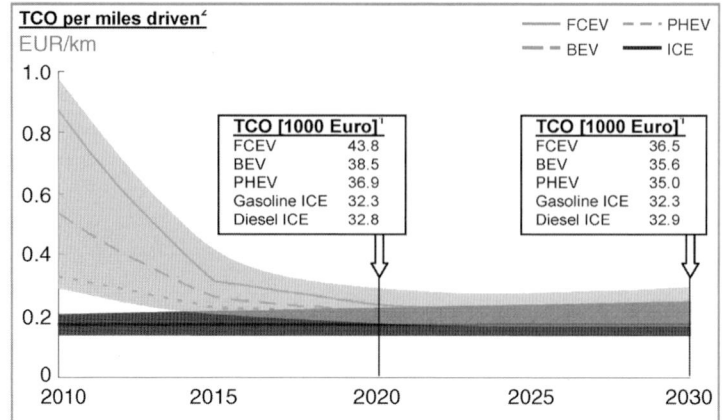

Figure 34.7 TCO of various future powertrains (European C/D segment): (1, 2020) TCO = purchase price + maintenance cost + fuel cost + infrastructure cost. (2, 2030) TCO per miles driven, with variance and sensitivities (fossil fuel prices varied by ±50%; learning rates varied by ±50%). Source: Reference [7].

Notably, social costs (e.g., subsidy, tax credits, or incentives) are also direct benefits for individuals and companies.

34.3.2.1 Social Costs
The major items for social costs are (i), subsidy, tax credits, or incentives to promote new powertrains and (ii) subsidy to build new infrastructure.

34.3.2.1.1 Subsidy, Tax Credits, or Incentives to Promote New Powertrains
Many countries have a support policy to promote new and alternative powertrains.

In Norway, BEVs (battery electric vehicles) are exempt from vehicle taxes, including purchase taxes, which are almost 100%, and value-added tax (VAT) of 25%. In addition, BEVs are allowed to drive in bus lanes and to park at public parking lots for free. These incentives help the sales of BEVs, and more than 10% of new car sales today are BEVs.

In Japan, hybrid cars enjoyed eco-car tax-reduction in 2009–2010 and the sales ratio reached more than 15% of new car sales. Today, the Japanese government offers a subsidy for BEVs, which is at a maximum ¥780 000 for ¥2 989 350 Nissan LEAF.

The German federal government levies a monthly automotive ownership tax based on catalog price, and battery-carrying vehicles can obtain a reduction of catalog price depending on battery size.

In the USA, federal government and state governments have several incentives for PHEVs (plug-in hydrogen electric vehicles) and BEVs. The federal government provides PHEVs and BEVs with a federal tax credit of up to $7500.

These subsidies and incentives are supposed to be applied also for FCEVs, when they are commercialized.

34.3.2.1.2 Subsidy to Build New Infrastructure

For new powertrains with non-conventional fuels, support policy to build new fuel infrastructure is mandatory, and the delay or absence of such infrastructure will become a disincentive for such vehicles.

Today, major leading countries and regions provide subsidies for hydrogen stations in preparation for FCEV launch in 2015–2020. A typical support policy is a subsidy to cover some of the CAPEX of a new fueling station building (Table 34.4). These counties are also considering the wider deployment of hydrogen stations beyond the initial stage, but support policies for the next stage are not clearly agreed at present (Table 34.5).

It is true that wider deployment of hydrogen stations requires certain investment, both from private sector and public sector. One example of the estimation of infrastructure deployment cost is McKinsey's 2010 study [7] on Europe over 40 years (Figure 34.8). Hydrogen infrastructure development requires €100 billion, assuming that hydrogen retail and distribution costs are €1000–2000 per vehicle. On the other hand, charging infrastructure needs €540 billion, assuming that charging stations cost €1500–2500 per vehicle. This estimate may justify hydrogen infrastructure building as the most promising and economical alternative powertrain.

34.3.2.2 Social Benefits

Compared with the social costs, the social benefits differ depending on analytical framework and policy scope and priorities of a government.

Table 34.4 Investments for initial deployment of hydrogen refueling network.

Country	Components
Germany	• Under the Clean Energy Partnership (CEP), a 50 refueling station network will be built to cover major cities. • In general, public funding ratio is 50%.
California	• According to the roadmap developed by the California Fuel Cell Partnership (CaFCP), a network of 68 stations is planned to cover Sacramento/San Francisco Bay area and Los Angeles region by 2016 [14]. • California government agencies, mainly California Energy Commission (CEC), provide the subsidy depending on the design criteria and scores of hydrogen stations, which sometimes become 80–70% of CAPEX.
Japan	• Japanese industries plan to install 100 stations in four major metropolitan cities: Tokyo, Nagoya, Osaka, and Fukuoka. • Japanese government, mainly Ministry of Economy, Trade and Industry (METI), offers subsidy for hydrogen stations, which is basically 50% of CAPEX (except for composite hydrogen storage vessels, which may get up to two-thirds).

Source: Technova Inc.

Table 34.5 Investments beyond initial deployment.

Country	Components
Germany	• In September 2013, H2 Mobility, German consortium with major automakers and fuel suppliers, announced their intention to build a refueling network of about 400 stations by 2023, and the overall investment seems to be €350 million. • It is generally assumed that around 40% of investment will come from the public sector.
California	• CaFCP announced that at least 100 stations are needed to make the refueling station business self-sustaining. • Public funding ratio has not been agreed.
Japan	• Fuel Cell Commercialization Conference of Japan (FCCJ), Japanese fuel cell/hydrogen industry association, published the roadmap in 2010, which indicates that nearly 1000 stations are needed by 2025 to support 2 million FCVs to achieve the CO_2 reduction target at transportation sector. • Public funding ratio has not been agreed.

Source: Technova Inc.

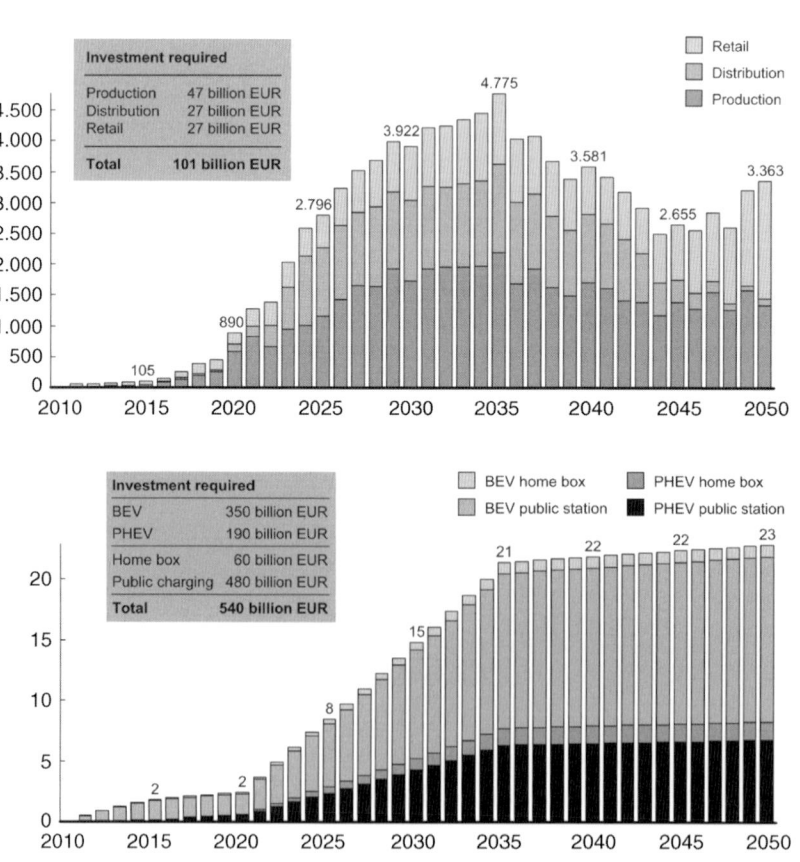

Figure 34.8 Required investment for hydrogen infrastructure and charging infrastructure. Source: Reference [7].

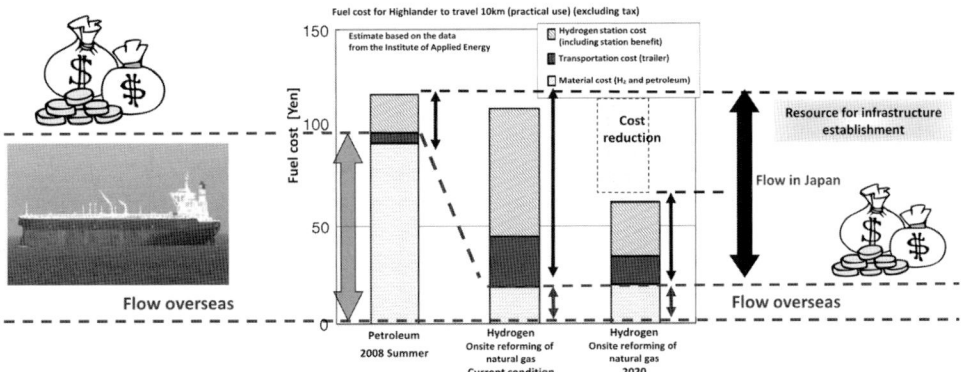

Figure 34.9 National benefits of hydrogen-FCEVs in Japan. Source: Toyota Motor Corporation.

The typical social benefits are the reduction of energy imports and the abated cost for public health. In addition, this may sometimes include the protected value of environment, the contribution to mitigating the impacts from climate change, as well as industrial competitiveness in the global economy.

The California state government enacted regulations to introduce zero emission vehicles (ZEVs), including BEVs and FCEVs. This ZEV mandate aims to improve California's air quality and then the public health of major cities, by reducing lung-related diseases. According to the American Lung Association, if California's vehicle fleet becomes 100% ZEV, the state will reduce smog-forming and fine particulate emissions by 95%, and also reduce public health, global warming, and societal costs by 90%.

Toyota estimates that the energy shift in the automotive sector from gasoline to natural-gas-delivered hydrogen will decrease the money out-flow to oil-producing countries, resulting in a reduction in trade deficit and the improvement of money flow within Japan (Figure 34.9). Of course, domestically-produced hydrogen will further help these improvements.

If 1 million FCEVs are on the road in Japan, Japanese oil imports will be cut by ¥35 billion ($350 million) annually. This is enough to support the building of an additional 100 hydrogen stations (¥350 million per station) annually, so the subsidy for hydrogen station building can be justified.

With these social benefits, the introduction and deployment of FCEVs can be justified socially and nationally. It is not discussed here, but support policy for the introduction of FCEVs and hydrogen infrastructure may help the FCEV/hydrogen industry be competitive with other suppliers in the world.

34.4 Conclusion

A hydrogen transportation system, representing FCEVs and hydrogen stations, has clear environmental benefits, as well as economic benefits for owners and society. The key factor for economic benefit and cost analysis is hydrogen cost, which can be divided into production cost and refueling cost (compression, storage, and dispensing cost), and delivery cost if necessary.

There are three approaches to defining the cost, bottom-up approach, top-down approaches, and total cost of ownership (TCO). With a bottom-up approach (today's cost), the hydrogen cost is in the range $10–20 kg^{-1} in Europe and Japan, whereas the US hydrogen cost is very low at around $4.5 kg^{-1} due to the low natural gas price in the US market. With a top-down approach (target cost), hydrogen cost is $5–9 kg^{-1} in Europe and Japan, with exception of DOEs target of $2–4 kg^{-1}. The TCO varies depending on assumptions of vehicle type and cost, fuel type and cost, as well as vehicle usage and lifetime.

Since the hydrogen/FCEV system has certain benefits (reduction of energy imports and reduction of lung-associated deceases), the introduction and deployment of FCEV can be justified socially and nationally.

References

1. WG2 FCV WG Report (March 2011) JHFC international seminar and roundtable. http://www.jari.or.jp/Portals/0/jhfc/data/seminor/fy2010/pdf/day1_J_10.pdf.
2. Wipke, K. (17 May 2012) Controlled hydrogen fleet and infrastructure analysis, 2012 DOE Annual Merit Review and Peer Evaluation Meeting. http://www.hydrogen.energy.gov/pdfs/review12/tv001_wipke_2012_o.pdf.
3. Ramsden, T. and Ruth, M. (14 May 2013) Pathway analysis: projected cost, well-to-wheels energy use and emissions of current hydrogen technologies, DOE Annual Merit Review. http://www.hydrogen.energy.gov/pdfs/review13/an036_ramsden_2013_o.pdf.
4. Dillich, S., Ramsden, T., and Melaina, M. (24 September 2012) DOE Hydrogen and Fuel Cells Program, Record #12024. http://hydrogen.energy.gov/pdfs/12024_h2_production_cost_natural_gas.pdf.
5. WG1 (March 2011) Hydrogen infrastructure WG Report, JHFC International Seminar and Roundtable. http://www.jari.or.jp/Portals/0/jhfc/data/seminor/fy2010/pdf/day1_J_09.pdf.
6. NEDO (2010) Hydrogen technology roadmap 2010. http://www.nedo.go.jp/content/100086294.pdf.
7. McKinsey & Company (2010) A portfolio of power-trains for Europe: a fact-based analysis. http://ec.europa.eu/research/fch/pdf/a_portfolio_of_power_trains_for_europe_a_fact_based__analysis.pdf.
8. GermanHy (2008) A study addressing the question: Where will the hydrogen in Germany come from by 2050? GermanHy Final Report, 2008. http://wupperinst.org/uploads/tx_wupperinst/GermanHy_Endbericht.pdf.
9. Ruth, M. and Joseck, F. (25 March 2011) Hydrogen threshold cost calculation. DOE Hydrogen and Fuel Cells Program Record #11007. http://www.hydrogen.energy.gov/pdfs/11007_h2_threshold_costs.pdf.
10. Randolph, K. (16 May 2013) Hydrogen production – session introduction, 2013 Annual Merit Review and Peer Evaluation Meeting. http://www.hydrogen.energy.gov/pdfs/review13/pd000_randolph_2013_o.pdf.
11. Sutherland, E. (15 May 2013) Hydrogen delivery – session introduction, 2013 Annual Merit Review and Peer Evaluation Meeting. http://www.hydrogen.energy.gov/pdfs/review13/pd00a_sutherland_2013_o.pdf.
12. Fuel Cells and Hydrogen Joint Undertaking (FCH JU) (2013) Annual implementation plan 2013. https://ec.europa.eu/research/participants/portal/

doc/call/fp7/fch-ju-2013-2/1585101-fch_ju_aip2013_en.pdf.
13. HyWays (2008) The European Hydrogen energy roadmap. http://www.hyways.de/docs/Brochures_and_Flyers/HyWays_Roadmap_FINAL_22FEB2008.pdf.
14. California Fuel Cell Partnership (CaFCP) (2012) A California road map: bringing hydrogen fuel cell electric vehicles to the Golden State. http://cafcp.org/sites/files/20120814_Roadmapv(Overview).pdf.
15. Hirose, K. (2013) Toyota's fuel cell development towards 2015. http://www.cenex-lcv.co.uk/presentations2013/day1/main/D1MPS2-Katsuhiko-Hirose.pdf.
16. American Lung Association (24 January 2012) Road to clean air II – a zero emission future. http://www.lung.org/associations/states/california/assets/pdfs/advocacy/clean-cars-campaign/zev-road-to-clean-air-stats.pdf.

35
Challenges and Opportunities of Hydrogen Delivery via Pipeline, Tube-Trailer, LIQUID Tanker and Methanation-Natural Gas Grid

Krishna Reddi, Marianne Mintz, Amgad Elgowainy, and Erika Sutherland

35.1
Introduction

Hydrogen provides a clean alternative to gasoline or other petroleum derivatives, especially when it is produced from renewable energy sources like wind and solar. For the successful commercialization of hydrogen fuel cell electric vehicles (HFCEVs), a reliable, safe, and convenient hydrogen delivery infrastructure is required. Currently, about 20 million [1] metric tons per year of hydrogen is produced, delivered, and used in the United States (USA), primarily as inputs to petroleum refining and ammonia production, and more recently to power fuel cells for stationary power and materials handling equipment (forklifts). The produced hydrogen is distributed via a limited infrastructure of pipelines to large industrial users, and liquid hydrogen tankers and tube-trailers to smaller customers. For deliveries up to 50 kg, hydrogen is transported in compressed gas cylinders, while for deliveries of more than 50 kg, hydrogen is shipped in gaseous tube-trailers and liquid tankers [2]. For large and consistent demand, hydrogen is delivered via dedicated pipeline. Today, there are about 1300 miles of such pipelines in the USA, from which hydrogen [2] is supplied mainly to petroleum refineries and fertilizer plants from hydrogen production facilities located in close proximity to these large customers.

Hydrogen delivery infrastructure is necessary only when hydrogen is produced at a central or semi-central location away from locations where HFCEVs are fueled (i.e., refueling stations). The delivery of hydrogen from the production plant to the vehicle's tank includes packaging, transport, and fueling operations [3]. The packaging operation includes the compression of hydrogen into pipelines or tube-trailers and the liquefaction of hydrogen for tanker deliveries. The transport operations consist of the transmission and distribution processes of hydrogen from the production plant to refueling stations, while the fueling operations include the processes used at the refueling station to dispense the hydrogen into the vehicle's tank at the desired fill rate and conditions. Currently, hydrogen is delivered in one of two physical forms: gaseous and liquid. The delivery infrastructure is defined by the physical and chemical forms of hydrogen

Hydrogen Science and Engineering: Materials, Processes, Systems and Technology, First Edition.
Edited by Detlef Stolten and Bernd Emonts.
© 2016 Wiley-VCH Verlag GmbH & Co. KGaA. Published 2016 by Wiley-VCH Verlag GmbH & Co. KGaA.

Table 35.1 Estimated delivery cost for distributed and centrally produced hydrogen from US DOE's FCTO [3].

Delivery cost (2007 US$)[a]	2011 Status	2020 Target
Distributed production		
Fueling station ($ gge^{-1})	2.50	<1.70
Centralized production		
Transmission and distribution ($ gge^{-1})	1.90–2.20	<1.30
Fueling station ($ gge^{-1})	1.60	<0.70

a) Does not include production cost.

(liquid, gas, or carrier material) being transported. Gaseous and liquid delivery pathways are mature, while the hydrogen carrier pathways are in technology development and laboratory testing phases with their scope for commercialization currently uncertain.

Delivery costs [3] associated with distributed and centralized hydrogen production as reported by the US Department of Energy's (DOE's) Fuel Cell Technology Office (FCTO) are given in Table 35.1. Costs using current delivery technologies are shown as "2011 Status," while costs using advanced technologies currently under development are shown as "2020 Targets." This nomenclature is used in several of the tables that follow.

35.2
Variation in Demand for Hydrogen

In the transportation sector, the demand for hydrogen varies with consumer driving and refueling patterns. From an analysis of gasoline filling events at approximately 400 Chevron gasoline stations, it has been observed that the daily demand for gasoline varies during the week with highest demand occurring on Friday (see Figure 35.1) [4]. Hourly demand also varies [4], exceeding the daily average

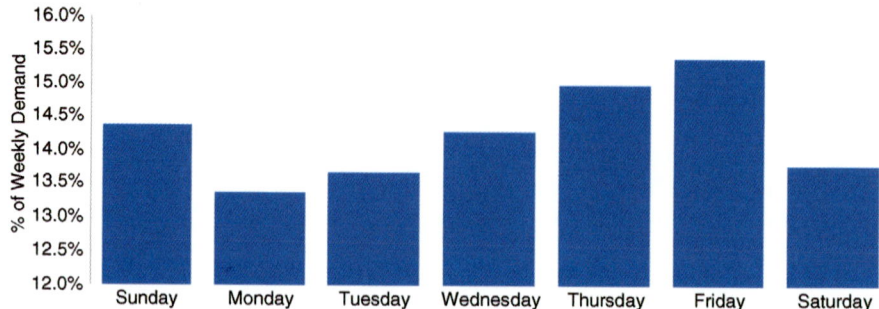

Figure 35.1 Profile of weekly fuel demand (by day).

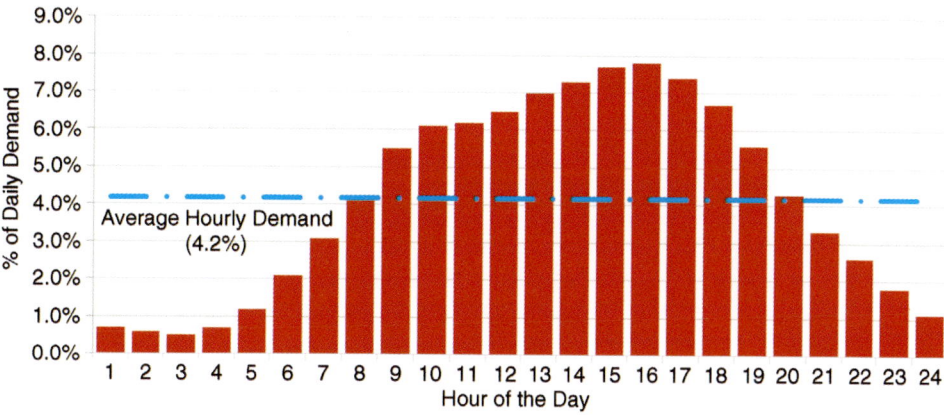

Figure 35.2 Profile of daily fuel demand (by hour).

between 9 a.m. and 7 p.m. (Figure 35.2). Additionally, demand during the summer driving season is about 10% higher than the annual average, while demand during the winter months is lower than the annual average (Figure 35.3) [4].

The hydrogen delivery infrastructure should address all these demand variations (seasonally, daily, and hourly) to ensure that consumers' hydrogen fueling experience is similar to what they have come to expect from the current gasoline infrastructure. The hydrogen delivery infrastructure should also provide sufficient storage capacity to accommodate planned production plant downtime associated with annual maintenance. To address these variations in the supply and demand for hydrogen, large storage systems are necessary at various stages of the delivery infrastructure. While the production plant is assumed to produce sufficient hydrogen to satisfy average daily demand, storage systems must be

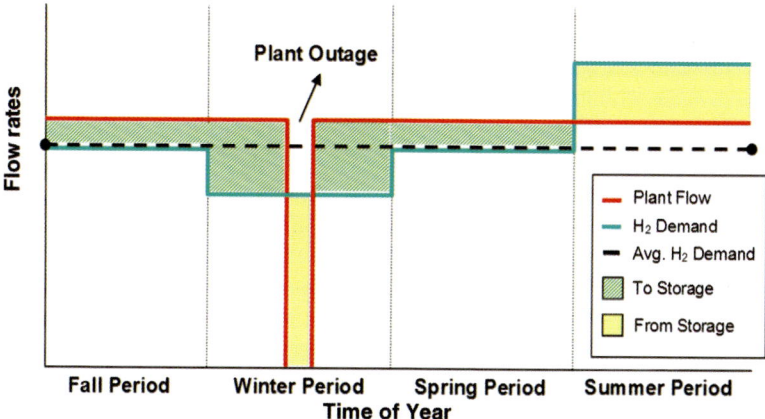

Figure 35.3 Seasonal fuel requirement due to typical consumer fueling demand and supply downtime.

filled when demand is low (below average) and withdrawn when demand is high (above average).

35.3
Refueling Station Components and Layout

Refueling stations normally contain a hydrogen supply source (in either liquid or gaseous state) from which the hydrogen is withdrawn, conditioning and dispensing equipment, and associated controls. Conditioning varies with the physical form of the supply source and the form of hydrogen required by the fuel cell vehicle. Hydrogen conditioning generally includes increasing its energy density through compression, cooling, or a combination of the two. In the following discussion refueling stations that receive hydrogen in gaseous form are described first, followed by a description of stations that receive hydrogen in liquid form.

35.3.1
Gaseous Hydrogen Refueling Station

At present, hydrogen in gaseous form is stored at either 350 or 700 bar in vehicle tanks. The former requires 350 bar dispensing capability, the current standard for refueling fuel cell buses and most fuel cell-powered lift trucks; the latter requires 700 bar and −40 °C dispensing capabilities. The cost to deliver hydrogen to a 350 bar vehicle tank is less than to a 700 bar tank since less compression and storage are needed, as well as no (or minimal) precooling. However, 700 bar tanks are needed for HFCEVs to achieve a driving range comparable to today's internal combustion engine vehicles (ICEVs), approximately 400 miles on a single tank of fuel. Though 350 bar hydrogen has relatively low delivery cost, the energy density of the hydrogen in the tank is also low, limiting driving range. As a result, all the major automobile Original Equipment Manufacturers (OEMs) have adopted the 700 bar [5] onboard hydrogen storage option. For this reason, we limit our discussion of gaseous hydrogen refueling stations to 700 bar fueling.

Typically, a gaseous hydrogen refueling station consists of a low pressure gaseous hydrogen source, a compressor with a high compression ratio (~45) [6] and throughput consistent with the maximum hourly demand, a high pressure buffer storage system to satisfy hourly variations in demand, a refrigeration unit to pre-cool the hydrogen to −40 °C, and a dispenser to regulate the flow of hydrogen from the high pressure buffer storage system into the vehicle's tank (Figure 35.4). The compressor brings inlet hydrogen from the low pressure hydrogen source (as low as 20 bar) to the high pressure storage buffer (at 950 bar for 700 bar fills). The amount of required compressor throughput and high pressure buffer storage are influenced mainly by the design capacity of the station and hourly refueling demand variations (usually represented by the number of required back-to-back vehicle fills). The high

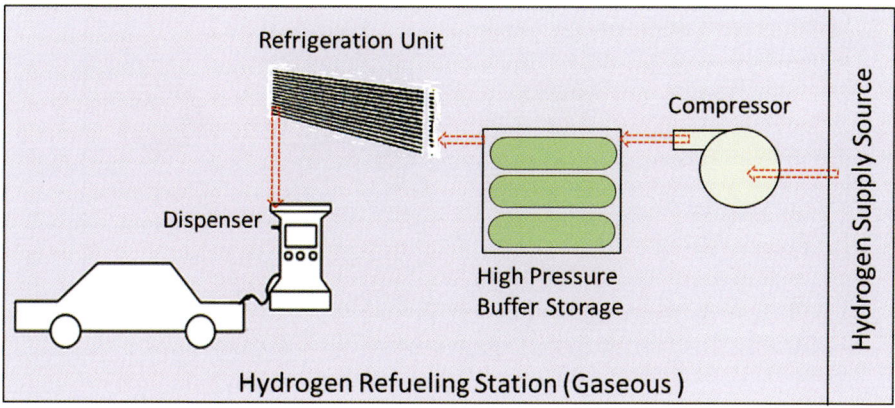

Figure 35.4 Schematic of a gaseous hydrogen dispensing refueling station.

pressure buffer storage system consists of multiple pressure vessel banks which are withdrawn by the dispenser and replenished by the compressor in a predefined order.

The hydrogen fueling protocol for surface transportation vehicles, J2601 [7], which has been developed by the Society of Automobile Engineers (SAEs), standardizes the fueling process by defining and establishing limits for various refueling parameters. The core parameters of the 700 bar fueling protocol are:

- The maximum hydrogen gas temperature in the vehicle tank cannot exceed 85 °C.
- The maximum hydrogen gas pressure in vehicle tank cannot exceed 875 bar at 85 °C.
- The maximum rate of fueling cannot exceed $60\,g\,s^{-1}$ (equivalent to 10 kg in 3 min).
- The target fueling time for a passenger car with capacity up to 5 kg hydrogen is 3 min.
- The 100% fueling target or state of charge, is defined as $40.2\,g\,l^{-1}$ (i.e., 700 bar at 15 °C; 875 bar at 85 °C).

The refueling station components and their operation should meet the SAE J2601 fueling protocol. Typically, the operation of refueling station includes a dispenser controlling the flow of precooled hydrogen into the vehicle tank from the high pressure buffer storage system, while the compressor replenishes the high pressure buffer storage system as required. The compressor at the refueling station requires a high compression ratio of about 45 (20 to 900 bar) and a throughput that meets the peak hourly demand. The installed refrigeration unit should be capable of precooling the hydrogen with a mass flow rate up to $60\,g\,s^{-1}$.

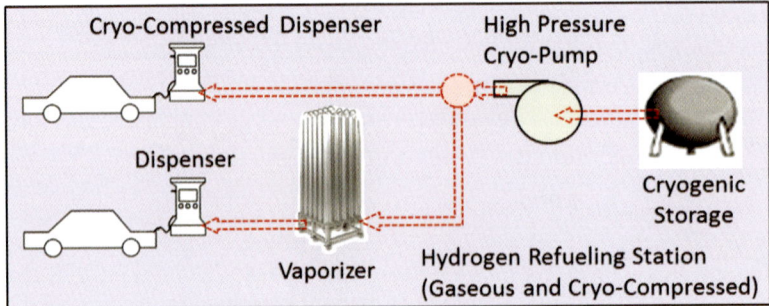

Figure 35.5 Schematic of a gaseous and cryo-compressed hydrogen dispensing refueling station.

35.3.2
Cryo-Compressed and Gaseous Refueling Station with Liquid Delivery

Some OEMs are pursuing cryo-compressed hydrogen as a potential onboard storage option for HFCEVs, since liquid hydrogen delivery from a production plant to refueling stations seems economical for some market scenarios [6].

Refueling stations with a liquid hydrogen supply can dispense hydrogen to vehicle tanks in either gaseous or cryo-compressed form. As shown in Figure 35.5, the latter type of hydrogen fueling station includes a cryogenic storage tank and a cryo pump. The pump raises the pressure of the hydrogen and permits it to be dispensed directly into the vehicle tank either in cryo-compressed form or 700 bar gaseous form. For 700 bar gaseous dispensing, the cryo-pump lifts the pressure of the liquid hydrogen to more than 700 bar, which is then vaporized at that pressure and dispensed into the vehicle's tank.

35.3.3
Refueling Station Challenges

- **Analysis:** The sizing of refueling station components is a function of such parameters as desired station capacity, expected hourly demand, onboard tank capacity and operating pressure, the size and configuration of cascade storage vessels (defined by the number of pressure banks and number of vessels in each pressure bank), and type of hydrogen source (gaseous, liquid, or cryo-compressed). These components are interdependent and must be well matched if the refueling station is to be optimized for low hydrogen refueling cost. Analysis of competing technologies is required to determine the relative advantages and disadvantages of each. Comparison of different technologies and combinations of technologies is required to identify the optimum combination of components for different sets of the influencing parameters defined above. In addition to identifying a complementary set of technologies across various station components, analyses should identify

optimum operation strategies for individual components to achieve further reductions in hydrogen cost.
- **Hydrogen quality:** Since fuel cells require high quality hydrogen, it must be produced at high purity and protected from contamination along the delivery chain to the point of use. Delivery of high quality hydrogen can be ensured by adopting contamination-free component designs or by purifying hydrogen before dispensing. Either of these two strategies would add to the accrued delivery cost.
- **Metering:** The 700 bar dispensers are expensive and manufactured by very few companies. The current metering technologies cannot meet the accuracy required by federal agencies. Development of reliable and durable flexible hoses capable of delivering 700 bar hydrogen and accurate metering are some of the challenges currently being addressed by the industry.
- **Refrigeration:** The SAE J2601 fueling protocol requires hydrogen to be cooled to $-40\,°C$ for dispensing into type 4 vehicle tanks. Compact, energy efficient, and cost-effective refrigeration technologies need to be developed to enable fast filling and meet the target delivery costs for hydrogen (see Table 35.1).
- **Storage costs:** High pressure storage is required at the refueling station for fast filling of HFCEVs. The real estate at the refueling station is premium, thus the high pressure storage should be compact and capable of storing enough hydrogen to address hourly demand variations. Steel and composite pressure vessels enable storage of hydrogen at high pressures and densities but are costly, highlighting the need for low cost storage systems. Improved steel and composite pressure vessel designs are currently being pursued by the DOE to reduce hydrogen storage cost.
- **Boil-off:** Boil-off is endemic to liquid hydrogen storage and strategies to reduce it should be implemented. Boil-off of hydrogen during tanker loading/unloading and refueling operations should also be minimized. The development of a low-cost cryo-compressor that recovers boil-off gas and conditions it to supplement the refueling service can overcome this challenge.
- **Land area:** The space at refueling stations is premium and land area requirements should be minimized. Present regulations require large setback distances for hydrogen storage which would increase the capital investment required for building and commissioning a refueling station. This should be addressed by developing appropriate code and standards to reduce setback distances while maintaining safety requirements. Developing liquid storage tanks for underground installation can minimize the required setback distances and reduce land area requirement and cost.
- **Compression:** The refueling station requires compressors with moderate throughput, high compression ratio (about 45 for pressurizing hydrogen from 20 to 900 bar), high reliability, and low maintenance. Current compression technologies require frequent maintenance and often necessitate installing a backup compressor to avoid costly station downtime. Special materials and seals should be employed to avoid contamination and leakage of the small hydrogen molecules in a cost effective manner.

35.4
Distributed Production of Hydrogen

During the initial stages of commercialization of HFCEVs, it may be practical to produce hydrogen in small quantities at refueling stations. Unless there is a sufficient number of HFCEVs to justify the construction of dedicated production facilities, it will be economical to produce and dispense hydrogen at refueling stations where there are no existing hydrogen production plants within a reasonable supply distance.

Figure 35.6 illustrates the layout of a fueling station with onsite hydrogen production. Components include the hydrogen producing equipment, a compressor, low-pressure buffer storage, high-pressure buffer storage, a refrigeration unit, and a dispenser. The hydrogen production equipment, typically a steam reformer that produces hydrogen from natural gas or an electrolyzer that splits water into hydrogen and oxygen, produces hydrogen at about 20 bar (or higher for electrolyzers). The low-pressure buffer is filled when hydrogen demand is below the average production rate and is drawn down when demand is above the average. The size of the low-pressure storage system should be greater than the difference between maximum daily demand and average daily demand. High-pressure buffer storage accommodates hourly demand variations and is sized to address the maximum hourly demand during the busiest day of the year. Typically, the compressor frequently replenishes the high-pressure buffer storage system by drawing from the on-site production (typically at 20 bar) or from the low-pressure buffer storage (at 150 to 250 bar) and pressurizing it to 950 bar for 700 bar vehicle tank fills. The high-pressure storage system is connected to the dispenser via a refrigeration unit, which precools the hydrogen to −40 °C for fast fills of 700 bar vehicle tanks. The dispenser permits the flow of hydrogen from the high-pressure buffer storage and regulates the flow to ensure a 5 kg fill in 3 min. The vehicle's tank pressure exceeds 700 bar (around 875 bar) at the end of the fill due to the increase in hydrogen temperature during the fill. Thus, the high-pressure buffer storage system should be rated above 875 bar (about 950 bar) to ensure 100% fills of vehicle tanks.

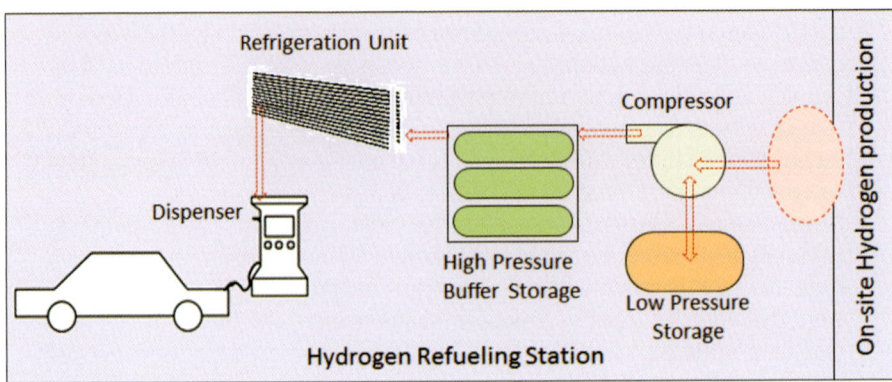

Figure 35.6 Schematic of a gaseous hydrogen dispensing refueling station with on-site production.

35.5
Central or Semi-central Production of Hydrogen

When hydrogen is produced in large quantities at central locations, a reliable hydrogen delivery infrastructure is needed to transport it from the production plant to refueling stations or other points of use. Hydrogen is currently transported in two physical forms, liquid and gaseous. It is delivered in gaseous form using tube-trailers and pipelines, and in liquid form using cryogenic tankers. Three major pathways that are considered commercially viable options for hydrogen delivery are discussed below.

35.5.1
Gaseous Hydrogen Delivery

35.5.1.1 Pipeline Delivery Pathway

Components Layout

The pipeline delivery pathway is economical when demand for hydrogen is large enough to justify the construction of dedicated transmission and distribution pipelines. Currently, pipelines are considered the lowest cost option to supply hydrogen to large refueling stations with daily demand greater than 1000 kg per day within a city/locality with a demand of more than 150 metric tons per day. As shown in Figure 35.7, the gaseous pipeline delivery pathway receives hydrogen from a production plant and includes a large geologic storage facility to provide backup

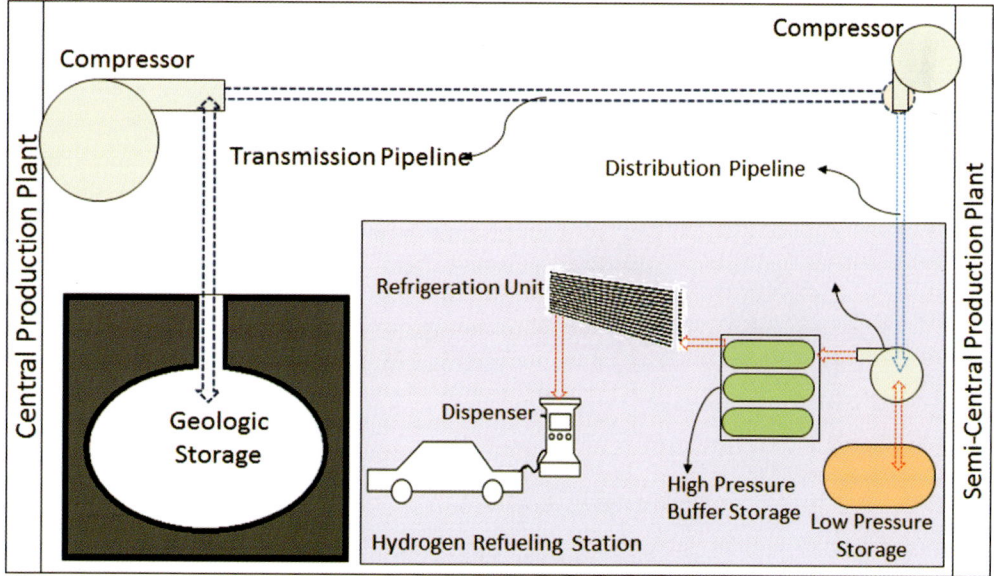

Figure 35.7 Schematic of gaseous hydrogen delivery pathway with pipeline supply.

Table 35.2 Relationships used to estimate US pipeline construction cost by component [8].

Cost component	Cost estimating equation (in 2009 dollars)[a)–c)]
Material	$63027 \times e^{0.0697D}$
Labor	$2.065 \times 24246 \times D^{0.9516}$
Right of way (ROW)	$2.302(1918.8D + 71347)$
Miscellaneous costs	37% of the sum of material, labor and ROW costs

a) D represents pipeline diameter and should be defined in inches.
b) The resulting cost estimates are in 2009 dollars.
c) Assumes a 10% cost premium for hydrogen pipelines versus natural gas pipelines.

supply during production plant outages and to buffer against seasonal demand variations; a compressor to pressurize hydrogen from the production plant or geological storage into the transmission pipeline; a transmission pipeline to transport hydrogen from the production plant to the city gate or distribution terminal; and a distribution pipeline network to bring hydrogen from the transmission pipeline to the refueling station. If hydrogen is produced at the city gate or distribution terminal, a compressor is needed to pressurize hydrogen from what is called a semi-central production plant to the distribution pipeline network.

Cost Estimates

Pipeline construction is a major investment and requires large and consistent demand to recover the cost. Pipeline construction can be broken down into material costs, labor costs, right-of-way (ROW) cost, and miscellaneous costs. A team from Pacific Northwest National Laboratory (PNNL) has developed equations for material, labor, and ROW costs, based on published data for natural gas pipeline construction. While the equations [8] have been developed for each region, average costs for the entire USA can be obtained from the relationships shown in Table 35.2 [8].

For hydrogen pipelines, cost is assumed to be about 10% [9] higher than the estimates shown in Table 35.2. Construction costs for distribution and transmission pipelines of different diameters are shown in Figures 35.8 and 35.9, respectively [9].

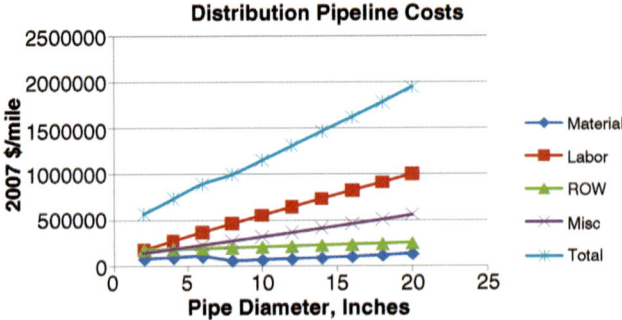

Figure 35.8 Distribution pipeline cost by expense type and pipe diameter.

35.5 Central or Semi-central Production of Hydrogen | 859

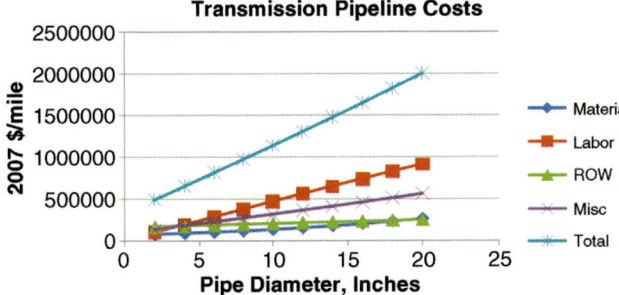

Figure 35.9 Transmission pipeline cost by expense type and pipe diameter.

Most pipelines used to transport industrial and domestic gas are of steel construction. Fiber-reinforced polymer (FRP), an alternative to steel, is estimated to cost about 25% more up front, but to save 40% in labor cost, resulting in an overall cost reduction of approximately 15% [10]. The major advantages of FRP are its flexibility, which enables spooling and simplifies pipe laying, and its resistance to chemical and corrosion resistance [11]. FRP pipe can be spooled for diameters up to about 6 inches; hence, it is more suitable for distribution pipelines than transmission pipelines, although laying several pipes in parallel can be used for transmission of large flow volumes.

The estimated costs of transmission and distribution pipelines [6] are shown in Table 35.3.

Challenges

- **High cost:** As mentioned above, the labor cost of pipeline construction is about 50% [6] of the total cost and is a major barrier to installing hydrogen pipelines. Labor costs are driven by the time required to lay and weld pipe sections. Steel is limited by its weight; hence, a given expanse of steel pipeline will require more pipe sections to be transported to the job site and more welds. Innovative packaging and joining techniques are required to bring down construction time and thus the labor cost. Efforts should be made to investigate alternative materials that can be more easily handled and joined, such as FRP.
- **Distribution:** Hydrogen has a low energy density and thus it is typically distributed at pressures of 20–30 bar to provide volumetric energy density

Table 35.3 Estimated costs for transmission and distribution pipelines.

Pipeline	2011 Status
Transmission ($ per mile for 6–40 inch pipelines, excluding ROW)	765–4500 K
Distribution ($ per mile for 4–8 inch pipelines, excluding ROW)	440–1200 K

comparable to other fuels. In contrast, natural gas distribution lines are within the 0.02–0.2 bar [6] range. As with any product, losses reduce revenue; hence, leaks must be avoided throughout the supply chain. Like natural gas, hydrogen is odorless and requires the addition of odorants to enable leak detection. Suitable odorants that can be separated easily from hydrogen and that do not endanger fuel cell performance or operator health and safety need to be developed. The problem is compounded by the low density and weight of the hydrogen molecule, which tends to leak more quickly than other molecules, including odorants.

- **Hydrogen and material interactions:** The interaction between hydrogen and pipe materials is not well understood at high operating pressures, especially when pressure cycling is involved. High pressures and pressure cycling also affect the durability of materials. Efforts should be made to develop new coatings to prevent embrittlement of steel pipelines.
- **Compression:** Reciprocating compressors are currently used to provide high throughput for transmission and distribution of hydrogen to industrial sites. The reciprocating compressors presently used for hydrogen service are costly and can contaminate the hydrogen with lubricants which may damage or degrade the performance of fuel cells. The embrittlement associated with hydrogen service equipment requires special materials. Possible solutions include new lubricant-free compression technologies that can provide high throughput and avoid contamination, or low-cost hydrogen purification processes that reduce the impact of purification on delivery cost.
- **Geologic storage:** Though hydrogen has been stored at low cost and in large quantities in geologic storage facilities, leakage and contamination are significant risks, high pressures may create operational challenges, and development costs can be high. Geologic storage needs a cushion gas (minimum amount of gas that needs to be left in storage), which for natural gas is about 15% [6] of the storage capacity. Leakage due to permeation into the surrounding rock may be unavoidable and may result in significant increases in storage cost. Contamination may require a post withdrawal purification step. The low energy density of hydrogen requires higher storage pressures than typically maintained for natural gas. The effects of those pressures per se, of cycling (i.e., filling and withdrawing hydrogen) at higher pressure, and of the reactivity of hydrogen with rock formations, are not known. Another major challenge is the lack of suitable geologic formations in certain regions, potentially limiting the feasibility of geologic storage in certain geographic locations.

35.5.1.2 Tube-Trailer Delivery Pathway

Components Layout

The tube-trailer delivery pathway is a viable option when the demand for hydrogen is low and a production plant is at a reasonable distance (less than 100 miles) (Figure 35.10). Hydrogen produced at a central plant may be transported to a tube-trailer loading terminal through transmission pipelines. In the case of semi-

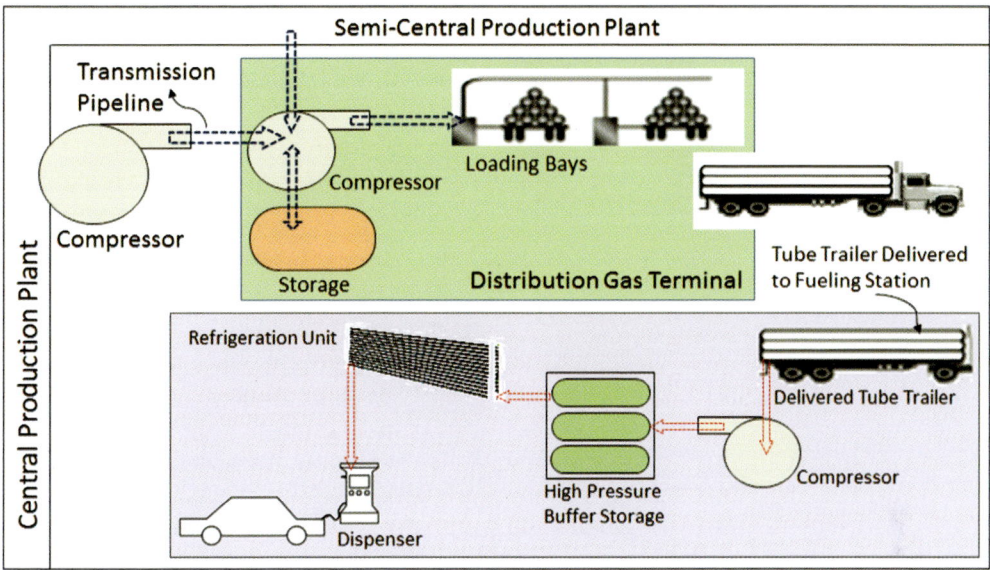

Figure 35.10 Schematic of gaseous hydrogen delivery pathway with tube-trailer supply.

central production, the production plant is co-located with the gas terminal. The gas terminal consists of loading bays with a connecting interface to the tube-trailers, a compressor that lifts the pressure of hydrogen from the production pressure to the tube-trailer operating pressure, and storage to smooth variations in demand. The tube-trailers are filled at the distribution gas terminal and dropped at refueling stations, where they are connected to the station's compressor. The compressor withdraws hydrogen from the tube-trailer to replenish the station's high-pressure buffer storage system, which is used to fill vehicle's tank. The dispenser establishes a connection between the high-pressure buffer storage and the vehicle's tank via a refrigeration unit, which precools the hydrogen to −40 °C. Supplying hydrogen by a tube-trailer eliminates the need for low-pressure storage which is typically used to buffer against supply and demand fluctuations.

Cost Estimates

The US Department of Transportation (DOT) regulates the size, weight, and operating pressure of gaseous cargo transported on US roadways. DOT regulations require pressure vessels to be mounted within an International Standard Organization (ISO) container measuring 40 feet long, 8 feet wide and 8 feet high [12]. Additionally, the DOT limits the total weight of the tube-trailer to 80 000 lbs [12]. Currently, steel pressure vessels mounted on trailers are used to transport hydrogen from production facilities to points of use. The steel pressure vessels are limited by relatively lower strength and higher density. The combination of steel material properties and DOT weight limitations restrict the maximum hydrogen payload to about 250 kg as shown in Figure 35.11. Currently,

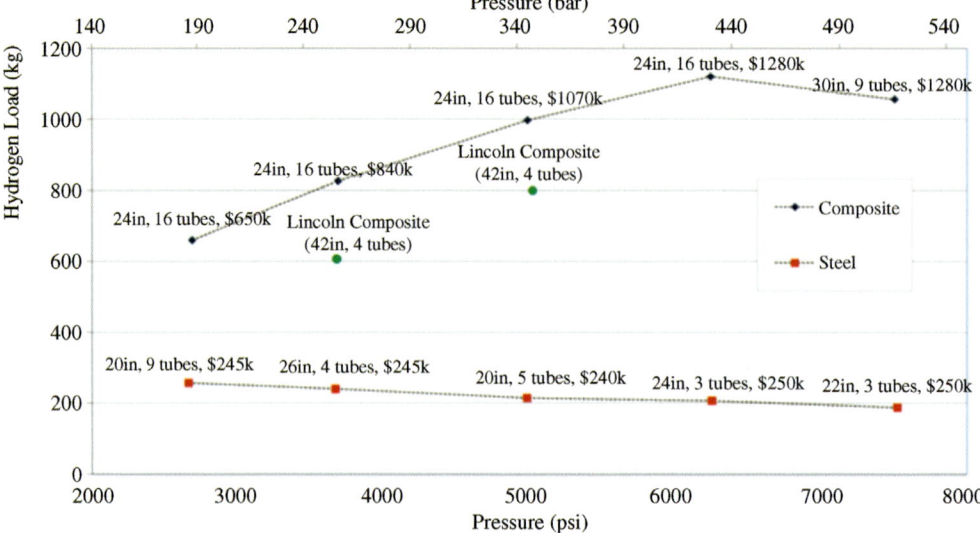

Figure 35.11 Maximum payload for various operating pressures, pressure vessel diameter, and count.

tube-trailer delivery is a viable option for distributing hydrogen to locations up to 200 miles from the production facility.

Pressure vessel thickness is directly proportional to diameter, operating pressure, and a safety factor [13] and inversely proportional to the strength of the material. A safety factor of 3.5 is required for steel pressure vessels used to transport gaseous fuels. For a pressure vessel of given material, diameter, and operating pressure, hydrogen payload can be maximized in compliance with DOT regulations. In comparison to steel, carbon fiber composites have higher strength, lower density, and a lower safety factor, 2.35 as compared to 3.5 [14]. Compared to steel pressure vessels, filament-wound pressure vessels are lighter and have thinner walls for any given operating pressure.

Composite and steel pressure vessels have been modeled for various outer diameters ranging from 10 to 30 inches, each 38 foot long for working pressures of 2700, 3700, 5100, 6600, and 7100 psia [15]. The maximum percentage of volume of the ISO container occupied by hydrogen for different operating pressures is shown in Figure 35.12 [15]. The difference in percentage of volume utilization between the steel and composite pressure vessels for each working pressure highlights the advantage of composites over steel.

The current maximum hydrogen payload that can be delivered for operating pressures of 2700, 3700, 5100, 6600, and 7100 psi using steel versus composite pressure vessels is shown as red and black data points in Figure 35.11 [15]. The operating pressure, vessel outer diameter, and maximum number of pressure vessels that can be mounted on the trailer in compliance with DOT regulations define the configuration of the tube-trailer. That configuration, along with the estimated

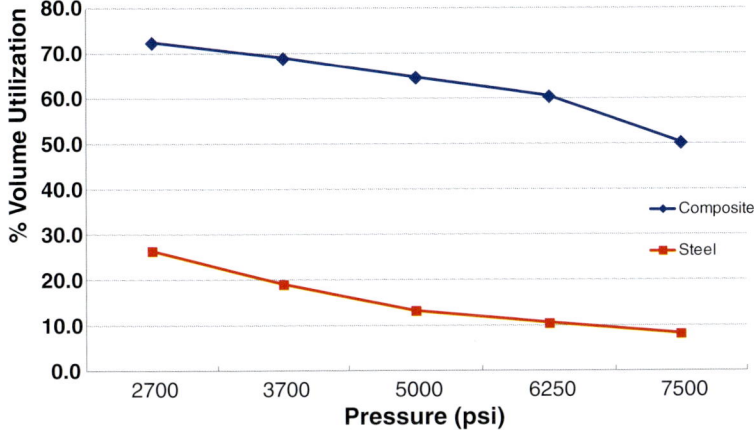

Figure 35.12 Decrease in volume utilization with service pressure.

manufacturing cost of the pressure vessels for each operating pressure, is shown in Figure 35.11 [15]. The payload of Hexagon Lincoln's Titan tube-trailer, which has been approved by DOT for 250 bar hydrogen deliveries, is also shown in green on Figure 35.11. It can be seen that composite tube-trailers can deliver more than 1000 kg payload with a working pressure of 5100 psi (350 bar) or higher. A complete techno-economic analysis is needed to quantify the costs of manufacturing, mounting, plumbing, valving, and so on, to estimate and identify the most economic combination of the number of tubes mounted on the trailer, the operating pressure, and material.

The US DOE has estimated the price of composite tube trailers for hydrogen storage at about $500 kg$_{H2}$$^{-1}$ by year 2010 and $300 kg$_{H2}$$^{-1}$ by year 2015. Currently, the cost of hydrogen storage is estimated to be $828 kg$_{H2}$$^{-1}$ for 250 bar and $783 kg$_{H2}$$^{-1}$ for 350 bar tube-trailers (Hexagon Lincoln Titan IV). Estimated costs in 2011 and the US DOE's 2020 cost targets [3] are shown in Table 35.4.

35.5.1.3 Challenges

- **Distribution issues:** The low density of the hydrogen molecule requires higher loading pressures to achieve the desired payload onboard a tube-trailer. The effect of pressure cycling on composite tanks needs to be investigated and better understood. Higher payloads are possible with higher loading pressures,

Table 35.4 Estimated cost and 2020 cost targets for gaseous tube trailers [3].

Tube-trailer	2011 Status	2020 Target
Operating pressure (bar)	250	520
Capacity (kg)	560	940
Capital cost (2007$)	470 K	540 K

but lengthy and expensive testing is required before DOT certification can be granted. Alternatives like delivery of cold gas using tube-trailers may enable the delivery of hydrogen at higher energy densities and lower pressures.
- **High storage cost:** Storage tanks for gaseous hydrogen are costly but are necessary at different locations along the delivery pathway (e.g., distribution gas terminal, refueling station, etc.). The 2011 cost of low-pressure (160 bar) storage is approximately $1000\,\text{kg}_{H2}^{-1}$, while the DOE targets are $850\,\text{kg}_{H2}^{-1}$ in 2015 and $700\,\text{kg}_{H2}^{-1}$ in $500 [3]. New technologies and materials need to be explored to achieve these targets.
- **Analysis:** Detailed analysis of distribution gas terminal operations is required to estimate required compression and storage and the optimum operational pressures. The minimum storage (empty) pressure and the compressor's throughput determine the time required to recharge an empty tube-trailer. Detailed models simulating the actual operations of distribution gas terminals would provide valuable data to understand the impact of component design parameters such as compressor throughput, required supply pressure, and so on. Analysis is also needed to study the relationship between loading time and temperature increase within the tubes so that limits can be set for safe recharging of tube-trailers. The effect of tube-trailer pressure on terminal and refueling station operations should be assessed so that optimum tube-trailer pressure can be determined for minimum delivery cost.
- **Compression:** Distribution gas terminals require high compression capacity in order to fill tube-trailers in the shortest time possible. Reliable compressors with large throughput are not commercially available. Commercial availability of these compressors is vital for the successful deployment of hydrogen infrastructure.

35.5.2
Liquid Hydrogen Delivery Pathway

35.5.2.1 Components Layout
Figure 35.13 shows that hydrogen produced at a central production plant may be transported to the distribution terminal via transmission pipelines, where it is liquefied and loaded into liquid tankers. In the case of semi-central production, hydrogen is produced and liquefied at the distribution terminal. The daily capacity of the liquefier equals or exceeds average daily demand. Cryogenic storage tanks usually hold 5–7 days of liquefier production. A pump draws liquid hydrogen from the cryogenic storage tank to the liquid tankers. The liquid tanker is transported to the refueling station where it is emptied into a cryogenic storage tank. The liquid hydrogen at the refueling storage tank is used to refuel vehicles as described in the Section 35.3.2 above.

35.5.2.2 Cost Estimates
Hydrogen exists in liquid state below 20 K at atmospheric pressure. Achieving such low temperatures is energy intensive and expensive. The liquefaction

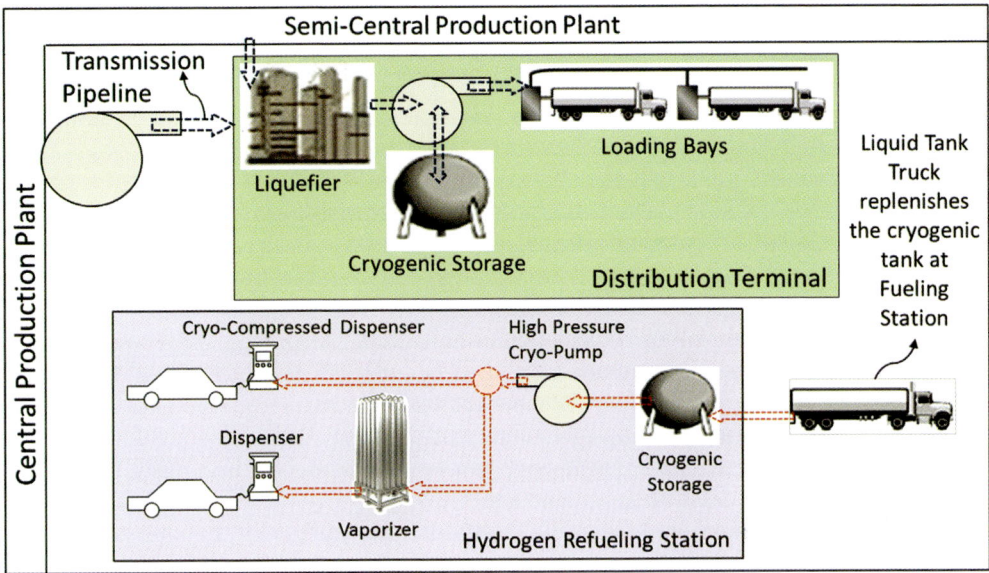

Figure 35.13 Schematic of the liquid delivery pathway.

energy represents about $\frac{1}{3}$ of the total energy content of the liquefied hydrogen. The liquefaction process accounts for more than $1 per kg of hydrogen cost [6]. The liquid hydrogen needs to be stored at 20 K in vacuum jacketed stainless steel tanks. Heat leak through the tank walls results in vaporization of liquid hydrogen (or boil-off). The boil-off can be recovered or vented to the atmosphere to avoid pressure buildup in the storage tank. The boil-off rate from the tank can be minimized with a high volume-to-surface ratio. Most stationary liquid hydrogen tanks are spherical in shape to minimize the boil-off rate. The boil-off rate varies from 0.4% to 0.06% per day for $50\,m^3$ and $20\,000\,m^3$ volume tanks, respectively [6]. Additionally, there is significant boil-off loss during unloading at the refueling station (pumping hydrogen from liquid tanker to on-site cryogenic tank). To minimize these boil-off losses, the number of deliveries should be limited and the delivery route should be planned accordingly.

Liquefying hydrogen increases its volumetric mass and energy densities many fold. The capacity of the liquid tanker is about 4 metric tons, which is 5–6 times the capacity of a composite tube-trailer and 15–20 times the capacity of a steel tube-trailer. DOE's 2011 estimate of the cost of liquid hydrogen tankers [9] are shown in Table 35.5. Notably, liquid tanker technology is considered mature and hence DOE sets no 2020 targets.

35.5.2.3 Challenges

- **Boil-off losses:** Boil-off is unavoidable and methods must be developed to either minimize it or capture the energy and utilize it elsewhere. Boil-off

Table 35.5 Estimated cost and key features of liquid hydrogen tankers [9].

Liquid tanker[a]	2011 Status
Operating pressure (bar)	5
Capacity (kg)	4300
Capital cost (2007$)	720 K

a) Cryogenic tank and trailer; excludes truck cab.

from pumps unloading fuel from the tanker to site storage tanks can account for an up to 5% loss (by volume) in the amount transferred [6]. Such losses should be minimized by more efficient component designs and improvements such as submerged pistons.

- **Underground storage:** Underground storage minimizes setback distances and is the preferred option. However, its high cost may offset the savings associated with lower land area.
- **Liquefaction:** During the liquefaction of hydrogen, it is cooled to 20 K in a multi-stage process, which is energy intensive. The exothermic conversion of the ortho form of hydrogen into the para form at low temperatures consumes a significant amount of energy as well. The total energy consumption to liquefy gaseous hydrogen from atmospheric conditions is approximately 12–15 kWh kg_{H2}^{-1} [6]. Technologies like magnetic or acoustic liquefaction may reduce the energy required and need to be investigated. Other storage options like high-pressure cryo-compressed storage tanks can reduce the energy demand, by avoiding the exothermic conversion of ortho into para hydrogen by allowing hydrogen storage at temperatures between 80 and 200 K [6].

35.6
Power-to-Gas Mass Energy Solution (Methanation)

35.6.1
Hydrogen Methanation Process

Hydrogen methanation, a well-known process that produces synthetic or substitute natural gas (SNG) by reacting CO_2 with H_2 over a metal catalyst, was originally reported by Sabatier in 1902. The reaction is reversible and exothermic:

$$CO_2 + 4H_2 \leftrightarrow CH_4 + 2H_2O \tag{36.1}$$

As shown in Figure 35.14, methane yield has been shown to peak in the vicinity of 350–400 °C in experiments using a Ru–TiO_2 catalyst [16]. Thus, the heat of reaction can be used for other processes.

Hydrogen methanation has been proposed as a means of reducing CO_2 emissions from fossil power plants. However, since most hydrogen is

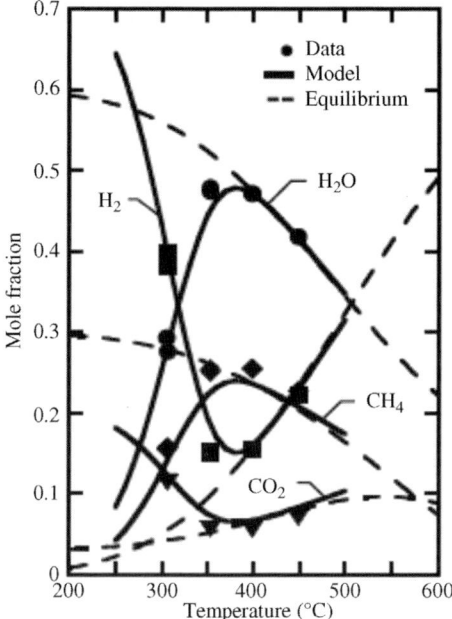

Figure 35.14 Gas composition as a function of temperature in hydrogen methanation.

produced by steam methane reforming (essentially the reverse reaction), using higher-valued hydrogen to produce natural gas only makes economic and environmental sense if the SNG product is superior to the commodity natural gas – having lower greenhouse gases (GHGs), for example – and thus able to command a higher price. For this reason, methanation projects typically propose using non-fossil sources for the input hydrogen and selling the SNG, a renewable fuel, in locations where it is highly valued. With its *Energiewende* program[1] and strong renewable energy incentives, Germany is home to most of the methanation or power-to-gas (P2G) projects under development today.

35.6.2
Current Applications

Methanation can produce SNG for power generation, energy storage, use as a chemical feedstock, or a fuel. As shown in Figure 35.15, methanation is a major component of the P2G value chain.

Several pilot and demonstration projects implementing portions of the P2G value chain are in various stages of development. The Center for Solar Energy and Hydrogen Research Baden-Wuerttemberg (ZSW), in partnership with ETO-GAS GmbH and the Fraunhofer Institute for Wind Energy and Energy System Technology (IWES), began operating a 25 kilowatt power-to-gas pilot plant in

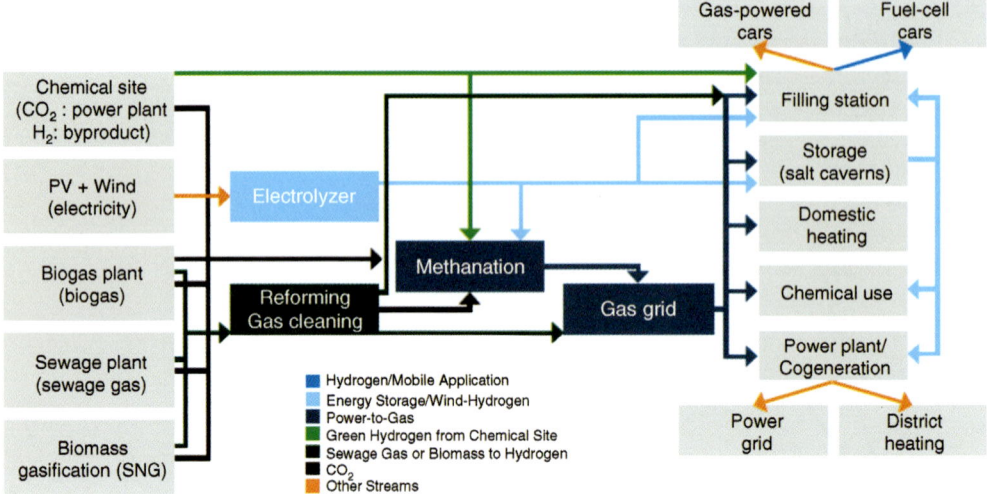

Figure 35.15 Power-to-gas value chain [17].

2009 and a 250 kW demonstration plant in 2012. The latter currently produces SNG for use by the ZSW research institute. Etogas has announced plans to begin series production of 1–20 MW modular commercial systems in 2015 [17].

Greenpeace Energy eG, the first German energy supplier to provide its customers with "green hydrogen" produced from electrolysis using wind power, is another participant in the P2G value chain. Their "proWindgas" product can be used for conventional heating and cooking or as a fuel for vehicles. Since October 2011, customers connected to the gas network have been able to take advantage of a "proWindgas" tariff which includes a 0.4 eurocent per kWh premium that is reinvested in further development of the wind–gas technology. Through its Planet energy subsidiary, Greenpeace Energy currently has eleven wind farms and three photovoltaic plants totaling 65 MW in operation. In the future, Greenpeace plans to build standardized electrolyzers, either along the gas grid or near wind farms, to further increase wind gas production [17].

In June 2013, Audi opened a new 6 MW power-to-gas plant in Wertle, Germany. Chemically binding some 2800 metric tons of CO_2 produced in an adjacent biogas plant with hydrogen produced from electrolysis using surplus electricity from off-shore wind turbines, the project is an integrated effort by Audi, ETOGAS GmbH, MT-BioMethan GmbH, and the energy company EWE AG. In addition to increasing energy efficiency by utilizing waste CO_2 from biogas production, the project also uses waste heat from methanation as process energy in biogas cleanup. The plant's output, known as Audi e-gas, is distributed to CNG fueling stations via the existing German natural gas network and annual output reportedly will power the equivalent of 1500 Audi A3 Sportback g-tron vehicles [18].

35.6.3
Challenges and Opportunities

By linking power and gas grids, P2G could be a mass solution to the intermittency of solar and wind power, leveraging some of the gas grid's extensive storage capacity in return for "greening" it with renewable power. Unlike the electric grid, which relies on a combination of generating reserves and short-term storage to balance hourly (or at most diurnal) supply and demand fluctuations, the natural gas network contains multiple sources of long-term storage to deal with major swings in seasonal demand. In the USA, some 400 storage facilities have 4.0 trillion cubic feet (Tcf) of working gas storage capacity, equivalent to nearly 20% of natural gas consumption [19]. In Germany, 48 underground storage facilities have a working capacity of 21.3 billion m^3 (752.2 bcf) of natural gas, equivalent to over 20% of natural gas use [20]. In the USA, much of this storage capacity is in depleted oil and gas wells and aquifers which can be cycled (i.e., filled and withdrawn) 1–2 times per year. On the other hand, most of Germany's geologic storage is in salt caverns which can be cycled multiple times per year. Salt caverns can hold significant quantities of SNG (in addition to fossil natural gas) and provide a means of utilizing excess renewable power from intermittent sources.

Owners of major wind assets are interested in storage solutions to prevent curtailment and increase available capacity. As shown in Figure 35.16, batteries and compressed air can be good choices for diurnal needs. But for multi-day and seasonal needs, pumped hydro, hydrogen, and SNG provide both longer discharge times and greater capacity [21]. While pumped hydro is an option limited to

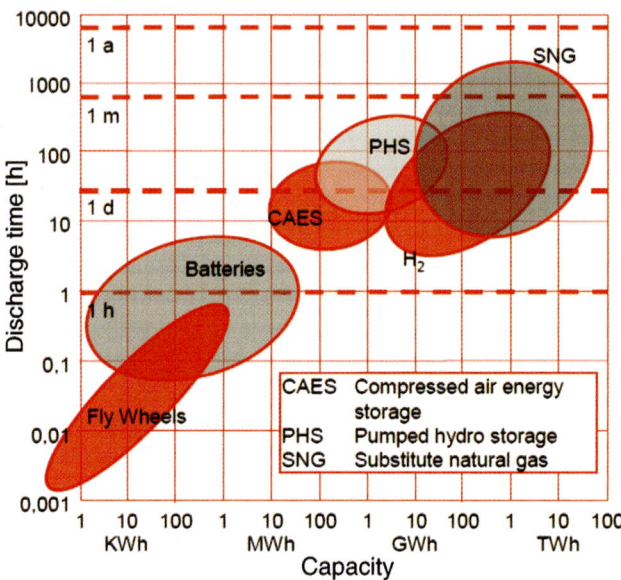

Figure 35.16 Capacity and discharge time of power storage technologies.

certain geographies, sites for hydrogen and SNG are more abundant. Note that the data in Figure 35.16 are plotted on a log-log scale. Per unit volume, hydrogen has ten times the energy density of compressed air storage; SNG has approximately three times the volumetric energy density of hydrogen.

As discussed in Sections 35.2, 36.5.1.1, and 35.5.1.3, storage is a critical element at several stages in the hydrogen supply chain, balancing hourly, daily, and seasonal fluctuations in demand. Unlike P2G, where SNG (or potentially "green hydrogen") is proposed to be stored in underground caverns in the vicinity of production or supply locations, geologic storage of hydrogen for use in HFCEVs is envisioned to be in the vicinity of demand locations. Such storage may not even need to be in geologic formations since a large network of hydrogen refueling stations could provide distributed capacity to store and utilize excess wind and solar energy at low demand periods with small on-site electrolyzers. At the project level, the location of geologic storage facilities will depend on local geology, which may be preferable near the supply source, near demand, or at some intermediate location between the two. Note that if hydrogen were stored closer to production locations, additional transmission infrastructure would be needed to bring it to refueling stations. In the long term, if a robust hydrogen infrastructure network were available, that would not be a problem. In the short-term, however, it would increase the cost of hydrogen delivery.

P2G is a means for utilizing excess wind and solar power during low electricity demand periods. Without the methanation step, "green hydrogen" could be injected into natural gas pipelines in volumetric percentages of up to 5%. Germany is exploring various aspects of doing so at scale. However, given North America's current over-supply of natural gas (and consequent depressed prices), combining high-priced hydrogen with low-priced natural gas is unlikely to be economic. Since low-cost hydrogen separation technologies do not yet exist, the stored hydrogen would likely be consumed along with the commodity natural gas, at significant loss to the producer. A more promising option would be to produce renewable methane where it can qualify for incentives to help it compete with low-priced natural gas, along with a small quantity of renewable hydrogen for use by early HFCEVs.

Finally, P2G is a potential solution to the problem of surplus generating reserves. If electricity can be stored in the megawatt range, generating reserves (particularly excess base- and medium-load power plants) can be reduced with consequent savings. P2G can also displace imported natural gas, an energy security issue outside North America, and may provide additional price competition for imported supplies.

35.7
Outlook and Summary

The hydrogen delivery infrastructure is expected to evolve over time, keeping pace with growth in the market penetration of HFCEVs and fuel cells in applications like stationary power and material handling equipment. During the initial stages of

HFCEV deployment, when existing hydrogen production plants may be at some distance from demand locations, it is envisioned that hydrogen will be produced in small quantities at the point of demand. Such distributed production [6] would not require a hydrogen delivery infrastructure. However, distributed production is only practical for supplying hydrogen in relatively small quantities. Distributed production lacks the economies of scale of centralized production facilities but offsets higher production cost by eliminating delivery cost (i.e., the cost incurred to package and transport hydrogen from the production facility to the dispensing nozzle). The delivery cost of hydrogen accounts for about half of total hydrogen cost [4] and must be minimized if hydrogen is to be competitive with petroleum derivatives.

Delivering hydrogen through a pipeline network is economical only when robust demand can justify the major capital investment. The labor for laying and welding steel pipeline sections is costly and can be reduced with the use of innovative joining techniques. Hydrogen embrittlement of steel pipes, especially due to pressure cycles, needs to be thoroughly investigated, as does the addition of suitable odorants to service pipelines to enable hydrogen leak detection. Delivering hydrogen via steel tube-trailers provides limited transport capacity (below 300 kg of H_2) due to weight restrictions imposed by regulatory authorities. Tube construction from light-weight materials such as carbon fiber composites can allow payloads exceeding 1 metric ton of hydrogen when charged at a pressure of 500 bar or above. Delivery of liquid hydrogen via tankers allows payloads that exceed 4 metric tons but requires an energy intensive liquefaction process that consumes significant amounts of electricity (approximately 12 kWh per kg of hydrogen). The carbon intensity of the electric grid may result in high greenhouse gas emissions for the liquid hydrogen pathway, thus negating the environmental benefit of using hydrogen compared to gasoline on a lifecycle basis. More efficient liquefaction technologies are needed to attain the cost benefits of delivering hydrogen in liquid form and the environmental benefits of using it in fuel cell vehicles.

Excess electric power that would otherwise be curtailed at times of high generation and low demand can be effectively utilized in producing hydrogen via electrolysis. Methanation of renewable hydrogen with concentrated CO_2 streams from other applications can prove vital to the economic viability of large-scale deployment of wind and solar power generation in various regions of the world. Methanation of hydrogen enhances the volumetric energy density threefold, which allows more effective utilization of energy storage in caverns. As the hydrogen fuel cell market develops, more hydrogen can be generated and stored onsite at points of use, providing additional options to store and utilize the renewable energy during low demand periods.

Note

[1] *Energiewende* is the term used to describe the set of policies, programs and projects intended to produce a transition to a sustainable economy by means of renewable energy, energy efficiency and sustainable development.

References

1 Garvey, M.D. (2011) The Hydrogen Report, CryoGas International, vol. 21(2), 32.
2 Bromaghim, G., Gibeault, K., Serfass, J., and Wagner, E. (2010) Hydrogen and Fuel Cells: The U.S. Market Report, National Hydrogen Association. Available at http://www.ttcorp.com/pdf/marketReport.pdf (accessed 27 January 2014).
3 U.S. Department of Energy, Fuel Cell Technology Office (2015) Hydrogen Delivery, in Multi-Year Research, Development and Demonstration Plan, Available at http://energy.gov/sites/prod/files/2015/08/f25/fcto_myrdd_delivery.pdf (accessed 13 August 2015).
4 Mintz, M., Elgowainy, A., and Gardiner, M. (2009) Rethinking Hydrogen Fueling: Insights from Delivery Modeling. Transportation Research Board of the National Academies, Transportation Research Record (2139).
5 Elgowainy, A. and Wang, M. (2011) Well-to-wheel analysis of sustainable vehicle fuels, in *Encyclopedia of Sustainability Science and Technology* (editor in chief R.A. Meyers), Springer, pp. 10502–10529.
6 US DRIVE Partnership (2013) Hydrogen Delivery Technical Team Roadmap. Available at http://www1.eere.energy.gov/vehiclesandfuels/pdfs/program/hdtt_roadmap_june2013.pdf (accessed 27 January 2014).
7 Society of Automotive Engineers (SAE) (2010) SAE International surface vehicle technical information report: Fueling Protocols for Light Duty Gaseous Hydrogen Surface Vehicles, SAE J2601, SAE International.
8 Brown, D., Cabe, J., and Stout, T. (2011) National lab uses OGJ data to develop cost equations. *Oil Gas J.*, **109** (1), 108.
9 Argonne National Lab, National Renewable Energy Laboratory and Pacific Northwest National Laboratory (2010) Hydrogen Delivery Scenario Analysis Model. 2.3 edn, Fuel Cell Technology Office, DOE.
10 Elgowainy, A., Mintz, M., and Brown, D., 2011 DOE Hydrogen Program Review: Hydrogen Delivery Infrastructure Analysis (10 May 2011). Available at http://www.hydrogen.energy.gov/pdfs/review11/pd014_mintz_2011_o.pdf (accessed 27 January 2014).
11 Rawls, G. and Adams, T., 05/15/2013, 2013 DOE Hydrogen Program Review: Fiber Reinforced Composite Pipelines. Available at http://www.hydrogen.energy.gov/pdfs/review13/pd022_adams_2013_o.pdf (accessed 27 January 2014).
12 U.S. Department of Transportation, Federal Highway Administration (1994) Federal Size Regulations for Commercial Motor Vehicles. Code of Federal Regulations (CFR), 23 CFR Part 658.
13 An International Code (2010) ASME Boiler & Pressure Vessel Code Section VIII Rules for Construction of Pressure Vessels – Division 3: Alternative Rules for Construction of High Pressure Vessels, ASME, 1 July 2011.
14 Baldwin, D. and Newhouse, N., (2011) Development of High Pressure Hydrogen Storage Tank for Storage and Gaseous Truck Delivery (10 May 2011). Available at http://www.hydrogen.energy.gov/pdfs/review11/pd021_baldwin_2011_o.pdf (accessed 27 January 2014).
15 Elgowainy, A., Reddi, K., Mintz, M., and Brown, D. (2013) 2013 DOE Hydrogen Program Review: Hydrogen Delivery Infrastructure Analysis (15 May 2013). Available at http://www.hydrogen.energy.gov/pdfs/review13/pd014_mintz_2013_o.pdf (accessed 27 January 2014).
16 Brooks, K., Hu, J., Zhu, H., and Kee, R. (2007) Methanation of carbon dioxide by hydrogen reduction using the Sabatier process in microchannel reactors. *Chem. Eng. Sci.*, **62** (4), 1161–1170.
17 Germany Trade & Invest (GTAI) (May 2012) Green Hydrogen and Power-to-Gas Technology. http://www.gtai.de/GTAI/Content/EN/Invest/_SharedDocs/Downloads/GTAI/Fact-sheets/Energy-environmental/fact-sheet-green-hydrogen-mass-energy-storage-for-future.pdf (accessed 27 January 2014).
18 Meza, E. (27 June 2013) Audi Opens 6 MW Power-to-Gas Facility. http://www.pv-magazine.com/news/details/beitrag/

audi-opens-6-mw-power-to-gas-facility_100011859/#axzz3Y7NFvaDP (accessed 27 January 2014).

19 U.S. Department of Energy, Energy Information Administration (EIA), Underground Natural Gas Storage (2XXX). http://www.eia.gov/pub/oil_gas/natural_gas/analysis_publications/ngpipeline/undrgrnd_storage.html (accessed 27 January 2014).

20 German Association of Energy and Water Industries (Bundesverband der Energie und Wasserwirtschaft – BDEW) (2XXX) Basic Data on Electricity, Gas and Heat. http://www.bdew.de/internet.nsf/id/99771DF985891796C1257A76003F897D (accessed 27 January 2014).

21 Linke, G., Steiner, K., Hoppe, M., and Mlaker, H. (2011) Power Storage in Smart Natural Gas Grids: Fiction or Fact? Presented at International Gas Union Research Conference, Seoul, 19–21 October 2011.

36
Pipelines for Hydrogen Distribution

Sabine Sievers[1] and Dennis Krieg[2]

36.1
Introduction

Pipelines in comparison to other distribution technologies offer several advantages. They are for instance by far less energy consuming per unit transported than trucks. In comparison to on-site production plants centralized production with pipeline distribution can benefit from improved economies of scope, that is, increased efficiency. All these aspects have a beneficial impact on the environment since less energy consumption reduces the related emissions. A second main advantage is the safety issue. In comparison to trucks pipelines only very seldom suffer from accidents and if they do then, normally, only very few people are impacted. Therefore pipelines are the most efficient and safest way to distribute hydrogen from the source to the sink.

36.2
Overview

36.2.1
Pipelines in Comparison to Other Transportation Possibilities

Hydrogen as a chemical product is used for several applications, for example, cooling within power plants, semiconductor production, refineries, and so on. Depending on the application different qualities and quantities are required. Especially the latter has a major influence on how the hydrogen is provided. Several possibilities exist (Table 36.1).

The minimum an industrial customer can buy is roughly 1 Nm^3. Smaller sizes, for example, for experiments in schools, are possible but rather seldom. Depending on the demand and the prevailing technology at the customer's facility, bottles or bulk distribution are possible for small to mid-size customers. If a very high quality is needed the hydrogen can be distributed in the liquid state. This guarantees a quality of 7.0 as well as a reasonable footprint. The density of liquid hydrogen is

Hydrogen Science and Engineering: Materials, Processes, Systems and Technology, First Edition.
Edited by Detlef Stolten and Bernd Emonts.
© 2016 Wiley-VCH Verlag GmbH & Co. KGaA. Published 2016 by Wiley-VCH Verlag GmbH & Co. KGaA.

Table 36.1 Comparison of existing supply possibilities for hydrogen.

Delivery		Quantity	Remarks
Trailer	Cylinder	1–10 Nm^3 per bottle	5 l, 10 l, and 50 l bottles available @ 200–300 bar
	Bulk (gaseous)	3000–5000 Nm^3 per truck	"Standard" for medium industrial customers
	Bulk (liquid)	20 000–30 000 Nm^3 per truck	Advantageous if very high quality, that is, 7.0, is needed, for example, for semiconductor production
On-site	Small	100–2000 Nm^3 h^{-1}	On-site: located at the customers plant; Over-the-fence: located close to but legally separated from the customer
	Large	1000 to >100 000 Nm^3 h^{-1}	
Pipelines	Gaseous	>500–1000 Nm^3 h^{-1}	>99% of the hydrogen transported in pipelines [1]; conjunction depends on the distance to the source and the throughput
	Liquid	—	Only one "pipeline" (500 m) in the world – rocket fuel for NASA

roughly five times higher than in gaseous state at 200 bar. Alternatively, pipelines can be used for the distribution. Since this requires a large investment, only big customers will be connected to a grid. Finally, the hydrogen can be produced on-site. Although not perfectly, it is possible to differentiate between small and big customers. The first are those who mainly want to be independent of truck transport and/or have access to cheap energy, for example, natural gas or electricity. Very big customers, for example, refineries or chemical plants, might choose to be connected to a pipeline grid if available or have a production plant on-site. Two different operating models exist. The first is the well-known on-site production. In this case customers normally introduce the hydrogen production into their process/facility. The second model is called "over-the-fence." The production plant is still close to the customer but a main legal difference exits. In the first case the plant is the property of the customer and contracts can provide operation and maintenance. The supplier is not responsible for anything outside of the scope of the contract, for example, increase of the price for the energy. In the second case the supplier will be paid at a fixed price for the hydrogen provided. This means that the supplier alone bares the risks of operation and maintenance.

36.2.2
An Overview of Existing Hydrogen Pipelines

Hydrogen is mainly produced and used in the USA, Europe, and Japan. The length of the pipeline grid, depending on the definition, in the USA is about 700–1300 km

36.2 Overview

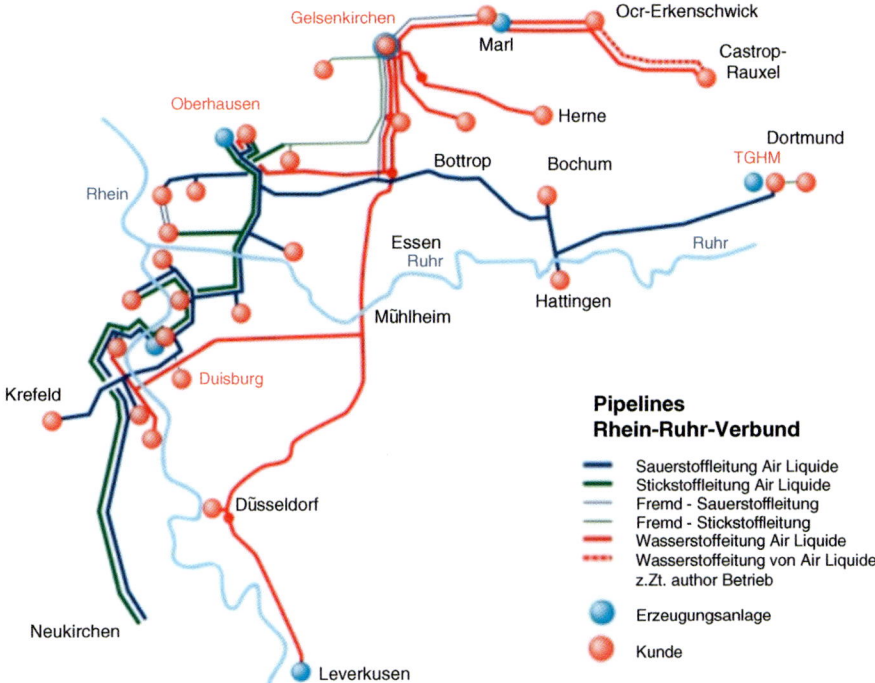

Figure 36.1 Pipeline grid of Air Liquide in the Rheine-Ruhr area. Internal communication of Air Liquide Deutschland GmbH.

and in Europe 1100–1800 km [2,3]. In the USA the main share is located at the Gulf Coast in order to supply hydrogen to the refineries and chemical plants. In Europe there are mainly three different hydrogen pipeline grids. The biggest is located in France, Belgium, and the Netherlands and is over 1000 km long. The other two are located in Germany and are 240 km (Figure 36.1) and 100 km long.

Most of the currently existing pipelines were not built and primarily used for hydrogen but for other chemicals, for example, ethylene. With growing demand new pipelines were taken into operation while old ones were refitted in order to be operated with hydrogen. Especially, fittings and sealings had to be replaced. Unlike ethylene, hydrogen is transported in dry mode, mainly due to material concerns. That caused accelerated degradation of the existing sealings and therefore might have caused trouble in terms of safety related issues. After the refitting of concerned parts no problems were reported.

36.2.3
Some Material Concerns

Hydrogen can penetrate conventional pipeline materials and then lead to loss of material strength. Existing pipelines therefore can suffer from accelerated

Table 36.2 Sources of hydrogen embrittlement and related effects [7].

Source	Effect	Remarks
Embrittlement due to chemical hydrogen reaction	Hydride formation	Materials absorb hydrogen, which leads to severe embrittlement, for example, Ti, Zr, V, Mg
	Hydrogen attack	Decarburization at elevated temperatures, that is, above 200 °C.
Internal hydrogen embrittlement due to metallic–physical processes	Hydrogen induced cracking	Chemical inter-reaction of material and water. Formation of metal oxides and hydrogen
	Blistering	Atomic hydrogen penetrates the material, gathers in material defects (inclusion, etc.) and finally recombines into molecular state, which causes an elevated internal pressure
Embrittlement due to environmental hydrogen	Hydrogen induced stress corrosion cracking	Combination of material exposed to a hydrogen atmosphere and external load

degradation and increased crack growth if operated with hydrogen. This phenomenon is called "hydrogen embrittlement." Although this expression is commonly used, it contains some very different physical effects that have rather little in common, except being related to hydrogen. Therefore, very few general statements about hydrogen embrittlement are possible. It is also not possible to conclude whether a material suffers from accelerated degradation in a hydrogen atmosphere only based on permeation data [4]. Hydrogen embrittlement is only of importance for metallic materials. Polymers are not impacted [5,6]. Table 36.2 shows the sources of hydrogen embrittlement and the effects related to them. The remarks give a brief explanation of the effect.

Nowadays pipelines are usually operated at fairly low pressure levels, that is, 20–30 bar, with very little variation, normally less than 10% of the operating pressure and only rather slowly, that is, within a few hours. This conservative conduct so far has allowed a safe operation. If hydrogen is to succeed as an energy carrier, for example, for fuel cell cars and in households, a new type of pipeline will be needed. These future pipelines will be operated at higher pressure levels, for example, 100 bar, and with much higher variation, since the pipeline will also serve as an intermediate energy storage system. Therefore future planning must take into consideration that negative impacts are much more likely to occur. To avoid negative impacts of hydrogen on a (future) pipeline, several options can be feasible, depending on the application and the requirements.

Table 36.3 provides a brief overview of potential possible ways hydrogen embrittlement might be avoided. Notably, it is not yet clear which process is the

Table 36.3 Possible ways to avoid hydrogen embrittlement in future pipelines [7].

Possibility	Virtue	Downside
Non-threatened material	Simple and safe	Expensive
Pipe-in-pipe (cladding)	Combination of mechanical stability of the outer pipe and chemical stability of the inner pipe	Longevity and production/welding not yet fully clarified for hydrogen
Over-dimensioning	Simple	Potentially expensive; threshold not clear
Coating	Combination of mechanical stability of the outer pipe and chemical stability of the coating	Best coating material and welding process not yet clear for hydrogen
Inhibitors	No modification of the pipeline needed	Purification of the hydrogen needed; threshold unclear

most promising one. The use of non-threatened materials, for example, Polyamide (PA), for instance is safe but also very expensive. PA gravimetrically costs ten times as much as steel and has a material strength that is roughly one order of magnitude lower. Therefore, far more material is needed, which is excessively expensive. Other possibilities like cladding, over-dimensioning, or coating have not yet been tested sufficiently, which makes it impossible to conclude whether, and to what extent, these are feasible options. Finally, the inhibition of hydrogen is also possible. In this case an added or preliminarily not-removed gas, like CO or O_2, will cover the surface and therefore prevent hydrogen entering the material. Unfortunately, it is not clear what the threshold, that is, the needed amount of impurities, for accelerated crack growing is. For fuel cell applications it is further more mandatory to purify the hydrogen in advance. Fuel cells can only withstand very small traces of impurities, for example, 5 ppm O_2 [8,9].

36.3
Brief Summary of Pipeline Construction

36.3.1
Planning and Approval

The first and most challenging step in the construction of a pipeline is the pipeline route. This must be suitable for obtaining approval, on the one hand, and economical on the other. Since a pipeline is only laid if a certain minimum viable scale exists, production and consumption locations are very precisely defined for the most part. Connecting these locations seems to be relatively uncomplicated in the first instance, but is subject to stringent legal, ecological, and economic criteria. These are considered in greater detail below.

The first important criterion is the type of pipeline. If it is a main supply line covering several hundred kilometers, for example, Hamburg–Frankfurt–Munich, it can be laid primarily in regions used for agriculture and forestry outside cities. If it is used to supply industrial zones in conurbations such as, for example, the Ruhr district, however, the outlay and the number of possible alternative routes on account of the space available usually increase significantly. Taking the infrastructure already in existence into consideration is a key factor in planning. Depending on the length and diameter, the project is subject to the UVPG (Gesetz über die Umweltverträglichkeitsprüfung – Environmental Impacts Assessment Act) in Germany.

Annex 1 of the UVPG lists the projects that fall within its scope. For the construction of a H2 pipe, no. 19.5 applies (unless the H2 pipe is subject to the Energy Act). Above a certain size, that is, DN>800 and more than 40 km, an environmental impact assessment (EIA) must always be carried out. With diameters of DN 300 or greater than DN 300, either general or location-related preliminary screening of the individual case is required, depending on the length, to determine whether the obligation to carry out an EIA exists. If the obligation to carry out an EIA exists, the project requires full planning permission; if no EIA obligation exists, the project requires planning approval. Planning approval is not required in cases of little importance. The EIA entails determining, describing, and evaluating the direct and indirect effects of a project on, for example, people, animals, plants, soil, air, climate, cultural assets, and interaction between the protected objects. Full planning permission has a collective effect with regard to all permits under public law. This means that the project is approved as a whole and no further applications have to be submitted. Full planning permission is usually the norm in the case of natural gas pipeline routes. If no full planning permission/planning approval is required, all public permits must be obtained by the client. In addition to the permits under public law, permission must be obtained under private law.

36.3.2
Assessment of Preferred Pipeline Routes and Alternative Routes

The selection of a potential preferred route and alternative routes takes account of different, sometimes competing aspects. These are, for example:

- Effects on people and the environment;
- use/laying in parallel to existing lines;
- transport routes;
- encroachments on the natural environment and landscape linked to the construction itself;
- economic efficiency.

If possible, the new route to be constructed should be oriented to existing energy pipeline routes and transport routes for both licensing and economic reasons. Notably, in this regard, in the case of federal roads, for example, a minimum distance of 20 m from the outer edge of the asphalted carriageway must be observed to facilitate subsequent development. Furthermore, a distance of 100 m should be aimed for in the case of federal motorways. Exceptions to this necessitate approval. Laying in parallel can make good sense, to combine routes. This is advantageous in particular in the case of longer projects. When high-voltage lines and electrified rail lines run in parallel, however, the influence of the electromagnetic field on the pipeline must be taken into consideration. To avoid negative effects on the active cathodic protection, action must be taken with regard to alternating current corrosion, for example, by earthing the pipeline or the jacket pipes up to active compensation.

Laying in service roads, roads, and so on results in higher construction costs for the most part. Less space is available and the civil engineering works must be carried out using manual digging in some cases to avoid damage to the existing infrastructure. In certain circumstances, however, such a route can offer clear advantages with regard to licensing. Depending on the region, this may even be the more cost-effective variant on the whole, because no compensation for agricultural land use is required.

Compensation consists of two elements:

1) One-off compensation for the entry in the land register; approx. 15–25% of the current market value.
2) Compensation for the loss of harvest and yield reduction in subsequent years.

The first element is a standard value and is always taken directly into account in planning. Since the prices for areas of agricultural/forestry land do not normally fluctuate significantly in a region, this element is not a basic determining factor for route mapping. The situation is different with regard to the second element, so-called compensation for agricultural land use. This sum compensates the farmer for lost crop yields. This compensation can vary substantially depending on the crop. At the least, the working strip required for the construction of the pipeline must always be compensated for. Depending on the crop, additional factors such as remaining areas, reduced yields for early harvesting, and so on come into play. A complete evaluation is only possible following completion of the construction works. Experience in this field shows that it is difficult to estimate a budget. It nearly always turns out to be higher, or in rare cases it is lower. While compensation is only given for the remaining areas for a maximum of two years (year of construction and following year), payments for the working strip are payable for longer:

- Year 1: 100%
- year 2: 50%
- year 3: 30%
- year 4: 10%.

Compensation levels differ substantially depending on the crop, and this has to be taken into account in the planning (Table 36.4). If a pipeline crosses a tree nursery, for example, for 100 m, with a working strip of just 6 m, costs of

Table 36.4 Land use compensation for different crop types at 100%.

Crop	Compensation (€ m^{-2})	Remarks:
Wheat	0.16	
Potatoes	0.35	Availability of headland
Maize	0.31	
Asparagus	12.50	Per row, three crops, 10 yr duration
Rhubarb	17	
Kitchen herbs	6.70	
Chinese cabbage	2.70	
Silver onions	10.00	
Tree nurseries	100–200	Single payment, no flat-rate statement possible

€60 000 to €120 000 can arise just for the agricultural land use compensation, and are thus even higher than the complete construction costs of the pipeline itself. The acquisition of route rights and compensation for agricultural land use can take up a not insignificant share of the overall budget.

A so-called headland is necessary for the cultivation of certain types of crop. These include potatoes and sugar beet, for example. It is used for turning the harvesting machine when harvesting. A pipeline that intersects the field unfavorably, as shown in Figure 36.2, can thus necessitate 100% compensation for the entire field in the worst case, depending on the construction period.

In addition to the costs for licensing and approvals, costs for compensatory measures, that is, measures to compensate for intervention in the ecosystem, must be budgeted for.

Pipelines should be laid underground where possible. Stations and a few special points, such as crossing the course of a stream, constitute an exception. In the case of sections above the ground, the use of stainless steels can be advantageous compared with untreated steel pipelines. The higher investment costs can work out thanks to lower maintenance and repair costs.

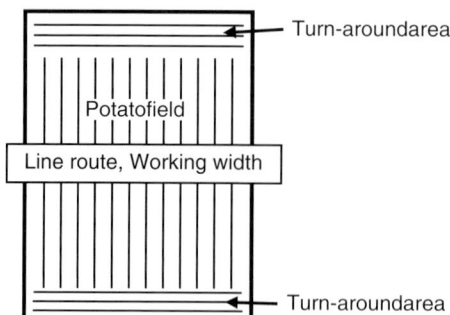

Figure 36.2 Potato field with headland. Internal communication of Air Liquide Deutschland GmbH.

Table 36.5 Schedule from start of planning to beginning of project execution.

Timescale	Step
2–5 yr	Procurement of route plans of other and supra-regional line operators
	Determining/recording the ecological sensitivity of a region
	Clarifying whether land-use plans, road modification measures, and so on are planned in the short or long term in the area of the "planned route"
	Groundwater and soil conditions
	Photographic documentation of the planned route
	Establishing possible locations for valve stations
	Ascertaining the owners; land registry documents
	Discussions with and visits to landscape custodians of the country with regard to assessing the impact. → Is the route feasible from the licensing viewpoint?
	Coordination of the course of the route and the location of the pressure reduction station on the premises of the potential customer
	In the case of "inner-city routes" and/or routes with difficult special points, visits to pipe construction and civil engineering companies in respect of technical feasibility and costs of planned route
	Project description, course of the route with photographic documentation
	"Advance discussions" with the authorities/associations
	Establishing all third-party lines, for example, sewers, telecommunications, electricity, water, gas, line operators, fiber optic cable routes. → Is the planned route technically feasible?
	Preliminary inquiry regarding material delivery times and prices to: – pipe supplier (up to 5 months delivery time) – fittings manufacturers (up to 6 months delivery time) – elbow supplier (up to 3 months delivery time)
	Planned route exists → Obtain unofficial quotation for construction
Start of project execution and construction	

The possible routes can be determined with reference to city plans, aerial photographs, and existing routes of other line operators. These are then finally evaluated with regard to their eligibility for licensing and from the viewpoint of construction costs and subsequent operating costs. Table 36.5 shows the necessary actions from the start of planning to the beginning of project execution and realization.

36.3.3
Project Execution, Construction and Commissioning

Once the planning is complete, project execution can commence. Various stages are gone through here. These are offset in time, but some also run in parallel:

- Final design of pipeline, that is, nominal width, pressure stage and choice of material, wall thickness;

- ordering of materials (pipes, elbows, fittings);
- production of documents in accordance with regulations (in Germany, for example: "Regulation on Pipelines") for commissioning/production of an expert opinion for the planned construction and operation of the pipeline;
- ordnance detection;
- production of documents for the responsible authority;
- production of an accompanying landscape conservation plan with regard to environmental impact;
- obtaining public law permits and approvals under private law (if no full planning permission or planning approval procedure is carried out);
- turnkey: putting the construction project out to tender and awarding the contract, taking account of all conditions from licenses and approvals.

The construction measures commence "locally" with the leasing and preparation of the construction and pipe storage sites. The number and size of these are naturally dictated by the size and length of the pipeline to be laid. The pipeline storage sites are always close to the route for logistical reasons, for example, every 2–4 km, and must be easily accessible to the delivery trucks. The individual pipes can be 16–18 m long. The construction depot and pipe storage areas can be procured by the client or by the contractor as part of the turnkey project.

As soon as the construction site has been set up, the equipment, pipes, and elbows are in situ and the route axis has been marked out, construction of the pipeline can begin. Figure 36.3 shows two photographs taken during construction work.

Even if good planning is extremely important for project realization, deviations from the master plan are to be expected in nearly every project (Table 36.6). These include unforeseeable obstacles, such as aerial bombs that have to be defused or realignments in a confined area. The list of things that can delay

(a)

(b)

Figure 36.3 Photographs of construction works: (a) excavator during a period of bad weather; (b) pipe trench with pipe string (DN 150) laid out in advance outside it. Internal communication of Air Liquide Deutschland GmbH.

Table 36.6 Schedule from start of project execution up to commissioning.

Timescale	Step
6–12 mo	Project planning, that is, design, licensing, approvals, and turnkey
	Pipeline and station construction
	Integration into monitoring and remote control
	Strength and leakage testing
	TÜV acceptance
	Time schedule coordinated with the customer and operating office with detailed plan for commissioning, including first filling
Commissioning of line and stations	

construction is long. In parallel to laying of the pipe and a data cable if necessary, stations including instrumentation and control technology are constructed and integrated into the telecontrol technology. Following completion of the construction work, which is monitored by an expert, the pipeline is subjected to strength and leakage testing (test pressure, e.g., 95% of yield strength; strength) and smart pigging if applicable. Acceptance by a recognized testing laboratory takes place prior to official commissioning. All planning and approval stages and all construction activities are documented extensively.

Relative to the time taken, it can be said that the planning and approval phase, without official planning permission, accounts for roughly $^3/_4$ of the overall time. In one Air Liquide project (Figure 36.4) two pipes (DN400) were laid with an overall length of approx. 50 km, 11 km of which was in an urban location. The pure construction time was approximately 8 months. The planning, including all the required coordination, licenses, and approvals, took 4.5 years to complete.

Figure 36.4 Pipe trench laid next to a federal road with two pipes DN 400. Internal communication of Air Liquide Deutschland GmbH.

36.4
Operation of an H_2 Pipeline

We often speak erroneously of the "operation" of a pipeline, but correctly speaking it should be termed at least as "operation, monitoring, maintenance, repair, troubleshooting." In addition to "operation and monitoring", a considerable number of activities/tasks arise in day-to-day business in relation to the pipeline (see Section 36.4.2 (Operations)). Compared with the investment the proportion of these costs is quite small and is currently roughly only 1–2% relative to the investment sum. Only a brief summary of "operation and monitoring" is given here. The following sectors can be broadly differentiated:

1) The control center
2) operations.

36.4.1
The Control Center

A pipeline must have at least one control center. This is where all the basic facilities for the safety of the pipeline, such as compressor stations, distribution, junction and transfer stations, pressure measuring points, and main shut-off elements are monitored. Figure 36.5 shows a couple of photographs of the construction of a small station. The basic facilities for the safety of the pipeline system must be capable of operation in sections or as a whole from the control center. The staff of the control center must be able to detect malfunctions at all times.

Figure 36.5 Station under construction. Internal communication of Air Liquide Deutschland GmbH.

According to the Pipeline Regulations, pipelines must be equipped with methods for detecting leaking substances. The information on leakage monitoring also goes to the control center, so that the necessary action, such as informing Operations/the emergency service up to shutting off the section affected and informing the authorities and customers, can be taken immediately. The control center is occupied around the clock, including during breaks in pumping.

As well as its original task of monitoring, the control center can also be responsible for "controlling" the pipeline. Control of the pipeline includes provision of the required hydrogen while adhering to the basic contractual conditions and operating the pipeline/pipeline network as efficiently as possible. For example, pressure, quality, and throughput must be ensured for the customer. At the same time, operation should be as efficient as possible from the point of view of energy, ecology, and economy. The more accurately the consumption profile of the customer can be anticipated, the simpler it is to organize operation. The differences in electricity prices on the stock exchange can be exploited and at the same time the installations can be run optimally energy-wise and, as a result, consume less power. Good planning is valuable, therefore, both economically and ecologically. Apart from controlling remotely operated valves and setting target values (pressure and quantity), the control center is not responsible for any operating tasks. All other work "on site" or on the route is carried out by Operations.

36.4.2
Operations

Apart from general documentation and management, Operations are responsible for all the necessary monitoring activities and measures according to the Technical Regulations on the pipeline "locally". These include:

- Regular route inspections, including emergency exercises;
- culvert soundings;
- recurring regular tests:
  - necessary maintenance/repair work;
  - monitoring of cathodic corrosion protection;
  - function testing of remotely operated valves and control valves;
  - calibration of quantity measurements/calibration with weights and measures authority;
  - filter testing;
  - testing of stations, external condition, tightness as well as electrical installations, lightning protection;
  - required commissioning and decommissioning;
  - planning and organization of pigging runs;
  - monitoring of third-party activities in the area of the pipeline;
  - documentation of monitoring measures.

36.5
Decommissioning/Dismantling/Reclassification

When a pipeline or part of a pipeline is finally decommissioned, the section concerned should always be partitioned off. To do this, it is emptied of hydrogen by flushing the pipeline with inert gas, for example, nitrogen, and the partial sections remaining are closed. Hazards to persons and the environment must be prevented. These measures are to be verified and certified by an expert.

Dismantling of the pipeline, similar to normal practice, for example, in power station construction, would mean encroaching on nature again as in construction. If the piece of pipe remaining in the ground might disrupt a further construction project, then the relevant partial section can be removed/dismantled without problems in the course of the construction work.

Backfilling of the decommissioned pipe can be carried out. This, though, must be examined in each individual case. It should be borne in mind in this regard, however, that backfilling is not reversible and thus another utilization, for example, as an empty cable pipe or as a jacket pipe, is explicitly ruled out.

Pipelines, even hydrogen pipelines, normally have a very long service life and can be used for several generations. Some sections in the Ruhr district, for example, have been in operation since the 1930s.

To ensure this, however, some things have to be considered:

- adherence to design parameters, especially with regard to the transport/pressure dynamics;
- selection of a suitable steel material;
- welding additive material;
- taking account of hydrogen-specific properties/requirements;
- dew point of at least approximately $-35\,°C$ to avoid internal corrosion;
- proper servicing and operation.

36.6
Conclusion

Hydrogen pipelines are considered the most reliable, efficient, energy saving, safe, economic, and environment saving way of transporting and distributing hydrogen at high penetration rates. In the long run pipelines for hydrogen will prevail as they do today for natural gas. There is much experience of such transportation. Nevertheless the transition of hydrogen as a chemical product to an energy carrier will impose some future challenges. These are more cost than technically driven, for example, initial investment and time taken for the permission and construction. Future projects will have to clarify the most efficient way of constructing and operating hydrogen pipelines but even more importantly they will need to standardize the permission and engineering process to avoid permission processes that take too long. The broad penetration and success of hydrogen for fuel cell cars and stationary applications is very much linked to the distribution infrastructure and in this respect pipelines are a key factor for the overall success.

References

1 Energy Information Administration (2008) The Impact of Increased Use of Hydrogen on Petroleum Consumption and Carbon Dioxide Emissions. www.eia.gov/analysis/requests/2008/sroiaf(2008)04.pdf (accessed 21 January 2013).

2 Mohitpour, M., Solanky, H., and Vinjamuri, G. (2004) Material selection and performance criteria for hydrogen pipeline transmission. *ASME/JSME Pressure Vessel & Piping Conference, 25–29 July 2004, San Diego, CA*, ASME, pp. 241–251. ISBN: 0-7918-4670-0.

3 Amos, W. (2007) Costs of Storing and Transporting Hydrogen, NREL/TP-570-25106, National Renewable Energy Laboratory, Golden, CO.

4 Rieke, E. and Grabke, H. (1994) Einfluss der Mikrostruktur von Stählen auf die Wasserstoffabsorption und die wasserstoffinduzierte Rissbildung. Technische Forschung Stahl – Abschlussbericht (1 July 1989 – 31 December 1993), Forschungsvertrag Nr. 7210-KE/122.

5 ISO (2004) (E): Basic considerations for the safety of hydrogen systems, ISO/TR 15916.

6 Azkarate, I. (2010) Materials and hydrogen; risk assessment and management of strategic energy technologies. Enero 3rd Scientific Workshop – 5 March 2010, Brussels.

7 Krieg, D. (2012) Konzept und Kosten eines Pipelinesystems zur Versorgung des deutschen Straßenverkehrs mit Wasserstoff. PhD-Thesis, RWTH Aachen and Forschungszentrum Jülich.

8 ISO (2008) Hydrogen fuel – Product specification – Part 2: PEM fuel cell applications for road vehicles, ISO/TS 14687-2.

9 Society of Automotive Engineers (2008) Information report on the development of a hydrogen quality guideline for fuel cells. SAE Surface Vehicle Information Report – J2719, pp. 1–14.

37
Refueling Station Layout
Patrick Schnell

37.1
Introduction

The finite nature of fossil fuels will bring about major structural changes in the energy and transport sector. Seen in a broad perspective, the use of fossil fuels must be reduced significantly and the energy supply has to be assured by the development of renewable, efficient, and low polluting energy systems. By 2012, the share of renewables in electricity generation in Germany already exceeded 20% [1]. Given Germany's commitment to the *Energiewende* (the transformation of national energy policy) and its decision to exit from nuclear energy by 2022, this share will rise significantly in the coming years. The German federal government's energy concept sets a target of 80% for the expansion of renewable energy by 2050 [2]. Ambitious energy policy objectives have also been set at the European level. European energy planners have set the goal of achieving a 20% share of renewable energy in total energy consumption by 2020 while at the same time reducing greenhouse gas emissions by 20% and increasing energy efficiency by 20% [3]. Seen in an international context, German and European energy policy efforts are playing a pioneering role.

In an energy supply system increasingly based on renewable energy sources, hydrogen has the potential of assuming a key role as an energy carrier. Hydrogen represents more than an alternative fuel source, because it can be used as a storable secondary energy carrier to balance fluctuating energy production from wind and sun and varying demand on the grid. Storing and making available surplus electricity according to need requires the development of efficient storage technologies, and represents one of the greatest challenges facing the energy revolution. There is a great deal that already needs to be done to ensure that the power grid can maintain network stability in the future.

Electrolysis and fuel cells represent existing technologies that make it possible to effectively use hydrogen as a secondary energy carrier. Stored hydrogen can be converted back into electricity as needed or used in various portable, mobile, or stationary applications. Using the principle of the electrochemical fuel cell, this conversion can be accomplished very efficiently and completely without harmful emissions. Produced from renewable sources, hydrogen thus makes possible a clean energy supply and mobility – independent of fossil resources.

One of the most promising areas of application for hydrogen and fuel cells is road transport, which accounts for about 16% of CO_2 emissions worldwide [4]. The use of hydrogen as a fuel in vehicles powered by fuel cells enables sustainable mobility without significant limitations compared to current driving habits. To accomplish this, there is an absolute need to systematically develop a hydrogen infrastructure, which represents an enormous challenge. To introduce a new fuel in the market and ensure universal service, a minimum number of refueling stations must be built. Germany alone would require 1000 hydrogen refueling stations (HRSs) to constitute a comprehensive network. To accomplish this will require major investments during the market introduction phase. The cost of constructing a comprehensive H_2-infrastructure in the European Union at high market penetration by 2050 has been estimated at between €700 and €2200 billion [5]. The enormous investments required on the infrastructure side coincide with major uncertainties on the side of revenue growth, since the market launch of fuel cell vehicles has been repeatedly postponed and future sales are very difficult to predict. Nevertheless, successful introduction of these vehicles presupposes the provision of an adequate infrastructure. Automobile manufacturers and refueling station operators are confronted with a classic chicken-and-egg dilemma. Therefore, the great challenge for the future is to develop the hydrogen infrastructure economically and in accordance with demand.

The Clean Energy Partnership (CEP) was established in December 2002 as a joint initiative by government and industry led by the German Ministry of Transport. The CEP is the largest demonstration project for hydrogen mobility in Europe. CEP partners include technology, oil and energy firms as well as most of the largest car manufacturers and two leading local public transport companies. The work of the CEP has generated valuable experience and insight about hydrogen vehicles and their associated refueling infrastructure through more than 20 000 refueling procedures for cars and buses. The CEPs experience has been continuously applied to the ongoing development of hydrogen refueling processes and technology. In 2012 the federal government and industry committed to building a network of 50 public hydrogen refueling stations in Germany by the end of 2015 [6]. The present chapter presents the current state of development in the design of hydrogen refueling stations and discusses possible approaches from the perspective of a station operator.

37.2
Basic Requirements for a Hydrogen Refueling Station

37.2.1
Car Refueling

The activities of the past few years have led car manufacturers, service station operators, and technology suppliers to jointly agree upon a technical guideline,

Table 37.1 Common industry requirements for fueling according to SAE TIR J2601.

Parameter	Value
Nominal working pressure (NWP):	70 or 35 MPa (at 15 °C)
Hydrogen storage capacity	From 1 up to 10 kg
Maximum pressure within the vehicle fuel system	87.5 or 43.75 MPa (125% NWP)
Gas temperature within the vehicle fuel system	$\leq$85 °C
Ambient temperature range at fueling station	$\geq$−40 and $\leq$+50 °C
Fuel flow rate at dispenser nozzle	$\leq$60 g-H$_2$ s^{-1} (3.6 kg min^{-1})
Fuel temperature at the dispenser nozzle	$\geq$−40 °C

SAE TIR J2601, for refueling cars with compressed gaseous hydrogen (CGH). The aim of the guideline, which is expected to be adopted as a standard in 2013, is to define the basic principles for safe and user-friendly hydrogen refueling that will approximate users' traditional refueling habits with conventional fuels. Targets include a refueling time of 3–5 min to achieve at least 90% of full capacity within the safety limits of the storage system. The guideline, published in 2010, defines basic requirements for the fueling process. The experiences gained from existing hydrogen refueling stations are being incorporated in the ongoing development of details of the standard.

SAE TIR J2601 establishes common industry performance requirements for refueling in accordance with Table 37.1 [7]:

- References to the standard SAE TIR J2799 that defines the mechanical connector geometry for fueling vehicles to 70 MPa and also provides specifications for the hardware for vehicle-to-station dispenser communication via an infrared interface with IrDA protocol.
- References to the standard SAE J2600 that provides connector hardware requirements for gaseous hydrogen fueling at 35 MPa.
- Definition of fueling tables for communication and non-communication fueling procedure that allow, dependent on ambient temperature measured at the station and state of charge of the tank, a safe refueling. The purpose is to optimize performance for each fueling type.
- The fueling procedure for communication fueling is specified to provide a fueling as close as possible to 100% state of charge (SOC) within the defined limits and specified fueling time.
- Fueling tables are provided for −40 °C (Type A), −20 °C (Type B), 0 °C (Type C), and ambient temperature (Type D) at 70 and 35 MPa.
- Definition of the fueling process: detecting of the communication interface, leak test, determination of the starting pressure in the vehicle tank.

Most vehicle manufacturers today are relying on 70 MPa (700 bar) quick refueling with −40 °C pre-cooling. The aim is to achieve a 3 min refueling time, which would differ only marginally from refueling with conventional fuels. Currently, typical tank sizes are 3.5–6 kg, which corresponds to driving distances of about 300–500 km. From a customer perspective, the use of a fuel cell vehicle – assuming the existence of an adequate fueling infrastructure – will differ very little from previous refueling habits.

Hydrogen station refueling is based on what is known as "average pressure ramp rate" methodology. A pressure ramp rate is selected depending on the starting pressure in the tank, as detected by a pressure pulse, and the starting temperature. The pressure ramp defines the rate of pressure rise. Thus, filling time is independent of tank size. The pressure ramp is selected in such a way that despite rapid refueling the heat of compression in the tank does not rise above a critical level. Pre-cooling the hydrogen to −40 °C also helps ensure a lower temperature in the tank.

The pressure ramps laid down in the guideline were determined through simulations and validated by practical testing. The refueling station is responsible for ensuring that fueling operates within the limits specified in the guidelines. The technical guideline differentiates between fueling where the refueling station communicates directly with the vehicle via an infrared interface (COM), and fueling without communication (NONCOM). Communication between the vehicle and refueling station makes it possible to determine important vehicle parameters, such as temperature and pressure. Without communication, the system has to assume the worst case and, consequently, filling is slower and reaches only 90% of full capacity. COM refueling makes it possible to go beyond the worst-case assumptions in the guideline, making possible a higher level of filling and a shorter refueling time.

37.2.2
Bus Refueling

Unlike the technology used for automotive hydrogen fueling, fuel cell powered buses have primarily used 350 bar (35 MPa) technology. In addition, no significant changes are anticipated in the next few years in this area. Since hydrogen tanks can be located on the roof of the buses, space concerns are not the same as with a car. Therefore, buses can dispense with high storage density and use cheaper 350 bar tanks instead. Overall, 350 bar refueling is less technically complex and has the advantage that it works without the need for pre-cooling. By eliminating the pre-cooling step and with lower efforts at compaction, refueling with 350 bar is much less energy intensive than the 700 bar technology. However, the buses require a separate dispenser with a special fueling nozzle. Due to the higher fueling quantities, bus-refueling stations need larger tanks and more powerful compressors. In designing the layout of the station, one must also consider that the buses require more space.

To date, no standard fueling protocols have been developed for buses. Before they are placed in operation, the hydrogen refueling stations must be harmonized to the individual bus type. Because of the small number of buses and local use by only one operator, this approach has not been a problem so far. In the medium term the plan is to expand the scope of the SAE J2601 regulation beyond automobiles and to define a common refueling protocol for buses. Since there is typically only a single local hydrogen refueling station available for today's small fleet of hydrogen-powered buses, it must be highly reliable. One way to achieve such reliability is through a redundant design of the main components.

Fuel cell buses in current operation have tank capacities up to 35 kg. Complete refueling at a high-performance hydrogen refueling station takes less than 10 min. Depending on the type of vehicle, fuel consumption is about 10 kg per 100 km. This enables the buses to drive distances up to 350 km with daily refueling.

37.3
Technical Concepts for Hydrogen Filling Stations

As part of the Clean Energy Partnership, several different refueling station concepts have been developed since 2002, and they have been comprehensively tested and validated in the demonstration phase. The refueling station concepts differ mainly in terms of different production and storage paths. Production is essentially divided into on-site and off-site procedures. In off-site production, hydrogen is produced in large, central generating plants or created as a byproduct, and is then delivered to the hydrogen refueling station. The alternative to off-site production is to produce hydrogen on a small scale on-site at the refueling station. Electrolysis is particularly well suited for this purpose. Using on-site production eliminates the need for hydrogen delivery. Figure 37.1 shows different existing models for hydrogen refueling stations, which are explained in more detail below.

37.3.1
Hydrogen Production

Currently, the total production of hydrogen in Germany is approximately 20 billion m^3 per year. Of this total, a large proportion is produced as a byproduct in the chemical industry and oil refining processes or as the principal product from reforming oil and gas. At present, less than 2% of Germany's hydrogen production results from the electrolysis of water. Today, the share of renewable energies used in the production of hydrogen, as anticipated in the vision of a sustainable hydrogen economy, remains low. Since hydrogen is a secondary energy carrier, its environmental equation can only be as positive as the equation for the primary energy source from which it is made. The Clean Energy Partnership is constantly striving to increase the proportion of certified green hydrogen (Certification by TUV SUD, or hydrogen produced from electrolysis with

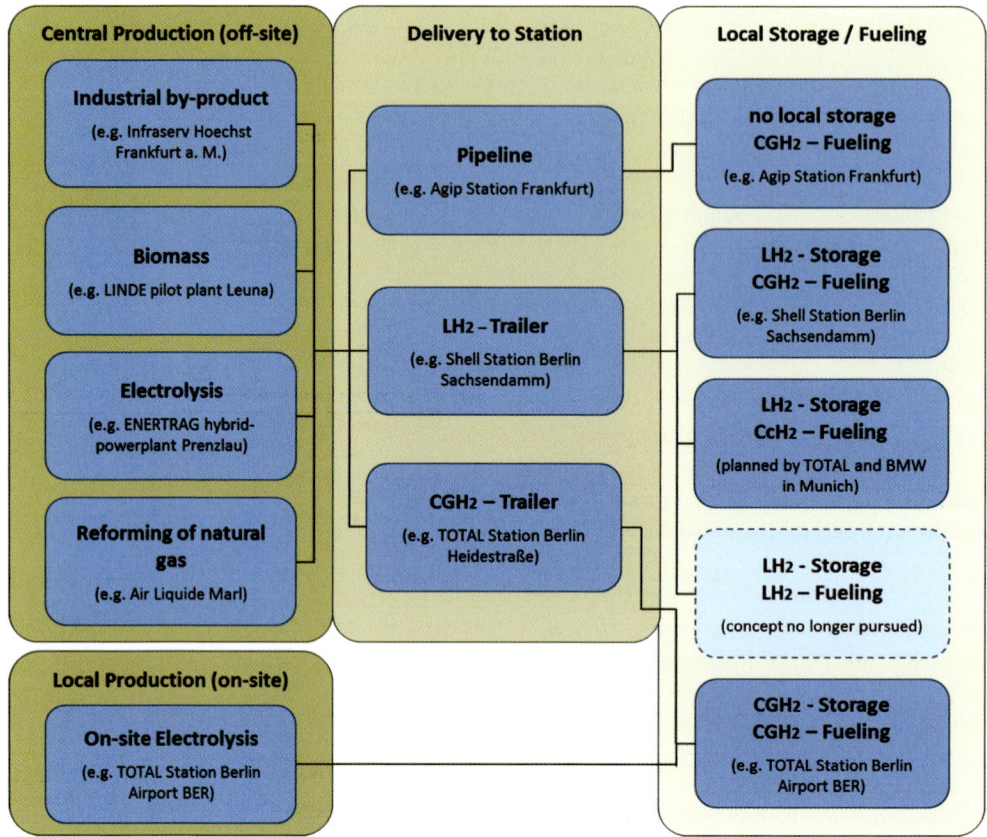

Figure 37.1 Scheme of station concepts.

certified green electricity) and to make hydrogen production independent of fossil fuels. Currently, at least 50% of the hydrogen sold at CEP refueling stations is certified as green [8].

37.3.1.1 Hydrogen Production from Biomass

In the future, biomass may play an important part in the development of sustainable sources for the production of hydrogen. There are several different methods for producing hydrogen from renewable organic materials. Currently, some problems of acceptance have arisen, since some of the plants that are processed for fuel can also serve as food sources ("food or fuel debate"). Of the procedures cited below, only glycerol reformation has been used to produce fuel in the context of hydrogen mobility to date. The most important processes are described below.

Glycerol Reforming

Crude glycerol obtained during the production of biodiesel from rapeseed is particularly well suited for the production of hydrogen. The potential efficiency of biomass per hectare is further increased by the conversion of glycerol into hydrogen for fuel production [9]. The Linde Company has built a demonstration plant to produce hydrogen from organic materials in Leuna, Germany. The plant for treatment, pyrolysis, and reforming of crude glycerol produces a hydrogen-rich gas, which is then purified and liquefied. In normal operation, the biomass-to-hydrogen (BTH) plant can produce up to 50 cubic meters of hydrogen per hour. One ton of crude glycerol yields about 100 kg of hydrogen. Compared to conventional hydrogen production by reforming natural gas, this process results in a 50% greater reduction in greenhouse gas (GHG) emissions. When scaled up to commercial scale production, the anticipated reduction in GHG emissions will increase to 80%. One problem with using glycerol is that fluctuations in its availability and price significantly affect production costs. Therefore, looking forward, there is a need to test alternative organic raw materials [10]. The certified green hydrogen generated at the BTH facility in Leuna is already being delivered to CEP refueling stations.

Biogas Reforming

Biogas generated in biogas plants can be converted into hydrogen by steam reforming in the same way as natural gas. Potential problems with this process are that biogas has a lower proportion of burnable gas than natural gas, and that its gas composition is variable over time. The technical concept has not been implemented yet in larger pilot plants, and, thus, no hydrogen made from biogas reforming has been supplied to hydrogen refueling stations to date.

Biomass Gasification

In the process of biomass gasification, biomass is gasified in gasification reactors, yielding a mixture of hydrogen, carbon monoxide, carbon dioxide, and methane. Usable hydrogen can be obtained by subsequently cleaning the gas mixture. Appropriate starting materials include organic waste, such as straw or wood chips. Biomass gasification has not been applied as yet for hydrogen mobility.

37.3.1.2 Electrolysis

Electrolysis has been known for decades as a technology for producing hydrogen, and is commercially available within certain technical parameters. Because it has been significantly less expensive to produce hydrogen from fossil fuels, research interest in modern electrolysis technology waned considerably over time, and only experienced a renaissance quite recently, driven by the search for storage capacity for fluctuating electricity production from renewable energies and the anticipated rise in the demand for non-fossil fuels for mobility. Over the past decade, electrolysis has become the focus of serious economic and scientific research. The key advantage of water electrolysis is the possibility of direct linkage to renewable energy sources (wind, PV, hydro). Electrolyzers can immediately respond to fluctuating power output, so electrolysis can be used for network power regulation as well as hydrogen production. Therefore, electrolysis is regarded as the most promising method for meeting future hydrogen demand.

The two electrolysis processes currently available on an industrial scale are alkaline electrolysis and PEM (polymer electrolyte membrane) electrolysis. Conventional alkaline electrolysis, with an efficiency of about 70% (in relation to the upper caloric value of hydrogen), has been used to produce hydrogen for many decades, either in settings where inexpensive electric current is available or when smaller amounts of pure hydrogen are needed for chemical processes. In recent years, several new concepts have been studied and technically developed to increase the efficiency of alkaline water electrolysis. Today, electrolysis systems are available in a wide range of performance capacity.

To date, PEM electrolysis has only been suitable for small system solutions (up to 100 kW) because of its unfavorable price/performance ratio. However, efforts are under way at CEP partner SIEMENS to apply this technology in larger-scale production.

Basically, electrolysis systems are classified into two groups: central (off-site) and decentralized (on-site) electrolysis systems. Central hydrogen production has the advantage that higher levels of efficiency can be achieved in larger installations. The reason for this increase in efficiency is that with increasing system size the end consumer's share of total energy consumption decreases. The disadvantage is that when hydrogen is produced centrally it must still be transported to the consumer. Decentralized production will become more significant in the specific context of increasing use of renewable energy, which is also produced in small, decentralized units. The current cost of producing hydrogen through electrolysis from wind power is still about five to six times greater than the cost of hydrogen production through steam reforming of natural gas [11]. However, it is anticipated that as fossil fuel prices rise and manufacturing costs for electrolysis systems fall as a result of technical progress, hydrogen production from electrolysis may prove profitable in the future. The production of wind-hydrogen by electrolysis is currently being researched and developed in a number of different demonstration projects.

An example of central hydrogen production by electrolysis is the hybrid power station located in Prenzlau Germany, operated by a partnership of ENERTRAG AG, TOTAL Deutschland GmbH, Vattenfall, and Deutsche Bahn. The hybrid power station uses surplus wind power to generate hydrogen that supplies a TOTAL station in Berlin with certified green hydrogen. It is delivered by means of a CGH_2 (compressed hydrogen) trailer [12].

An example of decentralized, on-site hydrogen production through electrolysis is the integrated hydrogen refueling station under construction at the Berlin-Brandenburg Airport (BER) (Figure 37.2). At this facility the hydrogen is produced by electrolysis using 100% wind power from a nearby wind farm constructed specifically for this purpose. In addition, all of the refueling station's other energy needs (electricity and heat) are covered by 100% renewable energy.

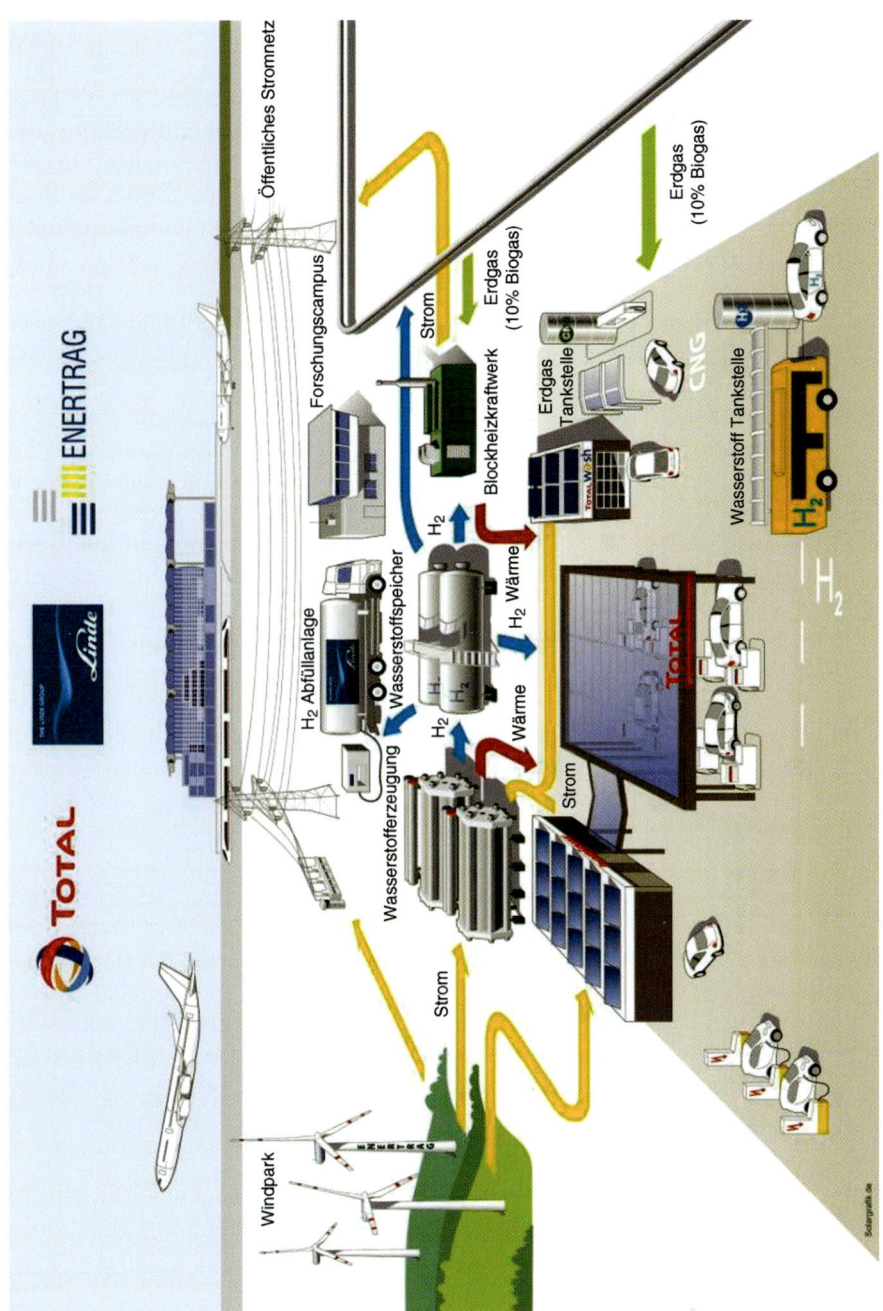

Figure 37.2 Scheme of the TOTAL refueling station at Berlin's new airport BER.

37.3.1.3 Steam Reforming

The steam reforming process produces hydrogen in two steps from naturally occurring hydrocarbon gases such as methane. Steam reforming is the most prevalent current method for producing hydrogen. Plants with capacities up to 100 000 m^3 h^{-1} have already been placed in operation. In the first step of steam reforming, the hydrocarbon is heated in the reformer to a temperature of 800–900 °C at a pressure of about 25 bar and reacted with water. The resulting carbon monoxide (CO) is separated out to increase the yield of hydrogen and in the second step, known as the water–gas shift reaction, it is converted into carbon dioxide and additional hydrogen.

Today's steam reforming process makes it possible to produce large quantities of hydrogen at relatively low cost [11]. Therefore, a portion of the hydrogen currently available at CEP stations is still produced by means of this process. For the future, the aim is to progressively shift to certified green hydrogen and gradually reduce the proportion of hydrogen obtained through steam reforming.

37.3.1.4 Byproduct Hydrogen

Many industrial processes, particularly those in the chemical industry, generate gaseous hydrogen as a byproduct. At least during the introductory phase, byproduct hydrogen can play a major role in the development of the hydrogen refueling infrastructure [13]. The (Agip) hydrogen refueling station in Frankfurt am Main is an example of the use of byproduct hydrogen for transport needs. The hydrogen refueling station is supplied through a 1000 bar (100 MPa) high-pressure hydrogen pipeline from the neighboring Höchst Industrial Park. Byproduct hydrogen is generated in large amounts at a chlorine production plant located in the industrial park [18]. Generally, as the H$_2$-infrastructure continues to expand, it makes sense to build refueling stations in close proximity to plants that generate hydrogen as an industrial byproduct.

37.3.1.5 Biological Hydrogen Production

Research on the production of hydrogen using biotechnological methods is still in its infancy. There are experiments under way with bacteria, micro-algae, and microbes that are known to release hydrogen under certain environmental conditions when stimulated by additional sunlight. This method relies on, among other things, the process of photosynthesis. To date, these production processes have been implemented only on a laboratory scale and require further basic research.

37.3.2
Hydrogen Delivery

When hydrogen is produced in central production plants, it must be delivered to the individual hydrogen refilling station sites. To date, three supply models have been implemented in practice: CGH_2 (compressed gaseous hydrogen) delivery, LH_2 (liquid hydrogen) delivery, and pipeline delivery. With the low current level of throughput at hydrogen refueling stations due to a dearth of vehicles, delivery and storage of gaseous hydrogen makes particular sense. As larger amounts of hydrogen are dispensed at the refueling stations, LH_2 delivery will gain in importance.

37.3.2.1 CGH_2 Delivery
Supplying refueling stations with compressed gaseous hydrogen (CGH_2) in trailers is a technically proven and simple form of hydrogen transport. Each CGH_2 trailer can carry about 500 kg of hydrogen at a pressure of 200–500 bar (20–50 MPa). In comparison to their total weight of 40 t they deliver a relatively small payload. Therefore, CGH_2 trailers are best suited for supplying smaller stations with low capacities at short distances (<200 km). At the hydrogen refueling station, the trailer either fills the station's low-pressure storage tank by overflow or the trailer itself serves as a low-pressure reservoir and remains on site. Both alternatives require one or more compression stages at the hydrogen refueling station to increase the pressure level to 350 or 700 bar (35 or 70 MPa) [14].

37.3.2.2 LH_2 Delivery
The large temperature difference between the cryogenic hydrogen in the tank ($-253\,°C$) and the ambient environment (-20 to $40\,°C$) makes specialized demands on the materials used to transport liquid hydrogen. At present, the storage technology has been adequately developed for many years to enable safe transport without any major difficulties. The significant advantage of transporting liquid hydrogen compared to gaseous transport is the nearly tenfold greater transport capacity of up to 3.5 t. Therefore, LH_2 delivery is particularly well suited for larger hydrogen refueling stations with greater turnover and for longer transport distances of 400–1000 km [14].

37.3.2.3 Pipeline Delivery
Supplying hydrogen refueling stations through a pipeline infrastructure is especially suitable for densely populated regions with high throughput. A pipeline infrastructure is the most economical method for transporting large quantities of hydrogen. Worldwide, there is already about 16 000 km of hydrogen pipeline in existence, which is used by the chemical industry and would meet the requirements for supplying hydrogen refueling stations with large hydrogen demands. One advantage of pipeline supply is its great flexibility in the case of variable quantities. The hydrogen is typically transported at a pressure of 20–100 bar (2–10 MPa) and must be further compressed at the hydrogen refueling station. In some cases it also makes sense to supply the hydrogen refueling station with a high-pressure pipeline. The high-pressure line eliminates the need for costly compression at the hydrogen refueling station.

37.3.3
Major Components of Hydrogen Refueling Stations

37.3.3.1 Production

Most currently existing hydrogen refueling stations are served by major gas suppliers that deliver hydrogen by trailer. Production takes place off-site at central generating plants. The supply is very reliable and responds flexibly to changing levels of demand. Of the 15 hydrogen refueling stations currently under construction or in operation in Germany, only five stations have on-site production capacity.

Until now, the only on-site production method employed as part of the CEP (Clean Energy Partnership) has been electrolysis. Alkaline electrolysis, in particular, has reached technical maturity and offers high reliability. The disadvantage of on-site generation is that a production failure may directly impair the operation of the hydrogen refueling station. Therefore, the latest generation of stations using on-site generation often employs a back-up option of hydrogen delivery.

37.3.3.2 Hydrogen Storage

CGH_2 Storage

Hydrogen refueling stations generally store gaseous hydrogen at three different pressure levels:

- 50 bar (5 MPa) low pressure storage;
- 150–500 bar (15–50 MPa) medium pressure storage;
- 700–1000 bar (70–100 MPa) high pressure storage.

In designing the layout for hydrogen refueling stations, it is important to take space requirements into consideration, especially when dealing with larger storage quantities. Two basic design options are to situate the reservoirs above ground or below ground. Underground storage will play a greater role in the future, both for considerations of space and for esthetic reasons.

Low-pressure storage is generally implemented in the form of a cylindrical steel tank with a pressure of up to 50 bar (5 MPa) and a storage capacity of up to 250 kg. These facilities are particularly well suited for the cost-efficient storage of larger amounts of hydrogen.

Medium-pressure reservoirs are implemented in either tubular form or as a bundle of cylinders. They accommodate pressures from 150 to 500 bar (15 to 50 MPa). These reservoirs are standard production items available from several different manufacturers. Smaller hydrogen refueling stations typically tend to avoid low-pressure storage and use only medium-pressure reservoirs for hydrogen storage.

Given the lack of a mass market, high-pressure hydrogen storage facilities designed for pressures up to 1000 bar (100 MPa) have only been available from a few manufacturers in the form of expensive custom-made solutions. Only a significant increase in production volumes will lead to significant cost reductions. At the present time, these facilities primarily use whole composite containers (Type IV) and steel composite containers (Type II) [15].

LH$_2$ Storage

Liquid hydrogen (LH$_2$) can be stored in stationary and mobile tanks with special insulation. The cylindrical gas tanks are similar to the low-pressure tanks and can each store 1–5 tons of hydrogen. The energy required for liquefaction is equivalent to about 20–30% of the energy content of the hydrogen [19]. Since it is impossible to completely prevent rewarming of the cryogenic hydrogen, about 0.5% of the stored hydrogen will evaporate every day. This boil-off loss can be used as an energy source. For example, at the Berlin Holzmarktstraße TOTAL station, a mini combined heat and power plant is fueled using the boil-off hydrogen.

In the worst-case situation, it may be necessary to simply discharge the unused hydrogen. In planning a hydrogen refueling station with LH$_2$ storage it is important to always make sure that the throughput is large enough to keep boil-off losses to a minimum.

Solid Hydrogen Storage

An alternative form of hydrogen storage is to chemically bind hydrogen and a metal to form a metal alloy. This approach, also referred to as metal hydride storage, is a safe and compact low-pressure storage method Compared to other storage methods, metal hydride storage systems offers the prospect of high energy efficiency. The technology is still in its early stages and has not yet demonstrated maturity in practice. As part of the CEP, a solid storage system with a capacity of up to 100 kg made by the manufacturer McPhy will be integrated for the first time as part of the hydrogen refueling station scheduled to begin operation in 2014 at the Berlin-Brandenburg Airport, and its performance will be tested in a demonstration project.

37.3.3.3 Compressors

Target pressures as high as 1000 bar (100 MPa) require high-performance compressors. The equipment must be compatible with hydrogen and not contaminate the extracted gas. Since these are highly specialized products, there are currently very few potential suppliers in this performance class. The compressors represent a significant cost factor for hydrogen refueling stations in terms of investment as well as operating costs. Therefore, compressors are key components in planning and operating hydrogen refueling stations.

Piston Compressors

Piston compressors typically compress the gas by means of a hydraulically driven piston. The heat generated by compression must be dissipated through a costly cooling process. A major challenge is to prevent contamination of the hydrogen at every stage, as this could damage the sensitive fuel cells in hydrogen-powered vehicles. Therefore, the piston must operate without any oil or lubricant, and this places a high demand on the seals. At the stations operated by the CEP equipped with piston compressors, problems partially arose as a result of impurities in the hydrogen. Piston compressors are also relatively large, so consideration has to be made for the additional space required when designing the layout of the hydrogen refueling station.

Diaphragm Compressors

In the diaphragm compressor, gas compression takes place in a double concave chamber by means of an oscillating diaphragm. The membrane hermetically seals and separates the gas chamber from the hydraulic drive. The membrane is set in motion on one side by the hydraulic drive. Through this deflective movement, the gas space between the diaphragm and the diaphragm cover is cyclically expanded and restricted, with the result that the gas is drawn in, compressed, and then discharged. Since the membrane separates the hydrogen from the hydraulic oil seal, contaminants and abrasive residues are excluded; however, the maximum flow rate is limited. Combinations of piston and diaphragm compressors are also available on the market.

Ionic Compressor

The ionic compressor has been developed by Linde, and has been used successfully at several CEP stations (e.g., TOTAL, Berlin Heidestraße and Vattenfall, Hamburg HafenCity). In place of a movable piston, ionic compressors employ a moving column of liquid that is shifted up and down by a pump to compress the gas. In this process the liquid is connected in two hydraulic pistons. As the column of fluid increases in one cylinder it decreases in the other cylinder. Since the compressor has few moving parts, mechanical wear is reduced, resulting in longer life and less need for maintenance. Another advantage is lower energy consumption when compared to other compressors.

LH$_2$ Pump

In the case of LH$_2$ storage at the hydrogen refueling station, compression can also be accomplished using an LH$_2$ pump (cryopump) that transforms the liquid hydrogen into compressed gaseous hydrogen. Greater efficiency can be achieved during compression as a result of the pressure build-up in the liquid phase. No separate cooling step is required for the hydrogen, since the liquid phase hydrogen provides adequate cooling. The pumps are also much more compact and have higher flow rates than traditional piston compressors. Cryopumps are being used at the Berlin-Sachsendamm Shell station, for example, which is the largest hydrogen refueling station within the CEP.

37.3.3.4 Pre-cooling

According to technical guideline SAE TIR J2601, hydrogen must be at a temperature between −33 and −40 °C at the time it is dispensed to the vehicle. When refueling at 700 bar (70 MPa), this is necessary so that the temperature in the tank remains below a critical level and to ensure rapid refueling. Thus, depending on the storage form of the hydrogen (gas or liquid), either cooling or an evaporator is necessary to provide the hydrogen at the specified temperature.

Hydrogen is pre-cooled by means of special compression chillers that consume large amounts of energy. When planning the layout of a hydrogen refueling station, consideration must be given to locating the pre-cooling system in close proximity to the dispenser so as to prevent energy loss. Stations with LH2 storage and cryogenic pumps do not require pre-cooling.

37.3.3.5 Dispenser

The dispenser is the part of the hydrogen refueling station that is seen and operated by the customer (Figure 37.3). It forms the interface between the hydrogen refueling station and the vehicle. The fueling nozzles are standardized according to SAE 2799 but, in other respects, no uniform design has been implemented as yet for dispensers. The CEP uses a coordinated service model to ensure that the customer always encounters a standardized user interface. Thus, for example, customer authorization is accomplished with a standardized H_2-card to ensure that only trained individuals carry out refueling. At integrated hydrogen refueling stations the dispenser is deliberately located alongside conventional petrol pumps so as to encourage customers to regard hydrogen as a normal fuel.

37.3.3.6 Controls

The control of hydrogen refueling stations basically consists of standard components typically used in industrial facilities. Various sensors, valves, and gauges are required for control, automation, and backup features. Appropriate material selection is required to assure that all components are compatible with hydrogen. The system is controlled by a programmable logic controller (PLC) and is

Figure 37.3 H_2-dispenser at the TOTAL station at Cuxhavener Strasse, Hamburg.

housed in control cabinets. Hydrogen filling stations are designed for automated operation. There is typically a provision made for remote diagnosis via data connection.

37.3.4
Integration of Hydrogen Refueling Stations

From the very outset, the CEP considered it important that customers regard hydrogen as a normal fuel, so in the interests of promoting acceptance the entire refueling process has been designed to resemble familiar conventional gas stations (Figure 37.4). The aim is to integrate hydrogen technology to the greatest extent possible into existing or newly built gas stations. In this way, hydrogen, still somewhat fraught with negative associations (explosive, flammable, dangerous) in the public mind, can instead be perceived as a normal fuel as part of a gas station's product portfolio. In this way, hydrogen refueling technology will gradually become integrated in the everyday life of a regular fuel customer, and future car buyers will have an easier time opting for this technology. First generation hydrogen refueling stations often had a distinctively temporary look to them and were clearly identifiable as research and development objects. In today's facilities, hydrogen technology will be fully integrated into the overall design of the station and scarcely noticeable to the customer.

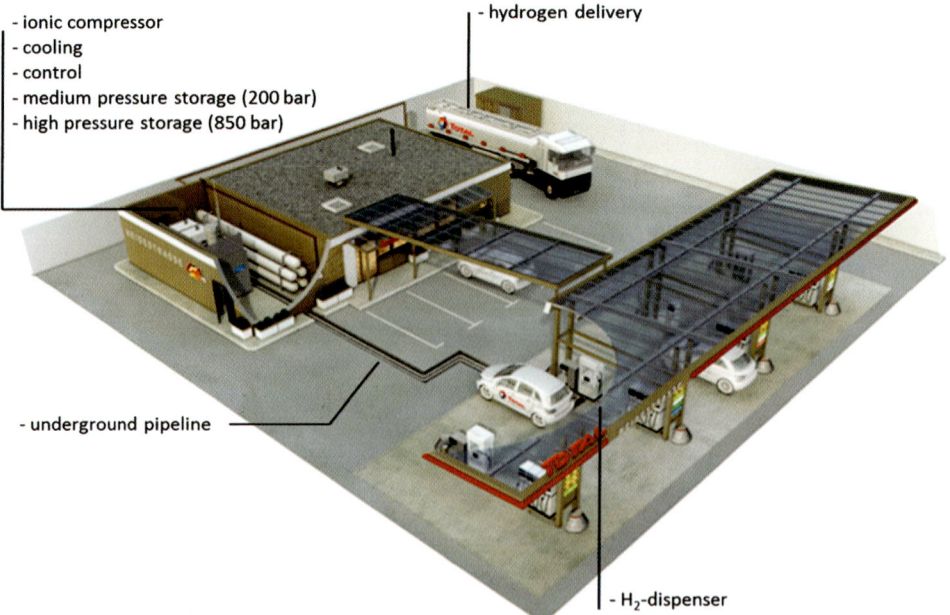

Figure 37.4 Berlin Heidestrasse TOTAL hydrogen refueling station.

37.3.5
Facility Size/Space Requirements

The space requirements of hydrogen refueling stations will pose a critical problem for the further development of the hydrogen infrastructure. With rising throughput rates, the space required for storage and compressors will also increase. Existing gas stations generally have very limited space to accommodate the subsequent integration of hydrogen. In addition, congested urban areas typically have very few open lots available for the construction of new stations and real estate costs are very high. To deal with these obstacles in the future, solutions will have to be found through technical innovations, such as improved space-saving storage concepts, as well as through organizational measures and targeted government support.

37.4
Challenges

Further efforts will be required to move hydrogen mobility along the path to market maturity. Over the course of the ongoing demonstration projects, several different problems were identified and the partners in the CEP have initiated efforts to find solutions for them. The following section describes some of the challenges that are current action priorities for working groups in the CEP.

37.4.1
Standardization

Until now, all of the hydrogen refueling stations that have been constructed have been custom made. Therefore, standardization is the most important current issue on the way to market readiness. Only by constructing larger numbers of stations with identical design will it be possible to resolve the problems that still remain. The following effects can be anticipated from a modular concept and series production of the main components:

- cost reduction due to higher volumes;
- more innovation through increased competition;
- increased level of technology maturity;
- increased availability and reliability;
- lower operating and maintenance costs;
- simplified procurement and shortening of the approval process.

One important step toward standardization has been the joint development of functional descriptions (specifications) for hydrogen refueling stations by the H2 Mobility Initiative, which define minimum standards for hydrogen refueling stations to be built in the future [16]. These specifications

Table 37.2 Performance specification of the different HRS sizes according to H2-Mobility HRS Functional Description.

	Very small HRS	Small HRS	Medium HRS	Large HRS
Number of refueling positions	1	1	2	4
Number of refueling per hour per position	2.5	6	6	10
Number of back to back refuelings per refueling position	0	1	1	10
Max. waiting time to fuel consecutive cars (min)	20	5	5	N.a.
Number of consecutive hours to meet the performance specifications	—	3	3	24/7
Average number of fuellings per day	10	30	60	125
Maximum number of fuellings per day	20	38	75	180
Maximum hydrogen hourly throughput (kg)	18	33.6	67.2	224
Average hydrogen throughput per day (kg)	56	168	336	700
Maximum hydrogen throughput per day (kg)	80	212	420	1000
Number of cars per station (approx.)	100	400	800	1600

incorporate many findings from the demonstration projects of the past decade. The specifications classify future hydrogen refueling stations into four size categories (Table 37.2).

In addition to setting performance requirements for hydrogen refueling stations, the specifications also include other points, such as safety requirements, gas purity, acceptance, and norms to be observed. Standardization is currently an iterative process, since there is continuous new input from additional practical experience, and the model is repeatedly corrected and revised accordingly. Larger numbers of identical stations would specifically serve to simplify the final technical acceptance test.

Opened in 2012, the Berlin Heidestrasse TOTAL station was built according to an earlier version of the H2-Mobility HRS Functional Description. In engineering and designing this "small" type hydrogen refueling station, the overall system design concept was always kept in focus. For the first time, all of the station's modules were designed with the aim of optimizing the system as a whole. This approach made it possible to significantly increase energy efficiency and capacity while at the same time minimizing space requirements for the system components. Another focus in the technical design of the facility was in the area of service and maintenance. German labor law guidelines were taken into consideration so that a single individual could perform future service and maintenance tasks. For example, the maximum weight of individual system components was kept below 15 kg. Furthermore, all parts of the system subject to wear can be changed on site and are freely accessible without requiring extensive disassembly of adjacent components or units. The principal components of the "small" type hydrogen refueling station are the hydrogen compressors, high-pressure hydrogen storage tank, and pre-cooling unit. In addition, the hydrogen refueling station includes a stand-alone dispenser for 700 bar (70 MPa) refueling.

37.4.2
Costs

Operators of hydrogen filling stations see the reduction of high investment and operating costs as a key challenge for building a comprehensive hydrogen infrastructure. The current investment costs for a "small" HRS are at €1.1 million including hardware and necessary preparatory construction works. In the current phase of demonstration operations and market preparation, it is not yet possible to operate hydrogen refueling stations economically. Due to uncertainties about the development of future demand, it is very difficult at this time to make serious economic calculations or develop reliable business models for future hydrogen refueling stations. Additional government support will be needed for the further development of the infrastructure. Looking forward, economies of scale in standardized systems will enable a reduction of investment costs. However, commercialization will only succeed if it is possible to achieve substantial financial returns from adequate sales volumes of hydrogen.

Operating costs are largely driven up by maintenance and energy costs. According to the experience within the CEP, the annual operating costs for a facility are around €110 000. The new technology is still relatively failure prone and requires short maintenance intervals. In the past, compressors have proven to be especially complex and costly components to maintain. Since existing hydrogen refueling stations have all been custom made, there has been a lack of a service infrastructure that could facilitate more rapid and cheaper repair and maintenance. For example, spare parts often take a long time to procure since they are not stocked as standard items. It is expected that maintenance costs can be significantly reduced as the number of stations grows and there is increased competition between manufacturers.

The principal causes of large energy costs are compression and pre-cooling. According to data in the literature, about 10% of the energy content of the hydrogen is needed for compression at 700 bar (70 MPa) [15]. Pre-cooling also uses large quantities of energy, especially at high ambient temperatures. Older facilities did not include an energy-optimized operation management system and, thus, the facility was constantly cooling, even in stand-by mode. In designing newer systems, attention was focused on making the operational model as energy sparing as possible. However, further work is required to monitor energy consumption and current developmental work on future facilities is focused on identifying potential ways for further energy optimization.

37.4.3
Reliability

The reliability of hydrogen refueling stations is an important element in public acceptance of the technology. Experience in the Clean Energy Partnership has shown that customers have a very negative reaction to closures of the hydrogen refueling stations related to breakdown or maintenance activities. Despite the fact that most hydrogen refueling stations achieved availability rates of 90% or more, a single wasted trip to a refueling station is annoying for the user. Because of the low number of available facilities, closure of a single hydrogen refueling station may affect many users or even an entire region. To respond to this issue, the CEP established a station availability monitoring system to let users check on the operational readiness of the station on the Internet before driving there.

There are many causes of operational failures in hydrogen refueling stations. In the past, it has been compressors and pre-cooling systems that have proven to be relatively failure-prone; in addition, several other problems arose that might be thought of as "diseases of childhood." Failure of a single component generally has led to failure of the entire refueling station. For larger systems, it is possible to include redundancy in designing the main components, but for smaller systems redundancy is not economically justifiable. New technology inevitably requires frequent maintenance, and this is often predictable. With increasing experience, however, maintenance cycles can be extended and maintenance can be optimized as needed.

In addition to frequent equipment failures, long downtimes are still a major problem. The reason is that, until now, technology suppliers have not set up local service infrastructures with standby service. Due to the small numbers of hydrogen refueling stations, there are few specialized technicians that are able to perform the service work. Service personnel often have to travel long distances. There is still no standard inventory for critical replacement parts, leading to long delivery times. In the past such circumstances have led hydrogen refueling stations to shut down for as long as several weeks for needed repairs.

Currently, the CEP has been considering the possibility of setting up a mobile hydrogen refueling station as a flexible back-up solution in the event of station outages as a way of assuring better availability during the network construction phase. In the event of a breakdown of a single hydrogen refueling station, this would enable faster replacement service for the customer and avoid depriving an entire region of its supply of hydrogen fuel. In terms of the market launch, system reliability needs to be further improved by appropriate technical measures, and downtime needs to be kept to a minimum by building a service infrastructure.

37.4.4
Approval Processes

From the perspective of the operator, the approval process for hydrogen refueling stations is still challenging, even though the experiences in Germany are more positive in comparison with other European countries. The approval process for such technically complex projects involves several regulators, including building authorities, environmental agencies, and fire departments. The smooth development of the necessary infrastructure requires the involvement of various stakeholders, and they often lack sufficient familiarity with the properties of hydrogen or the approval process for setting up hydrogen refueling stations. Earlier construction projects were hindered by enormous expenditures of time and effort in obtaining the necessary permits. For some projects, it was noted that the authorities demanded particularly stringent requirements due to their lack of understanding of hydrogen technology, and the approval process took unnecessarily long as the authorities attempted to eliminate every conceivable risk.

Based on previous experience in working with regulatory authorities, approval guidelines for station operators have been created with the aim of facilitating the approval process in the future [20]. A major impetus for streamlining future approval processes can be anticipated as a result of standardization and greater numbers of new hydrogen refueling stations. It has already been shown that in cities with a larger number of hydrogen refueling stations all of the involved parties learned a great deal from the projects and, as a result, the permitting procedure could be accomplished more easily and faster. Working together with the regulatory authorities and inspection bodies, the approval process should be streamlined and simplified [17]. The experiences of the CEP can become the model for unified international approaches.

37.4.5
Gauged H_2-Metering

Under existing laws for fuel sales, variations in volume measurement at new facilities must not exceed 1.5%. The equipment at the facilities must be calibrated at regular intervals. Until now, there has been no H_2-metering method available that achieves the required accuracy under the given constraints. The high pressures, large temperature gradients, and low partial mass flows are problematic for the Coriolis flow meters that have been used thus far. Small refueling amounts, in particular, generate large discrepancies. There is still further research and development required in the area of flow measurement instrumentation. In the CEP a working group devoted to this issue is currently working in cooperation with authorities and manufacturers of flow meters in an attempt to find a solution that ensures legal compliance pending commercial introduction.

37.4.6
Refueling According to Technical Guidelines

The technical guidelines described in Section 37.2 clearly define the requirements for pressure and temperature ramps during the hydrogen refueling process. Compliance on the part of the hydrogen refueling station ensures rapid and safe refueling. Previously, there had been no uniform hydrogen fueling acceptance tests that could check on the conformity of the fueling operation. Therefore, each new service station had to be analyzed by automobile manufacturers through individual acceptance tests. This was followed by a technical release that only applied to the specific vehicle type that had been tested. This enormous effort can be significantly reduced through a unified acceptance test. The CEP is currently looking for a test apparatus capable of verifying conformity to technical specifications for new hydrogen refueling stations. Ideally, future acceptance tests should be carried out or supervised by an independent approved inspection body.

37.4.7
Hydrogen Quality

At this time, there are still several unanswered questions about how to define, implement, and monitor uniform hydrogen quality at all hydrogen refueling stations. Quality standard ISO 14687 sets the expected standard for hydrogen quality for fuel cell vehicles and defines limit values for impurities in the hydrogen to which the hydrogen refueling stations must adhere (Table 37.3). Exceeding these

Table 37.3 Hydrogen quality according to ISO 14687.

Constituent	Chemical formula	ISO 14687 limits ($\mu mol\ mol^{-1}$)
Hydrogen fuel index	H_2	>99.97%
Total allowable non-hydrogen, non-helium, non-particulate constituents listed below		300
Water	H_2O	5
Total hydrocarbons	(C1 basis)	2
Oxygen	O_2	5
Helium	He	300
Nitrogen	N_2	100
Argon	Ar	
Carbon dioxide	CO_2	2
Carbon monoxide	CO	0.2
Total sulfur compounds		0.004
Formaldehyde	HCHO	0.01
Formic acid	HCOOH	0.2
Ammonia	NH_3	0.1
Total halogenated		0.05
Particulate concentration		$1\ mg\ kg^{-1}$

limit values may result in damage to sensitive fuel cells. Some of these limit values are below the ppm range and are thus extremely laborious to measure. Another problem is that, to date, no standardized sampling device has been developed for obtaining representative gas samples. To address this question, a working group within the CEP is devoting its efforts to develop appropriate sampling and analytic procedures for monitoring gas purity at hydrogen refueling stations. The challenge is to assure uniform quality at a manageable level of cost and effort. Meeting this challenge demands a broadly based economic analysis that includes consideration of both the vehicles and the infrastructure. The goal is to reach a compromise between the life expectancy of fuel cells and infrastructure-related expenditures. Based on vehicle-related experiences from the ongoing projects, it should be possible to re-evaluate the limit values for impurities. From the perspective of the infrastructure operator, such a re-evaluation could result in an adjustment of the fuel standard to be more consistent with practical realities.

37.5 Conclusion

The activities conducted to date have generated a wide range of new scientific understanding, and a good foundation exists to establish Germany as a leading market for hydrogen mobility. In the next step, strategic concepts for the construction of a basic nationwide infrastructure must be developed and implemented. Standardization is of particular importance at this stage. Standardization of design concepts for hydrogen refueling stations will significantly reduce costs, increase reliability and availability, and simplify approval processes and acceptance tests. We can anticipate that a growing market will ameliorate the current inadequacies in the availability of suppliers for certain components, including compressors, high-pressure storage units, refueling systems, and electrolyzers. Increased competition will also lead to lower costs and increased pressure for innovation on the part of manufacturers.

As the demand for hydrogen increases, the issue of sustainable hydrogen production will take on greater importance. Therefore, the inclusion of hydrogen mobility in current strategies for the energy transformation (*Energiewende*) should be encouraged. In this regard, electrolysis will have a critical part to play in both on-site and off-site hydrogen production.

To achieve full commercialization, the construction, acceptance testing, and operation of hydrogen refueling stations has to be consistently simplified and perfected. Although some problems were initially underestimated on the hydrogen refueling station side, and the emphasis was placed instead on the need to develop hydrogen-powered vehicles, the remaining challenges can surely be resolved through joint efforts on the part of industry, public authorities, and research institutions.

References

1 Bundesministerium für Umwelt, Naturschutz und Reaktorsicherheit BMU (2013) Zeitreihen zur Entwicklung der erneuerbaren Energien in Deutschland unter Verwendung von Daten der Arbeitsgruppe Erneuerbare Energien-Statistik (AGEE-Stat) in Deutschland, BMU, Berlin.

2 Bundesministerium für Wirtschaft und Technologie BMWi, Bundesministerium für Umwelt, Naturschutz und Reaktorsicherheit BMU (2010) Energiekonzept für eine umweltschonende, zuverlässige und bezahlbare Energieversorgung, BMWi, BMU, Berlin.

3 Bundesministerium für Wirtschaft und Technologie BMWi (2013) Europäische Energiepolitik. Available at http://www.bmwi.de/DE/Themen/Energie/Energiepolitik/europaeische-energiepolitik.html (accessed 13 May 2013).

4 International Energy Agency Statistics (2012) CO_2 Emissions from Fuel Combustion – Highlights 2012, IEA, Paris.

5 Tzimas, E., Castello, P., and Peteves, S. (2007) The evolution of size and cost of a hydrogen delivery infrastructure in Europe in the medium and long term. *Int. J. Hydrogen Energy*, **32**, 1369–1380.

6 Bundesministerium für Verkehr, Bau und Stadtentwicklung BMVBS (26 June 2012) Bundesregierung und Industrie errichten Netz von 50 Wasserstoff-Tankstellen, press release. Available at http://www.bmvbs.de/SharedDocs/DE/Pressemitteilungen/2012/125-ramsauer-wasserstofftankstellen.html (accessed 28 May 2013).

7 SAE International, Fuel Cell Standards Committee (2010) Fueling Protocols for Light Duty Gaseous Hydrogen Surface Vehicles, Product Code: J2601, SAE, Warrendale.

8 Clean Energy Partnership (2011) Wasserstoff bewegt – Ziele und Phasen der CEP. Available at http://www.bmvbs.de/cae/servlet/contentblob/70152/publicationFile/41671/wasserstoff-bewegt-broschuere-deutsch.pdf (accessed 22 May 2013).

9 Linde (2011) Wasserstoff: Von der Quelle in den Tank. Available at http://www.the-linde-group.com/de/clean_technology/clean_technology_portfolio/hydrogen_as_fuel/hydrogen/index.html (accessed 22 May 2013).

10 Tamhankar, S. (2012) Green hydrogen by pyroreforming of glycerol, presented at the 19th World Hydrogen Energy Conference– WHEC 2012 – 3–7 June 2012, Toronto, Canada. Available at http://www.whec2012.com/wp-content/uploads/2012/06/Tamhankar-WHEC2012-HPA11-2R.pdf (accessed 22. May 2013).

11 Trudewind, C.A. (2010) Vergleich von H2-erzeugungsverfahren, in: *Nachhaltigkeit von Energiesystemen – Kriterien und Bewertungen* (eds H.-J. Wagner and M.K. Koch), Lit Verlag, Münster.

12 ENERTRAG AG (18 January 2011) Wasserstoff-Hybridkraftwerk Prenzlau: ENERTRAG, TOTAL und Vattenfall vereinbaren Kooperation, Berlin, press release.

13 GermanHy (2004) Woher kommt der Wasserstoff in Deutschland bis 2050? Available at http://www.dena.de/fileadmin/user_upload/Publikationen/Verkehr/Dokumente/germanHy_Abschlussbericht.pdf (accessed 22 May 2013).

14 Ball, M., Weindorf, W., and Bünger, U. (2009) *Hydrogen Distribution in The Hydrogen Economy: Opportunities and Challenges* (eds M. Ball and M. Wietschel), Cambridge University Press, New York.

15 Smolinka, T. and Voglstätter, C. (2013) Wasserstoff-Infrastruktur für eine nachhaltige Mobilität – Entwicklungsstand und Forschungsbedarf, E-mobil BW GmbH, Stuttgart.

16 H2-Mobility (2010) 70 MPa Hydrogen Refuelling Station Standardization: Functional Description of Station Modules. Available at http://www.now-gmbh.de/fileadmin/user_upload/RE_Publikationen_NEU_2013/Publikationen_NIP/H2Mobility_HRS_Functional_Description.pdf (accessed 25 May 2015).

17 Wurster, R. (2011) Safe hydrogen infrastructure in Germany, NOW GmbH. Presented at the 19th World Hydrogen Energy Conference– WHEC 2012 – 3–7 June 2012, Toronto, Canada.

18 Zero Regio Project Consortium (2010) Demonstration of Fuel Cell Vehicles & Hydrogen Infrastructure: Summary of Results. Available at http://www.zeroregio.com/upload/Veroffentlichungen/list-of-reports/FinalSummaryofResults.pdf (accessed 14 January 2014).

19 Winter, C.-J. and Nitsch, J. (1988) *Wasserstoff als Energieträger: Technik, Systeme, Wirtschaft*, Springer Verlag, Heidelberg.

20 NOW GmbH (2013) Genehmigungsleitfaden für Wasserstoff-Stationen, Available at www.h2-genehmigung.de/ (accessed 14 January 2014).

Part 5
Handling of Hydrogen

Hydrogen Science and Engineering: Materials, Processes, Systems and Technology, First Edition.
Edited by Detlef Stolten and Bernd Emonts.
© 2016 Wiley-VCH Verlag GmbH & Co. KGaA. Published 2016 by Wiley-VCH Verlag GmbH & Co. KGaA.

38
Regulations and Codes and Standards for the Approval of Hydrogen Refueling Stations

Reinhold Wurster

38.1
Introduction

During the remainder of this decade, fuel cell electric vehicles (FCEVs) will become fully market ready towards 2020 and will be introduced by leading automotive companies (OEM) in key markets (e.g., China, Germany, Japan, Korea, UK, and USA). To ensure a smooth introduction of hydrogen fueled FCEVs in these markets, safe customer friendly to operate and affordable hydrogen refueling stations (HRSs) have to become available and need to be implemented in various locations under various existing local legal requirements. At present, unified procedures for the approval and permission of HRSs, allowing the use of hydrogen based on harmonized legislation, are missing on a global scale. It needs to be assured that this does not represent a roadblock for safe and widespread use of hydrogen (H_2) and FCEVs. The safety of hydrogen infrastructure is an essential prerequisite. A geographically area-wide coverage with harmonized requirements and efficient permitting procedures is therefore desirable if not indispensable.

The internationally applied term "Regulations, Codes and Standards," abbreviated as "RCS" sometimes is interpreted in an imprecise manner. Therefore, in the following the terms "regulation," "code" and "standard" are explained.

38.1.1
Explanation of the term "Regulations, Codes and Standards (RCS)"

38.1.1.1 Regulations
Regulations are legal requirements or laws. They have to be respected by all citizens. They can be treated as the highest level of coding, because they not only contain descriptions of the physical and operational features of the given technology or product they rule, but also performance definitions and limit values (tolerances) to be followed, and implicit restrictions for the use of non-standard or non-compliant items or systems. Usually, standards are enough for business use,

Hydrogen Science and Engineering: Materials, Processes, Systems and Technology, First Edition.
Edited by Detlef Stolten and Bernd Emonts.
© 2016 Wiley-VCH Verlag GmbH & Co. KGaA. Published 2016 by Wiley-VCH Verlag GmbH & Co. KGaA.

but regulations are needed to ensure that human safety and/or environmental compatibility are not unduly compromised by the given product or system [1].

In Europe, more precisely in the European Union, the main legislative acts come in two forms [2]: as Regulations and as Directives.

Regulations become law in all Member States the moment they come into force, without the requirement for any implementing measures or Member State per Member State adaptation, and automatically override conflicting domestic provisions.

Directives require Member States to achieve a certain result while leaving them a certain scope as to how they want to process and achieve the result. When the time limit for implementing directives passes, they may, under certain conditions, have direct effect in national law against Member States.

Decisions offer an alternative to the two above modes of legislation. They are legal acts that only apply to specified individuals or companies. They are often used in Competition Law, or on rulings on State Aid, but are also frequently used for procedural or administrative matters within the institutions.

As a result, EC Regulations and Directives need to be considered as a kind of supranational law, which has to be regarded by every Member State. In case of non-compliance fines will apply.

The above explanations are adapted from Reference [2]. This reference also provides a very detailed explanation of the European Union and Germany legal systems.

In the USA "codes," often referred to as legally binding, can be a collection of laws, regulations, ordinances, or other statutory requirements adopted by a government legislative authority that is involved in assuring the adequacy of the physical properties and healthy conditions of a product or equipment. Codes establish predictable, consistent minimum harmonized requirements that are to be applied, that is, criteria of being both "practical and adequate for protecting life, safety, and welfare of the public." Codes are adopted by a state or local government's legislative body, then enacted to regulate the manufacture or construction within a particular jurisdiction (city, county, or entire federal state). The primary purpose of a code is to regulate a new or proposed product, equipment, or construction. The earliest codes in the USA were for building construction and for fire protection.

38.1.1.2 Codes of Practice

Codes of practice usually explain the basic functions for safe handling and problem-preventive maintenance and are intended to guarantee trouble-free operation. In theory, those codes can be different for each product or at least for each manufacturer, but usually they share at least some basic elements that are built around the generic features and functionalities of the technology, and build a common understanding among relevant people of how to deal with this type of product or system. Good examples of this level of coding can be found around us everywhere, such as in automotive technology, where the order and functionality of the basic controls in a passenger car are not even written down in any code. Still a common philosophy is shared throughout the industry, and the pedals in all cars are in the same order from left to right, and the vehicle is steered with a wheel that turns the front wheels to the same direction as the driver turns the steering wheel. Even if these types of codes are often referred as "industry standards," they are not necessarily based on any formal agreement or even on any type of written document, but still exist. Nor are they produced by any deliberate development pathway, where the best, most efficient and safest practice is the target [1].

38.1.1.3 Standards

Should the technology in question interface with other systems or use common elements shared by different platforms, we need more precise descriptions of the codes that guarantee the compatibility of the related items. They need to be based on a written text, agreed in its scope and contents by those that have developed it and are using it to design and manufacture products and systems in that given field of technology. This type of coding is usually referred to as "standard," and such standards are usually developed and maintained by some specific organizations or interest groups built around the case or even the given branch of technology (so-called SDOs, Standards Drafting Organizations) [1].

In global markets, standards should be developed as international standards (not as national or regional ones – preferably UN-level like the ISO (International Organization for Standardization) and IEC (International Electrotechnical Commission) and not on the European regional or German, US or Japanese level). This will improve compatibility with all markets and avoid duplicate work and confusion [3].

ISO and IEC standards are widely adopted at the regional or national level and are used by all interested stakeholders, such as manufacturers, trade organizations, purchasers, consumers, certification bodies, testing laboratories and authorities. Since these standards reflect the best experience of industry, researchers, consumers and regulators worldwide and cover common needs in various countries, they constitute one of the important bases for the removal of unnecessary technical barriers to trade (Figure 39.1).

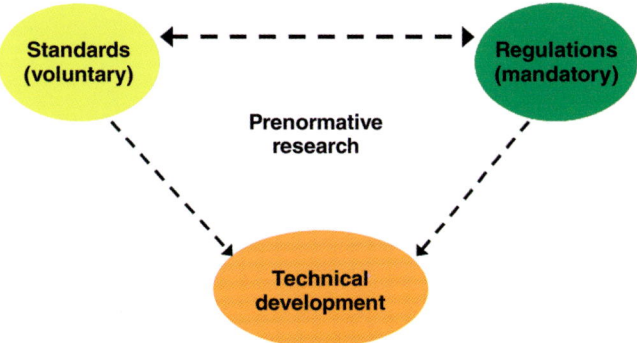

Figure 38.1 Relationship between standards, regulations, and products.

Standards as an instrument for harmonizing the requirements for interfaces in technical systems are usually agreed upon between private organizations and do not per se constitute a legal requirement but rather the state-of-the-art. In most countries they acquire legal status only if referenced from regulations or laws (legal documents). One exemption is China, where binding mandatory standards ("GB/" standards) also exist that have to be complied with like laws [4].

The incorporation of standards in legislative instruments by means of a reference constitutes a method of drafting a code or regulation in such a way that a detailed statement of technical requirements is replaced in the text of the code or regulation by a reference to one or more standards, or to the relevant parts thereof. The use in regulation of standards, and preferably of ISO and IEC standards, is an effective means of supporting national, regional, and global policies. It is already widely used in several regions of the world in concepts, agreements, and frameworks such as the New Approach in the European Union, Good Regulatory Practice developed by the Asia-Pacific Economic Cooperation, Subcommittee on Standards and Conformance, and the North American Free Trade Agreement. The use of international standards facilitates the elimination of unnecessary barriers to trade [4].

38.1.1.4 Referencing of standards

Direct referencing [4] means that the reference of a specific standard is directly quoted within a legal text using its identification number and title. This method often supports the mandatory use of a standard, so careful wording of the regulation will be necessary if the regulator wants the use of the ISO or IEC standard to remain optional (i.e., as one of a number of solutions to help comply with the regulation). By directly referencing standards in this way, regulators avoid reproduction of the standard in the legal text and eliminate the need to obtain permission for the use of copyright. Another advantage is that certain parts, or even single clauses of a standard, can be referenced where only a small part of a standard supports a regulation. There are two forms of direct referencing: dated and undated.

When using dated references in regulations any revisions of, or amendments to, ISO or IEC standards have to be taken into consideration by the relevant authority. The legal text will then need to be changed to note any amendments to, or revisions of, the standard.

In the case of an undated reference, the regulation quotes only the number and title of a specific standard and not the date. This method is therefore more flexible. In the case of a revision of a referenced standard, the regulation itself does not need to be adapted and the reference automatically corresponds to the latest edition of the standard and therefore the state of the art. In other words, the regulation allows the use of subsequent revised editions of the same standard.

Indirect referencing [4] involves recognizing and registering standards on an official information source external to the regulatory text. In this way, a list of standards deemed suitable by the regulator is compiled and published by an official process that the regulator controls. If a standard is revised or amended, no change is necessary to the legal text, only to the list. The list of standards may also include the dates of publication of the standards so as to ensure the legal certainty of a dated reference and to indicate when a particular edition is valid.

Such a list of recognized references needs to be kept up to date and made easily available to users through a web site or other means. This model has been applied in Europe where it is referred to as the "New Approach" (http://www.newapproach.org/) and is a joint effort of the three European Standards Organizations (CEN, CENELEC, and ETSI) together with both the European Commission and EFTA (European Free Trade Association).

38.1.1.5 Why Globally Harmonized Standards are Needed

The existence of non-harmonized standards for a given product in different countries or regions contributes to the so-called Technical Barriers to Trade (TBT). A fair competition needs to be based on clearly defined common references that are recognized from one country to the next and from one region to another. International standards, developed by consensus among trading partners, serve as the language of trade and represent a key ingredient of the World Trade Organization's (WTO's) Agreement on Technical Barriers to Trade (TBT).

Furthermore, the number of standardization bodies that have accepted the Code of Good Practice for the Preparation, Adoption and Application of Standards presented in Annex 3 to the WTO's TBT Agreement underlines the global importance and reach of this accord [5]. As of 2 February 2013, some 158 member countries have accepted this Code of Good Practice.

38.1.2
General Requirements for the Approval of Hydrogen Refueling Stations (HRSs)

The following general requirements have been taken from Reference [6], which has been compiled with the involvement of experts from China, Europe, Japan, and the United States of America. A more detailed break down and specification of requirements for HRSs can be found in ISO/DIS 2010 Gaseous hydrogen – Fueling stations [7].

HRS Design and Construction Recommendations: Recommendations for hydrogen plants (general design, hydrogen quality, piping, venting, pressure relief and venting discharge devices, material selection criteria, insulation, process control systems, instrumented safety systems), Safety distances (definition of safety distance, determination of safety distances, application to hydrogen refueling station, hazardous areas) [8], HRS plant level recommendations (site selection and HRS lay out, identification of and access to hazard zones, building, hydrogen gas detection and hydrogen fire detection, fire prevention and firefighting, HRS emergency shutdown, security), Specific recommendations (onsite production, hydrogen delivery by compressed gaseous hydrogen tube trailers or multi cylinder pack or by liquid hydrogen trailers or by pipeline, pipeline interface, hydrogen dispensing, gaseous hydrogen storage system, liquid hydrogen storage system, specific storage configurations).

Operation and Maintenance of an HRS: Operating requirements, Emergency safety plan, Maintenance requirements (general principle, typical maintenance program for a gaseous HRS).

Vehicle Interface Requirements: General requirements (compressed hydrogen refueling, liquid refueling), Grounding, Communication with the vehicle (gaseous hydrogen tank system, liquid hydrogen tank system), Refueling interface (nozzles and breakaway) for 35 MPa and 70 MPa gaseous hydrogen interface and for liquid hydrogen interface, Driver instructions (driver, station operator, vehicle manufacturer (OEM)), Acceptance of HRS based on OEM perspective (codes and approval authorities, passive safety compliance, functional safety, performance tests).

Risk Assessment Methodologies for HRS Approval: Concepts involved in risk assessment, Risk assessment process (risk management, risk acceptance criteria, risk assessment procedure, hydrogen accident databases), Risk assessment methods (HAZID – HAZard Identification, HAZOP – HAZard and OPerability study, SWIFT – Structured What-IF checklisT, FMECA – Failure Mode, Effect and Criticality Analysis, QRA – Quantitative Risk Assessment, RRR – Rapid Risk Ranking, Hazard consequence models used in QRA, Combination of risk assessment methods).

Country/location specific issues: Identification of stakeholders, Required information (laws/regulations applied, permits required, the required information by the authorities), External and occupational safety policy concerning hydrogen (target groups for a safety assessment, safety policy), Technical standards for the construction of an HRS, Methodologies and guidelines for the assessment of external (off-site) effects damage and risks, External safety (off-site safety) and land use planning, Inspection protocol, Contingency planning, Flow charts of the approval process.

38.2
European Union and Germany

38.2.1
Europe

Within the European Union of 27 national states (so-called Member States) a huge variety of procedures exist for the permitting of HRSs, which do not only differ between the Member States but also can differ within the single Member States. There are countries were requirements and procedures can be followed more easily or where more experience in permitting has already been obtained (e.g., Denmark, Germany, the Netherlands, Norway) and there are countries where this favorable situation does not exist (e.g., France and Italy). Therefore,

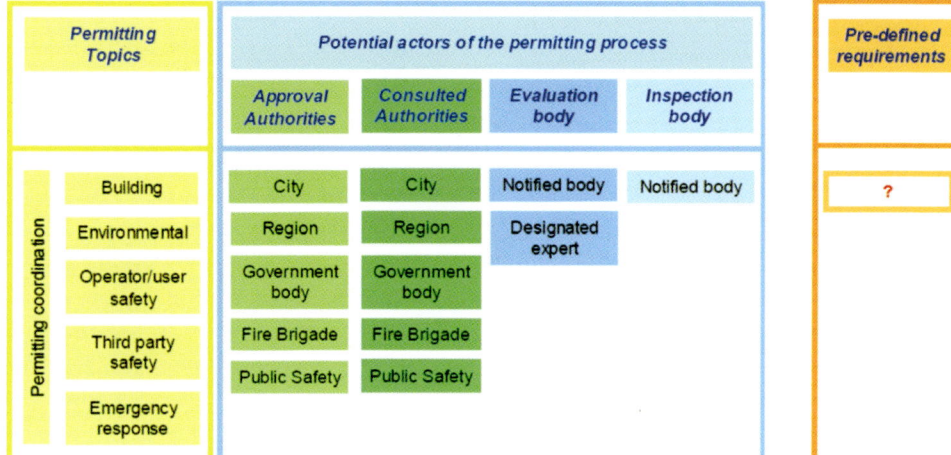

Figure 38.2 Potential permitting process complexity as identified for Europe by HyApproval [6].

it has long been obvious and recognized that this very fragmented situation needs improvement.

Hence, between 2005 and 2007, the European Commission co-funded project HyApproval [6] analyzed existing approval realities in six EU countries and sought to make best use of international experience, including in China, Japan, and the USA, and recommended improved approval practices.

The analysis has shown a comparatively complex approval practice for hydrogen refueling stations (HRS) by authorities in different countries worldwide (Figure 39.2).

This complex picture is already valid for Europe, where the approval process is handled by different authorities depending on the respective country, sometimes even differing within one country from province to province (Spain), region to region (Italy), department to department (France), or federal state to federal state (Germany). The approval process for a HRS in Italy or Belgium, for example, is coordinated by the fire brigades, whereas in Germany it is by 43 different labor safety authorities active in 16 federal states. Although the requirements according to which the approval process is performed, for example, in Germany are identical for all 16 federal states, the implementation of the process differs sometimes slightly from authority to authority due to the different level of awareness or knowledge available at the authorities.

Therefore, the HyApproval Handbook [6] already in 2008 published its key recommendation to develop an EC regulatory framework for hydrogen refueling stations based on the proven combination of Essential requirements, Harmonized standards, and Notified bodies. It was concluded that this could be achieved most efficiently through the development of an EC Regulation (as opposed to an EC Directive). Such a framework, which allows addressing the

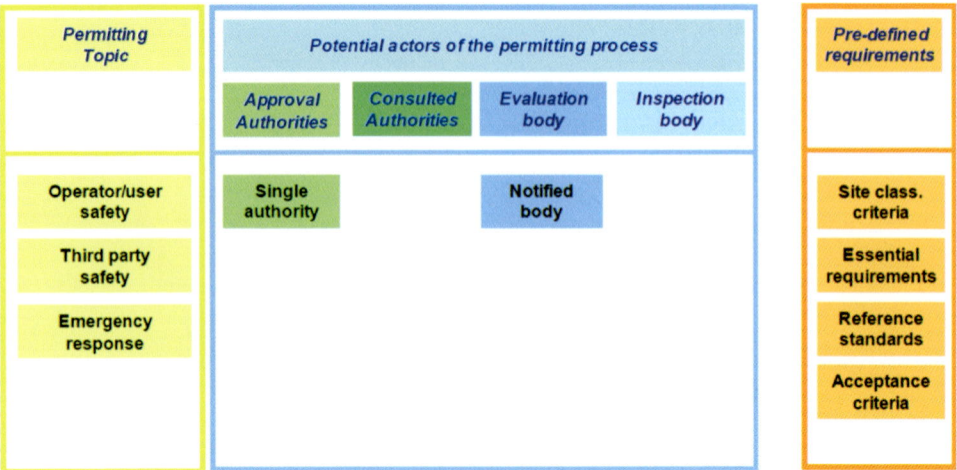

Figure 38.3 EU27 uniform permitting process suggested for EU27 by HyApproval [6].

key safety issues without impeding continued technological development, would establish a very streamlined EU 27 uniform permitting process as depicted in Figure 39.3.

After completion of HyApproval [6], in the meantime the European Union in 2009 published a regulation addressing the whole vehicle type approval of hydrogen propelled road vehicles [EC/79/2009] for EU27. In 2010, regulation EC/406/2010 [EC/406/2010] providing the implementing measures for EC/79/2009 was published, allowing the type approval of hydrogen vehicles in EU27. In preamble (16) of EC/79/2009, the European Commission (EC) was requested to "look into suitable measures to support the establishment of a Europe-wide filling-station network for hydrogen-powered vehicles."

Among others, as a consequence of this request, in January of 2013 the EC has issued a proposal for a Directive [9]. In Article 5 of this directive proposal regulating hydrogen supply for transport, the following requirements are stipulated:

1) Member States on the territory of which exist already at the day of the entry into force of this Directive hydrogen refueling points shall ensure that a sufficient number of publicly accessible refueling points are available, with distances not exceeding 300 km, to allow the circulation of hydrogen vehicles within the entire national territory by 31 December 2020 at the latest.
2) All hydrogen refueling points for motor vehicles shall be compliant with the technical specifications, set out in Annex III.2 by 31 December 2015 at the latest.
3) The Commission shall be empowered to adopt delegated acts in accordance with Article 8 concerning the updating of the technical specifications set out in Annex III.2.

Annex III.2 requires the following technical specifications to be complied with:

2. Technical specifications for hydrogen refueling points for motor vehicles.
2.1. Outdoor hydrogen refueling points dispensing gaseous hydrogen used as fuel on board land vehicles shall comply with the relevant EN standard, to be adopted by 2014, and, pending the publication of this standard, with the technical specifications of the ISO/TS 20 100:2008 Gaseous Hydrogen Fueling specification.
2.2. The hydrogen purity dispensed by hydrogen refueling points shall comply with the relevant EN standard, to be adopted by 2014, and, pending the publication of this standard, with the technical specifications included in the ISO 14 687-2 standard.
2.3. Hydrogen refueling points shall employ fueling algorithms and equipment complying with the relevant EN standard, to be adopted by 2014, and, pending the publication of this standard, with the ISO 20 100 Fueling Protocols for Light Duty Gaseous Hydrogen Surface Vehicles.[1)]
2.4. Connectors for vehicles for the refueling of gaseous hydrogen shall comply with the relevant EN standard, to be adopted by 2014, and, pending the publication of this standard, with the ISO 17 268 gaseous hydrogen land vehicle refueling connection devices standard.

This process clearly shows the importance of two RCS components: Regulations and therein referenced recognized Standards. It follows the New Approach [10] adopted by the European Commission, which recommends to reference standards from regulations or directives wherever appropriate to avoid duplicate work and/or confusion.

Whenever international standards exist, if they are regarded as appropriate and useful, the European standardization bodies CEN/CENELEC adopt these as EN standards then dubbed "EN ISO" or "EN IEC." The long procedure, which has carried on since the establishment of the ISO technical committee 197 (ISO/TC 197) in 1990, has resulted in various useful standards, draft industry standards, or technical specification, like international standards ISO 17 268 (which was adopted from SAE J2600), ISO 14 687-2, or draft standard ISO/DIS 20 100. CEN could adopt these as EN standards in 2014 if they were available as fully accepted ISO standards (which is not yet the case for ISO/DIS 20100). US standards regarded as reference by the global industry such as, for example, SAE J2601, which covers "Fueling Protocols for Light Duty Gaseous Hydrogen Surface Vehicles" and which was published in 2014 [11], for European regulatory use and cross-referencing would have to be transferred again into an ISO standard (similar to SAE J2600, which became ISO 17268).

---

1) Here a miss-referencing has definitely occurred as ISO 20 100 does not cover "Fueling Protocols for Light Duty Gaseous Hydrogen Surface Vehicles," which rather is covered in US SAE J2601 (a type of standard which cannot be referenced from an EU Regulation or Directive as it is not recognized as an international standard by EU).

In case no international ISO or IEC standard will come into force for HRSs in the foreseeable future, CEN/CENELC will have to draft a separate EN standard, which once in force replaces national standards in 33 European countries. For HRS to permit this the present way forward is the development of international standard ISO 19880-1, which is under way by ISO working group 24 and which aims to publish by the end of 2016. CEN could adopt this standard as EN standard in 2017. By the end of 2017, this adopted EN standard could be references as required in the AFID [9,12].

38.2.2
Germany

The approval process for an HRS in Germany only deals with the construction and the operation of an HRS. The manufacturing of the HRS equipment is already finalized. For this purpose, components such as pressure vessels have undergone successful tests and verifications, been accepted by a notified body, and received the relevant CE marking proving compliance with the applicable European Directive or Regulation.

As a general rule, for an HRS a construction permit has to be obtained and a permit according Section 13 BetrSichV (Regulation on Industrial Safety and Health) is required, once pressure equipment according to the Pressure Equipment Directive (PED) [13] with a CE marking has been installed in the plant. For the permit an expert opinion by a notified body has to be obtained, which assesses the safety relevant aspects of the HRS as well as the intended assembly of the pressure equipment within the HRS. During this process it will be checked if for all components to be installed a declaration of conformity has been obtained. The expert opinion by the notified body serves as a technical appraisal of the HRS for the approval authority. As soon as permission by the authority is obtained, the HRS before its first startup has to undergo an examination prior to startup according to Section 14 BetrSichV, which also has to be performed by a notified body.

BetrSichV furthermore rules risk assessment in Section 3, requirements for explosion protection in Section 6, and periodic inspection in Section 15.

The BetrSichV is the German law implementing EC directive 2009/104/EC. This directive requires the consideration of Reference [13], which in Germany is being implemented by the 14th GPSGV [14]. In its Annex I the requirements to be met are defined. These can be fulfilled by either following harmonized standards like Reference [15] or by applying the German regulation AD 2000.

In Germany the Equipment and Product Safety Act [16] ensures the implementation of the European directives applicable to the layout and approval of HRSs. These are 97/23/EC (Pressure Equipment Directive) by 14th GPSGV [14], 2006/42/EC (Machinery Directive) [17] by 9th GPSGV [18], 2006/95/EC (Voltage Limits Directive) [19] by 1st GPSGV [20], 2009/105/EC (Simple Pressure Vessel Directive) [21] by 6th GPSGV [22], 94/9/EC (ATEX Directive) [23] by 11th

GPSGV [24], and 2004/108/EC (Electromagnetic Compatibility Directive) [25] by EMVG 2008 [26].

If the plant is composed of parts that fall under the BImSchG (Federal Control of Pollution Law), for the HRS a corresponding permit by the relevant authority also has to be obtained. Which parts have to be considered by the BImschG is regulated in Section 1 of the 4th BImSchV (Federal Control of Pollution Decree), which deals with plants requiring approval. Such plants have to be mentioned in the decree, have to be operated for longer than 12 months at a given location, and are not to be considered as laboratory or pilot plant. Regarding HRSs this refers to onsite production of hydrogen, for example, by electrolysis or by compact natural gas steam reformer plants, or to the storage of hydrogen quantities larger than 3 t. The 9th BImSchV (Federal Control of Pollution Decree) rules the approval process itself.

The BImSchG permitting process stipulates further requirements to be complied with by the applicant. The documentation required for the permit according to the BetrSichV, however, is included in the BImSchG permitting process and has to be submitted together with all documentation handed in when applying for approval. At the end of the permitting process according to BImSchG an examination prior to startup according to Section 14 BetrSichV is required. The compliance with the requirements according to BImSchG has to be proven by a BImSchG expert. The BImSchG is the national implementation of Seveso II [27] and MAHD [28].

In 2010, in Germany a technical bulletin or code of practice was published that addresses the applicable minimum requirements to be respected in the approval of hydrogen refueling stations [29]. VdTÜV 514 is structured as: Scope, Definitions, Requirements for hydrogen fueling station components, Construction and installation requirements, Electrical requirements, Obligations under BetrSichV, Commissioning, Operation and maintenance, Regulations and standards. Safety distances are covered by TRBS 2152 [30], part 2 (Technical rules for operational safety/hazardous materials – Prevention or restriction of hazardous explosive atmospheres). The operational safety in the sense of the BetrSichV is evaluated according to TRBS 2141 [31] (Technical regulations for safety – Hazards by steam and pressure – General requirements).

The recognition of such VdTÜV bulletins is usually limited to use in Germany. For better harmonization of requirements on the European or international level, one report [32] has already recommended that it will be better to apply the more complete ISO/DIS 20 100 as soon as finalized. The advantage of this approach would be that a much more widely harmonized set of requirements covering 164 full ISO member countries globally would be applied. This would drastically reduce the need for national harmonization measures which presently have to be performed for local approval. National legislators could reference to ISO 20 100 and only request additional requirements if desired. This would allow HRS implementers to greatly better standardize their offering. Finally, this will also contribute to safety, reliability, and economics. In case of a non-availability of an ISO standard, an EN standard would be a fallback position for Europe.

References

1 HySociety (2004) WP3 Deliverable 10 – Measures for codes, standards and regulations for safe (and efficient) deployment of hydrogen in European Union, August 2004.
2 Garche, S., Mao, Z.Q., and Wolff, A. (2009) Analysis of European/German and Chinese Regulations regarding a hydrogen infrastructure for road traffic, German Chinese Sustainable Fuel Partnership (GCSFP). Available at http://www.dena.de/fileadmin/user_upload/Projekte/Verkehr/Dokumente/GCSFP-Study_H2-Regulations.pdf.
3 Dey, R. (2010) Advancing Commercialization of Hydrogen and Fuel Cell technologies through International Cooperation of Regulations, Codes and Standards (RCS). *18th World Hydrogen Energy Conference 2010 – WHEC 2010: Parallel Sessions Book 5: Strategic Analyses/Safety Issues/Existing and Emerging Markets WHEC, 16–21 May 2010, Essen* (eds D. Stolten and T. Grube), The Research Centre Jülich, Jülich.
4 Wurster, R. and He, Y. (2010) Analysis of the Situation of Standards and Codes for Hydrogen and Fuel Cell Vehicles and related Supply Infrastructures, German Chinese Sustainable Fuel Partnership (GCSFP). Available at http://www.dena.de/fileadmin/user_upload/Projekte/Verkehr/Dokumente/GCSFP-Study_H2-Codes_and_Standards.pdf
5 World Trade Organization (2013) http://www.wto.org/english/res_e/booksp_e/analytic_index_e/tbt_01_e.htm#p (accessed 2 February 2013).
6 HyApproval WP2 (2008) Handbook for Hydrogen Refuelling Station Approval - Version: 2.1, prepared by AIR LIQUIDE – DTA (AL DTA), Air Products PLC (APL), BP Gas Marketing Limited (BP), Chinese Academy of Sciences (CAS), Commissariat a l'Energie Atomique (CEA), Det Norske Veritas AS (DNV), Engineering Advancement Association of Japan (ENAA), ENI SpA (ENI), Forschungszentrum Karlsruhe (FZK), Health and Safety Executive (HSE/HSL), Hydrogenics Europe NV (HYGS), Institut National de l'Environnement Industriel et des Risques (INERIS), Instituto Nacional de Técnica Aeroespacial (INTA), Joint Research Centre – Institute for Energy (EC-JRC), Linde (Linde), National Center for Scientific Research Demokritos (NCSRD), National Renewable Energy Laboratory (NREL), Norsk Hydro ASA (Hydro), Shell Hydrogen BV (Shell), TNO - Netherlands Organisation for Applied Scientific Research (TNO), 4 June 2008. Available at http://www.hyapproval.org/Publications/The_Handbook/HyApproval_Final_Handbook.pdf
7 ISO (2008) ISO/DIS 20100 Gaseous hydrogen – Fuelling stations, 21 March 2011, which was under development until February 2013 and which is based on ISO/CD 20100 of 06 October 2009, which again is based on the published ISO 20100:2008 Gaseous hydrogen – Fuelling stations. http://www.iso.org/iso/home/search.htm?qt=20100&sort=rel&type=simple&published=on.
8 Barth, F. (2011) Risk informed separation distances for hydrogen refueling stations developed for ISO/DIS 20100 – Rationale and calculation basis, 10 July 2011.
9 COM (2013) 18/2 - 2013/0012 (COD) – Proposal for a Directive of The European Parliament and of The Council on the deployment of alternative fuels infrastructure, January 2013. Available at http://www.europarl.europa.eu/registre/docs_autres_institutions/commission_europeenne/com/2013/0018/COM_COM(2013)0018_EN.pdf.
10 New Approach Standardisation in the Internal Market in conjunction with: European Union (04/06 1985) Council Resolution of 7 May 1985 on a new

approach to technical harmonization and standards (85/C 136/01). *Official J. C*, 136, 0001–0009. Available at, www.newapproach.org/.

11 SAE (2014) standard J2601 "Fueling Protocols for Light Duty Gaseous Hydrogen Surface Vehicles", revision published: 2014-07-15. http://standards.sae.org/j2601_201407/.

12 European Parliament and European Council (28 October 2014) Directive 2014/94/EU of the European Parliament and of the Council of 22 October 2014 on the deployment of alternative fuels infrastructure. *Official J. Eur. Union*, L307. Available at http://eur-lex.europa.eu/legal-content/EN/TXT/PDF/?uri=CELEX:32014L0094&from=EN.

13 European Parliament and European Council (9 July 1997) Directive 97/23/EC of the European Parliament and of the Council of 29 May 1997 on the approximation of the laws of the Member States concerning pressure equipment. *Official J. Eur. Union*, L181, 1–55. Available at http://ec.europa.eu/enterprise/sectors/pressure-and-gas/documents/ped/index_en.htm

14 Anon (27 September 2002) Vierzehnte Verordnung zum Produktsicherheitsgesetz (Druckgeräteverordnung) (14. ProdSV) [14th decree on the equipment and product safety law (pressure equipment decree)]. Available at http://www.gesetze-im-internet.de/gsgv_14/BJNR380600002.html

15 CEN (2002) EN 13445. Unfired Pressure Vessels standard that provides rules for the design, fabrication, and inspection of pressure vessels, introduced in 2002 as a replacement for national pressure vessel design and construction codes and standards in the European Union and is harmonized with the Pressure Equipment Directive (97/23/EC or "PED"). Information available at http://en.wikipedia.org/wiki/EN_13445.

16 Anon (8 November 2011) Gesetz über die Bereitstellung von Produkten auf dem Markt (Produktsicherheitsgesetz - ProdSG), 08.11.2011. Available at http://www.gesetze-im-internet.de/bundesrecht/prodsg_2011/gesamt.pdf

17 European Parliament and European Council (9 June 2006) Directive 2006/42/EC of the European Parliament and of the Council of 17 May 2006 on machinery, and amending Directive 95/16/EC (recast). *Official J. Eur. Union*, L157, 24–86. Available at http://eur-lex.europa.eu/LexUriServ/LexUriServ.do?uri=OJ:L:2006:157:0024:0086:EN:PDF.

18 Anon (12 May 1993) Neunte Verordnung zum Produktsicherheitsgesetz (Maschinenverordnung) (9. ProdSV), 12.05.1993 [9th decree on the equipment and product safety law (decree on machinery in the market)]. Available at http://www.gesetze-im-internet.de/gsgv_9/BJNR070410993.html.

19 European Parliament and European Council (27 December 2006) Directive 2006/95/EC of the European Parliament and of the Council on the harmonisation of the laws of Member States relating to electrical equipment designed for use within certain voltage limits of December 2006. *Official J. Eur. Union*, L 374, 10–19. Available at http://eur-lex.europa.eu/legal-content/EN/TXT/PDF/?uri=CELEX:32006L0095&from=en.

20 Anon (11 June 1979) Erste Verordnung zum Produktsicherheitsgesetz (Verordnung über die Bereitstellung elektrischer Betriebsmittel zur Verwendung innerhalb bestimmter Spannungsgrenzen auf dem Markt) (1. ProdSV) [1st decree on the equipment and product safety law (decree on the provision of electrical equipment for the use within certain voltage limits in the market]. Available at http://www.gesetze-im-internet.de/techarbmgv_1/BJNR006290979.html.

21 European Parliament and European Council (8 October 2009) Directive 2009/105/EC of the European Parliament and of the Council of 16 September 2009 relating to simple pressure vessels. *Official J. Eur. Union*, L 264, 12–29. Available at http://eur-lex.europa.eu/LexUriServ/LexUriServ.do?uri=OJ:L:2009:264:0012:0029:en:PDF.

22 Anon (25 June 1992) Sechste Verordnung zum Produktsicherheitsgesetz (Verordnung über die Bereitstellung von einfachen Druckbehältern auf dem Markt)

(6. ProdSV) [6th decree on the equipment and product safety law (decree on the provision of simple pressure vessels in the market)]. Available at http://www.gesetze-im-internet.de/gsgv_6/BJNR111710992.html

23. European Parliament and European Council (19 April 1994) Directive 94/9/EC of the European Parliament and the Council of 23 March 1994 on the approximation of the laws of the Member States concerning equipment and protective systems intended for use in potentially explosive atmospheres. *Official J. Eur. Union*, L 100. Available at http://eur-lex.europa.eu/legal-content/EN/TXT/PDF/?uri=uriserv:OJ.C_.2014.445.01.0005.01.ENG

24. Anon (12 December 1992) Elfte Verordnung zum Produktsicherheitsgesetz (Explosionsschutzverordnung) (11. ProdSV) [11th decree on the equipment and product safety law (explosion protection decree)]. Available at http://www.gesetze-im-internet.de/bundesrecht/gsgv_11/gesamt.pdf.

25. European Parliament and European Council (31 December 2004) Directive 2004/108/EC of the European Parliament and the Council of 15 December 2004 on the approximation of the laws of the Member States relating to electromagnetic compatibility and repealing Directive 89/336/EEC. *Official J. Eur. Union*, L 390, 24–37. Available at http://eur-lex.europa.eu/legal-content/EN/TXT/PDF/?uri=CELEX:32004L0108&from=en.

26. Anon (26 February 2008) Gesetz über die elektromagnetische Verträglichkeit von Betriebsmitteln (EMVG), 26.02.2008. Available at http://www.bmwi.de/DE/Service/gesetze,did=21938.html.

27. European Parliament and European Council (14 January 1997) Directive 96/82/EC of the Council on the Control of Major Accident Hazards involving dangerous substances (SEVESO II). 09 December 1996. *Official J. Eur. Union*, L10, 13–33. Available at http://eur-lex.europa.eu/LexUriServ/LexUriServ.do?uri=CELEX:31996L0082:EN:HTML.

28. European Parliament and European Council (31 December 2003) Directive 2003/105/EC of the European Parliament and of the Council of 16 December 2003 amending Council Directive 96/82/EC on the control of major-accident hazards involving dangerous substances. *Official J. Eur. Union*, L 345, 97–105. Available at http://eur-lex.europa.eu/LexUriServ/LexUriServ.do?uri=OJ:L:2003:345:0097:0105:EN:PDF.

29. Verband der TÜV (2010) Compressed gases 514: Requirements for hydrogen fueling stations, MB DRGA 514-englisch, VdTÜV-Merkblatt: Compressed gases 514, Verband der TÜV e. V., Berlin, 04–2010. Available at http://www.vdtuev.de/shop/merkblaetter.

30. Committee on Operational Security (ABS) and The Committee on Hazardous Substances (2012) TRBS 2152 Part 2/TRGS 722: Technical rules for operational safety/hazardous materials - Prevention or restriction of hazardous explosive atmospheres TRBS 2152 Teil2/TRGS 722: Technische Regeln für Betriebssicherheit/Gefahrstoffe - Vermeidung oder Einschränkung gefährlicher explosionsfähiger Atmosphäre], GMBI Nr. 22, March 2012, pp. 398–410. Available at http://www.arbeitssicherheit.de/de/html/library/document/2201417,38.

31. Anon (2007) Technical regulations for safety TRBS 2141 – Hazards by steam and pressure – General requirements Technische Regeln für Betriebssicherheit TRBS 2141 Gefährdungen durch Dampf und Druck - Allgemeine Anforderungen -], GMBI Nr. 15, 23 March 2007, p. 327. Available at http://www.baua.de/de/Themen-von-A-Z/Anlagen-und-Betriebssicherheit/TRBS/TRBS-2141.html.

32. LBST, BASt, TÜV SÜD and CCS prepared for NOW (09 December 2011) Sichere Wasserstoffinfrastruktur (Safe Hydrogen Infrastructure), Report. Available at http://www.now-gmbh.de/fileadmin/user_upload/RE-Presse_Downloads/Sichere-H2-Infra_NOW-RCS_FR_final_02MAR2012.pdf.

39
Safe Handling of Hydrogen
William Hoagland

39.1
Introduction

Hydrogen has been used in industry for more than five decades, and the world currently produces more than 70 million metric tons per year for use in more than a dozen industries. As fuel cell technology has advanced and the use of hydrogen as a fuel has come to the forefront, the public is becoming more aware of its potential as a consumer automobile fuel, and there have emerged heightened concerns about its safety. Each of the major industrial gas companies have stated that hydrogen can be safely used as a consumer fuel for fuel cells, and the new generation of hydrogen vehicles and the first production vehicles are scheduled to debut in 2015.

However, hydrogen is, in many ways, different from the current petroleum based liquid fuels currently used to power our transportation system. Some of these differences pose additional hazards that must be considered, while other properties of hydrogen, such as it high diffusivity, may reduce potential hazards.

Hydrogen is safely handled in industries and manufacturing processes for petroleum refining, steel, chemicals, foods, electronics, and float glass. It is used to upgrade crude oil to produce cleaner burning gasoline and low sulfur diesel fuel. In the metallurgical industry, hydrogen is used to prevent oxidation when heat-treating certain metals and alloys. Hydrogen is also used by semiconductor manufacturers, and it is used in the chemical industry to synthesize ammonia and methanol [1].

Now, hydrogen fuels are set to assume a key role in the world's energy future. The rapid and dramatic improvements in fuel cell technologies have significantly enhanced the technical and economic viability in several areas: electricity to power for distributed power systems, utility generation, auxiliary and portable power systems, and various vehicle applications such as forklifts, specialty vehicles, automobiles, and buses.

But what about their use in consumer environments where untrained personnel refuel their own cars? Whereas industry has maintained an excellent safety record in controlled industrial environments, there is little historical experience on which to assess the risk of its use by the general public. For many years, the general public's knowledge of hydrogen has been limited to the Hindenburg disaster in 1937, the hydrogen bomb, and perhaps the Challenger space shuttle accident in 1986, and many perceive hydrogen as dangerous. Of course, none of these examples is an adequate comparison to the new markets for hydrogen and fuel cells; however, the memories of these events persist. The *New York Times*, as recently as October 2013, published a feature article in it Sunday magazine entitled "Image of Hydrogen Haunts Hydrogen Technologies" in which this reflexive fear of hydrogen was called "The Hindenburg Syndrome." On the brighter side, most experts are aware of the dramatic technical advances and excellent safety record in industry, and are not daunted by the unique properties and hazards posed by hydrogen. In fact several automakers have stated that their future is based on hydrogen fuel cell vehicles.

In support of this transition, there is developing significant ongoing cooperation among the world's safety experts concerning all aspects of using hydrogen in future energy systems. Notable are two international efforts where experts from government and industry share knowledge and experience to assess the risks and identify and resolve knowledge gaps where more information is needed. These organizations are the International Energy Agency Hydrogen Implementing Agreement Task on Hydrogen Safety (www.ieahia.org), and the International Association for Hydrogen Safety (www.hysafe.org). More information regarding safe handling of hydrogen may be found at their respective web sites. The International Conference on Hydrogen Safety has been held every two years since 2005, and is likewise a good resource for the state of the art and current research directions in hydrogen safety (e.g., www.ichs2013.com).

39.2
Hydrogen Safety and the Elements of Risk

Acceptable levels of safety may be determined by calculating risk which, in its simplest terms, may be defined as the probability of a hazardous event multiplied by the resulting cost:

$$R(\text{risk}) = P(\text{probability}) \times C(\text{consequence})$$

Therefore, to calculate risk, one needs to know, or estimate, the probability of a hazardous event occurring and the consequence in terms of costs or loss of life. It has become commonplace to measure the consequence in terms of loss of life, and the idea of not increasing the risk of fatality by more than one in a million (wiki Risk assessment).

In the case of hydrogen, there is usually not enough experience or data on which to base the probability or consequence of an accident or hazardous event occurring. For this reason, better predictive techniques are being developed to estimate the risk of these new systems in order to obtain their approval.

In many countries, the approval of a hydrogen energy project involves predominately one of two approaches to determine the maximum acceptable risk [2]:

1) "ALARP" – As low as reasonably practical. This criterion is based on the assumption that all available cost-effective measures should be applied to achieve the lowest risk feasible, and where the cost of further risk reduction measures is weighed against the benefits of their implementation.
2) "Do no harm" – This approach is based on the requirement that the presence of hydrogen systems not increase the generally accepted risk normally seen in a society. In most developed countries this value is usually considered to be 1×10^{-6} or a "one in one million." occurrence.

When a new technology such as hydrogen fueled automobiles is evaluated for risk, many assumptions about the overall system must be made to complete the assessment. As these new systems are deployed, historical data will be obtained with regard to failure probability, hazards, and consequences.

39.2.1
Assessing Risk

There are many methods for estimating the risks associated with hydrogen systems. Hydrogen safety researchers commonly use quantitative risk assessment (QRA), also known as probabilistic risk assessment (PRA), to assess the risk associated with a hydrogen system. QRA was first introduced about 40 years ago and applied to large technological systems, but it is less well suited for smaller, less complicated systems, since completing a full QRA can be quite costly. Many software programs have been introduced and offered for such hydrogen systems; however, quantitative risk assessments are based on specific systems and designs, and the lack of historical operating data upon which to estimate failure probabilities for each system component remains a problem.

Risk management is the identification, assessment, and prioritization of risks followed by coordinated and economical application of resources to minimize, monitor, and control the probability and/or impact of unfortunate events (Figure 39.1) [1].

39.2.2
Current Shortcomings of QRA for Hydrogen Safety

Human error – A major factor contributing to an accident is human error, but to date human error has not been explicitly included in hydrogen facility QRAs, which may result in a significant under estimation of the risk. Future QRA techniques must include it.

Ignition – Inadequate treatment of ignition probability, auto-ignition, and delayed ignition. Much work has been done but the knowledge has not been

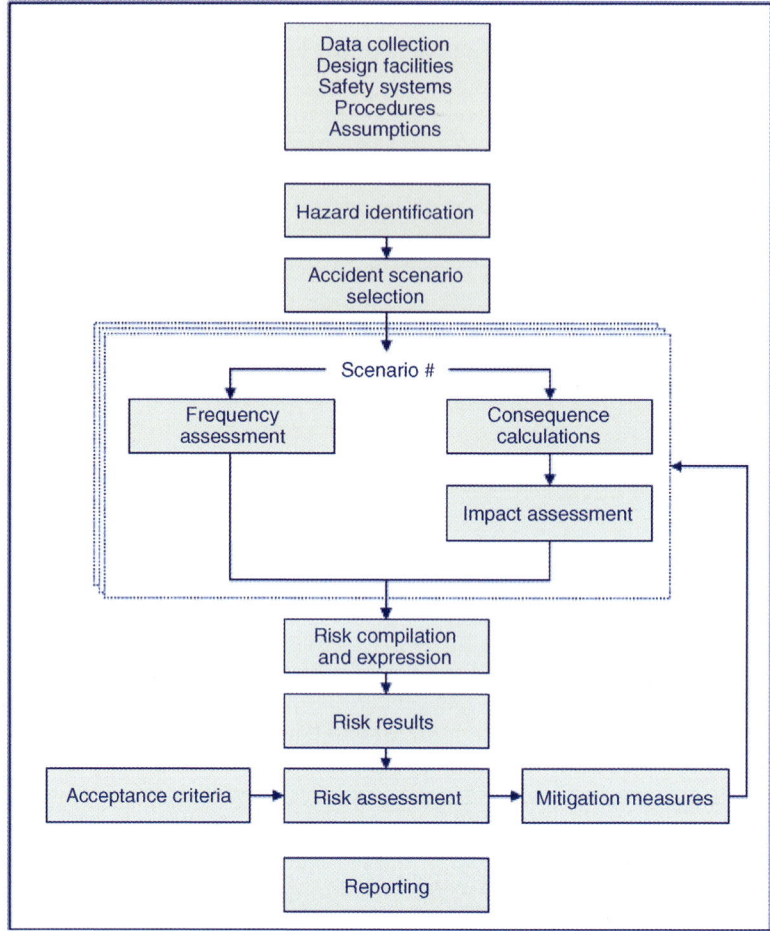

Figure 39.1 Quantitative risk analysis process schematic (DNV, 2008).

incorporated into probabilistic QRA models. Some work has been done based on available technical literature, empirical data, and limited experimental data; but more work is required.

Frequency probability of component failures – Little hydrogen specific data is available, although several data collection efforts have been initiated such as the HySafe Hydrogen Incident and Accident Database and the U.S. DOEs Hydrogen Incident Reporting and Lessons Learned database; but they are not yet sufficient for utilization in a QRA.

Simplified engineering models – Consequence calculations should predict the physical effects directly resulting from an accident (e.g., explosion overpressure and thermal radiation levels). For QRAs, where the consequences from a large number of scenarios need to be evaluated, simplified engineering models are needed.

Probability of causing damage – The results of consequence evaluations must be translated into a probability of causing damage to an individual, component or structure for use in a QRA. This can be done using Probit functions, which provide a statistical correlation between the magnitude of the consequence (e.g., thermal heat flux) and the resulting potential for an occurrence. Available Probit functions for predicting damage from a thermal heat flux are not hydrogen specific and thus introduce uncertainty in the risk results which can be evaluated using sensitivity studies.

Lack of uniform acceptance criteria – there are different types of acceptable citrieria for risk criteria and each country/locality is different.

Uncertainties in QRA models – A better understanding of the knowledge gaps is needed for reliably performing a QRA. Initially this information may be adapted from other industries.

39.3
The Unique, Safety-Related Properties of Hydrogen

The primary hazards associated with handling gaseous or liquid hydrogen are fire, explosion, asphyxiation, and exposure to extremely low temperatures. Virtually all of these are related to unintentional releases of hydrogen.

Hydrogen gas has certain physical, chemical, and materials/corrosion related characteristics that can affect safe handling, both positively and negatively:

1) *Physical properties* – Hydrogen molecules are the smallest of all and this makes it difficult to contain them as they are prone to leak through the smallest of openings in piping valves and fittings and storage containers. Hydrogen gas has an extremely low density at normal temperatures and pressures, and as a result the gaseous form is significantly lighter than air. When leaked, hydrogen gas rises rapidly and quickly diffuses, which can be a safety related benefit when outdoors. However, when the gas is released in a confined space (e.g., indoors), it will rise and accumulate at the ceiling or form pockets, potentially creating a flammable concentration and a potential hazard. Conversely, when not confined such as in an outdoor environment, hydrogen gas will rapidly disperse, reducing the size of any flammable cloud formed. In addition, its extremely low density reduces the energy content in the flammable cloud and thus the energy released in a fire or explosion.

2) *Chemical properties* – Hydrogen gas is highly flammable and will burn in air at a very wide range of concentrations between 4% and 75% by volume. The ignition energy of a flammable hydrogen–air mixture is extremely low, and mixtures may be ignited by spark, heat, or sunlight. Hydrogen will also spontaneously ignite in air above 500 °C (932 °F) [3] and when released at high pressures. Pure hydrogen–oxygen flames emit ultraviolet light and are

nearly invisible to the naked eye. The detection of a burning hydrogen leak may require a flame detector; such leaks can be very dangerous. Hydrogen flames under other conditions are blue, resembling blue natural gas flames. The destruction of the Hindenburg airship was an infamous example of hydrogen combustion; the cause is debated, but the visible orange flames were the result of a rich mixture of hydrogen to oxygen combined with carbon compounds from the airship skin.

H_2 reacts with every oxidizing element. It can react spontaneously and violently at room temperature with chlorine and fluorine to form the corresponding hydrogen halides, hydrogen chloride and hydrogen fluoride, which are also potentially dangerous acids.

3) *Corrosion and materials* compatibility – Hydrogen is non-corrosive. Typically used materials such as steel, copper, aluminum, and brass are suitable at normal temperatures. Steels with an ultimate tensile strength of less than 1000 MPa (~145 000 psi) or hardness of less than 30 HRC (Rockwell scale) are not generally considered susceptible to hydrogen embrittlement [4]. However, high-strength and low-alloy steels, nickel and titanium alloys as well as austempered iron are more susceptible [5].

39.4
General Considerations for the Safe Handling of Hydrogen

Although there are many sources of information on hydrogen safety (Figure 39.2), below are some generally accepted precautions to be followed. For specific locations and applications, the existing regulations, codes, and standard should be followed. A good resource is www.fuelcellstandards.com.

Chemical Formula	H_2
Molecular Weight	2.02
Boiling Point at 1 atm	−423°F (−252.9°C)
Freezing Point at 1 atm	−435°F (−259.2°C)
Critical Temperature	−400°F (−240°C)
Critical Pressure	186 psia (12.8 bar)
Density, Gas at 70°F (21°C), 1 atm	0.006 lb/ft^3 (0.1 g/l)
Specific Gravity, Gas (air=1) at 20°C, 1 atm	0.07
Specific Volume at 70°F (21°C), 1 atm	11.99 m^3/kg
Latent Heat of Vaporization	192 Btu/lb (446 kJ/kg)
Flammable Limits at 1 atm in air	4%–75% (by volume)

Figure 39.2 Safety related properties of hydrogen.

39.4.1
Gaseous Hydrogen

- Ensure adequate ventilations wherever hydrogen is used to reduce the risk of forming flammable mixtures in the event of a leak.
- Vent all pressure relief devices (PRDs) to a safe outside area away from ignition sources.
- Eliminate sources of ignition, such as open flames, sparks, cigarettes, and hot surfaces.
- Maintain adequate separation distances from exposures and other hazardous materials in accordance with applicable codes and local regulations.
- Use portable flammable gas detectors when performing maintenance on hydrogen systems.
- Ensure that electrical equipment near hydrogen systems meets the appropriate classification (NFPA 70, National Electric Code) or required separation distances are maintained.
- Purge equipment with an inert gas such as nitrogen before introducing hydrogen into the system.
- Use hydrogen or flammable gas detectors indoors where hydrogen gas may be released or accumulate in high areas.
- Storing hydrogen in confined areas where leaked hydrogen can accumulate should be avoided. Such spaces should be tested with a portable or continuous flammable gas analyzers prior to any maintenance being performed.
- As hydrogen flames are virtually invisible caution should be exercised to avoid walking into a hydrogen flame. For many years a "broom test" (using a straw broom places where a flames is suspected) was employed to test for hydrogen flames.

39.4.2
Liquid Hydrogen

- Storage and piping materials must be appropriately specified to prevent hydrogen and low temperature embrittlement. In addition, proper personal protective equipment must be worn when handling liquid hydrogen, including safety glasses with a full face shield, loose fitting, insulated gloves, flame resistant clothing, and high top safety shoes.
- An adequate pressure relief device must be provided in any confined space to guard against vapor expansion. A given volume of liquid hydrogen will vaporize and expand by about 850 times when warmed to ambient temperatures. This applies to piping systems and storage containers which may be subjected to overpressure.
- Liquid hydrogen can form two types of vapor clouds. The first occurs when exposed surfaces condense moisture in the surrounding air, which is generally harmless. However, a discharge of liquid hydrogen can produce a vapor cloud that is a fire hazard. Personnel should be alert to an abnormal vapor cloud which may indicate a leak. If a cloud is caused by a liquid hydrogen

leak, the flammable cloud may be larger than the visible cloud and the area should be sampled.
- More than one person should be present when liquid hydrogen is transferred, except when transferred by specially trained employees of a liquid hydrogen supplier. Withdrawal of a liquid from a tank, truck, or Dewar requires the use of a closed system with safety relief devices. The system should be purged with helium to eliminate any air from the system. Liquid transfer lines must be vacuum insulated to minimize vaporization. All equipment should be electrically grounded before beginning transfer.

Air will condense at liquid hydrogen temperatures and an oxygen enriched liquid may be formed due to the preferential evaporation of nitrogen from the mixture.

39.4.3 Handling Emergencies

The most important element of emergency preparedness is a detailed, well thought out Emergency Response Plan clearly delineating procedures in the event of hydrogen release and fire. The plan should include emergency notifications, evacuation procedures, and a schedule for periodic practice drills.

In addition, in the event of a hydrogen leak or when a hydrogen detector sounds, the following general guidelines should be followed:

- evacuate the area;
- eliminate all sources of ignition;
- if possible, safely shut off the hydrogen source to the leak;
- if possible, safely maximize the ventilation to the area;
- check the flammable gas concentration of all areas before entering.

39.5 Regulations, Codes, and Standards

The shift to a hydrogen based fueling infrastructure is significant, and involves a new set of hazards in addition to its flammability. In addition to the safety of new applications such as hydrogen vehicles, new infrastructure related systems and processes for hydrogen delivery, storage, refueling and dispensing will be needed.

Historically, the codes and standards development process has been lengthy and the body of existing codes and standards has been developed over a long period of time based on operating experience. The greater the experience, the more data will exist on the probability of equipment failures, accident scenarios, and the frequency and causes of mishaps. While it is beyond the scope of this chapter to provide a comprehensive review of current and under-development regulations, codes, and standards, the lack of such operating experience with hydrogen energy systems in consumer environments continues to be a significant barrier to the widespread adoption of these systems and the required infrastructure. In recent years, significant international efforts have been initiated for the development of codes and standards required for the introduction of these new systems. The lack of such long-term experience creates the natural tendency for codes and standards to be unnecessarily restrictive to ensure that an acceptable level of safety is maintained. One possible effect is to hinder the introduction of hydrogen systems and thus the operating experience upon which future infrastructure is developed. Besides unnecessarily restrictive regulations, codes, and standards, the commercial application of new hydrogen energy systems can be hindered by:

- The use of existing regulations, codes, and standards that have limited applicability to hydrogen (e.g., natural gas) and do not adequately take into account the unique properties of hydrogen that can affect its safe handling during production, storage, delivery, and use.
- High cost or lack of availability of insurance in the absence of actuarial accident and loss data.
- Public perceptions and sensitivities. The public has become comfortable with our current petroleum and natural gas infrastructure, and accept their hazards and consequences (automobile fires, natural gas explosions, carbon monoxide poisoning, environmental damage, etc.). However, the mention of hydrogen creates a concern for most people.
- Inadequate training of operators, first responders, and consumers.

In the United States, The National Fire Protection Association has published several documents related to the safe use of hydrogen. Of particular importance to hydrogen fuel cell vehicles is NFPA® 52, "Vehicular Gaseous Fuel Systems Code." The code is applicable to the following systems and systems' components:

- pressure relief devices (PRDs), including pressure relief valves,
- pressure gauges,
- pressure regulators,
- valves,
- hose and hose connections,
- vehicle fueling connections (nozzle),
- metal hydride storage,
- electrical equipment used with gaseous hydrogen systems,
- gas detection equipment and alarms,
- hydrogen generators,
- hydrogen dispensers,
- pressure switches,
- flow meters.

Another good resource is NFPA 2: "Hydrogen Technologies Code," a document that consolidates all the fire and life safety requirements applicable to generation, installation, storage, piping, use, and handling of hydrogen in compressed gas form or cryogenic liquid form into a single comprehensive resource. It includes fundamental requirements for hydrogen in both gaseous and liquid phases and contains additional use-specific categories, such as:

- vehicle fueling facilities;
- systems for fuel cell power and generation;
- applications involving combustion processes and special atmospheres;
- operations in the laboratory.

NFPA 2 applies to indoor fueling, hydrogen fueling stations, backup power, and electrolytic production of hydrogen. Annexes include a sample ordinance adopting NFPA 2, an example of a Class C furnace operational and maintenance checklist, and supplementary information on explosion hazards and protection in laboratories.

39.6
International Collaborations to Prioritize Hydrogen Safety Research

Several international collaborations and projects on hydrogen safety are working to share information and identify barriers to widespread use of hydrogen energy systems. The International Energy Agency Hydrogen Implementing Agreement Task 19 on Hydrogen Safety was initiated in 2004 and was succeeded by Task 31. While recognizing that overly restrictive or lacking codes and standard for hydrogen energy systems was a barrier, the experts agreed that their goal should be to provide a sound and technically credible basis for risk informed codes and standards. Since 2004, the Hydrogen Safety Task 19 and its successor Task 31 have identified research needs and conducted the necessary research and testing to develop a consensus on several important issues. In 2008, the task published three reports that helped to identify the following needs of the collaboration.

39.6.1
Survey of Hydrogen Risk Assessment Methods, 2008

This report, completed by Det Norske Veritas (DNV) of Norway is a survey of risk assessment methodologies for hydrogen production, storage, and refueling stations. The survey discusses how the approaches differ and how they are adaptable to the specifics of hydrogen risk.

39.6.2
Knowledge Gaps White Paper, 2008

The issue of knowledge gaps within Task 19 Hydrogen Safety was first raised in September 2005. The report recognized that more information was needed to validate the results of several models that were under development with the objective of reaching a consensus on what information was need that could be gained through research, testing, and modeling activities.

39.6.3
Comparative Risk Assessment Studies of Hydrogen and Hydrocarbon Fueling Stations, 2008

This report was based on the first three years of the Task 19 collaboration during which participating experts presented several risk assessment studies of refueling facilities. The objectives of this report were to describe the available relevant risk assessment studies, identify their key elements (approaches, methodologies, methods of analysis, key results, and recommendations) and review and compare the results to gain insights and identify research needs.

39.7 Current Directions in Hydrogen Safety Research [6]

There has been a significant research effort in Europe, North America, and Asia during the past ten years.

Below is a good representative sample, based primarily on the International Energy Agency's Hydrogen Implementing Agreement's collaboration on hydrogen safety in which participating researchers share their results, collectively interpret results, and identify additional research needs. A more comprehensive treatment of current research in the area may be found in the proceedings of the major hydrogen conferences, in particular the International Conference on Hydrogen Safety, the most recent of which was held in Brussels in September 2013 (www.ichs2013.com).

39.7.1
Research on the Physical Behavior of Leaked or Leaking Hydrogen

In the absence of historical operating and incident data, there is a need for better predictive methods of how hydrogen behaves during an unintended release. The goal of this research is to close the identified knowledge gap with respect to the source geometries, release phenomena, dispersion processes, ignition, and combustion modes. The thrust of this area of research includes experimental testing and simulation and model development and ultimately to compare them to validate the models. This information is important to the development of more reliable predictive risk assessment methods. Current areas of research include:

- the characterization of hydrogen releases from anticipated leak sources, such as pressure relief devices (PRD) and the effects of the geometry of orifices and cracks resulting from piping or equipment failures;

- a better understanding of the dispersion behavior of leaked hydrogen and the formation of flammable clouds – this includes simulations, modeling, and experiments for enclosed areas, large-scale releases, the effect of obstructions and dispersion, the effects of different release rates, and the validation of CFD (computational fluid dynamics) models;
- the effects of barriers and different geometries for both ignited and unignited releases;
- the characterization of jet releases of hydrogen with regard to surface effects and flammable cloud formation;
- the behavior of gas mixtures of hydrogen and other dangerous gases such as natural gas, carbon monoxide, and ammonia with respect to flammability limits, combustion and toxicity;
- liquid hydrogen release behavior with regard to flammable cloud formation, dispersion, and low temperature vapor clouds, and so on.;
- the ignition and combustion of leaked hydrogen to include ignition mechanisms, ignition probability, and auto-ignition;
- behavior of hydrogen jet flames, thermal radiation, crosswinds and surface effects and barriers;
- the effects of overpressure and explosions;
- assessment of quantitative tools for hydrogen safety and risk assessments and validation of CFD models and turbulence modeling.

39.7.2
Hydrogen Storage Systems Safety Research

Current research on the safety of hydrogen storage systems for mobile and stationary applications includes:

- materials research on on-board vehicular storage systems materials compatibility, hydrogen permeation, hydrogen embrittlement, hydrogen attack of metal liners and subsequent hydride embrittlement, stress corrosion cracking, liner delamination, thermal fatigue, mechanical fatigue, strain failure of vessel composite materials, low-temperature embrittlement;
- safety, reactivity, and risk mitigation of compressed hydrogen gas, liquid hydrogen and solid-state hydrogen storage in metal hydrides, chemical hydrides, such as NH_3BH_3, AlH_3, and adsorbents (such as AX-21 activated carbon and MOF-5);
- research on identifying and mitigating potential hazards such as material reactivity, mechanical impacts, fire safety and thermal impacts, thermally-activated pressure relief devices, fast depressurization, and so on;
- mathematical models (e.g., CFD, simulation, etc.) for storage vessel stresses, heat transfer during external fire scenarios (localized and engulfing fires), reaction kinetics between the stored hydride material and impurities such as moisture, oxygen intrusion, and so on;
- on-board fire suppression systems that are compatible with energetic storage materials such as metal and chemical hydrides.

39.7.3
Research that Supports Early Market Applications

The most advanced commercial applications for hydrogen technologies are backup power using polymer electrolyte membrane (PEM) fuel cells and forklifts. A good reference is "NFPA 2: Hydrogen Technologies Code" which will benefit permitting and enforcing officials when designing systems or reviewing permits for occupancies storing, handling, or using hydrogen or when inspecting existing facilities.

Early market applications for new hydrogen technologies for mobility, stationary, and materials handling applications will result in significant involvement of consumers. This will require adjustment to traditional approaches to risk characterization and hazard analysis that focused on operator and worker safety to a focus on the general public. Safety assessment methods, data, and use of prevention and mitigation features need to be tailored to address these early applications, for example, hydrogen forklifts/materials handling facilities, car and bus fleets, stationary power units, and so on. For these systems, efforts are being made to systematically collect failure and leak frequency data, and other information, to assess their safety and facilitate their commercial introduction.

Since new technologies are penetrating densely populated urban environments, focus is being placed on risk mitigation technologies and methods such as sensors, barriers/walls, and safety distances. These efforts aim to ensure that developers can produce codes and standards that are risk-informed and evidence-based.

39.7.4
Research on Risk Mitigation Measures

Mitigation includes measures that may be taken to reduce either the probability or consequences of an identified hazard. Such measures can be either a change to the system design or through operational procedures or training.

There are several ways of mitigating risk through safer system design or operational measures. Some are achieved at little or no cost and become recommended practice or are incorporated into regulations, codes, and standards. Others are achieved at a cost that may vary greatly according to the specifics of the application, system design, or location, and a cost–benefit analysis should be completed to choose the best combination of measures for any given hydrogen system installation.

Some ways hydrogen risks can be mitigated are:

- *separation distances* – maintaining safe distances that reduce or eliminate the exposure to personnel and property in the event of a hydrogen leak and possible ignition (Table 39.1);
- *barriers/walls* – to contain the propagation of a flammable cloud or ignited flame jet;

Table 39.1 Typical outdoor separation distances in feet from hydrogen systems (250–3000 psig, max piping diameter ¾-inch).[a]

	Distance (U.S. units)
Property boundaries	45[b]
Exposed persons other than those involved in servicing of the system	22[b]
Buildings and structures	
- **Combustible construction**	20[c]
- **Noncombustible non-fire-rated construction**	20[c]
- **Fire-rated construction with a fire resistance rating of not less than 2 h**	5[c]
Openings in buildings (doors, windows, and penetrations)	
- Opening	45[b]
- Non-opening	20[b]
Air intakes (HVAC, compressors, other)	45[d]
Fire barrier walls or structures used to shield the bulk system from exposures	5[c]
Unclassified electrical equipment	15[c]
Utilities (overhead), including electric power, building services, hazardous materials piping	20[d]
Ignition sources such as open flames and welding	45[d]
Parked cars	25[b]
Flammable gas storage systems, **above ground**	
< 400 scf for hydrogen	20[c]
> 400 scf for hydrogen	15[c]
Above-ground vents or exposed piping and components of **flammable gas storage systems**	20[c]
Hazardous materials (other than flammable gases) storage below ground	
Physical hazard materials or health hazard materials	20[c]
Hazardous materials storage above ground; physical hazard materials or health hazard materials	20[c]
Ordinary combustible material, **or slow burning solids**	20[c]

a) Based on NFPA 55 Compressed Gases and Cryogenic Fluids Code 2010 Table 10.3.2.2.1(a) Minimum Distance from Outdoor Gaseous Hydrogen Systems to Exposures (U.S. Units). Table 10.3.2.2.1(a) contains other pressure or diameter ranges.
b) Distances can be reduced by ½ of the distance shown if a fire barrier is installed between the system and the exposure. The fire barrier must have a minimum fire resistance rating of 2 h. The resultant distance should be measured from the downstream side of the fire barrier wall to the exposure.
c) All distances except for air intakes and overhead utilities, unless otherwise noted in note *a*, shall not apply if a fire barrier in installed between the system and the exposure. The fire barrier must have a minimum fire resistance rating of 2 h.
d) A fire barrier cannot be applied to air intakes and overhead utilities.

- *ventilation or open air design* to allow leaked hydrogen to rapidly diffuse in an unconfined manner;
- *hydrogen sensors and detectors* – hydrogen is not reliably detected by area sensors due to its high diffusivity and because leaked hydrogen, being lighter than air, rises so quickly; inexpensive passive leak detectors mounted close to potential leak sites have been developed that are extremely effective at detecting leaked hydrogen before it becomes a safety event;
- *safety education and training* to reduce the risk of unsafe conditions or procedures, and effective response to accidents or incidents;
- *safe designs* to reduce the probability and consequences of equipment or system failures;
- *systems safety analyses and safety inspections.*

39.7.5
Simplified Tools to Assess and Mitigate Risk

Several research organizations are directing efforts to develop simplified tools and engineering correlations that facilitate the assessment of risk. Such correlations will take advantage of available models and simulations, experimental testing data, and related research. One of the benefits to be derived from these efforts may be the ability to trade off various mitigation measures to arrive at the most cost-effective combination that meets established acceptable risk approval criteria.

Such simplified tools could be integrated to form a "toolkit" based on empirical correlations, approval criteria, historical data, and validated models. They could be used by systems designers, process engineers, codes and standards development experts, risk analysts and QRA professionals, and approving authorities. It is anticipated that such a toolkit would be based on a modular design with an agreed set of failure frequency data and ignition probabilities and would consist of open and free software that is well documented and tested [7].

39.8
Summary

Hydrogen has been safely used in industry for over five decades. All fuels are flammable by their very definition and hydrogen has different physical properties and behavior characteristics that must be considered when being handled. Nevertheless, most experts agree that it can be handled more safely than the current liquid fuels we now use; however, the lack of operating experience with hydrogen systems and mature regulations, codes, and standards is a barrier to widespread commercial adoption. Until more experience and the accompanying failure data is obtained, the risk posed by these systems will need to be assessed using models and experiences with other fuels. As fuel cell and hydrogen technologies develop and mature, the world will transition to them without much thought as to their relative safety.

References

1 Air Products (2014) Safetygram #4, Gaseous Hydrogen.
2 International Energy Agency (2008) Hydrogen Implementing Agreement, Task 19 Report, Survey of Hydrogen Risk Assessment Methods. Available at www.ieah2safety.com/reports.htm.
3 University of Southern Maine (2009) Dihydrogen. *O=CHem Directory*, University of Southern Maine. Retrieved 2009-04-06.
4 Jewett, R.P., Walter, R.J., Chandler, W.T., and Frohmberg, R.P. (1973) Hydrogen Environment Embrittlement of Metals, Report number NASA CR-2163.
5 Linde (2015) Safety Advice 13 – Handling of Hydrogen.
6 International Energy Agency (October 2010) Hydrogen Implementing Agreement Task 31-Hydrogen Safety Work Plan.
7 Kotchourko, A. *et al.* (2013) State of the art and research priorities in hydrogen safety. Presentation at the Fifth International Conference on Hydrogen Safety, Brussels, Belgium, 10 September 2013.

3 National Renewable Energy Laboratory (October 2013) Hydrogen Vehicle and Infrastructure Codes and Standards Citations. http://www.afdc.energy.gov/pdfs/48608.pdf.
4 Fifth International Conference on Hydrogen Safety, 9–11 September 2013. Conference presentations (2013) http://www.ichs2013.com/index.php/conference-presentations.
5 National Aeronautics and Space Administration (1997) Safety Standard For Hydrogen and Hydrogen Systems: Guidelines for Hydrogen System Design, Materials Selection, Operations, Storage, and Transportation, NSS 1740.16 (cancelled July 25 2005), http://www.hq.nasa.gov/office/codeq/doctree/canceled/871916.pdf.
6 Air Products (2104) Safetygram #9, Liquid Hydrogen. Available at http://www.airproducts.com/~/media/Files/PDF/company/safetygram-9.pdf.
7 Air Products (2104) Safetygram #4, Gaseous Hydrogen. Available at http://www.airproducts.com/~/media/Files/PDF/company/safetygram-4.pdf.
8 National Fire Protection Association (2010) NFPA 52, Vehicle Fuel Systems Code. Available at www.NFPA.org.
9 National Fire Protection Association (2011) NFPA 2, Hydrogen Technologies Code. Available at www.NFPA.org.

Bibliography

1 U.S. Department of Energy, Fuel Cell Technology Office (2015) Safety Codes and Standards. http://www1.eere.energy.gov/hydrogenandfuelcells/codes/.
2 H2 Safety Best Practices Online Manual (2015) A bibliography of cited references. h2bestpractices.org/bibliography.asp.

Part 6
Existing and Emerging Systems

40
Hydrogen in Space Applications
Jérôme Lacapere

40.1
Liquid Hydrogen for Access to Space

40.1.1
Liquid Storage

Hydrogen must be used in the liquefied phase as it is the most suitable option for storing large quantities of hydrogen in tanks with a low mass and optimized volume. Considering an upper stage of $50\,m^3$ containing 3500 kg of LH2, the tank mass could be less than 1500 kg, yielding a mass efficiency potentially greater than 70%:

$$\text{Mass efficiency is defined as } M_{H_2}/(M_{H_2} + M_{\text{tank}})$$

where:

M_{H_2} is the mass of hydrogen stored;

M_{tank} is the mass of the storage tank

To store the same mass of hydrogen in the pressurized vapor phase under 700 bar, (standard pressure considered today for hydrogen storage on automotive vehicles), the tank volume should be $87\,m^3$ (to be compared to the $50\,m^3$ considered above for liquid storage). For such on-ground storage, the mass efficiency is around 5%, yielding thus a tank mass of about 28 500 kg.

Even with an optimized design, this tank mass is obviously prohibitive for a space launcher.

Combined with liquid oxygen (LOX), liquid hydrogen (LH2) used as a fuel has a very large specific impulse (ISP) of around 450 s. This combination gives the larger ISP of different liquid propellant possibilities.

The ISP is defined as:

$$F_{\text{Thrust}} = \text{ISP}\dot{m}g_0$$

where:

F_{Thrust} is the thrust obtained from the engine, in Newton;

ISP is the specific impulse in seconds;

$\dot{m}$ is the mass flow rate in kg s^{-1};

$g_0 = 9.81$ ms^{-2} is the acceleration at the earth's surface.

Before using liquid hydrogen as a fuel, space rockets were already using LOX with storable propellants like kerosene (RP1). Compared with LOX/LH2, LOX/RP1 gives an ISP of around 353 s. That means to get the same thrust, a larger mass flow rate is needed with LOX/RP1 compared to LOX/LH2, with a ratio of $450/353 \approx 1.27$

40.1.2
Constraints Due to Liquid Hydrogen Use

The use of liquid hydrogen combined with liquid oxygen is very efficient but has some constraints due to the cryogenic storage of hydrogen around 20 K. Today's engines work with turbopumps to ensure large mass flow rates, which allow for large thrusts. However, liquid hydrogen is stored at low pressure (<4 bar) and the margins to avoid cavitation are quite low in turbopumps. Cavitation can be seen as a rapid phase-change process due to a sudden pressure evolution in the liquid. In turbopumps, liquid is accelerated, the static pressure thus decreases, and cavitation can occur if the static pressure becomes lower than the saturation pressure of the liquid. This saturation pressure is directly linked to the liquid temperature. The liquid temperature at the tank outlet must then be known very precisely (~0.1 K) to anticipate cavitation problems. Today, the temperature (linked to the pressure in the tank) at the turbopump inlet is specified by the engine designer. All the liquid hydrogen masses with a temperature larger than this specified temperature are called "thermal residuals" (Figure 40.1). These "thermal residuals" must be known very precisely to avoid having to take

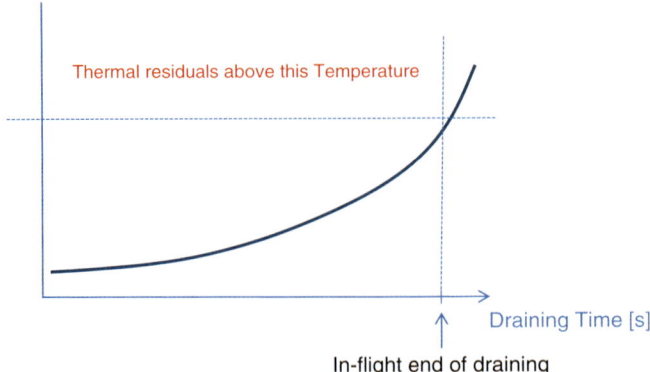

Figure 40.1 Evolution of liquid temperature during draining – thermal residual definition.

Figure 40.2 Illustration of computational results of liquid hydrogen temperature (expressed in kelvin) during propulsive phase.

additional margins and, consequently, reducing the payload mass. As illustrated in Figure 40.1, the draining is stopped before reaching the upper limit of temperature admissible by the engine system. To know precisely these thermal residuals, numerical simulations are performed to obtain a complete temperature map of liquid hydrogen in the tank during the whole launch phases (Figure 40.2).

Pressurization of a liquid hydrogen tank is normally ensured by helium on the ground thanks to on-ground storage and directly by hydrogen vapor during tank draining during flight. This hydrogen vapor is part of the liquid hydrogen feeding the cryogenic engine. Liquid hydrogen is then vaporized and heated up to be re-injected in the liquid hydrogen tank. In addition, significant pressure changes could appear in the cryogenic tanks during the propulsive phase. In fact, because of sloshing motions due to external perturbations, condensation of large quantities of hydrogen vapor could appear. This condensation due to de-stratification phenomenon and migration of subcooled liquid to the free surface [1] must be quantified precisely in order to design an adequate in-flight re-pressurization system.

40.1.3
Insulation

To store liquid hydrogen at 20.3 K (for 1 bar) without large mass losses by boiling, tank insulation must be very efficient. However, this is not really the case for space launchers: even with efficient insulation the tank walls, their thickness is not very large. In fact, one function of this insulation is to prevent air surrounding the launcher from cryo-condensation. Insulation will reduce the amount of boil-off masses but these masses will be very far from zero. Thus, very large quantities of liquid hydrogen are boiling off on the launch pad during the synchronized sequence. Due to these large boil-off quantities, the tanks are filled regularly and automatically up to the final on-ground pressurization phase.

40.2
To Go Beyond GTO

The cryogenic European launcher Ariane 5 ECA is currently releasing its two payloads in GTO (geostationary transfer orbit) after a single cryogenic upper stage propulsion phase. Then, the satellite should go to GEO orbit thanks to its own apogee engine. Future European launchers will be designed to perform several re-ignitions thanks to the Vinci engine. Several cryogenic upper stage propulsion phases will be performed to release payloads at more energetically favorable orbits called GTO+ or even directly into GEO (geostationary orbit) in some cases. Releasing the payload directly in GEO orbit, the satellite can be designed without an apogee engine, which means a significant gain in the functional mass of the satellite.

If the satellite is released in GTO+, the satellite has to include an apogee engine but the propellant mass needed to reach GEO will be lower than satellites released on classic GTO, which also represents a significant mass gain. To reach these GTO+ or GEO, the cryogenic upper stage needs to perform a new ignition in flight about 5.5 h after the first boost phase (this time corresponds to the coasting phase needed to reach the GTO apogee where a large thrust is needed to modify the orbit). To perform this re-ignition, the liquid hydrogen must be at the right temperature and pressure to be fed in the turbopump to avoid any cavitation risk. Furthermore, the turbopump and all volumes upstream of the combustion chamber of Vinci engine must be chilled before re-ignition. This is performed during the re-ignition preparation phase.

40.2.1
Coasting Phase

During all the coasting phase performed in a low gravity environment, the liquid hydrogen will not be submitted to any volumetric forces. The dominant forces are capillary forces, forcing the liquid to collect close to the tank walls. It is important to recall that the remaining liquid hydrogen mass is quite low (less than 30% of the initial mass) for this coasting phase since, at this stage, the first boost to reach GTO has already used a large quantity of hydrogen. Liquid hydrogen is then submitted to large heat fluxes inducing large boil-off masses after 5.5 h of coasting phases. To avoid these large quantities of boil-off masses, some additional techniques/systems have to be implemented keeping in mind the need to lower the global system mass. An interesting solution might consist, for example, in applying a thrust to settle liquid hydrogen at the bottom of the tank. This way, areas of tank walls wetted by liquid will be quite low, limiting at the same time the energy received by the liquid hydrogen. Numerous computations are running in different research and development entities to find the most relevant thrust values. Flight experiments are also needed to validate the computed values. In fact, a very low acceleration is needed to settle the liquid at the tank bottom. Theoretically, a Bond number above the critical Bond number is sufficient to maintain liquid hydrogen at the tank bottom. For general cylindro-spherical tanks, this critical Bond number can be taken as around 5 [2].

The Bond number (Bo) is the ratio between body forces (here linked to the low acceleration value) and surface tension forces:

$$Bo = \frac{\rho \gamma R^2}{\sigma}$$

where:

- ρ is the liquid density (or density difference between liquid and vapor) in kg m^{-3};
- γ is the acceleration in m s^{-2};
- R is the tank radius in m;
- σ is the liquid/vapor surface tension in N m^{-1}.

Considering a tank of 2 m diameter, a longitudinal acceleration of 1.5×10^{-4} m s^{-2} will then be theoretically sufficient to maintain the liquid hydrogen at the tank bottom. During past in-flight demonstrations, the Centaur rocket has demonstrated effective propellant control at 10^{-4} m s^{-2} [3]. In the case of upper stage with re-ignition possibilities, even with solutions to reduce the boil-off masses, the pressure will increase rapidly during the coasting phase up to the maximum allowable pressure in the tank (around 4 bar). When this maximum pressure is reached, a check valve is opened to release the additional vaporized mass. At the end of the coasting phase, the liquid is then saturated at the maximum allowable pressure in the tank. Other strategies can be implemented with different pressure levels during coasting phases. In any case, the liquid hydrogen is saturated and needs to be reconditioned for engine re-ignition.

40.2.2
Re-ignition Preparation Phase

To get margin with respect to turbopump NPSH (net pressure suction head), the liquid hydrogen needs to be in the subcooled state. A simple solution consists in performing a depressurization followed by helium pressurization. Additional helium capacities are then needed and they have to be quite large since the liquid fill ratio is quite low after the first boost of the cryogenic upper stage. But, before this depressurization phase, liquid hydrogen must be present at the tank bottom (if not already present thanks to continuous settling). A settling phase (or reorientation phase) is thus performed to get the liquid at the tank bottom [3,4]. This reorientation phase must also be predicted precisely to anticipate settling needs. An optimized value of thrust and duration have to be found to avoid geysering phenomenon occurring even with a low acceleration value; see Figure 40.3 showing a geyser occurring during a settling phase with a value of 10^{-3} m s^{-2}.

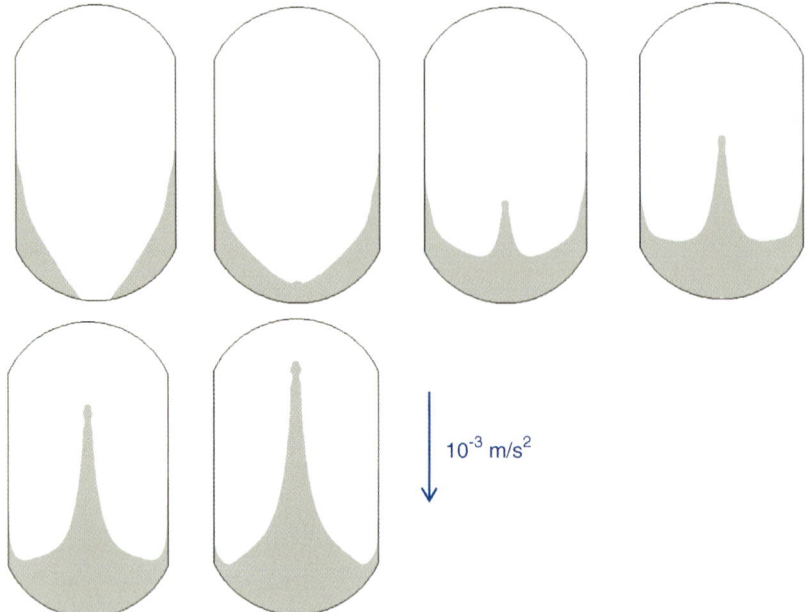

Figure 40.3 Geyser formation during a re-orientation phase with an acceleration of 0.001 m s^{-2} (images separated by a duration of 10 s).

To avoid this geyser formation, some solutions could consist in using pulse settling to limit velocities at the tank bottom. During depressurization, a low thrust is required to separate the vapor generated from the liquid phase. A large longitudinal acceleration will produce large bubble velocities and allow large depressurization rates. However, large longitudinal accelerations are not ideal from a system point of view (dry mass of the additional thrust system). Conversely, very small accelerations, which are easier to produce and not penalizing from the system mass point of view, will generate very low bubble velocities and large free surface deformation. In that case, the depressurization ratio will be low and this phase might be time consuming (several tens of minutes). Current research activities are looking for the optimum acceleration value allowing separation of vapor bubbles generated within a given time. Accelerations between 5×10^{-4} and 10^{-3} m s^{-2} could be considered for depressurization lasting about 10 min without too large free surface deformation. Figure 40.4 illustrates the formation of vapor bubbles in the liquid phase during a liquid hydrogen tank depressurization. Vapor bubbles are generated within the liquid phase and are separated from the liquid phase thanks to a small applied acceleration (5×10^{-4} m s^{-2} in this illustration case). This figure is a computational result.

Experimental activities with large tanks are required to validate these numerical results; Reference [5] has shown the formation of liquid "globules" ejected from the liquid bulk during the depressurization phase. Following this depressurization, a re-pressurization is necessary. This re-pressurization must

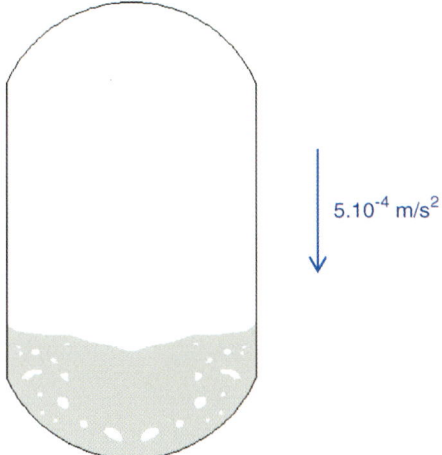

Figure 40.4 Liquid hydrogen tank depressurization under a small acceleration (5×10^{-4} m s^{-2}). Liquid hydrogen is represented in gray and vapor hydrogen is represented in white.

also be performed with a low longitudinal acceleration to avoid free surface deformation. Direct jet impinging the free surface must also be avoided [6]. During this re-ignition preparation phase, the turbopump, all the piping, and cavities upstream the combustion chamber must be chilled down thanks to liquid hydrogen mass injection. During this chill-down process performed under a low longitudinal acceleration (to evacuate bubbles), liquid hydrogen will boil at the walls. Boiling coefficients for liquid hydrogen are known on-ground but the impact of low gravity on boiling behavior is still not completely known. The critical heat flux seems to be lowered in a low gravity environment [7,8] (cf. boiling curves in 1g and in 0g in Figure 40.5) with film boiling occurring more easily under low gravity than on-ground. Thus, this chill-down process, which is also performed

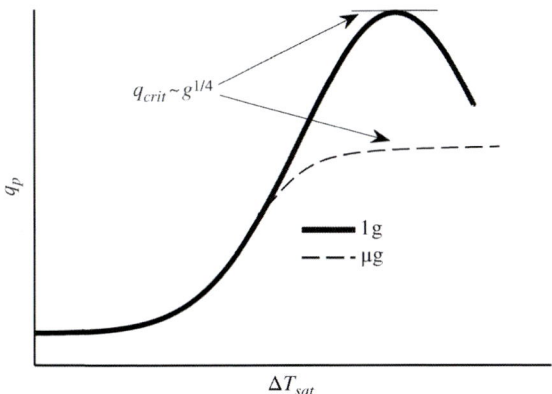

Figure 40.5 Boiling curves in 1g and 0g showing the low critical heat flux at low gravity. From Reference [7].

on-ground before launch is, however, not very well mastered in low gravity because of these uncertainties on boiling coefficient and bubble behavior in pipes [7,8]. Margins must thus be made on the time required to reach the desired value and experimental data are needed in a representative environment with representative geometries. The main risk lies in not chilling-down sufficiently parts of cavities due to poor heat transfer coefficients.

40.3
Relevant Tests in Low Gravity Environment

To validate all thermo-hydraulic behaviors and to reduce the safety margins [7,9], experiments in a low gravity environment are required. A multitude of validation tests performed with different similitude liquids will not be sufficient and some representative tests are mandatory to provide a final validation, that is, tests in a low gravity environment with liquid hydrogen must be performed. Some tests under low gravity were carried out in the 1960s accompanying the Mission to Moon program. The first in-flight experiments in the early 1960s [10] used several sounding rockets with a spherical liquid hydrogen tank (Figure 40.6).

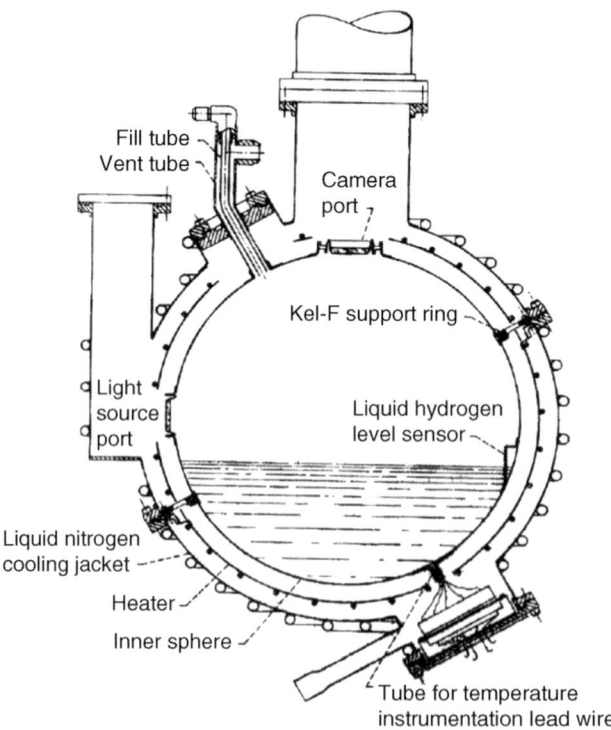

Figure 40.6 Hydrogen tank used in sounding rockets experiments [10].

A second important low gravity demonstration with liquid hydrogen was the AS-203 demonstration flight with S-IV demonstration cryogenic upper stage. Tanks were instrumented in order to obtain the temperature map evolution inside the liquid phase and the ullage. Pressure sensors were also included with diphasic sensors. A video camera located at the top of the tank allowed visualization of part of the free surface inside the tank. Different representative flight phases were performed during this demonstration in the course of several complete orbits: self-pressurization with temperature evolution, depressurization, venting, and propulsive venting. This demonstration flight was performed in 1966 [5] and remains today a reference for computational code validation and to understand various phenomena linked to liquid hydrogen behavior in a low gravity environment [11]. Since then, several additional experimental flights with liquid hydrogen were set-up in the USA but were later canceled for different reasons. These experimental flights were specified not only to obtain some data to validate numerical codes but also to increase the TRL (Technological Readiness Level) of additional devices dedicated to the Propellant Management Activities. For space activities, the common understanding of a new upper stage is that a specific technology should obtain a TRL of 6 (at least) in order to be able to engage the development phase of the tank. For the hydrogen tank of an upper stage launcher, a TRL of 6 corresponds to tests performed with liquid hydrogen under low gravity with a large-scale tank (diameter larger than 0.5 m in order to be representative of the large Rayleigh numbers encountered in real tanks even for low longitudinal acceleration). These kinds of tests to reach TRL 6 are thus very difficult and expensive to achieve.

A list of the most significant experiments is given hereafter. The reader is referred to the different references that provide an important source of analysis on the predicted liquid hydrogen behavior in space launcher hydrogen tanks:

- COLD-SAT [12,13],
- CRYOTE [14],
- CPST [15,16].

For all these US concepts, specific technologies to manage liquid hydrogen in space launcher tanks are referred as cryogenic fluid management (CFM) technologies. Additional experimental data could be obtained by adding instrumentation to current hydrogen tank launchers. The Japanese launcher H-IIA gave, for example, images of liquid hydrogen settling and depressurization in a liquid hydrogen tank thanks to a video camera located outside the tank and visualizing the free surface through an optical port. The drawbacks, however, of this interesting option remain the difficulty in obtaining precise and known boundary conditions and initial conditions. Furthermore, it is difficult to include new technologies in the recurrent tank without affecting its cost and the functional behavior. For long-term missions, managing liquid hydrogen during the whole flight requires the implementation of many new technologies, phase and experimental tests to reach TRL 6 are then mandatory [17]. Figure 40.7, from the

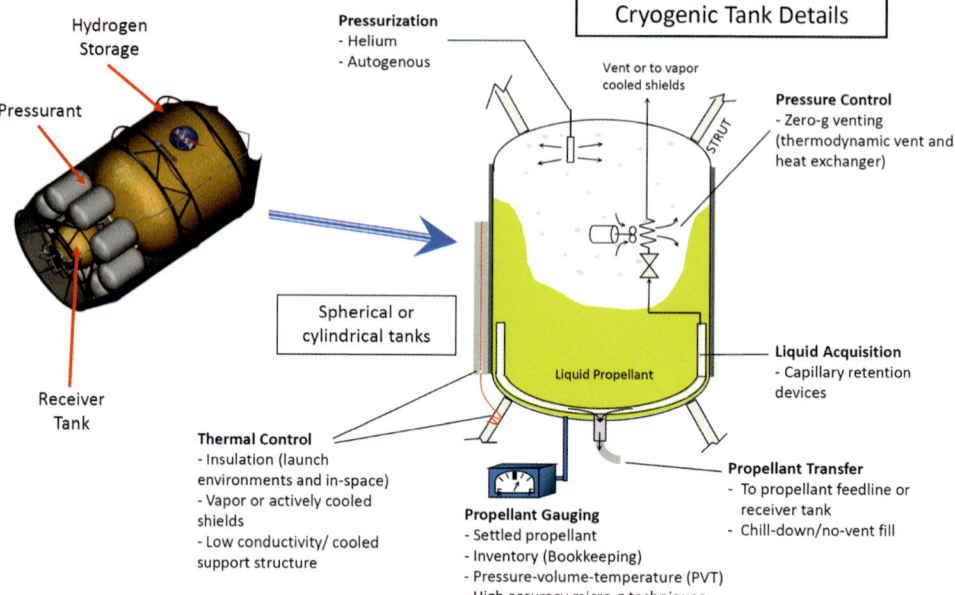

Figure 40.7 Illustration of the different technologies needed for long-term missions thanks to the CPST mission. From Reference [16].

CPST mission [15,16], illustrates the different technologies needed to master liquid hydrogen behavior in a low gravity environment.

The following approaches are today complementary:

- Use of specific experimental tests with liquid hydrogen in tanks of diameters less than flight tanks. Drawbacks: additional costs, feasibility, representativity.
- Use of instrumented recurrent tank. Drawbacks: unknown/uncertainties on initial and boundary conditions, difficulties in including new technologies without modification of functional behavior of recurrent tank.

40.4
In-Space Propulsion

Liquid hydrogen can also be used for future long-term missions like interplanetary missions: Mission to Mars for example. For this purpose, an efficient thermal protection of the liquid hydrogen tank becomes mandatory to reduce the overall boil-off masses, but is not sufficient. New methods to reduce the input heat fluxes need to be developed, such as:

- improved multi-layer insulation to reduce heat losses due to radiation;
- non-conductive struts to reduce conduction heat losses.

These new methods are generic for everyone concerned with cryogenic storage on-ground but need to be space launcher oriented, with specific concerns on the systems mass. Classical solutions can also consist of using the heat capacitance of hydrogen vapor. Once vaporized, hydrogen vapor is released from the tank and can be used to cool down different systems:

- the tank wall itself,
- a thermal shield,
- the structural struts,
- or all of the above.

All previous solutions are "passive" since they do not use compressors or circulators. Some "active" solutions can also be used to compensate the residual heat fluxes entering the liquid hydrogen tank. Furthermore, passive insulation methods need to be coupled with active systems for long missions: these active systems are based on cryocoolers on one side and a system to transfer energy from the tank to the cryocooler on the other side. Different solutions can also be envisioned to transfer the energy from the tank walls to the cryocoolers:

- direct connection of the cooler to the tank;
- liquid hydrogen subcooling and re-injection;
- use of cooling pipe around the tank walls [18,19].

Thanks to the relatively high storage temperature (about 90 K) of liquid oxygen, complete zero boil-off (ZBO) could be achieved by current cryocoolers without penalizing the dry mass. On the other hand, it is more difficult to remove completely boil-off masses of liquid hydrogen (stored about 20 K). Thus, on the liquid hydrogen side, reduce boil-off (RBO) concepts are replacing complete ZBO concepts in actual configuration studies. Figure 40.8 shows, for example, the direct cooling of liquid oxygen by cooling pipes (ZBO on LOX side) and the cooling of an external shield of the liquid hydrogen tank (RBO on LH2 side) by cooling pipes using very close temperatures and the same cryocooler [18].

In another concept, reference [19] uses vapor hydrogen generated to cool down the external shield and limit heat fluxes (Figure 40.9).

40.5 Conclusion

Even if it implies additional complexity and cost, the use of liquid hydrogen in space launchers is today the standard of every heavy launcher. The use of liquid hydrogen storage is now mastered for earth orbit access thanks to numerous on-ground and in-flight tests on the one hand and computations on the other hand. To increase these launch capacities and to use this very powerful hydrogen for missions longer than several hours, several technological developments are

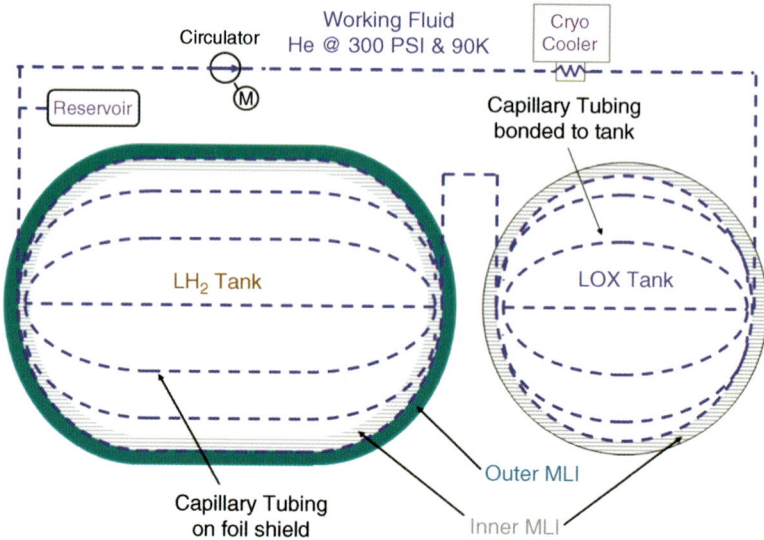

Figure 40.8 Direct tank cooling on liquid oxygen side and cooling of liquid hydrogen shield. From Reference [18].

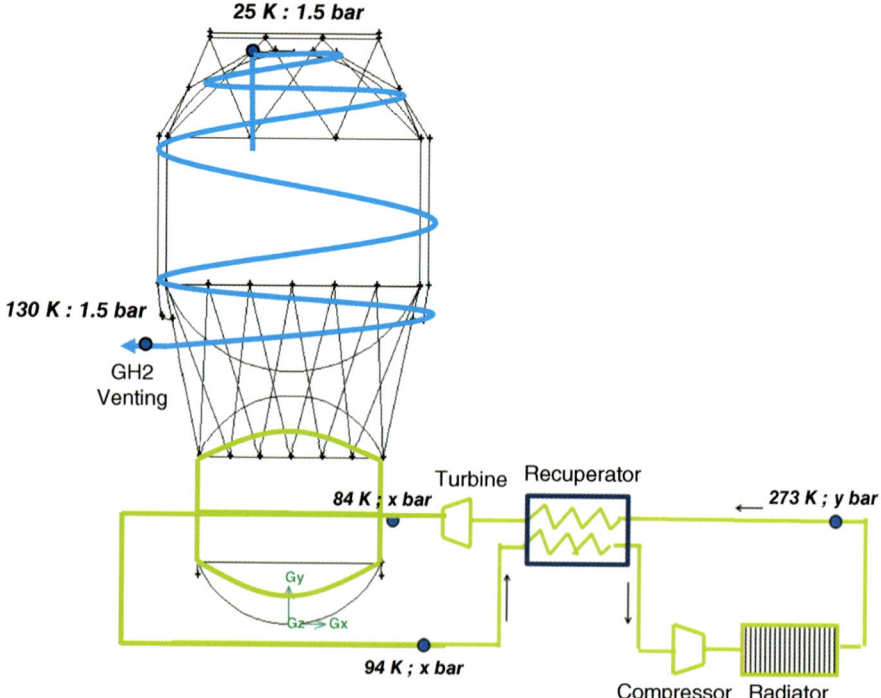

Figure 40.9 Concept of reduced boil-off for LH2 by vapor venting and active zero boil-off for liquid oxygen tank [19].

needed such as efficient multi-layer insulation and space cryocoolers. In return, such technological developments could have positive impacts on liquid hydrogen storage for on-ground mobility applications.

References

1. Lacapere, J. *et al.* (2009) Experimental and numerical results of sloshing with cryogenic fluids. *Prog. Propulsion Phys.*, **1**, 267–278.
2. Abramson, H.N. (1966) The Dynamic Behavior of Liquids in Moving Containers – with Applications to Space Vehicle Technology, NASA SP-106.
3. Kutter, B.F. *et al.* (2006) Settled Cryogenic Propellant Transfer, AIAA 2006-4436.
4. Dreyer, M. *et al.* (2009) Propellant behaviour in launcher tanks: an overview of the COMPERE program. *Progr. Propulsion Phys.*, **1**, 253–256.
5. Chrysler Corporation (13 January 1967) Evaluation of AS-203 Low Gravity Orbital Experiment.
6. Himeno, T. *et al.* (2002) Numerical investigation of liquid behaviour of the propellant tank of H-IIA. Presented at 38th AIAA/ASME/SAE/ASEE Joint Propulsion Conference and Exhibit, Indianapolis, 7–10 July 2002, AIAA-2002-3987.
7. Kannengieser, O. (2009) Etude de L'ébullition sur Plaque Plane en Microgravité, Application aux Réservoirs des Fusées Ariane V, Thesis, Université de Toulouse.
8. Lee, H.S. *et al.*, The pool boiling curve in microgravity. Presented at 34th Aerospace Sciences Meeting and Exhibit, Reno, NV, 15–18 January 1996, AIAA 96-0499.
9. Lacapere, J. *et al.* (2010) Cryogenic Propellant in Orbit: experimental data to validate numerical tools, CNES, French contribution to COSPAR 2010 (Committee on Space Research), Condensed Matter Physics, pp. 68–69.
10. Knoll, R.H. *et al.* (1962) Weightlessness Experiments with Liquid Hydrogen in Aerobee Sounding Rockets, Uniform Radiant Heat Addition – Flight 1, NASA, TM-X-484; Atmos, D. (1962), Flight 2, NASA TM-X-718; Corpas, E.L. (1963), Flight 3, NASA, TM-X-872; (1964), Flight 4, NASA, TM-X-873.
11. Grayson, G. *et al.* (2006) Cryogenic tank modeling for the Saturn AS-203 experiment. Presented at 42nd AIAA/ASME/ASEE Joint Propulsion Conference and Exhibit Sacramento, CA, 9–12 July 2006, AIAA 2006-5258.
12. Chuster, J.R. *et al.* (1989) COLD-SAT – An Orbital Cryogenic Hydrogen Technology Experiment, NASA TM 102303.
13. Kramer, E. (ed.) (1998) Cryogenic On-Orbit Liquid Depot-Storage, Acquisition and Transfer (COLD-SAT) – Experiment Conceptual Design and Feasibility Study, NASA TP 3523.
14. Piryk, D. *et al.* (2012) Thermal and fluid modeling of the CRYogenic Orbital TEstbed (CRYOTE) ground test article. Presented at Thermal and Fluid Analysis Workshop 2012, Pasadena, CA, 13–17 August. 2012.
15. Meyer, M.L. *et al.* (2012) Cryogenic Propellant Storage and Transfer Technology Demonstration for Long Duration In-Space Missions, NASA TM 2012–217642.
16. Chojnacki, K. (2013) Cryogenic Propellant Storage and Transfer (CPST) Technology Demonstration Mission (TDM), NASA.
17. Chato, D.J. (2006) The Role of Flight Experiments in the Development of Cryogenic Fluid Management Technologies, NASA/TM 2006-214261.
18. Christie, R.J. *et al.* (2011) Broad area cooler concepts for cryogenic propellant tanks. Presented at NASA, Thermal and Fluids Analysis Workshop (TFAWS), Newport News, VA, 15–19 August 2011.
19. Lacapere, J. *et al.* (2013) RBO/ZBO system for European long term cryogenic propulsion. Presented at 5th European Space Cryogenic Workshop, Noordwijk, The Netherlands, 16–18 December 2013.

41

Transportation/Propulsion/Demonstration/Buses: The Design of the Fuel Cell Powertrain for Urban Transportation Applications (Daimler)

Wolfram Fleck

41.1
Introduction

The PEM-fuel cell technology (FC) is a perfect energy conversion solution for highly efficient and low noise propulsion applications in a power range realistically of up to 300 kW. FC technology is supporting to 100% a sustainable energy approach via the use of hydrogen as a fuel, which can be derived from various alternative energy sources, hence supporting all CO_2 reduction initiatives known so far.

A powertrain based on fuel cell technology, using proton exchange membrane (PEM) technology and gaseous hydrogen as fuel on board is a promising solution for urban inner city transportation or utility vehicles. This chapter will mainly focus on bus applications. Other applications such as dump trucks, light duty trucks for mail delivery, and so on will be addressed briefly in Section 41.6 (Future Outlook). Due to the possibility of fleet operation, all those applications have an infrastructure advantage in common, thus resulting in an effective utilization of hydrogen infrastructure.

The FC-powertrain application for urban busses will be described, starting with its operational environment and requirements and specifications, including some certification related issues.

Section 41.4 (Design Solutions) describes the principal approach to a FC-powertrain lay out. How to deal with auxiliary power demands in a bus application will be evaluated as well as the evaluation of various solutions for the distribution of power to the traction system and to auxiliary systems. The simulation of the right choice of powertrain configuration with respect to hybridization is discussed and explained with examples. Operational strategy to minimize fuel consumption and maximize life time in conjunction with comfort requirements such as air-condition and heating is one of the key questions for a satisfactory powertrain lay out. Ideas for low cost design solutions are addressed in Section 41.6.

Hydrogen Science and Engineering: Materials, Processes, Systems and Technology, First Edition.
Edited by Detlef Stolten and Bernd Emonts.
© 2016 Wiley-VCH Verlag GmbH & Co. KGaA. Published 2016 by Wiley-VCH Verlag GmbH & Co. KGaA.

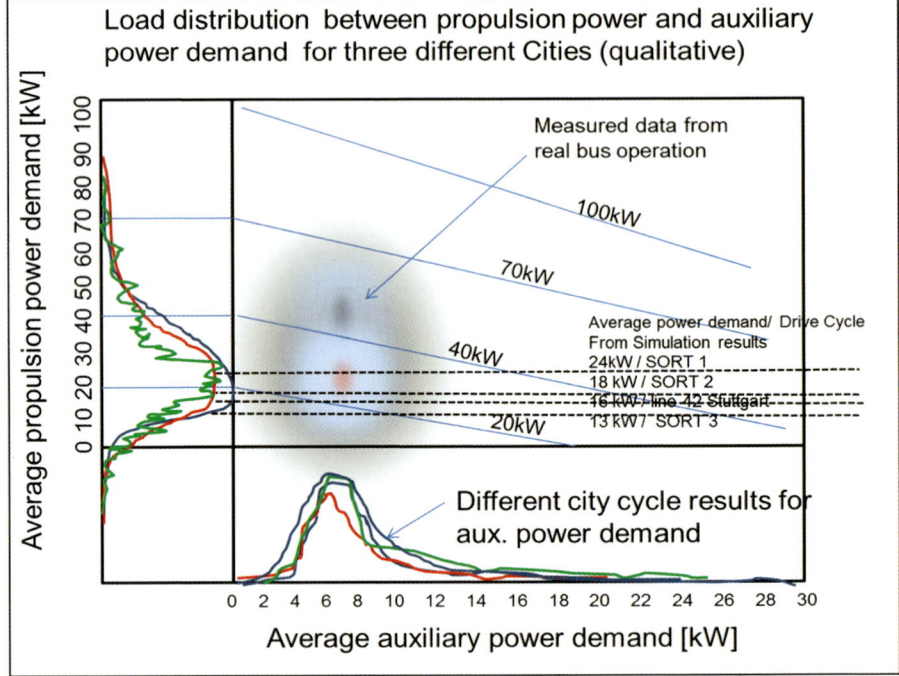

Figure 41.1 Determination of propulsion- and auxiliary power demand from measurements in various cities (schematic diagram). This information is necessary in order to balance the power needs between propulsion power and auxiliary power. This diagram is the result of detailed simulation calculations combined with real field data from fleet operation. The dotted horizontal lines are average power demand values for standard drive cycles as SORT and for an in-house comparison cycle [1].

41.2 Operational Environment

An inner city bus application is usually carefully adapted to the environment it is operated in. This is based on historical reasons, due to the major goal of transporting people safely as fast as possible by the lowest possible cost to the operator and hence to the passenger. There is some analogy to the mission profile planning of aircraft design in that buses are very much purpose designed compared to a long-haul truck or passenger car application.

The topology of a city and its street system infrastructure, as well as the bus line schedule, are the key parameters for the powertrain layout. Those key requirements determine maximum power demand, torque of drive train, range, and the time for refueling (Figure 41.1).

Specific maintenance and service requirements from the transit operator may influence the packaging of the powertrain in the bus coach as well as access hatches and lids for service.

Optimization of seating capacity depends strongly on an exact balanced weight distribution between all powertrain components relative to the bus axles.

Comfort requirements, mainly air conditioning (A/C) and heating followed by noise emissions and vibrations, influence the layout of auxiliary devices like cooling fans, water coolant pumps, heaters, oil pumps, compressors for suspension system, and many more.

Many communities are requesting city busses to fit into their individually defined city CO_2-reduction strategy, supporting local individual available sustainable energy sources such as wind power or low cost electric power availability at night. Some cities have very strict local excess restrictions, prohibiting the use of noisy vehicles in urban areas after 10 p.m.

Environmental friendly certificates are required to demonstrate responsibility for a clean urban transportation. During operation, service, and maintenance of the Fuel Cell bus, a vast number of regulations have to be fulfilled on top of the certification requirements of a bus.

The alternative FC powertrain has to fit into existing vehicle constructions with a strong request not to interfere and restrict with the existing passenger compartment. Other limitations include the maximum roof elevation defined by various cities.

41.3
Requirements

To determine the cross power demand of a bus propulsion system, three major areas of power consumption have to be defined:

- propulsion power to drive,
- power for auxiliary devices,
- power for heating and air condition.

41.3.1
Propulsion Power to Drive

A maximum vehicle velocity has to be defined considering legal requirements, customer requirements, and experience. This speed is usually defined as below $80\,\text{km}\,\text{h}^{-1}$ for European applications and may reach as high as $110\,\text{km}\,\text{h}^{-1}$ for US applications. The minimum design speed should be above $60\,\text{km}\,\text{h}^{-1}$ for legal operation on the highway (Autobahn). The maximum design speed requirement directly influences the licensing process and should be determined carefully [2]. For a successful design the lay out of the drive train components, for example, the electric motor and its inverters (power controllers) and the maximum speed and the average speed are important parameters (Figures 41.2 and 41.3). While the maximum velocity has a direct influence on the rotational speed requirements of the motor or any following gearbox, the average vehicle speed is

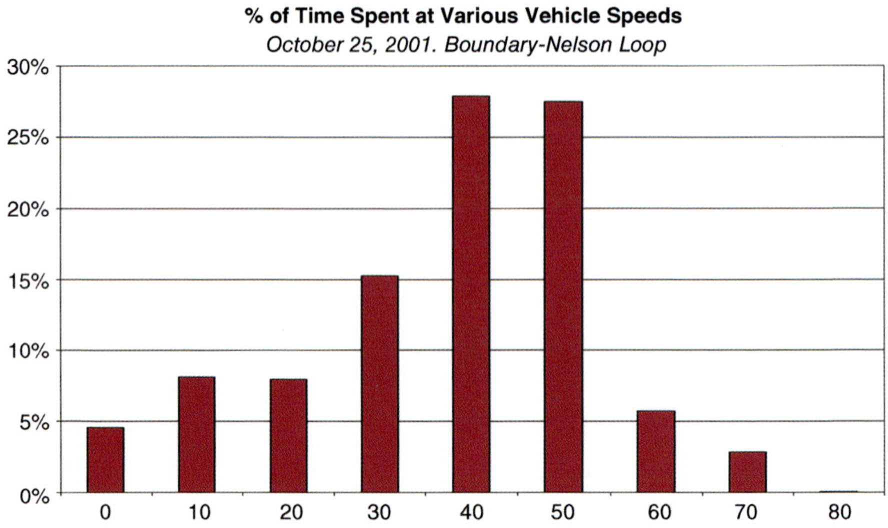

% of Time Spent at Various Vehicle Speeds
October 25, 2001. Boundary-Nelson Loop

Figure 41.2 Example for a typical time spent at various vehicle speeds for a cycle with little stop and go traffic. The average speed measured in ten cities over a period of six years varied between 9 and 19 km h^{-1}! Boundery-Nelson-Loop in Vancouver, B.C. Canada [4].

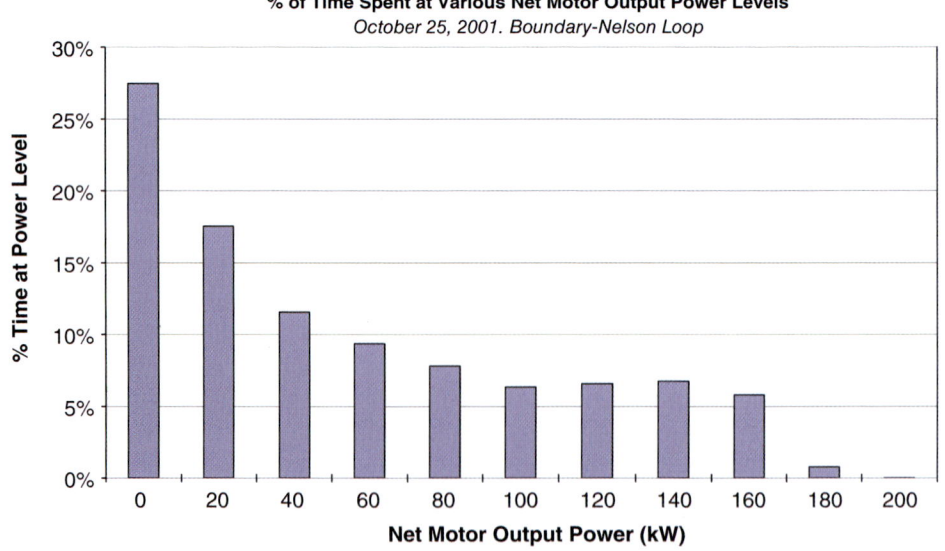

Figure 41.3 Example for a typical time spent at various power demands. Most time will be spent in idle power demand, while the maximum power is only required during acceleration or steep hill climbing. This particular test cycle included two hills: one with 10% grade over 1.5 km and one with 16% grade over approximately 500 m [4].

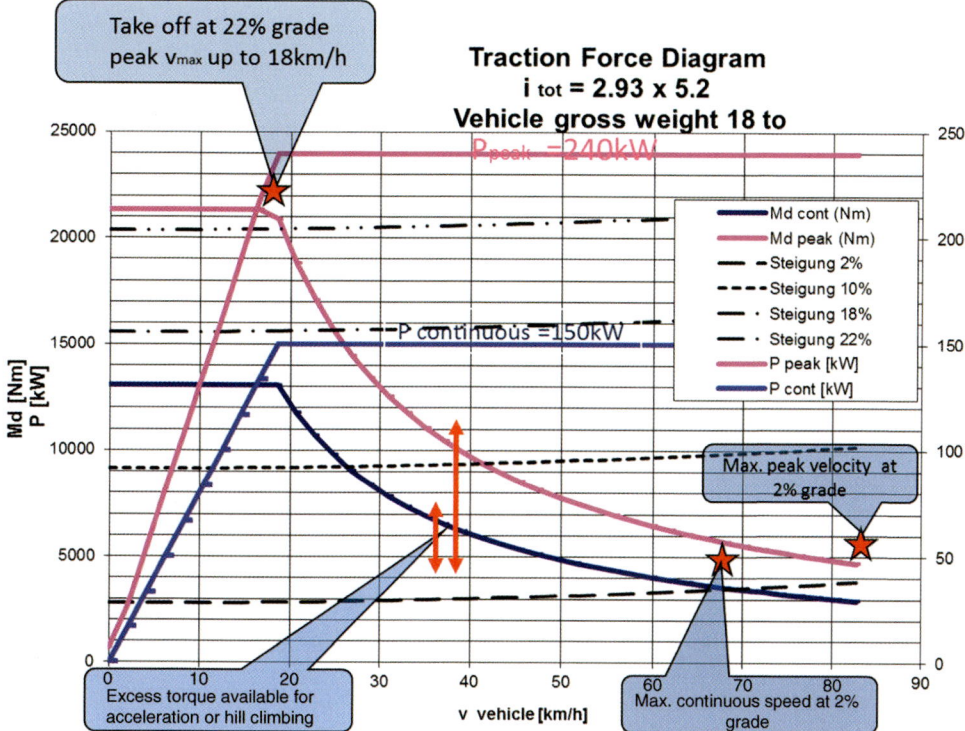

Figure 41.4 Schematic diagram showing motor torque and power in Nm and kW, respectively, against vehicle speed in km h^{-1}. Dotted lines are curves for constant grade in %. The red stars are key performance features, which have to be met to ensure sufficient power for usual traffic conditions [4].

important for location of the optimum efficiency point (sweet spot in a torque against speed diagram). In addition, the type of the electric drive may be decided at this point in time. From experience AC-motors are superior to PM-motors if the rotational speed is above approximately 6000 RPM [3].

The aerodynamic quality such as the Cw*A value and handling quality like rocking behavior around the roll axis are further criteria determining the maximum power requirement.

Hill climbing capability is required by several codes and standards and in addition by the transit authority. There main goal for this requirement is the avoidance of a bus becoming a traffic obstacle to the other traffic. The maximum design speed should be achieved for a grade of 2%. At 10% grade the minimum speed should not drop below 37 km h^{-1}. For a grade of 22% the powertrain must be strong enough to hold the vehicle in position, avoiding a backward rolling of the vehicle. All requested features will determine the specific torque requirement for the FC powered drive train (Figure 41.4).

Maximum acceleration will be defined as the time in seconds required to reach 50 km h^{-1} (Figure 41.5). For internal combustion engines this test is an

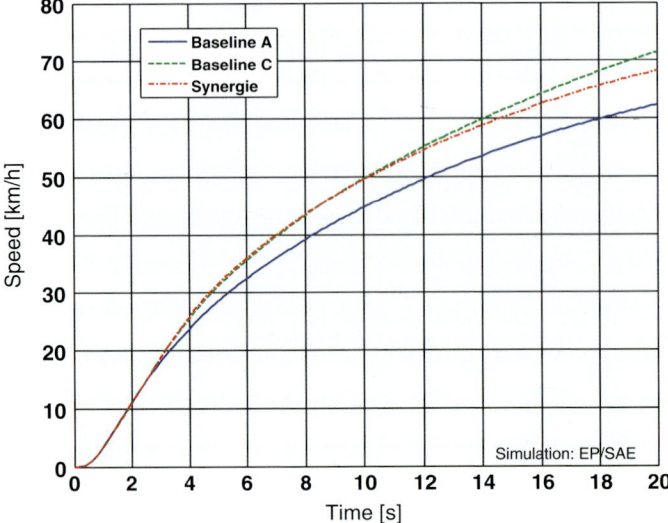

Figure 41.5 Diagram showing acceleration behavior of a 12 m bus application, simulation results. Test results show 10 s to the 50 km h^{-1} speed level for the CFCH and 12 s for the CUTE bus (all measured at empty vehicle weight). Remember a bus in normal service is only permitted to accelerate at <1.5 m s^{-2}, which results in acceleration times of >20 s to 50 km h^{-1}. Because electric motors develop an instantaneous maximum torque at nearly zero speed the acceleration of a fuel cell bus is superior to that of the diesel bus even considering the higher empty curb weight [12].

ultimate proof that the engine is still working acceptably and capable of developing its designed power output. For a FC-powertrain this test is of minor importance. However, bus drivers today still judge the agility of the vehicle by this test. A second acceleration requirement is based on experience. It ensures the capability of the coach to merge into traffic while leaving a bus stop located on an 8% hill. For an EvoBus product this may results in an acceleration requirement of 2 m s^{-2}. The third acceleration requirement for a city bus is due to safety reasons, protecting standing passengers. Acceleration of 1.2–1.5 m s^{-2} should not be exceeded.

41.3.2
Auxiliary Power Demand

Today's standard city busses are usually equipped with the following subsystems:

- power steering pump;
- brake compressor for pneumatic braking system;
- coolant water pumps;
- hydraulic pump for radiator fan operation if not managed via belt drive;
- alternators for 24 VDC onboard supply.

Table 41.1 Typical conventional auxiliary power demands for average bus applications (12 m coach, 18 ton GVW).

Subsystem	Power demand/source
Steering pump	5–7 kW, up to 12 kW if full steering power is required, belt or gear drive powered from ICE
Brake compressor for pneumatic braking system	Approximately 10 kW, operated in on/off mode, belt drive
Coolant water pump	Approximately 2–5 kW, operated from crankshaft or belt
Hydraulic pump for radiator fan operation	15–30 kW operated from crankshaft, some applications are belt driven
24 VDC supply via alternator	Up to 6 kW or 300 A @24 VDC operated via belt drive

In a conventional internal combustion engine (ICE) application auxiliary power is usually derived directly from the crankshaft of the engine (Table 41.1). The usual speed range for a standard bus diesel engine is between 600 and 2500 RPM depending on engine size. The so-called turn-down ratio, the ratio between idle speed and maximum speed, is approximately $1:4$. These values may change to $1:2.5$ for large engines and go up to $1:10$ for a passenger car application. All auxiliaries are propelled via belt or chain or gearboxes directly with a fix transmission ratio. Owing to this fact, not all auxiliary components are operated at the best point. Only some can be switched on and off via the use of an electromagnetic actuated clutch (e.g., A/C compressor or hydraulic pumps). This approach is well understood worldwide and works very reliably. This reliability and low cost approach comes with the disadvantage of poor efficiency. Most of the subsystems are designed for a typical diesel speed range from 800 to 2500 RPB.

A FC-powertrain offers the advantage of being able to propel all those auxiliaries decentralized, each on its optimum operational point. The NeBus technology demonstrator bus developed by Daimler research was the first proof that decentralized operation is a fundamental step towards better efficiency. However, in 1995 it was not possible to provide reliable power electronic devices to support this approach for a large fleet project; 15–20 years ago auxiliary components able to be operated by an electric motor were not readily available in terms of the required reliability and costs. The FC-powertrain design of the CUTE Bus fleet imitated a diesel engine application to overcome those reliability issues. At that time reliability was judged more important than efficiency, which led to the so-called "one Motor" approach.

41.3.3
Heating and Air Condition Power Demand

Heating and air conditioning of the bus passenger compartment is required by legal and comfort reasons. For the driver the wind shield defrosting and defogging is regulated by law, usually resulting in separate heating loops needing a superheating of the waste heat coming from any engine or fuel cell powertrain.

While the ICE engine provides enough waste heat at a temperature level well above 80–90 °C, the fuel cell system produces less waste heat at a temperature level between 55 and 75 °C. The lower operating temperature regime of the FC-engine requires higher surface areas and/or higher venting fan power demand to reject the waste heat either into atmosphere or into the passenger cabin heating system. Utilization of FC waste heat for cabin heating requires a superheating process for cold weather conditions if the coolant temperature is below the required preheating temperature for comfortable cabin heating. Special electrical heaters have to be installed to ensure a sufficient temperature gradient for the cabin heating system. This super-heater device is usually combined with the brake resistor in one apparatus. The brake resistor function is required to turn the negative torque or breaking moment of the drive train motor into usable heat. This occurs if the driver is using the brake pedal or the retarder function. The available electric brake power is roughly the inverse of traction power of the motor reduced by a certain factor [3]. While hydraulic retarder systems offer brake power up to 280 kW, the range of inverse electric brake power using the electric drive is limited to the lay out of the electric apparatus (depending on layout and size of bus application: 120–150 kW).

Up to 50 kW thermal power may be required to keep a 12 m three door city bus warm if ambient temperatures fall below −5 to −15 °C!

Even an ICE engine powered bus operated in central EU is usually equipped with a 35 kW diesel fuel powered heater system for cabin heating comfort. This heater system is necessary to compensate for cases when the ICE engine is either off or does not produce enough waste heat to keep the cabin temperature at the chosen comfort level. This auxiliary cabin heating system is also used to pre-heat the bus before operation in service. A fuel cell powered bus requires the same separate heating system, but uses the FC-power plant powering the brake resistor in order to generate the additional heat. An advantage is the low noise operation of the FC and no additional components.

Air condition (A/C) systems usually require between 15 and 25 kW shaft power with respect to an electric motor shaft power to propel an A/C compressor equipment providing 35–40 kW caloric chiller power.

Interestingly, in a 12 m coach application with three doors, the A/C system is only able to cool down the cabin from 30 °C ambient to 27 °C cabin inside temperature. The reduction in humidity is the major effect on passenger comfort in this case.

Obviously, the determination of the auxiliary loads and the power for heating and A/C is strongly dependent on drive condition, climate conditions, and drive cycle requirements. A proper prediction and simulation without a solid drive cycle input from test data is not possible.

Simulation performed so far separates A/C and heating demand from the auxiliaries required to drive and operate the bus. In real applications the fuel consumption is greatly influenced by heating and A/C effort. This may lead to the annoyance of customers in terms of expectations if not clearly communicated.

Figure 41.6 NeBus, May 1997, CUTE bus May 2003, and Citaro FuelCELL-Hybrid, October 2009 (from left to right). Picture documentation by the author, 3 October 2009.

41.4 Design Solutions

Fuel cell powertrains can be realized in various applications, from a straight FC-powertrain to a more complex hybrid configuration. Furthermore, the definition of how to propel the drive train and the many auxiliary devices needs to be solved to ensure successful bus operation.

41.4.1 NEBUS

The first technology demonstrator bus (NEBUS) was developed to address three important development goals (Figure 41.6):

1) show a fuel cell powertrain able to respond to sudden load changes without using a traction battery or ultra-caps for energy buffering;
2) demonstrate the capability of this technology in heavy duty applications as complex as a bus by the usage of as many components from the car application as possible;
3) show operation of all auxiliaries independent from drive train motor speed.

41.4.1.1 Design Solution

The fuel cell system was basically a scale up of the NECAR 2 [5] fuel cell engine, increasing the gross power from 50 to 250 kW. The design was based on Ballard

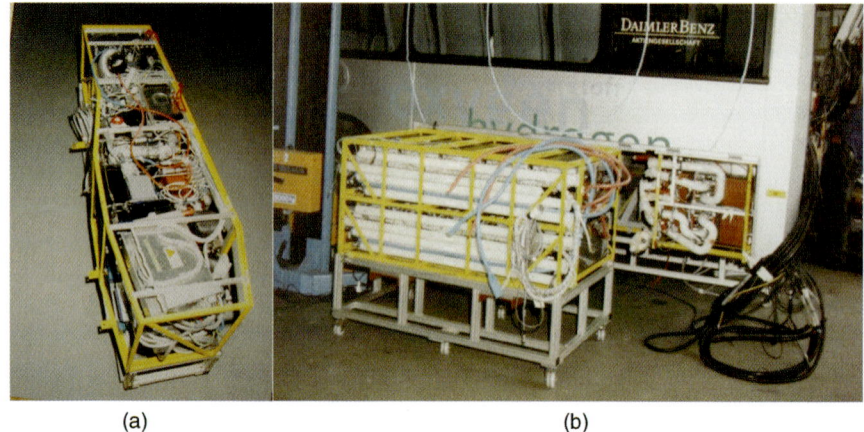

(a) (b)

Figure 41.7 The air module containing the two-stage compressor system with electric drive and converter plus the humidifaction apparatus for the anode loop, muffler, and exhaust system from the FC air supply system module (a). Still on its integration carrier the fuel cell stacks assembled to their water, air, and hydrogen manifolds just minutes before final integration into the bus engine bay. (b) The cooling and water purification system module is already installed. Power harness and 24 VDC harness is hanging out, on the very right-hand side of the picture, waiting for final integration and commissioning. Picture documentation from the author, early May 1997.

stacks with 132 cells MK 704-13 type, using DOW membranes. Ten stacks were electrically switched together so that two strings in parallel with five stacks in series worked to the power grid. Electrically the system worked in a range from 650 VDC at open circuit to 450 VDC at maximum load. The gross power output was 247 kW at 550 amp and 450 VDC; 190 kW could be used for the drive train. The delta power went into the supply of compressor, pumps, A/C, heating, and other auxiliary loads.

Aside from the stack itself everything else was a custom application. Humidification of the process gases, H_2, and air was based on membrane humidification. The air supply system was based on a two-stage compressor system with an electrical driven screw compressor at first stage and an exhaust powered second stage automotive turbo charger with variable nozzle turbine inlet (Figure 41.7). Hydrogen circulation was realized by the use of a special jet pump powered by the high pressure fuel, metered into the system via a special pressure regulator. The primary pure water coolant loop rejected its waste heat into a standard automotive glycol cooling loop. All pumps, fans, and actuators as three-way valves were operated by custom made pumps, valves actuators, and motors. The target for the system controller design was strongly focused to achieve the highest efficiency possible.

For the drive train a near-hub-wheel drive concept was chosen. Each wheel of the rear axle was powered via a separate motor and inverter (Figure 41.8). Each wheel was connected to a 50 kW rated AC-motor via a planetary reduction gear [6].

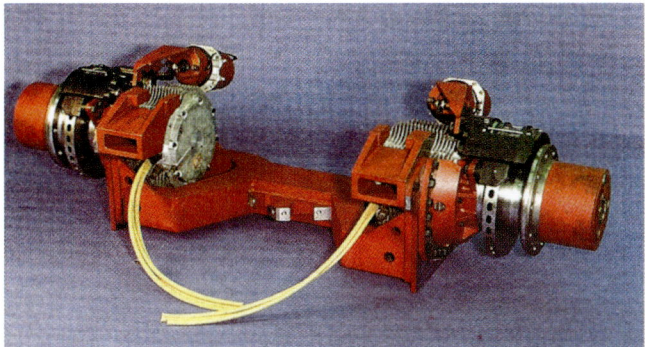

Figure 41.8 Air cooled ZF hub wheel drive axle Mark 1. This compact drive axle for low floor bus applications putting out 100 kW continuous and 150 kW peak power was a purpose design for hybrid applications and could be added into solo as well as articulated busses, with the latter using two drive axles to optimize traction and power supply [7].

The result was a very efficient powertrain but with poor reliability. This poor reliability was ascribed to the unavailability of electrified auxiliaries able to be operated in a voltage window defined by the fuel cell itself. Special custom made electric drives and their relevant inverters and controllers had to be custom developed supporting a voltage swing between 500 VDC at idle and 250 VDC at full load. None of those auxiliaries and drive train components went through a sound design validation program, which is the nature of a technology demonstrator!

The fuel storage system was based on Type3 Cylinder technology with a nominal operating pressure of 30 MPa. The tanks were made from an aluminum liner and a mix of glass and aramid fiber compound wrap. The storage system was realized with seven cylinders, 150 liters geometric volume each, resulting in 21 kg H_2 stored.

The performance of this bus was comparable to the diesel version. The additional weight of the FC-powertrain resulted in a significant reduction of passengers due to approximately 3200 kg added weight. The fuel consumption of 11–12 kg-H_2 per 100 km was about 15–20% lower compared to a standard diesel under assumption of equal gross vehicle weight (GVW).

41.4.2
CUTE and HyFleet:CUTE Program

The design of the fuel cell engine for this bus application was determined by the following targets and requirements:

- 12 m Citaro coach from EvoBus as platform vehicle;
- powertrain to serve the EU and NAFTA/DOT rules and regulation requirements;
- availability should be equal to the CNG Bus application;
- two-year operation under full warranty conditions, 4000 h life time to the stack.

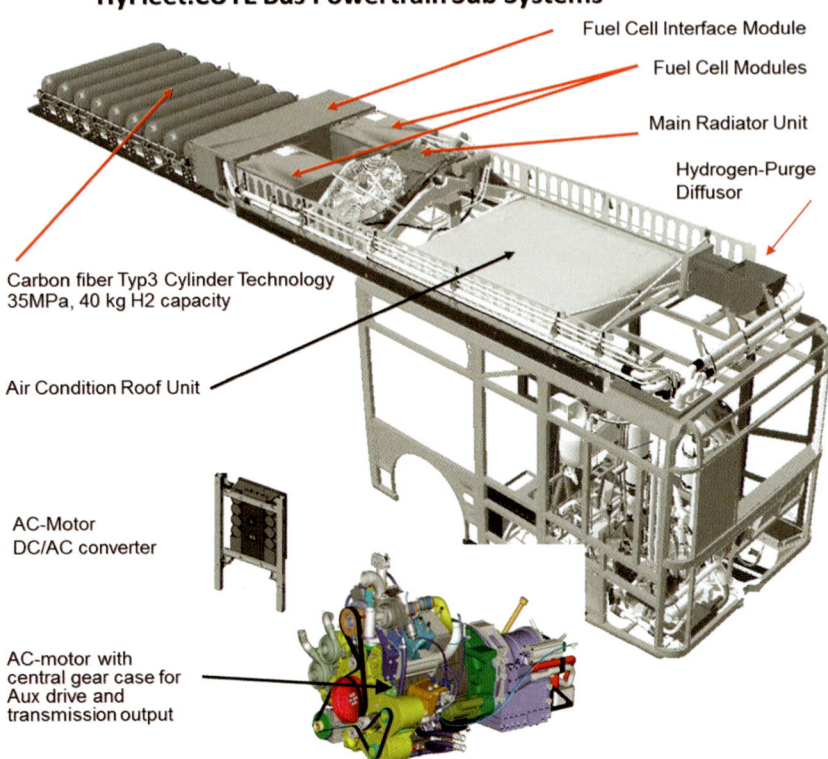

Figure 41.9 Packaging arrangement of the powertrain and other important sub-systems relative to the bus platform. Roof load was in the range of 2700 kg, which in turn required an active suspension system for driving comfort and safety during curve handling.

A new drive train and auxiliary power system had to be developed, because of the absence of a suitable drive motor solution like the ZF-hub-wheel-drive (Figure 41.9). The absence of sufficiently reliable auxiliary power systems for this specification significantly influenced the decision making process.

Interestingly, at the same time the Ballard Power Systems Corp. Vancouver and XCellsis Inc. intended to introduce six fuel cell buses onto the market, three in Chicago and three in Vancouver (test phase started in 1998 with CTA in Chicago). The powertrain concept was similar to the NEBUS with the difference being in the drive train axle arrangement. Instead of the innovative hub wheel drive axle from ZF a custom made electric motor with a reduction gear was fitted directly to a standard rear axle arrangement. All buses suffered from reliability problems on high voltage electric equipment similar to the problems experienced with the NeBus prototype from 1997 [8].

The concept approach of copying as much as possible of a standard ICE powertrain was steering a major discussion among experts at that time but was finally turned into realty, due to the existance of a much broader choice of

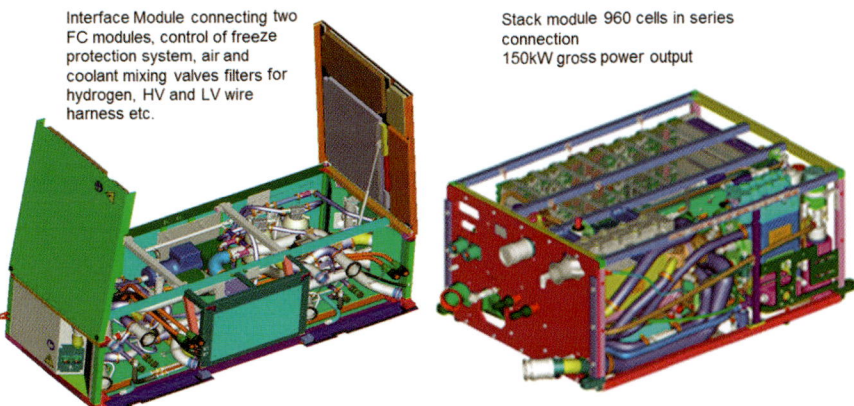

Figure 41.10 Interface module and one of the two Fuel Cell modules from Phase 5 design.

suitable suppliers, providing reliable and affordable electric components (2001). The use of a traction battery and hybridization control strategy was considered again as too risky for this fleet trial under full warranty conditions. The life time and reliability of the fuel cell itself was considered as the most important topic for this bus powertrain version.

After the realization of a feasibility demonstrator bus (so-called Phase 4 bus engine demonstrator) the concept was set as follows:

Fuel cell system based on MK 902-mod. stack design. Six cell rows with a total of 960 cells in series packaged in two separate fuel cell modules connected to an interface module that contained the heating, frost protection, and coolant water management (Figure 41.10).

The fuel storage system was based on Type3 Cylinder technology with a nominal operating pressure of 35 MPa. The tanks were made from aluminum and carbon fiber full wrap. The storage system was realized by use of nine cylinders, 205 liters geometric volume each, resulting in 45 kg H_2 stored.

The innovation in this FC-powertrain was a single electric AC-motor with the exact same torque and speed characteristics as an OM 906 bus diesel engine from EvoBus (Figure 41.11). On the one side the motor flange was attached to a custom made central gearbox. On the opposite flange a standard six-shift automatic transmission with a hydraulic retarder was bolted to the motor. The transmission was connected to a normal prop shaft to a standard bus T-axle with a differential gear ratio of 1 : 4.7.

The gearbox was able to propel all ten auxiliaries of the drivetrain and the bus. The automatic transmission was able to be adjusted by a transmission controller offering four different shift profiles. Together with the possibility to change the fixed rear axle transmission ratio at differential location, 28 different shift profiles could be selected to better adapt the powertrain to topological conditions.

The HyFleet:CUTE and CUTE-plus program was finally awarded as one of the most successful public funded programs operated simultaneously in Europe,

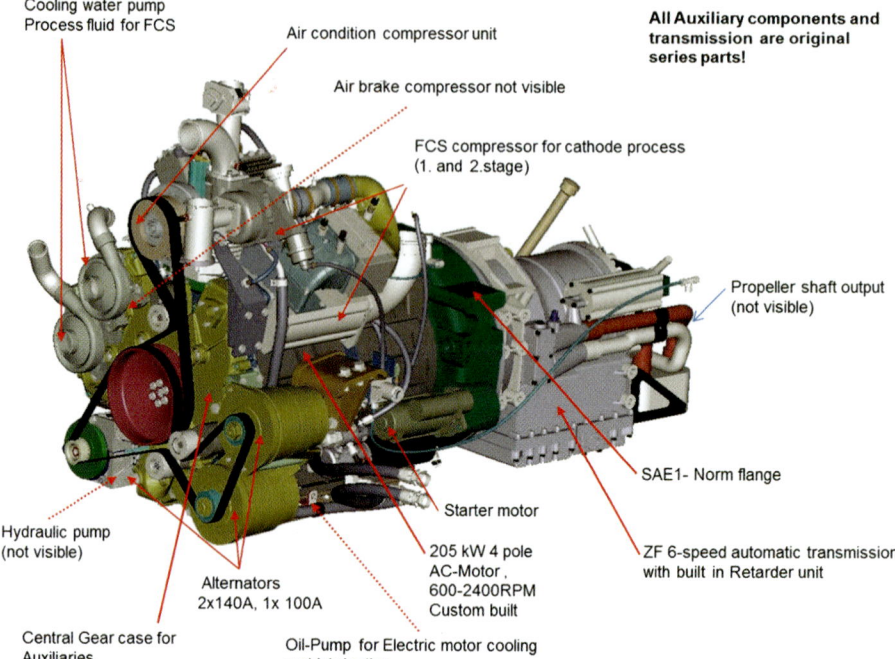

Figure 41.11 Heart of the "One-Motor"-design: Powertrain and auxiliary package for bus and fuel cell related components. The electric motor was able to develop 1100 Nm torque output in the range 600–2400 RPM.

Iceland, Australia and China. A total of 36 buses operated in ten locations worldwide, covering over 3 400 000 km of operation and field experience. The program was successfully operated from 2003 until 2009 [9].

41.4.3
Citaro FuelCELL-Hybrid

At the end of the HyFleet:CUTE program (2006) it was decided to develop a bus, which would incorporate all "lessons learned" from CUTE and HyFleet:CUTE. This program was called CUTE-plus program (Figure 41.12). Parallel to the execution of the HyFleet:CUTE program the Li-ion battery technology developed so rapidly that it became possible to consider a hybrid drive train lay out for the third generation of the FC bus (Figure 41.18). The major development targets for this fuel cell powertrain were defined as described below:

- The powertrain should match as much as possible into the current EvoBus Diesel Series-Hybrid program. Eventually, the FC-system may replace the Diesel Generator at a later point in time.
- To show improvements on the product cost side it was decided to utilize as much as possible the FC components and stacks from the FC car program.

Figure 41.12 Citaro FuelCELL-Hybrid (CFCH). Photograph: Daimler AG.

- Fully electrified auxiliary components were provided by the diesel hybrid project as well as the drive train concept. The drive train is based on the second generation of ZF hub wheel drive axle which is now water cooled.
- Fuel consumption should be 50% less compared to the CUTE powertrain.
- The life time of the FC system should be 8000 h or 5 year calendric service life.
- Maintenance should be able to be executed either by EvoBus standard service personnel or trained transit authority members.
- The bus total empty weight should support a passenger capacity of greater 72 people.
- The range should be >350 km, and the refueling time should be less than 10 min for the 35 kg H_2@35 MPa.
- The bus overall vehicle vertical extension should be reduced by 200 mm (from 3650 to 3450 mm).

The fuel cell system was based on the fuel cell power plant from the Mercedes Benz F-Cell passenger car using two times this very exact fuel cell system to provide a net power output of 140 kW to the high voltage (HV) grid of the bus (Figure 41.14a,b). The two F-Cell systems were packed into the roof in a twin arrangement (Figure 41.14a). The roof location was indicated to keep the effort for certification as low as possible and to avoid any costly repackaging of components.

The powertrain concept is designed to support the EvoBus series hybrid strategy as much as possible (Figure 41.15). The FC-twin system is linked into the series hybrid part of the powertrain by the use of a bi-directional DC/DC converter (Figure 41.13) matching the voltage swing of the FCS with the voltage swing governed by the HV-battery. Each of the two DC/DC converters is capable of transforming 100 kW from 470 to 280 VDC up to 550 to 750 VDC. The DC/DC technology used is completely new, compact, and avoids any heavy copper transformer coils for galvanic separation.

Figure 41.13 Each fuel cell system is connected to the traction power grid via a DC/DC converter. The compact and highly efficient design of this high voltage equipment is custom designed. The two lower wires are the plus and minus pole coming from the FCS, the two upper wires connect into the traction power grid. The device is water cooled. Photograph: documentary of RD/RFF and EvoBus Mannheim, 2009.

All HV components such as the hub wheel drive (Figure 41.16), inverters, multi converter for auxiliary power at 400 VDC and provision of 24 VDC board net, FC-DC/DC converter, and the HV-Battery (Figure 41.17) are water cooled (Figure 41.18).

The hydrogen fuel storage system is realized by a 35 MPa Type3 Cylinder technology using the valve technology and pressure regulators from the passenger car program. Due to the traction battery the storage system could be reduced from nine to seven cylinders. Each cylinder is capable of storing 5 kg H_2@35 MPa resulting in a total storage capacity of 35 kg.

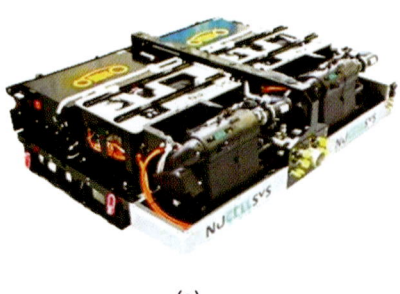
(a)

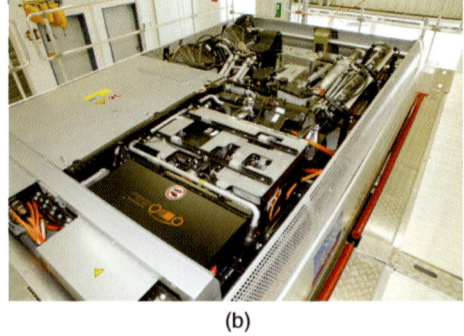
(b)

Figure 41.14 (a) Fuel Cell twin system designed and manufactured by Nucellsys GmbH [10]. FC-System: 120–160 kW output power, weight: 700 kg, volume: 1.8 m³, minimum ambient operation temperature −15 °C. (b) Final integration in the roof, rear location. In the top right-hand corner of the picture one can see the cooling management with the radiator package sucking in air from the center and venting it out towards the side of the bus.

41.4 Design Solutions

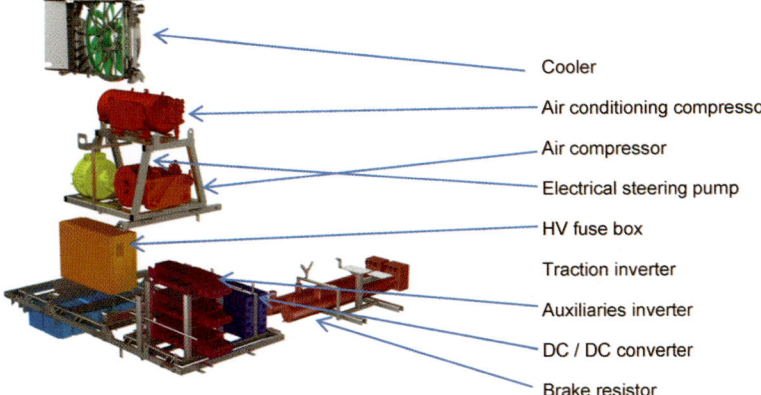

Figure 41.15 CAD view of the engine bay, giving an insight into the auxiliaries of the CFCH. Daimler AG.

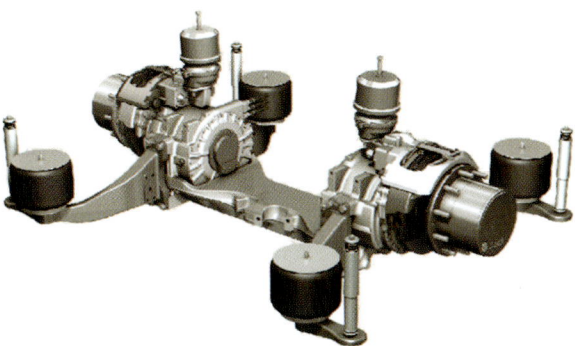

Figure 41.16 The new ZF-electric drive axle, AVE 130, is now water cooled, providing 21 000 Nm torque at a peak power of 240 kW and continuous rated output power of 130 kW. Reduction gear ratio $i = 22.85 : 1$. Motor maximum speed is 11 000 RPM with respect to 485 RPM at the wheel. Nominal current is 135 A, max is 350 A, while the nominal voltage range is between 350 and 420 VAC. Maximum axle loading capacity is 13 ton [11].

High Voltage Traction Battery

capacity: 26.1 kWh
peak power: 250 kW
continuous power: 120 kW
mass: 330 kg
Liquid cooled

Figure 41.17 The traction battery is a take over from the Diesel Hybrid bus application, based on Li-Fe technology. The box in front contains the A/C cooling system for the water cooled battery system (case located behind). Photograph from EvoBus GmbH.

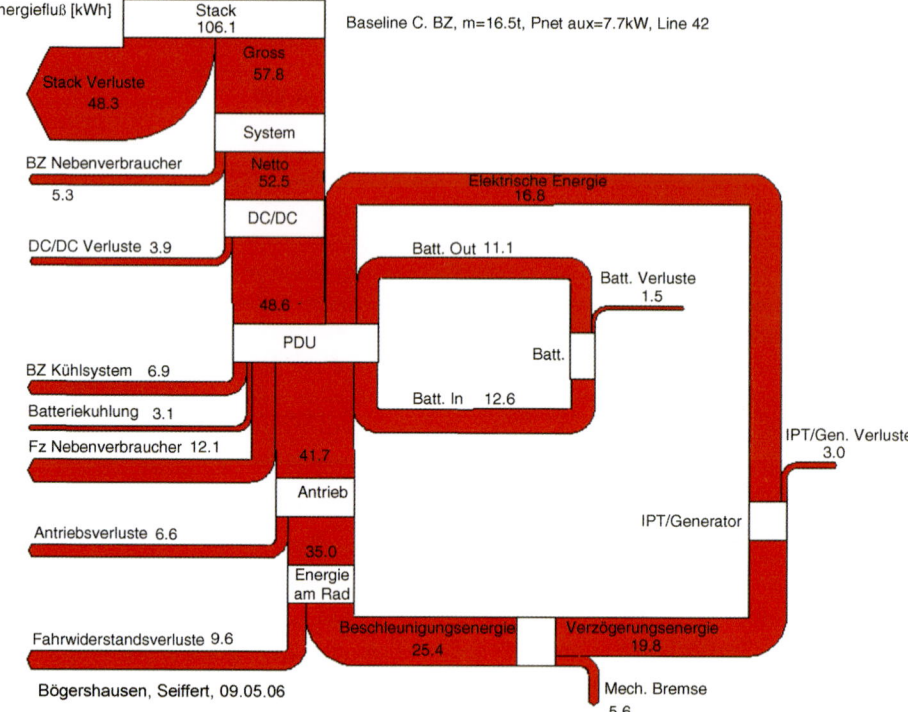

Figure 41.18 Sankey diagram of the series hybrid giving insight into the energy flow in kWh recovery and battery losses during acceleration and de-acceleration (braking). The energy flow is presented for a specific drive cycle and fuel cell configuration [12].

41.5
Test and Field Experience

41.5.1
NeBus

NeBus was designed as a pure research technology demonstrator vehicle, showing the great potential of the fuel cell technology (Table 41.2). It was not designed for lifetime or for use in active public transport. Nevertheless, after a detailed test program the bus was used as a presentation platform traveling all over the world. The test vehicle was operated for approximately 4–5 years until final termination of its active service and operation. Today it can be seen in the EvoBus Bus museum located in Neu-Ulm, Germany.

41.5.2
CTA, CMBC

Ballard Power Systems developed six busses for Chicago Transit Association (CTA) and Costal Mountain British Columbia (CMBC) a transit organization in

Table 41.2 Bus and key fuel cell drive train data for NeBus.

Parameter			
Fuel cell power plant	10× MK704-13 stacks, external membrane humidifaction, two-stage compressor system for cathode air supply, jet pump circulated anode loop	Gross power 250 kW Net power 190 kW	650–450 VDC
Auxiliary subsystems	No A/C, cabin heating via TRUMA gas heater powered by hydrogen, water pumps and air brake compressor at 380 V		
Drive train	ZF hub wheel drive, air cooled, 600 V Use for first time of water-cooled brake resistors in a bus application due to hydrogen storage on the roof. Typical ram air cooled resistors where not suitable due to operational temperature superseding 580 °C!	Rated power 100 kW	Peak power: 150 kW
Hybrid battery features	None		
Bus empty weight	15 100 kg		
Gross vehicle weight	19 000 kg		Experimental licence
Passenger capacity, legal	34 seating, 24 standing	58	
Hydrogen storage system	Type 3, 7 × 150 l. cylinders, NWP 30 MPa	21 kg	Cevlar glass fiber composite
Fuel consumption	11–12 kg per 100 km, no A/C system		Linie42 Stuttgart
Range	<200–250 km		Depending on topology and drive cycle

Vancouver Canada. One may ask why it is important to mention these busses in this chapter. It is due to their commonality of the fuel cell technology and the exact same deficiencies with respect to reliability and maturity of electric subsystems as motors, and high voltage inverter technology used in the NeBus. The six busses where based on third generation technology, called Phase 3 bus design, derived from the Phase 1 presented in 1987 and the Phase 2 bus presented in 1994. After Daimler-Benz had bought a 25% share of the Ballard

Company in 1995 it became liable for operation of the planned bus fleets for Chicago at CTA and Vancouver CMBC. The first three buses serving in Chicago showed a poor reliability record, failing 143 times in the 24 months of service due to fuel cell drive train failures of all kind. The Vancouver buses at CMBC started 6 months later, showing 96 failures in the same period of time. The number of road calls, meaning that the bus had to be towed home, was 23 or 1503 km distance between road calls [8]! The total mileage in service was for CTA approximately 48 000 km and for CMBC 60 000 km. These dramatic results turned into a reliability offensive that eventually resulted in the Phase 4 bus design (technology demonstrator for the one motor concept) and finally led to the Phase 5 design, which is named generation 2 at Daimler in respect of the EvoBus side. In 1997, XCELLSYS Canada was founded to improve the bus fuel cell applications.

41.5.3
CUTE and HyFleet:CUTE Program

The powertrain used in this international program was focused on reliability and lifetime (Table 41.3). Efficiency of the powertrain was a second priority. Starting in 2003 in Madrid, a total fleet of 36 buses operated for more than 162 000 h and transported more than 7 million passengers. The total travel distance up to 2009 was 3.40 million km. The bus availability was improved to 98% at real service conditions [13]!

The highest operation time of a bus was >5000 h. The durability of stack cells reached up to 4000 h. The lessons learned were:

- life time must come close to 15 000 h;
- fuel consumption of 20 kg per 100 km in the fleet average must be lowered by 50%;
- improved service and maintenance concepts;
- reduction of unit costs;
- hybridization required to recoup braking energy;
- development of bus specific auxiliaries;
- use of efficient traction motors for the drive train.

41.5.4
NaBuz DEMO and CHIC

The latest development, "Citaro FuelCELL-Hybrid," has been planned for operation in the following cities:

- Hamburg (Germany): 4× (in operation since October 2011);
- Brugg (Switzerland): 5× (in operation since January 2012);
- Karlsruhe (KIT, Germany): 2× (in operation since June 2013);

Table 41.3 Bus and key fuel cell drive train data for Citaro Fuel Cell bus (CUTE program).

Parameter			
Fuel cell power plant	2 × MK904 mod. Stacks, 960 cells in six cell rows in series, external spray humidifaction, two-stage compressor system for cathode air supply, two-stage jet pump circulating anode loop, first time for introduction of AL-alloy into pure water contact avoiding stainless steel where possible	Gross power 300 kW Net power 205 kW	980–550 VDC
Auxiliary subsystems	Standard A/C-compressor and cabin heating, all other auxiliaries propelled via central gear case with ten different power outputs	Capable of providing 170 kW simultaneous shaft power	
Drive train	Single motor from Reuland, Inc. AC motor with custom inverter from Saminco Inc. oil cooled motor, connected via SAE-1 flange to six speed automatic transmission from ZF	Rated power 205 kW @ 100 Nm	Peak power: 280 kW
Hybrid battery features	None		
Bus empty weight	14 200 kg		
Gross vehicle weight	18 000–19 000 kg (Spain only)		Series licence
Passenger capacity, legal	23 seating, 49 standing	72	
Hydrogen storage system	Type3, 9 × 205 l cylinders, NWP 35 MPa,	45 kg	Carbon fiber composite
Fuel consumption	18–21 kg per 100 km, average fleet results incl. heating and A/C		
Range	200–250 km		Depending on topology and drive cycle

- Milan and Bolzano (Italy): 8× (in operation since November 2013);
- Stuttgart and Fellbach (Germany): 4× (since 2014).

The most interesting achievement of this bus generation is its fuel efficient operation (Table 41.4). Even under real service operation the average fuel consumption is far below the predicted value of 10 kg-H_2 per 100 km (Figure 41.19). Refueling time in average varies between 6 and 10 min meeting the specification requirement successfully.

Table 41.4 Bus and key fuel cell drive train data for Citaro FuelCELL-Hybrid.

Parameter			
Fuel cell power plant	2× FC system from car application based on MK1100 stacks, 396 cells each, power reduced from 70 to 60 kW to achieve better durability and lifetime	Net power 120 kW	420–270 VDC
Auxiliary subsystems	Decentralized operation by use of three special converters. Each converter is able to power three separate electric motors at 400 V Two converters with 3× 400 VAC output each single converter operating as 12/24 VDC. Alternator replacement		
Drive train	ZF electric drive axle AVE 130	Rated power 130 kW@21 000 Nm	Peak power: 250 kW
Hybrid battery features	Li-Fe cells, 26.1 kWh energy content, water cooled	130 kW	250 kW
Bus empty weight	13 200 kg		
Gross vehicle weight	18 000–19 000 kg (Spain only)		Series licence
Passenger capacity, legal	26 seating, 50 standing	76	
Hydrogen storage system	Type3, 7 × 205 l cylinders, NWP 35 MPa, car components for valves and high pressure regulator, filling receptacle with car dimensions. Refueling time below 10 min	35 kg	First H_2-pressure system licenced to EC 79/2009
Fuel consumption	7–9 kg per 100 km, average fleet results incl. heating and A/C		Reference Test Line 42 Stuttgart
Range	>350 km		Depending on topology and drive cycle

41.6
Future Outlook

41.6.1
Transit Bus Applications

Fuel cell technology has matured and has been demonstrated to be a capable solution for heavy duty applications in the urban transportation business. The PEM FC

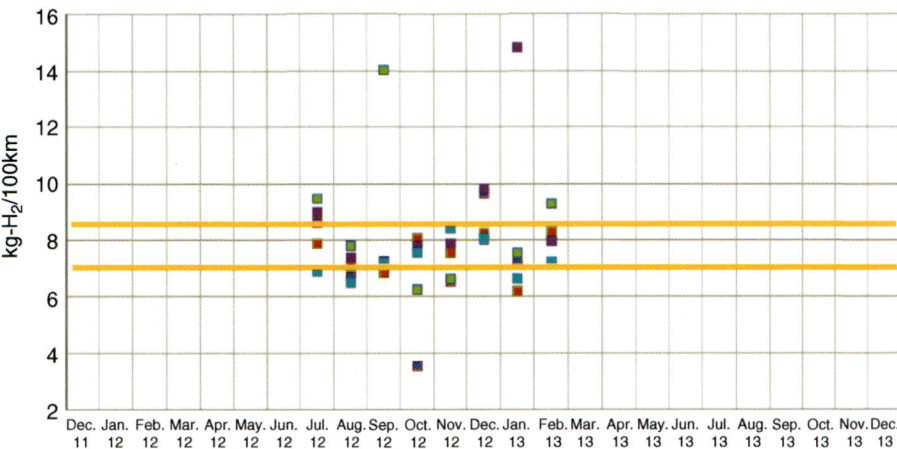

Figure 41.19 Actual fuel consumption data from bus operation in Brugg, Switzerland, 2012/2013 [14].

technology offers the possibility of linking multiple FC-power plant subsystem components together to increase the power output. The addition of a HV battery to the fuel cell powertrain enlarges the range of fuel saving and life time optimization strategies. The Li-ion battery technology combines the electrical features of a ultra-capacitor load bank with a light weight HV energy battery so that fast charging and discharging cycles during braking and acceleration will become possible.

Due to this battery type brake energy recovery is no longer limited by the battery technology.

Electronic brake paddles are available enabling the maximum utilization of brake energy recovery. This helps in saving brake pads on the mechanical side of the powertrain and increases overall fuel efficiency. The FC-technology supports start–stop strategies with much less stress to the components than an ICE.

The low noise operation of the powertrain enables the use of heavy duty application in urban areas even by night. This zero emission powertrain supports a clean urban transportation in city centers at the location of driving.

The next generation of fuel cell powertrain for city buses will fit into the currently available engine compartment (Figure 41.20). The whole power plant will fit in the area where maintenance can be performed as usual. The bus roof will only carry the fuel storage system, the heat management, and radiator package. More compact auxiliary components are needed such as for instance electric driven A/C compressors as demonstrated in the current Mercedes Benz Citaro FuelCELL-Hybrid bus.

41.6.2
Truck Applications

Daimler is investigating the use of this technology also for van and truck applications (Table 41.5). A recent investigation on truck applications powered by the

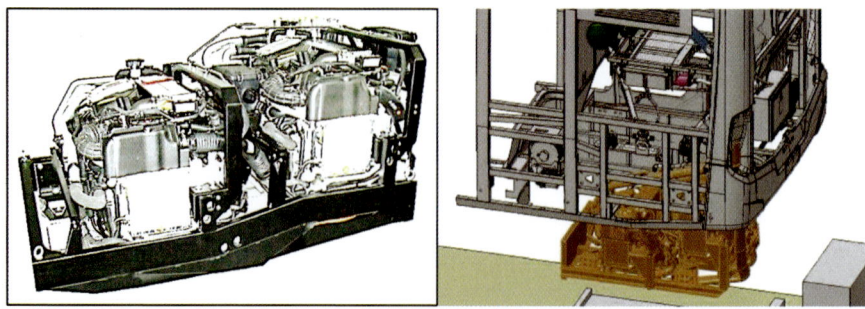

Figure 41.20 This most advanced 140 kW FCS is packaged for a Mercedes Benz Bus engine bay application. The total weight of this unit is approximately 450 kg and will fit into the exact same space as the ICE powertrain [15].

current fuel cell technology developed in-house for car applications shows high potential to fit into various investigated truck applications [16]. The study has proven that long haul truck applications are the most challenging ones due to power demand, fuel storage, lifetime requirements, and missing H_2-infrastructure. The most feasible applications are fleet operated vehicles and/or power demands that can be realized by the use of one or two car FCS (80–160 kW) and a H_2-storage demand below 20–25 kg. The application also depends on the size of the hybrid battery layout.

Table 41.5 Overview of truck platforms suitable for fuel cell applications derived from car applications.

	Platform			
	Light duty truck, delivery vans (3.5–7.5 ton GVW)	Small trucks (7.5–12 ton GVW)	Medium duty trucks: (12–18 ton GVW)	Heavy duty trucks long haul applications: (18–60 ton GVW)
H_2-storage capacity on board (kg)	10–16	15–25	30–60; requires long wheelbase for H_2 storage	>60; storage critical due to large volume required, switch to liquid H_2 indicated
FC-system (kW)	80	80	2 × 80	>280–300
Hybrid battery (kWh)	8 (80 kW)	8 (80 kW)	75	>75
Feasibility by use of car FCS	Yes	Yes	With restrictions	Requires new FCS power plant design and storage concepts based on LH2

Typical long haul applications require specific heavy duty FC-systems in the range of approximately 280–300 kW plus hybridization with a suitable HV battery. A purpose design FCS may become necessary to overcome packaging issues and lifetime requirements. The required amount of hydrogen of 80–100 kg on board is currently not possible to be packaged in diesel fuel tank spaces. The switch from compressed gas storage to liquefied cryogenic hydrogen storage containers is strongly indicated.

41.6.3
Life Time and Product Costs

Current fuel cell technology offers only a reasonable life time of 6000 h (respectively 8000–10 000 h for HD use). Customer expectations are located in areas beyond 20 000 h. To close this gap, fuel cell operational strategies and improved catalyst materials have to be developed. While the use of an optimized operational strategy is already successfully implemented into the current bus version, dedicated stack life time development is required to avoid costly replacement concepts in future.

For future development activities electrically powered auxiliaries, software controls, and high voltage (HV) component development such as DC/DC converters need more engineering effort and a broader market supplier base. Current activities to raise the low voltage on-board network from 12/24 to 48 VDC and more standardization on high voltage levels (300 or 600 VDC) will help improve the product cost situation. Product cost development is strongly dependent on production volume. Through a standardization of core sub-systems and components this challenge may be overcome in future. The biggest challenge for this technology to become competitive or superior to the current state of the art powertrain concepts is still the cost of the stack technology and the energy storage systems required (HV-battery and hydrogen pressure containers).

41.6.4
Summary

Today's PEM FC technology combined with powerful Li-ion hybrid batteries offers solutions for a wide range of bus and truck applications with power demands up to 250 kW (FCS and battery combined) and operational ranges below 350–500 km. The technology required to realize powertrains in this power range can be directly derived from high volume passenger car developments and its production processes. It is fair to say that with this approach the product cost challenge may be overcome for this market segment. Custom fuel cell system development for heavy duty applications may be indicated in order to solve future challenges of large and heavy truck applications such as intercity bus coaches and long-haul truck applications.

References

1 Jenewein, S. and Wepfer, K. (2013) EvoBus GmbH, Untersuchungen Simulation Konzeptauslegung CFCH Projekt 2013, internal report.
2 European Parliament and European Council (9 October 2007) Rahmenrichtlinie für kraftfahrzeuge und ihre anhänger 2007/46 EG. *Official J. Eur. Union*, L 263.
3 Schäfer, H. (2009) Wheel hub drive versus Axle Drives, Hofer EDS GmbH, efficiency map of different electrical machines, in *Hybrid Vehicles and Energy Management: 6th Symposium, 18 and 19 February 2009, Stadthalle Braunschweig*, ITS Niedersachsen, Braunschweig, p. 200.
4 Fleck, W. (2005–2006) Lecture series: Grobauslegung Brennstoffzellenantrieb am Beispiel Stadtbus, Vorlesungsunterlagen für FHTE Esslingen.
5 Dobler, K., Fleck, W., Sonntag, J. et al. (1997) Daimler Forschung, Testreport Necar2 Stack-Daten MK704, internal report.
6 Mercedes-Benz Omnibusse (1997) Technische Daten, NEBUS New Electric Bus, Mercedes-Benz Omnibusse, EvoBus GmbH, VM 6098.8003. 00-00/0597, 1997.
7 Mercedes-Benz Omnibusse (1995) EvoBus GmbH, Alternative Antriebskonzepte, brochure.
8 Ballard Power System (2001) Cleaning up: Zero Emission buses in real –world use, hand out brochure.
9 Sustainable Energy Europe Award (February 10th 2009) HyFLEET:CUTE was announced a winner of the 2009 Sustainable Energy Europe Award in the category of demonstration and dissemination. The awards honours Energy Best Practices – and are made for Outstanding Projects registered with the campaign.
10 Bunzel, G. (2009) Nucellsys GmbH, Technical report for Nabuz Prep.
11 EvoBus GmbH (2008) ZF Achsysteme Nutzfahrzeuge – Sonderkomponenten, Passau, techn. Dokumentation.
12 Bögershausen, C., Seiffert, S., and Anumu, P. (2005–2007) Simulation Study Results for concept evaluation, internal report Daimler AG.
13 Fleck, W. (Daimler AG) (2009) Current development activities within the Hyfleet: CUTE Program, in *Hybrid Vehicles and Energy Management: 6th Symposium, 18 and 19 February 2009, Stadthalle Braunschweig*, ITS Niedersachsen, Braunschweig.
14 Seraidou, N. (2013) Chic Test Results report from 24–25th April, 2013, Postbus Ltd., Brugg, CH.
15 Bunzel, Guido, Bauer, Rainer (2014) Nucellsys GmbH, Abschlussbericht "Heavy Duty Brennstoffzellensystem der 3. Generation für Anwendung im Bus", 2014, page 15, Nabuz Pre-Commercial.
16 Zuschlag, A., Fleck, W. (Daimler AG), Bunzel, G., Essling, R.-P. et al. internal study report.

42
Hydrogen and Fuel Cells in Submarines

Stefan Krummrich and Albert Hammerschmidt

42.1
Background

Non-nuclear submarines are generally based on a diesel-electric power supply system. A diesel generator in combination with lead acid batteries supplies the submarine's network system. The battery is charged during surfaced or snorkel operation of the submarine, because air is required for operation of the diesel engines. In submerged operation traditionally the entire power is taken from the lead acid batteries.

The limited capacity of the lead acid batteries results in a relatively short submerged operating period of several days, until resurfacing is required again to recharge the batteries. During this period of snorkeling the submarine is easily detectable – this is the main reason for the efforts that have been spent in recent decades to develop air-independent propulsion (AIP) systems for submarines.

The major requirements for such AIP systems are:

- high energy density (plant including storage of reactants);
- high efficiency ($\rightarrow$ low heat transfer to sea water);
- low noise level;
- low magnetic signature;
- small size;
- low weight;
- low effort for maintenance/no extra crew.

Many different technologies have been considered as AIP systems, but today fuel cell technology is by far the most successful technology in this area.

Hydrogen Science and Engineering: Materials, Processes, Systems and Technology, First Edition.
Edited by Detlef Stolten and Bernd Emonts.
© 2016 Wiley-VCH Verlag GmbH & Co. KGaA. Published 2016 by Wiley-VCH Verlag GmbH & Co. KGaA.

42.1.1
When it All Began . . .

Fuel cells have been under development by ThyssenKrupp Marine Systems for more than 20 years. At the beginning of the 1980s a land based test site was realized with a fuel cell system and a lead acid battery, comprising all supply systems for the reactants hydrogen and oxygen. The plant was built up with alkaline fuel cells, because polymer electrolyte membrane (PEM) FC technology was under development at Siemens, but not yet available at that time. After extensive testing of the plant, the system was installed onto the submarine U1, a Class 205 submarine that was still in operation with the German Navy. In 1988, first sea tests were performed. Thanks to the successful results of the testing, the HDW Class 212A submarine was designed from scratch based on fuel cell technology. The development of the PEM-fuel cells at Siemens was intensified. The result was the first PEM FC Module in 1994 with 30–40 kW power output, showing excellent operational performance.

Furthermore, system developments for the supply systems were performed to realize a safe and reliable system operation on board a submarine. Taking into consideration the very special requirements like operation in a closed atmosphere, shock loads, very strong acoustic requirements, and the need for a non-magnetic design, these tasks were very challenging. The type of hydrogen storage was also carefully evaluated, with the result that metal hydride storage was the best choice for submarines. The only problem regarding this choice was that metal hydride storage cylinders were available in finger-sized bottles for laboratories and so on, but not as the very large storage tanks needed to fulfill the storage capacity requirements for submarine applications. Therefore, these hydrogen storage cylinders had to be developed. This development was performed at ThyssenKrupp Marine Systems, from laboratory via the type approval testing by the German authorities up to the implementation of a reliable and quality optimized manufacturing process.

42.2
The HDW Fuel Cell AIP System

Figure 42.1 shows the system installed onboard the HDW Class 212A and 214 submarines.

The HDW fuel cell systems consists of the reactant storage of hydrogen and oxygen, the fuel cell modules developed by Siemens, a distillate cooling water system, and a system for handling the product water and residual gases.

The fuel cells are operated on the pure reactants hydrogen and oxygen. The reaction water is stored onboard to realize a closed system – no substances have to be sent overboard, no weight loss occurs by the consumption of the reactants, therefore no weight compensation is needed. A DC/DC converter regulates the output voltage of the system to fit with the actual boats network voltage.

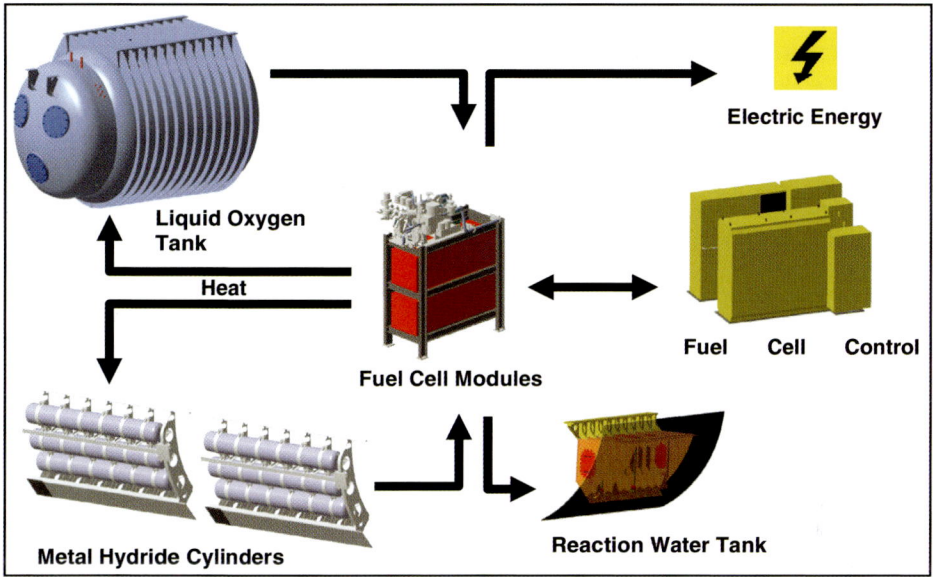

Figure 42.1 Overview of AIP system for submarines.

42.3
PEM Fuel Cells for Submarines

42.3.1
Introduction

The principal design and operation of PEM fuel cells are described in depth elsewhere in the present book.

PEM fuel cells convert chemical energy which is bound in hydrogen and oxygen via an electrochemical process into electrical energy. They are, in principle, well suited for use in a submarine environment: the energy conversion happens at a relatively low temperature (70–80 °C), the conversion is very effective (high electrical efficiency), and as an electrochemical process is – compared to combustions processes – silent. All these properties are favorable for the tactical needs of a modern submarine as described in the requirements above: the relatively low temperature and the highly efficient energy conversion are advantageous for low level heat dissipation around the submarine, keeping the thermal signature low.

Beyond these intrinsic properties of PEM fuel cells several additional and specific requirements should be fulfilled when using them as energy source in the underwater application. In addition to the major requirements for AIP systems (Section 42.1) the fuel cell modules should be characterized by:

- low volume of off-gases (residual gas hydrogen and oxygen) from fuel cell operation;

- high power density of the fuel cell module with a high degree of integration process equipment;
- low electrical stray field;
- shock resistance;
- compliance with submarine safety requirements.

42.3.2
The Oxygen/Hydrogen Cell Design

42.3.2.1 Constructive Features/Cell Design of Siemens PEM Fuel Cell

The basic constructional features of PEM fuel cells (Figure 42.2) are the proton conducting membrane (mostly based on a perfluorinated polymer, e.g., Nafion®), two catalyst layers (on the anode and cathode sides), a gas diffusion layer on both sides, and the bipolar or cooling plates. To achieve the power density metal-based bipolar plates are used that allow simple realization of a water-cooled and "thin" design.

The corrosive environment, which consists in the cathode compartment of pure oxygen and hot, deionized water, requires a high stability of the applied materials. Stainless, nickel-rich steel fulfils these stability requirements easily. This material can be embossed and welded to produce the structures for uniform coolant flow, gas supply, and residual gas product water removal.

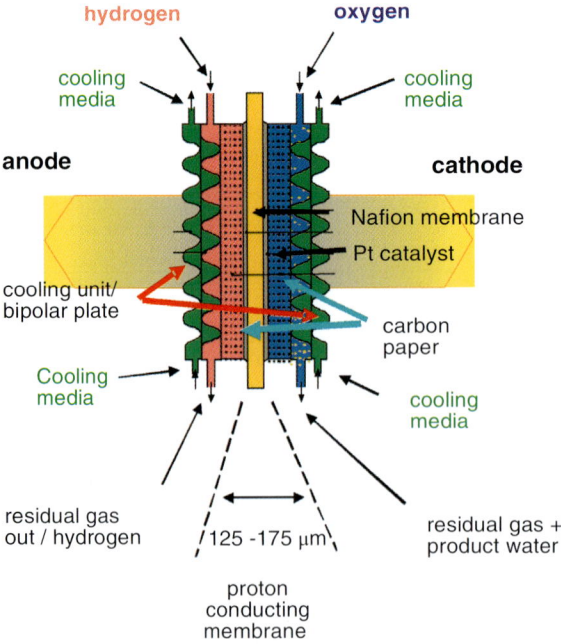

Figure 42.2 Schematic set-up of a PEM fuel cell [1].

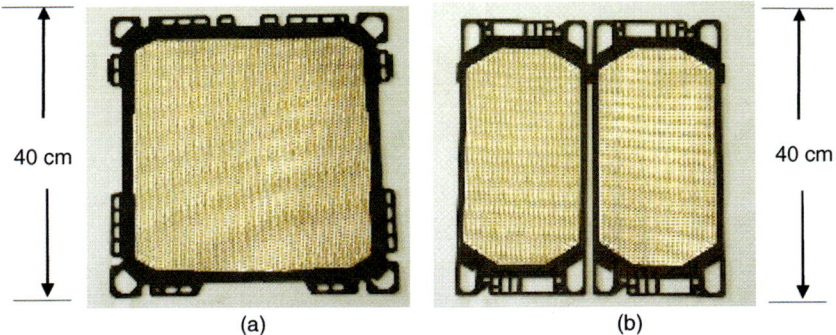

Figure 42.3 Metal-based bipolar plate, Design E4 (a) and Design D4 (b).

Figure 42.3 shows bipolar plates as used in the fuel cell module BZM 34 (design E4) and BZM 120 (design D4). The surface is gold-plated to ensure sufficient electrical contact between the metal and the carbon-paper based diffusion layer. The black gasket material, a fluoro-elastomer around the edge of the bipolar plate, is provided with axial gas channels conducting the media (coolant, reactant, product water) to and from the cells where they are consumed or released. The membrane electrode assembly (MEA), a five-layered component consisting of the membrane, catalyst, and gas diffusion layer, is placed in between two of these cooling units. An appropriate treatment of the diffusion layer is necessary to make it hydrophobic and to squeeze the product water out of the porous structure.

42.3.2.2 Results from Fuel Cell Operation

Figure 42.4 shows the significant difference of fuel cell operation between hydrogen/oxygen (blue) and hydrogen/air (green). The membrane basis for these investigations is DuPont Nafion® 115/117, respectively. In air operation a strong reduction of the cell voltage at a given current density can be observed even with the same membrane material.

The reason for this is the lower partial pressure of oxygen and transport phenomena at the reactive interface. But there is also a significant difference in pure oxygen operation between the thinner Nafion® 115 (nominal thickness @ dry state 125 μm) and the thicker Nafion 117® (175 μm) – in respect to power by nearly the factor of 1.5. The reason for this is the higher ohmic resistance of the thicker membrane.

42.3.3
Constructive Feature of Fuel Cell Module for Submarine Use

42.3.3.1 Preconditions

Fuel cells used in submarines are operated in a frequently closed environment with a confined volume and defined gas pressure. Both requirements must be

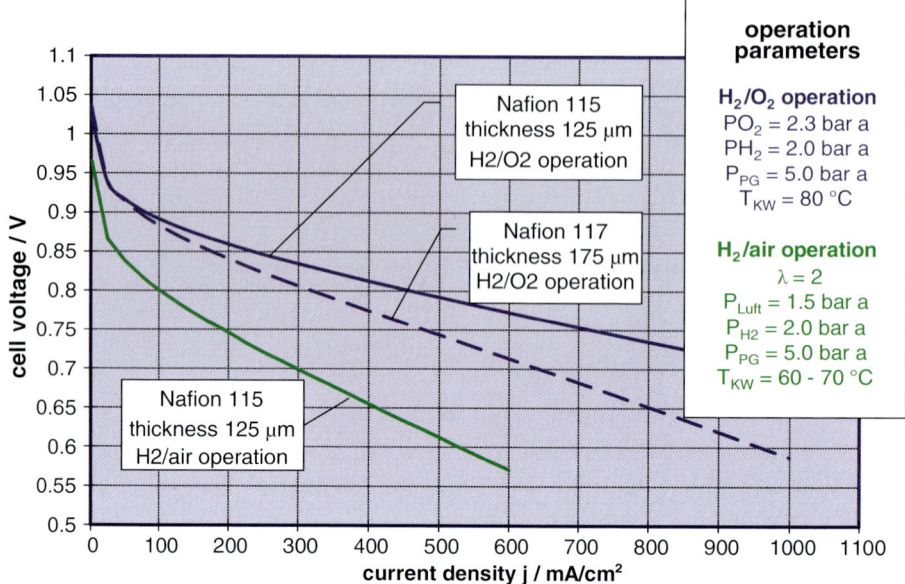

Figure 42.4 Voltage–current characteristic of a PEM fuel cell at different operation conditions.

considered in handling the off-gases. To avoid complicated or energy consuming residual gas treatment like compression and/or bringing it outboard it is important to know how to reduce the amount of residual gas as far as possible. A first step is to limit the impurities in the reactants: the quality of the oxygen used in the fuel cells (provided as liquid oxygen) is 99.5% purity with impurities like nitrogen or argon. Medical oxygen normally meets these requirements. Hydrogen is even much purer (99.999%) since it is stored in metal hydrides. These metal alloys act as purifiers: gases like CO_2, O_2, or water vapor are tightly adsorbed within the lattice structure of the metal alloys, reducing the storage capacity with time.

42.3.3.2 Cascaded Fuel Cell Stacks [2]

Figure 42.5 shows the construction principles of a cascaded fuel cells stack. To avoid unacceptably large hydrogen/oxygen quantities as residual gases a Siemens fuel cell module consists of several internal gas loops on the oxygen and on the hydrogen side, which physically do not need to be at the same location on both sides.

The dry reactants, which are released from the gas supply at pressures between 2 and 4 bar, pass through the membrane humidifier and the sections or cascades of fuel cells. Each section/cascade consists of a certain number of cells, which decreases from inlet to exit. The number of cells is adjusted corresponding to the depletion of reactants due to the electrochemical reaction within the cells in order to minimize the volume of the residual gas. After each loop the

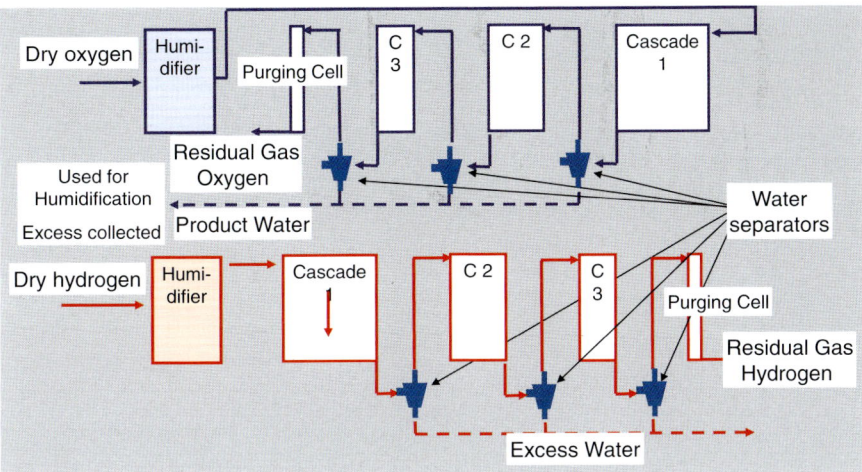

Figure 42.5 Schematic set-up of a cascaded fuel cell module with humidifier and purging cells.

reactants are separated from product water and conducted to the next cascade. The product or product water is collected separately. The last cascades of the anode and cathode side are the purging cells which define the volume and the quality (i.e., composition of reactant and inert gases) of the residual gas.

Figure 42.6 shows a typical purging behavior of a hydrogen purging cell. Whereas all fuel cells except the purging cells are operated with a continuous flow of reactants the purging cells are operated in the dead-ended mode. Due to the enrichment of inert gases and/or water and the operational current, the

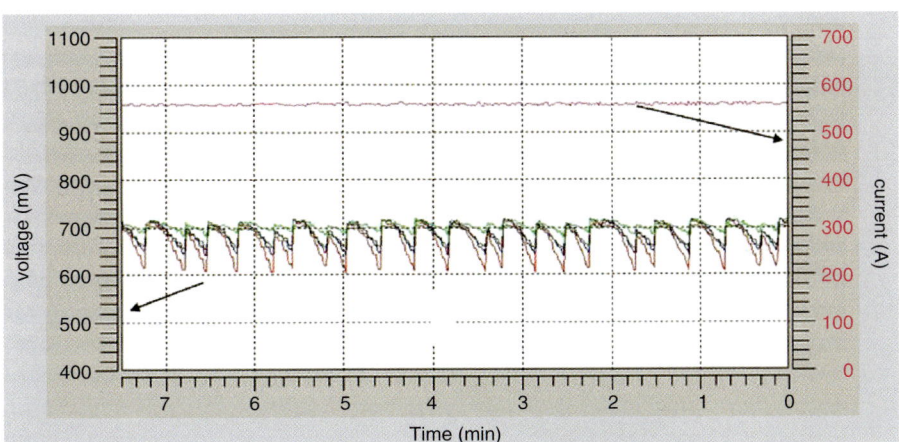

Figure 42.6 Typical behavior of a series of four hydrogen purging cells operating under full load.

voltage of the purging cells group shows a time-dependent behavior starting at a high voltage level with decay with time.

The trigger of the purging operation must be adjusted depending on acceptable purging volume or voltage levels [3]. The principle of the purging cells allows us to dispose the residual gases from both sides individually. Hydrogen can be brought into a hydrogen-rich environment, for example, a battery compartment, and is combusted there on recombination catalysts, whereas oxygen can be released into the boat atmosphere provided the amount of oxygen can be breathed away by the crew.

42.3.3.3 Pressure Cushion for Uniform Current Distribution [4]

The fuel cell modules are operated at current densities up to $1000\,\text{mA}\,\text{cm}^{-2}$; the stacks may consist of up to 200 single cells and more. It is important to guarantee a uniform contact pressure between the cooling units or bipolar plates and the membrane electrode units. Uneven current distribution has an impact on the potential distribution on the bipolar plate. As a result, one MEA may not operate homogenously without uniform water production and draining or inhomogeneous production of reaction heat. This may influence life time by partially drying out the membrane or inducing corrosion.

Fuel cell stacks with a rather high number of large area cells cannot simply be kept together by tie rods if extra boundary conditions must be considered like uniform pressure and sufficient gas tightness.

Using a pressure inducing element inside the cell stack has been considered to be the best solution for the Siemens fuel cell stacks (Figure 42.7).

The pressure cushion consists of an inflatable element that is filled with a flexible, electrically conductive material to ensure good electrical conductivity

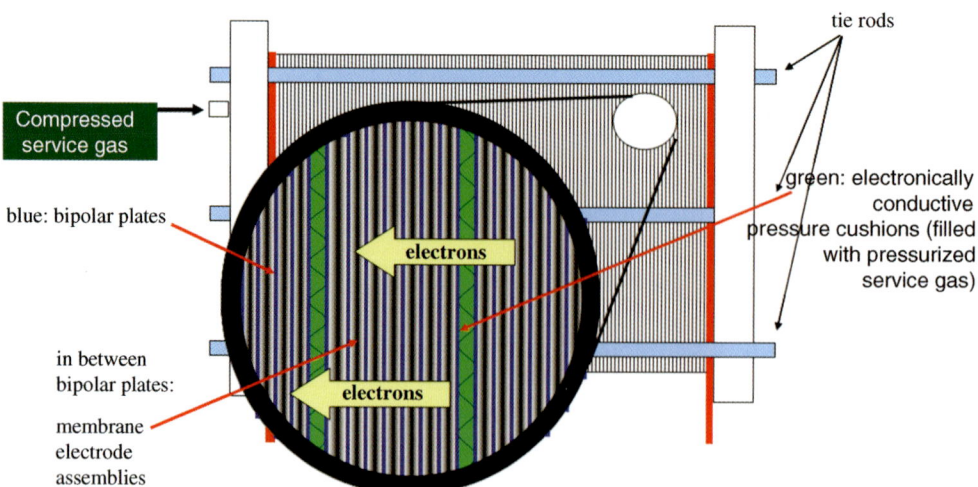

Figure 42.7 Fuel cell stack with integrated pressure cushions for uniform pressure distribution.

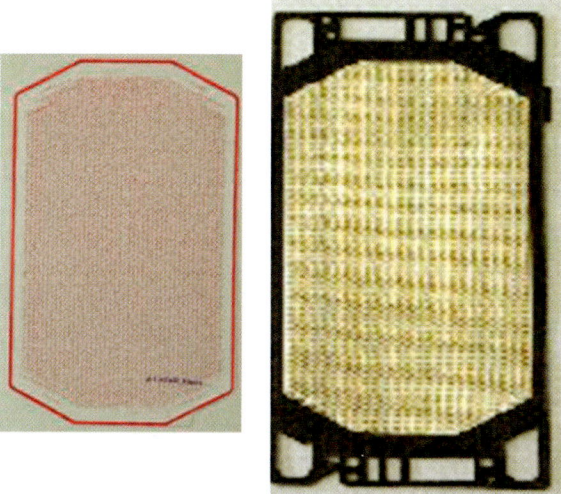

Figure 42.8 Measurement of uniform pressure application inside a fuel cell (design D4).

through the cushion. They ensure, together with the rigid tie rods and the stiff end plates, the necessary forces onto the fuel cell elements for good electrical conductivity. Figure 42.8 shows the quality of uniform pressing achieved by the pressure cushions as measured with a pressure sensitive paper material.

42.3.3.4 Fuel Cell Module

A schematic set-up of a Siemens PEM fuel cell stack is shown in Figure 42.9. The essential components of the stack are the hydrogen/oxygen humidifier, the cascaded cells stack within the pole plates, and the water separators to remove the product water from the gas stream, which are placed at the end plate.

The fuel cell module is supplied with dry reactants delivered from the gas storage (see below); the reactants are internally humidified and directly conducted into the fuel cell stack. The reaction water is employed to humidify the reactants and part of the reaction enthalpy produced during fuel cell operation is used as evaporation heat inside the humidifier.

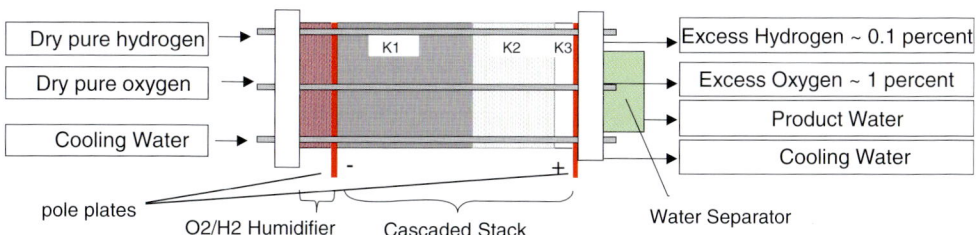

Figure 42.9 Schematic set-up of a cascaded Siemens PEM fuel cell stack with integrated humidifier, the cascaded stack and water separator.

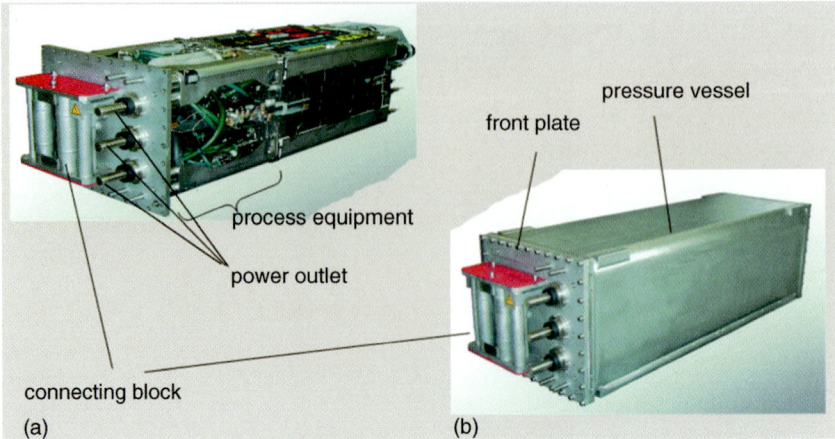

Figure 42.10 Siemens PEM fuel cell Module BZM 120 without (a) and with pressure container (b).

The advantage of the design is the simple use of dry reactants due to internal humidification.

The fuel cell module consists of the stack described above and, in addition, process equipment to handle the media like gases and cooling water (Figure 42.10).

Most of this process-related equipment is arranged in front of the fuel cell stack. It consists of valves and sensors for pressure, temperature, current, voltage, and flow measurements. The pipes and tube made of insulating material are used to conduct the media internally. The front plate of the module equipped with the electrical power connectors and a connecting block as interface for media supply is tightly connected to a container or pressure vessel. Due to its gas tightness the container is an active part of the safety system of the fuel cell plant. For this the pressure container is pressurized by nitrogen such that in case of an internal leak the critical media like hydrogen or oxygen are forced to remain inside the fuel cells stack. In case of a leakage no gas can leave the stack but nitrogen may enter and induces a pressure increase or voltage drop that shuts off the control system.

These constructional principles are valid for both BZM 34 and BZM 120 fuel cell modules (Figure 42.11). The essential difference between both modules is the cell area (see Figure 42.3, design E4 versus D4), the size of the active area, and the current density. The reason for the higher current density in BZM 120 is the use of Nafion® 115 instead Nafion® 117 in BZM 34 (see also Figure 42.4). The technical data of both modules are listed in Table 42.1.

42.3.3.5 Results from Fuel Cell Module Operation

Fuel cells modules can be easily operated when they are supplied by sufficient reactants, service gas, cooling water, and a load bench, provided they are connected to an appropriate control unit including the appropriate program. Since this is very product and process specific it will not reported here.

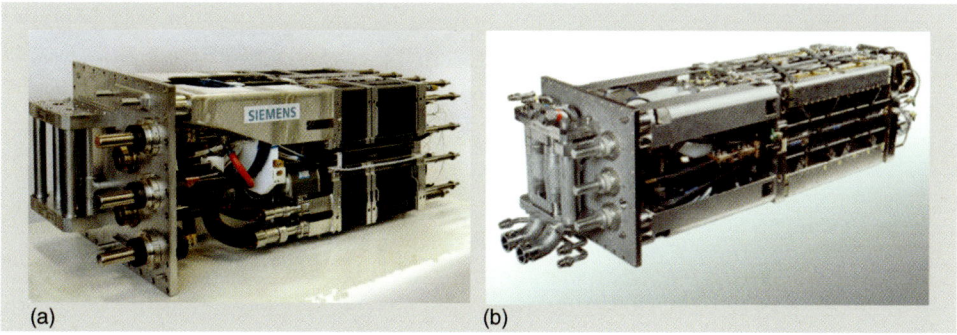

Figure 42.11 Inside of BZM 34 (a) and BZM 120 (b).

Table 42.1 Technical data of BZM 34 and BZM 120 [5].

	BZM 34	BZM 120
Rated power (kW)	30–40	120
Number of cells	~70	320
Rated current (A)		560
Rated voltage (V)	>50	215
Working temperature (°C)	70–80	70–80
Size (cm^3)	47 × 47 × 143	176 × 53 × 50
Weight (incl. pressure vessel) (kg)	650	900
Efficiency at full load (%)	>60	>55
Efficiency at 20% load (%)	>70	>65

Figure 42.12 presents a snapshot of the performance data of a BZM 120 operating onboard a submarine – mainly load steps between open circuit voltage and nominal load.

42.3.3.6 Safety Features of Submarine Fuel Cell Modules

The safe operation of PEM fuel cells onboard submarines is especially important due to the closed environment of the installation and the reactants hydrogen and oxygen, which are close together.

The fuel cell modules themselves are intrinsically safe. A basis of the safety philosophy is the pressure vessel concept discussed above. Another aspect is the process safety, provided by continuous measurement of physical data like temperatures, reactant pressure, flow, module current and voltage, and pairs of single cell voltages. These data are continuously compared with target and limit values and in case of a deviation appropriate measures are taken like the display of a warning signal, alarm, or automatic shut-off. This requires a fast PLC (programmable logic controller), which controls the automatic shut-down process,

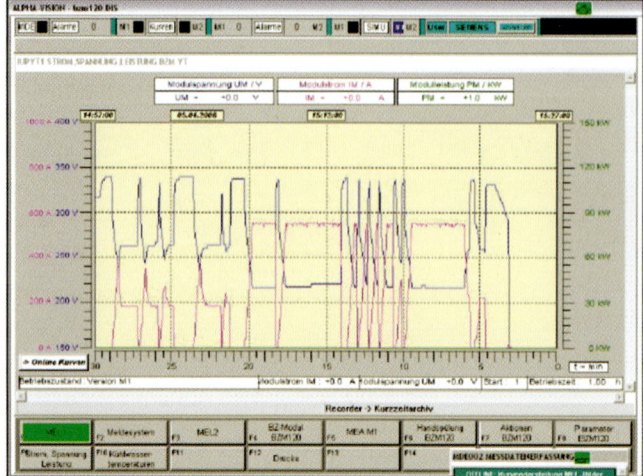

Figure 42.12 Snapshot of the performance of a BZM 120 operated under quick load changes (red: current, blue: voltage, x-axis: time in minutes).

consisting of the interruption of the media supply and the removal of reactants inside the fuel cells.

42.4
Hydrogen Storage

The hydrogen storage system for submarines is based on metal hydride storage cylinders, which have been developed by and are produced by ThyssenKrupp Marine Systems. This type of hydrogen storage ideally fulfills the specific requirements of this niche application. Compared to compressed hydrogen, the metal hydride storage offers much higher volumetric storage densities. The usage of liquid hydrogen was also considered, but because a cryogenic hydrogen tank has significant boil-off losses, and since the quantity of hydrogen required would result in a tank system with enormous influence on the entire submarine design, the metal hydride storage was chosen.

Today's metal hydride storage cylinders offer the highest safety combined with excellent operational features for the submarine. Waste heat from the fuel cells is used for discharging the tanks. As consequence, less waste heat is dissipated into the surrounding sea. During recharging of the metal hydride the cylinders should be cooled to enable a fast filling procedure.

The high weight of metal hydride compared to other storage technologies is no problem for the submarine application. Taking into consideration that non-nuclear submarines need lead in the keel area to achieve weight balance (principle of Archimedes), it is relatively easy to understand that the hydrogen storage cylinders can be installed by simply taking less or even no lead. Nevertheless, the

Figure 42.13 Metal hydride cylinder for submarines.

storage cylinders are installed outside the pressure hull, and therefore have to meet very harsh requirements like diving pressure, salt water environment, and maximum shock loads.

The storage cylinders are currently produced in several cassettes filled with metal hydride. These cassettes are then put into a pressure vessel. One cylinder is approximately 5 m long and has a diameter of approximately 500 mm (Figure 42.13). After the production procedure the storage cylinders have to be activated by several charging and discharging cycles. This is followed by the installation of two half shells made of glass fiber reinforced plastic around the cylinders. These shells are required for heating of the storage cylinders during operation with seawater that has been warmed up by the distillate cooling system.

Taking into consideration the usage of metal hydride, their behavior has to be understood in detail. Customers in most cases ask: How much hydrogen can be stored reversibly in the cylinders?

Figure 42.14 shows the influence of temperature and filling level on the pressure of the storage cylinders. It becomes obvious that the storage capacity is

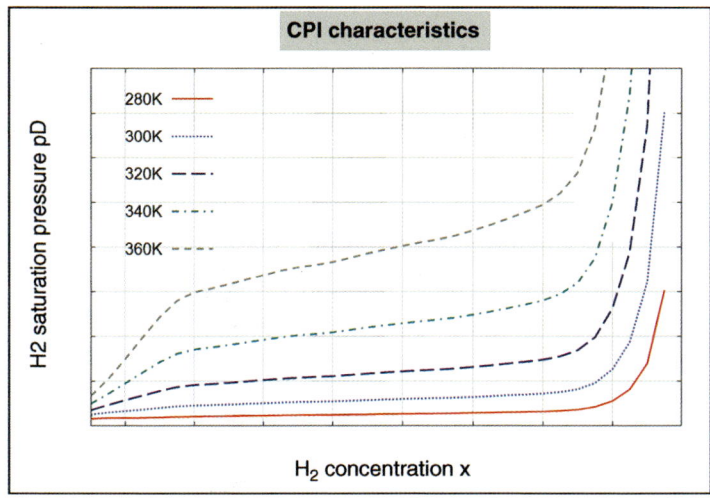

Figure 42.14 CPI (concentration–pressure isotherm) characteristics of metal hydride.

increased in cold conditions, provided that the pressure is the limiting factor for the hydride storage cylinder. Therefore, the storage capacity of the hydride storage system onboard submarines is directly influenced by the specified seawater temperature – the higher the temperature, the lower the storage capacity.

For refueling of the metal hydride ThyssenKrupp Marine Systems has developed a specific reactant filling station. This filling station controls the pressure and flow of hydrogen into the storage cylinders to achieve optimum conditions for maximum filling in a relatively short time. Nevertheless, as during the filling procedure heat has to be withdrawn from the storage cylinders, the procedure is a time-consuming action.

The number of metal hydride cylinders differs from submarine to submarine, and is related to the customer-specific requirements regarding the amount of AIP energy. To date, not even one single failure has occurred during operation in or at one of the metal hydride storage cylinders, with more than 500 of them produced so far.

42.5
The Usage of Pure Oxygen

Many challenges during the fuel cell development are caused by the usage of pure oxygen inside the fuel cells. The details are discussed above.

Storage of the required amount of oxygen inside the submarine is realized with cryogenic liquid oxygen tanks. In the HDW Class 212A submarines two liquid oxygen tanks are located outside the pressure hull (Figure 42.15), while in

Figure 42.15 HDW Class 212A submarine.

Figure 42.16 HDW Class 214 submarine.

HDW Class 214 boats (Figure 42.16) one large tank is located inside, hanging directly underneath the elastic platform deck, on which the HDW fuel cell system is located.

42.6
System Technology – Differences Between HDW Class 212A and Class 214 Submarines

The first submarines that have been developed from scratch based on fuel cell technology are the HDW Class 212A submarines for the German and Italian Navies. In these submarines the HDW fuel cell system is implemented electrically without a DC/DC-converter. This results in very harsh operating conditions for the fuel cells, because all dynamic load changes inside the ship's electrical network system directly affect the fuel cell modules. Thanks to the unique development of cascaded fuel cells, these harsh requirements can be fulfilled by the Siemens fuel cells.

Furthermore, the HDW Class 212A submarines have an entirely redundant system design – a failure of one single component will not lead to system degradation. Therefore, nine FC modules are installed, with only eight in operation, in order to have a spare module in case of a module-internal failure. On a system level, the HDW fuel cell system is the only alternative to the diesel engine, because only one single diesel generator is installed.

The HDW Class 214 submarines are equipped with the improved Siemens FC modules FCM120. Furthermore, a DC/DC converter is installed to adapt the FC output voltage to the ships network voltage. The system that has been developed for HDW class 214 can therefore also be integrated in other submarines, like for example in the HDW Class 209 submarine (Portugal and Greece) or the HDW Dolphin AIP Class submarine.

Both on HDW Class 212A and Class 214 submarines, the fuel cell control system realizes a fully automated operation of the fuel cell system. This means that no additional crew member is required onboard for the fuel cell system operation.

42.7
Safety Concept

The integration of a HDW fuel cell system working with pure hydrogen and oxygen inside the submarine's atmosphere is only possible when taking into consideration the technical safety at a very early design stage. Generally, any possibility of a hydrogen–oxygen mixture must be avoided. Furthermore, any uncontrolled outflow of one of these reactants into the submarine's atmosphere must be prohibited. These design criteria led to a system without any safety valves, which would usually be utilized in on-shore applications. The basic design principle is to ensure a double safety barrier between hydrogen/oxygen and the submarine's atmosphere. All actuators and sensors are designed according to the double-barrier principle. The fuel cell modules are encapsulated in a pressure container filled with nitrogen at a higher pressure than the corresponding hydrogen or oxygen pressures.

42.8
Developments for the Future – Methanol Reformer for Submarines

Although the existing HDW fuel cell system offers many advantages, for submarines larger than approximately 2000 t displacement with high AIP requirements a different solution is preferable. The system based on metal hydride storage cylinders is relatively heavy, resulting in the amount of hydrogen stored on board being limited by the capability of the submarine to carry them, having in mind the principle of Archimedes.

Generally, liquid fuels have high volumetric and gravimetric energy content and are easy to handle. These advantages in combination with fuel cell performance motivated ThyssenKrupp Marine Systems to initiate the development of a reformer system for onboard hydrogen production.

42.8.1
System Configuration

At the beginning of the development the requirements for the reformers were defined. A major requirement was operation based on the existing and proven

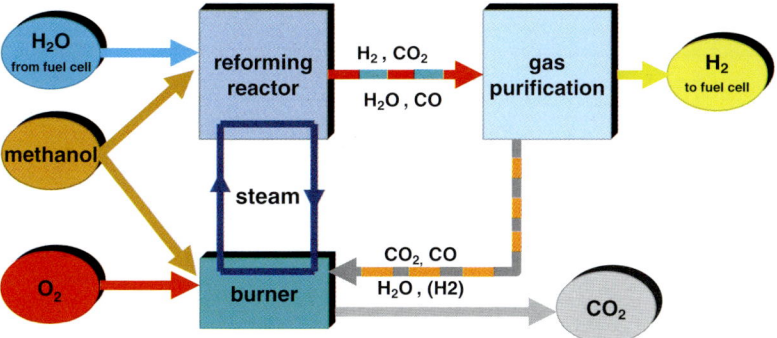

Figure 42.17 Overview of methanol reformer system.

Siemens fuel cells. Furthermore, the exhaust gas (CO_2) pressure should be high, to enable the discharge of exhaust gas into the surrounding seawater without the need for an additional exhaust gas compressor. Of major importance were the overall system efficiency and the reliability and availability of the system.

Based on these requirements, the choice was a methanol–steam reformer system operated at elevated pressure. Hydrogen purification is performed with a membrane purification unit. The required thermal energy is produced in a high-pressure oxygen burner. Figure 42.17 shows an overview of the process.

The methanol is mixed with water, evaporated and fed to the steam reformer. The reforming reactor is heated by a boiling water cycle operating between 60 and 100 bar. The methanol–water mixture is converted into a hydrogen-rich gas mixture at a temperature between 250 and 300 °C. This reformate gas is further processed in a gas purification unit based on hydrogen-permeable membranes. The major fraction of hydrogen passes through the membrane and can be fed directly to the Siemens fuel cell. The rest of the reformate gas is burned with pure oxygen in the burner, under the addition of methanol, to provide the required heat for the reforming process. The only product gas from the reformer is CO_2; the H_2O in the exhaust gas is condensed and reused internally.

The methanol reformer itself will be operated in an enclosure with several safety features, to protect the crew from any harmful gases and liquids.

42.8.2
Challenges of the Methanol Reformer Development

The methanol reformer development was started based on the procurement of a functional demonstrator. This plant was operated with the process conditions that were foreseen for the submarine plant. The process itself worked well, but many components showed weaknesses regarding the special onboard requirements.

Nevertheless, after showing that the methanol reforming process works well under the special process conditions that had to be chosen, a major effort has to be made to achieve a submarine-suitable layout of the system.

42.8.3
Hydrogen Purification Membranes

One major milestone was a successful shock test of the palladium membranes used for the separation of hydrogen from the reformate gas stream. This shock test was performed with membranes that were saturated with hydrogen, the operational differential pressure was applied, and the operating temperature was applied to the membrane. This kind of shock testing is a specific requirement for systems and components that are foreseen to be integrated onboard a submarine.

Because the existing Siemens FC modules are foreseen to be coupled with the reformer, the purity requirements are extremely high. Therefore, any leakage from the reformate (feed) to the hydrogen (permeate) must be avoided. These unique requirements led to the choice of a thick film membrane. This kind of membrane shows highest stability, although the costs are comparatively high. No membrane based on a support structure, for example, sintered metal with a thin palladium film, showed the required stability for operating cycles (temperature as well as load changes), lifetime, and the special submarine requirements.

42.8.4
High Pressure Catalytic Oxidation

The burner inside the methanol reformer is designed as a multi-stage catalytic oxidation device, because the catalytic oxidation does not require an ignition system or a flame monitoring system. These two mentioned systems are not available in a submarine-suitable design. Furthermore, the oxygen control of a multi-stage burner is beneficial compared to a flame burner, and shows only minor deviation in the oxygen concentration in the exhaust gas.

The burner is a patented unique design, operating at each stage at sub-stoichiometric conditions. The temperature in each stage should be limited by only partial oxidation. Even if this technology seems to be unusual and complicated related to the control system of the burner, the control can be performed by simple temperature sensors and an oxygen sensor. The catalytic oxidation device shows best exhaust gas quality and very high modulation ability.

42.8.5
Integration on Board a Submarine

The conceptual integration of a reformer system on board requires the consideration of several special requirements derived from the special environment. To meet all requirements for onboard operation, a special safety concept has been worked out. This safety concept led to the solution of an encapsulated reformer system. A view of the encapsulated reformer is shown in Figure 42.18. Inside the encapsulation, a special ventilation system is installed for cooling purposes. Furthermore, the encapsulated system is continuously monitored for leakages

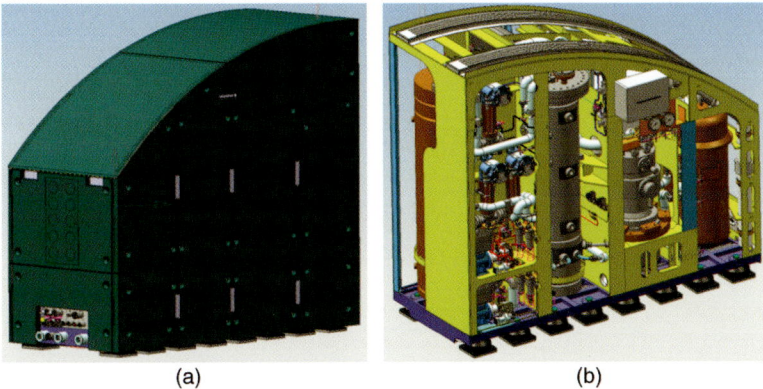

Figure 42.18 Methanol reformer system including (a) and excluding (b) encapsulation.

(CO, H_2, CH_3OH). In the unlikely event of the detection of any of these substances, a special air purification device (catalytic oxidation) is operated to convert these substances into harmless gases. The escape of these substances into the submarine's atmosphere is effectively avoided.

The methanol reformer system itself is installed on an elastic platform inside the submarine. The LOX (liquid oxygen) tank is located below the platform. Generally, the size of the LOX tank is the dominant factor for the submarine design. Because methanol is a liquid energy carrier, it can be stored on board in tanks that are part of the ship's structure.

The dynamic behavior of fuel cells is much faster than the dynamics of the hydrogen production inside the reformer. As submarines always have a very large battery installed, the dynamic requirements can be met by the entire energy system, consisting of fuel cell system (+ reformer) and battery. To control the load of the fuel cell system in relation to the reformer dynamics, the DC/DC-converter controls the load of the fuel cell in accordance with the actual hydrogen output of the reformer.

42.9
Conclusion

The usage of fuel cells onboard submarines is one of the most successful applications for fuel cells in the world. The HDW fuel cell system based on metal hydride storage of hydrogen fits best the unique requirements of a submarine. In combination with the Siemens fuel cells with their cascaded design the system is absolutely quiet and highly efficient. Today, submarine fuel cell technology can be considered as a mature technology that enables navies to operate submerged submarines for several weeks.

The reformer as an additional option offers further increased underwater endurance. All advantages and challenges derived from the usage of fuel cells, metal hydride storage, and pure oxygen have been addressed – resulting in a proven system for every day submarine use.

The experience gained in the entire submarine fuel cell development that began in the 1980s shall contribute to the development and market penetration of fuel cell systems also in other, for example, civilian applications. In addition, the success story of submarine fuel cells shows that customer benefit is of utmost importance during the implementation of a new technology.

References

1 Hammerschmidt, A. (2007) Fuel cell propulsion of submarines. *Naval Forces*, special issue, **28**, 132.
2 Mehltretter, I. (2010) EP 2122737.
3 Strasser, K. (1997) EP 0596366 A1.
4 Hartnack, H., Lersch, J., and Mattejat, A. (2010) EP 1627445.
5 Hammerschmidt, A. and Scholz, D. (2011) Electrical platform and fuel cell based AIP systems for submarines. Presented at Submarine Institute of Australia, Inc. Technology Conference, Adelaide, Australia, 8–10 November 2011.

43
Gas Turbines and Hydrogen
Peter Griebel

43.1
Introduction

The motivation for burning hydrogen in gas turbines comes from the need to decarbonize the fuels used in the power generation sector in order to lower the carbon dioxide (CO_2) emissions and their impact on climate change. Hydrogen (H_2) and H_2-rich fuels, for example, synthesis gas or "syngas" (a mixture mainly consistent of hydrogen and carbon monoxide), originate from production routes like gasification of coal and biomass and other sources. These production technologies have already been discussed in detail in Chapter 2. If a pre-combustion, CO_2 capture and storage (CCS) process is applied, the syngas can be decarbonized and a fuel that mainly consists of hydrogen is derived. Another source of hydrogen is the nowadays intensively discussed power-to-gas route (see Chapter 2).

Hydrogen as such is a noble fuel that should be converted into power with the highest achievable system efficiency. One candidate for such a highly efficient conversion technology is the gas turbine (GT), in which an efficiency of up to 42% in single cycle and up to 61% in combined cycle mode (gas + steam turbine) is nowadays achieved.

Another option instead of centralized production, storage, and direct energy conversion of hydrogen into power is to feed hydrogen to the natural gas pipeline grid, thereby using the natural gas grid as a "decentralized" storage. As a consequence the gas turbines would have to deal with H_2-enriched natural gas blends.

However, burning hydrogen-rich fuels or even pure hydrogen poses several challenges to the gas turbine components, especially to the combustor. Today's gas turbines have to be highly load-flexible to compensate for the increasing share of renewables in the power sector. In combination with the demand of a high fuel flexibility (gas turbines should always also run on natural gas as a back-up fuel) these requirements are even more challenging.

The main challenges are related to the gas turbine combustor. Because of a much higher reactivity of hydrogen in comparison with natural gas, the basic

Hydrogen Science and Engineering: Materials, Processes, Systems and Technology, First Edition.
Edited by Detlef Stolten and Bernd Emonts.
© 2016 Wiley-VCH Verlag GmbH & Co. KGaA. Published 2016 by Wiley-VCH Verlag GmbH & Co. KGaA.

combustion properties like ignition delay time, flame speed, and flame temperature are significantly more challenging, causing a higher pre-ignition and flashback risk. In addition, for the same thermal power, the volumetric fuel flow rate of hydrogen is much higher compared to natural gas because of the lower heating value of hydrogen per volume. As a consequence today's most common combustion technology in stationary gas turbines "lean premixed combustion" cannot be easily adapted to burn hydrogen in a safe and environmentally friendly way. Major burner or combustor re-designs or completely new combustion technologies have to be developed to avoid pre-ignition and flashback.

Burning hydrogen in gas turbines also poses challenges for the hot path materials and turbo machinery components. The moisture content in the exhaust gas is higher than for natural gas operation causing a higher heat transfer to the combustor walls and turbine blade material in the hot section, as well as a higher potential for hot corrosion of the hot path material. Overall, the materials challenge is focused on solving the component life-limiting problems of overheating and enhanced corrosion resulting from higher temperatures.

The challenges in turbo machinery are related to the higher fuel flow rate of hydrogen leading to turbine choking. Consequently, the demand for air from the compressor is reduced leading to increased compressor back pressure pushing it to instability. The higher water vapor content in the exhaust gas leads to changes in enthalpy drop per stage in the turbine and also changes the temperature distributions on blades. This requires careful expander geometry and cooling flow modifications.

The following section focuses on combustion issues when burning hydrogen in a gas turbine because these are the major challenges.

43.2
Combustion Fundamentals of Hydrogen relevant for Gas Turbines

Here the fuel and combustion properties of hydrogen are discussed in comparison to methane and natural gas. From Table 43.1 in which the main fuel and combustion properties are listed, it is obvious that hydrogen has completely different properties and is a much more reactive fuel than natural gas or methane.

Hydrogen is by a factor of about 8 lighter than methane and the lower heating value (LHV) per unit mass is more than two-times higher than for methane. However, on a per unit volume basis the lower heating value of hydrogen is about three-times lower. As a consequence, for the same thermal power the required volumetric fuel flow rate of hydrogen is more than two-times higher than the one of natural gas. Hydrogen has much wider flammability limits than methane or natural gas. Lean extinction occurs at an equivalence ratio of $\Phi = 0.1$, a much lower equivalence ratio than for natural gas. The wide flammability limits are related to the chemical kinetics and the large diffusion coefficient of hydrogen. The chemical kinetics of the hydrogen/air system are well known and a detailed discussion can be found in, for example, References [6,7]. The fast

Table 43.1 Fuel properties of hydrogen, natural gas, and methane [1–5].

Fuel and combustion properties	Hydrogen	Natural gas	Methane
Density at 273 K, 1.1013 bar (kg m^{-3})	0.09	0.7–0.9	0.72
Flammability limits (vol.% in air)	4–75	4.5–13.5	5–15
Flammability limits (Φ)	0.1–7.1		0.4–1.6
Lower heating value (LHV) (MJ kg^{-1})	120	38.9–47.1	50.0
LHV (MJ m^{-3})	10.8	31–41	35.9
Maximum laminar flame speed (in air, atmospheric conditions) (m s^{-1})	3.25 ($\Phi \approx 1.80$)		0.45 ($\Phi \approx 1.08$)
Adiabatic flame temperature ($\Phi = 1$) (K)	2370		2226
Lower Wobbe Index (MJ m^{-3})	40.7	46.5–48	47.9

reaction rates of the H_2/O_2 system are the reason for the higher laminar flame speed of hydrogen compared to methane (maximum laminar flame speed is about a factor 7 higher than for methane, see also Figure 43.3 below).

The adiabatic flame temperature of a hydrogen flame is higher than for a hydrocarbon flame (at stoichiometric conditions about 150 K higher) which will result in higher nitrogen oxide (NO_x) emissions if the stoichiometry is not adapted.

Hydrogen has a significantly lower Wobbe Index (WI), which is the most commonly used parameter for specifying the acceptability of a gaseous fuel in a combustion system. The lower Wobbe Index is typically defined as:

$$\text{WI} = \frac{\text{LHV}}{\sqrt{\text{relative density}}} \quad (43.1)$$

The Wobbe Index values in Table 43.1 are based on the lower heating value expressed in MJ m^{-3} and the relative density (ratio of the fuel gas density to the air density). The significance of the Wobbe Index is that for a given fuel supply and combustor conditions (temperature and pressure) and a given control valve positions two gases with different compositions, but the same Wobbe Index, will give the same energy input to the combustion system. Thus the greater the change in Wobbe Index the greater the degree of flexibility in the control of the combustion systems that is needed to achieve the design heat input. Gas turbine manufacturers typically specify that their turbines are capable of operating over a range of Wobbe Index values. Ranges in excess of ±10% of mid-range values are normal. However, hardware changes may be required to cover all the range and individual gas turbines are unlikely to be able to accommodate all the range without re-tuning. For a particular gas turbine installation a range of ±5% of the commissioned/tuned value of Wobbe Index and/or heating value is typical. For some gas turbines an even smaller range as low as of ±2% of the commissioned/tuned Wobbe Index has been specified [4].

All the discussed differences in the fuel properties of hydrogen result as a consequence of significantly different combustion properties like ignition delay and flame speed, which are discussed in more detail in the following sections.

43.2.1
Ignition Delay

One crucial parameter for the proper design of a combustion system is the ignition delay time. This variable determines how much time is available for fuel/air premixing prior to the onset of ignition and combustion. On one hand, a high level of premixing is required to achieve low emissions. On the other hand, the premixing section has to be designed to avoid unwanted ignition in this section in order not to overheat the walls. Due to the high inlet temperature and pressure in gas turbine combustors this is not an easy task and it is even more challenging for highly reactive fuels like hydrogen.

The ignition delay time of hydrogen and hydrogen/natural gas mixtures are shown in Figure 43.1. In these shock tube measurements, the composition of a typical natural gas was simulated by a reference gas (RG: a mixture of 92 vol.% methane and 8 vol.% ethane). Ignition delay times of diluted hydrogen/reference gas/oxygen/Ar mixtures with hydrogen contents of 0, 40, 80 and 100 vol.% were measured. It can be seen that the ignition delay time is strongly dependent on temperature. For all temperature conditions the ignition delay times are significantly shorter with increasing hydrogen content. For more details see Reference [8].

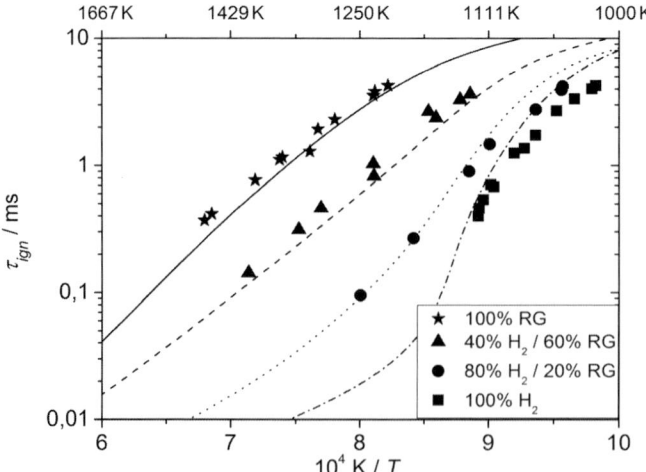

Figure 43.1 Measured and calculated ignition delay times as a function of temperature for mixtures of hydrogen/reference gas/O_2/Ar ($\Phi = 0.5$, dilution 1:5) at pressures of 16 bar. Experimental results are illustrated with symbols. Lines represent simulation results using the RDv06i8 reaction mechanism.

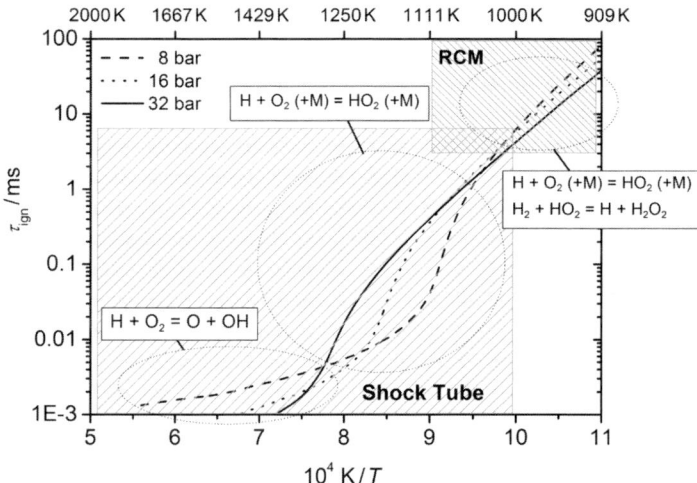

Figure 43.2 Pressure and temperature influence on the ignition delay time of hydrogen; main kinetic reactions as a function of the temperature regime for a $H_2/O_2/Ar$ mixture tested with an updated kinetic mechanism (see Reference [9]) at 8, 16, and 32 bar.

Figure 43.2 shows the ignition behavior of hydrogen at low, intermediate, and high temperature regimes for 8, 16, and 32 bar. The governing kinetic reactions of each temperature regime as well as the testing devise (rapid compression machine (RCM) or shock tube) used in the measurements are also indicated in the figure.

In general, increasing the pressure results in shorter ignition delay times in the high-temperature ($T > 1350$ K) and the low-temperature regime ($T < 1000$ K). The opposite dependence is observed in the temperature range between 1000 and 1350 K.

Under low- to intermediate-temperature conditions (in the temperature range investigated in RCMs), hydrogen oxidation is governed by reaction (43.2):

$$H + O_2 (+M) = HO_2(+M) \qquad (43.2)$$

which leads to the production of HO_2 radicals. The hydroperoxyl (HO_2) radical reacts with H_2 leading to the formation of H_2O_2 which decomposes into two OH radicals.

At higher temperatures (in the temperature range investigated in shock tubes), the competition between the reaction:

$$H + O_2 = O + OH \qquad (43.3)$$

and reaction (43.2) leads to an unusual pressure dependence of the ignition delay times. Depending on the pressure, at higher temperatures the oxidation process is mainly governed by reaction (43.3). Due to the pressure dependence of reaction (43.2), the temperature range at which the competition between (43.3) and (43.2) occurs depends on the pressure. For more details see Reference [9].

43.2.2
Flame Speed

The bulk of laminar flame studies found in the literature were performed at atmospheric pressure with non-preheated mixtures. Experimental laminar flame speed data of hydrogen or natural gas/H_2 blends at gas turbine relevant conditions (high pressure, high pre-heating temperature, lean mixtures) is scarce. Only a few studies (e.g., References [10–13]) were performed at higher pressures.

The general trends are discussed with the help of Figure 43.3. Again, as in Figure 43.1, the natural gas is simulated by a reference gas (RG) mixture of 92 vol.% methane and 8 vol.% ethane. From Figure 43.3 it is obvious that the laminar flame speed of hydrogen is significantly higher and depends more strongly on the equivalence ratio (Φ) than the flame speeds of natural gas (simulated by RG). The flame speed of hydrogen peaks on the fuel rich side (at $\Phi \approx 1.8$) whereas for natural gas the maximal value is reached at $\Phi \approx 1.1$. Note that up to a H_2 content in natural gas of about 30 vol.% the laminar flame speed values are only moderately higher than those of natural gas. This is in a good agreement with kinetic calculations performed by Brower et al. [14]. For higher amounts of hydrogen the flame speeds are significantly higher than the ones of natural gas (for 50 vol.% H_2 about a factor of 2, for 80 vol.% almost a factor of 4).

Laminar flame speeds at gas turbine relevant conditions as a function of adiabatic flame temperatures and equivalence ratios are presented in Figure 43.4. For a constant H_2 content in the fuel the flame speed increases with equivalence ratio because the flame temperature increases. Again, as in Figure 43.3, up to a H_2 content in the fuel of about 30 vol.% the flame speed values on the fuel lean

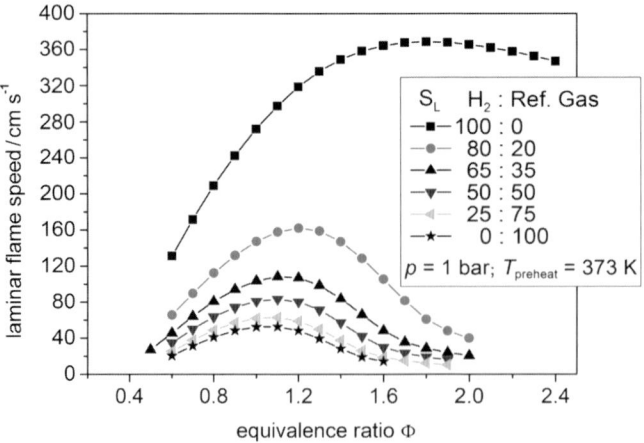

Figure 43.3 Laminar flame speeds of hydrogen and hydrogen/reference gas blends as a function of equivalence ratio (1 bar, 373 K air preheating temperature). Kinetic simulations were performed with the RDv06i8 reaction mechanism (Naumann, C. personal communications).

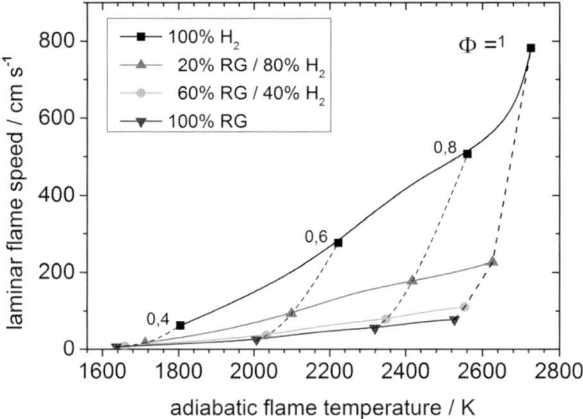

Figure 43.4 Laminar flame speeds of hydrogen and hydrogen/natural gas blends at typical gas turbine conditions (16 bar, 723 K air preheating temperature) as a function of equivalence ratio and flame temperature. Kinetic simulations were performed with the RDv06i8 reaction mechanism (Naumann, C. personal communications).

side are almost the same than for natural gas. For pure hydrogen however, they are significantly, about a factor of 10, higher.

From a practical point of view the turbulent flame speed is the more relevant quantity for the proper design of gas turbine combustors than the laminar flame speed. In the presence of turbulence the flame speed is significantly increased due to enhanced mixing and wrinkling of the flame surface thereby enlarging the flame surface area and the reaction rate. It is very common to express the turbulent flame speed (S_T) correlation as a function of the corresponding quantities of the un-stretched laminar flame speed (S_L) and the turbulence intensity (u') (e.g., References [15,16]):

$$\frac{S_T}{S_L} = 1 + A \left(\frac{u'}{S_L} \right)^B \tag{43.4}$$

In Eq. (43.4) the factor A is expected to depend on the ratio of the characteristic length scales of the turbulent (l_T) and the laminar flame (l_L) and B is an exponent between 0.5 and 1. This simplified flame speed correlation has been extended to account for additional effects like flame stretch, Lewis number effects (the diffusivity of hydrogen is much higher than for methane resulting in a different Lewis number), and so on. For more detailed information the reader is referred to additional literature [17–19].

Figure 43.5 gives an example of how hydrogen in the fuel changes the flame speed. In this figure, the measured turbulent flame speed values, which can be interpreted as global consumption rates, are compared to calculated un-stretched laminar flame speeds for methane and methane/hydrogen blends at lean conditions, 673 K preheating temperature and a pressure of 5 bar [20].

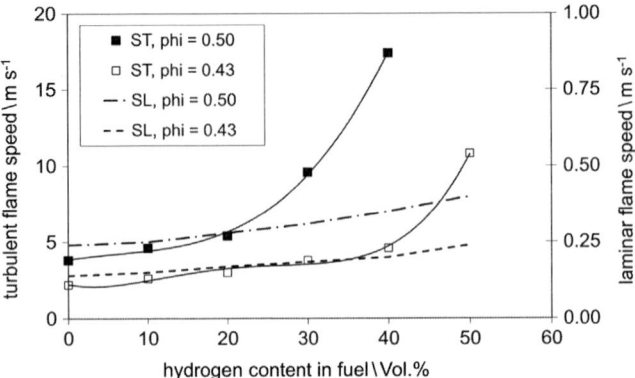

Figure 43.5 Influence of hydrogen content in the fuel on experimentally determined turbulent flame speed and calculated laminar flame speed.

Up to a hydrogen content in the fuel of 20 vol.% the results for both equivalence ratios show approximately the same increase in turbulent flame speed. This trend is quite linear up to this point and continues up to 35 vol.% hydrogen substitution for the leaner case ($\Phi = 0.43$). The relative increase of the calculated laminar flame speeds is moderate and the same for both equivalence ratios over the entire range of hydrogen contents studied. However, for the turbulent flame speeds, the increase is significantly higher. As can be noted from Figure 43.5, for H_2-contents higher than 20 vol.% (at $\Phi = 0.5$) or 35 vol.% (at $\Phi = 0.43$) the turbulent flame speed rapidly increases nonlinearly towards much higher values. For $\Phi = 0.50$ and 40 vol.% H_2 the turbulent flame speed increases by a factor of about 4 compared to pure methane. The effect of hydrogen in the fuel mixture is to consistently increase the turbulent flame speed over the increase of the laminar flame speed. Multiple effects contribute to this observed trend. First, the presence of hydrogen in the fuel mixture increases the OH radicals being supplied to the reaction zone. As a consequence the global consumption rate and the flame propagation increase. Because hydrogen has a Lewis number less than unity, under fuel lean conditions its mass diffusivity is greater than its thermal diffusivity. This has the tendency of increasing the hydrogen concentration in the reaction zone and thereby promoting fuel consumption and flame propagation. Additionally, fluid dynamic effects can contribute because flames with high hydrogen contents stabilize at different axial locations with different effective turbulent quantities than pure methane flames. The higher turbulent flame speed of hydrogen flames compared to natural gas flames is likely a consequence of the above-mentioned effects.

43.2.3
Flame Temperature, Stability, and Emissions

The adiabatic flame temperature of a hydrogen flame is about 150 K higher than a natural gas flame of the same stoichiometry (Figure 43.4). Without adapting the operating conditions (e.g., shifting towards a leaner equivalence ratio) this would result in higher NO_x emissions of the hydrogen flame. However, because of the wider flammability limits of hydrogen, especially on the fuel lean side, lean extinction of a hydrogen flame occurs at a lower equivalence ratio than for a natural gas flame. A comparison of extinction limits of natural gas and hydrogen-enriched natural gas flames is given in Griebel *et al.* [21]. Thus a hydrogen flame can be operated at much leaner conditions with lower flame temperature compared to a natural gas flame, which results in lower nitrogen oxide (NO_x) emissions.

The extension of the lean extinction limit of hydrogen is due to a higher strain resistance. Hydrogen flames are less sensitive to flame stretch because of a higher diffusivity of hydrogen, which results in a higher effective fuel concentration in the flame front [18,22,23].

Emissions of hydrogen flames are only briefly discussed here because this topic has already been described in detail in the literature (e.g., References [24,25]). In a hydrogen flame, due to the absence of carbon in the fuel, no carbon monoxide is formed during fuel conversion and nitrogen monoxide (NO) is formed only via the thermal NO route [26] and the nitrous oxide pathway [25]. Thermal NO is one of the main sources of NO_x in many gas turbines. The rate-limiting step is the reaction of the molecular nitrogen with oxygen atoms at high temperatures. This NO formation route is exponentially dependent on temperature and high formation rates are observed at temperatures higher than 1800 K. At lean premixed conditions at higher pressures, for example, typical for gas turbines, the nitrous oxide (N_2O) pathway can also significantly contribute to the total NO_x formation.

43.3
State-of-the-art Gas Turbine Technology for Hydrogen

Gas turbines for power generation are commonly classified according to their power range, starting with the lowest power level of micro gas turbines (MGT) in the kW range, all the way to heavy-duty GT with up to 500 electrical power. Typical values of the power range, the operating conditions (pressure, combustor air inlet temperature (T_{inlet}), turbine inlet temperature (T_{t3})), as well as some technological characteristics of each GT class are listed in Table 43.2. Today's most widely used combustion technology for modern stationary gas turbines fired with natural gas is the lean premixed combustion (LPC) technique. It allows for highly efficient turbine operation and very low emissions such as carbon monoxide and NO_x. With current state-of-the-art hot gas temperatures of up to 1873 K and combustor operating pressure of up to 30 bar of heavy-duty GTs, NO_x emission limits of less than 25 ppm NO_x (at 15% O_2) are achieved. Such a performance can usually only be achieved for gaseous fuels of a certain specification (pipeline natural gas quality) and is limited to an operational load range close to full load. At low load conditions flame stability issues (e.g., lean

Table 43.2 Classification of gas turbines for power generation.

GT type	Electrical power	p (bar)	T_{Inlet} (K)	T_{t3} (K)	Efficiency (single cycle) (%)	Technological characteristics turbo-machinery, combustion
MGT	1–500 kW	1–5	873–973	≤1273	25–30	Uncooled turbine blades, diffusion, partially premixed, LPC recuperator
Industrial GT	5–70 MW	8–15	573	≤1373	≈35	Partially premixed, LPC
Heavy duty GT	70–500 MW	10–30	673–923	≤1873	≤42	Heavily cooled turbine blades, LPC
Aero-derivatives	≤100 MW	20–40	≤1023	≤1923		Compact, highly sophisticated blade cooling technology

blow-out) arise and measures have to be taken to stabilize combustion by deviating from the premix combustion mode (diffusion flame pilots, fuel staging, etc.).

These problems are heightened with the use of highly reactive fuels like hydrogen which exhibit strongly different combustion properties than hydrocarbon fuels (see Section 43.2). For this reason current gas turbine combustion systems for hydrogen-rich fuels are very often based on diffusion burner systems with high levels of nitrogen or steam dilution. In addition, these turbines are specifically designed for a particular fuel composition and often tailored towards a moderate gas turbine operating design point (i.e., moderate pressure and firing temperature). Pre-ignition or flashback and subsequent overheating of burner components for operation with hydrogen-rich fuels in current gas turbine combustor designs can only be controlled by dilution with inert species like nitrogen or steam. With this fuel dilution the turbulent burning rates are lower and the ignition delay times increase in a way to meet similar values to those for natural gas. Local flame temperatures in the diffusion flame zone can be dampened by the increased thermal capacity provided by the diluent species and thus thermal NO_x formation can be significantly reduced in order to achieve NO_x emission levels similar to natural gas fired units. However, this requires significant amounts of nitrogen or steam (up to about 50 vol.%) to achieve a sufficiently high fuel dilution. As a consequence, this results in a higher complexity of the fuel supply system and it also increases the capital and operational costs because of the infrastructure needed for nitrogen and steam production.

Many of the large gas turbine manufacturers have developed turbines for operation on syngas in the last two decades. Most of these machines were installed in integrated gasification combined cycle (IGCC) plants for power generation or cogeneration of electrical power, steam, fuels, and chemicals. A broad variety of feedstocks (coal, pet coke, oil, waste, wood, etc.) were used for these IGCC installations in the USA, Europe, and Asia. Over the past 15–20 years of operation IGGC plants have faced a couple of technical and economic problems. IGCC plants are of a high technical complexity, which results in high capital investments. In addition, the availability and reliability of these plants are not as high as for other fossil fuel based power generation technologies like natural gas combined cycle (NG-CC) or coal plants. Because of the high technical complexity – an IGCC plant consists of many components like gasifier, air-separation unit, syngas cleaning unit, heat recovery steam generator, and so on – the likelihood that parts of the plant are out of service due to repair or maintenance is high and this dramatically decreases the plant availability. Together with high capital investment and maintenance costs this results in high costs of electricity.

Some of the major gas turbine developments with respect to the use of hydrogen-rich fuels are described in the following.

Alstom has developed and commissioned a heavy-duty gas turbine for syngas in the 200 MW range (GT13E2) that can handle hydrogen-rich fuels up to 45 vol.% H_2. Nitrogen dilution up to 55 vol.% was used to control the flame parameters and the NO_x emissions [27].

GE has gained experience with burning hydrogen-rich fuels in various gas turbines in the power range 10–280 MW, mostly from IGCC plants. Combustors based on diffusion flames with nitrogen or steam dilution of the syngas were applied to achieve the NO_x emission targets. Successful operation on syngas containing up to 45 vol.% H_2 is reported for F-class machines (e.g., Reference [28]).

Siemens has investigated syngas operation in different gas turbines in the power range from 10 up to 250 MW. Successful operation of these turbines installed in IGCC plants in the USA and Europe with hydrogen contents of up to 41 vol.% has been reported (e.g., Reference [29]). Steam (up to 22 vol.%) or nitrogen (up to 30 vol.%) dilution were applied to achieve the NO_x emission targets.

A similar operating experience with H_2-rich fuels (syngas) in a gas turbine has been reported by Ansaldo Energia (e.g., Reference [30]). The reference reported that a 170 MW gas turbine (V94.2K) was tested with syngas (H_2 content up to 45 vol.%) and NO_x emissions were controlled by steam dilution (up to $\approx$50 vol.%).

From the above mentioned syngas applications, much experience was gained with respect to gas turbine operation on highly reactive fuels (fuels with high amounts of hydrogen). The state-of-the-art combustor technology used in these applications is based on diffusion flames with nitrogen or steam dilution to control NO_x emissions. Up to now, no gas turbine has been installed that can handle 100 vol.% hydrogen without the need of diluting the fuel with nitrogen or steam. Of course, many manufacturers are working on further developments towards hydrogen fired GTs in their R&D programs and a couple of promising new combustor technologies for hydrogen are being investigated in the academic research community. This will be discussed in the next section.

43.4
Research and Development Status, New Combustion Technologies

Future gas turbines have to guarantee a safe and reliable power generation with a high efficiency and low emissions. In addition, GT have to be highly fuel and load flexible to compensate for strong and fast supply fluctuations in the power generation due to the increasing contribution of renewables. Due to their fast startup, current GTs already provide peak capacity and high load flexibility. Nevertheless, this capability has to be extended to cope with additional requirements of the future energy supply system, for example, operation on decarbonized fuels. The development of decentralized power generation or combined heat and power (CHP) systems (e.g., based on micro gas turbines) is becoming increasingly important, demanding also a high fuel and load flexibility of these systems.

In the centralized power generation sector IGCC power plants based on coal or biomass together with a carbon capture and storage (CCS) process can be a solution for decarbonized power generation. Applying a pre-combustion CCS step would shift the syngas (H_2/CO mixture) to an almost pure hydrogen fuel delivered to the combustor with the consequence of the technical challenges described in Section 43.1. However, besides technical problems, economic, political, and social hurdles have to be resolved for a successful implementation of this technology. IGCC plants have a very high complexity because they consist of several sub-systems like gasifier, air-separation unit, syngas cleaning unit, CCS step, gas turbine, steam turbine, heat recovery steam generator, and so on. The risk of a technical failure of plant sub-systems is high resulting in a lower plant availability compared to current combined cycle gas turbine (CC-GT) plants. Together with high investment costs this results in a higher cost of electricity (CoE) compared to CC-GT or coal based power plants. This is partly caused by the low fossil fuel prices, for example, of coal or shale gas, and the very low CO_2-certificate price at the moment. In addition, the public acceptance of CCS technology in some countries, for example, Germany, is also not very high, acting as a block to further development of this technology.

Nevertheless, GT related industry and academia with support by public funding have initiated R&D programs to develop gas turbines that can handle H_2-rich fuels in order to be prepared when the political and market conditions have changed. A brief summary of the R&D activities in the GT industry are described in the following paragraphs.

Alstom is investigating the development of a gas turbine for hydrogen-rich fuel based on the GT24 and GT26. This gas turbine is characterized by a two stage combustion technology. Research work is concentrated on the second, the reheat combustor that operates at high pressures (about 20 bar) and temperatures (about 1273 K). The work aims at achieving low NO_x emissions in the reheat combustor for hydrogen-rich fuels up to 70 vol.% H_2 [31]. This is a typical fuel composition of an IGCC-CCS plant with about 90% carbon capture [32]. This work has been carried out within the European project DECARBit [33]. In parallel, applied research on autoignition of hydrogen-rich fuels at reheat conditions is performed at the DLR Institute of Combustion Technology in Stuttgart. In an optically accessible generic reheat combustor that mimics the relevant operating conditions with respect to pressure, temperature, and flue gas composition (exhaust gas from the first combustor) very detailed investigations on the autoignition process are performed with sophisticated optical and laser diagnostics [34–37].

GE Power & Water is developing a new premixing fuel injector for heavy-duty GTs and hydrogen-rich fuels. The design is based on small-scale jet-in-crossflow mixing similar to lean direct injection (LDI) which will be discussed in the next section. Single nozzle tests as well as test of a full-scale multi-nozzle combustor of 10 MW were performed and reported (e.g., Reference [38]). These tests were performed at F-class operating conditions and for various fuels including hydrogen contents up to 90 vol.%. Results indicate a reliable low NO_x solution for H_2-rich fuels. GE Oil & Gas have tested a 10 MW prototype GT at Enel's coal-fired power plant. The GT was equipped with a diffusive flame combustor and NO_x emissions were controlled by steam injection. The gas turbine was successfully operated with pure hydrogen [39].

Siemens is developing GTs for syngas and hydrogen-rich fuels for integration in future IGCC plants based on their F- and G-class machines. Part of this work was performed in Siemens development programs and the European projects "High Efficient Gas Turbine with Syngas Application" (HEGSA) [40], ENhanced CAPture of CO_2 (ENCAP) [32], and the US-DOE funded "Advanced Hydrogen Development Turbine" programme. A syngas/hydrogen prototype combustor with a diffusion flame and steam or nitrogen dilution was investigated at high pressure and successful testing was reported [41]. A more recent development focuses on a fuel flexible lean premixed combustor for natural gas and hydrogen-rich fuels. Successful operation on syngas has been reported for H-class firing temperatures. Operation with hydrogen-rich fuels containing up to 70 vol.% H_2 is reported [42]. A scaled staged GT burner has been tested on natural gas and NG/H_2 blends with up to 40 vol.% hydrogen at the DLR Stuttgart [43]. In the industrial gas turbine sector Siemens is developing fuel flexible combustors based on the SGT-700 and SGT-800 technology in the 30–50 MW range. Test results for different fuels including hydrogen-rich fuels with a hydrogen content of the fuel of up to 32 vol.% are reported in References [44,45].

Several technology development programs were initiated worldwide to enhance state-of-the-art GT technology with respect to operation on hydrogen-rich fuels.

In Europe a few European Commission and industry co-funded R&D projects, for example, "New Combustion Systems for Gas Turbines (NGT)" [46], "ENhanced CAPture of CO_2" (ENCAP) [32], and DECARBit [33,47], enabling advanced pre-combustion capture techniques and plants were established within the 5th, 6th, and 7th Framework programmes. The recently finished European project "Low Emission Gas Turbine Technology for Hydrogen-rich Syngas (H_2-IGCC)" [48] focused on technical solutions for the use of state-of-the-art GTs in the next generation of IGCC plants. The main objective was to enable GT combustors to burn undiluted hydrogen-rich syngas with low NO_x emissions, combined with high fuel flexibility. In this project, combustion, materials, and turbo-machinery aspects were investigated and system analysis was performed. Two development pathways, conventional lean premixed combustion and innovative combustion concepts, for example, FLOX®, were investigated. Ansaldo Energia was developing a gas turbine combustor for hydrogen-rich fuels within this project.

In the USA, the Department of Energy (DoE) launched a three-phase development program for a hydrogen turbine. The objective of this "Advanced Hydrogen Turbine Development" programme is to develop a fuel flexible (natural gas, syngas, hydrogen) gas turbine for future IGCC plants. The overall DOE Advanced Power System goal is to conduct the research and development necessary to produce CO_2 sequestration ready coal-based IGCC power systems with high efficiency (45–50%), near zero emissions (<2 ppm NO_x @ 15% O_2), and competitive plant capital cost (<$1000 kW^{-1}). The achievements of the developed technology will be validated in by testing a pre-commercial prototype planned for 2016/2017 [42,49].

43.4.1
New Combustion Technologies

The well-established lean premixed combustion (LPC) technology for stationary gas turbines operated on natural gas has led to a significant reduction of NO_x emissions. In most cases, the flames in LPC combustors are aerodynamically stabilized by swirl, resulting in a vortex breakdown generated central recirculation zone. Liability to thermoacoustic instabilities, risk of flashback, and sensitivity with respect to changes of the fuel composition are limitations for a reliable, safe, and low-emission operation of lean premixed swirl stabilized burner systems. Therefore, alternative combustion concepts for gas turbines are desirable that can handle the challenging requirements of a high fuel and load flexibility together with low emissions, a high reliability, and a safe operation. In this respect MILD Combustion or Flameless Oxidation (FLOX® is a registered trademark of WS Wärmeprozess technik GmbH, Renningen, Germany), lean direct injection (LDI), multi-injection, and micro-mixing are attractive alternatives with a high potential. A summary of the R&D activities in industry and academia with respect to these combustion technologies is given in the following paragraphs.

43.4.1.1 MILD Combustion or Flameless Oxidation
The common characteristics of MILD combustion and flameless oxidation (FLOX®) are that the reactants are highly diluted by inert gas species like N_2, CO_2, H_2O and as a consequence the flame reactions are delayed and combustion

43.4 Research and Development Status, New Combustion Technologies

is distributed over a larger volume and not concentrated in a thin reaction zone (flame front). Thereby, high peak temperatures causing high NO formation rates (thermal NO) are avoided. The name MILD combustion was proposed by Cavaliere [50]: "A combustion process is named Mild when the inlet temperature of the reactant mixture is higher than mixture self-ignition temperature whereas the maximum allowable temperature increase with respect to inlet temperature during combustion is lower than mixture self-ignition temperature (in Kelvin)."

Under special flow and temperature conditions, for example, formed by a large and intense internal recirculation zone, Wünning [51,52] found that stable combustion takes place without any visible or audible flame. For that reason the combustion mode was named "flameless oxidation." The first industrial applications of FLOX® combustion were in atmospheric, low caloric furnaces at low power density [53,54]. In gas turbine combustors with high power densities and high flame temperatures, combustion does not fully meet this MILD combustion characterization [55]. Nevertheless, the name FLOX® denotes a certain GT burner type that is characterized by high momentum fuel/air jets, discharged into the combustor through several orifices arranged on a circle (Figure 43.6a). The high axial momentum jets induce a distinct recirculation in the combustion chamber (see Figure 43.6b; streamlines indicate flow field) leading to an intense mixing of burned gas with fresh fuel/air mixture. High flashback resistance is obtained through the absence of low velocity zones, which favors this concept for multi-fuel applications, especially for hydrogen-rich fuels.

An experimental study of flameless combustion for GT has been performed at atmospheric conditions [56].

More recent developments of FLOX® combustors at gas turbine operating conditions have been performed at the DLR Institute of Combustion Technology in Stuttgart. The proof of principle of this technology for gas turbine application is described elsewhere [57]. Successful tests of a combustor with optical access for laser diagnostics operated with natural gas and hydrogen at pressures up to 30 bar are reported. In the following years several FLOX® model combustors

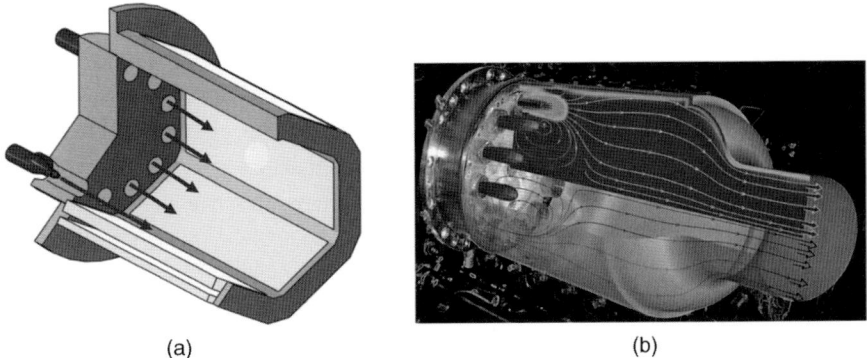

(a)   (b)

Figure 43.6 FLOX® combustor characteristics (Lammel, O. and Schütz, H. personal communication).

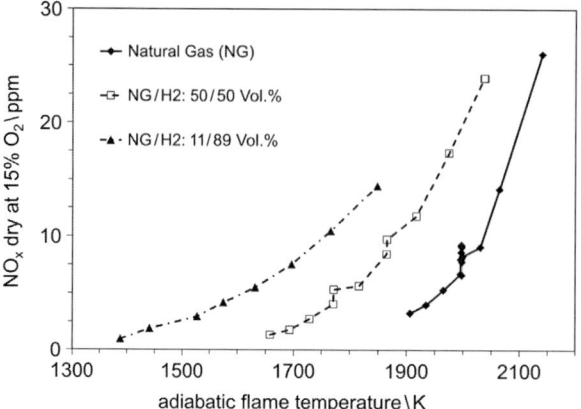

Figure 43.7 Measured NO_x values in a FLOX burner for different fuels. Data extracted from (Lammel, O. and Schütz, H. personal communication).

with high power densities, similar to current GT combustors, were investigated and low NO_x emissions for various fuels (NG, NG + H_2, NG + propane) were achieved [58]. Typical results of NO_x emissions as a function of adiabatic flame temperature are shown in Figure 43.7. With increasing flame temperature NO_x emissions increase due to higher thermal NO formation, which exponentially depends on temperature. For a constant flame temperature, higher hydrogen contents in the fuel result in higher NO_x emissions. Because of the shorter ignition delay time and the higher flame speed of the hydrogen/NG mixtures the flame stabilizes closer to the fuel/air jet outlet in comparison to the natural flame. Thereby, less time for fuel/air premixing is available, causing higher NO_x emissions. In addition, because of short and attached flames in the hydrogen/NG cases the residence time in the post flame region is longer, contributing thereby also to a higher NO formation. To improve the part load performance, special patented burner designs (HiPerMix® [59], EZEE® [60]) have been developed and tested [55,61].

FLOX®-based combustors have proven to be an attractive alternative for reliable, fuel and load flexible, low-emission gas turbine combustors [55,58,61].

43.4.1.2 Lean Direct Injection, Multi-injection, and Micro-mixing

Lean direct injection (LDI), multi-injection, and micro-mixing are discussed together here because they have some similarities. The fuel and air flow is injected into the combustor in a multiple jet arrangement. The number of jets is much higher than in a FLOX® combustor and they are also not arranged on a circle. As a result, the combustor flow field and the reaction zone are different (no large and intense inner recirculation zone and no volumetric combustion, both typical for FLOX®).

The LDI concept was originally developed as a low NO_x combustion concept for aeroengines. In these combustors the liquid fuel is directly injected into the flame zone. However, a fine atomization and rapid mixing of fuel and air is very important to achieve low-NO_x emissions [62,63]. This concept has a low auto-ignition and flashback risk, which makes it attractive for application in stationary gas turbines operated on hydrogen-rich fuels.

Successful tests of a multi-point fuel injection, lean burn concept have been reported by Tacina et al. [64]. The combustor was operated with Jet-A fuel at air inlet temperatures of 810 K and pressures up to ≈28 bar. Each injector had an air swirler to achieve rapid fuel/air mixing and to provide a small recirculation zone for flame stabilization.

Various testing of LDI combustors has been performed in the past to demonstrate the low NO_x potential also for pure hydrogen. Marek et al. [65] performed flame tube experiments of LDI injectors for pure hydrogen and at air inlet temperatures up to ≈800 K and pressures up to ≈14 bar. The results show that NO_x emissions are comparable to Jet-A, LDI combustor concepts. However, some of the tested injectors result in high NO_x levels. Nitrogen dilution might be an option to achieve lower NO_x emissions.

Whether it is better to add the diluent to the air or fuel stream was studied by Weiland et al. [66]. For the investigated LDI burner set-up (co-axial air flame), fuel stream dilution was preferable because an operation with higher fuel and air velocities was possible, thereby reducing the residence time and consequently the NO_x emissions.

The conclusion from the above-described studies is that more fuel injection ports per air jet reduce NO_x emissions due to higher fuel jet momentum and thereby better fuel/air mixing. A higher number of air jets with a smaller size reduce NO_x by reducing the residence time in the flame due to a shorter length of the combustion zone. Of course, the penalty of this is a higher combustor pressure loss.

A multiple-injection prototype combustor for hydrogen-rich fuels has been developed by Hitachi [67,68]. The burner consists of multiple fuel nozzles and a perforated plate with inclined air holes. The air is injected in such a way that each fuel jet is shielded by the air flow. The individual fuel/air jets join each other inside the combustor forming a swirling flow motion with converging-and-diverging streamlines and a weak inner recirculation zone. A lifted flame is formed which is beneficial for flashback resistance and low NO_x emissions. Tests with hydrogen-rich fuels up to 65 vol.% H_2 at air inlet temperatures of 660 K and a pressure of 6 bar indicate a stable operation with low NO_x emissions.

Test results of a full-scale micro-mixing lean-premixed injector for hydrogen-rich fuels have been reported by Parker Hannifin and UCI [69]. The multi-point injection concept was adopted from previous work at NASA [64,70]. This design is based on an annular injection scheme consisting of small injection points that are distributed over the entire combustor dome. The fuel and air mixing channels are integrated into the dome. The full-scale injector consisting of sixteen micro-mixing cups was tested at a pressure of 12 bar and at flame temperatures up to 1750 K. Very low NO_x emissions are reported for nitrogen or CO_2 diluted hydrogen-rich fuels up to 100 vol.% H_2.

43.5
Concluding Remarks

The use of hydrogen in gas turbines is feasible, although a couple of technical and economic hurdles have to be resolved. The main technical challenge of using hydrogen as a gas turbine fuel is related to the high flashback and auto-ignition risk because of the significantly higher flame speed and shorter ignition delay time of hydrogen compared to natural gas. In addition, without adapting the operating conditions (e.g., shifting to leaner equivalence ratios) due to a higher flame temperature the NO_x emissions of a GT fired with hydrogen would be higher than for natural gas. On the other hand, hydrogen has wider flammability limits than natural gas. Therefore, shifting the operating conditions towards leaner equivalence ratios is not a concern. However, the above-mentioned challenges will require design modifications of current state-of-the-art gas turbine combustors.

Current gas turbine combustion systems for hydrogen-rich fuels are based on diffusion burners that require a high level of inertization (nitrogen or steam dilution) in order to reduce the high fuel reactivity and to achieve the NO_x emission targets. This necessary fuel dilution is disadvantageous because it increases the complexity of the fuel supply system and causes higher capital and operational costs. In addition, these turbines are specifically designed for a particular fuel composition and often tailored towards a moderate gas turbine operating design point. Therefore, alternative combustion concepts for hydrogen fired gas turbines are desirable that can handle the challenging requirements of a high fuel and load flexibility together with low emissions, high reliability, and safe operation.

Studies of new, innovative combustion concepts for hydrogen fired gas turbines have shown promising results in laboratory-scale and model combustor high-pressure testing but these concepts need to be further investigated and enhanced to a higher technology readiness level before they can be implemented in gas turbines.

On the economic side, power plant technologies for the use of hydrogen-rich fuels like syngas or pure hydrogen (e.g., IGCC plants) need to be further developed and enhanced in order to be cost-competitive with state-of-the-art power generation technology.

References

1 Glassman, I. (ed.) (1996) *Combustion*, 3rd edn, Academic Press, San Diego, CA.
2 Dunn-Rankin, D. (ed.) (2008) *Lean Combustion*, Academic Press, Burlington.
3 Lackner, M., Palotás, Á.B., and Winter, F. (eds) (2013) *Combustion*, 1st edn, Wiley-VCH Verlag GmbH, Weinheim.
4 Abott, D.J., Maunand, J., Deneve, M., and Bastiaans, R. (2009) The Impact of Natural Gas Quality on Gas Turbine Performance. European Turbine Network, Position Paper.
5 Hydrogen data. Presented by HyWeb, http://www.h2data.de/ (accessed 12 August 2015).

6 Warnatz, J., Maas, U., and Dibble, R.W. (2006) *Combustion: Physical and Chemical Fundamentals, Modeling and Simulation, Experiments, Pollutant Formation*, 4th edn, Springer, Berlin.

7 Burke, M.P., Chaos, M., Ju, Y., Dryer, F.L., and Klippenstein, S.J. (2012) Comprehensive H2/O2 kinetic model for high-pressure combustion. *Int. J. Chem. Kinet.*, **44** (7), 444–474.

8 Herzler, J. and Naumann, C. (2009) Shock-tube study of the ignition of methane/ethane/hydrogen mixtures with hydrogen contents from 0% to 100% at different pressures. *Proc. Combust. Inst.*, **32** (1), 213–220.

9 Kéromnès, A., Metcalfe, W.K., Heufer, K.A., Donohoe, N., Das, A.K., Sung, C., Herzler, J., Naumann, C., Griebel, P., Mathieu, O., Krejci, M.C., Petersen, E.L., Pitz, W.J., and Curran, H.J. (2013) An experimental and detailed chemical kinetic modeling study of hydrogen and syngas mixtures oxidation at elevated pressures. *Combust. Flame*, **160**, 995–1011.

10 Davis, S.G., Joshia, A.V., Wanga, H., and Egolfopoulos, F. (2005) An optimized kinetic model of H2/CO combustion. *Proc. Combust. Inst.*, **30**, 1283–1292.

11 Bradley, D., Lawes, M., Liu, K., Verhelst, S., and Woolley, R. (2007) Laminar burning velocities of lean hydrogen–air mixtures at pressures up to 1.0 MPa. *Combust. Flame*, **149** (1–2), 162–172.

12 Burke, M.P., Dryer, F.L., and Ju, Y. (2011) Assessment of kinetic modeling for lean H2/CH4/O2/diluent flames at high pressures. *Proc. Combust. Inst.*, **33** (1), 905–912.

13 Krejci, M.C., Mathieu, O., Vissotski, A.J., Ravi, S., Sikes, T.G., Petersen, E.L., Kéromnès, A., and Metcalfe, W.K. (2013) Laminar flame speed and ignition delay time data for the kinetic modeling of hydrogen and syngas fuel blends. *J. Eng. Gas Turb. Power*, **135** (2), 021503-1–0215030-9.

14 Brower, M., Petersen, E.L., Metcalfe, W., Curran, H.J., Füri, M., Bourque, G., Aluri, N., and Güthe, F. (2013) Ignition delay time and laminar flame speed calculations for natural gas/hydrogen blends at elevated pressures. *J. Eng. Gas Turbines Power*, **135** (2), 021504-1–021504-10.

15 Bradley, D. (1992) How fast can we burn. *Proc. Combust. Inst.*, **24**, 247–262.

16 Peters, N. (2000) *Turbulent Combustion*, Cambridge University Press, Cambridge.

17 Lipatnikov, A.N. and Chomiak, J. (2002) Turbulent flame speed and thickness: phenomenology, evaluation, and application in multi-dimensional simulations. *Prog. Energ. Combust.*, **28** (1), 1–74.

18 Lipatnikov, A.N. and Chomiak, J. (2005) Molecular transport effects on turbulent flame propagation and structure. *Prog. Energ. Combust.*, **31** (1), 1–73.

19 Filatyev, S.A., Driscoll, J.F., Carter, C.D., and Donbar, J.M. (2005) Measured properties of turbulent premixed flames for model assessment, including burning velocities, stretch rates, and surface densities. *Combust. Flame*, **141** (1–2), 1–21.

20 Boschek, E., Griebel, P., and Jansohn, P. (2007) Fuel variability effects on turbulent, lean premixed flames at high pressures, in *ASME Turbo Expo 2007: Power for Land, Sea and Air, Montreal, Canada, 14–17 May 2007*, vol. **2**, paper GT-2007-27496, ASME, pp. 373–383.

21 Griebel, P., Boschek, E., and Jansohn, P. (2007) Lean blowout limits and NOx emissions of turbulent, lean premixed hydrogen-enriched methane/air flames at high pressure. *J. Eng. Gas Turbines Power*, **129** (2), 404–410.

22 Gaucheau, J.L., Denet, B., and Searby, G. (1998) A numerical study of lean CH4/H2/air premixed flames at high pressure. *Combust. Sci. Technol.*, **137** (1–6), 81–99.

23 Jackson, G.S., Sai, R., Plaia, J.M., and Boggs, C.M.K.T.K. (2003) Influence of H2 on the response of lean premixed CH4 flames to high strained flows. *Combust. Flame*, **132** (3), 503–511.

24 Lefebvre, A.H. and Ballal, D.R. (2010) *Gas Turbine Combustion*, 3rd edn, CRC Press, Boca Raton, FL.

25 Lieuwen, T.C. and Yang, V. (2013) *Gas Turbine Emissions* (eds T.C. Lieuwen and V. Yang), Cambridge University Press, New York.

26 Zeldovich, Y.B., Frank-Kamenetskii, D., and Sadovnikov, P. (1947) *Oxidation of Nitrogen in Combustion*, House of the Academy of Sciences of USSR.

27 Reiss, F., Griffin, T., and Reyser, K. (2002) The ALSTOM GT13E2 medium BTU gas turbine, in *ASME Turbo Expo 2002: Power for Land, Sea and Air, Amsterdam, The Netherlands, 3–6 June 2002*, vol. **1**, paper GT2002-30108, ASME, pp. 705–712.

28 Payrhuber, K. (2007) Gas turbines for syngas. Presented at the 2nd International Freiberg Conference on IGCC and ItL Technologies, 8–12 May 2007, Freiberg, Germany.

29 Gadde, S., Wu, J., Gulati, A., McQuiggan, G., Koestlin, B., and Prade, B. (2006) Syngas capable combustion systems development for advanced gas turbines, in *ASME Turbo Expo 2006: Power for Land, Sea and Air, Barcelona, Spain, 8–11 May 2006*, vol. **4**, paper GT2006-90970, ASME, pp. 547–554.

30 Bonzani, F. and Gobbo, P. (2007) Operating experience of high flexibility syngas burner for IGCC power plant, in *ASME Turbo Expo 2007: Power for Land, Sea and Air, Montreal, Canada, 14–17 May 2007*, vol. **2**, paper GT2007-27114, ASME, pp. 65–71.

31 Poyyapakkam, M., Wood, J., Mayers, S., Ciani, A., Guethe, F., and Syed, K. (2012) Hydrogen combustion within a gas turbine reheat combustor, in *ASME Turbo Expo 2012: Power for Land, Sea and Air, Copenhagen, Denmark, 11–15 June 2012*, vol. **2**, paper GT2012-69165, ASME, pp. 847–854.

32 CORDIS (2010) ENCAP Project: Enhanced Capture of CO2 (ENCAP), 6th Framework Program, Contract No: SES6-CT-2004-502666, http://www.encapco2.org/ (accessed 12 August 2015).

33 CORDIS (2015) DECARBit Project: Enabling Advanced Pre-combustion Capture Techniques and Plants (DECARBit), 7th Framework Programme, FP7-211971, http://www.sintef.no/Projectweb/DECARBit/ (accessed 12 August 2015).

34 Fleck, J.M., Griebel, P., Steinberg, A.M., Stöhr, M., Aigner, M., and Andrea Ciani, A. (2010) Experimental investigation of a generic, fuel flexible reheat combustor at gas turbine relevant operating conditions, in *ASME Turbo Expo 2010: Power for Land, Sea and Air, Glasgow, UK, 14–18 June 2010*, vol. **2**, paper GT2010-22722, ASME, pp. 583–592.

35 Fleck, J., Ciani, A., Griebel, P., Steinberg, A., Stöhr, M., and Aigner, M. (2012) Autoignition limits of hydrogen at relevant reheat combustor operating conditions. *J. Eng. Gas Turbines Power*, **134** (4), 041502–041502-8.

36 Fleck, J.M., Griebel, P., Steinberg, A.M., Arndt, C.M., Naumann, C., and Aigner, M. (2013) Autoignition of hydrogen/nitrogen jets in vitiated air crossflows at different pressures. *Proc. Combust. Inst.*, **34**, 3185–3192.

37 Fleck, J.M., Griebel, P., Steinberg, A.M., Arndt, C.M., and Aigner, M. (2013) Autoignition and flame stabilization of hydrogen/natural gas/nitrogen jets in a vitiated cross-flow at elevated pressure. *Int. J. Hydrogen Energy*, **38** (36), 16441–16452.

38 York, W.D., Zminsky, W.S., and Yilmaz, E. (2013) Development and testing of a low NOx hydrogen combustion system for heavy-duty gas turbines. *J. Eng. Gas Turbines Power*, **135** (2), 022001-1–022001-8.

39 Cocchi, S. and Sigali, S. (2010) Development of a low NOx hydrogen-fuelled combustor for 10 MW class gas turbines, in *ASME Turbo Expo 2010: Power for Land, Sea and Air, Glasgow, UK, 14–18 June 2010*, vol. **2**, paper GT2010-23348, ASME, pp. 1025–1035.

40 CORDIS, HEGSA (2008) HEGSA Project: High Efficient Gas Turbine with Syngas Application (HEGSA), 5th Framework Programme, EC Project NNE5/644/2001. http://cordis.europa.eu/project/rcn/86914_en.html (accessed 12 August 2015).

41 Wu, J., Brown, P., Diakunchak, I., Gulati, A., Lenze, M., and Koestlin, B. (2007) Advanced gas turbine combustion system development for high hydrogen fuels, in *ASME Turbo Expo 2007: Power for Land, Sea and Air, Montreal, Canada, 14–17 May 2007*, vol. **2**, paper GT2007-28337, ASME, pp. 1085–1091.

42 Bradley, T. and Marra, J. (2012) Advanced hydrogen turbine development update, in *ASME Turbo Expo 2012: Power for Land, Sea and Air, Copenhagen, Denmark, 11–15 June 2012*, vol. **3**, paper GT2012-68169, ASME, pp. 537–545.

43 Lückerath, R., Lammel, O., Stöhr, M., Boxx, I., Stopper, U., Meier, W., Janus, B., and Wegner, B. (2011) Experimental investigations of flame stabilization of a gas turbine combustor, in *ASME Turbo Expo 2010: Power for Land, Sea and Air, Vancouver, Canada, 6–10 June 2011*, vol. **2**, paper GT2011-45790, ASME, pp. 725–736.

44 Larfeldt, J., Larson, A., and Andersson, M. (2012) SGT-700 and SGT-800 fuel flexibility testing activities, the future of gas turbine technology. Presented at the 6th International Gas Turbine Conference (IGTC), Brussels, Belgium, 17–18 October 2012.

45 Lörstad, D., Wang, L., Axelsson, S., and Björkman, M. (2013) Siemens Gas Turbine SGT-800 enhanced to 50 MW: design modifications, validation and operation experience, Power-Gen Europe, Vienna, Austria. http://pennwell.websds.net/2013/vienna/pge/papers/T3S5O2-paper.pdf (accessed 12 August 2015).

46 CORDIS (2005) NGT Project: New Combustion Systems for Gas Turbines (NGT), 5th Framework Programme, Contract No. ENK5-CT-2001-00564. http://cordis.europa.eu/project/rcn/59861_en.html (accessed 12 August 2015).

47 DECARBit (2011) Project Overview. http://www.sintef.no/Projectweb/DECARBit/Project-Overview/ (accessed 12 August 2015).

48 H2-IGCC Project (2009–2014) Low Emission Gas Turbine Technology for Hydrogen-rich Syngas, 7th Framework Programe, FP7-239349. http://cordis.europa.eu/projects/rcn/98465_en.htm, www.h2-igcc.eu/default.aspx.

49 Office of Fossil Energy: Hydrogen Turbine Program (2012) 2012 Portfolio: Turbines for Coal Based Systems that Capture Carbon. Annual Report, DOE/NETL-2012/1587.

50 Cavaliere, A. and de Joannon, M. (2004) Mild combustion. *Prog. Energy Combust. Sci.*, **30** (4), 329–366.

51 Wünning, J.A. and Wünning, J.G. (1992) Burners for flameless oxidation with low NOx formation even at maximum air preheat. *Gas Wärme Int.*, **41** (10), 438–444.

52 Wünning, J.A. and Wünning, J.G. (1997) Flameless oxidation to reduce thermal NO-formation. *Prog. Energy Combust. Sci.*, **23** (1), 81–94.

53 Weber, R., Verlaan, A., Orsino, S., and Lallemant, N. (1999) On emerging furnace design methodology that provides substantial energy savings and drastic reductions in CO2, CO and NOx emissions. *J. Energy Inst.*, **72**, 77–83.

54 Coelho, P.J. and Peters, N. (2001) Numerical simulation of a mild combustion burner. *Combust. Flame*, **124** (3), 503–518.

55 Schütz, H., Lammel, O., Schmitz, G., Rödiger, T., and Aigner, M. (2012) EZEE®: a high power density modulating FLOX® combustor, in *ASME Turbo Expo 2012: Power for Land, Sea and Air, Copenhagen, Denmark, 11–15 June 2012*, vol. **2**, paper GT2012-68997, ASME, pp. 701–712.

56 Li, G., Gutmark Ephraim, J., and Stankovic, D. (2006) Experimental study of flameless combustion in gas turbine combustors, 44th AIAA Aerospace Sciences Meeting and Exhibit, 9–12 January 2006, Reno, Nevada, USA.

57 Lückerath, R., Meier, W., and Aigner, M. (2008) FLOX® combustion at high pressure with different fuel compositions. *J. Eng. Gas Turbines Power*, **130** (1), 011505-1–011505-7.

58 Lammel, O., Schütz, H., Schmitz, G., Lückerath, R., Stöhr, M., Noll, B., Aigner, M., Hase, M., and Krebs, W. (2010) FLOX® combustion at high power density and high flame temperatures. *J. Eng. Gas Turbines Power*, **132** (12), 121503-1–121503-10.

59 Schütz, H. and Schmitz, G. (2009) Patent DE 10 2007 036 953 B3. PCT/EP2008/059744.

60 Schütz, H., Schmitz, G., and Lammel, O. (2010) Patent DE 10 2008 032 265 B4.

61 Rödiger, T., Lammel, O., Aigner, M., Beck, C., and Krebs, W. (2013) Part-load operation of a piloted FLOX® combustion system. *J. Eng. Gas Turbines Power*, **135** (3), 031503-1–031503-9.

62 Tacina, R., Wey, C., Liang, P., and Mansour, A. (2001) A low NOx lean-direct injection, multipoint integrated module combustor concept for advanced aircraft gas turbines, Sixth International Conference on Technologies and Combustion for a Clean Environment (Clean Air VI), Porto, Portugal, 9–12 July 2001.

63 Tacina, R., Wey, C., and Choi, K.J. (2001) Flame tube NOx emissions using a lean-direct-wall-injection combustor concept, 37th Joint Propulsion Conference and Exhibit, Salt Lake City, Utah, USA, 8–11 July 2001, AIAA-2001-3271.

64 Tacina, R., Partelow, L., Mansour, A., and Wey, C. (2004) Experimental sector and flame-tube evaluations of a multipoint integrated module concept, in *ASME Turbo Expo 2004: Power for Land, Sea and Air, Vienna, Austria 14–17 June 2004*, vol. **1**, paper GT2004-53263, ASME, pp. 107–119.

65 Marek, C.J., Smith, T.D., and Kundu, K. (2005) Low emission hydrogen combustors for gas turbines using lean direct injection, 41st Joint Propulsion Conference and Exhibit, Tucson, Arizona, USA 10–13 July 2005, AIAA-2005-3776.

66 Weiland, N.T. and Strakey, P.A. (2010) NOx reduction by air-side versus fuel-side dilution in hydrogen. *J. Eng. Gas Turbines Power*, **132** (7), 071504-1–071504-9.

67 Dodo, S., Asai, T., Koizumi, H., Takahashi, H., and Yoshida, S.I.H. (2011) Combustion characteristics of a multiple-injection combustor for dry low-NOx combustion of hydrogen-rich fuels under medium pressure, in *ASME Turbo Expo 2010: Power for Land, Sea and Air, Vancouver, Canada, 6–10 June 2011*, vol. **2**, paper GT2011-45459, ASME, pp. 467–476.

68 Asai, T., Dodo, S., Koizumi, H., Takahashi, H., and Yoshida, S.I.H. (2011) Effects of multiple-injection-burner configuration on combustion characteristics for dry low-NOx combustion of hydrogen-rich fuels, in *ASME Turbo Expo 2010: Power for Land, Sea and Air, Vancouver, Canada, 6–10 June 2011*, vol. **2**, paper GT2011-45295, ASME, pp. 311–320.

69 Hollon, B., Steinthorsson, E., and Mansour, A. (2011) Ultra-low emission hydrogen/syngas combustion with a 1.3 MW injector using a micro-mixing lean-premix system, in *ASME Turbo Expo 2010: Power for Land, Sea and Air, Vancouver, Canada, 6–10 June 2011*, vol. **2**, paper GT2011-45929, ASME, pp. 827–834.

70 Tacina, R., Wey, C., Laing, P., and Mansour, A. (2002) Sector tests of a low NOx, lean-direct-injection, multipoint integrated module combustor concept, in *ASME Turbo Expo 2010: Power for Land, Sea and Air, Amsterdam, The Netherlands, 3–6 June 2002*, vol. **1**, paper GT2002-30089, ASME, pp. 533–544.

44
Hydrogen Hybrid Power Plant in Prenzlau, Brandenburg

Ulrich R. Fischer, Hans-Joachim Krautz, Michael Wenske, Daniel Tannert, Perco Krüger, and Christian Ziems

44.1
Introduction

In an environmentally concerned world hydrogen is considered to become one of the most important future energy carriers. With pressure to reduce the emission of carbon dioxide especially the production of hydrogen from plentiful renewable energy sources such as wind, solar, biomass, or hydropower offers an extraordinary low carbon footprint. The increase of renewable energies could promote the production of green hydrogen and initiate future markets like green mobility, large-scale renewable energy storage, and green hydrogen for industries. In recent years in Germany there has been a rapid rise of production capacity of renewable energies especially from wind turbines, photovoltaic installations, and biomass plants. The installed nominal power of wind turbines is actually (June 2013) 32.6 GW and of photovoltaic installations 32.9 GW [1]. This is a consequence of the successful guaranteed power-energy-tariff for feeding in electricity from renewable sources. The renewable electric energy production in Germany is about 136 TWh per year, which represents a share of 22% of the yearly demand [1]. Due to the fluctuating character of renewable energy sources there is often a time shift between production and consumption. Furthermore, we are faced with a strong local discrepancy between production of renewable energy and demand. In the southern and western regions of Germany with dense population and industry there is a high electricity demand. In contrast, especially in the northern or eastern parts of Germany, which have more installed wind power but less population and industry, there is often a surplus of electricity.

This temporal and local imbalance has a serious impact on the requirements for grid stability. As a consequence there is a demand for additional construction of power transmission lines. The situation in terms of concentration of renewable energy production capacity is extraordinary in the Federal State Brandenburg, with an actual installed wind power of (June 2013) 4.9 GW and photovoltaic power of 2.6 GW [2]. The renewable electricity production is 14.2 TWh per year, which represents 75% of the yearly energy demand of the Federal State Brandenburg.

Hydrogen Science and Engineering: Materials, Processes, Systems and Technology, First Edition.
Edited by Detlef Stolten and Bernd Emonts.
© 2016 Wiley-VCH Verlag GmbH & Co. KGaA. Published 2016 by Wiley-VCH Verlag GmbH & Co. KGaA.

To meet the requirements for grid stability different measures have been considered. First of all, further grid expansion is necessary. This is true for local medium voltage transmission lines with up to 110 kV as well as for the high voltage power grid with 220/380 kV. Furthermore, the grid could be relieved by demand side management. Eventually, a very powerful measure for grid balance is the intermediate storage of excess renewable energy. The necessary maximum positive balance power in Germany is estimated to be in the range of 8.0–10.5 GW and the negative balance in the range of 6.8–8.6 GW in 2025 [3].

Driven by these demands, as far back as 2007 the German company Enertrag AG started to develop the concept of the hybrid power plant for the intermediate storage of wind energy as hydrogen. The hybrid power plant project was started with involvement of the companies Total Deutschland GmbH, Vattenfall Europe Innovation GmbH, and Deutsche Bahn AG and with additional funding support by the Federal State Brandenburg. The cornerstone for the plant was laid in Dauerthal in the northeast of Berlin in April 2009 and operation started in October 2011.

The development of the hybrid power plant by Enertrag AG was based on long-term experience especially in alkaline water electrolysis and wind turbine–electrolyzer coupling. Indeed, scientific cooperation between Enertrag AG and the University of Applied Sciences Stralsund started in this field several years ago. At the Stralsund University the first German wind–photovoltaics–electrolyzer system was installed in 1995 [4]. The main components are a 100 kW wind turbine, a 10 kW photovoltaic installation, a 20 kW alkaline pressure electrolyzer at 25 bar, and a 200 Nm^3 hydrogen storage unit. The operational experience of this experimental scale system was conveyed to the upscaled and further expanded hybrid power plant of Enertrag AG.

The main target of the hybrid power plant is to transform wind power into hydrogen, which is much easier to store, and to provide predictable feed-in of renewable energy into the electric grid. In this way renewable energy can be 100% flexible and can be used when and where it is needed.

Another important target is the future demand for carbon-free fuel for mobility. In this respect, the market for fuel cell vehicles is also addressed by the concept of a hybrid power plant. According to Reference [5] hydrogen could deliver up to 40% of the energy demand for mobility in Germany in 2050 and up to 60% of this hydrogen could be produced from renewable energy sources. With this approach a cost competitive hydrogen mobility and a distinct carbon dioxide reduction in this sector is feasible.

A main issue with hydrogen fuel cell vehicles is the infrastructure required to deliver the hydrogen. Actually, the hybrid power plant is a direct hydrogen supplier for the first CO_2 free hydrogen filling station of Total in Berlin and therefore a first step for the emergence of a green hydrogen filling station infrastructure. The wind-produced hydrogen will also be available at the filling station at the new Berlin-Brandenburg International Airport due to open in the near future.

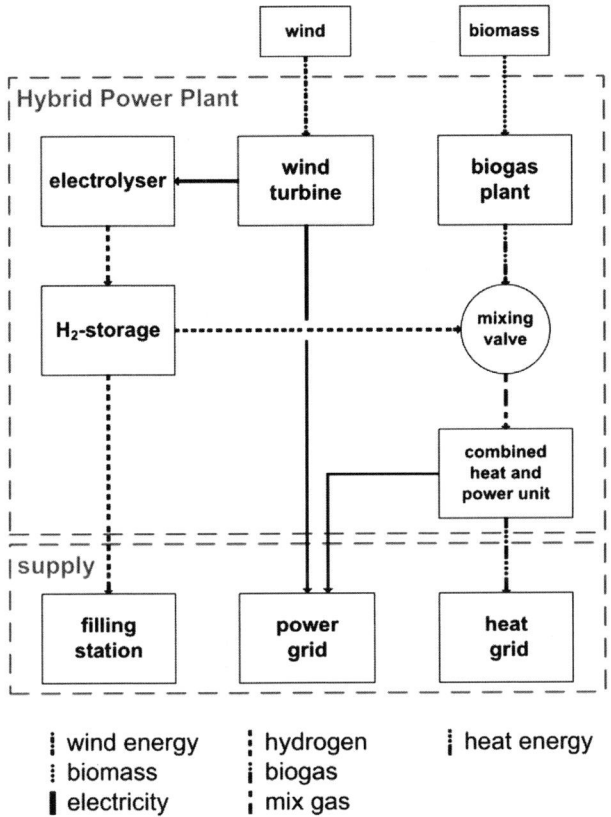

Figure 44.1 Main components of the hybrid power plant.

44.2
Description of the Concept of the Hybrid Power Plant at Prenzlau

44.2.1
Overview

The configuration of the entire system (hybrid power plant) ensures the following functions:

- production of hydrogen by electrolysis;
- generating electrical and thermal energy;
- predictable and homogenized energy supply to the electrical grid;
- increased reliability of forecasts of energy supply to the electrical grid.

The hybrid power plant consists of the main components (Figure 44.1).

A view of the hybrid power plant is given in Figure 44.2. Three wind turbines, 2.3 MW each, are directly electrically connected via a medium-voltage cable

Figure 44.2 Hybrid power plant Prenzlau.

(20 kV) to the electrolyzer. This medium-voltage cable is integrated into the medium-voltage grid, which is directly connected to the 220 kV high-voltage transmission grid of 50 Hertz Transmission GmbH by the transformer station Bertikow.

The electrolyzer operates on the principle of atmospheric alkaline electrolysis. Its installed power is 600 kW with a production capacity of 120 Nm^3 h^{-1} of hydrogen and 60 Nm^3 h^{-1} of oxygen. The hydrogen is compressed by a low-pressure reciprocating compressor to 50 bar. A detailed description of the electrolyzer is given in the next section.

The hydrogen storage consists of three stationary tanks with a maximum pressure of 42 bar. The total capacity of the storage is 1350 kg. A high-pressure diaphragm compressor allows a trailer to be filled at 200 bar for transport to a hydrogen filling station.

The biogas plant will be equipped with 33 tons of corn silage per day and produces about 350 Nm^3 h^{-1} of biogas. The chemical output of the biogas amounts to 1540 kW.

For the conversion of biogas and hydrogen, two CHP-units were installed and connected to the power and heat grid. Each CHP-unit delivers a maximum of 366 kW electrical power and 409 kW heat. The heat of one CHP-unit is fed into the district heating pipeline of the city Prenzlau. The maximum possible hydrogen-share in the CHP fuel gas is 70%. Table 44.1 summarizes the main specifications.

44.2.2
Alkaline Electrolyzer

A photograph of the atmospheric electrolyzer with the stack and gas separators visible is shown in Figure 44.3. The electrolyzer system consists of several components, which are depicted in Figure 44.4. The stack was designed according to the filter press principle with a bipolar cell configuration and includes 70 single cells. The electrodes were coated by the VPS process. Potassium hydroxide is

44.2 Description of the Concept of the Hybrid Power Plant at Prenzlau

Table 44.1 Specifications of the main components.

Integral components	Specification
Electrical grid	• Local medium voltage grid: 20 kV
Wind turbines	• Cumulated power: 6.9 MW (each 2.3 MW)
Electrolyzer	• Power input: 600 kW; • gas output: 120 Nm^3 h^{-1} hydrogen, 60 Nm^3 h^{-1} oxygen
H$_2$-storage tank	• Storage volume: 1350 kg (at 42 bar)
Combined heat and power (CHP)	• Two CHP-units; • co-firing of biogas and hydrogen: up to 70% hydrogen; • maximum power output (each unit): 366 kW$_{el}$/ 409 kW$_{therm}$
Biogas plant	• 33 t per day corn silage; • 350 m^3 h^{-1} biogas

used as electrolyte at operation temperatures between 75 and 80 °C. The maximum hydrogen production is 122 Nm^3 h^{-1} at a current density of 2.8 kA m^{-2}. The electrical power is delivered to the electrolyzer via a transformer and rectifier unit, which is connected to the medium-voltage grid. The rectifier can be controlled between 0 and 4200 A.

Applying a DC voltage causes water splitting into hydrogen and oxygen. Both gases rise into two single separators, where they are separated from the liquid potassium hydroxide. The oxygen is vented to the atmosphere. The hydrogen is

Figure 44.3 Photograph of the alkaline electrolyzer.

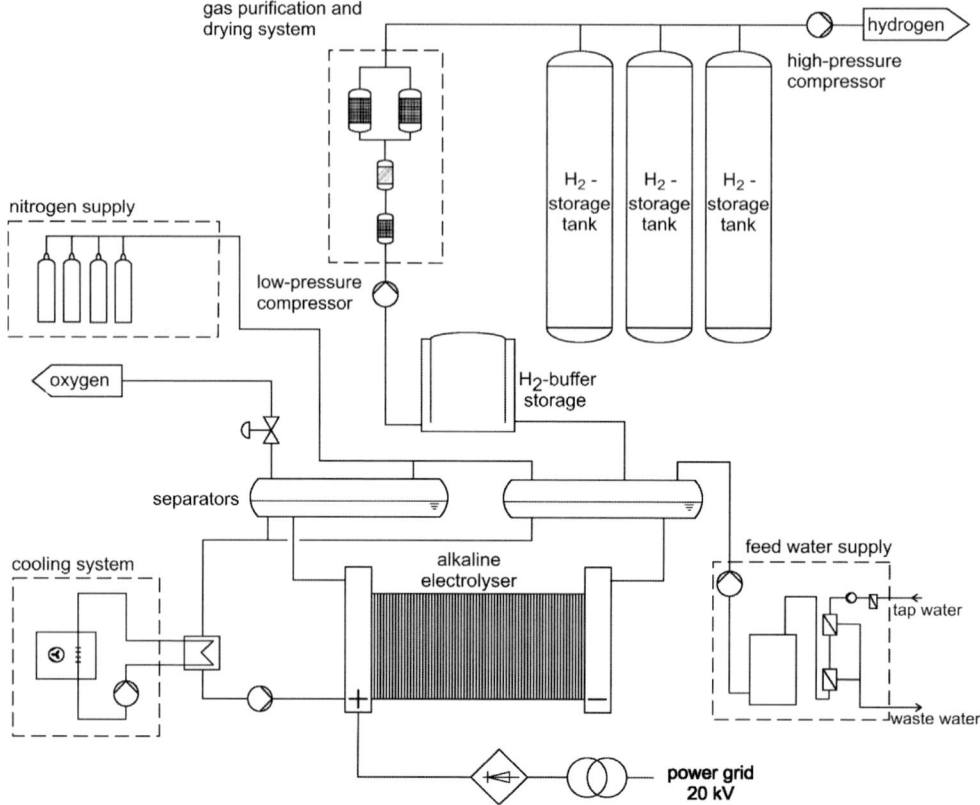

Figure 44.4 Main components of the alkaline electrolyzer.

fed into a buffer tank and subsequently compressed by a low-pressure compressor. At a pressure of 42 bar$_g$ the hydrogen flows through the gas purification and drying system and is cleaned up to a purity of 5.0 (99.999%). The purified hydrogen is stored in hydrogen tanks and can be compressed with a high-pressure membrane compressor for transport.

The feed water treatment is carried out in three steps. First, possible suspended solids are removed by filters from the tap water. In the second step, the water passes through a reverse osmosis process. The filtered water is separated into solutions with high and low concentrations by pressure and a semi-permeable membrane. Finally, the permeate passes through an ion-exchanger to remove remaining anions and cations via two different ion-exchange resins. With a yield of 50%, conductivities of less than 1 µS are achieved.

The cooling water system provides cooling capacity for several circuits. The main electrolyte cooling circuit upholds the electrolyzer stack operating temperature. Further cooling circuits are used to cool the oxygen and hydrogen gases after they leave the separators. The transformer and rectifier also require a cooling circuit.

Table 44.2 Specifications of the alkaline electrolysis system.

Integral components	Specification
Rectifier	• Nominal current: 0–4200 A
Cell stack and separators	• 70 Single cells; • filter press design with bipolar cell configuration; • maximum hydrogen output: 122 Nm^3 h^{-1}; • current density: 2.8 kA m^{-2}; • gas purity: 99.9 vol.% (without additional gas conditioning); • working temperature: 75–80 °C; • electrolyte: potassium hydroxide (KOH); • electrode activation (cathode): VPS-coating
Gas scrubber	• Pressure limitation and compensation oxygen versus hydrogen; • water trap: potassium hydroxide removal
Buffer tank	• Nominal volume: 10 m^3; • operating pressure: 17 mbar; • maximum pressure: 45 mbar
Low-pressure compressor	• Reciprocating compressor (2pce, each 60 Nm^3 h^{-1}); • product gas pressure: up to 42 bar$_g$
Gas purification and drying system (hydrogen)	• Potassium-hydroxide filter, catalytic oxygen removal, heat-generated adsorber; • output gas quality: 5.0 (99.999 vol.% pure hydrogen)
Storage tank	• Storage capacity: 1350 kg hydrogen; • storage pressure: 42 bar$_g$
High-pressure compressor	• Membrane compressor • product gas pressure: up to 230 bar • capacity about 600 Nm^3 h^{-1}

The nitrogen supply is used for inertization of the piping system before restart of the electrolyzer after long plant shutdowns or emergency stops. Table 44.2 shows the main specifications of the alkaline electrolyzer.

44.2.3
Safety Engineering

Since the hybrid power plant could be exposed to explosive gas mixtures, the German "Regulation on Industrial Safety and Health" (BetrSichV) requires the analysis of hazards and the creation of an explosion protection document to avoid these risks or to deal with them.

The overall concept of the hybrid power plant is still in the demonstration stage. Therefore, an actual overall analysis integrating the surrounding infrastructure and including considerations for operational safety will take place after detailed operating experience has been gained and the analysis of all documentation will then be completed. Operating experience and resulting data sets enable a realistic view on electrolyzer based risk analysis. This analysis can be derived from the probability of occurrence and quantifiable periods of availability of individual system components.

In the case of atmospheric operated alkaline electrolyzers, the potential hazards are mainly related to the corrosive properties of the potassium hydroxide solution and the formation of flammable and explosive gas mixtures, for example by permeation of the respective gases through the diaphragm. Recent research efforts at the Brandenburg Technical University Cottbus-Senftenberg (BTU C-S) aimed to use the internal pressure caused by the process in the alkaline electrolysis technology to make use of product gases at pressures up to 58 bar without additional compressors [6]. By shifting the explosion limits of hydrogen mixtures, the pressure in pressurized alkaline electrolyzers is another important process variable in safety assessment [7]. The explosion limits of hydrogen in air under standard conditions are 4–75 vol.%. The advantage of atmospheric pressure alkaline electrolysis technology compared to high-pressure alkaline electrolyzers is the less complex construction and sealing.

The operating pressure of the installed alkaline electrolysis system is 30 mbar above atmospheric pressure in order to prevent the intrusion of foreign gas contents into the plant and thus counteracting the formation of explosive mixtures.

In addition, any leaks that cause a pressure drop in the system are easier to detect. During an inspection of the pipes for leaks, the excess pressure is also helpful in finding leaks in detachable pipe connections.

For the startup of the electrolyzer, some adapted procedures were implemented. Thus, the system pressure before starting the system is checked and is verified to be below a predefined threshold. It is mandatory that the electrolysis system is purged with nitrogen. In the initial phase a 15 min production period of hydrogen and oxygen with a current up to 1800 A is started to release contaminated gas mixtures to the atmosphere. During the hydrogen-purge, the gas stream is discharged through the gas scrubber and the connected vent lines to the atmosphere. After 15 min, the gas quality is checked by the sensor. The production process is started when the hydrogen gas stream reaches a contamination below 1.8 vol.% oxygen. After reaching this value the valve to the buffer tank and the subsequent compression and storage path are opened.

Within the electrolysis system, the formation of explosive gas mixtures is prevented by separating the anolyte from the catholyte through the membrane. The volatile nature of renewable energies with minimum load feeds leads to an increase in foreign gas concentrations in the product gases. To detect unwanted cross overs of the respective gases, in both the path of oxygen production and of hydrogen production, sensors are installed. Any contaminants will usually first become noticeable in the oxygen path by an increased hydrogen concentration resulting from the high diffusion coefficient of hydrogen. If the threshold value is exceeded, shutdown of the production process is started. Subsequently, the detected foreign gases are purged from the system by means of an adjusted procedure.

The gas produced in the electrolysis cells is a two-phase mixture of gas and electrolyte and flows owing to the buoyancy force of the gas bubbles [8] to the gas separators located above the stack. The separators are half filled with electrolyte in normal operation mode. According to the stoichiometric ratios double the hydrogen volume compared to the oxygen volume is fed by the gas–electrolyte stream from the cells to the hydrogen separator. The different volumes of gases in the respective separators cause a pressure increase, and in turn a reduction of the lye levels in the ratio of approximately 1 : 2. Because the separator pressures are directly transmitted to the cells, large level shifts and a high differential pressure should be avoided to prevent a diaphragm gas breakthrough and thus the formation of explosive gas mixtures. The response of the filling levels must therefore be continuously monitored and compensated by means of process control by partial pressure reduction. After passing a gas cooling unit to remove residual aerosols, the product gases flow into the scrubber. Here, residual traces of potassium hydroxide are removed from the gas. Furthermore, balancing of the pressure difference between the two gas channels is ensured by special system design and control procedures. The control concept is therefore linked to both the levels in the separators and the levels in the scrubber.

Regarding the design of the electrolyzer, the arrangement of the membrane electrode assembly was chosen based on the proven principle of the filter press configuration. Here, the single cells are connected electrically in series and by means of tie rods are pressed between two end plates. The anode and cathode spaces are air-tightly separated with this system by means of gasket seals and the pressure of the end plates. To avoid the establishment of an explosion protection zone in the entire area, a large sensor system and ventilation system with an integrated flow monitoring was installed. Using this system, the necessary air exchange rates can be realized both in the case of operation and an accident. For this, fresh air is fed into the explosion protection zones inside. In addition, passive ventilation roof openings are installed, which ensure a constant air exchange and discharge any accumulation of hydrogen in the roof area.

The safety evaluation of the compressors in the machine hall was made in accordance with the German standard BGR 104 Section 1.2.4.3.1 (b2). Through adequate protection mechanisms according to TRBS 2152 Part 2, the machine hall is expected to have explosion protection zone II. Explosion protection zone I designates occasionally hazardous areas whereas the less hazardous zone II designates only rarely or temporarily hazardous areas. The room fail-safe system is equipped with sensors and alarm devices.

In the case of a leakage in the explosion protected areas, the hydrogen sensor at 20% of the lower explosive limit (LEL) of hydrogen activates a pre-alarm. At a concentration of 40% of the LEL an emergency stop of the electrolysis system with activation of the forced ventilation will be initiated. Moreover, the subsequent compressors are switched off, the gas supply is interrupted, and a pressure reduction would be carried out through the vent lines.

In the exterior of the operation building, there are openings of vent lines and in the field of hydrogen storage, the explosion protection zones I and II are defined. In the field of the biogas fermenter tank, the designation of an explosion protection zone II is sufficient.

44.3
Operating Modes of the Hybrid Power Plant

The hybrid power plant is capable of operating in four different modes depending on the demand for hydrogen, predictable feed in of electricity, or with respect to economic issues of the electrical power market. A superior level central control is to monitor the single component controls of the wind-turbine, the electrolyzer, the mixing valve, and the CHP-plant to ensure the proper functioning of the modes. The four different modes are described in the following subsections.

44.3.1
Hydrogen Production Mode

In operating mode "hydrogen production" (Figure 44.5) the hybrid power plant is operated as a hydrogen production plant. The aim is to produce a maximum amount of hydrogen or at least a guaranteed amount of hydrogen with the available energy determined by the wind profile. The hydrogen can then be used as fuel in the area of mobility or for industrial processes. Figure 44.5 illustrates this mode with a real plot of the input power of the wind-turbines (performance WP) and the almost constant hydrogen production power. The residual wind energy is fed directly to the electric grid.

The advantage of this mode is CO_2-neutral hydrogen production. In addition, the hydrogen can be provided locally with associated benefits, such as saving expensive transport by water, road, or rail.

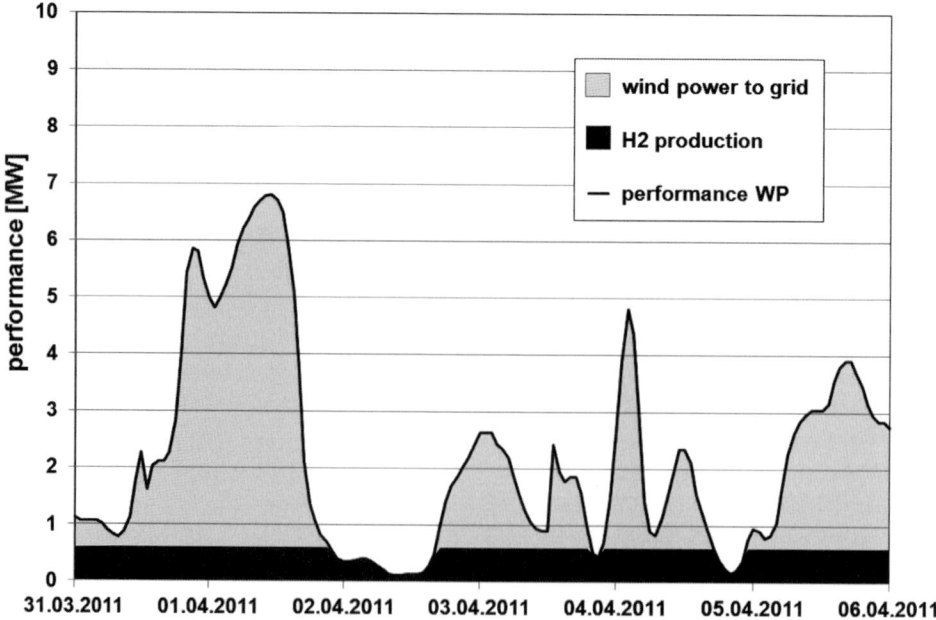

Figure 44.5 Example of hydrogen production mode.

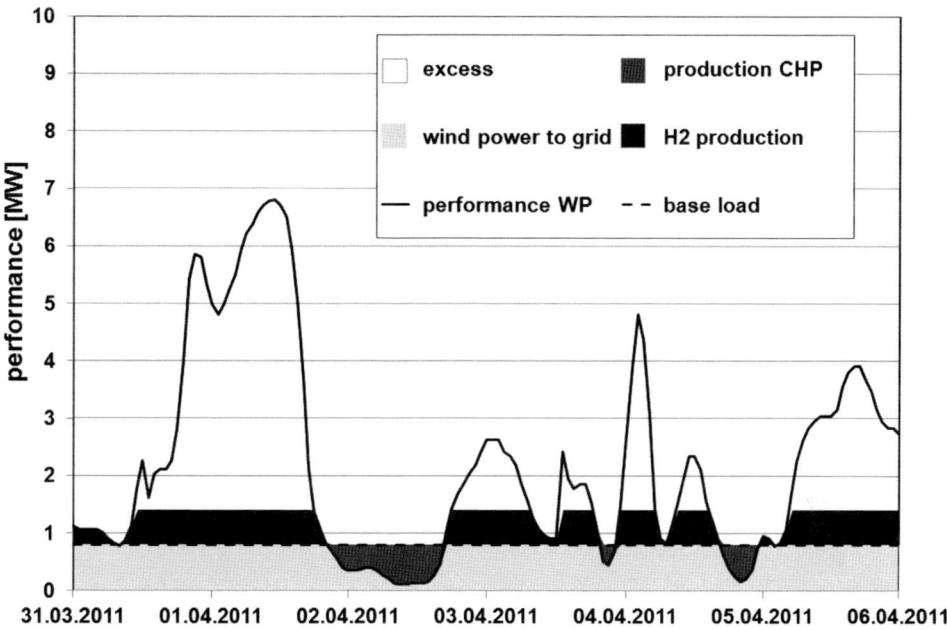

Figure 44.6 Example of base load mode.

This mode of operation is characterized by the following points:

- constant daily supply of hydrogen;
- the main decision criteria is the level of hydrogen storage;
- maximum utilization of the electrolyzer;
- the electrolyzer is operated with the lowest output, if the tank is almost filled;
- at maximum tank level the electrolyzer is on stand-by;
- CHP is not used with mixed hydrogen–natural gas.

44.3.2
Base Load Mode

The "base load" operation mode (Figure 44.6) is intended to guarantee a constant electric power (base load) irrespective of the input wind power. The hybrid power plant mode compensates the fluctuations in the wind profile. Wind energy exceeding the demand is converted into hydrogen and stored. During low power wind periods, hydrogen and biogas are mixed and used for electricity production by CHP-engines. With respect to the electric power grid, the hybrid power plant works as a base load power plant. As shown in Figure 44.6 only a constant base load (dotted line) is supplied to the electrical grid.

The main benefit of this mode is the decoupling of the electric power supply network from the fluctuating wind conditions. The power supply from wind

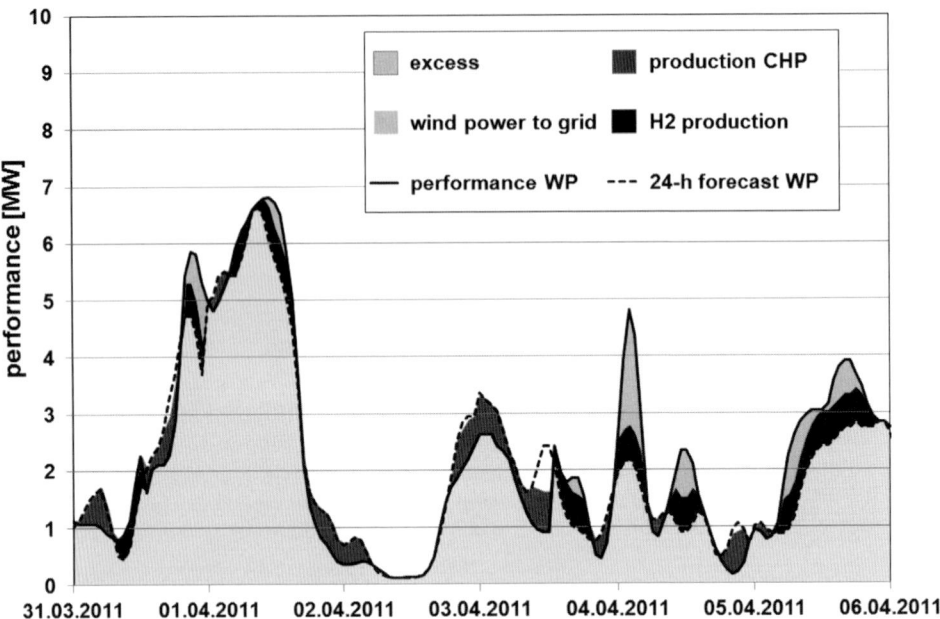

Figure 44.7 Example of forecast mode.

turbines becomes predictable by a connected hybrid power plant and it results in an effective utilization of the electrical grid capacity.

This mode of operation is characterized by the following points:

- Constant power is delivered over a defined period;
- The main decision criteria is the guaranteed electrical output power;
- If the current wind turbine performance exceeds the guaranteed output power, the excess power is fed into the electrolyzer;
- Further excess power can be limited by pitching (power reduction of wind turbines by adjusting the rotor blades);
- If the current wind turbine performance falls below the guaranteed power output, the power deficit is balanced by CHP.

44.3.3
Forecast Mode

When operating in "forecast" mode (Figure 44.7) an accurate match of the forecasted power of the hybrid power plant is the main target. For a definite period of up to 24 h in advance a forecast value is given for the power of the hybrid power plant. The values are based on the wind forecast values and the maximum control potential of the hybrid power plant. In the following period the plant tries to reach these default values. Due to the almost hundred percent assurance of the forecasted performance, the network operation can be planned.

Figure 44.7 gives an impression of the forecast mode with real wind data and simulated hybrid power plant components. Plotted are the forecast power to be exactly delivered by the hybrid power plant (dashed line) and the real wind-turbine power (continuous line). If the forecast exceeds the real wind power, the deviation is compensated by CHP. If vice versa the wind power exceeds the forecast, the electrolyzer produces hydrogen. As shown in Figure 44.7 the hybrid power plant is capable of following the forecast for almost the whole time. There are some time periods with residual excess energy from the wind turbine (see, for example, 04.04.2011) and some rare time periods with insufficient power to cover the forecast (see, for example, 03.04.2011, around noon).

This mode of operation is characterized by the following points:

- main criteria is the exact power output according to the wind power forecast;
- forecast is issued for up to 24 h in advance;
- performance forecast of the hybrid power plant takes into account the internal losses and the current hydrogen storage level.

44.3.4
EEX Mode

This mode takes into account the achievable prices for electricity, for example at the European Energy Exchange (EEX). When a high price can be obtained for wind power, the hybrid power plant feeds into the grid. If the price for wind power is under a minimum price, electricity will be used for hydrogen production and fed directly into the grid when the hydrogen storage capacity is exceeded.

Figure 44.8 illustrates the situation with assumed high prices during low wind power periods and low prices during high wind power periods. The CHP-systems operate at high price periods (see, for example, 02 April 2011) in this example simulation.

This mode of operation is characterized by the following points:

- main criteria is the market price for electricity;
- forecast is given 24 h in advance.

44.4
Operational Management and Experiences

44.4.1
Dynamic Load Operation

A prerequisite for wind–electrolyzer coupling is a good dynamic behavior of the electrolyzer. Therefore, long-term operation tests concerning the dynamic response of the electrolyzer were performed. Figure 44.9 gives an example of

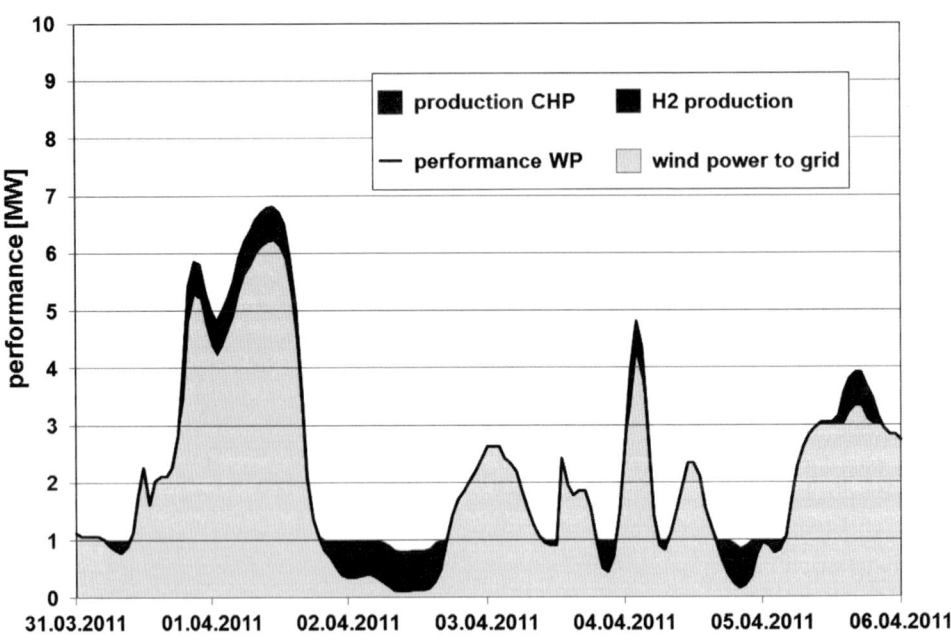

Figure 44.8 Example of EEX mode.

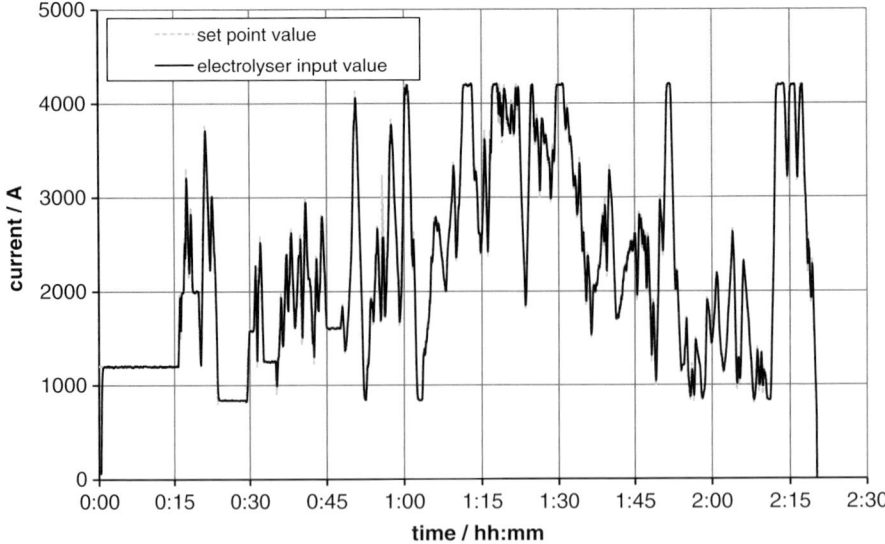

Figure 44.9 Dynamic load performance – set point value and electrolyzer input current against time.

the load characteristic of the electrolyzer against time during direct wind turbine coupling. The graphs show the set point value determined by the excess energy of the three wind turbines (wind power exceeds the forecast) and the input value of the electrolyzer.

As illustrated in Figure 44.9, the electrolyzer is capable of following input transients with a current slope of 70 A s^{-1}. The maximum current is 4200 A, which corresponds to a current density of 2.8 kA m^{-2}. A further increase of the current slope is feasible and under test. During the transient operation mode a continuous monitoring of the gas quality of both oxygen and hydrogen was performed. As depicted in Figure 44.4 the electrolyte streams from the separators are joined before recirculating into the stack to compensate KOH concentration differences. Due to the dissolved oxygen and hydrogen in the confluent electrolyte stream we get hydrogen impurities at the anode sides and oxygen impurities on the cathode sides of the cells. In case of partial load operation with lower currents and lower gas production the ratio of impurity gas to produced gas increases. Hence the partial load operation is critical, because we have to limit the hydrogen content in oxygen to well below the lower explosion limit of 4.0 vol.%. With respect to this critical limit an electrolyzer partial load of 20% was reached with a hydrogen content below 2.0 vol.%. During the experimental electrolyzer operation of about 2100 h no significant degradation of the stack voltage was measured.

44.4.2
Temperature Influence on Stack Voltage

For further comparison with measurement results some equations concerning the temperature dependence of the cell voltage will be provided. The overall effective single cell voltage (V_{cell}) of an alkaline electrolyzer is the sum of the reversible cell voltage (V_{rev}), the activation overpotentials V_{an} and V_{cath} at the anode and the cathode, respectively, and the voltage drops across the ohmic resistances of the cell (V_{ohm}):

$$V_{cell} = V_{rev} + V_{an} + V_{cath} + V_{ohm} \tag{44.1}$$

At sufficiently high current densities there is theoretically an approximately linear dependence of the overpotential on the current density. The voltage V_{ohm} itself is represented by the sum of the voltage drops due to the ohmic resistances of the electrolyte with bubbles ($V_{ohm,KOH}$) at anode and cathode side and the voltage drop at the diaphragm ($V_{ohm,dia}$):

$$V_{ohm} = V_{ohm,KOH} + V_{ohm,dia} \tag{44.2}$$

The decrease of the reversible cell voltage (V_{rev}) with increasing temperature (T) is theoretically determined by the temperature dependence of the change of the Gibbs free energy of reaction (ΔG) with:

$$\frac{\partial V_{rev}}{\partial T} = \frac{1}{2F}\frac{\partial \Delta G}{\partial T} = -\frac{\Delta S}{2F} \tag{44.3}$$

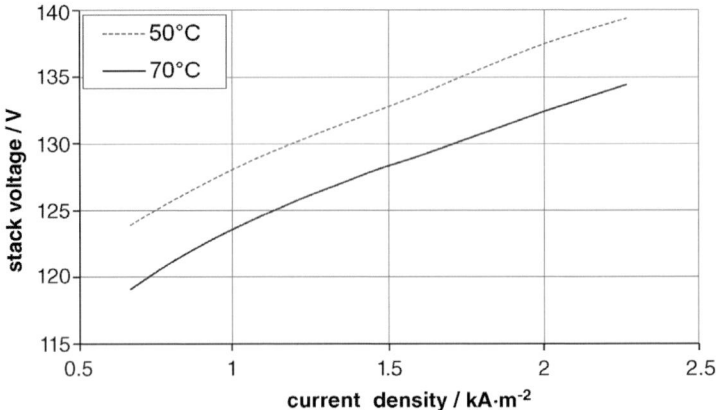

Figure 44.10 Temperature dependence of stack voltage. Stack with 70 single cells.

With the entropy of reaction (ΔS) and the Faraday constant (F) it is possible to calculate the temperature dependence of the reversible cell voltage in the case of water electrolysis:

$$\frac{\partial V_{rev}}{\partial T} = -\frac{163.25 \text{ J K}^{-1} \text{ mol}^{-1}}{2 \cdot 96485 \text{ A s mol}^{-1}} = -0.845 \frac{\text{mV}}{\text{K}} \quad (44.4)$$

The sum of the voltage drops ($V_{ohm,KOH}$) due to the resistance of the electrolyte can be estimated with the specific conductivity of the electrolyte ($\kappa(T,c)$), the current density (j), and the thickness (d) of the electrolyte layers between the electrodes and the diaphragm according to:

$$V_{ohm,KOH} = \frac{jd}{\kappa(T,c)} \quad (44.5)$$

The specific conductivity of the electrolyte ($\kappa(T,c)$) is a function of temperature (T) and concentration (c). In our special case the concentration of the electrolyte is 28 wt%. With this concentration we obtain a specific conductivity of the KOH solution of about $\kappa = 0.6 \, \Omega^{-1} \text{cm}^{-1}$ at 25 °C and $\kappa = 1.5 \, \Omega^{-1} \text{cm}^{-1}$ at 100 °C using data extracted from References [9] and [10]. A linear approximation yields values within this temperature range.

As shown in Figure 44.10 an overall stack voltage decrease of 4.5–5 V was measured corresponding to a temperature increase of 20 K. This corresponds to a voltage decrease for the single cell (ΔV_{cell}) of 3.4–3.6 mV K^{-1}. Other authors [11,12] reported a decrease of 2.8 mV K^{-1} in the temperature range 15–65 °C but for a pressure electrolyzer at 25 bar and a current density of 0.4 A m^{-2}. As shown above, the temperature dependence of the change of the Gibbs free energy is responsible for only 0.845 mV K^{-1}. The temperature dependence of the specific conductivity of the electrolyte accounts for about 0.9 mV K^{-1} at a current density of $j = 0.2$ A cm^{-2} and 0.45 mV K^{-1} at $j = 0.1$ A cm^{-2} according to formula (44.5). The residual part of the voltage decrease ΔV_{cell} is therefore caused by the temperature dependence of the activation overpotentials at the electrodes and the ohmic resistance of the diaphragm. The temperature dependent measurements confirm the advantage of increasing the electrolyzer operating temperature with a voltage efficiency increase of actually 0.14% per degree temperature increase at 0.2 A cm^{-2} in the range above 70 °C.

44.4.3
Influence of Activated Electrodes

A significant decrease of the activation overpotentials of the electrodes and therefore of the cell voltages is possible by enhancing the catalytic activity of the electrodes. For this purpose cathodes coated with Raney-nickel by vacuum plasma spraying (VPS) [13] are used. The main components within this alloy are nickel, molybdenum, and aluminium. To gain a high porous nickel layer with a high specific surface area an activation treatment is necessary prior to the first electrolyzer operation to remove the aluminium content. This activation process was performed with a solution of potassium hydroxide and potassium sodium tartrate tetrahydrate as described in Reference [14]. After this activation treatment a sponge-like nickel layer was obtained.

To evaluate the influence of both the overpotential of the electrodes and the ohmic overpotential due to the resistance of the KOH solution several measurement series were performed at different temperatures. For comparative purposes the electrolyzer stack was equipped partly with activated Raney nickel and non-activated plain nickel cathodes. The stack consists of a total of 70 single cells with 35 cells activated and 35 cells non-activated.

The voltage measurement was performed separately on both different stack parts. As shown in Figure 44.11 a significant decrease in single cell voltages was observed due to the cathode activation process.

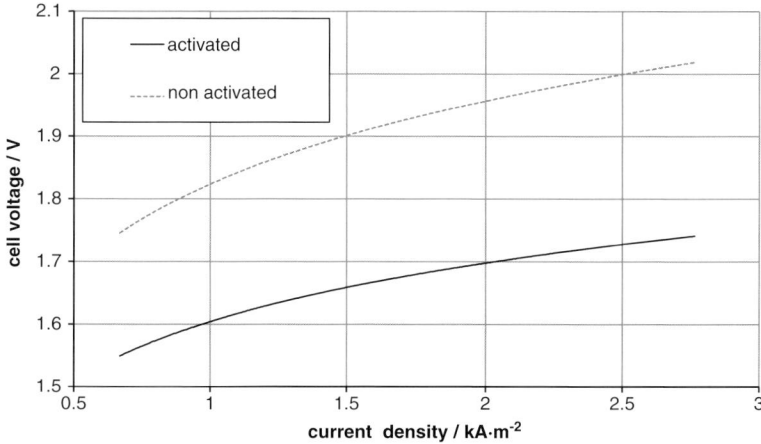

Figure 44.11 Cell voltage versus current density with activated and non-activated cells.

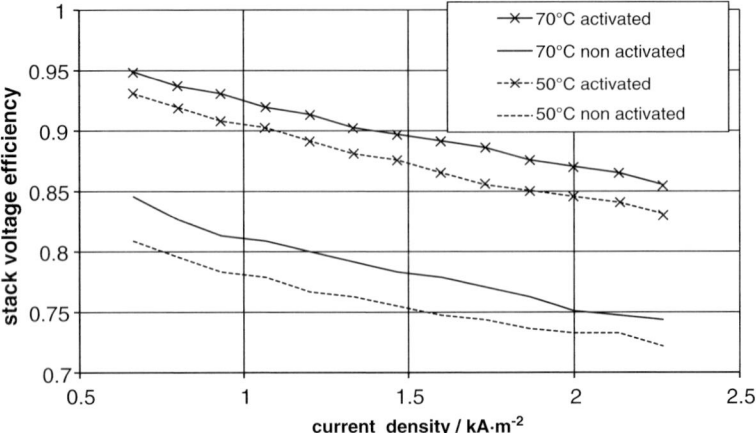

Figure 44.12 Stack voltage efficiency dependence on current density, temperature, and cathode activation process.

44.4.4
Voltage Efficiency of the Electrolyzer

With the voltage measurement data of the two stack parts, a calculation of the voltage efficiency in dependence of the current density, stack temperature, and cathode activation process is possible. The voltage efficiency (η_V) is defined as:

$$\eta_V = \frac{V_{th}}{V_{cell}} = \frac{1.48 V}{V_{cell}} \tag{44.6}$$

Here, V_{th} is the thermoneutral cell voltage at STP and V_{cell} the cell voltage. With formula (44.6) the voltage efficiencies shown in Figure 44.12 are calculated. For the activated stack part a voltage efficiency of 85.5% was attained at 70 °C and a current density of $2.3 \, kA \, m^{-2}$.

44.5
Outlook

After 20 months of operational experience the hybrid power plant has proven to fulfill properly the main function of generating hydrogen by direct coupling with wind turbines. Further attention will be focused on the long-term availability of the system with enhanced reliability and minimal maintenance and service effort. Long-term degradation processes especially of the electrolyzer system are another issue under investigation. An ongoing development is being carried out to refine and optimize the superior level central control of the plant to make the operation modes of the hybrid power plant work satisfactory. This has to be accompanied by long-term tests and evaluation of the operating modes.

The next future plan is the direct injection of hydrogen into a natural gas pipeline operated by ONTRAS. A further development of the hybrid power plant is on the way. A 60 bar pressurized alkaline electrolyzer will be investigated with respect to dynamic behavior and minimal and maximal load. The investigation results will contribute to the development of next electrolyzer models optimized for dynamic operation.

The experience drawn from the operation of the hybrid power plant is intended to be used in subsequent up-scaled projects with an electrolyzer in the megawatt range coupled with a wind farm as well as additional hydrogen underground cavern storage and natural gas pipeline injection.

References

1 DGS (2013) Die Karte der Erneuerbaren Energien – EnergyMap: Bundesrepubik Deutschland http://www.energymap.info/energieregionen/DE/105.html (accessed 8 August 2013).
2 DGS (2013) Die Karte der Erneuerbaren Energien – EnergyMap: Bundesland Brandenburg http://www.energymap.info/energieregionen/DE/105/108.html (accessed 8 August 2013).
3 Hannig, F., Smolinka, T., Bretschneider, P., Nicolai, S., Krüger, S., Meißner, F., and Voigt, M. (2009) *Stand und Entwicklungspotenziale der Speichertechniken für Elektroenergie – Ableitung von Anforderungen an und Auswirkungen auf die Investi-tionsgüterindustrie*, BMWi-Auftragsstudie.
4 Gamallo, F., Lehmann, J., Luschtinetz, O.u.T., Miege, A., and Sponholz, C. (2007) Hydrogen production as a suitable alternative for grid-balancing. Sixth World Wind Energy Conference, Mar del Plata, Argentina, 2–4 October 2007.
5 Joest, St., Fichtner, M., Wietschel, M., Bünger, U., Stiller, C., Schmidt, P., and Merten, F. (2009) Woher kommt der Wasserstoff in Deutschland bis 2050? Studie im Auftrag des BMVBS.
6 Ziems, C., Tannert, D., and Krautz, H.-J. (2012) Project presentation: design and installation of advanced high pressure alkaline electrolyzer-prototypes. *Energy Procedia*, **29**, 744–753.
7 Schröder, V., Emonts, B., Janßen, H., and Schulze, H.-P. (2003) Explosionsgrenzen von Wasserstoff/Sauerstoff-Gemischen bei Drücken bis 200bar. *Chem. Ing. Tech.*, **75** (7), 914–918.
8 Mat, M.D., Aldas, K., and Ilegbusi, O.J. (2004) A two-phase flow model for hydrogen evolution in an electrochemical cell. *Int. J. Hydrogen Energy*, **29**, 1015–1023.
9 Häussinger, P., Lohmüller, R., and Watson, A.M. (2009) Hydrogen, in *Ullmann's Encyclopedia of Industrial Chemistry*, vol. 7, Wiley-VCH Verlag GmbH, Weinheim, p. 41.
10 Bayer, W. (2000) Entwicklung alternativer Elektroden und Aktivierungskonzepte für die alkalische Hochleistungselektrolyse. Thesis, University of Ulm.
11 Ursúa, A. and Sanchis, P. (2012) Static-dynamic modeling of the electrical behavior of a commercial alkaline water electrolyser. *Int. J. Hydrogen Energy*, **37**, 18598–18614.
12 Diéguez, P.M., Ursúa, A., Sanchis, P., Sopena, C., Guelbenzu, E., and Gandia, L.M. (2008) Static-dynamic modeling of the electrical behavior thermal performance of a commercial alkaline water electrolyzer: experimental study and mathematical modeling. *Int. J. Hydrogen Energy*, **33**, 7338–7354.
13 Schiller, G. and Borck, V. (1992) Vacuum plasma sprayed electrodes for advanced alkaline water electrolysis. *Int. J. Hydrogen Energy*, **17**, 261–273.
14 Schiller, G., Henne, R., and Borck, V. (1995) Vacuum plasma spraying of high-performance electrodes for alkaline water electrolysis. *J. Therm. Spray Technol.*, **4**, 185–194.

45
Wind Energy and Hydrogen Integration Projects in Spain

Luis Correas, Jesús Simón, and Milagros Rey

45.1
Introduction

Wind power installed in Spain at the end of 2012 reached 22 785 MW, making it the third power generation technology after coal fired thermal power plants and nuclear energy, with a production of 48 156 GWh and electricity demand coverage of 17.41%. Such power is very well accepted by society as this wind electricity avoids emissions of 22 million of CO_2 tons, contributes directly or indirectly to the GDP (M€2623 or 0.24%), gives employment to 27 000 people, and is an export amounting to M€1893. This international leadership is mainly due to development of the national wind industry with companies such as Gamesa, Acciona, the former Ecotecnia, or MTorres, who have taken the lead in the manufacture and development of wind turbines, while utilities such as Iberdrola and many IPPs promote new wind farms both in Spain and abroad (Asociación Empresarial Eólica, http://www.aeeolica.org/es/sobre-la-eolica/la-eolica-en-espana/) [1].

According to the National Renewable Energy Plan in Spain 2005–2010 [2], the foreseen wind power installed should have reached 20 155 MW at the end of the period. This objective was achieved almost one year in advance. The revised Plan for the timeframe 2011–2020 [3] sets targets of 27 869 MW in 2015 and 35 750 MW in 2020 with on-shore as well as off-shore wind power. The various regional plans already intended by 2008 accumulated to 40 968 MW [4]. Between 40 and 42.5 GW of wind power capacity are expected by 2020 in Spain according to the low and high scenarios of EWEA [5], 1–1.5 GW of which could be off-shore capacity. These predictions match well with the objectives of the Spanish Government. However, this big potential for wind power generation can lead to issues with power integration into the grid.

Red Eléctrica de España, the Spanish electricity transport operator, stated in 2008 a feasibility study for the expected 44 000 MW of wind energy in Spain by 2020, provided enough grid reinforcement was put in place and changes in the regulatory framework were undergone [6]. Above all it is absolutely necessary to satisfy the requirements against voltage dips according to the Operational Procedures 12.3 and 12.2 defined by the transport grid operator Red Eléctrica

Hydrogen Science and Engineering: Materials, Processes, Systems and Technology, First Edition.
Edited by Detlef Stolten and Bernd Emonts.
© 2016 Wiley-VCH Verlag GmbH & Co. KGaA. Published 2016 by Wiley-VCH Verlag GmbH & Co. KGaA.

de España (http://www.ree.es/es/actividades/operacion-del-sistema/procedimientos-de-operacion). Moreover, wind energy must provide grid services to ensure the stability and security of the electrical system.

To follow the demand, fast-response and safe reserve power generation systems are needed to ensure the consumption coverage every time. Wind power is hardly able to satisfy this requirement, as sometimes its production profile is opposite to the ramp of the demand curve. Therefore, the availability and reliability to cover peak loads are low. Its variable behavior, sometimes with sharp ramps, the small reduction of the peak in summer and winter, and forecast errors on the order of 20% rising with the prediction horizon time, makes it very difficult to consider wind power as a firm power generation system. Instead it is an energy resource.

To briefly reflect the evolution of these issues some recent data are available. Wind power in 2012 several times exceeded its previous records. On the 18th of April at 16:41 p.m. a new record in instantaneous power was reached with 16 636 MW (12% higher than the previous record on 9th November 2010). That same day the maximum hourly and daily energy was also exceeded with 16 455 MW and 334 850 MWh, respectively. April 2012 showed a new monthly production record with 5362 GWh wind power (8.8% higher than the previous peak in December 2009) and in September a new record in demand coverage with wind power was set when, on the 24th at 3.03 p.m., 64.25% of mainland demand was covered, surpassing the previous high of 61.06% on 19th April of the same year at 1:37 p.m. (these data can be obtained from demanda.ree.es/demanda.html).

The heavy storm winds that occurred in early 2013 together with the heavy rains of March caused complex situations in managing the national grid. An example occurred during the Easter holidays, where the decline in industrial consumption for the holiday period dramatically forced a stop to wind production that the network could not absorb and a reduction by almost 20% of the power output of nuclear power plants in the country. This had not happened for 15 years when installation of the first combined cycle took place and hence nuclear power ceased to be modular.

Difficulties arise when large forecast errors occur. An example in Spain occurred on 11th November 2008 when wind generation fell by more than 4000 MW in 8 h and reached a minimum from which demand started growing. These situations do not occur very often, but the system should be prepared for them. In other examples of extreme ramp rates recorded during storms, an installed power of 11 GW could increase by 800 MW (7%) in 45 min (ramp rate of 1067 MW h^{-1}, 9% of capacity) and decrease by 1000 MW (9%) in 1 h 45 min (ramp rate -570 MW h^{-1}, 5% of capacity).

The fact that the peninsular electricity system remains an electrical island with little interconnection with Europe through France does not help with this type of congestion in the network. Spain, with nearly 20% of its energy coming from wind, has only round 2.5% of interconnection capacity with the UCTE through France, and thereby has to solve issues due to wind on its own (in some cases, Spain and Portugal are considered together as a single system, both sufficiently interconnected and both with a high wind penetration).

Wind energy has some special characteristics that may have a significant effect on the power system, the most important of which are its intermittency and variability. The operation and capabilities of wind energy should be adapted to the system and one should recognize the specific characteristics of the wind energy and adapt its operation and management rules to better integrate this new major energy source, as the following section will analyze in more detail.

45.2
The Role of Hydrogen in Wind Electricity Generation

Wind promoters and manufacturers have worked hard to overcome most of the requirements stated by Power System Operators. Thus, wind turbines have fault ride through capabilities, there are forecasting tools that allow wind to participate in markets assuming the balancing costs, the visibility and controllability of wind power generation have increased, and progress is expected on providing wind farms with voltage and frequency control capabilities or ancillary services. While current technology can, apparently, deal with all the expected improvements they depend, of course, on the availability of wind. The impact of wind energy on the power system affects to the operational security, reliability and efficiency and could be analyzed from a technical and economical point of view. The impacts can be related to three focus areas: balancing, adequacy of power and grid as the IEA WIND task 25 group has shown in some results obtained from this international collaboration [7].

In this sense, hydrogen production, storage, and use in the context of wind–hydrogen systems can involve a wide range of elements to comply with various roles. Figure 45.1 provides a generalized schematic. Re-electrification can take place in gas turbines, fuel cells, or internal combustion engines. Other end uses are possible, such as water desalination. If hydrogen generation is not located

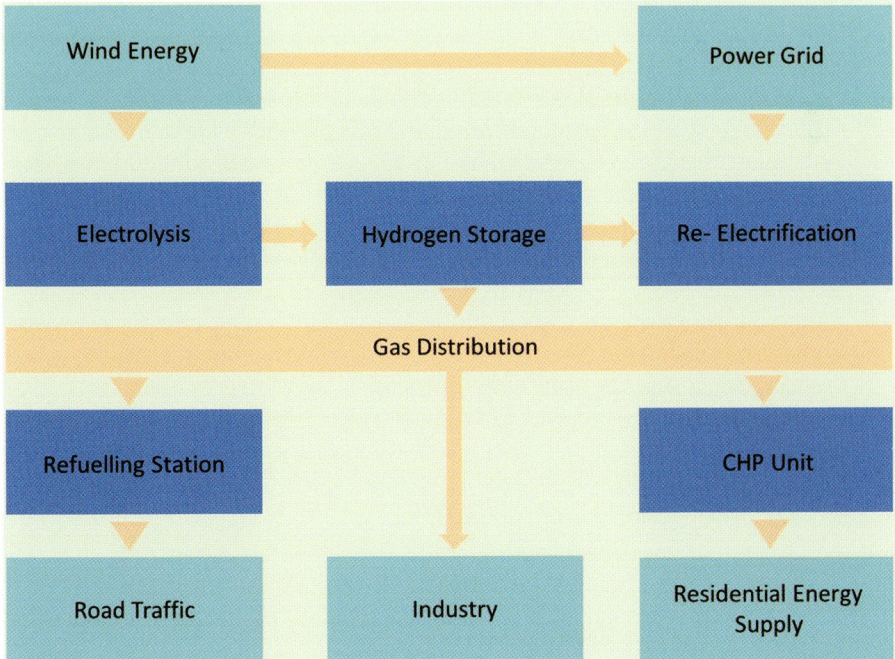

Figure 45.1 Generalized schematic of generating hydrogen from wind energy and using it (CHP, combined heat and power).

next to a wind farm but further "downstream," electrical energy for electrolysis will be taken from the grid, unlike as indicated in Figure 45.1.

IEA HIA Task 24 findings [8] established three main categories of wind–hydrogen systems, which are discussed in the following subsections:

- mini grids
- electricity storage
- fuel production.

45.2.1
Mini Grids

The mini grids category refers to fully or partially islanded systems that include wind energy and typically other (decentralized) power generation. Partially islanded systems in this context are those that have significant constraints with respect to their link to the main grid. The main purpose of hydrogen production is the storage of temporary surpluses of energy from renewable energy and the provision of a demand side management solution for energy supply (the electrolyzer serving as a controllable or dispatchable load). The hydrogen thus produced could be used for any purpose, including as a transport fuel, for re-electrification at times of low renewable generation, water purification, and in industry.

Mini grids serve as a solution to the perennial problem of reliable energy supply in remote areas (matching of supply and demand) and are, therefore, typically on a small scale. Fully islanded systems in remote locations will face a less challenging economic environment thanks to high fuel prices (costs for delivery), high grid connection costs, and, in some cases, tolerance to lower availability. They will also be less affected by energy price variation in markets.

Environmental legislation related to landscape conservation areas or remote regions in general can be a both driver in favor of wind–hydrogen systems and also against them (due to possible bird or bat collisions with wind turbines, etc.). Environmental benefits include less pollution and less noise, elimination of fuel delivery miles, avoiding risk of fuel spillages, and low-carbon energy supply.

Among the competing technologies are diesel generators, battery systems, and pumped hydro. Diesel systems and batteries could also serve as complementary systems.

In Spain two categories of mini-grid systems could be developed. One is related to the peninsula area, where very few electrically isolated areas exist, but where they do energy supplies are usually provided by diesel generators or very small hybrid renewable installations (wind/photovoltaic) with batteries for storage. Islands are the other application for which hydrogen mini-grids are appropriate, an example of which is the RES2H2 project situated in the Canary Islands (see below).

45.2.2
Electricity Storage

The main purpose of systems under this category is to "smooth out" short-term fluctuations in wind power, that is, negative and positive balancing, and produce hydrogen at times of surplus power production and re-electrifying it during periods of underproduction. Such devices could facilitate wind power integration on a large scale, independent of support from fossil-fuel power stations.

Increasing penalties for deviations could create market opportunities here. A wind–hydrogen system will have to compete with conventional (fossil) power plants. However, unlike these, it can provide both negative and positive balancing. Large amounts of hydrogen could also be stored in underground caverns to bridge longer-term (seasonal) variations of wind power production.

Balancing power plants have a low number of annual full-load hours (i.e., a low capacity factor). This is typical of conventional units, usually fueled by natural gas. Simulations carried out in the HyWindBalance project indicate a similar situation for the "matching a day-ahead forecast" case [9]. This will have a negative effect on economic viability. Costs for electricity fed back into the grid could be higher than €1 kWh^{-1}. The electrical efficiency of a power-to-hydrogen-to-power cycle (round trip efficiency) is unlikely to exceed 40% in the future.

Combining "electricity storage" with "fuel production" (see next section) could help to overcome the issue of low capacity factor. On the one hand, the operating strategy of such a plant will become more complex, since it includes additional functionality and hardware. On the other hand, the "tightrope walk" of grid balancing is somewhat eased by the flexibility of operation that the addition of fuel production can offer. Carefully designed criteria will be required for allocating priority to re-electrification or to fuel supply under a given set of circumstances. Therefore, it is hardly feasible to define "standard systems" at present.

The relevance of (large) electricity storage systems awaits the development and maturity of components such as appropriate electrolysers and gas turbines that are suitable for hydrogen. With respect to electrolysers, this refers to cost reduction, efficiency improvement (regarding both stack and balance of plant, including AC/DC conversion), and meeting the requirements of a high degree of intermittency (low capacity factor, stand-by losses), and potentially a requirement to follow rapid power changes in the sub-second range. The latter could be solved or mitigated by combination with ultracapacitors or battery systems.

Technically and economically optimal sites will depend on the structure of the grid that the wind–hydrogen system is intended to support (i.e., to stabilize). In a closely meshed system, electrolysis and re-electrification could take place virtually anywhere. In the case of significant bottlenecks between renewable generation and centers of consumption, energy buffering would have to take place upstream of such points.

Among the competing technologies (adiabatic) compressed air systems and pumped hydro appear to be the most and probable ones to contend with (see Reference [10]). Around 2400 MW reversible hydro power plants and, as of 2010, 20 000 MW wind power capacity are installed in Spain. The interconnections with France are equivalent to 2.5% of total grid capacity, whereas a figure of around 10% would be desirable. The National Renewable Energy Plan in Spain 2011–2020 [3] sets targets for nearly 36 GW of wind power and also 8.8 GW of reversible pumped hydro, with a clear goal of achieving short-term grid balancing. According to more recent studies [11], Spain has the biggest potential in the EU for pumped hydro, amounting to 17.6 TWh under some realizable assumptions.

Examples of this "electricity storage" configuration are both the Hidrólica project and Sotavento, both of which are situated upstream of the network, directly at the wind farm.

45.2.3
Fuel Production

The main purpose of facilities under this category is in supplying hydrogen fuel to road vehicles. The simplest mode of electrolysis operation would be to produce and store hydrogen continuously on a 24 h a day/7 days per week basis to satisfy the average fuel demand. However, this plays no role in the management of wind power, as it does not respond to the variable output of either local or distant wind turbines.

In terms of energy security and climate change, significant benefits are gained in operating electrolysers in a more responsive, grid-balancing mode that enables more effective and wider deployment of renewable energy, particularly wind power. In this case, the electrolysers are switched on and off or modulated to respond to the supply and demand balance of the grid. This should be rewarded through flexible tariffs and price signals on the spot market, for example, at times of surplus renewable power. Due to the considerably lower capacity factors that this entails, a larger electrolysis plant and hydrogen storage facility is needed, as well as a more complex operational control. From a commercial point of view, the electrolyzer must be much cheaper than one operating at high capacity factors, since the return on investment will be much slower.

In practice, a mixed approach that incorporates both re-electrification and fuel production may be advantageous, as is outlined in the preceding section, an example of which is the ITHER project. One option, or niche market, in this context could be that of "hydrogen cities" or "hydrogen communities," where hydrogen is also used in CHP units and for industrial purposes, with the fuel transported in local hydrogen pipeline networks.

Reference [13] has determined the costs per kilometer driven when employing hydrogen derived from wind power in a $60\,000\,\text{Nm}^3\,\text{h}^{-1}$ plant. They amount to about three times those of using a corresponding diesel vehicle (prices as of 2004, including taxes). This difference can be considered as non-prohibitive for the introduction of "green" hydrogen as a fuel for road vehicles. In the light of rising crude oil prices on the one hand (although currently mitigated by the financial and economic crisis, and the probability of increased oil production from non-conventional sources in the midterm) and technological progress and cost-reduction potential in hydrogen and fuel cells on the other hand, cost parity in terms of kilometers driven appears to be a realistic perspective in the not too distant future.

Emerging technologies that potentially compete with wind–hydrogen include battery-electric vehicles, biofuels, and synthetic liquid fuels.

45.2.4
Comparison of the Three Configurations

Facilities under the "electricity storage" category are the simplest. They comprise just the upper two rows in Figure 45.1, that is, hydrogen is produced, stored, and re-electrified. "Mini grids" can involve any of the elements displayed in Figure 45.1. In the case of "fuel production," hydrogen may be produced and stored at the garage forecourt (small units) or compressed or liquefied for transportation to the individual refueling stations on trailers, or even injected to pipelines provided there is infrastructure in place or a long-term solution in a developed hydrogen economy.

The main findings of this section can be abstracted as:

- System under the "mini grids" category are likely to become economically feasible once the challenges in the technical domain have been met, given the market attraction.
- As for "electricity storage" and "fuel production," actual and future systems are likely to serve both purposes to some extent, due to economic and technical benefits.
- Despite the low round-trip efficiency, but given the high energy density compared to competing technologies, "electricity storage" is expected to be most promising in the domain of long-term (several weeks up to seasonal) bulk storage of hydrogen in underground salt caverns.
- "Fuel production" opens up new markets for energy harvested by wind turbines. With respect to this opportunity, additional opportunities have been discussed recently in the light of rising fossil energy prices. This includes the use of hydrogen in industrial processes, as raw material, as a substitute for natural gas, and even its methanization, provided CO_2 as raw material becomes available in the amount required.

Under the headline "power to gas" the injection of hydrogen into transmission pipelines of the natural gas system has also been considered as an alternative to re-electrification. Given the vast amounts of energy that are stored and transported in such networks, this could be a simple option in technical terms while staying within the boundaries set by the established codes and standards (grid codes) for natural gas systems (as regards energy density and Wobbe Index, in particular).

45.3
Description of Wind–Hydrogen Projects

Given the blooming wind energy industry in Spain during the last decade – now very dependent on the upcoming reforms in the electricity market – and the special conditions of isolation and a powerful transport network a lot of interest in hydrogen has been shown in terms of coupling hydrogen production to wind farms. Several studies and projects were conducted by energy companies and R&D centers. The main highlights and outputs of the projects are shown under this section.

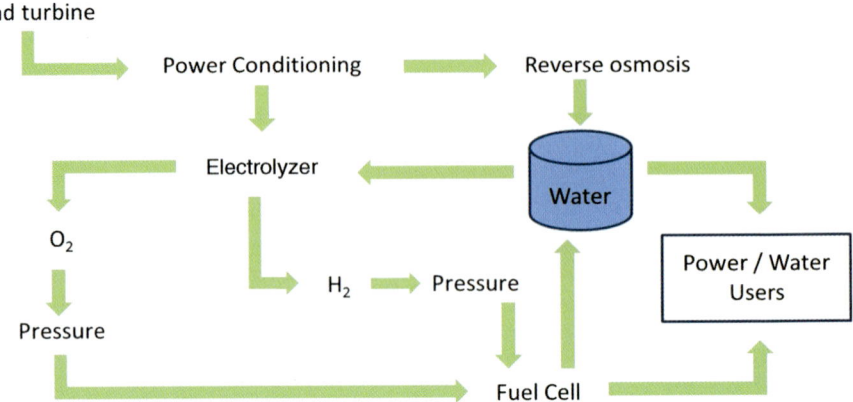

Figure 45.2 Schematic of RES2H2 site in Spain.

45.3.1
RES2H2 Project

The RES2H2 project implemented two wind–hydrogen prototype systems, one in Greece and the second in Spain, at the Instituto Tecnológico de Canarias premises, in Gran Canaria Island (Figures 45.2 and 45.3). Only the latter is discussed here. The project was financed by the Fifth Framework Programme of the European Commission. The system was put into operation in 2007 [12].

Figure 45.3 RES2H2 test site at ITC (Spain).

The prototype installed in Spain was designed to satisfy the electricity and water needs of a theoretical isolated village (mini grid type). When the electricity supply exceeded the theoretical demand, a 100 kW alkaline electrolyzer used excess electricity to produce 0.99 kg of hydrogen per hour at 25 bar (5500 Nm^3 of storage) and a reverse osmosis plant (40 kWe of nominal power) also used this excess electricity to produce a maximum of 110 m^3 per day of desalinated water. When power from the wind turbine did not cover the demand, the stored hydrogen was used in six 5 kW PEM fuel cells to produce electricity.

The system was nevertheless not directly connected to a wind turbine. Power production data of a 225 kW wind turbine installed close to the prototype were recorded and scaled appropriately to generate times series that were used as input power signal to the system for "hardware simulations." Stationary and dynamic tests of all components were performed to obtain the efficiency, range of operation, power consumption, transient response, and especially the operating curves of fuel cells and electrolyzer. The response of the components to power variation typical from wind energy resources and the interaction of each component with the rest of the system were analyzed. The performance data would allow definition of the appropriate wind turbine capable of satisfying the demands of the prototype and the load profiles in stand-alone operation. An initial topology of the stand-alone system was designed. More insight into technical details and conclusions can be found in Reference [13].

45.3.2 Hidrólica Project

This project involved a wind–hydrogen pilot plant located near Cádiz (Andalucía) which is a part of a research project led by ENDESA Generation with Green Power Technologies, AICIA, and INERCO as partners. After a preliminary study of the electricity production of the wind farm where the pilot plant was located (by comparing the production and prediction curves of the last three years), simulations were made to optimize wind energy generation by means of an integrated system of hydrogen and electricity generation. This system, the main components of which are an electrolyzer, a fuel cell, and a hydrogen tank, allowed the generation of hydrogen by using part of the energy produced by a variable-speed wind turbine.

The system was located on site at the Tahivilla 80 MW farm, an equivalent of 1900 production hours per year equipped with MADE 800 wind turbines. Located in the south of Spain, it was composed of a PEM electrolyzer with a maximum electricity consumption of 41 kWe. The resulting oxygen was vented and the hydrogen stored at intermediate pressure, 15 bar, in a storage tank. The system also had a compressor for storage at 200 bar, and a PEM fuel cell capable of generating 12 kWe (Figures 45.4 and 45.5).

The wind–hydrogen plant was tested and its technical viability demonstrated. As main conclusions, the time response of the equipment was quite good and more than enough to absorb the wind farms variable output, despite a too low overall efficiency of around 14% in the round trip. Due to this fact, a major conclusion is that the plant systems must be designed to keep a high efficiency working at all loads. In addition, some lessons learned point to a very high complexity for the overall control system, failures in ancillary equipment (deionised water feed pump or frequency transformer), and high total harmonic distortion (THD) in wind generator electrical power input.

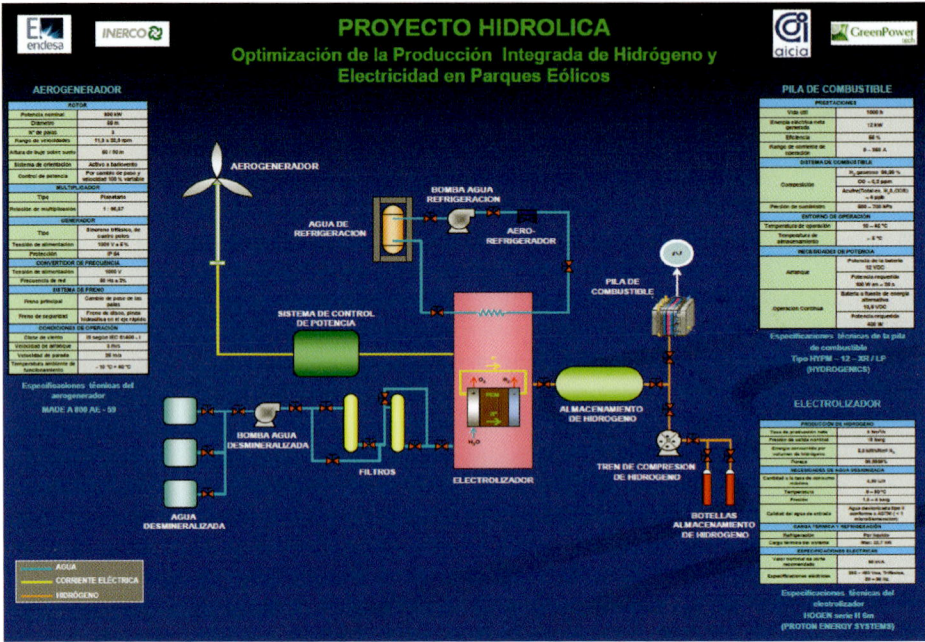

Figure 45.4 Layout of the Hidrólica Project.

Figure 45.5 Equipment container.

45.3.3
ITHER Project

The Foundation for the Development of New Hydrogen Technologies in Aragon is a private non-profit organization promoted by the Regional Government, other public bodies, and private companies. At its premises located at Walqa Technology Park, a complete showcase of the whole hydrogen chain is displayed as a full scale demo and R&D test bench, consisting of renewable generation, hydrogen production by 10 $Nm^3 h^{-1}$ alkaline electrolysis, 120 $Nm^3 h^{-1}$ @ 30 bar buffer storage, 25 kg @ 350 bar high pressure storage, filling station capable of delivery at 200 bar and 350 bar, and several different PEM fuel cells operating on a daily basis integrated in the building: 1.7 kWe back-up power for the IT servers and computers, 1.2 kWe back-up power for lighting, 10 kW fuel cell for distributed generation, and 12 kW CHP fuel cell. The first cornerstone of this whole showcase, called the ITHER project, consisted of the erection of a wind farm and the installation of an alkaline electrolyzer capable of producing H_2 from wind and photovoltaic [14].

The aim of the project was to allow testing of hydrogen generation by electrolysis with electricity obtained from renewable sources, with the most diverse available technologies. The secondary goal of the installation was to learn about the integration at real scale of renewable sources and buffer hydrogen storage, as well as dissemination, training, and public awareness, "green" hydrogen generation from renewable sources for their consumption in stationary or portable applications, and evaluation of the efficiency and functionality of the whole system.

The generation part consists of three wind turbines (80, 225, and 330 kW) and 100 kW PV integrated in a parking roof (60 kW) and in four double axis trackers (20 kW + 10 kW + 5 kW + 5 kW), featuring different kinds of panels. Electricity produced can be sent to the local grid or to the pressurized alkaline electrolyzer (10 $Nm^3 h^{-1}$ capacity, manufactured by IHT, formerly Lurgi technology, Figure 45.6). The system features customized power electronics due to the high intensity used in the alkaline electrolyzer. Recent improvements in the cell efficiency have allowed better interconnection with an intermittent resource, wind and solar, increasing the maximum intensity at the rectifier. The system also includes a control system and supervision (SCADA), connection to the electrical grid to harness the excess production, and additional devices to improve the electrical power quality, including a stabilization systems based on batteries and supercapacitors. The elements of the balance of plant must also be considered: water demineralization, hydrogen purification, compression and storage, both by metal hydride (capacity of 100 kWh energy) and compressed hydrogen gas at 30 bar in a buffer tank and at 350 bar for dispensing to vehicles. The annual renewable production with the renewable resources rounds 700 MWh per year. The maximum hydrogen production is limited by the electrolyzer, and is higher than 7000 kg H_2 per year, with an efficiency of 60%, which has been improved recently using new membrane materials, although the actual hydrogen consumption by the fuel cell vehicles and the stationary fuel cells on site are much lower. Hydrogen not delivered to vehicles can be used in various applications at the Foundation's main premises, such as an off-grid hybrid

Figure 45.6 Alkaline electrolyzer at the ITHER project.

application (PV + batteries + 1 kWe fuel cell), distributed generation (10 kWe), a back-up power system for the IT server, and a residential CHP fuel cell. The system represents a test bench that can provide a flexible energy storage system for management of the renewable resources at FHa premises. The main objective of the project was the improvement of a system able to provide flexibility to the power grid, mainly with high penetration of renewable resources.

The infrastructure put in place by ITHER has been used as a basis for further R&D projects, such as Elygrid (FCH JU Grant no 278824), spanning till 2014. This project aims to reduce the total cost of hydrogen produced via electrolysis coupled to renewable energy sources, mainly wind turbines, and focuses on megawatt size electrolysers (from 0.5 MW and up). The objectives are to improve the efficiency related to the complete system by 20% and to reduce capital costs by 25%. The work is structured in three different parts, namely, cell improvements, power electronics, and balance of plant (BOP).

45.3.4
Sotavento Project

This project took place at the premises of the Experimental Wind Farm Sotavento, consisting of a storage plant of wind energy using hydrogen at a scale that, without being one that solves the variability of generation, allows experience to be obtained in real operations that may easily be extrapolated to design full scale solutions.

Sotavento Galicia, S.A. was established in 1997 as an Experimental Wind Farm. It has a rated capacity of 17.56 MW with 24 turbines of five different technologies. The Sotavento Wind Farm was selected as the location of the hydrogen project, as one of its purposes is to promote R&D and to be a center for training and dissemination of renewable energy.

The surplus electricity generated by the wind farm is used in an alkaline Hydrogenics electrolyzer. The electrolyzer is built with four stacks of electrolyte cells, from where H_2 and O_2 are obtained separately. The O_2 is released into the atmosphere whereas the H_2, produced at a maximum rate of 60 Nm^3 h^{-1} and at a pressure of 10 bar, undergoes a process of purification and drying through heat exchangers, a filter coalescer, an O_2 removal column, and two drying columns (one in operation and one in regeneration) to obtain a purity of >99.99% (Figure 45.7).

To increase storage capacity, the H_2 generated is compressed into two piston compressors, Bauer HFS 15.4-13-DUO II, supporting up 61.8 Nm^3 h^{-1} from 4 bar compressing up to 200 bar. Between the electrolyzer and the compressor there is a buffer capacity of 1000 l, to stabilize the compressor suction. The high pressure H_2 at 200 bar is stored in a set of seven blocks of 28 cylinders each, providing a maximum storage capacity of 1725 Nm^3. These blocks are

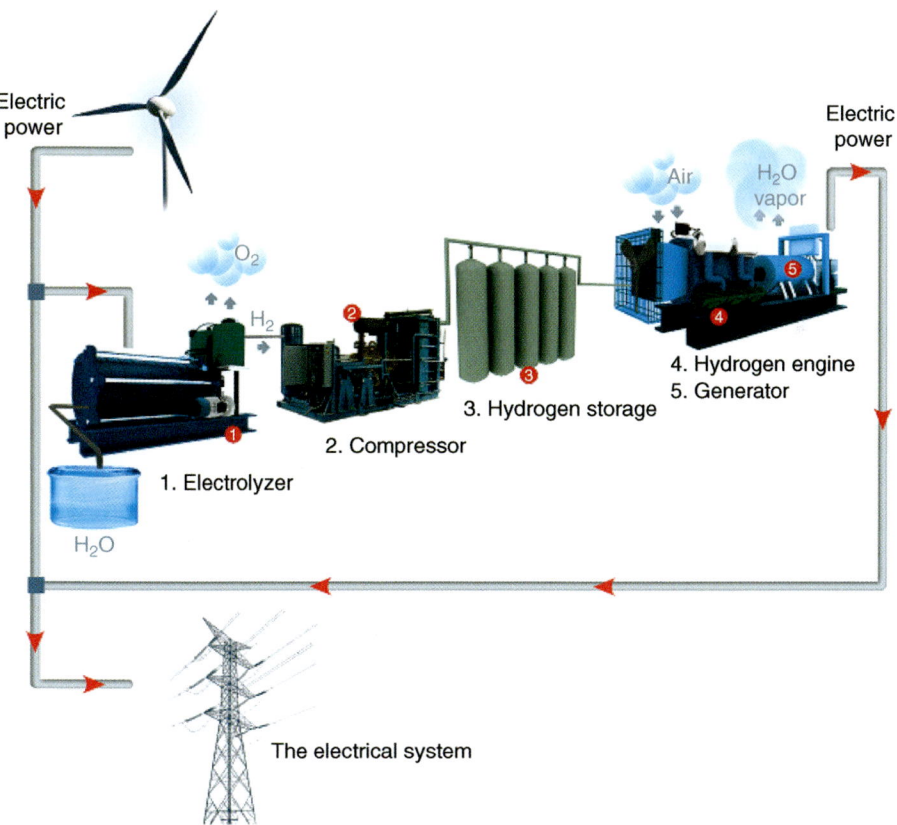

Figure 45.7 Schematics of the Sotavento wind–hydrogen project.

interconnected so that they form two groups of H_2 storage, with the possibility of isolation of each group.

The H_2 stored can be consumed in the case of energy deficit when the amount of energy produced by the wind farm is lower than the generation forecast. To carry out this function a genset Continental Energy Systems Group of 55 kW has been installed (Scania engine and Stamford generator). To feed the engine with the hydrogen stored a two stage decompression panel is needed, so that the hydrogen stored at 200 bar is decompressed in a first stage down to 14 bar and then to the suction pressure to the engine at the second. The genset has a consumption of up to 70 $Nm^3 h^{-1}$ of H_2 at a pressure of 25–60 mbar. In conclusion, the facility allows the use of H_2 as an energy storage system.

Each component has its own control and an integrated communications system has been developed. Data acquisition is performed by a stand-alone controller SNAP-PAC-S supported by a periphery of inputs and outputs. A MODBUS communication protocol was implemented and measurements from the engine, compressor, electrolyzer, and network analyzers integrated into a common server. The control system consists of a monitoring application developed under the Opto22 PACDisplay environment. This program allows commands to be sent from the operator and the reading of data from the controller, presenting information, disk storage, and generation of graphs and trends of relevant historical data.

The control system also contains information about electricity market prices and operating parameters of the wind farm. Electricity price data are obtained directly from the web site of OMEL, the national market operator. The controller connects to the web site, downloads the data, and makes them available to be represented in the graphical interface to help the operator take decisions.

45.4
Operation Strategies Tested in the Sotavento Project

The test campaign had initially two main objectives. The first was the characterization and study of all the equipment making up the system and developing semi-empirical models to describe their behavior. The second was the characterization of the plant in continuous and automated operation, following different possible strategies of wind farm management.

The electrolyzer is a system designed to meet a demand of constant hydrogen production, as traditionally this kind of equipment has been used to supply hydrogen to industries that consume it as a raw material in a manufacturing process. Its operation is based on the strategy, therefore, to meet demand at any time, keeping a constant process pressure (about 9.5 bar). However, if the goal is to manage renewable energy the operating strategy should be based on the energy the electrolyzer is supposed to consume, that is, the energy available in the wind farm, which in turn determines the flow of hydrogen produced. Another feature of the electrolyzer control worth considering is that, in order to

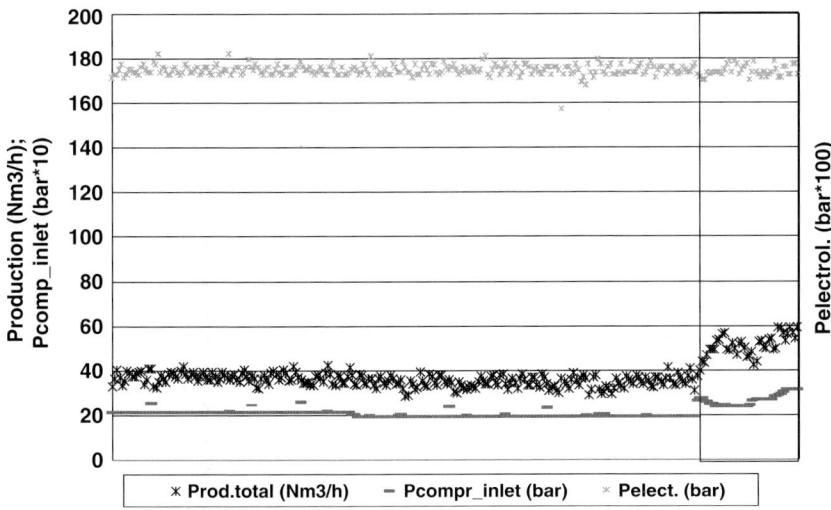

Figure 45.8 Representation of increased production of the electrolyzer.

satisfy a demand for hydrogen and operate in a stable way, it must maintain its downstream pressure within a certain range (7.5–9.5 bar).

Figure 45.8 shows that while the compressor suction pressure remains constant (dashed line), the production (bold crosses) and the pressure in the electrolyzer (light crosses) do, too, hovering around $40\,\text{Nm}^3\,\text{h}^{-1}$ and 9.5 bar, respectively. At the end of the graph the compressor suction pressure increases, and shows how the electrolyzer reacts by producing more in order to maintain the pressure in the process constant, about 9.5 bar.

To avoid instabilities in electrolyzer and compressor and prevent venting, which implies a loss of hydrogen and therefore electric power, downstream pressure to the electrolyzer is maintained within this interval. To achieve this it is necessary to define combinations of stacks of cells and compressor units that adapt production flows and hydrogen compression to optimize system performance, achieving maximum efficiency.

Four different management strategies are discussed below. For more insight, recommended texts are References [15–17].

45.4.1
Peaking Plant Strategy

By using H_2 as energy management of a wind farm production, the operator can choose to manage the farm by consuming wind power produced during low price hours (usually 5–6 h daily during the night) and injecting electricity to the grid from the previously stored H_2 at peak prices (1–2 h per day). This strategy will be most appropriate the higher the difference between the price per kWh

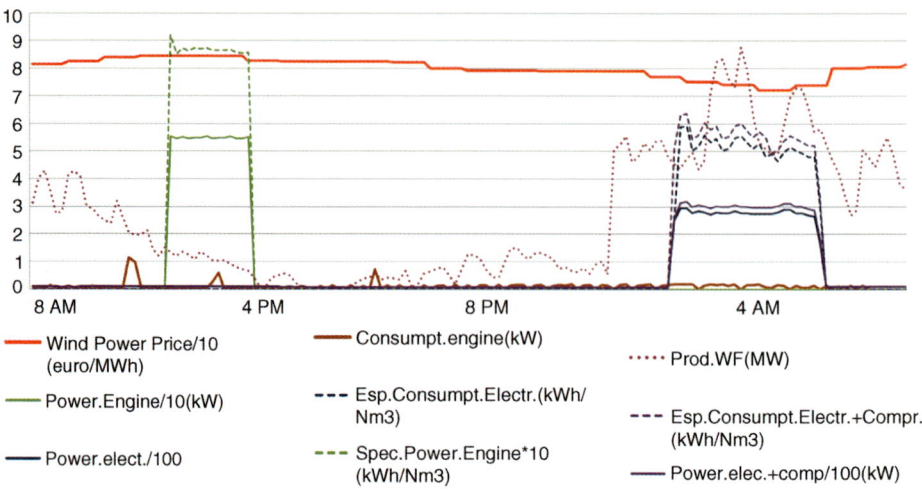

Figure 45.9 Results for a peaking strategy in summer.

between valley and peak time, depending exclusively on the electricity market (Figure 45.9). However, the high costs of equipment and the low conversion efficiency of electricity–hydrogen–electricity make this type of facility still economically unfavorable. Even without considering the investment required in the installation, for this type of facility managed with a "peaking" strategy to be economically viable in the market there must be a high difference between extreme prices to compensate for the losses due to the low efficiency.

The red line in Figure 45.9 represents the daily evolution of prices in the pool for the day of testing, with €25.8 MWh^{-1} being the minimum price and €68 MWh^{-1} the peak price. The purple dots represent the wind farm power performance on the test day. In blue and purple (dashed lines) the behavior of the electrolyzer and compressor, respectively, are represented; they operate in the valley hours between 3 and 7 a.m. In shades of brown and green the behavior of the engine is displayed, which operates during peak hours, from 19:00 to 22:00 h. On this particular day, the production of the wind farm is higher during valley and lower during peak, which means this strategy definitely helps in increasing a low production of the wind farm during peak hours.

In more detail, the lines in dark blue and purple, respectively, represent the average power of the electrolyzer and electrolyzer + compressor assembly (Figure 45.9). The average value of power of the electrolyzer during the 4 h of operation was 283.94 kW, while the average power of the assembly has risen to 305.61 kW, that is, the average power of the compressor was 21.66 kW, equivalent to 7.1% of the total power consumed. Regarding the operation of the engine, power is represented by a green line. The power output remains almost constant throughout the 3 h of operation with an average value of 54.95 kW. An overall efficiency of less than 20% was obtained.

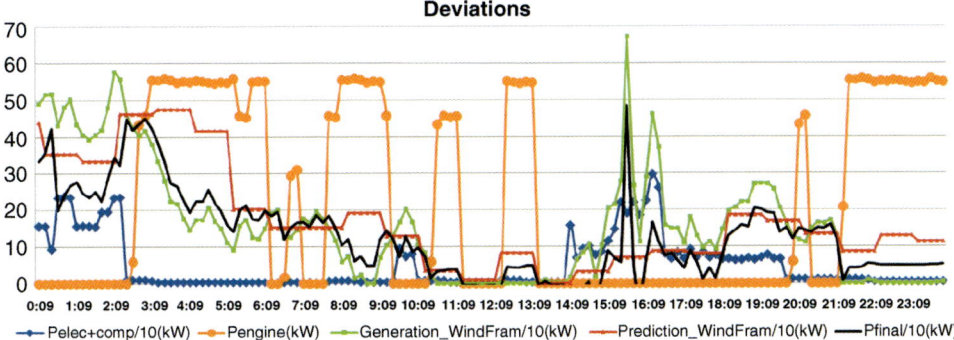

Figure 45.10 Test results for one day for the deviation correction strategy.

45.4.2
Strategy of Deviation Correction

Another management strategy using H_2 storage in wind farms is to avoid deviations from forecasted production. In this section we analyze the case of the Spanish electricity market considering the treatment that current legislation makes of those deviations. This strategy includes the "balancing" in the short term to ensure grid stability and power quality (voltage and frequency) in addition to power output trying to match the prediction offered by the wind farm at the connection point of the network.

The results analyzed correspond to days of low production of the wind farm, with power outputs of the order of hundreds of kWs (Figure 45.10).

Figure 45.10 shows the prediction of production of the wind farm in red; the actual output is represented in green, which was analyzed at all times throughout the day; genset power is displayed in orange; the power consumed by the compressor and electrolyzer is in blue; and the black line shows the net power produced by the wind farm including the H_2 facility.

It can be observed that when the production is higher than the prediction of the farm, the electrolyzer comes into operation, thus consuming the excess production over the prediction.

On the other hand, when production of the wind farm is less than what was committed, the dark red line representing the genset power starts to increase electricity export, trying to deliver the committed power to the grid.

For instance, the first two and a half hours in Figure 45.10 represent a positive deviation eliminated by the operation of the electrolyzer. Depending on the size of the deviation, the power of the electrolyzer is regulated. As stated previously, the set point available for matching compressor and electrolyzer power is the compressor suction pressure. Notably, therefore, the available control of the electrolyzer does not fulfill the requirements for adjusting the power of it to a certain value. Other situations of negative or positive deviations can be seen along the figure and interpreted accordingly.

45.4.3
Strategy for Increasing the Capacity Factor of the Wind Farm

This third strategy aims to improve the capacity factor of the wind farm, that is, to increase the hours of operation of the turbine, enabling its operation during times with technical limitations or when generation exceeds the rated power of the network connection. During these periods the wind energy produced is consumed in the electrolyzer.

This situation occurred in the first quarter of 2010 in Spain on numerous occasions. These shutdowns caused a loss of 0.6% of wind electricity produced in 2010. In future scenarios in which wind power capacity is higher, these situations will occur just as often as there will be more wind farms producing during peak hours where electricity demand is very low.

In 2010 the Sotavento wind farm experienced limitations to its production on two days in January, four in February, seven in March, three in April, one in May, one in June, one in September, three in October, and two in November, totaling 157 h with an equivalent production loss of 1293.42 MWh. Some or all of this energy can be used to generate hydrogen at these times of limitations.

This strategy can also be applied in repowered wind farms where less powerful older machines are replaced by higher capacity wind turbines. The limiting factor in these situations is usually the connection capacity of the existing network. By taking advantage of the occasions when the wind farm exceeds the connection capacity of the network the operation of wind farms can be enhanced (Figure 45.11).

Figure 45.11 shows that the electrolyzer has been used at full power, simulating moments of technical restrictions of the wind farm or production exceeding its connection capacity. Power consumed by the electrolyzer is shown by dotted blue crosses, the total power (electrolyzer + compressor assembly) by lilac crosses, the power consumed by the stacks in the process of electrolysis by

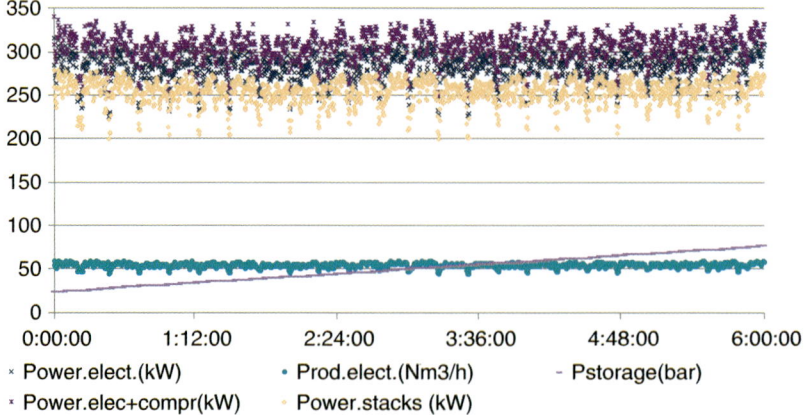

Figure 45.11 Test results for increasing the wind farm capacity factor.

orange dots, the production of H_2 in the electrolyzer by light blue dots; the pressure of hydrogen storage is shown as a lilac line, representative of the state of filling of storage, and a green-blue line represents the compressor suction pressure. Throughout the night the electrolyzer is operated at its maximum output with specific consumption of $5.21\,\mathrm{kWh\,Nm^{-3}\,H_2}$ generated.

45.4.4
Load Leveling Enabling Distributed Generation or Island

The load leveling strategy at wind farm level or substation level (for a local group of wind farms) enables the management of wind power according to the power demand profile at that point, resulting in a complete matching of generation and consumption at local level. This means that when demand is low and there is excess wind power hydrogen is produced and stored. When demand is high, it is covered by wind–hydrogen production if necessary (see Reference [18]). Such systems are often isolated small-scale systems, like a small island. This strategy can also be used in distributed generation systems mainly based on renewable energy and connected to distribution networks with multi-MW peak loads.

Load leveling at large-scale, for distributed generation systems, could be described as a demand management with electrolysers, which would mean controllable loads distributed in different locations and of different sizes, such as for hydrogen refueling stations for fuel cell vehicles, in network substations, at homes or in centralized plants (see Reference [19]). Taking into account local generation and consumption, depending on wind generation, other renewable and conventional generation of electricity and the existing demand, electrolysers would work (nominal power, partial loads or switched off) by absorbing the surplus energy to produce hydrogen. In this case, the hydrogen gas would be used later to produce electricity but more probably it would be used as fuel for transport applications or as a supplement or by natural gas piping. The final application of this hydrogen would depend on the prices of competitor fuels, such as gasoline or natural gas.

Considering the tests performed, a hydrogen storage facility could not participate in the market for primary regulation, requiring response times of less than 15 s. The secondary regulation requires response times in the range of 30 s, so this kind facility would also not be suitable for this type of regulation, as the response time of the engine is about 30 s. This kind of wind–hydrogen concept would match the specifications of tertiary control systems, whose response time is set at 15 min.

45.5
Conclusions

Wind energy is an established technology, which has seen some rather impressive developments in the last 30 years, as a result of environmental concerns and the need for energy self-sufficiency. Overall, wind energy is the most advanced renewable energy technology for generating electricity. Further wind power deployment in countries such as Spain, with strong limitations for network connection to other countries, will surely lead to increased complexity in power management, technical restrictions, and the need for reserve power or higher curtailments.

The wind–hydrogen alternative appears to be an attractive storage option for overcoming the major drawback of wind power, that is, its intermittent availability, which affects both stand-alone and grid-connected applications. In stand-alone wind-power applications such storage can be used to stabilize the availability of electricity by converting wind into hydrogen during periods of surplus capacity and, during periods of low or no wind availability, the hydrogen can be used to generate electricity.

The hydrogen option is also important in grid-connected applications, where it can be used to manage the flow of electricity to the grid. Of course, in the long anticipated "hydrogen economy" of the future, hydrogen would be a candidate for transportation fuel.

The technologies associated with hydrogen production, storage, and distribution are fairly well established and, largely as a result of environmental concerns and the anticipated hydrogen economy, are benefiting from continuing research and development. Demonstrations of associated wind–hydrogen technologies have taken place in Spain in recent years, leading to extensive experience of some companies and R&D centers. These projects have demonstrated that there are no insurmountable problems with the use of wind power to produce hydrogen by the electrolysis of water.

Moreover, not only do the sizing and choice of equipment and layout design have a great influence on the results of a wind–hydrogen facility, but also the operation strategy, which can range from a peaking plant, making the economic optimum, a deviation correction from the wind forecast, increasing the capacity factor (interesting in terms of deferring investments in grid reinforcement), load leveling to allow for distributed generation, to even the island concept.

Most of the wind power management applications and strategies that have been discussed result in a rather low number of annual operating hours. This fact puts the economic feasibility of wind–hydrogen systems today into question since the facilities still require in addition high specific costs. Further R&D and industrialization for the main components along with more stringent regulation and grid codes could allow in the near future the commercial deployment of the concepts discussed.

References

1 Deloitte for Asociación Empresarial Eólica (2012) Impacto Macroeconómico del Sector Eólico en España. Available at http://www.aeeolica.org/uploads/documents/Estudio_macro_AEE_2012.pdf.

2 Instituto para la Diversificación y Ahorro Energético (2005) Plan de Energías Renovables en España 2005–2010. Available at http://www.idae.es/index.php/mod.pags/mem.detalle/id.14/relmenu.12.

3 Instituto para la Diversificación y Ahorro Energético (2011) Plan de Energías Renovables 2011–2020. Available at http://www.idae.es/uploads/documentos/

documentos_11227_PER_2011-2020_def_93c624ab.pdf.
4 Ministerio de Industria, Turismo y Comercio (2008) Planificación de los sectores de electricidad y gas 2008–2016. Desarrollo de las redes de transporte. Available at http://www.minetur.gob.es/energia/planificacion/Planificacionelectricidadygas/desarrollo2008-2016/DocTransportes/planificacion2008_2016.pdf.
5 European Wind Energy Association (2011) Pure Power - Wind energy targets for 2020 and 2030. Available at http://www.ewea.org/fileadmin/ewea_documents/documents/publications/reports/Pure_Power_Full_Report.pdf.
6 Red Eléctrica de España (2012) La integración de la generación de régimen especial en el sistema eléctrico. Propuestas normativas y estudios de capacidad zonal. Available at http://www.ree.es/sites/default/files/01_ACTIVIDADES/Documentos/AccesoRed/Estudios_capacidad_zonal_v3_26sep12.pdf.
7 Holttinen, H. *et al.* (2008) Impacts of large amounts of wind power on design and operation of power systems. Presented at WindPower 2008, 1–4 June 2008, Houston, Texas, NREL/CP-500-43540.
8 Hoskin, A. *et al.* (2013) *IEA-HIA Task 24 Wind Energy & Hydrogen Integration* (ed. L. Correas), IEA HIA. ISBN: 978-0-9815041-5-5.
9 Stolzenburg, K. *et al.* (2008) Hydrogen as a means of controlling and integrating wind power into electricity grids – The HyWindBalance Project. Presented at the International Conference "Hydrogen on Islands", 22–25 October 2008, Bol, Croatia.
10 VDE Verband der Elektrotechnik Elektronik Informationstechnik e.V., Power Engineering Society (2008) Energy storage in power supply systems with a high share of renewable energy sources, significance, state of the art, need for action.
11 Gimeno-Gutiérrez, M. and Lacal-Arántegui, R. (2013) *Assessment of the European potential for pumped hydropower energy storage - A GIS-based assessment of pumped hydropower storage potential*. Joint Research Centre, Report EUR 25940 EN, Luxembourg: Publications Office of the European Union. ISBN 978-92-79-29511-9.
12 European Commission (2007) EESD RES2H2 Project: Cluster Pilot Project for the Integration of renewable Energy Sources into European Energy Sectors using Hydrogen. Publishable Final Report. European Commission EESD contract N°: ENK5-CT-2001-00536. Available at http://www.docstoc.com/docs/163298576/103966621EN6.
13 Linnemann, J. and Steinberger, R. (2007) Realistic costs of wind-hydrogen vehicle fuel production. *Int. J. Hydrogen. Energy*, **32**, 1492–1499.
14 Correas, L. and Aso, I. (2010) Task 24: Wind Energy and Hydrogen Integration, in *Proceedings of the 18th World Hydrogen Energy Conference 2010 - WHEC 2010, May 16–21 2010, Essen* (ed. D. Stolten and T. Grube), Forschungszentrum Jülich GmbH, Zentralbibliothek, Verlag, Jülich, pp. 419–427. ISBN: 978-3-89336-654-5.
15 Rey, M., Aguado, M., Garde, R., García, G., and Carretero, T. (2012) H_2 production from wind power in a wind farm in Spain. *J. Energy Power Eng.*, **6**, 49–59.
16 Rey, M., Carretero, T., Aguado, M., and Garde, R. (2010) H2 production in Sotavento Wind Farm. Presented at 18th World Hydrogen Energy conference 2010 – WHEC, 2010, 16–21 May Essen, Germany.
17 Rey, M. (2012) Desarrollo de una metodología de caracterización para instalaciones integradas de hidrógeno y energía eólica según sus estrategias de gestión. Ph.D. Thesis, Universidad Pública de Navarra.
18 Jørgensen, C. and Ropenus, S. (2008) Production price of hydrogen from grid connected electrolysis in a power market with high wind penetration. *Int. J. Hydrogen Energy*, **33** (20), 5335–5344.
19 Troncoso, E. and Newborough, M. (2007) Implementation and control of electrolysers to achieve high penetrations of renewable power. *Int. J. Hydrogen Energy*, **32** (13), 2253–2268.

46
Hydrogen Islands – Utilization of Renewable Energy for an Autonomous Power Supply

Frano Barbir

46.1
Introduction

The World is facing an imminent shift from the present unsustainable energy system, primarily based on utilization of fossil fuels, to a new sustainable energy system based on renewable energy sources. This transition may have already begun, as solar PV and wind energy installations are growing at unprecedented rates. Renewable energy sources are more than sufficient to satisfy all the energy needs of today's world and technologies for their utilization are available. It is therefore possible to envision an energy system solely based on utilization of renewable energy (Figure 46.1) [1]. Various forms of renewable energy, primarily solar, wind, and hydro can be converted into electricity and heat. This transformation may be accomplished in centralized power plants (probably a better word would be energy plants), but also a significant portion may be accomplished locally or even individually. In some cases economy of scale may favor larger units (e.g., large wind turbines cost less per kW then small ones) but in some cases mass produced small units may also result in lower cost.

In such a system there are obvious needs: (i) to store energy – to overcome fluctuations in availability of renewable energy sources; (ii) to transport energy over mid and long distances; and (iii) to fuel the transportation sector (land, sea, and air). Hydrogen could satisfy those needs. Electricity and hydrogen are the only non-carbon currencies that, together, can supply the full menu of energy services. A vision of an energy system of the future presented here in which hydrogen plays a significant role is logical and sustainable. However, such a system should not be called a "hydrogen economy," but rather "hydricity" or "hydrogen-electricity economy" [2].

While a future energy system based only on renewable energy sources with electricity and hydrogen as the main energy carriers is technically feasible, a more difficult task is to define a path of how to get there from here. One problem is the inherently high cost of renewable energy. Hydrogen makes the renewable energy even more expensive. An energy system based primarily on renewable energy sources would be impossible without the storage, transport, and fuel features that hydrogen offers. However, renewable energy sources can

Hydrogen Science and Engineering: Materials, Processes, Systems and Technology, First Edition.
Edited by Detlef Stolten and Bernd Emonts.
© 2016 Wiley-VCH Verlag GmbH & Co. KGaA. Published 2016 by Wiley-VCH Verlag GmbH & Co. KGaA.

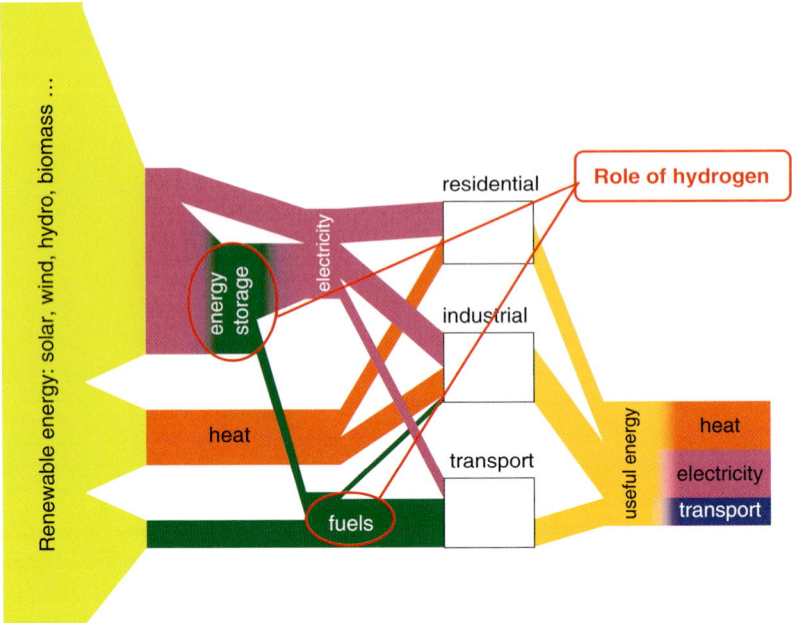

Figure 46.1 Hypothetical energy system of the future based on renewable energy and hydrogen's role in it [1].

start penetrating the energy markets without those features, and so at this stage they do not need hydrogen.

On the other hand, hydrogen without renewable energy does not make much sense, although it can be produced from fossil fuels. Hydrogen produced from fossil fuels, particularly from natural gas and coal, as a fuel cannot compete in today's market with the very fuels it is produced from, and therefore hydrogen production from fossil fuels makes sense only in a transition period to help establish hydrogen supply infrastructure and to help commercialize hydrogen utilization technologies. Because coal has by far the largest reserves of fossil fuels, hydrogen production from coal would make sense in this transition period, but only providing carbon capture and sequestration is applied.

Another difficulty in the market penetration of hydrogen as a fuel and hydrogen technologies including fuel cells is the non-existence of hydrogen infrastructure. This infrastructure can be built, but it would take decades to do so. The key problem is that hydrogen technologies for production, storage, and utilization are so interdependent and can hardly be commercialized individually. Technologies for hydrogen utilization, such as fuel cells, cannot be introduced and commercialized without technologies for hydrogen production and storage, and vice versa. (The high temperature fuel cells that can run on natural gas are the exception – they can be commercialized without hydrogen infrastructure, providing they offer economic advantages.)

Implementation of hydrogen energy technologies may be easier on a smaller scale, but still within a complete energy system. Islands represent the perfect platforms for such efforts as they offer a defined project scope, including sources of energy and demand, as well as physical boundaries. The choice of conventional energies on the islands is often limited and their cost may be several times higher than on the mainland. At the same time, most islands do not have energy intense activities, have historically learned how to conserve energy, and most importantly have plenty of clean renewable energy available, such as solar, wind, geothermal, waves, tides, ocean current, and ocean thermal gradient. These sources may be used to generate electricity and hydrogen, which then can satisfy all the energy needs of the island communities.

46.2
Existing Hydrogen Projects on Islands

Islands are therefore ideal polygons and showcases for the demonstration of individual hydrogen technologies as well as the entire mini hydrogen, that is, hydricity economies [3]. There are already several ongoing or proposed demonstrations of the integrated hydrogen systems on islands all over the world. Table 46.1 summarizes studied, planned, and completed hydrogen installations on islands [4–21]. They vary in size from kW to MW and in scope from a refueling station to a complete autonomous island energy supply.

The PURE (Promoting Unst Renewable Energy) project is situated on the island of Unst in the Shetland Islands in the most northerly part of Scotland. The area has some of the best wind and wave energy sources in Europe. The project was commissioned by the Unst Partnership Ltd., a community development agency established by the Unst Community Council to support local economic development and regeneration. The system consists of two 6 kW wind turbines which provide electric power to directly heat five business units. Excess electricity from the wind turbines is used to produce hydrogen which can then be stored and is also used to fuel a fuel cell/battery hybrid vehicle and for a back-up power unit consisting of a 5 kW fuel cell and an inverter [4,5].

Utsira is a windswept island 20 km off the west coast of Norway. A wind hydrogen demonstration project was started in 2004. The project was launched by StatoilHydro and Enercon. Local authorities, members of the local community, and representatives of local trade and commerce have been involved throughout the development and operation phases and the project reports that this was vital for achieving success. The system is designed to meet the entire energy demand of ten households using wind energy. Using excess wind power to create hydrogen, energy can be stored for use at peak hours or when the wind is not blowing. The system consists of a 600 kW wind farm, a 10 Nm^3 h^{-1} alkaline electrolyzer, an 11 Nm^3 h^{-1} hydrogen compressor, a 2400 Nm^3 hydrogen gas storage at 200 bar, a 55 kW hydrogen engine generator system, and a 10 kW fuel cell [4,6].

Table 46.1 Studied, planned, and completed hydrogen projects on islands.

Island/project	Status	Energy source	Equipment/capacities	Use	Reference
Unst, UK/PURE Promoting Unst Renewable Energy	In service	Wind	2×6 kW wind turbines 5 kW fuel cell	Fuel cell and car	[4,5]
Utsira Wind Hydrogen Energy Project, Norway	In service	Wind	600 kW wind turbines 10 Nm^3 h^{-1} electrolyzer 2400 Nm^3 H$_2$ storage @ 200 bar 55 kW hydrogen genset, 10 kW fuel cell	Community power	[4,6]
ECTOS Hydrogen Refueling station, Iceland	In service	Hydro, geothermal	60 Nm^3 h^{-1} electrolyzer 200 kg day^{-1} refueling station	Buses, vehicles, boat	[4,7]
Gran Canaria, Canary Islands, Spain	In service	Wind	11 Nm^3 h^{-1} electrolyzer 500 Nm^3 H$_2$ storage @ 25 bar 6×5 kW fuel cell	Desalination plant	[4]
Gran Canaria, Canary Islands, Spain	In service	Solar PV + wind	3 kW PV, 10 kW wind turbine 1.16 Nm^3 h^{-1} electrolyzer 50 Nm^3 H$_2$ storage @ 200 bar 5 kW fuel cell	Fuel cell and vehicle	[4]
Totara Valley, North Island, New Zealand	In service	Wind	300 W wind turbine 400 W electrolyzer 18 mm 2 km H$_2$ pipeline 1 kW fuel cell	Fuel cell and H$_2$ combustor	[4,8]
EcoIsland project, Isle of Wight, UK	Planned	Wind	80 kg day^{-1} H$_2$ generation unit	Hydrogen vehicles and boat	[9]
Hawaii Hydrogen Initiative, Oahu, Hawaii, USA	Planned	Syngas	PSA H$_2$ generator 65 kg day^{-1} electrolyzer	Fleet of vehicles	[9]
Veskenskov village, Lolland Island, Denmark	In service	Wind	16 Nm^3 h^{-1} electrolyzer 25 Nm^3 H$_2$ storage @ 6 bar H$_2$ pipeline up to 35 1.5 kW FC µCHP	Combined heat and power	[9]
PEI Wind Hydrogen Village, Prince Edward Island, Canada	Experiment	Wind	Electrolyzer 125 kW hydrogen genset	Experiment	[10]
Matiu/Somes, Wellington Harbor, NZ	Installed	Wind, solar PV	6 kW wind turbine 4 kW PV 1.2 kW hydrogen system	Water heating, cooking	[11]

Location	Status	Energy source	System details	Purpose	Ref.
UNIDO ICHET Bozcaada Wind Hydrogen Project, Bozcaada, Turkey	Installed	Wind, solar	30 kW wind turbine, 20 kW PV 50 kW electrolyzer 35 kW H_2 genset 20 kW fuel cell	Demonstration, community power	[12]
Ramea Island, Canada	Installed	Wind	90 $Nm^3 h^{-1}$ electrolyzer 2000 Nm^3 H_2 storage @ 450 bar 4 × 62.5 kW H_2 engine gensets	Community power	[13]
La Reunion, France	Study	Solar PV	333 kW PV 200 kW electrolyzer 46 $Nm^3 h^{-1}$ compressor 200 bar H_2 storage 25 kW fuel cell	Community power	[14]
Yakushima, Japan	Installed	Hydro	1.25 $Nm^3 h^{-1}$ electrolyzer 126 Nm^3 H_2 storage @ 35 MPa 30 $Nm^3 h^{-1}$ refueling station	Vehicle	[15]
Mljet, Croatia	Study	Solar PV, wind	Varying	Energy storage, shuttle buses	[16]
Karpathos, Greece	Study	Wind	600 kW wind turbine 450 kW electrolyzer 2300 kg H_2 storage 100 kW fuel cell	Community power	[17]
Milos, Greece	Study	Wind	Wind turbines, diesel generators, H_2 system	Community power	[18]
Porto Santo, Madeira, Portugal	Study	Wind solar PV	25 MW wind or 11 MW wind and 20 MW PV 11 MW electrolyzer 5.5 MW fuel cell	Community power 100% renewable island	[19]
Corvo, Azores, Portugal	Study	Wind	2 × 100 kW wind turbines 120 + 160 kW diesel gensets 19 $Nm^3 h^{-1}$ (80 kW) electrolyzer 200 kg H_2 storage 50 kW fuel cell	Community power	[20]
Terceira, Azores, Portugal	Study	Wind, wave, geothermal, biogas	2400 $Nm^3 h^{-1}$ hydrogen generation	Hydrogenopolis	[21]

Iceland is located in the North Atlantic and is famous for its geothermal renewable energy. In fact, in 2008, 84% of Iceland's energy needs were met with hydro and geothermal power, with the remaining 16% coming from imported oil. Therefore, the only element of Iceland's energy use not supplied domestically is the transportation sector, although coal is imported as raw material for aluminum smelters. Iceland has a population of approximately 300 000 people and plans to be completely energy independent. Iceland's project, the ECTOS Hydrogen Fueling Station in Reykjavik, consists of production, storage, and a fueling station and three FC Citaro test buses. The fueling station has the capacity to deliver 30 kg of hydrogen per day at 440 bar and the total production and capacity of the station is 200 kg day^{-1}. The buses had a 250 kW PEM system from Ballard, 360 bar storage on the roof and an operating range of 150–240 km. The station has been in continuous use since 2003. Since the conclusion of the bus test it serves private and rental vehicles and a fuel cell boat [4,7].

Gran Canaria is one of the Canary Islands located off the west coast of Africa and is a very popular tourist destination. The Canary Islands consist of seven small island systems with only two being electrically interconnected. The rest of the islands, given the distance separating them and the depth of the ocean in this area, remain isolated electric systems. Gran Canaria has abundant solar and wind resources but has few fresh water resources and must operate desalination facilities. There are two systems in operation at The Instituto Tecnológico de Canarias (ITC). The first has been designed for stationary applications to meet electricity and water demands in stand-alone operation. The system includes a 55 kW alkaline electrolyzer with a nominal production of 11 Nm^3-H$_2$ h^{-1}, 500 Nm^3 H$_2$ storage at 25 bar, six 5 kW fuel cells, and inverters to supply AC power to electrical loads. The system also includes a wind turbine and a reverse osmosis desalination plant, which takes excess electric energy from the wind turbine when the hydrogen storage tank is full [4]. The second system is a small-scale hydrogen production system for transportation which consists of a 10 kW wind turbine, a 3 kWp photovoltaic system, a small PEM electrolyzer (1.16 Nm^3-H$_2$ h^{-1}), a compressor, and a 5 kW fuel cell. Hydrogen produced is used to supply a small vehicle. The vehicle has a 4 kW fuel cell and has an estimated range of 60 km. It has 7 Nm^3 of hydrogen storage at 200 bar [4].

The Totara Valley site involves a small remote farming community in the Wairarapa hill country on the North Island of New Zealand. The community expressed an interest in local renewable energy as an alternative to grid supply. Initial technologies included small roof top PV and solar hot water systems and a 1 kW micro-hydro system. Unfortunately, the best wind resources were 2 km from the farms, so a small-scale pilot project was then launched to demonstrate a low cost, high reliability pipeline based hydrogen energy network. One half of the system was thus located at the wind site and the other at a farm. The system at the wind site consists of a 300 W wind turbine, a supercapacitor to smooth the very fast fluctuations in the winds output, and a PEM electrolyzer. The hydrogen produced is fed through a plastic polymer pipeline running 2 km underground to the farm. This pipeline can also act as a store for 6 kWh of hydrogen (2000 l at

4 bar). A 1 kW PEM fuel cell provides heat and power loads at the farm. In the future an additional hydrogen burner will also provide hot water. At the farm, there are also an inverter and buffer batteries. The combined heat and power (CHP) fuel cell and direct water heater conversion at the residences maximizes the hydrogen link efficiency [4,8].

The EcoIsland project aims to make the UK's Isle of Wight, population circa 150 000, a net energy exporter by 2020 through the integration of its increasing wind, tidal, and photovoltaic renewable energy supplies via a smart grid, battery and hydrogen storage for grid balancing and the provision of clean transportation [9].

One island that has already harnessed its renewable energy potential and become a net energy exporter to the mainland is the Danish island of Lolland, population circa 70 000. In 2007 the island generated 1 150 000 MWh of electricity and consumed just 750 000 MWh. Approximately 90% of the island's electrical generation comes from the largely unobstructed wind that sweeps the land; the remaining generation is also green, coming from biomass, waste, and to a lesser extent biogas. The Lolland Hydrogen Community in the village of Vestenskov is the EUs first full-scale residential fuel cell combined heat and power demonstration, with installations in 35 homes. Hydrogen for the community is produced through excess wind-powered electrolysis; the by-product oxygen created in the process is used to accelerate biological reactions in a nearby municipal water treatment plant [9].

One of the largest island demonstrations of hydrogen and fuel cell solutions is Hawaii. With a 90% dependence on imported oil, Hawaii, population ca 1.4 million, is America's most fossil fuel dependent state resulting in exceptionally high fuel and electricity prices. This has driven both islanders and the US Government to actively pursue sustainable energy solutions for Hawaii and a roadmap aiming to achieve 70% clean energy use by 2030 has been set. After a handful of previous hydrogen projects, a consortium of twelve companies, agencies, and universities launched the Hawaii Hydrogen Initiative (H2I). H2I aims to make Hawaii the first US "hydrogen state" beginning by installing 20 to 25 hydrogen stations at strategic locations across Hawaii's most populous island, Oahu. Hydrogen will be produced by a pressure swing adsorption (PSA) process from industrial syngas, and plans to include renewable biogas from plant oils and animal fats as syngas feedstocks have been announced. Oahu is also to see the construction of a 65 kg per day Proton OnSite hydrogen generator (PEM electrolyzer) for installation at the Marine Corps Base Hawaii to fuel a fleet of GM Equinox FCEV [9].

The PEI Wind-Hydrogen Village has been developed in North Cape, Prince Edward Island as Canada's first grid-independent Sustainable Energy Supply System for northern and remote communities. The PEI Wind-Hydrogen Village is intended to demonstrate an effective and sustainable means for addressing the intermittency of wind power in stand-alone applications. The system is designed to operate on the basis that when the wind is blowing the wind turbines supply power to connected loads including a hydrogen production system. When there is low or no wind conditions, stored hydrogen is used to fuel a back-up generator that keeps the electricity flowing without any disruption of supply [10].

To evaluate the real-world potential for hydrogen energy storage, a small pilot-scale system has been developed and installed at Matiu/Somes Island in Wellington Harbor. This "remote" island is powered with a hybrid renewable energy system including both wind and solar PV generation, backed up by a diesel genset. An electrolyzer based hydrogen storage system has been installed to evaluate the technology readiness level (TRL) and commercial potential for this type of technology. The hydrogen gas is currently used to substitute for LPG in cooking applications, and will also be used in the future for instant water heating [11].

The UNIDO-ICHETs Bozcaada Wind-Solar Hydrogen project, on a Turkish Bozca Island, is a facility aimed at studying how hydrogen and renewable energies can be integrated in stand-alone applications for the powering of island communities. This experimental power plant supplied by AccaGen produces electricity thanks to a 20 kW solar photovoltaic array and a 30 kW set of wind turbines and stores an equivalent amount of energy as hydrogen via a 50 kW electrolyzer. At times of grid failure or peak demand, the stored hydrogen can be converted back into electricity using a 20 kW fuel cell and 35 kW hydrogen engine, allowing uninterrupted electricity supply to the equivalent of 20 households for up to 24 h [12].

46.3
System Design/Configuration

In general, a system for autonomous energy supply (Figure 46.2) consists of devices for utilization of available renewable energy sources, such as photovoltaic

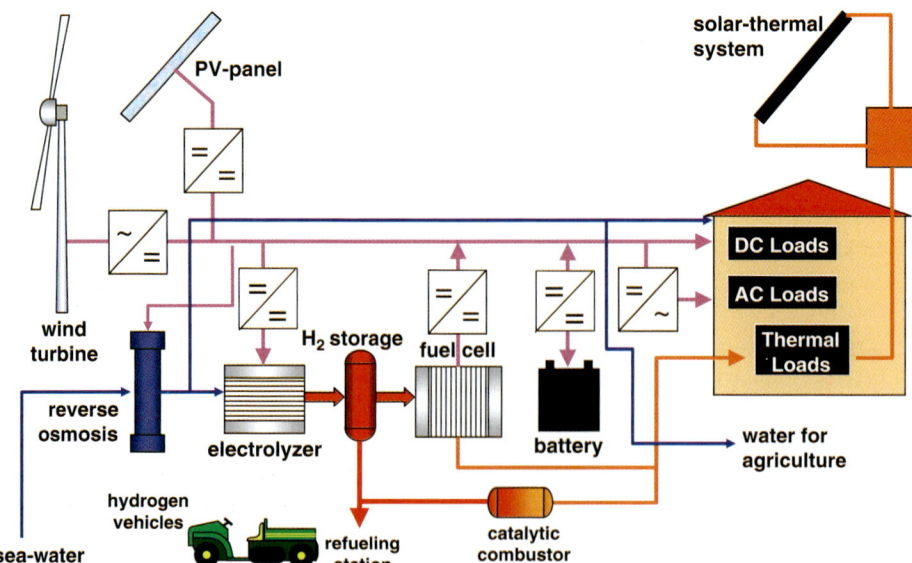

Figure 46.2 Schematic diagram of an autonomous island energy system based on renewable energy sources with a significant role for hydrogen.

panels or wind turbines, a subsystem for energy storage, inverters and converters necessary for matching the voltage and type of current between power sources, energy storage and the load, and a control and power management subsystem. A subsystem where hydrogen serves for energy storage usually consists of an electrolyzer, hydrogen storage, which may or may not include a compressor, and a fuel cell. Sometimes this entire subsystem is called a regenerative fuel cell, although this name better suites a device that can act both as an electrolyzer and a fuel cell (so-called unitized regenerative fuel cell) [22]. Usually, such a system also includes batteries for short-term energy storage, as a fuel cell may not react fast enough to sudden changes of load.

Basically, all the energy conversion devices are connected to a common bus from which power is supplied to the load. Each device may require its own DC/DC inverter to match its voltage to the bus voltage and to enable a desired control strategy. The load typically requires a DC/AC converter, unless it is a purely DC load [23–25].

46.4
Key Technologies

46.4.1
Electrolyzer

There are currently two types of electrolyzers commercially available, namely, alkaline and proton exchange membrane (PEM) electrolyzers. Alkaline electrolyzers are more mature, while PEM electrolyzers offer several advantages such as the possibility for higher operating pressure, higher current density and higher system efficiency, simpler system (no need for electrolyte management), and thus less corrosion [26]. Of particular interest for use in conjunction with the highly intermittent renewable energy sources is the capability of operation with rapid changes in current. Another important feature is the minimum power required for safe operation of the electrolyzer. Yet another problem related to operation with a highly variable power source is thermal management. The electrolyzer takes time to reach its normal operating temperature, but due to intermittent operation it may operate most of the time at a temperature below nominal, which results in a lower efficiency.

In electrolysis processes, the hydrogen generation rate is proportional to current. Nominal capacity of an electrolyzer is usually given in $Nm^3\,h^{-1}$ or $kg\,h^{-1}$ ($1\,Nm^3$ of hydrogen $= 0.0893\,kg$). The performance of an electrolyzer may be given by the power required to generate $1\,Nm^3$ of hydrogen ($kWh\,Nm^{-3}$) or by its efficiency, using either the lower ($33.3\,kWh\,kg^{-1}$ or $2.98\,kWh\,Nm^{-3}$) or higher heating value ($39.4\,kWh\,kg^{-1}$ or $3.52\,kWh\,kg^{-1}$) of hydrogen. Usually, electrolyzers require 4–6 kWh of electrical energy to generate $1\,Nm^3$ of hydrogen (corresponding to efficiencies between 0.88 and 0.59, based on the hydrogen higher heating value).

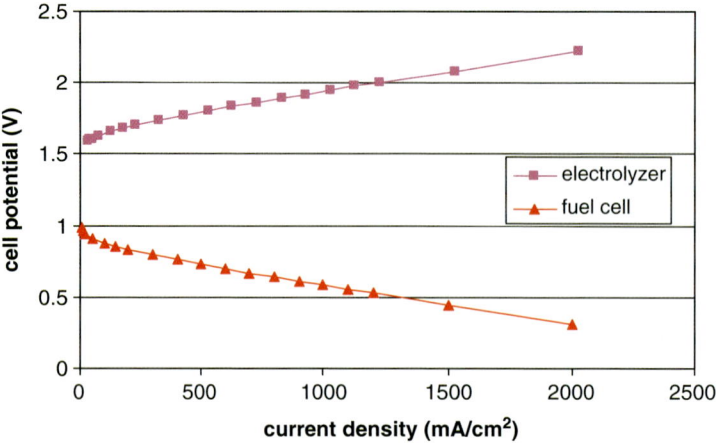

Figure 46.3 Typical electrolyzer and fuel cell polarization curves [22].

The efficiency of an electrolyzer is inversely proportional to the cell potential, which in turn is determined by the current density, which in turn directly corresponds to the rate of hydrogen production per unit of electrode active area. Cell voltage and current (or current density) are related by a polarization curve (Figure 46.3). A higher voltage would result in more hydrogen production, but at a lower efficiency. Typically, the cell voltage is selected at about 2 V, but a lower nominal voltage (as low as 1.6 V) may be selected, if the efficiency is more important than the size (and the capital cost) of the electrolyzer.

Another "source" of inefficiency is hydrogen permeation (loss) through the polymer membrane. This is typically insignificant at low operating pressures, but it may significantly affect the overall efficiency at very high pressures (>100 bar).

In addition, there are power losses in voltage regulation and some power is needed for the auxiliary equipment (pumps, fans, solenoid valves, instrumentation, and controls). The electrolyzer efficiency is therefore [22]:

$$\eta_{EL} = \frac{1.482}{V_{cell}} \frac{i - i_{loss}}{i} \frac{\eta_{DC}}{1 + \xi} \quad (46.1)$$

where:

1.482 = potential corresponding to the hydrogen higher heating value ($\Delta H/nF$);
V_{cell} = individual (average) cell potential (V);
i = operating current density (A cm^{-2});
i_{loss} = current density loss corresponding to hydrogen crossover and internal currents (A cm^{-2});
η_{DC} = efficiency of the DC/DC voltage regulator;
ξ = ratio between parasitic power and net power consumed by the electrolyzer.

The nominal power required to run an electrolyzer is simply a product of its nominal capacity and its performance characteristics (or a ratio between its nominal capacity and its efficiency, multiplied by the corresponding hydrogen heating value). The efficiency of an electrolyzer is not constant throughout its operating range. It reaches a maximum value at a certain hydrogen generation rate, which depends on hydrogen (and oxygen) crossover, which in turn is a linear function of pressure and an exponential function of temperature.

An electrolyzer, depending on its design and construction may operate at elevated pressure (up to 340 bar has been reported [27], but the commercial units rarely exceed 14 bar). Obviously, operation at elevated pressure simplifies the connection to the hydrogen storage, that is, it eliminates the need for a compressor. Of course, pressurization of hydrogen requires slightly higher operating voltage (as per Nernst equation), and the excess power actually corresponds to isothermal compression (plus activation losses on both electrodes and ohmic losses). Oxygen may also be pressurized and stored for later use in a fuel cell, although problems and additional cost related to oxygen storage often exceed benefits of fuel cell operation with pure oxygen versus operation with air (higher voltage and efficiency). Oxygen is therefore usually just vented from the system.

Recently, high temperature solid oxide electrolysis has attracted attention for large-scale generation of hydrogen from renewable (or nuclear) power. At high temperatures, electricity requirements for the water splitting process decrease, so it is possible to operate such electrolyzers at thermoneutral potential (corresponding to 100% stack efficiency) [28].

46.4.2
Hydrogen Storage

There are different types of hydrogen storage, namely, compressed gas, liquid hydrogen, metal hydrides, chemical hydrides, and so on [29]. However, the simplest method, and thus most frequently applied in the systems for autonomous power supply, is to store hydrogen as compressed gas. Hydrogen has a very low density (0.08376 kg m^{-3} at room conditions) so its storage requires large volumes (11.93 m^{-3} for 1 kg of hydrogen). These can be somewhat reduced by compressing hydrogen, so at 200 bar 1 kg of hydrogen would require 0.068 m^3 of storage space. Table 46.2 shows the volume required to store 1 kg of hydrogen at 20 °C. Pressures up to 700 bar have been applied in automotive applications. However, pressurization to such high pressures requires special storage tank materials and significant energy (about 10% of the hydrogen lower heating value) and it can be justified only where the space is really limited (such as in the automobiles). Pressurization may be accomplished by the electrolyzer (if a pressurized electrolyzer is available) or a hydrogen compressor may be needed. In the latter case a buffer tank may be required between the electrolyzer and the compressor. A usual way of storing hydrogen at pressures up to 200 bar is in steel cylinders which then may be connected in packs (usually of six cylinders in a pack). Note that a typical cylinder at 200 bar contains only 0625 kg of hydrogen and weighs about 50 kg.

Table 46.2 Volume required to store 1 kg of hydrogen at 20 °C.

Pressure (bar)	Volume (m³)
1	11.9344
10	1.2235
100	0.1288
200	0.0684
300	0.0484
350	0.0427
450	0.0349
700	0.0257

46.4.3
Fuel Cell

The most common type of fuel cells used in autonomous power supply systems is the PEM fuel cells. PEM fuel cell technology is the most advanced for low power operation (<100 kW), has quick startup, due to low operating temperatures (typically from 45 to 80 °C), and also has relatively fast response to sudden changes of load, and has a relatively simple supporting system. The fuel cell supporting system includes air supply, and pressure and temperature controls.

Fuel cell efficiency is proportional to its voltage, but the fuel cell system efficiency must also take into account the parasitic losses (i.e., losses in voltage regulation and power needed for auxiliary components in the fuel cell system, such as pumps, fans, relays, controller, etc.). The fuel cell system efficiency (based on the hydrogen higher heating value) is [22]:

$$\eta_{FC} = \frac{V_{cell}}{1.482} \frac{i}{i + i_{loss}} (\eta_{DC} - \xi) \quad (46.2)$$

where:

V_{cell} = individual/average cell potential (V);
i = current density corresponding to V_{cell} (A cm^{-2});
i_{loss} = current density loss corresponding to hydrogen crossover and internal currents (A cm^{-2});
η_{DC} = efficiency of DC/DC voltage regulator;
ξ = ratio between parasitic power and fuel cell gross power output;

Usually, fuel cell efficiency is expressed based on the hydrogen lower heating value (in that case the number 1.482 in Eq. (46.2) must be replaced by 1.253, which is the voltage that corresponds to hydrogen lower heating value). In commercial systems the fuel cell efficiency at nominal power is available from 40% to 60%. Fuel cell efficiency is not constant throughout the operating range, so at lower power levels it may be even higher than the rated efficiency at nominal power. However, it is not recommended to operate a fuel cell system at very low power levels as the efficiency could be very low.

46.5
System Issues

46.5.1
Capacity Factor

An electrolyzer may be sized to receive all the power generated from a PV array, but it in that case it would operate with the same capacity factor as the PV array or lower, which is determined by the sun availability. The capacity factor, defined as a ratio between average and nominal power, for solar energy only in the best locations can exceed 25% and in most cases is about 20% or lower (depending on the latitude and climate conditions). For wind power plants the capacity factor may exceed 40% or even 50% for good windy locations and in most cases is about 30%. The capacity factor represents a coefficient of utilization of installed capital, and therefore it is an important factor in determining the economics of any power generating or energy conversion device. A more economical option may be to size the electrolyzer at a power lower than the PV maximum power output. In that case some of the power from the PV could be unutilized, or stored in the batteries, but the electrolyzer would operate with a higher capacity factor. For any combination of PV availability and load profiles, there is an optimum electrolyzer capacity. Similar consideration is valid for sizing the fuel cell. The economics of PV–hydrogen or wind–hydrogen systems greatly depends on the configuration of the system and its application, in addition to the available solar insolation and/or wind.

46.5.2
Coupling Efficiency

Coupling of a source such as a PV array with an electrolyzer may result in somewhat lower efficiency, due to the losses related to power/voltage matching. DC/DC voltage regulators may be designed to operate with an efficiency as high as 93–95%, but this high efficiency may be achieved only in a very narrow power range. In a highly variable mode of operation such as with the input from a PV array, this efficiency may be considerably lower. If there is a good match between the polarization curves of PVs and electrolyzers, experience from a handful of PV/electrolysis pilot plants shows that they can be matched directly (with no power tracking electronics) with relatively high efficiency, >93% coupling efficiency [30,31].

46.5.3
Intermittent Operation

Direct coupling of an electrolyzer to a PV array or to a wind turbine implies intermittent operation with highly variable power output. The problem is that at very low loads the rate at which hydrogen and oxygen are produced (which is proportional to current density) may be lower than the rate at which these gases permeate through electrolyte, and mix with each other. This may create hazardous conditions inside the electrolyzer. Hydrogen flammability limits in oxygen are between 4.6% and 93.9%, but the alarms and automatic shutdown of the electrolyzer are set at much safer concentrations. The hydrogen permeation rate at 80 °C through Nafion 117, typically used in PEM electrolyzers, should be less than $1.25 \times 10^{-4}\,cm^3\,s^{-1}\,cm^{-2}$ at atmospheric pressure, corresponding to a current density of $0.002\,A\,cm^{-2}$ [32], which is rather negligible compared to $1\,A\,cm^{-2}$, a typical current density in PEM electrolyzers at full power. However, hydrogen permeation rate increases linearly with pressure, which means that at 200 bar the hydrogen loss would be $\sim 0.4\,A\,cm^{-2}$. The oxygen permeation rate through hydrated Nafion membrane is considerably (an order of magnitude) lower [33].

46.5.4
Water Consumption

Theoretically, from 1 liter of water $1.24\,Nm^3$ of hydrogen can be produced. Actual water consumption is about 25% higher, since both gases leave the electrolyzer wet, and some water is lost from the system due to oxygen exhaust and periodic hydrogen purge. This is particularly important for systems that are supposed to work in a closed loop with a fuel cell. The fuel cell produces water but in the systems that use air instead of pure oxygen most water is lost due to air exhaust from the fuel cell.

46.5.5
Performance Degradation with Time

When sizing an electrolyzer for operation in an autonomous power system, one must take into account voltage increase with time due to performance degradation. If the efficiency or the hydrogen generation rate are important, then the end-of-life electrolyzer performance must be taken into account rather than beginning-of-life performance. The voltage increase rate in typical electrolyzers is in the range $2\text{--}10\,\mu V\,h^{-1}$ [27]. Similar degradation occurs in fuel cells, but results in voltage decrease over time.

46.6
Sizing

For a given load and given resources profiles (solar insolation, wind availability) it is necessary to size the key system components, namely, nominal power of the power source, nominal capacity (or nominal power) of the electrolyzer, size of hydrogen storage, capacity of the batteries, and fuel cell nominal power. A fuel cell should be sized to meet the power need at any instant. However, by combining with other power technologies, such as batteries or ultracapacitors, it is possible to reduce the fuel cell installed/nominal power. Obviously, there is room for optimization and an optimal configuration is a function of fluctuations in both renewable energy availability and power demand, on one side, and the unit prices of individual equipment (namely, fuel cell, batteries, hydrogen storage, etc.) on the other.

The power source must be sized so that produced electricity on an annual basis is sufficient not only to satisfy the load but also to include all the losses related to energy storage and energy management (battery charging/discharging, electrolyzer and fuel cell operation, voltage regulators/inverters). The efficiency of the entire autonomous system defined as a ratio between delivered energy and incoming/captured renewable energy is:

$$\eta_{SYS} = \eta_{PS}[\zeta + \eta_{ST}(1-\zeta)]\eta_{DC/AC} \qquad (46.3)$$

where:

η_{PS} = efficiency of the power source;

ζ = ratio between energy directly delivered from the power source to the load and energy generated by the power source;

η_{ST} = efficiency of energy storage system, defined as the ratio between energy coming out of the energy storage and energy coming in the energy storage (for electrolyzer/hydrogen storage/fuel cell path this can be from 30% to 40% and for batteries it may exceed 80%);

$\eta_{DC/AC}$ = efficiency of DC/AC converter.

The electrolyzer must be sized so that excess power (power from the power source minus load) may be converted into hydrogen. The fuel cell must be sized so that the required power to the load is provided at any given moment. Hydrogen storage must be sized so that the amount of hydrogen is sufficient for fuel cell operation. These three components (electrolyzer, fuel cell, and hydrogen storage) must be sized in conjunction with a battery bank, which is usually a part of the system, as part of the energy is also stored in the batteries. The amount of energy to be stored (in batteries and in hydrogen) must be determined from the dynamics of power availability and power demand. For any given load and resources profile there is an infinite number of combinations of the component sizes that would satisfy the requirements of an autonomous power supply system. The selected/designed system must satisfy the power requirements at any given moment during the entire year. The sizing usually requires simulation and optimization procedure with minimum total cost (capital and operating) as the optimization objective. However, the sizing procedure is tightly connected to controls and energy management, as a different controls and energy management strategy may affect the electrolyzer and fuel cell efficiencies and energy to be stored in batteries as opposed in hydrogen, and thus affect the required size of the individual equipment.

46.7
Energy Management

Several different energy management strategies may be employed in an autonomous energy system [23–25]. In such a system power from the power source and power needed by the load are uncontrollable inputs. The supervisory control system makes decisions on where the power is going or where the power is coming from based on power from the power source and power needed by the load at any given moment, and based on the pre-programmed set of rules. The variables needed to make decisions are battery state of charge and hydrogen storage pressure (as a measure of hydrogen content in the storage). An optimal control strategy in making decisions assigns higher priority to the power route with higher transmission efficiency. The route with the highest efficiency is obviously a direct route from the source to the end user load. The next highest efficiency is a route through the batteries (charging and discharging process). The next highest efficiency is a route through the electrolyzer, hydrogen storage, and the fuel cell. The power routes that include power from the batteries to the electrolyzer and from the fuel cell to the batteries are the least efficient and should be used only when necessary. Therefore, such a strategy will operate the system as follows:

1) When the instantaneous electrical power from the power source exceeds the demand of the end-use load power is transferred directly from the source to the end-use load. At the same time, if the battery is under-charged (i.e., battery state-of-charge is low), the excess power is used to charge the battery. If the battery is fully charged (i.e., battery state-of-charge is high), the excess electricity would be inputted into the electrolyzer to produce hydrogen.
2) When power from the power source is insufficient for the demand of the end user load and battery state-of-charge is high, power is also directly sent to the end user and the battery provides the electricity to compensate the deficit power.
3) When power from the power source is insufficient for the demand of the end user load and battery state-of-charge is low, the fuel cell is switched on to utilize the stored hydrogen to generate electricity and compensate the deficit power.

The on/off-switching of the electrolyzer and fuel cell in an autonomous H_2 system that uses a battery as a short-term energy buffer can be based on the state of charge (SOC) of the battery in a double hysteresis loop control scheme [23,24] as shown in Figure 46.4. The hysteresis loops are used to prevent the electrolyzer and the fuel cell from being switched on/off too frequently. The on/off switching actions of the electrolyzer and the fuel cell will be determined

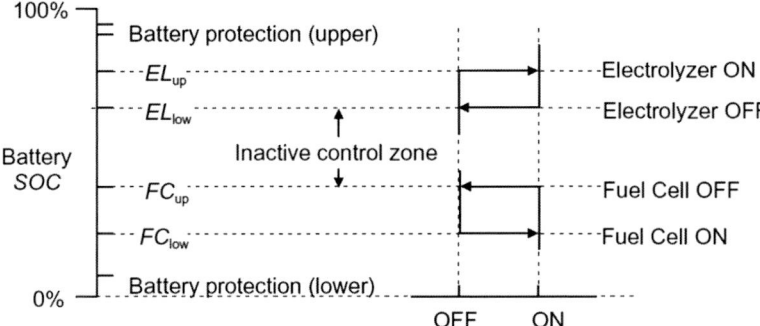

Figure 46.4 Control strategy using battery state-of-charge for an autonomous H_2 system (Reprinted from [23] with permission from Elsevier).

by the battery state-of-charge, where EL_{up}, EL_{low}, FC_{up}, and FC_{low} are the key control parameters. Usually, EL_{up}, $EL_{low} = (0.9, 0.8)$ or $(0.85, 0.75)$ and FC_{up}, $FC_{low} = (0.45, 0.35)$ or $(0.5, 0.4)$ are the appropriate parameters in practical applications. The basic control settings for the on/off-switching of the electrolyzer and fuel cell can also be made seasonally dependent [23,24].

The electrolyzer can be operated in either the fixed or the variable input current/power mode. Simulation results in Reference [23] have demonstrated that the electrolyzer in the variable power mode leads to a higher energy efficiency of the PV-H_2 system than the electrolyzer in fixed power mode does. The fuel cell can also be operated in either the fixed or the variable input current/power mode; however, most simulation studies consider it as a constant current/power source and its current/power as yet another control variable. The following general recommendations can be made [23]:

- The electrolyzer should operate in a variable power mode, rather than in a fixed power mode.
- The set point for switching on the electrolyzer should be as high as possible.
- The fuel cell should operate at the lowest power possible power level.
- The set point for switching off the fuel cell should be as low as possible.

Performance of an autonomous H_2 system can be significantly affected by relatively small changes made in the control strategy. The hydrogen storage can be used as a key system performance indicator. Figure 46.5 shows the hydrogen storage level and battery state of charge during a typical year for an autonomous PV-H_2 system [23].

However, real life control strategies should also take into account other important parameters such as performance degradation of the various electrochemical components (electrolyzer, fuel cell, and battery). It is necessary to avoid operating schemes that are likely to reduce the lifetime (due to degradation). Hence, other system performance parameters, such as the number of starts and stops and minimum allowable run times for the electrolyzer and the fuel cell,

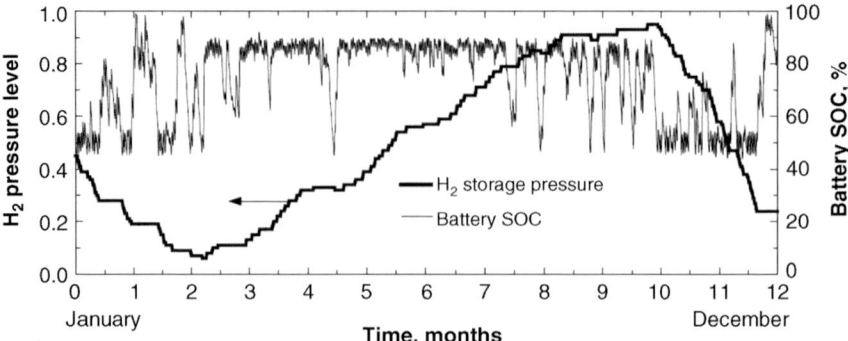

Figure 46.5 Hydrogen pressure level (normalized) and battery state of charge during a typical year for an autonomous PV-H_2 system (Reprinted from [23] with permission from Elsevier).

and battery state of charge levels, must also be considered. A more detailed system design and optimization of system operation should take into consideration more factors, such as advanced control strategy (an intelligent control algorithm, which is based on solar radiation, wind and load forecasting), thermal impacts on the operation efficiency of critical components (e.g., PV, electrolyzer, battery, and fuel cell), and optimum operation pressures for system components (e.g., electrolyzer, storage tank, and fuel cell). Moreover, to have an accurate techno-economic optimization, the overall system cost should be calculated on the basis of investment costs, operating costs, and life times of the main components.

46.8
Other Uses/System Configurations

46.8.1
Demand Side Management

An autonomous power supply system operates in regimes where either there is excess available power which then must be stored or there is insufficient power which then must be drawn from the energy storage. In such a system, demand side management, including energy conservation and rational use of energy, is an absolute necessity. With adequate demand side management it is possible to achieve significant savings in size of the power generation equipment and particularly in energy storage. Therefore, demand side management may be incorporated within the supervisory system control.

46.8.2
Seawater Desalination

In most cases, islands do not have water wells, but water may be produced by desalination of seawater. However, that process requires energy (around

$3\,kWh\,m^{-3}$). One of the most common processes is reverse osmosis, which happens to be the same process that is used to prepare water for the electrolysis process. It is therefore possible to integrate the water generation plant with the energy supply system. In that case, excess electricity from a renewable energy source may be used in the seawater desalination process. Of course, in sizing the power source one must take into account the energy needed for desalination.

46.8.3
Oxygen Use

The water electrolysis results in both hydrogen and oxygen being produced on the opposite sides of the membrane. The electrolyzer produces stoichiometric amounts of oxygen, that is, for every two moles of hydrogen a mole of oxygen is produced as well. Oxygen is typically vented, but if needed oxygen may be captured and stored for later use (e.g., in a fuel cell). Storage and handling of oxygen, particularly at elevated pressures, requires special care. Oxygen produced in the electrolysis process as a byproduct may be used in wastewater treatment plants on the islands. The township Kalyves on the island of Crete uses a seawater electrolysis process to generate oxygen and ozone which are then used in wastewater treatment (and hydrogen in that case is just a byproduct).

46.8.4
Fuel for Vehicles

To make an island completely energy self-sustained the energy supply system must also produce fuel for transportation. Hydrogen, of course, is an ideal fuel, although electric vehicles may be competitive because of the limited driving range. Hydrogen has the advantage because it can be stored, which means it can be produced during periods when there is excess power from the renewable source. Fuel cell vehicles usually store hydrogen at 350 or 700 bar. Therefore, compressors are needed to raise the pressure from the hydrogen storage tank. Hydrogen fuel cell vehicles are already on the brink of commercialization by major automobile companies. Various utility vehicles may also be used on the islands. Several of the existing island hydrogen demonstration projects involve a refueling station for hydrogen vehicles.

46.9
Conclusions

Islands are ideal polygons and showcases for demonstration of individual hydrogen technologies as well as entire mini hydrogen, that is, hydricity, economies. Experience from the demonstration projects will be valuable in designing and implementing larger hydrogen systems on national and regional levels. There are already several ongoing or proposed demonstrations of the integrated hydrogen systems on islands all over the world. They vary in size from kW to MW, and in scope from a refueling station to a complete autonomous island energy supply.

A survey of some existing islands energy supply projects involving hydrogen [4] have pointed out the main barriers such as:

- lack of mass produced, reliable, off-the shelf systems;
- funding and permitting for systems;
- challenges in design, installation, and maintenance of these projects due to location and a lack of local expertise.

To address these barriers the International Association for Hydrogen Energy launched the Hydrogen Islands Initiative about a decade ago, which was carried out by the UNIDO-International Center for Hydrogen Energy Technologies (2006–2012) [3]. This Initiative has the following goals:

- facilitate the exchange of technical and organizational information between the hydrogen related demonstration projects on the islands all over the world through a web site, newsletter, and periodic conferences;
- provide technical and organizational assistance for starting new projects;
- represent a link between the users, suppliers, financiers, local governments, and international organizations;
- encouraging partnerships and clustering in projects, rather than one-of-a-kind approaches;
- facilitating international business opportunities for communities by actively supporting communities in exporting their systems or expertise;
- popularize hydrogen energy and its applicability to the island communities worldwide.

There is still a need, and even urgency, to bring this initiative to the forefront of national, international, and business energy agendas. Island hydrogen projects may create a significant market for equipment manufacturers (electrolyzers, fuel cells, hydrogen vehicles) who are struggling to introduce their products. With self-sufficient renewable energy systems, in which hydrogen serves as energy storage medium and fuel, islands worldwide will become energy self-sufficient, and serve as the showcases for development of the global sustainable energy system.

References

1 Barbir, F. (2011) Future of hydrogen and fuel cells, in: *Proceedings HYCELTEC 2011, III Iberian Symposium on Hydrogen Fuel Cells and Advanced Batteries* (eds F. Barreras, R. Mustata, A. Lozano, L. Valino, and J.A. Picazo), Spanish National Research Council, Zaragoza, pp. 71–74.

2 Scott, D.S. (2007) *Smelling Land: The Hydrogen Defense Against Climate Catastrophe*, Enhanced Edition, Queen's Printer of British Columbia, Victoria, B.C.
3 Lymberopoulos, N., Barbir, F., and Fjermestadt Hagen, E. (2008) Islands as niche markets for hydrogen energy technologies, in *17th World Hydrogen Energy Conference 2008 (WHEC 2008)*, 15–19 June 2008, Brisbane, Australia, Australian Association for Hydrogen Energy (AAHE), vol. 1, pp. 347–348.
4 Miles, S. and Gillie, M. (2009) Benefits and Barriers of the Development of Renewable/Hydrogen Systems in Remote and Island Communities, International Energy Agency, Hydrogen Implementation Agreement, Task 18 Evaluation of Integrated Hydrogen Systems. http://ieahia.org/pdfs/RemoteIslandBenefits.pdf (accessed: 27 December 2013).
5 Gazey, R., Aklil, D., and Salman, S. (2006) A field application experience of integrating hydrogen technology with wind power in a remote island location. *J. Power Sources*, 157 (2), 841–847.
6 Ulleberg, Ø., Nakken, T., and Eté, A. (2010) The wind/hydrogen demonstration system at Utsira in Norway: evaluation of system performance using operational data and updated hydrogen energy system modeling tools. *Int. J. Hydrogen Energy*, 35 (5), 1841–1852.
7 Sigfusson, T.I. (2007) Hydrogen island: the story and motivations behind the Icelandic hydrogen society experiment. *Mitigation Adapt. Strateg. Global Change*, 12 (3), 407–418.
8 IPHE (March 2011) HYLINK: Distributed Hydrogen Energy System for Remote Areas. http://www.iphe.net/docs/Renew_H2_HYLINK.pdf (accessed: 27 December 2013).
9 Wing, J. (2011) Islands as Hydrogen Infrastructure Demonstrators. http://www.cleantechinvestor.com/portal/islands/10029-islands-as-hydrogen-infrastructure-demonstrators.html (accessed: 27 December 2013).
10 Department of Transportation, Infrastructure and Energy (2011) PEI Wind Hydrogen Village. http://www.gov.pe.ca/energy/index.php3?number=1007450&lang=E (accessed: 27 December 2013).
11 Gardiner, A. (2013) Matiu/Somes Island – a remote site for distributed hydrogen. Presented at HYPOTHESIS X, Edinburgh, 11–12 June 2013. Available at http://www.hypothesis.ws/HYPOTHESIS_X/Presentazioni/HYP_012.pdf (accessed: 5 January 2014).
12 Yazici, M.S. and Hatipoglu, M. (2012) Hydrogen and fuel cell demonstrations in Turkey. *Energy Proc.*, 29, 683–689.
13 Oprisan, M. (2007) Introduction of hydrogen technologies to Ramea Island. Presented at the IEA Wind – KWEA Joint Workshop, April 2007. Available at http://www.ieawind.org/wnd_info/KWEA_pdf/Oprisan_KWEA_.pdf (accessed: 6 January 2014).
14 Avril, S., Arnaud, G., Florentin, A., and Vinard, M. (2010) Multi-objective optimization of batteries and hydrogen storage technologies for remote photovoltaic systems. *Energy*, 35 (12), 5300–5308.
15 Kai, T., Uemura, Y., Takanashi, H., Tsutsui, T., Takahashi, T., Matsumoto, Y., Fujie, K., and Suzuki, M. (2007) A demonstration project of the hydrogen station located on Yakushima Island – operation and analysis of the station. *Int. J. Hydrogen Energy*, 32 (15), 3519–3525.
16 Krajačić, G., Martins, R., Busuttil, A., Duić, N., and Carvalho, M.G. (2008) Hydrogen as an energy vector in the islands' energy supply. *Int. J. Hydrogen Energy*, 33 (4), 1091–1103.
17 Ntziachristos, L., Kouridis, C., Samaras, Z., and Pattas, K. (2005) A wind-power fuel-cell hybrid system study on the non-interconnected Aegean islands grid. *Renew. Energy*, 30 (10), 1471–1487.
18 Tzamalis, G., Zoulias, E.I., Stamatakis, E., Parissis, O.-S., Stubos, A., and Lois, E. (2013) Techno-economic analysis of RES & hydrogen technologies integration in remote island power system. *Int. J. Hydrogen Energy*, 38 (26), 11646–11654.
19 Duić, N. and Carvalho, M.G. (2004) Increasing renewable energy sources in island energy supply: case study Porto Santo. *Renew. Sustain. Energy Rev.*, 8 (4), 383–399.

20 Parissis, O.-S., Zoulias, E., Stamatakis, E., Sioulas, K., Alves, L., Martins, R., Tsikalakis, A., Hatziargyriou, N., Caralis, G., and Zervos, A. (2011) Integration of wind and hydrogen technologies in the power system of Corvo island, Azores – a cost-benefit analysis. *Int. J. Hydrogen Energy*, 36 (13), 8143–8151.

21 Alves, M. (2008) Hydrogen energy: Terceira Island demonstration facility. *Chem. Ind. Chem. Eng. Q.*, 14 (2), 77–85. http://www.doiserbia.nb.rs/img/doi/1451-9372/2008/1451-93720802077A.pdf (accessed 7 January 2014).

22 Barbir, F., Molter, T., and Dalton, L. (2005) Efficiency and weight trade-off analysis of regenerative fuel cells as energy storage for aerospace applications. *Int. J. Hydrogen Energy*, 30 (4), 231–238.

23 Ulleberg, O. (2004) The importance of control strategies in PV–hydrogen systems. *Solar Energy*, 76 (1–3), 323–329.

24 Zhou, K., Ferreira, J.A., and de Haan, S.W.H. (2008) Optimal energy management strategy and system sizing method for stand-alone photovoltaic-hydrogen systems. *Int. J. Hydrogen Energy*, 33 (2), 477–489.

25 Castaneda, M., Cano, A., Jurado, F., Sanchez, H., and Fernandez, L.M. (2013) Sizing optimization, dynamic modeling and energy management strategies of a stand-alone PV/hydrogen/battery-based hybrid system. *Int. J. Hydrogen Energy*, 38 (10), 3830–3845.

26 Ayers, K. (2013) PEM Electrolysis R&D Webinar http://www1.eere.energy.gov/hydrogenandfuelcells/pdfs/webinarslides052311_pemelectrolysis_ayers.pdf (accessed: 2 January 2014).

27 Barbir, F. (2005) PEM electrolysis for production of hydrogen from renewable energy sources. *Solar Energy*, 78 (5), 661–669.

28 Petipas, F., Brisse, A., and Bouallou, C. (2013) Model-based behaviour of a high temperature electrolyser system operated at various loads. *J. Power Sources*, 239, 584–595.

29 Sherif, S.A., Barbir, F., Veziroglu, T.N., Mahishi, M., and Srinivasan, S.S. (2007) Hydrogen energy technologies, in *Handbook of Energy Efficiency and Renewable Energy* (eds F. Kreith and D.Y. Goswami), CRC Press, Boca Raton, pp. 27.1–27.16.

30 Steeb, H.A., Brinner, A., Bubmann, H., and Seeger, W. (1990) Operation experience of a 10 kW PV-electrolysis system in different power matching modes, in *Hydrogen Energy Progress VIII*, vol. 2 (eds T.N. Veziroglu and P.K. Takahashi), Pergamon Press, New York, pp. 691–700.

31 Atlam, O., Barbir, F., and Bezmalinović, D. (2011) A method for optimal sizing of an electrolyzer directly connected to a PV module. *Int. J. Hydrogen Energy*, 36 (12), 7012–7018.

32 Kocha, S., Plasse, P., Onishi, L., Wheeler, D., and Bett, J. (2002) In-situ electrochemical measurement of hydrogen crossover currents in PEMFC, in *Proceedings Fuel Cell Technology: Opportunities and Challenges, 2002 AIChE Spring National Meeting*, 10–14 March 2002, New Orleans, LA, AIChE, pp. 307–312.

33 Sakai, T., Takenaka, H., and Torikai, E. (1986) Gas diffusion in dried and hydrated Nafions. *J. Electrochem. Soc.*, 133 (1), 88–92.

Index

a

acetate, 116, 420–422, 471, 475
acetic acid, 34, 116, 117, 139, 203, 421
– decarboxylation of, 116
acetylene, 135, 137, 141–142, 145
achievable hydrogen prices per sector., 275
acid gases, 499
acidic electrolysis, 331. *see also* PEM electrolysis
activated alumina, 557
activated carbon, 494
– adsorption equilibrium isotherms of CO_2 on, 500
– adsorption of CO_2 on, 500
– CO_2 diffusion, 500
– efficiency for CO_2 removal, 500
– isotherms of water on, 499
– layer of, 500
– loading of CO, 500
– mechanical resistances of, 500
– order of adsorption in, 500
– utilization of, 499
activated/non-activated cells, 1049
additives, 115, 302–303, 339, 779–780
– titanium-based, 779
adiabatic expansion energy, 713, 730
adiabatic reactors, 180–181, 185–185, 206
adsorbents, 494
– activated carbon, 494
– – capillary filling, 494
– – water isotherms, 494
– adsorption equilibrium properties of, 496
– alumina, 494
– – adsorption isotherms, 494
– – silica gel, 494
– – water isotherm, 494
– selective, 499
– zeolites, 494
– – utilization of, 494

adsorption, 16, 316, 492
– energies of, 492
– equilibrium isotherms, 492
– – shape of, 492
– – type I, 492
– – of CO_2, CH_4, 493
advanced hydrogen development turbine programme, 1023
aeroengines, 1026
agitated granular sludge bed reactor (AGSBR), 427
agricultural residues, 423–424
AIP. *see* air-independent propulsion (AIP) systems
air condition (A/C) systems, 972
aircraft design, planning of, 966
air-independent propulsion (AIP) systems, 991
– fuel cell technology in, 991
– HDW fuel cell, 992
– – class 214, 992
– – class 212A, 992
– overview of, 993
– requirements for, 991
air pollutants, 545
alanates, 774
alcohols, 116, 146, 179, 181, 182, 360, 427, 473
Alder reaction, 115
algae
– biomass production, high, 473
– [FeFe]-hydrogenase activation, 466
– ferredoxin, 472
– genetically modified, 473
– H_2 production in, 465
– hydrogenase, 466
– physiology and biochemistry of, 465
alkaline earth metal oxides, 153
alkaline electrolysis, 27, 283, 311–312, 331
– characteristic cell performance for, 313
– compression, 314

Hydrogen Science and Engineering: Materials, Processes, Systems and Technology, First Edition.
Edited by Detlef Stolten and Bernd Emonts.
© 2016 Wiley-VCH Verlag GmbH & Co. KGaA. Published 2016 by Wiley-VCH Verlag GmbH & Co. KGaA.

– current–voltage characteristics and operating ranges for, 332
– gas separation in, 313–314
– schematic of, 314
– technology, 1040
alkaline electrolyzer, 317, 1036, 1037, 1061
– based risk analysis, 1039
– cooling water system, 1038
– design of, 1041
– feed water treatment in, 1038
– main components of, 1038
– safety evaluation of compressors, 1041
– specifications of, 1039
alkaline water electrolysis cell
– anode and cathode polarization, voltage of, 291
– compositions of typical cell voltage, 292
alkylation, 10
alumina, 7, 499
– mechanical properties of, 499
– utilization of, 499
– volumetric adsorption equilibrium, higher, 499
aluminum, 143
aluminum plate fin heat exchangers, 758
– advantages, 760
– – high flexibility, 760
– – high specific heat exchange surface, 760
– – low pressure losses, 760
– – low specific costs, 760
– parameters, 758
aluminum powder, 799
ambient temperature, 796
American Lung Association, 845
amides, 775
amines like diethanolamine (DEA), 558
amino acids, 115
ammonia, 5, 7, 10, 19, 491, 544, 814, 933
– coal-based production, 26
– global production of, 19, 32
– Haber–Bosch process, 32
– production, 283
– production, based on steam reforming of natural gas, 32–33
– – schematic depiction of, 33
– production data based on, 21
– syngas production for, 32
– syngas production from natural gas
– – composition of, 240
– – methane steam reforming, 240
– – steps, 239
– – water–gas shift reaction

– – Influence of temperature on equilibrium constant, 241
– synthesis based on hydrogen
– – produced from water electrolysis, 33–34
– synthesis reactor, 33
anaerobic digestion technology, 398
anaerobic filter (AF) reactor, 429
anaerobic sequencing batch reactor (ASBR), 427
anode–electrolyte interface, 448
anoxygenic photosynthesis, 465
aqueous hydrogen peroxide, 659
argon, 134, 142
argon–methane mixtures, 141
Arthrospira platensis, 470
artificial leaf, 472
– development of, 472
– silicone, 472
– technical difficulties, development of, 472
asbestos, 41
Asia-Pacific Economic Cooperation, 922
Aspen Custom Modeler, 516
asphaltene molecule, 10
ASTERIX project, 168
atmospheric alkaline electrolysis, 1036
ATR. see autothermal reforming (ATR)
ATR reactors, 220
Au–CeO_2 interaction, 208
automotive hydrogen storage, 712
automotive pressurized fuel cell system, 793
automotive technology, 920
autonomous energy system, 1090
– energy management strategies for, 1090–1092
autonomous H_2 system, 1091
– control strategy using battery state-of-charge for, 1091
autonomous power supply, 1075
– converters, 1083
– fuel cells used in, 1086
– schematic diagram of an, 1082
– system configurations for, 1092–1093
– – demand side management, 1092
– – fuel for vehicles, 1093
– – oxygen use, 1093
– – seawater desalination, 1092–1093
– system design/Configuration for, 1082–1083
autonomous PV-H_2 system, 1092
– hydrogen pressure level, and battery state of charge, 1092
autothermal reformer, 204

autothermal reforming (ATR), 25, 35, 131, 133, 134, 181, 190, 195, 197, 199, 202, 220, 221, 234
autotrophs, 471
auxiliary cabin heating system, 972
auxiliary power systems, 976
auxiliary power unit (APU), 133
– gliding arc unit, components, 133
Azospirillum brasilense, 470
Azospirillum lipoferum, 470

b

balance equations, 179
balancing mechanism (BM), 261
balancing power plants, 1057
band gap, 443–448, 453, 454
– necessary to split water, 444
– positions, for semiconductors, 443, 444
batteries, 265
– and charging technology, 711
– quick charging stations, 711
battery electric vehicles (BEVs), 842
battery-powered vehicles, 827
battery type brake energy recovery, 987
bauxite, 53
benzene–water systems, 659
Berlin Heidestrasse TOTAL hydrogen refueling station, 906
BEVs. *see* battery electric vehicles (BEVs)
biodesulfurization, 16
biodiesel, 178, 473
biofilm, 429
biofuels, 65–66, 202, 393, 402, 417, 473, 476, 1058
– environmental performance of, 403
– production by other organisms, 473
– – biodiesel, 473
– – petroleum products, 473
biogas, 224, 401, 427, 476, 543
– fermenter tank, 1041
– reforming, 398–399
biohydrogen, 393, 417, 476
– advantages of, 477
– application of, 476
– approaches, 395
– – state of technology and research, and industrial players, 395–397
– – technical characteristics, 395–397
– economic aspects of, 477
– production, 427
– – BOD, 427
– – by dark fermentation using bioreactor, 428

– – from organic residues by dark fermentation, 427
– – palm oil mill effluent (POME), 427
bio-hydrogenation processes
– economic analyses of, 430
– future prospects, 431
– to make economically viable, 431
– reactor costs, 431
biological H_2 production, 463
– advantages of, 463
– classification, 464
– designing of, 471
– – Calvin cycle, 471
– – genetically modified organisms, 471
– endergonic reaction, 469
– factors affecting, 470
– factors influencing, 466
– hydrogenase for, 466
– methods to enhance, 474
– – dark fermentative H_2 Production, 475
– – genetic engineering for, 476
– – photobioreactors, 474
– – photolytic H_2 Production by microorganisms, 474
– – photosynthetic bacterial H_2 production, 474
– – water electrolysis, 474
– nitrogenase, 463, 466
– nitrogen fixation, 463
biological hydrogen production, 419–420
biomass, 109, 116, 237
– catalysts, coke produced from, 119
– cheaper, conversion into hydrogen, 431
– concentration of typical intermediates for degradation of phytomass glucose and, 117
– -derived fuels, 419
– hydrogen production from, 237
– hydrothermal carbonization (HTC) of, 110
– hydrothermal gasification, 111
– influence of lignin, 118–119
– influence of proteins, 118
– influence of salts, 118
– liquefaction, 110
– methane, nickel and noble metal catalysts, 119
– reaction scheme of hydrothermal degradation, 117
– solubilization of, 114
– sources, as substrate for, 423
biomass gasification product gas
– carbon dioxide removal from, 534, 535
– – separation performance for, 535
biomass production and supply, 393

biomimetic approaches, using inorganic systems, 441
biomimetics, 471
biophotolysis, 464
– of water, 420
bioreactor, 427
– biohydrogen production, by dark fermentation using, 428
– choice of, 427
– prevent corrosion, 429
bipolar cell configuration, 1036
bitumen, from tar sand, 3
boundary conditions, 515
– definition of, 515
Bradyrhizobium japonicum, 467
brake resistor function, 972
Brandenburg, concentration of renewable energy production capacity in, 1033
Brandenburg Technical University Cottbus-Senftenberg (BTU C-S), 1040
Brunauer–Emmett–Teller (BET) model, 781
BTU C-S. *see* Brandenburg Technical University Cottbus-Senftenberg (BTU C-S)
bubble management, 302–303
bubble phenomena, 293
– advancing and receding angles, 294
– bubble departure diameter predictions, 293–296
– critical diameter, stagnant model approximated equation, 295
– gas bubble on an electrode surface, schematic diagram, 294
– model predictions with experimental observations
– – comparison of, 296–297
– prediction of critical diameter, parameters used in, 296
bubble resistance, 293
buffer capacity, 796
buffer hydrogen, 796
buffer pressure, 797
bulk density, 800
bus applications, auxiliary power demands for, 970, 971
bus propulsion system, 967
– acceleration behavior, 970
– aerodynamic quality of, 969
– auxiliary power demand, 970
– heating and air condition power demand, 971
– hill climbing capability, 969
– internal combustion engines, 969
– maximum acceleration, definition of, 969

– propulsion power to drive, 967
bus refueling, 894
– 350 bar tanks, 894
– fueling nozzle, 894
– high storage density, 894
– pre-cooling step, 894
– protocols, 894
Butler–Volmer kinetics, 596

c

Ca-compounds, 101
CAES. *see* compressed air energy storage; compressed air energy storage (CAES)
Caldanaerobacter subterraneus, 420
Caldicellulosiruptor owensensis, 429
Caldicellulosiruptor saccharolyticus, 420
California Energy Commission (CEC), 843
Canada
– energy framework and relevant policies, 70
– – National Energy Board role, 70
– hydrogen related energy policy strategies, 74
– – Canadian Hydrogen and Fuel Cell Association (CHfCa), 74
– hydrogen research, development, demonstration, and deployment activities, 74–75
– – hydrogen and fuel cells progress, 75
– – RD&D expenditure, 74, 75
– – refueling infrastructure, 74
– – viability of hydrogen production and purification, 74
Canary Islands, 1080
– Instituto Tecnológico de Canarias (ITC), 1080
capacity factor, 1087
CAPEX. *see* capital expenditure (CAPEX)
capital expenditure (CAPEX), 276, 406, 408, 733, 751, 843
carbohydrates, 475
carbon capture and storage (CCS) process, 1022
carbon-containing byproducts, 138
carbon dioxide (CO_2), 190, 417, 517, 547, 629, 635
– emissions, 1011
– – man-made, 283
– platinum, 547
– thermodynamic equilibrium calculations, 547
carbon fiber (CF), 713, 794
carbon footprint
– of refinery operations, 4

carbon footprints
– of different energy sources, 418
carbon formation, 153, 179
– during cracking, 7
carbon monoxide (CO), 186, 190, 206, 237, 544, 546, 635, 900
carbon nanostructures growth, 785
carbon number, 116
carbon oxides, 139, 491
carbon source, 470
carbon suboxide, 140
car refueling, 892
– with compressed gaseous hydrogen (CGH), 893
– with conventional fuels, 894
– user-friendly, 893
carrier induced granular sludge bed reactor (CIGSB), 427
CASE. see compressed air energy storage (CAES)
case study, 800
– reverse engineering, 800
– – metal hydride refueling, 802
– – reference targets, 804
– – sensitivity study, 805
– – study parameters, 801
– – system analysis model, 803
catalyst poison, 546
– carbon monoxide, 548
catalysts, 119
– choice, 201
– -coated membrane (CCM), 300, 335, 338, 339
– development, 200–201
– screening, 186
catalytic combustion, 188
catalytic cracking, 7, 9, 11
catalytic dehydrogenation, 819
catalytic hydrolysis, 783
catalytic oxidation, 41
catalytic partial oxidation (CPOX), 193–198, 220
– advantages, 193
catalytic reactors, 180–181
– catalytic combustion, 188
– hydrogen generation, 181–184
– preferential oxidation, 186–187
– selective CO methanation, 187–188
– water–gas shift reaction, 184–186
catalytic reduction, 557–558
cathode activation process, 1050
caustic soda, 41, 43, 46
– economy of, 53–55

CcH_2. see cryo-compressed hydrogen (CcH_2)
CCM. see catalyst, -coated membrane (CCM)
CCS. see carbon capture and storage (CCS) process; CO_2 capture and storage (CCS)
CEC. See California Energy Commission (CEC)
cell density, 429
cellulose, 115
– hydrothermal biomass degradation, reaction scheme of, 117
centralized hydrogen production, 835
CeO_2/Ce_2O_3 redox system, 95
CEP. see Clean Energy Partnership (CEP)
ceramic honeycomb receivers, 156
– development, 156–157
ceramic membranes, 510
– molecular sieving, 511
– separation of hydrogen, 510
Ceria, 95
cerium-based oxides, 96
cerium oxide, 95
CF. see carbon fiber (CF)
CGH. see compressed gaseous hydrogen (CGH)
CGH_2. see compressed gaseous hydrogen storage (CGH_2)
chemical batteries, 264. see also batteries
chemical energy, 441
chemical engineering thermodynamics, 175–180
chemical equilibrium, 175, 178–180, 199
– limiting cases, 178
chemical industry, 19
chemical kinetics, 797
chemical reactions, 178
chemoheterotrophs, 471
Chernobyl, 88
Chicago Transit Association (CTA), 982
– test and field experience, 982
Chlamydomonas reinhardtii mutant Stm6Glc4, 466
chlorine, 41
– area of research, focused on, 49
– diaphragm electrolyzers, 45–46
– economy of, 53–55
– electrochemistry of production, 42
– energy saving in chlorine industry, 55–56
– ion exchange membrane electrolyzers, 46–49
– liquefaction systems, 49
– mercury electrolyzer technology, 43–45
– parameters for chlorine production cost, 43
– production, industrial technologies, 42
– – based on common principles, 42
– usage, 52–53

– – data from World Chlorine Council, 52
– World chlorine production, 54
chlorine-alkali production
– with oxygen depolarized cathode (ODC), 49–51 (*see also* oxygen depolarized cathode (ODC))
chlorine production, 28–29
– daphragm cell process, 31
– hydrogen as by-product of, 32
– membrane cell process, 29–30
– mercury cell process, 30–31
– new developments, 31–32
chloroplast, 465
– thylakoid membrane of, 465
chloroplasts, 441
CH_4/O_2 ratio, 142
CHP. *see* combined heat and power (CHP) systems
CHP fuel cell, 1064
Citaro FuelCELL-hybrid, 984
city bus, acceleration requirement for, 970
class 205 submarine, 992
clathrate hydrates, 659
clathrates, 597
– structure of the cubic clathrates, 597
– tetrahydrofuran (THF), 597
– tetra-*n*-butylammoniumfluoride (TBAF), 597
Clean Energy Partnership (CEP), 61, 843, 892
Clean Power Net, 64
clean-up technology, 230–231
clean urban transportation, 967
climate change, 1011
climate protection, 417
Clostridium thermocellum, 420
CMBC. *see* Costal Mountain British Columbia (CMBC)
CO. *see* carbon monoxide (CO)
coal, 3, 543, 1076
– gasification of, 543
– partial oxidation of, 32
coal-based production
– of ammonia, 26
coal gasification, 242, 244, 331, 393
– in different reactor types, conditions of, 243
cobalt, 7, 91
cobalt-molybdenum-alumina, 7
cobalt-molybdenum catalysts, 7
cobalt oxide, 7
CO_2 capture and storage (CCS), 1011
CoE. *See* cost of electricity (CoE)
CO_2 emissions, 64, 109
– from small-scale reforming, 231

CO_2 footprint, 231
CO/H_2 ratio, 168
CO_2 hydrogenation, 783
coke formation, 5
coke-forming reactions, 5
cold gas clean-up, 552
– adsorption processes, 552
– downstream synthesis, 552
– liquid scrubbing technology, 552
combined cycle power plant, 244
combined heat and power (CHP) systems, 1022
combustion technologies, 1012, 1021, 1024
– flameless oxidation, 1024
– – lean direct injection, 1026
– – micro-mixing, 1026
– – multi-injection, 1026
CO_2 (or dry) methane reforming (DMR), 151
complete gas cleaning system, design of, 205–208
complex hydrides, 591–592, 773
– alanates, 774
– amides, 591, 775
– borohydrides, 779
– hydrido alanates, 591
– hydrido borates, 591
– imides, 591
compressed air energy storage (CAES), 265, 622
compressed gas, 441
compressed gaseous hydrogen (CGH), 893
compressed gaseous hydrogen storage (CGH_2), 712–714, 812
– refueling, 717–719
– validation
– – challenges, and opportunities, 729
– – procedure, 726–729
compressed gas storage, special forms of, 704
compressed gas systems for vehicles, specific requirements, 699
– common filling stations, types, 703
– filling procedure, standardized, 701
– filling receptacle, 701
– legal and normative requirements, 699–700
– refueling, 700–704
– temperature curve, filling vehicle tank, 702
compressed hydrogen storage, 659, 662
– hydrogen compression, 662–663
– hydrogen pressure vessels, 663
– – conventional small-scale bulk storage, 663–665
– – current and potential small- to medium-scale bulk storage, 665–669

– – medium-scale bulk storage, new ideas for, 669
– large quantities, 659
compressibility factor, 606
compression, 314–316, 752
– compressors, application ranges for, 754
– cost optimum for, 315
– energy demand
– – isothermal/adiabatic, 315
– – multi-step adiabatic, 315
– of helium gas, 752
– of hydrogen gas, 752
– ideal isentropic compression process, 752
– inlet temperature (T_{in}), 752
– isentropic exponent, 752
– – impact of, 752
– low density, 753
– Lysholm screw rotors, 754
– outlet temperature (T_{out}), 752
– pressure ratio (p_{out}/p_{in}), 752
– reciprocating compressors, 755
– screw compressor, 753
compressors, 643, 903
– diaphragm, 904
– ionic, 904
– LH_2 pump, 904
– piston, 903
concentrated solar power (CSP), 151
– technology, development and perspective, 168
concentrated solar power (CSP) plants, 825
concentrating receiver system (CRS), 88
concentration polarization, 515
condensation, 5
conduction band (CB), 442
conductivity, 447
CO_2-neutral hydrogen production, 1042
conformable containers, 705–707
continuous stirred tank reactors (CSTRs), 427, 429
– disadvantage with, 430
– hydrogen production using, 429
control strategy using battery state-of-charge, 1091
conventional hydrogen production, 419
conventional vehicles, 834
coolant, 791
– temperature, 794
copper, 544
CO_2-reduction strategy, 967
corona-based plasma, 142
corrosion, 98, 100, 429, 447, 450, 635, 888, 944, 1012, 1083

– resistance, 859
Costal Mountain British Columbia (CMBC), 982
– test and field experience, 982
cost of electricity (CoE), 1022
coupling hydrogen production, 1059
coupling reforming with solar energy, 154
covalent bonding, 565–566
– atomic orbitals, 565
– diatomic molecules, 565
– natural isotopes, 566
– wave functions, 565
covalent hydrides, 590–591
– bond, 590
– hydrogen bonds, 591
– London forces, weak, 591
– nonpolar bonds, 590
– oxidizing agents, 591
cracking quality, 7
crude glycerol, 897
crude oil, chemical nature, 3
cryoadsorbing storage systems, 659
cryo-compressed gas, 711
cryo-compressed hydrogen (CcH_2), 713
– heat receptivity, 716
– hydrogen safety validation, 730–731
– refueling, 718, 719, 724
– – storage density change during, 725
– storage in a density–temperature diagram of hydrogen, operating range, 715
– storage options, heat receptivity, 716
– storage system
– – operating principles, 722–725
– system design, 720–722
– – prototype 2013, 720–722
– thermodynamic opportunities, 714–717
– validation challenges, and opportunities, 729
– validation opportunities, 729
– validation procedure, for storage system, 726–729
– vehicle storage, 713
– – operating range, 726
– – validation challenges, 725–726
– volumetric efficiency, 714
– – storage options, 714
cryo-compressed refueling unit, 718
cryocooler, 961
cryogenic air separation, 237
cryogenic fluid management (CFM) technologies, 959
cryogenic hydrogen, 903
– tank, 1002
cryogenic liquid, 601

cryogenic liquid hydrogen storage, 670
– hydrogen liquefaction, 670–671
– – physical and chemical properties relevant for, 670
– liquid hydrogen storage tanks, 671–675
– – boil-off rate of storage tanks, 675
– – horizontal and vertical tanks, technical data of, 672
– – spherical storage tank of NASA, 673
– – tanks depending on their size, technical data, 673
cryogenic liquid oxygen tanks, 1004
cryogenic storage tanks, 864
cryopumps, 904
CSP. *see* concentrated solar power (CSP) plants
CSS. *see* cyclic steady state
CTA. *see* Chicago Transit Association (CTA)
Cu catalysts, 201
cumulated non-renewable energy demand, 405
CUTE bus fleet, FC-powertrain design of, 971
CUTE-plus program, 977
CUTE program, 975
– test and field experience, 984
Cu/ZnO catalysts, 139
cyanobacteria and microorganisms, 420
– culture of, 471
– genetic modification of gene in, 473
– for H_2 production, 465
– [NiFe]-hydrogenase, 466
– nitrogen fixing/non-nitrogen fixing, 465
– Nostoc species, 467
– oxidative stress, 470
– photosynthetic electron transport pathway, 468
– physiology of, 465
– uptake hydrogenase
– – *Bradyrhizobium japonicum*, 467
– – cyanothece, 468
– – *Pyrodictium brockii*, 467
– uptake hydrogenase (HUP), 467
cyclic steady state (CSS), 495
cyclohexane, 203

d

Dalton's law, 316
Danish island of Lolland, 1081
– biogas, 1081
– Hydrogen Community of, 1081
dark fermentation, at thermophilic temperatures, 420
dark fermentative H_2 production, 475
– acetate, 475
– butyrate, 475
– carbohydrates, 475
– glucose dehydrogenase, 476
– hydrogen partial pressure, 476
– metal ions, 476
DBD reactor, 139, 141
DC/AC converter, 1083
DC/DC inverter, 1083
DEA. *see* amines like diethanolamine
de-carbonization, 256
– of energy sectors and chemical industry, 275
– strategies, 59
dehydriding reaction, 774
dehydrogenation, 135
demand-side management (DSM), 262, 263
demand–supply matching, 258
density
– current, 289, 291–292, 319, 348, 452, 1048, 1050, 1084, 1088
– energy, 271, 670, 852
– flux, 580, 583
– gravimetric, 763
– hydrogen and methane, 632
– material, 800
– power, 368, 381, 639, 994, 1025
– storage, 694, 715, 724
Department of Energy (DoE), 1024
derived equilibrium hydrogen pressure for electrochemical hydrogenation, 596
desorption, 496
– enthalpy, 792
– kinetics, 797
destructive hydrogenation, 5
desulfurization, 11, 16, 201, 218, 237
– developments in routes to, 16
detonation, 645
deuterated water, 566
deuterium, 566, 764
Deutsche Bahn, 898
diaphragm cell process, for production of chlorine, 31
diaphragm compressors, 904
diaphragm electrolyzers, 45–46
diatomic molecules, 565, 764
dielectric barrier discharge (DBD) plasma apparatus, 139
dielectric heating, 140
dielectric material (quartz), 139
Diels–Alder reaction, 115
diesel-electric power supply system, 991
diesel engines, 991

diesel fuels, 3, 143, 201, 419
– fuel processor subsystem, for autothermal reforming, 196
– low sulfur, 933
– potential design, fuel processor subsystem for steam reforming, 197
diesel generator, 991
diesel oil, 16
diesel systems, 1056
diffusion in gasses, 579–580
– mass flux density, 579
– self-diffusion coefficients of pure H_2, 581
dimerization, 115
4,6-dimethyldibenzothiophene (DMDBT), 201
di-olefins, 8
DIPPR project, 177
direct current (DC), 317
– -spark discharge plasma, 143
directly irradiated reactor/receiver, 155
directly irradiated reactors, 163
– non-structured reactors based on
– – solid particles, 167
– structured reactors based on
– – ceramic foams, 163–166
– – ceramic honeycombs, 163
– – fins, 166
directly irradiated receivers (DIRs), 156
– with stationary absorbers, 156
directly irradiated volumetric receiver–reactors (DIVRRs), 157, 168
direct methods, 611
direct tank cooling, on liquid oxygen side, 962
discharge time, 264
– electricity storage systems, 277
distillate hydrotreating, 5, 7
distillery industry waste, 425
dodecane, 176
– autothermal reforming
– – heat of reaction and heat for steam supply for, 198
– – mapping of hydrogen yield for, 200
– bond energies for, 178
dodecane ($C_{12}H_{26}$), 146
DOE. see US Department of Energy (DOE)
doping, 300
DOT. see US Department of Transportation (DOT)
double-barrier principle, 1006
DSM. see demand-side management (DSM)
Dutch energy system, 260

e

EC. see European Commission
EcoIsland project, 1081
eddy currents, 643
EEX. See European Energy Exchange (EEX)
E_F. See Fermi energy (E_F)
effective medium theory, 769
EFTA. see European Free Trade Association
EIA. see environmental impact assessment (EIA)
electrical and transport resistances, 292
electrical energy, 331, 629
electrical grid loads, 350
electrical grid modernization, 262
electrical power grid, 284
electrical resistances, 292–293
electric discharge systems, 141
electricity, 89
– demand, 284
electricity energy storage, installed capacities by technology, 267
electricity, from the German grid, 403
electricity grid operation
– overview of services, energy storage solutions provide relating to different aspects of, 267
electricity grid system, 258
electricity load management
– potential of DSM for, 264
electricity network, 255
electricity peak shaving, 284
electricity price, 406
electricity shedding, 261
– annual, 260
electricity shortage, 255
electricity storage, 264, 1057, 1058
– large-scale, 265
electric vehicles, 833
electrocatalysts, 448
– electrode potential, as a function of current density, 449
– and overvoltage, 448
electro-catalysts, 298–300
electrochemical devices, 543
electrochemical environment, 441
electrochemical equations, 443
electrochemical splitting, of water, 310
electrochemistry
– chlorine production, 42
– – with ODC technology, 49
– of NaCl electrolysis with an ODC, 50
electrode–electrolyte interfaces, 441– 442, 447, 448
electrode kinetics, 287–288

– cell overpotential, 291–292
– electrical and transport resistances, 292
– hydrogen generation overpotential, 288–290
– oxygen generation overpotential, 290–291
electrodes, 297–298
– hybrid, 447
– hydrogen overpotential, 301
– for a mass market, 441
– metal, 297
– Ni electrodes, 298
– oxygen overpotential, 301
– photoactive, 445
– surface modifications, 298
electrolysis, 41, 217, 226, 544, 897, 1083
– high temperature solid oxide, 1085
– hydrogen generation rate, 1083
– oxygen produced in, 1093
– water, 1093
electrolyte, 50, 206, 289, 296, 302, 312, 362, 376, 441, 450, 544, 1041, 1065, 1088
electrolyzers, 41, 461, 629, 814, 1083–1085
– alkaline, 1083
– capacity factor, 1087
– commercial, 331
– coupling efficiency, 1087
– direct coupling of, 1088
– dynamic operation of, 317–318
– dynamic system response, 320–321
– efficiency of an, 1084
– electronic control for, 457
– gas drying, 316–317
– for hydrogen production, 331
– intermittent operation, 1088
– operation range of, 318–320
– performance degradation with time, 1088
– polymer membrane, 1084
– power required to run, 1085
– power supply technologies for, 317
– proton exchange membrane (PEM), 1083
– role in energy transition, 270–274
– sizing of, 1088–1090
– standby mode, 322
– startup and shutdown process, 322
– system design criterion, 322
– system efficiency, 322–325
– technology, 835
– thyristor-based power supply, 317
– transistor-based power supplies, 317
– typical, 1084
– water consumption, 1088
electrostatic forces, 554
electrostatic precipitators (ESPs), 554

ENCAP. *See* ENhanced CAPture of CO_2 (ENCAP)
energetic application, 393
energetic efficiency, 401
– free, 402
energy conversion solution, 965
energy demand, 309, 401, 402
– for water electrolysis, 365
energy losses
– caused by reaction resistances, 293
energy management, 261
energy production, fluctuated, 309
energy security, 1058
energy storage, 711
– capabilities of UK power and gas network., 276
– evolution of demand for, 268–270
– facilitates, 262
– globe drive demand for, 256
energy supply, 309
– by renewable energy sources, 310
– system, 811
energy systems, features, 360
energy transition, 255
energy, Worldwide demand for, 417
ENERTRAG AG, 898
ENhanced CAPture of CO_2 (ENCAP), 1023
enthalpy, 132, 174, 175, 177, 199
enthalpy–entropy relationship, 794
entrained-flow gasifier, 243
entropy, 177, 445
environmental friendly certificates, 967
environmental impact assessment (EIA), 880
environmental regulations, 3
equation of state (EOS)
– for gas/fluid mixtures, 631
equilibrium constant, 179, 241, 574, 580, 595
equilibrium pressure, 595–596, 766, 794, 796–799, 807
Escherichia coli, 473
– genetic engineering of, 473
ESPs. *see* electrostatic precipitators (ESPs)
ethane, 6, 135, 202, 203, 545, 1016
ethanol, 134, 135, 139, 174, 182, 422, 473, 840
– steam reforming, 201
ethene, 202, 203
ethylene, 8, 201, 877
ethylene glycol, 792
EU27 member states
– hydrogen production, 21
– uniform permitting process suggested for, 926

European Association for Storage of Energy (EASE), 266
European chlorine production capacity, 30
European Commission (EC), 926
– directive proposal regulating hydrogen supply for transport, 926–927
European Energy Exchange (EEX), 1045
European energy policy, 891
European Free Trade Association (EFTA), 923
European Hydrogen Association (EHA), 81
– Project HyER, 81
European hydrogen production, 21
European launchers, 954
– Ariane 5 ECA, 954
– coasting phase, 954–955
– GEO/GTO orbits, 954
– re-ignition preparation phase, 954–958
– – boiling curves, 957
– – formation of vapor bubbles, 956
– – geyser formation during re-orientation phase, 956
– satellite, with apogee engine, 954
– Vinci engine, 954
European project DECARBit, 1021
European standardization bodies, 927
European Union, 920
– CO_2 emission reduction target, 256
– Commission
– – adopted communication "Energy Roadmap 2050," 256
– – de-carbonization objective, 256
– energy framework and relevant policies, 68–69
– – cut carbon emissions by 2050, key goals, 68
– – objectives for 2020, 68
– hydrogen related energy policy strategies, 69
– hydrogen research, development, demonstration, and deployment activities, 69–70
– intermittent renewable energy sources (RES), 258–259
– legislative acts, 920
– – directives, 920
– – regulations, 920
– renewable electricity in, 263
– survey of FCH JU support activities, 70
– survey of FCH JU supported projects of call for proposals made in 2011, 70–73
– uniform permitting process suggested for EU27, 926
evaporators, 192
EvoBus product, 970
EvoBus vehicle, 975

exhaust gas, 1012
expanded natural graphite, 799
expansion turbine, 755
– gas bearing, 757
– oil bearing, 756

f

facultative photoautotrophic microalgae, 465
failure mode, effect and criticality analysis (FMECA), 924
Faraday constant, 442
FC bus, 978
FCC. *see* fluid catalytic cracking
FCCJ. *see* Fuel Cell Commercialization Conference of Japan (FCCJ)
FCEVs. *see* fuel cell electric vehicles (FCEVs)
FCH JU. *see* Fuel Cells and Hydrogen Joint Undertaking (FCH JU)
FC-power plant, 972
FC-powertrain, 970
FCSs. *see* fuel cell systems (FCSs)
FCTO. *see* Fuel Cell Technology Office (FCTO)
FC-twin system, 979
Fe-compounds, 101
Federal Ministry of Education and Research (BMBF), 64
feedstock options, 223
feedstock purification, 238
Fe_3O_4/FeO, as redox material, 90
Fermi energy (E_F), 769
fermions, 566
ferredoxin (PetF), 465
ferrites, 91–93
– cobalt and nickel ferrites, favorable thermodynamic properties, 93
– demonstrated within the European projects, 91
– – Hydrosol pilot plant, 92
– – solar HYDROSOL reactor, 92
– – two-step thermochemical cycle, 91
– hydrogen production via a redox cycle using, 93
– utilization of a rotary kiln, 93
– work on fluidized bed reactors, 93
fiber-reinforced polymer (FRP), 859
FICFB gasification technology, 398
Fick's law, 512
filter, 553
– types of, 553
filter press principle, 1036
first law of thermodynamics, 618
Fischer-Tropsch reactions, 23, 36, 151, 225, 246, 519, 814

fixed-bed bioreactor, 427
fixed bed gasifier (Lurgi reactor), 242
flameless oxidation (FLOX®), 1024
– combustor characteristics, 1025
– measured NO_x values in, 1026
flame monitoring system, 1008
flat sheet modules, 526
– baffle plates, 527
– envelope type membrane, 526
– hydrogen selective membranes, 526
– poly(ethylene oxide), 526
– polysep membranes, 526
– spiral wound membrane element, 526
Flory–Huggins model, 512
fluid catalytic cracking (FCC), 557
fluidised-bed gasifier (Winkler process), 243
fluidized-bed bioreactor, 427
flywheels, 264, 265
FMECA. see failure mode, effect and criticality analysis
food industry waste, 425, 427
formic acid, 814
fossil-based CO_2 emission, 217
fossil diesel, 403
– in agricultural machines, 405
fossil energy, 109, 309, 419
fossil fuels, 22, 88, 109, 140, 151, 168, 201, 237, 283, 359, 405, 417, 419, 517, 629, 811, 891, 1075
– catalyst development for, 201
– combustion of, 403
– hydrogen production from, 1076
– for rapid industrialization and urbanization, 417
– synthesis gas (syngas) produced from, 237
– technologies, 256
– used for electricity generation, 417
– utilization of, 1075
FRP. see fiber-reinforced polymer (FRP)
fuel cell, 1086–1087
– bus, 967
– efficiency of, 1086
– module, 977, 996, 1000, 1001
– polarization curves, 1084
– regenerative, 1083
– supporting system, 1086
fuel cell/battery hybrid vehicle, 1077
Fuel Cell Commercialization Conference of Japan (FCCJ), 844
fuel cell electric vehicles (FCEVs), 61, 64, 274, 711–712, 833
– components, 834
– functions, 834

fuel cell modules, 977, 1000
– BZM 34, 1000, 1001
– BZM 120, 1000, 1001
fuel cell powertrains, 973
– automotive glycol cooling loop, 974
– Citaro fuel cell-hybrid, 978, 979
– CUTE program, 975
– design solutions, 973
– – NEBUS, 973
– hydrogen circulation in, 974
– NEBUS, 973
– – design solution, 973
fuel cells, 284, 543
Fuel Cells and Hydrogen Joint Undertaking (FCH JU), 841
fuel cell systems (FCSs), 204, 791
– in mobile applications, 173
fuel cell technology, 982
– life time of, 989
– product costs, 989
– start–stop strategies, 987
fuel cell technology office (FCTO), 850
fuel choice, 173
fuel consumption, minimization strategies, 965
fuel gases, 543
– potential sources of, 543
fueling operations, 849
fueling station, 217, 660
fuel/pollutants, 545
– carbonyl sulfide, 545
– gas composition of untreated product gas, 546
– gasification synthesis gas, 545
– hydrogen sulfide, 545
– main gas components, 545–547
– – carbon dioxide, 547
– – carbon monoxide, 546
– – hydrogen, 546
– – methane, 546
– – nitrogen, 547
– – steam, 547
– mercaptans, 545
– trace gas components, 547–550
– – alkali compounds, 549
– – halogen compounds, 550
– – nitrogen compounds, 550
– – other potential contaminants, 550
– – particulate matter, 549
– – siloxanes, 550
– – sulfur compounds, 549
– – tars, 547
fuel power, 134

fuel processing, 174
- arrangement of subsystems
- - with no connecting heat exchangers, 193
- current challenges in, 201–203
- subsystems of, 192–193
fuel processor subsystem, 191
- for autothermal reforming of diesel as a draft, 196
- consisting of autothermal reforming of diesel and a gas cleaning unit for PEFC systems, 191
- for steam reforming of diesel as a potential design, 197
fuel production, 1057, 1059
fugacity coefficients, 176, 177
fullerenes, 145
future energy carriers, 1033

g

gas cleaning subsystem, as a possible design for a PEFC system, 207
gas clean-up for fuel cell systems, 543
gas drying, 316–317
- specific gas constants, 316
gaseous hydrogen, 812
- safety related precautions, 939
gaseous hydrogen delivery, 857
- pipeline delivery pathway, 857
- tube-trailer delivery pathway, 860
gas hourly space velocity (GHSV)., 181
gasification, 5, 26
- of alcohols and organic acids, 116
- glucose, 113–115
- processes, 237
- of solid fuels, hydrogen production by, 242–244
gas oils, 7
gasoline, 3, 5, 134, 143, 201, 419
- -based cars, 827
- dispenser, 711
- reformed, effects of sulfur in, 202
- refueling stations, 711
- yields, 15
gas permeation, 190
gas permeation, in polymers, 580–583
- coefficients of hydrogen for, 582
- Fick's first law, 580
- Henry's constant, 580
- permeation coefficients of some selected polymer, 584
- polymer membrane, 580
gas separation, 313–314, 509
- KOH-solution level in vessels, 314

- membrane process, 510,
gas–solid reactions, 101
gas storage system, 617
- energy balance of, 618
- storage temperature, 619
- temporal change of mass, 617
gas transport, 642
gas turbine (GT), 1011
- burning hydrogen in, 1012
- classification of, 1020
- combustion fundamentals of hydrogen, 1012
- - emissions, 1018
- - flame speed, 1016
- - flame temperature, 1018
- - ignition delay, 1014
- - stability, 1018
- combustor, 1011
- new combustion technologies, 1022, 1024
- - flameless oxidation, 1024
- re-electrification of, 1055
- research and development status, 1022
gas turbines, 309
genetically modified organisms, 471
genetic biomarkers, 476
genetic engineering, to enhance hydrogen production, 476
- genetic biomarkers, 476
- renewable biofuel, 476
- *Rhodobacter sphaeroides*, 476
GEO (geostationary orbit), 954
geological hydrogen storage, 629
- fundamental aspects, 631
- geological storage option, general types of, 635–636
- hydrogen and microbial inventories, interaction between, 634–635
- hydrogen storage, in caverns, 640–642
- hydrogen storage, in porous rocks, 636–640
- integrated power-to-gas concept, schematic drawing of, 630
- operational hydrogen storage projects, 649–650
- - mixed hydrogen–natural gas, 650
- - pure hydrogen sites, 650–651
- - town gas storages, 651–654
- physicochemical properties, of hydrogen and hydrogen mixtures, 631–634
- process engineering, importance in, 642–649
- storage options, according to storage capacity and discharge time, 630
geothermal power, 1080

geothermal renewable energy, 1080
German Renewable Energy Law, 406
German wind–photovoltaics–electrolyzer system, 1034
Germany, 60, 601, 891
– copper plate model, 261
– electricity grid, 261
– energy framework and relevant policies, 60
– – Energy Package, 60
– – master plan, 60
– energy management, 261
– hydrogen as road transport fuel in, 601
– hydrogen related energy policy strategies, 60–61
– – National Hydrogen and Fuel Cell Technology Innovation Programme (NIP), 60–61
– hydrogen research, development, demonstration, and deployment activities, 61
– – hydrogen production, 63
– – industry activity, 64–65
– – goal of, 64
– – "H2Mobility" group, 64
– – multi-energy filling station, 61–63
– – special markets, 64
– – stationary and residential applications, 63–64
– – transportation, 61, 63
– PEM-fuel cells development at Siemens, 992
– renewable electric energy production in, 1033
– renewable energy, fuel capacity of, 1033
– town gas storage project, 652–653
– wind energy shedding, 261
GHGs. see greenhouse gases (GHGs)
Gibbs energy, 90, 178, 179
Gibb's free energy, 145, 442, 445
gliding arc, 135
– discharge plasma assisted methane reforming with CO_2, 135
– plasma, 132–136
global annual hydrogen production, 393
global warming, 255, 417
glucose, 113–115
– aldol-splitting, 113
– conversion under hydrothermal conditions, 114
– degradation products, 113
– gasification, 114
glycerol, 471
glycine, 115
glycosidic bonds, 115

graphene coatings, 352
gravimetric energy density, 812
green algae, 420, 465, 469
green energy, 423
greenhouse effect, 188
greenhouse gases (GHG), 132, 402, 417, 629, 867
– emissions, 59, 394, 403, 417
– – for hydrogen concepts, 405
– mitigation potential, 404
green hydrogen filling station infrastructure, 1034
green hydrogen, production of, 1033
grid-related services, 266
grid-scale renewable energy storage, 276–277
grid stability, 255, 309
gross vehicle weight (GVW), 975
growth conditions, 470
GTO (geostationary transfer orbit), 954
gums, 5
GVW. see gross vehicle weight (GVW)

h

Haber–Bosch process, 32
hardware simulations, 1061
harmonization, 218
harvesting system, 441
Hawaii, 1081
– animal fats, 1081
– energy solutions for, 1081
– plant oils, 1081
– pressure swing adsorption (PSA) process, 1081
– syngas feedstocks, 1081
HAZard and OPerability study (HAZOP), 924
HAZard Identification (HAZID), 924
HAZID. see HAZard Identification
HDW class 212A submarine, 992, 1004–1005
HDW class 214 submarine, 992, 1005
HDW fuel cell system, 1005
– safety concept of, 1006
heat conduction, 204
heat exchangers, 192, 758
– aluminum plate fin, 758
heat extraction, 156
heat generation, 349
heating–cooling cycles, 156
heat loss, 430
heat-of-combustion, 177
heat of formation, 178
heat of reaction, 193
heat of vaporization, 156
heat pipe receiver, 156

heat pumps, 785
heat receptivity, of cryo-compressed and liquid hydrogen storage, 716
heat transfer, 155, 799
– equations, 617, 619
heat transfer fluid, 155, 156
heavy aromatic feedstock, 9
heavy duty transport, 173
heavy metal
– impurities, 237
heavy oils, 16, 237
– feedstock and raw syngas composition
– – of Texaco process, 246
– gasification of, 245
– syngas by partial oxidation of, 244
HEC. see hydrogen evolving catalyst (HEC)
HEGSA. see high efficient gas turbine with syngas application (HEGSA)
helium, 134
Helmholtz free energy, 603
Henry's constant, 580
Henry's law, 512
hercynite cycle, 93–94
heterocyst, 465
heterogeneous catalysis, 174
heterotrophs, 471
hexadecane, 140
hexane, 202
Heyrovski reaction, 596
HFCEVs. see hydrogen fuel cell electric vehicles (HFCEVs)
H_2–H_2 interaction, 764
Hidrólica project, 1058
high efficient gas turbine with syngas application (HEGSA), 1023
high-frequency pulsed plasma (HFPP), 143
high-temperature electrolysis, 331
Hindenburg syndrome, 934
hollow fiber modules, 525
– Evonik Industries design, 525
– high packing density, 525
– schematic representation of, 526
hot gas clean-up, 552
– industrial-scale applications, 552
HRSs. see hydrogen refueling stations (HRSs)
HRU. see hydrogen, release unit (HRU)
HSER. see hydrogen sorption-enhanced reforming (HSER)
HUP. see uptake hydrogenase
HV components, 980
– hub wheel drive, 980
– HV battery, 980
– inverters, 980

– multi converter for auxiliary power, 980
hybrid copper chlorine cycle, 100–102
– heat required for reaction, 101
– idealized efficiency, 101–102
– thermal reactions, 101
hybrid electric passenger vehicles, 711
hybrid electric vehicle (HEV), 712
hybrid electrodes, 447
hybrid power plant, operating modes of, 1042
– base load, 1043
– – example of, 1043
– EEX, 1045
– forecast, 1044
– hydrogen production, 1042
– – advantage of, 1042
– – example of, 1042
hybrid power plant project, 1034
– alkaline water electrolysis, 1034
– main components of, 1035
– at Prenzlau, 1035, 1036
– wind turbine–electrolyzer coupling, 1034
hybrid SMR-POX catalytic system, 147
hybrid sulfur cycle, 89, 96–98
hybrid thermochemical cycles, 89
hydraulic pistons, 904
hydraulic retarder, 972, 977
hydraulic retention time (HRT), 429
hydrides, 588–597
– classification of, 588
– – complex hydrides (hydrido complexes), 591–592
– – covalent hydrides, 590–591
– – interstitial (metallic) hydrides, 592–593
– – ionic hydrides, 588–590
– conversion with complex, 593–594
– conversion with other, 593
– formation of, 593–597
– hydrolysis, 594
– properties of, 588
– – covalent compounds, 588
hydrido complexes. see complex hydrides
hydrobromic acid, 98
hydrocarbon cracking, 141
hydrocarbon fuels, 140, 237, 1020
– hydrogen production from, 237
hydrocarbon membranes, 350
hydrocarbons, 134, 143, 151, 237, 565
– autothermic reforming of, 249
– small-scale reforming, 217
hydrochloric acid, 41
hydrocracking, 5, 8–11, 16, 518
– homogeneous catalysts based, 10
hydrodesulfurization, 3, 5, 16, 238

hydrodesulfurized products, 7
hydrofining, 7
hydrogen, 19, 135, 491, 546, 566, 601, 763, 891
– absorption, 784
– adsorption of, 493
– alloys interactions, 764
– applications, 891
– – and production processes in a refinery, 5
– automotive industry, 491
– based on biomass, 393
– bomb, 934
– bonds, 115, 591
– cars, 219
– cities, 1058
– communities, 1058
– compressed gaseous, 601
– compressibility factor of, 606
– compression work of, 614
– containing pressure vessels, design, 715
– conversion, main challenge for, 278
– as coupled stream
– – in electrolytic production of chlorine, 28–29
– data reported by Eurostat, 20–21
– density of, 569
– desorbstion, 791
– desorption
– – from sodium alanate, 774
– diffusion, 771
– dissociation–recombination reaction, 771
– distribution
– – to a filling station, 401
– – of produced hydrogen, 399–400
– distribution costs, 410–411
– – distribution cost for a mean transport distance, 410
– economy, 283, 1075
– energy applications of, 784
– energy industries, 491
– energy relevant properties, 764
– European production, 20
– fractionation by gas permeation, 516–518
– fuel cell vehicles, 217, 1034
– as a fuel, importance and applications, 419
– fuel properties of, 1013
– fuel storage system, 980
– gas, 1011
– – enthalpy of combustion, 461
– – storage, 1077
– – turbine technology, 1019
– generation processes, indicated by educt mixture, 194
– heat flow of, 624

– H_2-PSA unit, 502
– – improvement of, 502
– hydricity, 1075
– ideal fuel, 1093
– ignition delay time of, 1014
– implementing agreement, 59, 81
– inorganic solids interaction, 764
– Islands Initiative, 1094
– isotopes, 566, 764
– – deuterium, 764
– – tritium, 764
– isotopes storage, 785
– Joule–Thomson coefficient of, 608
– laminar flame speeds of, 1016
– lean molecules, 814
– liberatation, 791
– liquid hydrogen storage, 601
– management, 14–16
– – network, 16
– metal interactions, 763
– molecule, 565
– network optimization, 3
– ortho, 606
– para, 606
– pressures, 7, 1085
– production
– – in nature, diverse pathway for, 464
– – route, 491
– properties of, 461
– purging cell, 997
– purification, 491
– – membranes, 1008
– purity, 221–223, 491, 495
– – upgrade, 15
– quality, 912
– real gas behavior of, 622
– real isentropic exponent of, 610
– recovery of, 492
– in refineries, 4, 491
– refueling stations, 217
– refueling with compressed, 601
– release unit (HRU), 827
– requirements worldwide, 331
– reversible orption of, 765
– -rich fuels, 1011
– as road transport fuel, 601
– roles along chemical value chain, 19
– safety and the elements of risk, 934–935
– as secondary energy carrier, 891
– selective membranes, 526
– solubility, effect of salt on, 631
– in space applications, 951
– speed data of, 1016

– storage, 601, 767, 1085–1086
– – compressor, 1083
– – fuel cell, 1083
– – options, 712
– – systems, 173
– – tank, 1080
– supply infrastructure, 393, 1076
– technologies, under research, development, 59
– temperature, 614
– – change during expansion of, 616
– thermal conductivity of, 579
– in thermal processes, 4
– thermodynamic properties with methane, comparison of, 606–610
– uses and applications of, 36
– utilization, 15, 16
– – across different sectors, 20
– – technologies, commercialization of, 1076
– variation of storage tank
– – pressure of, 621
– – temperature of, 621
– vehicles, 1093
– volume required to store, 1086
– yield, 202
hydrogen and Fuel Cells in the Economy (IPHE), 81
– mission of the International Partnership, 81
hydrogenases, 463, 465, 466
– active-site structures of, 467
– carbon monoxide, 467
– catalytic activity of, 466
– classes, identified, 466–467
– for H_2 production, 466
– hydrogen uptake, 467
hydrogenation, 4, 7, 10, 237, 815
– of aromatics, 11
– selective, 11
hydrogen/carbon monoxide separation, 517
– methanol production, 517
– purge gas, 517
hydrogen combustion properties of, 1012
– flame speed, 1012
– flame temperature, 1012
– ignition delay time, 1012
hydrogen cost
– calculation, 836
– – bottom-up approach, 836
– – top-down approach, 836
– – total cost of ownership (TCO) approach, 836
– in Europe, 838
– in Germany, 838

– in Japan, 837
– in USA, 836
hydrogen delivery, 849, 900
– CGH_2 delivery, 901
– challenges, 849
– costs, 850
– infrastructure, 849
– LH_2 delivery, 901
– opportunities, 849
– pipeline delivery, 901
hydrogen demand, 850
– daily, 850
– hourly, 850
– seasonally, 851
– variation, 850
– weekly, 850
hydrogen diffusion, in metals, 583–587
– electrochemical formation, 595–597
– fluorite, 583
– metallic membranes, 585
– permeation coefficient for selected transition metal membrane, 586
– rock salt, 583
– Sievert's law, 583
– wurzite, 583
hydrogen embrittlement, 878
– chemical hydrogen reaction, 878
– environmental hydrogen, 878
– metallic–physical processes, 878
hydrogen energy project, 935
– ALARP approach, 935
– Do no harm approach, 935
hydrogen energy storage, 1082
– potential for, 1082
hydrogen energy systems
– NFPA 2: "Hydrogen Technologies Code," 941
– NFPA® 52, 941
– regulations, codes, and standards, 940–942
– Vehicular Gaseous Fuel Systems Code, 941
hydrogen evolving catalyst (HEC), 448, 450
hydrogen filling stations, 895
– technical concepts for, 895
– – facility size, 907
– – hydrogen delivery, 900
– – hydrogen production, 895
– – hydrogen refueling stations components, 902
– – hydrogen refueling stations, integration of, 906
– – space requirements, 907
hydrogen fuel cell electric vehicles (HFCEVs), 849

hydrogen/hydrocarbon separation, 518
– boiler fuel, 518
– cold train distillation, 518
– Huntsman Advanced Materials, 514
– hydrocracker, 518
– hydrotreater product gases, 518
hydrogenics electrolyzer, 1065
Hydrogen Implementing Agreement (HIA) Task 30
– on Global Hydrogen Resource Analysis, 59
hydrogen in chemical industry, sources of, 22
– electrolytic processes, 26
– – alkaline electrolysis, 27
– – high-temperature electrolysis, 27
– – PEM electrolysis, 27
– hydrogen as coupled stream
– – in electrolytic production of chlorine, 28–29
– hydrogen production
– – steam reforming *vs.* electrolysis, 28
– process variations, 25
– – autothermal reforming, 25
– – gasification, 26
– – other waste and coupled streams, 26
– – partial oxidation, 25
– – pre-reforming, 25
– – water–gas shift conversion, 26
– steam reforming, 23–25
– synthesis gas-based processes, 22–23
hydrogen infrastructure, 835, 907, 965
– compression, 835
– delivery cost, 835
– dispensing cost, 835
– incentives for promotion, 842
– production cost, 835
– refueling cost, 835
– social benefits, 843
– storage, 835
– subsidy for, 843
– tax credits, 842
hydrogen interacting materials, energy applications of, 784
– chemical functions, 784
– energy storage, 784
hydrogen liquefaction, 733
– conventional process, 740
– – compression, 740
– – cooling, 740
– – expansion, 741
– economics of, 733
– hardware components, 751
– – compression, 752
– – expansion turbine, 755

– – heat exchangers, 758
– history of, 734
– principles of, 739
– – final cooling, 749
– – hydrogen liquefaction processes, evolution of, 745
– – nitrogen Brayton Cycle precooling, 746
– – nitrogen precooling, Joule-Thomson process, 742
– – power requirements, 739
– – precooling part, process design of, 745
– – precooling purposes, mixed gas refrigeration for, 748
hydrogen/nitrogen separation, 517
– ammonia production, 517
hydrogen peroxide (H_2O_2), 574
hydrogen physics, 565
– clathrates, 597
– hydrides, 588–597
– molecular hydrogen, 565
– – covalent bonding, 565–566
– – molecular orbitals, 565–566
– – natural isotopes, 566
– – nuclear spin-states, 566–569
– – thermodynamic properties, 569–573
– reaction kinetics, 574–578
– transport kinetics, 578–587
hydrogen pipelines, 876
– construction, 879
– – commissioning, 883
– – design parameters, adherence to, 888
– – dew point, 888
– – planning and approval, 879
– – project execution, 883
– – routes assessment, 880
– – servicing and operation, 888
– – steel material, selection of, 888
– – welding additive material, 888
– decommissioning, 888
– – backfilling, 888
– dismantling, 886
– investment costs, 882
– operations, 886
– – control center, 886
– – culvert soundings, 887
– – maintenance, 886
– – monitoring, 886
– – recurring regular tests, 887
– – regular route inspections, 887
– – repair, 886
– – troubleshooting, 886
– reclassification, 888
– system, 284

hydrogen producing countries, 876
– Belgium, 877
– France, 877
– Germany, 877
– Japan, 876
– Netherlands, 877
– USA, 876
hydrogen production, 12–14, 20, 88, 895, 1042
– biological, 463, 900
– from biomass, 896
– – biogas reforming, 897
– – biomass gasification, 897
– – glycerol reforming, 897
– from biomass as feedstock, 422–423
– byproduct hydrogen, 900
– by other organisms, 473
– by photosynthetic organisms, 461, 463
– – in the presence of sunlight, 464
– – source of renewable energy, 463
– plant, 1042
– power, 1042
– processes, and impact on environment, 418
– from solid fuels (coal, biomass), 242
– – basic principles and reactions of syngas production from solid fuels, 242
– – hydrogen production by gasification of solid fuels, 242–244
– by steam reforming, 900
– through electrolysis, 897
– via methane reforming, 152
hydrogen production costs, 406, 408–410
– cost parameters at regional frame conditions/ time horizons, 406
– – capital expenditures, 406
– – fixed OPEX, 406
– – revenues, 406
– – variable operational expenditures, 406
– and distribution costs, 407
– economic assessment, assumptions for, 407
– guideline VDI number 6025, 406
– sensitivity analysis, 409
hydrogen projects on islands, 1077–1082
– fuel cell/battery hybrid vehicle, 1077
– hydrogen compressor, 1077
– hydrogen gas storage, 1077
– Promoting Unst Renewable Energy (PURE), 1077
– studied, planned, and completed projects on, 1078
hydrogen, properties at low temperature, 735
– ortho and para modifications, 735
– thermodynamic properties, 735
hydrogen refueling stations (HRSs), 892, 923

– approval process for Germany, 928–930
– average pressure ramp rate methodology, 894
– basic requirements for, 892
– – bus refueling, 894
– – car refueling, 892
– challenges, 907
– – approval processes, 910
– – costs, 909
– – gauged H_2-metering, 911
– – hydrogen quality, 912
– – refueling, technical guidelines, 911
– – reliability, 910
– – standardization, 907
– codes of practice for, 920–921
– construction recommendations for, 923–924
– country/location specific issues related with, 924
– design/construction recommendations for, 923–924
– design of, 923
– globally harmonized standards, need for, 923
– – Technical Barriers to Trade (TBT), 923
– – World Trade Organization's (WTO's), 923
– maintenance of, 924
– major components of, 902
– – compressors, 903
– – controls, 905
– – gauges, 905
– – sensors, 905
– – valves, 905
– – dispenser, 905
– – hydrogen storage, 902
– – pre-cooling, 904
– – production, 902
– operation of, 924
– procedures for the permission for, 924–930
– – in European Union, 924–930
– referencing of standards, 922–923
– regulations for, 919–920
– – decisions, 920
– – directives, 918
– requirements for the approval of, 923–924
– risk assessment methodologies for approval of, 924
– technology standards for, 921–922
– – direct referencing, 922
– – indirect referencing, 922–923
– vehicle interface requirements, 924
hydrogen, role in wind electricity generation, 1055
– electricity storage, 1056

– fuel production, 1058
– mini grids, 1056
– three configurations, comparison of, 1059
hydrogen safety research, 943
– hydrogen storage systems safety research, 944
– research on physical behavior of leaked or leaking hydrogen, 943–944
– research supporting early market applications, 945
– risk mitigation measures, research on, 945–947
– simplified tools to assess and mitigate risk, 947
hydrogen separation, 516–519
– carbon monoxide separation, 517
– energy generation processes, 516
– fractionation applications in hydrogen production, 519
– hydrocarbon separation, 518
– from hydrocracker flash gas, 532
– – gas permeation unit for, 532
– – hydrogen content in retentate, 533
– by membrane technology, 531
– – from hydrocracker flash gas, 532–533
– – from purge streams in ammonia production, 531–532
– nitrogen separation, 517
– with polymeric membranes, 509
– from purge streams, 531
– – purge gas treatment in ammonia production, 532
hydrogen sorption-enhanced reforming (HSER), 229
hydrogen stations, 835
– off-site, 835
– on-site, 835
hydrogen storage, 781, 791, 811, 902
– CGH_2 storage, 902
– in complex hydrides, 773
– gravimetric energy density, 813
– high open-porosity structures for, 781
– in intermetallic compounds, 767
– LH_2 storage, 903
– for light-duty fuel cell vehicles, 792
– metal hydrides based requirements, 792
– – buffer hydrogen, 796
– – containment tank, 794
– – desorption kinetics, 797
– – material compaction and heat transfer, 799
– – material thermodynamics, 794
– – operating temperatures, 792
– – sorption kinetics, 798

– on-board efficiency, 792
– physisorption, 781
– solid hydrogen storage, 903
– for submarines, 1002
– volumetric energy density, 813
hydrogen storage, in vehicles, 691–693
– design of a tank system, 695–697
– – container, for compressed gas storage, 697–699
– – flow diagram, for compressed hydrogen, 696
– pressurized storage, advantages of, 693–695
hydrogen storage projects, 650
– pure hydrogen storage projects, 650–651
– town gas storages, 651–652
hydrogen sulfide, 5, 6, 7, 10
hydrogen sulfite, 635
– impacts on the energy balance, 635
hydrogen transportation, 811, 833, 875
– dry mode, 877
– economics, 836
– – cost, 836
– – social cost and benefits, 841
– embrittlement, 878
– fuel cell electric vehicles (FCEVs), 833
– hydrogen infrastructure, 835
– low pressure levels, 878
– pipelines refitting, 877
– safety related issues, 877
hydrogen transport kinetics, 578–587
– diffusion in gasses, 579–580
– diffusion in metals, 583–587
– diffusion in polymers, 580–583
– permeation in metals, 583–587
– permeation in polymers, 580–583
– thermal conductivity, 578–579
hydrogen utilization, in chemical industry, 32
– ammonia, 32–34
– methanol, 34–36
hydroisomerization, 5
hydrophilic electrode, 303
hydroprocesses, in refinery, 4
hydroprocessing operation, 15
hydroprocessing parameters, 6
hydro storage, 265
hydrothermal biomass conversion, 109, 113
– and saturation pressure curve of water, 110
hydrothermal carbonization, 113
hydrothermal degradation
– of sodium acetate, 116
hydrotreater product gases, 518
hydrotreating, 7, 8, 10, 11
– catalysts, 11

- processes, 6–8
- technology, 16
hypothetical energy system of future, 1076
HyWindBalance project, 1057

i
ICE. *see* internal combustion engine (ICE)
Iceland, 1080
- ECTOS Hydrogen Fueling Station, 1080
- geothermal renewable energy, 1080
ICEVs. *see* internal combustion engine vehicles (ICEVs)
ideal Carnot system, 91
ideal gases, 514, 567, 601, 602–603
- binary mixture of, 514
- enthalpy of, 602
- gas law, 602
- hydrogen, 567
- internal energy of, 602
- isentropic relations for, 609
- isobaric heat capacity, 602
- isochoric heat capacity, 602
- self diffusion coefficient of, 579
- state equation of, 572
- thermodynamic properties of, 601
ideal gas law, 176
IGCC. *see* integrated gasification combined cycle (IGCC)
ignition delay time, 1014
- pressure and temperature influence on, 1015
ignition temperature, 188
immobilization, 429
indirectly heated reactors, 159
- heat transfer medium, 159
- - air, 159–160
- - molten salts, 160–162
- - sodium (Na) vapors, 160
- - solid particles, 162–163
indirectly irradiated allothermal reformer, 155
indirectly irradiated integrated tubular reactor/receiver, 155
indirectly irradiated reactors, 168
indirectly irradiated receivers (IIR), 155
indirect method, 611
indium oxide, 447
industrial reformers, typical
- schematics of operation of, 153
industrial reformer unit, components, 218
industrial residues, 424–425
industry standards, 921
inner city bus application, operational environment of, 966
InP photoelectrode, 450

insoluble materials, 5
in-space propulsion, 960–961
insulation, 713
insulation materials, 430
integrated gasification combined cycle (IGCC), 1021
intelligent control algorithm, 1092
intermetallic compounds, 767
- AB type alloy TiFe, 772
- AB_5 type compounds, 771
- AB_2 type hydrogen absorbing alloys, 771
- hydrides, 773
intermittent renewable energy sources, 261
- integration, options for, 261
internal combustion engine (ICE), 971
internal combustion engine vehicles (ICEVs), 852
internal energy, 618
- rate of change of, 618
International Association for Hydrogen Energy, 82
International Association for Hydrogen Safety, 934
international collaborations, to prioritize hydrogen safety research, 942
- comparative risk assessment studies
- - of hydrogen and hydrocarbon fueling stations, 2008, 943
- Knowledge Gaps White Paper, 2008, 942–943
- survey of hydrogen risk assessment methods, 2008, 942
International Energy Agency (IEA), 59
International Networks, 80–81
- European Hydrogen Association (EHA), 81
- - Project HyER, 81
- Hydrogen and Fuel Cells in the Economy (IPHE)
- - mission of the International Partnership, 81
- Hydrogen Implementing Agreement of, 81
- International Association for Hydrogen Energy, 82
International Organization for Standardization (ISO), 921
International Standard Organization (ISO), 861
interstitial (metallic) hydrides, 592–593
- direct reaction of the elements, 594
- pressure isotherms for the hydrogen, 595
- properties of ternary interstitial hydrides, 593
- ternary Interstitial hydrides, 592

– thermally activated reaction, 594
inverters, 1083
investment cost, for power and gas networks, 276
ion exchange membrane, 47
– cross section, 47
– electrolyzers, 46
ionic compressor, 904
ionic hydrides, 588–590
– BeH_2, 590
– Brønsted acids, weak, 588
– crystal structures for, 590
– direct reaction of the elements, 593
– nucleophiles, 588
– overview of the (stable) binary hydrides of all elements, 589
– thermally activated reaction, 593
ionic liquids (ILs), 302
ionic resistance, 293
iridium oxide (IrO_2), 340
iridium utilization, 351
iron based ammonia synthesis catalyst, 238
iron oxide (Fe_3O_4), 185, 634
iron sulfide (FeS), 185
isenthalpic expansion, 613
– direct method for, 615
– indirect method for, 615
isentropic compression, 611–615
isentropic expansion, 613
isentropic exponent, 609–610
– relations for ideal and real gases, 609
– thermodynamic properties, definitions of, 604
islanded systems, 1056
ISO. see International Standard Organization (ISO)
iso-butane, 10
isobutyraldehyde, 473
isocyanates, 52
isomerization, 10
isooctane, 203
isothermal compression, 611–615
isothermal refueling, 799
isotope separation, 785
ITHER project, 1058

j

jacket cooling, 611
Japan
– energy framework and relevant policies, 78–79
– – energy policy, 79
– – Energy White Paper, approval for, 78

– hydrogen related energy policy strategies, 79
– – goal for 2025, 79
– hydrogen research, development, demonstration, and deployment activities, 80
– – METI, subsidiary program for, 80
– – NEDO role, 80
– Japanese commercialization scenario for hydrogen and fuel cells, 79
– JHFC3 outline of a demonstration project, 80
Japanese Atomic Energy Research Institute, 88
jet fuels, 3, 5, 201
Joule–Thomson coefficient, 606, 607
– effect, 607
– expansion, 607
Joule–Thomson effect, 515, 572–573, 602
– coefficient of hydrogen, 573
– real gas, 572
Joule–Thomson Process, 742

k

kerosene, 5, 8, 143
– fraction, 8
– -range material, 10
Koppers–Totzek process, 243
the Kyoto Protocol, 87–88

l

Landolt-Börnstein series, 771
landscape conservation, environmental legislation to, 1056
lanthanum-strontium-vanadium-oxide (LSV), 545
large specific impulse (ISP), 951
Laves phase alloys, 771–772
law of mass conservation, 617
LCA (life cycle assessment) methodology, 403
– steps and assumptions for biohydrogen concepts, 404
LDI. see lean direct injection (LDI)
lead acid, 265
– batteries, 991
leading automotive companies (OEM), 919
leaf H_2 Production, 472
– artificial leaf, 472
– artificial photosynthesis, 472
– electrocatalyst, 472
– semiconductor photoelectrochemical materials, modern, 472
lean direct injection (LDI), 1023
lean premixed combustion (LPC) technique, 1012, 1019, 1024

LEL. *see* lower explosive limit (LEL)
Lewis number, 1017
LH$_2$. *see* liquefied hydrogen (LH2); liquid hydrogen (LH$_2$)
LHV. *see* lower heating value (LHV)
light intensity, 470
lignin, 115
– degradation of, 115
– hydrothermal biomass degradation
– – simplified reaction scheme of, 117
lignocelluloses, 114, 420
lignocellulosic feedstock, 431
Li-ion battery technology, 978
liquefaction, 764, 866
liquefied hydrogen (LH$_2$), 441, 733
– properties, 733
– – high energy density, 733
– – light weight, 733
– – low environmental impact, 733
– pump, 904
liquid hydrocarbon fuels, 143
liquid hydrogen (LH$_2$), 812, 903, 951, 1085
– advantages of compression, 717
– based hydrogen stations with cryogenic compression, 717
– compression, advantage of, 717
– cryogenic compression, 717
– heat receptivity of, 716
– refueling, 717
– – unit, 718
– storage, 713, 715–716, 719, 724
– tank insulation, 953
– thermodynamic opportunities, 714
– usage, constraints due to, 952–953
liquid hydrogen delivery pathway, 864
– challenges, 865
– – boil-off losses, 865
– – liquefaction, 866
– – underground storage, 866
– components layout, 864
– cost estimates, 864
liquid hydrogen, for access to space, 951
liquid hydrogen, safety related precautions, 939–940
liquid hydrogen storage systems, 173
liquid hydrogen tankers, 849
liquid organic hydrogen carrier (LOHC) systems, 814
– alicyclic compounds, 814
– applications of, 822
– – in energy transport over long distances, 824
– – in mobile sector, 827

– – in storage of renewable energy equivalents, 823
– development of, 819
– ecotoxicological properties, 815
– heteroalicyclic compounds, 814
– liquid aromatic or heteroaromatic compound, 814
– toxicological properties, 815
Liquid oxygen, 996
liquid oxygen (LOX), 951–952
– tank, 1009
liquid-petroleum gas (LPG), 209
liquid storage, 951
– larger ISP, of liquid propellant, 951
– mass efficiency, 951
lithium ion [Li-ion], 265
LOHC. *see* liquid organic hydrogen carrier (LOHC) systems
London forces, weak, 591
low carbon technologies, 256
lower explosive limit (LEL), 1041
lower heating value (LHV), 318, 614, 1012
low gravity environment, relevant tests in, 958–960
– CPST mission, illustrating technologies, 960
– demonstration with liquid hydrogen, 959
– source of analysis
– – on predicted liquid hydrogen behavior, 959
– spherical liquid hydrogen tank, 958
low-temperature metal hydride (LTMH) systems, 791
low-temperature water–gas shift reactor (LT-WGS), 176, 185
LOX. *see* liquid oxygen (LOX), tank
LPC. *see* lean premixed combustion (LPC) technique
LPG (liquefied pressurized gas), 24–25
LSV. *see* lanthanum-strontium-vanadium-oxide
LTMH. *see* low-temperature metal hydride (LTMH) systems
Lurgi-MegaMethanol Process®, 35

m

magnetite, 91
manganese, 91
manganese ferrite plus activated sodium carbonate system, 93–94
Mark 13A cycle, 98
markets, energy price variation in, 1056
Mark 13 V2 , as hybrid cycle, 98
mass fractions, 800
mass transport, 174

material compaction, 799
material densities, 800
material fill factor, 800
MATLAB Simulink software, 398
maximum vehicle velocity, 967
MCFC. *see* molten carbonate fuel cells
MDEA. *see* methyl-diethanolamine
MEA. *see* membrane electrode assembly (MEA)
medical oxygen, 996
medium-temperature metal hydride (MTMH) system, 791
mega-ammonia technology, 34
membrane bioreactor (MBR), 427
membrane cell process, for production of chlorine, 29
membrane classification, 519
– free volume, quality of the, 520
– hydrogen, properties of, 520
– Kuhn segments, 519
– rubbery polymers, 519
membrane electrode assembly (MEA), 338, 995
membrane electrolysis process, 47
membrane gas separation, basics of, 510–516
– chemical potential, 511
– – difference in, 510
– coating technique, 511
– gas separation membrane process, 510
– isenthalpic throttling, 515
– model for membrane modules, 516
– permeate, 510
– permeate pressure, applied, 516
– permeation mechanism, 513
– pressure ratio, definition of, 513
– retentate, 510
– solution diffusion mechanism, 511
– sorption coefficient, 513
– thermodynamic equilibrium, 512
– transport resistances in gas permeation, 511
– types of membranes, 510
– – carbonous membranes, 510
– – ceramic membranes, 510
– – metal alloy membranes, 510
– – polymeric membranes, 510
membrane humidification, 974
membrane material classification, 519–520, 528–531
– based on glassy polymer, 528
– – chemical structure of Matrimid®, 531
– – diffusion selectivity, 529
– – industrial membrane gas separation, 530
– – permeability selectivity of polymer, 530
– – polymers of intrinsic microporosity (PIM), 530
– – Robeson plot, 530
– – scheme of polyimide conversion, 531
– based on rubbery polymers, 527
– – chemical structure of Polyactive^{TM}, 528
– – gas transport properties of selected polymers, 529
– – poly(ethylene glycol) (PEG), 528
membrane materials, 519
– classification of, 519–520
– defect curing, 524
– – Poiseuille mechanisms, 524
– geometrical classification, 523–524
– – Loeb–Sourirajan integral asymmetric membranes, 523
– morphological classification, 520–523
membrane modules, 510, 519
– classification of, 523
– – flat sheet modules, 526–528
– – hollow fiber modules, 525–526
membrane morphological classification, 519
– cross-section of gas separation membrane, 520
– cross-section of polyacrylonitrile porous membrane, 522
– gas separation membranes, 520
– Knudsen flow, 520
– plasticization of pores walls, 520
– surface porosity, 522
– TFC membrane, 521
membrane reformer (MRF), 230
membrane technology, 41, 46, 517
Mercedes Benz F-cell passenger car, 979
mercury
– cell process, for production of chlorine, 30
– cell room, 44
– electrolyzer, 44
– – technology, 43–45
– environmental and health problems, 41
mesophiles, 429
metal alloy membranes, 511
– palladium, 511
metal alloys, 448
metal-based catalysts, 10
metal chips, 143
metal electrodes, 297
metal hydrides (MH), 441, 791, 1085
– CPI characteristics of, 1003
– -filled pressure container, 705
– formation, thermodynamics of, 765
– for stationary storage, 659
– storage

– – cylinders, 1002
– – devices, 704
– – systems, thermal integration of, 792
metallic membranes, 585
metallic piping, 545
metalloprotein, catalytic, 463
metal membranes, 511
– palladium, 511
metal/metal oxide thermochemical cycles, 89, 90
– CeO_2/Ce_2O_3, 95–96
– Fe_3O_4/FeO as redox material, 90–91
– ferrites, 91–93
– hercynite cycle, 93–94
– manganese ferrite plus activated sodium carbonate, 94
– production of hydrogen based on, 90
– Zn/ZnO, as promising redox system, 94–95
metal nitrides (M_3N_x), 775
metal-on-metal sealing, 645
metal–organic frameworks (MOFs), 499
– lower density of, 499
metal oxides (MOs), 89, 90, 447
– catalyst, 7
– cycles, 89
metal pnictides, 443
metering, 644
methanation, 179, 238, 867
– with CO_2, 255
methane, 110, 143, 544, 546, 606, 814
– as carrier gas, 140
– catalyst poison, 546
– comparison of thermodynamic properties with hydrogen, 606–610
– compressibility factor of, 606
– compression work of, 610
– conversion
– – into acetylene and hydrogen, 142
– – into hydrogen and C_2 hydrocarbons, 141
– – main products, 134
– – rate, 136
– decomposition, 153
– direct methods for, 617
– flow rate, 136
– formation, 179
– fuel properties of, 1013
– heat flow of, 624
– indirect methods for, 617
– isothermal compression of, 615
– Joule–Thomson coefficient of, 608
– nickel catalysts, 546
– oxidation, 188
– positive effect on, 134

– pyrolysis, 134
– reactivity in a microwave plasma, 137
– real gas behavior of, 622
– real isentropic exponent of, 610
– reforming process, 151
– solar thermal reforming of, 157–159
– temperature, 614
– – change during expansion of, 616
– variation of the storage tank
– – pressure of, 621
– – temperature of, 621
methanogenesis, 429
methanogenic bacteria, 429
methanol, 20, 139, 174, 177, 179, 491, 814, 933
– conversion, 139–140
– diaphragm technology, 41
– Lurgi-MegaMethanol Process®, 35
– process schematics for, 36
– production data based on, 21
– reactions of syngas to, 35
– reformer, 138
– – development, 1007
– – high pressure catalytic oxidation, 1008
– – multi-stage catalytic oxidation device, 1008
– – for submarines, 1006
– – system, 1007, 1009
methylcyclohexane, 203
methyl–cyclohexane–toluene–hydrogen cycle, 659
methyl-diethanolamine (MDEA), 558
methylnaphthalene, 203
MGT. see micro gas turbines (MGT)
MH. see metal hydrides (MH)
microalgae, 473
– biodiesel, 473
microbial-based biofilms, 427
microbial biomass, 427
microbial immobilization, 427
microbial inventories, 635
microbial metabolism, 635
micro fuel cells, 64
micro gas turbines (MGT), 1019, 1022
microorganisms, 634
micro-sized gliding arc discharge reactor, 135
microwave plasma, 136–138
microwave plasma source, types of, 137
microwave (2.45 GHz) "tornado"-type plasma, 138
mixed oxides, 152
mixed oxides (MnO_2-CuO), 187
MK 904-mod stack design, 977
mobile applications, 309

MODBUS communication protocol, 1066
MOFs. *see* metal-organic frameworks
molar concentrations, 179
molar specific enthalpy, 178
molar specific entrop, 178
molecular hydrogen, 19, 583
– diffusion coefficients of, 583
molecular orbitals, 565–566
– antibonding MO, 565
– bonding, antibonding molecular orbitals of H_2 molecule, 566
– heteronuclear molecules, 566
– nuclear spin-states, 566–569
molten carbonate fuel cell (MCFC), 64, 544
molybdenum, 11
– oxides, 7
Mo–Ni/Al_2O_3 catalyst, 141
monolithic ceramic structures
– explored as volumetric receivers, 155
monoolefins, 8
Monte Carlo program, 782
MOs. *see* metal oxides (MOs)
MTMH. *see* medium-temperature metal hydride (MTMH) system
multi-column PSA, 502
– large flows of gas, approach to handle, 502
– twelve-column PSA cycle, 503
multi-fuel reforming, 174
multivalent metals, 89
municipal sewage, 431
mutual solubility, of hydrogen and water, 631

n

NaAlH$_4$. *see* sodium alanate (NaAlH$_4$)
Nafion ionomer, 339
Nafion membrane, hydrated, 1088
– oxygen permeation rate through, 1088
nano- and microstructured photoelectrodes, 455–457
nanosecond pulsed DBD system, 141
nanostructures, 300
naphthas, 7, 218, 224
– hydrotreater, 5
naphthenes, 8
National Bureau of Standards (NBS), 734
National Fire Protection Association, 941
National Hydrogen and Fuel Cell Technology Innovation Programme (NIP), 60–61
– defined goals for, 61
– financial requirements for, 61
– first/second phase of, 61
– research, development and demonstration projects within, 61
– total budget within, 61
National Renewable Energy Laboratory (NERL), 836
natural gas, 3, 201, 218, 237, 545, 601, 1076
– autothermic reforming of, 249
– combined cycle (NG-CC), 1021
– composition of, 601
– fuel properties of, 1013
– pipeline grid, 1011
– price projections, 227
– syngas by steam reforming, 246–249
– – equilibrium content of methane, 247
– – evolution of methane conversion, 249
– – main reaction, 246
– – primary reformer, 248
– – secondary reformer, 250
– – temperature profile and heat fluxes, 248
– typical process data for steam reforming/ secondary reforming, 250
natural isotopes, of H_2, 566
– Au-electrodes, 566
– deuterated water, 566
– deuterium, 566
– mass difference, high, 566
– neutron, 566
– protium, 566
– Pt-electrodes, 566
– tritium, 566
– water electrolysis, 566
natural polymers, for immobilization, 429
NBS. *see* National Bureau of Standards (NBS)
NeBus prototype, 976
NeBus technology, 971
– electronic brake paddles, 987
– fuel cell drive train data for, 983
– test and field experience, 982
NEC. *see* n-ethyl-carbazole (NEC)
NECAR 2 fuel cell engine, 973
NEDO. *see* New Energy and Industrial Technology Development Organization (NEDO)
NERL. *see* National Renewable Energy Laboratory (NERL)
Nernst equation, 1085
n-ethyl-carbazole (NEC), 816
new combustion systems for gas turbines (NGT), 1023
New Energy and Industrial Technology Development Organization (NEDO), 837
NG-CC. *see* natural gas, combined cycle (NG-CC)
NGOs, 81

NGT. *see* new combustion systems for gas turbines (NGT)
n-hexadecane, 141, 143
Ni alloys
– tafel slopes of, 299
Ni/Al$_2$O$_3$ catalysts, 188
Ni-based catalysts, 152
– inhibit carbon formation, 153
Ni catalysts, 136, 201
NiCGO. *see* nickel-ceria-gadolinium-oxide
nickel, 11, 48, 91
– catalysts, 546
nickel-ceria-gadolinium-oxide (NiCGO), 545
nickel hydroxide cathode, 771
nickel-metal hydride (NiMH), 265, 771
nickel oxide-silica-alumina, 7
Ni electrode, 299
Ni-gadolinium doped ceria (NiGDC), 545
Ni/γ-Al$_2$O$_3$ catalysts, 139
NiGDC. *see* Ni-gadolinium doped ceria
NiMH. *see* nickel-metal hydride (NiMH)
nitric acid, 32
nitrogen, 5, 134, 224, 403, 547, 888
nitrogenase, 465, 466
– active site of, 469
– endergonic reaction, 469
– nitrogen fixation by, 469
nitrogen-based fertilizers, 19
nitrogen fertilizers, 32
nitrogen monoxide (NO), 1019
nitrogen oxide (NOx), 1013
nitrogen source, 470
nitrogen starvation, 471
nitrous oxide (N$_2$O), 1019
NiYSZ. *see* Ni-yttrium stabilized zirconia
Ni-yttrium stabilized zirconia (NiYSZ), 545
NO. *see* nitrogen monoxide (NO)
noble metal-free catalysts, 441
noble metal oxides, 448
non-catalytic partial oxidation, 25, 237
nondestructive hydrogenation, 5
non-nuclear submarines, 991
non-thermal plasmas, 132
– gliding-Arc plasma, 132
– types, in hydrogen production from hydrocarbons, 133
– used in hydrogen production, 133
North American Free Trade Agreement, 922
Norway
– energy framework and relevant policies, 65
– – climate and energy fund, 65
– – CO$_2$ emissions, 65–66

– hydrogen related energy policy strategies, 65–66
– – carbon capture and storage (CCS) program, 65
– – zero emission vehicles, 66
– hydrogen research, development, demonstration, and deployment activities, 66–67
– – HyNor Project, 67
– – Norwegian R&D institutions related to hydrogen, 66
– – Transnova Hydrogen Projects, 68
– – ZEG Power, 67
n-TiO$_2$ layer, internal resistivity, 447
n-tridecane, 143
nuclear energy, 1053
nuclear reactors, 87, 88
nuclear spin-states, 566–569
– antiparallel, 567
– diatomic molecules, 568
– equilibrium composition of (normal) n-H$_2$, 567
– fermions, 566
– molar heat capacity of p-H$_2$, 568
– normal H$_2$, 567
– ortho-H$_2$, 567
– para-H$_2$, 567
– parallel, 567
– paramagnetic materials, 567
– quantum number, 568
– wave function, total, 568
nuclear very high temperature reactors (VHTRs), 96
nucleophiles, 588

O

octane, 203
– number, 8, 218
ODC. *see* oxygen depolarized cathode (ODC)
OEMs. *see* leading automotive companies; Original Equipment Manufacturers (OEMs)
ohmic resistance, 293
oil refining, 153
OM 908 bus diesel engine, 977
one motor approach, 971
one-motor-design, 978
on-site hydrogen supply, 217
on-site reforming, 217
operating expenditure (OPEX), 733
operational management and experiences, 1045
– activated electrodes, influence of, 1049
– dynamic load, 1045
– temperature influence on stack voltage, 1047

– voltage efficiency of electrolyzer, 1050
OPEX. *see* operating expenditure (OPEX)
optimization
– of power system operation with intermittent generators, 277
organic acids, 427, 635
organic fertilizers, 140
Original Equipment Manufacturers (OEMs), 852
ortho-H_2, 566–569
osmosis, 509
Ostwald process, 32
overvoltage, 448
oxalate, 116
oxidant, 133
oxidation, of hydrochloric acid, 41
oxidative desulfurization, 16
oxinitrides, 447
oxygen, 180, 1093
oxygenates, 139
oxygen depolarized cathode (ODC)
– chlorine-alkali production with, 49–51
– cross section of NaCl ODC based electrolysis halfshell, 51
– electrochemistry of chlorine production with ODC technology, 49
– – reactions, 49
– electrochemistry of NaCl electrolysis with, 50
– structure, 50
oxygen sensor, 1008
oxygen-to-carbon ratio (O/C), 200, 204

p

PA. *see* polyamide (PA)
PACDisplay environment, 1066
Pacific Northwest National Laboratory (PNNL), 858
palladium, 511
paper industry, 53
para-H_2, 566–569
– heat capacity (c_V), 568
– thermal conductivity of, 568, 569
parallel compressed gas, 704
parallel hydride/compressed gas storage, 704
paramagnetic materials, 567
partial methanol flux, 138
partial oxidation, 25, 133, 220, 237, 331
– of methane to syngas, 139
– of propane to syngas, 134
partial oxidation (POX), 5, 131, 133, 136, 142, 146, 206
particle size, 351

particulate matter, 552
– electrostatic forces, 554
– electrostatic interaction, 552
– electrostatic precipitators (ESPs), 554
– gravitational settling, 553
– inertial separation, 552
– separation efficiencies, 553
– technologies for particulate removal, 553
passenger car application, 966
passenger car program, 980
PCT. *see* pressure–composition–temperature (PCT) isotherm
Pd/Ag membranes, 190
PDMS. *see* polydimethylsiloxane
Pd/ZnO catalysts, 201
PED. *see* pressure equipment directive
PEFC. *see* proton exchange fuel cell (PEFC) system
PEFCs (polymer electrolyte fuel cells), 173
– demand highest fuel quality with, 205
– fuel processor subsystem, consisting of autothermal reforming of diesel and a gas cleaning unit for, 191
– gas cleaning subsystem as a possible design for, 207
– principle scheme of gas cleaning processes, for carbon monoxide and hydrogen sulfide in, 206
PEG. *see* poly(ethylene glycol)
PEM. *see* polymer electrolyte membrane (PEM); polymer electrolyte membrane (PEM) electrolysis; proton exchange membrane; proton exchange membrane (PEM) technology
PEM electrolyzer, 1081
PEM FC module, 992
PEMFCs. *see* polymer electrolyte membrane fuel cells
PEM fuel cells, 1086
PEM fuel cells for submarines, 993
– constructive feature of fuel cell module, 995
– – cascaded fuel cell stacks, 996, 997
– – preconditions, 995
– – pressure cushion, 998
– oxygen/hydrogen cell design, 994
– – constructive features, 994
– – fuel cell operation results, 995
– safety features of, 1001
– voltage–current characteristic of, 996
PEM-fuel cell technology, 965
permeate mole fraction, 514
petroleum
– fractions, 5

– products, 5
– refining, 933
PFSA (perfluorosulfonic acid) membranes, 350
p-GaP electrode, 447
phase 4 bus engine demonstrator, 977
phenols, 115
– from carbohydrates, 115
– from lignin, 115
PHEVs. see plug-in hydrogen electric vehicles (PHEVs)
Phosphoric acid fuel cell (PAFC), 64
photocathode, 450
photocurrent, 441
photodecomposition, 464
photoelectrochemical (PEC) cells, 442, 443
– characteristic of photovoltaic component
– – of water splitting PEC, 445
– energy balance, 457
– with single photoelectrode, 443–446
– STH efficiency, 450–451
– with a tandem photocathode immersed in acidic electrolyte, 449
– with two photoelectrodes, 446–448
– for water splitting device, 446
photoelectrochemical devices, 461
– water splitting, 442
photoelectrochemical systems, 457
– economic aspects, 457
photoelectrochemical water splitting, 442
photoelectrode preparation, 457
photoelectrodes, 447
photofermentation, 420
photofermentative hydrogen production, 420
photofermentative processes, 465
photoheterotrophic microalgae, 465
photoheterotrophs, 471
photolytic H_2 production, by microorganisms, 474
– antenna system, reduction of, 475
– biophotolysis process, 474
photosynthesis, 441
photosynthetic bacterial H_2 production, 475
– metal ion, 475
– photofermentative bacteria, 475
– purple non-sulfur bacteria, 475
photosynthetic microorganisms, 476
– genetic study of, 476
photosynthetic organisms, 461, 463
– catalytic metalloprotein, 463
– cyanobacteria, 463
– green algae, 463
– hydrogenase, 463

photosystem II (PSII), 461
– $CaMn_4O_5$ cluster in, 462
– repair cycle, 469
– water oxidation, 472
photosystems, 465
– cytochrome b6f, 470
– electron transport chain, 465
– LHCII, 469
– light harvesting complexes, 465
– multiprotein complexes, 469
– plastoquinones, 470
– repair cycle, 469
photovoltages, 441, 457
photovoltaics (PV), 59
– efficiency, 446
– panels, 1083
– systems, 309
physisorption, 781
PID (proportional-integral-derivative) controller, 232
PIM. see polymers of intrinsic microporosity
pipeline delivery pathway, 857
– challenges, 859
– – compression, 860
– – distribution, 859
– – geologic storage, 860
– – high cost, 859
– – hydrogen and material interactions, 860
– components layout, 857
– cost estimates, 858
pipeline grid, 877
pipelines, 642
piston compressors, 903
plasma, 131
– –catalyst interactions, 139
– -catalytic methanol-steam reforming, 139
– characteristics, 131
– cracking, 141
– decomposition of hydrocarbons co-produces carbon, 131
– non-thermal, 131–132
– reactor, 134, 143
– thermal, 132
plasmagen gas, 134
plasmatron device, 133, 137
plasma type, preferred
– for hydrogen production from hydrocarbon conversion, 147
plastoquinones, 470
platinum, 448
platinum-alumina, 7
PLC. see programmable logic controller (PLC)

plug-in hydrogen electric vehicles (PHEVs), 842
PNNL. *see* Pacific Northwest National Laboratory (PNNL)
policies promoting increased penetration of renewables
– global view of, 257
pollutant level requirements, 550–551
– contaminant limits suggested, 551
– current density, 551
pollutants, 132
polyamide (PA), 879
polycyclic aromatic compounds, 10
polycyclic aromatic content, 9
polydimethylsiloxane (PDMS), 514
poly(ethylene glycol) (PEG), 528
polymer, 580
– permeation coefficient of, 580
polymer electrolyte membrane (PEM), 898, 992
– electrolysis systems, 309–310, 812
polymer electrolyte membrane (PEM) electrolysis, 27, 312
– cell operated at temperatures range, 349
– cell under varying flows on anode side with pin-type flow field, 347
– challenges, 334–335
– characteristic cell performance for, 313
– compression, 314
– control of temperature in, 348
– cost reduction, 350
– – catalyst coated membranes, 350–351
– – cell design, 354
– – modular system design, 354–355
– – porous transport layer, 351–352
– – separator plates, 352–354
– current-voltage characteristics and operating ranges for, 332
– future trends in, 349–350
– gas separation in, 313–314
– general principles of, 335
– historical perspective, 333–334
– schematic of, 314
– stack design, 338
– – catalyst coated membranes (CCMs), 338–340
– – flow field designs, 344
– – porous transport layer, 340–343
– – separator plate, 343–344
– stack operation, 345
– – balanced pressure operation, 345
– – differential pressure operation, 345–346
– – thermal management, 348–349
– – water circulation, 346–348
– stacks commercial, developers/manufacturers of, 336–337
– state-of-the-art, 335
– – stack, specifications/components of, 335–336
– with and without cathode water circulation, 348
polymer electrolyte membrane fuel cells (PEMFCs), 544
polymeric membranes, 509, 511
polymeric membranes classification, 519
– membrane classification, 519
– – glass transition temperature (Tg), 519
– – Kuhn segments, 519
– – morphological classification, 520–523
– – solubility exclusion mechanisms, 519
polymerization, 5, 113
polymer membrane, 1084
polymers of intrinsic microporosity (PIM), 530
polytrophic coefficients, 191
porous material, 492
– activated carbon, 493
– activated zeolites, 493
porous reservoirs, 631, 634
– geological storage options for hydrogen in, 637
porous support layer, 515
– concentration gradients in, 515
porous transport layers
– gas flow from, 341
– SEM micrographs, 342–343
Porphyidium cruentum, 473
potassium chloride, 41
potassium hydroxide, 41, 1036, 1037
power plants, characteristics, 263
power quality, and reliability, 265
power rating, 264
power storage technologies, 869
– capacity, 869
– discharge time of, 869
power system
– demand-side management (DSM), 263
– flexibility resources and flexibility needs in, 262
– IEA listing categories of flexibility sources, 262
– operators, 1055
– supply-side management, 263
power technologies, 1089
"power-to-gas" concept, 255, 1059
power-to-gas mass energy solution, 866
– challenges and opportunities, 869

– current applications, 867
– hydrogen methanation process, 866
power-to-gas (PtG) plants, 274–275
power-to-gas technology, 309
powertrain layout, 966
– maximum power demand, 966
– range, 966
– time for refueling, 966
– torque of drive train, 966
preferential CO oxidation, 206
Prenzlau, hybrid power plant project in, 1035, 1036
– alkaline electrolyzer, 1036, 1037
– electrolyzer operation, 1034
– overview of, 1035
pressure–composition isotherms, 765
pressure–composition–temperature (PCT) isotherm, 797
pressure equalization, 497
pressure equipment directive (PED), 928
pressure swing adsorption (PSA), 15, 24, 221, 230, 238
pressure–temperature phase diagram, 569–571
– atomic plasma, 571
– cubic closed-packed structures (fcc), 571
– phase regions for molecular/atomic hydrogen, 570
– supercritical fluid, 570
– triple point, 570
pressure–volume phase diagram, 571–572
– critical point (p_{crit}), 572
– *p–V_m pase diagram of hydrogen,* 572
pressurized gas storage, thermodynamics of, 601
– compression/expansion processes, analysis of, 610–617
– – isenthalpic expansion, 615–617
– – isentropic compression, 611–615
– – isentropic expansion, 615–617
– – isothermal compression, 611–615
– expansion processes, analysis of
– – isothermal compression, 611–615
– gas storage process, thermodynamic modeling of, 617–619
– – application examples of, 620–624
– – – refueling of a vehicle storage tank, 620–622
– – – salt cavern, 622–22
– – heat transfer equations, 619
– thermodynamic properties of gases, comparison of, 606–610
– – compressibility factor, 606–607

– – isentropic exponent, 609–610
– – Joule–Thomson coefficient, 607–609
– thermodynamic state variables, calculation of, 602–606
– – ideal gases, for, 602–603
– – real gases, for, 603–606
probabilistic risk assessment (PRA), 935
process control instruments, 175
process engineering, 642
product gas composition, for various hydrogen generation processes, 195
product gases, 544
production-to-demand-approach, 811
programmable logic controller (PLC), 905, 1001
Promoting Unst Renewable Energy (PURE), 1077
propane, 134, 143, 203
– conversion, 142
– conversion depended on, 145
propane–butane mixture (LPG), 173
propanoic acid, 139
propene, 203
propulsion and auxiliary power demand, determination of, 966
protium, 566
proton channel, 471
– biochemical modifications of, 471
– genetic modifications of, 471
proton conducting membrane, 994
proton exchange fuel cell (PEFC) system, 792
proton exchange membrane (PEM), 1083
– technology, 965
Proton OnSite hydrogen generator, 1081
PSA cycle, improving, 501
– adsorption processes, 501
– large flows of gas, approach to handle, 502
– multi-column PSA, 502
– numerical simulation of, 502
– purge gas, 502
– rapid pressure swing adsorption (RPSA) units, 503
– rotary valves, 503
– total cycle time, 502
– total time of, 501
– twelve-column PSA cycle, 503
– unit productivity, calculation of the, 501
PSA technology, basics, 492
– activated carbon, 493
– activated zeolites, 493
– adsorbent, 492
– adsorbent productivity, 495
– adsorption equilibrium isotherms, 492
– adsorption processes, 492

- black-box design, 495
- black-box scheme of, 495
- blowdown step, 497
- cost of the H_2 separation, 496
- cyclic adsorption–desorption operation, 496
- cyclic loading, 496
- cyclic steady state, 495, 496
- depressurization, 497
- desorption, 496
- energy consumption, 495
- equilibrium state, 492
- feed gas, 495
- fundamental operation of, 495
- gas mixture, 496
- high H_2 recovery, 495
- hydrogen recovery, 497
- inlet stream of, 499
- isotherms of a contaminant, 495
- light-recycle, 497
- multicomponent gas phase, 492
- pressure equalization, 497
- pressure history, in a two-column PSA process, 498
- pressure swing adsorption, 493
- provide purge, 498
- Skarstrom cycle, 496
- steam–methane reforming off gases, 493
- tail gas, 495
- temperature swing adsorption (TSA), 493
PSA technology, for H_2 separation, 491
- basics of, 492
- improving the PSA Cycle, 501
- operating costs of, 492
PSA units, 502
- cycle times of, 502
PSI, 465
Pt and Ru catalysts, 141
Pt/CeO_2 WGS catalyst, 207
PTFE coating, 352
Pt-Ru catalysts, 187
pulsed corona, 142
pulsed plasmas, 142–144
pumped hydro storage [PHS], 265
PURE. *see* Promoting Unst Renewable Energy
pure hydrogen storage projects, 650
- Clemens Dome, Texas, USA, 650
- Moss Bluff, Texas, USA, 650
- Teesside Project, Yorkshire, UK, 651
pure oxygen, use of, 1004
purple photosynthetic bacteria, 469
- biological nitrogen fixation in, 469
- FeMo cofactor, 469
PV array, 1087

PV–hydrogen systems, 1087
- economics of, 1087
Pyrodictium brockii, 467
pyrolysis, 8, 153, 195
pyrolytic carbon, 143
pyrophoricity, 229

q
QRA. *see* quantitative risk assessment
quantitative risk assessment (QRA), 924, 935
- for hydrogen safety, shortcomings, 935
- - frequency probability of component failures, 936
- - human error, 935
- - ignition, 935
- - lack of uniform acceptance criteria, 937
- - probability of causing damage, 937
- - simplified engineering models, 936
- - uncertainties in QRA models, 937
- schematic, 936
quantum number, 568
quick charging stations, 711

r
rapeseed methyl ether (RME), 556
rapid compression machine (RCM), 1015
rapid pressure swing adsorption (RPSA), 503
rapid risk ranking (RR), 924
RCM. *see* rapid compression machine (RCM)
reactant filling station, 1004
reaction kinetics, 574–578
- reaction with O_2
- - explosion limits, 577–578
- - microscopic mechanisms, 574–577
- - thermodynamics, 574
reaction number, defined, 178
reaction of H_2 with O_2
- explosion limits, 577
- - autoignition temperature of H_2/air-mixtures, 578
- - isothermal explosion, 577
- - for mixtures of H_2 and O_2, 577
- mechanism of, 574
- - chain reaction (unbranched), 576
- - hydrogen peroxide (H_2O_2), 574
- - hydroperoxyl radical, 575
- - hydroxyl radical, 575
- - isothermal explosion, 576
- - oxygen radical, 575
- - reaction between hydrogen and oxygen, 576
- - thermal explosion, 576
- thermodynamics of, 574

Index

– – binary hydrides, 574
– – equilibrium constant (K), 574
– – free formation enthalpy of H_2O vapor, 575
– – hydrogen fluoride, 574
– – mass action law, 574
reactor configurations, 427
– and scale-up challenges, 427
reactor improvements, and design aspects, 230
reactor operating conditions, 7
real gases, 572, 601, 603–606
– caloric equations of, 615
– direct method for, 615
– final states of, 617
– Helmholtz free energy, 603
– indirect method for, 615
– isentropic relations for, 609
– temperature of, 572
– thermal equation of state for, 603
– thermodynamic properties in terms Helmholtz free energy, 604
– thermodynamic properties of, 601
reduce boil-off (RBO) concept, 961
reduced boil-off
– for LH_2 by vapor venting and, 962
refining heavy feedstocks, 11–12
reformate gas stream, 1008
reformer technologies
– technology and pros/cons, 221
reforming reaction, 491
reforming technologies, 219, 491
refueling, 711, 713, 717
– and infrastructure perspectives, 717–719 (*see also cryo-compressed gases*)
refueling station, 852
– challenges, 854
– – analysis, 854
– – boil-off, 855
– – compression, 855
– – hydrogen quality, 855
– – land area, 855
– – metering, 855
– – refrigeration, 855
– – storage costs, 855
– cryo-compressed, 854
– gaseous hydrogen, 852
Regulation on Industrial Safety and Health, 1039
Regulations Codes and Standards (RCS), 919
renewable-based hydrogen economy, 152
renewable electricity, 59, 255
renewable energy, 262, 477, 811, 1075
– alternative sources of, 477
– cost of, 1075

– import, from Iceland, 826
– – aluminum production, 826
– – hydrogen-rich LOHC systems, liquid transport of, 826
– storage of, 813
– transport of, 813
– utilization of, 1075
renewable energy integration, 264
– non-storage options for, 264
renewable energy sources, 151, 255, 309, 417, 711, 891
– grid integration of, 309
renewable generators, 276
renewable heat source, 87
"renewable" hydrogen production route, 151
reservoir rocks, 634
resistance model, 509
retentive membranes, 427
reverse engineering, 800
– back pressure, 805
– bulk density, 803
– charge kinetic rate, 803
– discharge kinetic rate, 803
– heat exchanger tubes, 806
– heat transfer tubes, 806
– refueling pressure, 803
– refueling temperature, 803
– state-of-charge conversion, 802
– temperature profile, 802
– thermal conductivity, 805
reversible solid oxide fuel cell (RSOFC) technology, 359, 379–383
– based sustainable energy system, 382
– cell and stack design flexibility, 362
– efficient and stable cyclic operation, 380–383
– efficient production of hydrogen and chemicals in electrolysis mode, 364–366
– electrode performance, 379–380
– features, 361–366
– key requirements for, 379
– materials for, 361
– multi-fuel capability in fuel cell mode, 364
– operating principles, 359–360, 359–361
– performance curve of YSZ based cell, 380
– simplified system schematic, 382
– stability for reversible operation, 379–380
– ten-cell RSOFC stack and performance, 381
– voltage–current density curves, YSZ-based cells with, 380
Rhizobium, uptake hydrogenase (HUP), 464
rhodium nanoparticles, 448
Rhodobactor sphaeroides, 476

– light harvesting antenna complexes, deletion of, 476
– mutation of, 476
risk management, 935
RME. see rapeseed methyl ether
RPSA. see rapid pressure swing adsorption
RR. see rapid risk ranking
RSOFC. see reversible solid oxide fuel cell (RSOFC) technology
RSOFC technology. see reversible solid oxide fuel cell (RSOFC) technology
Ru/Al$_2$O$_3$ catalysts, 188
RuO$_x$ film, 446
ruthenium dioxide (RuO$_2$), 299
ruthenium oxide (RuO$_2$), 340

s

Saccharomyces cerevisiae, 473
safe handling, of hydrogen, 938
– gaseous hydrogen, 939
– handling emergencies, 940
– liquid hydrogen, 939–940
safe reserve power generation systems, 1054
safety engineering, 1039
safety-related properties, of hydrogen, 937, 938
– chemical properties, 937–938
– corrosion and materials compatibility, 938
– physical properties, 937
saline aquifers, 635
salt cavern, 622
– compressed air energy storage (CAES), 622
– diagram/input parameters of, 623
– variation of the temperature, 623
salting-out effect, 632
Sankey diagram, 982
SCWG. see supercritical water gasification (SCWG)
SDOs. see standards drafting organizations
seawater, 1092
– desalination, 1092–1093
secondary battery system, 834
selective adsorbents, 499
– alumina, 499
– capillary condensation, 499
– heavy hydrocarbons, 499
– HKUST-1 adsorbent, 499
– linear isotherm, 501
– mechanical properties of, 499
– mechanical stability, 499
– metal–organic frameworks (MOFs), 499
– particle movement, to avoid, 499
– pressure equalizations, 501
– PSA unit, 499
– silica gel, 499
– type V isotherm, 499
– vapor–liquid equilibrium, 499
– water loading of a basic gamma alumina, 499
– zeolite layer, 501
selective CO methanation, 206
self-immobilized microbes, 427
self-sustaining cooling, 713
semiconductors, 443, 447
– band gap positions, 443, 444
– materials, 441
– photoelectrochemical materials, modern, 472
– surfaces, 19
– valence and conduction band
– – related to electrochemical redox potentials, 444
separation devices, 189
– adsorption process, 189
– balance-of-plant (BOP) components, 191
– – heat exchangers and evaporators, 192
– – pumps and compressors, 191–192
– membrane process, 189–190
sewage sludge, 424
SHE. see standard hydrogen electrode
shock tube measurements, 1014
short term operating reserve (STOR), 277
Sievert's law, 190
silica gel, 499
– water adsorbed in, 499
silicone, 472
siloxanes, 550
simulated grid load, 310
single-wall nanotubes (SWNTs), 784
sintering, 232
Skarstrom cycle, 492
– steps of, 496
– – adsorption, 496
– – counter-current blowdown, 496
– – counter-current purge, 497
– – pressurization, 497
SLFs. see synthetic liquid fuels (SLFs)
slurry hydrocracking, 10
small-scale CO$_2$ capture, 231
small-scale reformer systems, 219
– categorized by defining sizes, 219
small-scale reforming, for on-site hydrogen supply, 217
– emerging technologies, 228–229
– – clean-up technology, 230–231
– – material development, 229–230
– – reactor improvements and design aspects, 230

– – small-scale CO_2 capture, 231
– feedstock options, 223
– – biodiesel, 224–225
– – biomethanol and bioDME, 224
– – biopropane, 225
– – upgraded biogas, 223–224
– industrial reformer unit, schematics of components in, 218
– process control, 232
– reforming technologies, 219
– – hydrogen purity, 221–223
– – partial oxidation, 220–221
– – steam methane reforming (SMR), 219–220
– small-scale reformer systems
– – categorized by defining sizes, 219
– suppliers and products, 225
– – cost trends, 225
   – – central plant production, 225–226
   – – economics, on-site vs. central plant production, 227–228
   – – on-site SR, 226–227
small scale water electrolyzers, 284
smart grid, 1081
SMES. see superconducting magnetic energy storage (SMES)
SMRs. see steam-methane reformers
SNG. see substitute natural gas (SNG)
SOC. see state-of-charge (SOC)
sodium alanate (NaAlH$_4$), 774
– dehydrogenation of, 774
– suspensions of, 774
sodium alanates, 791
sodium chloride, 41
sodium hydroxide
– usage, 53
sodium hypochlorite, 53
sodium sulfur, 265
SOEC. see solid oxide electrolysis (SOEC)
SOEC technology. see solid oxide electrolysis cell (SOEC) technology
SOFCs. see solid oxide fuel cells
SOFC technology. see solid oxide fuel cell (SOFC) technology
solar
– -aided chemistry, 155
– -aided methane reforming, 154
– chemical receiver/reactors, 157
– concentration systems, 154
– -driven PEC water splitting device, 457
– energy, 87, 89, 152, 267, 461
– – import, from Northern Africa, 825
– furnaces, 90
– -heated tubular receivers, 156

– H_2 production, 472
– installations, 256
– irradiation, 155, 156
– light, 450
– power, 157, 260
– – plants, 266, 711
– PV systems, 258
– radiation, 91, 156
– reactors, 90
– receivers, 155
– – operational principles, 155
– – -reactor concepts, 154
– thermal power, 785
– – plants, 87
– thermal reforming, 151
– – current development status and future prospects, 167–169
– – of methane, 157–159
– thermal water decomposition, 87, 88
– thermochemical cycles, 89
– thermochemical reactor, 96
solid fuels (coal, biomass), 242
– basic principles and reactions of syngas production, 242
– hydrogen production from, 242
– – gasification, 242–244
solid–gas reactions, 91
solid oxide electrolysis (SOEC), 309
solid oxide electrolysis cell (SOEC) technology, 372–379
– applications, 378–379
– Co-electrolysis of H_2O and CO_2, 378
– current densities at thermoneutral voltage, 374
– efficient production of hydrogen and chemicals in electrolysis mode, 364–366
– materials for, 361
– operation of multi-cell stacks, 373
– performance (for hydrogen production from steam), 373
– performance degradation and life, 375–376
– performance degradation of 25-cell stack in terms of ASR, 377
– performance of a ten-cell stack for electrolysis, 378
– selected YSZ-based SOECs at thermoneutral voltage, current densities for, 374
– ten-cell SOEC stack, 375
– voltage as a function of time for ten-cell stack, 377
solid oxide fuel cells (SOFCs), 544
solid oxide fuel cell (SOFC) technology
– applications, 371–372

– area specific resistance (ASR) breakdown, for a planar SOFC stack, 368
– catalysts for, 201
– cathode/metallic interconnect contact evolution, 370
– cell and stack design flexibility, 362
– cell-to-cell voltage variation, 367
– chromium poisoning of SOFC cathodes, potential steps in, 369
– fuel utilization, 370–371
– long-term performance of 96-cell planar stack, 371
– market sectors and applications, 372
– materials for, 361
– micrograph of Co-Mn spinel coated interconnect, 370
– multi-fuel capability in fuel cell mode, 364
– multiple fabrication options, 362
– Ni-YSZ anode, 368
– operating temperature choices, 362, 364
– performance, 366
– performance degradation and life, 368
– power systems, 373
– single cell (anode-supported) performance, 366
– single cells and microstructures, 362
– stacks and systems, operation and durability, 370
– state-of-the-art SOFC single cells, 367
solid oxide fuel cell (SOFCs) technology, 366–371
– market sectors and applications, 372
solution-diffusion theory, 509
sorbent-enhanced reforming (SER), 229
sorption kinetics, 798
Sotavento wind–hydrogen project, 1065
– operation strategies tested in, 1066
– – capacity factor of wind farm, increasing of, 1070
– – deviation correction, 1069
– – load leveling enabling distributed generation, 1071
– – peaking plant, 1067
Spain
– electricity transport operator, 1053
– National Renewable Energy Plan, 1053
– RES2H2 site, 1060
– wind power, 1053
spark discharge, 142
special ventilation system, 1008
specific heat capacity, 175
spherical liquid hydrogen tank, 958
Spirulina maxima, 473

Spirulina platensis, 470
SSM. *see* supply-side management (SSM)
stable compounds, 5
stack voltage efficiency dependence, 1050
stack voltage, temperature dependence of, 1048
standard bus diesel engine, 971
standard hydrogen electrode (SHE), 596
Standards Drafting Organizations (SDOs), 921
standards, regulations, and products, relationship between, 921
standard systems, 1057
Starch, 470
– catabolism of, 470
state-of-charge (SOC), 797, 1090
stationary application areas, and requirements, 660
– chemical usage, as a versatile energy carrier, 660
– compressed gaseous tanks, 661
– delivery infrastructure, 660–661
– renewable energy production, 661
– small-scale storage facilitie, 661
– storage facilities of different sizes, 661
– underground storage facilities, 660
stationary hydrogen storage
– costs for bulk storage, 675
– economic data and targets
– – of DOE, 677
– – of EC, 677
– investment cost, based on storage size and targets of EC for, 676
– methods for, 659
– technical and economic data, of spherical pressure vessels, 676
– technical assessment of bulk hydrogen storage technologies, 678–682
steam-methane reformers (SMRs), 5, 15, 493
steam methane reforming, 218, 231
steam methane reforming (SMR), 131, 151, 219–220
– steam-to-carbon ratios, 220
– water–gas shift, 220
steam reforming, 237, 331, 398
– advantage of, 194
– of natural gas, 23
storage capacity, 630
– electricity storage systems, 277
storage efficiency, 660
storage heat receptivity
– in cold operation, 713
storage parameters, 661–662
– gravimetric and volumetric capacity, 661

– gross and net storage capacity, 662
– lower heating value (LHV), 662
– standard temperature and pressure (STP), 662
street system infrastructure, 966
"structured" catalytic systems, 155
Structured What-IF checklisT (SWIFT), 924
submarine
– hydrogen and fuel cells in, 991
– – historical perspectives, 992
– – – ThyssenKrupp Marine Systems, 992
– – overview, 991
– network system, 991
– reformer system conceptual integration, 1008
– snorkel operation of, 991
substitute natural gas (SNG), 866
subsystem reforming. see also fuel processing
– optimization of, 193
– – catalyst development, 200–201
– – process selection and optimization, 193–200
– technical outlook, 204–205
sugars, 471
sulfate compounds, 634
sulfur, 3, 5, 237
– impurities, 237
sulfur components, 558
– amines like diethanolamine (DEA), 558
– methyl-diethanolamine (MDEA), 558
– removal of, 558
– sodium hydroxide (NaOH), 558
sulfur compounds, 152
sulfur-containing compounds, 19
sulfur-containing hydrocarbons, 190
sulfur cycles, 96
sulfuric acid, 89, 97
sulfur–iodine cycle, 88
sulfur–iodine process, 98
sulfur–iodine water-splitting cycle, 98–99
– basic reactions, 98
– efficiency, 100
sulfur tolerance, 201
supercapacitors, 264
superconducting magnetic energy storage (SMES), 264
superconductivity, 785
supercritical steam reforming, 195
supercritical water gasification (SCWG), 110
– biomass used for, 109–119
– catalysts, 119
– limitation, 119
– – catalyst stability, 122

– – heating-up, 119
– – heat recovery, 120
– – material choice, 121–122
– – salt deposition, 121
– – yield, 120–121
– new developments, 122–123
– – feedstock, 122–123
– – new catalysts, based on Ni-Mg-Al/Ru, 122
– – pretreatment methods, 122
– scale-up and technical application, 122
– – plant VERENA, 122
super-heater device, 972
supply-side management (SSM), 262, 263
surface tension, 303
sustainable energy system, 1075
SWNTs. see single-wall nanotubes (SWNTs)
Synechococcus elongatus, 470
syngas, 139, 237, 1011. see also synthesis gas
– ammonia syngas production from natural gas
– – steps, 239
– production, 153
– – based on fossil fuels and biomass, 238
– – by partial oxidation of heavy oils/gasification, 241, 244–246
– – by steam reforming of natural gas, 246–249
– – and treatment, depending on final use, 239
– treatment steps, to remove unwanted impurities, 237
synthesis gas, 237, 543
– ratio, adjustment of the, 509
synthetic diamond growth, 785
synthetic fuels, 36
– producer of, 36
synthetic liquid fuels (SLFs), 151

t
Tafel slope, 299
tail gas, 495
tars, 556
– separation of, 556
tars/higher hydrocarbons, 555
– catalytic reduction, 557
– – Al_2O_3, 557
– – fluid catalytic cracking (FCC), 557
– – synthetic catalysts, 557
– catalytic reforming, 556
– downstream processes, 556
– hot gas tar reforming, 556
– rapeseed methyl ether (RME), 556
– thermal cracking, 556
– zeolites, 557
TBAF. see tetra-*n*-butylammoniumfluoride
TBT. see Technical Barriers to Trade

TCA. *see* temperature cycling adsorption (TCA)
TCIs. *see* total capital investments (TCIs)
TCO. *see* total cost of ownership (TCO)
Technical Barriers to Trade (TBT), 923
technical data of, 1001
– BZM 34, 1000, 1001
– BZM 120, 1000, 1001
technologies, to remove pollutants, 551–558
– alkali components, 555
– – Al_2O_3, 555
– – getter materials, 555
– – overview of alkali removal options, 555
– – SiO_2, 555
– cold gas clean-up (CGC), 551, 552
– hot gas clean-up (HGC), 551, 552
– particulate matter, 552–555
– sulfur components, 558–559
– tars/higher hydrocarbons, 555–558
technology options, for integrating variable renewables, 262
technology readiness level (TRL), 1082
Teflon-sealed reactor, 450
temperature cycling adsorption (TCA), 189
temperature swing adsorption (TSA), 493
tetrahydrofuran (THF), 597
tetra-*n*-butylammoniumfluoride (TBAF), 597
TFC membrane, 522
– discontinuous process, 523
– resistance model of, 524
THD. *see* total harmonic distortion (THD)
thermal-based electricity storage, 265
thermal conductivity, 578–579, 800
– definition of, 578
– fourier's law, 578
thermal cracking, 556–557
thermal decomposition, of water, 88
thermal dissociation, of water, 87
thermal efficiencies, 101, 136
thermal plasmas
– DC-RF plasma, 145–146
– DC torch plasma, 144–145
– three-phase AC plasma, 145
– used in hydrogen production, 144
thermal power plants, 1053
thermal reforming, 218
Thermoanaerobacterium thermosaccharolyticum, 420
thermochemical cycles, 87
– decompose water into, 89
– historical perspective, 88
– *T–S* diagram of two-step cycle to split water, 90

thermochemical hydrogen production, 87
thermochemical processes, characterization, 393
thermochemistry, 152
thermodynamics, 88
– limitations, 174
– of methane reforming, 152
– properties of H_2, 569–573
– – Joule–Thomson effect, 572–573
– – pressure–temperature phase diagram, 569–571
– – pressure–volume phase diagram, 571–572
– properties relations, 175–178
– quantities, decision tree for determination of, 176
– state variables, calculation of, 602–606
thermolysis reaction, 87
thermoneutral cell voltage at STP, 1050
thermophiles, 420, 429
thermophilic biohydrogen production, 420
– via dark fermentation, biochemistry, 420–421
thermophilic dark fermentation, 429
– costs of operation, reduction, 431
– problem associated with, 430
thermophilic hydrogen producing bacteria, 420
– microbial characteristics, 421–422
thermophilic temperature, 430
Thermotoga elfii, 420
THF. *see* tetrahydrofuran
thin film composite membranes, 511
thylakoid membrane, 441, 465
ThyssenKrupp Marine Systems, 992
TiO_2 based catalyst, 141
titanium, 352
Tokyo motor show, 833
toluene, 203, 205
total capital investments (TCIs)
– calculation, 406
– of hydrogen production, 407–408
– – cost fractions, representation, 409
– for similar technology devices, 406
total cost of ownership (TCO), 836
total European chlorine production capacity, 28
total harmonic distortion (THD), 1062
town gas storages, 651–652
– Beynes, Ile de France, France, 654
– town gas storage at ketzin, Germany, 652–653
– town gas storage at lobodice, Czech Republic, 652

trace gas components, 547
- alkali compounds, 549
- biomass gasification synthesis gas, 547
- contaminants in biomass synthesis gas, 548
- halogen compounds, 550
- - C_2Cl_4, 550
- - halides of catalyst metals, 550
- - HCl, 550
- - HF, 550
- nitrogen compounds, 550
- - HCN, 550
- - NO_x, 550
- other potential contaminants, 550
- - metal–organic compounds, 550
- - phosphorus, 550
- particulate matter (particles), 549
- - particle content in untreated product gas, 549
- siloxanes, 550
- - deposition of silicon compounds, 550
- - wastewater treatment, 550
- sulfur compounds, 549
- tars, 547
- - content in untreated product gas, 548
- - cyclic hydrocarbons, 547
- - naphthalene, 548
transit bus applications, 986
transition metal compounds, 448
transition metal oxides, 443
transmembrane flux
- calculation of, 515
transparent hydrocarbon fuels, 143
transport costs, 406
transport of molecules, 509
trimethylbenzene, 203
2,2,4-trimethylpentane, 143
triple point, 570
tritium, 566, 764
TRL. see technology readiness level (TRL)
truck applications, 987
TSA. see temperature swing adsorption (TSA)
tube-trailer delivery pathway, 860
- challenges, 863
- - analysis, 864
- - compression, 864
- - distribution issues, 863
- - high storage cost, 864
- components layout, 860
- cost estimates, 861
tubing, 642
tubular reactors, 141, 168
tubular reformers, 154
tungsten-nickel sulfide, 7

turbulent flame speed, 1017
Turkish Bozca Island, 1082
- UNIDO-ICHETs Bozcaada Wind-Solar Hydrogen project, 1082
two-stage compressor system, 974
type3 cylinder technology, 975
typical time spent
- at power demands, 968
- at vehicle speeds, 968

u

UASB reactor, 429
UK H2 mobility program, 217
ultra-low-sulfur (ULS) gasoline, 3
United States of America, 76
- energy framework and relevant policies, 76
- - state and local government activities, 76
- hydrogen related energy policy strategies, 76–77
- - Hydrogen and Fuel Cells Program, coordination, 76
- - US DOE Hydrogen and Fuel Cells Program Annual Progress Report, 76
- - zero emission vehicles, 76–77
- hydrogen research, development, demonstration, and deployment activities, 77–78
- - Hydrogen and Fuel Cells Program, key targets/collaborations, 77–78
- - US DoE Hydrogen and Fuel Cells Program overview, 77
unit operations, 180
upflow anaerobic sludge blanket bioreactor (UASB), 427, 429
UPS (uninterruptable power supply), 64
uptake hydrogenase (HUP), 467
uranium–europium cycle, 100, 102
urea production
- International Fertilizer Organization data, 32
US Department of Energy (DOE), 850
US Department of Transportation (DOT), 861
UT-3 cycle, 88, 100, 101
- as competing process in hydrogen production, 101
- gas–solid reactions, 101
- research focused on, 101
- thermal efficiency, 101
UV irradiation, 476

v

vacuum plasma spraying (VPS), 1049
value-added tax (VAT), 842

vanadium redox (VRB), 265
Van't Hoff plots, 765
VAT. *see* value-added tax (VAT)
Vattenfall, 898
vehicle batteries, 711
vehicle onboard storage, 711
vehicle storage tank, 620
– refueling of a, 620–622
– – time of, 620
viscosity, 296, 340, 634, 639
volatile compounds, 242
volatile solid (VS), 429
voltage efficiency, 1050
volumetric hydrogen density, 770
"volumetric" receivers, 156

w

waste products, 403
wastewater, 431
– treatment, 1093
water
– consumption rate, 347
– dielectric constant, 112
– hydrogen bonds, 112
– polarity, 112
– properties of, 111–112
– – as function of temperature, 112
– role in different reaction pathways, 112
– at supercritical conditions, 112
– – ionic product, 112
– thermal decomposition, 88
water electrolysis, 283, 309, 457, 566
– conceptual distributed energy system with, 284
– gas drying, 316–317
– hydrogen production cost, 326
– investment costs, 325
– lifetime of, 326
– operating principles of different types of, 332
– power supply, 317
– process steps and system integration of, 310
– system design and operation strategy, 327
– thermodynamic consideration, 285
– – electrode kinetics, 287–292
– – theoretical cell voltages, 285–287
– unit, 255
water-gas shift (WGS) reaction, 26, 118, 152, 154, 179, 181, 184–186, 231, 238
water oxidation, 461
water oxidizing photoelectrodes, 447
water splitting devices
– design of, 448–455
water splitting mechanism, 461

– other methods of, 462
– – electrolysis, 462
– – gasification, 463
– – photobiological splitting, 463
– – photochemical splitting, 463
– – steam electrolysis, 462
– – steam reforming, 462
– – thermal splitting of water, 463
water splitting membrane, 442
– solar-to-hydrogen (STH) efficiency of, 445
wave speed, 646
Westinghouse cycle, 89
WGS reactor, 185
WI. *see* Wobbe Index (WI)
wind capacity, 260
wind constraints, 261
wind–electrolyzer coupling, 1045
wind energy, 260
– as hydrogen, storage of, 1034
wind farm production, 1067
– capacity factor of, 1070
– H_2 storage, 1069
wind generator electrical power input, 1062
wind–hydrogen plant, 1061
wind–hydrogen projects, description of, 1059
– Hidrólica project, 1061
– ITHER project, 1063
– RES2H2 project, 1060
– Sotavento project, 1064
wind–hydrogen systems, 1056, 1087
wind load factors, 256
wind penetration, 260
wind power, 261, 1054
– integration on energy system, 256
– management, 1058
– for the Netherlands, 260
wind power plants, 1087
– capacity factor, 1087
wind promoters, 1055
wind turbines, 255, 309, 317, 1083
Wobbe Index (WI), 1013
wood chip gasification, 398
– energy demand, 405
World energy consumption, 359
World Trade Organization's (WTO's), 923
wurtzite, 91

y

yttria-doped ceria (YDC), 367

z

zeolites, 493, 557, 784
– cost of, 501

– density of, 501
– pores in, 501
zero boil-off (ZBO), 961
zero emission vehicles (ZEVs), 349, 711, 845
– air quality, 845
– global warming reduction, 845
– public health, 845
– societal costs reduction, 845

ZEVs. *See* zero emission vehicles (ZEVs)
zinc, 13, 91, 95, 283
zinc bromine [Zn Br], 265
zirconium oxides, 189
ZnO bed, 153
ZnO/Cu_2O photocathode, 447
ZnO, solar dissociation of, 95
Zn/ZnO system, 94–95
Zr-supported Rh/Pt catalyst, 201